Cosine Identities

$\cos(u + v) = \cos u \cos v - \sin u \sin v$

$\cos(u - v) = \cos u \cos v + \sin u \sin v$

$\cos 2t = \cos^2 t - \sin^2 t = 1 - 2\sin^2 t = 2\cos^2 t - 1$

$\cos\left(\dfrac{t}{2}\right) = \pm\sqrt{\dfrac{1 + \cos t}{2}}$

$m\cos(kt - c) = a\cos kt + b\sin kt$, provided $m = \sqrt{a^2 + b^2}$, $\tan c = \dfrac{b}{a}$

Sine Identities

$\sin\left(\dfrac{\pi}{2} - t\right) = -\cos t$

$\cos\left(\dfrac{\pi}{2} + t\right) = \sin t$

$\sin(u + v) = \sin u \cos v + \cos u \sin v$

$\sin(u - v) = \sin u \cos v - \cos u \sin v$

$\sin 2t = 2\sin t \cos t$

$\sin\left(\dfrac{t}{2}\right) = \pm\sqrt{\dfrac{1 - \cos t}{2}}$

Tangent Identities

$\tan(u + v) = \dfrac{\tan u + \tan v}{1 - \tan u \tan v}$

$\tan(u - v) = \dfrac{\tan u - \tan v}{1 + \tan u \tan v}$

$\tan 2t = \dfrac{2\tan t}{1 - \tan^2 t}$

$\tan\left(\dfrac{t}{2}\right) = \dfrac{1 - \cos t}{\sin t}$

Identities for Sums, Differences, and Products of Sines and Cosines

$\sin u \cos v = \dfrac{\sin(u + v) + \sin(u - v)}{2}$

$\sin u \sin v = \dfrac{\cos(u - v) - \cos(u + v)}{2}$

$\cos u \cos v = \dfrac{\cos(u + v) + \cos(u - v)}{2}$

$\cos u + \cos v = 2\cos\left(\dfrac{u + v}{2}\right)\cos\left(\dfrac{u - v}{2}\right)$

$\sin u + \sin v = 2\sin\left(\dfrac{u + v}{2}\right)\cos\left(\dfrac{u - v}{2}\right)$

$\cos u - \cos v = -2\sin\left(\dfrac{u + v}{2}\right)\sin\left(\dfrac{u - v}{2}\right)$

$\sin u - \sin v = 2\cos\left(\dfrac{u + v}{2}\right)\sin\left(\dfrac{u - v}{2}\right)$

Applied Technical Mathematics
with Calculus

Merwin J. Lyng
Mayville State University

L. J. Meconi
University of Akron

Earl J. Zwick
Indiana State University

Wm. C. Brown Publishers

Book Team

Editor *Earl McPeek*
Developmental Editor *Theresa Grutz*
Production Editor *Renée A. Menne*
Designer *Heidi J. Baughman*
Visuals/Design Freelance Specialist *Barbara J. Hodgson*

 Wm. C. Brown Publishers

President *G. Franklin Lewis*
Vice President, Publisher *George Wm. Bergquist*
Vice President, Operations and Production *Beverly Kolz*
National Sales Manager *Virginia S. Moffat*
Group Sales Manager *Vincent R. Di Blasi*
Vice President, Editor in Chief *Edward G. Jaffe*
Executive Editor *Earl McPeek*
Marketing Manager *Elizabeth Robbins*
Advertising Manager *Amy Schmitz*
Managing Editor, Production *Colleen A. Yonda*
Manager of Visuals and Design *Faye M. Schilling*
Production Editorial Manager *Julie A. Kennedy*
Production Editorial Manager *Ann Fuerste*
Publishing Services Manager *Karen J. Slaght*

WCB Group

President and Chief Executive Officer *Mark C. Falb*
Chairman of the Board *Wm. C. Brown*

Cover photo © A. M. Rosario/The Image Bank
All line art prepared by PC&F, Inc.

Copyright © 1992 by Wm. C. Brown Publishers. All rights reserved

Library of Congress Catalog Card Number: 90–85163

ISBN 0-697-05970-7

No part of this publication may be reproduced, stored in a retrieval system, or transmitted, in any form or by any means, electronic, mechanical, photocopying, recording, or otherwise, without the prior written permission of the publisher.

Printed in the United States of America by Wm. C. Brown Publishers, 2460 Kerper Boulevard, Dubuque, IA 52001

10 9 8 7 6 5 4 3 2 1

Contents

		Preface	xi
CHAPTER 1		**The Real Number System**	**1**
	1.1	Numbers, Symbols, and Operations	2
	1.2	Properties of Real Numbers	10
	1.3	Exponents	14
	1.4	Roots and Radicals	23
	1.5	Rational Exponents	31
	1.6	Addition and Subtraction of Polynomials	33
	1.7	Multiplication of Polynomials	36
	1.8	Division of Polynomials	39
	1.9	Solving Linear Equations	43
	1.10	Ratio and Proportion	48
	1.11	Variation	56
	1.12	Summary of Terms, Formulas, Rules, and Procedures	62
	1.13	Chapter 1 Review Exercises	64
	1.14	Chapter 1 Test	68
CHAPTER 2		**Basic Algebraic Skills**	**69**
	2.1	Special Products	70
	2.2	Factoring	74
	2.3	Algebraic Fractions	85
	2.4	Multiplication and Division of Fractions	91
	2.5	Addition and Subtraction of Fractions	95
	2.6	Fractional Equations	101
	2.7	Summary of Terms, Rules, and Procedures	105
	2.8	Chapter 2 Review Exercises	106
	2.9	Chapter 2 Test	108
	2.10	Cumulative Review—Chapters 1 and 2	109

CHAPTER 3 Graphing — 111

3.1	The Coordinate Plane	112
3.2	Graphing Linear Equations	116
3.3	Graphing Quadratic Equations	123
3.4	Graphing Nonlinear Equations	127
3.5	Functions	130
3.6	Summary of Terms, Formulas, Rules, and Procedures	135
3.7	Chapter 3 Review Exercises	136
3.8	Chapter 3 Test	137

CHAPTER 4 Systems of Linear Equations — 139

4.1	Introduction: Linear Equations	140
4.2	Solving Systems of Two Linear Equations by Graphing	142
4.3	Solving Systems of Two Linear Equations by Substitution	145
4.4	Solving Systems of Two Linear Equations by Addition or Subtraction	150
4.5	Solving Systems of Two Linear Equations by Determinants	156
4.6	Solving Systems of Three Linear Equations in Three Unknowns	161
4.7	Summary of Terms, Formulas, Rules, and Procedures	168
4.8	Chapter 4 Review Exercises	170
4.9	Chapter 4 Test	171

CHAPTER 5 Solving Quadratic Equations — 173

5.1	Solving Quadratic Equations by Graphing	174
5.2	Solving Quadratic Equations by Factoring	177
5.3	Solving Quadratic Equations by Completing the Square	181
5.4	Solving Quadratic Equations with the Quadratic Formula	186
5.5	Solving Quadratic Inequalities	192
5.6	Summary of Terms, Formulas, Rules, and Procedures	196
5.7	Chapter 5 Review Exercises	197
5.8	Chapter 5 Test	198
5.9	Cumulative Review—Chapters 3, 4, and 5	198

CHAPTER 6 Numerical Trigonometry — 201

6.1	Angles and Triangles	202
6.2	The Tangent Ratio	212
6.3	Trigonometric Ratios	220
6.4	The Law of Sines	229

6.5	The Law of Cosines	238
6.6	Summary of Terms, Formulas, Rules, and Procedures	246
6.7	Chapter 6 Review Exercises	248
6.8	Chapter 6 Test	250

CHAPTER 7 Analytic Trigonometry — 253

7.1	Radian Measure	254
7.2	The Circular Functions	262
7.3	The Graphs of the Sine and Cosine Functions	273
7.4	The Graphs of $y = a\sin(bt + c)$ and $y = a\cos(bt + c)$	286
7.5	The Graphs of the Other Trigonometric Functions	294
7.6	Composite Curves	301
7.7	Inverse Trigonometric Functions	308
7.8	Summary of Terms, Formulas, Rules, and Procedures	318
7.9	Chapter 7 Review Exercises	324
7.10	Chapter 7 Test	326

CHAPTER 8 Trigonometric Identities and Equations — 329

8.1	Trigonometric Identities	330
8.2	Special Formulas for the Cosine Function	336
8.3	Special Formulas for the Sine and Tangent Functions	346
8.4	Sums, Differences, and Products of Sines and Cosines	357
8.5	Trigonometric Equations	363
8.6	Summary of Terms, Formulas, Rules, and Procedures	370
8.7	Chapter 8 Review Exercises	372
8.8	Chapter 8 Test	374
8.9	Cumulative Review—Chapters 6, 7, and 8	375

CHAPTER 9 Vectors and Complex Numbers — 377

9.1	Vectors and Vector Operations	378
9.2	Operations with Complex Numbers	388
9.3	Geometric Representation of Complex Numbers	395
9.4	Powers and Roots of Complex Numbers	403
9.5	Summary of Terms, Formulas, Rules, and Procedures	409
9.6	Chapter 9 Review Exercises	410
9.7	Chapter 9 Test	412

CHAPTER 10 — Exponential and Logarithmic Functions — 413

10.1	Exponential Functions	414
10.2	Logarithmic Functions	419
10.3	Properties of Logarithms	423
10.4	Natural Logarithms and Common Logarithms	428
10.5	Exponential and Logarithmic Equations	431
10.6	Summary of Terms, Formulas, Rules, and Procedures	436
10.7	Chapter 10 Review Exercises	437
10.8	Chapter 10 Test	439

CHAPTER 11 — Some Nonlinear Equations — 441

11.1	Graphic Solutions of a System of Equations	442
11.2	Algebraic Solutions of a System of Equations	446
11.3	Radical Equations	452
11.4	Summary of Terms, Rules, and Procedures	456
11.5	Chapter 11 Review Exercises	457
11.6	Chapter 11 Test	458
11.7	Cumulative Review—Chapters 9, 10, and 11	458

CHAPTER 12 — Inequalities — 461

12.1	Linear Inequalities	462
12.2	Nonlinear Inequalities	468
12.3	Absolute Value Equations and Inequalities	475
12.4	Linear Inequalities in Two Variables	482
12.5	Graphical Linear Programming	487
12.6	Summary of Terms, Rules, and Procedures	495
12.7	Chapter 12 Review Exercises	496
12.8	Chapter 12 Test	497

CHAPTER 13 — Matrices and Determinants — 499

13.1	Addition and Scalar Multiplication of Matrices	500
13.2	Solving Systems of Equations Using Matrices	507
13.3	Multiplication of Matrices	517
13.4	The Inverse of a Square Matrix	529
13.5	The Determinant of a Square Matrix	538
13.6	Properties of Determinants	542

13.7	Summary of Terms, Rules, and Procedures	549
13.8	Chapter 13 Review Exercises	550
13.9	Chapter 13 Test	552
13.10	Cumulative Review—Chapters 12 and 13	553

CHAPTER 14 Theory of Equations 555

14.1	Introduction	556
14.2	Synthetic Division	557
14.3	Rational Roots of Polynomial Equations	564
14.4	Approximate Roots of Polynomial Equations	575
14.5	Summary of Terms, Rules, and Procedures	580
14.6	Chapter 14 Review Exercises	581
14.7	Chapter 14 Test	582

CHAPTER 15 Sequences and Series 583

15.1	Sequences	584
15.2	Series	589
15.3	Arithmetic Progressions	592
15.4	Geometric Progressions	596
15.5	Infinite Series	601
15.6	The Binomial Theorem	605
15.7	Summary of Terms, Formulas, Rules, and Procedures	612
15.8	Chapter 15 Review Exercises	613
15.9	Chapter 15 Test	614
15.10	Cumulative Review—Chapters 14 and 15	615

CHAPTER 16 Analytic Geometry 617

16.1	Introduction	618
16.2	Linear Equations	619
16.3	The Circle	627
16.4	The Parabola	632
16.5	The Ellipse	643
16.6	The Hyperbola	653
16.7	Polar Equations	664
16.8	Summary of Terms and Formulas	672
16.9	Chapter 16 Review Exercises	676
16.10	Chapter 16 Test	677

CHAPTER 17 Statistics — 679

17.1	Organizing Data	680
17.2	Measures of Central Tendency	690
17.3	Measures of Dispersion	699
17.4	The Least Squares Line	711
17.5	Nonlinear Curve Fitting	720
17.6	Summary of Terms, Formulas, Rules, and Procedures	729
17.7	Chapter 17 Review Exercises	731
17.8	Chapter 17 Test	733
17.9	Cumulative Review—Chapters 16 and 17	735

CHAPTER 18 Differential Calculus — 737

18.1	The Limit of a Function	738
18.2	Continuity	742
18.3	Limits and Infinity	745
18.4	The Slope of a Tangent Line to a Curve	752
18.5	Definition of Derivative	756
18.6	Rules for Finding Derivatives	759
18.7	The Chain Rule	768
18.8	The Derivative as a Rate of Change	772
18.9	Differentials	776
18.10	Implicit Differentiation and Higher-Order Derivatives	780
18.11	Summary of Terms and Theorems	786
18.12	Chapter 18 Review Exercises	787
18.13	Chapter 18 Test	789

CHAPTER 19 Derivatives of Transcendental Functions — 791

19.1	Derivatives of Logarithmic Functions	792
19.2	Derivatives of Exponential Functions	797
19.3	Derivatives of Trigonometric Functions	801
19.4	Derivatives of Inverse Trigonometric Functions	807
19.5	Summary of Terms, Rules, and Theorems	813
19.6	Chapter 19 Review Exercises	814
19.7	Chapter 19 Test	816

CHAPTER 20 — Applications of Derivatives — 817

20.1	Tangent Lines and Normal Lines	818
20.2	The First Derivative Test	821
20.3	The Second Derivative Test	827
20.4	Curve Sketching	832
20.5	Maxima and Minima Problems	838
20.6	Velocity and Acceleration	842
20.7	Related Rates	845
20.8	Summary of Terms and Rules	849
20.9	Chapter 20 Review Exercises	849
20.10	Chapter 20 Test	851
20.11	Cumulative Review—Chapters 18, 19, and 20	852

CHAPTER 21 — Integral Calculus — 855

21.1	Antiderivatives	856
21.2	Basic Rules	861
21.3	Variations of the Power Rule	866
21.4	Indefinite and Definite Integrals	872
21.5	Integrals Involving Exponential and Logarithmic Functions	877
21.6	Integrals of the Trigonometric Functions	882
21.7	Summary of Terms and Rules	890
21.8	Chapter 21 Review Exercises	892
21.9	Chapter 21 Test	894

CHAPTER 22 — Techniques of Integration — 895

22.1	Integration by Parts	896
22.2	Trigonometric Substitution	901
22.3	Partial Fractions	909
22.4	Integral Tables	916
22.5	Summary of Terms and Rules	922
22.6	Chapter 22 Review Exercises	925
22.7	Chapter 22 Test	926

CHAPTER 23	**Applications of Integration**	**927**
23.1	The Area under a Curve	928
23.2	The Area between Two Curves	936
23.3	The Volume of a Solid of Revolution by Cylindrical Solids	941
23.4	The Volume of a Solid of Revolution by Cylindrical Shells	947
23.5	Work	952
23.6	Centroids	955
23.7	Moments of Inertia	960
23.8	Other Applications	964
23.9	Summary of Terms and Rules	973
23.10	Chapter 23 Review Exercises	974
23.11	Chapter 23 Test	976
23.12	Cumulative Review—Chapters 21, 22, and 23	977
	Answer Key	981
	Index	1137

Preface

A solid background in mathematics is fundamental for success in many technical fields. This background involves a knowledge of algebra, an understanding of elementary functions and their properties, a working knowledge of trigonometry and trigonometric functions and their properties, and, for some technical courses, an introduction to analytic geometry and calculus. This book is designed for students in technical programs that require an understanding of these basic mathematical concepts. The topics included and their treatment reflect the opinions of a broad spectrum of experts in the technical mathematics field.

Text Organization

This book has the flexibility to be used in many different contexts. Chapters 1 and 2 provide a review of basic algebra for those students with limited mathematical preparation. An instructor who is teaching a course designed for these students may want to proceed at a relatively slow pace through these two chapters. For better-prepared students, a faster pace will provide a quick review of basic algebraic concepts.

Chapters 3, 4, 5, 10, 11, and 12 develop those algebraic concepts that are necessary for technical courses. Chapters 6, 7, 8, and 9 develop the basic concepts of trigonometry, vectors, and complex numbers.

The trigonometry chapters are placed together so trigonometry can be treated as a cohesive unit by those instructors who desire to develop it in this manner. However, those instructors who wish to treat each trigonometry chapter independently may do so as long as the prerequisites are observed (see chart).

Chapter 13 (Matrices and Determinants), chapter 14 (Theory of Equations), chapter 15 (Sequences and Series), chapter 16 (Analytic Geometry), and chapter 17 (Statistics) provide supplementary material for courses in which these topics are appropriate.

Chapters 18, 19, 20, 21, 22, and 23 provide the essential concepts from differential and integral calculus that are necessary for more advanced technical courses. The theory of calculus is kept to a minimum, but important rules, formulas, and theorems are carefully developed. A wide variety of technical applications of the calculus are presented in each chapter.

The following chart indicates which chapters are **review chapters,** which are **core chapters,** and which are **optional.** The numbers in parentheses indicate the **prerequisite chapters.**

Review	Core	Optional
1	3	13(5)
2	4(3)	14(5)
	5(4)	17(5)
	6(5)	
	7(6)	
	8(7)	
	9(8)	
	10(5)	
	11(10)	
	12(11)	
	15(5)	
	16(8)	
	18(16)	
	19(18)	
	20(19)	
	21(20)	
	22(21)	
	23(22)	

Approach

The book is written in an informal style. The use of extensive lists of axioms and formal proofs is kept at a minimum. Some proofs are included to enhance the students' understanding of the materials. In early chapters of the text considerable step-detail and clarifying step-comments are included in the examples. These details save time for the reader, review basic procedures, and give the students confidence as they proceed through the early chapters of the text. The amount of step-detail is gradually decreased in later sections in an effort to encourage the student to think independently.

Chapter Introductions

Most chapters are introduced with a brief historical sketch or with nontrivial applications that suggest the past and current relevance or value of the material in the chapter. In addition to motivating students, the introductions provide students with important historical background information on key figures in the development of algebraic, trigonometric, and calculus concepts.

Examples

The text contains approximately 700 examples that help to develop the concepts and motivate the students. These examples are selected from many trade, vocational, technical, and industrial areas.

Trial Problems

Before students attempt section exercise sets, they can practice using the skills and concepts just learned by working the **Trial Problems.** Three to six problems precede each section exercise set. Answers are given in the Answer Key to ensure that students have learned the correct method of solving each type of problem. Trial problems can also be used as in-class examples for discussion or as informal quizzes to assess student understanding of specific concepts and skills.

Exercises

There are approximately 6300 exercises in the text. More than 4400 of these are in the section exercise sets. The exercise sets are designed both to provide practice in routine calculations and to challenge the students. The first portion of each exercise set allows the student to practice the concepts that are developed in the section. The next portion applies the concepts learned to situations from the industrial and technical world. Some of these applications are word problems and are intended to develop problem-solving strategies, while others extend ideas that were developed in the section. Many of these applications are intended to be a challenge to the student, but are deliberately not identified as such, so that neither the student nor the instructor is prejudiced. More exercises than are necessary for understanding the concepts are included in each set. This, together with the fact that the even-numbered exercises are duals of the odd-numbered exercises, provides flexibility in making assignments. The answers to the odd-numbered exercises appear at the end of the book. The answers to all exercises are provided in the instructor's manual.

Applications

Nontrivial applications are used extensively in the examples and exercises, not only to enhance the development of the mathematical concepts, but also to relate the mathematics to the real world. The applications are chosen so that neither the student nor the instructor needs to have an extensive background in a technical field to understand them. There are more than 1100 applications in the exercise sets of the text and more than 100 applications in the examples throughout the text.

Boxed Formulas, Rules, and Procedures

Rules, properties, procedures, and strategies are boxed for easy identification. They are also listed in the chapter summary and are keyed to the section in which they were first developed.

Calculators

This book is written under the assumption that the students have access to scientific calculators. The development of these calculators and the recent reduction in their cost have made them a necessity for students in technical fields. Many of the solutions to the examples are followed by a listing of the calculator keystrokes needed to solve the problem. Calculators allow the exclusion of tables of trigonometric function values, logarithmic tables, and tables for the values of exponential functions. They also allow the inclusion of exercises and applications that do not have to be oversimplified to fit paper-and-pencil calculations. Those exercises in which the use of a calculator is specifically recommended are identified by a calculator icon.

Chapter Summaries

A chapter summary is included at the end of each chapter. The summary identifies the terms, formulas, rules, and procedures that were introduced in the chapter. To enhance student review, these are each identified according to the page or section in which they first appeared.

Review Exercises

There are nearly 1200 review exercises in the text. Of these, nearly 200 are real-world applications. A set of review exercises is placed at the end of each chapter to provide the students with practice problems in a setting outside of the particular section in which the concepts were introduced. The sections to which odd-numbered review exercises apply are identified in the Answer Key.

Chapter Tests

There are over 400 chapter test items in the text. A chapter test is provided at the end of each chapter. These tests are suitable for a 50 minute period. All of the chapter test answers in the Answer Key are coded to the section number to which the problem applies.

Cumulative Reviews

There are nearly 350 cumulative review exercises in the text. Cumulative review exercises are placed at the end of chapters 2, 5, 8, 11, 13, 15, 17, 20, and 23. These sets of exercises provide the students with practice problems on the mathematical concepts outside the chapter in which they were introduced. All answers and chapter reference numbers are included in the Answer Key.

For the Instructor

Numerous ancillary materials are available for instructors who adopt this text. These include:

- **Instructor's Manual,** containing an introduction to the manual, several suggested course plans with reproducible final exams for each plan, three reproducible tests for each chapter, blank answer forms for the tests, several forms for final exams, answers for all of the reproducibles, and answers for the even-numbered exercises in *Applied Technical Mathematics with Calculus*.

- **wcb TestPak,** a computerized testing service, provides instructors with either a mail-in/call-in testing program or the complete test item file on diskette for use with the IBM PC, Apple, or Macintosh computer. **wcb** TestPak requires no programming experience. Tests can be generated randomly, by selecting specific test items, or by objective. In addition, new test items can be added and existing test items can be edited.

- **wcb QuizPak,** a part of TestPak 3.0, which provides students with true/false, multiple choice, and matching questions for each chapter in the text. Using this portion of the program will help students to prepare for examinations. Also included with the **wcb** QuizPak is an on-line testing option, which allows professors to prepare tests for students to take using the computer. The computer will automatically grade the test and update the gradebook file.

- **wcb GradePak,** also a part of TestPak 3.0, is a computerized grade management system for instructors. This program tracks student performance on examinations and assignments. It will compute each student's percentage and corresponding letter grade, as well as the class average. Printouts can be made utilizing both text and graphics. The items in the Test Item File and Quiz Item File are different from the items in the prepared tests in the Instructor's Manual. Hence, you will have even more items to choose from for your tests.

- **Test Item File/Quiz Item File,** a printed version of the computerized testing service, which allows instructors to choose test items based on chapter, section, or objective.

For the Student

For students who use this text, a student study guide with solutions to selected problems is available. In addition to providing extra practice with the key concepts presented in *Applied Technical Mathematics with Calculus,* comprehensive chapter summaries and interesting extension problems are also included. All answers to additional problems are provided in the appendix.

Acknowledgments

A text of this magnitude is the combined effort of many people. Our immediate families have been a positive influence on each of us. Our colleagues have offered their constructive criticism and encouragement and our students have been enthusiastic about the development of the text as it evolved from a rough draft to its present form. Specific mention must be made of those people who carefully reviewed the manuscript at its various stages of development and made positive comments for its improvement. These people are listed below.

G. Don Benson	Virginia Western Community College
Glenn R. Boston	Catawba Valley Community College
Debra Lee Gallo	Kent State University–Salem Campus
Patricia L. Hirshy	Delaware Technical & Community College
Maryann E. Justinger	Erie Community College, South
Jon C. Luke	Indiana University–Purdue University at Indianapolis
Marcel Maupin	O.S.U. Technical Branch
Marilyn L. Peacock	Tidewater Community College
Phyllis R. Schott	Catonsville Community College
Darrel H. Schwartz	St. Cloud Technical College
Arlene M. Sherburne	Montgomery College
Ralph W. Spence	Columbus Technical Institute (Retired)
James Wolfe	Hocking Technical College
Kelly Wyatt	Umpqua Community College

Merwin J. Lyng

L. J. Meconi

Earl J. Zwick

Chapter 1

The Real Number System

Two thousand years ago the systems of numeration pictured to the right were adequate for the society of the day. However, today's technological society demands a much greater understanding of numbers.

In chapter 1, the real numbers, the operations on these numbers, and the resulting properties are studied. This knowledge is basic in the technological and industrial world where mathematics is used.

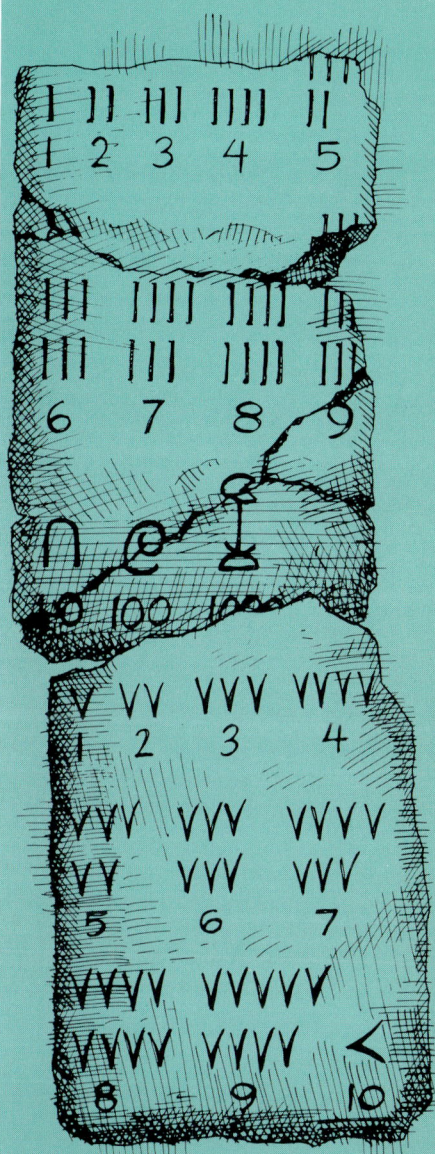

CHAPTER 1 THE REAL NUMBER SYSTEM

1•1 Numbers, Symbols, and Operations

In the scientific, technical, and industrial world, many kinds of numbers are used for many purposes. Probably the most fundamental use of numbers is that of counting.

Natural Numbers
Counting Numbers
Positive Integers

The numbers used for counting are called **natural numbers** and are represented by the symbols 1, 2, 3, 4, 5, ... These numbers are also called **counting numbers** or **positive integers.**

Notice that zero is not included in the list of natural numbers. If we include zero, the symbols 0, 1, 2, 3, . . . represent **whole numbers.**

Whole Numbers

The whole numbers together with the "opposites" of the counting numbers, or negative integers (. . . −4, −3, −2, −1), make up the set of numbers we call the **integers.** The integers can be symbolized as . . . −4, −3, −2, −1, 0, 1, 2, 3, 4. . . .

Integers

Figure 1.1 illustrates how the positive integers, negative integers, and zero combine to form the integers and how these integers can be placed on a number line.

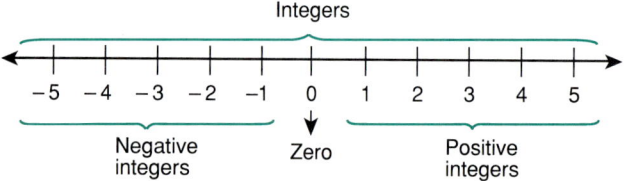

Figure 1.1

The negative integers are necessary when we discuss such things as temperature below zero, distance below sea level, deficit spending, or the time before blast-off in rocket launches.

Any number that can be expressed in the form $\dfrac{p}{q}$, where p and q are both integers and q is not zero, is called a **rational number.** The rational numbers can also be represented as **repeating** or **terminating decimals,** as is illustrated in examples 1.1 and 1.2.

Rational Number
Repeating Decimals
Terminating Decimals

• *Example 1.1:*   Express $\dfrac{3}{7}$ as a repeating decimal.

Solution: $\dfrac{3}{7} = 3 \div 7 = 0.428571428571. \ldots$

In this example the sequence of digits 428571 is repeating.

Section 1.1 Numbers, Symbols, and Operations

- *Example 1.2:*   Express $\dfrac{1}{2}$ as a terminating decimal.

  Solution: $\dfrac{1}{2} = 1 \div 2 = 0.5$.

  Note that the terminating decimal 0.5 can be expressed as a repeating decimal as 0.5000 . . . with zeros repeating after the 5.

Hence, rather than saying that the rational numbers can be expressed as repeating *or* terminating decimals, we could simply say that the rational numbers can be expressed as repeating decimals.

It is sometimes convenient to use a bar over those digits that repeat rather than to use the three dots as we have done in the examples. Then 0.428571428571 . . . becomes $0.\overline{428571}$ and 0.5 . . . becomes $0.5\overline{0}$.

Since integers can be expressed as rational numbers and rational numbers can be expressed as repeating decimals, it must be concluded that integers can also be expressed as repeating decimals. It may not be apparent that equations such as

$$1.0 = 0.999 \ldots$$
$$2.0 = 1.999 \ldots$$
$$3.0 = 2.999 \ldots$$

are true. However, this must be the case as is demonstrated in example 1.3.

- *Example 1.3*   Show that $1.0 = 0.999. \ldots$

  Solution: If it is agreed that $\dfrac{1}{3} = 0.333 \ldots$, then it must be agreed that $\dfrac{3}{3} = 0.999. \ldots$ (Multiply both sides by 3.) But since $\dfrac{3}{3} = 1.0$, it follows that

  $$1.0 = 0.999. \ldots$$

  By adding 1, 2, 3, etc., to both sides of this last equation, we get

  $$2.0 = 1.999 \ldots$$
  $$3.0 = 2.999 \ldots$$
  $$4.0 = 3.999 \ldots, \text{ etc.}$$

4 CHAPTER 1 THE REAL NUMBER SYSTEM

 Knowing that integers can be expressed as repeating decimals, as described previously, is sometimes useful when interpreting the results of calculations done on hand-held calculators. Do the following keystrokes on different hand-held calculators.

You will note that on some calculators the display will indicate 0.999 . . . and on others the display will indicate 1.0. You should be aware that in this case both of these answers represent the same value.

Irrational Numbers

Real Numbers

The rational numbers can be expressed as repeating decimals. Consequently, the nonrepeating decimals are identified as **irrational numbers.** Numerals such as $\sqrt{2}$, $\sqrt{3}$, and π represent irrational numbers. Now if we combine the rational numbers and the irrational numbers we have created the **real numbers.** Figure 1.2 illustrates the development of the real numbers from the natural numbers.

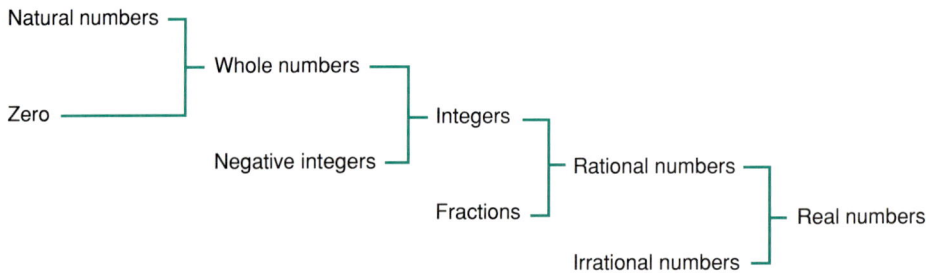

Figure 1.2

We can represent numbers as points on a number line. Figure 1.3 illustrates a number line with various numbers represented as points on the line. When we represent a number as a point on a line we say that the point is a **graph** of the number.

Graph

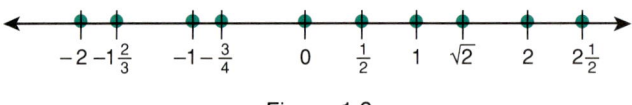

Figure 1.3

Real Number Line

Every real number can be represented as a point on such a line and every point on such a line can be represented by a real number. We say that the real numbers can be placed in one-to-one correspondence with the points on this line. Therefore, we refer to this line as the **real number line.**

The real numbers are adequate for solving all of the problems we encounter in the first 11 chapters of this book. In chapter 12 we extend the real number system to form the complex number system.

SECTION 1.1 NUMBERS, SYMBOLS, AND OPERATIONS   5

Literal Number
Variables
Constants

Is Less Than
Is Greater Than
Is Equal To

When dealing with numbers it is often necessary to represent the numbers with letters from our alphabet or with symbols from other alphabets such as the Greek alphabet. A number represented by a letter is called a **literal number.** Literal numbers that may vary in a given situation are called **variables.** Literal numbers that do not vary in a given problem are called **constants.**

It is also useful to be able to discuss how two numbers are related. If a and b are two numbers, and a is to the left of b on the number line, then a **is less than** b and is symbolized by $a < b$. If a is to the right of b on the number line, then a **is greater than** b and we write $a > b$. Of course, if a and b represent the same point, then a **is equal to** b and we write $a = b$. The following examples illustrate this concept.

$2 < 3$ since 2 is to the left of 3 on the real number line.

$\frac{1}{4} < \frac{3}{4}$ since $\frac{1}{4}$ is to the left of $\frac{3}{4}$ on the real number line.

$-\frac{1}{3} > -\frac{1}{2}$ since $-\frac{1}{3}$ is to the right of $-\frac{1}{2}$ on the real number line.

$0.5 < 0.6$ since 0.5 is to the left of 0.6 on the real number line.

$\frac{7}{8} > \frac{3}{4}$ since $\frac{7}{8}$ is to the right of $\frac{3}{4}$ on the real number line.

$0.67 > \frac{2}{3}$ since 0.67 is to the right of $\frac{2}{3}$ (which is $0.\overline{6}$) on the real number line.

$-2 > -3$ since -2 is to the right of -3 on the real number line.

$\sqrt{5} > 2.23$ since $\sqrt{5}$ (which is a little more than 2.236) is to the right of 2.23 on the real number line.

$0.\overline{1} = \frac{1}{9}$ since $0.\overline{1}$ is the repeating decimal that is equal to the rational number $\frac{1}{9}$.

You may have noticed that many of the numerical expressions that have occurred in this section could have been approximated on a hand-held calculator. In this section, and in the remainder of this text, you are encouraged to use your calculator. Those problems in which the use of a hand-held calculator is specifically recommended are identified by the symbol . You should consult the manual that accompanies your calculator for specific operating instructions.

If the numbers that are being used in a calculation are the result of some process of measurement, they are usually approximations. When numbers have been arrived at as a result of a counting process or from a definition, the numbers are said to be

Significant Digits

exact. When calculating with approximate numbers, there are three terms that are important to understand. These are accuracy, precision, and **significant digits** (or significant figures). A digit is *significant* if it satisfies any of the following rules.

1. All nonzero digits are significant. For example, the number 4.57 contains three significant digits.

2. All zeros between significant digits are significant. The number 3.082 contains four significant digits.

3. Leading zeros are never significant. The number 0.0037 contains only two significant digits.

4. Trailing zeros are significant only if they lie to the right of a decimal point. For example, the numbers 3.60, 0.400, and 0.00820 each have three significant digits.

5. Trailing zeros are not significant on integers. Hence, the numbers 8, 70, and 200 each have one significant digit. But if trailing zeros on integers are meant to be significant, this can be indicated by writing the integer in scientific notation. For example, the integer 60,000, when written as 6.00×10^4, means that it has three significant digits.

Accuracy
Precision

The **accuracy** of a number is defined as the number of significant digits it contains. The **precision** of a number refers to the decimal position of the last significant digit. Thus, 43.9 is accurate to three significant digits, but is precise to only one decimal place (tenths). Whereas, the number 0.007 is accurate to only one significant digit, but is precise to three decimal places (thousandths).

When performing calculations that involve more than one of the basic operations, certain rules for the **order of operations** must be followed. These are

Order of Operations

ORDER OF OPERATIONS

1. Perform the operations within grouping symbols.

2. Perform all of the multiplications and divisions, working from left to right.

3. Perform all of the additions and subtractions, working from left to right.

Absolute Value

When operating with real numbers it is often necessary to deal with the **absolute value** of a number. The absolute value of a number is the distance of that number from zero on the number line. If a number is positive, the absolute value is the number itself. If the number is negative, the absolute value is the opposite of that number. The opposite of a number is obtained by changing its sign. Thus, the absolute value of 5, which can be symbolized as $|5|$, is 5. The absolute value of $-5 = |-5| = -(-5) = 5$ (that is, the opposite of -5). Note that the absolute value of a real number is never negative. Consider the following examples.

$$|3| = 3, \quad |-2| = -(-2) = 2, \quad |5 - 8| = |-3| = -(-3) = 3,$$
$$|9 - 3| = |6| = 6, \quad -|4| = -4, \quad -|-8| = -(8) = -8.$$

SECTION 1.1 NUMBERS, SYMBOLS, AND OPERATIONS   **7**

The laws that govern the addition, subtraction, multiplication, and division of real numbers are of importance in all of our future work with these numbers. These laws are stated and illustrated with the following statements and examples.

> To add real numbers whose signs are alike, add the numbers and use the sign of the original numbers.

For example, $(+8) + (+4) = +12$, or $(-12) + (-4) = -16$.

> To add real numbers whose signs are unlike, subtract the smaller absolute value from the larger, and attach the sign of the number with the larger absolute value.

For example, $(-8) + (+3) = -(8 - 3) = -5$ (-8 has larger absolute value)
and $(+8) + (-3) = +(8 - 3) = +5$.   ($+8$ has larger absolute value)

> To subtract one real number from another, change the sign of the second number and proceed as in addition.

For example, $(-8) - (-3) = (-8) + (+3) = -5$ (Change the sign and add)
and $(-2) - (-6) = (-2) + (+6) = +4$.   (Change the sign and add)

MULTIPLICATION RULES FOR REAL NUMBERS

1. If the signs of all of the numbers are positive, multiply the numbers and attach a positive sign to the answer.
2. If there is an odd number of negative numbers to be multiplied, multiply the numbers and attach a negative sign to the answer.
3. If there is an even number of negative numbers to be multiplied, multiply the numbers and attach a positive sign to the answer.

For example, $(+3) \times (+6) = +18$ (All numbers are positive)
$(+3) \times (-8) = -24$ (Odd number of negative numbers)
$(-9) \times (-4) = +36$.   (Even number of negative numbers)

The rules for dividing two real numbers are similar to those for multiplying real numbers.

8 CHAPTER 1 THE REAL NUMBER SYSTEM

> When dividing two real numbers, the answer is positive if both numbers have the same sign. The answer is negative if the two numbers have opposite signs.

For example, $(+8) \div (+2) = +4,$
$(-6) \div (-3) = +2,$
$(-12) \div (+4) = -3,$
$(+14) \div (-2) = -7.$

Trial Problems 1.1

Before you begin the section exercises, warm up with these problems. Complete answers are included in the answer key.

Place the proper symbol, $<$, $>$, or $=$, between the two numbers.

1. $-5 \quad -2$
2. $\dfrac{3}{4} \quad \dfrac{5}{8}$
3. $4.00\ldots \quad 3.99\ldots$

Simplify.

4. $(-3) + (-5)$
5. $(-12) \div (-3)$

Exercises 1.1

In exercises 1–50 place the correct symbol ($<$, $>$, or $=$) between the two numbers.

1. $3 \quad 5$
2. $8 \quad 9$
3. $-5 \quad -2$
4. $-6 \quad -3$
5. $-2 \quad 9$
6. $-3 \quad 8$
7. $\dfrac{3}{4} \quad \dfrac{7}{4}$
8. $\dfrac{3}{5} \quad \dfrac{4}{5}$
9. $35 \quad 32$
10. $38 \quad 34$
11. $52 \quad -60$
12. $57 \quad -50$
13. $-\dfrac{2}{3} \quad 4$
14. $-\dfrac{5}{8} \quad 3$
15. $\dfrac{3}{4} \quad -1$
16. $\dfrac{7}{8} \quad -2$
17. $\dfrac{9}{10} \quad 1$
18. $\dfrac{1}{2} \quad 3$
19. $\dfrac{3}{4} \quad \dfrac{5}{8}$
20. $\dfrac{2}{3} \quad \dfrac{5}{6}$
21. $\dfrac{2}{3} \quad 0.66$
22. $\dfrac{1}{3} \quad 0.33$
23. $0.44 \quad 0.444$
24. $0.33 \quad 0.333$
25. $\dfrac{1}{3} \quad 0.333\ldots$
26. $\dfrac{2}{3} \quad 0.666\ldots$
27. $2.0 \quad 1.999\ldots$
28. $3.0 \quad 2.999\ldots$
29. $4.0 \quad 3.999\ldots$
30. $5.0 \quad 4.999\ldots$
31. $-\dfrac{2}{3} \quad -\dfrac{3}{4}$
32. $-\dfrac{7}{8} \quad -\dfrac{5}{6}$

| 33. -2.3 $\quad$ -3.2 | 34. -2.4 $\quad$ -4.2 | 35. -0.5 $\quad$ -0.6 | 36. -0.7 $\quad$ -0.8 |

37. 0.66 $\quad$ $0.\overline{6}$ 38. $0.\overline{9}$ $\quad$ 0.99 39. $0.\overline{2}$ $\quad$ $\dfrac{2}{9}$ 40. $0.\overline{3}$ $\quad$ $\dfrac{3}{9}$

41. -0.67 $\quad$ $-\dfrac{2}{3}$ 42. -1.67 $\quad$ $-\dfrac{1}{6}$ 43. $\sqrt{2}$ $\quad$ $\sqrt{3}$ 44. $\sqrt{5}$ $\quad$ $\sqrt{6}$

45. $\sqrt{7}$ $\quad$ 2.64 46. $\sqrt{8}$ $\quad$ 2.82 47. -2.33 $\quad$ -2.333 48. -5.22 $\quad$ -5.222

49. $-\sqrt{5}$ $\quad$ $-\sqrt{6}$ 50. $-\sqrt{3}$ $\quad$ $-\sqrt{2}$

Simplify.

51. $(+2) + (+8)$
52. $(-6) + (-7)$
53. $(-3) - (+5)$
54. $(+7) - (-3)$
55. $(+6) + (+5) + (+3)$
56. $(-9) + (-2) + (-5)$
57. $(+2) - (+3) + (-6) - (-4)$
58. $(-5) - (-2) + (+6) - (+8)$
59. $(+7) \times (-8)$
60. $(-5) \times (+3)$
61. $(-4) \times (-6)$
62. $(-2) \times (-8)$
63. $(+8) \times (-3) \times (-2)$
64. $(-7) \times (+3) \times (-5)$
65. $(+4) \times (-4) \times (-3) \times (+3)$
66. $(-9) \times (+9) \times (+2) \times (-2)$
67. $(+20) \div (-4)$
68. $(-36) \div (+9)$
69. $(-15) \div (-3)$
70. $(-18) \div (-6)$

List in order, starting with the smallest.

71. $-2, 8, \sqrt{6}, |-7|, -|-4|, -12, \dfrac{5}{8}$

72. $\dfrac{3}{4}, 7, |-5|, \sqrt{7}, -|-5|, -13, -6$

Simplify.

73. $\dfrac{(+3)(-8) - 3(-2)}{9 - 2}$
74. $\dfrac{(+7)(-2) - 2(-4)}{2 - 5}$
75. $|-7|$
76. $|-8|$
77. $|+4|$
78. $|+3|$
79. $-|-6|$
80. $-|-3|$
81. $-|+2|$
82. $-|+5|$
83. $|3 - 7|$
84. $|2 - 6|$
85. $-|5 - 9|$
86. $-|7 - 12|$
87. $-|-24 + 32|$
88. $-|-13 + 17|$
89. $-|35 - 27|$
90. $-|42 - 37|$

91. If Pike's Peak is 14,108 feet above sea level and Death Valley is 280 feet below sea level, what is the change in elevation from the top of Pike's Peak to the floor of Death Valley?

92. A rocket launch was last put on hold 15 seconds before blast-off. The rocket was destroyed by the range safety-officer 27 seconds after blast-off. How much time elapsed from the last hold position to the time of destruction?

93. The highest temperature ever recorded in North Dakota was 121 degrees on July 6, 1936. The lowest temperature ever recorded in North Dakota was −60 degrees on February 15, 1936. Find the change in temperature from the high to the low.

94. The driest three-month period in the Twin Cities was in 1894, when 1.72 inches of rain fell. This compares to a normal of 11.22 inches of rain for the same period. How much below normal was the driest three-month period?

95. The quality control department of a factory reduced the number of defective parts from 16 per 1000 to 9 per 1000. Find the reduction and express the result as parts per 1000.

96. The efficiency expert on an assembly line determined that that particular line produced 275 pieces per 1000 man hours. By using robots to do some of the tasks on the line the efficiency was improved to 305 pieces per 1000 man hours. Find the increase in efficiency and express the result in pieces per 1000 man hours.

97. The temperature of a chemical was 21° C. The chemical was cooled to −114° C. What was the change in temperature of the chemical?

98. A defective refrigerator caused the temperature in a storage facility to decrease 2° C per day. Ten days passed before the defect was discovered. If the original temperature was −5° C, what was the temperature when the defect was discovered?

99. The diagonal of a rectangular piece of metal calculates to be $4\sqrt{5}$ inches in length. How does this length compare with 8.95 inches?

100. The diagonal of a square piece of metal calculates to be $5\sqrt{2}$ centimeters. How does this length compare with 7.07 centimeters?

1 • 2 Properties of Real Numbers

The operation of addition on the real numbers and the operation of multiplication on the real numbers result in a variety of properties. These properties are summarized in table 1.1.

ADDITION PROPERTY OF EQUALITY	If $a = b$, then $a + c = b + c$.

This property suggests that we can add the same quantity to both sides of an equation without altering the nature of the equation. For example, if $2 = 2$, then $2 + 3 = 2 + 3$.

MULTIPLICATION PROPERTY OF EQUALITY	If $a = b$, then $ac = bc$.

This expression indicates that we can multiply both sides of an equation by the same quantity without altering the nature of the equation. For example, if $4 = 4$, then $(4)(7) = (4)(7)$. Both of these properties will become useful when we begin to simplify equations.

When we multiply a quantity by zero, the result is zero. This property is called

Zero Factor Property the **zero factor property.**

SECTION 1.2 PROPERTIES OF REAL NUMBERS

TABLE 1.1 AXIOMS FOR REAL NUMBERS

$a + b$ is a unique real number.	Closure for addition
$a + b = b + a$.	Commutative property of addition
$(a + b) + c = a + (b + c)$.	Associative property of addition
ab is a unique real number.	Closure for multiplication
$ab = ba$.	Commutative property of multiplication
$(ab)c = a(bc)$.	Associative property of multiplication
$a(b + c) = (ab) + (ac)$.	Distributive property
There exists a unique number 0 with the property $$a + 0 = a \quad \text{and} \quad 0 + a = a.$$	Identity element for addition
There exists a unique number 1 with the property $$a \cdot 1 = a \quad \text{and} \quad 1 \cdot a = a.$$	Identity element for multiplication
For each real number a, there exists a unique real number $-a$ (called the *negative* of a) with the property $$a + (-a) = 0$$ and $(-a) + (a) = 0$.	Negative, or additive-inverse, property
For each real number a, except 0, there exists a unique real number $\dfrac{1}{a}$ (called the *reciprocal* of a) with the property $$a\left(\frac{1}{a}\right) = 1 \quad \text{and} \quad \left(\frac{1}{a}\right)a = 1.$$	Reciprocal, or multiplicative-inverse, property

As a consequence of these assumptions, the real numbers have other properties.

ZERO FACTOR PROPERTY

$$(a)(0) = 0.$$

Double Negative Property

We can also prove that the negative of a negative of a number is the number itself. This is called the **double negative property.**

DOUBLE NEGATIVE PROPERTY

$-(-a) = a$, where the negative of a is $-a$.

Additive Inverse

The negative of a is also called the **additive inverse** of a.

Similarly, the reciprocal of the reciprocal of a number is the number itself. This can be called the **double reciprocal property.**

Double Reciprocal Property

CHAPTER 1 THE REAL NUMBER SYSTEM

DOUBLE RECIPROCAL PROPERTY	$\dfrac{1}{\frac{1}{a}} = a$, where $\dfrac{1}{a}$ is the reciprocal of a.

Multiplicative Inverse   The reciprocal of a is also called the **multiplicative inverse** of a.

We often define subtraction in terms of addition as follows:

$$a - b \text{ means } a + (-b).$$

For example, $9 - 3$ means $9 + (-3)$ or $(8) - (-5)$ means $(8) + (5)$.

Division (by a nonzero number) can be defined in terms of multiplication as follows.

$$a \div b \text{ means } (a)\dfrac{1}{b}.$$

• **Example 1.4:**   Use the definition of subtraction to express $3 - 5$ as a sum.

     Solution: $3 - 5 = 3 + (-5)$, since (-5) is the negative of 5.

• **Example 1.5:**   Use the definition of division to express $8 \div 3$ as a product.

     Solution: $8 \div 3 = 8\left(\dfrac{1}{3}\right)$ since $\dfrac{1}{3}$ is the reciprocal of 3.

Trial Problems 1.2   Before you begin the section exercises, warm up with these problems. Complete answers are included in the answer key.

Which property, commutative, associative, distributive, identity, or inverse, describes each of the following?

1. $2 + 3 = 3 + 2$
2. $5 + (3 + 4) = (3 + 4) + 5$
3. $a + (b + c) = (a + b) + c$
4. $2(r + s) = 2r + 2s$
5. $3 + (-3) = 0$

Exercises 1.2

In exercises 1–20 fill in the blank to make the statement an application of the stated property.

1. $5 + x = x +$ _____   Commutative property of addition
2. $(2r)s = 2($ _____ $)$ Associative property of multiplication
3. $(2)(4) =$ _____   Closure property of multiplication

4. $(2 + a) + 3 = 2 + (\underline{})$ Associative property of addition
5. $8 + \underline{} = 0$ Additive inverse property
6. $(3)(b) = \underline{}$ Commutative property of multiplication
7. $(5)\left(\dfrac{1}{5}\right) = \underline{}$ Multiplicative inverse property
8. $t + \underline{} = t$ Identity element for addition
9. $a + (b + 2) = (b + 2) + \underline{}$ Commutative property for addition
10. $(w)(\underline{}) = w$ Identity element for multiplication
11. $3(x + y) = \underline{} + \underline{}$ Distributive property
12. $(8)\left(\dfrac{1}{8}\right) = \underline{}$ Commutative property of multiplication
13. If $y = 5$, then $y + 4 = 5 + \underline{}$ Addition property of equality
14. $-(-6) = \underline{}$ Double negative property
15. If $z = 6$, then $-3(z) = \underline{} (6)$ Multiplication property of equality
16. $(12)(\underline{}) = 0$ Zero-factor property
17. $-(-3a) = \underline{} a$ Double negative property
18. If $a = b$, then $a + 5 = b + \underline{}$ Addition property of equality
19. $(2)(3 + 4) = (3 + 4)(\underline{})$ Commutative property of multiplication
20. $x + (y + 3) = x + (3 + \underline{})$ Commutative property of addition

In exercises 21–29 rewrite each subtraction problem as an addition problem and each division problem as a multiplication problem.

21. $12 - 3$
22. $6 - (-7)$
23. $-3 - 5$
24. $-7 - 3$
25. $-4 - (-6)$
26. $3 \div 5$
27. $\dfrac{2}{3} \div 4$
28. $10 \div 2$
29. $10 \div \dfrac{1}{2}$

Simplify by expressing as a single numeral.

30. $3 + 7$
31. $4 + (-9)$
32. $-11 - (-7)$
33. $8 - 3$
34. $-4 + (-12)$
35. $-5 - (-9)$
36. $4 - 3 - 11$
37. $-4 - 3 - (-5)$
38. $8 - (-3) - 2$
39. $-8 - 3 + 2$
40. $7 - (5 - 2)$
41. $(4 - 2 + 7) - 3$

42. $(5 - 9) + (4 - 3)$

43. $(3 - 5 + 4) - (8 - 10)$

44. $(27 + 3 - 6) - (18 - 12 + 1)$

45. $(-5) - \dfrac{-12}{3} - 4(2)$

46. $\dfrac{(-4)(9) - 3(-2)}{3 - 18}$

47. $\dfrac{5 - (-7)}{(-2)(-3)}$

48. $\dfrac{(-6)(30) \div (4 + 5)}{(-5)(4) \div (-6 + 4)}$

49. $0 - 6(-7) + (-5)$

50. $\dfrac{(+3)(-8)(+2)}{0 - 12}$

51. In a certain electrostatic field, the voltage at one point is 1580 volts, and the voltage at another point is -760 volts. Find the potential difference from the first point to the second pont.

52. A paper supply company had an inventory of 1450 cases of #43 continuous feed computer paper. What is the inventory after selling 250 cases and restocking the warehouse with 580 cases?

53. A Brinell gauge reading of 230 is set for testing the hardness of steel plates. Readings above or below this setting are identified by using signed numbers. A reading of -5 means the hardness test indicates a durability of $230 + (-5)$ or 225. What is the durability if the reading on the Brinell gauge is $+35$?

54. The current in amperes for a certain electric circuit is determined by calculating the value of this expression

$$\dfrac{-5 + (-10)(14)}{+11 + (-6)}.$$

What is the current?

1 • 3 Exponents

The addition expression $x + x$ makes sense, but it may be more convenient to write $x + x$ as $2x$. Similarly we abbreviate $x + x + x$ as $3x$ and $x + x + x + x$ as $4x$, etc.

• *Example 1.6:* The formula for the perimeter of a square with side s is expressed as $P = s + s + s + s$ or $P = 4s$.

There are corresponding abbreviations for expressions involving multiplication. We agree to write the expression $x \cdot x$ as x^2 and $x \cdot x \cdot x$ as x^3 and $x \cdot x \cdot x \cdot x$ as x^4. In each of these cases the letter x is called the **base** and the natural numbers 2, 3, and 4 are called **exponents**. In general we write $a^n = a \cdot a \cdot a \cdot a \cdots a$ (n factors), where a is a real number, $a \neq 0$, and n is a positive integer.

Base
Exponents

We read a^n as "a to the nth power" or "a to the nth." The number a^n is called the nth **power** of a.

nth Power

This notation can also be used when the base is a number or a combination of numbers and letters as illustrated in example 1.7.

• *Example 1.7:*
$(5)(5)(5) = 5^3$
$(10)(10) = 10^2$
$(3b)(3b)(3b)(3b) = (3b)^4$
$(r + s)(r + s)(r + s) = (r + s)^3$.

What happens when we multiply two or more numbers having the same base?

- *Example 1.8:* $(4^2)(4^3) = [(4)(4)][(4)(4)(4)] = 4^5$
  $(a^3)(a^4) = (aaa)(aaaa) = a^7.$

  Note that the exponent on the result is the sum of exponents on the two factors.

In general, when we multiply two numbers having the same base, we simply add the exponents and keep the base the same. This can be symbolized by

$$(a)^m(a^n) = a^{m+n}.$$

What happens when we divide two numbers having the same base?

- *Example 1.9:* $\dfrac{4^5}{4^3} = \dfrac{(4)(4)(4)(4)(4)}{(4)(4)(4)} = (4)(4) = 4^2 = 4^{5-3}$

  $\dfrac{a^7}{a^4} = \dfrac{aaaaaaa}{aaaa} = a^3 = a^{7-4}.$

  Note that the exponent on the result is the exponent on the dividend (top number) minus the exponent on the divisor (bottom number).

In general, when we divide two numbers having the same base we subtract the exponent of the divisor from the exponent of the dividend and keep the base the same. This property is called the quotient of like bases. This we symbolize by

$$\frac{a^m}{a^n} = a^{m-n}, \text{ when } m > n.$$

So far we have used exponents that were positive integers. We can expand the properties of exponents to include zero and negative exponents. Consider examples 1.10 and 1.11.

- *Example 1.10:* Compute $\dfrac{a^n}{a^n}$.

  Solution: $\dfrac{a^n}{a^n} = 1$, but $\dfrac{a^n}{a^n} = a^{n-n} = a^0.$

  Therefore we must define

  $a^0 = 1,$

  when $a \neq 0$.

Using the fact that $a^0 = 1$, the following examples must be true.
$$m^0 = 1, \quad (4s)^0 = 1, \quad 4s^0 = 4, \quad 2x^2(3y)^0 = 2x^2.$$
The symbol 0^0 is undefined.

Example 1.11 suggests the meaning of negative exponents.

- **Example 1.11:** Compute $\dfrac{a^3}{a^5}$.

    Solution: $\dfrac{a^3}{a^5} = \dfrac{aaa}{aaaaa} = \dfrac{1}{aa} = \dfrac{1}{a^2}.$

    But, if we subtract exponents when we divide we get
    $$\frac{a^3}{a^5} = a^{3-5} = a^{-2},$$
    therefore $a^{-2} = \dfrac{1}{a^2}.$

In general we define
$$a^{-n} = \frac{1}{a^n}.$$

Note that $a^{-1} = \dfrac{1}{a^1} = \dfrac{1}{a}$. We also know that $a\left(\dfrac{1}{a}\right) = 1$. Hence,
$$a(a^{-1}) = 1.$$

We also note that $(a^{-n})(a^n) = 1$ so a^{-n} is the multiplicative inverse of a^n. Symbolically this can be written
$$a^{-n} = (a^n)^{-1}.$$

Consequently, a^{-n} can be written in several ways.
$$a^{-n} = (a^n)^{-1} = (a^{-1})^n = \left(\frac{1}{a}\right)^n = \frac{1}{a^n}.$$

Examples 1.12 through 1.16 suggest further properties of exponents.

- **Example 1.12:** Compute $(a^4)^3$.

    Solution: $(a^4)^3 = (a^4)(a^4)(a^4) = a^{4+4+4} = a^{12} = a^{(4)(3)}.$

• *Example 1.13:* Compute $(a^4)^{-3}$.

Solution: $(a^4)^{-3} = \dfrac{1}{(a^4)^3} = \dfrac{1}{a^{12}} = a^{-12}$.

Note that the exponent on the result is the product of the two exponents involved. This property is called the power of a power.

In general, we write

$$(a^m)^n = a^{mn}.$$

The property related to the exponent on a product, often referred to as the power of a product, is illustrated by examples 1.14 and 1.15.

• *Example 1.14:* Compute $(ab)^3$.

Solution:
$$(ab)^3 = (ab)(ab)(ab)$$
$$= (aaa)(bbb) = a^{1+1+1}b^{1+1+1}$$
$$= a^3b^3.$$

Note that an expression with no exponent is understood to have an exponent of one (1).

• *Example 1.15:* Compute $(ab)^{-3}$.

Solution: $(ab)^{-3} = \dfrac{1}{(ab)^3} = \dfrac{1}{a^3b^3} = a^{-3}b^{-3}$.

Note that the power on the product can be written as a power on each of the factors in the product.

In general, we write

$$(ab)^n = a^n b^n.$$

The property of exponents related to the power of a quotient is illustrated by example 1.16.

CHAPTER 1 THE REAL NUMBER SYSTEM

• *Example 1.16:* Compute $\left(\dfrac{a}{b}\right)^4$.

Solution:
$$\left(\dfrac{a}{b}\right)^4 = \left(\dfrac{a}{b}\right)\left(\dfrac{a}{b}\right)\left(\dfrac{a}{b}\right)\left(\dfrac{a}{b}\right)$$
$$= \dfrac{aaaa}{bbbb}$$
$$= \dfrac{a^4}{b^4}.$$

Note that the power of a quotient can be written as a power of the dividend and as a power of the divisor separately.

In general, we write
$$\left(\dfrac{a}{b}\right)^n = \dfrac{a^n}{b^n}.$$

We have defined a^n for every integer exponent n when a is any real number except zero. We have also illustrated some of the rules of exponents. For convenience we summarize these rules.

RULES OF EXPONENTS

If a and b are nonzero real numbers and if m and n are integers, then

1. $a^m a^n = a^{m+n}$
2. $\dfrac{a^m}{a^n} = a^{m-n}$
3. $a^{-n} = (a^n)^{-1} = (a^{-1})^n = \left(\dfrac{1}{a}\right)^n = \dfrac{1}{a^n}$
4. $a^0 = 1$
5. $a(a^{-1}) = 1$
6. $(a^m)^n = a^{mn}$
7. $(ab)^n = a^n b^n$
8. $\left(\dfrac{a}{b}\right)^n = \dfrac{a^n}{b^n}$

• *Example 1.17:* Use the rules of exponents to simplify
$$(xy)^2(x^2y^3)^{-1}.$$

Solution 1:
$(xy)^2(x^2y^3)^{-1} = (xy)^2 \dfrac{1}{x^2y^3}$ (Express $(x^2y^3)^{-1}$ as $\dfrac{1}{x^2y^3}$)

$= \dfrac{x^2y^2}{x^2y^3}$ (Express $(xy)^2$ as x^2y^2 and multiply)

$= \dfrac{1}{y}$ (Simplify)

$= y^{-1}.$ (Express with negative exponent if requested to do so)

Solution 2: $(xy)^2(x^2y^3)^{-1} = (x^2y^2)(x^{-2}y^{-3}) = x^{2-2}y^{2-3}$
$= x^0y^{-1} = y^{-1}.$

• **Example 1.18:** Express $\dfrac{(2^{-3})(16^5)}{(4^3)(32)}$ as a power of 2.

Solution: $\dfrac{(2^{-3})(16^5)}{(4^3)(32)} = \dfrac{(2^{-3})(2^4)^5}{(2^2)^3(2^5)}$ (Express 16 as 2^4, 4^3 as $(2^2)^3$ and 32 as 2^5)

$= \dfrac{(2^{-3})2^{20}}{(2^6)(2^5)}$ (Apply rules of exponents)

$= \dfrac{2^{17}}{2^{11}}$ (Apply rules of exponents)

$= 2^6.$ (Simplify)

• **Example 1.19:** Use the rules of exponents to compute $\dfrac{(2^6)(5^7)}{(50)(10^3)}$.

Solution: $\dfrac{(2^6)(5^7)}{(50)(10^3)} = \dfrac{(2^6)(5^7)}{(2)(25)[(2)(5)]^3}$ (Express 50 as $(2)(25)$ and 10 as $(2)(5)$)

$= \dfrac{(2^6)(5^7)}{(2)(5^2)(2^3)(5^3)}$ (Apply rules of exponents)

$= \dfrac{(2^6)(5^7)}{(2^4)(5^5)}$ (Apply rules of exponents)

$= (2^2)(5^2)$ (Apply rules of exponents)

$= (4)(25)$ (Expand)

$= 100.$ (Multiply)

In scientific work we often encounter very large numbers or very small numbers. For example, the distance light travels in one year, 5,872,000,000,000 miles, is called a light year. The mass of an oxygen molecule equals 0.0000000000000000000005313 grams. A laser beam may have energy of as much as 10,000,000,000,000 watts, or the mass of an electron equals 0.00054875 atomic mass units.

One important application of exponents appears in scientific notation in which we have a very efficient method of expressing very large or very small numbers.

Scientific Notation

In **scientific notation** we express all positive numbers as a number between one and ten times a power of ten. Symbolically we express all positive numbers in the form $c \times 10^n$, where $1 \leq c < 10$ and n is an appropriate integer.

The following examples illustrate this form.

$$130 = 1.3 \times 100 = 1.3 \times 10^2$$
$$5250 = 5.25 \times 10^3$$
$$1{,}768{,}000 = 1.768 \times 10^6$$
$$54{,}000{,}000{,}000 = 5.4 \times 10^{10}$$
$$2.54 = 2.54 \times 10^0$$

$$0.04 = 4/100 = 4/10^2 = 4.0 \times 10^{-2}$$
$$0.056 = 5.6 \times 10^{-2}$$
$$0.00768 = 7.68 \times 10^{-3}$$
$$0.0000000000007 = 7 \times 10^{-13}.$$

The mass of a neutron is approximately

$$0.00000000000000000000000016.$$

In scientific notation this is 1.6×10^{-24} gram.

• *Example 1.20:* Express the mass of five million neutrons in scientific notation.

Solution: $(5{,}000{,}000)(1.6 \times 10^{-24})$
$= (5 \times 10^6)(1.6 \times 10^{-24})$ (Express 5,000,000 in scientific notation)
$= (5)(1.6)(10^6)(10^{-24})$ (Commute)
$= 8.0 \times 10^{-18}$ gram.   (Multiply)

• *Example 1.21:* Compute using scientific notation. Leave your answer in scientific notation.

$$\frac{(14{,}000)(0.00003)(8{,}800{,}000)}{(1200)(0.000002)}$$

Solution:
$$\frac{(1.4 \times 10^4)(3 \times 10^{-5})(8.8 \times 10^6)}{(1.2 \times 10^3)(2 \times 10^{-6})}$$ (Express each factor in scientific notation)

$$= \frac{(1.4)(3)(8.8)}{(1.2)(2)} \times 10^{4 - 5 + 6 - 3 + 6}$$ (Commute and multiply powers of ten)

$= 15.4 \times 10^8$ (Multiply)
$= 1.54 \times 10^9.$ (Express in scientific notation)

A scientific calculator is very useful for doing problems of this type.

Very large or very small numbers can be displayed in scientific notation on a scientific calculator. The calculator has a key such as $\boxed{\text{EXP}}$ or $\boxed{\text{EE}}$ that allows you to enter suitable powers of 10 (usually up to 99 or down to -99).

• *Example 1.22:*   Key in the number 3.90816×10^{57} on your calculator.

Solution: Press **3.90816** $\boxed{\text{EXP}}$ **57** .

On the display you should now have 3.90816 57

• • • • • • • • • •

• *Example 1.23:*   Key in the number 1.5×10^{-27}.

Solution: Press **1.5** $\boxed{\text{EXP}}$ **27** $\boxed{+/-}$.

The key $\boxed{+/-}$ (or similar key) changes the exponent from 27 to -27. On the display you should now have $1.5 - 27$.

• • • • • • • • • •

On a calculator or computer, scientific notation is also called *floating point notation.*

Note the key $\boxed{y^x}$ on your calculator. It can be used for the calculation (approximately) of powers. To calculate y^x, we first key in y, then $\boxed{y^x}$, then the exponent x followed by $=$.

• *Example 1.24:*   Calculate $(5.32)^7$.

Solution: **5.32** $\boxed{y^x}$ **7**
$\boxed{=}$ 120609.475.

• • • • • • • • • •

• *Example 1.25:*   Calculate $(153.4)^{-13}$.

Solution 1:   **153.4** $\boxed{y^x}$ **13** $\boxed{+/-}$ $\boxed{=}$
$3.8394615 - 29$.

Solution 2:   **1** $\boxed{\div}$ **153.4** $\boxed{y^x}$ **13** $\boxed{=}$
$3.8394615 - 29$.

• • • • • • • • • •

It is important to note that there may be more than one way to arrive at a correct answer by using a calculator. It is also important to realize that the sequence of steps that gives a correct answer on one calculator may not give the same answer on another type of calculator. This is due to the different kinds of logic that may be used in the manufacture of the calculator. Furthermore, the number of digits that can be displayed on different calculators may vary. The manual that accompanies each type of calculator should be consulted for operating instructions that may be unique to that calculator.

Chapter 1 The Real Number System

Trial Problems 1.3

Before you begin the section exercises, warm up with these problems. Complete answers are included in the answer key.

Compute.

1. $(-2)^3$
2. $\left(\dfrac{1}{3}\right)^4$
3. $\left(\dfrac{1}{4}\right)^{-3}$
4. $2^{-1}\left(\dfrac{1}{3}\right)^2$
5. $\dfrac{(x^2)(y^{-2})}{(x^{-3})(y^{-3})}$

Exercises 1.3

Compute.

1. 2^3
2. 5^2
3. $(-3)^2$
4. $(-2)^4$
5. 3^{-2}
6. 2^{-3}
7. $\left(\dfrac{1}{5}\right)^2$
8. $\left(\dfrac{1}{3}\right)^2$
9. $\left(\dfrac{1}{2}\right)^{-3}$
10. $\left(\dfrac{1}{4}\right)^{-3}$
11. $(4)^0$
12. $(5)^0$
13. $\left(-\dfrac{3}{2}\right)^3$
14. $\left(-\dfrac{2}{3}\right)^3$
15. $3^{-2}\left(\dfrac{1}{2}\right)^3$
16. $4^{-2}\left(\dfrac{1}{3}\right)^2$
17. $\dfrac{(a^5)(b^3)}{(a^3)(b^{-2})}$
18. $\dfrac{(x^2)(y^{-2})}{(x^{-3})(y^{-3})}$
19. $(x^3)(y^{-3})(xy)^2(x^{-3})$
20. $(y^3)(x^{-3})(xy)^2(x^{-2})$
21. $(xy)^3(x^{-4})(y^4)$
22. $(xy)^2(y^{-2})(x^8)$
23. $\dfrac{2^3}{3^{-2}}$
24. $\dfrac{3^2}{2^{-3}}$

Express as powers of 2.

25. 4^2
26. 8^2
27. 16^{-3}
28. 8^{-2}
29. $\left(\dfrac{1}{2}\right)^6$
30. $\left(\dfrac{1}{2}\right)^5$
31. $[(2^3)(8^4)]^3$
32. $[(2^4)(8^3)]^2$
33. $2^4\left(\dfrac{1}{4}\right)^3 8^2\left(\dfrac{1}{16}\right)$
34. $2^3\left(\dfrac{1}{4}\right)^2 4^3\left(\dfrac{1}{16}\right)$
35. $\left(\dfrac{1}{2}\right)\left(\dfrac{1}{4}\right)\left(\dfrac{1}{8}\right)^{-1}$
36. $\left(\dfrac{1}{2}\right)\left(\dfrac{1}{4}\right)^{-1}\left(\dfrac{1}{8}\right)$

In exercises 37–54, express each value in scientific notation.

37. 53.2
38. 23.5
39. 0.4
40. 0.6
41. 2.48
42. 4.82
43. 0.8016
44. 0.6018
45. 0.00067
46. 0.00076
47. 29
48. 37
49. 5280
50. 2850
51. 300,000
52. 400,000
53. 452,000,000,000
54. 326,000,000,000

SECTION 1.3 EXPONENTS

In exercises 55–66, compute and express the answer in scientific notation.

55. $(130)(18{,}000{,}000)(0.00021)$

56. $\dfrac{(16{,}000)(0.0002)}{(0.000002)(204{,}000)}$

57. $(0.003)^2(0.00004)(0.00006)(5{,}000{,}000{,}000)$

58. $\dfrac{800(0.00001)^2}{(200{,}000)^4}$

59. $(300{,}000)^5$

60. $5(40)(300)(2000)(100{,}000)$

61. $\dfrac{(0.0000005)^2}{(2000)^4}$

62. $(3000)(80{,}000)^3(0.00002)^6$

63. The number of inches in a mile.

64. The number of cubic centimeters in a cubic meter.

65. The total mass in grams of 4,000,000 neutrons.

66. The length in kilometers of a light year (the distance light travels in one year). The speed of light is approximately 300,000 km/sec and one year is approximately 365 days.

In exercises 67–72, use your calculator to find approximate values.

67. $(3.5)^9$

68. $(1.327)^5$

69. $(1.09)^{69}$

70. $(1.09)^{-120}$

71. $(1.7085 \times 10^7)^{-8}$

72. $(2.817 \times 10^{-12})^5$

Solve the following problems using scientific notation and your calculator where appropriate.

73. If the volume of a gas is expressed in cm³ and the pressure on the same gas is measured in Pa (pascal), then the product of the volume and the pressure is expressed as Pa · cm³. For a certain gas the product of the pressure and volume is 25.6 Pa · cm³ and the pressure is 0.00207 Pa. What is the volume?

74. The sun weighs approximately 1.8×10^{27} tons. The earth weighs approximately 1.32×10^{25} pounds. How many times heavier is the sun than the earth?

75. The diameter of the universe is about 2,000,000,000 light years. A light year is approximately 9,460,000,000,000 kilometers. Calculate the diameter of the universe in kilometers.

76. If a spacecraft is in synchronous orbit about the earth and the angular radius to its horizon circle is 30° then its distance h from the earth can be determined by the equation

$$h = \dfrac{6378(1 - 0.8660)}{0.8660}.$$

Calculate h to three significant digits.

1 • 4 Roots and Radicals

Square Root A **square root** of a positive number x is a number whose square is x. A square root of 16 is 4, since $4^2 = 16$. Another square root of 16 is -4, since $(-4)^2 = 16$. Every positive number has two square roots, one a positive number, and the other a negative number. Hence, the two square roots are opposites of one another.

24 CHAPTER 1 THE REAL NUMBER SYSTEM

Radical
Principal Square Root
Radicand

The symbol "$\sqrt{}$" is called a **radical.** It is used to indicate the positive or **principal square root** of a number. Thus $\sqrt{16} = 4$ and $\sqrt{25} = 5$. The number under the radical is called the **radicand.** Since we have agreed that $\sqrt{16} = 4$, we also agree that $-\sqrt{16} = -4$ and that $+\sqrt{16} = +4$.

Most of our work with roots deals with square roots, but we do have occasion to consider other roots of a number. Thus we designate the **principal nth root** of a as $\sqrt[n]{a} = x$ where $x^n = a$. The letter n is called the **index** of the radical.

Principal nth Root
Index

If $n = 2$, the 2 is usually not written, and we call it a square root ($\sqrt[2]{a} = \sqrt{a}$). We have already seen examples of this. If a is positive, x is positive. If a is negative, and n is odd, x is negative. We do not consider cases where a is negative and n is an even integer greater than 2. Examples 1.26 and 1.27 illustrate these points.

• *Example 1.26:* $\sqrt[3]{64} = 4$, since $4^3 = 64$ (a is pos., x is pos.)
  $\sqrt[3]{-64} = -4$, since $(-4)^3 = -64$ (a is neg., n is odd, x is neg.)
  $\sqrt[4]{81} = 3$, since $(3)^4 = 81$ (a is pos., x is pos.)
  $\sqrt[5]{32} = 2$, since $(2)^5 = 32$ (a is pos., x is pos.)
  $\sqrt[5]{-32} = -2$, since $(-2)^5 = -32$ (a is neg., n is odd, x is neg.)

If a minus sign precedes the radical sign, we express the result as the negative of the principal nth root.

• *Example 1.27:* $-\sqrt[3]{64} = -4$
  $-\sqrt[3]{-64} = -(-4) = 4$
  $-\sqrt[4]{16} = -(2) = -2$
  $-\sqrt[5]{-32} = -(-2) = 2$.

Perfect Square

The square of a rational number is called a **perfect square.** For example, $\left(\dfrac{2}{3}\right)^2 = \dfrac{4}{9}$, $\left(\dfrac{3}{4}\right)^2 = \dfrac{9}{16}$, $7^2 = 49$, $9^2 = 81$, and $12^2 = 144$. So $\dfrac{4}{9}$, $\dfrac{9}{16}$, 49, 81, and 144 are perfect squares. The square root of a perfect square is a rational number. Hence,

$$\sqrt{\dfrac{4}{9}} = \sqrt{\left(\dfrac{2}{3}\right)^2} = \dfrac{2}{3}$$

$$\sqrt{\dfrac{9}{16}} = \sqrt{\left(\dfrac{3}{4}\right)^2} = \dfrac{3}{4}$$

$$\sqrt{49} = \sqrt{7^2} = 7$$

$$\sqrt{81} = \sqrt{9^2} = 9$$

$$\sqrt{144} = \sqrt{12^2} = 12.$$

If a number is not a perfect square, its square root can only be approximated. The $\sqrt{2}$ is approximately 1.4142135 . . . , and the $\sqrt{5} = 2.2360679$. . . . These decimal numbers do not repeat. Therefore we call them nonrepeating decimals or *irrational numbers.* Remember that we discussed repeating and nonrepeating decimals in section 1.1.

SECTION 1.4 ROOTS AND RADICALS

The approximate square root of some positive integers can be found by using the $\sqrt{}$ key on most calculators.

- **Example 1.28:** Find the square root of 7.

  Solution: Press 7 $\sqrt{}$. The display should read 2.645751311. Thus $\sqrt{7} = 2.645751311$ (approximately). The number of decimal places may vary depending on the type of calculator you have.

Radical Form

Most mathematicians prefer that numbers like the square root of 7 be left in **radical form** like $\sqrt{7}$ since this is an exact value. However, in industrial and technical applications the approximate decimal form may be more desirable. The calculator is used if a decimal approximation of the radical is required.

Square roots obey the following algebraic rules:

RULES FOR SQUARE ROOTS:

Let a and b be positive real numbers. Then

1. $(\sqrt{a})^2 = a$
2. $\sqrt{a^2} = a$
3. $\sqrt{ab} = \sqrt{a}\sqrt{b}$
4. $\sqrt{\dfrac{a}{b}} = \dfrac{\sqrt{a}}{\sqrt{b}}$

Rules 1 and 2 follow directly from our definition of square roots. We illustrate rule 3 with an example:

$$\sqrt{9 \cdot 4} = \sqrt{36} = 6$$
$$\sqrt{9} \cdot \sqrt{4} = 3 \cdot 2 = 6.$$

Thus,

$$\sqrt{9 \cdot 4} = \sqrt{9} \cdot \sqrt{4}.$$

We can illustrate rule 4 with a similar example.

$$\sqrt{\frac{144}{36}} = \sqrt{4} = 2$$

$$\frac{\sqrt{144}}{\sqrt{36}} = \frac{12}{6} = 2.$$

Thus,

$$\sqrt{\frac{144}{36}} = \frac{\sqrt{144}}{\sqrt{36}}.$$

Rule 3 can be used to simplify radicals that are not perfect squares.

- *Example 1.29:* Find $\sqrt{500}$.

  Solution: $\sqrt{500} = \sqrt{(100)(5)}$ (Factor)
  $= \sqrt{100}\sqrt{5}$ (Express as product of square roots)
  $= 10\sqrt{5}.$ (Extract roots where possible)

- *Example 1.30:* Find $\sqrt{r^2 b}$, $(r > 0)(b > 0)$.

  Solution: $\sqrt{r^2 b} = \sqrt{r^2}\sqrt{b}$ (Express as product of square roots)
  $= r\sqrt{b}.$ (Extract roots where possible)

- *Example 1.31:* Find $\sqrt{\dfrac{108 x^3}{75 x^2}}$

  Solution: $\sqrt{\dfrac{108 x^3}{75 x^2}} = \sqrt{\dfrac{(3)(36) x^3}{(3)(25) x^2}}$ (Factor)
  $= \sqrt{\dfrac{36 x}{25}}$ (Simplify)
  $= \dfrac{\sqrt{36}\sqrt{x}}{\sqrt{25}}$ (Express as product/quotient of square roots)
  $= \dfrac{6\sqrt{x}}{5}.$ (Extract roots where possible)

- *Example 1.32:* Simplify the radical $\sqrt[4]{16 z^5}$.

  Solution: $\sqrt[4]{16 z^5} = \sqrt[4]{2^4 (z^4) z}$ (Express as powers of four where possible)
  $= 2z\sqrt[4]{z}.$ (Extract roots where possible)

- *Example 1.33:* Compute $(4 + \sqrt{7})(5 - 2\sqrt{7})$.

  Solution: $(4 + \sqrt{7})(5 - 2\sqrt{7})$
  $= (4)(5) + (\sqrt{7})(5) - (4)(2\sqrt{7})$
  $\quad - (\sqrt{7})(2\sqrt{7})$ (Distributive property)
  $= 20 + 5\sqrt{7} - 8\sqrt{7} - 2\sqrt{7^2}$ (Multiply)
  $= 20 - 3\sqrt{7} - 2(7)$ (Simplify)
  $= 20 - 3\sqrt{7} - 14$ (Multiply)
  $= 6 - 3\sqrt{7}.$ (Subtract)

SECTION 1.4 ROOTS AND RADICALS

Rationalizing the Denominator

It is often desirable to eliminate expressions containing radicals from the denominator of a quotient. This process is called **rationalizing the denominator.**

To rationalize expressions whose denominators contain radicals we must multiply both the numerator and the denominator by the same value. We choose this value so that the exponent on each factor in the denominator will be a multiple of the index of the radical of the denominator. This is illustrated in examples 1.34 through 1.36.

• *Example 1.34:* Rationalize the denominator in $\dfrac{5}{4\sqrt{3}}$.

Solution:
$$\dfrac{5}{4\sqrt{3}} = \dfrac{5}{4\sqrt{3}} \cdot \dfrac{\sqrt{3}}{\sqrt{3}} \quad \text{(Multiply both numerator and denominator by } \sqrt{3}\text{)}$$
$$= \dfrac{5\sqrt{3}}{4\sqrt{3^2}} \quad \text{(Multiply)}$$
$$= \dfrac{5\sqrt{3}}{(4)(3)} \quad \text{(Extract roots where possible)}$$
$$= \dfrac{5\sqrt{3}}{12}. \quad \text{(Multiply)}$$

••••••••••

• *Example 1.35:* Rationalize the denominator in $\dfrac{\sqrt[3]{a}}{\sqrt[3]{b}}$.

Solution:
$$\dfrac{\sqrt[3]{a}}{\sqrt[3]{b}} = \dfrac{\sqrt[3]{a}\,\sqrt[3]{b^2}}{\sqrt[3]{b}\,\sqrt[3]{b^2}} \quad \text{(Multiply both numerator and denominator by } \sqrt[3]{b^2}\text{)}$$
$$= \dfrac{\sqrt[3]{ab^2}}{\sqrt[3]{b^3}} \quad \text{(Multiply)}$$
$$= \dfrac{\sqrt[3]{ab^2}}{b}. \quad \text{(Extract roots where possible)}$$

••••••••••

• *Example 1.36:* Rationalize the denominator in $\dfrac{\sqrt[4]{x}}{\sqrt[4]{y^5}}$.

Solution:
$$\dfrac{\sqrt[4]{x}}{\sqrt[4]{y^5}} = \dfrac{\sqrt[4]{x}\,\sqrt[4]{y^3}}{\sqrt[4]{y^5}\,\sqrt[4]{y^3}} \quad \text{(Multiply both numerator and denominator by } \sqrt[4]{y^3}\text{)}$$
$$= \dfrac{\sqrt[4]{xy^3}}{\sqrt[4]{y^8}} \quad \text{(Multiply)}$$
$$= \dfrac{\sqrt[4]{xy^3}}{y^2}. \quad \text{(Extract roots where possible)}$$

••••••••••

28 CHAPTER 1 THE REAL NUMBER SYSTEM

If the denominator is of the form $\sqrt{a} + \sqrt{b}$, we multiply the numerator and the denominator by the expression $\sqrt{a} - \sqrt{b}$. This eliminates the radicals in the denominator and leaves a denominator of the form $a - b$.

• **Example 1.37:** Rationalize the denominator in $\dfrac{4}{\sqrt{5} + \sqrt{2}}$.

Solution:

$$\dfrac{4}{\sqrt{5} + \sqrt{2}} = \dfrac{4}{\sqrt{5} + \sqrt{2}} \cdot \dfrac{\sqrt{5} - \sqrt{2}}{\sqrt{5} - \sqrt{2}} \quad \text{(Multiply both numerator and denominator by } \sqrt{5} - \sqrt{2}\text{)}$$

$$= \dfrac{4(\sqrt{5} - \sqrt{2})}{5 - 2} \quad \text{(Multiply)}$$

$$= \dfrac{4\sqrt{5} - 4\sqrt{2}}{3}. \quad \text{(Simplify)}$$

• **Example 1.38:** Rationalize $\dfrac{1}{\sqrt{3} - 1}$.

Solution:

$$\dfrac{1}{\sqrt{3} - 1} = \dfrac{1}{\sqrt{3} - 1} \cdot \dfrac{\sqrt{3} + 1}{\sqrt{3} + 1} \quad \text{(Multiply both numerator and denominator by } \sqrt{3} + 1\text{)}$$

$$= \dfrac{\sqrt{3} + 1}{\sqrt{3^2} - 1} \quad \text{(Multiply)}$$

$$= \dfrac{\sqrt{3} + 1}{3 - 1} \quad \text{(Extract roots where possible)}$$

$$= \dfrac{\sqrt{3} + 1}{2}. \quad \text{(Simplify)}$$

• **Example 1.39:** Rationalize the denominator in $\dfrac{\sqrt{2} + \sqrt{3}}{\sqrt{5} - \sqrt{7}}$.

Solution:

$$\dfrac{\sqrt{2} + \sqrt{3}}{\sqrt{5} - \sqrt{7}}$$

$$= \dfrac{\sqrt{2} + \sqrt{3}}{\sqrt{5} - \sqrt{7}} \cdot \dfrac{\sqrt{5} + \sqrt{7}}{\sqrt{5} + \sqrt{7}} \quad \text{(Multiply both numerator and denominator by } \sqrt{5} + \sqrt{7}\text{)}$$

$$= \dfrac{\sqrt{10} + \sqrt{14} + \sqrt{15} + \sqrt{21}}{5 - 7} \quad \text{(Multiply)}$$

$$= \dfrac{\sqrt{10} + \sqrt{14} + \sqrt{15} + \sqrt{21}}{-2}. \quad \text{(Simplify)}$$

Section 1.4 Roots and Radicals

• **Example 1.40:** Rationalize the denominator in $\dfrac{\sqrt{r}}{-2\sqrt{r}-3\sqrt{s}}$.

Solution:

$$\dfrac{\sqrt{r}}{-2\sqrt{r}-3\sqrt{s}}$$

$$= \dfrac{\sqrt{r}}{-2\sqrt{r}-3\sqrt{s}} \cdot \dfrac{-2\sqrt{r}+3\sqrt{s}}{-2\sqrt{r}+3\sqrt{s}} \quad \text{(Multiply both numerator and denominator by } -2\sqrt{r}+3\sqrt{s}\text{)}$$

$$= \dfrac{-2r+3\sqrt{rs}}{4r-9s} \quad \text{(Multiply and simplify)}$$

Trial Problems 1.4

Before you begin the section exercises, warm up with these problems. Complete answers are included in the answer key.

Find the indicated principal root.

1. $\sqrt{49}$
2. $\sqrt[3]{-64}$
3. $\sqrt[3]{0.027}$
4. Simplify $\sqrt[3]{64a^4}$
5. Rationalize $\dfrac{\sqrt{5}}{\sqrt{2}}$

Exercises 1.4

Find the indicated principal root of each of the following.

1. $\sqrt{36}$
2. $\sqrt{81}$
3. $-\sqrt{144}$
4. $-\sqrt{169}$
5. $\sqrt{0.16}$
6. $\sqrt{0.49}$
7. $\sqrt[3]{8}$
8. $\sqrt[3]{64}$
9. $\sqrt[3]{-8}$
10. $\sqrt[3]{-64}$
11. $-\sqrt[3]{125}$
12. $-\sqrt[3]{27}$
13. $\sqrt[3]{0.027}$
14. $\sqrt[3]{0.064}$
15. $\sqrt[4]{81}$
16. $\sqrt[4]{625}$
17. $\sqrt[5]{-32}$
18. $\sqrt[5]{-243}$
19. $\sqrt[6]{64}$
20. $\sqrt[6]{729}$

Express the given radicals in simplest form. Do not rationalize.

21. $-\sqrt[3]{-8}$
22. $-\sqrt[3]{-27}$
23. $\sqrt{25a^4}$
24. $\sqrt{36x^4}$
25. $\sqrt{9x^2y^4}$
26. $\sqrt{16x^2y^4}$
27. $\sqrt{x^4y^6z}$
28. $\sqrt{x^6y^2z}$
29. $\sqrt{12a^2b^3x}$
30. $\sqrt{8a^2b^5x}$
31. $\sqrt{\dfrac{16a^2}{25b^2}}$
32. $\sqrt{\dfrac{9a^2}{36b^2}}$
33. $\sqrt{\dfrac{72x^3}{50x}}$
34. $\sqrt{\dfrac{18y^3}{2y}}$
35. $\sqrt{\dfrac{2x^2y}{5a^3}}$
36. $\sqrt{\dfrac{3xy^5}{8a^3}}$

Use your calculator to approximate the values of the given square roots.

37. $\sqrt{500}$
38. $\sqrt{600}$
39. $\sqrt{7800}$
40. $\sqrt{8700}$
41. $\sqrt{0.04}$
42. $\sqrt{0.09}$
43. $\sqrt{459.23}$
44. $\sqrt{954.32}$

Simplify these radicals.

45. $\sqrt[3]{16}$
46. $\sqrt[3]{54}$
47. $\sqrt[3]{27a^4}$
48. $\sqrt[3]{64b^4}$
49. $\sqrt[4]{81x^9}$
50. $\sqrt[4]{16y^5}$
51. $\sqrt[5]{64x}$
52. $\sqrt[5]{96y^3}$
53. $\dfrac{\sqrt[3]{54a^6b^7c^8}}{16a^3bc^2}$
54. $\dfrac{\sqrt[3]{162x^9y^{10}z^{11}}}{16x^3yz^2}$

Rationalize the denominators.

55. $\dfrac{\sqrt{2}}{\sqrt{3}}$
56. $\dfrac{\sqrt{3}}{\sqrt{2}}$
57. $\dfrac{\sqrt[3]{32}}{\sqrt[3]{2}}$
58. $\dfrac{\sqrt[3]{18}}{\sqrt[3]{2}}$
59. $\dfrac{\sqrt{50}}{2\sqrt{3}}$
60. $\dfrac{\sqrt{98}}{2\sqrt{3}}$
61. $\dfrac{\sqrt{3}}{\sqrt{5b}}$
62. $\dfrac{\sqrt{5}}{\sqrt{3b}}$
63. $\dfrac{1}{\sqrt{2}+\sqrt{3}}$
64. $\dfrac{2}{\sqrt{2}+\sqrt{3}}$
65. $\dfrac{2+\sqrt{5}}{\sqrt{3}-\sqrt{2}}$
66. $\dfrac{2-\sqrt{5}}{\sqrt{3}-\sqrt{2}}$
67. $\dfrac{\sqrt{a}}{\sqrt{a}+2\sqrt{b}}$
68. $\dfrac{\sqrt{a}}{\sqrt{a}+3\sqrt{b}}$
69. $\dfrac{\sqrt{x}}{-3\sqrt{x}+2\sqrt{y}}$
70. $\dfrac{\sqrt{x}}{-2\sqrt{x}-3\sqrt{y}}$

71. The ratio of the rates of diffusion of two different gases is given by the equation $\dfrac{r_1}{r_2} = \dfrac{\sqrt{m_2}}{\sqrt{m_1}}$, where m_1 and m_2 are the masses of the molecules of the gases. Find the ratio $\dfrac{r_1}{r_2}$ if $m_1 = 49$ and $m_2 = 64$.

72. Work exercise 71 if $m_1 = 64$ and $m_2 = 81$.

73. A cubical container holds 729 cu ft. What is the length of an edge of the container?

74. Work exercise 73 if the container holds 343 cu ft.

75. A square piece of plastic used in the fabrication of a computer base has an area of 540 in.² Find the length of the side of this square.

76. A square cathode ray tube has an area of 612 in.² Find the length of a side of this tube.

77. In the theory of waves in wires we encounter the expression

$$\dfrac{\sqrt{d_1}-\sqrt{d_2}}{\sqrt{d_1}+\sqrt{d_2}},$$

where each d represents the wave length. Write this expression in simplified form, then evaluate the expression if $d_1 = 10$, and $d_2 = 3$.

78. Work exercise 77 if $d_1 = 7$ and $d_2 = 5$.

79. In Kepler's problem dealing with orbiting satellites the expression

$$\sqrt{\dfrac{1+e}{1-e}}$$

is encountered when calculating the true anomaly of an ellipse with eccentricity e. Simplify this expression and evaluate the expression if $e = 0.25$.

80. When calculating the velocity of a body in an elliptical orbit at a distance r from the focus, in terms of the semimajor axis a, we encounter the expression

$$\sqrt{\dfrac{2}{r}-\dfrac{1}{a}}.$$

Find a common denominator, rationalize the denominator, and evaluate the expression when $r = 10{,}260$ km and $a = 14{,}460$ km.

1•5 Rational Exponents

Thus far, the only numbers we have used as exponents have been integers. Fractions can also be used as exponents. Many formulas in science and technology use fractional exponents in their expressions. For example, the formula

$$A = 0.5^{t/h}$$

gives the portion A of a radioactive material that remains after t years, given that the half-life of the material is h years.

Fractional exponents are also very convenient for converting expressions containing radicals into expressions containing fractional exponents, and vice versa. For fractional exponents to be useful and meaningful, they must satisfy all of the basic laws of exponents. The significance of fractional exponents can be established with example 1.41.

• *Example 1.41:* What meaning can we give to $a^{1/3}$?

Solution: In section 1.4 we established that $(\sqrt{a})^2 = a$. Similarly, $(\sqrt[3]{a})^3 = a$. However, using the rules of exponents,

$$(a^{1/3})^3 = a^{(1/3)(3)} = a.$$

Since

$$(\sqrt[3]{a})^3 = a$$

and

$$(a^{1/3})^3 = a,$$

it follows that

$$\sqrt[3]{a} = a^{1/3}.$$

If we generalize on the basis of the example, we must define $a^{1/n}$ to be the principal nth root of a.

THE PRINCIPAL NTH ROOT OF a

$$a^{1/n} = \sqrt[n]{a}.$$

Example 1.42 illustrates this generalization.

• *Example 1.42:*
$$8^{1/3} = \sqrt[3]{8} = 2$$
$$32^{1/5} = \sqrt[5]{32} = 2$$
$$(-32)^{1/5} = \sqrt[5]{-32} = -2$$
$$16^{1/2} = \sqrt{16} = 4.$$

When $a^{1/n}$ is raised to the m power, the basic laws of exponents should still hold. Again we will look at an example and make a generalization.

• **Example 1.43:** $a^{2/3} = (a^{1/3})^2 = (a^2)^{1/3}$. (Use properties of exponents)

Thus,

$$(\sqrt[3]{a})^2 = \sqrt[3]{a^2}.$$ (Convert to radical form using principal nth root property)

In general, we write

$$a^{m/n} = (\sqrt[n]{a})^m = \sqrt[n]{a^m}.$$

• **Example 1.44:**
$8^{2/3} = (8^{1/3})^2 = 2^2 = 4$
$32^{3/5} = (32^{1/5})^3 = 2^3 = 8$
$27^{4/3} = (27^{1/3})^4 = 3^4 = 81$
$(y^4)^{3/2} = [(y^4)^{1/2}]^3 = (y^2)^3 = y^6$
$(-64)^{2/3} = [(-64)^{1/3}]^2 = (-4)^2 = 16.$

Trial Problems 1.5

Before you begin the section exercises, warm up with these problems. Complete answers are included in the answer key.

1. Change to radical form $x^{2/5}$.
2. Change to exponential notation $\sqrt[3]{6}$.
3. Evaluate $25^{1/2}$.
4. Evaluate $(16)^{-1/4}$.
5. Evaluate $(-64)^{1/3}$.

Exercises 1.5

Change the given expression to radical form.

1. $3^{1/2}$
2. $2^{1/2}$
3. $x^{1/4}$
4. $y^{1/5}$
5. $b^{3/5}$
6. $a^{2/5}$
7. $r^{5/3}$
8. $s^{7/3}$

Change the given expression to exponential form and simplify if possible.

9. $\sqrt{3}$
10. $\sqrt{2}$
11. $\sqrt[3]{5}$
12. $\sqrt[3]{6}$
13. $\sqrt[4]{7^2}$
14. $\sqrt[4]{9^2}$
15. $\sqrt[5]{a^2 b}$
16. $\sqrt[5]{ab^2}$

Evaluate the given expressions.

17. $9^{1/2}$
18. $4^{1/2}$
19. $(16)^{1/2}$
20. $(36)^{1/2}$
21. $(27)^{1/3}$
22. $(64)^{1/3}$
23. $(-32)^{1/5}$
24. $(-27)^{1/3}$

25. $8^{4/3}$ 26. $9^{3/2}$ 27. $(-8)^{2/3}$ 28. $(-27)^{5/3}$

29. $(16)^{-1/4}$ 30. $(25)^{-1/2}$ 31. $\dfrac{(25)^{-3/2}}{(81)^{3/4}}$ 32. $\dfrac{(64)^{1/2}}{(64)^{1/3}}$

33. The side x of a square, in terms of the area A, is given by the equation $x = \sqrt{A}$. Write this equation using fractional exponents.

34. An expression that arises in Einstein's theory of relativity is $\sqrt{\dfrac{(c^2 - v^2)}{c^2}}$, where v is the velocity of an object and c is the velocity of light. Write this expression using fractional exponents.

35. One equation for the efficiency of an internal combustion engine is $E = 100\left(1 - \dfrac{1}{R^{2/5}}\right)$, where E is the efficiency in percent and R is the compression ratio of the engine. Express this equation in radical form.

36. The foreshortening factor k from Kepler's laws of orbiting bodies is established by the relationship

$$k = \sqrt{1 - e^2},$$

where e is the eccentricity of the elliptical orbit. Express this equation using fractional exponents.

1 • 6 Addition and Subtraction of Polynomials

Monomial   Algebraic expressions like x^2, $4x$, and $5x^2y$ are called monomials. A **monomial** is an expression that consists of the product of a real number and one or more variables
Term   whose exponents are nonnegative integers. A monomial may also be called a **term**. The
Polynomial   indicated sum (or difference) of more than one monomial is called a **polynomial**. A
Binomial   polynomial with two terms is called a **binomial**, one with three terms is called a **tri-**
Trinomial   **nomial**. Thus the following are polynomials.

$$\begin{array}{ll} 4x^3 + 7x^2 - 5x + 2 & \text{(Polynomial)} \\ 10x + 4 & \text{(Binomial)} \\ 16x^2 + 5x^4 + 7x^3y^2 & \text{(Trinomial)} \end{array} \right\} \text{(Polynomials)}$$

In some cases, a monomial is considered to be a special case of a polynomial.

Degree of a Monomial in a Variable   The **degree of a monomial in a variable** is the exponent of the variable in the monomial. Thus, $6xy^4z^2$ is of degree 1 in x, degree 4 in y, and degree 2 in z.

Degree of a Monomial   The **degree of the monomial** is the sum of the degrees in each of the variables. The monomial $6xy^4z^2$ has degree $1 + 4 + 2 = 7$.

Degree of a Polynomial   The **degree of a polynomial** is the degree of the term that has the highest degree. Thus, $5x^2 + 4x^3 + 6x + 7$ has degree 3, $12x - 9$ has degree 1, and $12x^5y + 9xy^3 + 7x^2y^3$ has degree 6.

Similar or Like Terms   If two monomials have the same degree in each of their variables, the monomials are called **similar terms** or **like terms**. Thus, $4x^2y^3$ and $6x^2y^3$ are similar terms, $3x$ and $5x$ are similar terms, $9xy^4$ and $3xy^4$ are similar terms, and $5xyz$ and $-7xy^2z$ are not similar terms.

To add polynomials, we add the coefficients of the similar terms. We illustrate this with example 1.45.

- *Example 1.45:* Add $5x^3 - 4x^2 - 6x + 3$ and $x^3 + x^2 - 4$.

    Solution:
    $$\begin{array}{r} 5x^3 - 4x^2 - 6x + 3 \\ \underline{x^3 + x^2 - 4} \\ 6x^3 - 3x^2 - 6x - 1 \end{array}$$

For certain values of x, the polynomials in example 1.46 may represent the lengths of the sides of a triangle.

- *Example 1.46:* Find an expression for the perimeter of a triangle if the lengths of the sides of the triangle are represented by the polynomials $2x^2 + x + 3$, $x^2 + 2x + 4$, and $3x^2 + 4x + 5$.

    Solution: The perimeter is the sum of the lengths of the sides of the triangle. Thus, the perimeter is the sum of the three polynomials.

    $$\begin{array}{r} 2x^2 + x + 3 \\ x^2 + 2x + 4 \\ \underline{3x^2 + 4x + 5} \\ 6x^2 + 7x + 12 \text{ is the perimeter.} \end{array}$$

In section 1.2 we learned that $a - b = a + (-b)$. That is, to subtract one number from another, we add the negative (or opposite) of the second number to the first number. We use the same rule when we subtract polynomials.

- *Example 1.47:* Subtract $x^2 - 4x + 6$ from $2x^2 - 3x + 4$.

    Solution: The negative of $x^2 - 4x + 6$ is $-x^2 + 4x - 6$.

    Hence, $2x^2 - 3x + 4$
    Plus $\underline{-x^2 + 4x - 6}$
    Gives $x^2 + x - 2.$

- *Example 1.48:* Find the area of the shaded region of the rectangle in figure 1.4 if the area of the large rectangle is represented by the polynomial $4x^3 - 2x^2 + 7x$ and the area of the small rectangle is represented by the polynomial $2x^2 + 3x - 5$.

Section 1.6 Addition and Subtraction of Polynomials

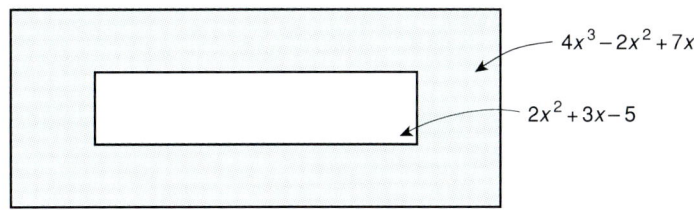

Figure 1.4

Solution: The negative of $2x^2 + 3x - 5$ is $-2x^2 - 3x + 5$. If we add this polynomial to $4x^3 - 2x^2 + 7x$ we get

$$
\begin{array}{r}
4x^3 - 2x^2 + 7x \\
-2x^2 - 3x + 5 \\
\hline
4x^3 - 4x^2 + 4x + 5
\end{array}
$$

(The area of the shaded region)

Trial Problems 1.6

Before you begin the section exercises, warm up with these problems. Complete answers are included in the answer key.

1. Determine the degree of the polynomial $2x^3 + 4x^2 + 5x - 1$.
2. Add $2x^2 + 4x - 6$ to $3x^2 + 2x + 1$.
3. Write the negative of the polynomial $x^3 + 3x^2 - 4x + 2$.
4. Subtract $2x^2 + 3x - 2$ from $3x^2 - 4x - 5$.
5. Subtract $2x^3 - 3x^2 - x + 5$ from $x^4 - 4x^2 - 2x$.

Exercises 1.6

Determine the degree of each monomial.

1. $5x^2y$
2. $3x^2y^2$
3. $12xy^2$
4. $11abc$

Determine the degree of each polynomial.

5. $4x^3 + 3x^2 - 7x + 5$
6. $5x^3 + 4x^2 + 6x - 3$
7. $3xy + x$
8. $8xy + y$
9. $2x^3y^2 + 3x^2y^2 + 5xy$
10. $6x^2y^3 + 5x^2y^2 + 3xy$
11. $5x^6 + 1$
12. $6x^5 - 1$

Add the following polynomials.

13. $3x^2 + 4x + 6$
    $\underline{2x^2 + 2x - 1}$

14. $5x^3 + 6x^2 - 2x + 7$
    $\underline{ 3x^2 + 4x - 2}$

15. $4x^3y - 7x^2y^2 + 2xy^3$
    $\underline{2x^3y - 3x^2y^2 - 4xy^3}$

16. $5x^3y - 7x^2y^2 + 6xy^3$
    $\underline{4x^3y + 5x^2y^2 + 3xy^3}$

17. $5x^6 + 1$
    $\underline{3x^6 - 1}$

18. $4y^6 + 3$
    $\underline{5y^6 - 3}$

Write the negative of each polynomial.

19. $x^3 + 3x^2 - 5x + 6$

20. $4x^3 - 5x^2 + 7x - 2$

21. $3x^6 + 5$

22. $4x^6 - 8$

Subtract.

23. $x^2 + 3x + 2$ from $3x^2 + 4x + 5$

24. $x^2 - 2x - 3$ from $8x^2 - 6x + 7$

25. $3x^2y - 4xy^2$ from $6x^2y + 7xy^2$

26. $4x^2y - 3xy^2$ from $7x^2y + 6xy^2$

27. $8x^3 + 4x^2 + 3$ from $x^4 + 5x^3 + 3x - 8$

28. $2x^3 - 3x^2 + x - 5$ from $x^5 + 4x^3 + x^2$

29. A rectangle has a length represented by the polynomial $x^2 - 3x + 6$ and a width represented by the polynomial $2x + 1$. Find an expression for the perimeter of the rectangle.

30. A rectangle has a length represented by the polynomial $2x^2 + 4x + 7$ and a width represented by the polynomial $x^2 + x$. Find an expression for the perimeter of the rectangle.

31. The fuel consumption of an engine is given by the expression $3.7x^2 + 0.4y + 7.2$. After a turbocharger was added to the engine, the fuel consumption was $4.3x^2 + 0.3y + 7.2$. Write an expression for the increase in consumption caused by the turbocharger.

32. The fuel consumption of an engine is given by the expression $1.9x^2 + 3.4y + 3.7$. After a turbocharger was added to the engine, the fuel consumption was $1.3x^2 + 3.5y + 3.7$. Write an expression for the increase in consumption caused by the turbocharger.

33. The area of a rectangular piece of fiberglass is represented by the expression $4x^3 - 2x^2 + 7x$. The area of a square region that has been punched out of the center of the given piece of fiberglass is represented by the expression $2x^2 + 3x - 8$. Find the area of the piece of fiberglass that remains after the center piece has been removed.

34. The perimeter of a triangular piece of aluminum is found by adding the lengths of the three sides of the triangle. If the lengths of the three sides are $3x^2 + 4x - 5$, $5x + 6$, and $x^2 - 8$, find the perimeter of this piece of aluminum.

1 • 7 Multiplication of Polynomials

To multiply two monomials, we must use the properties of real numbers and the properties of exponents. Example 1.49 uses the commutative property and the property that we add the exponents when we multiply factors with like bases.

• *Example 1.49:* Simplify $(2x^3y)(3x^2y^4)$.

Solution:
$$(2x^3y)(3x^2y^4) = (2)(3)(x^3x^2)(yy^4) \quad \text{(Commute)}$$
$$= 6x^{3+2}y^{1+4} \quad \text{(Multiply)}$$
$$= 6x^5y^5. \quad \text{(Simplify)}$$

We note that the product of two monomials is a monomial.

To multiply a polynomial by a monomial, we use the distributive property and multiply each term of the polynomial by the monomial. This is illustrated in examples 1.50 and 1.51.

Section 1.7 Multiplication of Polynomials

- *Example 1.50:* Simplify $(3x^2)(x^2 + 4x - 3)$.

  Solution:
  $(3x^2)(x^2 + 4x - 3)$
  $= (3x^2)(x^2) + (3x^2)(4x) + (3x^2)(-3)$ (Distribute)
  $= (3x^2x^2) + 12x^2x + [3(-3)x^2]$ (Multiply)
  $= 3x^4 + 12x^3 - 9x^2$ (Simplify)

- *Example 1.51:* Simplify $(3x^2y^2)(x^2 - 2xy + y^2)$.

  Solution:
  $3x^2y^2(x^2 - 2xy + y^2)$
  $= 3x^2y^2(x^2) + (3x^2y^2)(-2xy) + (3x^2y^2)(y^2)$ (Distribute)
  $= (3x^2x^2y^2) + [3(-2)x^2xy^2y] + (3x^2y^2y^2)$ (Multiply)
  $= 3x^4y^2 - 6x^3y^3 + 3x^2y^4$. (Simplify)

To multiply a polynomial by another polynomial, we multiply each term in the first polynomial by each term in the second polynomial. Then we combine the similar or like terms. Example 1.52 is solved by two different techniques. The first solution uses a vertical technique much like that used in arithmetic. The second uses the distributive law. There are other techniques, such as the FOIL method, that can be used when each polynomial has just two terms. This method is discussed in detail when we consider special products in chapter 2.

- *Example 1.52:* Multiply $x^2 + 3x - 2$ by $x + 5$.

  Solution 1: (Using the vertical method.)

  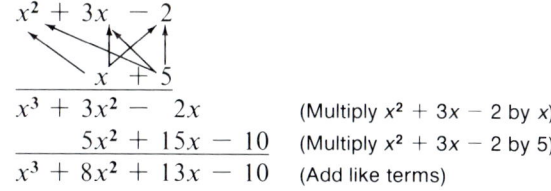

  $x^3 + 3x^2 - 2x$ (Multiply $x^2 + 3x - 2$ by x)
  $\underline{ 5x^2 + 15x - 10}$ (Multiply $x^2 + 3x - 2$ by 5)
  $x^3 + 8x^2 + 13x - 10$ (Add like terms)

  Solution 2: (Using the distributive law.)

  $(x + 5)(x^2 + 3x - 2)$ (Multiply x times x^2
  $= x(x^2 + 3x - 2) + 5(x^2 + 3x - 2)$ $+ 3x - 2$ then multiply
  5 times $x^2 + 3x - 2$)
  $= x(x^2) + x(3x) + x(-2)$ (Distribute)
  $ + 5(x^2) + 5(3x) + 5(-2)$
  $= x^3 + 3x^2 - 2x + 5x^2 + 15x - 10$ (Multiply)
  $= x^3 + 8x^2 + 13x - 10$. (Simplify)

CHAPTER 1 THE REAL NUMBER SYSTEM

Trial Problems 1.7

Before you begin the section exercises, warm up with these problems. Complete answers are included in the answer key.

Multiply.

1. $x^3 x^4$
2. $(2x^3 y^2)(-3x^2 y)$
3. $5x(3x + 2)$
4. $(x + 3)(x - 3)$
5. $(x + 2)(x^2 + x - 1)$

Exercises 1.7

Perform the indicated calculations.

1. $x^5 x^4$
2. $y^3 y^6$
3. $(4x^2 y)(-3xy^2)$
4. $(3x^2 y^2)(-4xy^2)$
5. $x^4 x$
6. yy^5
7. $(2x^2 y^3)(x^3)$
8. $(3x^2 y^3)(x^3)$
9. $(6x^2 y^3) \dfrac{xy^2}{6}$
10. $(5x^2 y^3) \dfrac{xy^2}{5}$
11. $3(x^2 + 2x + 3)$
12. $4(x^2 + 3x + 2)$
13. $5x(2x + 3)$
14. $5x(3x + 2)$
15. $4x^3(x^2 + x + 2)$
16. $5x^3(x^3 + 2x + 1)$
17. $-2x(x^3 - 4x^2 + 5x - 3)$
18. $-3x(x^3 - 2x^2 + 4x - 5)$
19. $(x + 2)(x - 2)$
20. $(x - 3)(x + 3)$
21. $(2x + 5y)^2$
22. $(3x + 2y)^2$
23. $(x + 1)(x^2 - x + 1)$
24. $(x + 2)(x^2 - 2x + 4)$
25. $(x^2 + y^2)(3x - y)$
26. $(x^2 + y^2)(2x + y)$
27. $(3x^2 - 4y^2)(2x^2 + y^2)$
28. $(4x^2 - 3y^2)(2x^2 + y^2)$

Solve the following problems.

29. When determining the focal length of a lens we often encounter the expression

$$(n - 1)\left(\dfrac{1}{R_1} - \dfrac{1}{R_2}\right),$$

where n is the index of refraction, R_1 is the radius of curvature of the surface upon which the light is incident, and R_2 is the radius of curvature of the surface through which the light leaves the lens. Rewrite this expression by multiplying the two factors.

30. Under certain conditions, the velocity of an object in terms of time t is given by the formula

$$v = 5(t - 2)(t - 4).$$

Simplify the right-hand side of this equation by multiplying the factors.

31. If v_0 is the initial velocity, v the final velocity, and t the time, the following expressions occur in the study of motion.

$$\dfrac{v_0 - v}{t} \qquad \dfrac{v_0 + v}{2}$$

Find the product of these two expressions.

32. When calculating the energy of electrons with mass m and velocities v_1 and v_2, we may encounter the expression

$$m(v_2 - v_1)(v_2 + v_1).$$

Simplify the expression by multiplying.

33. Multiply to show that

$$(x + y)(x^2 - xy + y^2) = x^3 + y^3.$$

Then show by substitution that the same value is obtained on each side of the equation when $x = 3$ and $y = 2$.

34. Solve exercise 33 when $x = -2$ and $y = 3$.

1·8 Division of Polynomials

To divide one monomial by another we must use the properties of real numbers and the properties of exponents much like we did for multiplication.

• *Example 1.53:*   Simplify $\dfrac{12a^5}{3a^2}$.

Solution: $\dfrac{12a^5}{3a^2} = \dfrac{(3)(4)a^2 a^3}{3a^2}$ (Write the numerator and the denominator in factored form, then remove the common factors.)

$= 4a^3$.

Note that when we divide one monomial by another monomial the result is a monomial.

To divide a polynomial by a monomial we divide each term in the polynomial by the monomial.

• *Example 1.54:*   Simplify $(28a^3b^3 + 35a^2b^2) \div (7ab^2)$.

Solution: $\dfrac{28a^3b^3 + 35a^2b^2}{7ab^2} = \dfrac{28a^3b^3}{7ab^2} + \dfrac{35a^2b^2}{7ab^2}$ (Divide each term by $7ab^2$)

$= 4a^2b + 5a.$ (Simplify)

• *Example 1.55:*   Simplify $(axy^2 + ax^3 - 4a^3x^2) \div (-ax)$.

Solution: $\dfrac{axy^2 + ax^3 - 4a^3x^2}{-ax}$

$= \dfrac{axy^2}{-ax} + \dfrac{ax^3}{-ax} - \dfrac{4a^3x^2}{-ax}$ (Divide each term by $-ax$)

$= -y^2 - x^2 + 4a^2x.$ (Simplify)

We divide polynomials by using a method similar to the long division process used in arithmetic. To divide one polynomial by another we arrange both the dividend and the divisor in descending powers of the same variables. Then we divide the first term of the dividend by the first term of the divisor to produce the first term of the quotient. Next we multiply the entire divisor by the first term of the quotient and subtract this product from the dividend. Next we divide the first term of the difference just obtained by the first term of the divisor to obtain the second term of the quotient. Finally we multiply the entire divisor by the second term of the quotient and subtract this product from the first difference. This process is repeated until the difference is either zero or is an expression whose degree is less than the degree of the divisor. This long division process is illustrated in examples 1.56, 1.57, and 1.58. Follow the process described very carefully as you work through each example.

• *Example 1.56:* Divide $7x + 3x^2 - 4$ by $x + 3$.

Solution: First write the dividend in descending order: $3x^2 + 7x - 4$.

$$
\begin{array}{r}
3x \phantom{{}+ 7x - 4} \\
x + 3 \overline{\smash{)}\, 3x^2 + 7x - 4} \\
3x^2 + 9x \\
\hline
-2x - 4
\end{array}
$$

(Divide: $3x^2 \div x = 3x$. Write $3x$ in the quotient)
(Multiply: $(3x)(x + 3) = 3x^2 + 9x$)
(Subtract: $(3x^2 + 7x) - (3x^2 + 9x) = -2x$)
(Bring down next term, -4)

We now repeat these basic steps by dividing $-2x$ by x.

$$
\begin{array}{r}
3x - 2 \\
x + 3 \overline{\smash{)}\, 3x^2 + 7x - 4} \\
3x^2 + 9x \\
\hline
-2x - 4 \\
-2x - 6 \\
\hline
+ 2
\end{array}
$$

(Divide: $-2x \div x = -2$. Write -2 in the quotient)
(Multiply: $(-2)(x + 3) = -2x - 6$)
(Subtract: $(-2x - 4) - (-2x - 6) = 2$)

This process ends when the degree of the remainder is less than the degree of the divisor. Hence, we stop since the degree of the remainder 2 is zero and the degree of the dividend $x + 3$ is one. Thus, our quotient is $3x - 2$ and our remainder is 2. To check the result we must multiply the quotient times the divisor and then add the remainder.

$(3x - 2)(x + 3) = 3x^2 + 7x - 6$ (Multiply quotient times divisor)

$(3x^2 + 7x - 6) + (2) = 3x^2 + 7x - 4$ (Add the remainder)

Since we got $3x^2 + 7x - 4$, which is equal to our dividend, the answer checks.

Examples 1.57 and 1.58 leave some of the details to the reader.

- **Example 1.57:** $(x^3 + 3x^2 - 4x - 12) \div (x + 2) = \underline{?}$

  Solution:

  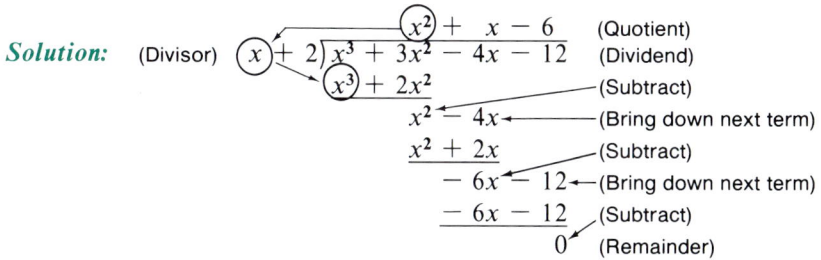

- **Example 1.58:** $(6x^2 + 4x^3 + 1) \div (2x - 1) = \underline{?}$

  Solution: First we must arrange the terms of the dividend so that the exponents on the variable x are in descending order. Note that we must leave a space for the first power of x, or insert $0x$, since there was no such term in the dividend.

  $$\begin{array}{r}
  2x^2 + 4x + 2 \quad \text{(Quotient)} \\
  (Divisor) \quad 2x - 1 \overline{\smash{\big)}\, 4x^3 + 6x^2 + 0x + 1} \quad \text{(Dividend)} \\
  \underline{4x^3 - 2x^2} \quad \text{(Subtract)} \\
  8x^2 + 0x \quad \text{(Bring down next term)} \\
  \underline{8x^2 - 4x} \quad \text{(Subtract)} \\
  4x + 1 \quad \text{(Bring down next term)} \\
  \underline{4x - 2} \quad \text{(Subtract)} \\
  3 \quad \text{(Remainder)}
  \end{array}$$

  Since we have a remainder, we may express it as part of the quotient as follows:

  $$2x^2 + 4x + 2 + \frac{3}{2x - 1}.$$

Trial Problems 1.8

Before you begin the section exercises, warm up with these problems. Complete answers are included in the answer key.

Divide.

1. $6x^4 \div 2x^3$
2. $a^2x \div (-a)$
3. $(-8x^2y^3z) \div (-24x^2z^5)$
4. $(x^2 + 3x - 10) \div (x - 2)$
5. $\dfrac{3t^4 - t^3 - 14t^2 + 4t + 8}{3t + 2}$

Exercises 1.8

In exercises 1–32, perform the indicated operations and simplify the answers.

1. $8m^4 \div 2m$
2. $10m^5 \div 5m$
3. $\dfrac{14s^3}{2s^2}$
4. $\dfrac{16t^3}{2t^2}$
5. $(a^2x) \div (-a)$
6. $(b^2y) \div (-b)$
7. $(-x^3y^4) \div (xy^2)$
8. $(-x^4y^5) \div (x^2y^3)$
9. $(-4ac^2t^4) \div (-2ac^3)$
10. $(-12c^3p^4y) \div (-9cpy^6)$
11. $\dfrac{-7t^3uy}{-6ty}$
12. $\dfrac{9abc^4ds}{15bc^4d^2}$
13. $(a^3x^4 - a^2x^3) \div (ax^2)$
14. $(x^3y^4 - 6x^2y^5) \div (xy^2)$
15. $(-2x^3y^5 - 4x^2y^6) \div (2xy^2)$
16. $(-8x^3y^5 - 10x^2y^6) \div (2x^2y)$
17. $\dfrac{a^2b^2c^3 - a^3b^4c^6 + a^2b^2c}{a^2b^2c}$
18. $\dfrac{r^3s^2t^2 - r^2s^2t^3 + r^2s^3t^3}{r^2st^3}$
19. $\dfrac{a^4b^3 - a^3b^2 + a^2b - a^3b^4}{-a^2b}$
20. $\dfrac{s^3t^4 - s^4t^2 + s^2t^3 - s^3t}{-s^2t}$
21. $(x^2 - 2x - 3) \div (x + 1)$
22. $(x^2 + 3x + 2) \div (x + 1)$
23. $(y^2 - 7y + 10) \div (y - 2)$
24. $(y^2 + 3y - 10) \div (y - 2)$
25. $(2s^2 - s - 6) \div (s - 2)$
26. $(3s^2 + 8s + 4) \div (s + 2)$
27. $(5t - 5t^2 + 2t^3 - 6) \div (t - 2)$
28. $(4t - 5t^2 + 3t^3 - 2) \div (t - 1)$
29. $\dfrac{3x^4 + 5x^3 + x^2 + 5x - 2}{3x - 1}$
30. $\dfrac{3x^4 + 2x^3 + 5x^2 - 8x + 2}{3x - 1}$
31. $\dfrac{6x^4 + 8x^3 - 4x^2 - 8x - 5}{2x - 2}$
32. $\dfrac{3x^4 - x^3 - 14x^2 + 4x + 10}{3x + 2}$

33. In the study of electricity, the three resistances R_1, R_2, and R_3 combine to form the expression
$$\dfrac{R_1R_2 + R_2R_3 + R_1R_3}{R_1R_2R_3}.$$
Simplify this expression by dividing.

34. When dealing with certain electronic coils we may encounter the expression
$$\dfrac{30r^2}{3r + 0.6}.$$
Simplify this expression by dividing.

35. The sum of the first four terms of a geometric progression may be expressed as
$$\dfrac{a(1 + r + r^2 + r^3) - a(r + r^2 + r^3 + r^4)}{1 - r}.$$
Simplify this expression.

36. When we study optics we may find the expression
$$\dfrac{lD}{d}\left(n + \dfrac{1}{2}\right) - \dfrac{lD}{d}\left(n - \dfrac{1}{2}\right).$$
Simplify this expression.

1 • 9 Solving Linear Equations

Linear Equation in One Variable

A **linear equation in one variable,** x, is an equation of the form

$$ax + b = 0,$$

where a and b are constants and $a \neq 0$, or an equation that can be brought into that form by transposing terms. **Transposing** is another expression for adding the same quantity to both sides of the equation. For example,

Transposing

$$2x + 3 = -5, \qquad x - \frac{3}{2} = 2x + 1, \qquad x^2 + 2x = x^2 + 5$$

are linear equations, since

$2x + 3 = -5$	becomes	$2x + 8 = 0$	(Add 5 to both sides)
$x - \dfrac{3}{2} = 2x + 1$	becomes	$x + \dfrac{5}{2} = 0$	(Add $\dfrac{3}{2} - x$ to both sides)
$x^2 + 2x = x^2 + 5$	becomes	$2x - 5 = 0.$	(Add $-x^2 - 5$ to both sides)

We discuss linear equations in two variables later.

To solve the general linear equation $ax + b = 0$ ($a \neq 0$) for the variable x means to isolate the variable x on one side of the equation. We do this by undoing what has been done. That is, we must subtract b from both sides, then divide both sides by a. Subtracting b and dividing by a is the same as adding $(-b)$ and multiplying by $\dfrac{1}{a}$ as we discussed in section 1.2.

$$ax + b = 0,$$
$$ax = -b, \qquad \text{(Subtract } b \text{ from both sides)}$$
$$x = -\frac{b}{a}. \qquad \text{(Divide both sides by } a\text{)}$$

We see that $ax + b = 0$ has only one possible solution and that this solution checks, since

$$ax + b = 0,$$
$$a\left(-\frac{b}{a}\right) + b = 0, \qquad \text{(Substitute } -\frac{b}{a} \text{ for } x\text{)}$$
$$-b + b = 0, \qquad \text{(Multiply)}$$
$$0 = 0. \qquad \text{(Simplify)}$$

Therefore, we say that the linear equation $ax + b = 0$ ($a \neq 0$) has exactly one solution, namely $x = -\dfrac{b}{a}$.

When this solution is graphed on the real number line, the graph is a single point, namely $-\frac{b}{a}$. Example 1.59 illustrates the solution of a specific linear equation.

• **Example 1.59:** Solve $3x + 1 = 8$.

Solution:
$$3x + 1 = 8$$
$$3x = 7 \quad \text{(Subtract 1 from both sides)}$$
$$\left(\frac{1}{3}\right)(3x) = \left(\frac{1}{3}\right)(7) \quad \text{(Multiply both sides by } \frac{1}{3}\text{)}$$
$$x = \frac{7}{3}.$$

Graphed on the real number line we get the point $\frac{7}{3}$. This is illustrated in figure 1.5.

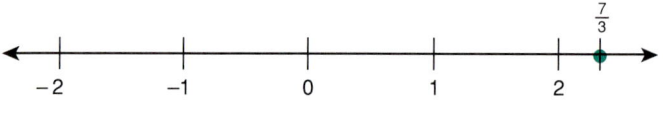

Figure 1.5

To solve some linear equations we must first simplify the expressions on either side of the equality. This is illustrated in examples 1.60, 1.61, and 1.62.

• **Example 1.60:** Solve $4(x - 3) = 2x + 1$.

Solution:
$$4(x - 3) = 2x + 1$$
$$4x - 12 = 2x + 1 \quad \text{(Expand left side)}$$
$$4x = 2x + 13 \quad \text{(Add 12 to both sides)}$$
$$2x = 13 \quad \text{(Subtract 2x from both sides)}$$
$$\frac{2x}{2} = \frac{13}{2} \quad \text{(Divide both sides by 2)}$$
$$x = \frac{13}{2}.$$

• *Example 1.61:* Solve $4\left(\dfrac{x}{4} - 1\right) = 4\left(\dfrac{2x}{4} - 6\right)$.

Solution: $4\left(\dfrac{x}{4} - 1\right) = 4\left(\dfrac{2x}{4} - 6\right)$

$\begin{aligned} x - 4 &= 2x - 24 & \text{(Expand both sides)} \\ x - 2x &= 4 - 24 & \text{(Transpose } 2x \text{ and transpose } 4) \\ -x &= -20 & \text{(Simplify both sides)} \\ x &= 20. & \text{(Divide both sides by } -1) \end{aligned}$

• *Example 1.62:* Solve $x^3 + 2x - 13 = x^2(x - 3) + 3(x^2 - 2x + 1)$.

Solution:
$\begin{aligned} x^3 + 2x - 13 &= x^2(x - 3) + 3(x^2 - 2x + 1) \\ x^3 + 2x - 13 &= x^3 - 3x^2 + 3x^2 - 6x + 3 & \text{(Expand right side)} \\ 2x - 13 &= -6x + 3 & \text{(Combine terms)} \\ 2x + 6x &= 13 + 3 & \text{(Transpose } -6x \text{ and transpose } -13) \\ 8x &= 16 & \text{(Combine terms)} \\ x &= 2. & \text{(Divide both sides by 8)} \end{aligned}$

Linear Equation in Two Variables

A **linear equation in two variables** x and y is an equation of the form

$$ax + by + c = 0, \quad (a \text{ and } b \text{ not both } 0)$$

or one that can be brought into this form by appropriate algebraic manipulation.

We can solve for either variable in terms of the other. To solve for x in terms of y, we transpose the terms by and c, then divide by a. Remember, the word "transpose" means that we add the same amount to both sides of an equation or we subtract the same amount from both sides of the equation.

$\begin{aligned} ax + by + c &= 0 \\ ax &= -by - c & \text{(Transpose)} \\ x &= -\dfrac{b}{a}y - \dfrac{c}{a}. & \text{(Divide by } a) \end{aligned}$

Similarly, if we solve for y in terms of x, we get

$\begin{aligned} ax + by + c &= 0 \\ by &= -ax - c & \text{(Transpose)} \\ y &= -\dfrac{a}{b}x - \dfrac{c}{b}. & \text{(Divide by } b) \end{aligned}$

When either of these expressions is graphed on a rectangular coordinate system, the graph is a straight line. Hence the terminology "linear" equation. We discuss the rectangular coordinate system and graph linear equations of this type in chapter 3.

Examples 1.63 and 1.64 illustrate how a linear equation in two variables can be solved for either of the variables.

• **Example 1.63:** Solve $2x + 3y - 6 = 0$ for x and for y.

   Solution: Solving for x, we get

   $$2x + 3y - 6 = 0$$
   $$2x = -3y + 6 \quad \text{(Transpose)}$$
   $$x = -\frac{3y}{2} + \frac{6}{2} \quad \text{(Divide)}$$
   $$x = -\frac{3y}{2} + 3. \quad \text{(Simplify)}$$

   Solving for y, we get

   $$2x + 3y - 6 = 0$$
   $$3y = -2x + 6 \quad \text{(Transpose)}$$
   $$y = -\frac{2x}{3} + \frac{6}{3} \quad \text{(Divide)}$$
   $$y = -\frac{2x}{3} + 2. \quad \text{(Simplify)}$$

   • • • • • • • • • •

• **Example 1.64:** The relation between Fahrenheit and Celsius temperatures is

   $$\frac{C}{5} = \frac{F - 32}{9}.$$

Solve this linear equation for C and for F.

   Solution: Solving for C we get

   $$\frac{C}{5} = \frac{F - 32}{9}$$
   $$\frac{5(C)}{5} = \frac{5(F - 32)}{9} \quad \text{(Multiply by 5)}$$
   $$C = \frac{5(F - 32)}{9}. \quad \text{(Simplify)}$$

Solving for F, We get

$$\frac{C}{5} = \frac{F-32}{9}$$

$$(9)(5)\frac{C}{5} = (9)(5)\frac{(F-32)}{9} \quad \text{(Multiply by 9 and 5)}$$

$$9C = 5(F-32) \quad \text{(Simplify)}$$

$$9C = 5F - 160 \quad \text{(Distribute)}$$

$$-5F = -9C - 160 \quad \text{(Transpose)}$$

$$\frac{-5F}{-5} = \frac{-9C}{-5} - \frac{160}{-5} \quad \text{(Divide by } -5\text{)}$$

$$F = \frac{9C}{5} + 32. \quad \text{(Simplify)}$$

Trial Problems 1.9

Before you begin the section exercises, warm up with these problems. Complete answers are included in the answer key.

Solve for the variable.

1. $4y - 1 = 23$
2. $\frac{x}{2} - 5 = 3$
3. $4(2x + 1) = 2(3x + 8)$
4. $1 + \frac{y}{9} = 3$
5. $(2a - 3)^2 = 4a^2 - 8$

Exercises 1.9

In exercises 1–8, solve for the variable and graph the solution on a number line.

1. $4x + 1 = 25$
2. $3x + 1 = 25$
3. $\frac{y}{2} - 4 = 3$
4. $\frac{y}{2} - 3 = 4$
5. $6s + 3 = 2s - 5$
6. $5t + 2 = 3t + 8$
7. $-3(2x + 1) = 4(4x - 3)$
8. $-3(3x + 1) = 4(5x - 4)$

In exercises 9–12, solve for x.

9. $3\left(\frac{x}{3} - 1\right) = 3\left(\frac{2x}{3} + 6\right)$
10. $2\left(\frac{x}{2} + 1\right) = 2\left(\frac{3x}{2} + 5\right)$
11. $(2x - 3)(x + 2) = 4 - x + 2x^2$
12. $(2x - 3)^2 = 4x^2 - 8$
13. $5x + 2y = 5$. Solve for x in terms of y.
14. $4x + 3y = 6$. Solve for y in terms of x.

In exercises 15–22, solve for x.

15. $1 + \dfrac{x}{9} = 3$

16. $4 + \dfrac{x}{5} = 34$

17. $(x - 1)^2 = (x - 2)^2 + 5$

18. $(x + 1)^2 = (x + 2)^2 - 5$

19. $x^3 + 5x + 3 = x^2(x - 2) + 2(x^2 - 4x - 1)$

20. $x^3 - 6x + 2 = x^2(x - 3) + 3(x^2 + 2x + 1)$

21. $ax + b = cx + d, (a \neq c)$

22. $ax - b = cx - d, (a \neq c)$

The formula relating distance d, rate r, and time t is $d = rt$. This formula is the basis for exercises 23–26.

23. In problems involving distance, rate, and time, this equation involving t may arise

$$110 = 55t.$$

Solve for t.

24. In problems involving distance, rate, and time, this equation involving rate may arise

$$110 = 2r.$$

Solve for r.

25. In problems involving distance, rate, and time, this equation relating distance and time may arise

$$d = 55t.$$

Solve for t in terms of d.

26. In problems involving distance, rate, and time, this equation involving distance and rate may arise

$$\dfrac{d}{4} = r.$$

Solve for d in terms of r.

27. A board 30 ft long is cut into two pieces. One piece has length x and the other has length y. Express this information as a linear equation in variables x and y.

28. A trust fund of $50,000 is divided among two sons and a daughter so that each gets an equal share. Express this information as a linear equation in variables s and d.

29. The potential difference V (in volts) between two points in a circuit when it starts at 115 volts and increases at a rate of 20 volts per second, where t represents the time in seconds, is given by the formula

$$V = 115 + 20t.$$

Find the value of V when $t = 12$ seconds.

30. In the study of motion, the equation $v = v_0 - gt$ relates the initial velocity v_0, the time t, and the acceleration due to gravity g to the final velocity v. Solve this equation for g.

31. According to Hooke's law, the elongation of a spring due to stretching is proportional to the tension. Hooke's law is expressed as

$$E = cT,$$

where E is the elongation, T is the tension, and c is a constant. Solve this equation for c.

32. The magnetomotive force F around a path in a 10-amp circuit is calculated with the formula

$$F = 4\pi N,$$

where N is the number of times the electric circuit links the path. Solve the formula for N.

1 • 10 Ratio and Proportion

Ratio A **ratio** is a comparison of two numbers by division. The two numbers must represent quantities that have the same units. Ratios can be written in many forms. For example, the ratio of 3 to 7 can be written as $3/7$, $3 \div 7$, $\dfrac{3}{7}$, or $3:7$.

Although a ratio compares two quantities that have the same units, a ratio has no units of measurement and is simply a number.

SECTION 1.10 RATIO AND PROPORTION

- **Example 1.65:** One gear has 72 teeth and a second gear has 32 teeth. Find the ratio of the number of teeth in the first gear to the number of teeth in the second gear.

    Solution: The ratio is $\dfrac{72 \text{ teeth}}{32 \text{ teeth}} = \dfrac{72}{32} = \dfrac{9}{4}$. Note that the ratio has no units and is usually written in simplest form (reduced form).

- **Example 1.66:** The length of a rectangular template is 2 feet and the width is 18 inches. Find the ratio of the length to the width.

    Solution: Before the ratio can be found the measurements used must be in the same units. Two feet is 24 inches, so the ratio is $\dfrac{24 \text{ in.}}{18 \text{ in.}}$

    $= \dfrac{4(6)}{3(6)} = \dfrac{4}{3}$, or 4:3.

Rate A **rate** is a comparison of two quantities that have different units.

- **Example 1.67:** If a person travels 165 miles in three hours, what is the rate in miles per hour?

    Solution: The miles-to-hour rate is $\dfrac{165 \text{ mi}}{3 \text{ hr}} = \dfrac{55 \text{ mi}}{\text{hr}}$.
    Note that the rate is also written in its simplest form, and that the denominator should be in terms of one unit.

Proportion A **proportion** is an equation that states the equality of two ratios or rates. An example of a proportion that equates two rates might be

$$\dfrac{30 \text{ mi}}{4h} = \dfrac{15 \text{ mi}}{2h}.$$

A proportion that equates two ratios might look like this:

$$\dfrac{4}{6} = \dfrac{8}{12} \quad \text{or} \quad \dfrac{5}{10} = \dfrac{x}{20}.$$

If the ratio $\dfrac{a}{b}$ is equal to the ratio $\dfrac{c}{d}$, then

$$\dfrac{a}{b} = \dfrac{c}{d}$$

Extremes is a proportion and is read *a is to b as c is to d*. In this proportion, *a* and *d* are called
Means **extremes** and *b* and *c* are called **means**.

Since a proportion is also an equation, we note that all of the rules that relate to equations hold for proportions as well.

If we consider the general proportion $\dfrac{a}{b} = \dfrac{c}{d}$ and multiply both sides by bd we get

$$bd\left(\dfrac{a}{b}\right) = bd\left(\dfrac{c}{d}\right).$$

Simplifying, this becomes

$$ad = bc.$$

Conversely, if we begin with $ad = bc$ and divide both sides by bd we get

$$\dfrac{ad}{bd} = \dfrac{bc}{bd}.$$

Simplifying, this becomes

$$\dfrac{a}{b} = \dfrac{c}{d}.$$

Thus we have the property that the product of the extremes is equal to the product of the means. That is, if we cross multiply the products should be equal, or the cross-products should be equal. We will call this the **cross-product rule**.

CROSS-PRODUCT RULE

If $\dfrac{a}{b} = \dfrac{c}{d}$, then $ad = bc$ and if $ad = bc$, then $\dfrac{a}{b} = \dfrac{c}{d}$ where $b, d \neq 0$.

Examples 1.68 and 1.69 illustrate this property.

• *Example 1.68:* Is $\dfrac{3}{5} = \dfrac{12}{20}$?

   Solution: Since the product of the extremes, $(3)(20) = 60$ and the product of the means, $(5)(12) = 60$, or, since the cross products are both equal, the proportion is true.

• *Example 1.69:* Is $\dfrac{5}{6} = \dfrac{16}{18}$?

   Solution: Since the product of the extremes, $(5)(18) = 90$ and the product of the means, $(6)(16) = 96$, or, since the cross products are *not* equal, the proportion is *not* a true proportion.

Examples 1.70 through 1.73 illustrate some additional properties of proportions. Example 1.70 illustrates that the means may be interchanged or the extremes may be interchanged and the proportion remains true.

- *Example 1.70:* If $\dfrac{6}{9} = \dfrac{2}{3}$, then $\dfrac{3}{9} = \dfrac{2}{6}$, or $\dfrac{6}{2} = \dfrac{9}{3}$

    Solution: $(6)(3) = 18$ $(3)(6) = 18$ $(6)(3) = 18$
    $(9)(2) = 18$ $(9)(2) = 18$ $(2)(9) = 18$

    Since the cross products are all equal, each proportion is true.

Note that in each proportion the ratios change but the equality remains true. We generalize this example into a rule, which we call the **extreme-means rule**.

Extreme-Means Rule

EXTREMES-MEANS RULE

If $\dfrac{a}{b} = \dfrac{c}{d}$, then $\dfrac{d}{b} = \dfrac{c}{a}$, or if $\dfrac{a}{b} = \dfrac{c}{d}$, then $\dfrac{a}{c} = \dfrac{b}{d}$.

Example 1.71 illustrates that a proportion remains true after the ratios have been inverted.

- *Example 1.71:* If $\dfrac{2}{9} = \dfrac{4}{18}$, then $\dfrac{9}{2} = \dfrac{18}{4}$.

    Solution: $(2)(18) = 36$ $(9)(4) = 36$
    $(9)(4) = 36$ $(2)(18) = 36$

    Since the cross products are equal in every case, both proportions are true.

Inverted Ratios Rule

We generalize this example into a rule, which we call the **inverted ratios rule**.

INVERTED RATIOS RULE

If $\dfrac{a}{b} = \dfrac{c}{d}$, then $\dfrac{b}{a} = \dfrac{d}{c}$.

Solve a Proportion

To **solve a proportion** means to find the missing number or missing numbers, if possible.

• **Example 1.72:** Solve the proportion $\dfrac{x}{4} = \dfrac{3}{2}$.

Solution: $\dfrac{x}{4} = \dfrac{3}{2}$

$(2)(x) = (4)(3)$ (Cross multiply)

$2x = 12$ (Simplify)

$x = 6.$ (Divide by 2)

Mean Proportional   If the means of a proportion are equal, that number is called the **mean proportional** between the extremes.

For example, in the proportion

$$\dfrac{2}{4} = \dfrac{4}{8},$$

4 is a mean proportional between 2 and 8.

Example 1.73 illustrates this idea.

• **Example 1.73:** Solve the proportion $\dfrac{2}{x} = \dfrac{x}{8}$ for the mean proportional x.

Solution: $\dfrac{2}{x} = \dfrac{x}{8}$

$x^2 = (2)(8)$ (Cross multiply)

$x^2 = 16$ (Multiply)

$x = \pm 4.$ (Extract roots)

Note that there are two solutions for this mean proportion.

Ratios and proportions are useful for finding missing parts of congruent or similar geometric figures.

Congruent   Geometric figures are said to be **congruent** if they have the same size and shape. The triangles in figure 1.6 are congruent.

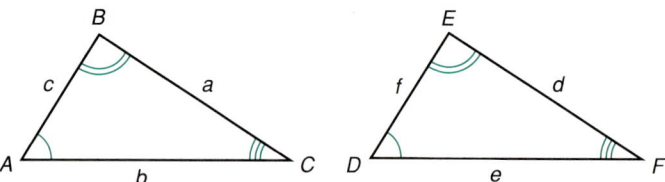

Figure 1.6

SECTION 1.10 RATIO AND PROPORTION

Notice that the measures of the corresponding angles are equal as well as the measures of the corresponding sides. That is,

$$\sphericalangle A = \sphericalangle D \quad \sphericalangle B = \sphericalangle E, \quad \sphericalangle C = \sphericalangle F,$$
$$a = d, \quad b = e, \quad c = f.$$

Similar   We give a special name to geometric figures that have the same shape, but not necessarily the same size. The triangles in figure 1.7 are **similar.**

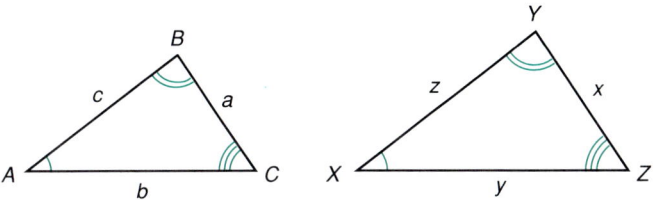

Figure 1.7

Similar figures such as the triangles in figure 1.7 have some important properties.

PROPERTIES OF SIMILAR TRIANGLES

1. The measures of corresponding angles are equal.
2. The measures of corresponding sides are in proportion. That is,

$$\sphericalangle A = \sphericalangle X, \quad \sphericalangle B = \sphericalangle Y, \quad \sphericalangle C = \sphericalangle Z,$$
$$\frac{a}{x} = \frac{b}{y}, \quad \frac{a}{x} = \frac{c}{z}, \quad \frac{b}{y} = \frac{c}{z}.$$

If we know that two geometric figures are similar, and if we know the lengths of some of the sides, we can often use proportions to find the lengths of some of the missing sides.

• *Example 1.74:* In figure 1.8 △ABC is similar to △XYZ. Use proportions to find the lengths of y and x.

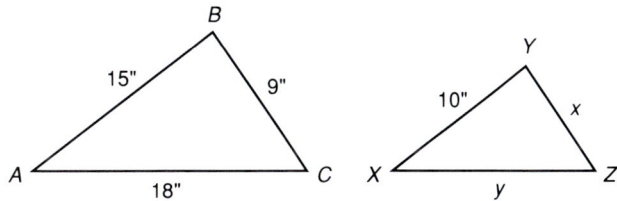

Figure 1.8

Solution 1:
$$\frac{y}{18} = \frac{10}{15} \qquad \frac{x}{10} = \frac{9}{15}$$
$$15y = (10)(18) \qquad 15x = (9)(10)$$
$$y = \frac{(10)(18)}{15} \qquad x = \frac{(9)(10)}{15}$$
$$y = 12''. \qquad x = 6''.$$

Solution 2: Since the means can be interchanged without losing equality, the values of y and x could have been found as follows.
$$\frac{y}{10} = \frac{18}{15} \qquad \frac{x}{9} = \frac{10}{15}$$
$$15y = (18)(10) \qquad 15x = (10)(9)$$
$$y = \frac{(18)(10)}{15} \qquad 15x = \frac{(10)(9)}{15}$$
$$y = 12'' \qquad x = 6''.$$

Trial Problems 1.10

Before you begin the section exercises, warm up with these problems. Complete answers are included in the answer key.

Express each as a ratio or a rate.

1. 20 to 15
2. 2 yards to 3 feet
3. 110 miles in 2 hours
4. $600 in 3 years

Solve for x.

5. $2:3 = 6:x$

Exercises 1.10

Express the following as ratios in simplest form.

1. 15 to 20
2. 18 to 27
3. $\frac{1}{2}$ to $\frac{3}{10}$
4. $\frac{1}{3}$ to $\frac{4}{9}$
5. 3 yards to 2 feet
6. 2 yards to 8 feet
7. 20 cm to 30 mm
8. 15 cm to 50 mm
9. 6 lb 2 oz to 4 lb 8 oz
10. 4 lb 2 oz to 2 lb 1 oz
11. 2 ft 7 in. to 7 ft 9 in.
12. 1 ft 10 in. to 3 ft 8 in.

SECTION 1.10 RATIO AND PROPORTION

Express each rate in terms of the first quantity to the second quantity. Simplify.

13. 30 mi in 4 hrs
14. 25 mi in 5 hrs
15. 2 pounds of sugar in 6 gallons of water
16. 3 pounds of sugar in 12 gallons of water
17. $500 in 4 years
18. $400 in 5 years
19. 16 tiles for each 9 sq ft of floor area
20. 48 tiles for each 27 sq ft of floor area
21. Three ounces of medication for each 150 lbs of weight.
22. Two ounces of medication for each 140 lbs of weight.

Solve the following proportions.

23. $2:3 = 4:x$
24. $3:9 = 6:x$
25. $\dfrac{8}{x+3} = \dfrac{4}{x}$
26. $\dfrac{2}{x+3} = \dfrac{6}{5x+5}$

27. Sixteen asphalt tiles are required to tile 9 sq ft. How many tiles are necessary to tile 234 sq ft?
28. Sixteen asphalt tiles are required to tile 9 sq ft. How many square feet can be tiled with 256 tiles?
29. Statistics show that two out of five people will vote in a particular election. At this rate, how many people would be expected to vote in a city of 250,000?
30. Statistics show that three out of seven people will vote in a particular election. At this rate, what is the population of a city if 9000 people voted?
31. In figure 1.9 △ABC is similar to △DEF and $f = 12'$. Find the length of the e and d.
32. Find f and d in figure 1.9 if $e = 30'$.

33. A pattern for a rectangular end table is printed in a book. The pattern measures 6″ by $9\frac{1}{2}$″. If the width of the finished table is to be 24″, what will be the finished length? The height of the table is 32″.
34. Find the finished width of the end table in exercise 33 if the finished length is 28.5″ and the height is 30″.
35. If an American National Standard thread has 12 threads per inch, how many threads are there on 6 inches of threaded rod that is $\frac{3}{4}$″ thick?
36. If an Acme thread has 16 threads per inch, how many threads are needed on 2 inches of threaded bolt? Assume that the bolt is $\frac{5}{8}$″ thick.

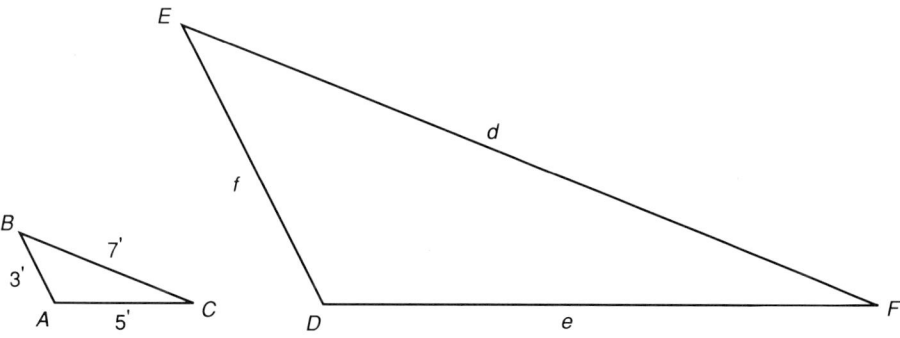

Figure 1.9

1 • 11 Variation

In industry, science, and technology, many formulas are established through observation and experience. In many applied situations, the ratio of one quantity to another remains the same and is the basis for establishing many important formulas. Consider examples 1.75, 1.76, and 1.77.

• *Example 1.75:* In a given measurement, the ratio of the length in inches i to the length in feet f is 12. This may be written as

$$\frac{i}{f} = 12.$$

• *Example 1.76:* The ratio of the electrical resistance R of a given wire to the length L of the wire is always the same. This may be written as

$$\frac{R}{L} = k \qquad (k \text{ is a constant}).$$

• *Example 1.77:* If we are traveling at 55 miles per hour, the rate of the distance d traveled to the time t in hours is constant. This may be written as

$$\frac{d}{t} = 55.$$

Each of these examples illustrates the same basic idea, that certain ratios, or rates, remain constant. In general, we express this as

$$\frac{y}{x} = k,$$

Constant of Proportionality

where x and y are variables and k is a constant called the **constant of proportionality**. If we solve this equation for y, we get

$$y = kx.$$

Direct Variation

This type of relation is called a **direct variation**. There are many ways of expressing this equation in words. Some are:

"y is proportional to x"

"y is directly proportional to x"

"y varies directly as x."

SECTION 1.11 VARIATION

In any variation there are at least two variables and a constant of proportionality. If we know all but one of these values we may use them to find the missing value. Very often we are given a set of values for the variables and are asked to find the constant of proportionality. Then with this constant and another set of values for all but one of the variables, we are asked to find the value of the missing variable.

This procedure can be summarized in a three-step process, which we call the **proportion solving strategy**.

Proportion Solving Strategy

PROPORTION SOLVING STRATEGY

1. Set up an equation representing the variation.
2. Use this equation and the first set of values to find k.
3. Solve for the value of the missing or required variable.

We illustrate this strategy with examples 1.78 and 1.79.

• **Example 1.78:** If y varies directly as x, and $x = 3$ when $y = 12$, find the value of y when $x = 5$.

Solution 1: Since y varies directly as x, the relation between x and y is

$$y = kx. \quad \text{(Set up the equation)}$$

Substituting the given values for x and y, we get

$$y = kx$$
$$12 = k(3) \quad \text{(Substitute for } x \text{ and } y\text{)}$$
$$k = 4. \quad \text{(Find the value of } k\text{)}$$

Now with $k = 4$ and $x = 5$ (the second value of x), we get

$$y = kx$$
$$y = (4)(5) \quad \text{(Substitute for } k \text{ and } x\text{)}$$
$$y = 20. \quad \text{(Solve for the required value)}$$

Solution 2: Since y varies directly as x, the ratios $\frac{12}{3}$ and $\frac{y}{5}$ must be equal. Hence,

$$\frac{y}{5} = \frac{12}{3}$$
$$3y = (5)(12) \quad \text{(Cross multiply)}$$
$$3y = 60 \quad \text{(Simplify)}$$
$$y = 20. \quad \text{(Divide by 3)}$$

In solution 2 we found y without first finding the constant of proportionality.

• **Example 1.79:** The distance d that an object falls under the influence of gravity is directly proportional to the square of the time t of the fall. If an object falls 64 ft in 2 secs, how far does it fall in 5 secs?

Solution: Since d varies directly as the square of t we write

$$d = kt^2. \quad \text{(Set up the equation)}$$

Now, $d = 64$ when $t = 2$, so we write

$$d = kt^2$$
$$64 = k(2)^2$$
$$64 = 4k$$
$$16 = k. \quad \text{(Solve for } k\text{)}$$

When $t = 5$ and $k = 16$, we write

$$d = kt^2$$
$$d = 16(5)^2$$
$$d = 16(25)$$
$$d = 400 \text{ ft.} \quad \text{(Solve for the required value)}$$

• • • • • • • • • •

Inverse Variation

A second type of variation that occurs quite frequently in technical and industrial problems is the **inverse variation,** which we express by the relation

$$y = \frac{k}{x},$$

where x and y are variables and k is again the constant of proportionality. We say:

"y varies inversely as x"

"y is inversely proportional to x."

• **Example 1.80:** If r varies inversely as t, and $r = 55$ when $t = 2$, find r when $t = 5$.

Solution: Since r varies inversely as t, we write

$$r = \frac{k}{t}. \quad \text{(Set up the equation)}$$

Substituting the given values for r and t, we get

$$r = \frac{k}{t}$$

$$55 = \frac{k}{2} \quad \text{(Substitute for } r \text{ and } t\text{)}$$

$$110 = k. \quad \text{(Solve for } k\text{)}$$

With this value of k for the constant of proportionality, and the second value of r, we write

$$r = \frac{k}{t}$$

$$r = \frac{110}{5} \quad \text{(Substitute for } k \text{ and } t\text{)}$$

$$r = 22. \quad \text{(Solve for the required value)}$$

Combined or Joint Variation

Finally, it is possible to relate one variable to more than one other variable. When such is the case we say that we have a **combined variation** or a **joint variation.** For example, if y varies directly as x and inversely as z, we write

$$y = \frac{kx}{z}.$$

• *Example 1.81:* The force F between two electrically charged particles with charges q_1 and q_2 varies directly as the product of the charges. Express this as a combined variation.

Solution: $F = kq_1q_2$.

• *Example 1.82:* The electrical resistance R of a wire varies directly as the length l and inversely as the square of the diameter d of the wire. If a wire 400 feet long with diameter 0.16 inch has a resistance of 128 ohms, how much resistance will there be if the wire is only 50 feet long?

Solution: Since R varies directly with length l and inversely with the square of the diameter we write

$$R = \frac{kl}{d^2}. \quad \text{(Set up the equation)}$$

Substituting the given values of R, l, and d, we get

$$R = \frac{kl}{d^2}$$

$$128 = \frac{k(400)}{(0.16)^2} \quad \text{(Use your calculator)}$$

$$k = 0.008192. \quad \text{(Solve for } k\text{)}$$

With this value of k and the new value for l we write

$$R = \frac{kl}{d^2}$$

$$R = \frac{(0.008192)(50)}{(0.16)^2} \quad \text{(Use your calculator)}$$

$$R = 16 \text{ ohms.} \quad \text{(Solve for the required value)}$$

Trial Problems 1.11

Before you begin the section exercises, warm up with these problems. Complete answers are included in the answer key.

Express each statement as an equation.

1. r varies directly as s.
2. u is proportional to v.
3. t varies inversely as r.
4. p varies inversely as the square of q.
5. y varies directly as the square of x and inversely as the square root of z.

Exercises 1.11

In exercises 1–22 express the statement as an equation.

1. s varies directly as t.
2. p varies directly as r.
3. y is proportional to z.
4. t is proportional to s.
5. p is inversely proportional to q.
6. u is inversely proportional to v.
7. s varies inversely as t.
8. t varies inversely as s.
9. r varies directly as the square of d.
10. s varies directly as the square of t.
11. p varies inversely as the square of q.
12. r varies inversely as the square of s.
13. y varies directly as s and inversely as t.
14. x varies directly as p and inversely as q.
15. v varies directly as the cube of s and inversely as the square of t.
16. x varies directly as the product of y and z and inversely as d squared.

In the following exercises, determine the constant of proportionality.

17. y varies directly as x, and $y = 20$ when $x = 5$.
18. s varies directly as t, and $s = 30$ when $t = 6$.
19. r varies directly as s and inversely as t squared, and $s = 150$ when $t = 5$ and $r = 12$.
20. p varies directly as q and inversely as r squared, and $q = 81$ when $r = 3$ and $p = 27$.
21. t is directly proportional to n and varies inversely as p, and $t = 20$ when $n = 15$ and $p = 3$.
22. y varies directly as x and is inversely proportional to z, and $y = 27$ when $x = 18$ and $z = 6$.

In exercises 23–28 find the required value.

23. s varies directly as t, and $s = 30$ when $t = 6$. Find s when $t = 2$.

24. u varies directly as v, and $u = 20$ when $v = 5$. Find u when $v = 3$.

25. s varies inversely as the square of t, and $s = 2$ when $t = 3$. Find s when $t = 6$.

26. p varies inversely as the square of q, and $p = \frac{1}{2}$ when $q = 4$. Find p when $q = 3$.

27. r is directly proportional to s cubed and inversely proportional to t squared, and $r = 16$ when $s = 2$ and $t = 3$. Find r when $s = 3$ and $t = 9$.

28. v is directly proportional to t squared and inversely proportional to u cubed, and $v = 18$ when $t = 3$ and $u = 2$. Find v when $t = 3$ and $u = 4$.

29. The pressure exerted by a liquid at a given point varies directly as the depth of the point beneath the surface of the liquid. If a certain liquid exerts a pressure of 50 pounds per square foot at a depth of 10 feet, find the pressure at a depth of 40 feet.

30. The volume v of a gas varies directly as its temperature T and inversely as its pressure P. A gas occupies 20 cubic feet at a temperature of 300 K (kelvin) and a pressure of 30 pounds per square inch. Find the volume if the temperature is raised to 360 K and the pressure is decreased to 20 pounds per square inch.

31. The maximum safe uniformly distributed load L for a horizontal beam varies jointly as the breadth b and the square of the depth d and inversely as the length l. An 8 foot beam with $b = 3$ in. and $d = 4$ in. will support a uniformly distributed load up to 720 pounds. How many uniformly distributed pounds will an 8 foot beam support if $b = 3$ inches and $d = 6$ inches?

32. The intensity of illumination I of a light source varies inversely as the square of the distance d from the source, and $I = 20$ units when $d = 5$ feet. Find I when $d = 10$.

33. The kinetic energy of an object varies directly as the square of its velocity. If an object with a velocity of 24 meters per second has a kinetic energy of 19,200 joules, what will be the kinetic energy of the same object when its velocity is 48 meters per second?

34. The break horsepower (bhp) of an engine, as measured on a dynamometer, is directly proportional to the product of the length (R) of the breaking arm in feet, the rpm (N) of the engine, and the force (F) in pounds required to break the engine. Find the constant of proportionality if the bhp is 57.12, R is 3 feet, F is 100 pounds, and N is 1000 rpm. Express your answer to four significant digits.

35. In aerodynamics, the two flat-plate areas f and f', the drag D, and the dynamic pressure q are related by the formulas

$$f' = \frac{D}{1.28q} \quad \text{and} \quad f = \frac{D}{q}.$$

Solve these formulas for f' in terms of f.

36. The common flexure formula dealing with the strength of a beam uses the following variables.

P = working unit stress of the metal.
I = moment of inertia of the girder about its axial base.
M = bending moment.
Y = distance from the neutral axis at the most stressed fiber of the girder.

Express this formula if M is directly proportional to the product of I and P and inversely proportional to Y.

1·12 Summary of Terms, Formulas, Rules, and Procedures

Terms

Absolute Value (p. 6)
Accuracy (p. 6)
Additive Inverse (p. 11)
Base (p. 14)
Binomial (p. 33)
Combined Variation (p. 59)
Congruent (p. 52)
Constant of Proportionality (p. 56)
Constants (p. 5)
Counting Numbers (p. 2)
Degree of a Monomial (p. 33)
Degree of a Monomial in a Variable (p. 33)
Degree of a Polynomial (p. 33)
Direct Variation (p. 56)
Double Negative Property (p. 11)
Double Reciprocal Property (p. 11)
Exponents (p. 14)
Extreme-Means Rule (p. 51)
Extremes (p. 49)
Graph (p. 4)
Index (p. 24)
Integers (p. 2)
Inverse Variation (p. 58)
Inverted Ratios Rule (p. 51)

Irrational Numbers (p. 4)
Is Equal To (p. 5)
Is Greater Than (p. 5)
Is Less Than (p. 5)
Joint Variation (p. 59)
Like Terms (p. 33)
Linear Equation in One Variable (p. 43)
Linear Equation in Two Variables (p. 45)
Literal Number (p. 5)
Mean Proportional (p. 52)
Means (p. 49)
Monomial (p. 33)
Multiplicative Inverse (p. 12)
Natural Numbers (p. 2)
nth Power (p. 14)
Order of Operations (p. 6)
Perfect Square (p. 24)
Polynomial (p. 33)
Positive Integers (p. 2)
Precision (p. 6)
Principal nth Root (p. 24)
Principal Square Root (p. 24)
Proportion (p. 49)
Proportion Solving Strategy (p. 57)

Radical (p. 24)
Radical Form (p. 25)
Radicand (p. 24)
Rate (p. 49)
Ratio (p. 48)
Rational Number (p. 2)
Rationalizing the Denominator (p. 27)
Real Number Line (p. 4)
Real Numbers (p. 4)
Repeating Decimals (p. 2)
Scientific Notation (p. 19)
Significant Digits (p. 6)
Similar (p. 53)
Similar Terms (p. 33)
Solve a Proportion (p. 51)
Square Root (p. 23)
Term (p. 33)
Terminating Decimals (p. 2)
Transposing (p. 43)
Trinomial (p. 33)
Variables (p. 5)
Whole Numbers (p. 2)
Zero Factor Property (p. 10)

Formulas

Direct Variation: $y = kx$ (k is the constant of proportionality) (1.11)

Inverse Variation: $y = \dfrac{k}{x}$ (k is the constant of proportionality) (1.11)

Joint Variation: $y = \dfrac{kx}{z}$ (k is the constant of proportionality) (1.11)

Rules and Procedures

- To add real numbers whose signs are alike, add the numbers and use the sign of the original numbers. (1.1)

- To add real numbers whose signs are unlike, subtract the smaller absolute value from the larger, and attach the sign of the number with the larger absolute value. (1.1)

- To subtract one real number from another, change the sign of the second number and proceed as in addition. (1.1)

- Multiplication rules for real numbers.
  1. If the signs of all of the numbers are positive, multiply the numbers and attach a positive sign to the answer.

2. If there is an odd number of negative numbers to be multiplied, multiply the numbers and attach a negative sign to the answer.
3. If there is an even number of negative numbers to be multiplied, multiply the numbers and attach a positive sign to the answer. (1.1)

- When dividing two real numbers, the answer is positive if both numbers have the same sign. The answer is negative if the two numbers have opposite signs. (1.1)

- Axioms for real numbers (1.2)

$a + b$ is a unique real number.	Closure for addition
$a + b = b + a$.	Commutative property of addition
$(a + b) + c = a + (b + c)$.	Associative property of addition
ab is a unique real number.	Closure for multiplication
$ab = ba$.	Commutative property of multiplication
$(ab)c = a(bc)$.	Associative property of multiplication
$a(b + c) = (ab) + (ac)$.	Distributive property
There exists a unique number 0 with the property $a + 0 = a$ and $0 + a = a$.	Identity element for addition
There exists a unique number 1 with the property $a \cdot 1 = a$ and $1 \cdot a = a$.	Identity element for multiplication
For each real number a, there exists a unique real number $-a$ (called the negative of a) with the property $a + (-a) = 0$ and $(-a) + (a) = 0$.	Negative, or additive-inverse, property
For each real number a, except 0, there exists a unique real number $\dfrac{1}{a}$ (called the reciprocal of a) with the property $a\left(\dfrac{1}{a}\right) = 1$ and $\left(\dfrac{1}{a}\right)a = 1$.	Reciprocal, or multiplicative-inverse, property

- ADDITION PROPERTY OF EQUALITY
  If $a = b$, then $a + c = b + c$. (1.2)
- MULTIPLICATION PROPERTY OF EQUALITY
  If $a = b$, then $ac = bc$. (1.2)
- ZERO FACTOR PROPERTY
  $(a)(0) = 0$ (1.2)

- **DOUBLE NEGATIVE PROPERTY** (1.2)
  $-(-a) = a$, where the negative of a is $-a$
- **DOUBLE RECIPROCAL PROPERTY** (1.2)
  $\dfrac{1}{\frac{1}{a}} = a$, where $\dfrac{1}{a}$ is the reciprocal of a.
- **DEFINITION OF SUBTRACTION** (1.2)
  $a - b$ means $a + (-b)$.
- **DEFINITION OF DIVISION** (1.2)
  $a \div b$ means $a\left(\dfrac{1}{b}\right)$
- Rules of Exponents. If a and b are nonzero real numbers and if m and n are integers, then (1.3)
  1. $a^m a^n = a^{m+n}$
  2. $\dfrac{a^m}{a^n} = a^{m-n}$
  3. $a^{-n} = (a^n)^{-1} = (a^{-1})^n = \left(\dfrac{1}{a}\right)^n = \dfrac{1}{a^n}$
  4. $a^0 = 1$
  5. $a(a^{-1}) = 1$
  6. $(a^m)^n = a^{mn}$
  7. $(ab)^n = a^n b^n$
  8. $\left(\dfrac{a}{b}\right)^n = \dfrac{a^n}{b^n}$
- Rules for Square Roots. Let a and b be positive real numbers. Then (1.4)
  1. $(\sqrt{a})^2 = a$
  2. $\sqrt{a^2} = a$
  3. $\sqrt{ab} = \sqrt{a}\sqrt{b}$
  4. $\sqrt{\dfrac{a}{b}} = \dfrac{\sqrt{a}}{\sqrt{b}}$

- **THE PRINCIPAL nTH ROOT OF a** (1.5)
  $a^{1/n} = \sqrt[n]{a}$
- **CROSS-PRODUCT RULE** (1.10)
  If $\dfrac{a}{b} = \dfrac{c}{d}$, then $ad = bc$ and if $ad = bc$, then $\dfrac{a}{b} = \dfrac{c}{d}$ where $b, d \neq 0$.
- **EXTREMES-MEANS RULE** (1.10)
  If $\dfrac{a}{b} = \dfrac{c}{d}$, then $\dfrac{d}{b} = \dfrac{c}{a}$, or
  if $\dfrac{a}{b} = \dfrac{c}{d}$, then $\dfrac{a}{c} = \dfrac{b}{d}$.
- **INVERTED RATIOS RULE** (1.10)
  If $\dfrac{a}{b} = \dfrac{c}{d}$, then $\dfrac{b}{a} = \dfrac{d}{c}$.
- **PROPERTIES OF SIMILAR TRIANGLES** (1.10)
  1. The measures of corresponding angles are equal.
  2. The measures of corresponding sides are in proportion. That is,
  $\angle A = \angle X$, $\angle B = \angle Y$, $\angle C = \angle Z$,
  $\dfrac{a}{x} = \dfrac{b}{y}$, $\dfrac{a}{x} = \dfrac{c}{z}$, $\dfrac{b}{y} = \dfrac{c}{z}$.
- **PROPORTION SOLVING STRATEGY** (1.11)
  1. Set up an equation representing the variation.
  2. Use this equation and the first set of values to find k.
  3. Solve for the value of the missing or required variable.

1·13 Chapter 1 Review Exercises

Place the correct symbol ($<$, $>$, or $=$) between the two numbers.

1. 3   5
2. -5 -3
3. 4 $\dfrac{8}{2}$
4. $\dfrac{6}{3}$ 3
5. $\dfrac{2}{3}$ $\dfrac{5}{6}$
6. $\dfrac{3}{4}$ $\dfrac{7}{8}$

Find the additive inverse (opposite) of each number.

7. 4
8. 6
9. -5
10. -8
11. $-(-3)$
12. $-(-2)$

Fill in the blank to make the statement an application of the stated property.

13. $5 + 3 = 3 + \underline{}$ Commutative property of addition
14. $(9 + 8) + 2 = 9 + (\underline{})$ Associative property of addition
15. $6(4 + 7) = (4 + 7)\underline{}$ Commutative property of multiplication
16. $3(5 + 2) = (3)(5) + \underline{}$ Distributive property
17. $-(-6) = \underline{}$ Double negative property
18. $8 + \underline{} = 8$ Additive identity property

Simplify each expression.

19. $4 + (-9)$
20. $-5 - (-8)$
21. $(4 - 2 + 7) - 3$
22. $(4 - 2) + (7 - 3)$

Compute.

23. $(-3)^2$
24. $\left(-\dfrac{2}{3}\right)^2$
25. $\left(-\dfrac{3}{2}\right)^{-1}$
26. $\left(\dfrac{1}{2}\right)^{-3}$
27. $\dfrac{2^3}{3^{-2}}$
28. $\dfrac{3^2}{2^{-3}}$

Express as powers of two.

29. 16^{-3}
30. $\left(\dfrac{1}{2}\right)^6$
31. $[(2^3)(8^4)]^3$
32. $(2)(4^3)(8^2)$
33. $\left(\dfrac{1}{2}\right)\left(\dfrac{1}{4}\right)\left(\dfrac{1}{8}\right)^{-1}$
34. $\left[\left(\dfrac{1}{2}\right)\left(\dfrac{1}{4}\right)\left(\dfrac{1}{8}\right)\right]^{-1}$

Express in scientific notation.

35. 543.2
36. 23.45
37. 0.00678
38. 0.00087

Approximate with your calculator.

39. $(2.03)^{-90}$
40. $(3.02)^{60}$
41. $(1.8705 \times 10^8)^{-7}$
42. $(2.178 \times 10^5)^{-12}$

Find the indicated principal root.

43. $\sqrt{169}$
44. $\sqrt[3]{0.027}$
45. $\sqrt[5]{-243}$
46. $\sqrt[4]{625}$

Express in simplest radical form.

47. $-\sqrt[3]{-27}$
48. $\sqrt{25a^4}$
49. $\sqrt{16x^2y^4}$
50. $\sqrt{12a^2b^3x}$
51. $\sqrt[3]{27a^4}$
52. $\sqrt[5]{96y^3}$
53. $\sqrt[4]{16y^5}$
54. $\sqrt[3]{\dfrac{54a^6b^7}{16a^3b}}$

Approximate with your calculator.

55. $\sqrt{599}$
56. $\sqrt{0.09}$
57. $\sqrt{459.32}$
58. $\sqrt{0.00054}$

Rationalize the denominator.

59. $\sqrt{\dfrac{18}{2}}$
60. $\sqrt{\dfrac{5}{3}}$
61. $\dfrac{1}{\sqrt{2}+\sqrt{3}}$
62. $\dfrac{2+\sqrt{5}}{\sqrt{3}-\sqrt{2}}$

Change the given exponential expression to radical form. Do not simplify.

63. $3^{1/2}$
64. $2^{1/3}$
65. $a^{2/5}$
66. $b^{5/2}$
67. $(-8)^{2/3}$
68. $(16)^{-1/4}$
69. $(25)^{-3/2}/(81)^{3/4}$
70. $(64)^{1/2}/(64)^{1/3}$

Change the given expression to exponential form.

71. $\sqrt{3}$
72. $\sqrt[3]{2}$
73. $\sqrt[3]{5}$
74. $\sqrt[4]{7^2}$
75. $\sqrt[5]{a^2 b}$
76. $\sqrt[5]{ab^2}$

Determine the degree of each polynomial.

77. $4x^3 + 3x^2 - 7x^2 + 5$
78. $8xy + y$

Add.

79. $3x^2 + 4x + 6$ and $2x^2 - 2x + 1$
80. $5x^3y + 7x^2y^2 + 6xy^3$ and $4x^3y + 5xy^3 - 4x^2y^2$

Subtract.

81. $3x^2y - 4xy^2$ from $6x^2y + 7xy^2$
82. $2x^3 - 3x^2 + x - 5$ from $x + 4x^3 + x^2$

Multiply.

83. $2x + 3$ by $3x - 5$
84. $3x^2 + x - 7$ by $x - 2$

Divide.

85. $x^2 + 5x + 6$ by $x + 3$
86. $x^3 + 3x^2 + 3x + 1$ by $x + 1$

Solve.

87. $2x + 3y + 60 = 0$ for x
88. $3x - 2y - 40 = 0$ for y

Solve the proportion.

89. $\dfrac{x}{6} = \dfrac{9}{18}$
90. $\dfrac{9}{x} = \dfrac{x}{16}$

Express as a rate.

91. $240 in 6 years
92. 32 tiles for 18 square feet

Express as a ratio in simplest form.

93. 2 yards to 4 feet

94. 14 ounces to 2 pounds

Express as equations.

95. s varies directly as t and inversely as u squared.

96. p is directly proportional to the product of q and r.

Find the required value.

97. u varies directly as v squared, and $u = 144$ when $v = 3$. Find u when $v = 2$.

98. s is directly proportional to r and inversely proportional to t, and $s = 5$ when $r = 2$ and $t = 6$. Find s when $r = 2$ and $t = 3$.

99. The electric resistance R of a wire varies directly as the length l and inversely as the square of the diameter d of the wire. Express this relation as an equation. Then solve for the variable l.

100. In aerodynamics, the flat-plate area f is directly proportional to the drag D and inversely proportional to the dynamic pressure q. Express this relationship as an equation.

101. In aerodynamics, the aspect ratio AR of a nonrectangular wing is directly proportional to the square of the span S and inversely proportional to the wing area W. Express this relationship as an equation.

102. The indicated horsepower (ihp) of an engine is based on the average pressure P on each square inch of piston, the number N of power strokes per minute, the length L of each stroke in feet, and the area A of each piston in square inches. If the formula for ihp is

$$\text{ihp} = kPLAN,$$

find k if $P = 125$ lb per sq in.

$L = 0.23$ ft
$A = 28.3$ sq. in.
$N = 6000$ strokes per min
ihp $= 148$

Express k to five significant digits.

103. If the temperature in Mayville, North Dakota at 5 A.M. was $-28°$ F and the temperature is rising at a rate of $4°$ F per hour, what will the temperature be at 3 P.M.?

104. Light travels at a rate of about 186,000 miles per second. The average distance from the sun to the earth is 93,000,000 miles. Calculate the time it takes light to reach the earth from the sun.

105. The diagonal d of a rectangle is determined by the formula

$$d = \sqrt{l^2 + w^2},$$

where l is the length of the rectangle and w is the width. Calculate d if $l = 40$ meters and $w = 30$ meters.

106. The compound amount A that results when P dollars is invested at a rate r per period for 3 periods is given by the formula

$$A = P(1 + r)(1 + r)(1 + r).$$

Write the expanded expression for the amount A by multiplying the factors on the right-hand side.

107. Boyle's law dealing with the compressibility of gasses relates the original volume V_0, and the original pressure P_0, to the final volume V and the final pressure P in the following equation

$$P_0 V_0 = P V.$$

Solve this equation for P.

108. The ratio of gear A to gear B is given by the relationship

$$\frac{S_a}{S_b} = \frac{N_a}{N_b},$$

where S_a is the speed of gear A, S_b is the speed of gear B, N_a is the number of teeth in gear A, and N_b is the number of teeth in gear B. Solve this formula for N_b.

1·14 Chapter 1 Test

True or false.

1. $\dfrac{2}{3} > \dfrac{5}{6}$

2. $0.66 = 0.\overline{6}$

3. $\sqrt{3} > \sqrt{2}$

Simplify.

4. $\dfrac{(-2) + 3(4)}{-5}$

5. $-|-13 + 17|$

6. $\dfrac{(-6)(3) \div (4 + 5)}{(-5)(4) \div (-6 + 4)}$

7. $(x^3)(y^{-3})(xy)^2(x^{-3})$

8. $2^3\left(\dfrac{1}{4}\right)^2 4^3\left(\dfrac{1}{16}\right)$

Express in scientific notation.

9. 2850

10. 0.816

Use your calculator to evaluate the following.

11. $(3.5)^9$

12. $(1.09)^{-15}$

Compute.

13. $(4 + \sqrt{7})(5 - 2\sqrt{7})$

Rationalize.

14. $\dfrac{5}{4\sqrt{3}}$

15. $\dfrac{4}{\sqrt{5} + \sqrt{2}}$

16. Express in radical form: $r^{5/3}$.

17. Express in exponential form: $\sqrt[5]{ab^2}$.

18. Subtract $x^2 + 3x + 2$ from $3x^2 + 4x - 5$.

19. Multiply $(x + 1)$ times $(x^2 - x + 1)$.

20. Divide $(3s^2 + 8s + 4)$ by $(s + 2)$.

21. Solve for x: $3\left(\dfrac{x}{3} - 1\right) = 3\left(\dfrac{2x}{3} + 6\right)$.

22. Solve for x: $(x + 1)^2 = (x + 2)^2 - 5$.

23. Express as a ratio: 2 yards to 8 feet.

24. Solve for x in $2:3 = 4:x$.

25. If y varies directly as x, and $x = 3$ when $y = 12$, find the value of y when $x = 5$.

26. Express as an equation: x varies directly as the product of y and z and inversely as d squared.

Chapter 2: Basic Algebraic Skills

The origin of the word algebra can be traced to a treatise written by an Arab mathematician, al-Khowarizmi. The name of the treatise was *al-jabr w'al-muqabalah*. Literally translated this title means the "science of the reunion and the opposition." It is more freely translated as the "science of reduction and cancellation."

al-Khowarizmi's text became known in Europe through its Latin translation and made the word *al-jabr*, or algebra, synonymous with the science of equations. Since the middle of the nineteenth century, algebra has come to mean a great deal more.

In chapter 2 we develop some basic algebraic skills. These skills are necessary for the successful development and understanding of the mathematics that follows in the remaining chapters of this text. Indeed, we will be developing algebraic manipulation and the "science of equations."

2•1 Special Products

In section 1.7 we learned how to multiply general polynomials. There are certain products that appear so frequently in algebra that it is worthwhile to look at them separately.

We have noted that the multiplication of polynomials depends on repeated use of the distributive law. This is still the case with special products. However, we will attempt to identify patterns that can be generalized in order to simplify the multiplication process.

First, let us consider the product of two binomials:

$$(a + b)(x + y).$$

Multiplying as we did in the last chapter, we see that

$$(a + b)(x + y) = ax + ay + bx + by.$$

The arrows indicate the various products that, when added, produce the answer. Notice that we have multiplied the first terms (F), the outer terms (O), the inner terms (I), and the last terms (L). This procedure is often called the FOIL (first, outer, inner, last) method.

• **Example 2.1:** Multiply and simplify $(4x + y)(3x + 2y)$.

Solution: $(4x + y)(3x + 2y)$

$= (4x)(3x) + (4x)(2y) + (y)(3x) + (y)(2y)$ (Expand using FOIL)
$= 12x^2 + 8xy + 3xy + 2y^2$ (Simplify)
$= 12x^2 + 11xy + 2y^2.$ (Add like terms)

Square of a Binomial

Another common product is the multiplication of two identical binomials. This is called the **square of a binomial**, since

$$(x + y)(x + y) = (x + y)^2.$$

To square a binomial means to use the binomial as a factor twice. Using the method just developed, we see that

$$\begin{align*}(x + y)^2 &= (x + y)(x + y) \\ &= (x)(x) + (x)(y) + (y)(x) + (y)(y) \quad \text{(FOIL)} \\ &= x^2 + xy + xy + y^2 \quad \text{(Simplify)} \\ &= x^2 + 2xy + y^2. \quad \text{(Add like terms)}\end{align*}$$

This can be generalized to form a rule for squaring a binomial.

SQUARE OF A BINOMIAL

The square of a binomial is equal to the square of the first term plus twice the product of the first and last terms plus the square of the last term.

$$(x + y)^2 = x^2 + 2xy + y^2 \quad \text{or} \quad (x - y)^2 = x^2 - 2xy + y^2.$$

We use this rule to square the binomial in example 2.2.

• **Example 2.2:** Expand and simplify $(3a + 4b)^2$.

Solution: $(3a + 4b)^2 = (3a)^2 + 2(3a)(4b) + (4b)^2$ (Square the binomial)
$= 9a^2 + 24ab + 16b^2.$ (Multiply and simplify)

As described by the above rule, notice that the first and last terms of the product are the square of $3a$ (the first term of the binomial) and $4b$ (the last term of the binomial). The middle term of the product ($24ab$) is twice the product of the first and last terms of the binomial, $3a$ and $4b$.

This rule applies to differences also.

• **Example 2.3:** Expand and simplify $(2m - 3n)^2$.

Solution: $(2m - 3n)^2$
$= (2m)^2 + 2(2m)(-3n) + (-3n)^2$ (Square the binomial)
$= 4m^2 - 12mn + 9n^2.$ (Multiply and simplify)

• **Example 2.4:** Expand and simplify $(4r + 3s^2)^2$.

Solution: $(4r + 3s^2)^2 = (4r)^2 + 2(4r)(3s^2) + (3s^2)^2$ (Square the binomial)
$= 16r^2 + 24rs^2 + 9s^4.$ (Multiply and simplify)

Another special product is the product of the sum of two terms and the difference of the same two terms. That is, the terms in each binomial are identical except for the signs between them. Consider

$$(a + b)(a - b).$$

Using the methods previously developed, we may write

$(a + b)(a - b)$
$= (a)(a) + (a)(-b) + (b)(a) + (b)(-b)$ (FOIL)
$= a^2 - ab + ab - b^2$ (Multiply)
$= a^2 - b^2.$ (Simplify)

Notice that the sum of the outer and inner products is zero. The answer is called the **difference of two squares** and is summarized as the rule for multiplying the sum and the difference of two terms.

Difference of Two Squares

PRODUCT OF THE SUM AND DIFFERENCE OF TWO TERMS

The product of the sum of two terms and the difference of the same two terms is the square of the first term minus the square of the last term:

$$(a + b)(a - b) = a^2 - b^2.$$

Use this rule in example 2.5.

• *Example 2.5:* Multiply and simplify $(2x + 3y)(2x - 3y)$.

Solution: $(2x + 3y)(2x - 3y) = (2x)^2 - (3y)^2$ (Square the first and last terms)

$= 4x^2 - 9y^2.$ (Simplify)

The product is the square of the first term, $(2x)^2$, minus the square of the second term, $(3y)^2$.

Use the rule to multiply the binomials in example 2.6.

• *Example 2.6:* Multiply and simplify $(xy + 5z)(xy - 5z)$.

Solution: $(xy + 5z)(xy - 5z)$
$= (xy)^2 - (5z)^2$ (Square the first and last terms)
$= x^2y^2 - 25z^2.$ (Simplify)

The product is the square of the first term, $(xy)^2$, of the binomial minus the square of the second term, $(5z)^2$, of the binomial.

Trial Problems 2.1

Before you begin the section exercises, warm up with these problems. Complete answers are included in the answer key.

Find the following products.

1. $(a + b)(r + s)$
2. $(x + y)^2$
3. $(2x - 3y)^2$
4. $(a + 2x)(a - 2x)$
5. $(30 - 4)(30 + 4)$

Exercises 2.1

Find the following products.

1. $(x + y)(r + s)$
2. $(a + b)(c + d)$
3. $(x + y)(u - r)$
4. $(a + b)(m - n)$
5. $(x - y)(2x - y)$
6. $(x - y)(3x - y)$
7. $(a - b)(x + y)$
8. $(c - d)(x + y)$
9. $(a + 2b)(a + 2b)$
10. $(2x + y)(2x + y)$
11. $(a + b)^2$
12. $(c + d)^2$
13. $(2x + y)^2$
14. $(3x + y)^2$
15. $(2a - 3b)^2$
16. $(3a - 2b)^2$
17. $(x + y)(x - y)$
18. $(a + b)(a - b)$
19. $(a - 2b)(a + 2b)$
20. $(3a + 2b)(3a - 2b)$
21. $(x^2 + 1)(x^2 - 1)$
22. $(3 - y^2)(3 + y^2)$
23. $(5 + 3t)(5 - 3t)$
24. $(3r + s)(3r - s)$
25. $(2a + 5)(2a - 5)$
26. $(3b + 4)(3b - 4)$
27. $(30 + 2)(30 - 2)$
28. $(50 + 1)(50 - 1)$

29. In the physics of elastic collision, the expression $m_1(V_a + V_b)(V_a - V_b)$ may appear. Simplify this expression by multiplication.

30. The deflection of a beam might involve the expression $w(d^2 - x^2)^2$. Simplify this expression by multiplication.

31. The echo range R measured by radar and the distance to the ground echo point x lead to the expression $(2R - x)^2 - x^2 - R^2$. Simplify this expression.

32. The expression for the maximum power in an electric circuit can be written as $(R + r)^2 - 2r(R + r)$. Simplify this expression.

33. When finding the reactance in an electrical circuit, the impedance Z and the resistance R are combined in the following expression

$$(Z + R)(Z - R).$$

Simplify this expression by multiplying.

34. When calculating the present value of an investment we may encounter the expression

$$(1 + i)^2,$$

where i is the rate of interest. Expand this expression by squaring the binomial.

35. When calculating the goodness of fit to test a statistical hypothesis we encounter the expression

$$(f_0 - f_t)^2,$$

where f_0 is the observed frequency and f_t is the theoretical frequency. Expand the expression by squaring the binomial.

36. A square lot that measured 50 feet on each side was altered by adding 4 feet to one side and subtracting 4 feet from the other side. The area of the new lot can now be expressed as

$$(50 + 4)(50 - 4).$$

Evaluate this expression by multiplying the two binomials. Check your result with your calculator.

37. A piece of sheet metal measures 3 feet by 4 feet. If an amount x is cut off the length and an amount y is cut off the width, the new area can be expressed as

$$(4 - x)(3 - y).$$

Express this product as a polynomial.

38. To determine the number of permutations of n different objects taken three at a time we must evaluate the expression

$$n(n - 1)(n - 2).$$

Rewrite this expression by multiplying.

2•2 Factoring

Factoring

Factoring is the process of expressing a polynomial as a product of polynomials of lower degree (factors). Factoring is useful for simplifying algebraic expressions, for solving certain types of equations, and for writing numbers as products.

Removing a Common Factor

The first type of factoring problem we discuss is that of **removing a common factor** from an expression. This is always the first procedure to consider during the factoring process.

Recall from section 1.7 that to multiply a monomial times a polynomial we multiplied each term of the polynomial by the monomial. That is,

$$a(x + y + z) = (a)(x) + (a)(y) + (a)(z)$$
$$= ax + ay + az.$$

To factor the answer, we reverse the process. That is,

$$ax + ay + az = a(x + y + z),$$

where a is one of the factors and $x + y + z$ is the other. The terms x, y, and z share no other common factor. Hence we have removed the highest (or greatest) common factor.

Remove a common factor from the expressions in examples 2.7, 2.8, and 2.9.

• *Example 2.7:* Factor $3x^4 - 6x^2 - 5x^9$ by removing the highest common factor.

Solution: The numerical coefficients 3, 6, and 5 share no common factor. The variable factors x^4, x^2, and x^9 have x^2 as their highest common factor. Removing the common factor, x^2, from the given polynomial yields

$$3x^4 - 6x^2 - 5x^9 = x^2(3x^2 - 6 - 5x^7).$$

• *Example 2.8:* Factor $2x^2y + 24xy + 6y$.

Solution: The greatest common factor of the numerical coefficients 2, 24, and 6 is 2. The literal factors, x^2y, xy, and y have y as their greatest common factor. Hence, $2y$ is the greatest common factor. Removing this common factor, $2y$, produces

$$2x^2y + 24xy + 6y = 2y(x^2 + 12x + 3).$$

• *Example 2.9:* Factor $24a^3b^2c - 18a^5bc^4 + 12a^6b^4$.

Solution: The greatest common factor of the numerical coefficients 24, 18, and 12 is 6. The literal factors a^3b^2c, a^5bc^4, and a^6b^4 have a^3b as their greatest common factor. Hence, $6a^3b$ is the greatest common factor. Removing this common factor produces

$$24a^3b^2c - 18a^5bc^4 + 12a^6b^4 = 6a^3b(4bc - 3a^2c^4 + 2a^3b^3).$$

To check that the factors on the right-hand side produce the expression on the left-hand side, simply multiply the factors to see if the expression on the left-hand side is obtained. Be careful, since it is possible to have the product of the factors yield the expression without one of them being the *greatest* common factor.

The next type of factoring involves quadratic (second degree) binomials or trinomials. We consider four forms of factoring quadratic polynomials.

FACTORING QUADRATIC POLYNOMIALS

(1) $\quad x^2 + (a + b)x + ab = (x + a)(x + b)$

(2) $\quad x^2 + 2ax + a^2 = (x + a)^2$

(3) $\quad x^2 - a^2 = (x + a)(x - a)$

(4) $\quad acx^2 + (ad + bc)xy + bdy^2 = (ax + by)(cx + dy)$

Each of these forms can be checked by multiplying the factors on the right-hand side of each equation. The product, when simplified, will yield the polynomial on the left-hand side.

The discussion of these quadratic expressions is limited to those whose coefficients are integers and whose exponents are nonnegative integers.

As an application of form (1), consider the trinomial

$$x^2 + 5x + 6.$$

If we compare this polynomial with the left-hand side of form (1)

(1) $\quad x^2 + (a + b)x + ab = (x + a)(x + b)$
$\quad\quad\, x^2 + \quad\quad 5x + 6,$

we conclude that the factored form of our polynomial should match the right-hand side of form (1). Remembering our FOIL method, we see from form (1) that a and b are integers such that $a + b = 5$ and $ab = 6$. That is, the sum of a and b must result from the outer and inner products, and the product of a and b must result from the last product.

Following is a list of factors whose product is $+6$, and their corresponding sums.

Factors	Sums
$+6$ and $+1$	$+7$
-6 and -1	-7
$+3$ and $+2$	$+5$
-3 and -2	-5

The only sum of any of these pairs of factors that will produce $+5$ is

$$+3 \quad \text{and} \quad +2.$$

Therefore, a and b must be $+3$ and $+2$ so that

$$x^2 + 5x + 6 = (x + 3)(x + 2). \quad \text{(Check by multiplying)}$$

Trial and Error This process is often called the **trial and error** process. We try certain factors. If we make an error we try other factors. Since there are only a finite number of possible factors, we must ultimately find the correct factors, if the trinomial is factorable.

Another example further illustrates this concept.

• *Example 2.10:* Factor $x^2 + 2x - 8$.

Solution: Again we would like to find, if possible, two binomial factors whose product is the given trinomial. Again, remembering our FOIL method, the only integer factors whose product is -8, with their corresponding sums, are listed below.

Factors	Sums
$+8$ and -1	$+7$
-8 and $+1$	-7
$+4$ and -2	$+2$
-4 and $+2$	-2

The only sum of any of these pairs of factors that produces $+2$ is

$$+4 \quad \text{and} \quad -2.$$

Therefore, we see that

$$x^2 + 2x - 8 = (x + 4)(x - 2).$$

Example 2.11 illustrates how the trial and error method fails to find the desired factors.

• *Example 2.11:* Factor $x^2 - 7x - 6$.

Solution: The only integer factors whose product is -6, with their corresponding sums, are listed below.

Factors	Sums
$+6$ and -1	$+5$
-6 and $+1$	-5
$+3$ and -2	$+1$
-3 and $+2$	-1

Since the sum of any pair of these factors never equals -7, we can find no factors using just integers as coefficients.

• • • • • • • • •

Form (2) is a special case of form (1). Notice that the right-hand side of form (2) is a trinomial and the left-hand side is the square of a binomial. For example, the trinomial $x^2 + 6x + 9$ was the result of squaring the binomial $x + 3$. That is, $(x + 3)^2 = x^2 + 6x + 9$, so

$$x^2 + 6x + 9 = (x + 3)(x + 3)$$
$$= (x + 3)^2.$$

Examples 2.12 and 2.13 further illustrate this form of factoring.

• *Example 2.12:* Factor $x^2 + 14x + 49$.

Solution: Since $7^2 = 49$ and $7 + 7 = 14$, we get

$$x^2 + 14x + 49 = (x + 7)(x + 7)$$
$$= (x + 7)^2.$$

• • • • • • • • •

• *Example 2.13:* Factor $x^2 + 10x + 25$.

Solution: Since $5^2 = 25$ and $5 + 5 = 10$, we get

$$x^2 + 10x + 25 = (x + 5)(x + 5)$$
$$= (x + 5)^2.$$

• • • • • • • • •

Form (3) is also a special case of form (1). In this case, the coefficient of the linear or first degree term is zero. Consider

$$x^2 - 25 = (x + 5)(x - 5)$$

or

$$x^2 - 5^2 = (x + 5)(x - 5).$$

We see that the quadratic expression is the difference of the squares of two terms and that it factors into the sum and the difference of the two terms.

Example 2.14 further illustrates this special case.

• *Example 2.14:* Factor $x^2 - 36$.

Solution: Since $6^2 = 36$, and since the first degree, or middle term, is zero we must have

$$x^2 - 36 = x^2 - 6^2$$

or

$$x^2 - 6^2 = (x + 6)(x - 6).$$

If the expression we wish to factor was the result of multiplying factors such as $(3x + 5y)$ and $(3x - 5y)$, then the expression would be $9x^2 - 25y^2$. So if we wish to factor $9x^2 - 25y^2$ we should think of it as $(3x)^2 - (5y)^2$. Now it is expressed as the difference of two squares and factors into $(3x + 5y)(3x - 5y)$. Consider example 2.15.

• *Example 2.15:* Factor $16x^2 - 81y^2$.

Solution: $16x^2 - 81y^2 = (4x)^2 - (9y)^2 = (4x + 9y)(4x - 9y)$

You should note that expressions that are the *sum* of two squares like $x^2 + 5^2$ do not factor in the real numbers.

Form (4) is a generalization of form (1). That is, in form (4) we have a quadratic trinomial in which the coefficient of the second degree term of the variable is other than 1. The factoring of such a trinomial can sometimes be done by trial and error. Consider the trinomial.

$$8x^2 - 21x - 9.$$

1. First we consider possible combinations of factors of the first term. The factors of 8 are 8 and 1 or 4 and 2. Again, note the use of the FOIL method.

$$\overset{F}{\overbrace{}}$$
$(8x\ \ )(x\ \ )$
$(4x\ \ )(2x\ \ )$

2. Next we consider possible combinations of factors (L) of the last terms together with those of the first terms. The factors of the last term 9 are 9 and 1 or 3 and 3.

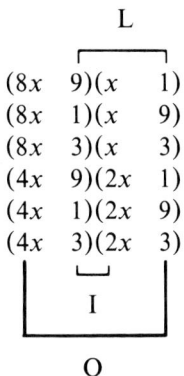

3. Select the combination(s) of products (O) and (I) whose sum(s) *could* equal the middle term $(-21x)$. All of the possible combinations are listed below.

$$(8x \quad 9)(x \quad 1) \qquad (4x \quad 9)(2x \quad 1)$$
$$\pm 9x \qquad\qquad \pm 18x$$
$$\pm 8x \qquad\qquad \pm 4x$$

$$(8x \quad 1)(x \quad 9) \qquad (4x \quad 1)(2x \quad 9)$$
$$\pm 1x \qquad\qquad \pm 2x$$
$$\pm 72x \qquad\qquad \pm 36x$$

$$(8x \quad 3)(x \quad 3) \qquad (4x \quad 3)(2x \quad 3)$$
$$\pm 3x \qquad\qquad \pm 6x$$
$$\pm 24x \qquad\qquad \pm 12x$$

4. The only sum that produces $-21x$ is $-24x$ plus $+3x$. Therefore the factors are

$$(8x + 3)(x - 3).$$

5. Check by multiplying the binomials:

$$(8x + 3)(x - 3) = 8x^2 - 21x - 9.$$

Use the trial and error method to factor example 2.16.

* Example 2.16: Factor $5x^2 + 8x - 4$.

Solution: Possible combinations of factors for $5x^2$ are

$$(5x\quad)(x\quad).$$

Possible combinations of factors for 4, together with the factors for the first term, are

$$(5x\quad 4)(x\quad 1)$$
$$(5x\quad 1)(x\quad 4)$$
$$(5x\quad 2)(x\quad 2).$$

Selecting the combination of products whose sum could equal our middle term $(+8x)$, we get

$$(5x\quad 2)(x\quad 2)$$
$$\pm 2x$$
$$\pm 10x.$$

Inserting the proper signs gives

$$(5x - 2)(x + 2).$$

Checking, we see that

$$(5x - 2)(x + 2) = 5x^2 + 8x - 4.$$

Example 2.17 further illustrates this concept.

* Example 2.17: Factor $3x^2 - 10x - 8$.

Solution: Possible combinations of factors for $3x^2$ are

$$(3x\quad)(x\quad).$$

Possible combinations of factors of 8, together with the factors for the first term, are

$$(3x\quad 8)(x\quad 1)$$
$$(3x\quad 1)(x\quad 8)$$
$$(3x\quad 2)(x\quad 4)$$
$$(3x\quad 4)(x\quad 2).$$

Selecting the combination of products whose sum could equal our middle term $(-10x)$, we get

$$(3x \quad 4)(x \quad 2) \quad \text{or} \quad (3x \quad 2)(x \quad 4)$$

$$\pm 4x \qquad\qquad\qquad \pm 2x$$

$$\pm 6x. \qquad\qquad\qquad \pm 12x.$$

To get $-10x$ we could have used $-6x$ and $-4x$, which would mean that both factors would have a negative sign and be $(3x - 4)(x - 2)$. Multiplying these we get $3x^2 - 10x + 8$, which is not correct. To get $-10x$ we could also have used $+2x$ and $-12x$. This would mean our factors are $(3x + 2)(x - 4)$. If we multiply these we see that

$$(3x + 2)(x - 4) = 3x^2 - 10x - 8.$$

The polynomials that we have factored so far either had a common factor, or factored by one method or another into the product of two binomials.

There are other types of polynomials that occur in technical applications that are worth considering. When some of these polynomials are factored they result in special factoring formulas.

SPECIAL FACTORING FORMULAS

(5) $\quad ax + ay + bx + by = (a + b)(x + y)$

(6) $\quad x^3 + y^3 = (x + y)(x^2 - xy + y^2)$

(7) $\quad x^3 - y^3 = (x - y)(x^2 + xy + y^2)$

These can be readily verified by multiplying the right-hand member of each equation.

Factoring by Grouping

Example 2.18 further illustrates form (5), which is called **factoring by grouping.**

• *Example 2.18:* Factor by grouping $3x^2y + 2y + 3xy^2 + 2x$.

Solution 1: Group the first term with the last term $(3x^2y + 2x)$ and the third term with the second term $(3xy^2 + 2y)$ by using the commutative and associative properties of addition. Now we have

$$3x^2y + 2y + 3xy^2 + 2x = (3x^2y + 2x) + (3xy^2 + 2y).$$

Next, remove the common factor x from $(3x^2y + 2x)$ and the common factor y from $(3xy^2 + 2y)$. We get

$$(3x^2y + 2x) + (3xy^2 + 2y) = x(3xy + 2) + y(3xy + 2).$$

The factor $(3xy + 2)$ appears in both terms on the right-hand side of our last equation. Removing this common factor produces

$$x(3xy + 2) + y(3xy + 2) = (x + y)(3xy + 2).$$

Thus, the original polynomial expressed in factored form is

$$3x^2y + 2y + 3xy^2 + 2x = (x + y)(3xy + 2).$$

This solution is summarized below.

$$3x^2y + 2y + 3xy^2 + 2x$$
$$= (3x^2y + 2x) + (3xy^2 + 2y) \quad \text{(Regroup)}$$
$$= x(3xy + 2) + y(3xy + 2) \quad \text{(Remove common factors } x \text{ and } y\text{)}$$
$$= (x + y)(3xy + 2). \quad \text{(Remove common factor } 3xy + 2\text{)}$$

Solution 2: Group the first term with the third term $(3x^2y + 3xy^2)$ and the last term with the second term $(2x + 2y)$ by using the commutative and associative properties of addition. Now we have

$$3x^2y + 2y + 3xy^2 + 2x = (3x^2y + 3xy^2) + (2x + 2y).$$

Next, remove the common factor $3xy$ from $(3x^2y + 3xy^2)$ and the common factor 2 from $2x + 2y$. This produces

$$(3x^2y + 3xy^2) + (2x + 2y) = 3xy(x + y) + 2(x + y).$$

The factor $(x + y)$ appears in both terms on the right-hand side of our last equation. Removing this common factor gives

$$3xy(x + y) + 2(x + y) = (3xy + 2)(x + y).$$

Thus the original polynomial expressed in factored form is

$$3x^2y + 2y + 3xy^2 + 2x = (x + y)(3xy + 2).$$

Notice that each solution gave us the same answer, although we grouped the polynomial differently in each case. Solution 2 is summarized below.

$$3x^2y + 2y + 3xy^2 + 2x$$
$$= (3x^2y + 3xy^2) + (2x + 2y) \quad \text{(Regroup)}$$
$$= 3xy(x + y) + 2(x + y) \quad \text{(Remove common factors)}$$
$$= (x + y)(3xy + 2). \quad \text{(Remove common factor } x + y\text{)}$$

Example 2.19 illustrates the use of form (5).

• Example 2.19: The interface stress in the glass member of a glass-metal cylindrical seal is

$$c_m T_0 - c_g T_0 + c_m T_1 - c_g T_1.$$

Factor this expression.

Solution: Grouping the first two terms and the last two terms gives

$$c_m T_0 - c_g T_0 + c_m T_1 - c_g T_1$$
$$= (c_m T_0 - c_g T_0) + (c_m T_1 - c_g T_1).$$

Removing the common factor T_0 from $(c_m T_0 - c_g T_0)$ and common factor T_1 from $(c_m T_1 - c_g T_1)$ gives

$$(c_m T_0 - c_g T_0) + (c_m T_1 - c_g T_1) = (c_m - c_g) T_0 + (c_m - c_g) T_1.$$

The factor $(c_m - c_g)$ appears in both terms on the right-hand side of the last equation. Removing this common factor produces

$$(c_m - c_g) T_0 + (c_m - c_g) T_1 = (c_m - c_g)(T_0 + T_1).$$

Thus, the original polynomial expressed in factored form is

$$c_m T_0 - c_g T_0 + c_m T_1 - c_g T_1 = (c_m - c_g)(T_0 + T_1).$$

∙∙∙∙∙∙∙∙∙∙

Sum and Difference of Two Cubes

Forms (6) and (7) are verified directly. They are called the **sum and difference of two cubes,** respectively. Examples 2.20 and 2.21 illustrates the use of form (7).

• *Example 2.20:* The volume of material needed to build a spherical container with inside diameter r and outside diameter R can be expressed as

$$\frac{4}{3}\pi R^3 - \frac{4}{3}\pi r^3.$$

Factor this expression.

Solution: The factor $\frac{4}{3}\pi$ appears in both terms of the expression.

Removing this common factor yields

$$\frac{4}{3}\pi R^3 - \frac{4}{3}\pi r^3 = \frac{4}{3}\pi(R^3 - r^3).$$

The factor $(R^3 - r^3)$ is the difference of two cubes. Factoring it using form (7) produces

$$R^3 - r^3 = (R - r)(R^2 + Rr + r^2).$$

Thus, the original expression in factored form is

$$\frac{4}{3}\pi R^3 - \frac{4}{3}\pi r^3 = \frac{4}{3}\pi(R - r)(R^2 + Rr + r^2).$$

∙∙∙∙∙∙∙∙∙∙

• *Example 2.21:* Factor completely $4x^3 + 500$.

Solution: First look for a common factor:
$$4x^3 + 500 = 4(x^3 + 125).$$
The factor, $x^3 + 125$, can now be written as the sum of two cubes. Hence, we get
$$4x^3 + 500 = 4(x^3 + 5^3)$$
$$= 4(x + 5)(x^2 - 5x + 25).$$

Trial Problems 2.2

Before you begin the section exercises, warm up with these problems. Complete answers are included in the answer key.

Factor.

1. $ay + by + cy$
2. $4x^3 + 8x^2 + 12x$
3. $x^2 - 10x + 16$
4. $16x^2 - 9y^2$
5. $x^3 - 8$

Exercises 2.2

Factor.

1. $ax + bx + cx$
2. $ay + by + cy$
3. $3x^3 + 6x^2 + x$
4. $3x^4 - 5x^9 + 6x^2$
5. $ab + a^2b^2 + a^3b^3$
6. $xy + x^2y^2 + x^3y^3$
7. $2^2a + 2^3a^2 + 2^4a^3$
8. $3^2b + 3^3b^2 + 3^4b^3$
9. $4x^2yz + 16x - 12z$
10. $3xy^2z + 9x - 27y$
11. $7x^2y + 14xy^2 + 21xy$
12. $5a^2b + 10ab^2 + 15ab$
13. $a^4b^3c^2 + a^2b^4c^3 + a^2b^3c^4$
14. $x^3y^5z^4 + x^5y^3z^4 + x^4y^3z^5$

Factor completely.

15. $a^2 + 2a + 1$
16. $x^2 + 2x + 1$
17. $x^2 - 9x + 14$
18. $x^2 - 3x + 2$
19. $x^2 - 5x - 14$
20. $x^2 - 7x - 18$
21. $5x^2 - 80$
22. $3x^2 - 48$
23. $4x^2 - 9y^2$
24. $9x^2 - 16y^2$
25. $x^2 + 16x + 64$
26. $x^2 + 10x + 25$
27. $2x^2 - 5x - 12$
28. $3x^2 - 15x + 18$
29. $10x^2 - 3x - 18$
30. $3x^2 - 7x + 2$
31. $9x^2 - 6ax + a^2$
32. $25m^2 - 10mn + n^2$
33. $3z^3 - 4z^2 + z$
34. $a^2x + 2ax + x$

Factor by grouping.

35. $ab + a + b + b^2$
36. $ab + bc + ad + cd$
37. $2x^2 - x + 2xy - y$
38. $mn + m^2 + mn + n^2$

Factor the sums and differences of cubes.

39. $x^3 + 8$
40. $y^3 + 27$
41. $8x^3 - y^3$
42. $27a^3 - y^3$
43. $54a^3 + 128b^3$
44. $16a^3 + 250b^3$
45. $375x^3 - 81y^3$
46. $192x^3 - 648y^3$

47. The cross sectional area of a rectangular metal tube is given by the expression $36x - 2x^2$. Factor this expression.

48. The area of a plastic disc can be represented by the expression $\pi R^2 - \pi r^2$. Factor this expression.

49. The energy radiated by an electric light at two different temperatures is expressed as $kt_2^4 - kt_1^4$. Factor this expression.

50. Under certain conditions in the physics of motion, the displacement of an object for time t can be expressed as $16t^2 - 32t - 240$. Factor this expression.

51. In the study of electricity the impedance Z and the resistance R are combined in the expression

$$Z^2 - R^2.$$

Factor this expression.

52. If from a large circular disc of radius R, nine smaller circular discs of radius a are punched out, the area of the remaining portion of the large disc is given by the expression

$$\pi R^2 - 9\pi a^2.$$

Factor this expression.

53. If from a square with length x, a smaller square of length y is removed from each corner, the area of the remaining piece is given by the expression

$$x^2 - 4y^2.$$

Factor this expression.

54. The difference of the expressions for the volume of a cube with edge e and the surface area of the same cube produces the expression

$$e^3 - 6e^2.$$

Factor this expression.

55. From a square of side $2r$ a circle of radius r is cut. The wasted portion of the square can be expressed as

$$4r^2 - \pi r^2.$$

Factor this expression.

56. After adding x feet to each side of an existing parking lot the area of the new lot could be represented by the expression

$$20{,}000 + 300x + x^2.$$

Factor this expression to obtain expressions for the length and the width of the new lot.

2•3 Algebraic Fractions

Rational Expression
Algebraic Fraction

A fraction is an expression denoting a quotient. If the numerator (dividend) and the denominator (divisor) are polynomials, then the fraction is called a **rational expression** or an **algebraic fraction**. For example,

$$\frac{5}{z}, \quad \frac{y}{y+1}, \quad \frac{x^2 + 2x - 1}{x}, \quad \frac{x}{x^2 - 1}, \quad \text{and} \quad \frac{x^2 + 1}{2x - 1}$$

are algebraic fractions. Any polynomial can be considered an algebraic fraction, since it is the quotient of itself and 1. Thus,

$$3x, \quad 2y + y^2, \quad \text{and } 6$$

are algebraic fractions since

$$3x = \frac{3x}{1}, \quad 2y + y^2 = \frac{2y + y^2}{1}, \quad \text{and } 6 = \frac{6}{1}.$$

At times the variables in a rational expression are replaced with numerical values and the fraction can be evaluated. In this case, the rational expression represents a real number as long as the denominator is not zero. For any value of the variables for which the denominator becomes zero, the fraction does not represent a real number and is said to be undefined. For example, the algebraic fraction

$$\frac{x + 2}{x - 3}$$

is undefined when $x = 3$, since

$$\frac{3 + 2}{3 - 3} = \frac{5}{0}.$$

Similarly, $\frac{x^2}{y}$ is undefined when $y = 0$.

The expression

$$\frac{x + 5}{a^2 - 3a + 2},$$

which factors into

$$\frac{x + 5}{(a - 2)(a - 1)},$$

is undefined when $a = 2$ or when $a = 1$. When $a = 2$, the expression becomes

$$\frac{x + 5}{(2 - 2)(2 - 1)}$$

or

$$\frac{x + 5}{(0)(1)}.$$

When $a = 1$, the expression becomes

$$\frac{x + 5}{(1 - 2)(1 - 1)}$$

or

$$\frac{x + 5}{(-1)(0)}.$$

Since algebraic fractions represent quotients of real numbers, they satisfy the rules of algebra for quotients.

SECTION 2.3 ALGEBRAIC FRACTIONS

QUOTIENT RULES

$$\frac{a}{b} = \frac{c}{d} \text{ if and only if } ad = bc, \quad (b, d \neq 0) \quad (2.1)$$

$$\frac{a}{b} = \frac{ka}{kb} \quad (k \neq 0, b \neq 0) \quad (2.2)$$

Equivalent Fractions

Rule (2.1) establishes a means by which we can identify fractions that represent the same number. We call such fractions **equivalent fractions**. For example,

$$\frac{2}{3} = \frac{4}{6} \text{ because } (2)(6) = (3)(4).$$

Rule (2.2) asserts that an equivalent fraction is obtained if the numerator and the denominator of a fraction are each multiplied or divided by the same nonzero number. For example,

$$\frac{1}{2} = \frac{1 \cdot 2}{2 \cdot 2} = \frac{2}{4} \text{ or } \frac{6}{8} = \frac{6 \div 2}{8 \div 2} = \frac{3}{4}.$$

To reduce a fraction, to write a fraction in lowest terms, or to simplify a fraction all mean the same thing. Rule (2.2) is useful when we wish to simplify a fraction.

• **Example 2.22:** Simplify $\frac{x^2 - 4}{x^2 - 2x - 8}$.

Solution: To apply rule (2.2), we need to express the numerator and denominator as products. We do this by factoring to produce

$$\frac{x^2 - 4}{x^2 - 2x - 8} = \frac{(x - 2)(x + 2)}{(x - 4)(x + 2)}.$$

We can now divide both the numerator and the denominator by $(x + 2)$. That is, cancel the factors $(x + 2)$. This leaves

$$\frac{(x - 2)\cancel{(x + 2)}}{(x - 4)\cancel{(x + 2)}} = \frac{x - 2}{x - 4}.$$

There are three signs associated with a fraction: a sign for the numerator, a sign for the denominator, and a sign for the fraction itself.

The following properties suggest that there are four different possible symbols for $\frac{a}{b}$ and four different possible symbols for its additive inverse, $-\frac{a}{b}$.

88 CHAPTER 2 BASIC ALGEBRAIC SKILLS

ADDITIVE INVERSE PROPERTIES

$$\frac{a}{b} = \frac{-a}{-b} = -\frac{a}{-b} = -\frac{-a}{b}, \quad (b \neq 0) \qquad (2.3)$$

$$-\frac{a}{b} = \frac{-a}{b} = \frac{a}{-b} = -\frac{-a}{-b}, \quad (b \neq 0). \qquad (2.4)$$

These properties follow directly from the application of rule (2.1).

The forms $\frac{a}{b}$ and $\frac{-a}{b}$, in which the sign of the fraction and the sign of the denominator are both positive, are usually most convenient. We refer to these forms as

Standard Forms **standard forms.** For example,

$$-\frac{2}{5}, \quad \frac{4}{5}, \quad \text{and} \quad \frac{6}{7}$$

are in standard form, while

$$\frac{2}{-5}, \quad \frac{-4}{-5}, \quad \text{and} \quad -\left(\frac{6}{-7}\right)$$

are not in standard form.

Recall that

$$-(a - b) = -a + b = b - a.$$

This fact, together with the forms we have just discussed, is very useful when we simplify certain fractions.

• *Example 2.23:* Rewrite the fraction $\frac{-(2 - x)}{x + 1}$ in standard form.

Solution:
$$\frac{-(2 - x)}{x + 1} = \frac{-2 + x}{x + 1} \qquad \text{(Remove parentheses in numerator)}$$

$$= \frac{x - 2}{x + 1} \qquad \text{(Commute terms in numerator)}$$

• *Example 2.24:* Replace the "?" mark with the appropriate expression to make the fractions equal:

$$\frac{-x}{y - x} = \frac{?}{x - y}.$$

Solution: To get the denominator in the desired form we must change the signs of the denominator. Hence, in order to maintain equality we must also change the sign of the numerator. Thus, we get

$$\frac{-x}{y-x} = \frac{-(-x)}{-(y-x)} = \frac{x}{x-y}.$$

• **Example 2.25:** Simplify $\dfrac{10 + 3x - x^2}{x^2 - 4x - 5}$.

Solution:
$$\frac{10 + 3x - x^2}{x^2 - 4x - 5} = \frac{(5-x)(2+x)}{(x-5)(x+1)} \quad \text{(Factor)}$$

$$= -\frac{-(5-x)(2+x)}{(x-5)(x+1)} \quad \text{(Change two signs)}$$

$$= -\frac{(x-5)(2+x)}{(x-5)(x+1)} \quad (-(5-x) = x-5)$$

$$= -\frac{2+x}{x+1} \quad \text{or} \quad -\frac{x+2}{x+1}. \quad \text{(Cancel and commute)}$$

• **Example 2.26:** Simplify $\dfrac{3x^2 + 6x - 24}{5x^2 + 20x - 60}$.

Solution:
$$\frac{3x^2 + 6x - 24}{5x^2 + 20x - 60} = \frac{3(x^2 + 2x - 8)}{5(x^2 + 4x - 12)} \quad \text{(Remove common factors)}$$

$$= \frac{3(x+4)(x-2)}{5(x+6)(x-2)} \quad \text{(Factor each trinomial)}$$

$$= \frac{3(x+4)}{5(x+6)}. \quad \text{(Cancel)}$$

Trial Problems 2.3

Before you begin the section exercises, warm up with these problems. Complete answers are included in the answer key.

Make the fraction on the right equivalent to the fraction on the left.

1. $\dfrac{2}{2-x} = \dfrac{}{x-2}$

2. $\dfrac{-b}{b-a} = \dfrac{}{a-b}$

Express in lowest terms.

3. $\dfrac{15a^4b^3}{45ab^5}$

4. $\dfrac{5x^2 - 20}{5x - 10}$

5. $\dfrac{2b^2 + b - 6}{b^2 + b - 2}$

Exercises 2.3

Write in standard form and specify any real values of the variables for which the fraction is undefined.

1. $-\dfrac{2}{3}$
2. $-\dfrac{3}{4}$
3. $\dfrac{3}{-4}$
4. $\dfrac{2}{-3}$
5. $-\dfrac{-5}{7}$
6. $-\dfrac{-3}{7}$
7. $\dfrac{4}{-7}$
8. $\dfrac{5}{-6}$
9. $\dfrac{-2}{-5}$
10. $\dfrac{-3}{-7}$
11. $-\dfrac{-3}{-5}$
12. $-\dfrac{-2}{-3}$
13. $\dfrac{3x}{-y}$
14. $\dfrac{2x}{-y}$
15. $-\dfrac{2y}{x-3}$
16. $-\dfrac{2x}{y-3}$
17. $-\dfrac{-2x}{-3y^2}$
18. $-\dfrac{-3y}{-2x^2}$
19. $\dfrac{-x^2}{-y^2}$
20. $\dfrac{-y^3}{-x^2}$

Make the fraction on the right-hand side equivalent to the fraction on the left-hand side by replacing the "?" mark with the appropriate expression. Assume that no denominator equals zero.

21. $\dfrac{-1}{2-x} = \dfrac{?}{x-2}$
22. $\dfrac{-a}{b-a} = \dfrac{?}{a-b}$
23. $-\dfrac{4}{3-y} = \dfrac{?}{y-3}$
24. $-\dfrac{6}{2-x} = \dfrac{?}{x-2}$
25. $\dfrac{x-2}{3-x} = \dfrac{?}{x-3}$
26. $\dfrac{x-5}{3-y} = \dfrac{?}{y-3}$
27. $\dfrac{-x}{x-y} = \dfrac{?}{y-x}$
28. $\dfrac{-y}{x-y} = \dfrac{?}{y-x}$
29. $-\dfrac{x}{2x-3y} = \dfrac{?}{3y-2x}$
30. $-\dfrac{a}{3a-2b} = \dfrac{?}{2b-3a}$

Express in lowest terms.

31. $\dfrac{16x^4y^3}{48xy^5}$
32. $\dfrac{17a^4b^3}{51ab^5}$
33. $\dfrac{3x^2-12}{3x-6}$
34. $\dfrac{3a^2-12}{3a-6}$
35. $\dfrac{a-b}{a^2-b^2}$
36. $\dfrac{x-y}{x^2-y^2}$
37. $\dfrac{x-4}{x^2-16}$
38. $\dfrac{x-5}{x^2-25}$
39. $\dfrac{a-b}{b^2-a^2}$
40. $\dfrac{x-y}{y^2-x^2}$
41. $\dfrac{a^3-3a^2+2a}{a}$
42. $\dfrac{3x^3-6x^2+3x}{3x}$
43. $\dfrac{4x^2-9y^2}{2x^2+xy-6y^2}$
44. $\dfrac{2y^2+y-6}{y^2+y-2}$
45. $\dfrac{2a^2-ab-6b^2}{a^2+ab-6b^2}$
46. $\dfrac{2r^2-rs-15s^2}{6r^2+17rs+5s^2}$

47. The length of a rectangular metal plate under a certain temperature T can be expressed as
$$\frac{2000 + 90T + T^2}{2500 - T^2}.$$
Simplify this expression.

48. The width of a rectangular metal plate under a certain temperature T can be expressed as
$$\frac{16,000 - 120T - T^2}{400 + 2T}.$$
Simplify this expression.

49. When using a steam pile driver, the safe load P per pile may be computed by simplifying the right-hand side of the formula
$$P = \frac{2wh(s + k) + 2amh(s + k) - 2bh(s + k)}{(s + k)^2},$$

where h = stroke, or height of the fall in feet,
a = effective area of the piston in square inches,
m = mean effective pressure of the steam on the downward stroke, in pounds per square inch,
b = total back pressure,

and k = a constant, sometimes taken from 0.1 to 0.3.

50. The safe load p in pounds, when using a drop hammer pile driver, can be determined by simplifying the right-hand side of the formula
$$p = \frac{6whs + 6wh}{3(s + 1)^2},$$

where h = fall of the hammer in feet,
s = penetration in inches under the last blow,

and w = weight of the hammer in pounds.

2•4 Multiplication and Division of Fractions

The product of two fractions is a fraction whose numerator is the product of the numerators of the two fractions and whose denominator is the product of the denominators of the two fractions.

PRODUCT OF FRACTIONS	$\dfrac{a}{b} \cdot \dfrac{c}{d} = \dfrac{ac}{bd}.$	(2.5)

• **Example 2.27:** Multiply $\dfrac{2}{3} \cdot \dfrac{4}{5}$.

Solution: $\dfrac{2}{3} \cdot \dfrac{4}{5} = \dfrac{(2)(4)}{(3)(5)} = \dfrac{8}{15}.$ (Multiply numerators)
(Multiply denominators)

• **Example 2.28:** Multiply $\dfrac{3x}{y} \cdot \dfrac{2}{z}$.

Solution: $\dfrac{3x}{y} \cdot \dfrac{2}{z} = \dfrac{(3x)(2)}{yz} = \dfrac{6x}{yz}.$ (Multiply numerators and simplify)
(Multiply denominators)

• **Example 2.29:** Multiply $\dfrac{x-2}{3} \cdot \dfrac{2}{x+2}$.

Solution: $\dfrac{x-2}{3} \cdot \dfrac{2}{x+2} = \dfrac{(x-2)(2)}{3(x+2)} = \dfrac{2x-4}{3x+6}$.   (Multiply numerators)
(Multiply denominators)

In chapter 1 we defined division by a real number in terms of multiplication by the reciprocal of that real number. Since algebraic fractions are real numbers, the definition holds for the division of algebraic fractions, or rational expressions, as we have sometimes called them.

QUOTIENT OF FRACTIONS

$$\frac{a}{b} \div \frac{c}{d} = \frac{a}{b} \cdot \left(\frac{c}{d}\right)^{-1}$$

or

$$\frac{a}{b} \div \frac{c}{d} = \frac{a}{b} \cdot \frac{d}{c}. \tag{2.6}$$

• **Example 2.30:** Divide $\dfrac{3}{4}$ by $\dfrac{5}{7}$.

Solution: $\dfrac{3}{4} \div \dfrac{5}{7} = \dfrac{3}{4} \cdot \dfrac{7}{5}$ (Invert divisor)

$= \dfrac{21}{20}.$ (Multiply)

• **Example 2.31:** Divide $\dfrac{2x}{y}$ by $\dfrac{a^2}{2b^2}$.

Solution: $\dfrac{2x}{y} \div \dfrac{a^2}{2b^2} = \dfrac{2x}{y} \cdot \dfrac{2b^2}{a^2}$ (Invert divisor)

$= \dfrac{4b^2 x}{a^2 y}.$ (Multiply)

Section 2.4 Multiplication and Division of Fractions

- **Example 2.32:** $\dfrac{4x^2y^3}{15a^2b^3} \div \dfrac{6xy}{5a^3b^5}$.

    Solution:
    $$\dfrac{4x^2y^3}{15a^2b^3} \div \dfrac{6xy}{5a^3b^5} = \dfrac{4x^2y^3}{15a^2b^3} \cdot \dfrac{5a^3b^5}{6xy} \quad \text{(Invert divisor)}$$
    $$= \dfrac{(4)(5)a^3b^5x^2y^3}{(15)(6)a^2b^3xy} \quad \text{(Multiply)}$$
    $$= \dfrac{2ab^2xy^2}{9}. \quad \text{(Simplify)}$$

- **Example 2.33:** $\dfrac{x^2 - 16}{2x^2 - 7x - 4} \cdot \dfrac{4x^2 + 6x + 2}{6x^2 - 12x} \div \dfrac{x^2 + 5x + 4}{3x^3 - 5x^2 - 2x}$.

    Solution:
    $$\dfrac{x^2 - 16}{2x^2 - 7x - 4} \cdot \dfrac{4x^2 + 6x + 2}{6x^2 - 12x} \div \dfrac{x^2 + 5x + 4}{3x^3 - 5x^2 - 2x}$$
    $$= \dfrac{x^2 - 16}{2x^2 - 7x - 4} \cdot \dfrac{4x^2 + 6x + 2}{6x^2 - 12x} \cdot \dfrac{3x^3 - 5x^2 - 2x}{x^2 + 5x + 4}$$
    $$= \dfrac{(x + 4)(x - 4)}{(2x + 1)(x - 4)} \cdot \dfrac{2(2x + 1)(x + 1)}{6x(x - 2)} \cdot \dfrac{x(3x + 1)(x - 2)}{(x + 4)(x + 1)}$$
    $$= \dfrac{3x + 1}{3}.$$

Trial Problems 2.4

Before you begin the section exercises, warm up with these problems. Complete answers are included in the answer key.

Multiply. Simplify where possible.

1. $\dfrac{2}{5} \cdot \dfrac{3}{7}$
2. $\dfrac{8a^2}{4b^3} \cdot \dfrac{3b^2}{4a^2}$
3. $\dfrac{4x - 8}{5x - 20} \cdot \dfrac{10x - 40}{36x - 72}$

Divide. Simplify where possible.

4. $\dfrac{4}{3} \div \dfrac{2}{5}$
5. $\dfrac{6a - 12}{8a + 32} \div \dfrac{18a - 36}{10a + 40}$

Exercises 2.4

Multiply. Simplify where possible.

1. $\dfrac{2}{3} \cdot \dfrac{4}{5}$

2. $\dfrac{3}{5} \cdot \dfrac{2}{7}$

3. $\dfrac{8x^2}{4y^3} \cdot \dfrac{3y^2}{4x^3}$

4. $\dfrac{6a^2}{3b^3} \cdot \dfrac{4b^2}{5a^3}$

5. $\dfrac{3x - 6}{5x - 20} \cdot \dfrac{10x - 40}{27x - 54}$

6. $\dfrac{8x - 12}{14x + 7} \cdot \dfrac{42x + 21}{32x - 48}$

7. $\dfrac{x^2 + 5x + 4}{x^3 y^2} \cdot \dfrac{x^2 y^2}{x^2 + 2x + 1}$

8. $\dfrac{x^2 + x - 2}{xy^2} \cdot \dfrac{x^3 y}{x^2 + 5x + 6}$

9. $\dfrac{2x^2 - 5x}{2xy + y} \cdot \dfrac{2xy^2 + y^2}{5x^2 - 2x^3}$

10. $\dfrac{3x^3 + 4x^2}{5xy - 3y} \cdot \dfrac{3y^3 - 5xy^3}{3x^2 + 4x}$

11. $\dfrac{x^2 - 2x - 24}{x^2 + 6x + 8} \cdot \dfrac{x^2 + 5x + 6}{x^2 - 5x - 6}$

12. $\dfrac{x^2 - 8x + 7}{x^2 - 9x + 14} \cdot \dfrac{x^2 + 3x - 10}{x^2 + 3x - 4}$

13. $\dfrac{x^2 - 6x - 27}{x^2 - 10x - 24} \cdot \dfrac{16 + 6x - x^2}{x^2 - 17x + 72}$

14. $\dfrac{20 - x - x^2}{x^2 + 7x + 10} \cdot \dfrac{x^2 - 13x + 42}{x^2 - 11x + 28}$

Divide. Simplify where possible.

15. $\dfrac{2}{3} \div \dfrac{4}{5}$

16. $\dfrac{3}{4} \div \dfrac{2}{3}$

17. $\dfrac{ax^2}{by} \div \dfrac{bx}{ay^2}$

18. $\dfrac{rx}{sy^2} \div \dfrac{sx^2}{ry^3}$

19. $\dfrac{4x^3 y^3}{15a^2 b^3} \div \dfrac{6xy}{5a^3 b^5}$

20. $\dfrac{9x^3 y^4}{16a^4 b^3} \div \dfrac{45x^4 y^2}{14a^7 b}$

21. $\dfrac{6x - 12}{8x + 32} \div \dfrac{18x - 36}{10x + 40}$

22. $\dfrac{28x + 14}{45x - 30} \div \dfrac{14x + 7}{30x - 20}$

Perform the indicated operations. Simplify where possible.

23. $\dfrac{x^2 - 9}{3x^2 - 8x - 3} \cdot \dfrac{6x^2 + 8x + 2}{6x^2 - 12x} \div \dfrac{x^2 + 4x + 3}{2x^3 - 3x^2 - 2x}$

24. $\dfrac{3y - 3}{4y^2 - 4} \cdot \dfrac{2y^2 - 2y}{3y^2} \div \dfrac{1 - y}{1 + y}$

25. $\dfrac{a^2 - a}{a - 3} \cdot \dfrac{a + 1}{a^2 + 4a} \div \dfrac{a^2 - 3a - 4}{a^2 - 16}$

26. $\dfrac{u^2 - 4u + 3}{u^2} \cdot \dfrac{u^2 + u}{u^2 - 6u + 9} \div \dfrac{u^2 - 2u - 3}{u^2 - u - 6}$

27. The ratio of the volume of a sphere with radius r to the surface area of the same sphere is given by the expression

$$\dfrac{4}{3}\pi r^3 \div 4\pi r^2.$$

Simplify this expression.

28. The volume of a cone of radius r and height h when compared with the volume of the top half of the same cone yields the ratio

$$\dfrac{\pi r^2 h}{3} \div \dfrac{\pi r^2 \left(\dfrac{h}{2}\right)}{3}.$$

Simplify this ratio.

29. The volume of a cube with edge $2r$ compared to the volume of a sphere of radius r can be expressed as

$$(2r)^3 \div \dfrac{4\pi r^3}{3}.$$

Simplify this expression.

2·5 Addition and Subtraction of Fractions

Fractions represent real numbers for permissible real number replacements of any variables involved. Hence, we can use the properties of real numbers that we developed in chapter 1 to simplify sums involving fractions. For example, since

$$\frac{a}{c} = a(c)^{-1} = a\left(\frac{1}{c}\right), \qquad (c \neq 0) \qquad \text{(Multiply by reciprocal)}$$

and

$$\frac{b}{c} = b(c)^{-1} = b\left(\frac{1}{c}\right), \qquad \text{(Multiply by reciprocal)}$$

then

$$\begin{aligned}
\frac{a}{c} + \frac{b}{c} &= a\left(\frac{1}{c}\right) + b\left(\frac{1}{c}\right) & \text{(Add equals to equals)} \\
&= (a + b)\left(\frac{1}{c}\right) & \text{(Remove common factor } \frac{1}{c}\text{)} \\
&= \frac{a + b}{c}. & \text{(Multiply)}
\end{aligned}$$

Hence, we have the following property for the sum of two fractions with the same denominator.

SUM OF TWO FRACTIONS WITH THE SAME DENOMINATORS

$$\frac{a}{c} + \frac{b}{c} = \frac{a + b}{c}, \qquad (c \neq 0). \qquad (2.7)$$

That is, to add two fractions with a common denominator add the numerators and write the sum over the common denominator.

If the fractions to be added do not have the same denominator, we simply change them into equivalent fractions that do have the same denominators, then apply the property that we just developed.

In general, for the sum of two fractions with different denominators, we have

$$\begin{aligned}
\frac{a}{b} + \frac{c}{d} &= \left(\frac{a}{b}\right)\left(\frac{d}{d}\right) + \left(\frac{c}{d}\right)\left(\frac{b}{b}\right) & \text{(Find common denominators)} \\
&= \frac{ad}{bd} + \frac{bc}{bd} & \text{(Multiply)} \\
&= \frac{ad + bc}{bd}, & (b, d \neq 0).
\end{aligned}$$

Thus, the rule for finding the sum of two fractions with different denominators can be summarized as follows.

SUM OF TWO FRACTIONS WITH DIFFERENT DENOMINATORS

$$\frac{a}{b} + \frac{c}{d} = \frac{ad + bc}{bd}, \quad (b, d \neq 0). \tag{2.8}$$

To subtract one fraction from another, we simply add to the first fraction the additive inverse of the second. Thus,

$$\frac{a}{b} - \frac{c}{d} = \frac{a}{b} + \frac{-c}{d} \quad \text{(Add additive inverse)}$$

$$= \frac{a}{b} \cdot \frac{d}{d} + \frac{-c}{d} \cdot \frac{b}{b} \quad \text{(Find common denominators)}$$

$$= \frac{ad - bc}{bd}. \quad \text{(Multiply)}$$

This results in a rule for finding the difference of two fractions with different denominators.

DIFFERENCE OF TWO FRACTIONS WITH DIFFERENT DENOMINATORS

$$\frac{a}{b} - \frac{c}{d} = \frac{ad - bc}{bd}, \quad (b, d \neq 0). \tag{2.9}$$

We illustrate these two rules with examples 2.34 and 2.35.

• *Example 2.34:* $\frac{2}{3} + \frac{4}{5} = \underline{?}.$

Solution:
$$\frac{2}{3} + \frac{4}{5} = \frac{2}{3} \cdot \frac{5}{5} + \frac{4}{5} \cdot \frac{3}{3} \quad \text{(Find common denominators)}$$

$$= \frac{10}{15} + \frac{12}{15} \quad \text{(Multiply)}$$

$$= \frac{10 + 12}{15} \quad \text{(Add)}$$

$$= \frac{22}{15}. \quad \text{(Simplify)}$$

• *Example 2.35:* $\frac{4}{5} - \frac{2}{3} = \underline{?}.$

Solution: $\dfrac{4}{5} - \dfrac{2}{3} = \dfrac{4}{5} + \dfrac{-2}{3}$ (Add additive inverse)

$= \dfrac{4}{5} \cdot \dfrac{3}{3} + \dfrac{-2}{3} \cdot \dfrac{5}{5}$ (Find common denominators)

$= \dfrac{12}{15} + \dfrac{-10}{15}$ (Multiply)

$= \dfrac{12 - 10}{15}$ (Add)

$= \dfrac{2}{15}.$ (Simplify)

Least Common Denominator

Least Common Multiple

Forms (2.8) and (2.9) cannot be used directly if we have more than two fractions to be added or subtracted. In such cases we must first find a common denominator. In fact, if we find the **least common denominator,** the process of simplifying the result will be easier.

The least common denominator (LCD) is the **least common multiple** (LCM) of the denominators. The LCM of two or more natural numbers is the smallest natural number that is exactly divisible by each of the given numbers. For example, 24 is the LCM of 3 and 8, since 24 is the smallest natural number that is exactly divisible by 3 and 8.

By factoring each number completely, we can develop a convenient method for finding the LCM. This is illustrated in example 2.36.

• *Example 2.36:* Find the LCM of 12 and 18.

Solution: Write each number in factored form:

$$12 = (2)(2)(3)$$
$$18 = (2)(3)(3).$$

Use each factor the greatest number of times it appears in the factored form of any of the given numbers. In this problem, the 2 appears twice in the factored form of the first number and once in the factored form of the second number. Hence, we use it twice. The 3 appears once as a factor of the first number and twice as a factor of the second. Hence, we use it twice also. Thus,

$$(2)(2)(3)(3)$$

is the factored form of the LCM of 12 and 18. Therefore (2)(2)(3)(3), or 36, is the LCM of 12 and 18.

• **Example 2.37:** Find the LCM of 9, 12, and 15.

Solution: Writing each number in factored form, we get

$$9 = (3)(3)$$
$$12 = (2)(2)(3)$$
$$15 = (3)(5).$$

Thus, the factored form of the LCM of 9, 12, and 15 must be

$$(2)(2)(3)(3)(5),$$

which is equal to 180. So the LCM of 9, 12, and 15 is 180.

We can define the LCM of polynomials in a manner analogous to that just described for natural numbers. Thus, the LCM of a set of polynomials is the polynomial of lowest degree that, when divided by each of the given polynomials, yields a polynomial quotient. This is illustrated in example 2.38.

• **Example 2.38:** Find the LCM of x^2, $x^2 - 9$, and $x^3 - x^2 - 6x$.

Solution: Writing each polynomial in factored form, we get

$$x^2 = xx$$
$$x^2 - 9 = (x + 3)(x - 3)$$
$$x^3 - x^2 - 6x = x(x - 3)(x + 2).$$

Thus, the factored form of the LCM is

$$xx(x + 3)(x - 3)(x + 2).$$

Examples 2.39 and 2.40 illustrate how we add or subtract rational expressions with different denominators.

• **Example 2.39:** Compute $\dfrac{7}{5x - 10} + \dfrac{4}{3x - 6}$.

Solution: Factoring the denominators makes

$$\frac{7}{5x - 10} + \frac{4}{3x - 6} = \frac{7}{5(x - 2)} + \frac{4}{3(x - 2)}.$$

The LCM of the denominators is $(3)(5)(x - 2)$. To create a common denominator in each fraction we multiply the first fraction by $\dfrac{3}{3}$ and the second fraction by $\dfrac{5}{5}$. This produces

$$\left(\frac{3}{3}\right)\frac{(7)}{(5)(x - 2)} + \left(\frac{5}{5}\right)\frac{4}{(3)(x - 2)}.$$

Multiplying we get

$$\frac{21}{(3)(5)(x-2)} + \frac{20}{(3)(5)(x-2)}.$$

Adding these fractions with common denominators gives

$$\frac{21 + 20}{(3)(5)(x-2)} \quad \text{or} \quad \frac{41}{15(x-2)}.$$

• **Example 2.40:** Compute $\dfrac{5}{x^2 - x - 2} - \dfrac{3}{x^2 + 2x + 1}$.

Solution: Factoring the denominators makes

$$\frac{5}{x^2 - x - 2} - \frac{3}{x^2 + 2x + 1}$$
$$= \frac{5}{(x-2)(x+1)} - \frac{3}{(x+1)(x+1)}.$$

The LCM of the denominators is

$$(x-2)(x+1)(x+1) \quad \text{or} \quad (x-2)(x+1)^2.$$

To create a common denominator in each fraction we multiply the first fraction by $\dfrac{(x+1)}{(x+1)}$ and the second fraction by $\dfrac{(x-2)}{(x-2)}$. This produces

$$\frac{5}{(x-2)(x+1)} \cdot \frac{(x+1)}{(x+1)} - \frac{3}{(x+1)(x+1)} \cdot \frac{(x-2)}{(x-2)}.$$

Multiplying we get

$$\frac{5x + 5}{(x-2)(x+1)^2} - \frac{3x - 6}{(x-2)(x+1)^2}.$$

Subtracting these fractions with common denominators gives

$$\frac{5x + 5 - 3x + 6}{(x-2)(x+1)^2} \quad \text{or} \quad \frac{2x + 11}{(x-2)(x+1)^2}.$$

Note that it is often more convenient to leave the answers in factored form.

Trial Problems 2.5

Before you begin the section exercises, warm up with these problems. Complete answers are included in the answer key.

Find the least common multiple.

1. 12, 15, 10
2. $3x, 6x, 9xy^2$

Add or subtract.

3. $\dfrac{4}{27} + \dfrac{5}{18}$
4. $\dfrac{5}{8} - \dfrac{7}{12}$
5. $\dfrac{x}{a} + \dfrac{y}{b}$

Exercises 2.5

Find the LCM.

1. 24, 30, 20
2. 12, 18, 24
3. $2a, 4b, 6ab^2$
4. $6xy, 8x^2, 3xy^2$
5. $2ab, 6b^2$
6. $12xy, 24x^2y^3$
7. $2a - 2, a - 1$
8. $x^2 - 1, 2(x - 1)^2$
9. $x^2 - 3x + 2, (x - 1)^2$
10. $x^2 + 3x - 4, (x - 1)^2$

Perform the indicated operations.

11. $\dfrac{2}{27} + \dfrac{5}{18}$
12. $\dfrac{5}{12} + \dfrac{7}{18}$
13. $\dfrac{7}{8} - \dfrac{5}{12}$
14. $\dfrac{5}{9} - \dfrac{7}{15}$

15. $\dfrac{a}{x} + \dfrac{b}{y}$
16. $\dfrac{c}{x} + \dfrac{d}{y}$
17. $\dfrac{12}{x} - \dfrac{5}{2x}$
18. $\dfrac{5}{a} - \dfrac{3}{2a}$

19. $\dfrac{1}{2x} - \dfrac{2}{4x} + \dfrac{3}{6x}$
20. $\dfrac{1}{2a} + \dfrac{2}{4a} - \dfrac{3}{6a}$
21. $\dfrac{x + 4}{8x} + \dfrac{x - 3}{6x}$
22. $\dfrac{2x - 3}{2x} + \dfrac{x + 3}{3x}$

23. $\dfrac{2x + 9}{9x} - \dfrac{x - 5}{5x}$
24. $\dfrac{3y - 2}{12y} - \dfrac{y - 3}{18y}$
25. $\dfrac{4}{x - 2} + \dfrac{5}{x + 3}$
26. $\dfrac{2}{x - 3} + \dfrac{5}{x + 4}$

27. $\dfrac{6}{x - 7} - \dfrac{4}{x + 3}$
28. $\dfrac{3}{y + 6} - \dfrac{4}{y - 3}$
29. $\dfrac{2a}{a - 7} + \dfrac{5}{7 - a}$
30. $\dfrac{4x}{6 - x} + \dfrac{5}{x - 6}$

31. $\dfrac{2}{y + 2} + \dfrac{3}{y + 3}$
32. $\dfrac{1}{2x + 1} + \dfrac{3}{x - 2}$
33. $\dfrac{1}{b^2 - 1} - \dfrac{1}{b^2 + 2b + 1}$

34. $\dfrac{y}{y^2 - 16} - \dfrac{y + 1}{y^2 - 5y + 4}$
35. $\dfrac{3a}{a^2 + 3a - 10} + \dfrac{2a}{a^2 + a - 6}$
36. $\dfrac{x + 4}{x^2 - x - 2} + \dfrac{2x - 3}{x^2 + 2x + 1}$

37. $\dfrac{1}{(s - t)(t - u)} + \dfrac{1}{(t - u)(u - s)} + \dfrac{1}{(u - s)(s - t)}$
38. $\dfrac{1}{(s + t)(t + u)} - \dfrac{1}{(s + u)(t + u)} - \dfrac{1}{(s + t)(s + u)}$

39. $\dfrac{1}{r^2 - 7r + 12} + \dfrac{2}{r^2 - 5r + 6} - \dfrac{3}{r^2 - 6r + 8}$
40. $\dfrac{4}{p^2 - 4q^2} + \dfrac{2}{p^2 + 3pq + 2q^2} + \dfrac{4}{p^2 - pq - 2q^2}$

41. A magnetic field can be described by the expression

$$\frac{m}{x^2 + y^2} - \frac{2mx^2}{(x^2 + y^2)^2}.$$

Simplify this expression.

42. An analysis of the forces acting on a plastic material gives the expression

$$1 - \frac{4c}{5d} + \frac{c^3}{3d^3}.$$

Simplify this expression.

43. When condensers are connected in series the capacity may be found with the expression

$$\frac{1}{\frac{1}{C_1} + \frac{1}{C_2} + \frac{1}{C_3}},$$

where each C represents an individual capacity. Simplify this expression.

44. When four resistances are connected in a series-parallel circuit the expression

$$\frac{1}{R_1 + R_2} + \frac{1}{R_3 + R_4}$$

must be evaluated to find the total resistance. If each R represents a resistance, simplify the expression.

2·6 Fractional Equations

In science and technology we encounter many important equations that contain fractions. The solution of these equations involves the use of the basic properties and operations that were developed in chapter 1. However, certain care must be exercised to assure that the solutions check when substituted for the variable in the original equation.

When solving equations that contain fractions, we can use a basic procedure to eliminate the fractions.

If we multiply every term of the equation by the lowest common denominator, the resulting equation does not involve fractions and can be solved by the methods outlined in chapter 1. This is illustrated in examples 2.41, 2.42, and 2.43.

• **Example 2.41:**   Solve $1 + \dfrac{x}{9} = \dfrac{4}{3}$ for x.

Solution: To clear fractions, multiply both sides by the LCD. The lowest common denominator of the fractions in this equation is 9, so we multiply each term by 9.

$$1 + \frac{x}{9} = \frac{4}{3}$$

$$(9)(1) + (9)\frac{x}{9} = (9)\frac{4}{3} \quad \text{(Multiply every term by 9)}$$

$$9 + x = 12 \quad \text{(Cancel)}$$

$$x = 12 - 9 \quad \text{(Subtract 9)}$$

$$x = 3. \quad \text{(Simplify)}$$

Checking in the original equation, we have

$$1 + \frac{x}{9} = \frac{4}{3}$$

$$1 + \frac{3}{9} = \frac{4}{3} \quad \text{(Substitute 3 for } x\text{)}$$

$$1 + \frac{1}{3} = \frac{4}{3} \quad \text{(Reduce)}$$

$$\frac{4}{3} = \frac{4}{3}. \quad \text{(Simplify)}$$

Since the substitution of 3 for x gives a true statement, $\frac{4}{3} = \frac{4}{3}$, $x = 3$ checks.

• **Example 2.42:** Solve $\frac{2}{x} + \frac{9}{2} = 4$ for x.

Solution: The lowest common denominator of the fractions in the equation is $2x$. Therefore, we multiply each term in the equation by $2x$.

$$\frac{2}{x} + \frac{9}{2} = 4$$

$$(2x)\frac{2}{x} + (2x)\frac{9}{2} = (2x)4 \quad \text{(Multiply by } 2x\text{)}$$

$$4 + 9x = 8x \quad \text{(Cancel)}$$

$$9x - 8x = -4 \quad \text{(Transpose } 8x \text{ and } 4\text{)}$$

$$x = -4. \quad \text{(Simplify)}$$

Checking, we get

$$\frac{2}{x} + \frac{9}{2} = 4$$

$$\frac{2}{-4} + \frac{9}{2} = 4 \quad \text{(Substitute } -4 \text{ for } x\text{)}$$

$$\frac{-1}{2} + \frac{9}{2} = 4 \quad \text{(Reduce)}$$

$$\frac{8}{2} = 4 \quad \text{(Add)}$$

$$4 = 4. \quad \text{(Simplify)}$$

Hence, $x = -4$ is a solution.

SECTION 2.6 FRACTIONAL EQUATIONS

• **Example 2.43:** Solve $\dfrac{3}{x} + \dfrac{2}{x-2} = \dfrac{4}{x^2 - 2x}$ for x.

Solution: The lowest common denominator of the terms in the equation is $x(x-2)$. Next we multiply each term in the equation by $x(x-2)$.

$$\dfrac{3}{x} + \dfrac{2}{x-2} = \dfrac{4}{x^2 - 2x}$$

$$\dfrac{3}{x}x(x-2) + \dfrac{2}{x-2}x(x-2) = \dfrac{4}{x^2-2x}x(x-2) \quad \text{(Multiply by } x(x-2)\text{)}$$

In factored form, this can be written as

$$\dfrac{3}{x}x(x-2) + \dfrac{2}{x-2}x(x-2) = \dfrac{4}{x(x-2)}x(x-2).$$

$$\begin{aligned}
3(x-2) + 2x &= 4 &&\text{(Cancel)} \\
3x - 6 + 2x &= 4 &&\text{(Distribute)} \\
5x - 6 &= 4 &&\text{(Simplify)} \\
5x &= 4 + 6 &&\text{(Transpose } -6\text{)} \\
5x &= 10 &&\text{(Add)} \\
x &= 2. &&\text{(Divide by 5)}
\end{aligned}$$

Checking, we get

$$\dfrac{3}{x} + \dfrac{2}{x-2} = \dfrac{4}{x^2 - 2x}$$

$$\dfrac{3}{2} + \dfrac{2}{2-2} = \dfrac{4}{(2)^2 - 2(2)} \quad \text{(Substitute 2 for } x\text{)}$$

$$\dfrac{3}{2} + \dfrac{2}{0} = \dfrac{4}{0}. \quad \text{(Simplify)}$$

Since division by zero is undefined, the value $x = 2$ cannot be a solution. Therefore there is no solution to this equation.

• • • • • • • • • •

Whenever an equation is solved by multiplying both sides of the equation by a factor involving the variable, an extraneous root (solution) may be introduced. Since we do not know the value of the variable, multiplying by a factor involving the variable may be the same as multiplying both sides of the equation by zero. Multiplying both sides of an equation by zero causes both sides of the equation to be zero. This is the reason we sometimes get "solutions" that do not check.

Example 2.44 illustrates how an equation that contains many variables can be solved for one of the variables.

Given the equation from optics,

$$\frac{1}{p} + \frac{1}{q} = \frac{1}{f},$$

solve the equation for f.

$$\frac{1}{p}(pqf) + \frac{1}{q}(pqf) = \frac{1}{f}(pqf) \quad \text{(Multiply each term by the least common denominator } pqf\text{)}$$
$$qf + pf = pq \quad \text{(Simplify)}$$
$$(q + p)f = pq \quad \text{(Factor } f \text{ on left side)}$$
$$f = \frac{pq}{q + p}. \quad \text{(Divide both sides by } p + q\text{)}$$

Trial Problems 2.6

Before you begin the section exercises, warm up with these problems. Complete answers are included in the answer key.

Solve the equations and check the results.

1. $\dfrac{a}{2} + \dfrac{a}{5} = 7$
2. $\dfrac{b}{2} - \dfrac{b}{3} = 3$
3. $\dfrac{t}{5} - \dfrac{t-5}{10} = 2$
4. $\dfrac{3x-1}{4x+1} = \dfrac{2}{3}$
5. $\dfrac{1}{2} = \dfrac{3}{4y-2}$

Exercises 2.6

Solve the equations and check the results.

1. $\dfrac{x}{5} + \dfrac{x}{2} = 7$
2. $\dfrac{x}{5} + \dfrac{x}{3} = 8$
3. $2 - \dfrac{x}{3} = \dfrac{x}{4}$
4. $9 + \dfrac{x}{2} = \dfrac{x}{5}$
5. $\dfrac{6}{7} - \dfrac{y}{3} = \dfrac{5}{42}$
6. $\dfrac{y}{5} - \dfrac{5}{6} = \dfrac{4}{15}$
7. $\dfrac{r}{2} - \dfrac{r-5}{6} = 4$
8. $\dfrac{r}{6} + \dfrac{1-5r}{10} = 3$
9. $\dfrac{2}{s} = \dfrac{6}{s-4}$
10. $\dfrac{3}{s} = \dfrac{5}{s-8}$
11. $\dfrac{3x-1}{3x+1} = \dfrac{1}{2}$
12. $\dfrac{2x-1}{2x+1} = \dfrac{1}{4}$
13. $\dfrac{1}{x} = \dfrac{4}{3x+1}$
14. $\dfrac{1}{x} = \dfrac{5}{3x-1}$
15. $\dfrac{1}{s} + \dfrac{3}{s+2} = \dfrac{14}{s^2+2s}$
16. $\dfrac{1}{t} + \dfrac{5}{t-3} = \dfrac{9}{t^2-3t}$

17. In optics, an important equation is

$$\frac{1}{p} + \frac{1}{q} = \frac{1}{f},$$

where p is the distance of the object from the lens, q is the distance of the image from the lens, and f is the focal length of the lens. Solve this equation for q.

18. Solve the equation in exercises 17 for p.

19. An equation that relates the number of teeth N of a gear, the outside diameter D_o of the gear, and the pitch diameter D_p is

$$D_p = \frac{D_o N}{N + 2}.$$

Solve for N.

20. Solve the equation in exercise 19 for D_o.

21. In the thermodynamics of refrigeration, we may find the equation

$$\frac{W}{Q_1} = \frac{T_2}{T_1} - 1,$$

where W is the work input, Q_1 is the heat absorbed, T_1 is the temperature inside the refrigerator, and T_2 is the temperature outside the refrigerator. Solve the equation for T_1.

22. Solve the formula in exercise 21 for T_2.

23. A formula relating the depth h of a gear tooth to the major diameter D of the gear and the minor diameter d of the gear may be expressed as

$$\frac{h}{D - d} = 2.$$

Solve for D.

24. Solve the formula in exercise 23 for d.

25. For a beam fixed at one end and unsupported at the other, the maximum bending moment at the point of support is determined with the formula

$$M = PL + \frac{WL}{2},$$

where M = maximum bending moment,
P = load concentrated at the free end,
L = length of span,
and W = dead load.
Solve the equation for P.

26. Solve the equation in exercise 25 for L.

27. The safe load in pounds when a drop hammer is used is

$$p = \frac{2wh}{s + 1},$$

where w = weight of the hammer in pounds,
h = fall of the hammer in feet,
s = penetration in inches under the last blow,
and p = safe load in pounds.
Solve this equation for w.

28. The power needed to lift an object is determined with the formula

$$p = \frac{fd}{t},$$

where p = power,
f = force,
d = distance,
and t = time.
Solve this equation for d.

2·7 Summary of Terms, Rules, and Procedures

Terms

Algebraic Fraction (p. 85)
Difference of Two Squares (p. 72)
Equivalent Fractions (p. 87)
Factoring (p. 74)
Factoring by Grouping (p. 81)

Least Common Denominator (p. 97)
Least Common Multiple (p. 97)
Rational Expression (p. 85)
Removing a Common Factor (p. 74)

Square of a Binomial (p. 70)
Standard Forms (p. 88)
Sum and Difference of Two Cubes (p. 83)
Trial and Error (p. 76)

Rules and Procedures

- **SQUARE OF A BINOMIAL.**
  The square of a binomial is equal to the square of the first term plus twice the product of the first and last terms plus the square of the last term.

  $$(x + y)^2 = x^2 + 2xy + y^2$$
  or
  $$(x - y)^2 = x^2 - 2xy + y^2. \quad (2.1)$$

- **PRODUCT OF THE SUM AND DIFFERENCE OF TWO TERMS.**
  The product of the sum of two terms and the difference of the same two terms is the square of the first term minus the square of the last term.

  $$(a + b)(a - b) = a^2 - b^2. \quad (2.1)$$

- **FACTORING QUADRATIC POLYNOMIALS.**
  (1) $x^2 + (a + b)x + ab = (x + a)(x + b)$
  (2) $x^2 + 2ax + a^2 = (x + a)^2$
  (3) $x^2 - a^2 = (x + a)(x - a)$
  (4) $acx^2 + (ad + bc)xy + bdy^2$
  $\quad = (ax + by)(cx + dy)$
  $$(2.2)$$

- **SPECIAL FACTORING FORMULAS.**
  (5) $ax + ay + bx + by = (a + b)(x + y)$
  (6) $x^3 + y^3 = (x + y)(x^2 - xy + y^2)$
  (7) $x^3 - y^3 = (x - y)(x^2 + xy + y^2)$
  $$(2.2)$$

- **QUOTIENT RULES.**

  $\dfrac{a}{b} = \dfrac{c}{d}$ if and only if $ad = bc$, $\quad (b, d \neq 0)$

  $\dfrac{a}{b} = \dfrac{ka}{kb} \quad (k \neq 0, b \neq 0).$ $\quad (2.3)$

- **ADDITIVE INVERSE PROPERTIES.**

  $$\frac{a}{b} = \frac{-a}{-b} = -\frac{a}{-b} = -\frac{-a}{b}, \quad (b \neq 0)$$

  $$-\frac{a}{b} = \frac{-a}{b} = \frac{a}{-b} = -\frac{-a}{-b}, \quad (b \neq 0). \quad (2.3)$$

- **PRODUCT OF FRACTIONS.**

  $$\frac{a}{b} \cdot \frac{c}{d} = \frac{ac}{bd}. \quad (2.4)$$

- **QUOTIENT OF FRACTIONS.**

  $$\frac{a}{b} \div \frac{c}{d} = \frac{a}{b} \cdot \left(\frac{c}{d}\right)^{-1}$$

  $$\frac{a}{b} \div \frac{c}{d} = \frac{a}{b} \cdot \frac{d}{c}. \quad (2.4)$$

- **SUM OF TWO FRACTIONS WITH THE SAME DENOMINATORS.**

  $$\frac{a}{c} + \frac{b}{c} = \frac{a + b}{c}, \quad (c \neq 0). \quad (2.5)$$

- **SUM OF TWO FRACTIONS WITH DIFFERENT DENOMINATORS.**

  $$\frac{a}{b} + \frac{c}{d} = \frac{ad + bc}{bd}, \quad (b, d \neq 0). \quad (2.5)$$

- **DIFFERENCE OF TWO FRACTIONS WITH DIFFERENT DENOMINATORS.**

  $$\frac{a}{b} - \frac{c}{d} = \frac{ad - bc}{bd}, \quad (b, d \neq 0). \quad (2.5)$$

2·8 Chapter 2 Review Exercises

Perform the indicated operations.

1. $(a + b)(x + y)$
2. $(c + d)(r + s)$
3. $(a + b)(x - y)$
4. $(c + d)(r - 2)$
5. $(x - y)(2x - y)$
6. $(x - y)(3x - y)$
7. $(2a + b)(2a + b)$
8. $(x + 2y)(x + 2y)$
9. $(x + y)^2$
10. $(a + b)^2$
11. $(2a + 3b)^2$
12. $(3a + 2b)^2$
13. $(a - 2b)^2$
14. $(3a - b)^2$
15. $(a + 5b)(a - 5b)$
16. $(3b + 4)(3b - 4)$

Factor completely.

17. $ax + ay + az$
18. $ay + by + cy$
19. $x^2 + 2x + 1$
20. $x^2 + 5x + 6$
21. $x^2 - 5x - 14$
22. $x^2 - 2x - 15$
23. $y^2 - 16$
24. $b^2 - 25$
25. $9r^2 - 49s^2$
26. $64u^2 - 25v^2$
27. $2x^2 - 5x - 12$
28. $3x^2 - 15x + 18$
29. $xy + x + y + y^2$
30. $rs + r + s + s^2$
31. $3x^3 + 24$
32. $2y^3 - 54$

Simplify.

33. $\dfrac{x^4 y^3}{xy^5}$
34. $\dfrac{r^2 s^3}{rs^4}$
35. $\dfrac{3t^2 - 12}{t - 2}$
36. $\dfrac{4s^2 - 16}{s + 2}$
37. $\dfrac{x^3 - 3x^2 + 2x}{x}$
38. $\dfrac{2a^3 - 4a^2 + 2a}{2a}$
39. $\dfrac{2b^2 + b - 6}{b^2 + b - 2}$
40. $\dfrac{4a^2 - 9b^2}{2a^2 + ab - 6b^2}$
41. $\dfrac{8a^2}{4b^3} \cdot \dfrac{3b^2}{4a^3}$
42. $\dfrac{5x^2}{6y^3} \cdot \dfrac{3y^2}{4x^3}$
43. $\dfrac{r^2 + 5r + 4}{r^3 s^2} \cdot \dfrac{r^2 s^3}{r^2 + 2r + 1}$
44. $\dfrac{t^2 + t - 2}{st^2} \cdot \dfrac{t^3 s}{t^2 + 5t + 6}$
45. $\dfrac{a^2 - 2a - 24}{a^2 + 6a + 8} \cdot \dfrac{a^2 + 5a + 6}{a^2 - 5a - 6}$
46. $\dfrac{u^2 - 8u + 7}{u^2 - 9u + 14} \cdot \dfrac{u^2 + 3u - 10}{u^2 + 3u - 4}$
47. $\dfrac{4a^2 b^3}{15x^2 y^3} \div \dfrac{6ab}{5x^3 y^5}$
48. $\dfrac{14 r^3 s^4}{45 t^4 u^3} \div \dfrac{9 r^4 s^2}{16 t^7 u}$
49. $\dfrac{6m - 12}{8m + 32} \div \dfrac{18m - 36}{10m + 40}$
50. $\dfrac{28n + 14}{45n - 30} \div \dfrac{14n + 7}{30n - 20}$
51. $\dfrac{x}{a} + \dfrac{y}{b}$
52. $\dfrac{r}{s} + \dfrac{t}{u}$
53. $\dfrac{3}{2a} - \dfrac{2}{4a} + \dfrac{1}{6a}$
54. $\dfrac{3}{r} - \dfrac{2}{2r} + \dfrac{1}{3r}$
55. $\dfrac{5}{r + 4} + \dfrac{2}{r - 3}$
56. $\dfrac{5}{s + 3} + \dfrac{4}{s - 2}$
57. $\dfrac{3}{t + 6} - \dfrac{4}{t - 3}$
58. $\dfrac{6}{a - 7} - \dfrac{4}{a + 3}$
59. $\dfrac{2x}{x - 7} + \dfrac{5}{7 - x}$
60. $\dfrac{4s}{6 - s} + \dfrac{5}{s - 6}$
61. $\dfrac{a}{a^2 - 16} - \dfrac{a + 1}{a^2 - 5a + 4}$
62. $\dfrac{1}{s^2 - 1} - \dfrac{1}{s^2 + 2s + 1}$

Solve for the variable.

63. $\dfrac{x}{2} + \dfrac{x}{5} = 7$
64. $\dfrac{r}{3} + \dfrac{r}{5} = 8$
65. $\dfrac{x}{2} - \dfrac{x - 5}{6} = 4$
66. $\dfrac{t}{6} + \dfrac{1 - 5t}{10} = 3$
67. $\dfrac{5}{3x + 9} = \dfrac{2}{2x - 6}$
68. $\dfrac{1}{r} + \dfrac{1}{r - 1} = \dfrac{1}{r(r - 1)}$

69. In hydrodynamics we may encounter the equation

$$\frac{p}{d} = \frac{P}{d} - \frac{m^3}{2\pi^2 r^2}.$$

Solve for d.

70. A projectile is fired. The height y above ground after t seconds may be represented by the equation

$$y = \frac{v_0 t}{2} - 16t^2.$$

Solve for v_0.

71. The thickness T of tubing with inside diameter d and outside diameter D is expressed with the formula

$$T = \frac{D - d}{2}.$$

Solve this equation for d.

72. The time t necessary for an object to fall 136 feet if it is thrown downward with an initial velocity of 120 feet per second is represented in the expression

$$16t^2 + 120t - 136.$$

Factor this expression.

73. The volume of a square pyramid whose base has side s and whose height is h, when compared with the top half of this same pyramid yields the ratio

$$\frac{s^2 h}{3} \div \frac{s^2 \left(\frac{h}{2}\right)}{3}.$$

Simplify this ratio.

2·9 Chapter 2 Test

Find the following products.

1. $(a + b)(x + y)$
2. $(u - v)(r + s)$
3. $(x - y)(3x - y)$
4. $(a + 2b)(a + 2b)$
5. $(c + d)^2$
6. $(3a + 2b)(3a - 2b)$
7. $(50 + 1)(50 - 1)$

Factor completely.

8. $2ax + 4ay + 6az$
9. $x^2 - 7x - 18$
10. $5x^2 - 80$
11. $x^2 + 16x + 64$
12. $a^2 x + 2ax + x$
13. $16a^3 - 250b^3$

Simplify these fractional expressions.

14. $\dfrac{3x^2 - 12}{3x - 6}$
15. $\dfrac{x - 9}{x^2 - 81}$
16. $\dfrac{a^3 - 3a^2 + 2a}{a}$
17. $\dfrac{2y^2 + y - 6}{y^2 + y - 2}$
18. $\dfrac{2r^2 - rs - 15s^2}{6r^2 + 17rs + 5s^2}$

Perform the indicated operations. Simplify where possible.

19. $\dfrac{8x^2}{4y^2} \cdot \dfrac{3y^2}{4x^3}$
20. $\dfrac{8x - 12}{14x + 7} \cdot \dfrac{42x + 21}{32x - 48}$
21. $\dfrac{x^2 - 2x - 24}{x^2 + 6x + 8} \cdot \dfrac{x^2 + 5x + 6}{x^2 - 5x - 6}$
22. $\dfrac{a^2 - a}{a - 2} \cdot \dfrac{a + 1}{a^2 + 4a} \div \dfrac{a^2 - 3a - 4}{a^2 - 16}$
23. $\dfrac{12}{x} - \dfrac{5}{2x}$
24. $\dfrac{x + 4}{8x} + \dfrac{x - 3}{6x}$

25. $\dfrac{y}{y^2 - 16} - \dfrac{y + 1}{y^2 - 5y + 4}$ 26. $\dfrac{x}{5} + \dfrac{x}{2} = 7$ 27. $\dfrac{2}{s} = \dfrac{6}{s - 4}$

28. $\dfrac{1}{t} + \dfrac{5}{t - 3} = \dfrac{9}{t^2 - 3t}$

2·10 Cumulative Review—Chapters 1 and 2

True or False.

1. $-5 > -2$ 2. $\dfrac{2}{3} = 0.66$ 3. $\sqrt{2} < 1.42$

4. $\dfrac{2}{3} > \dfrac{5}{8}$ 5. $3(x + y) = 3x + y$

Simplify by expressing as a single numeral.

6. $\dfrac{5 - (-7)}{(-2)(-3)}$ 7. $\dfrac{(-6)(30) \div (4 + 5)}{(-5)(4) \div (-6 + 4)}$

Express as powers of two.

8. $2^3 \left(\dfrac{1}{4}\right)^2 4^3 \left(\dfrac{1}{16}\right)$ 9. $\left(\dfrac{1}{2}\right)\left(\dfrac{1}{4}\right)\left(\dfrac{1}{8}\right)^{-1}$

Express in scientific notation.

10. 0.00076 11. 2594

Use your calculator to find approximate values.

12. $(1.09)^{69}$ 13. $(2.817 \times 10^{-12})^5$

Rationalize the denominators.

14. $\sqrt{\dfrac{3xy^5}{8a^3}}$ 15. $\dfrac{\sqrt{a}}{\sqrt{a} + 3\sqrt{b}}$

Change the given expression to radical form.

16. $a^{2/5}$ 17. $s^{7/3}$

Change the given expressions to exponential form and simplify where possible.

18. $\sqrt[3]{6}$ 19. $\sqrt[4]{9^2}$

Simplify the given expressions.

20. $(-27)^{5/3}$ 21. $\dfrac{25^{-3/2}}{81^{3/4}}$

Perform the indicated operations.

22. $(3x^2 + 4x + 6) + (2x^2 + 2x - 1)$

23. $(8x^2 - 6x + 7) - (x^2 - 2x - 3)$

24. $(x + 1)(x^2 - x + 1)$

25. $(4x^2 - 3y^2)(2x^2 + y^2)$

26. $(-12c^3p^4y) \div (-9cpy^6)$

27. $(5t - 5t^2 + 2t^3 - 6) \div (t - 2)$

Solve for the variable.

28. $-3(2x + 1) = 4(4x - 3)$

29. $(x - 1)^2 = (x - 2)^2 + 5$

Solve the proportions.

30. $2:3 = 8:x$

31. $x:4 = 16:x$

Find the required values.

32. s varies directly as t, and $s = 30$ when $t = 6$. Find s when $t = 2$.

33. v is directly proportional to t^2 and inversely proportional to u^3, and $v = 18$ when $t = 3$ and $u = 2$. Find v when $t = 3$ and $u = 4$.

Find the products.

34. $(x - y)(2x - y)$

35. $(2a - 3b)^2$

36. $(3 - y^2)(3 + y^2)$

Factor completely.

37. $2x^2 - 5x - 12$

38. $9x^2 - 6ax + a^2$

39. $16a^3 - 2y^3$

Express in lowest terms.

40. $\dfrac{4x^2 - 9y^2}{2x^2 + xy - 6y^2}$

41. $\dfrac{2r^2 - rs - 15s^2}{6r^2 + 17rs + 5s^2}$

Perform the indicated operations. Simplify where possible.

42. $\dfrac{x^2 - 8x + 7}{x^2 - 9x + 14} \cdot \dfrac{x^2 + 3x - 10}{x^2 + 3x - 4}$

43. $\dfrac{28x - 14}{45x - 30} \div \dfrac{14x + 7}{30x - 20}$

44. $\dfrac{x + 4}{8x} + \dfrac{x - 3}{6x}$

45. $\dfrac{y}{y^2 - 16} - \dfrac{y + 1}{y^2 - 5y + 4}$

Solve for the variable.

46. $\dfrac{3x - 1}{3x + 1} = \dfrac{1}{2}$

47. $\dfrac{1}{t} + \dfrac{5}{t - 3} = \dfrac{9}{t^2 - 3t}$

CHAPTER

3 *Graphing*

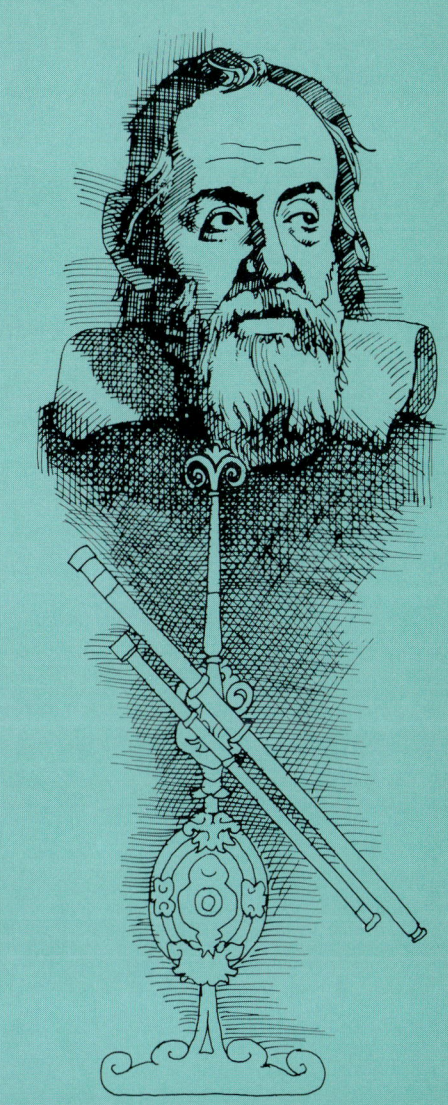

*T*here are many situations in everyday life in which one thing depends on another. When Galileo dropped an object from the Leaning Tower of Pisa he observed that the velocity of the object depended on the amount of time the object had been falling. The owner of a baseball team might observe that the number of people who attend a home game depends on (or is a function of) who the pitchers are. The amount of money it takes to fill the tank of an automobile with gasoline is a function of the number of gallons needed.

In mathematics there are many situations in which one variable depends on (or is a function of) another variable. In this chapter we introduce the concept of a function and look at the vocabulary and notation of functions. Later we consider different types of functions such as linear functions, quadratic functions, exponential functions, logarithms functions, trigonometric functions, and many others.

3•1 The Coordinate Plane

In chapter 1 we indicated how the real numbers could be represented on a horizontal line. A beginning point was chosen, and the number zero was assigned to it. An arbitrary unit of length was selected to determine an appropriate scale. Then we graphed various real numbers by identifying the point on the line that corresponded to each of the real numbers.

We also noted in chapter 1 how the solutions to certain linear equations in one variable could be graphed as points on the real number line.

Now we are ready to consider how we can graph the values that satisfy equations in two variables. Examples of equations in two variables are

Equation	Variables
$2x + 3y = 6$	x, y
$C = \dfrac{5}{9}(F - 32)$	C, F
$6F_1 + 5F_2 = 300$	F_1, F_2

Ordered Pair Since the equations have two variables, there are pairs of values that satisfy each equation. We use the term **ordered pair** to indicate a pair of numbers where the *order* in which the numbers are written is important. For example, in section 1.9 we considered the formula $F = \dfrac{9}{5}C + 32$, which relates temperatures on the Fahrenheit and Celsius scales. By letting $C = -40, 0, 20,$ and 100 in the formula, corresponding values of F can be found.

If $C = -40, F = \dfrac{9}{5}(-40) + 32 = -40.$

If $C = 0, F = \dfrac{9}{5}(0) + 32 = 32.$

If $C = 20, F = \dfrac{9}{5}(20) + 32 = 68.$

If $C = 100, F = \dfrac{9}{5}(100) + 32 = 212.$

These pairs of values for F and C are represented in the following table:

C	−40	0	20	100
F	−40	32	68	212

SECTION 3.1 THE COORDINATE PLANE   113

The data in this table can also be represented as a set of ordered pairs $\{(-40,-40),(0,32),(20,68),(100,212)\}$, where the first number in each pair represents a Celsius temperature and the second number represents the corresponding Fahrenheit temperature.

Coordinate System
Rectangular or Cartesian Coordinate System
Coordinates
x-coordinate
Abscissa
y-coordinate
Ordinate

To represent a set of ordered pairs on a graph, we introduce the idea of a **coordinate system**. We call the coordinate system a **rectangular** or **Cartesian coordinate system**.

The ordered pairs of numbers are called **coordinates** of a point. The first number in the pair is called the **x-coordinate** or **abscissa** of the point. The second number in the pair is called the **y-coordinate** or **ordinate** of the point. For example, in the ordered pair (5,7), the x-coordinate, or abscissa, is 5, and the y-coordinate, or ordinate, is 7.

We make the coordinate system by drawing a pair of perpendicular lines, as in figure 3.1.

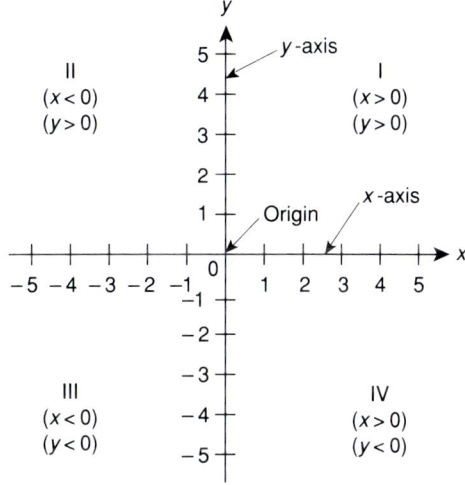

Figure 3.1

Origin
x-axis
y-axis

We select an arbitrary scale for each of the lines beginning at the point where the two lines intersect. We call this point of intersection the **origin**. The horizontal line is usually called the **x-axis** and the vertical line is usually called the **y-axis**. It is customary to scale the x-axis increasingly from left to right and the y-axis increasingly from bottom to top. The size of the unit used to scale the axis is completely arbitrary, and usually depends on the magnitude or size of the numbers encountered in the problems.

Quadrants

The x- and y-axes divide the plane into four regions called **quadrants**. Quadrants I, II, III, and IV are illustrated in figure 3.1.

Note that in the first quadrant (I) both the x-coordinate and the y-coordinate are positive. In quadrant II the x-coordinate is negative and the y-coordinate is positive. In the third quadrant (III) the x-coordinate and the y-coordinate are both negative and in quadrant IV the x-coordinate is positive and the y-coordinate is negative.

Once we have drawn a coordinate system, we can plot the graph of any ordered pair of real numbers on this system. In fact, there is a one-to-one correspondence between the ordered pairs of real numbers and the points on the rectangular coordinate system.

In figure 3.2, point A represents the ordered pair (2,3). That is, the point is two units to the right of the y-axis and three units above the x-axis. Point B represents the ordered pair (5,4); point C, the ordered pair (−4,1); point D, the ordered pair (−3,−2); point E, the ordered pair $\left(4, -\frac{7}{2}\right)$; point F, the ordered pair (−7,0); and point G, the ordered pair (0,6).

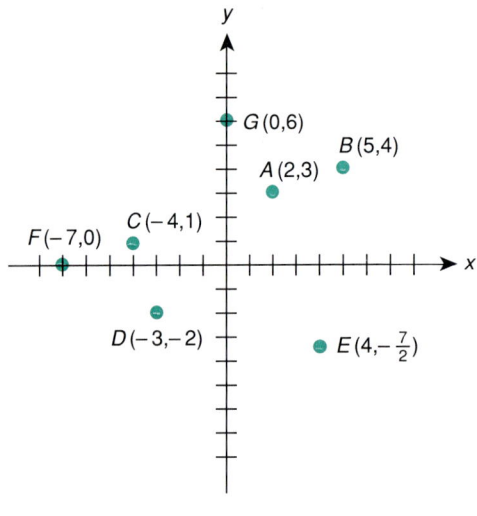

Figure 3.2

Note that the first number in the ordered pair indicates how far the point is to the right or left of the y-axis, and the second number in the ordered pair indicates how far the point is above or below the x-axis.

Trial Problems 3.1

Before you begin the section exercises, warm up with these problems. Complete answers are included in the answer key.

Draw a rectangular coordinate system and plot the ordered pairs.

1. (5,3),(2,−1),(−3,−3),(−2,4)
2. (4,0),(0,3),(−3,0),(0,−4)

Write the coordinates of the points in the figure below.

3. A, B, C, D
4. E, F, G, H

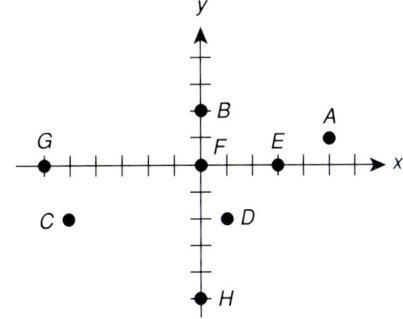

Given the formula $y = 3x - 5$.

5. Determine the set of ordered pairs when $x = 0, 1, 2,$ and 3.

Exercises 3.1

1. Draw a rectangular coordinate system and plot the ordered pairs $(-3,-5)$, $(0,4)$, $(6,0)$, $(2,-7)$, $(-3,8)$, $(-4,6)$, $(5,0)$ and $(0,0)$.

2. Draw a rectangular coordinate system and plot the ordered pairs $(2,3)$, $(4,-5)$, $(-3,6)$, $(0,5)$, $(-4,0)$, and $(0,0)$.

3. Write the coordinate of points A, B, C, D, and E in figure 3.3.

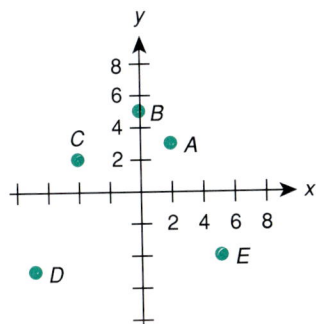

Figure 3.3

4. Write the coordinates of the points P, Q, R, S, and T in figure 3.4.

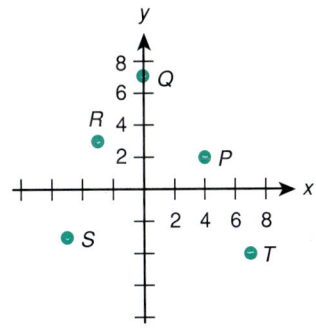

Figure 3.4

5. Draw a Cartesian coordinate system, plot the given points, and connect the points in the order given. $A(0,0)$, $B(4,0)$, $C(4,6)$, $D(2,7)$, and $E(0,6)$. Also draw segments EA, CE, BE, and AC.

6. Draw a Cartesian coordinate system, plot the given points, and connect the points in the order given. $A(0,0)$, $B(0,6)$, $C(8,4)$, and $D(2,4)$. Also draw segments AD, BD, and AC.

7. Given the formula $C = 0.4x + 50$, determine the set of ordered pairs when $x = 10, 20, 30, 40,$ and 50. Plot the ordered pairs on an appropriate coordinate system.

8. Given the formula $C = 0.6x + 30$, determine the set of ordered pairs when $x = 10, 20, 30, 40,$ and 50. Plot the ordered pairs on an appropriate coordinate system.

9. For American National Standard screw thread the formula $D = \dfrac{0.6495}{N}$ relates the depth D of thread to the number N of threads per inch. Determine the set of ordered pairs when $N = 13, 18, 20,$ and 24. Plot the ordered pairs on an appropriate coordinate system.

10. For Acme screw thread the formula $D = \dfrac{0.500}{N} + 0.01$ relates the depth D of thread to the number N of threads per inch. Determine the set of ordered pairs when $N = 16, 18, 20,$ and 24. Plot the ordered pairs on an appropriate coordinate system.

11. The formula that relates pressure P, force F, and area A is written as
$$P = \frac{F}{A}.$$
Graph the relation between P and F if A is 5 square inches. Let $F = 50, 40, 30, 20,$ and 10.

12. Volts E, amperes I, and ohms of resistance R are related by the expression
$$E = IR.$$
Graph this relation for $R = 40$ and $E = 220, 200, 140, 120,$ and 100.

13. The rule for pulleys states that the force P multiplied by the number of moving strands n equals the weight that can be raised. This can be symbolized as
$$W = Pn.$$
Graph this relation if $n = 5$ and $W = 50, 45, 40, 35,$ and 30.

3·2 Graphing Linear Equations

In section 1.9 we defined linear equations in one variable ($ax + b = 0$) and linear equations in two variables ($ax + by + c = 0$). The graph of a linear equation in one variable was a point on the real number line. We also stated that the graph of a linear equation in two variables is a straight line when graphed on a rectangular coordinate system. We are now ready to consider the graphs of equations of the type

$$ax + by + c = 0, \text{ (}a \text{ and } b \text{ not both } 0\text{)};$$

that is, linear equations in two variables. Note that the two variables in this case are x and y and that each is to the first power.

In chapter 1 and again in section 3.1, we considered the formula $F = \frac{9}{5}C + 32$, which describes the relation between Fahrenheit and Celsius temperature scales. If we write this formula in the form $F - \frac{9}{5}C - 32 = 0$, we see that it fits the definition of a linear equation with variables F and C. That is, it matches the form of

$$ax + by + c = 0$$

as

$$ax + by + c = 0,$$
$$1F - \frac{9}{5}C - 32 = 0.$$

Note that $a = 1$, $b = -\frac{9}{5}$, and $c = -32$.

Figure 3.5 illustrates the graph of $F - \frac{9}{5}C - 32 = 0$ for the ordered pairs that were determined in section 3.1.

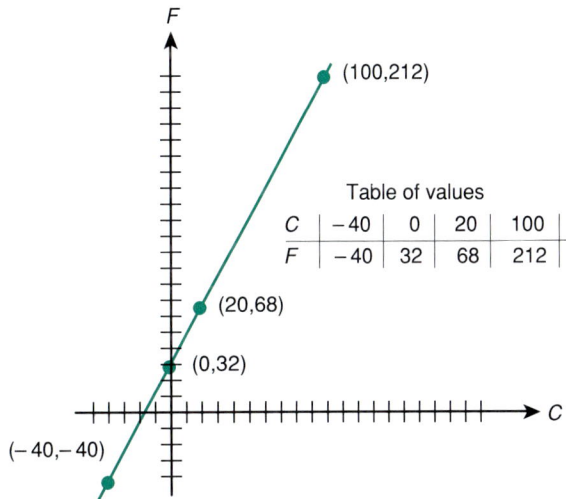

Figure 3.5

Examples 3.1, 3.2, and 3.3 illustrate effective techniques for graphing linear equations.

• **Example 3.1:** Graph the equation $3x - 2y = 6$.

Solution: When $3x - 2y = 6$ is written as $3x - 2y - 6 = 0$, we see that it fits the definition of a linear equation and will graph as a straight line. It is necessary to find only two ordered pairs, since two points will determine the position of the straight line on the rectangular coordinate plane. It is often desirable to determine a third point to ensure that we have not made a mistake.

To find an ordered pair, arbitrarily select a value for one variable, substitute for that variable, and solve for the second variable. For example, find y when $x = 0$.

$$3x - 2y = 6$$
$$3(0) - 2y = 6 \quad \text{(Let } x = 0, \text{ substitute)}$$
$$-2y = 6 \quad \text{(Simplify)}$$
$$y = -3. \quad \text{(Divide by } -2)$$

When $x = 0$, $y = -3$, thus, $(0,-3)$ is a point on the graph.
Find x when $y = 0$.

$$3x - 2y = 6$$
$$3x - 2(0) = 6 \quad \text{(Let } y = 0, \text{ substitute)}$$
$$3x = 6 \quad \text{(Simplify)}$$
$$x = 2. \quad \text{(Divide by 3)}$$

When $y = 0$, $x = 2$, hence, $(2,0)$ is a point on the graph.

To ensure that the points lie on the proper straight line find a third point. Find y when $x = 6$.

$$3x - 2y = 6$$
$$3(6) - 2y = 6 \quad \text{(Let } x = 6, \text{ substitute)}$$
$$18 - 2y = 6 \quad \text{(Multiply)}$$
$$-2y = 6 - 18 \quad \text{(Transpose)}$$
$$-2y = -12 \quad \text{(Subtract)}$$
$$y = 6. \quad \text{(Divide by } -2)$$

So $(6,6)$ should also be a point on the graph.

These ordered pairs $(0,-3)$, $(2,0)$, and $(6,6)$ are graphed in figure 3.6.

••••••••••

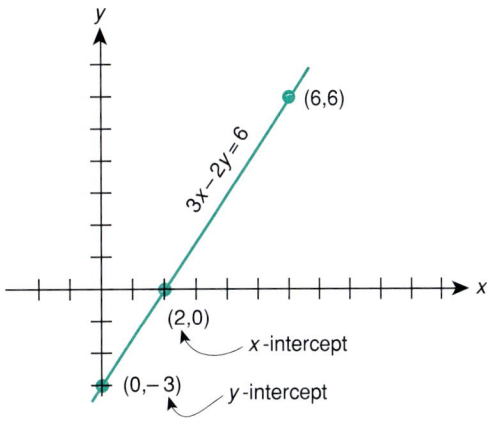

Figure 3.6

The x-coordinate of the point where the line crosses the x-axis is called the **x-intercept** and the y-coordinate of the point where the line crosses the y-axis is called the **y-intercept**. The intercepts are usually very easy to determine. For instance, in example 3.1, we found the y-intercept by letting $x = 0$ and then solving the resulting equation for y. The y-intercept was -3. We found the x-intercept by letting $y = 0$ then solving the resulting equation for x. The x-intercept was 2.

x-intercept
y-intercept

• *Example 3.2:* The cost of producing a particular device is described by the equation $C = 800 + 1.5x$, where 800 represents the fixed cost of operating and $1.5x$ the variable cost of producing x of these devices. Graph the cost equation for x from 0 to 1000.

Solution: Select three values for x and find the corresponding values for C. When $x = 0$ we get

$$C = 800 + 1.5x$$
$$C = 800 + 1.5(0)$$
$$C = 800.$$

So (0,800) is one of the points on the graph. When $x = 500$ we get

$$C = 800 + 1.5x$$
$$C = 800 + 1.5(500)$$
$$C = 800 + 750$$
$$C = 1550.$$

So (500,1550) is also a point on the graph.
To find a third (insurance) point, let $x = 1000$ and get

$$C = 800 + 1.5x$$
$$C = 800 + 1.5(1000)$$
$$C = 800 + 1500$$
$$C = 2300.$$

So (1000,2300) is the third point.

Now we graph these ordered pairs (0,800), (500,1550), and (1000,2300) on an appropriate coordinate system and draw the line segment containing them. Figure 3.7 represents an appropriate graph.

Note that the x-values we selected were nonnegative, since it does not make sense to produce a negative number of items. Also note that the graph is a line segment rather than a line since x is limited to values from 0 to 1000, inclusive.

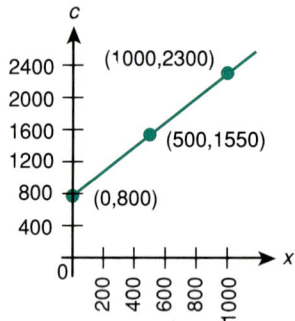

Figure 3.7

In the general linear equation $ax + by + c = 0$, if $c = 0$, the equation reduces to the form $ax + by = 0$. The graph of such an equation passes through the origin (0,0) since when $x = 0$, then $y = 0$. Consider example 3.3.

• **Example 3.3:** Graph $3x + 4y = 0$.

Solution: Select three values for x and find the corresponding values for y.
When $x = 0$ we get

$$3x + 4y = 0$$
$$3(0) + 4y = 0 \quad \text{(Let } x = 0, \text{ substitute)}$$
$$4y = 0 \quad \text{(Multiply and simplify)}$$
$$y = 0. \quad \text{(Divide by 4)}$$

So (0,0) is one of the points on the graph.
When $x = 8$ we get

$$3x + 4y = 0$$
$$3(8) + 4y = 0 \quad \text{(Let } x = 8, \text{ substitute)}$$
$$24 + 4y = 0 \quad \text{(Multiply)}$$
$$4y = -24 \quad \text{(Transpose 24)}$$
$$y = -6. \quad \text{(Divide by 4)}$$

So $(8,-6)$ is a point on the graph.
To find a third (insurance) point, let $y = 6$ and get

$$3x + 4y = 0$$
$$3x + 4(6) = 0 \quad \text{(Let } y = 6, \text{ substitute)}$$
$$3x + 24 = 0 \quad \text{(Multiply)}$$
$$3x = -24 \quad \text{(Transpose 24)}$$
$$x = -8. \quad \text{(Divide by 3)}$$

So $(-8,6)$ is a third point on the graph.

Plot the ordered pairs (0,0), (8,−6), and (−8,6) on an appropriate coordinate system and draw the line passing through the points. Figure 3.8 represents an appropriate graph. Note that the line passes through the origin.

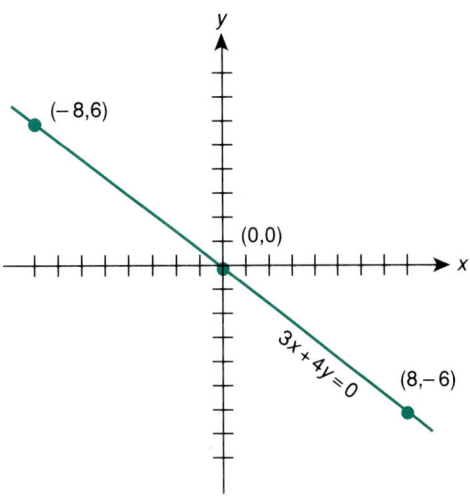

Figure 3.8

The equations $x = r$ and $y = s$, where r and s are constants, are special cases of linear equations. The graph of the equation $x = r$ is a straight line parallel to the y-axis. If $r > 0$, the line is r units to the right of the y-axis. If $r < 0$, the line is r units to the left of the y-axis. The graph of the equation $y = s$ is a straight line parallel to the x-axis. If $s > 0$, the line is s units above the x-axis and if $s < 0$, the line is s units below the x-axis. The graphs of these two special cases are shown in figure 3.9.

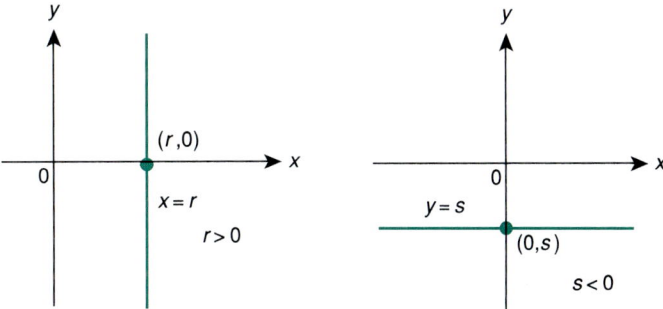

Figure 3.9

122 CHAPTER 3 GRAPHING

Although straight lines can be graphed by selecting points on the coordinate system, there are other techniques for graphing straight lines that are very useful. These are developed in chapter 16.

Trial Problems 3.2

Before you begin the section exercises, warm up with these problems. Complete answers are included in the answer key.

Draw the graphs of these linear equations on an appropriate coordinate system.

1. $x - 2y - 8 = 0$
2. $2x + y + 6 = 0$
3. $x = y$
4. $x = 3$
5. $y = -2$

Exercises 3.2

Draw the graphs of these linear equations on an appropriate coordinate system.

1. $2x - y - 8 = 0$
2. $x + 2y - 6 = 0$
3. $x + y = 0$
4. $x - y = 0$
5. $2x + y = 0$
6. $x - 2y = 0$
7. $3x - 3y - 2 = 0$
8. $2x - 2y - 3 = 0$
9. $x = -3$
10. $y = -2$

Graph the following for the values suggested.

11. A contractor buys a machine for $24,000 and depreciates it over a 10-year period to a value of $6,000. The value v of the machine at any given time t is given by the equation $v = 24{,}000 - 1800t$. Graph the equation as t goes from 0 to 10.

12. Work exercise 11 if the value equation is $v = 30{,}000 - 2400t$.

13. The current through a 10-ohm resistor varies according to the number of volts across the resistor. This relation is described by the equation $v = 10i$, where i represents the current in amperes and v represents the voltage in volts. Graph the equation as v varies from 0 to 110.

14. Work exercise 13 as v varies from 110 to 220.

15. The **mechanical advantage** of an inclined plane is the ratio of the length of the plane to its height. This can be expressed by the equation $M = L/h$. If the height of a given plane is 8 feet, plot the graph of relation between M and L.

16. Work exercise 15 if the height of the inclined plane is 6 feet.

17. A circuit has a resistance R of 200 ohms. The applied voltage E varies with the current A according to the formula

$$E = 200A.$$

Graph this relation as A ranges from 0.0 to 10.0 amperes.

18. Work W, force F, and distance D are related by the formula

$$W = FD.$$

Graph this equation for a force of 100 pounds.

19. Under certain conditions, the speed of pulley A and the speed of pulley B will be related by the equation

$$A = \frac{B}{5}.$$

Graph this equation.

20. Two forces, F_1 and F_2, act on a supporting beam. The equation relating these two forces is

$$6F_1 + 5F_2 = 300.$$

Graph this equation.

3•3 Graphing Quadratic Equations

Now let us consider a quadratic equation in two variables of the form

$$y = ax^2 + bx + c, \quad (a \neq 0),$$

or one that can be made to match this form through appropriate algebraic manipulation.

Some examples of quadratic equations of this type might be

$$y = 3x^2 - 2x + 1$$
$$y = -x^2 + 3$$
$$A = 5x^2 + 2x$$
$$d = 16t^2.$$

Parabola The graph of a quadratic equation of this type is a **parabola**. Figure 3.10 illustrates
Opens Up the graph of a parabola that **opens up**. A parabola that **opens down** is illustrated by
Opens Down the graph in figure 3.11.

Curves of this general type appear quite frequently in science and technology. Satellite receiving stations are parabolic. Reflectors on flashlights and automobile headlights tend to be parabolic. Parabolic mirrors are used on many of the largest telescopes, and projectiles travel in parabolic paths. As you continue your study of mathematics, science, technology, and industry, you will constantly become aware of quadratic equations in two variables and their corresponding parabolic curves.

The following examples illustrate how quadratic equations of the type $y = ax^2 + bx + c$ can be graphed. Since we know that the graphs will be parabolas and not straight lines, it will be desirable for us to plot many points in order to establish a smooth curve.

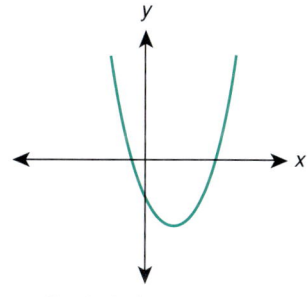

Parabola that opens up

Figure 3.10

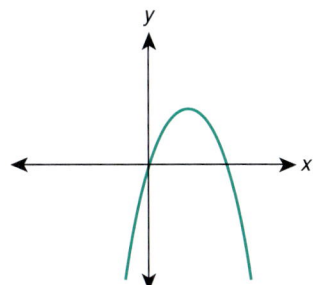

Parabola that opens down

Figure 3.11

• *Example 3.4:* Graph $y = x^2 - 2x - 3$.

Solution: Find several solutions of the equation. Choose appropriate values for x and find the corresponding values for y. For example when $x = 3$,

$$\begin{aligned} y &= (3)^2 - 2(3) - 3 & \text{(Let } x = 3\text{, substitute)} \\ y &= 9 - 6 - 3 & \text{(Multiply)} \\ y &= 0. & \text{(Simplify)} \end{aligned}$$

It might be appropriate to list solutions in table form, then plot these ordered pairs on a rectangular coordinate system. The table and graph should look like those in figure 3.12.

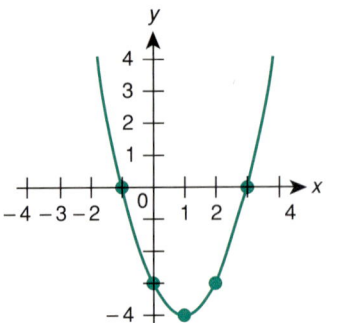
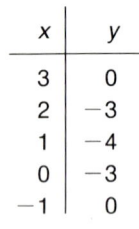

Figure 3.12

Notice that this parabola crosses the x-axis at two points, (3,0) and (−1,0). Note also that the graph of this parabola appears to have an extreme point at (1,−4). In fact, all parabolas have an extreme point. This point is called the **vertex** of the parabola. If this extreme point is a minimum point for the parabola, the parabola opens upward. If the extreme point is a maximum point, the parabola opens downward.

It can be shown that the graph of a quadratic equation of the type

$$y = ax^2 + bx + c$$

will have its extreme point when $x = \dfrac{-b}{2a}$. In example 3.4, the extreme point occurs when $x = \dfrac{-(-2)}{2(1)}$ or when $x = 1$. When $x = 1$, $y = -4$, so the extreme point is indeed at (1,−4) as the graph indicated. Finding the extreme point first helps us to choose other points so that we get a smooth curve quickly.

The concept of the extreme point is used as further concepts are developed in examples 3.5, 3.6, and 3.7.

• **Example 3.5:** Graph $y = x^2 - 2x + 1$.

Solution: Choose several appropriate values for x and find the corresponding values for y. The extreme point occurs when $x = \dfrac{-b}{2a} = \dfrac{-(-2)}{2(1)} = 1$, so the extreme point is at (1,0). Place this point in a table and choose other values of x on either side of this value, then plot these points on a coordinate system. The graph should look like the one in figure 3.13.

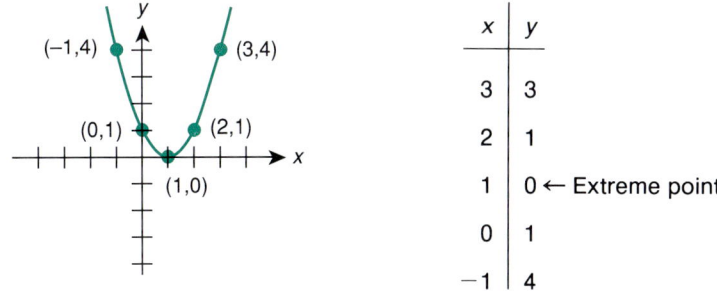

Figure 3.13

Notice that the parabola crosses (touches) the x-axis at only one point, namely the extreme point $(1,0)$.

• **Example 3.6:** Graph $y = x^2 - 2x + 5$.

Solution: Choose several appropriate values for x and find the corresponding values for y. List these values in a table. If we plot these ordered pairs on a rectangular coordinate system, the graph should look like the one in figure 3.14.

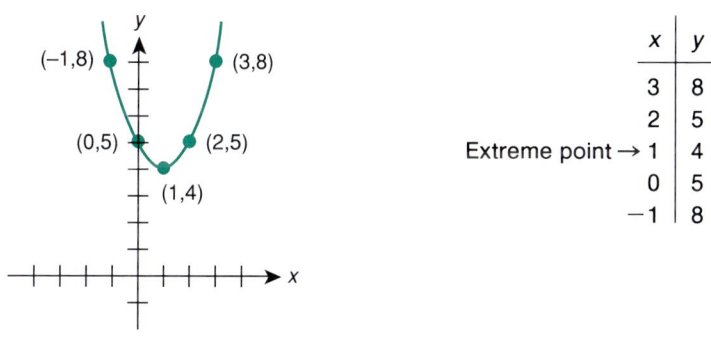

Figure 3.14

Notice that this parabola does not cross the x-axis at all.

The number of times a parabola intersects the x-axis is a significant concept. We discuss this concept in chapter 5 when we solve quadratic equations.

Example 3.7 illustrates the nature of the graph of the parabola if the coefficient of the second degree term is negative.

• **Example 3.7:** $y = -2x^2 + 1$

Solution: Choose several appropriate values for x and find the corresponding values for y. List these values in a table and plot them on an appropriate coordinate system, as in figure 3.15. The extreme point occurs when $x = \dfrac{-b}{2a}$ or when $x = \dfrac{-(0)}{2(-2)}$. That is, when $x = 0$. When $x = 0$, $y = 1$, so the extreme point is $(0,1)$.

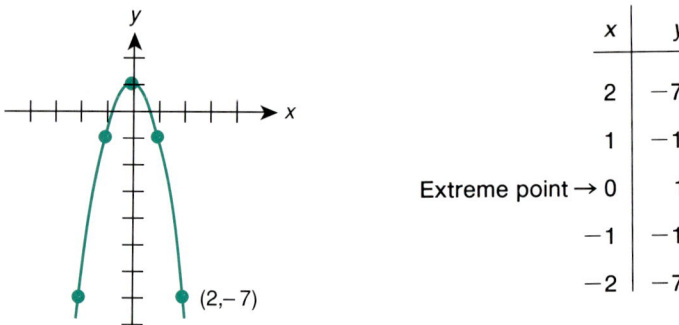

x	y
2	−7
1	−1
Extreme point → 0	1
−1	−1
−2	−7

Figure 3.15

Notice that this parabola opens downward, whereas examples 3.4, 3.5, and 3.6 yielded parabolas that opened upward. This is caused by the negative coefficient (-2) of x^2. That is, if $a < 0$ the parabola opens downward and the extreme point is a maximum point. If $a > 0$, the parabola opens upward and the extreme point is a minimum point.

Trial Problems 3.3

Before you begin the section exercises, warm up with these problems. Complete answers are included in the answer key.

Draw the graphs of these equations on an appropriate coordinate system.

1. $y = 3x^2$
2. $y = x^2 - 2$
3. $y = 2x^2 + 3$
4. $y = \dfrac{1}{2}x^2$
5. $y = -3x^2$

Exercises 3.3

Graph.

1. $y = x^2$
2. $y = 2x^2$
3. $y = x^2 - 1$
4. $y = x^2 + 1$
5. $y = -x^2$
6. $y = -2x^2$
7. $y = \dfrac{x^2}{2}$
8. $y = \dfrac{x^2}{3}$
9. $y = x^2 - 4x + 3$
10. $y = x^2 - 2x - 3$
11. $y = -x^2 + 2x + 3$
12. $y = -x^2 - 2x + 3$

13. The width of a rectangular piece of metal is x and its length is $28 - 2x$. Express the area A of this rectangle as a quadratic equation in x.

14. The width of a rectangular piece of metal is x and its length is $36 - 2x$. Express the area A of this rectangle as a quadratic equation in x.

15. Under certain conditions the distance d that an object is above the ground is given by $d = 96t - 16t^2$, where t is the time in seconds. Graph this equation for $t = 0, 1, 2, 3, 4, 5,$ and 6.

16. Under certain conditions the distance d that an object is above the ground is given by $d = 48t - 8t^2$, where t is the time in seconds. Graph this equation for $t = 0, 1, 2, 3, 4, 5,$ and 6.

17. The pressure p, measured as a head necessary to give a fluid, originally at rest, a velocity v, is determined by the formula

$$p = \dfrac{v^2}{2g},$$

where g is the acceleration due to gravity, approximately $\dfrac{32 \text{ ft}}{\sec^2}$. Graph p as a function of v as $v = 5, 10, 15,$ and 20.

18. The maximum velocity v (in miles per hour) at which an automobile can safely negotiate a curve of radius r (in feet) is given by the formula $r = 0.48v^2$. Graph r as a function of v as v goes from 0 to 55 miles per hour.

19. The equation $p = -s^2 + 50s - 150$ describes the relation between the profit p and the selling price s of an item. Graph p as a function of s for s between 5 and 45.

20. The demand for a product can be described by the equation

$$d = 400 - 5p^2,$$

where d represents the demand for the product at price p. Graph d as a function of p for values of p between 1 and 8.

3•4 Graphing Nonlinear Equations

Many relations in the industrial world cannot be represented by linear equations. We saw some of these in the last section. In this section we look at others. Consider examples 3.8 and 3.9.

• *Example 3.8:*   The unit cost of producing electronic signs in a small plant is described by the equation

$$c = \dfrac{900 + 300x}{x},$$

where c represents the cost per unit of producing x units per day. However, the plant cannot produce more than ten units per day. Graph the relation.

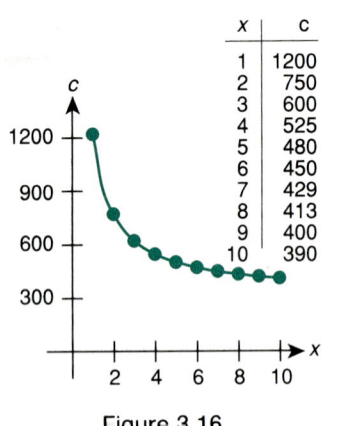

x	c
1	1200
2	750
3	600
4	525
5	480
6	450
7	429
8	413
9	400
10	390

Figure 3.16

Solution: Determine the coordinates of several points on the graph as $x = 1, 2, \ldots, 10$. Place these values in a table and plot the points on a suitable coordinate system. Then connect the points with a smooth curve. The graph should look like the one in figure 3.16.

• *Example 3.9:* This table shows the relation between the diameter D in inches of standard copper wire and the resistance R in ohms per 1000 feet of the wire. Graph the relation.

D	R
0.035	10.2
0.072	3.2
0.134	1.0
0.220	0.3
0.300	0.1

Solution: Graph the points on an appropriate coordinate system. Then connect the points with a smooth curve. The resulting graph should look like that in figure 3.17.

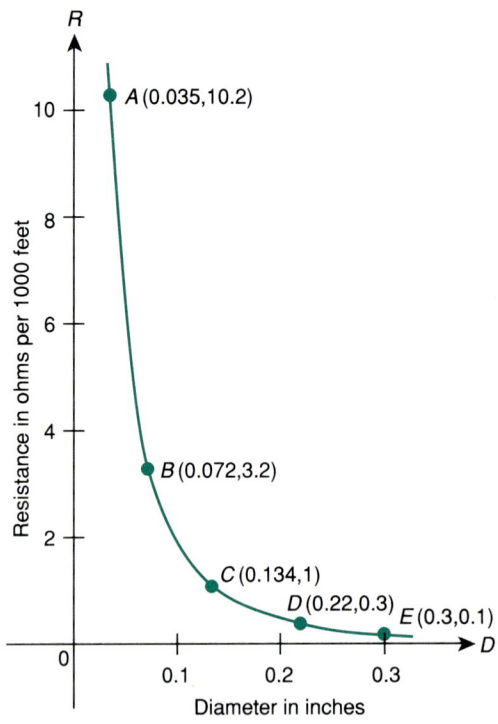

Figure 3.17

SECTION 3.4 GRAPHING NONLINEAR EQUATIONS 129

Notice that in example 3.8 we were given the equation plus some additional information. From the equation and the given information, we were able to construct the table of values and then plot the graph.

In example 3.9 we were not given an equation, but we were given a table of values from which we constructed the graph. As you see, it is not always necessary to have the equation as long as we have a table of values.

Trial Problems 3.4

Before you begin the section exercises, warm up with these problems. Complete answers are included in the answer key.

Graph the relation shown in the table.

1.

x	-2	-1	0	1	2
y	5	3	2	$\frac{3}{2}$	$\frac{6}{5}$

2.

x	1	2	3	4	5
y	10	5	$\frac{10}{3}$	$\frac{5}{2}$	2

For the values given, graph the relation determined by the equation.

3. $y = \dfrac{10}{x+2}$ for $x = 0, 1, 2,$ and 3.

4. $y = \dfrac{15}{x+1}$ for $x = 0, 2, 4,$ and 6.

5. $y = \dfrac{3}{x}$ for $x = -1, -2, -3,$ and -4.

Exercises 3.4

1. Graph the relation shown in the table.

x	0	10	20	30	40	50
y	1000	500	330	250	200	167

2. Graph the relation shown in the table.

p	1	2	3	4	5	6	8	10
y	15	7.5	5	3.75	3	2.5	1.875	1.5

3. Graph the relation determined by the equation $y = \dfrac{30}{x+1}$ for x from 0 to 9.

4. Graph the relation determined by the equation $y = \dfrac{25}{x+1}$ for x from 0 to 9.

5. A contractor has determined that the hourly cost of operating a certain piece of equipment is described by the equation $C = \dfrac{800}{r} + r$, where r represents the speed in miles per hour that the equipment is run. Graph the relation for r from 10 to 60. Estimate the most economical speed of operation.

6. Do exercise 5 using the equation $C = \dfrac{1500}{r} + \dfrac{r}{2}$.

7. The table shows the weight w in pounds per foot for square steel bars with a width of d inches. Graph the relation.

d	0.5	1.0	1.5	2.0	2.5	3.0
w	0.85	3.40	7.65	13.60	21.25	30.60

8. The table shows the weight w in kilograms per meter for square steel bars with a width of d centimeters. Graph the relation.

d	1	2	3	4	5	6	7
w	8	31	69	123	193	277	377

9. The electrical resistance R of a particular 20-foot length of wire varies inversely as its cross-sectional area A according to the formula

$$R = \frac{20}{A}.$$

Graph R as a function of A as A varies from 0.1 to 1.0.

10. The area A of a footing required to support a building of weight W depends on the bearing capacity B of the soil. The formula that relates these three values is

$$A = \frac{W}{B}.$$

Graph A as a function of B as $B = 2000, 2100, 2200, 2300, 2400,$ and 2500 if $W = 60,000$.

11. The Brinell number is the ratio of the load on a sphere used to indent the material to be tested to the area of the spherical indentation produced. The formula for the Brinell number is

$$\text{Bn} = \frac{P}{\pi t D},$$

where P equals the load in kg, t equals the depth of indentation, and D equals the diameter of the ball. All lengths are expressed in millimeters. Graph the Brinell number Bn as a function of the diameter D if $t = 3.14$, $P = 9.86$, and D varies from 1.0 to 10.0.

12. Under certain conditions, Boyle's law dealing with the compressibility of gases relates the original pressure P_0 to the original volume V_0 by the formula

$$P_0 = \frac{200}{V_0}.$$

Graph P_0 as a function of V_0, as V_0 varies from 50 to 400.

3•5 Functions

In this chapter we introduced the rectangular coordinate system and graphed various equations. We observed that any set of ordered pairs of real numbers can be graphed on the rectangular coordinate system.

Now we consider certain sets of ordered pairs that we define as functions.

Function

A **function** is a set of ordered pairs such that no two pairs have the same first component. Thus, the set of pairs $\{(1,2),(3,2),(4,5)\}$ is a function, but the set of pairs $\{(1,2),(1,3),(2,4)\}$ is not a function, since $(1,2)$ and $(1,3)$ have the same first component.

Figure 3.18 and figure 3.19 represent the graphs of these sets of ordered pairs.

In figure 3.19 notice that the pairs $(1,2)$ and $(1,3)$ lie on the same vertical line, or that one point is directly above the other. Wherever this happens we do not have a function. In fact, one way of testing to see if a graph represents a function is to pass a vertical line across the graph. If the vertical line (as it passes across the graph) intersects the graph in more than one point, we do not have a function. Otherwise we **Vertical Line Test**. do. We call this the **vertical line test**.

Since a function is a set of ordered pairs, we often find it convenient to identify and discuss the numbers that are first components of these pairs and the numbers that **Domain** are second components of these pairs. The set of first components is called the **domain** **Range** and the set of second components is called the **range.** Thus, the function represented in figure 3.18 has the set $\{1,3,4\}$ as its domain and the set $\{2,5\}$ as its range.

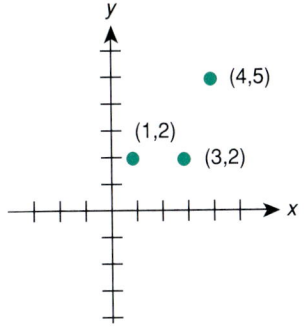

Figure 3.18

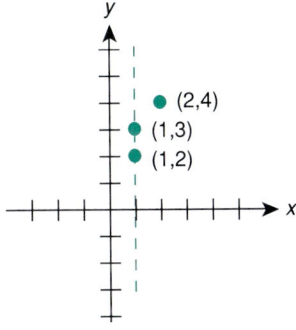

Figure 3.19

Sometimes we find it necessary to specify the domain of a function, and other times we simply assume that the domain includes all of those values that make sense. For example, in the function represented by the equation

$$y = \sqrt{x - 1},$$

we see that x must be greater than or equal to 1 in order that $x - 1$ be nonnegative so that $\sqrt{x - 1}$ is a real number. Thus, we simply assume that the domain includes all real numbers x where $x \geq 1$.

Since functions are sets of ordered pairs satisfying certain conditions, they play an important role in all of mathematics. For this reason it is desirable to discuss some of the notation that is associated with functions. We have already discussed how a set of ordered pairs can represent a function. Functions can also be represented by formulas and equations.

When representing a function by an equation, it is often convenient to use the expression "y is a function of x." Symbolically, we write this expression as $y = f(x)$. For example, for the equation $y = x^2 + 3$, we may write $y = f(x)$, where $f(x) = x^2 + 3$. Thus, $y = x^2 + 3$ and $f(x) = x^2 + 3$ are different ways of expressing the same function. The notation $f(x)$ does not indicate multiplication. It represents a variable. In the examples above, $f(x)$ represents the variable y. We are not restricted to using $f(x)$ for the variable. The expressions $g(x)$, $h(x)$, $\phi(x)$, etc., can also be used for functional notation.

An important use of the functional notation $f(x)$ is to designate the value of a function for a specified value of x.

• **Example 3.10:** Given $f(x) = 5x - 4$, find the value of the function when $x = 2$; that is, find $f(2)$ by substituting 2 for x in the given equation.

Solution: Since

$$f(x) = 5x - 4$$

then

$$f(2) = 5(2) - 4 \quad \text{(Substitute 2 for } x\text{)}$$

or

$$f(2) = 10 - 4 \quad \text{(Multiply)}$$

so

$$f(2) = 6. \quad \text{(Subtract)}$$

Hence, the pair (2,6) is in the function or would represent a point on the graph of the function.

- **Example 3.11:** Given $f(x) = 3x^2 - 5x + 6$, find the value of the function when $x = h$.

    Solution: Since
    $$f(x) = 3x^2 - 5x + 6,$$
    then
    $$f(h) = 3h^2 - 5h + 6. \quad \text{(Substitute } h \text{ for } x\text{)}$$

- **Example 3.12:** Given $f(x) = 7x + 4$, find $\dfrac{f(x + h) - f(x)}{h}$.

    Solution: Since
    $$f(x) = 7x + 4,$$
    then
    $$f(x + h) = 7(x + h) + 4 \quad \text{(Substitute } x + h \text{ for } x\text{)}$$
    $$= 7x + 7h + 4. \quad \text{(Simplify)}$$
    So
    $$\begin{aligned} f(x + h) - f(x) &= (7x + 7h + 4) - (7x + 4) && \text{(Subtract)} \\ &= 7x + 7h + 4 - 7x - 4 && \text{(Remove parentheses)} \\ &= 7h. && \text{(Simplify)} \end{aligned}$$
    Hence,
    $$\dfrac{f(x + h) - f(x)}{h} = \dfrac{7h}{h} \quad \text{(Substitute)}$$
    $$= 7. \quad \text{(Cancel)}$$
    Therefore,
    $$\dfrac{f(x + h) - f(x)}{h} = 7.$$

- **Example 3.13:** Given $f(x) = 3x^2$, find $\dfrac{f(x + h) - f(x)}{h}$.

Solution: If
$$f(x) = 3x^2,$$
then
$$\begin{aligned}f(x + h) &= 3(x + h)^2 \\ &= 3(x^2 + 2hx + h^2) \quad \text{(Square the binomial)} \\ &= 3x^2 + 6hx + 3h^2. \quad \text{(Multiply by 3)}\end{aligned}$$

Subtracting the value of $f(x)$ from the value of $f(x + h)$ gives,
$$\begin{aligned}f(x + h) - f(x) \\ &= (3x^2 + 6hx + 3h^2) - 3x^2 \quad \text{(Substitute)} \\ &= 3x^2 + 6hx + 3h^2 - 3x^2 \quad \text{(Remove parentheses)} \\ &= 6hx + 3h^2. \quad \text{(Simplify)}\end{aligned}$$

Dividing this expression by h gives,
$$\begin{aligned}\frac{f(x + h) - f(x)}{h} &= \frac{6hx + 3h^2}{h} \quad \text{(Substitute)} \\ &= 6x + 3h. \quad \text{(Divide)}\end{aligned}$$

Therefore,
$$\frac{f(x + h) - f(x)}{h} = 6x + 3h.$$

• • • • • • • • • •

There are numerous examples in everyday life and in science and technology in which one quantity, or variable, is a function of (or depends on) another quantity or variable.

Many of these may be stated verbally, such as:

The circumference of a circle is a function of its radius.

The area of a square is a function of the length of a side.

The volume of a cube is a function of its edge.

The voltage across a resistor is a function of the current.

The distance traveled at 55 miles per hour is a function of the time traveled.

Often we are faced with the problem of converting the verbal statement into a symbolic statement. Consider examples 3.14 and 3.15.

• *Example 3.14:* Express the area of a circle as a function of its radius.

Solution: Since the area of a circle can be expressed as
$$A = \pi r^2,$$
we can write $f(r) = \pi r^2$ where $A = f(r)$.

• • • • • • • • • •

• **Example 3.15:** Express the distance s in miles traveled in 3 hours as a function of the velocity v in miles per hour.

Solution: The formula for distance in terms of velocity and time is

$$s = vt.$$

Since $t = 3$ hrs, we write

$$s = v(3) \quad \text{(Substitute 3 for } v\text{)}$$

or

$$s = 3v. \quad \text{(Commute)}$$

Thus, we can express distance as a function of velocity as

$$f(v) = 3v, \quad \text{where} \quad s = f(v).$$

Trial Problems 3.5

Before you begin the section exercises, warm up with these problems. Complete answers are included in the answer key.

Express the equation in functional notation.

1. $y = x - 3$
2. $y = 5x^2$

Find the indicated values of the given functions.

3. $f(x) = x - 2, f(3) = $ _____
4. $f(t) = 2t + 3, f(-4) = $ _____
5. $f(x) = x^2 - 3x + 1, f(a) = $ _____

Exercises 3.5

Express each of the equations by using functional notation.

1. $y = x + 3$
2. $y = x + 4$
3. $v = 16t^2$
4. $v = -16t^2$
5. $A = s^2$
6. $A = 5s$
7. $p = 2(5 + w)$
8. $p = 2(3 + w)$
9. $y = 3x^2 - 2$
10. $y = 3x - x^3$

Find the indicated values of the given functions.

11. $f(x) = x + 2, f(3) = $ _____
12. $f(x) = x + 3, f(4) = $ _____
13. $f(t) = 2t - 1, f(3) = $ _____
14. $f(t) = 3t - 2, f(4) = $ _____
15. $f(p) = p^3 + 2p - 1, f(3) = $ _____
16. $f(p) = p^2 - 3p + 1, f(3) = $ _____
17. $f(s) = s^3, f(4) = $ _____
18. $f(s) = 4s, f(3) = $ _____
19. $f(x) = 4^x, f(2) = $ _____
20. $f(x) = 3^x, f(4) = $ _____
21. $f(x) = x^3 + 2x - 1, f(a) = $ _____
22. $f(x) = x^2 + 3x - 1, f(b) = $ _____
23. $f(x) = x^2 - 4x + 5, f(x + h) = $ _____
24. $f(x) = x^2 - 3x + 2, f(x + h) = $ _____

In exercises 25–30, find $\dfrac{f(x+h)-f(x)}{h}$ for the given values of $f(x)$.

25. $f(x) = x + 2$
26. $f(x) = x - 3$
27. $f(x) = 2x + 5$
28. $f(x) = 3x - 4$
29. $f(x) = 4x^2 - 3x$
30. $f(x) = 5x^2 + 4x$

Express each as indicated.

31. Express the circumference of a circle as a function of its radius.

32. Express the perimeter of a square as a function of its side.

33. Express a taxi fare F as a function of the distance d (miles) traveled if the fee for the first $\tfrac{1}{4}$ mile is 50¢ and 10¢ for each $\tfrac{1}{4}$ mile thereafter.

34. Express the simple interest I on $100 at an interest rate of 18% per year as a function of the number of years t.

35. A printer will print a certain brochure for $200 plus 16 cents per copy. Express the cost as a function of the number of copies x.

36. Express the number of seconds as a function of the number of hours h.

37. An electric pump can pump 1500 liters more per hour than a gasoline powered pump. Express the rate R_1 of the first pump as a function of the rate R_2 of the second pump.

38. The stopping distance d of a vehicle varies directly as the square of its velocity v. Express d as a function of v if $d = 120$ feet when $v = 40$ miles per hour.

3·6 Summary of Terms, Formulas, Rules, and Procedures

Terms

Abscissa (p. 113)
Cartesian Coordinate System (p. 113)
Coordinate System (p. 113)
Coordinates (p. 113)
Domain (p. 130)
Function (p. 130)
Mechanical Advantage (p. 122)
Opens Down (p. 123)

Opens Up (p. 123)
Ordered Pair (p. 112)
Ordinate (p. 113)
Origin (p. 113)
Parabola (p. 123)
Quadrants (p. 113)
Range (p. 130)
Rectangular Coordinate System (p. 113)

Vertex (p. 124)
Vertical Line Test (p. 130)
x-axis (p. 113)
x-coordinate (p. 113)
x-intercept (p. 119)
y-axis (p. 113)
y-coordinate (p. 113)
y-intercept (p. 119)

Formulas

An equation of the type $ax + by + c = 0$, (a and b not both zero) is called a linear equation in two variables, x and y, and will graph on the rectangular coordinate system as a straight line.   (3.2)

An equation of the type $y = ax^2 + bx + c$, ($a \neq 0$) is called a quadratic equation in the variable x. The graph of such an equation on the rectangular coordinate system will be a parabola.   (3.3)

Rules and Procedures

- Equations whose graphs are not straight lines are called nonlinear equations. (3.4)
- A function is a set of ordered pairs such that no two pairs have the same first component. (3.5)
- The notation $y = f(x)$ means y is a function of x. (3.5)

3·7 Chapter 3 Review Exercises

1. Plot the following points on an appropriate rectangular coordinate system: $(5,100), (-6,300), (-4,-200)$, and $(2,-100)$.

2. Plot the following points on an appropriate rectangular coordinate system: $(100,5), (300,-6), (-200,-4)$, and $(-100,2)$.

3. Plot the given points and connect the points in the order given: $(5,3), (-5,-3), (0,5), (5,-3), (-5,3)$, and $(5,3)$.

4. Plot the given points and connect the points in the order given: $(-5,0), (3,5), (-3,-5), (-3,5), (3,-5)$, and $(-5,0)$.

5. Given the formula $C = 4x + 30$, determine the set of ordered pairs when $x = 1, 2, 3, 4$, and 5. Plot the ordered pairs on an appropriate coordinate system.

6. Given the formula $C = 5x + 40$, determine the set of ordered pairs when $x = 1, 2, 3, 4$, and 5. Plot the ordered pairs on an appropriate coordinate system.

7. Graph the equation $3x - 2y = 6$.
8. Graph the equation $3x + 4y = 12$.
9. Graph the equation $y = 2x^2 + 7x + 3$.
10. Graph the equation $y = 2x^2 - 7x + 3$.
11. Graph the equation $y = -3x^2 + 5x + 2$.
12. Graph the equation $y = -2x^2 + x + 3$.
13. Plot the values in this table on a suitable coordinate system.

x	1	2	3	4	5	6	7	8
y	1	0.5	0.3	0.25	0.2	0.16	0.14	0.12

14. Plot the values in this table on a suitable coordinate system.

x	0	1	2	3	4	5
y	1	2	4	8	16	32

Express each of these equations using functional notation.

15. $y = x^2 + 3$
16. $y = x^2 + 5$
17. $p = 2(w + 5)$
18. $p = 2(w + 3)$

Find the indicated value of the given function.

19. $f(x) = x - 3, f(4) = $ _____
20. $f(x) = x + 2, f(3) = $ _____
21. $f(t) = 3t^2 - t, f(3) = $ _____
22. $f(t) = 2t^3 + t, f(2) = $ _____
23. $f(x) = 3x^2 - 2x + 4, f(a) = $ _____
24. $f(x) = x^2 + 3x - 5, f(b) = $ _____

Express each as indicated.

25. Express the perimeter of an equilateral triangle as a function of a side.

26. Express the perimeter of an isosceles triangle as a function of one of the congruent sides if the third side is 8.

27. An appliance repair company charges $35.00 plus $25.00/hour for labor whenever it makes a service call. Express the cost C of a service call as a function of the number of hours h spent on the call.

28. The cutting speeds of lathes and milling machines is represented by the formula

$$C = \frac{\pi RD}{12},$$

where C = cutting speed,
R = revolutions per minute,
D = diameter of work, or diameter of cutter in inches,

and π = 3.1416.
Express C as a function of D if R is 360 rpm.

29. The law of conductance may be stated as

$$Q = UA\Delta T,$$

where U = the conductance,
A = area perpendicular to the heat flow through which it is passing,
and ΔT = the temperature difference between the hot and cold sides of the substance through which the heat is being transferred.

Express Q as a function of U if A = 10 square inches and ΔT = 25° F.

30. A rectangular piece of sheet metal has a length of x and a width of $x - 2$. Express the area A of this rectangle as a function of x.

3·8 Chapter 3 Test

1. Draw a rectangular coordinate system and plot the ordered pairs $(-3,5)$, $(-2,-4)$, $(1,-3)$, $(2,4)$, and $(0,-5)$.

2. Given the formula $C = 0.5x + 40$. Determine the ordered pairs when x = 10, 20, 30, 40, and 50.

3. Write the coordinates of the points $A, B, C, D, E,$ and F in this figure.

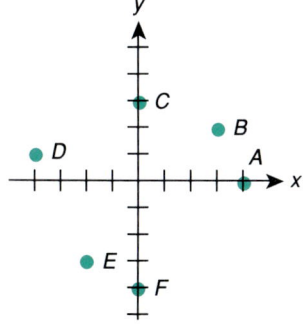

4. Draw the graph of $x + 2y - 6 = 0$ on an appropriate coordinate system.

5. Work W, force F, and distance D are related by the formula $W = FD$. Graph this equation for a force of 10 grams.

6. Graph the quadratic equation $y = x^2 - 1$.

7. Under certain conditions the distance d that an object is above the ground is given by $d = 96t - 16t^2$, where t is the time in seconds. Graph this equation for t from 0 to 5 inclusive.

8. Graph the relation determined by the equation

$$y = \frac{25}{x + 1}$$

for x from 0 to 9 inclusive.

9. The area A of a footing required to support a building of weight W depends on the bearing capacity B of the soil. The formula that relates these three variables is $A = \dfrac{W}{B}$. Graph A as a function of B as $B = 2000$, 2100, 2200, 2300, 2400, and 2500 if $W = 60{,}000$.

10. Express $y = x + 4$ using functional notation.

11. Express $P = 2(5 + w)$ using functional notation.

12. If $f(x) = x^2 + 3x - 1$, find $f(2)$.

13. Express the circumference of a circle as a function of its radius.

14. If $f(x) = x^2 + 3x - 1$, find $f(x + h)$.

15. If $f(t) = 2t - 1$, find $f(5) - f(-3)$.

16. If $f(x) = 4x^2 - 3x$, find $\dfrac{f(x + h) - f(x)}{h}$.

Chapter 4: Systems of Linear Equations

*F*rancois Viete (1540–1603) was a lawyer and a member of the French parliament. He is considered to be the greatest French mathematician of the sixteenth century, although he devoted only his spare time to mathematics.

Among his many publications is one called *In Artem*. In this text, Viete introduced the practice of using the vowels in the alphabet to represent unknown quantities and the consonants to represent known quantities. Viete did not have a symbol for equality, nor did he have exponents. Instead he used words to convey the idea of an equation or the power of a variable.

Our present practice of using the later letters of the alphabet for unknowns and the early letters for constants was introduced by another French mathematician, René Descartes (1596–1650). In an algebra book called *The Whetstone of Witte*, published in 1557 by the English mathematician Robert Recorde, the modern symbol for equality was used for the first time.

Today, because of the widespread use of mathematics in technical and industrial areas, nearly any letter or symbol might be used as a variable or a constant. With symbols for equality and exponents we can very efficiently express ideas from many disciplines in the form of mathematical equations. This chapter continues our study of some of these equations, namely linear equations in two variables.

4•1 Introduction: Linear Equations

It is not unusual, in trade, technical, and industrial applications, to encounter problems involving two different relationships between two variables, that is, two equations in two unknowns. When this happens we have a **system of equations.** To solve a system of equations we need values that satisfy both equations at the same time. In this chapter we look at various techniques that can be used to solve systems of equations.

System of Equations

Linear Equation

If the relationship between the two variables results in a first degree equation ($ax + by + c = 0$), the equation is called a **linear equation,** since its graph on the rectangular coordinate system would be a straight line.

If we have a pair of linear equations graphed on the same coordinate system, three possibilities result. Figure 4.1 illustrates these three possibilities.

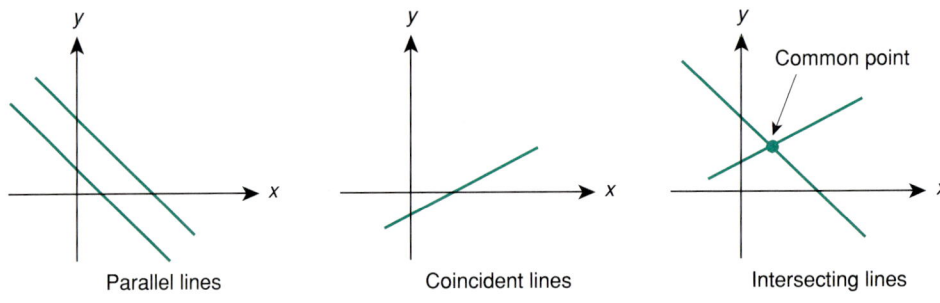

Figure 4.1

Inconsistent

Dependent

Consistent and Independent
Simultaneous Equations

In the first of these possibilities, the lines are parallel and the equations have no common solution. Such a system is said to be **inconsistent.** In the second of these possibilities, the lines coincide and the equations have an infinite number of common solutions. This system is said to be **dependent.**

The third possibility is the one that is of greatest interest to us. Here the lines intersect at a common point and the lines have a single ordered pair as a common solution. Such a system is said to be **consistent and independent.**

All such pairs of equations are called **simultaneous equations.**

• *Example 4.1:* The graph of the system of equations $3x + y = 3$ and $3x + y = 6$ is shown in figure 4.2. The system is inconsistent, since there is no common solution. In such a system the variable terms are the same, or can be made the same, by removing common factors, but the constants are different.

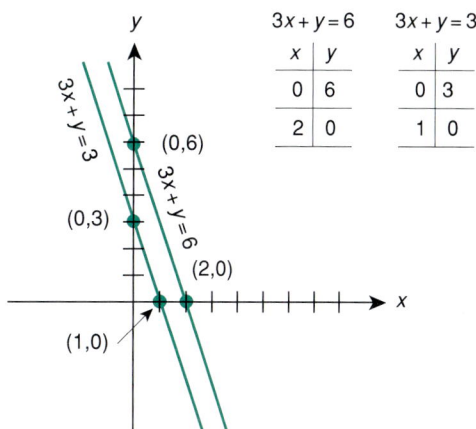

Figure 4.2

- *Example 4.2:* The graph of the system $x + y = 3$ and $4x + 4y = 12$ is a single line, as shown in figure 4.3. Every point satisfying the first equation also satisfies the second, and every point satisfying the second also satisfies the first. The system is dependent. In such a system, if the common factors are removed from the equations, the resulting equations are identical. By removing common factors in this example,

$$\begin{cases} x + y = 3 \\ 4x + 4y = 12 \end{cases} \text{becomes} \begin{cases} x + y = 3 \\ 4(x + y) = 4(3) \end{cases}$$

$$\text{becomes} \begin{cases} x + y = 3 \\ x + y = 3. \end{cases}$$

••••••••••

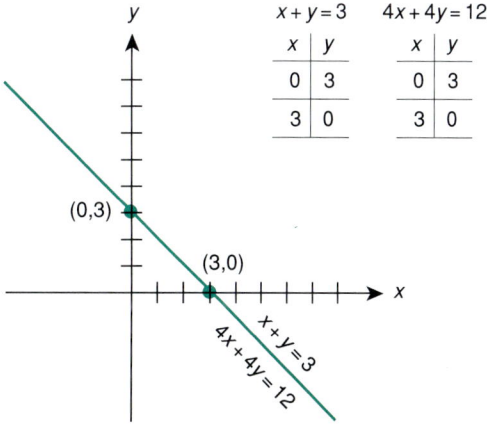

Figure 4.3

- **Example 4.3:** The system of equations $x + y = 3$ and $3x - y = 1$ has a common point $(1,2)$. Since the point $(1,2)$ checks in both equations, as

$$x + y = 3 \qquad\qquad 3x - y = 1$$
$$(1) + (2) = 3 \qquad\qquad 3(1) - (2) = 1$$
$$3 = 3 \qquad\qquad 1 = 1,$$

this system is consistent and independent. The graph of this system is illustrated in figure 4.4.

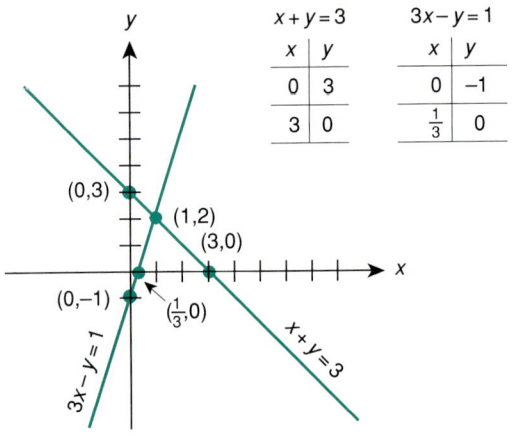

Figure 4.4

In the remaining sections of this chapter we discuss various methods that can be used to solve systems of equations simultaneously.

4•2 Solving Systems of Two Linear Equations by Graphing

One method of solving a system of two equations in two different variables is to sketch the graphs of the two equations. We then determine their common solution, if there is one, which is the point of intersection on the graph. Examples 4.4 and 4.5 illustrate this technique.

- **Example 4.4:** Graph the equations $x + y = 7$ and $x - y = 5$ and estimate the coordinates of the point that is common to both lines.

SECTION 4.2 SOLVING SYSTEMS OF TWO LINEAR EQUATIONS BY GRAPHING

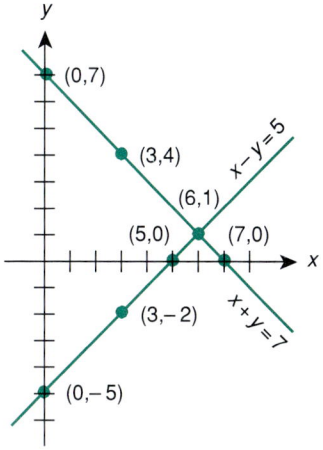

Figure 4.5

Solution: First, determine a table of values for each equation. The intercepts provide the most convenient points. A third point helps to check the graph.

$$x + y = 7 \qquad x - y = 5$$

x	y		x	y
0	7		0	−5
7	0		5	0
3	4		3	−2

Now draw the graph of each equation. The point of intersection is approximately $(6,1)$, as illustrated in figure 4.5.

The point $(6,1)$ should be the common solution. By letting $x = 6$ and $y = 1$ in each of the given equations, we can check to see that the approximate solution is correct.

$$\begin{array}{ll} x + y = 7 & x - y = 5 \\ (6) + (1) = 7 & (6) - (1) = 5 \\ 7 = 7 & 5 = 5. \end{array}$$

••••••••••

• **Example 4.5:** Graph the equations $x = 5$ and $x + 3y = 11$ and estimate the coordinates of the point that is common to both lines.

Solution: First determine a table of values for each equation. The intercepts might be convenient points. A third (insurance) point helps us to check our work.

$$x + 0y = 5 \qquad x + 3y = 11$$

x	y		x	y
0	?		0	$\frac{11}{3}$
5	0		11	0
5	6		−1	4

Note that in the first equation, when $x = 0$, y did not have a value. Also, for any specified value of y, x is always five. Now draw the graph of each equation. The point of intersection is approximately $(5,2)$, as is illustrated in figure 4.6. The point $(5,2)$ should be a common solution. By letting $x = 5$ and $y = 2$ in each of the given equations, we can check to see that the approximate solution is correct.

$$\begin{array}{ll} x + 0y = 5 & x + 3y = 11 \\ (5) + 0(2) = 5 & (5) + 3(2) = 11 \\ 5 + 0 = 5 & 5 + 6 = 11 \\ 5 = 5 & 11 = 11. \end{array}$$

••••••••••

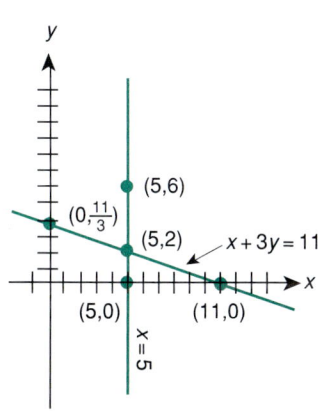

Figure 4.6

Solving Systems of Equations by Graphing

1. Graph the first equation on a rectangular coordinate system.
2. Graph the second equation on the same coordinate system.
3. If the graphs meet at one point, estimate the coordinates of that point. This is the solution. Such a system is consistent and independent.
4. If the graphs of the two equations are parallel lines there is no common solution to the system. Such a system is called inconsistent.
5. If the graphs of the two equations result in the same line there are an infinite number of solutions. This system is called dependent.
6. If there appears to be a unique solution, this solution should be checked in both of the given equations.

Solving systems of equations graphically does not always yield a point that can be easily read from the graph. In some cases the graphic solution can only be estimated.

In the next sections we consider more efficient methods of solving simultaneous linear equations.

Trial Problems 4.2

Before you begin the section exercises, warm up with these problems. Complete answers are included in the answer key.

Graph each pair of equations on a coordinate system. Then estimate the coordinates of the points (if any) that are common to the graphs.

1. $x + y = 6$, $x - y = 2$
2. $x + y = 2$, $x - y = 4$
3. $x + y = -4$, $2x + 2y = 4$
4. $x = 6$, $y = 1$
5. $4x + 3y = -12$, $8x + 6y = -24$

Exercises 4.2

Graph each pair of equations on a coordinate system. Then estimate the coordinates of the points (if any) that are common to the graphs.

1. $x + y = 7$, $x - y = 1$
2. $x + y = 5$, $x - y = 3$
3. $x + y = 4$, $x - y = 4$
4. $x + y = 6$, $x - y = 6$
5. $x + y = 3$, $2x + y = 4$
6. $x + y = 2$, $2x + y = 3$
7. $x = 2$, $y = 6$
8. $y = 2$, $2x = 10$
9. $x = -1$, $3x - y = -5$
10. $x - y = -1$, $y = 5$
11. $x + y = 6$, $x + 3y = 2$
12. $x + y = 3$, $x + 3y = -1$
13. $4x + 3y = 26$, $3x - y = 13$
14. $2x + 3y = 5$, $3x - 4y = 2$
15. $x + y = 1$, $3x + 3y = 3$
16. $x + y = 2$, $2x + 2y = 4$
17. $2x + y = 1$, $2x + y = 5$
18. $3x + y = 1$, $3x + y = 6$
19. $3x = y - 3$, $6x = y + 3$
20. $2x = y - 10$, $3x = y - 13$
21. $y = 6x + 4$, $2y = 8x + 12$
22. $3y = 4x + 7$, $4y = 5x + 10$

23. Determine the approximate **break-even point** for a cost equation $y = 300 + 2x$ and a revenue equation $y = 5x$. The break-even point is the common solution for this system of equations.

24. Determine the approximate break-even point for a cost equation $y = 500 + 3x$ and a revenue equation $y = 4x$. The break-even point is the common solution for this system of equations.

25. An electrician and his apprentice received $120 for working 4 hours on a job. They received a total of $108 when the electrician worked 4 hours and the apprentice worked 3 hours on another job. Their hourly rates e and a can be found by solving the equations $4e + 4a = 120$ and $4e + 3a = 108$. Determine the hourly rates by solving this system of equations.

26. A plumber and an apprentice received $160 for 8 hours on a job. They received a total of $96 when the plumber worked 6 hours and the apprentice worked 3 hours on another job. Their hourly rates p and a can be found by solving the equations $8p + 8a = 160$ and $6p + 3a = 96$. Determine the hourly rates by solving this system of equations.

27. When a current of 2 amps passes through a resistor R_1 and a current of 3 amps passes through a resistor R_2, the total voltage across the resistors is 8 volts. If the current in the first resistor is changed to 4 amps and that in the second resistor is changed to 1 amp, the total voltage is 11 volts. The resistances (in ohms) can be found by solving the equations $2R_1 + 3R_2 = 8$ and $4R_1 + R_2 = 11$. Determine the resistances by solving this system of equations.

28. The perimeter of a rectangle is 40 cm. The length is 4 cm longer than the width. The dimensions l and w can be found by solving these equations: $2l + 2w = 40$; $l - w = 4$. Determine the dimensions by solving this system of equations.

4 • 3 Solving Systems of Two Linear Equations by Substitution

In section 4.2 we learned to estimate the solution of a pair of linear equations by graphing. As we noticed, the graphic solutions are not always exact.

If the exact solutions are required, we must often use other methods to find them. Solving linear equations by substitution is one method that gives us exact solutions. To solve a system by substitution, we first solve one of the equations for one of the variables in terms of the other. Next we substitute the expression for this variable into the second equation. This process eliminates one of the variables so that we can find the remaining variable. Once the value of one variable has been found, the value of the remaining variable can be found by substituting the known value into one of the equations and solving this equation for the remaining variable. Example 4.6 illustrates this process.

• *Example 4.6:* Solve, by substitution, the system of linear equations

$$2x + y = 4, \quad 3x - y = 1.$$

Solution: First we must solve one of the equations for one of the unknowns. Observation indicates that solving the first equation for y would require the least effort and manipulation. Thus, we have

$2x + y = 4$ or $y = 4 - 2x$. This value is substituted for y in the second equation, giving

$$3x - y = 1$$
$$3x - (4 - 2x) = 1 \quad \text{(Substitute } 4 - 2x \text{ for } y)$$
$$3x - 4 + 2x = 1 \quad \text{(Remove parentheses)}$$
$$3x + 2x - 4 = 1 \quad \text{(Commute)}$$
$$5x - 4 = 1 \quad \text{(Add)}$$
$$5x = 1 + 4 \quad \text{(Transpose } -4)$$
$$5x = 5 \quad \text{(Add)}$$
$$x = 1. \quad \text{(Divide by 5)}$$

The value of y that corresponds to $x = 1$ is found by substituting $x = 1$ into either equation. Letting $x = 1$ in the first equation gives

$$2x + y = 4$$
$$2(1) + y = 4 \quad \text{(Substitute for } x)$$
$$2 + y = 4 \quad \text{(Multiply)}$$
$$y = 4 - 2 \quad \text{(Transpose)}$$
$$y = 2. \quad \text{(Simplify)}$$

Thus, the solution is $x = 1$ and $y = 2$.

We can check the result by substituting both values into the second equation. This gives

$$3x - y = 1$$
$$3(1) - (2) = 1 \quad \text{(Substitute for } x \text{ and } y)$$
$$1 = 1. \quad \text{(Simplify)}$$

Thus the solution checks. The graph of this system and its solution are illustrated in figure 4.7.

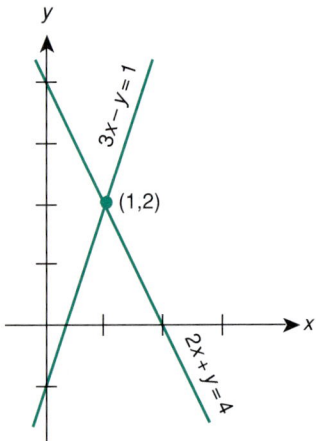

Figure 4.7

• *Example 4.7:* Solve, by substitution, the system of equations

$$4x + 6y = -9, \quad 2x - 4y = 13.$$

Solution: It is not obvious which variable would be the easiest to solve for, or which equation we should use. Therefore, we simply solve the first equation for x. We get

$$4x + 6y = -9$$
$$4x = -9 - 6y \quad \text{(Transpose)}$$
$$x = \frac{-9 - 6y}{4}. \quad \text{(Divide by 4)}$$

Substituting this value for x in the second equation and then solving for y, we get

$$2x - 4y = 13$$
$$2\left(\frac{-9 - 6y}{4}\right) - 4y = 13 \quad \text{(Substitute for } x\text{)}$$
$$\frac{-9 - 6y}{2} - 4y = 13 \quad \text{(Cancel)}$$
$$-9 - 6y - 8y = 26 \quad \text{(Multiply by 2)}$$
$$-14y = 35 \quad \text{(Simplify)}$$
$$y = \frac{35}{-14} \quad \text{(Divide by } -14\text{)}$$
$$y = -\frac{5}{2}. \quad \text{(Reduce)}$$

Substituting the value $y = -\frac{5}{2}$ into the expression $x = \frac{-9 - 6y}{4}$, we get

$$x = \frac{-9 - 6\left(-\frac{5}{2}\right)}{4} \quad \text{(Substitute for } y\text{)}$$
$$x = \frac{-9 - 3(-5)}{4} \quad \text{(Cancel)}$$
$$x = \frac{-9 + 15}{4} \quad \text{(Multiply)}$$
$$x = \frac{6}{4} \quad \text{(Simplify)}$$
$$x = \frac{3}{2}. \quad \text{(Reduce)}$$

Thus, the solution is $x = \frac{3}{2}$, $y = -\frac{5}{2}$.

Checking this solution in the second equation gives

$$2x - 4y = 13$$
$$2\left(\frac{3}{2}\right) - 4\left(-\frac{5}{2}\right) = 13 \quad \text{(Substitute for } x \text{ and } y\text{)}$$
$$3 - 2(-5) = 13 \quad \text{(Cancel)}$$
$$3 + 10 = 13 \quad \text{(Multiply)}$$
$$13 = 13. \quad \text{(Simplify)}$$

Thus, the solution checks.

• • • • • • • • • •

When we try to solve a dependent system by substitution, one of the equations reduces to $0 = 0$. This means that there are an infinite number of solutions.

• Example 4.8: Solve, by substitution, the system of equations

$$2x + y = 3, \quad 4x + 2y = 6.$$

Solution: Solving the first equation for y gives

$$y = 3 - 2x. \quad \text{(Solve for } y \text{ in terms of } x\text{)}$$

Substituting this value for y into the second equation gives

$$\begin{aligned}
4x + 2y &= 6 \\
4x + 2(3 - 2x) &= 6 & \text{(Substitute for } y\text{)} \\
4x + 6 - 4x &= 6 & \text{(Multiply)} \\
4x - 4x &= 6 - 6 & \text{(Transpose and simplify)} \\
0x &= 0 & \text{(Note: } x \text{ can be any value)} \\
0 &= 0.
\end{aligned}$$

In the equation $0x = 0$ we can see that any value of x works since zero times x is always zero. Hence there are an infinite number of solutions. Therefore the system is dependent.

• • • • • • • • • •

Solving an inconsistent system by substitution leads us to an equation that cannot be true. This is illustrated in example 4.9.

• Example 4.9: Solve, by substitution, the system

$$3x - 2y = 4, \quad 9x - 6y = 2.$$

Solution: Solving the first equation for x gives

$$x = \frac{2y + 4}{3}. \quad \text{(Solve for } x \text{ in terms of } y\text{)}$$

Substituting this expression into the second equation gives

$$\begin{aligned}
9x - 6y &= 2 \\
9\left(\frac{2y + 4}{3}\right) - 6y &= 2 & \text{(Substitute for } x\text{)} \\
3(2y + 4) - 6y &= 2 & \text{(Cancel)} \\
6y + 12 - 6y &= 2 & \text{(Multiply)} \\
12 &= 2 & \text{(Simplify)} \\
10 &= 0. & \text{(Subtract 2)}
\end{aligned}$$

This cannot be true, therefore the system is inconsistent. The graph of this system would be two parallel lines.

• • • • • • • • • •

SECTION 4.3 SOLVING SYSTEMS OF TWO LINEAR EQUATIONS BY SUBSTITUTION

SUBSTITUTION METHOD

1. Solve one equation for one of the variables in terms of the other variable.
2. Substitute the expression for the variable found in step 1 into the second equation.
3. Solve the resulting equation for the remaining variable.
4. Substitute the value of the variable found in step 3 into one of the original equations to find the remaining variable.
5. Check the results by substituting the values of the variables into each of the given equations.
6. If the system is either dependent or inconsistent, the process will fail.

Trial Problems 4.3

Before you begin the section exercises, warm up with these problems. Complete answers are included in the answer key.

Solve the following systems by substitution.

1. $2x + y = 3$, $x - y = 3$
2. $y = 1 - x$, $2x + y = 7$
3. $x - 3y = 1$, $3x - 4y = 3$
4. $x = 3 - y$, $x = y + 11$
5. $4x - 2y = 6$, $2x + 6 = 4y$

Exercises 4.3

Solve the following systems of equations by substitution.

1. $y = 1 - x$, $2y - 8x = 1$
2. $y = 2 - x$, $3y - 7x = -4$
3. $y = 5 + x$, $2x - y = 1$
4. $y = 6 + x$, $x - 2y = -11$
5. $x - 4y = 3$, $2x - 3y = 9$
6. $x - 3y = 2$, $3x - 4y = 16$
7. $x + 2y = 3$, $3x + 5y = 10$
8. $x - 2y = 3$, $2x - y = 6$
9. $2x + y = 0$, $3x + 2y = 1$
10. $3x - y = 0$, $4x - 2y = -2$
11. $2x + 5y = 4$, $3x - 5y = 0$
12. $5x + 2y = 4$, $5x - 3y = 0$
13. $x - y = 4$, $2x + 3y = 6$
14. $x - y = 5$, $3x + 2y = 20$
15. $y = x - 1$, $x - y = 5$
16. $y = x - 2$, $x - y = -2$
17. $2x + 3y = 6$, $4x - 6 = -6y$
18. $4x - 2y = 1$, $2x + 7 = 4y$

19. Given the *demand equation* $y + 20p = 740$ and the *supply equation* $y - 30p = 200$, find the coordinates of the point where the graphs of the demand and supply equations cross. This point is called the point of **market equilibrium**.

20. Given the demand equation $y + 30p = 690$ and the supply equation $y - 20p = 240$, find the point of market equilibrium, the point where the two graphs cross.

21. The sum of two resistors in a series circuit is 320 ohms. The smaller resistor has a resistance $\frac{3}{5}$ that of the larger resistor. What are the resistances of the two resistors?

22. The sum of two resistors in a series circuit is 210 ohms. The smaller resistor has a resistance $\frac{3}{4}$ that of the larger resistor. What are the resistances of the two resistors?

23. The perimeter of a rectangle is 48 meters and the length of the rectangle is 2 meters longer than the width. Find the length and the width of the rectangle.

24. A corporation has found that the cost c of producing n items is represented by the equation $c = 800 + 3n$. The revenue c from the sale of n items is represented by the equation $c = 3.5n$. The point (c,n) that satisfies both equations is called the break-even point. Find the break-even point for this corporation.

25. A pair of metal rods has a combined length of 24.27 cm. The longer rod is 13.12 cm longer than the shorter rod. Find the lengths of the rods.

26. A certain spring of length L stretches P cm for each kilogram of weight that is attached to the spring. Experimentation reveals that the spring is 13 cm long when a weight of 3 kg is attached and is 19 cm long when a weight of 5 kg is attached. Solve the system of equations for L and P.

4 • 4 Solving Systems of Two Linear Equations by Addition or Subtraction

In sections 4.2 and 4.3 we solved systems of equations graphically and by substitution. The graphic method did not always give exact answers. The method of substitution was sometimes cumbersome if fractions resulted. Solving systems of equations by addition or subtraction sometimes eliminates these problems.

To solve a system of two linear equations by addition or subtraction we simply add one equation to the other or subtract one equation from the other. Or we add one equation to a nonzero multiple of the other or subtract one equation from a nonzero multiple of the other. We can do this since we are adding equals to equals. This process eliminates one of the variables so that we can solve for the remaining variable. Example 4.10 further illustrates this technique.

• *Example 4.10:* Solve the system of equations

$$2x + y = 4, \quad 3x - y = 1.$$

Note that this is the same system that we solved by substitution in example 4.6.

Solution: If we place one equation over the other and then add them, we see that the y terms are eliminated and we can solve for x.

$$\left. \begin{array}{r} 2x + y = 4 \\ 3x - y = 1 \end{array} \right\} \text{ (Each equation has both } x \text{ and } y.)$$
$$5x = 5 \quad \text{(The equation has only the variable } x.)$$
$$x = 1.$$

Section 4.4 Solving Systems of Two Linear Equations by Addition or Subtraction

To find y, we substitute $x = 1$ into either of the given equations. If we select the first equation, we get

$$
\begin{aligned}
2x + y &= 4 \\
2(1) + y &= 4 \quad &\text{(Substitute for } x\text{)} \\
2 + y &= 4 \quad &\text{(Multiply)} \\
y &= 4 - 2 \quad &\text{(Transpose)} \\
y &= 2. \quad &\text{(Simplify)}
\end{aligned}
$$

The desired solution is $x = 1$ and $y = 2$.

•••••••••

If neither of the variables can be eliminated as a result of addition or subtraction, we must first multiply either one or both of the equations by a nonzero constant. Example 4.11 illustrates this process.

• *Example 4.11:* Solve the system of equations

$$3x - 3y = 6, \quad 5x + 2y = 10.$$

Solution: If we multiply the first equation by 2 and the second equation by 3, both equations have a $6y$ term.

The new system is

$$
\begin{aligned}
6x - 6y &= 12 \quad &\text{(Multiply first equation by 2)} \\
15x + 6y &= 30. \quad &\text{(Multiply second equation by 3)}
\end{aligned}
$$

We now add these two equations and eliminate the y-terms.

$$
\begin{aligned}
6x - 6y &= 12 \\
\underline{15x + 6y} &= \underline{30} \\
21x &= 42. \quad \text{(Add the two equations)}
\end{aligned}
$$

The resulting equation is $21x = 42$.

We now solve this equation for x and get $x = 2$.

To find y, we substitute $x = 2$ into either of the two given equations. If we select the second equation, we get

$$
\begin{aligned}
5x + 2y &= 10 \\
5(2) + 2y &= 10 \quad &\text{(Substitute 2 for } x\text{)} \\
10 + 2y &= 10 \quad &\text{(Multiply)} \\
2y &= 0 \quad &\text{(Subtract 10)} \\
y &= 0. \quad &\text{(Divide by 2)}
\end{aligned}
$$

The desired solution is $x = 2$, $y = 0$.

•••••••••

Addition or Subtraction Method

1. If necessary multiply each equation by a constant chosen so that one of the variables can be eliminated from both equations by adding or subtracting the equations.
2. Obtain an equation in one variable by adding or subtracting the equations resulting from step 1.
3. Solve the resulting equation.
4. Substitute this value in one of the original equations and solve for the other variable.
5. Check the results by substituting the values of the variables into each of the given equations.
6. If the system is either dependent or inconsistent, the process will fail.

Examples 4.12, 4.13, and 4.14 further illustrate the solution of systems of two equations in two unknowns by the addition or subtraction method.

• *Example 4.12:* Solve the pair of equations
$$y = 1200 + 2x, \quad y = 3.5x.$$

Solution:
1. Since both equations contain the same y term, we do not need to multiply by a constant.
2. Subtract the equations,
$$y = 3.5x$$
$$y = 2x + 1200$$
$$0 = 1.5x - 1200.$$
3. Solve this equation
$$1.5x = 1200$$
$$x = \frac{1200}{1.5} = 800.$$
4. Substitute $x = 800$ into the equation $y = 3.5x$, as
$$y = 3.5(800) = 2800.$$

The desired solution is $x = 800$ and $y = 2800$.

SECTION 4.4 SOLVING SYSTEMS OF TWO LINEAR EQUATIONS BY ADDITION OR SUBTRACTION

• *Example 4.13:* Solve the system

$$2x + 4y = 13, \qquad 3x - 5y = 3.$$

Solution: 1. If we multiply the first equation by 5 and the second equation by 4, both of the resulting equations will have a $20y$ term:

$$10x + 20y = 65$$
$$12x - 20y = 12.$$

2. Add the equations:

$$22x = 77.$$

3. Solve this equation:

$$x = \frac{77}{22} = \frac{7}{2}.$$

4. Substitute $x = \frac{7}{2}$ in the equation $2x + 4y = 13$, and solve for y.

$$2\left(\frac{7}{2}\right) + 4y = 13$$
$$7 + 4y = 13$$
$$4y = 6$$
$$y = \frac{6}{4} = \frac{3}{2}.$$

The solution is $x = \frac{7}{2}$, $y = \frac{3}{2}$.

•••••••••••

• *Example 4.14:* Solve a demand equation $y + 20p = 800$, where y represents the demand for a product at a price p, and a supply equation $y - 30p = 200$, where y represents the supply of the product that manufacturers are willing to furnish at a price p. The point of *market equilibrium* is the point $E(p, y)$ where the graphs of the supply and demand equations cross. Find the coordinates of this point.

Solution: $y + 20p = 800$
$y - 30p = 200.$

In this system of equations the coefficients of y are both one. Hence, we can subtract one equation from the other to eliminate the y terms.
  Subtract the equations and solve for p:

$$50p = 600$$
$$p = 12.$$

Substitute $p = 12$ into the equation $y = 20p + 800$:

$$y + 20(12) = 800$$
$$y + 240 = 800$$
$$y = 800 - 240 = 560.$$

The coordinates of the equilibrium point E are $(12, 560)$. Market equilibrium is attained with a price of 12 and a supply and demand of 560 units of the product. This system is graphed in figure 4.8.

• • • • • • • • • •

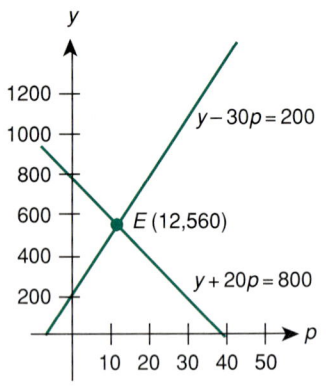

Figure 4.8

Example 4.15 indicates what happens when we attempt to solve a dependent system by addition or subtraction.

• **Example 4.15:** Solve the system

$$x + y = 3$$
$$2x + 2y = 6.$$

Solution: If we multiply the first equation by 2 then subtract the second equation from the first we get

$2x + 2y = 6$	(Multiply first equation by 2)
$2x + 2y = 6$	(Second equation)
$0x + 0y = 0.$	(Subtract second equation)

This suggests that there are an infinite number of solutions to the system since this last equation is true for any values of x and y.

• • • • • • • • • •

When we attempt to solve an inconsistent system by addition or subtraction, an inconsistency develops. Example 4.16 illustrates this.

• **Example 4.16:** Solve the system

$$x + y = 7$$
$$x + y = 3.$$

Solution: If we subtract the second equation from the first, we get

$x + y = 7$	(First equation)
$x + y = 3$	(Second equation)
$0x + 0y = 4.$	(Subtract second equation)

This suggests that there are no solutions to the system since there are no values of x and y that, when multiplied by zero and added, equal 4.

• • • • • • • • • •

Section 4.4 Solving Systems of Two Linear Equations by Addition or Subtraction

Trial Problems 4.4

Before you begin the section exercises, warm up with these problems. Complete answers are included in the answer key.

Solve by addition or substitution.

1. $x - y = 3$, $x + y = 7$
2. $x = y + 7$, $x + y = 5$
3. $2x + 5y = -26$, $3x - 5y = 11$
4. $x - 2y = -7$, $y + 2x = -9$
5. $5x - 2y = -2$, $7x - 3y = -3$

Exercises 4.4

Solve by addition or subtraction.

1. $x + y = 4$, $x - y = 2$
2. $x + y = 7$, $x - y = 3$
3. $3x + 4y = 24$, $x + 2y = 11$
4. $2x + 4y = 10$, $x + y = 3$
5. $x - 4y = 3$, $2x - 7y = 9$
6. $x - 3y = 1$, $2x - 5y = 3$
7. $2x + 5y = 4$, $3x - 5y = 0$
8. $2x + 5y = 1$, $3x - 5y = 14$
9. $2x = 7 - y$, $2x = 1 + y$
10. $3x = 9 - y$, $3x = 21 + y$
11. $x - 2y = 5$, $2x + y = 20$
12. $x - 2y = 1$, $2x + y = 12$
13. $5x - 2y = 7$, $7x - 3y = 5$
14. $5x - 2y = 11$, $7x - 3y = 15$
15. $y = -100x + 400$, $y = 300x - 300$
16. $y = 100x + 100$, $y = 300x - 100$
17. $0.1x + 0.3y = 2.4$, $0.5x - 1.5y = 12$
18. $0.1x + 0.3y = -0.7$, $0.5x - 1.5y = 2.5$

19. A rectangular lot with length x and width y has a perimeter of 720 ft. The lot is 60 ft longer than it is wide. Find the length and the width.

20. A rectangular lot with length x and width y has a perimeter of 270 meters. The lot is 15 meters longer than it is wide. Find the length and width.

21. A rocket is launched with an average velocity of 2000 mi/hr. An hour later an interceptor rocket, with an average velocity of 2500 mi/hr, is launched along the same path. Find the times of flights of the rockets when the interceptor rocket overtakes the first rocket by solving the following system of equations: $2000t_1 = 2500t_2$, $t_1 = t_2 + 1$.

22. Solve exercise 21 if the first rocket has an average velocity of 3200 km/hr and the interceptor rocket has an average velocity of 4000 km/hr.

23. Two electric currents (in amps) can be related as indicated by these two equations: $3I_1 + 4I_2 = 3$, $3I_1 - 5I_2 = -6$. Solve these equations for I_1 and I_2.

24. Solve exercise 23 for I_1 and I_2 if the two currents are related by these two equations: $5I_1 + 8I_2 = 36$, $4I_1 - I_2 = 14$.

25. Given the demand equation $y + 15p = 700$ and the supply equation $y - 20p = 280$, solve by addition or subtraction to find the values of y and p that produce the point of market equilibrium.

26. Given the demand equation $y - 15p = 105$ and the supply equation $y - 12p = 180$, solve by addition or subtraction to find the values of y and p that produce the point of market equilibrium.

4 • 5 Solving Systems of Two Linear Equations by Determinants

In section 4.4 we solved systems of linear equations by addition or subtraction. In this section we solve systems of linear equations using determinants.

Matrix A **matrix** is a rectangular array of symbols. Matrices are enclosed by parentheses or brackets and are usually denoted by capital letters. A matrix **B** with two rows and two columns is called a 2×2 matrix and may be expressed as

$$\mathbf{B} = \begin{pmatrix} 4 & 1 \\ 2 & 3 \end{pmatrix} \quad \text{or} \quad \mathbf{B} = \begin{bmatrix} 4 & 1 \\ 2 & 3 \end{bmatrix}.$$

The concept of a matrix is introduced here so that we can define a determinant of a matrix. The theory of matrices is further developed in chapter 13.

A matrix that has an equal number of rows and columns is called a square matrix.

Determinant Associated with each square matrix there is a real number called the **determinant** of that matrix. For the matrix $\mathbf{A} = \begin{pmatrix} a & b \\ c & d \end{pmatrix}$ we define the determinant of **A** to be

$$\begin{vmatrix} a & b \\ c & d \end{vmatrix} = ad - bc.$$

Order of the Determinant The **order of the determinant** is the number of entries in any row or column. This determinant is of order two. The determinant of a 2×2 matrix is obtained by taking the product of the entries on the diagonal from upper left to lower right

Primary Diagonal (**primary diagonal**) and subtracting the product of the entries on the other diagonal
Secondary Diagonal (**secondary diagonal**).

• *Example 4.17:* $\begin{vmatrix} 4 & 1 \\ 3 & 2 \end{vmatrix} = (4)(2) - (1)(3) = 8 - 3 = 5.$

• *Example 4.18:* $\begin{vmatrix} 7 & -1 \\ 3 & 2 \end{vmatrix} = (7)(2) - (-1)(3) = 14 + 3 = 17.$

We now use determinants to solve two linear equations in two unknowns.
If we take the two general linear equations in this form

$$ax + by = e$$
$$cx + dy = f,$$

Section 4.5 Solving Systems of Two Linear Equations by Determinants

and solve them by addition or subtraction, we get

$$x = \frac{ed - bf}{ad - bc}$$

$$y = \frac{af - ec}{ad - bc}.$$

The following manipulations show just how we solved the system for x:

$a(d)x + b(d)y = e(d)$	(Multiply the first equation by d)
$(b)cx + (b)dy = (b)f$	(Multiply the second equation by b)
$adx - bcx = ed - bf$	(Subtract second equation from first)
$(ad - bc)x = ed - bf$	(Remove common factor x from left-hand side)
$x = \dfrac{ed - bf}{ad - bc}.$	(Divide by $ad - bc$)

Similarly, to solve for y we manipulate as follows:

$a(c)x + b(c)y = ec$	(Multiply the first equation by c)
$(a)cx + (a)dy = (a)f$	(Multiply the second equation by a)
$ady - bcy = af - ec$	(Subtract first equation from second)
$(ad - bc)y = af - ec$	(Remove common factor y from left side)
$y = \dfrac{af - ec}{ad - bc}.$	(Divide by $ad - bc$)

Since $ed - bf = \begin{vmatrix} e & b \\ f & d \end{vmatrix}$, $af - ec = \begin{vmatrix} a & e \\ c & f \end{vmatrix}$, and $ad - bc = \begin{vmatrix} a & b \\ c & d \end{vmatrix}$, we can substitute these values and obtain

$$x = \frac{ed - bf}{ad - bc} = \frac{\begin{vmatrix} e & b \\ f & d \end{vmatrix}}{\begin{vmatrix} a & b \\ c & d \end{vmatrix}}$$

$$y = \frac{af - ec}{ad - bc} = \frac{\begin{vmatrix} a & e \\ c & f \end{vmatrix}}{\begin{vmatrix} a & b \\ c & d \end{vmatrix}}.$$

Note how the constants e and f are placed in the first column of the numerator when we are solving for the first variable x, and in the second column of the numerator when solving for the second variable y. The remaining column in each numerator is like the corresponding column in the denominator.

The denominator in each case is formed by placing the coefficients of *x* in the first column and the coefficients of *y* in the second column. Since the denominators are the same in each case we need to evaluate the denominator only once. Occasionally we let the Greek letter delta (Δ) stand for the denominator.

Cramer's Rule This process of solving a system of equations using determinants is called **Cramer's rule.** We summarize this process and then illustrate it with examples 4.19, 4.20, and 4.21.

SOLUTION OF SYSTEM OF TWO EQUATIONS IN TWO UNKNOWNS BY DETERMINANTS

1. Write the equations in the form
$$ax + by = e$$
$$cx + dy = f.$$

2. Form the determinant in the denominator by using the coefficients of the variables:
$$\begin{vmatrix} a & b \\ c & d \end{vmatrix}.$$

3. Form the determinant in the numerator for the *first* variable by placing the constants *e* and *f* in the *first* column of the determinant.

4. Make the remaining column in the determinant of the numerator identical to the corresponding column in the determinant of the denominator.

5. Form the determinant in the numerator of the *second* variable by placing the constants *e* and *f* in the *second* column of the determinant.

6. Make the remaining column in the determinant of this numerator identical to the corresponding column in the determinant of the denominator.

7. Evaluate the determinants and simplify the results.

8. Check the answers.

9. If the determinant in the denominator is zero and the determinant in the numerator is not zero, then the system is inconsistent and there is no solution.

10. If the determinant in the denominator is zero and the determinant in the numerator is also zero, then the system is dependent and there are many solutions.

Section 4.5 Solving Systems of Two Linear Equations by Determinants

• **Example 4.19:** Solve, with determinants, the system

$$2x + 3y = 7$$
$$x + 4y = 6.$$

Solution:
$$x = \frac{\begin{vmatrix} 7 & 3 \\ 6 & 4 \end{vmatrix}}{\begin{vmatrix} 2 & 3 \\ 1 & 4 \end{vmatrix}} = \frac{(7)(4) - (3)(6)}{(2)(4) - (3)(1)} = \frac{28 - 18}{8 - 3} = \frac{10}{5} = 2$$

$$y = \frac{\begin{vmatrix} 2 & 7 \\ 1 & 6 \end{vmatrix}}{\begin{vmatrix} 2 & 3 \\ 1 & 4 \end{vmatrix}} = \frac{(2)(6) - (7)(1)}{(2)(4) - (3)(1)} = \frac{12 - 7}{8 - 3} = \frac{5}{5} = 1.$$

The solution $x = 2$ and $y = 1$ can be checked in either of the original equations.

....

If the denominator is zero the system is either inconsistent or dependent. Example 4.20 illustrates an inconsistent system.

• **Example 4.20:** Solve, with determinants, the system

$$3x + 1y = 6$$
$$3x + 1y = 3.$$

Solution:
$$x = \frac{\begin{vmatrix} 6 & 1 \\ 3 & 1 \end{vmatrix}}{\begin{vmatrix} 3 & 1 \\ 3 & 1 \end{vmatrix}} = \frac{(6)(1) - (1)(3)}{(3)(1) - (1)(3)} = \frac{6 - 3}{3 - 3} = \frac{3}{0}$$

$$y = \frac{\begin{vmatrix} 3 & 6 \\ 3 & 3 \end{vmatrix}}{\begin{vmatrix} 3 & 1 \\ 3 & 1 \end{vmatrix}} = \frac{(3)(3) - (6)(3)}{(3)(1) - (1)(3)} = \frac{9 - 18}{3 - 3} = \frac{-9}{0}.$$

Since division by zero is not allowed, there is no solution. This system was graphed in figure 4.2 and was called inconsistent. Notice that the left-hand side of each of the original equations was the same but that the right-hand sides were different. This suggests an inconsistency. Hence the name, inconsistent system. Inconsistent systems have no solution.

....

160 CHAPTER 4 SYSTEMS OF LINEAR EQUATIONS

Example 4.21 illustrates a dependent system.

• *Example 4.21:* Solve, with determinants, the system
$$x + y = 3$$
$$4x + 4y = 12.$$

Solution:
$$x = \frac{\begin{vmatrix} 3 & 1 \\ 12 & 4 \end{vmatrix}}{\begin{vmatrix} 1 & 1 \\ 4 & 4 \end{vmatrix}} = \frac{(3)(4) - (1)(12)}{(1)(4) - (1)(4)} = \frac{12 - 12}{4 - 4} = \frac{0}{0}$$

$$y = \frac{\begin{vmatrix} 1 & 3 \\ 4 & 12 \end{vmatrix}}{\begin{vmatrix} 1 & 1 \\ 4 & 4 \end{vmatrix}} = \frac{(1)(12) - (3)(4)}{(1)(4) - (1)(4)} = \frac{12 - 12}{4 - 4} = \frac{0}{0}.$$

Again we have division by zero but in this case the numerators are also zero. This system was graphed in figure 4.3 and was called dependent since one equation is a multiple of the other or "depends" on the other. Dependent systems have many solutions.

Trial Problems 4.5

Before you begin the section exercises, warm up with these problems. Complete answers are included in the answer key.

Evaluate the determinants.

1. $\begin{vmatrix} 3 & 1 \\ 2 & 2 \end{vmatrix}$ 2. $\begin{vmatrix} 1 & 3 \\ 2 & -1 \end{vmatrix}$ 3. $\begin{vmatrix} -1 & 1 \\ 1 & -1 \end{vmatrix}$

Solve with determinants, the following pairs of linear equations.

4. $3x + y = 7, 2x + 2y = 6$ 5. $x + 3y = 3, 2x - y = -8$

Exercises 4.5

Solve, with determinants, the following pairs of linear equations.

1. $2x - 3y = 7, 3x + 5y = 1$
2. $2x - 3y = 1, 3x + 5y = 11$
3. $2x + 3y = 13, 5x - 4y = -2$
4. $2x + 3y = 5, 5x - 4y = 1$
5. $3x - y = 10, 4x + 2y = 10$
6. $3x - y = 0, 4x + 2y = 2$
7. $5x - y = 9, 4x = 2y$
8. $5x - y = -7, 4x = 2y - 8$
9. $2x + 3y = 6, x = 4 - 3y$
10. $2x + 3y = 6, 4x = 1 - 3y$

Use Cramer's rule to solve these systems of equations. From the results state whether the systems are dependent or inconsistent.

11. $1.5x + 2.1y = 7.3$
    $4.5x + 6.3y = 3.7$

12. $1.6x + 2.1y = 4.0$
    $4.8x + 6.3y = 6.0$

13. $2.7x - 3.1y = 1.6$
    $8.1x - 9.3y = 4.8$

14. $1.8x - 2.6y = 1.0$
    $2.7x - 3.9y = 1.5$

Solve for the specified variables.

15. Stresses in an aluminum bar connected to a prestressed steel sleeve are related by
$$S_S + 0.5S_A = 0$$
$$S_S - 2S_A = 80{,}000 \text{ MPa.}$$
Use Cramer's rule to find stresses S_S and S_A in MPa.

16. Forces on a beam produce the equilibrium equations
$$R_1 + R_2 = 1000$$
$$5R_1 - 3R_2 = 2990.$$
User Cramer's rule to find the reaction forces R_1 and R_2.

17. When two forces F_1 and F_2 act on a certain lever supported by a fulcrum, the first of the following equations results. When the forces are moved to different locations on the lever, the second equation results. Solve the system of equations using Cramer's rule:
$$50F_1 - 70F_2 = 0$$
$$55F_1 - 65F_2 = 60.$$

18. If the sum of two forces F_1 and F_2 is 48 and the F_2 is eight more than F_1 the following two equations will result. Use Cramer's rule to solve this system of equations:
$$F_1 + F_2 = 48$$
$$F_2 = F_1 + 8.$$

19. An investor wishes to invest \$10,000. P dollars will be invested at 12% and Q dollars will be invested at 16%. The return on these investments is \$1360 per year. The following equations result:
$$P + Q = 10{,}000$$
$$0.12P + 0.16Q = 1360.$$
Solve these equations using Cramer's rule.

20. If x kilograms of alloy containing 25% tin and y kilograms of alloy containing 45% tin are combined, the result is 50 kilograms of alloy containing 39% tin. This information can be expressed with the following equations:
$$x + y = 50$$
$$0.25x + 0.45y = (0.39)(50).$$
Solve these equations using Cramer's rule.

4•6 Solving Systems of Three Linear Equations in Three Unknowns

The solution of three equations in three unknowns can be determined by using the same techniques that were used with two equations in two unknowns. The techniques used with two equations were graphing, addition-subtraction, substitution, and Cramer's rule. Sketching the graphs of equations in three unknowns requires three-dimensional graphing, which is beyond the scope of this chapter. Graphing equations in three unknowns often results in approximate solutions as was the case when we graphed two equations in two unknowns.

In example 4.22 we solve the system of three equations in three unknowns using the addition-subtraction process. Our goal is to add or subtract pairs of equations so that one of the variables is eliminated. This results in two equations in the same two unknowns. Then we use the techniques of the previous section to solve this pair of equations for each of these variables. With the value of two variables known we can substitute these values into any of the three given equations and find the value of the third variable.

• *Example 4.22:* Solve the following system of equations by addition-subtraction:

$$2x + y - z = 4, \qquad (4.1)$$
$$x - y - z = -3, \qquad (4.2)$$
$$3x - 2y + z = 9. \qquad (4.3)$$

Solution: If we add equation (4.1) to equation (4.3) we get

$$5x - y = 13. \qquad (4.4)$$

We can get another equation in the *same two* unknowns if we add equation (4.2) to equation (4.3). Doing so, we get

$$4x - 3y = 6. \qquad (4.5)$$

Now we have reduced the problem of solving three equations in three unknowns to that of solving a system of two equations in two unknowns. This system of two equations is

$$5x - y = 13 \qquad (4.4)$$
$$4x - 3y = 6. \qquad (4.5)$$

If we multiply equation (4.4) by 3, we get

$$15x - 3y = 39. \qquad (4.6)$$

If we subtract equation (4.5) from equation (4.6), we get

$$15x - 3y = 39 \qquad (4.6)$$
$$\underline{4x - 3y = 6} \qquad (4.5)$$
$$11x = 33$$
$$x = 3.$$

Substituting $x = 3$ into equation (4.4), we get

$$
\begin{aligned}
5x - y &= 13 & &\text{(4.4)}\\
5(3) - y &= 13 & &\text{(Substitute 3 for } x\text{)}\\
15 - y &= 13 & &\text{(Multiply)}\\
-y &= 13 - 15 & &\text{(Transpose 15)}\\
-y &= -2 & &\text{(Subtract)}\\
y &= 2. & &\text{(Divide by } -1\text{)}
\end{aligned}
$$

We have now found values for x and y. Substituting these values into any one of the given equations will determine z.

Substituting $x = 3$ and $y = 2$ into equation (4.2) gives

$$
\begin{aligned}
x - y - z &= -3 &\\
(3) - (2) - z &= -3 & &\text{(Substitute 3 for } x \text{ and 2 for } y\text{)}\\
3 - 2 - z &= -3 & &\text{(Remove parentheses)}\\
1 - z &= -3 & &\text{(Add)}\\
-z &= -3 - 1 & &\text{(Transpose 1)}\\
-z &= -4 & &\text{(Subtract)}\\
z &= 4 & &\text{(Divide by } -1\text{)}
\end{aligned}
$$

Section 4.6 Solving Systems of Three Linear Equations in Three Unknowns

The solution is $x = 3$, $y = 2$, and $z = 4$. This solution can also be expressed as (3,2,4).
This solution checks since,

$$2x + y - z = 4$$
$$2(3) + (2) - (4) = 4$$
$$6 + 2 - 4 = 4$$
$$4 = 4,$$

$$x - y - z = -3$$
$$(3) - (2) - (4) = -3$$
$$3 - 2 - 4 = -3$$
$$-3 = -3,$$

$$3x - 2y + z = 9$$
$$3(3) - 2(2) + (4) = 9$$
$$9 - 4 + 4 = 9$$
$$9 = 9.$$

Although we can use the method of substitution to solve a system of three equations in three variables, we usually do not use this technique when we have more than two equations since the process becomes quite cumbersome. We do not illustrate the substitution method as it applies to three equations.

If we wish to solve three linear equations in three unknowns by using Cramer's rule, we must first learn how to evaluate a determinant of order three. If the matrix under consideration has dimensions 3×3, then its determinant is of **order three** and is defined as follows:

Order Three

$$\begin{vmatrix} a_1 & b_1 & c_1 \\ a_2 & b_2 & c_2 \\ a_3 & b_3 & c_3 \end{vmatrix} = a_1b_2c_3 + a_2b_3c_1 + a_3b_1c_2 - a_1b_3c_2 - a_2b_1c_3 - a_3b_2c_1.$$

A convenient scheme for obtaining this result is to copy the first two columns of the determinant to the right of the last column. Then to evaluate, we compute the sum of the products on each diagonal from upper left to lower right. Then we subtract the sum of the products on each diagonal from upper right to lower left. These products are indicated below:

$$\begin{vmatrix} a_1 & b_1 & c_1 \\ a_2 & b_2 & c_2 \\ a_3 & b_3 & c_3 \end{vmatrix} = \begin{vmatrix} a_1 & b_1 & c_1 \\ a_2 & b_2 & c_2 \\ a_3 & b_3 & c_3 \end{vmatrix} \begin{matrix} a_1 & b_1 \\ a_2 & b_2 \\ a_3 & b_3 \end{matrix} = a_1b_2c_3 + a_2b_3c_1 + a_3b_1c_2 - a_1b_3c_2 - a_2b_1c_3 - a_3b_2c_1.$$

Practice this technique in examples 4.23 and 4.24.

• *Example 4.23:* Evaluate

$$\begin{vmatrix} 1 & 0 & 2 \\ 2 & 1 & 2 \\ 1 & 3 & 4 \end{vmatrix}.$$

Solution: 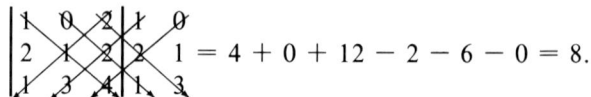 $= 4 + 0 + 12 - 2 - 6 - 0 = 8.$

• *Example 4.24:* Evaluate

$$\begin{vmatrix} 1 & -2 & 2 \\ 2 & 3 & 1 \\ -1 & 4 & 2 \end{vmatrix}.$$

Solution: 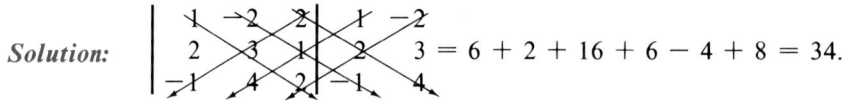 $= 6 + 2 + 16 + 6 - 4 + 8 = 34.$

Suppose we are given these three equations with these three unknowns:

$$a_1 x + b_1 y + c_1 z = d_1$$
$$a_2 x + b_2 y + c_2 z = d_2$$
$$a_3 x + b_3 y + c_3 z = d_3.$$

If we wish to solve this system using Cramer's rule, the solution would look like this:

$$x = \frac{\begin{vmatrix} d_1 & b_1 & c_1 \\ d_2 & b_2 & c_2 \\ d_3 & b_3 & c_3 \end{vmatrix}}{\begin{vmatrix} a_1 & b_1 & c_1 \\ a_2 & b_2 & c_2 \\ a_3 & b_3 & c_3 \end{vmatrix}} \qquad y = \frac{\begin{vmatrix} a_1 & d_1 & c_1 \\ a_2 & d_2 & c_2 \\ a_3 & d_3 & c_3 \end{vmatrix}}{\begin{vmatrix} a_1 & b_1 & c_1 \\ a_2 & b_2 & c_2 \\ a_3 & b_3 & c_3 \end{vmatrix}} \qquad z = \frac{\begin{vmatrix} a_1 & b_1 & d_1 \\ a_2 & b_2 & d_2 \\ a_3 & b_3 & d_3 \end{vmatrix}}{\begin{vmatrix} a_1 & b_1 & c_1 \\ a_2 & b_2 & c_2 \\ a_3 & b_3 & c_3 \end{vmatrix}}.$$

The determinant of the denominator is formed by placing the coefficients of the first variable x in the first column, the coefficients of the second variable y in the second column, and the coefficients of the third variable z in the third column. The Greek letter (Δ) is often used to represent the denominator.

Note how the three constants d_1, d_2, and d_3 are placed in the first column of the numerator when we are solving for the first variable x, in the second column of the numerator when we are solving for the second variable y, and in the third column of the numerator when we are solving for the third variable z. The remaining columns in each numerator are like the corresponding columns in the denominator.

Note that the denominator in each case is the same. Hence, we must evaluate only four different determinants. Example 4.25 illustrates the solution of three equations in three unknowns using Cramer's rule.

• **Example 4.25:** Solve by Cramer's rule:

$$\begin{aligned} 3x - z &= 15 \\ 2x + y - 2z &= 0 \\ 4x - 3y + z &= 5. \end{aligned}$$

Solution:

$$\Delta = \begin{vmatrix} 3 & 0 & -1 \\ 2 & 1 & -2 \\ 4 & -3 & 1 \end{vmatrix} \begin{matrix} 3 & 0 \\ 2 & 1 \\ 4 & -3 \end{matrix}$$

$$= 3 + 0 + 6 + 4 - 18 + 0 = -5$$

$$x = \frac{\begin{vmatrix} 15 & 0 & -1 \\ 0 & 1 & -2 \\ 5 & -3 & 1 \end{vmatrix} \begin{matrix} 15 & 0 \\ 0 & 1 \\ 5 & -3 \end{matrix}}{\Delta}$$

$$= \frac{15 + 0 + 0 + 5 - 90 - 0}{-5} = \frac{-70}{-5} = 14$$

$$y = \frac{\begin{vmatrix} 3 & 15 & -1 \\ 2 & 0 & -2 \\ 4 & 5 & 1 \end{vmatrix} \begin{matrix} 3 & 15 \\ 2 & 0 \\ 4 & 5 \end{matrix}}{\Delta}$$

$$= \frac{0 - 120 - 10 + 0 + 30 - 30}{-5} = \frac{-130}{-5} = 26$$

$$z = \frac{\begin{vmatrix} 3 & 0 & 15 \\ 2 & 1 & 0 \\ 4 & -3 & 5 \end{vmatrix} \begin{matrix} 3 & 0 \\ 2 & 1 \\ 4 & -3 \end{matrix}}{\Delta}$$

$$= \frac{15 + 0 - 90 - 60 - 0 - 0}{-5} = \frac{-135}{-5} = 27.$$

The solution is $x = 14$, $y = 26$, $z = 27$.

We should check these solutions by substituting them into each of the three given equations.

• • • • • • • • • •

Solution of a System of Three Equations in Three Unknowns Using Determinants

1. Write the equations in proper form with the variables appearing in order on the left-hand side of each equation and the constants appearing on the right-hand side of each equation.

2. Create the determinant of the denominator by placing the coefficients of the first variable into the first column of the determinant, the coefficients of the second variable into the second column, and the coefficients of the third variable into the third column.

3. Create the determinant of the numerator for each variable by placing the constants in the first column when seeking the first variable, by placing the constants in the second column when seeking the second variable, and by placing the constants in the third column when seeking the third variable.

4. Complete the determinant of each numerator by making the remaining columns in each numerator identical to the corresponding column in the denominator.

5. Evaluate each of the determinants and simplify the results.

6. Check the answer in each of the given equations.

7. If the determinant of the denominator is zero, then the system is either dependent or inconsistent as it was in the case of two equations in two unknowns.

In this chapter we have solved two simultaneous linear equations by various methods. First we looked at graphic solutions. Then we used the method of substitution. Later we solved two equations simultaneously by addition or subtraction and finally by determinants.

There are many other methods that can also be used to solve two equations simultaneously. Furthermore, some of the methods that we have discussed can be used to solve three, four, or more equations simultaneously. For example, we have just seen how three equations can be solved simultaneously with determinants.

However, if we encounter situations in industry and technology requiring the solution of large numbers of simultaneous equations, we would probably not solve them by the methods that have been illustrated in this chapter. We would turn instead to electronic computers, which, when programmed properly, would give us accurate results much more quickly.

Section 4.6 Solving Systems of Three Linear Equations in Three Unknowns

Trial Problems 4.6

Before you begin the section exercises, warm up with these problems. Complete answers are included in the answer key.

Solve this system of equations by addition, subtraction, or substitution.

1. $x + y + z = 6$
   $x + y - z = 2$
   $x - y + z = 4$

Evaluate these determinants.

2. $\begin{vmatrix} 0 & 3 & -1 \\ 1 & 2 & -2 \\ -3 & 4 & 1 \end{vmatrix}$

3. $\begin{vmatrix} 3 & 2 & 1 \\ 1 & 2 & 3 \\ 3 & 1 & 2 \end{vmatrix}$

4. $\begin{vmatrix} 2 & 0 & 0 \\ -2 & 1 & 2 \\ 3 & 6 & 0 \end{vmatrix}$

Solve this system using Cramer's rule.

5. $x + y = 5$
   $y + z = 4$
   $x + z = 3$

Exercises 4.6

Evaluate these determinants.

1. $\begin{vmatrix} 1 & 2 & 3 \\ 3 & 2 & 1 \\ 2 & 1 & 3 \end{vmatrix}$

2. $\begin{vmatrix} 3 & 1 & 2 \\ 2 & 3 & 1 \\ 1 & 2 & 3 \end{vmatrix}$

3. $\begin{vmatrix} 4 & 5 & 6 \\ 5 & 6 & 4 \\ 6 & 4 & 5 \end{vmatrix}$

4. $\begin{vmatrix} 5 & 6 & 4 \\ 4 & 5 & 6 \\ 6 & 4 & 5 \end{vmatrix}$

5. $\begin{vmatrix} -1 & 1 & 2 \\ 3 & -1 & 3 \\ 4 & 0 & 2 \end{vmatrix}$

6. $\begin{vmatrix} -1 & 2 & -1 \\ 3 & 1 & 3 \\ 0 & 4 & 2 \end{vmatrix}$

7. $\begin{vmatrix} 3 & 6 & 0 \\ 2 & 1 & -2 \\ 0 & 0 & 2 \end{vmatrix}$

8. $\begin{vmatrix} 2 & 4 & 0 \\ 2 & -1 & 2 \\ 0 & 0 & 3 \end{vmatrix}$

9. $\begin{vmatrix} 4 & -3 & 1 \\ 2 & 1 & -2 \\ 3 & 0 & -1 \end{vmatrix}$

10. $\begin{vmatrix} 3 & 0 & -1 \\ 2 & 1 & -2 \\ 4 & -3 & 1 \end{vmatrix}$

Solve by Cramer's rule.

11. $x + y - z = 3$
    $x - y + z = 2$
    $-x + y + z = 1$

12. $x + y - z = 4$
    $x - y + z = 2$
    $-x + y + z = 0$

13. $2y - 3z = 0$
    $x + y + z = 1$
    $3y + 5z = 0$

14. $y - 3z = 8$
    $x + y = 3$
    $x + z = 1$

15. $2x + y + 2z = 1$
    $4y + 2z = x$
    $3x + y = -1 - z$

16. $3x + y + z = -1$
    $4y + 2z = x$
    $2x + y = 1 - 2z$

17. The node equations for a circuit are

$$13v_1 - 3v_2 = 1000$$
$$-9v_1 + 29v_2 - 20v_3 = 0$$
$$5v_2 - 8v_3 = 0.$$

Determine the node voltages v_1, v_2, and v_3.

18. An alloy is composed of three metals, A, B, and C. The percentages of each metal are related by the equations

$$A + B + C = 100$$
$$A - 2C = 0$$
$$-3A + B = 0.$$

Determine the percentage of metals A, B, and C.

19. Three forces F_1, F_2, and F_3 are acting on a beam. The equations that represent these forces are

$$F_1 + F_2 + F_3 = 30$$
$$2F_1 + 3F_2 + F_3 = 70$$
$$3F_1 - F_2 + 2F_3 = 25.$$

Solve this system with determinants.

20. Three currents I_1, I_2, and I_3 are acting in a given circuit. The three equations that represent these currents are

$$-I_1 + I_2 - I_3 = 0$$
$$2I_1 + 4I_3 = 16$$
$$3I_2 + 4I_3 = 26.$$

Solve this system with determinants.

4·7 Summary of Terms, Formulas, Rules, and Procedures

Terms

Break-Even Point (p. 145)
Consistent and Independent (p. 140)
Cramer's Rule (p. 158)
Dependent (p. 140)
Determinant (p. 156)

Inconsistent (p. 140)
Linear Equation (p. 140)
Market Equilibrium (p. 149)
Matrix (p. 156)
Order of the Determinant (p. 156)

Order Three (p. 163)
Primary Diagonal (p. 156)
Secondary Diagonal (p. 156)
Simultaneous Equations (p. 140)
System of Equations (p. 140)

Formulas

Determinant of a matrix of order two. (4.5)

$$\begin{vmatrix} a & b \\ c & d \end{vmatrix} = ad - bc.$$

Determinant of a matrix of order three. (4.6)

$$\begin{vmatrix} a_1 & b_1 & c_1 \\ a_2 & b_2 & c_2 \\ a_3 & b_3 & c_3 \end{vmatrix} = a_1b_2c_3 + a_2b_3c_1 + a_3b_1c_2 - a_1b_3c_2 - a_2b_1c_3 - a_3b_2c_1.$$

Rules and Procedures

- Summary of the steps used to solve systems of equations by graphing. (4.2)
  1. Graph the first equation on a rectangular coordinate system.
  2. Graph the second equation on the same coordinate system.
  3. If the graphs meet at one point, estimate the coordinates of that point. This is the solution. Such a system is consistent and independent.
  4. If the graphs of the two equations are parallel lines there is no common solution to the system. Such a system is called inconsistent.

5. If the graphs of the two equations result in the same line there are an infinite number of solutions. This system is called dependent.
  6. If there appears to be a unique solution, this solution should be checked in both of the given equations.

- Summary of the steps used to solve systems of equations by the substitution method. (4.3)
  1. Solve one equation for one of the variables in terms of the other variable.
  2. Substitute the expression for the variable found in step one into the second equation.
  3. Solve the resulting equation for the remaining variable.
  4. Substitute the value of the variable found in step three into one of the original equations to find the remaining variable.
  5. Check the results by substituting the values of the variables into each of the given equations.
  6. If the system is either dependent or inconsistent, the process will fail.

- Summary of the steps used to solve systems of equations by the method of addition or subtraction. (4.4)
  1. If necessary multiply each equation by a constant chosen so that one of the variables can be eliminated from both equations by adding or subtracting the equations.
  2. Obtain an equation in one variable by adding or subtracting the equations resulting from step 1.
  3. Solve the resulting equation.
  4. Substitute this value in one of the original equations and solve for the other variable.
  5. Check the results by substituting the values of the variables into each of the given equations.
  6. If the system is either dependent or inconsistent, the process will fail.

- Summary of the steps used in the solution of a system of two equations in two unknowns by determinants. (4.5)
  1. Write the equations in the form

  $$ax + by = e$$
  $$cx + dy = f$$

  2. Form the determinant of the denominator by using the coefficients of the variables:

  $$\begin{vmatrix} a & b \\ c & d \end{vmatrix}.$$

  3. Form the determinant of the numerator for the *first* variable by placing the constants e and f in the *first* column of the determinant.
  4. Make the remaining column in the determinant of the numerator identical to the corresponding column in the determinant of the denominator.
  5. Form the determinant of the numerator of the *second* variable by placing the constants e and f in the *second* column of the determinant.
  6. Make the remaining column in the determinant of this numerator identical to the corresponding column in the determinant of the denominator.
  7. Evaluate the determinants and simplify the results.
  8. Check the answers.
  9. If the determinant in the denominator is zero and the determinant in the numerator is not zero, then the system is inconsistent and there is no solution.
  10. If the determinant in the denominator is zero and the determinant in the numerator is also zero, then the system is dependent and there are many solutions.

- Summary of the steps used in the solution of a system of three equations in three unknowns by determinants. (4.6)
  1. Write the equations in proper form with the variables appearing in order on the left-hand side of each equation and the constants appearing on the right-hand side of each equation.
  2. Create the determinant of the denominator by placing the coefficients of the first variable into the first column of the determinant, the coefficients of the second variable into the second column, and the coefficients of the third variable into the third column.
  3. Create the determinant of the numerator for each variable by placing the constants in the first column when seeking the first variable, by placing the constants in the second column when seeking the second variable, and by placing the constants in the third column when seeking the third variable.
  4. Complete the determinant of each numerator by making the remaining columns in each numerator identical to the corresponding column in the denominator.
  5. Evaluate each of the determinants and simplify the results.
  6. Check the answer in each of the given equations.
  7. If the determinant of the denominator is zero, then the system is either dependent or inconsistent as it was in the case of two equations in two unknowns.

4·8 Chapter 4 Review Exercises

Solve the following systems by graphing each pair of equations on a coordinate system. Then estimate the coordinates of the points (if any) that are common to the graphs.

1. $x + y = 7, x - y = 3$
2. $x + y = 8, x - y = 2$
3. $x + y = 5, x - y = 5$
4. $x + y = 7, x - y = 7$
5. $x - y = 2, x - y = -3$
6. $2x + y = 13, 2x + y = 9$
7. $3x + 4y = 10, 6x + 8y = 20$
8. $2x + 3y = 9, 4x + 6y = 18$

Solve the following systems of equations by substitution.

9. $y = 1 - x, 2y - 8x = 1$
10. $y = 2 - x, 3y - 7x = -4$
11. $x + 2y = 3, 3x + 5y = 10$
12. $x - 2y = 3, 2x - y = 6$
13. $2x + 5y = 4, 4x + 10y = 4$
14. $5x + 2y = 4, 10x + 4y = 8$

Solve by addition or subtraction.

15. $x + y = 4, x - y = 2$
16. $x + y = 7, x - y = 3$
17. $2x + 5y = 4, 3x - 5y = 0$
18. $2x + 5y = 1, 3x - 5y = 14$
19. $5x - 2y = 7, 7x - 3y = 5$
20. $5x - 2y = 11, 7x - 3y = 15$

Solve with determinants.

21. $2x - 3y = 7, 3x + 5y = 1$
22. $2x - 3y = 1, 3x + 5y = 11$
23. $5x - y = 9, 4x - 2y = 0$
24. $5x - y = -7, 4x - 2y = -8$

Evaluate these determinants.

25. $\begin{vmatrix} 1 & 2 & 3 \\ 3 & 2 & 1 \\ 2 & 1 & 3 \end{vmatrix}$

26. $\begin{vmatrix} 3 & 1 & 2 \\ 2 & 3 & 1 \\ 1 & 2 & 3 \end{vmatrix}$

27. $\begin{vmatrix} 4 & 5 & 6 \\ 6 & 5 & 4 \\ 5 & 4 & 6 \end{vmatrix}$

28. $\begin{vmatrix} 5 & 4 & 6 \\ 6 & 5 & 4 \\ 4 & 5 & 6 \end{vmatrix}$

Solve by Cramer's rule.

29. $x + y = 7$
    $x - y = 3$

30. $x + y = 8$
    $x - y = 2$

31. $x + y - z = 3$
    $x - y + z = 2$
    $-x + y + z = 1$

32. $y - 3z = 8$
    $x + y = 3$
    $x + z = 1$

Solve.

33. Determine the break-even point for a cost equation $y = 300 + 2x$ and a revenue equation $y = 5x$.

34. Determine the break-even point for a cost equation $y = 500 + 3x$ and a revenue equation $y = 4x$.

35. Given the demand equation $y + 20p = 740$ and the supply equation $y - 30p = 200$, find the coordinates of the point where the graphs of the demand and supply equations cross.

36. Given the demand equation $y + 30p = 690$ and the supply equation $y - 20p = 240$, find the point of market equilibrium.

37. A rectangular lot with length x and width y has a perimeter of 720 ft. The lot is 60 ft longer than it is wide. Find the length and the width.

38. A rectangular lot with length x and width y has a perimeter of 270 meters. The lot is 15 meters longer than it is wide. Find the length and the width.

39. Three unknown forces F_1, F_2, and F_3 are acting on a structure. The system of equations that relates these forces is

$$0.5F_1 + F_2 = 6.5$$
$$F_1 - F_3 = -5$$
$$4F_2 - 3F_3 = -4.$$

Solve this system with determinants. The forces are in kilograms.

40. The displacement of an object moving in a straight line under constant acceleration for three different experimental times results in this system of equations. Solve this system with determinants.

$$s_0 + 2v_0 + 2a = 23$$
$$s_0 + 3v_0 + 4.5a = 41$$
$$s_0 + 5v_0 + 12.5a = 95$$

$s_0 = $ distance in ft.

$v_0 = $ velocity in ft/sec.

$a = $ acceleration in ft/sec².

4·9 Chapter 4 Test

1. Graph this pair of equations on a coordinate system. Then estimate the coordinates of the point (if any) that will be a common solution.

$$x + y = 4$$
$$x - y = 3.$$

2. Solve this system by substitution.

$$y = 2 - x$$
$$3y - 7x = -4.$$

3. Solve this system by substitution.

$$5x + 2y = 4$$
$$5x - 3y = 0.$$

4. Solve by addition or subtraction.

$$x + y = 7$$
$$x - y = 3.$$

5. Solve by addition or subtraction.

$$2x + 5y = 4$$
$$3x - 5y = 0.$$

6. Solve by addition or subtraction.

$$5x - 2y = 7$$
$$7x - 3y = 5.$$

7. Use Cramer's rule to solve this system.

$$2x - 3y = 1$$
$$3x + 5y = 11.$$

8. Solve this system using determinants.

$$2x + 3y = 6$$
$$x = 4 - 3y.$$

9. Solve the system by any method. Is the system dependent, inconsistent, or consistent and independent?

$$2x + 3y = 6$$
$$4x = 1 - 3y.$$

10. Evaluate the determinant

$$\begin{vmatrix} -1 & 1 \\ 3 & -1 \end{vmatrix}.$$

11. Evaluate the determinant

$$\begin{vmatrix} 3 & 6 & 0 \\ 2 & 1 & -2 \\ 0 & 0 & 2 \end{vmatrix}.$$

12. Solve by Cramer's rule.

$$x + y - z = 3$$
$$x - y + z = 2$$
$$-x + y + z = 1.$$

Chapter 5
Solving Quadratic Equations

Equations of the second degree, often called quadratic equations, were solved arithmetically by the Egyptians (2160–1700 B.C.), geometrically by Euclid and his followers (300 B.C.), and algebraically by the Hindus (A.D. 510–630). The ninth century Arab writer, al-Khowarizmi, gave arithmetic rules, whose validity was demonstrated geometrically, for the solution of quadratics. These methods were in use in Europe until the close of the sixteenth century, when writers began to consider the solution of the general quadratic equation with literal coefficients.

Today we solve quadratic equations by many methods. In this chapter we solve quadratic equations by graphing, by factoring, by extracting roots, by completing the square, and by the quadratic formula.

The process of solving quadratic equations by completing the square is important for two reasons. First, the process of completing the square is necessary so that we can later solve the general quadratic equation by completing the square and produce the quadratic formula. Second, the process of completing the square is valuable in later chapters in this text, when our goal is to write an expression in a particular form.

Quadratic equations occur quite frequently in applied problems. The formula for the area of a circle ($A = \pi r^2$) is a quadratic equation. Formulas dealing with falling bodies, such as $d = \frac{1}{2}gt^2$, are represented by quadratic equations. These are just a few of the examples that you might encounter in this chapter.

5 • 1 Solving Quadratic Equations by Graphing

In section 1.9 we discussed the solution of linear equations in one variable. In this section we discuss the solution of quadratic equations in one variable, that is, equations of the type $ax^2 + bx + c = 0$.

Many equations of this type are found in various technical fields. An equation describing the acceleration of an object is $s_0 + v_0 t + \frac{1}{2}at^2 = s$. This is a quadratic equation in the variable t. The equation $i^2 R + iE - 8000 = 0$ is a quadratic equation in variable i (current) in terms of resistance R and voltage E.

What we are really doing when we solve a quadratic equation in one variable is taking a quadratic equation in two variables, namely $y = ax^2 + bx + c$, and setting $y = 0$.

Roots
Zeros

The solutions to such equations are sometimes called **roots** or **zeros**, since we are finding the value of x that makes $y = 0$.

In section 3.3 we graphed quadratic equations of the type $y = ax^2 + bx + c$. We noted that the graphs were parabolas. We also noted that some parabolas crossed the x-axis at two distinct points, some crossed (touched) at one point, and some parabolas didn't intersect the x-axis at all. These graphs are illustrated in figures 5.1, 5.2, and 5.3.

Since any point on the x-axis has a y-coordinate equal to zero, the x-coordinate of each of these points is a solution to the equation. That is, to solve a quadratic equation by graphing, all we have to do is graph the parabola and determine the x-coordinate of the points where the parabola intersects the x-axis. If the parabola does not intersect the x-axis, there are no real-valued solutions. There are nonreal solutions, which are discussed later. Recall that the solutions to an equation consist of all of those values that, when substituted into the equation, make the equation true. There are two

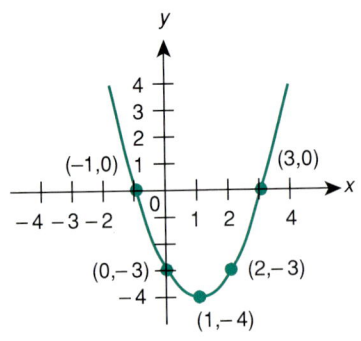

Figure 5.1

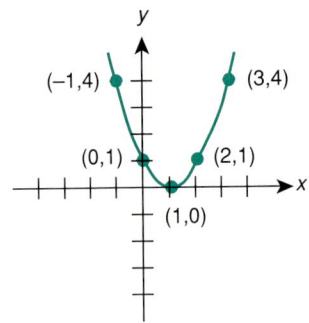

Figure 5.2

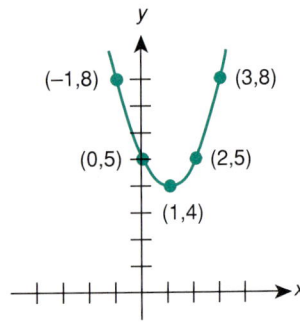

Figure 5.3

such solutions (roots) for a quadratic equation. For example, the quadratic equation $2x^2 + 5x - 3 = 0$ has solutions $x = \frac{1}{2}$ and $x = -3$ since both of these values make the equation true as is indicated below.

$$2x^2 + 5x - 3 = 0$$
$$2\left(\frac{1}{2}\right)^2 + 5\left(\frac{1}{2}\right) - 3 = 0$$
$$2\left(\frac{1}{4}\right) + \frac{5}{2} - 3 = 0$$
$$\frac{2}{4} + \frac{5}{2} - 3 = 0$$
$$\frac{1}{2} + \frac{5}{2} - 3 = 0$$
$$\frac{6}{2} - 3 = 0$$
$$3 - 3 = 0$$
$$0 = 0$$

$$2x^2 + 5x - 3 = 0$$
$$2(-3)^2 + 5(-3) - 3 = 0$$
$$2(9) + (-15) - 3 = 0$$
$$18 - 15 - 3 = 0$$
$$3 - 3 = 0$$
$$0 = 0.$$

Double Roots

Occasionally the roots are both the same value. In such cases they are called **double roots.** Examples 5.1, 5.2, and 5.3 illustrate the number of solutions that result when we solve quadratic equations by graphing.

• *Example 5.1:* Solve by graphing $x^2 - 2x - 3 = 0$.

Solution: The equation $y = x^2 - 2x - 3$ is graphed in figure 5.1. The parabola intersects the x-axis at $(-1,0)$ and $(3,0)$. Hence when $y = 0$, $x = -1$ or $x = 3$. Therefore, -1 and 3 are solutions to the quadratic $x^2 - 2x - 3 = 0$.

We can check these solutions by replacing x by -1 to get

$$x^2 - 2x - 3 = 0$$
$$(-1)^2 - 2(-1) - 3 = 0$$
$$1 + 2 - 3 = 0$$
$$0 = 0,$$

and replacing x by 3 to get

$$x^2 - 2x - 3 = 0$$
$$(3)^2 - 2(3) - 3 = 0$$
$$9 - 6 - 3 = 0$$
$$0 = 0.$$

Hence, both solutions check.

- **Example 5.2:** Solve by graphing $x^2 - 2x + 1 = 0$.

  Solution: The equation $y = x^2 - 2x + 1$ is graphed in figure 5.2. The parabola intersects the x-axis at only one point, namely (1,0). Hence, when $y = 0$, $x = 1$. Therefore, 1 is the only solution. It is a double root. We can check this solution by replacing x by 1 to get

$$x^2 - 2x + 1 = 0$$
$$(1)^2 - 2(1) + 1 = 0$$
$$1 - 2 + 1 = 0$$
$$0 = 0.$$

  Hence, the solution checks.

- **Example 5.3:** Solve by graphing $x^2 - 2x + 5 = 0$.

  Solution: The equation $y = x^2 - 2x + 5$ is graphed in figure 5.3. The parabola does not intersect the x-axis. Hence, there is no real number x for which this equation equals zero. Thus, there are no real number solutions to this equation. However, there are solutions to this equation that are not real numbers. Such solutions, which are called complex numbers, are discussed later.

Trial Problems 5.1

Before you begin the section exercises, warm up with these problems. Complete answers are included in the answer key.

Solve by graphing.

1. $y = x^2 - 5x + 6$
2. $y = x^2 - 4$
3. $y = x^2 + 6x + 9$
4. $y = x^2 + 6x + 10$
5. $y = x^2$

Exercises 5.1

Solve by graphing.

1. $y = x^2 - 4x + 3$
2. $y = x^2 - 2x - 3$
3. $y = -x^2 + 2x + 3$
4. $y = -x^2 - 2x + 3$
5. $y = x^2 - 1$
6. $y = x^2 - 4$
7. $y = \frac{1}{2}x^2$
8. $y = \frac{1}{3}x^2$
9. $y + x^2 = 0$
10. $y + 2x^2 = 0$

Solve the following exercises by graphing on an appropriate coordinate system.

11. When a ball rolls down an inclined plane, it travels a distance $d = 6t + \dfrac{t^2}{2}$ feet in t seconds. Graph this equation to illustrate how d depends on t by plotting time t along the horizontal axis. How long will it take the ball to travel 14 feet?

12. The current in a circuit flows according to the equation $i = 12 - 12t^2$, where i is the current and t is the time in seconds. Graph this equation to illustrate how i depends on t by plotting time t along the horizontal axis.

13. If a projectile is fired vertically into the air with an initial velocity of 96 ft/sec, the distance s in feet above the ground in t seconds is given by $s = 96t - 16t^2$. Graph this equation to illustrate how s depends on t by plotting time t along the horizontal axis. Determine the maximum height s and the value of t when the projectile strikes the ground.

14. An object is dropped from the top of the Empire State Building. The distance s that the object is from the ground in t seconds is given by the equation $s = 1250 - 16t^2$. Graph this equation to illustrate how s depends on t by plotting time t along the horizontal axis. What is the value of t when the object strikes the ground?

5•2 Solving Quadratic Equations by Factoring

In section 2.1 we considered special products and in section 2.2 we factored various polynomials. The skills we developed in those sections are useful when we wish to solve quadratic equations by factoring. Examples 5.4 through 5.9 illustrate how we solve quadratic equations by factoring.

• **Example 5.4:** Solve $x^2 + 6x - 27 = 0$ by factoring.

Solution: Since the left-hand side of this equation is a quadratic, the trial-and-error method of factoring yields

$$x^2 + 6x - 27 = 0$$
$$(x - 3)(x + 9) = 0.$$

The left-hand side is now a product. Since this product is equal to zero, we conclude that either $(x - 3) = 0$ or $(x + 9) = 0$. This is true, because in the real number system, when the product of two factors is zero either one or both of the factors must be zero. Thus, if $x - 3 = 0$, then $x = 3$. If $x + 9 = 0$, then $x = -9$. So the solutions are 3 and -9. We often put solutions within set braces such as $\{3, -9\}$ and refer to the set as a **solution set.**

Solution Set

Checking these results, we get

$$x^2 + 6x - 27 = 0$$
$$(3)^2 + 6(3) - 27 = 0$$
$$9 + 18 - 27 = 0$$
$$0 = 0,$$

and

$$x^2 + 6x - 27 = 0$$
$$(-9)^2 + 6(-9) - 27 = 0$$
$$81 - 54 - 27 = 0$$
$$0 = 0.$$

Therefore, both solutions check.

• **Example 5.5:** Find the solution set for $x^2 + 5x = -6$.

Standard Form

Solution: First we must write the equation so that the right-hand side is equal to zero. This is sometimes called **standard form**. Then we proceed as we did in example 5.4:

$$x^2 + 5x = -6$$
$$x^2 + 5x + 6 = 0 \quad \text{(Transpose (add 6))}$$
$$(x + 2)(x + 3) = 0 \quad \text{(Factor left-hand side)}$$

$x + 2 = 0$ or $x + 3 = 0$ (Set each factor equal to zero)

$x = -2$ $x = -3$. (Solve each equation for x)

Therefore the solution set is $\{-2, -3\}$. You should check the solutions.

• **Example 5.6:** Solve $x^2 - 8x + 16 = 0$ by factoring.

Solution:
$$x^2 - 8x + 16 = 0$$
$$(x - 4)(x - 4) = 0 \quad \text{(Factor left-hand side)}$$

$x - 4 = 0$ $x - 4 = 0$ (Set each factor equal to zero)

$x = 4$ $x = 4$.   (Transpose the 4)

Checking this solution, we get

$$x^2 - 8x + 16 = 0$$
$$(4)^2 - 8(4) + 16 = 0$$
$$16 - 32 + 16 = 0$$
$$0 = 0.$$

Therefore the only solution checks. It is a double solution, or a double root.

Section 5.2 Solving Quadratic Equations by Factoring

- **Example 5.7:** Solve $x^2 = 16$ by factoring.

  Solution:
  $$x^2 = 16$$
  $$x^2 - 16 = 0 \quad \text{(Transpose 16)}$$
  $$(x + 4)(x - 4) = 0 \quad \text{(Factor left-hand side)}$$
  $$x + 4 = 0 \quad \text{or} \quad x - 4 = 0 \quad \text{(Set each factor equal to zero)}$$
  $$x = -4 \qquad\qquad x = 4. \quad \text{(Solve for } x\text{)}$$

  The solution set is $\{-4, 4\}$.

Extracting Roots

Note that we could have solved this equation by **extracting roots** much like we developed in section 1.4. That is, since $x^2 = 16$, $x = \sqrt{16}$ or $x = -\sqrt{16}$, so $x = 4$ or $x = -4$.

- **Example 5.8:** Find the solution set by factoring $2a^2 = 5a$.

  Solution:
  $$2a^2 = 5a$$
  $$2a^2 - 5a = 0 \quad \text{(Transpose 5a)}$$
  $$a(2a - 5) = 0 \quad \text{(Factor left-hand side)}$$
  $$a = 0 \quad \text{or} \quad 2a - 5 = 0 \quad \text{(Set each factor equal to zero)}$$
  $$2a = 5 \quad \text{(Transpose 5)}$$
  $$a = \frac{5}{2}. \quad \text{(Solve for } a\text{)}$$

  The solution set is $\{0, \frac{5}{2}\}$.

  A word of caution may be appropriate at this point. Had we solved the equation $2a^2 = 5a$ by first dividing by a, we would have $2a = 5$ or $a = \frac{5}{2}$. Thus, we would have lost the solution $a = 0$. This happens since dividing by a is the same as dividing by zero in this case, and division by zero is not defined.

- **Example 5.9:** Solve by factoring $7x^2 - 28 = 0$.

  Solution:
  $$7x^2 - 28 = 0$$
  $$7(x^2 - 4) = 0 \quad \text{(Factor the common factor)}$$
  $$7(x + 2)(x - 2) = 0 \quad \text{(Factor the left-hand side)}$$
  $$x + 2 = 0 \quad \text{or} \quad x - 2 = 0 \quad \text{(Set each factor equal to zero)}$$
  $$x = -2 \qquad\qquad x = 2. \quad \text{(Solve for } x\text{)}$$

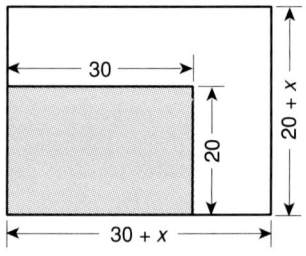

Figure 5.4

Although every quadratic equation has two solutions, occasionally one of those solutions has no meaning in real-life situations. Example 5.10 illustrates this point.

• *Example 5.10:* A rectangular parking lot is 30 units long and 20 units wide. An equal number of units are to be added to the length and to the width so that the area of the lot is doubled. Find the dimensions of the new parking lot.

Solution: It might be desirable to draw a diagram like that in figure 5.4. If we let x equal the number of units to be added to the length and to the width, the new length becomes $30 + x$ and the new width becomes $20 + x$. Since the original area of the lot was 20 units times 30 units, or 600 square units, the expanded lot must have an area of 1200 square units. Since the formula for area is length times width equals area we have

$$\begin{aligned} (\text{length})(\text{width}) &= \text{area} \\ (30 + x)(20 + x) &= 1200 \\ 600 + 50x + x^2 &= 1200 \quad \text{(Multiply left-hand side)} \\ x^2 + 50x + 600 - 1200 &= 0 \quad \text{(Transpose and commute)} \\ x^2 + 50x - 600 &= 0 \quad \text{(Subtract 1200)} \\ (x - 10)(x + 60) &= 0 \quad \text{(Factor left-hand side)} \\ x - 10 = 0 \quad x + 60 &= 0 \quad \text{(Set each factor equal to zero)} \\ x = 10 \quad x &= -60. \end{aligned}$$

Note that the length of the lot cannot be increased by -60 units. Therefore the only value of x that makes sense is $x = 10$. Thus, the new dimensions of the lot are $30 + 10$ or 40 units and $20 + 10$ or 30 units.

Trial Problems 5.2

Before you begin the section exercises, warm up with these problems. Complete answers are included in the answer key.

Factor these quadratics.

1. $x^2 - 5x + 6$
2. $x^2 + 8x + 16$
3. $6x^2 + x - 15$

Solve these quadratic equations by factoring.

4. $x^2 + 7x + 12 = 0$
5. $6x^2 - x - 15 = 0$

Exercises 5.2

Solve by factoring. Check exercises 1–10.

1. $x^2 - 7x + 12 = 0$
2. $x^2 + 6x - 16 = 0$
3. $a^2 + 14a + 49 = 0$
4. $y^2 + 16y + 64 = 0$
5. $a^2 - 11a = 12$
6. $b^2 - 14b = 15$
7. $t^2 - 32 = 4t$
8. $s^2 - 27 = 6s$
9. $r^2 = 36 - 9r$
10. $p^2 = 28 - 3p$

Find the solution set.

11. $2x^2 - 7x - 9 = 0$
12. $2y^2 - y - 3 = 0$
13. $6a^2 - 5a + 1 = 0$
14. $3b^2 + 10b - 8 = 0$
15. $r^2 = 25$
16. $s^2 = 49$
17. $2x^2 = 32$
18. $3y^2 = 27$
19. $5x^2 - 45 = 0$
20. $8y^2 - 18 = 0$
21. $x(x + 3) = -2$
22. $2y(2y + 6) = -8$
23. $2(x^2 - 6) = -5x$
24. $5(x^2 - 5) = 20x$

Solve.

25. Under certain conditions, the motion of an object suspended by a helical spring yields the equation $D^2 + 8D + 12 = 0$. Solve for D.

26. Work exercise 25 if the equation is $D^2 + 8D + 15 = 0$.

27. The length of a rectangle is 8 inches more than the width w. The area of this rectangle is 48 square inches. The relation between the length, width, and area is a quadratic equation. Determine the equation. Solve the equation to find the length and width.

28. The length of a rectangle is 13 inches more than the width w. The area of this rectangle is 48 square inches. The relation between the length, width, and area is a quadratic equation. Determine the equation. Solve the equation to find the length and width.

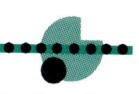

5 • 3 Solving Quadratic Equations by Completing the Square

Perfect Square

If a trinomial is the square of some binomial, then the trinomial is called a **perfect square.** For example, $x^2 + 2x + 1$ and $x^2 - 6x + 9$ are perfect squares because,

$$x^2 + 2x + 1 = (x + 1)^2,$$

and

$$x^2 - 6x + 9 = (x - 3)^2.$$

If one side of an equation is a perfect square involving the variable, and the other side of the equation is a constant, we can solve the equation by taking the square root of both sides of the equation. We did this in example 5.7 in section 5.2 and said that we had solved the equation by extracting roots. Example 5.11 illustrates this idea.

CHAPTER 5 SOLVING QUADRATIC EQUATIONS

• *Example 5.11:* Solve the equation $x^2 - 6x + 9 = 16$ by extracting roots.

Solution:
$x^2 - 6x + 9 = 16$
$(x - 3)^2 = 16$ (Express the left-hand side as a perfect square)
$(x - 3)^2 = 4^2$ (Express the right-hand side as a constant squared)
$x - 3 = \pm 4$ (Extract square roots)
$x - 3 = 4$ or $x - 3 = -4$ (Express as separate equations)
$x = 7$ or $x = -1$.   (Solve each equation for x)

It is possible to take a quadratic equation whose left-hand side is not a perfect square and change it into one that is. Then we can solve the resulting equation by extracting roots. The technique of solving quadratic equations in this manner is called **Completing the Square** completing the square. This technique is illustrated in examples 5.12, 5.13, and 5.14.

• *Example 5.12:* Solve $x^2 + 6x - 27 = 0$ by completing the square.

Solution: The left-hand side of this equation is not a perfect square, so we write the equation as

$$x^2 + 6x = 27. \quad \text{(Transpose (add 27))}$$

Notice that the coefficient of x^2 is 1 and the coefficient of x is 6.

To complete the square on the left-hand side, we take half of six, and get $\frac{6}{2}$ or 3, square it to get 3^2, then add 3^2 to both sides of the equation to get

$$x^2 + 6x + 3^2 = 27 + 3^2. \quad \text{(Add } 3^2 \text{ to both sides)}$$

Next we rewrite the left-hand member as the square of a binomial and simplify the right-hand side to get

$$(x + 3)^2 = 36. \quad \text{(Rewrite and simplify)}$$

As you see, the left-hand side is now a perfect square. If we take the square root of each side, we get

$$x + 3 = 6 \quad \text{or} \quad x + 3 = -6. \quad \text{(Extract square roots)}$$

Solving each of these equations gives us the solutions

$$x = 3 \quad \text{or} \quad x = -9. \quad \text{(Solve for } x\text{)}$$

You may have noticed that the equation $x^2 + 6x - 27 = 0$ could have been solved more quickly by factoring. We solved the equation by completing the square because we are interested in developing an alternate method of solution, not just a quick solution.

SECTION 5.3 SOLVING QUADRATIC EQUATIONS BY COMPLETING THE SQUARE   183

• *Example 5.13:*   Solve $4t^2 - 8t + 3 = 0$ by completing the square.

Solution: To get the left-hand side in the desired form, it is necessary to subtract 3 from both sides, and it is also necessary to divide both sides by 4. Hence, we get

$$4t^2 - 8t + 3 = 0$$
$$4t^2 - 8t = -3 \quad \text{(Transpose (subtract 3))}$$
$$t^2 - 2t = -\frac{3}{4}. \quad \text{(Divide both sides by 4)}$$

To complete the square on the left-hand side, we take half of 2 and get 1, square it to get 1^2, then add 1^2 to both sides of the equation to get

$$t^2 - 2t + 1^2 = -\frac{3}{4} + 1^2. \quad \text{(Add 1^2 to both sides)}$$

Next we rewrite the left-hand member as a square of a binomial and simplify the right-hand side:

$$(t - 1)^2 = \frac{1}{4}. \quad \text{(Rewrite and simplify)}$$

Extract the square roots and get

$$t - 1 = \frac{1}{2} \quad \text{or} \quad t - 1 = -\frac{1}{2}. \quad \text{(Extract square roots)}$$

Solving each of these equations gives us the solutions $t = \frac{3}{2}$ or $t = \frac{1}{2}$.

••••••••••

• *Example 5.14:*   Solve $x^2 - 3x - 1 = 0$ by completing the square.

Solution: Adding 1 to both sides of the equation gives

$$x^2 - 3x = 1. \quad \text{(Transpose (add 1))}$$

Next we take half of three to get $\frac{3}{2}$, square it to get $\left(\frac{3}{2}\right)^2$, then add it to both sides to get

$$x^2 - 3x + \left(\frac{3}{2}\right)^2 = 1 + \left(\frac{3}{2}\right)^2. \quad \text{(Add $\left(\frac{3}{2}\right)^2$ to both sides)}$$

Next we rewrite the left-hand member as the square of a binomial and simplify the right member:

$$\left(x - \frac{3}{2}\right)^2 = \frac{13}{4}.$$ (Rewrite and simplify)

Extract the square roots and get

$$x - \frac{3}{2} = \sqrt{\frac{13}{4}} \quad \text{or} \quad x - \frac{3}{2} = -\sqrt{\frac{13}{4}}.$$ (Extract roots)

Solving these two equations for x gives us the two solutions

$$x = \frac{3 + \sqrt{13}}{2} \quad \text{or} \quad x = \frac{3 - \sqrt{13}}{2}.$$

••••••••••

The basic idea behind this method of solution is to create the square of a simple binomial on the left-hand side of the equation. Since $(x + a)^2 = x^2 + 2ax + a^2$, we see that the square of half of the coefficient of x, namely $\left(\frac{2a}{2}\right)^2$ or a^2, added to $x^2 + 2ax$ gives $x^2 + 2ax + a^2$, which is a perfect square. We summarize this process as follows. Consider the equation

$$3x^2 - 5x - 1 = 0.$$

1. Divide both sides of the equation by the coefficient of x^2 (if it is not already 1):

$$\frac{3x^2}{3} - \frac{5x}{3} - \frac{1}{3} = \frac{0}{3}$$

$$x^2 - \frac{5x}{3} - \frac{1}{3} = 0.$$

2. Transpose the constant term to the right-hand side of the equation and all terms containing the variable to the left-hand side:

$$x^2 - \frac{5x}{3} = \frac{1}{3}.$$

3. Complete the square on the left-hand side by adding the square of one half the coefficient of the first degree term to both sides of the equation:

$$x^2 - \frac{5x}{3} + \left[\frac{1}{2}\left(\frac{5}{3}\right)\right]^2 = \frac{1}{3} + \left[\frac{1}{2}\left(\frac{5}{3}\right)\right]^2$$

$$x^2 - \frac{5}{3}x + \left(\frac{5}{6}\right)^2 = \frac{1}{3} + \left(\frac{5}{6}\right)^2.$$

Section 5.3 Solving Quadratic Equations by Completing the Square

4. Factor the left-hand side and simplify the right-hand side:

$$\left(x - \frac{5}{6}\right)^2 = \frac{37}{36}.$$

5. Extract the square root from both sides of the equation:

$$x - \frac{5}{6} = \pm\frac{\sqrt{37}}{6}.$$

6. Solve for x:

$$x = \frac{5}{6} \pm \frac{\sqrt{37}}{6}.$$

The roots of the equation $3x^2 - 5x - 1 = 0$ are

$$\frac{5 + \sqrt{37}}{6} \quad \text{and} \quad \frac{5 - \sqrt{37}}{6}.$$

We summarize this procedure then we use this technique in the following exercises.

Solving a Quadratic Equation by Completing the Square

1. Divide both sides of the equation by the coefficient of the second degree term.
2. Transpose the constant term to the right-hand side of the equation and all terms containing the variable to the left-hand side.
3. Complete the square on the left-hand side by adding the square of one half the coefficient of the first degree term to both sides of the equation.
4. Factor the left-hand side of the equation and simplify the right-hand side.
5. Extract the square root from both sides of the equation.
6. Solve the resulting equations for the variable.

Trial Problems 5.3

Before you begin the section exercises, warm up with these problems. Complete answers are included in the answer key.

Solve these quadratic equations by completing the square.

1. $x^2 - 5x + 6 = 0$
2. $x^2 - x - 20 = 0$
3. $2x^2 = 12 - 2x$
4. $3x^2 = -7x - 2$
5. $2x^2 = 3x$

Exercises 5.3

Solve by completing the square.

1. $x^2 + 4x - 12 = 0$
2. $x^2 + 3x + 2 = 0$
3. $x^2 + 4x + 4 = 0$
4. $x^2 + 9x + 20 = 0$
5. $x^2 - x - 6 = 0$
6. $x^2 - x - 20 = 0$
7. $2x^2 = 6 - 5x$
8. $2x^2 = 3 - 4x$
9. $3x^2 + x = 4$
10. $2x^2 - 3x = 5$
11. $4x^2 - 3 = 2x$
12. $5x^2 - 3 = 2x$

13. A metal bar is divided into two parts so that one part is 4 inches longer than the other part. If the sum of the squares of the two lengths is 208 square inches, find the two lengths.

14. One surface of a rectangular solid has a width w that is 8 millimeters shorter than the length l. If the area A of the surface is 105 square millimeters, find the length and the width ($A = lw$).

15. The length of a rectangular piece of metal is 8 centimeters more than the width. If the area A is 153 square centimeters, find the length and width ($A = lw$).

16. P dollars is invested at r percent compounded annually. At the end of two years it will grow to an amount $A = P(1 + r)^2$. What is the rate r if $500 grows to $577.81 in two years?

5•4 Solving Quadratic Equations with the Quadratic Formula

If we take the general quadratic equation and solve it by completing the square, we get a formula for solving any quadratic equation.

Take $ax^2 + bx + c = 0$. Subtract c from both sides and divide both sides of the equation by a to get

$$ax^2 + bx = -c \quad \text{(Transpose } c\text{)}$$

$$x^2 + \frac{bx}{a} = \frac{-c}{a}. \quad \text{(Divide both sides by } a\text{)}$$

To complete the square on the left-hand side, we take half of $\dfrac{b}{a}$ to get $\dfrac{b}{2a}$, square it to get $\left(\dfrac{b}{2a}\right)^2$, then add $\left(\dfrac{b}{2a}\right)^2$ to each side of the equation to get

$$x^2 + \frac{b}{a}x + \left(\frac{b}{2a}\right)^2 = \left(\frac{b}{2a}\right)^2 - \frac{c}{a}. \quad \left(\text{Add } \left(\frac{b}{2a}\right)^2 \text{ to both sides}\right)$$

Rewrite the left-hand member of the equation as the square of a binomial:

$$\left(x + \frac{b}{2a}\right)^2 = \left(\frac{b}{2a}\right)^2 - \frac{c}{a}. \quad \text{(Rewrite left-hand side)}$$

Expand and simplify the right-hand side of the equation:

$$\left(x + \frac{b}{2a}\right)^2 = \frac{b^2 - 4ac}{4a^2}. \quad \text{(Expand and simplify right-hand side)}$$

Section 5.4 Solving Quadratic Equations with the Quadratic Formula

Next we extract the square roots and get

$$x + \frac{b}{2a} = +\frac{\sqrt{b^2 - 4ac}}{2a} \quad \text{or} \quad x + \frac{b}{2a} = -\frac{\sqrt{b^2 - 4ac}}{2a}. \quad \text{(Extract square roots)}$$

Solving these two equations for x gives us two solutions for the general quadratic equation:

$$x = \frac{-b + \sqrt{b^2 - 4ac}}{2a} \quad \text{or} \quad x = \frac{-b - \sqrt{b^2 - 4ac}}{2a}. \quad \text{(Solve for } x\text{)}$$

We often combine these two equations into one expression,

$$x = \frac{-b \pm \sqrt{b^2 - 4ac}}{2a},$$

Quadratic Formula and call this equation the **quadratic formula**. This is a formula for the roots of a quadratic equation expressed in terms of the coefficients a, b, and c. We summarize the procedure for solving quadratic equations with the quadratic formula, then we use the procedure in examples 5.15–5.18.

Solving a Quadratic Equation Using the Quadratic Formula

1. If the equation has fractional coefficients, multiply both sides of the equation by the common denominator.
2. If necessary, transpose terms to get the equation in the form $ax^2 + bx + c = 0$.
3. Write the quadratic formula.
4. Write the quadratic formula leaving spaces for the values of a, b, and c.
5. Substitute the proper values for a, b, and c.
6. Calculate, simplify, and extract roots.
7. Solve the resulting equations for the variable.
8. Check the results for accuracy and reasonableness.

• **Example 5.15:** Solve $x^2 + 6x - 27 = 0$ using the quadratic formula.

Solution: Write the quadratic formula. Next, write the formula again, leaving places for the values of a, b, and c. Then substitute 1 for a, 6 for b, and -27 for c. This procedure is desirable because it gives you practice writing the formula and reduces the chances for errors.

$$x = \frac{-b \pm \sqrt{b^2 - 4ac}}{2a}$$ (Write the quadratic formula)

$$x = \frac{-(\ ) \pm \sqrt{(\ )^2 - 4(\ )(\ )}}{2(\ )}$$ Leave spaces for the values of a, b, and c

$$x = \frac{-(6) \pm \sqrt{(6)^2 - 4(1)(-27)}}{2(1)}.$$ (Substitute values for a, b, and c)

Simplifying, we get

$$x = \frac{-6 \pm \sqrt{36 + 108}}{2}$$ (Multiply)

$$x = \frac{-6 \pm \sqrt{144}}{2}$$ (Add)

$$x = \frac{-6 \pm 12}{2}$$ (Extract square roots)

$$x = \frac{-6 + 12}{2} \quad \text{or} \quad x = \frac{-6 - 12}{2}$$ (Express as separate equations)

$$x = 3 \qquad\qquad\qquad x = -9.$$ (Solve for x)

These are the same solutions we got when we solved the equation $x^2 + 6x - 27 = 0$ by factoring in example 5.4 and by completing the square in example 5.12. Of course, there is little point in using the quadratic formula to solve an equation that could be readily factored. The quadratic formula is valuable when we can't conveniently solve the equation by some other means. Example 5.16 illustrates this situation.

• **Example 5.16:** Solve $2x^2 - x - 2 = 0$ using the quadratic formula.

Solution: Using the procedure we developed in the previous example, we get

$$x = \frac{-b \pm \sqrt{b^2 - 4ac}}{2a}$$ (Write the quadratic equation)

$$x = \frac{-(\ ) \pm \sqrt{(\ )^2 - 4(\ )(\ )}}{2(\ )}$$ (Leave spaces for the values of a, b, and c)

$$x = \frac{-(-1) \pm \sqrt{(-1)^2 - 4(2)(-2)}}{2(2)}$$ (Substitute values for a, b, and c)

$$x = \frac{1 \pm \sqrt{1 + 16}}{4}$$ (Multiply)

$$x = \frac{1 \pm \sqrt{17}}{4}.$$ (Add)

SECTION 5.4 SOLVING QUADRATIC EQUATIONS WITH THE QUADRATIC FORMULA

Thus,

$$x = \frac{1 + \sqrt{17}}{4} \quad \text{or} \quad x = \frac{1 - \sqrt{17}}{4}. \quad \text{(Extract roots)}$$

It would have been quite difficult for us to find these solutions so quickly by any other means.

• • • • • • • • • •

Example 5.17 can be solved easily by factoring, but we use the quadratic formula to illustrate an important point.

• **Example 5.17:** Solve $x^2 - 6x + 9 = 0$ using the quadratic formula.

Solution: Using the procedure suggested in examples 5.15 and 5.16, we have

$$x = \frac{-b \pm \sqrt{b^2 - 4ac}}{2a} \quad \text{(Write the quadratic equation)}$$

$$x = \frac{-(\;) \pm \sqrt{(\;)^2 - 4(\;)(\;)}}{2(\;)} \quad \text{(Leave spaces for the values of } a, b, \text{ and } c\text{)}$$

$$x = \frac{-(-6) \pm \sqrt{(-6)^2 - 4(1)(9)}}{2(1)} \quad \text{(Substitute values for } a, b, \text{ and } c\text{)}$$

Simplifying, we get

$$x = \frac{6 \pm \sqrt{36 - 36}}{2} \quad \text{(Multiply)}$$

$$x = \frac{6 \pm \sqrt{0}}{2} \quad \text{(Extract square roots)}$$

$$x = \frac{6 + 0}{2} \quad \text{or} \quad x = \frac{6 - 0}{2} \quad \text{(Express as separate equations)}$$

$$x = 3 \quad \text{or} \quad x = 3. \quad \text{(Solve for } x\text{)}$$

Since the value under the radical is zero, both roots are equal. We say that we have two equal solutions, or simply one solution. Such a solution is also called a *double root*.

• • • • • • • • • •

Discriminant In the quadratic formula, the expression under the radical, $b^2 - 4ac$, is called the **discriminant**. The value of the discriminant affects the nature of the solutions in the following ways:

1. If $b^2 - 4ac = 0$, there is one real solution.

2. If $b^2 - 4ac > 0$, there are two unequal real solutions.

3. If $b^2 - 4ac < 0$, there are two nonreal solutions. (This case is discussed in later chapters).

If the quadratic equation that we wish to solve has any fractional coefficients, it is usually desirable to eliminate the fractions first. By eliminating the fractional coefficients, the calculations usually become easier. This is illustrated in example 5.18.

• **Example 5.18:** Solve $\dfrac{x^2}{4} + x = \dfrac{5}{4}$ using the quadratic formula.

Solution: To eliminate the fractional coefficients, we must multiply both sides of the equation by 4, then simplify:

$$4\left(\dfrac{x^2}{4} + x\right) = 4\left(\dfrac{5}{4}\right) \quad \text{(Multiply both sides by 4)}$$

$$x^2 + 4x = 5 \quad \text{(Simplify)}$$

$$x^2 + 4x - 5 = 0. \quad \text{(Transpose (subtract 5))}$$

Now that we have eliminated the fractional coefficients, we can solve the equation using the quadratic formula or any other convenient method. We use the quadratic formula.

$$x = \dfrac{-b \pm \sqrt{b^2 - 4ac}}{2a} \quad \text{(Write the quadratic equation)}$$

$$x = \dfrac{-(\ ) \pm \sqrt{(\ )^2 - 4(\ )(\ )}}{2(\ )} \quad \text{(Leave spaces for the values of } a, b, \text{ and } c)$$

$$x = \dfrac{-(4) \pm \sqrt{(4)^2 - 4(1)(-5)}}{2(1)} \quad \text{(Substitute values for } a, b, \text{ and } c)$$

$$x = \dfrac{-4 \pm \sqrt{16 + 20}}{2} \quad \text{(Multiply)}$$

$$x = \dfrac{-4 \pm \sqrt{36}}{2} \quad \text{(Add)}$$

$$x = \dfrac{-4 \pm 6}{2} \quad \text{(Extract roots)}$$

$$x = \dfrac{-4 + 6}{2} \quad \text{or} \quad x = \dfrac{-4 - 6}{2} \quad \text{(Express as separate equations)}$$

$$x = 1 \quad\quad\quad x = -5. \quad \text{(Solve for } x)$$

Trial Problems 5.4

Before you begin the section exercises, warm up with these problems. Complete answers are included in the answer key.

Identify the values of a, b, and c in exercises 1, 2, and 3.

1. $4x^2 + 3x + 5 = 0$
2. $3x^2 - 7x = 0$
3. $\dfrac{x^2}{2} - 8 = 0$

4. Find the value of the discriminant for the quadratic equation $3x^2 + 2x - 5 = 0$.

5. Use the quadratic formula to solve the equation $3x^2 - 5x = -1$.

Exercises 5.4

Solve using the quadratic formula.

1. $x^2 - 5x + 4 = 0$
2. $x^2 + 3x - 4 = 0$
3. $y^2 - 5y = 6$
4. $y^2 - 3y = -2$
5. $z^2 = 3z - 1$
6. $2z^2 = 7z - 6$
7. $x^2 - \dfrac{5}{3}x = -\dfrac{1}{3}$
8. $x^2 - \dfrac{1}{2}x = -\dfrac{1}{2}$
9. $r^2 - 5r = 0$
10. $r^2 + 3r = 0$
11. $t^2 - 10t = -25$
12. $t^2 - 14t = -49$

Solve using the quadratic formula and your calculator.

13. $3.41x^2 - 2.57x - 0.23 = 0$
14. $2.05x^2 + 3.56x + 0.94 = 0$
15. $0.092x^2 - 5.77x + 2.91 = 0$
16. $0.35x^2 + 0.12x + 0.011 = 0$

17. Under certain conditions, the distance s that a projectile is above the ground is given by $s = 96t - 16t^2$, where t is the time in seconds. After what amount of time is the object 144 feet above the ground?

18. Under certain conditions, the distance s that a projectile is above the ground is given by $s = 64t - 16t^2$, where t is the time in seconds. After what amount of time is the object 64 feet above the ground?

19. The power P developed in a particular electric circuit is
$$P = EI - RI^2,$$
where E is the voltage and R the resistance. Assuming P, E, and R are constants, solve this equation for I.

20. The current m in an alternating electric circuit is related to an inductance L, a resistance R, and a capacitor C. This relation can be expressed by the equation
$$LCm^2 + RCm + 1 = 0.$$
Solve this equation for m.

Conditional Inequalities

5•5 Solving Quadratic Inequalities

An inequality that involves a variable may be true for some values of the variable and false for others. Such inequalities are called **conditional inequalities**. To solve a conditional inequality, we must find the set of values for which the inequality is true.

To solve a quadratic inequality, we use many of the techniques that we have learned in previous chapters. We also have to develop some new strategies that are unique to this section. Examples 5.19–5.22 illustrate some of these strategies.

• *Example 5.19:* For what values of x is

$$x^2 - x - 12 > 0?$$

Solution: First we factor the left hand side of this inequality, which is a quadratic:

$$x^2 - x - 12 > 0$$
$$(x - 4)(x + 3) > 0.$$

Now we must find values of x for which the product of these two factors is greater than zero; that is, the product must be positive. In order for the product of two factors to be positive, both factors must be of the same sign; that is, either both must be positive, or both must be negative.

Figure 5.5 illustrates a device that is very useful for determining the values that satisfy the inequality. On the top real number line we identify those values that make the first factor $(x - 4)$ equal to zero, those values that make $(x - 4)$ positive, and

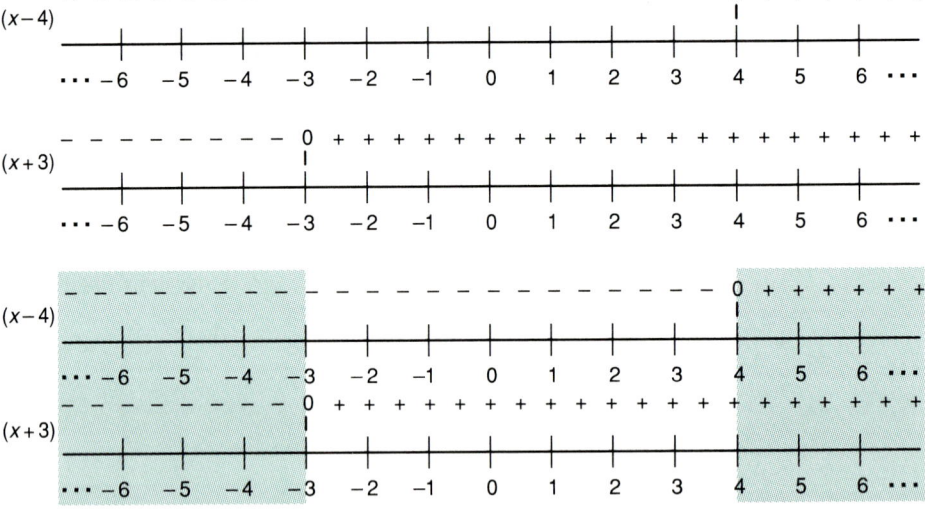

Figure 5.5

those that make it negative. On the next real number line we identify those values that make the second factor $(x + 3)$ zero, positive, or negative.

When the two number lines are combined, it is apparent from the shaded regions that both factors are positive when $x > 4$ and both factors are negative when $x < -3$. Thus, the solution to the inequality is $x > 4$ or $x < -3$.

• *Example 5.20:* For what values of x is

$$x^2 - x - 12 < 0?$$

Solution: Let us use the technique that we developed in example 5.19. First, factor the quadratic

$$x^2 - x - 12 < 0$$
$$(x - 4)(x + 3) < 0.$$

Next, indicate on real number lines those values that make each factor negative, zero, or positive. Figure 5.6 illustrates these values.

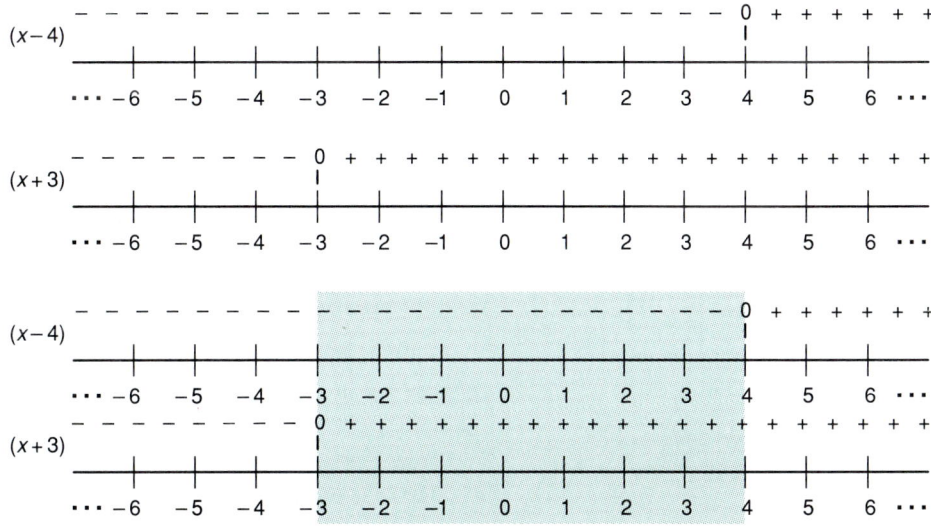

Figure 5.6

In order for the product of two factors to be less than zero (negative), one of the factors must be positive and the other must be negative. By studying figure 5.6 we see that this happens when x is between -3, and 4 (shaded region). Therefore, we write the solution as $-3 < x < 4$.

• *Example 5.21:* For what values of x is
$$x^2 - x \le 2?$$

Solution: We must add -2 to both members of this inequality so that the left-hand member is in general quadratic form and the right-hand member is zero. Now we have
$$x^2 - x - 2 \le 0.$$

Next we factor the left-hand member, then place each factor on a real number line to determine the values that make the factor positive, negative, or zero. The factored form is
$$(x - 2)(x + 1) \le 0.$$

Figure 5.7 illustrates these values.

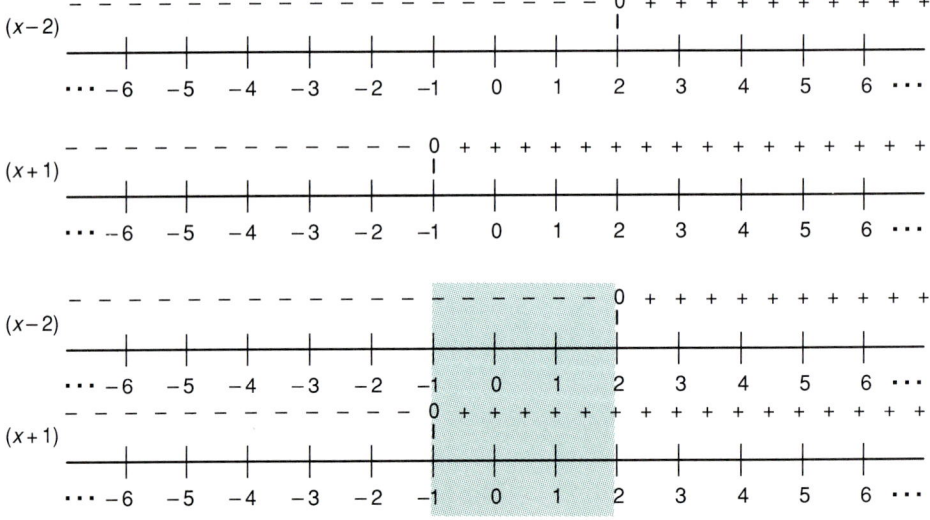

Figure 5.7

In order for the product to be less than zero, the factors must have opposite signs. In order for the product to equal zero, at least one of the factors must be zero. Thus, we write the solution as
$$-1 \le x \le 2.$$

Notice that the equality indicates that the endpoints of this interval are also solutions.

• *Example 5.22:* For what values of x is
$$-x^2 - x + 12 > 0?$$

Solution: Factoring this inequality we get
$$(-x - 4)(x - 3) > 0.$$

Next indicate on real number lines those values that make each factor negative, zero, or positive. Figure 5.8 illustrates these values.

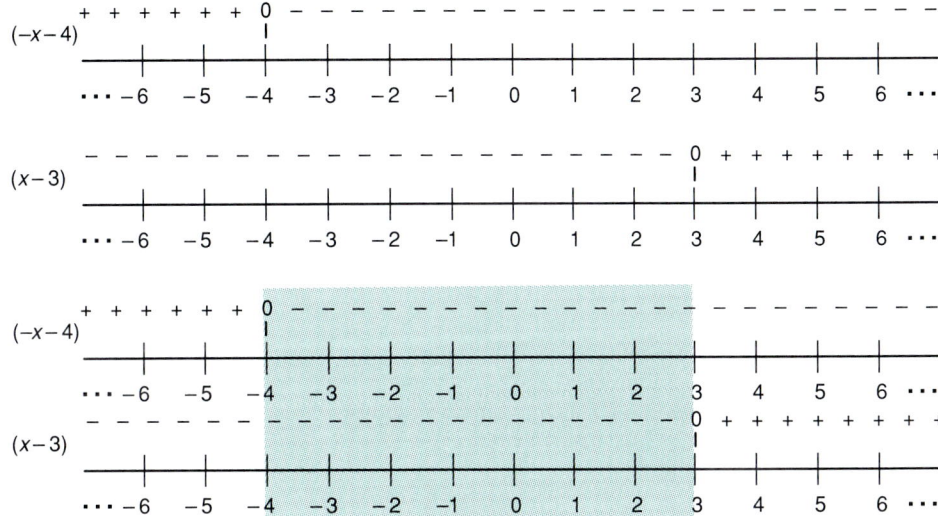

Figure 5.8

Notice that the values that make the factor $(-x - 4)$ negative are to the right, while those values that make the factor positive are to the left. In order for the product to be greater than zero, the factors must have the same sign. By studying figure 5.8 we see that this happens when x is between -4 and 3 (shaded region). Therefore we write the solution as $-4 < x < 3$.

We could also have solved this inequality by first multiplying both sides of the inequality by -1. This would reverse the sense of the inequality and the example would be much like the previous examples.

Trial Problems 5.5

Before you begin the section exercises, warm up with these problems. Complete answers are included in the answer key.

1. For what values of x is $x - 3$ negative?
2. For what value of x is $2x + 5$ equal to zero?
3. For what values of x is $-x + 4$ positive?
4. For what values of x is the product $(x + 3)(x + 2)$ negative?
5. For what values of x is the quadratic inequality $x^2 - x - 6 < 0$ true?

Exercises 5.5

Solve the inequalities.

1. $x^2 + 2x - 15 < 0$
2. $x^2 - x - 6 < 0$
3. $x^2 - 9x + 20 > 0$
4. $x^2 + 9x + 20 > 0$
5. $2x^2 + 9x + 9 < 0$
6. $3x^2 + 8x + 4 < 0$
7. $-x^2 - x + 12 > 0$
8. $-x^2 + x + 12 > 0$
9. $x^2 + 3x < 5(x + 3)$
10. $x^2 + 7x < 6(x + 1)$
11. $x^2 - 8x + 23 \leq 8$
12. $4x^2 - 13x + 5 < 2$
13. $x - \dfrac{8}{x} \leq 7$
14. $x + \dfrac{15}{x} \leq 8$

15. The force F, in newtons, acting on a certain object varies according to the time t, in seconds, and is given by the formula $F = 2t^2 - 5t - 1$. Find the positive values of t for which the force is greater than 4 newtons.

16. A skylight designed for a particular location must have a length that is 30 cm more than its width. For what values of the width will the area be greater than 2800 cm²?

5·6 Summary of Terms, Formulas, Rules, and Procedures

Terms

Completing the Square (p. 182)
Conditional Inequalities (p. 192)
Discriminant (p. 189)
Double Roots (p. 175)

Extracting Roots (p. 179)
Perfect Square (p. 181)
Quadratic Formula (p. 187)
Roots (p. 174)

Solution Set (p. 177)
Standard Form (p. 178)
Zeros (p. 174)

Formulas

The Quadratic Formula: $x = \dfrac{-b \pm \sqrt{b^2 - 4ac}}{2a}$ (5.4)

Rules and Procedures

- Procedure for solving a quadratic equation by completing the square. (5.3)
  1. Divide both sides of the equation by the coefficient of the second degree term.
  2. Transpose the constant term to the right-hand side of the equation and all terms containing the variable to the left-hand side.
  3. Complete the square on the left-hand side by adding the square of one half the coefficient of the first degree term to both sides of the equation.
  4. Factor the left-hand side of the equation and simplify the right-hand side.
  5. Extract the square root from both sides of the equation.
  6. Solve the resulting equations for the variable.

- Procedure for solving quadratic equations using the quadratic formula. (5.4)
  1. If the equation has fractional coefficients, multiply both sides of the equation by the common denominator.
  2. If necessary, transpose terms to get the equation in the form $ax^2 + bx + c = 0$.
  3. Write the quadratic formula.
  4. Write the quadratic formula leaving spaces for the values of a, b, and c.
  5. Substitute the proper values for a, b, and c.
  6. Calculate, simplify, and extract roots.
  7. Solve the resulting equations for the variable.
  8. Check the results for accuracy and reasonableness.

5·7 Chapter 5 Review Exercises

Solve by graphing.

1. $y = x^2 - 4x - 5$
2. $y = x^2 + 4x - 5$
3. $y = -x^2 - 4x + 5$
4. $y = -x^2 + 4x + 5$
5. $y = x^2 - 16$
6. $y = x^2 - 9$

Solve by factoring.

7. $x^2 - 8x + 12 = 0$
8. $x^2 + 6x - 16 = 0$
9. $a^2 + 10a + 25 = 0$
10. $b^2 + 16b + 64 = 0$
11. $t^2 - 27 = 6t$
12. $s^2 - 32 = 4s$

Solve by completing the square.

13. $x^2 - 7x + 12 = 0$
14. $x^2 - 3x - 4 = 0$
15. $x^2 + 3x - 1 = 0$
16. $x^2 + 4x - 2 = 0$
17. $5x^2 - 3 = 2x$
18. $4x^2 - 3 = 2x$

Solve with the quadratic formula.

19. $x^2 - 6x + 5 = 0$
20. $x^2 - 5x + 6 = 0$
21. $x^2 + 5x + 5 = 0$
22. $x^2 + 6x + 7 = 0$
23. $3x^2 - 18x + 15 = 0$
24. $4x^2 - 20x + 24 = 0$

Solve the inequalities.

25. $x^2 + 2x - 16 < 0$
26. $x^2 - 8x + 12 < 0$
27. $a^2 - a - 6 > 0$
28. $b^2 + 2b - 15 > 0$
29. $x^2 - 9x + 24 \leq 4$
30. $x^2 + 9x + 16 \leq -4$

31. A projectile is fired upward over level ground with an initial velocity such that the distance s (in feet) of the object above the ground is given by the formula

$$s = 48t - 16t^2,$$

where t is the time in seconds. For what period of time is the projectile above the ground?

32. The electrical resistance R (in ohms) of a certain piece of material depends on the temperature T according to the relation

$$R = -75 + 10T + T^2.$$

For what values of T ($T > 0°$ C) is the resistance greater than 125 ohms?

5·8 Chapter 5 Test

1. Solve $x^2 + x - 6 = 0$ by graphing.
2. Solve $x^2 = 16$ by graphing.
3. Solve $2x^2 - 7x - 9 = 0$ by factoring.
4. Solve $a^2 = 49$ by factoring.
5. Solve $x^2 + 6x - 27 = 0$ by completing the square.
6. Solve $5x^2 - 1 = 2x$ by completing the square.
7. Solve $2z^2 - 7z = 6$ with the quadratic formula.
8. Solve $x^2 - \dfrac{5}{3}x = -\dfrac{1}{3}$ with the quadratic formula.
9. Solve the quadratic inequality $x^2 - x > 2$.
10. Solve the quadratic inequality $-x^2 + x + 12 > 0$.
11. A projectile is fired upward over level ground with an initial velocity such that the distance s (in feet) of the object above the ground is given by the formula $s = 32t - 16t^2$, where t is the time in seconds. For what period of time is this projectile above the ground?
12. The force F, in newtons, acting upon a certain object varies according to the time t, in seconds, and is given by the formula $F = 2t^2 - 5t - 1$. Find the positive values for t for which the force is greater than 6.

5·9 Cumulative Review—Chapters 3, 4, and 5

1. Draw the rectangular coordinate system and plot the ordered pairs $(2,3)$, $(4,-5)$, $(-3,6)$, $(0,5)$, $(-4,0)$, and $(0,0)$.
2. Volts E, amperes I, and resistance R are related by the formula

$$E = IR.$$

Graph the relation when $R = 40$ and $E = 220, 200, 140, 120,$ and 100.

3. Graph the equation $4x - 3y = 12$.
4. Graph the equation $3x + 2y = 0$.
5. Graph the equation $y = x^2 - 2x - 3$.
6. Graph the equation $y = -2x^2 + 1$.
7. Graph the relation shown in this table.

x	0	10	20	30	40	50
y	800	500	330	250	200	167

8. Graph the relation determined by the equation

$$y = \frac{10}{x+1} \text{ for } x = 0, 1, 2, 3, 4, \text{ and } 5.$$

9. Express $y = 3x^2 - 2$ using functional notation.
10. If $f(s) = s^3$, find $f(4)$.
11. If $f(x) = x^2 - 4x + 5$, find $f(x + h)$.
12. Graph the system

$$x + y = 3$$
$$3x - y = 1.$$

Determine a common point if one exists.

13. Determine the appropriate break-even point for a cost equation $y = 500 + 3x$ and a revenue equation $y = 4x$.

14. Solve the system by substitution.
$$y = x - 1$$
$$x - y = 5.$$

15. The perimeter of a rectangle is 24 cm and the length is 2 cm longer than the width. Find the length and the width of the rectangle.

16. Given the demand equation $y = 10p + 800$ and the supply equation $y - 5p = 900$. Solve by addition or subtraction to find the values of y and p that will produce market equilibrium.

17. Solve with determinants.
$$2x - 3y = 1$$
$$3x + 5y = 11.$$

18. Solve using Cramer's rule.
$$1.5x + 2.1y = 9.3$$
$$4.5x + 6.3y = 27.9.$$

19. Evaluate the determinant.
$$\begin{vmatrix} 4 & -3 & 1 \\ 2 & 1 & -2 \\ 3 & 0 & -1 \end{vmatrix}.$$

20. Solve using Cramer's rule.
$$3x + y + z = -1$$
$$4y + 2z = x$$
$$2x + y = 1 - 2z.$$

21. Solve $y = x^2 + 4x + 3$ by graphing.

22. Solve $8x^2 - 32 = 0$ by factoring.

23. Find the solution set for $2y^2 + y = 3$.

24. Solve $4x^2 - 3 = 2x$ by completing the square.

25. Solve $2z^2 = 7z - 6$ using the quadratic formula.

26. Solve the quadratic inequality $3x^2 - 8x + 4 > 0$.

27. A certain rectangular opening must have a length that is 60 cm greater than its width. For what values of the width will the area of the rectangular opening be greater than 1600 cm²?

Chapter 6
Numerical Trigonometry

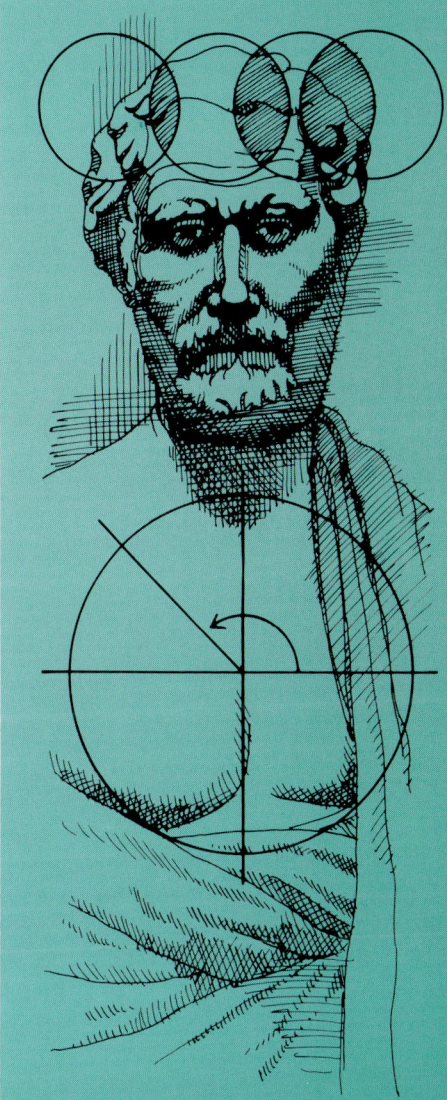

Numerical trigonometry can be traced to obscure beginnings as early as 1900 B.C. An ancient Babylonian cuneiform tablet contains a table that is equivalent to a modern trigonometry table. The early Greek astronomers used celestial observations to calculate interplanetary distances, and they may have been the first to divide a circle into 360 degrees around the second century B.C.

Some historians refer to the Greek astronomer Hipparchus as the "Father of Trigonometry." He has been credited with constructing his *Table of Chords,* which a successor Claudius Ptolemy used to construct a table that gives the lengths of the chords of all central angles of a circle in half-degree intervals from 0.5° to 180°. The table is equivalent to a modern table of sines.

In this chapter the trigonometry of angles and triangles is considered. Later chapters consider the trigonometric functions using the set of real numbers as the domain of the functions.

We assume that the students have access to a scientific calculator, and their use is encouraged in this and in succeeding chapters.

6•1 Angles and Triangles

Angle An **angle** is defined as the union of two distinct rays having a common endpoint. A ray $\overrightarrow{MN}$ is the set of points on the half-line having M as the endpoint and that contains point N. Thus, angle BAC ($\measuredangle BAC$) in figure 6.1 consists of the set of points that are on ray AB or on ray AC.

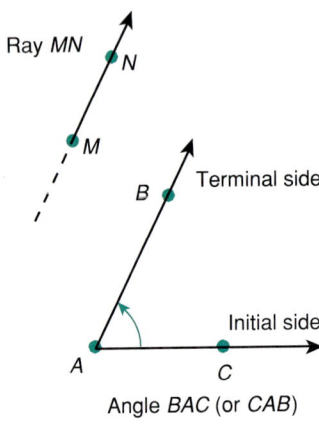

Angle BAC (or CAB)

Figure 6.1

We measure an angle in degrees based on a circle having 360° (° = degrees). Thus, if a circle is divided into 360 equal parts, each part represents one degree.

The angles shown in figure 6.2a are angles that measure 90°, 45°, and 135°.

Angle in Standard Position When an angle is represented on a coordinate system with the vertex at the origin and the initial side of the angle as the positive horizontal axis, it is said to be an **angle in standard position**. These angles have positive measure when measured in a counterclockwise direction and negative measure when measured in a clockwise direction. Some angles in standard position are shown in figure 6.2b.

Just as time is measured in hours, minutes, and seconds, angles are measured in degrees, minutes, and seconds. A minute is $\frac{1}{60}$ of a degree, and a second is $\frac{1}{60}$ of a minute.

An angle measuring 36 degrees, 45 minutes, and 15 seconds is written as

$$36°45'15''. \quad ('=\text{minutes},\ ''=\text{seconds})$$

Since we depend on scientific calculators to determine trigonometric functions of angles, we need to change the measures of angles from this form to decimal form. The following relationships are useful.

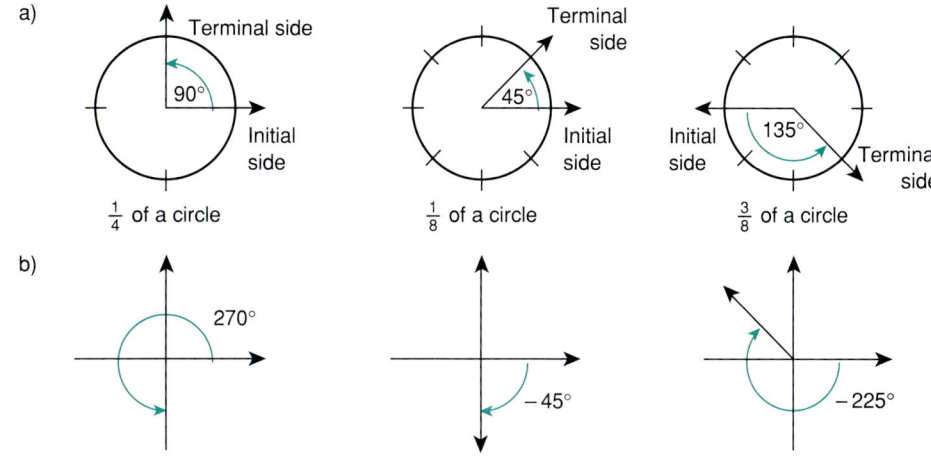

Figure 6.2

$$1 \text{ minute} = \frac{1}{60} \text{ degrees}$$

$$1 \text{ second} = \frac{1}{3600} \text{ degrees} = \frac{1}{60} \text{ minute}$$

$$k \text{ minutes} = \frac{k}{60} \text{ degrees}$$

$$k \text{ seconds} = \frac{k}{3600} \text{ degrees}$$

To change the angle measuring $36°45'15''$ to decimal form, we proceed as follows:

$$15'' = \left(\frac{15}{3600}\right)° = 0.00417°$$

$$45' = \left(\frac{45}{60}\right)° = 0.75°$$

$$36°45'15'' = (36 + 0.75 + 0.00417)° = 36.75417°.$$

15 ÷ 60 + 45 = ÷ 60 + 36 = .

Display: 36.754167

Many calculators have buttons that allow entering the angle in degrees, minutes, and seconds, making these calculations unnecessary.

For example, on a Texas Instruments model TI-55III calculator, changing 36°45′15″ to degrees is accomplished by the following sequence of steps:

$$36.4515 \boxed{\text{2ND}} \boxed{\text{DMS/DD}}.$$

The display shows 36.754167.

Conversely, if we wish, for example, to change 52.8714° to degrees, minutes, and seconds, we use the following sequence:

$$52.8714 \boxed{\text{INV}} \boxed{\text{2ND}} \boxed{\text{DMS/DD}}.$$

The display shows 52.521704. The first 52 indicates the number of degrees. The next 52 indicates the number of minutes and the 1704 indicates that there are 17.04 seconds. Thus, 52.8714° = 52°52′17.04″.

Coterminal Angles
Angles that have the same initial and terminal sides are called **coterminal angles**. Figure 6.3 shows three angles in standard position that are coterminal.

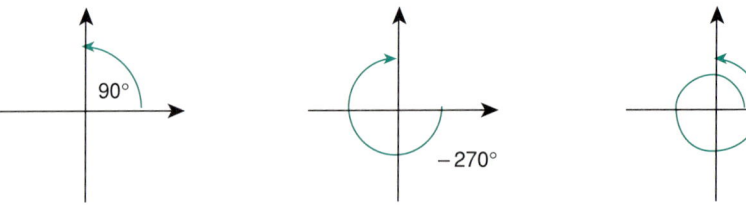

Figure 6.3

Right Angle
An angle that measures 90° is called a **right angle**. The small square inserted at the vertex of the angle is the conventional way of indicating that the angle is a right angle. This is illustrated in figure 6.4.

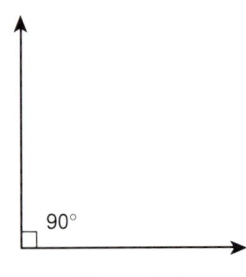

Figure 6.4

Two other ideas that are useful in the study of the trigonometry of angles and triangles are the ideas of complementary angles and supplementary angles.

Complementary Angles
If the sum of the measures of two angles equals 90°, the angles are called **complementary angles**. Two pairs of complementary angles are shown in figure 6.5.

Supplementary Angles   If the sum of the measures of two angles equals 180°, the angles are called **supplementary angles**. The third pair of angles in figure 6.5 is supplementary.

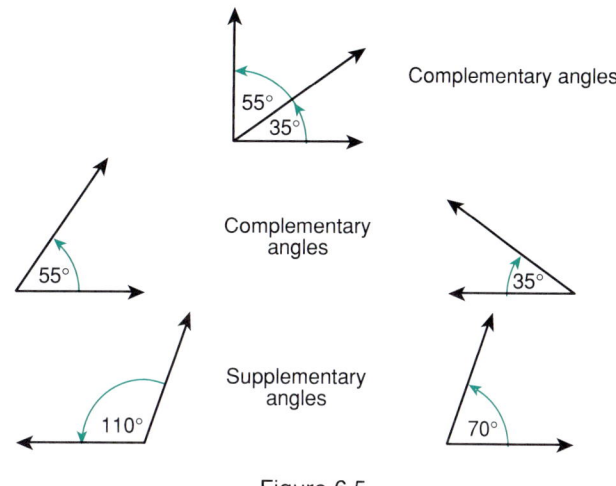

Figure 6.5

Since we are studying the trigonometry of triangles, some definitions are necessary.

> Given three noncollinear points *A*, *B*, and *C*, triangle *ABC* ($\triangle ABC$) is the union of segments *AB*, *BC*, and *CA*.

For convenience, the length of the side opposite $\angle A$ is designated as *a*, the length of the side opposite $\angle B$ is designated as *b*, and the length of the side opposite $\angle C$ is designated as *c*. Figure 6.6 illustrates a typical triangle with its sides and angles properly labeled.

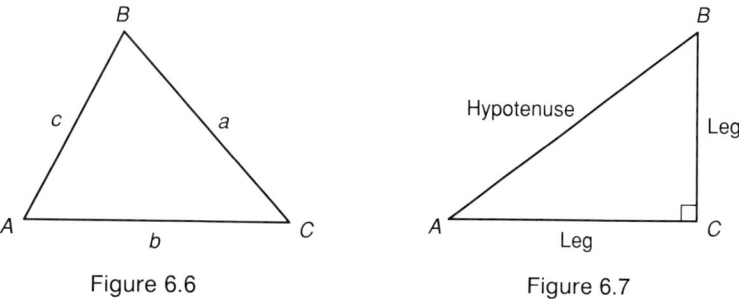

Figure 6.6   Figure 6.7

Right Triangle   For a **right triangle** (a triangle that has a right angle), we call the side opposite
Hypotenuse   the right angle the **hypotenuse**. The other two sides are called the **legs** of the triangle.
Legs   Figure 6.7 illustrates a right triangle.

One of the relationships for right triangles that is useful in solving problems with
Theorem of Pythagoras   right triangles is the **Theorem of Pythagoras**.

Theorem of Pythagoras

In the right triangle the square of the hypotenuse is equal to the sum of the squares of the legs. For $\triangle ABC$ in figure 6.8,

$$c^2 = a^2 + b^2 \quad \text{or} \quad c = \sqrt{a^2 + b^2}.$$

Given a pair of sides of a right triangle, we can use this theorem to determine the length of the other side.

• *Example 6.1:* Given $a = 11.0$ and $b = 12.0$ in figure 6.8, determine c, the length of the hypotenuse.

Solution: Use the Pythagorean theorem.

$$c^2 = a^2 + b^2$$
$$c^2 = (11)^2 + (12)^2$$
$$c^2 = 121 + 144$$
$$c^2 = 265$$
$$c = \sqrt{265} = 16.3 \text{ (approx.)}.$$

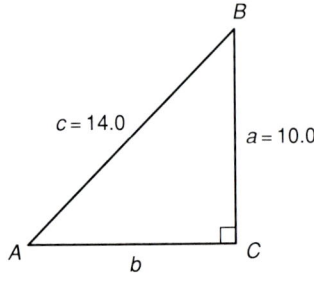

Figure 6.8

We can also use this relationship to determine the length of one of the legs of a right triangle when the lengths of the hypotenuse and the other leg are known.

• *Example 6.2:* If the hypotenuse of the right triangle in figure 6.9 is 14.0 cm long and one of the legs is 10.0 cm long, determine the length of the other leg.

Solution: Use the pythagorean theorem $a^2 + b^2 = c^2$, where $a = 10$, $c = 14$:

$$a^2 + b^2 = c^2$$
$$(10)^2 + b^2 = (14)^2.$$

Solve the resulting equation for b:

$$100 + b^2 = 196$$
$$b^2 = 96$$
$$b = \sqrt{96} = 9.8 \text{ (approx.)}.$$

The length of the other leg is 9.8 cm.

A useful relationship from elementary geometry concerning the angles of a triangle is the following one. Refer to the triangle in figure 6.10.

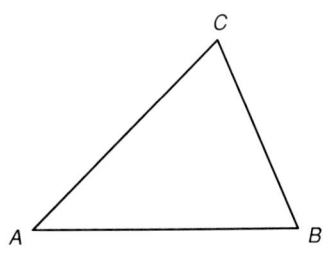

Figure 6.10

The sum of the measures of the angles of a triangle equals 180°.

$$\angle A + \angle B + \angle C = 180°.$$

Section 6.1 Angles and Triangles

- *Example 6.3:* In △ABC angles A and B measure 46°20' and 84°45', respectively. Determine the measure of the third angle (∡C).

Solution:
$$\angle A + \angle B + \angle C = 180°$$
$$46°20' + 84°45' + \angle C = 180°$$
$$130°65' + \angle C = 180°.$$

We substitute 1°5' for 65':

$$131°5' + \angle C = 180°$$
$$\angle C = 180° - 131°5'$$
$$= 179°60' - 131°5'$$
$$= 48°55'.$$

The final relationship of triangles that we consider in this section is the relationship among the lengths of the sides of similar triangles.

Similar Triangles

△ABC is **similar** to △A'B'C' provided.

$$\angle A = \angle A'$$
$$\angle B = \angle B'$$
$$\angle C = \angle C'.$$

Figure 6.11 illustrates a pair of similar triangles.

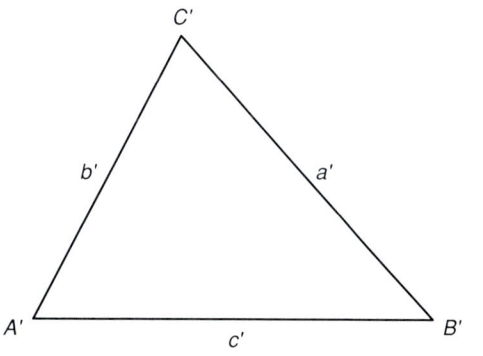
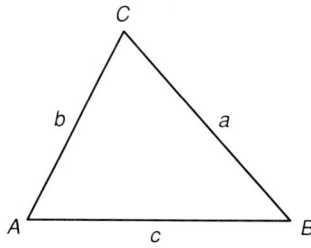

Figure 6.11

The particular property of similar triangles that is useful in the study of trigonometry is the relationship involving the ratios of the sides of similar triangles. Briefly stated, the relationship is

For a pair of similar triangles the ratio of corresponding sides is constant.

Note that for the similarity relationship between triangles ABC and $A'B'C'$ ($\triangle ABC \sim \triangle A'B'C'$), angles A and A', B and B', and C and C' are pairs of corresponding angles. The pairs of sides a and a', b and b', and c and c' are pairs of corresponding sides.

In particular, if $\triangle ABC$ is similar to $\triangle A'B'C'$, where $\angle A = \angle A'$, $\angle B = \angle B'$, and $\angle C = \angle C'$, then

$$\frac{a}{a'} = \frac{b}{b'} = \frac{c}{c'}.$$

Examples 6.4 and 6.5 illustrate how we can use this relationship to solve problems involving similar triangles.

• *Example 6.4:* In figure 6.12, given that $b = 3.2$, $b' = 4$, $a' = 6$ and $c' = 8$, determine the lengths of the remaining sides in $\triangle ABC$.

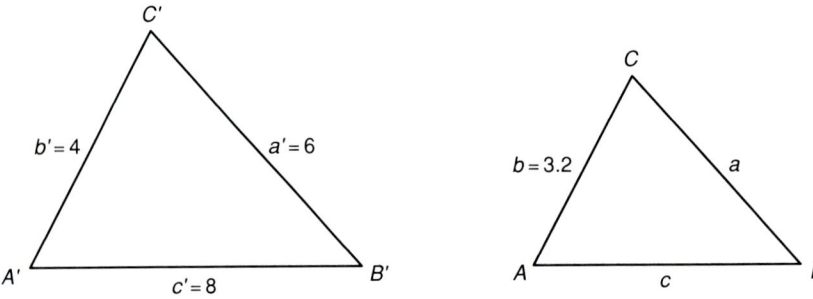

Figure 6.12

Solution: 1. Determine a in $\triangle ABC$ using the relationship $\dfrac{a}{a'} = \dfrac{b}{b'}$, where $b' = 4$, $a' = 6$, and $b = 3.2$:

$$\frac{a}{6} = \frac{3.2}{4}.$$

Solving the resulting equation for a gives

$$a = \frac{(3.2)(6)}{4} = 4.8.$$

2. Determine c in $\triangle ABC$ using the relationship $\dfrac{c}{c'} = \dfrac{b}{b'}$, where $c' = 8$, $b' = 4$, and $b = 3.2$

$$\frac{c}{8} = \frac{3.2}{4}$$

Solving the resulting equation for c gives

$$c = \frac{(3.2)(8)}{4} = 6.4.$$

• *Example 6.5:* In figure 6.13, the city plans to extend 54th Street from Route 431 to Allisonville Road. For this project they need to find the distance from B' to C'. It is known that $AB' = 1995.0$ yd, $BC = 1050.0$ yd, $AC = 1840.0$ yd, and $AC' = 2430.0$ yd.

Solution: $\triangle ABC$ is similar to $\triangle AB'C'$. Do you see why? We can use the following relationship to determine $B'C'$:

$$\frac{BC}{B'C'} = \frac{AC}{AC'}$$

$$\frac{1050}{B'C'} = \frac{1840}{2430}.$$

Solving the equation for $B'C'$ gives

$$(1050)(2430) = 1840\, B'C'$$

$$B'C' = \frac{(1050)(2430)}{1840} = 1387.$$

The length of $B'C'$ is 1387 yd.

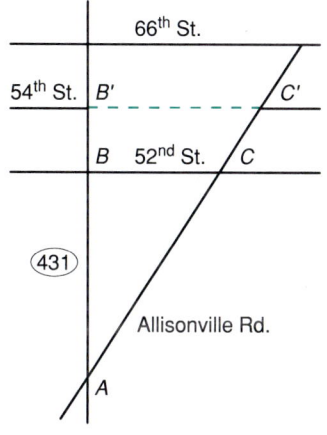

Figure 6.13

The following steps summarize the procedure for solving similar triangles.

SOLVING SIMILAR TRIANGLES

1. Select a pair of equal ratios that contain the three known quantities and the unknown quantity.
2. Substitute the known quantities.
3. Solve the resulting equation.

Trial Problems 6.1

Before you begin the section exercises, warm up with these problems. Complete answers are included in the answer key.

1. Convert $26°25'$ to decimal form.
2. In a triangle ABC, angle $A = 93°$ and angle $B = 62°$. Find the measure of angle C.
3. The legs of a right triangle measure 16.5 cm and 18.3 cm. Find the length of the hypotenuse.
4. A square has a diagonal 14.3 in. long. Find the lengths of the sides.
5. In triangle ABC, $a = 11.0$ and $b = 14.8$. If triangle $A'B'C'$ is similar and $a' = 18.2$, find b'.

Exercises 6.1

In exercises 1–6 convert the measures of the angles to decimal form. Round the answers to four decimal places.

1. $30°20'$
2. $28°35'$
3. $68°22'$
4. $69°22'$
5. $118°23'15''$
6. $113°17'45''$

7. An angle measures $16.3°$. Determine the measure of the complementary angle.
8. An angle measures $28.6°$. Determine the measure of the complementary angle.
9. An angle measures $63°20'$. Determine the measure of the supplementary angle.
10. An angle measures $125°35'$. Determine the measure of the supplementary angle.
11. Two angles of a triangle measure $68°$ and $72°$. Determine the measure of the third angle.
12. Two angles of a triangle measure $80°$ and $42°$. Determine the measure of the third angle.
13. Two angles of a triangle measure $68°40'$ and $55°30'$. Find the measure of the third angle.
14. Two angles of a triangle measure $50°45'$ and $82°35'$. Find the measure of the third angle.
15. One angle of a right triangle measures $46°23'$. Find the measure of the other angles.
16. One angle of a right triangle measures $49°12'$. Find the measures of the other angles.
17. An **isosceles triangle** is a triangle that has a pair of angles with equal measure. If the equal angles each measure $40°$, find the measure of the third angle.
18. If the third angle of an isosceles triangle measures $30°$, find the measures of each of the equal angles.
19. In figure 6.14, $\triangle ABC$ is isosceles ($AB = AC$). If angle A measures $32°$, find the measures of angles B and C.
20. In figure 6.14, $\triangle ABC$ is isosceles ($AB = AC$). If angle B measures $74°$, find the measures of angles A and C.

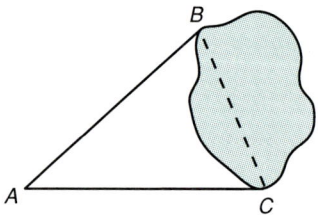

Figure 6.14

Use figure 6.15 as a guide for exercises 21–24.

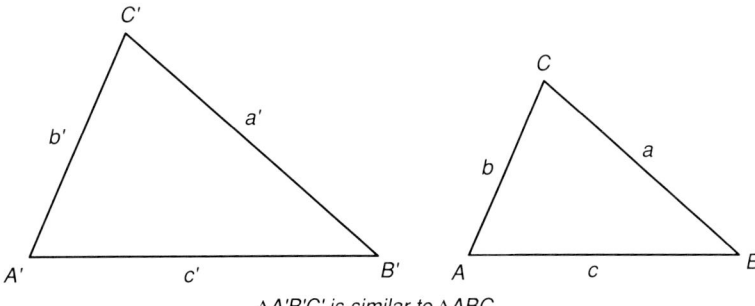

△A'B'C' is similar to △ABC

Figure 6.15

21. Given $a = 14$ in., $b = 12$ in., and $c = 16$ in., determine b' and c' if $a' = 24$ in.

22. Given $a = 10$ cm, $b = 12$ cm, and $c = 15$ cm, determine b' and c' if $a' = 20$ cm.

23. Given $a = 5.2$ in., $b = 4.1$ in., $c = 6.2$ in., and $b' = 8.1$ in., determine a' and c'.

24. Given $a = 10.7$ cm, $b = 6.8$ cm, $c = 12.3$ cm, and $b' = 8.1$ cm, determine a' and c'.

Use figure 6.16 as a guide for exercises 25–28.

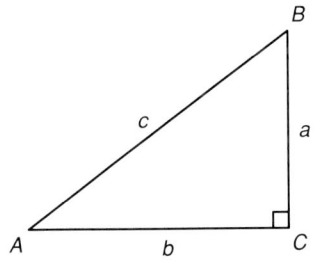

Figure 6.16

25. Given $a = 120$ and $b = 119$, find c.

26. Given $a = 73$ and $b = 65$, find c.

27. Given $a = 60$ in. and $c = 75$ in., find b.

28. Given $a = 90$ ft and $c = 106$ ft, find b.

29. A square has sides 17 in. long. Determine the length of the diagonal.

30. A square has sides 12.3 cm long. Determine the length of the diagonal.

31. One of the ways to determine whether a figure is a rectangle is to measure the diagonals to check whether they are equal. A sheet of stainless steel is supposed to be rectangular, and the lengths of adjacent sides are 47.2 in. and 19.8 in. How long should the diagonals be for the sheet to be rectangular?

32. A sheet of stainless steel has adjacent sides that are 80.5 cm and 42.4 cm long. If the diagonals are the same length, then the sheet is rectangular. How long should the diagonals be?

33. A square bar 1 in. on a side is to be milled from a circular rod (see figure 6.17). If the available rods are 1 in., $1\frac{1}{4}$ in., and $1\frac{1}{2}$ in. diameters, which rod should be used?

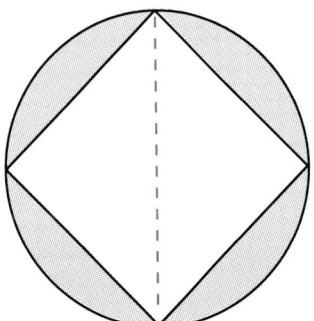

Figure 6.17

34. A square bar 2.5 cm on a side is to be milled from a circular rod (see figure 6.17). If the available rods are 3.2 cm, 3.4 cm, and 3.6 cm in diameter, which rod should be used?

35. The county wants to extend Dean Road from 62nd Street until it intersects with Allisonville Road in figure 6.18. Given $CC' = 1637$ yd, $B'C' = 993$ yd, and $BC = 243$ yd, determine the length of AC. Hint: $AC' = AC + CC'$.

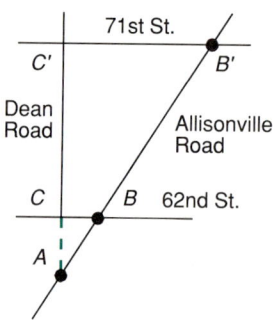

Figure 6.18

6 • 2 The Tangent Ratio

A surveyor needs to find distance BC in figure 6.19, when the measure of angle A and distance AC are known. To solve this problem and similar problems, we introduce a ratio called the **tangent** ratio.

Tangent

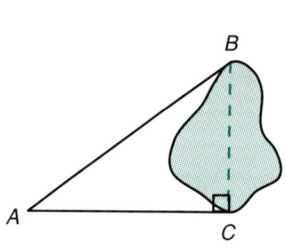

Figure 6.19

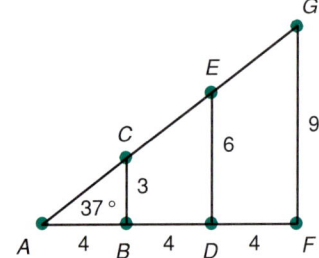

Figure 6.20

Figure 6.20 shows three similar right triangles, ABC, ADE, and AFG, each of which contains a 37° angle at A. Using the blocks of the graph paper as a measuring device, $AB = 4$, $AD = 8$, $AF = 12$, $BC = 3$, $DE = 6$, and $FG = 9$. Since triangles ABC, ADE, and AFG are similar, the ratios between corresponding sides are equal. Thus,

$$\frac{BC}{AB} = \frac{DE}{AD} = \frac{FG}{AF},$$

or

$$\frac{3}{4} = \frac{6}{8} = \frac{9}{12} = 0.75.$$

This ratio is called the *tangent of A,* and we write

$$\tan A = \tan 37° = 0.75 \text{ (approximately)}.$$

SECTION 6.2 THE TANGENT RATIO

In general, the tangent of an acute angle (measure $< 90°$) of a right triangle is defined as follows (refer to $\triangle ABC$ in figure 6.21):

$$\tan A = \frac{\text{length of side opposite } \angle A}{\text{length of side adjacent to } \angle A} = \frac{a}{b}.$$

Similarly,

$$\tan B = \frac{\text{length of side opposite } \angle B}{\text{length of side adjacent to } \angle B} = \frac{b}{a}.$$

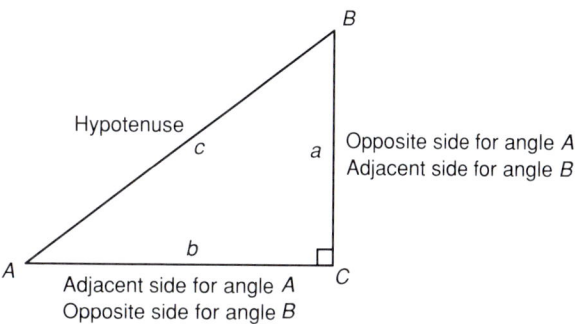

Figure 6.21

Figure 6.22

We assume that the students have access to a scientific calculator for calculating tangents and other trigonometric ratios of angles. For the scientific calculator in figure 6.22, we can determine the tangent of an angle by entering 37 and pressing the button marked $\boxed{\text{TAN}}$. Make sure the calculator is in degree mode. The display reads 0.753554. We usually round off the ratios to four decimal places. Hence, $\tan 37° = 0.7536$.

• **Example 6.6:** Determine the tangent of an angle that measures $72°35'$.

Solution: Since the calculator must have the angle in decimal form, we first must convert $72°35'$ to this form:

$$35' = \left(\frac{35}{60}\right)° = (0.583333)°$$

$$72°35' = 72.583333°$$
$$\tan 72°35' = 3.1878 \text{ (rounded to 4 places)}.$$

The most efficient way to get this result is by following this sequence. The sequence may vary depending on the logic of the calculator.

Display: 3.18775.
We can also use the calculator to find angle measures when the tangent value is known.

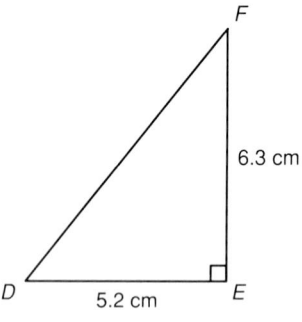

Figure 6.23

Find the measure of angles D and F in figure 6.23.

$$\tan D = \frac{\text{length of opposite side}}{\text{length of adjacent side}}$$
$$= \frac{6.3}{5.2} = 1.2115385.$$

Note that so far we have tan D. We want the measure of D.

$$D = 50.5°$$

 6.3 ÷ 5.2 = INV TAN .

Display: 50.4638.

$$\tan F = \frac{\text{length of opposite side}}{\text{length of adjacent side}}$$
$$= \frac{5.2}{6.3} = 0.8253968$$
$$F = 39.5°$$

5.2 ÷ 6.3 = INV TAN .

Display: 39.536157.

NOTE: The INV button on the calculator stands for INVERSE. In the equation tan $x = y$, the sequence of steps

y INV TAN

determines x. The sequence of steps x TAN determines y. On some calculators the button is labeled ARC instead of INV .

For example,

 1.567 INV TAN

gives 57.455524.

57.455524 TAN

gives 1.567.

The tangent ratio can also be used to determine unknown lengths of sides of right triangles.

SECTION 6.2 THE TANGENT RATIO

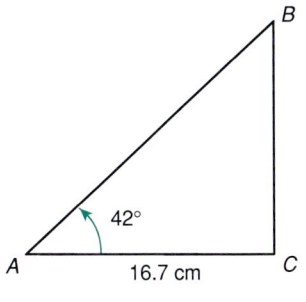

Figure 6.24

• **Example 6.8:** Find the length of BC in figure 6.24.

Solution: Notice that BC is the side opposite $\angle A$, that $\angle A$ is known, and the adjacent side AC is known. We can use the tangent ratio for $\angle A$ to determine the length of BC:

$$\tan A = \frac{BC}{AC}$$

$$\tan 42° = \frac{BC}{16.7}.$$

Solving the equation for BC gives, $BC = (16.7)\tan 42° = 15.0$ cm.

Calculator solution:

 42 [TAN] × 16.7 = .

Display: 15.036747.

Example 6.9 illustrates the use of the tangent ratio to solve an electrical problem.

• **Example 6.9:** An AC circuit contains a coil and a resistor in series. The phase angle, θ, is the angle by which the current leads the impressed voltage. The angle can be found by representing the resistance of the resistor as the adjacent side and the reactance of the coil as the opposite side of a right triangle containing the phase angle θ (see figure 6.25). Suppose the reactance $X_L = 74.8$ ohms and the resistance $R = 30$ ohms. Find the measure of the phase angle.

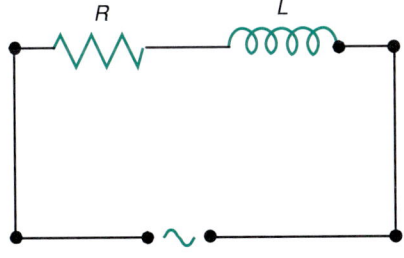 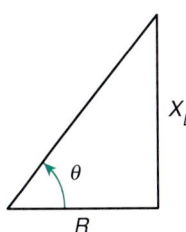

Figure 6.25

Solution: $\tan \theta = \dfrac{X_L}{R}$.

Substitute $X_L = 74.8$ and $R = 30$.

$$\tan \theta = \frac{74.8}{30}$$

$$\theta = 68.1°.$$

This sequence of steps should produce the solution.

 74.8 ÷ 30 = INV TAN .

Display: 68.145783.

• *Example 6.10:* An object that decreases gradually in diameter so that it assumes a conical shape is said to be **tapered**. The angle included between the sides (or the extended sides) of the object is called the **taper angle**. In figure 6.26, angle *CEB* is the taper angle. Given the dimensions shown in the figure, determine the taper angle.

Tapered
Taper Angle

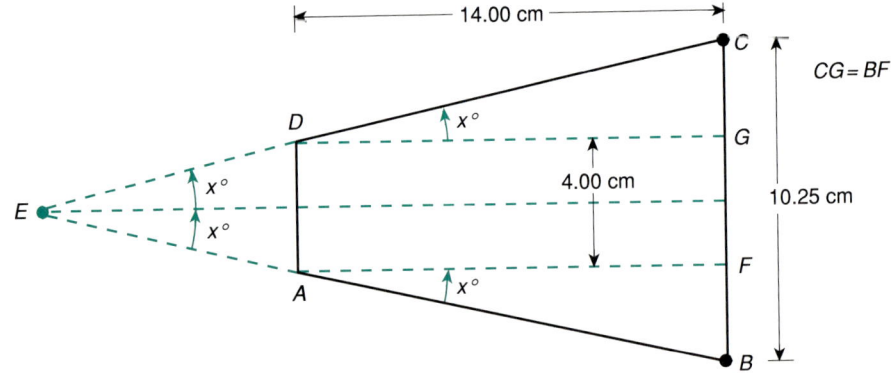

Figure 6.26

Solution: The measure of ∡*DEA* is twice the measure of ∡*CDG*. Let *x* represent the measure of ∡*CDG*. To use the tangent ratio in △*CDG*, we must first determine the length of *CG*:

$$CG = \frac{1}{2}(BC - AD) = \frac{1}{2}(10.25 - 4) = 3.125.$$

We can now use the tangent ratio in △ *GDC*:

$$\tan x = \frac{CG}{DG} = \frac{3.125}{14} = 0.2232143$$

$$x = 12.582963°$$

$$\angle DEA = 2x = 25.2°.$$

This sequence of steps should produce the solution:

 3.125 ÷ 14 = INV TAN × 2 = .

Display: 25.16592.

SECTION 6.2 THE TANGENT RATIO

Trial Problems 6.2

Before you begin the section exercises, warm up with these problems. Complete answers are included in the answer key.

1. If $A = 26°$ and $a = 18.3$, find b.
2. If $A = 48°$ and $b = 16.1$, find a.
3. If $a = 11.3$ and $b = 22.4$, find A.
4. If $B = 67°$ and $a = 0.43$, find b.
5. If $a = 11.0$ and $b = 16.7$, find B.

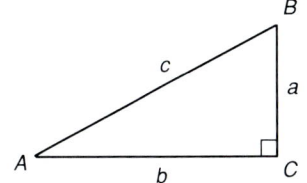

Exercises 6.2

Using figure 6.27 as a guide, find the missing dimensions and angles.

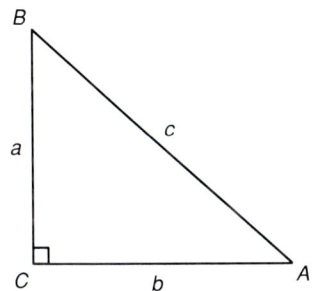

Figure 6.27

	Angle A	Angle B	a	b
1.	67°	23°	11 in.	?
2.	42°	48°	14 in.	?
3.	62°	38°	?	16.5 in.
4.	55°	35°	?	13.2 cm
5.	22.3°	67.7°	?	140 yd
6.	31.6°	58.4°	?	11.3 m
7.	35.7°	54.3°	62.3 in.	?
8.	55.6°	34.4°	21.2 ft	?
9.	54°20′	35°40′	23.2 cm	?
10.	27°45′	62°15′	34.6 cm	?
11.	28°11′	61°49′	3.71 m	?
12.	12°23′	77°37′	6.85 in.	?
13.	16°12′45″	?	?	146 m
14.	70°27′40″	?	?	18 ft

	Angle A	Angle B	a	b
15.	?	?	150 cm	82 cm
16.	?	?	42 in.	63 in.

17. How tall is the tree in figure 6.28 if its shadow is 72 feet long and the angle to the sun from the horizontal plane is 38°?

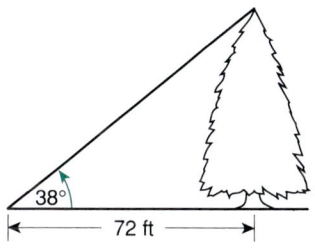

Figure 6.28

18. Solve exercise 17 given that the angle is 31.5° and the length of the shadow is 62.5 feet.

19. Find the height of a building in figure 6.29 if the angle of sight from the plane of the base to the top of the building is 40°.

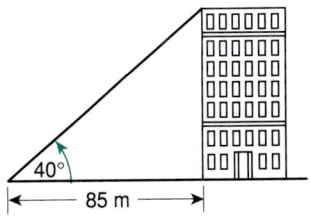

Figure 6.29

20. Solve exercise 19 given that the angle of sight is 38.5° and the distance from the building is 37 feet.

21. In figure 6.30 a surveyor wishes to determine the distance between points A and B, given that $BC = 50$ feet and the angle C measures 47°. Determine AB.

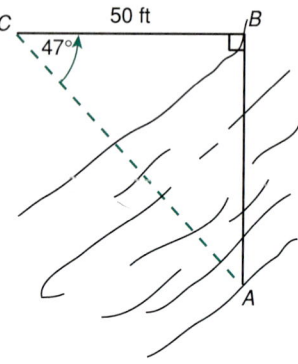

Figure 6.30

22. Solve exercise 21 given that $BC = 62$ feet and angle $C = 43°30'$.

23. Determine the taper angle, the measure of angle DEA in figure 6.31, if $AD = 2.3$ in., $BC = 5.3$ in., and $DG = 7.1$ in.

24. Solve exercise 23 if $AD = 2.7$ cm, $BC = 7.4$ cm, and $DG = 9.5$ cm.

25. The roof in figure 6.32 has a rise of 4 feet and a run of 12 feet. Find the angle that a rafter makes with the horizontal.

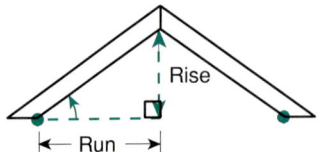

Figure 6.32

26. Solve exercise 25 if the rise is 5.2 feet and the run is 16.5 feet.

27. A guy wire on an antenna tower in figure 6.33 is attached to a point that is 12 feet from the base of the tower. The tower is 50 feet high.
    a. What angle does the wire make with the ground?
    b. What angle does the wire make with the tower?
    c. How long is the wire? (Use the Pythagorean relationship.)

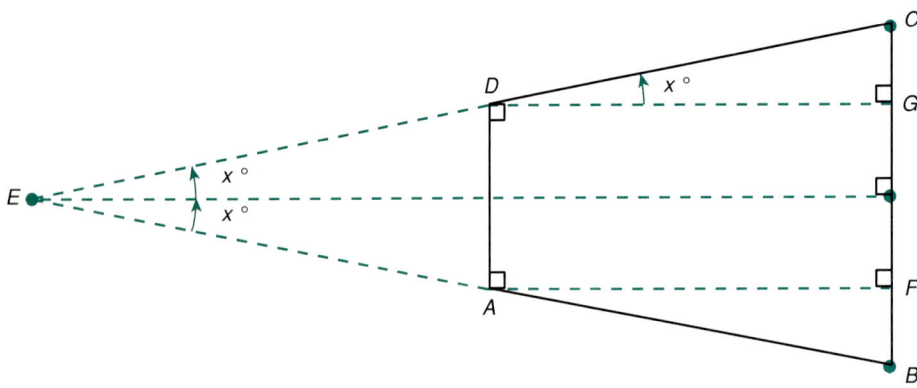

Figure 6.31

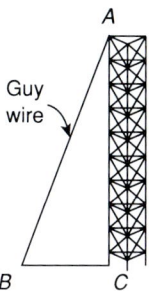

Figure 6.33

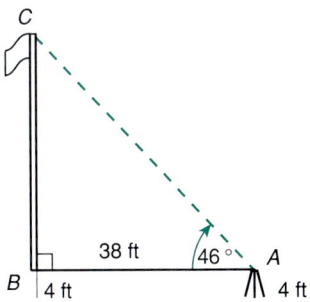

Figure 6.35

28. Solve exercise 27 given that point B is 9.8 feet from the base of the tower and AB is 48.3 feet.

29. A flagpole in figure 6.34 is located on the roof at the edge of a building. The angles of elevation from a point 40 feet from the base of the building to the base and to the top of the pole are 60° and 67°, respectively. Find the height of the flagpole.

32. Solve exercise 31 given that the transit height is 4 feet, $AB = 41.2$ feet, and the angle at A is 42°35′.

33. For the hexagon in figure 6.36, $AB = 2.2$ cm and angle $BAC = 30°$. Determine the length of AC.

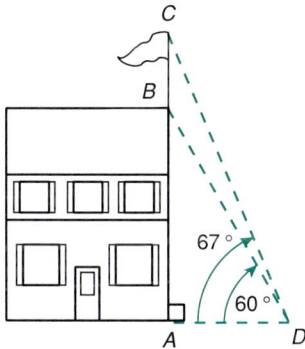

Figure 6.34

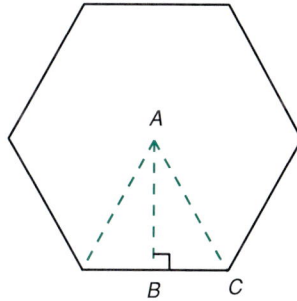

Figure 6.36

34. Solve exercise 33 given that angle $BAC = 30°$ and $AB = 3.25$ in.

35. An AC circuit contains a coil and a resistor in series. Determine the phase angle of the circuit, given that the reactance of the coil is 32.7 ohms and the resistance is 42.3 ohms (see example 6.9).

36. Solve exercise 35 given that the reactance is 27.2 ohms and the resistance is 20 ohms.

30. Solve exercise 29 if $AD = 38$ feet, angle $ADB = 59°20′$, and angle $ADC = 66°30′$.

31. A surveyor, using a transit that is located 4 feet above the ground at point A, at a distance of 38 feet from the base of a pole, sights an angle of 46° to the top of the pole. Find the height of the pole in figure 6.35.

37. The coefficient of friction of an object moving with a constant velocity down an inclined plane is calculated by the formula $k = \tan \theta$, where θ represents the angle between the inclined plane and the corresponding horizontal plane, and k represents the coefficient of friction. If $k = 0.105$, determine the angle of inclination.

38. Solve exercise 37, given that $k = 0.223$.

39. When light is reflected from a surface, the angle of incidence i is equal to the angle of reflection r (see figure 6.37). Suppose a beam of light reflecting from a surface is intercepted at point P, which is 12.3 cm horizontally and 16.1 cm vertically from the point of reflection. Determine the angle of incidence.

40. Solve exercise 39 given that $y = 29.2$ in. and $x = 22.3$ in.

41. Given an AC circuit containing a capacitor, a resistor, and an inductor in series, the phase angle is determined by the formula

$$\tan \theta = \frac{X_L - X_C}{R},$$

where θ is the phase angle, R is the resistance, X_L is the inductive reactance, and X_C is the capacitive reactance. Given that $X_L = 7$ ohms, $X_C = 4$ ohms, and $\theta = 45°$, determine R.

42. Solve exercise 41, given that $X_L = 8.5$ ohms, $X_C = 6.5$ ohms, and $\theta = 60°$.

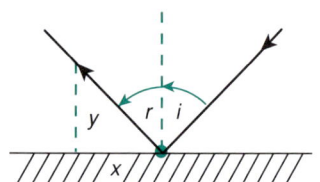

Figure 6.37

6•3 Trigonometric Ratios

Figure 6.38 shows a corner of a metal piece that must be rounded off according to the given specifications. To do this, it is necessary to determine the distance AC. To solve this and similar problems, we introduce two more trigonometric ratios, the sine and cosine ratios.

Sine   The **sine** of $\angle A$ (written sin A) is the ratio of the length of the side opposite $\angle A$
Cosine to the length of the hypotenuse of a right triangle containing $\angle A$. The **cosine** of $\angle A$ (written cos A) is the ratio of the length of the side adjacent to $\angle A$ to the length of the hypotenuse of a right triangle that contains $\angle A$. Figure 6.39 will help you to see these ratios.

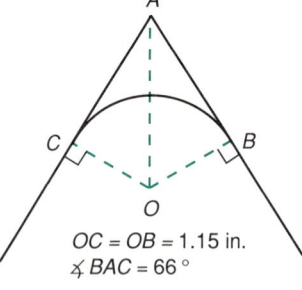

$OC = OB = 1.15$ in.
$\angle BAC = 66°$

Figure 6.38

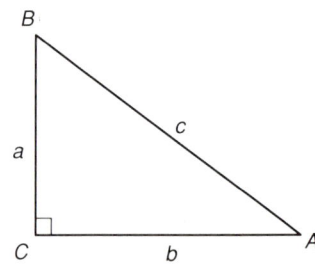

Figure 6.39

Right Triangle Ratios

$$\sin A = \frac{\text{length of side opposite } A}{\text{length of hypotenuse}} = \frac{a}{c}$$

$$\cos A = \frac{\text{length of side adjacent to } A}{\text{length of hypotenuse}} = \frac{b}{c}$$

$$\sin B = \frac{\text{length of side opposite } B}{\text{length of hypotenuse}} = \frac{b}{c}$$

$$\cos B = \frac{\text{length of side adjacent to } B}{\text{length of hypotenuse}} = \frac{a}{c}$$

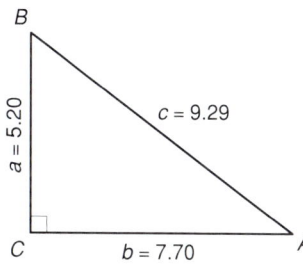

Figure 6.40

• *Example 6.11:* In triangle ABC of figure 6.40, determine sin A, cos A, and the measure of angle A.

Solution: $\sin A = \dfrac{a}{c} = \dfrac{5.20}{9.29} = 0.5597$

$\cos A = \dfrac{b}{c} = \dfrac{7.70}{9.29} = 0.8288.$

To determine the measure of angle A, we can use either the $\boxed{\text{SIN}}$ or the $\boxed{\text{COS}}$ button on the calculator along with the $\boxed{\text{INV}}$ button (or the $\boxed{\text{ARC}}$ button).
Either of these two sequences calculates the measure of angle A.

$\boxed{5.20}\ \boxed{\div}\ \boxed{9.29}\ \boxed{=}\ \boxed{\text{INV}}\ \boxed{\text{SIN}}$.

Display: 34.037934.

$\boxed{7.70}\ \boxed{\div}\ \boxed{9.29}\ \boxed{=}\ \boxed{\text{INV}}\ \boxed{\text{COS}}$.

Display: 34.019396.
The measure of angle A is 34.0°.

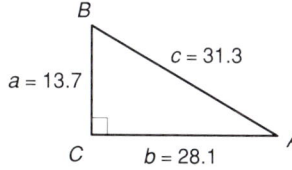

Figure 6.41

• *Example 6.12:* In triangle ABC of figure 6.41, determine the measures of angles A and B.

Solution: $\sin A = \dfrac{\text{length of side opposite } \angle A}{\text{length of hypotenuse}} = \dfrac{a}{c}$

$= \dfrac{13.7}{31.3} = 0.4377$

$A = 26.0°.$

13.7 ÷ **31.3** = **INV** **SIN**.

Display: 25.957204.

$$\cos B = \frac{\text{length of side adjacent to } \angle B}{\text{length of hypotenuse}} = \frac{a}{c}$$

$$= \frac{13.7}{31.3} = 0.4377$$

$$B = 64.0°$$

 13.7 ÷ **31.3** = **INV** **COS**.

Display: 64.042796.

Note that since angles A and B are complementary angles, we could have found the measure of angle B as follows:

$$B = 90.0° - 26.0° = 64.0°$$

•••••••••

The sine and cosine ratios can be used to determine unknown lengths of sides of right triangles in a manner similar to the way the tangent ratio was used in section 6.2.

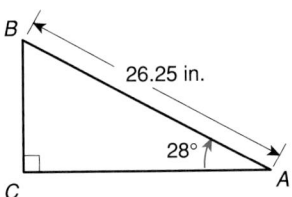

Figure 6.42

For the triangle in figure 6.42, find the lengths of AC and BC.

Several different ratios could be used, since it is very easy to find the measure of angle B (the complement of angle A). Since we already have the measure of angle A, we use the sine and cosine ratios for angle A to calculate the lengths of BC and AC.

1. $\sin A = \dfrac{\text{length of side opposite } \angle A}{\text{length of hypotenuse}} = \dfrac{BC}{AB}$

$$\sin A = \frac{BC}{AB}$$

$$\sin 28° = \frac{BC}{26.25}$$

$$BC = (\sin 28°)(26.25) = 12.32 \text{ in.}$$

 28 **SIN** × **26.25** =.

Display: 12.323629.

2. $\cos A = \dfrac{\text{length of side adjacent to } \angle A}{\text{length of hypotenuse}} = \dfrac{AC}{AB}$

$$\cos A = \dfrac{AC}{AB}$$

$$\cos 28° = \dfrac{AC}{26.25}$$

$$AC = (\cos 28°)(26.25) = 23.18 \text{ in.}$$

28 [COS] [×] 26.25 [=].

Display: 23.177374.

Examples 6.14, 6.15, and 6.16 illustrate how the sine and cosine ratios can be used to solve problems in applied areas.

• *Example 6.14:* A metal plate is bent according to the specifications shown in figure 6.43. Calculate the lengths of AB and AC.

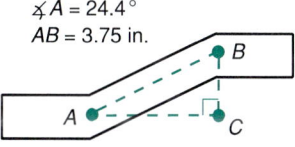

$\angle A = 24.4°$
$AB = 3.75$ in.

Figure 6.43

Solution: 1. $\sin A = \dfrac{\text{length of side opposite } \angle A}{\text{length of hypotenuse}} = \dfrac{BC}{AB}$

$$\sin A = \dfrac{BC}{AB}$$

$$\sin 24.4° = \dfrac{BC}{3.75}$$

$$BC = (\sin 24.4°)(3.75) = 1.55 \text{ in.}$$

24.4 [SIN] [×] 3.75 [=].

Display: 1.5491416.

2. $\cos A = \dfrac{\text{length of side adjacent to } \angle A}{\text{length of hypotenuse}} = \dfrac{AC}{AB}$

$$\cos A = \dfrac{AC}{AB}$$

$$\cos 24.4° = \dfrac{AC}{3.75}$$

$$AC = (\cos 24.4°)(3.75) = 3.42 \text{ in.}$$

24.4 [COS] [×] 3.75 [=].

Display 3.4150637.

224 CHAPTER 6 NUMERICAL TRIGONOMETRY

• *Example 6.15:* In electrical theory, the power P in a particular circuit is given by the equation

$$P = EI \cos \theta,$$

where I is the current in amperes, E is the voltage in volts, and θ is the phase angle between E and I. Find the power in a 110 volt circuit if the current is 20 amperes and $\theta = 60°$.

Solution: We substitute $E = 110$, $I = 20$, and $\theta = 60°$ into the equation:

$$\begin{aligned} P &= EI \cos \theta \\ &= 110(20)(\cos 60°) \\ &= 110(20)(0.5) = 1100. \end{aligned}$$

The power delivered is 1100 watts.
This sequence of steps should produce the solution.

$$60 \;\boxed{\text{COS}}\; \boxed{\times}\; 110 \;\boxed{\times}\; 20 \;\boxed{=}.$$

Display: 1100.

Return to the problem of figure 6.38 at the beginning of section 6.3. We now have the mathematics necessary for the solution of the problem.

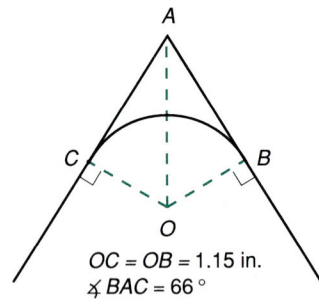

Figure 6.44

• *Example 6.16:* Determine the lengths of segments OA and CA in figure 6.44.

Solution: We attack the problem in three stages:

1. Find the measure of angle OAC that is half the measure of angle A.
2. Find the length of segment OA.
3. Find the length of segment AC

1. $\angle OAC = \dfrac{1}{2} \angle A = \dfrac{1}{2}(66°) = 33°.$

2. In triangle OCA

$$\sin \angle OAC = \frac{\text{length of side opposite } \angle OAC}{\text{length of hypotenuse}} = \frac{OC}{OA}$$

$$\sin \angle OAC = \frac{OC}{OA}$$

$$\sin 33° = \frac{1.15}{OA}$$

$$OA = \frac{1.15}{\sin 33°} = 2.11 \text{ in.}$$

SECTION 6.3 TRIGONOMETRIC RATIOS

33 [SIN] [1/x] [×] 1.15 [=].

Display: 2.1114902.

3. In triangle OAC

$$\cos \angle OAC = \frac{\text{length of side adjacent to } \angle OAC}{\text{length of hypotenuse}} = \frac{AC}{OA}$$

$$\cos \angle OAC = \frac{AC}{OA}$$

$$\cos 33° = \frac{AC}{2.11}$$

$$AC = (\cos 33°)(2.11) = 1.77 \text{ in.}$$

33 [COS] [×] 2.11 [=].

Display: 1.7695949.

There are three other trigonometric ratios that are considered in more detail in a later chapter. They are the reciprocals of the sine, cosine, and tangent functions. We define these functions in terms of the lengths of the sides of right triangles. The reciprocal ratios can be used to calculate their numerical values for particular angles. Referring to figure 6.45, the definitions of the cotangent, secant, and cosecant ratios are as follows.

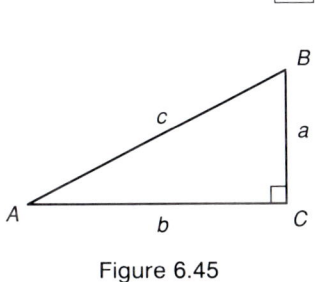

Figure 6.45

COTANGENT, SECANT, AND COSECANT

$$\text{Cotangent } A = \cot A = \frac{\text{length of adjacent side}}{\text{length of opposite side}} = \frac{b}{a} = \frac{1}{\tan A}$$

$$\text{Secant } A = \sec A = \frac{\text{length of hypotenuse}}{\text{length of adjacent side}} = \frac{c}{b} = \frac{1}{\cos A}$$

$$\text{Cosecant } A = \csc A = \frac{\text{length of hypotenuse}}{\text{length of opposite side}} = \frac{c}{a} = \frac{1}{\sin A}.$$

• *Example 6.17:* For an angle $A = 62°15'$, calculate $\cot A$, $\sec A$, and $\csc A$.

Solution: $62°15' = 62.25°$

$$\cot A = \frac{1}{\tan A} = \frac{1}{\tan 62.25°} = 0.5261$$

62.25 [TAN] [1/x].

Display: 0.5261255.

$$\sec A = \frac{1}{\cos A} = \frac{1}{\cos 62.25°} = 2.1477$$

 62.25 COS 1/x .

Display: 2.1476993.

$$\csc A = \frac{1}{\sin A} = \frac{1}{\sin 62.25°} = 1.1300$$

62.25 SIN 1/x .

Display: 1.1299593.

Trial Problems 6.3

Before you begin the section exercises, warm up with these problems. Complete answers are included in the answer key.

1. If $a = 17.9$ and $A = 65.3°$, find c.
2. If $a = 17.9$ and $c = 25.8$, find B.
3. If $b = 14.3$ and $c = 25.9$, find A.
4. If $A = 22.6°$, find cot A.
5. If $P = EI \cos \theta$, $\theta = 60°$, $E = 110$, and $I = 20$, find P.

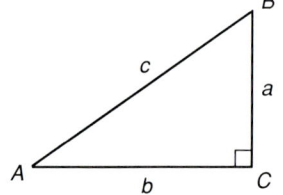

Exercises 6.3

Using figure 6.46 as a guide, find the missing lengths and angles.

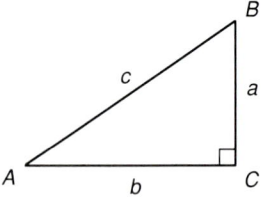

Figure 6.46

	Angle A	Angle B	a	b	c
1.	?	?	3.2 in.	4.6 in.	5.6 in.
2.	?	?	12.4 cm	13.3 cm	18.2 cm
3.	32°	?	?	?	16.2 in.
4.	47°	?	?	?	18.7 cm
5.	61°	?	147 m	?	?
6.	38°	?	19 ft	?	?
7.	?	?	120 ft	119 ft	?
8.	?	?	72 m	65 m	?

	Angle A	Angle B	a	b	c
9.	?	32°25′	?	?	8.14 cm
10.	?	66°35′	?	?	17.23 in.
11.	16°30′	?	8.25 in.	?	?
12.	27°45′	?	6.17 in.	?	?

13. For each angle determine cot θ, sec θ, and csc θ.
    a. $\theta = 37°$ **b.** $\theta = 84.6°$ **c.** $\theta = 62°35′$

14. For each angle determine cot θ, sec θ, and csc θ.
    a. $\theta = 53°$ **b.** $\theta = 5.4°$ **c.** $\theta = 27°25′$

15. Using the formula $P = EI \cos \theta$, find the power P in a 110 volt circuit if the current is 15 amps and the phase angle is θ.
    a. $\theta = 45°$ **b.** $\theta = 30°$ **c.** $\theta = 65°$

16. Solve exercise 15, given the following phase angles.
    a. $\theta = 60°$ **b.** $\theta = 35°$ **c.** $\theta = 62.5°$

17. In figure 6.47, calculate the measures of angles XYZ, WZY, and ZXY if $YZ = 24$ ft, $XZ = 10$ ft, and $ZW = 9.2$ ft.

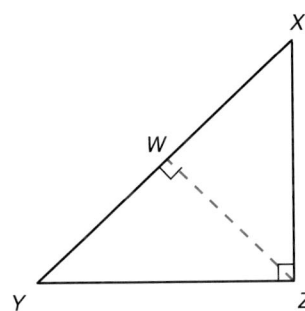

Figure 6.47

18. In figure 6.47, calculate the measures of angles XYZ, WZY, and ZYX if $YZ = 7.4$ m, $XZ = 3.0$ m, and $ZW = 2.8$ m.

19. For the machine part shown in figure 6.48, calculate the measure of angle ECD.

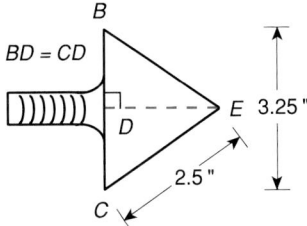

Figure 6.48

20. For the machine part shown in figure 6.49, calculate the measure of angle ECD.

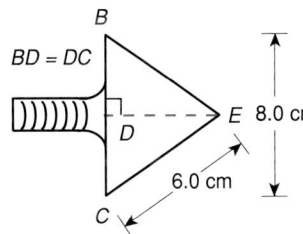

Figure 6.49

21. Calculate the measure of angle ABC in figure 6.50.

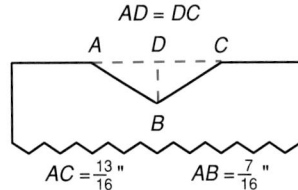

Figure 6.50

22. Calculate the measure of angle ABC in figure 6.51.

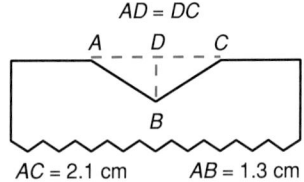

Figure 6.51

23. Calculate the length of AB in figure 6.52, if $AC = \dfrac{3}{4}$ in. and angle $BAC = 48°$.

24. Calculate the length of AB in figure 6.52, if $AC = 1.8$ cm and angle $BAC = 48°$.

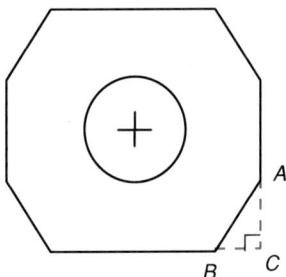

Figure 6.52

25. A bridge is being built across a canyon that is 6550 feet wide. If one side of the canyon is 240 feet higher than the other side, determine the angle of elevation of the bridge. (The angle of elevation is the angle between a line drawn from a point on the edge of the lower side of the canyon to a point on the edge of the higher side, and a line drawn from the original point horizontally across the canyon.)

26. Solve exercise 25 given that the canyon is 8230 feet wide and the difference in elevations is 460 ft.

27. When light passes from one medium to another, the bending is called *refraction*. In figure 6.53, θ_1 is called the angle of incidence and θ_2 is called the angle of refraction. When the first medium is air the *index of refraction* is determined by the formula $n \sin \theta_2 = \sin \theta_1$. Given that $n = 1.7$ for a particular second medium and $\theta_1 = 42°$, determine the measure of θ_2.

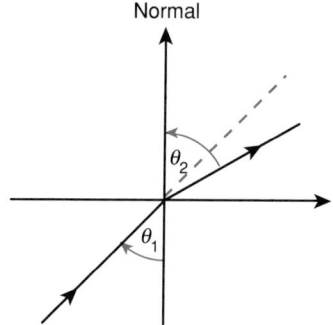

Figure 6.53

28. Solve exercise 27, given that $n = 2.42$ and $\theta_1 = 41.50°$.

29. The average power P delivered to a particular AC circuit is determined by the formula

$$P = \left(\dfrac{V_m I_m}{2}\right) \cos \theta,$$

where P represents the average power in watts, V_m represents the maximum voltage in volts, I_m represents the maximum current in amps, and θ represents the phase angle.
Suppose $\theta = 30°$, $V_m = 110$, and $I_m = 15$. Determine P.

30. Determine θ in the formula of exercise 29 given that $P = 210$, $V_m = 150$, and $I_m = 3$.

For exercises 31–34, three equally spaced holes are drilled on the circumference of a circle. Find the center-to-center distance between the holes, given the diameter d of the circle.

31. $d = 6.0$ in.   32. $d = 0.875$ in.   33. $d = 2.2$ cm   34. $d = 15.0$ cm

35. In figure 6.54, if $AB = 1.25$ in. and $BC = 2.60$ in., determine the measure of the taper angle (angle ACB).

36. In figure 6.54, if $AB = 1.15$ cm and $BC = 2.60$ cm, determine the measure of the taper angle (angle ACB).

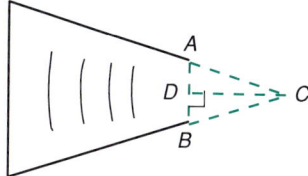

Figure 6.54

37. The Pentagon Building has a distance of 784 ft from the center of the building to each of its corners (see figure 6.55). Find the lengths of the sides of the building.

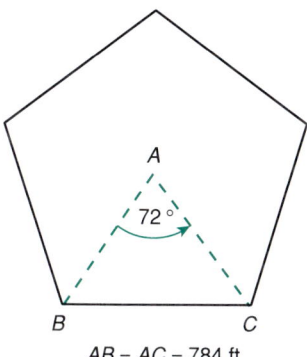

Figure 6.55

38. Solve exercise 37 for a pentagonal building where $AB = 239$ m.

39. The measurements in figure 6.56 were used to determine the length of a tunnel to be built through a hill. Find x, the length of the tunnel.

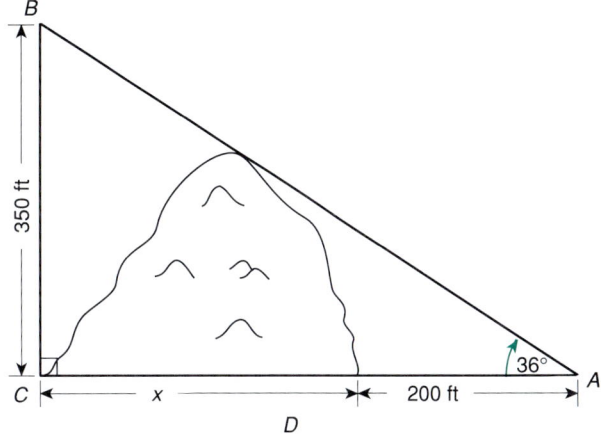

Figure 6.56

40. Solve exercise 39 given that $AD = 250$ m, $BC = 260$ m, and $A = 36°$.

6•4 The Law of Sines

So far in this chapter the problems have involved situations that could be modeled using right triangles. In section 6.4 and in section 6.5 problems that can be modeled with general triangles are considered.

Obtuse Angles Since some triangles contain **obtuse angles** (angles that have measures between 90° and 180°), it is necessary to calculate the trigonometric ratios for these angles. The following formulas can be used.

Trigonometric Ratios for Obtuse Angles

For $90° < \theta < 180°$

$$\sin \theta = \sin(180° - \theta)$$
$$\cos \theta = -\cos(180° - \theta)$$
$$\tan \theta = -\tan(180° - \theta)$$

For example,

$$\sin 130° = \sin(180° - 130°) = \sin 50° = 0.7660$$
$$\cos 120° = -\cos(180° - 120°) = -\cos 60° = -0.5$$
$$\tan 160° = -\tan(180° - 160°) = -\tan 20° = -0.3640.$$

These formulas are necessary only if the values of the ratios must be obtained from a table. Using a calculator, you can obtain the values of the ratios in the usual manner.

For $\sin 130°$: 130 [SIN]   Display: 0.7660444

For $\cos 120°$: 120 [COS]   Display: -0.5

For $\tan 160°$: 160 [TAN]   Display: -0.3639702.

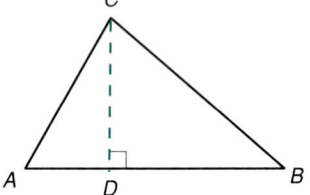

Figure 6.57

Consider the acute triangle (all angles are acute angles) in figure 6.57. Let CD be the altitude to side AB. (The altitude is the perpendicular from a vertex to the opposite side.)

Using triangle ADC,

$$\sin A = \frac{CD}{AC} \quad \text{or} \quad CD = (AC)\sin A.$$

Using triangle BDC,

$$\sin B = \frac{CD}{BC} \quad \text{or} \quad CD = (BC)\sin B.$$

Thus,

$$(AC)\sin A = (BC)\sin B.$$

Divide both sides of the equation by $(AC)(BC)$

$$\frac{\sin A}{BC} = \frac{\sin B}{AC}.$$

It can be shown in a similar manner that

$$\frac{\sin B}{AC} = \frac{\sin C}{AB}$$

These results yield the following rule.

SECTION 6.4 THE LAW OF SINES

LAW OF SINES

$$\frac{\sin A}{BC} = \frac{\sin B}{AC} = \frac{\sin C}{AB}.$$

The ratio between the sine of an angle of a triangle and the length of the opposite side is constant.

Using figure 6.58 we can restate this law in a more convenient form.

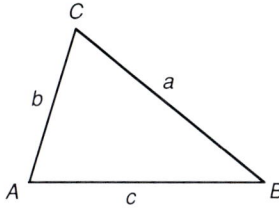

Figure 6.58

LAW OF SINES (ALTERNATE)

If a, b, c represent the lengths of the sides of a triangle ABC that are opposite the angles A, B, and C, respectively, of the triangle, then

$$\frac{\sin A}{a} = \frac{\sin B}{b} = \frac{\sin C}{c}.$$

For the case of the obtuse triangle in figure 6.59, the measure of angle C is greater than $90°$.

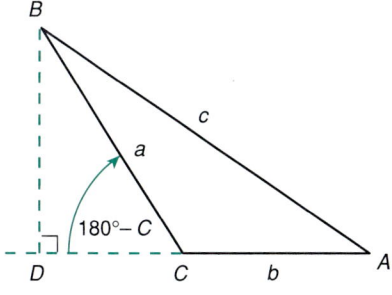

Figure 6.59

$$\sin A = \frac{BD}{c} \quad \text{and} \quad \sin(180° - C) = \frac{BD}{a}$$

$$BD = c \sin A = a \sin(180° - C)$$

$$\frac{\sin A}{a} = \frac{\sin B}{b} = \frac{\sin(180° - C)}{c}.$$

Since $\sin(180° - C) = \sin C$, the formula for the law of sines does not have to be changed if the triangle is an obtuse triangle.

The law of sines can be used to solve two cases of the general triangle.

Case I   Two angles and a side of the triangle are known.

Case II   Two sides and an angle opposite one of the given sides are known.

Let us consider Case I first.

• *Example 6.18:*   If $A = 56°$, $B = 72°$, and $b = 10$ cm, find a and c.

Solution: I. Determine a.

Since we know A, B, and b, we select the proportion that contains these values and the value a, which we want to find:

$$\frac{\sin A}{a} = \frac{\sin B}{b}.$$

Substitute 56° for A, 72° for B, and 10 for b in the proportion and solve for a:

$$\frac{\sin 56°}{a} = \frac{\sin 72°}{10}$$

$$a = \frac{(10)(\sin 56°)}{\sin 72°} = 8.7 \text{ cm.}$$

Use this sequence of calculator steps.

10 $\boxed{\times}$ 56 $\boxed{\text{SIN}}$ $\boxed{=}$ $\boxed{\div}$ 72 $\boxed{\text{SIN}}$ $\boxed{=}$.

Display: 8.7170169.

II. Determine c.

We select the proportion containing B, b, C, and c. The measure of C can be found as follows:

$$C = 180° - (A + B)$$
$$= 180° - (56° + 72°)$$
$$= 52°.$$

Substitute the values for *C*, *B*, and *b* into this proportion:

$$\frac{\sin B}{b} = \frac{\sin C}{c}$$

$$\frac{\sin 72°}{10} = \frac{\sin 52°}{c}.$$

Solve the resulting equation for *c*:

$$c = \frac{(10)\sin 52°}{\sin 72°} = 8.3 \text{ cm}.$$

10 $\boxed{\times}$ 52 $\boxed{\text{SIN}}$ $\boxed{=}$ $\boxed{\div}$ 72 $\boxed{\text{SIN}}$ $\boxed{=}$.

Display: 8.2856354.

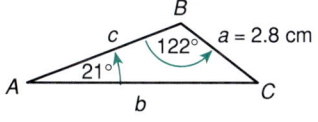

Figure 6.60

• *Example 6.19:* A blueprint for a circuit board calls for resistors to be located at points *A*, *B*, and *C* in figure 6.60. Find the length of *AC*.

Solution: Since we are given angles *A* and *B* and the length of the side opposite angle *A*, we use the proportion containing *A*, *B*, *BC*, and *AC*. Let *b* represent the length of *AC* and *a* represent the length of *BC*:

$$\frac{\sin A}{a} = \frac{\sin B}{b}.$$

Solve this proportion for *b* and substitute the values for *A*, *B*, and *a*:

$$b = \frac{(a)\sin B}{\sin A} = \frac{(2.8)\sin 122°}{\sin 21°} = 6.6 \text{ cm}.$$

Calculator solution

2.8 $\boxed{\times}$ 122 $\boxed{\text{SIN}}$ $\boxed{=}$ $\boxed{\div}$ 21 $\boxed{\text{SIN}}$ $\boxed{=}$.

Display: 6.6259683.

Examples 6.18 and 6.19 illustrated how to determine the remaining parts of a general triangle when two angles and a side are known (Case I). Example 6.20 demonstrates how to use the law of sines to solve a triangle when the lengths of two sides and the measure of an angle opposite one of the given sides are known. There may be some difficulty, however, since some problems have more than one solution.

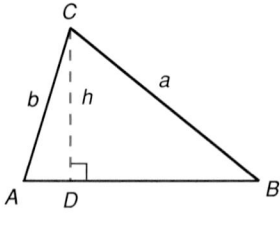

Figure 6.61

Suppose we are given the lengths of two sides a and b of $\triangle ABC$ in figure 6.61. Further suppose that the measure of angle A, the angle opposite side BC is given. Let h represent the length of the altitude CD to the other side AB. Then, in $\triangle ADC$,

$$\sin A = \frac{h}{b} \quad \text{or} \quad h = b \sin A.$$

We can determine h from the given information. Since a is the hypotenuse of $\triangle BDC$ and h is a leg of the triangle,

$$a \geq h \quad \text{and} \quad a \geq b \sin A.$$

There are four possibilities for the relationship between h, a, and b. These are illustrated in figure 6.62.

i. If $a = h$, then $\triangle ABC$ is a right triangle, and there is one solution to the problem.

ii. If $h < a < b$, there are two possible solutions. $\triangle AB_1C$ and $\triangle AB_2C$ in (ii) of figure 6.62 fit the required conditions.

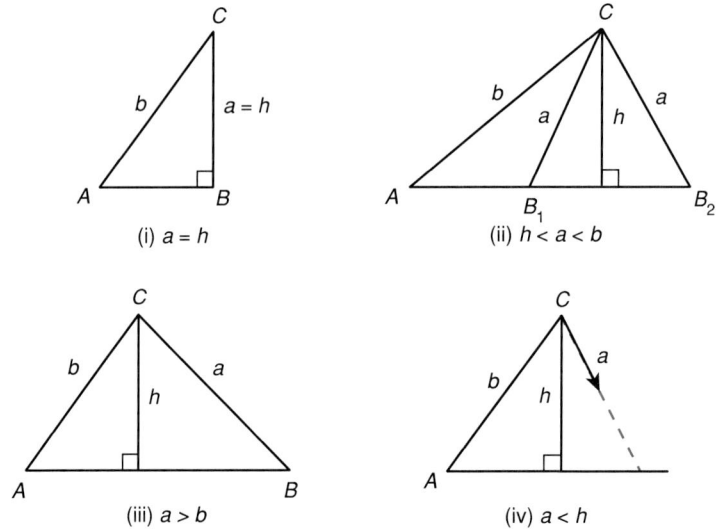

Figure 6.62

iii. If $a > b$, there is one solution.

iv. If $a < h$, there is no solution to the problem. Note that in reality we eliminated this case by showing that $a \geq h$ in our previous discussion.

Example 6.20 considers the case (ii), where there are two possible solutions to a problem.

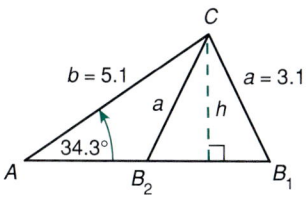

Figure 6.63

• **Example 6.20:** Given a triangle ABC with $a = 3.1$ cm, $b = 5.1$ cm, and $A = 34.3°$, determine the measures of angles B and C and the length of c.

Solution: We first check to see how many solutions we have to the problem.

$$h = b \sin A = (5.1) \sin 34.3° = 2.9$$

Since $2.9 < 3.1 < 5.1$

$$h < a < b,$$

we can expect to have two solutions to the problem. We solve the problem in three stages. First we find the two possible measures for angle B (B_1 and B_2 in figure 6.63). Then we take each of these possibilities for angle B and find the corresponding values of C and c.

1. Find the measures of B_1 and B_2:

$$\frac{\sin B}{b} = \frac{\sin A}{a}$$

$$\sin B = \frac{(b) \sin A}{a}$$

$$\sin B = \frac{(5.1) \sin 34.3°}{3.1} = 0.9270912.$$

The first possibility for B is $B_1 = 68.0°$. The other possibility for B is

$$B_2 = (180° - B_1) = 180° - 68° = 112°.$$

2. Use $B_1 = 68°$ and determine the corresponding C and c:

$$C = 180° - (A + B_1)$$
$$= 180° - (68° + 34.3°)$$
$$= 77.7°.$$

Use the law of sines to find c:

$$\frac{\sin C}{c} = \frac{\sin B}{b}$$

$$c = \frac{b \sin C}{\sin B} = \frac{(5.1) \sin 77.7°}{\sin 68°} = 5.4 \text{ cm}.$$

3. Use $B_2 = 112°$ and determine the corresponding values of C and c:

$$C = 180° - (A + B_2)$$
$$= 180° - (34.3° + 112°)$$
$$= 33.7°.$$

Use the law of sines to find c:

$$c = \frac{a \sin C}{\sin A} = \frac{(3.1)\sin 33.7°}{\sin 34.3°} = 3.1 \text{ cm}.$$

We can summarize these results by listing the dimensions of the two triangles that fit the conditions in the problem.

Triangle 1	Triangle 2
$a = 3.1$ cm	$a = 3.1$ cm
$b = 5.1$ cm	$b = 5.1$ cm
$c = 3.1$ cm	$c = 5.4$ cm
$A = 34.3°$	$A = 34.3°$
$B_2 = 112°$	$B_1 = 68°$
$C = 33.7°$	$C = 77.7°$

●●●●●●●●●●

Trial Problems 6.4

Before you begin the section exercises, warm up with these problems. Complete answers are included in the answer key.

1. If $\theta = 122°$, find $\sin \theta$ and $\csc \theta$.
2. If $A = 43°$, $B = 68°$, and $b = 14$, find a.
3. If $B = 62.3°$, $A = 66.8°$, and $a = 16.5$, find b.
4. If $C = 85.3°$, $c = 18.6$, and $B = 62.3°$, find b.
5. If $C = 105°$, $a = 19.6$, and $c = 28.3$, find A.

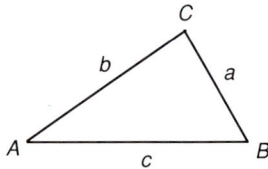

Exercises 6.4

For each of the given angles, determine the sine, cosine, tangent, cotangent, secant, and cosecant.

1. $\theta = 105°$
2. $\theta = 127°$
3. $\theta = 98°$
4. $\theta = 105°45'$

Using figure 6.64 as a guide, find the missing parts for each triangle.

Angle A	Angle B	Angle C	a	b	c
5. 80°	70°	?	10 in.	?	?
6. 80°	70°	?	25 cm	?	?

Angle A	Angle B	Angle C	a	b	c
7. 41.3°	65.2°	?	?	6.3 cm	?
8. 16.7°	62.3°	?	?	32.5 in.	?
9. ?	55°20′	35°30′	?	?	142 ft
10. ?	66°35′	72°15′	?	?	120 m
11. 65°	?	?	14 cm	11 cm	?
12. 65°	?	?	6.3 in.	4.1 in.	?
13. 55.2°	?	?	4.3 in.	6.7 in.	?
14. 45.7°	?	?	10 cm	13 cm	?

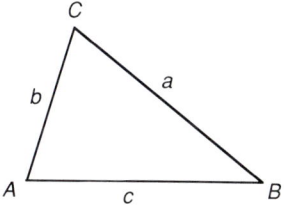

Figure 6.64

15. In figure 6.65 holes to be drilled at points B, C, D, E, and F are equally spaced around a circle with a radius of 4 in. Calculate the length of BC.

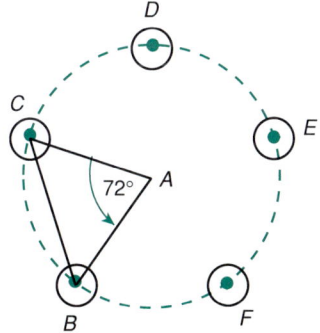

Figure 6.65

16. Solve exercise 15 if the radius of the circle is 10 cm.

17. Find the lengths of the sides of the plot of land shown in figure 6.66, given that $AB = 1250$ feet.

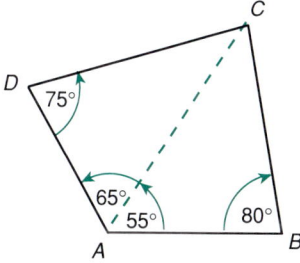

Figure 6.66

18. Solve exercise 17, given that $AB = 450$ m.

19. Two sides of a parallelogram in figure 6.67 intersect at an angle of 110°. One of the sides of the parallelogram is 25 inches long. The longer diagonal makes an angle of 30° with one of the 25 inch sides. Find the length of the other side.

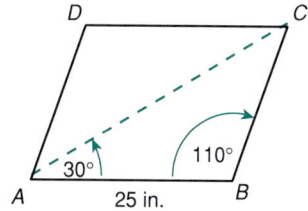

Figure 6.67

20. Solve exercise 19, given that the sides intersect at an angle of 120°, the side is 32 cm long, and the angle of intersection of the given side and the longer diagonal is 20°.

21. A surveyor sets up a transit at points A and B in figure 6.68 and measures the indicated angles. If $AB = 50$ ft, determine the lengths of AC, AD, BC, and BD.

22. Solve the problem in exercise 21 using the given angles and $AB = 30$ m.

23. Compute the chlorine (Cl) to chlorine bond length in figure 6.69 given that the carbon (C) radius is 0.77 A (angstrom), the chlorine radius is 0.99 A, and the bonding angle (C) is $120°$.

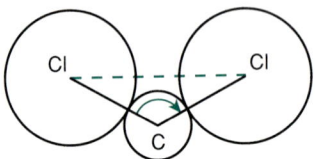

Figure 6.69

24. Solve exercise 23 given that the bonding angle is $109.5°$.

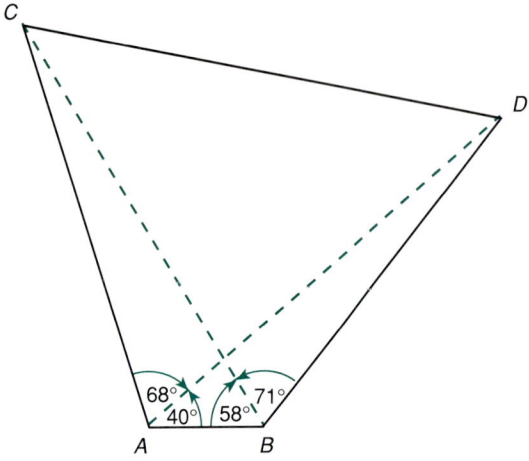

Figure 6.68

6 • 5 The Law of Cosines

In section 6.4 we solved triangles in which two angles and a side of a triangle were given. We now introduce a method for determining the parts of a triangle when two sides and the included angle are given, or when the lengths of the three sides are given.

In triangle ADC in figure 6.70

$$\cos A = \frac{x}{b} \quad \text{or} \quad x = b \cos A.$$

Using the Pythagorean theorem on triangles ADC and BDC, we get

$$h^2 = b^2 - x^2 \quad \text{and} \quad h^2 = a^2 - (c - x)^2.$$

Setting the two results for h^2 equal,

$$b^2 - x^2 = a^2 - (c - x)^2.$$

We solve this equation for a^2:

$$b^2 - x^2 = a^2 - c^2 + 2cx - x^2$$
$$a^2 = b^2 + c^2 - 2cx.$$

We substitute $b(\cos A)$ for x in this equation:

$$a^2 = b^2 + c^2 - 2bc(\cos A).$$

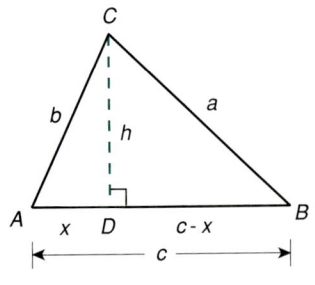

Figure 6.70

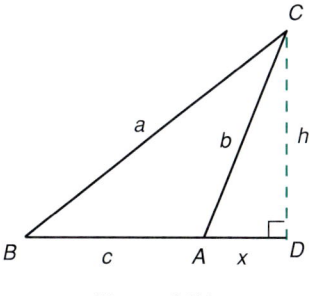

Figure 6.71

If angle A is an obtuse angle as in figure 6.71, the development of the formula is similar:

$$\cos A = -\cos(180° - A) = -\frac{x}{b} \quad \text{or} \quad x = -b \cos A.$$

We use the Pythagorean theorem on triangles ADC and BDC:

$$h^2 = b^2 - x^2 \quad \text{and} \quad h^2 = a^2 - (c + x)^2$$
$$b^2 - x^2 = a^2 - (c + x)^2.$$

We solve the resulting equation for a^2:

$$b^2 - x^2 = a^2 - c^2 - 2cx - x^2$$
$$a^2 = b^2 + c^2 + 2cx.$$

However, $x = -b \cos A$:

$$a^2 = b^2 + c^2 + 2c(-b \cos A)$$
$$a^2 = b^2 + c^2 - 2bc(\cos A).$$

This is the same result that we obtained when angle A was an acute angle. Since the choice of the positions of the angles in the triangle was arbitrary, we can obtain similar formulas for determining b^2 and c^2. The results are called the **law of cosines**.

Law of Cosines

LAW OF COSINES

For a triangle ABC, with a, b, and c the lengths of the sides opposite angles A, B, and C, respectively, the law of cosines is

$$a^2 = b^2 + c^2 - 2bc(\cos A)$$
$$b^2 = a^2 + c^2 - 2ac(\cos B)$$
$$c^2 = a^2 + b^2 - 2ab(\cos C).$$

The law of cosines can be used to solve triangles of the following types:

1. Two sides and the included angle are known.
2. Three sides are known.

Examples 6.21 and 6.22 illustrate problems of the first type and example 6.23 illustrates a problem of the second type.

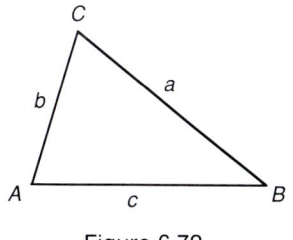

Figure 6.72

- **Example 6.21:** For $\triangle ABC$ in figure 6.72, suppose $\angle C$ measures $65°$, $a = 6.5$ cm, and $b = 14.5$ cm. Determine the length of c and the measures of angles A and B.

240 CHAPTER 6 NUMERICAL TRIGONOMETRY

Solution: We solve the problem in three stages.

1. Use the law of cosines to determine c.
2. Use the law of sines to determine $\angle A$.
3. Find B using $A + B + C = 180°$.

1. $c^2 = a^2 + b^2 - (2ab)\cos C$.

Substitute the given values of a, b, and C into the formula.

$$c^2 = (6.5)^2 + (14.5)^2 - 2(6.5)(14.5) \cos 65°$$
$$c^2 = 42.25 + 210.25 - 2(6.5)(14.5)(0.4226)$$
$$c^2 = 42.25 + 210.25 - 79.6635$$
$$c^2 = 172.8365$$
$$c = 13.15 \text{ cm.}$$

6.5 $\boxed{x^2}$ $\boxed{+}$ 14.5 $\boxed{x^2}$ $\boxed{-}$ 2 $\boxed{\times}$ 6.5 $\boxed{\times}$ 14.5 $\boxed{\times}$ 65 $\boxed{\text{COS}}$ $\boxed{=}$ $\boxed{\sqrt{x}}$.

Display: 13.146728.

2. $\dfrac{\sin C}{c} = \dfrac{\sin A}{a}$

$$\sin A = \dfrac{(a)\sin C}{c}$$

$$\sin A = \dfrac{(6.5)\sin 65°}{13.15} = 0.4480$$

$$A = 27°.$$

A calculator sequence for this solution is

6.5 $\boxed{\times}$ 65 $\boxed{\text{SIN}}$ $\boxed{=}$ $\boxed{\div}$ 13.15 $\boxed{=}$ $\boxed{\text{INV}}$ $\boxed{\text{SIN}}$.

Display: 26.614467.

3. $A + B + C = 180°$.
Substitute $A = 27°$ and $C = 65°$ and solve for B:

$$27° + B + 65° = 180°$$
$$B = 88°.$$

• *Example 6.22:* Two forces of 140 pounds and 180 pounds act on an object at an angle of 56° with each other. Find the magnitude of the resultant force in figure 6.73.

SECTION 6.5 THE LAW OF COSINES

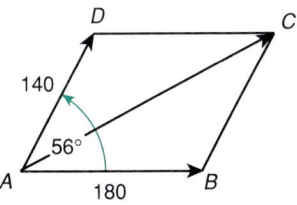

Figure 6.73

Solution: The resultant force is determined by the length of the diagonal AC of the parallelogram in which $AD = 140$ and $AB = 180$. In a parallelogram the sum of the measures of adjacent angles is $180°$:

$$\angle A + \angle B = 180°$$
$$\angle B = 180° - 56° = 124°.$$

We use the law of cosines to determine the length of AC:

$$(AC)^2 = (AB)^2 + (BC)^2 - 2(AB)(BC)\cos B$$
$$(AC)^2 = (180)^2 + (140)^2 - 2(180)(140)\cos 124°$$
$$(AC)^2 = 32{,}400 + 19{,}600 - 2(180)(140)(-0.5592)$$
$$(AC)^2 = 32{,}400 + 19{,}600 + 28{,}183.68 = 80{,}183.68$$
$$AC = 283.2 \text{ pounds.}$$

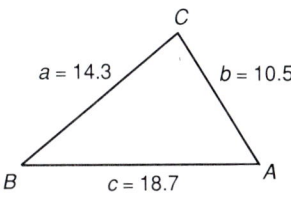

Figure 6.74

• *Example 6.23:* The triangle in figure 6.74 has the three sides given: $a = 14.3$ cm, $b = 10.5$ cm, and $c = 18.7$ cm. Determine the measures of the three angles of the triangle.

Solution: Choose any angle, say angle C, and use the law of cosines to determine the measure of C:

$$c^2 = a^2 + b^2 - 2ab \cos C.$$

Solve this equation for $\cos C$:

$$\cos C = \frac{a^2 + b^2 - c^2}{2ab}$$

Substitute the values of a, b, and c into the equation:

$$\cos C = \frac{(14.3)^2 + (10.5)^2 - (18.7)^2}{2(14.3)(10.5)}$$
$$= \frac{204.49 + 110.25 - 349.69}{300.3}$$
$$= \frac{-34.95}{300.3} = -0.11638.$$

Note that $\cos C$ is negative, so $\angle C$ is $> 90°$. However, the calculator takes care of this automatically:

$$\text{angle } C = 96.7°.$$

A calculator sequence for this solution is

14.3 $\boxed{x^2}$ $\boxed{+}$ 10.5 $\boxed{x^2}$ $\boxed{-}$ 18.7 $\boxed{x^2}$ $\boxed{=}$ $\boxed{\div}$ 2 $\boxed{\div}$ 14.3 $\boxed{\div}$ 10.5 $\boxed{=}$ $\boxed{\text{INV}}$ $\boxed{\text{COS}}$.

Display: 96.683436.

We can determine the measures of angles A and B by using either the law of sines or the law of cosines. Since the law of sines requires fewer calculations, we select this method:

$$\frac{\sin A}{14.3} = \frac{\sin 96.7°}{18.7}$$

$$\sin A = \frac{(14.3)(0.99317)}{18.7} = 0.75948$$

$$A = 49.4°$$
$$B = 180° - (A + C)$$
$$ = 180° - (49.4° + 96.7°)$$
$$ = 33.9°.$$

•••••••••

Example 6.24 illustrates a method for determining the distance between two points using a surveyor's transit. What makes the problem interesting is that the surveyor does not need to make any measurements from either of the points.

• *Example 6.24:* A surveyor is able to set up a transit at points A and B in figure 6.75 and take the following measurements:

$$AB = 100 \text{ ft}, \qquad a_1 = 79°40',$$
$$a_2 = 34°35', \qquad b_1 = 95°5',$$
$$b_2 = 25°20'.$$

Determine the length of CD.

Solution: We attack the problem in three stages. First we use triangle ABD and the law of sines to find AD. Then we use triangle ABC and the law of sines to find AC. Finally, we use triangle ACD and the law of cosines to find CD.

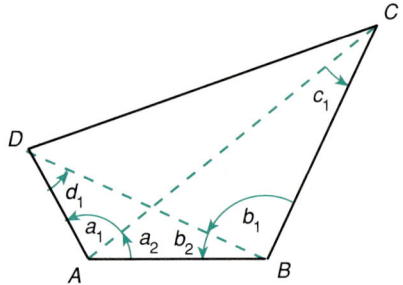

Figure 6.75

C
D
d_1
a_1
a_2
b_2
A
B
b_2= 25° 20'
d_1 = 40° 25'

Figure 6.76

C
c_1
D
b_1
a_2
b_2
A
B
c_1 = 25°
b_1 = 95° 5'
b_2= 25° 20'

Figure 6.77

1. In triangle ABD in figure 6.76,

$$\begin{aligned} d_1 &= 180° - (a_1 + a_2 + b_2) \\ &= 180° - (79°40' + 34°35' + 25°20') \\ &= 40°25'. \end{aligned}$$

Use the law of sines to find AD:

$$\frac{AD}{\sin b_2} = \frac{AB}{\sin d_1}$$

$$\begin{aligned} AD &= \frac{(AB)\sin 25°20'}{\sin 40°25'} \\ &= \frac{(100)(0.42788)}{0.64834} \\ &= 66.0 \text{ ft.} \end{aligned}$$

2. In triangle ABC in figure 6.77,

$$\begin{aligned} c_1 &= 180° - (25°20' + 34°35' + 95°5') \\ &= 25°. \end{aligned}$$

Use the law of sines to find AC:

$$\frac{AC}{\sin(b_1 + b_2)} = \frac{AB}{\sin c_1}$$

$$\begin{aligned} AC &= \frac{(AB)\sin(25°20' + 95°5')}{\sin 25°} \\ &= \frac{(100)(0.86240)}{0.42262} \\ &= 204.06 \text{ ft.} \end{aligned}$$

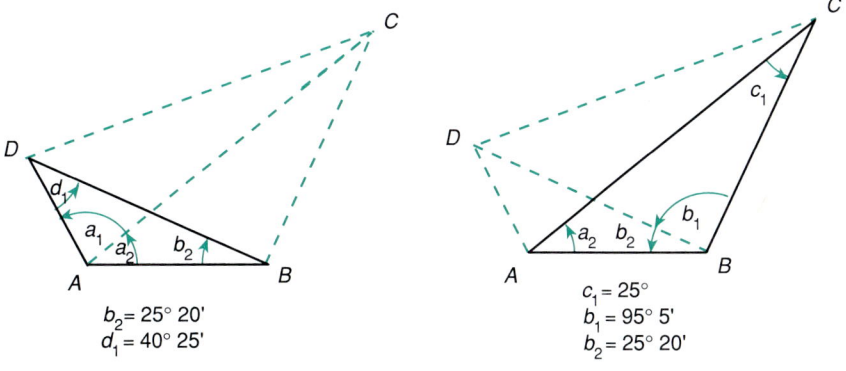

Chapter 6 Numerical Trigonometry

Figure 6.78

3. Use the law of cosines on triangle ACD to find CD in figure 6.78:

$$(CD)^2 = (AD)^2 + (AC)^2 - 2(AD)(AC)\cos a_1$$
$$= (66.0)^2 + (204.06)^2 - 2(66.0)(204.06)(0.179)$$
$$= 4356.0 + 41,640.484 - 4821.52968$$
$$= 41,174.954$$
$$CD = \sqrt{41,174.954} = 202.92 \text{ ft.}$$

Trial Problems 6.5

Before you begin the section exercises, warm up with these problems. Complete answers are included in the answer key.

1. If $a = 17$, $b = 14$, and $C = 98°$, find c.

2. If $c = 18.6$, $a = 13.3$, and $B = 46.3°$, find b.

3. If $c = 0.45$, $b = 0.86$, and $A = 25°30'$, find a.

4. If $a = 14.3$, $b = 16.7$, and $c = 20.4$, find A.

5. If $a = 0.72$, $b = 0.85$, and $c = 0.98$, find C.

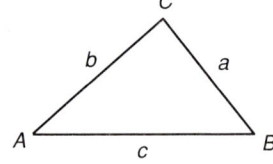

Exercises 6.5

Using figure 6.79 as a guide, determine the values indicated in the table.

	Angle A	Angle B	Angle C	a	b	c
1.	80°	?	?	?	10 in.	14 in.
2.	40°	?	?	?	12.1 in.	6.3 in.
3.	?	60°	?	26 ft	?	16 ft
4.	?	35°	?	$\frac{7}{8}$ in.	?	$\frac{1}{2}$ in.
5.	?	?	67°	45 ft	180 ft	?

SECTION 6.5 THE LAW OF COSINES

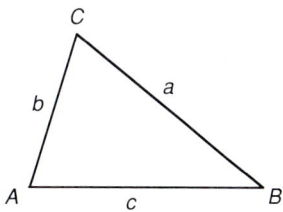

Figure 6.79

Angle A	Angle B	Angle C	a	b	c
6. ?	?	108°	16 in.	18 in.	?
7. 80°25'	?	?	?	25 cm	21 cm
8. 40°30'	?	?	?	12.5 cm	15 cm
9. ?	60°25'	?	9 m	?	5 m
10. ?	35°35'	?	22 mm	?	12 mm
11. ?	?	?	150 m	90 m	142.1 m
12. ?	?	?	24 cm	27 cm	41.4 mm

13. A surveyor wishes to determine the distance BC across the lake in figure 6.80. Determine BC if angle A measures 52°, $AB = 600$ ft, and $AC = 800$ ft.

16. Two forces AB, with a magnitude of 26 lb, and AC, with a magnitude of 42 lb, act at point A. The angle between the two forces is 48°. Find the magnitude of the resultant force.

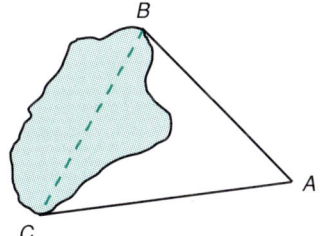

Figure 6.80

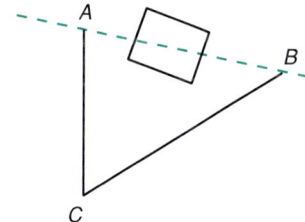

Figure 6.81

14. Solve exercise 13 if angle A measures 52°, $AB = 89.4$ m, and $AC = 119.2$ m.

15. Two planes are approaching an airport at the same altitude. One plane approaches from due west at a distance of 14.3 mi from the runway. The other plane approaches at an angle of 21° north of due west from a distance of 17.1 mi. How far apart are the two planes?

17. A water line passes under a building. The company plans to replace section AB in figure 6.81. It can be determined that angle C measures 67°, $AC = 120$ m, and $BC = 140$ m. How much pipe will the company need?

18. Solve exercise 17, given angle $C = 61°25'$, $AC = 87.2$ ft, and $BC = 98.3$ ft.

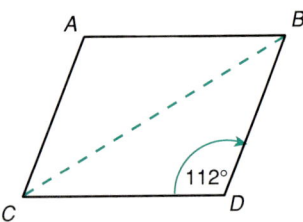

Figure 6.82

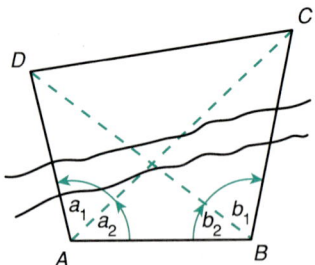

Figure 6.83

19. Adjacent sides of a parallelogram in figure 6.82 are 12 in. and 7 in. long. One of the larger angles contains 112°. Find the lengths of the diagonals.

20. Solve exercise 19, given that the lengths of adjacent sides are 14.3 cm and 8.7 cm.

21. A surveyor sets up a transit at points A and B in figure 6.83 and makes the following measurements. $AB = 100$ ft, $a_1 = 53°10'$, $a_2 = 34°20'$, $b_1 = 87°10'$, and $b_2 = 53°40'$.
    a. Determine the length of AC.
    b. Determine the length of AD.
    c. Use the results of parts a and b to determine the length of DC.

22. In exercise 21, use the following measurements: $AB = 40$ m, $a_1 = 117°$, $a_2 = 31°40'$, $b_1 = 113°$, and $b_2 = 20°20'$.
    a. Determine the length of AC.
    b. Determine the length of AD.
    c. Use the results of parts a and b to determine the length of DC.

6·6 Summary of Terms, Formulas, Rules, and Procedures

Terms

Angle (p. 202)
Angle in Standard Position (p. 202)
Complementary Angles (p. 204)
Cosecant of an Angle (p. 225)
Cosine of an Angle (p. 220)
Cotangent of an Angle (p. 225)
Coterminal Angles (p. 204)
Hypotenuse (p. 205)

Isosceles Triangle (p. 210)
Law of Cosines (p. 239)
Law of Sines (p. 231)
Legs of a Right Triangle (p. 205)
Obtuse Angles (p. 229)
Right Angle (p. 204)
Right Triangle (p. 205)
Secant of an Angle (p. 225)

Similar Triangles (p. 207)
Sine of an Angle (p. 220)
Supplementary Angles (p. 205)
Tangent (p. 212)
Taper Angle (p. 216)
Tapered (p. 216)
Theorem of Pythagoras (p. 205)

Formulas

Theorem of Pythagoras: In a right triangle, the square of the hypotenuse is equal to the sum of the squares of the legs, that is, $c^2 = a^2 + b^2$ in figure 6.84.   (6.1)

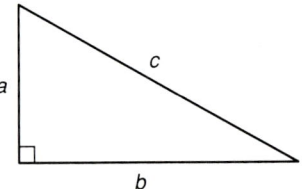

Figure 6.84

Section 6.6 Summary

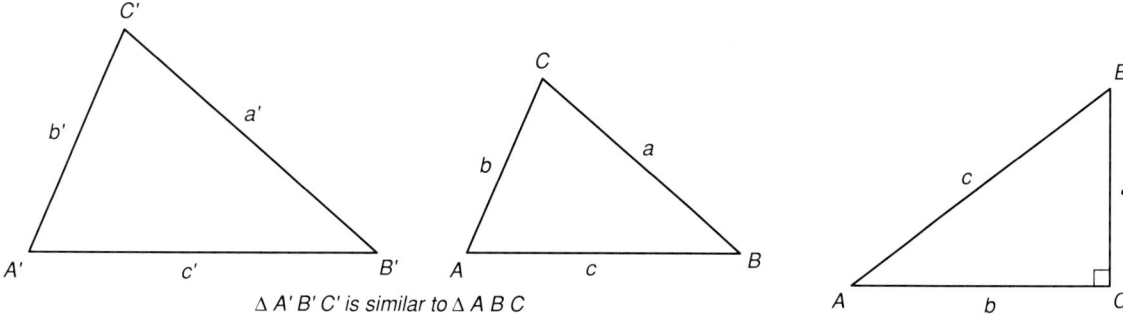

△ A'B'C' is similar to △ ABC

Figure 6.85

Figure 6.86

Similar Triangles: Triangle ABC is similar to triangle $A'B'C'$ if $\angle A = \angle A'$, $\angle B = \angle B'$, and $\angle C = \angle C'$ (figure 6.85). (6.1)

For similar triangles ABC and $A'B'C'$, $\dfrac{a}{a'} = \dfrac{b}{b'} = \dfrac{c}{c'}$.

Trigonometric Ratios: For a right triangle ABC (figure 6.86):

$\sin A = \dfrac{a}{c}$ (6.3) $\qquad \sin B = \dfrac{b}{c}$ (6.3)

$\cos A = \dfrac{b}{c}$ (6.3) $\qquad \cos B = \dfrac{a}{c}$ (6.3)

$\tan A = \dfrac{a}{b}$ (6.2) $\qquad \tan B = \dfrac{b}{a}$ (6.2)

$\cot A = \dfrac{b}{a}$ (6.3) $\qquad \cot B = \dfrac{a}{b}$ (6.3)

$\sec A = \dfrac{c}{b}$ (6.3) $\qquad \sec B = \dfrac{c}{a}$ (6.3)

$\csc A = \dfrac{c}{a}$ (6.3) $\qquad \csc B = \dfrac{c}{b}$ (6.3)

Rules and Procedures

- The sum of the measures of complementary angles equals 90°. (6.1)
- The sum of the measures of supplementary angles equals 180°. (6.1)
- The sum of the measures of the angles of a triangle equals 180°. (6.1)

- SOLVING RIGHT TRIANGLES (6.1)
  1. Select a trigonometric ratio that contains the two known quantities and the unknown quantity.
  2. Substitute the known quantities.
  3. Solve the resulting equation.

- LAW OF SINES. For any triangle ABC (figure 6.87), (6.4)

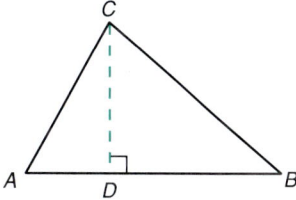

Figure 6.87

$$\dfrac{\sin A}{a} = \dfrac{\sin B}{b} = \dfrac{\sin C}{c}.$$

Use the law of sines to solve triangles when
I. Two angles and a side of a triangle are known.
II. Two sides and an angle opposite one of the given sides are known.

For case II listed above, if a, b, and angle A are given, then $h = b \sin A$ and

a. If $a = h$, then the triangle is a right triangle and there is one solution to the problem.
b. If $h < a < b$, there are two solutions to the problem.
c. If $a > b$, there is one solution to the problem.

- LAW OF COSINES. For any triangle ABC (figure 6.88), (6.5)

$$a^2 = b^2 + c^2 - 2bc \cos A$$
$$b^2 = a^2 + c^2 - 2ac \cos B$$
$$c^2 = a^2 + b^2 - 2ab \cos C.$$

Use the Law of Cosines to solve triangles when:
I. Two sides and the included angle are known.
II. Three sides are known.

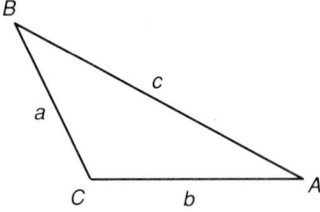

Figure 6.88

6·7 Chapter 6 Review Exercises

Determine $\sin \theta$, $\cos \theta$, $\tan \theta$, $\cot \theta$, $\sec \theta$, and $\csc \theta$ for each of the given angles.

1. $\theta = 37°$
2. $\theta = 41°$
3. $\theta = 32°18'$
4. $\theta = 62°55'$

Using figure 6.89 as a guide, use the Pythagorean theorem to determine the lengths of the missing sides.

5. $a = 7.2$ cm, $b = 6.3$ cm
6. $a = 115.2$ in., $b = 137.2$ in.
7. $a = 7.83$ in., $c = 8.4$ in.
8. $b = 141.6$ cm, $c = 184.6$ cm

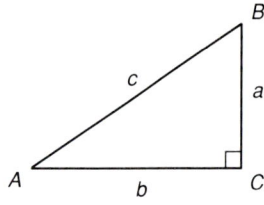
Figure 6.89

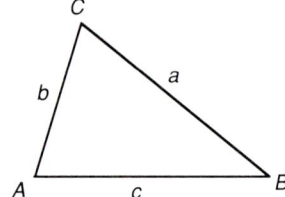
Figure 6.90

Using figure 6.89 as a guide, determine the indicated missing part for each triangle.

9. $a = 3.5$, $b = 7.8$, find $\angle A$
10. $a = 14.2$, $b = 6.3$, find $\angle A$
11. $a = 7.8$, $c = 14.5$, find $\angle B$
12. $a = 6.1$, $c = 17.5$, find $\angle B$
13. $\angle A = 63°10'$, $a = 11.3$, find c
14. $\angle A = 41°50'$, $a = 7.6$, find c

Using figure 6.90 as a guide, determine the missing indicated parts for each $\triangle$.

15. $\angle A = 33°$, $\angle C = 104°$, $a = 6.3$ in., determine $\angle B$, b, c.
16. $\angle A = 37°$, $\angle C = 96°$, $a = 7.8$ cm, determine $\angle B$, c.
17. $a = 3.1$ cm, $b = 5.1$ cm, $\angle A = 36.1°$, determine $\angle B$, $\angle C$, c.
18. $a = 1.2$ in., $b = 2.0$ in., $\angle A = 33.4°$, determine $\angle B$, $\angle C$, c.
19. $\angle C = 65°20'$, $a = 6.4$ cm, $b = 14.2$ cm, determine $\angle A$, $\angle B$, c.
20. $\angle C = 62°30'$, $a = 16.3$ in., $b = 36.1$ in., determine $\angle A$, $\angle B$, c.
21. $a = 14$ cm, $b = 10$ cm, $c = 18$ cm, determine $\angle A$, $\angle B$, $\angle C$.
22. $a = 5.5$ in., $b = 3.9$ in., $c = 7.1$ in., determine $\angle A$, $\angle B$, $\angle C$.

23. Determine the length of a diagonal of a square having sides that are 14.3 cm long.

24. A square has a diagonal that is 18.7 in. long. Determine the lengths of the sides.

25. Using the formula $P = EI \cos \theta$, determine the power in a circuit given that $\theta = 60°$, $E = 110$ volts, and $I = 15$ amperes.

26. Using the formula $P = EI \cos \theta$, determine the voltage E in a circuit, given that $P = 825$ watts, $I = 15$ amperes, and $\theta = 60°$.

27. One side of a canyon is 115 ft higher than the other side. The angle of elevation from the lower rim to the higher rim is $14°20'$. Determine the width of the canyon.

28. The horizontal distance across a canyon is 825 ft. The angle of elevation from the lower rim to the higher rim is $11°35'$. How much higher is the higher rim than the lower rim?

29. Two forces of 80 lb and 100 lb act on an object at an angle of 60° with each other. Find the magnitude of the resultant force.

30. Two forces of 60 kg and 80 kg act on an object at an angle of 64° with each other. Find the magnitude of the resultant force.

31. A surveyor makes the following measurements in figure 6.91. $AB = 50$ ft, $a_1 = 79°$, $a_2 = 34°$, $b_1 = 26°$, $b_2 = 91°$. Determine the lengths of AD, AC, BD, BC, and CD.

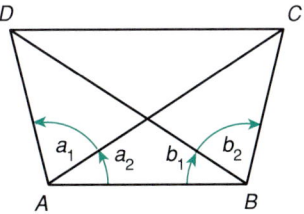

Figure 6.91

32. A surveyor makes the following measurements in figure 6.91: $AB = 32$ m, $a_1 = 80°$, $a_2 = 36°$, $b_1 = 29°$, and $b_2 = 88°$. Determine the lengths of AD, AC, BC, BD, and CD.

33. In figure 6.92, $\angle A = 62°$, $\angle BDC = 78°$, and $AD = 35$ ft. Determine the length of BC.

34. In figure 6.92, $\angle A = 56°$, $\angle BDC = 72°$, and $AD = 35$ m. Determine the length of BC.

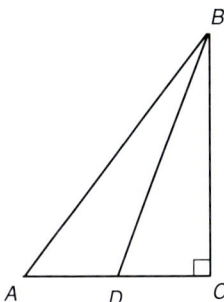

Figure 6.92

35. For the AC series circuit in figure 6.93, V_R, V_L, and V_C represent the voltages across the resistor, the inductor, and the capacitor. These are related to the phase angle θ by the formula

$$\tan \theta = \frac{V_L - V_C}{V_R}.$$

Given that $\theta = 22.5°$, $V_R = 11.3$, and $V_L = 22.3$, determine V_C.

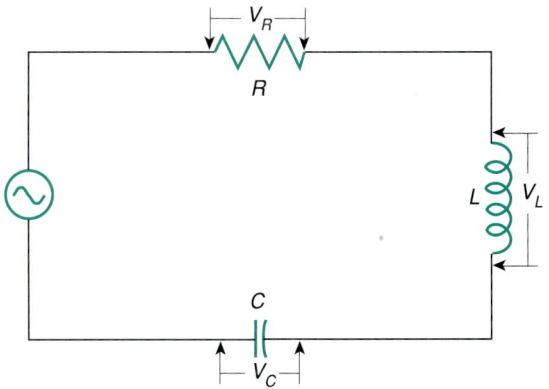

Figure 6.93

36. Solve exercise 35, given that $\theta = 60°$, $V_R = 11.8$, and $V_L = 30.7$.

37. Three different types of atoms are bonded as indicated in figure 6.94. Determine the distance AC.

38. Solve exercise 37 if the angle is changed to 29°.

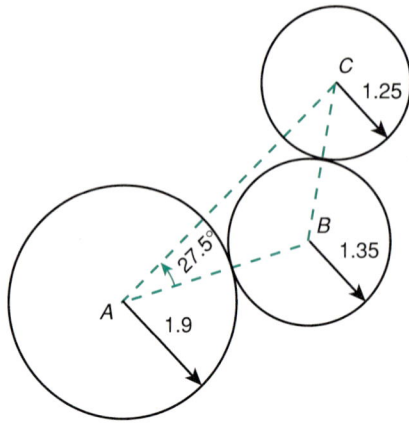

Figure 6.94

The identity $(a - b)\cos\dfrac{C}{2} = c\sin\dfrac{(A - B)}{2}$ is often used to check the final solution of a general triangle since all six parts of the triangle are involved in the identity. Use this identity to check whether the following solutions to triangles are valid.

39. $\angle A = 73°40'$, $\angle B = 60°$, $\angle C = 46°20'$, $a = 4$, $b = 3.61$, and $c = 3$

40. $\angle A = 127°$, $\angle B = 29.43°$, $\angle C = 23.57°$, $a = 26$, $b = 16$, and $c = 13$

41. $\angle A = 77°$, $\angle B = 41°$, $\angle C = 62°$, $a = 110$, $b = 74$, and $c = 80$

42. $\angle A = 76.1°$, $\angle B = 54°$, $\angle C = 49.9°$, $a = 150$, $b = 125$, and $c = 92$

6·8 Chapter 6 Test

1. If two angles of a triangle measure 65°20' and 75°15', find the measure of the third angle.

2. Find the measure of the angle that is the supplement of an angle that measures 108°55'.

3. Find the length of a leg of a right triangle if the hypotenuse is 8.9 cm long and the other leg measures 5.3 cm.

4. Find the length of the hypotenuse of a right triangle that has legs that measure 114.3 in. and 132.5 in.

5. A triangle has sides with lengths 3.5 in., 4.6 in., and 5.8 in. A second triangle, which is similar to the first triangle, has a side of length 10.5 in. This side corresponds to the 4.6 in. side in the first triangle. Find the lengths of the remaining sides of the second triangle.

6. If the tangent of an angle is 0.7002, find the sine and the cosine of the angle.

7. Given that the sides of a right triangle measure 5.3 cm, 8.6 cm, and 10.1 cm, find the sine, cosine, and tangent of the angle that is opposite the shortest side of the triangle.

8. The angle of elevation to the top of a tower at a point that is 120 ft from the base of the tower is 65.3°. Find the height of the tower.

9. A square has sides that are 14.8 cm long. Find the length of the diagonal of the square.

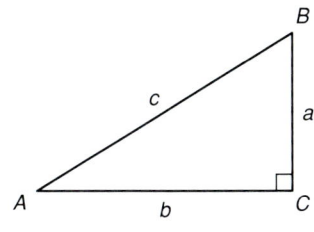

Figure 6.95

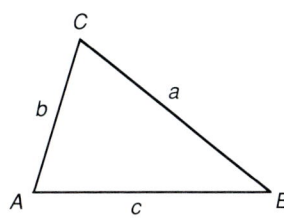

Figure 6.96

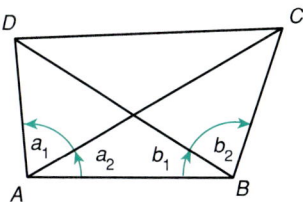

Figure 6.97

Use figure 6.95 as a guide for problems 10–13.

10. $a = 1.7$, $b = 2.3$, find A.

11. $a = 143$, $b = 263$, find B.

12. $A = 55°$, $a = 16.3$, find c.

13. $B = 45°20'$, $b = 18.3$, find c.

Use figure 6.96 as a guide in solving problems 14–17.

14. $A = 56°$, $C = 94°$, $a = 8.3$ cm, find b and c.

15. $a = 1.5$ in., $b = 2.3$ in., $A = 42.3°$, find C and c.

16. $a = 14$ cm, $b = 16$ cm, $c = 12$ cm, find A and B.

17. $C = 60°25'$, $a = 18.3$, $b = 14.6$, find A and c.

18. Using the formula $P = EI \cos \theta$, determine the voltage (E) in a circuit if $P = 725$ watts, $I = 10$ amps, and $\theta = 60°$.

19. Two forces of 25 kg and 65 kg act on an object. If the angle between the forces is 56°, find the magnitude of the resultant force.

20. In figure 6.97, $AB = 75$ ft, $a_1 = 72°$, $a_2 = 26°$, $b_1 = 28°$, and $b_2 = 95°$. Calculate the lengths of BD, BC, and DC.

Chapter 7

Analytic Trigonometry

Johannes Müller (1436–1476), one of the most influential mathematicians of the fifteenth century, was more widely known as Regiomontanus, the latinized form of his birthplace, Köenigsberg, Germany. His *De Triangulis Omnimodus* (*On Triangles of All Kinds*) was one of the first publications to develop trigonometry as a separate branch of mathematics independent of astronomy. He also devised procedures to determine a triangle that satisfies a given set of conditions, for example, determine a triangle given a side, the altitude on this side, and the ratio of the other two sides. He later computed a table of tangents and finally, published an almanac, *Ephemerides Astronomicae*. Some historians claim that Columbus carried the almanac with him on his four voyages to the west and used Regiomontanus' prediction of a lunar eclipse to frighten natives into providing provisions for his ships.

Chapter 6 provided an introduction to the trigonometric ratios based on the lengths of the sides of right triangles. Many natural phenomena repeat over definite periods of time, for example, alternating electrical current, sound waves, and the angular displacement of pendulums. The trigonometric functions are also repeating functions, and they are useful in analyzing these and other phenomena. In analytic trigonometry, the trigonometric functions are defined in terms of real numbers rather than in terms of measures of angles.

7 • 1 Radian Measure

Figure 7.1 shows a circle with radius r. If an arc with length r is marked off on the circumference of the circle from the x-axis in a counterclockwise direction, the central angle that intercepts this arc has a measure of one **radian**.

Radian

Referring to figure 7.2, the radian measure of an angle θ is defined as follows.

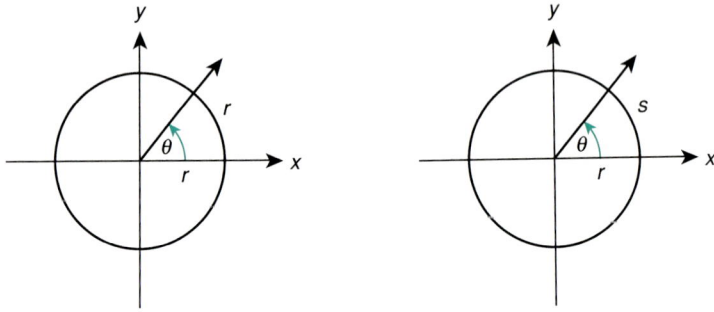

Figure 7.1   Figure 7.2

RADIAN MEASURE OF AN ANGLE

$$\theta \text{ (radians)} = \frac{s \text{ (arc length)}}{r \text{ (radius of the circle)}}.$$

Note that the radian measure of an angle is independent of the radius of the circle.

Since the circumference of a circle of radius r is $C = 2\pi r$, and the circle contains 360°, 2π radians = 360° and π radians = 180°. This relationship leads to the following formulas for converting between degrees and radians.

CONVERTING BETWEEN DEGREES AND RADIANS

Degrees to Radians:

$$\theta \text{ degrees} = \frac{\theta \pi}{180} \text{ radians}.$$

Radians to Degrees:

$$t \text{ radians} = \frac{180t}{\pi} \text{ degrees}.$$

• *Example 7.1:*   a. 60° = ? radians

b. 135° = ? radians

c. $\frac{\pi}{4}$ radians = ? degrees

d. 2.7 radians = ? degrees

Solution: **a.** $60° = \dfrac{60\pi}{180}$ radians $= \dfrac{\pi}{3}$ radians

b. $135° = \dfrac{135\pi}{180}$ radians $= \dfrac{3\pi}{4}$ radians

c. $\dfrac{\pi}{4}$ radians $= \left(\dfrac{180(\pi/4)}{\pi}\right)° = 45°$

d. 2.7 radians $= \left(\dfrac{180(2.7)}{\pi}\right)° = \left(\dfrac{180(2.7)}{3.1416}\right)° = 154.7°$ (approx.)

Example 7.2 illustrates the use of the formula

$$\theta = \dfrac{s}{r}$$

to solve a problem involving a pendulum.

• **Example 7.2:** A pendulum swings through an arc of $30°$. The arc formed by a swing of the pendulum is 28 in. long. Find the length of the pendulum.

Solution: **1.** Change $30°$ to radians:

$$30° = 30\left(\dfrac{\pi}{180}\right) \text{ radians} = \dfrac{\pi}{6} \text{ radians.}$$

2. Use the formula $\theta = \dfrac{s}{r}$ with $s = 28$ and $\theta = \dfrac{\pi}{6}$ to calculate the length of the pendulum:

$$\dfrac{\pi}{6} = \dfrac{28}{r}$$

$$r = \dfrac{(28)(6)}{\pi} = 53 \text{ in.}$$

Sector of a Circle

The area of a sector of a circle can be found when the measure of the central angle and the radius of the circle are known. Figure 7.3 shows a sector of the circle as the shaded area. A **sector of a circle** is a region bounded by the sides of a central angle and the circle.

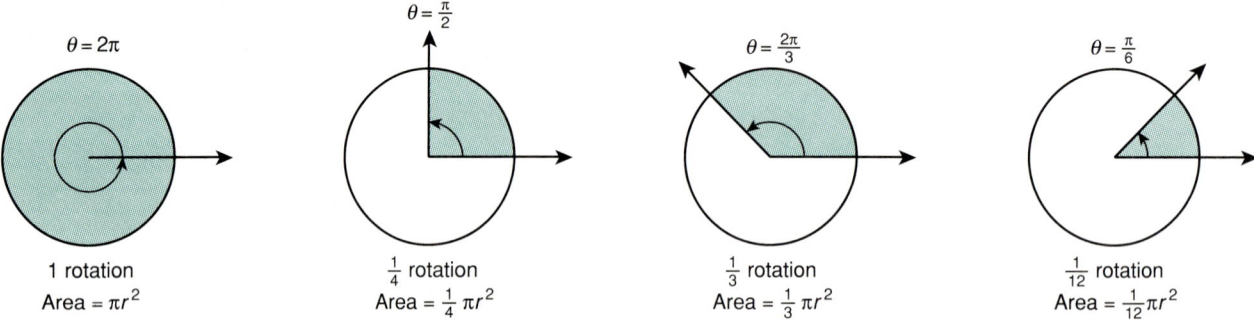

Figure 7.3

Since we know that one revolution (2π radians) results in the area of the entire circle (πr^2), we can set up a proportion to determine the area of a sector using the measure of the central angle in radians:

$$\frac{\text{Area of sector}}{\text{Area of circle}} = \frac{\text{Central angle in radians}}{\text{One complete revolution in radians}}$$

$$\frac{A}{\pi r^2} = \frac{\theta \text{ rad}}{2\pi \text{ rad}}$$

$$A = \frac{\theta}{2\pi}(\pi r^2) = \frac{1}{2}r^2\theta.$$

AREA OF A SECTOR OF A CIRCLE

$$A = \frac{1}{2}r^2\theta,$$

where r = radius of the circle and θ = measure of the central angle in radians.

• *Example 7.3* Find the area of the sector of a circle that has a radius of 14.3 ft and an angle of $46°22'$.

Solution: First change $46°22'$ to decimal degrees, then to radians, and second, find the area of the sector.

1. $46°22' = \left(46 + \frac{22}{60}\right)° = 46.36667°$

$$46.36667° = (46.36667)\left(\frac{\pi}{180}\right) \text{ radians}$$

$$= 0.809251 \text{ radians}.$$

2. Use the formula $A = \frac{1}{2}r^2\theta$ with $r = 14.3$ and $\theta = 0.809251$.

$$A = \frac{1}{2}(14.3)^2(0.809251) = 82.7.$$

The area is 82.7 sq. ft.
The key strokes for this solution are

14.3 $\boxed{x^2}$ $\boxed{\times}$ 0.809251 $\boxed{\div}$ 2 $\boxed{=}$.

Display: 82.741868.

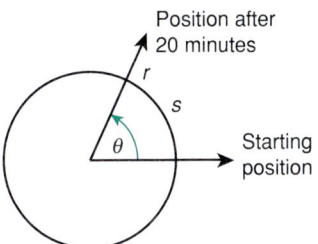

Figure 7.4

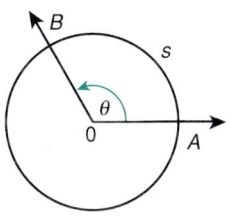

Figure 7.5

Linear Velocity

• *Example 7.4:* The radial arm in an irrigation system is 80.0 ft long. It makes one revolution every 2 hours (120 min).

a. How far does the end of the arm travel in 20 minutes?

b. How much area is irrigated in 20 minutes? (See figure 7.4.)

Solution: **a.** In 20 minutes the arm makes $\frac{20}{120} = \frac{1}{6}$ of a revolution. The circumference of the circle is

$$2\pi r = 2(\pi)80 = 160\pi.$$

Hence, in 20 minutes the arm travels

$$(160\pi)\left(\frac{1}{6}\right) \text{ ft} = 83.8 \text{ ft}.$$

b. Since the arm makes $\frac{1}{6}$ of a revolution, $\theta = 2\pi\left(\frac{1}{6}\right) = \frac{\pi}{3}$.

$$A = \frac{1}{2}r^2\theta = \frac{1}{2}(80)^2\left(\frac{\pi}{3}\right) = 3350 \text{ sq ft}.$$

From a starting point (point A in figure 7.5), a particle moves along an arc of a circle to point B during a time t. The **linear velocity** of the particle is given by the following formula.

CHAPTER 7 ANALYTIC TRIGONOMETRY

LINEAR VELOCITY

$$v = \frac{s}{t}$$

where $s =$ length of the arc and $t =$ time.

Angular Velocity

If the central angle generated by the motion of the particle measures θ radians, the **angular velocity** (the rate of change in θ with respect to the time t) is given by this formula.

ANGULAR VELOCITY

$$\omega = \frac{\theta}{t}$$

where

$\theta =$ measure of the central angle in radians
$t =$ time

(ω is the Greek letter omega).

The relationship between linear and angular velocity can be developed from the formulas $v = \dfrac{s}{t}$ and $\omega = \dfrac{\theta}{t}$ as follows:

$$v = \frac{s}{t}$$

$$v = \frac{r\theta}{t} \quad \text{(Substitute } r\theta \text{ for } s.\text{)}$$

$$v = r\left(\frac{\theta}{t}\right)$$

$$v = r\omega. \quad \left(\text{Substitute } \omega \text{ for } \frac{\theta}{t}.\right)$$

$v = r\omega$

$$v = r\omega$$

where v is the linear velocity, r is the radius, and ω is the angular velocity.

Examples 7.5, 7.6, and 7.7 illustrate some of the applications of linear and angular velocity.

• **Example 7.5:** For the irrigation system in example 7.4, find the linear and angular velocities of the end of the radial arm.

Solution: **1.** The linear velocity is calculated using the formula $v = \dfrac{s}{t}$, where $s = 84$ ft and $t = 20$ min,

$$v = \frac{84}{20} \text{ ft/min} = 4.2 \text{ ft/min}.$$

2. The angular velocity is calculated using the formula $\omega = \dfrac{\theta}{t}$ where $\theta = \dfrac{\pi}{3}$ and $t = 20$ min,

$$\omega = \frac{\frac{\pi}{3}}{20} \text{ rad/min} = 0.05 \text{ rad/min (approx.)}.$$

Note that we could have calculated the angular velocity using the formula $v = r\omega$, where $r = 80$ and $v = 4.2$ ft/min,

$$4.2 = 80\omega$$

$$\omega = \frac{4.2}{80} = 0.05 \text{ rad/min}.$$

• *Example 7.6:* A wheel is rotating with an angular velocity of 6.2 rad/sec. If the linear velocity of the wheel is 99.2 in./sec, what is the radius of the wheel?

Solution: Use the formula $v = \omega r$, where $v = 99.2$ and $\omega = 6.2$:

$$99.2 = 6.2r$$

$$r = \frac{99.2}{6.2} = 16 \text{ in}.$$

• *Example 7.7:* A motor is rotating at a speed of 2100 rpm (revolutions per minute). A grinding wheel with a radius of 6 in. is attached to the motor. Find the linear velocity of the wheel.

Solution: **1.** First convert 2100 rpm to rad/min. Since each revolution is equivalent to 2π radians,

$$2100 \text{ rpm} = (2100)(2\pi)$$
$$= 4200\pi \text{ rad/min}.$$

2. Given $r = 6$ in, use the formula $v = \omega r$:

$$v = 4200\pi(6) = 25{,}200\pi \text{ in./min}.$$

A more realistic way of expressing the answer is in ft/sec rather than in in./min. Divide by 60 to convert minutes to seconds:

$$25{,}200\pi \text{ in./min} = \frac{25{,}200\pi}{60} = 420\pi \text{ in/sec.}$$

Divide by 12 to convert inches to feet:

$$420\pi \text{ in./sec} = \frac{420\pi}{12} = 35\pi = 110 \text{ ft/sec.}$$

The linear velocity is about 110 ft/sec.

Trial Problems 7.1

Before you begin the section exercises, warm up with these problems. Complete answers are included in the answer key.

1. $41°40' = $ _____ radians.
2. 6.17 radians $= $ _____ degrees.
3. Given a circle with a radius of 6.3 and an arc length of 7.2, find the measure of the central angle in radians.
4. An angular velocity of 3.2 rad/min along a circle with a radius of 14.2 ft represents what linear velocity?
5. Find the area of a sector of a circle with a radius of 3.5 in. if the central angle of the sector is $\frac{\pi}{8}$ radians.

Exercises 7.1

1. $35° = $ __?__ radians
2. $55° = $ __?__ radians
3. $66.3° = $ __?__ radians
4. $78.3° = $ __?__ radians
5. $53°15' = $ __?__ radians
6. $65°45' = $ __?__ radians
7. $43°12'25'' = $ __?__ radians
8. $98°45'35'' = $ __?__ radians
9. $\frac{\pi}{8}$ radians $= $ __?__ degrees
10. $\frac{\pi}{16}$ radians $= $ __?__ degrees
11. $\frac{5\pi}{12}$ radians $= $ __?__ degrees
12. $\frac{7\pi}{12}$ radians $= $ __?__ degrees
13. 2.61 radians $= $ __?__ degrees
14. 1.32 radians $= $ __?__ degrees
15. 3.42π radians $= $ __?__ degrees
16. 7.28π radians $= $ __?__ degrees

In the following exercises, calculate the missing angle (θ), arc length (s), or radius (r). Use the formula $s = r\theta$.

17. $s = 18$ in., $r = 9.2$ in.
18. $s = 14$ in., $r = 7.3$ in.
19. $s = 8.7$ cm, $r = 2.3$ cm
20. $s = 9.1$ cm, $r = 2.7$ cm
21. $s = 8$ in., $\theta = 3$ radians
22. $s = 4$ in., $\theta = 1.5$ radians
23. $s = 4$ in., $\theta = 35°$
24. $s = 8$ in., $\theta = 67°$
25. $s = 4$ in., $\theta = 62°25'$
26. $s = 8$ in., $\theta = 74°35'$
27. $r = 3.7$ in., $\theta = 2.35$ radians
28. $r = 6.2$ cm, $\theta = 5.17$ radians

In the following exercises, calculate the missing linear velocity (*v*), angular velocity (ω), or radius (*r*). Use the formula $v = r\omega$.

29. $\omega = 2.1$ rad/min, $r = 3.7$ cm
30. $\omega = 2.9$ rad/min, $r = 5.3$ cm
31. $v = 14.3$ ft/sec, $r = 2.3$ ft
32. $v = 16.7$ m/sec, $r = 3.7$ m
33. $\omega = 2.9$ rad/sec, $v = 33$ ft/sec
34. $\omega = 3.2$ rad/sec, $v = 15$ ft/sec
35. $v = 88$ ft/sec, $\omega = 1200$ rpm
36. $v = 44$ ft/sec, $\omega = 600$ rpm

37. An ammeter has an indicator that is 4.2 cm long. The indicator swings through an arc that has a central angle of 55°. How long is the scale at the end of the indicator?

38. Solve exercise 37 if the indicator is 3.9 cm and the central angle measures 62°.

39. A person standing on the earth's moon observes the earth (approximate diameter 8000 mi). Calculate the approximate angle measure that the earth makes to the observer. Use 240,000 miles as the distance from the observer to the center of the earth.

40. A person standing on the earth observes the moon (approximate radius 1740 km). Calculate the approximate angle measure that the moon makes to the observer. Use 384,000 km as the distance from the observer to the center of the moon.

41. A pendulum 32 in. long swings through an arc with a central angle of 13°15′. Calculate the length of the arc through which the pendulum swings.

42. Solve exercise 41 using a pendulum of length 28 and a central angle of 14°30′.

43. Two cities are located on the same meridian (see figure 7.6). One city is located at *M* (16°20′ north latitude) and the second city is located at *N* (22°35′ north latitude). Calculate the approximate distance between the cities. Use 4000 mi as the radius of the earth.

44. Solve exercise 43 if *M* and *N* are located at 25°20′ and 42°35′ north latitude, respectively.

45. A windshield wiper is 14 in. long and it rotates through an angle of $\dfrac{3\pi}{4}$. If the tip of the blade is 18 in. from the base of the supporting arm, find the area that is swept by the blade.

46. Solve exercise 45 using a 16 in. blade, an angle of $\dfrac{3\pi}{4}$, and a distance of 20 inches from the tip of the blade to the base of the supporting arm.

47. A radar beam has an effective range of 50 mi and sweeps through an angle of 135°. Find the area swept by the beam.

48. A directional light has an illumination range of 50 ft through an angle of 65°. Find the area that is illuminated.

49. Find the area of the shaded region in figure 7.7. *Hint:* The region consists of a sector less a triangle.

50. Find the area of the shaded region in figure 7.8. *Hint:* The region consists of a sector less a triangle.

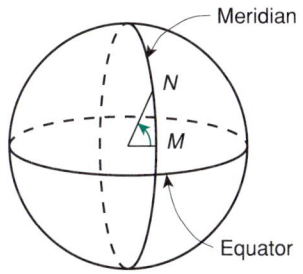

Figure 7.6

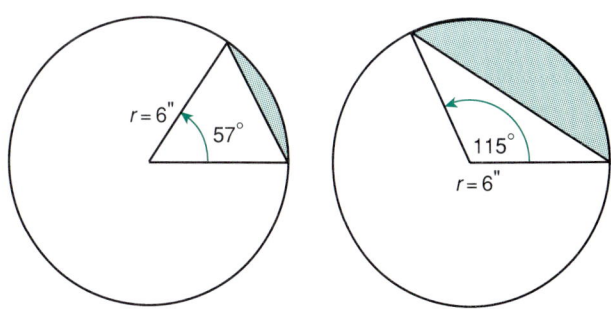

Figure 7.7

Figure 7.8

51. A satellite is in circular orbit around a planet. The radius of the orbit is 4750 mi. If a complete orbit requires 3 hours, determine the linear and angular velocities of the satellite.

52. Solve exercise 51 given a radius of 6300 mi and a 4 hour time for a complete orbit.

53. A tire has an outside diameter of 28 in. If the linear velocity is 55 mph, calculate the angular velocity and the number of rpm's.

54. A truck tire has an outside diameter of 36 in. The tire is turning at 400 rpm. Determine the linear and angular velocities of a point on the tread of the tire.

55. A vehicle moves at 50 mph on a curve that is a circular arc with a 1200 ft radius. What is the angular velocity of the vehicle?

56. A vehicle is moving on a curve that is a circular arc with a 1000 ft radius. If the angular velocity is 0.09 rad/sec, find the linear velocity in mph.

57. A 4 in. ammeter indicator moves through an angle of 55° in 0.25 sec. Find the angular velocity of the indicator in rad/sec.

58. In exercise 57, find the linear velocity of the tip of the indicator in ft/sec.

7 • 2 The Circular Functions

In this section the trigonometric functions are defined using the real numbers as the domain of the functions.

Unit Circle Consider a point moving steadily in a counterclockwise direction along a **unit circle** (a circle with the center at the origin of the coordinate system with a radius of 1). See figure 7.9. The equation of the unit circle is $x^2 + y^2 = 1$. At any given time, the point occupies a position on the circle. The point is then associated with an ordered pair (x,y). If the distance along the circle from the point $(1,0)$ to the point (x,y) is called t, we can associate the real number t with the ordered pair (x,y). Thus, for each positive real number t, there is a unique pair (x,y) associated with t. Similarly, if we consider the clockwise direction along the circle from the point $(1,0)$ as negative, there is associated with each negative real number t a unique ordered pair (x,y). For $t = 0$, the associated ordered pair is $(1,0)$.

The trigonometric functions can now be defined using the unit circle, where t is the length of the arc from the point $(1,0)$ to the point (x,y) on the circle. Using the unit circle results in simpler expressions for the functions that would be obtained using a general circle.

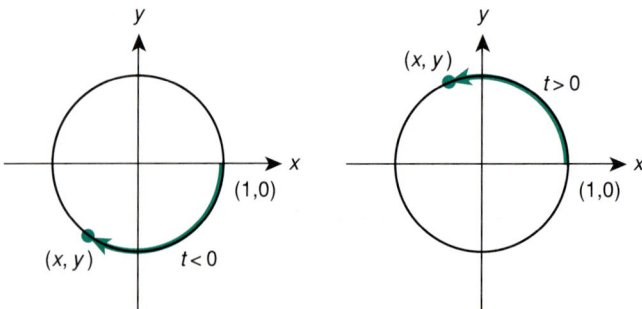

Figure 7.9

SECTION 7.2 THE CIRCULAR FUNCTIONS 263

TRIGONOMETRIC FUNCTIONS

$$\sin t = y \qquad \tan t = \frac{y}{x} \qquad \sec t = \frac{1}{x}$$

$$\cos t = x \qquad \cot t = \frac{x}{y} \qquad \csc t = \frac{1}{y}.$$

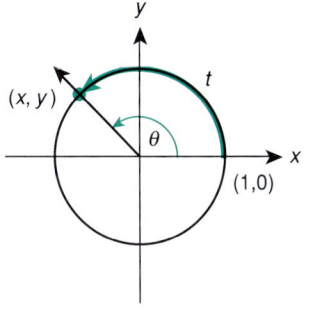

Figure 7.10

Note that these functions are very similar to the angular trigonometric functions defined in chapter 6. In figure 7.10, the angle θ in radians is numerically equal to the arc length t, that is,

$$\theta \text{ (radians)} = t \text{ (real number)}.$$

Normally, the units of measure are included only if the central angle is in degrees. Thus,

$$\sin(45°) = 0.7071,$$

while

$$\sin(45) = 0.8509$$

means the sine of 45 radians or the sine of the real number 45.

We can calculate exact values for the trigonometric functions for $t = \frac{\pi}{6}$, $t = \frac{\pi}{4}$, and $t = \frac{\pi}{3}$, since these are the values for which the central angle is respectively 30°, 45°, and 60°. In figure 7.11a we have a 30°-60°-90° triangle. From elementary geometry, in a 30°-60°-90° triangle the side opposite the 30° angle is $\frac{1}{2}$ the

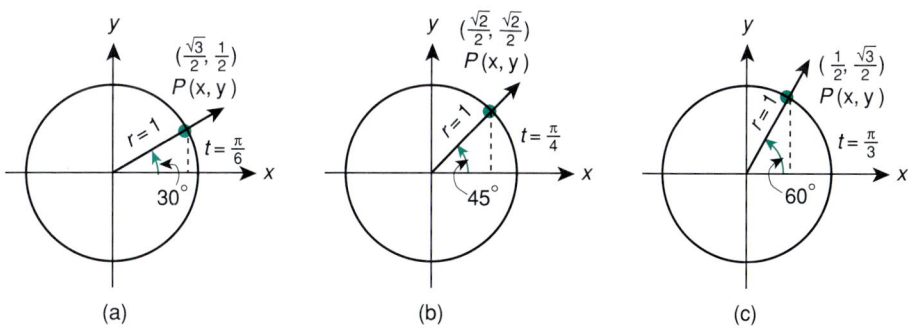

Figure 7.11

hypotenuse or $\frac{1}{2}(1) = \frac{1}{2}$. Using the Theorem of Pythagoras, $x^2 + y^2 = 1$ with $y = \frac{1}{2}$.

$$x^2 + \left(\frac{1}{2}\right)^2 = 1$$

$$x = \frac{\sqrt{3}}{2}.$$

$$\sin\frac{\pi}{6} = y = \frac{1}{2} \qquad \tan\frac{\pi}{6} = \frac{y}{x} = \frac{1}{\sqrt{3}} \qquad \sec\frac{\pi}{6} = \frac{1}{x} = \frac{2\sqrt{3}}{3}$$

$$\cos\frac{\pi}{6} = x = \frac{\sqrt{3}}{2} \qquad \cot\frac{\pi}{6} = \frac{x}{y} = \sqrt{3} \qquad \csc\frac{\pi}{6} = \frac{1}{y} = 2.$$

In figure 7.11b, we have a 45°-45°-90° triangle. Since $x = y$ and $x^2 + y^2 = 1$, $2x^2 = 1$ and $x = y = \frac{\sqrt{2}}{2}$.

$$\sin\frac{\pi}{4} = y = \frac{\sqrt{2}}{2} \qquad \tan\frac{\pi}{4} = \frac{y}{x} = 1 \qquad \sec\frac{\pi}{4} = \frac{1}{x} = \sqrt{2}$$

$$\cos\frac{\pi}{4} = x = \frac{\sqrt{2}}{2} \qquad \cot\frac{\pi}{4} = \frac{x}{y} = 1 \qquad \csc\frac{\pi}{4} = \frac{1}{y} = \sqrt{2}.$$

In figure 7.11c, we again have a 30°-60°-90° triangle, so $x = \frac{1}{2}$ and $y = \frac{\sqrt{3}}{2}$, so the trigonometric functions for $t = \frac{\pi}{3}$ are as follows.

$$\sin\frac{\pi}{3} = \frac{\sqrt{3}}{2} \qquad \tan\frac{\pi}{3} = \sqrt{3} \qquad \sec\frac{\pi}{3} = 2$$

$$\cos\frac{\pi}{3} = \frac{1}{2} \qquad \cot\frac{\pi}{3} = \frac{1}{\sqrt{3}} \qquad \csc\frac{\pi}{3} = \frac{2\sqrt{3}}{3}.$$

Some of the expressions above were simplified using the properties of radicals. For example, $\sec\frac{\pi}{4} = \frac{1}{\cos\left(\frac{\pi}{4}\right)} = \frac{1}{\frac{\sqrt{2}}{2}} = \frac{2}{\sqrt{2}} = \frac{2\sqrt{2}}{\sqrt{2}\sqrt{2}} = \frac{2\sqrt{2}}{2} = \sqrt{2}.$

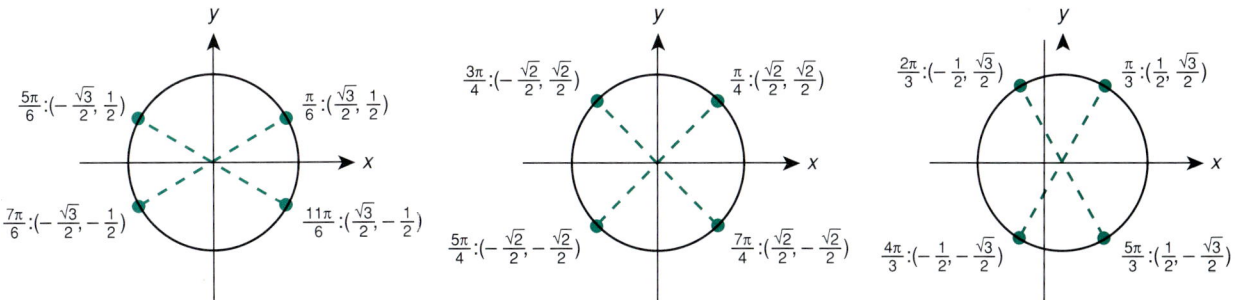

Figure 7.12 Figure 7.13

These values of the trigonometric functions can be used, along with the symmetry properties of a circle, to obtain exact values of the trigonometric functions for other values of t. For example, using figure 7.12 as a guide, the following sine and cosine values can be obtained.

$$\sin \frac{5\pi}{6} = \sin \frac{\pi}{6} = \frac{1}{2} \qquad \cos \frac{5\pi}{6} = -\cos \frac{\pi}{6} = -\frac{\sqrt{3}}{2}$$

$$\sin \frac{7\pi}{6} = -\sin \frac{\pi}{6} = -\frac{1}{2} \qquad \cos \frac{7\pi}{6} = -\cos \frac{\pi}{6} = -\frac{\sqrt{3}}{2}$$

$$\sin \frac{11\pi}{6} = -\sin \frac{\pi}{6} = -\frac{1}{2} \qquad \cos \frac{11\pi}{6} = \cos \frac{\pi}{6} = \frac{\sqrt{3}}{2}.$$

Using figure 7.13 and the similarity properties of the circle, the following sines and cosines are obtained.

$$\sin \frac{3\pi}{4} = \frac{\sqrt{2}}{2} \qquad \cos \frac{3\pi}{4} = -\frac{\sqrt{2}}{2}$$

$$\sin \frac{5\pi}{4} = -\frac{\sqrt{2}}{2} \qquad \cos \frac{5\pi}{4} = -\frac{\sqrt{2}}{2}$$

$$\sin \frac{7\pi}{4} = -\frac{\sqrt{2}}{2} \qquad \cos \frac{7\pi}{4} = \frac{\sqrt{2}}{2}$$

$$\sin \frac{2\pi}{3} = \frac{\sqrt{3}}{2} \qquad \cos \frac{2\pi}{3} = -\frac{1}{2}$$

$$\sin \frac{4\pi}{3} = -\frac{\sqrt{3}}{2} \qquad \cos \frac{4\pi}{3} = -\frac{1}{2}$$

$$\sin \frac{5\pi}{3} = -\frac{\sqrt{3}}{2} \qquad \cos \frac{5\pi}{3} = \frac{1}{2}$$

Similar calculations can be made for the other trigonometric functions.

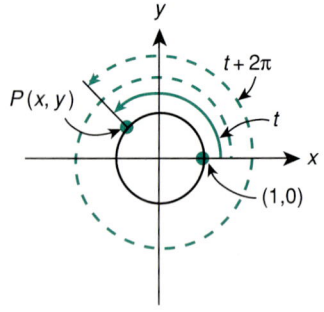

Figure 7.14

If $t \geq 2\pi$, the values of the trigonometric functions can be determined using these relationships:

$$\sin(t + 2\pi) = \sin t$$
$$\cos(t + 2\pi) = \cos t.$$

As figure 7.14 indicates, these are valid relationships, since going around the circle once ($t = 2\pi$) gets us back to the starting point (1,0). For $t > 2\pi$, we go around the circle as many times as necessary for $t = 2\pi, 4\pi, 6\pi, \ldots, 2k\pi, \ldots$ and calculate the values of the functions at the terminal point. More generally,

$$\left. \begin{array}{l} \sin(t + 2\pi k) = \sin t \\ \cos(t + 2\pi k) = \cos t \end{array} \right\} \quad k = 1, 2, 3, \ldots.$$

We can use these relationships to determine the values of the trigonometric functions for $t > 2\pi$.

• *Example 7.8:* Determine the following values.

    a. $\sin \dfrac{9\pi}{4}$ **b.** $\cos 23\pi$ **c.** $\tan \dfrac{45\pi}{4}$

Solution: **a.** $\sin \dfrac{9\pi}{4} = \sin\left(2\pi + \dfrac{\pi}{4}\right) = \sin\dfrac{\pi}{4} = \dfrac{\sqrt{2}}{2}$

b. $\cos 23\pi = \cos(22\pi + \pi) = \cos\pi = -1$
($t = \pi$ gives us the terminal point $(-1,0)$ on the unit circle)

c. $\tan \dfrac{45\pi}{4} = \dfrac{\sin\left(\dfrac{45\pi}{4}\right)}{\cos\left(\dfrac{45\pi}{4}\right)}$

$\dfrac{45\pi}{4} = \dfrac{(40 + 5)\pi}{4} = 10\pi + \dfrac{5\pi}{4}.$

We need to determine $\sin \dfrac{5\pi}{4}$ and $\cos \dfrac{5\pi}{4}$.

Since $P_2 \left(\dfrac{-\sqrt{2}}{2}, \dfrac{-\sqrt{2}}{2}\right)$ in figure 7.15 is directly opposite point $P_1 \left(\dfrac{\sqrt{2}}{2}, \dfrac{\sqrt{2}}{2}\right)$ on the circle, $\sin \dfrac{5\pi}{4} = \cos \dfrac{5\pi}{4} = \dfrac{-\sqrt{2}}{2}$ and $\tan \dfrac{5\pi}{4} = 1.$

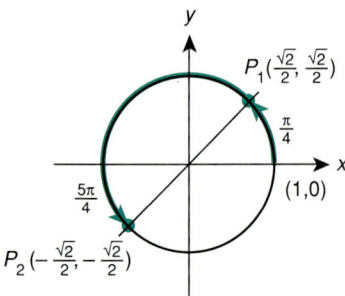

Figure 7.15

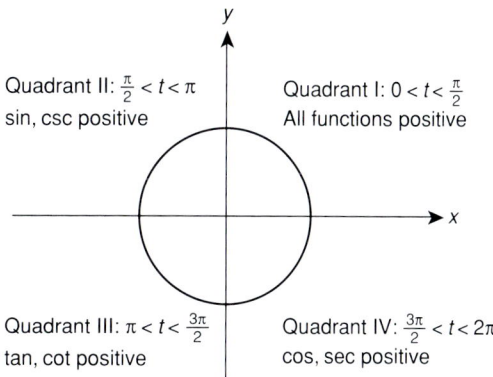

Figure 7.16

The following chart indicates the signs of the six trigonometric functions in the four quadrants for $0 \leq t \leq 2\pi$. These relationships are illustrated in figure 7.16.

1. In quadrant I, both x and y are positive, so all six functions are positive.
2. In quadrant II, $x < 0$ and $y > 0$, so only the sine and cosecant are positive.
3. In quadrant III, $x < 0$ and $y < 0$, so only the tangent and cotangent are positive.
4. In quadrant IV, $x > 0$ and $y < 0$, so only the cosine and secant are positive.

For negative numbers, the values of the trigonometric functions are determined by the following formulas:

TRIGONOMETRIC FUNCTIONS FOR NEGATIVE NUMBERS

$$\sin(-t) = -\sin t$$
$$\cos(-t) = \cos t$$
$$\tan(-t) = -\tan t$$
$$\cot(-t) = -\cot t$$
$$\sec(-t) = \sec t$$
$$\csc(-t) = -\csc t.$$

The validity of these formulas can be established by considering figure 7.17. For a particular value of $t > 0$, we obtain a terminal point $P_1(x,y)$. For $-t < 0$, we obtain the terminal point $P_2(x,-y)$.

Using the definition of the sine function:

$$\left. \begin{array}{c} \sin t = y \\ \sin(-t) = -y \end{array} \right\} \sin(-t) = -\sin t.$$

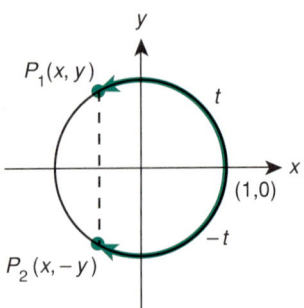

Figure 7.17

Using the definition of the tangent function:

$$\left. \begin{array}{l} \tan t = \dfrac{y}{x} \\[1em] \tan(-t) = \dfrac{-y}{x} \end{array} \right\} \quad \tan(-t) = -\tan t.$$

The other four relationships can be established in a similar manner.

Note that the trigonometric functions are examples of functions that are classified as **even functions** or as **odd functions.**

Even Function
Odd Function

In general, a function f is *even* if $f(-t) = f(t)$ for all t in the domain of f. That is, if (t,y) is a point on the graph of f, then $(-t,y)$ is also a point on the graph of f. Geometrically, this means that the graph of f is symmetric with respect to the y-axis. Thus, $y = \cos t$ and $y = \sec t$ and are even functions.

A function g is *odd* if $g(-t) = -g(t)$ for all t in the domain of g. That is, if (t,y) is a point on the graph of g, then $(-t,-y)$ is also a point on the graph of g. Geometrically, this means that the graph of g is symmetric with respect to the origin. Thus, $y = \sin t$, $y = \tan t$, $y = \cot t$, and $y = \csc t$ are odd functions.

We use these even/odd properties when we construct the graphs of the trigonometric functions.

We can use these relationships to calculate the values of the trigonometric functions for $t < 0$ or we can use the relationships like $\sin(t + 2\pi) = \sin(t)$.

• *Example 7.9:* Determine the following values.

a. $\sin\left(-\dfrac{\pi}{2}\right)$

b. $\cos\left(-\dfrac{3\pi}{4}\right)$

c. $\tan(-27\pi)$.

SECTION 7.2 THE CIRCULAR FUNCTIONS

Solution 1: (Using the negative value formulas.)

a. $\sin\left(-\dfrac{\pi}{2}\right) = -\sin\left(\dfrac{\pi}{2}\right) = -1$

b. $\cos\left(-\dfrac{3\pi}{4}\right) = \cos\left(\dfrac{3\pi}{4}\right) = -\dfrac{\sqrt{2}}{2}$

c. $\tan(-27\pi) = \tan(-26\pi - \pi) = \tan(-\pi) = \dfrac{\sin(-\pi)}{\cos(-\pi)}$

$= \dfrac{-\sin \pi}{\cos \pi} = \dfrac{-0}{-1} = 0.$

Solution 2: (Using the $t + 2k\pi$ relationships.)

a. $\sin\left(-\dfrac{\pi}{2}\right) = \sin\left(2\pi - \dfrac{\pi}{2}\right) = \sin\left(\dfrac{3\pi}{2}\right) = -1$

b. $\cos\left(-\dfrac{3\pi}{4}\right) = \cos\left(2\pi - \dfrac{3\pi}{4}\right) = \cos\left(\dfrac{5\pi}{4}\right) = -\dfrac{\sqrt{2}}{2}$

c. $\tan(-27\pi) = \tan(28\pi - 27\pi) = \tan(\pi) = \dfrac{\sin \pi}{\cos \pi} = \dfrac{0}{-1}$
$= 0.$

Many applications in electrical circuits lead to expressions that contain trigonometric functions. Example 7.10 illustrates such a problem.

• **Example 7.10:** Under certain conditions the current i in the circuit in figure 7.18, expressed as a function of time t, is represented by the equation

$$i = 2\cos\dfrac{500}{3}t - \dfrac{1}{5}\sin\dfrac{500}{3}t,$$

where i is in amps and t is in seconds.

a. Determine i, when $t = \dfrac{\pi}{2}$.

b. Determine i, when $t = \dfrac{3}{5}$.

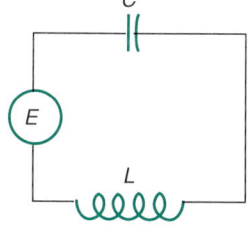

Figure 7.18

Solution: a. $i = 2\cos\left(\dfrac{500}{3}\right)\left(\dfrac{\pi}{2}\right) - \dfrac{1}{5}\sin\left(\dfrac{500}{3}\right)\left(\dfrac{\pi}{2}\right)$

$= 2\cos\dfrac{250\pi}{3} - \dfrac{1}{5}\sin\dfrac{250\pi}{3}.$

We need to calculate $\cos\left(\dfrac{250\pi}{3}\right)$ and $\sin\left(\dfrac{250\pi}{3}\right)$.

You can either set your calculator on radian measure or calculate as follows:

$$\cos\left(\frac{250\pi}{3}\right) = \cos\left(82\pi + \frac{4\pi}{3}\right) = \cos\left(\frac{4\pi}{3}\right) = -\frac{1}{2},$$

$$\sin\left(\frac{250\pi}{3}\right) = \sin\left(82\pi + \frac{4\pi}{3}\right) = \sin\left(\frac{4\pi}{3}\right) = -\frac{\sqrt{3}}{2}.$$

Therefore,

$$i = 2\left(\frac{-1}{2}\right) - \frac{1}{5}\left(\frac{-\sqrt{3}}{2}\right) = -1 + 0.173 = -0.827$$

b. $i = 2\cos\left(\dfrac{500}{3}\right)\left(\dfrac{3}{5}\right) - \dfrac{1}{5}\sin\left(\dfrac{500}{3}\right)\left(\dfrac{3}{5}\right)$

$= 2\cos 100 - \dfrac{1}{5}\sin 100.$

The easiest way to get cos 100 and sin 100 is with a hand calculator. Set the calculator in radian mode, enter 100 and press the cos key or sin key:

$$i = 2(0.8623) - \frac{1}{5}(-0.5064) = 1.83 \text{ (approx.)}.$$

The key strokes for $i = 2\cos 100 - \dfrac{1}{5}\sin 100$ are

$$2 \;\boxed{\times}\; 100 \;\boxed{\text{COS}}\; \boxed{=}\; \boxed{-}$$
$$0.2 \;\boxed{\times}\; 100 \;\boxed{\text{SIN}}\; \boxed{=}.$$

Display: 1.8259109.

Trial Problems 7.2

Before you begin the section exercises, warm up with these problems. Complete answers are included in the answer key.

1. Use your calculator to find $\sec\left(-\dfrac{\pi}{6}\right)$.

2. Find an exact expression for $\cos\dfrac{17\pi}{4}$.

3. Evaluate $y = 30\sin\left(25t + \dfrac{\pi}{3}\right)$ for $t = \dfrac{\pi}{6}$.

4. Evaluate $y = 10{,}000 + 1400\sin\dfrac{\pi t}{12}$ for $t = 18$.

5. Evaluate $y = \left(\dfrac{v_0}{32}\right)^2 \sin 2\theta$ for $v_0 = 240$ and $\theta = 0.632$.

Exercises 7.2

1–2. Complete the following table.

TABLE 7.1

t	sin t	cos t	tan t	cot t	sec t	csc t
0	0	1	0	undefined	1	undefined
$\dfrac{\pi}{6}$	$\dfrac{1}{2}$	$\dfrac{\sqrt{3}}{2}$				
$\dfrac{\pi}{4}$	$\dfrac{\sqrt{2}}{2}$	$\dfrac{\sqrt{2}}{2}$	1	1	$\sqrt{2}$	$\sqrt{2}$
$\dfrac{\pi}{3}$	$\dfrac{\sqrt{3}}{2}$	$\dfrac{1}{2}$				
$\dfrac{\pi}{2}$	1	0	undefined	0	undefined	1
$\dfrac{2\pi}{3}$						
$\dfrac{3\pi}{4}$						
$\dfrac{5\pi}{6}$						
π						
$\dfrac{7\pi}{6}$						
$\dfrac{5\pi}{4}$						
$\dfrac{4\pi}{3}$						
$\dfrac{3\pi}{2}$						
$\dfrac{5\pi}{3}$						
$\dfrac{7\pi}{4}$						
$\dfrac{11\pi}{6}$						

Use your calculator to approximate the following values.

3. $\sin 3.7$
4. $\cos 6.2$
5. $\tan 1.7$
6. $\cot(-1.7)$
7. $\sec(-5.2)$
8. $\csc 5.2$
9. $\sin\dfrac{3}{\pi}$
10. $\cos\dfrac{5}{\pi}$

Use table 7.1 and the formulas developed in this section to calculate an exact expression for each problem.

11. $\sin\dfrac{100\pi}{4}$
12. $\cos\dfrac{100\pi}{4}$
13. $\tan\dfrac{11\pi}{4}$
14. $\cot\dfrac{11\pi}{4}$
15. $\sin(-22\pi)$
16. $\cos(-17\pi)$
17. $\sec\dfrac{15\pi}{4}$
18. $\csc\left(-\dfrac{15\pi}{4}\right)$

19. The voltage across a resistor in an electrical circuit is indicated by the expression $v = 100 \sin 50t$, where t is the time. Determine the voltage when $t = \dfrac{\pi}{4}$ seconds.

20. The current across a resistor in a circuit is indicated by the expression $i = 10 \sin 50t$, where t represents the time. Determine the current when $t = \dfrac{\pi}{4}$ seconds.

21. The current across a resistor is represented by the expression $i = 40\left(\sin 50t + \dfrac{\pi}{6}\right)$. Determine the current when $t = \pi$.

22. The voltage across a resistor is represented by the expression $v = 200\left(\sin 50t + \dfrac{\pi}{6}\right)$. Determine the voltage when $t = \pi$.

23. A weight is attached to a spring. If the weight is pulled y feet below the equilibrium point and released, the equation for the motion is $y = \dfrac{1}{2}\cos 20t$, where t is the time in seconds. Determine the position of the weight after $\dfrac{\pi}{10}$ seconds (see figure 7.19).

24. Solve exercise 23 if the equation for the motion of the weight is $y = \dfrac{1}{4}\sin 16t$ and $t = \dfrac{\pi}{32}$.

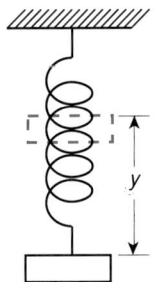

Figure 7.19

25. The population of a particular species of animal in a region is approximated by the equation $y = 5000 + 1200 \sin\left(\dfrac{\pi t}{12}\right)$, where y represents the population at any time t in months. Estimate the population for $t = 13$ months.

26. Solve exercise 25 given that $y = 20{,}000 + 5{,}500 \sin\left(\dfrac{\pi t}{12}\right)$ and $t = 15$ months.

27. If an object with weight W (lbs) is dragged across a horizontal floor by a force of F lbs directed at an angle of θ radians with the plane of the floor, the force is given by the equation

$$F = \dfrac{kW}{k \sin\theta + \cos\theta},$$

where k is the coefficient of friction. Given $k = 0.5$, $W = 120$ lb, and $\theta = \dfrac{\pi}{5}$, determine F.

28. Solve exercise 27 given $k = 0.5$, $W = 220$ lb, and $\theta = \dfrac{2\pi}{11}$.

29. A projectile is fired at an angle of θ radians, an initial velocity of v_0 ft/sec, and y is the range of the projectile. The following relationship holds,

$$y = \dfrac{v_0^2}{32} \sin 2\theta.$$

Determine the range if $v_0 = 360$ ft/sec and $\theta = 0.628$ radians.

30. Solve exercise 29 given $v_0 = 450$ ft/sec and $\theta = 0.524$ radians.

7•3 The Graphs of the Sine and Cosine Functions

In this section we consider the graphs of the functions $y = \sin t$ and $y = \cos t$, where the domains of the functions are the real numbers. While we could use any values for t to plot the functions on a coordinate system, we have already calculated the values of the trigonometric functions for $t \in \left\{0, \dfrac{\pi}{6}, \dfrac{\pi}{4}, \dfrac{\pi}{3}, \dfrac{\pi}{2}\right\}$ and for various multiples of these values. Table 7.2 lists the values of $\sin t$ and $\cos t$ for $0 \le x \le 2\pi$.

TABLE 7.2

t	$y = \sin t$	$y = \cos t$	t	$y = \sin t$	$y = \cos t$
0	0	1	π	0	-1
$\dfrac{\pi}{6}$	$\dfrac{1}{2}$	$\dfrac{\sqrt{3}}{2}$	$\dfrac{7\pi}{6}$	$-\dfrac{1}{2}$	$-\dfrac{\sqrt{3}}{2}$
$\dfrac{\pi}{4}$	$\dfrac{\sqrt{2}}{2}$	$\dfrac{\sqrt{2}}{2}$	$\dfrac{5\pi}{4}$	$-\dfrac{\sqrt{2}}{2}$	$-\dfrac{\sqrt{2}}{2}$
$\dfrac{\pi}{3}$	$\dfrac{\sqrt{3}}{2}$	$\dfrac{1}{2}$	$\dfrac{4\pi}{3}$	$-\dfrac{\sqrt{3}}{2}$	$-\dfrac{1}{2}$
$\dfrac{\pi}{2}$	1	0	$\dfrac{3\pi}{2}$	-1	0
$\dfrac{2\pi}{3}$	$\dfrac{\sqrt{3}}{2}$	$-\dfrac{1}{2}$	$\dfrac{5\pi}{3}$	$-\dfrac{\sqrt{3}}{2}$	$\dfrac{1}{2}$
$\dfrac{3\pi}{4}$	$\dfrac{\sqrt{2}}{2}$	$-\dfrac{\sqrt{2}}{2}$	$\dfrac{7\pi}{4}$	$-\dfrac{\sqrt{2}}{2}$	$\dfrac{\sqrt{2}}{2}$
$\dfrac{5\pi}{6}$	$\dfrac{1}{2}$	$-\dfrac{\sqrt{3}}{2}$	$\dfrac{11\pi}{6}$	$-\dfrac{1}{2}$	$\dfrac{\sqrt{3}}{2}$
			2π	0	1

Plotting the approximate position of the pairs (t, sin t) on a coordinate system and connecting the points with a smooth curve yields the graph of $y = \sin t$ for $0 \leq x \leq 2\pi$. This graph is pictured in figure 7.20.

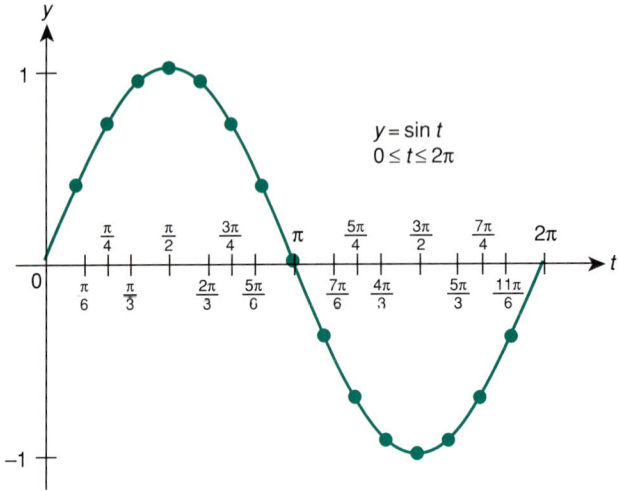

Figure 7.20

From section 7.1 we know that for any real number t, $\sin(t \pm 2\pi k) = \sin t$, $k \in \{1, 2, 3, \ldots\}$. Therefore, to complete the graph of $y = \sin t$, simply continue the graph in figure 7.20 in both directions repeating the graph for the domain $0 \leq t \leq 2\pi$ for each multiple of 2π. The resulting graph is shown in figure 7.21.

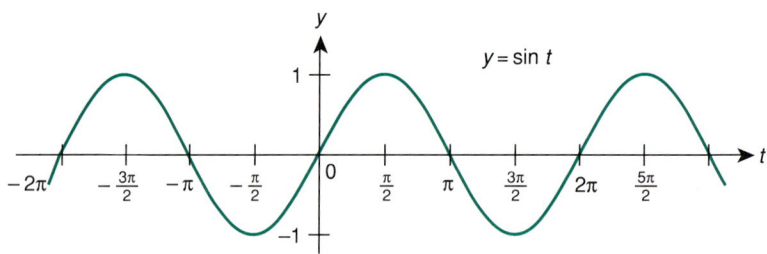

Figure 7.21

Period of a Function

Since the graph of $y = \sin t$ repeats for multiples of 2π, we say that the function has a **period** of 2π. This suggests the following definition:

PERIODIC FUNCTION

A function f is **periodic** if there is a positive real number k so that $f(t + k) = f(t)$. The smallest such number k (provided it exists) is called the **period** of f.

These properties of the sine function can be summarized as follows.

PROPERTIES OF THE SINE FUNCTION

1. $y = \sin t$ is defined for any real number t (domain).
2. $-1 \leq \sin t \leq 1$ for any t (range).
3. The sine function is an odd function, that is, the graph is symmetric with respect to the origin.
4. $\sin(t + \pi) = -\sin t$.
5. $\sin(t + 2\pi) = \sin t$ (periodic with period 2π).

If the points $(t, \cos t)$ from table 7.2 are plotted on a coordinate system and the points connected with a smooth curve, the graph of $y = \cos t$ for $0 \leq t \leq 2\pi$ is obtained. The graph is shown in figure 7.22.

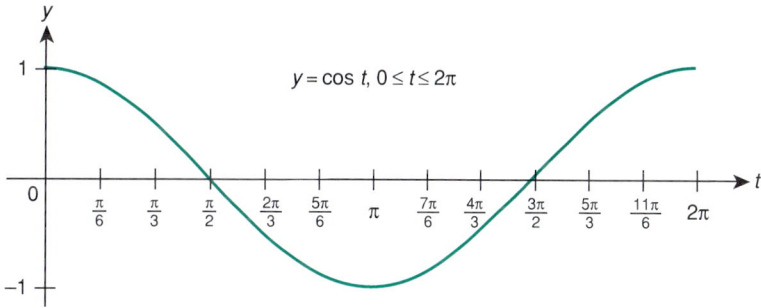

Figure 7.22

Since we know that $\cos(t \pm 2\pi k) = \cos t$, the function $y = \cos t$ also has period 2π. The curve can be extended to obtain a curve for $y = \cos t$ for any t. The graph is shown in figure 7.23. From figure 7.23 and from previous calculations, we can see that the function $y = \cos t$ has the following properties.

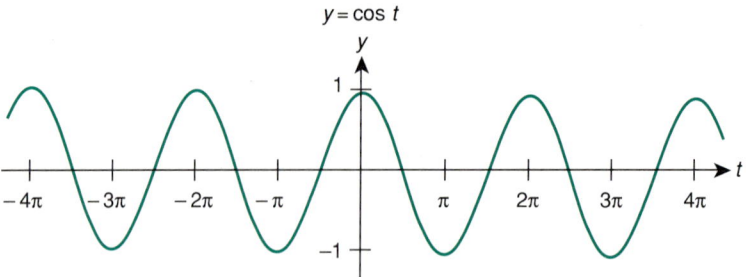

Figure 7.23

PROPERTIES OF $y = \cos t$	1. $y = \cos t$ is defined for any real number t (domain). 2. $-1 \leq \cos t \leq 1$ for any t (range). 3. The cosine function is an even function, that is, the graph is symmetric to the y-axis. 4. $\cos(t \pm \pi) = -\cos t$. 5. $\cos(t \pm 2\pi) = \cos t$ (period $= 2\pi$).

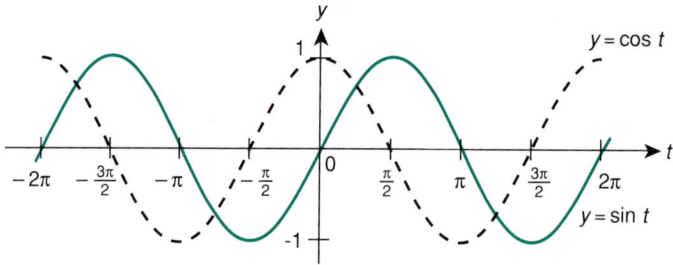

Figure 7.24

Figure 7.24 shows the graphs of $y = \sin t$ and $y = \cos t$ plotted on a coordinate system. Notice that the graph of the cosine function can be obtained by shifting the sine function $\dfrac{\pi}{2}$ units to the left, and the sine function can be obtained by shifting the cosine function $\dfrac{\pi}{2}$ units to the right. This leads to the following relationships.

$$\cos t = \sin\left(t + \frac{\pi}{2}\right)$$

$$\sin t = \cos\left(t - \frac{\pi}{2}\right).$$

Sinusoidal Curve
Cycle

Any curve that has the same shape as the graph of the sine function is called a *sinusoidal curve*. Since the graph of the cosine function fits this definition, it is a sinusoidal curve. One period of such a curve is called a **cycle**.

We now consider the graphs of some variations of the sine and cosine functions. The first such variations to be considered are the graphs of functions of the form $y = a \sin t$ and $y = a \cos t$, where a is an arbitrary constant. They have the same period (2π) as the sine and cosine functions. However, the range of the functions is not the same as the ranges of the sine and cosine functions.

• **Example 7.11:** Graph the equation $y = 3 \sin t$.

Solution: The "3" in the equation triples the y-values for the graph of $y = \sin t$, as indicated in table 7.3.
Graph $y = \sin t$ and triple all of the y-values. See figure 7.25.
Notice that range of the function $y = 3 \sin t$ is $-3 \leq y \leq 3$.

TABLE 7.3

t	-2π	$-\dfrac{3\pi}{2}$	$-\pi$	$-\dfrac{\pi}{2}$	0	$\dfrac{\pi}{2}$	π	$\dfrac{3\pi}{2}$	2π
$\sin t$	0	1	0	-1	0	1	0	-1	0
$3 \sin t$	0	3	0	-3	0	3	0	-3	0

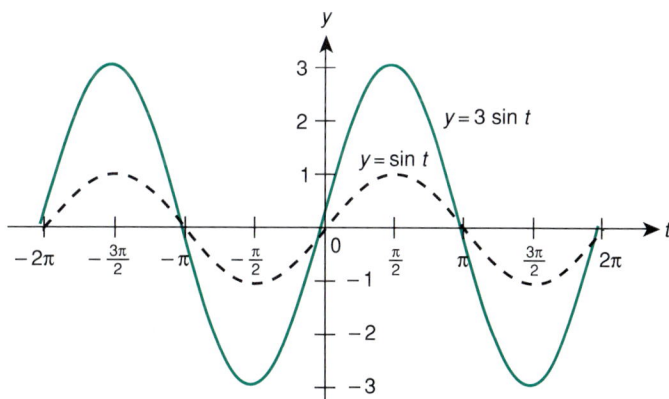

Figure 7.25

278 CHAPTER 7 ANALYTIC TRIGONOMETRY

Amplitude The **amplitude** of a sinusoidal function is the number calculated by the following formula.

> Amplitude $= \frac{1}{2}(M - m)$ where $M =$ largest value of the function and $m =$ smallest value of the function.

Thus, the amplitude of $y = 3 \sin t$ is calculated as follows:
$$\text{Amplitude} = \frac{1}{2}(M - m) = \frac{1}{2}(3 - (-3)) = \frac{6}{2} = 3.$$

> In general, if $y = a \sin t$ or $y = a \cos t$,
> $$\text{Amplitude} = |a|.$$

• **Example 7.12:** Graph the function $y = -\frac{1}{2} \sin t$ and find the amplitude.

Solution: The amplitude is $\left| -\frac{1}{2} \right| = \frac{1}{2}$. This means that the minimum and maximum values for the function are $-\frac{1}{2}$ and $+\frac{1}{2}$, respectively. The graph of the function can be found by first graphing the function $y = \frac{1}{2} \sin t$. Then reflect the graph about the t-axis. This means that if a point (t,y) is on the graph of $y = \frac{1}{2} \sin t$, then the point $(t,-y)$ is on the graph of $y = -\frac{1}{2} \sin t$. The results of this procedure are shown in table 7.4 and in figure 7.26.

TABLE 7.4

t	-2π	$-\frac{3\pi}{2}$	$-\pi$	$-\frac{\pi}{2}$	0	$\frac{\pi}{2}$	π	$\frac{3\pi}{2}$	2π
$\sin t$	0	1	0	-1	0	1	0	-1	0
$\frac{1}{2} \sin t$	0	$\frac{1}{2}$	0	$-\frac{1}{2}$	0	$\frac{1}{2}$	0	$-\frac{1}{2}$	0
$-\frac{1}{2} \sin t$	0	$-\frac{1}{2}$	0	$\frac{1}{2}$	0	$-\frac{1}{2}$	0	$\frac{1}{2}$	0

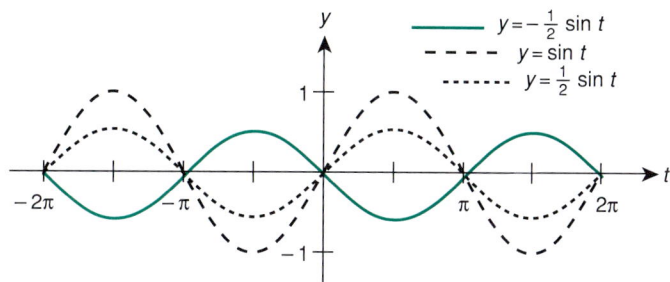

Figure 7.26

The second types of sinusoidal curves to be considered are those that can be written in the form $y = a \sin bt$ or $y = a \cos bt$, where a and b are arbitrary constants. For functions of these forms, the amplitude is $|a|$. The coefficient of t (b) changes the period of the graph. Since bt ($b > 0$) varies from 0 to 2π as t varies from 0 to $\frac{2\pi}{b}$, the period becomes $\frac{2\pi}{b}$.

$y = a \sin bt$ or $y = a \cos bt$,

$$\text{Amplitude} = |a|$$

$$\text{Period} = \left| \frac{2\pi}{b} \right|.$$

• **Example 7.13:** Graph the function $y = \cos 3t$ and determine the period of the function.

Solution: The "3" in the equation changes the period from 2π to $\frac{2\pi}{3}$. Table 7.5 shows the coordinates of some points on the graph that are illustrated in figure 7.27.

TABLE 7.5

t	$-\frac{2\pi}{3}$	$-\frac{\pi}{2}$	$-\frac{\pi}{3}$	$-\frac{\pi}{6}$	0	$\frac{\pi}{6}$	$\frac{\pi}{3}$	$\frac{\pi}{2}$	$\frac{2\pi}{3}$
$3t$	-2π	$-\frac{3\pi}{2}$	$-\pi$	$-\frac{\pi}{2}$	0	$\frac{\pi}{2}$	π	$\frac{3\pi}{2}$	2π
$\cos 3t$	1	0	-1	0	1	0	-1	0	1

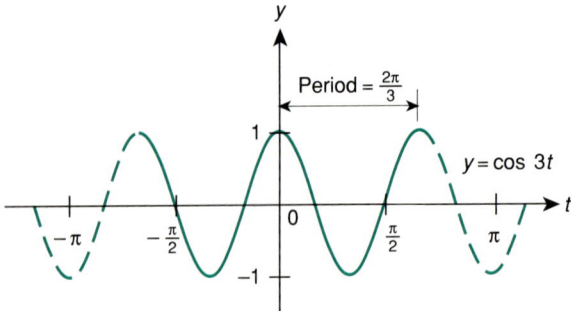

Figure 7.27

The choices of $t = 0, \frac{\pi}{6}, \frac{\pi}{3}, \frac{\pi}{2}$, and $\frac{2\pi}{3}$ for Table 7.5 were made to obtain one complete period of the graph. The period of the function is $\frac{2\pi}{3}$. If we divide the interval $\left[0, \frac{2\pi}{3}\right]$ into 4 equal parts, we obtain the t-values for which the function $y = \cos 3t$ takes on the values $1, 0, -1, 0$, and 1 in succession. The t-values are

$$\left\{0, 0 + \frac{1}{4}\frac{(2\pi)}{3}, 0 + \frac{2}{4}\frac{(2\pi)}{3}, 0 + \frac{3}{4}\frac{(2\pi)}{3}, 0 + \frac{4}{4}\frac{(2\pi)}{3}\right\}$$

or

$$\left\{0, \frac{\pi}{6}, \frac{\pi}{3}, \frac{\pi}{2}, \frac{2\pi}{3}\right\}.$$

Perhaps a more intuitive way of listing these values is in this form

$$\left\{0, \frac{\pi}{6}, \frac{2\pi}{6}, \frac{3\pi}{6}, \frac{4\pi}{6}\right\}.$$

Note that the numerators of the fractions are successive multiples of π.

• • • • • • • • • •

The following summary combines the procedures of examples 7.12 and 7.13 into the form of a set of procedures for constructing graphs for equations of the forms $y = a \sin bt$ and $y = a \cos bt$ (see figure 7.28).

SECTION 7.3 THE GRAPHS OF THE SINE AND COSINE FUNCTIONS 281

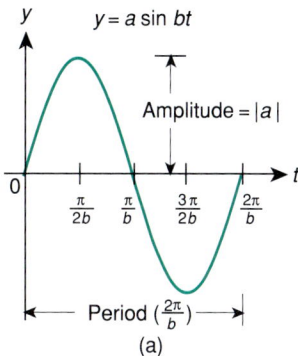
(a)

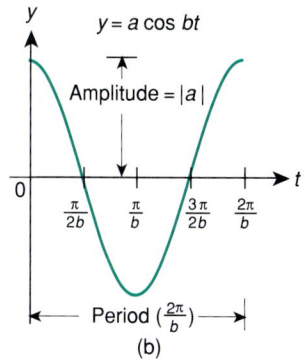
(b)

Figure 7.28

GRAPHING
$y = a \sin bt$
AND, $y = a \cos bt$

1. Find the period $\dfrac{2\pi}{b}$ and locate the point $\left(\dfrac{2\pi}{b}, 0\right)$ on the t-axis.

2. Divide the interval from 0 to $\dfrac{2\pi}{b}$ into four equal parts using the t-values 0, $\dfrac{\pi}{2b}, \dfrac{\pi}{b}, \dfrac{3\pi}{2b}, \dfrac{2\pi}{b}$.

3. Make a table of y-values using the t-values obtained in step 2 and plot these points.

4. Plot as many additional points as needed and sketch a smooth curve.

5. Sketch as many additional periods of the curve that are needed.

• *Example 7.14:* The voltage across a resistor in an electrical circuit expressed as a function of time is represented by the equation $v = 10 \sin 50t$. Graph the function for two periods with $t \geq 0$.

Solution: Use the procedure outlined above. (See figure 7.29).

1. The period is $\dfrac{2\pi}{b} = \dfrac{2\pi}{50} = \dfrac{\pi}{25}$. Plot the point $\left(\dfrac{\pi}{25}, 0\right)$ on the t-axis.

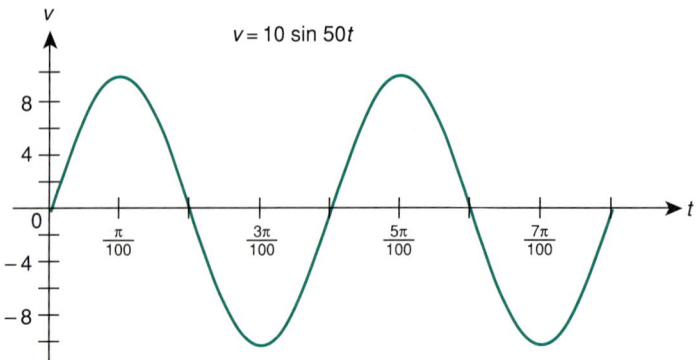

Figure 7.29

TABLE 7.6

t	0	$\dfrac{\pi}{100}$	$\dfrac{\pi}{50}$	$\dfrac{3\pi}{100}$	$\dfrac{\pi}{25}$
$50t$	0	$\dfrac{\pi}{2}$	π	$\dfrac{3\pi}{2}$	2π
$\sin 50t$	0	1	0	-1	0
$v = 10 \sin 50t$	0	10	0	-10	0

2. The interval from $t = 0$ to $t = \dfrac{\pi}{25}$, divided into four equal parts, yields the values that are needed. $\dfrac{\pi}{25} \div 4 = \dfrac{\pi}{100}$, so the values are

$$0 + \dfrac{\pi}{100}, \quad 0 + 2\left(\dfrac{\pi}{100}\right), \quad 0 + 3\left(\dfrac{\pi}{100}\right), \quad \text{and } 0 + 4\left(\dfrac{\pi}{100}\right)$$

or

$$\left\{0, \dfrac{\pi}{100}, \dfrac{\pi}{50}, \dfrac{3\pi}{100}, \dfrac{\pi}{25}\right\}.$$

3. Make a table of v-values using the t-values obtained in step 2. See table 7.6. Plot these points.

4. Suppose we plot these additional points, which are selected as the midpoints of the intervals in step 2:

$$\left(\dfrac{\pi}{200}, 7.1\right), \quad \left(\dfrac{3\pi}{200}, 7.1\right), \quad \left(\dfrac{5\pi}{200}, -7.1\right), \quad \left(\dfrac{7\pi}{200}, -7.1\right).$$

SECTION 7.3 THE GRAPHS OF THE SINE AND COSINE FUNCTIONS

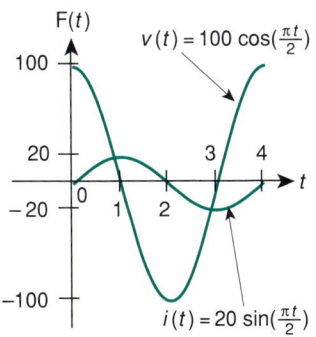

Figure 7.30

5. Sketch one more period for $t > \dfrac{\pi}{25}$.

•••••••••

• **Example 7.15:** The current i in a circuit is represented by the equation $i = 20 \sin \dfrac{\pi t}{2}$ and the voltage v by $v = 100 \cos \dfrac{\pi t}{2}$. Graph these equations on the same set of axes (see figure 7.30).

Solution: **A.** Graph the current function.

1. The period is $\dfrac{2\pi}{b} = \dfrac{2\pi}{\dfrac{\pi}{2}} = 4$. Plot the point $(4,0)$.

2. Dividing the interval $(0,4)$ into four equal parts yields the t-values $0, 1, 2, 3, 4$.

3. Make a table of i-values using $t \in \{0, 1, 2, 3, 4\}$. See table 7.7. Plot the points.

4. Plot some additional points. Suppose we use the midpoints $t = \dfrac{1}{2}, \dfrac{3}{2}, \dfrac{5}{2}, \dfrac{7}{2}$. The points are as follows:

$$\left(\dfrac{1}{2}, 14.1\right), \left(\dfrac{3}{2}, 14.1\right), \left(\dfrac{5}{2}, -14.1\right), \left(\dfrac{7}{2}, -14.1\right).$$

Sketch the curve.

TABLE 7.7

t	0	1	2	3	4
$\dfrac{\pi t}{2}$	0	$\dfrac{\pi}{2}$	π	$\dfrac{3\pi}{2}$	2π
$\sin \dfrac{\pi t}{2}$	0	1	0	-1	0
$20 \sin \dfrac{\pi t}{2}$	0	20	0	-20	0

TABLE 7.8

t	0	1	2	3	4
$\dfrac{\pi t}{2}$	0	$\dfrac{\pi}{2}$	π	$\dfrac{3\pi}{2}$	2π
$\cos \dfrac{\pi t}{2}$	1	0	-1	0	1
$100 \cos \dfrac{\pi t}{2}$	100	0	-100	0	100

B. Graph the voltage function.

1–2. Steps 1 and 2 are identical to steps 1 and 2 in part A.

3. Make a table of v-values using $t = 0, 1, 2, 3, 4$. See table 7.8. Plot these points.

4. Plot additional points using $t = \dfrac{1}{2}, \dfrac{3}{2}, \dfrac{5}{2}, \dfrac{7}{2}$. These points are $(\dfrac{1}{2}, 70.7), (\dfrac{3}{2}, -70.7), (\dfrac{5}{2}, -70.7), (\dfrac{7}{2}, 70.7)$. Sketch the curve.

Trial Problems 7.3

Before you begin the section exercises, warm up with these problems. Complete answers are included in the answer key.

Find the amplitude and the period for each function.

1. $y = \sin \dfrac{3t}{2}$ **2.** $y = \cos 3.5t$ **3.** $y = 3.5 \cos t$

4. $y = -1.4 \cos 3t$ **5.** $y = 3.8 \sin \dfrac{\pi t}{4}$

Exercises 7.3

Draw one cycle of each curve and state the amplitude and the period.

1. $y = 2 \sin t$ **2.** $y = 2 \cos t$ **3.** $y = \dfrac{1}{2} \sin t$ **4.** $y = \dfrac{1}{2} \cos t$

5. $y = -2 \cos t$ **6.** $y = -2 \sin t$ **7.** $y = -\dfrac{1}{3} \cos t$ **8.** $y = -\dfrac{1}{3} \sin t$

9. $y = \cos 2t$ **10.** $y = \sin 2t$ **11.** $y = \sin \dfrac{t}{3}$ **12.** $y = \cos \dfrac{t}{3}$

13. $y = \sin \dfrac{t}{\pi}$ 14. $y = \cos \dfrac{t}{\pi}$ 15. $y = 2 \sin 2t$ 16. $y = 2 \cos 2t$

17. $y = -\dfrac{1}{2} \cos 3t$ 18. $y = -\dfrac{1}{2} \sin 3t$ 19. $y = 2 \cos \dfrac{\pi t}{2}$ 20. $y = 3 \sin \dfrac{\pi t}{2}$

21. Draw the graphs of $y = \sin t$ and $y = \sin 4t$ on the same set of axes. For each period of $y = \sin t$, how many periods of $y = \sin 4t$ are there?

22. Draw the graphs of $y = \cos t$ and $y = \cos \dfrac{t}{3}$ on the same set of axes. For each period of $y = \cos \dfrac{t}{3}$, how many periods of $y = \cos t$ are there?

23. The voltage across a resistor in a circuit is described by the equation $v = 20 \sin 60t$. Determine the period of amplitude of the function and graph one period of the function.

24. The voltage in a circuit is represented by the equation $v = 100 \sin 300t$. Determine the amplitude and the period and graph one period of the function.

25. If the voltage across a 5-ohm resistor in a circuit is expressed by the equation $v = 150 \sin 300t$ and the current is expressed by the equation $i = \dfrac{v}{r}$, write an equation for the current and graph the current and voltages on the same set of axes.

26. Do exercise 25 given that $v = 80 \sin 200t$ and $r = 10$ ohms.

27. Use a calculator to find values of $y = \sin t$ for $t = 0$ to $t = 0.2$ in increments of $t = 0.04$. Compare these values with $y = t$ over the same domain. This shows why the approximation formula $\sin x \approx x$ is used for small values of x.

28. Do exercise 27 for $y = \tan t$ for $t = 0$ to $t = 0.2$ using increments of $t = 0.04$.

29. The voltage across a resistor in an AC circuit is described by the equation $v = 100 \sin 360t$. The current is described by the equation $i = 10 \sin 360t$. Graph the equations on the same set of axes and state the period and amplitude for each equation.

30. Solve exercise 29 given that $v = 100 \sin 240t$ and $i = 10 \sin 240t$.

31. A pendulum swings with a harmonic motion according to the equation

$$y = a \sin \sqrt{\dfrac{g}{L}} \, t,$$

where a = amplitude in feet,
g = 32 ft/sec^2,
t = time in seconds,
and L = length of pendulum in feet.
If $L = 4$ and $a = 0.5$, graph the function for 2 cycles.

32. Solve exercise 31 given that $L = 6$, $a = 0.75$, and $g = 32$.

33. A projectile is fired at an angle of θ radians, an initial velocity of v_0 ft/sec and y represents the range of the projectile. The following relationship holds:

$$y = \dfrac{v_0^2}{32} \sin 2\theta.$$

Given $v_0 = 360$ ft/sec, graph the equation for $\theta = 0$ to $\dfrac{\pi}{2}$.

34. Solve exercise 33 given $v_0 = 450$ ft/sec.

7·4 The Graphs of $y = a \sin(bt + c)$ and $y = a \cos(bt + c)$

In section 7.3 graphs of functions of the form $y = a \sin bt$ and $y = a \cos bt$ were considered. This section considers the graphs of more general types of sinusoidal curves.

The first curve to be considered is the graph of an equation of the form $y = \sin(t + k)$.

• **Example 7.16:** Graph the function $y = \sin(t + \pi)$.

Solution: As t increases from 0 to 2π, the graph of $y = \sin t$ is one period of the function.

Now consider what happens to t as $t + \pi$ increases from 0 to 2π. If $t + \pi = 0$, then $t = -\pi$. If $t + \pi = 2\pi$, then $t = \pi$. Thus, t increasing from $-\pi$ to π produces one period of the function $y = \sin(t + \pi)$.

Table 7.9 shows several points on the graph.

TABLE 7.9

$t + \pi$	0	$\dfrac{\pi}{2}$	π	$\dfrac{3\pi}{2}$	2π
t	$-\pi$	$-\dfrac{\pi}{2}$	0	$\dfrac{\pi}{2}$	π
$\sin(t + \pi)$	0	1	0	-1	0

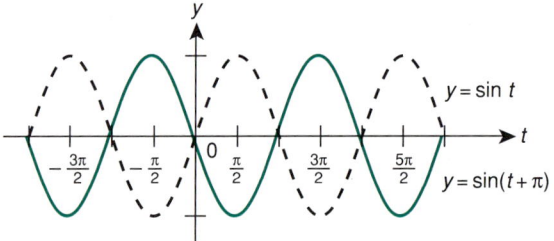

Figure 7.31

Figure 7.31 illustrates the graph of the function.

Note that the graph of $y = \sin(t + \pi)$ is, in effect, the same as the graph of $y = \sin t$ with the y-axis moved π units to the right. This is equivalent to shifting the graph of $y = \sin t$ π units to the left. The graph shows that the function $y = \sin(t + \pi)$ has the same amplitude and period as the sine function (amplitude 1,

SECTION 7.4 THE GRAPHS OF $y = a\sin(bt + c)$ AND $y = a\cos(bt + c)$ 287

Phase Shift

and period 2π). The fact that the graph of $y = \sin(t + \pi)$ was obtained by shifting π units to the left (negative direction) is expressed by saying that the **phase shift** of the function is $-\pi$.

> The phase shift for an equation of the form $y = \sin(t + k)$ or $y = \cos(t + k)$ is $-k$, where a negative shift is to the left and a positive shift is to the right.

Example 7.17 illustrates the procedure for graphing an equation of the form $y = a\sin(bt + c)$.

• **Example 7.17:** Graph the function $y = 2\sin\left(4t + \dfrac{\pi}{2}\right)$.

Solution: We write the equation in a slightly different form. Factor 4 from the expression $\left(4t + \dfrac{\pi}{2}\right)$, resulting in the expression $4\left(t + \dfrac{\pi}{8}\right)$. The reason for this change will become apparent. The function now has the form

$$y = 2\sin 4\left(t + \dfrac{\pi}{8}\right).$$

Graph the function $y = 2\sin 4t$ and do a phase shift of $-\dfrac{\pi}{8}$ units. As $4t$ increases from 0 to 2π, t increases from 0 to $\dfrac{\pi}{2}$. Thus, the period of $y = 2\sin 4t$ is $\dfrac{\pi}{2}$ and the amplitude is 2.

Table 7.10 illustrates some of the values for $y = 2\sin 4t$. These values are graphed in figure 7.32.

TABLE 7.10

$4t$	0	$\dfrac{\pi}{2}$	π	$\dfrac{3\pi}{2}$	2π
t	0	$\dfrac{\pi}{8}$	$\dfrac{\pi}{4}$	$\dfrac{3\pi}{8}$	$\dfrac{\pi}{2}$
$2\sin 4t$	0	2	0	-2	0

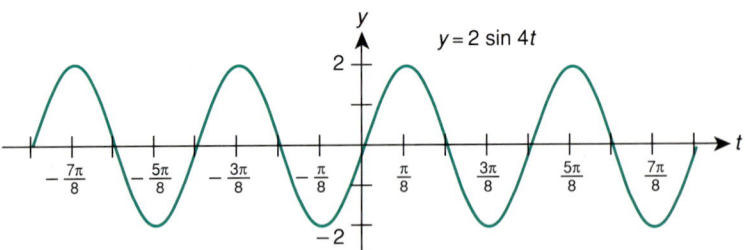

Figure 7.32

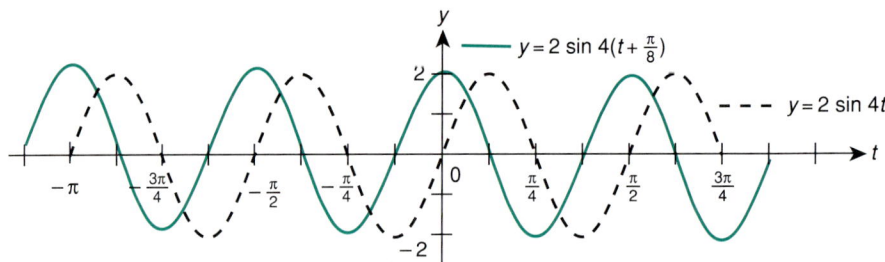

Figure 7.33

We now perform a phase shift of $-\dfrac{\pi}{8}$ on the graph in figure 7.32 to obtain the graph of $y = 2 \sin 4\left(t + \dfrac{\pi}{8}\right)$, which is illustrated in figure 7.33.

In general, the graph of $y = a \sin(bt + c)$ has the following properties:

$$y = a \sin(bt + c) = a \sin b\left(t + \dfrac{c}{b}\right).$$

1. amplitude $= |a|$
2. period $= \dfrac{2\pi}{|b|}$
3. phase shift $= -\dfrac{c}{|b|}$.

SECTION 7.4 THE GRAPHS OF $y = a \sin(bt + c)$ AND $y = a \cos(bt + c)$

The graph of $y = a \cos(bt + c)$ has similar properties.

$$y = a \cos(bt + c) = a \cos b\left(t + \frac{c}{b}\right)$$

1. amplitude $= |a|$
2. period $= \dfrac{2\pi}{|b|}$
3. phase shift $= -\dfrac{c}{|b|}$.

We now use these properties to simplify the process of graphing a function of the type $y = a \cos(bt + c)$.

• **Example 7.18:** Graph the function $y = 4 \cos\left(2t - \dfrac{\pi}{2}\right)$ for one period.

Solution: Rewrite the equation in the form $a \cos b\left(t + \dfrac{c}{b}\right)$.

This means factoring 2 from the expression $2t - \dfrac{\pi}{2}$.

$$y = 4 \cos 2\left(t - \dfrac{\pi}{4}\right)$$

1. amplitude $= |4| = 4$
2. period $= \dfrac{2\pi}{2} = \pi$
3. phase shift $= -\left(-\dfrac{\pi}{4}\right) = \dfrac{\pi}{4}$.

We now graph the function using these facts. Suppose we start and end the period when $4 \cos 2\left(t - \dfrac{\pi}{4}\right) = 4$:

$$4 \cos 2\left(t - \dfrac{\pi}{4}\right) = 4$$

$$\cos 2\left(t - \dfrac{\pi}{4}\right) = 1.$$

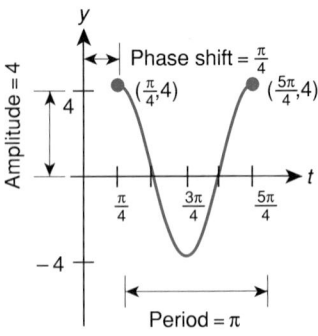

Figure 7.34

One of the values for which this equation holds is

$$2\left(t - \frac{\pi}{4}\right) = 0$$

$$t = \frac{\pi}{4}.$$

We begin with the point $\left(\frac{\pi}{4}, 4\right)$. Since the period is π, we end the period with the value $t = \frac{\pi}{4} + \pi = \frac{5\pi}{4}$. Therefore, the right endpoint is $\left(\frac{5\pi}{4}, 4\right)$. This is illustrated in figure 7.34.

••••••••••

Pure Waves
Simple Harmonic Motion

Frequency

Equations of the form $y = a \sin(bt + c)$ and $y = a \cos(bt + c)$, where t represents time, describe periodic motion, which occurs in many applied areas. These equations describe what is usually referred to as **pure waves** or **simple harmonic motion**. If t is measured in seconds, the period for such equations is $\frac{2\pi}{|b|}$ seconds. The number of periods per second is called the **frequency** f of the motion and $f = \frac{|b|}{2\pi}$ cycles/sec (or $\frac{|b|}{2\pi}$ hertz). Example 7.19 describes one such application.

• **Example 7.19:** Consider a capacitor with a capacitance of C farads and a maximum charge of y_0 coulombs. If the capacitor is connected in series with a coil having an inductance of L henries, the charge y on the capacitor after t seconds is given by the following function:

$$y = y_0 \sin\left(\frac{t}{\sqrt{LC}} - \frac{\pi}{2}\right) = y_0 \sin\left(\frac{1}{\sqrt{LC}}\right)\left(t - \frac{\sqrt{LC}\,\pi}{2}\right).$$

Assuming that the resistance of the coil is negligible, if $L = 0.5$ henry and $C = 0.00005$ farad, determine the period and the frequency of the circuit and graph the function given that $y_0 = 0.00025$ coulombs.

Solution: $\frac{1}{\sqrt{LC}} = \frac{1}{\sqrt{(0.5)(0.00005)}} = \frac{1}{\sqrt{0.000025}} = \frac{1}{0.005} = 200.$

The period is $\frac{2\pi}{|b|} = \frac{2\pi}{\frac{1}{\sqrt{LC}}} = \frac{2\pi}{200} = 0.031$ sec.

The frequency is $\frac{\frac{1}{\sqrt{LC}}}{2\pi} = \frac{100}{\pi} = 31.8$ hertz.

The equation for the charge is

$$y = y_0 \sin\left(\frac{1}{\sqrt{LC}}\left(t - \frac{\sqrt{LC}\pi}{2}\right)\right)$$

or

$$y = 0.00025(\sin(200(t - 0.0025\pi)))).$$

The amplitude of the function is $|y_0| = 0.00025$. Setting $200(t - 0.0025\pi) = 0$ yields $t = 0.008$. Setting $200(t - 0.0025\pi) = 2\pi$ yields $t = 0.039$. One period of the graph extends from $t = 0.008$ to $t = 0.039$. Using an appropriate scale, plot one period of the graph and extend the graph in both directions.

t	0.008	0.016	0.024	0.031	0.039
y	0	0.00025	0	−0.00025	0

The results are shown in figure 7.35.

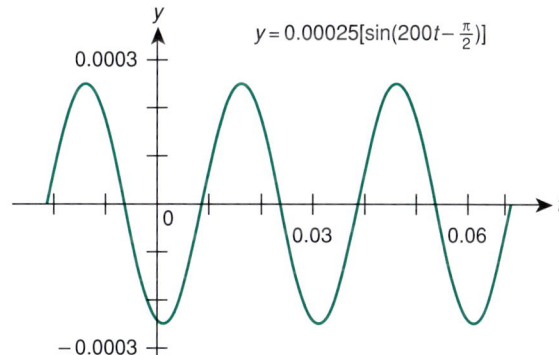

Figure 7.35

Electrical circuits contain many examples in which the concept of phase shift is important. A simple alternating current circuit is shown in figure 7.36. The relationship between the voltage V and the current I in the circuit is also shown.

Out of Phase From the graph we can see that the current and the voltage are **out of phase** by $\frac{\pi}{2}$. In a pure capacitance circuit, the current **leads** the voltage by $\frac{\pi}{2}$ or the voltage **lags** lags the current by $\frac{\pi}{2}$.

Leads

Lags

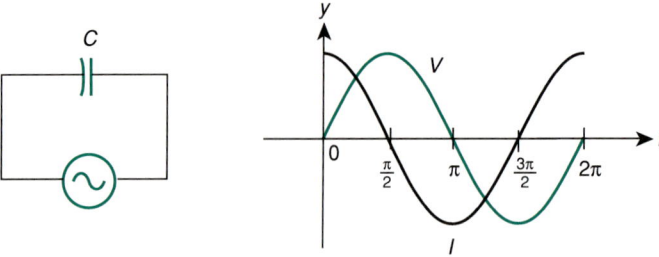

Figure 7.36

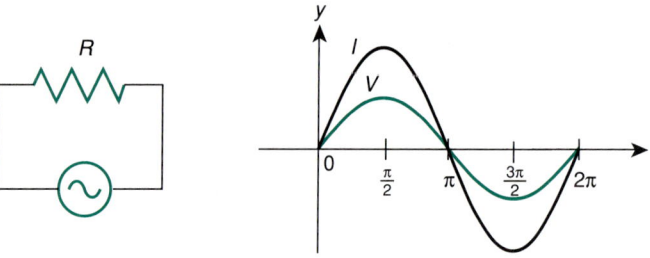

Figure 7.37

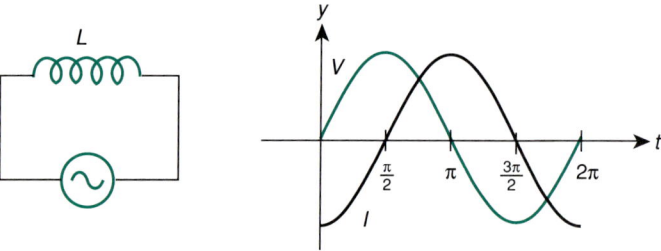

Figure 7.38

In Phase In a purely resistive AC circuit, the voltage and current are **in phase,** that is, there is no lag between the current and voltage. Such a circuit is illustrated in figure 7.37.

For a purely inductive AC circuit, the voltage leads the current by $\dfrac{\pi}{2}$ or the current lags the voltage by $\dfrac{\pi}{2}$. This is illustrated in figure 7.38.

Trial Problems 7.4

Before you begin the section exercises, warm up with these problems. Complete answers are included in the answer key.

Find the period and the phase shift for each function.

1. $y = 10 \sin\left(t - \dfrac{\pi}{8}\right)$
2. $y = 20 \sin\left(t + \dfrac{\pi}{3}\right)$
3. $y = 1.8 \sin\left(2t - \dfrac{\pi}{6}\right)$
4. $y = 2.9 \sin\left(-3t - \dfrac{\pi}{12}\right)$
5. $y = 40 \cos\left(40t + \dfrac{\pi}{6}\right)$

Exercises 7.4

Graph each function for one period.

1. $y = \sin\left(t + \dfrac{\pi}{4}\right)$
2. $y = \cos\left(t + \dfrac{\pi}{4}\right)$
3. $y = \cos\left(t + \dfrac{\pi}{6}\right)$
4. $y = \sin\left(t + \dfrac{\pi}{6}\right)$
5. $y = \sin\left(t - \dfrac{\pi}{2}\right)$
6. $y = \cos\left(t - \dfrac{\pi}{2}\right)$
7. $y = \cos\left(t - \dfrac{\pi}{3}\right)$
8. $y = \sin\left(t - \dfrac{\pi}{3}\right)$
9. $y = 2 \sin\left(t + \dfrac{\pi}{4}\right)$
10. $y = 2 \cos\left(t + \dfrac{\pi}{4}\right)$
11. $y = \dfrac{1}{2} \cos\left(t + \dfrac{\pi}{6}\right)$
12. $y = \dfrac{1}{2} \sin\left(t + \dfrac{\pi}{6}\right)$
13. $y = 3 \sin\left(t - \dfrac{\pi}{2}\right)$
14. $y = 3 \cos\left(t - \dfrac{\pi}{2}\right)$
15. $y = 1.3 \cos\left(t - \dfrac{\pi}{3}\right)$
16. $y = 2.3 \sin\left(t - \dfrac{\pi}{3}\right)$
17. $y = 2 \sin\left(3t + \dfrac{\pi}{4}\right)$
18. $y = 2 \cos\left(3t + \dfrac{\pi}{4}\right)$
19. $y = \dfrac{1}{2} \cos\left(2t + \dfrac{\pi}{6}\right)$
20. $y = \dfrac{1}{2} \sin\left(2t + \dfrac{\pi}{6}\right)$
21. $y = 3 \sin\left(5t - \dfrac{\pi}{2}\right)$
22. $y = 3 \cos\left(5t - \dfrac{\pi}{2}\right)$
23. $y = 1.5 \cos\left(2.3t - \dfrac{\pi}{3}\right)$
24. $y = 2.3 \sin\left(1.5t - \dfrac{\pi}{3}\right)$
25. $y = 50 \sin\left(200t - \dfrac{\pi}{2}\right)$
26. $y = 50 \cos\left(200t - \dfrac{\pi}{2}\right)$

27. The circuit in figure 7.39 has a coil with inductance L henries and a capacitor with a capacitance of C farads. If the capacitor is charged and the switch is closed, the value of E in volts oscillates by a cosine curve with amplitude E_0, frequency $\dfrac{1}{2\pi \sqrt{LC}}$, and phase shift 0. Given that $E_0 = 200$ volts, $L = 8$ henries, and $C = 0.0004$ farad, determine the period, the frequency, and the equation for the voltage at any time t. Graph the equation.

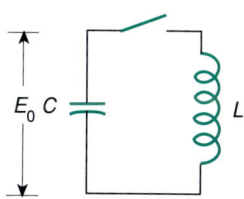

Figure 7.39

28. Solve exercise 27 given that $E_0 = 100$ volts, $L = 5$ henries, and $C = 0.00004$ farad.

29. The current through a resistor in an AC circuit is given by the equation $I = 40 \sin\left(360t + \dfrac{\pi}{6}\right)$ and the voltage by the equation $v = 200 \sin\left(360t + \dfrac{\pi}{6}\right)$. Graph the equations on the same set of axes and determine the amplitude and period for each equation.

30. Solve exercise 29 using the equations
$I = 20 \sin\left(200t + \dfrac{\pi}{3}\right)$ and $v = 80 \sin\left(200t + \dfrac{\pi}{3}\right)$.

31. The current through a coil in an AC circuit is given by the equation $I = 10 \sin 200t$ and the voltage by the equation $v = 200 \sin 200\left(t + \dfrac{\pi}{2}\right)$. Find the period and phase shift for each function. Determine the maximum values of the current and voltage.

32. Solve exercise 31 given that $I = 7 \sin\left(360t - \dfrac{\pi}{4}\right)$ and $v = 252 \sin\left(360t + \dfrac{\pi}{4}\right)$.

33. The pressure of a moving sound wave is given by the equation
$$p = 1.4 \sin \pi (k - 300t),$$
where k is a constant in meters,
t is the time in seconds,
and P is the pressure in newtons/m².
Determine the amplitude and the frequency of the sound wave.

34. Solve exercise 33 given that
$$P = 1.6 \sin\pi(k - 250t).$$

7 • 5 The Graphs of the Other Trigonometric Functions

In the previous sections, we graphed the sine and cosine functions. In this section, we graph the four remaining trigonometric functions.

We first consider the graph of $y = \tan t$ for $-\dfrac{\pi}{2} < t < \dfrac{\pi}{2}$.

Table 7.11 shows some values of $\tan t$ obtained from a calculator, rounded off to two decimal places. We have used the customary values of t with the additional values of $\dfrac{5\pi}{12}$ and $-\dfrac{5\pi}{12}$, since $\dfrac{5\pi}{12}$ is midway between $\dfrac{\pi}{3}$ and $\dfrac{\pi}{2}$. Note that $\tan\left(-\dfrac{\pi}{2}\right)$ and $\tan\dfrac{\pi}{2}$ are undefined, since $\tan t = \dfrac{\sin t}{\cos t}$ and $\cos t = 0$ for these values.

TABLE 7.11

t	$\tan t$	t	$\tan t$	t	$\tan t$
$-\dfrac{\pi}{2}$	undefined	$-\dfrac{\pi}{6}$	-0.58	$\dfrac{\pi}{3}$	1.73
$-\dfrac{5\pi}{12}$	-3.73	0	0	$\dfrac{5\pi}{12}$	3.73
$-\dfrac{\pi}{3}$	-1.73	$\dfrac{\pi}{6}$	0.58	$\dfrac{\pi}{2}$	undefined
$-\dfrac{\pi}{4}$	-1	$\dfrac{\pi}{4}$	1		

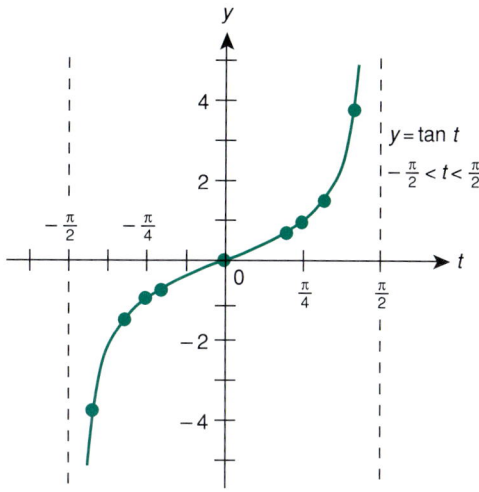

Figure 7.40

We plot the approximate positions of these points on a coordinate system and connect the points with a smooth curve. The graph of the tangent function for $-\frac{\pi}{2} < t < \frac{\pi}{2}$ is illustrated in figure 7.40.

The tangent function is a periodic function like the sine and cosine functions. However, the values of tan t repeat whenever t increases or decreases by π. Hence, the period of $y = \tan t$ is π. Thus,

$$\tan(t + \pi) = \tan t.$$

This equality can be shown algebraically. Recall that

$$\cos(t + \pi) = -\cos t \quad \text{and} \quad \sin(t + \pi) = -\sin t.$$

Hence,

$$\tan(t + \pi) = \frac{\sin(t + \pi)}{\cos(t + \pi)} = \frac{-\sin t}{-\cos t} = \tan t.$$

Figure 7.41 shows the graph of the tangent function for several periods. Note that the graph is symmetric with respect to the origin. Thus,

$$\tan(-t) = -\tan t.$$

Vertical Asymptote

Also note that the graph of $y = \tan t$ approaches the vertical lines $t = -\frac{\pi}{2}$ and $t = \frac{\pi}{2}$ but does not intersect the lines. The lines $t = -\frac{\pi}{2}$ and $t = \frac{\pi}{2}$ are called **vertical asymptotes** of the graph of $y = \tan t$.

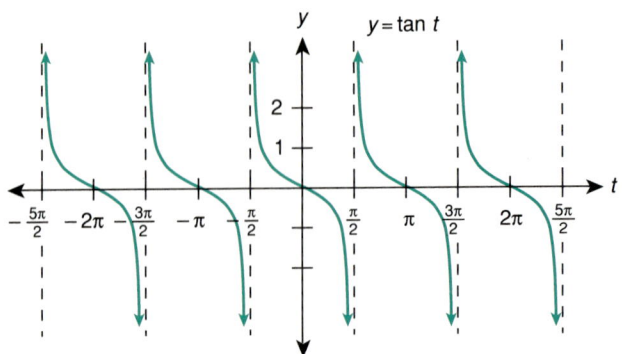

Figure 7.41

Also note that where the cosine function is zero, the tangent function is undefined. Thus, there are asymptotes for the tangent curve where the cosine is zero, that is, for $t = \dfrac{\pi}{2} + k\pi$.

For the function $y = \cot t$, we proceed in a manner similar to the one we used to graph the tangent function. This time, however, we graph one period for $0 < t < \pi$. We use this interval, since $\cot t = \dfrac{\cos t}{\sin t}$ is undefined for $\sin t = 0$ at $t = 0$ and $t = \pi$.

Table 7.12 shows several values of $(t, \cot t)$ for $0 < t < \pi$. The $t = \dfrac{\pi}{12}$ value was selected to obtain a point for t halfway between 0 and $\dfrac{\pi}{6}$. We plot these points on a coordinate system and connect them with a smooth curve. The graph of the cotangent function for $0 < t < \pi$ is illustrated in figure 7.42.

TABLE 7.12

t	$\cot t$	t	$\cot t$	t	$\cot t$
0	undefined	$\dfrac{\pi}{3}$	0.58	$\dfrac{5\pi}{6}$	-1.73
$\dfrac{\pi}{12}$	3.73	$\dfrac{\pi}{2}$	0	$\dfrac{11\pi}{12}$	-3.73
$\dfrac{\pi}{6}$	1.73	$\dfrac{2\pi}{3}$	-0.58	π	undefined
$\dfrac{\pi}{4}$	1	$\dfrac{3\pi}{4}$	-1		

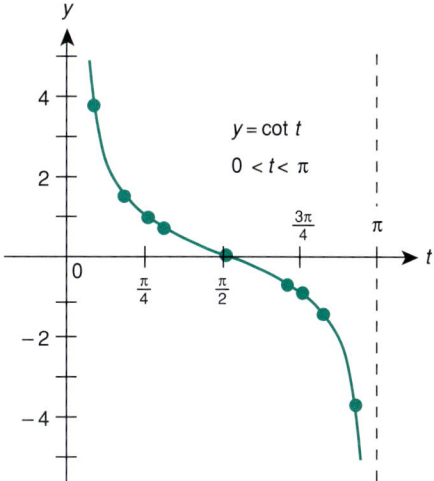

Figure 7.42

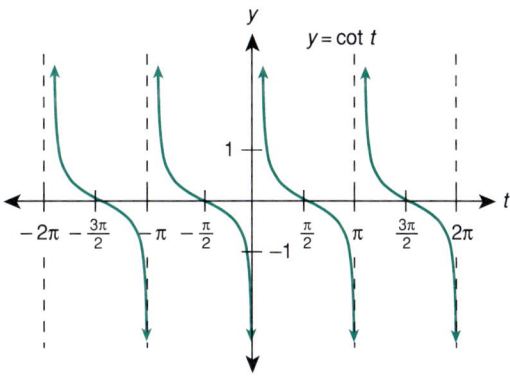

Figure 7.43

Like the graph of the tangent function, the graph of the cotangent function has a period of π. Figure 7.43 shows several periods of the graph of $y = \cot t$. Note that where the sine function is zero, the cotangent function is undefined. Thus, there are asymptotes for the cotangent where the sine is zero, namely for $t = \pm k\pi$.

Since the graph of the cotangent is symmetric with respect to the origin,

$$\cot(-t) = -\cot t.$$

To graph the secant and cosecant functions, we make use of the facts that the secant function is the reciprocal of the cosine function, and that the cosecant function is the reciprocal of the sine function. To graph the secant function we first sketch the graph of the cosine function and plot the reciprocal values of the y-values to obtain the y-values of the secant function.

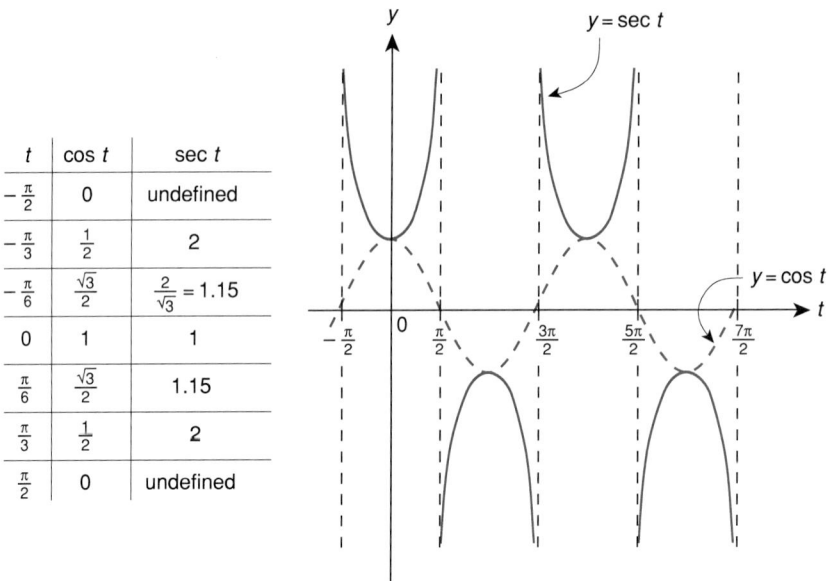

t	$\cos t$	$\sec t$
$-\frac{\pi}{2}$	0	undefined
$-\frac{\pi}{3}$	$\frac{1}{2}$	2
$-\frac{\pi}{6}$	$\frac{\sqrt{3}}{2}$	$\frac{2}{\sqrt{3}} = 1.15$
0	1	1
$\frac{\pi}{6}$	$\frac{\sqrt{3}}{2}$	1.15
$\frac{\pi}{3}$	$\frac{1}{2}$	2
$\frac{\pi}{2}$	0	undefined

Figure 7.44

Figure 7.44 shows several periods of the function $y = \sec t$. Note that the period of the function is 2π. Note also that the graph is symmetric with respect to the y-axis. Thus,

$$\sec(-t) = \sec t.$$

Note that where the graph of the cosine curve crosses the t-axis, the secant function is undefined. Thus, there are asymptotes for the secant curve where the cosine is zero, namely for $t = \frac{\pi}{2} \pm k\pi$.

Similarly, for the graph of $y = \csc t$, we first sketch a graph of the sine function and plot the reciprocal y-values.

Figure 7.45 shows two periods of the function $y = \csc t$. Note that the period is 2π and that the graph is symmetric with respect to the origin. Thus,

$$\csc(-t) = -\csc t.$$

Also note that where the graph of the sine curve crosses the t-axis, the cosecant function is undefined. Thus, there are asymptotes for the cosecant curve where the sine is zero, namely for $t = \pm k\pi$.

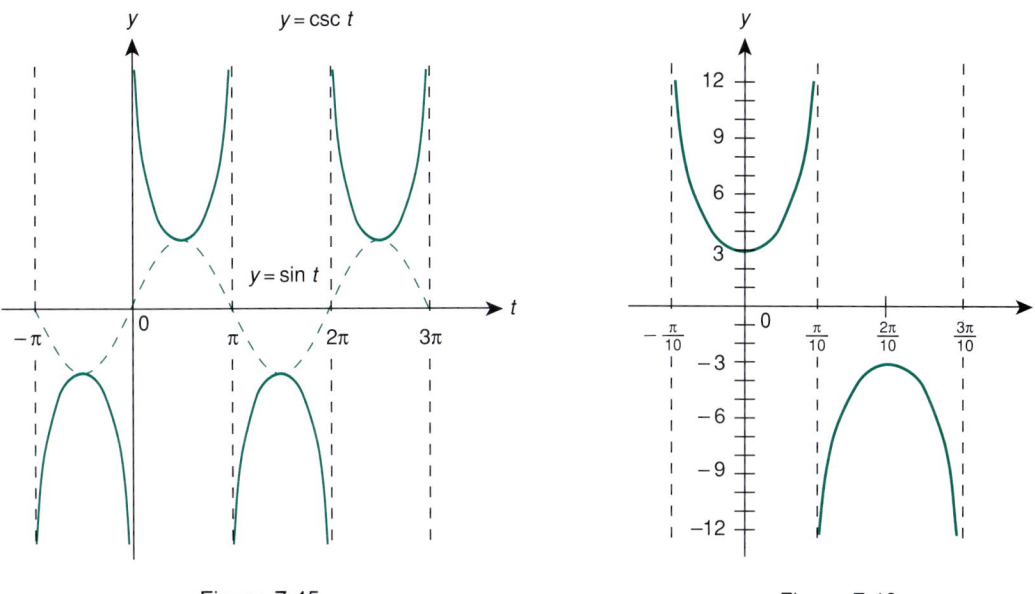

Figure 7.45

Figure 7.46

Examples 7.20 and 7.21 illustrate how to graph some variations of these functions.

- **Example 7.20** Graph one period of the function $y = 3 \sec 5t$.

  Solution: Since the period of the secant function is 2π, the function $y = 3 \sec 5t$ generates one cycle as

  $$5t \text{ varies from } -\frac{\pi}{2} \text{ to } \frac{3\pi}{2},$$

  or

  $$t \text{ varies from } -\frac{\pi}{10} \text{ to } \frac{3\pi}{10}.$$

  One way to graph the function $y = 3 \sec 5t$ is to sketch the graph of $y = \sec 5t$ and multiply the y-values by 3 (see figure 7.46).

  Another method of graphing this function is to sketch the curve $y = 3 \cos 5t$, draw the asymptotes for $y = 3 \sec 5t$ (the values for which $\cos 5t$ is zero and $\sec 5t$ is undefined), and plot the reciprocal values of $3 \cos 5t$. Then sketch the curve.

• Example 7.21: Sketch the graph of one period of the function $y = \tan \dfrac{t}{2}$.

Solution: The graph of $y = \tan t$ contains one period as t varies from $-\dfrac{\pi}{2}$ to $\dfrac{\pi}{2}$. The graph of $y = \tan \dfrac{t}{2}$ contains one period as $\dfrac{t}{2}$ varies from $-\dfrac{\pi}{2}$ to $\dfrac{\pi}{2}$, or as t varies from $-\pi$ to π. Thus, the period of $y = \tan \dfrac{t}{2}$ is 2π.

Table 7.13 shows the coordinates of several points on the graph. This table is an adaptation of table 7.12. Figure 7.47 shows the graph of $y = \tan \dfrac{t}{2}$.

In effect, we could have changed the scale on the t-axis of figure 7.40 to obtain the graph in figure 7.47.

TABLE 7.13

$\dfrac{t}{2}$	$-\dfrac{\pi}{2}$	$-\dfrac{5\pi}{12}$	$-\dfrac{\pi}{3}$	$-\dfrac{\pi}{4}$	$-\dfrac{\pi}{6}$	0	$\dfrac{\pi}{6}$	$\dfrac{\pi}{4}$	$\dfrac{\pi}{3}$	$\dfrac{5\pi}{12}$	$\dfrac{\pi}{2}$
t	$-\pi$	$-\dfrac{5\pi}{6}$	$-\dfrac{2\pi}{3}$	$-\dfrac{\pi}{2}$	$-\dfrac{\pi}{3}$	0	$\dfrac{\pi}{3}$	$\dfrac{\pi}{2}$	$\dfrac{2\pi}{3}$	$\dfrac{5\pi}{6}$	π
$\tan \dfrac{t}{2}$	undefined	-3.73	-1.73	-1	-0.58	0	0.58	1	1.73	3.73	undefined

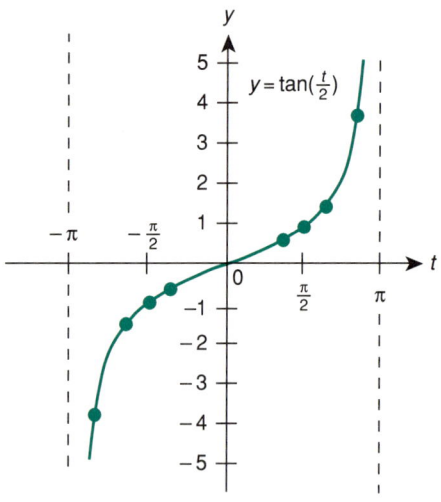

Figure 7.47

Trial Problems 7.5

Before you begin the section exercises, warm up with these problems. Complete answers are included in the answer key.

Find the lowest three nonnegative values of t for which the given function is undefined.

1. $y = \tan 3t$
2. $y = \cot \dfrac{t}{2}$
3. $y = \dfrac{1}{2} \sec 2t$
4. $y = 3 \csc \dfrac{t}{3}$
5. $y = 1.5 \tan 4.5t$

Exercises 7.5

Draw one cycle of each graph and state the period of each function.

1. $y = \tan 2t$
2. $y = \cot 2t$
3. $y = 3 \tan t$
4. $y = 3 \cot t$
5. $y = \sec \dfrac{t}{2}$
6. $y = \csc \dfrac{t}{2}$
7. $y = \dfrac{1}{2} \sec t$
8. $y = \dfrac{1}{2} \csc t$
9. $y = 3 \tan 2t$
10. $y = 3 \cot 2t$
11. $y = \dfrac{1}{2} \sec \dfrac{t}{2}$
12. $y = \dfrac{1}{2} \csc \dfrac{t}{2}$
13. $y = \dfrac{1}{2} \cot 3t$
14. $y = \dfrac{1}{2} \tan 3t$
15. $y = 3 \csc 2t$
16. $y = 3 \sec 2t$

7•6 Composite Curves

Composite functions are "composed" of two or more functions. At times these functions are added or subtracted. In this case we can add or subtract the function values at corresponding domain values.

• **Example 7.22:** Graph the function $y = 2 + \sin t$.

Solution: We first graph the functions $y_1 = 2$ and $y_2 = \sin t$ on the same axes. We now get the points on the graph of $y = 2 + \sin t$ by adding the y-values of these two graphs at the key values of t. For example, at $t = \dfrac{\pi}{2}$, $y_2 = \sin t$ gives $y_2 = 1$ and $y = 2$ gives the $y_1 = 2$. So $y = 2 + \sin t$, evaluated at $t = \dfrac{\pi}{2}$ gives $y = y_1 + y_2 = 2 + 1 = 3$. These values are illustrated in table 7.14.

TABLE 7.14

t	0	$\frac{\pi}{2}$	π	$\frac{3\pi}{2}$	2π
$y_2 = \sin t$	0	1	0	-1	0
$y_1 = 2$	2	2	2	2	2
$y = 2 + \sin t$	2	3	2	1	2

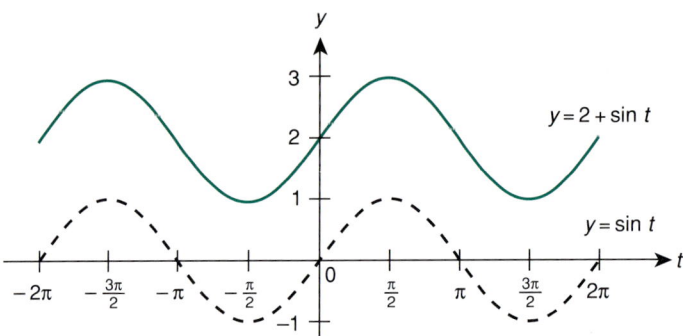

Figure 7.48

These results are shown in figure 7.48.

In effect, we have moved the curve $y = \sin t$ up two units on the coordinate system. Example 7.23 illustrates a more complex case.

• **Example 7.23:** Graph the function $y = \sin t + \cos t$ for $0 \leq t \leq 2\pi$.

Solution: The function can be graphed by letting $y_1 = \sin x$ and $y_2 = \cos x$. We draw both graphs on the same set of axes and add the corresponding y-coordinates, that is, $y = y_1 + y_2$ for any value of t. Table 7.15 contains some values for this function, which is graphed in figure 7.49.

TABLE 7.15

t	0	$\dfrac{\pi}{4}$	$\dfrac{\pi}{2}$	$\dfrac{3\pi}{4}$	π	$\dfrac{5\pi}{4}$	$\dfrac{3\pi}{2}$	$\dfrac{7\pi}{4}$	2π
$y_1 = \sin t$	0	$\dfrac{\sqrt{2}}{2}$	1	$\dfrac{\sqrt{2}}{2}$	0	$-\dfrac{\sqrt{2}}{2}$	-1	$-\dfrac{\sqrt{2}}{2}$	0
$y_2 = \cos t$	1	$\dfrac{\sqrt{2}}{2}$	0	$-\dfrac{\sqrt{2}}{2}$	-1	$-\dfrac{\sqrt{2}}{2}$	0	$\dfrac{\sqrt{2}}{2}$	1
$y_1 + y_2$	1	$\sqrt{2}$	1	0	-1	$-\sqrt{2}$	-1	0	1

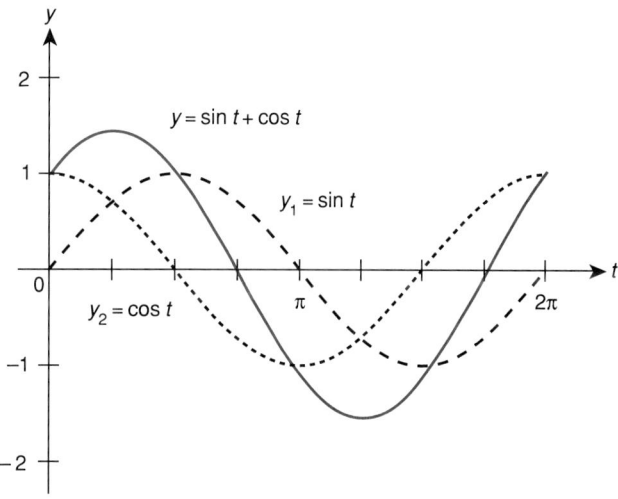

Figure 7.49

From the graph in figure 7.49 we can see that the maximum value of $y = \sin t + \cos t$ occurs at $t = \dfrac{\pi}{4}$, and the minimum value occurs at $t = \dfrac{5\pi}{4}$:

$$\text{at} \quad t = \dfrac{\pi}{4}, \quad y = \dfrac{\sqrt{2}}{2} + \dfrac{\sqrt{2}}{2} = \sqrt{2},$$

$$\text{at} \quad t = \dfrac{5\pi}{4}, \quad y = \dfrac{-\sqrt{2}}{2} - \dfrac{\sqrt{2}}{2} = -\sqrt{2}.$$

Therefore, the amplitude of the function is $\sqrt{2}$.

Example 7.24 shows an application of the method of addition of coordinates to an electrical problem.

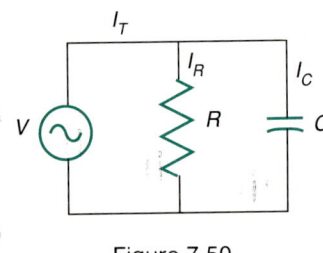

Figure 7.50

- **Example 7.24:** In a parallel AC circuit containing a resistance R and a capacitance C as in figure 7.50, the current through the elements is given by these equations:

$$I_C = 60 \cos\left(60t + \frac{\pi}{2}\right) \quad \text{(The current through } C\text{)}$$

$$I_R = 100 \cos 60t \quad \text{(The current through } R\text{)}$$

$$I_T = I_C + I_R. \quad \text{(The total current)}$$

Graph the function for I_T for $\dfrac{-\pi}{120} \leq t \leq \dfrac{5\pi}{120}$.

Solution: $I_T = 60 \cos\left(60t + \dfrac{\pi}{2}\right) + 100 \cos 60t$.

1. Graph $I_R = 100 \cos 60t$. Table 7.16 contains some of the values for this function.

2. Graph $I_C = 60 \cos\left(60t + \dfrac{\pi}{2}\right) = 60 \cos 60\left(t + \dfrac{\pi}{120}\right)$.
   Table 7.17 contains some of the values for this function.

3. Graph $I_T = I_C + I_R$ by addition of I-coordinates. This graph is pictured in figure 7.51.

TABLE 7.16

cos 60t	0	1	0	−1	0	1	0
60t	$\dfrac{-\pi}{2}$	0	$\dfrac{\pi}{2}$	π	$\dfrac{3\pi}{2}$	2π	$\dfrac{5\pi}{2}$
t	$\dfrac{-\pi}{120}$	0	$\dfrac{\pi}{120}$	$\dfrac{\pi}{60}$	$\dfrac{\pi}{40}$	$\dfrac{\pi}{30}$	$\dfrac{5\pi}{120}$
100 cos 60t	0	100	0	−100	0	100	0

SECTION 7.6 COMPOSITE CURVES

TABLE 7.17

$\cos 60\left(t + \dfrac{\pi}{120}\right)$	1	0	-1	0	1	0	-1
$60\left(t + \dfrac{\pi}{120}\right)$	0	$\dfrac{\pi}{2}$	π	$\dfrac{3\pi}{2}$	2π	$\dfrac{5\pi}{2}$	3π
$t + \dfrac{\pi}{120}$	0	$\dfrac{\pi}{120}$	$\dfrac{\pi}{60}$	$\dfrac{\pi}{40}$	$\dfrac{\pi}{30}$	$\dfrac{5\pi}{120}$	$\dfrac{\pi}{20}$
t	$\dfrac{-\pi}{120}$	0	$\dfrac{\pi}{120}$	$\dfrac{\pi}{60}$	$\dfrac{\pi}{40}$	$\dfrac{\pi}{30}$	$\dfrac{5\pi}{120}$
$60\cos 60\left(t + \dfrac{\pi}{120}\right)$	60	0	-60	0	60	0	-60

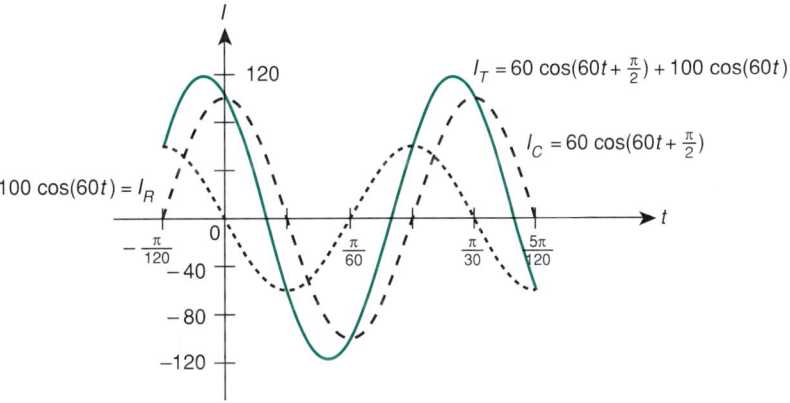

Figure 7.51

In examples 7.22, 7.23, and 7.24, functions have been added or subtracted to form a new function. In example 7.25, the composite function has been formed by multiplying two functions together.

• **Example 7.25:** Graph the function $y = t \sin t$ for $-3\pi \le t \le 3\pi$.

  Solution: The function is *not* periodic, since multiplying $\sin t$ by t magnifies the value of the function as t increases. We construct a table of y-values for some of the critical t-values. Table 7.18 contains some of these values.

TABLE 7.18

t	-3π	$-\dfrac{5\pi}{2}$	-2π	$-\dfrac{3\pi}{2}$	$-\pi$	$-\dfrac{\pi}{2}$	0	$\dfrac{\pi}{2}$	π	$\dfrac{3\pi}{2}$	2π	$\dfrac{5\pi}{2}$	3π
$\sin t$	0	-1	0	1	0	-1	0	1	0	-1	0	1	0
$t \sin t$	0	$\dfrac{5\pi}{2}$	0	$-\dfrac{3\pi}{2}$	0	$\dfrac{\pi}{2}$	0	$\dfrac{\pi}{2}$	0	$-\dfrac{3\pi}{2}$	0	$\dfrac{5\pi}{2}$	0

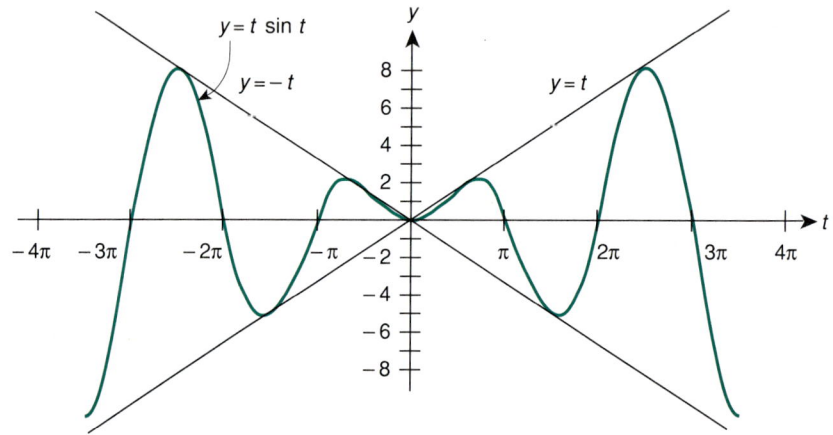

Figure 7.52

We can see from the table that the function is always zero when t is a multiple of π and that $t \sin t = t$ when t is a multiple of $\dfrac{\pi}{2}$. To aid in the graphing, we first graph the lines $y = t$ and $y = -t$ on the coordinate system (see figure 7.52).

We now plot the points listed in the table and connect them with a smooth curve. Note that the curve is symmetric with respect to the y-axis. This can be shown by the following calculations:

$$y = f(t) = t \sin t = (-t)(-\sin t) = (-t)(\sin(-t)) = f(-t).$$

That is,

$$f(t) = f(-t).$$

Trial Problems 7.6

Before you begin the section exercises, warm up with these problems. Complete answers are included in the answer key.

Find the maximum and the minimum value for each function.

1. $y = 2 - \sin t$
2. $y = 3 + \cos t$
3. $y = t + \cos 2t$
4. $y = 8000 + 2000 \cos \dfrac{\pi t}{12}$
5. $y = 2t \cos 3t$

Exercises 7.6

Graph the given functions.

1. $y = 2 + \sin t$
2. $y = 2 + \cos t$
3. $y = -3 + \cos t$
4. $y = -3 + \sin t$
5. $y = 1 + \tan t$
6. $y = 1 + \cot t$
7. $y = \sin t - \cos t$
8. $y = \cos t - \sin t$
9. $y = 3 + \cos 2t$
10. $y = 3 + \sin 2t$
11. $y = 2 \cos t + 3 \sin t$
12. $y = 3 \cos t + 2 \sin t$
13. $y = \sin t + \sin 2t$
14. $y = \cos t + \cos 2t$
15. $y = \sin t + \cos 2t$
16. $y = \cos 2t + 2 \cos t$
17. $y = t + \sin t$
18. $y = t + \cos t$
19. $y = t \cos t$
20. $y = t \sin t$

21. Solve example 7.24 of this section given that
$I_C = 40 \cos\left(50t + \dfrac{\pi}{2}\right)$ and $I_R = 100 \cos 50t$ for
$\dfrac{-\pi}{100} \leq t \leq \dfrac{\pi}{20}$.

22. Solve example 7.24 of this section given that
$I_C = 20 \cos\left(40t + \dfrac{\pi}{2}\right)$ and $I_R = 50 \cos 40t$ for
$\dfrac{-\pi}{80} \leq t \leq \dfrac{3\pi}{80}$.

23. The population of a species of animal in a region fluctuates according to the equation

$y = 20,000 + 5000 \cos \dfrac{\pi t}{12}$, where y is the population after t months. Graph the function for $0 \leq t \leq 48$. What are the maximum and minimum sizes of the population?

24. Solve exercise 23 given that
$y = 10,000 + 300 \cos \dfrac{\pi t}{12}$.

25. The instantaneous power in a particular circuit is given by the equation

$$p = \dfrac{V_m I_m}{2} - \dfrac{V_m I_m}{2} \cos 60t,$$

where V_m = the maximum voltage,
and I_m = the maximum current.
Given that $V_m = 50$ volts and $I_m = 5$ amps, graph the power function for two cycles. What is the maximum power?

26. Solve exercise 25, given that $V_m = 60$ volts and $I_m = 12$ amps.

27. The monthly sales y (in thousands of units) of a product are approximated by the formula

    $y = 53.7 + 22.1 \sin \dfrac{\pi t}{6}$, where t represents the time t in months and $t = 0$ represents July, the first month of the fiscal year. Plot the graph of the function for $t = 0$ to $t = 12$.

28. Solve exercise 27 for $y = 63.2 + 31.2 \sin \dfrac{\pi t}{6}$.

29. An indicator y of the level of economic activity is related to the time t in years by the equation

    $y = 200 + 50 \sin \dfrac{\pi t}{6}$. Graph the function for $t = 0$ to $t = 12$.

30. Solve exercise 29 for the equation $y = 100 + 25 \sin \dfrac{\pi t}{12}$ for $t = 0$ to $t = 24$.

7•7 Inverse Trigonometric Functions

Inverse Function

In this section we consider the inverses of the trigonometric functions. By the **inverse** of a function $f(t)$ we mean the function $g(t)$ that satisfies the following condition:

$$f(g(t)) = g(f(t)) = t.$$

Practically speaking, an inverse function g can be constructed from the graph of a function f using the following rule.

> If (a,b) is a point on f, then (b,a) is a point on g.

Figure 7.53 shows the graphs of $f(t) = t^3$ and $g(t) = \sqrt[3]{t}$, which are inverse functions. Note that the functions are symmetric with respect to the line $y = t$.

One-to-One Function

In order to have an inverse, a function must be a **one-to-one function.** To be one to one, a function must have a different range element for each domain element. Practically speaking, this means that a function is one to one if every horizontal line cuts the graph of the function in at most one point.

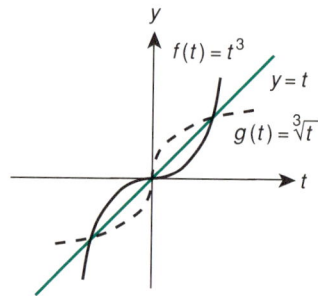

Figure 7.53

Figure 7.54a shows the graph of a function that is not one to one, since the range element, k, is assigned to three different domain elements. Figure 7.54b shows a one-to-one function, since each range element is assigned to only one domain element.

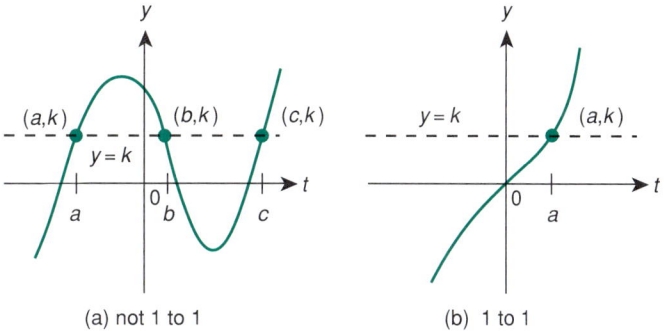

(a) not 1 to 1   (b) 1 to 1

Figure 7.54

Clearly, the trigonometric functions do not fit the conditions for being one to one. The tangent function shown in figure 7.55 is not one to one, since every horizontal line cuts the graph in an infinite number of points.

However, if we restrict the domains of the trigonometric functions, we can construct a one-to-one function from these functions. For example, consider the function $f(t) = \sin t$.

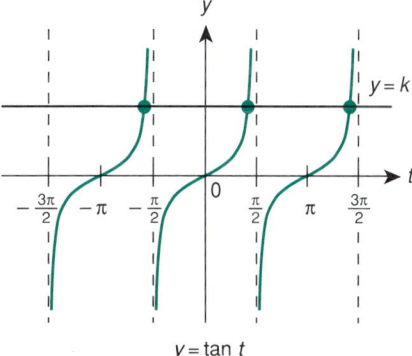

Figure 7.55

310   CHAPTER 7 ANALYTIC TRIGONOMETRY

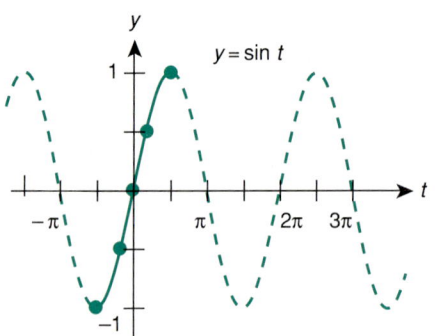

Figure 7.56

If we restrict the domain of $f(t)$ to $\dfrac{-\pi}{2} \leq t \leq \dfrac{\pi}{2}$, as in figure 7.56, the function is indeed one to one. We now define the inverse sine function $\sin^{-1} t$ as follows:

$y = \sin^{-1} t$ provided $t = \sin y$, where $-1 \leq t \leq 1$ and $-\dfrac{\pi}{2} \leq y \leq \dfrac{\pi}{2}$.

In effect, the graph of $y = \sin^{-1} t$ can be constructed using the following property:

If (a,b) is a point on the graph of $y = \sin t$, then (b,a) is a point on the graph of $y = \sin^{-1} t$.

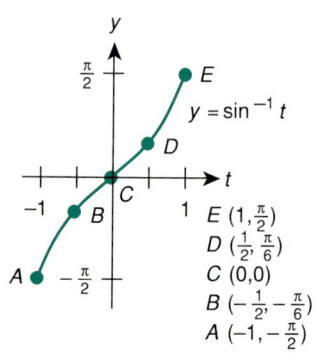

Figure 7.57

The set of points $\left\{\left(-\dfrac{\pi}{2}, -1\right), \left(-\dfrac{\pi}{6}, -\dfrac{1}{2}\right), (0,0), \left(\dfrac{\pi}{6}, \dfrac{1}{2}\right), \left(\dfrac{\pi}{2}, 1\right)\right\}$ is on the graph of $y = \sin t$. The set of points $\left\{\left(-1, -\dfrac{\pi}{2}\right), \left(-\dfrac{1}{2}, -\dfrac{\pi}{6}\right), (0,0), \left(\dfrac{1}{2}, \dfrac{\pi}{6}\right), \left(1, \dfrac{\pi}{2}\right)\right\}$ is, therefore, on the graph of $y = \sin^{-1} t$. See figure 7.57.

SECTION 7.7 INVERSE TRIGONOMETRIC FUNCTIONS   311

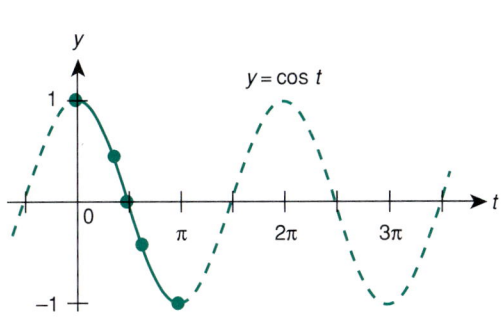

Figure 7.58

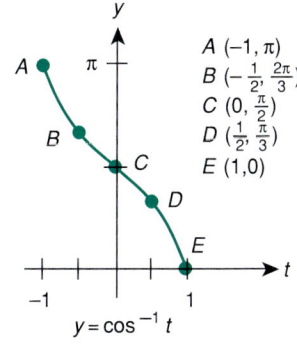

Figure 7.59

For the cosine function, we can construct an inverse function if we restrict the domain of $y = \cos t$ to the interval $0 \leq t \leq \pi$, as in figure 7.58. The set of points $\left\{(0,1), \left(\dfrac{\pi}{3}, \dfrac{1}{2}\right), \left(\dfrac{\pi}{2}, 0\right), \left(\dfrac{2\pi}{3}, -\dfrac{1}{2}\right), (\pi,-1)\right\}$ on the graph of $y = \cos t$ gives us the set of points $\left\{(1,0), \left(\dfrac{1}{2}, \dfrac{\pi}{3}\right), \left(0, \dfrac{\pi}{2}\right), \left(-\dfrac{1}{2}, \dfrac{2\pi}{3}\right), (-1, \pi)\right\}$ on the graph of $y = \cos^{-1} t$.

We graph these on a set of axes and connect them with a smooth curve (figure 7.59).

For the tangent function, we can construct an inverse function in a similar manner by restricting the domain of the tangent to $-\dfrac{\pi}{2} < t < \dfrac{\pi}{2}$.

The range of $y = \tan^{-1} t$ is $-\dfrac{\pi}{2} < y < \dfrac{\pi}{2}$ for the domain that consists of all real numbers. Figure 7.60 contains the restricted tangent function and its inverse function.

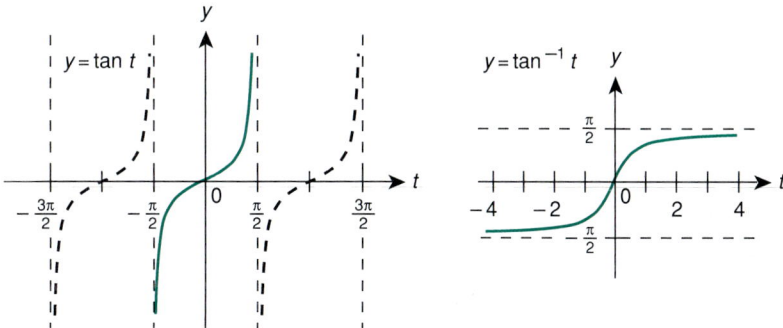

Figure 7.60

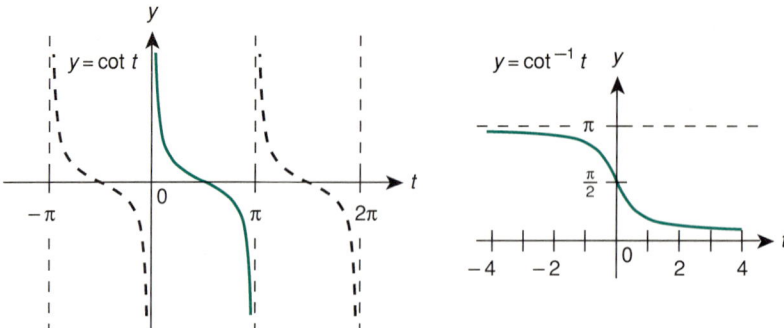

Figure 7.61

For the cotangent function, we can construct an inverse function by restricting the domain of $y = \cot t$ to $0 < t < \pi$.
The domain of $y = \cot^{-1} t$ is the set of all real numbers and the range is $0 < y < \pi$. Figure 7.61 contains these functions.

To construct an inverse for the secant function, we restrict the domain of $y = \sec t$ to $0 < t < \pi$.
The function $y = \sec^{-1} t$ has no t-values between -1 and $+1$, so the domain is $t \leq -1$ or $t \geq 1$. The range of the function is $0 \leq y \leq \pi$, $y \neq \dfrac{\pi}{2}$. Figure 7.62 contains these functions. Note: the definitions of $\sec^{-1}$ and $\csc^{-1}$ are not universally agreed upon.

The student is invited to construct the graph of $y = \csc^{-1} t$ using the graph of $y = \csc t$ over the domain $-\dfrac{\pi}{2} \leq t \leq \dfrac{\pi}{2}$, $t \neq 0$. Check your results with table 7.21 in the chapter review.

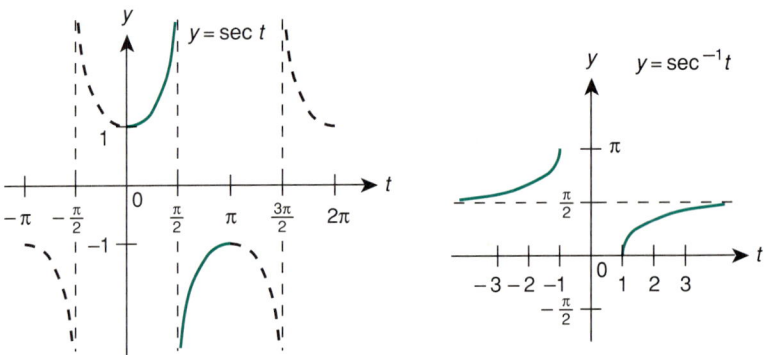

Figure 7.62

Briefly stated, an inverse function of a trigonometric function is the function $f^{-1}(t)$ such that $f(f^{-1}(t)) = t$ and $f^{-1}(f(t)) = t$. For example,

$$\sin(\sin^{-1} t) = t \quad \text{for } -1 \leq t \leq 1$$

$$\sin^{-1}(\sin t) = t \quad \text{for } -\frac{\pi}{2} \leq t \leq \frac{\pi}{2}.$$

Similar results can be stated for the other trigonometric functions.

INVERSE TRIGONOMETRIC FUNCTIONS

$$\cos(\cos^{-1} t) = t \quad \text{for } -1 \leq t \leq 1$$
$$\cos^{-1}(\cos t) = t \quad \text{for } 0 \leq t \leq \pi$$
$$\tan(\tan^{-1} t) = t \quad \text{for any } t$$
$$\tan^{-1}(\tan t) = t \quad \text{for } \frac{-\pi}{2} < t < \frac{\pi}{2}$$
$$\cot(\cot^{-1} t) = t \quad \text{for any } t$$
$$\cot^{-1}(\cot t) = t \quad \text{for } 0 < t < \pi$$
$$\sec(\sec^{-1} t) = t \quad \text{for } t \leq -1 \text{ or } t \geq 1$$
$$\sec^{-1}(\sec t) = t \quad \text{for } 0 \leq t \leq \pi, t \neq \frac{\pi}{2}$$
$$\csc(\csc^{-1} t) = t \quad \text{for } t \leq -1 \text{ or } t \geq 1$$
$$\csc^{-1}(\csc t) = t \quad \text{for } -\frac{\pi}{2} \leq t \leq \frac{\pi}{2}, t \neq 0$$

We can calculate approximate values for the inverse trigonometric functions using a scientific calculator that has an [INV] or [ARC] and [SIN], [COS], and [TAN] buttons. We usually get an ERROR in the display if we enter a number that is not in the domain of the inverse function and try to get the inverse.

Examples (set the calculator for radians):

a. $\sin^{-1}(0.314) = 0.319$ (approx.)

0.314 [INV] [SIN].

Display: 0.3194032.

b. $\cos^{-1}(-0.327) = 1.904$ (approx.)

0.327 [+/−] [INV] [COS].

Display: 1.9039236.

c. $\cos^{-1}(1.32)$ does not exist

$$1.32 \; \boxed{\text{INV}} \; \boxed{\text{COS}}.$$

Display: ERROR or E.
This happens because it is not possible to find a number t such that $\cos t = 1.32$. Note the restrictions on the domain of $y = \cos^{-1} t$.

d. $\tan^{-1}(18.732) = 1.517$ (approx.)

$$18.732 \; \boxed{\text{INV}} \; \boxed{\text{TAN}}.$$

Display: 1.5174624.
There are some difficulties with using a calculator to determine the values for an inverse trigonometric function. Example 7.26 illustrates this.

• *Example 7.26:* Use a calculator to calculate $\sin^{-1}(\sin 1.94)$.

Solution: Using the formula $\sin^{-1}(\sin t) = t$, it would appear that

$$\sin^{-1}(\sin 1.94) = 1.94.$$

However, if we use the calculator set on radian measure to check this, we get the following steps.

$$1.94 \; \boxed{\text{SIN}} \; \boxed{\text{INV}} \; \boxed{\text{SIN}}.$$

Display: 1.2015927.
  What went wrong? An examination of the graph of the function $y = \sin t$ in figure 7.63 shows the difficulty.
  When we calculated $\sin(1.94)$, the calculator did indeed get the correct value, 0.9326 (approx.). However, when we calculated $\sin^{-1}(0.9326)$, the calculator assumed that we meant the 0.9326 that was in the domain of the inverse sine function, and it correctly reported the $\sin^{-1}(0.9326)$ as 1.20 (approx.). Hence, in such exercises we must take care to observe the domain restrictions on the inverse functions.

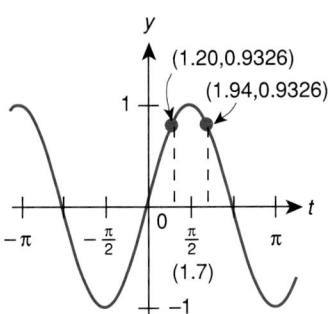

Figure 7.63

Here are some examples that illustrate how to evaluate expressions that contain inverse functions.

• **Example 7.27:** Use a calculator to calculate the value of each expression in radians to four decimal places.

   a. $\cos(\tan^{-1} 0.43)$

   b. $\sin(\tan^{-1} 0.326)$

   c. $\sin^{-1}(\cos 0.842)$

   d. $\tan^{-1}(\cos 3.42)$

Solution: **a. 0.43** $\boxed{\text{INV}}$ $\boxed{\text{TAN}}$ $\boxed{\text{COS}}$

Display: 0.9186692.
$\cos(\tan^{-1} 0.43) = 0.9187$.

b. 0.326 $\boxed{\text{INV}}$ $\boxed{\text{TAN}}$ $\boxed{\text{SIN}}$

Display: 0.3099459.
$\sin(\tan^{-1} 0.326) = 0.3099$.

c. 0.842 $\boxed{\text{COS}}$ $\boxed{\text{INV}}$ $\boxed{\text{SIN}}$

Display: 0.7287963.
$\sin^{-1}(\cos 0.842) = 0.7288$.

d. 3.42 $\boxed{\text{COS}}$ $\boxed{\text{INV}}$ $\boxed{\text{TAN}}$

Display: -0.7657699.
$\tan^{-1}(\cos 3.42) = -0.7658$.

Examples 7.28 and 7.29 illustrate the procedure for graphing variations of the inverse trigonometric functions.

• **Example 7.28:** Sketch the graph of $y = \dfrac{3\pi}{4} + \tan^{-1} t$

Solution: Graph the functions $y = \tan^{-1} t$ and $y = \dfrac{3\pi}{4}$ on the same set of axes and construct the graph of $y = \dfrac{3\pi}{4} + \tan^{-1} t$ by addition of y-coordinates. The results are shown in figure 7.64.

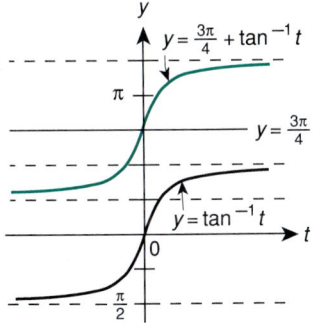
Figure 7.64

316 CHAPTER 7 ANALYTIC TRIGONOMETRY

• **Example 7.29:** Sketch the graph of the function $y = 3 \sin^{-1} t$.

TABLE 7.19

t	$\sin^{-1} t$	$3 \sin^{-1} t$
-1	$-\dfrac{\pi}{2}$	$-\dfrac{3\pi}{2}$
$-\dfrac{1}{2}$	$-\dfrac{\pi}{6}$	$-\dfrac{\pi}{2}$
0	0	0
$\dfrac{1}{2}$	$\dfrac{\pi}{6}$	$\dfrac{\pi}{2}$
1	$\dfrac{\pi}{2}$	$\dfrac{3\pi}{2}$

Solution: Since the domain of $y = \sin^{-1} t$ is $-1 \le t \le 1$, the function $y = 3 \sin^{-1} t$ has the same domain. Table 7.19 shows the coordinates of five points on the graph. Plot the points on a coordinate system and sketch a smooth curve (figure 7.65).

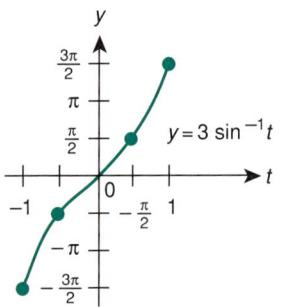

Figure 7.65

Trial Problems 7.7

Before you begin the section exercises, warm up with these problems. Complete answers are included in the answer key.

Use a calculator to evaluate the expressions correct to four decimal places.

1. $\cos^{-1}(-0.4632)$
2. $\sin^{-1}(0.5617)$
3. $\tan^{-1}\left(\dfrac{2\pi}{3}\right)$
4. $\tan\left(\sin^{-1}\left(\dfrac{\pi}{4}\right)\right)$
5. $\sin(\sin^{-1}(0.5))$

Exercises 7.7

Use a calculator to find the following in radians, correct to four decimal places.

1. $\sin^{-1}(0.87)$
2. $\cos^{-1}(0.87)$
3. $\cos^{-1}(-0.4642)$
4. $\sin^{-1}(-0.4642)$
5. $\sin^{-1}(1.324)$
6. $\cos^{-1}(1.324)$
7. $\tan^{-1}(1.872)$
8. $\tan^{-1}(2.631)$
9. $\tan^{-1}(-37.62)$
10. $\tan^{-1}(62.36)$
11. $\sin^{-1}\left(\dfrac{\pi}{12}\right)$
12. $\cos^{-1}\left(\dfrac{\pi}{7}\right)$
13. $\sin(\cos^{-1}(0.862))$
14. $\cos(\sin^{-1}(0.862))$
15. $\sin^{-1}(\cos(0.378))$
16. $\cos^{-1}(\sin(0.378))$
17. $\tan(\sin^{-1}(0.463))$
18. $\sin(\tan^{-1}(1.32))$
19. $\sin\left(\cos^{-1}\left(\dfrac{\pi}{6}\right)\right)$
20. $\cos^{-1}\left(\sin\left(\dfrac{\pi}{6}\right)\right)$
21. $\tan\left(\sin^{-1}\left(\dfrac{\pi}{4}\right)\right)$
22. $\tan\left(\cos^{-1}\left(\dfrac{\pi}{4}\right)\right)$
23. $\sin\left(\tan^{-1}\left(\dfrac{\pi}{3}\right)\right)$
24. $\cos\left(\tan^{-1}\left(\dfrac{\pi}{3}\right)\right)$

25. Graph the functions $y = \sin t$, $-\dfrac{\pi}{2} \le t \le \dfrac{\pi}{2}$, $y = t$, and $y = \sin^{-1} t$, $-1 \le t \le 1$ on the same set of axes, using the property that the graphs of a function and its inverse are symmetric about the line $y = t$.

26. Graph the functions $y = \cos t$, $0 \le t \le \pi$, $y = t$, and $y = \cos^{-1} t$, $-1 \le t \le 1$ on the same set of axes, using the property that the graphs of a function and its inverse are symmetric about the line $y = t$.

Graph the given functions.

27. $y = 1 + \sin^{-1} t$
28. $y = 2 + \cos^{-1} t$
29. $y = 2 \cos^{-1} t$
30. $y = 2 \sin^{-1} t$
31. $y = \pi + \tan^{-1} t$
32. $y = \pi + \cot^{-1} t$

33. In the 1800s an English mathematician used the following relationship to calculate a value of π to over 100 decimal places:
$$\pi = 16 \tan^{-1}\left(\dfrac{1}{5}\right) - 4 \tan^{-1}\left(\dfrac{1}{70}\right) + 4 \tan^{-1}\left(\dfrac{1}{90}\right).$$
Use a calculator to verify that this is correct (approx.).

34. In the 1800s a German mathematician used the following relationship to calculate a value of π to 200 decimal places:
$$\pi = 4\left[\tan^{-1}\left(\dfrac{1}{2}\right) + \tan^{-1}\left(\dfrac{1}{5}\right) + \tan^{-1}\left(\dfrac{1}{8}\right)\right].$$
Use a calculator to verify that this is correct (approx.).

Under certain conditions, if the leads to an oscilloscope are connected at the points shown in the circuit of figure 7.66, the phase angle between the two voltages can be calculated according to the formula $t = \sin^{-1}\left(\dfrac{AA'}{BB'}\right)$ where AA' and BB' are the distances shown on the figure on the screen of the oscilloscope. (Note: If the phase angle is 0, the figure is a line, and if the phase angle is $\dfrac{\pi}{2}$, the figure is a circle.) In the following exercises calculate the phase angle using the given measurements.

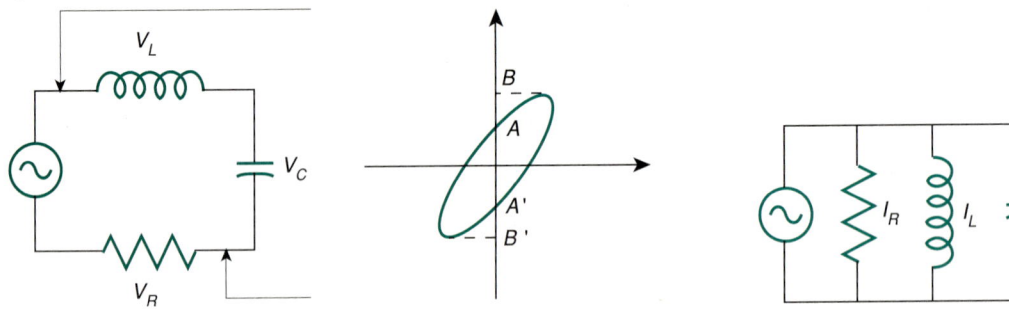

Figure 7.66

Figure 7.67

35. $AA' = 1.4 \quad BB' = 2.3$
37. $AA' = 0.37 \quad BB' = 0.74$
36. $AA' = 0.6 \quad BB' = 0.8$
38. $AA' = 0.42 \quad BB' = 0.96$

Under certain conditions, the phase angle t in figure 7.67 can be calculated according to the formula

$$t = \tan^{-1}\left(\frac{I_c - I_L}{I_R}\right).$$

In the following exercises, calculate the phase angle using the given measurements.

39. $I_c = 0.5, I_L = 0.2, I_R = 0.3$
40. $I_c = 0.6, I_L = 0.3, I_R = 0.3$
41. $I_c = 0.73, I_L = 0.52, I_R = 0.66$
42. $I_c = 1.32, I_L = 0.85, I_R = 1.03$

7·8 Summary of Terms, Formulas, Rules, and Procedures

Terms

Amplitude (p. 278)
Angular Velocity (p. 258)
Cycle (p. 277)
Even Function (p. 268)
Frequency (p. 290)
In Phase (p. 292)
Inverse Function (p. 308)
Lags (p. 291)

Leads (p. 291)
Linear Velocity (p. 257)
Odd Function (p. 268)
One-to-one Function (p. 308)
Out of Phase (p. 291)
Period of a Function (p. 274)
Periodic Function (p. 275)
Phase Shift (p. 287)

Pure Waves (p. 290)
Radian (p. 254)
Sector of a Circle (p. 255)
Simple Harmonic Motion (p. 290)
Sinusoidal Curve (p. 277)
Unit Circle (p. 262)
Vertical Asymptote (p. 295)

Formulas

Radian Measure (7.1)

$$\theta \text{ (radians)} = \frac{s \text{ (arc length)}}{r \text{ (radius of the circle)}}$$

$$\theta° = \left(\frac{\theta\pi}{180}\right) \text{ radians}$$

$$t \text{ radians} = 180\left(\frac{t}{\pi}\right) \text{ degrees.}$$

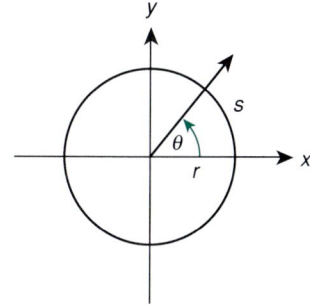

Figure 7.68

Area of a sector of a circle (A) (7.1)

$$A = \frac{1}{2}r^2\theta$$

where r = radius of the circle and θ = measure of the central angle in radians.

Linear Velocity (v) (7.1)

$$v = \frac{s \text{ (arc length)}}{t \text{ (time)}}.$$

Angular velocity (ω) (7.1)

$$\omega = \frac{\theta \text{ (measure of the central angle in radians)}}{t \text{ (time)}}.$$

Linear velocity (v) (7.1)

$v = r\omega$, where r = radius of the circle and ω = angular velocity.

Trigonometric functions for t in radians (7.2)

$$\sin t = y \qquad \tan t = \frac{y}{x} \qquad \sec t = \frac{1}{x}$$

$$\cos t = x \qquad \cot t = \frac{x}{y} \qquad \csc t = \frac{1}{y}.$$

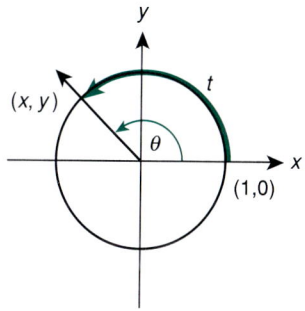

Figure 7.69

Trigonometric Properties (7.2)

$\sin(t + 2\pi k) = \sin t$, where k is an integer
$\cos(t + 2\pi k) = \cos t$, where k is an integer
$\sin(-t) = -\sin t \qquad \cos(-t) = \cos t$
$\tan(-t) = -\tan t \qquad \cot(-t) = -\cot t$
$\sec(-t) = \sec t \qquad \csc(-t) = -\csc t.$

Rules and Procedures

- Graphs of the trigonometric functions (7.3, 7.5)

TABLE 7.20

Function	Domain	Range	Period	Graph $0 \leq t \leq 2\pi$
$y = \sin t$	all real numbers	$-1 \leq y \leq 1$	2π	
$y = \cos t$	all real numbers	$-1 \leq y \leq 1$	2π	
$y = \tan t$	all real numbers except $t = \dfrac{\pi}{2} + k$, k an integer	all real numbers	π	

SECTION 7.8 SUMMARY

TABLE 7.20—*Continued*

Function	Domain	Range	Period	Graph $0 \leq t \leq 2\pi$
$y = \cot t$	all real numbers except $t = k$, k an integer	all real numbers	π	
$y = \sec t$	all real numbers except $t = \dfrac{\pi}{2} + k$, k an integer	$y \leq -1$ or $y \geq 1$	2π	
$y = \csc t$	all real numbers except $t = k$, k an integer	$y \leq -1$ or $y \geq 1$	2π	

- Graphs of the inverse trigonometric functions (7.7)

TABLE 7.21

Function	Domain	Range	Graph
$y = \sin^{-1} t$	$-1 \leq t \leq 1$	$-\dfrac{\pi}{2} \leq y \leq \dfrac{\pi}{2}$	
$y = \cos^{-1} t$	$-1 \leq t \leq 1$	$0 \leq y \leq \pi$	
$y = \tan^{-1} t$	all real numbers	$-\dfrac{\pi}{2} < y < \dfrac{\pi}{2}$	

TABLE 7.21—*Continued*

Function	Domain	Range	Graph
$y = \cot^{-1} t$	all real numbers	$0 < y < \pi$	
$y = \sec^{-1} t$	$t \leq -1$ or $t \geq 1$	$0 \leq y \leq \pi$ $y \neq \dfrac{\pi}{2}$	
$y = \csc^{-1} t$	$t \leq -1$ or $t \geq 1$	$-\dfrac{\pi}{2} \leq y \leq \dfrac{\pi}{2}$ $y \neq 0$	

- Graph of $y = a \sin(bt + c) = a \sin b\left(t + \dfrac{c}{b}\right)$ (7.4)

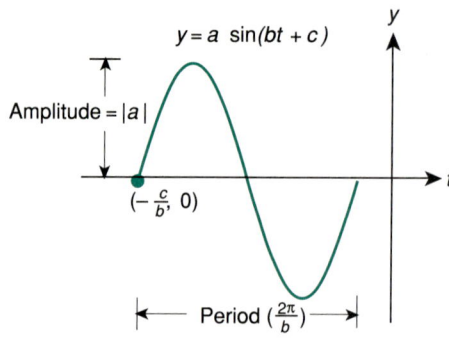

Figure 7.70

- Graph of $y = a \cos(bt + c) = a \cos b\left(t + \dfrac{c}{b}\right)$ (7.4)

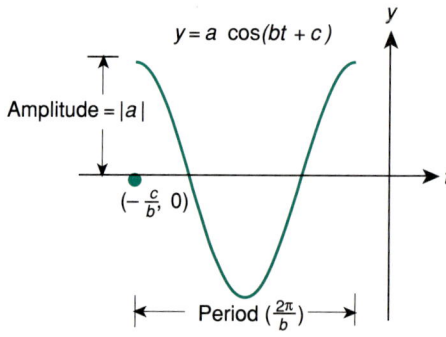

Figure 7.71

7·9 Chapter 7 Review Exercises

1. $61°25' = \underline{}$ radians
2. $42°25' = \underline{}$ radians
3. $\dfrac{\pi}{3}$ radians $= \underline{}$ degrees
4. $\dfrac{4\pi}{3}$ radians $= \underline{}$ degrees
5. 3.24 radians $= \underline{}$ degrees
6. 1.72 radians $= \underline{}$ degrees

In the following exercises, calculate the missing angle (θ), arc length (s), or radius (r).

7. $s = 3.7$ cm, $r = 2.6$ cm
8. $s = 12.0$ in., $r = 6.5$ in.
9. $s = 3.5$ cm, $\theta = 1.4$ radians
10. $s = 2.7$ mm, $\theta = 0.5$ radians

In the following exercises find the missing linear velocity (v), angular velocity (ω), or radius (r).

11. $v = 66$ ft/sec, $\omega = 600$ rpm
12. $\omega = 2.6$ rad/sec, $v = 16.5$ ft/sec
13. $v = 14.3$ m/sec, $r = 3.6$ m
14. $\omega = 2.7$ rad/sec, $r = 5.1$ in.

SECTION 7.9 CHAPTER 7 REVIEW EXERCISES

Find the exact values for each of the following expressions.

15. $\sin \dfrac{31\pi}{2}$
16. $\cos \dfrac{31\pi}{2}$
17. $\tan \dfrac{11\pi}{4}$
18. $\cot \dfrac{11\pi}{4}$

Use a calculator to approximate each value.

19. $y = \dfrac{1}{2} \sin 16t$, where $t = \dfrac{\pi}{10}$
20. $y = \dfrac{1}{4} \cos 20t$, where $t = \dfrac{\pi}{10}$
21. $y = 40\left(\sin\left(50t + \dfrac{\pi}{6}\right)\right)$, where $t = \dfrac{\pi}{2}$
22. $y = 200\left(\sin\left(40t + \dfrac{\pi}{6}\right)\right)$, where $t = \dfrac{\pi}{2}$
23. $y = 20\left(\cos \dfrac{\pi}{5} + \sin \dfrac{\pi}{5}\right)$
24. $y = 30\left(\cos \dfrac{\pi}{8} + \sin \dfrac{\pi}{8}\right)$

Graph each function for the given domain.

25. $y = 3 \sin t,\ 0 \le t \le 2\pi$
26. $y = 3 \cos t,\ 0 \le t \le 2\pi$
27. $y = 3 \cos \dfrac{\pi t}{3},\ 0 \le t \le 6$
28. $y = 3 \sin \dfrac{\pi t}{3},\ 0 \le t \le 6$
29. $y = \tan \dfrac{t}{3},\ -\dfrac{3\pi}{2} \le t \le \dfrac{3\pi}{2}$
30. $y = \cot \dfrac{t}{3},\ 0 \le t \le 3\pi$

Graph each function for one period.

31. $y = \sin\left(t + \dfrac{\pi}{3}\right)$
32. $y = \cos\left(t + \dfrac{\pi}{3}\right)$
33. $y = 3 \cos\left(t + \dfrac{\pi}{3}\right)$
34. $y = 3 \sin\left(t + \dfrac{\pi}{3}\right)$
35. $y = 2 \cos\left(3t - \dfrac{\pi}{4}\right)$
36. $y = 2 \sin\left(3t - \dfrac{\pi}{4}\right)$
37. $y = 50 \sin\left(100t + \dfrac{\pi}{2}\right)$
38. $y = 50 \cos\left(100t + \dfrac{\pi}{2}\right)$
39. $y = 1.5 + \cos t$
40. $y = 1.5 + \sin t$
41. $y = \sin 2t + 2 \cos t$
42. $y = 2 \cos t + \sin t$

Use your calculator to find each value.

43. $y = 3 \cos^{-1} t$, where $t = 0.45$
44. $y = \sin^{-1} t$, where $t = 0.67$
45. $y = \tan^{-1} t$, where $t = 3.22$
46. $y = \tan^{-1} t$, where $t = -3.81$
47. $y = \tan(\sin^{-1} t)$, where $t = -0.57$
48. $y = \cot(\sin^{-1} t)$, where $t = 0.82$
49. $y = \sin^{-1}(\cos t)$, where $t = \dfrac{\pi}{4}$
50. $y = \cos^{-1}(\sin t)$, where $t = \dfrac{\pi}{4}$

Graph each function.

51. $y = \sin^{-1}(2t)$ for $-\dfrac{1}{2} \le t \le \dfrac{1}{2}$

52. $y = \cos^{-1}(2t)$ for $-\dfrac{1}{2} \le t \le \dfrac{1}{2}$

53. The population (y) of a species of animal in a region at a time (t) in months is approximated by the equation $y = 10{,}000 + 1200 \sin \dfrac{\pi t}{12}$. Find the maximum and minimum populations for the species.

54. The population (y) of a species of animal in a region at a time (t) in months is approximated by the equation $y = 20{,}000 + 2000 \sin \dfrac{\pi t}{12}$. Find the maximum and minimum populations for the species.

55. The current across a resistor is given the equation $i = 30 \sin\left(300t + \dfrac{\pi}{6}\right)$ and the voltage by the equation $v = 150 \sin\left(300t + \dfrac{\pi}{6}\right)$. Graph both equations on the same set of axes and determine the amplitude and period for each equation.

56. Solve exercise 55 using the equations $i = 30 \sin\left(200t + \dfrac{\pi}{3}\right)$ and $v = 90 \sin\left(200t + \dfrac{\pi}{3}\right)$.

57. The power in watts delivered in a particular circuit is determined by the equation

$$P = \sqrt{3}\,VI \cos\theta,$$

where V is the voltage in watts, I is the current in amperes, and t is the phase angle between V and I. Given $P = 1530$ watts, $V = 440$ volts, and $I = 2.4$ amps, determine the phase angle.

58. Solve exercise 57 given that $P = 1200$ watts, $V = 440$ volts, and $I = 2.2$ amps.

7·10 Chapter 7 Test

1. $53°20' = $ _?_ radians

2. 1.88 radians $=$ _?_ degrees

In the following exercises, calculate the missing angle (θ), arc length (s), or circle radius (r).

3. $s = 4.2$ cm, $\theta = 1.62$ radians

4. $s = 11.3$ in., $r = 3.8$ in.

In the following exercises find the missing linear velocity (v), angular velocity (ω), or circle radius (r).

5. $\omega = 2.4$ rad/sec, $v = 33$ ft/sec

6. $v = 18.4$ m/sec, $r = 0.4$ m

Find an exact value.

7. $\cot \dfrac{9\pi}{4}$

8. $\cos \dfrac{19\pi}{4}$

Use a calculator to approximate each value.

9. $y = \dfrac{1}{4} \sin 20t$, where $t = \dfrac{\pi}{12}$

10. $y = 40\left(\sin\left(40t + \dfrac{\pi}{3}\right)\right)$, where $t = \dfrac{\pi}{6}$

11. $y = 40\left(\cos \dfrac{\pi}{10} + \sin \dfrac{\pi}{10}\right)$

12. $y = 1.8\left(\cos \dfrac{\pi}{20} + \sin \dfrac{\pi}{20}\right)$

Graph each function for the given domain.

13. $y = 2.5 \cos t, \ 0 \le t \le 2\pi$

14. $y = 3 \cos \dfrac{\pi t}{2}, \ 0 \le t \le 4$

15. $y = \tan \dfrac{t}{2}, \ -\pi < t < \pi$

16. $y = \cos\left(t + \dfrac{\pi}{6}\right), \ 0 \le t \le 2\pi$

Graph each function for one period.

17. $y = 2 \cos\left(t + \dfrac{\pi}{2}\right)$

18. $y = 3 \cos\left(2t - \dfrac{\pi}{2}\right)$

19. $y = 40 \sin\left(50t + \dfrac{\pi}{2}\right)$

20. $y = 2(\sin t + \cos t)$

21. $y = t + \cos t$

22. $y = \cos^{-1}(3t)$ for $-\dfrac{1}{3} \le t \le \dfrac{1}{3}$

23. The current through a resistor is represented by the equation $i = 40 \sin\left(60t + \dfrac{\pi}{3}\right)$. Find the current when $t = \pi$.

24. The motion of a weight attached to a spring is represented by the equation $y = 0.25 \cos 25t$, where t is the time in seconds and y is the displacement of the weight in inches. Find the position of the weight after $\dfrac{\pi}{8}$ seconds.

CHAPTER 8

Trigonometric Identities and Equations

A sixteenth century French mathematician, François Viète (1540–1603), in his publication *Canon Mathematicus Seu ad Triangula*, made several important contributions to trigonometry. He was one of the first mathematicians to develop methods of solving triangles using all six of the trigonometric functions. He also developed some of the trigonometric identities still used today. Some examples are

$$\sin \theta = \sin(60° + \theta) - \sin(60° - \theta)$$

$$\sin x + \sin y = 2 \sin \frac{x + y}{2} \cos \frac{x - y}{2}$$

$$\csc \theta + \cot \theta = \cot \frac{\theta}{2}.$$

Viète is also credited with developing expressions for $\cos n\theta$, where n is a positive integer less than or equal to ten, and for using trigonometric substitution to solve certain types of cubic equations.

8 • 1 Trigonometric Identities

In previous chapters we developed some of the relationships between the different trigonometric functions. These relationships, along with some similar ones, are listed below.

TRIGONOMETRIC IDENTITIES

I. Reciprocal Identities

$$\sin t = \frac{1}{\csc t} \qquad \cos t = \frac{1}{\sec t} \qquad \tan t = \frac{1}{\cot t}$$

$$\cot t = \frac{1}{\tan t} \qquad \sec t = \frac{1}{\cos t} \qquad \csc t = \frac{1}{\sin t}.$$

II. Quotient Identities

$$\tan t = \frac{\sin t}{\cos t} \qquad \cot t = \frac{\cos t}{\sin t}.$$

III. Sign Identities

$$\sin(-t) = -\sin t \qquad \cos(-t) = \cos t \qquad \tan(-t) = -\tan t$$
$$\cot(-t) = -\cot t \qquad \sec(-t) = \sec t \qquad \csc(-t) = -\csc t.$$

IV. Pythagorean Identities

$$\sin^2 t + \cos^2 t = 1 \qquad \tan^2 t + 1 = \sec^2 t$$
$$\cot^2 t + 1 = \csc^2 t.$$

The new identities like $\sin^2 t + \cos^2 t = 1$ can be easily verified. Consider the unit circle shown in figure 8.1. The Pythagorean theorem yields the relationship

$$y^2 + x^2 = 1.$$

Substitute $y = \sin t$ and $x = \cos t$ into the equation:

$$\sin^2 t + \cos^2 t = 1.$$

We can verify the identity $\tan^2 t + 1 = \sec^2 t$ using the above identity. Begin with the identity $\sin^2 t + \cos^2 t = 1$. Dividing each member of the equation by $\cos^2 t$ yields the equation

$$\frac{\sin^2 t}{\cos^2 t} + \frac{\cos^2 t}{\cos^2 t} = \frac{1}{\cos^2 t}.$$

Substitute $\tan^2 t$ for $\frac{\sin^2 t}{\cos^2 t}$ and $\sec^2 t$ for $\frac{1}{\cos^2 t}$:

$$\tan^2 t + 1 = \sec^2 t.$$

This procedure illustrates one of the methods of verifying a trigonometric identity. The following suggestions are useful in proving these identities.

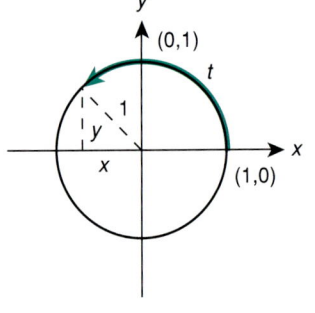

Figure 8.1

Verifying a Trigonometric Identity

1. If possible, (a) simplify one side of the equation until it is the same as the other side, or (b) simplify both sides of the equation until they are the same.

2. If the identity contains fractions, (a) combine the sum or the difference of the two fractions into a single fraction, or (b) split a single fraction into the sum or difference of two fractions.

3. Write all of the expressions in the equation in terms of sines and cosines, then simplify the equation.

4. Usually it is best to work with the most complex side of the equation.

Examples 8.1 and 8.2 illustrate the use of these suggestions.

• *Example 8.1:* Write the given expression in terms of sine and cosine and simplify the expression.

$$\frac{1 + \tan t}{\sin t} - \sec t.$$

Solution: Substitute $\tan t = \dfrac{\sin t}{\cos t}$ and $\sec t = \dfrac{1}{\cos t}$ into the given expression to get

$$\frac{1 + \tan t}{\sin t} - \sec t = \frac{1 + \dfrac{\sin t}{\cos t}}{\sin t} - \frac{1}{\cos t}.$$

Next determine the common denominator to be $(\sin t)(\cos t)$, and express both of the terms using this denominator.

$$= \frac{\left(1 + \dfrac{\sin t}{\cos t}\right)}{\sin t} \cdot \frac{\cos t}{\cos t} - \frac{1}{\cos t} \cdot \frac{\sin t}{\sin t}$$

$$= \frac{\cos t + \sin t}{\sin t \cos t} - \frac{\sin t}{\sin t \cos t}$$

$$= \frac{\cos t + \sin t - \sin t}{\sin t \cos t} \qquad \text{(Simplify the numerator and combine fractions with a common denominator)}$$

$$= \frac{\cos t}{\sin t \cos t} \qquad \text{(Simplify the numerator)}$$

$$= \frac{1}{\sin t}. \qquad \text{(Reduce to lower terms)}$$

Often there is more than one way to simplify an expression. Examples 8.2–8.5 give one method of simplification. If you see another possible approach, give it a try and see if it works.

• **Example 8.2:** Write the given expression in terms of sines and cosines and simplify.

$$\frac{1 + \cot t \sec t}{\tan t + \sec t}.$$

Solution: Substitute $\cot t = \dfrac{\cos t}{\sin t}$, $\sec t = \dfrac{1}{\cos t}$, and $\tan t = \dfrac{\sin t}{\cos t}$ into the expression:

$$\frac{1 + \cot t \sec t}{\tan t + \sec t} = \frac{1 + \left(\dfrac{\cos t}{\sin t}\right)\left(\dfrac{1}{\cos t}\right)}{\dfrac{\sin t}{\cos t} + \dfrac{1}{\cos t}}.$$

We can simplify the expression considerably if we multiply the numerator and denominator by $(\sin t \cos t)$, the common denominator of the fractions in the numerator of the expression. Hence,

$$\frac{\left(1 + \left(\dfrac{\cos t}{\sin t}\right)\left(\dfrac{1}{\cos t}\right)\right)}{\left(\dfrac{\sin t}{\cos t} + \dfrac{1}{\cos t}\right)} \cdot \frac{(\sin t \cos t)}{(\sin t \cos t)} = \frac{\sin t \cos t + \cos t}{\sin^2 t + \sin t}.$$

By factoring the numerator and denominator and then simplifying the expression we get

$$\frac{\cos t(\sin t + 1)}{\sin t(\sin t + 1)} = \frac{\cos t}{\sin t}.$$

••••••••••

We now consider some examples in which we prove trigonometric identities.

• **Example 8.3:** Prove that $\sec t = \sec t \sin^2 t + \cos t$.

Solution: Since the left-hand side of the equation contains only the secant function, the best way to proceed is to start with the right-hand side of the equation and make substitutions for each term using only secants.

$$\sec t \sin^2 t + \cos t = \sec t(1 - \cos^2 t) + \cos t \quad \text{(Substitute } 1 - \cos^2 t \text{ for } \sin^2 t\text{)}$$

$$= \sec t\left(1 - \frac{1}{\sec^2 t}\right) + \frac{1}{\sec t} \quad \left(\text{Substitute } \frac{1}{\sec^2 t} \text{ for } \cos^2 t\right)$$

$$= \sec t - \frac{1}{\sec t} + \frac{1}{\sec t} \quad \text{(Multiply)}$$

$$= \sec t. \quad \text{(Simplify)}$$

• **Example 8.4:** Prove that $\sec t - \tan t = \dfrac{\cos t}{1 + \sin t}$.

Solution: Since the right-hand side of the equation contains only the sine and cosine functions, we start with the left-hand side and make substitutions in terms of sine and cosine.

$$\sec t - \tan t$$

$$= \frac{1}{\cos t} - \frac{\sin t}{\cos t} \quad \left(\text{Substitute } \frac{1}{\cos t} \text{ for } \sec t \text{ and } \frac{\sin t}{\cos t} \text{ for } \tan t\right)$$

$$= \frac{1 - \sin t}{\cos t}. \quad \text{(Add the fractions)}$$

The expression is still not in the form of the right-hand side. However, if we multiply the numerator and denominator by $(1 + \sin t)$ and use appropriate substitutions we get

$$\frac{(1 - \sin t)(1 + \sin t)}{\cos t(1 + \sin t)}$$

$$= \frac{1 - \sin^2 t}{\cos t(1 + \sin t)} \quad \text{(Multiply in the numerator)}$$

$$= \frac{\cos^2 t}{\cos t(1 + \sin t)} \quad \text{(Replace } 1 - \sin^2 t \text{ with } \cos^2 t, \text{ Pythagorean identity)}$$

$$= \frac{\cos t}{1 + \sin t}. \quad \text{(Reduce the fraction)}$$

Often it is more convenient to work with both sides of an identity and construct a proof using the results. Example 8.5 illustrates this technique.

• **Example 8.5:** Prove the given identity:
$$(\csc t - \sin t)(\csc t + \sin t) = \csc^2 t \,(1 - \sin^4 t).$$

Solution: We attack the problem by replacing the expressions on both sides of the equation with expressions that contain only sines and cosines.

$(\csc t - \sin t)(\csc t + \sin t) \qquad \csc^2 t (1 - \sin^4 t).$ (Replace $\csc t$ with $\dfrac{1}{\sin t}$)

$\left(\dfrac{1}{\sin t} - \sin t\right)\left(\dfrac{1}{\sin t} + \sin t\right) \qquad \dfrac{1}{\sin^2 t}(1 - \sin^4 t)$ (Multiply)

$\dfrac{1}{\sin^2 t} - \sin^2 t \qquad\qquad \dfrac{1 - \sin^4 t}{\sin^2 t}$ (Add fractions)

$\dfrac{1 - \sin^4 t}{\sin^2 t}$

Since we obtained the same expression for the left- and right-hand sides of the equation, we can construct a proof by starting with the initial expression on the left-hand side, proceeding down the left column and up the right column. (Follow the arrows). This produces the following proof:

$$(\csc t - \sin t)(\csc t + \sin t)$$
$$= \left(\dfrac{1}{\sin t} - \sin t\right)\left(\dfrac{1}{\sin t} + \sin t\right)$$
$$= \dfrac{1}{\sin^2 t} - \sin^2 t$$
$$= \dfrac{1 - \sin^4 t}{\sin^2 t}$$
$$= \dfrac{1}{\sin^2 t}(1 - \sin^4 t)$$
$$= \csc^2 t (1 - \sin^4 t).$$

••••••••••

Some pairs of expressions look as if they may be equal, but in fact are not valid identities. To prove that an "identity" is not valid, we need only find one instance in which the expressions on the left- and right-hand sides do not give the same values. Example 8.6 shows how to handle this type of problem.

• **Example 8.6:** Prove that the given expression is *not* an identity:
$$\sin(t + u) = \sin t + \sin u.$$

Solution: We need to find one pair of values for t and u that produce different values for the expressions
$$\sin(t + u) \text{ and } \sin t + \sin u.$$

Let $t = \dfrac{\pi}{3}$ and $u = \dfrac{\pi}{4}$.

$$\sin(t + u) = \sin\left(\frac{\pi}{3} + \frac{\pi}{4}\right) = \sin\frac{7\pi}{12} = 0.9659$$

$$\sin t + \sin u = \sin\frac{\pi}{3} + \sin\frac{\pi}{4}$$
$$= 0.8660 + 0.7071 = 1.5731.$$

Therefore, $\sin(t + u) \neq \sin t + \sin u.$

Trial Problems 8.1

Before you begin the section exercises, warm up with these problems. Complete answers are included in the answer key.

Simplify each expression.

1. $(\tan t)(\csc t)(\sec t)$
2. $\dfrac{\cos t}{\cot t}$
3. $\dfrac{\tan^2 t}{1 + \tan^2 t}$
4. $\dfrac{\sin t + \sin t \cot^2 t}{\csc^2 t}$

Exercises 8.1

Simplify each expression.

1. $\sin t \sec t$
2. $\cos t \csc t$
3. $\cot t \sec t$
4. $\tan t \csc t$
5. $\dfrac{\cos t}{1 - \sin^2 t}$
6. $\dfrac{\sin t}{1 - \cos^2 t}$
7. $\cot t \tan^2 t$
8. $\tan t \cot^2 t$
9. $\csc t \sec t - \tan t$
10. $\csc t \sec t - \cot t$
11. $\dfrac{\tan t + \cot t}{\sec t \csc t}$
12. $\dfrac{\sin t}{\csc t} + \dfrac{\cos t}{\sec t}$
13. $\dfrac{\sec t}{\tan t + \cot t}$
14. $\dfrac{\csc t}{\tan t + \cot t}$
15. $(\sec t + \tan t)(1 - \sin t)$
16. $\sin t(\csc t - \sin t)$
17. $\dfrac{\sin t}{\csc t} + \dfrac{\cos t}{\sec t}$
18. $\dfrac{\csc t}{\sin t} + \dfrac{\sec t}{\cos t}$
19. $\dfrac{1 + \sin t}{\cos t} + \dfrac{\cos t}{1 + \sin t}$
20. $\dfrac{\sin t}{1 + \cos t} + \dfrac{1 + \cos t}{\sin t}$

Prove the given identities.

21. $\sin t = \dfrac{\tan t}{\sec t}$

22. $\cos t = \dfrac{\cot t}{\csc t}$

23. $\csc t = \dfrac{\cot t}{\cos t}$

24. $\sec t = \dfrac{\tan t}{\sin t}$

25. $\sec^2 t + \tan^2 t = 2\tan^2 t + 1$

26. $\csc^2 t + \cot^2 t = 2\cot^2 t + 1$

27. $\sec t - \cos t = \sin t \tan t$

28. $\tan t + \cot t = \sec t \csc t$

29. $\dfrac{1}{\sec t + \tan t} = \sec t - \tan t$

30. $\dfrac{1}{\csc t - \cot t} = \csc t + \cot t$

31. $\dfrac{\cos t + \sin t}{\cos t - \sin t} + \dfrac{\cot t - 1}{\cot t + 1} = \dfrac{2}{\cos^2 t - \sin^2 t}$

32. $\dfrac{1 + \cos^2 t}{1 - \cos t} + \dfrac{\sec t - 1}{\sec t + 1} = 2(\cot^2 t + \csc^2 t)$

33. $\dfrac{1 - \cot^2 t}{1 + \cot^2 t} = 2\sin^2 t - 1$

34. $\dfrac{1 - \tan^2 t}{\sec^2 t} = \dfrac{\cot t - \tan t}{\cot t + \tan t}$

35. $\dfrac{\cot t + \csc t}{\sin t + \tan t} = \cot t \csc t$

36. $\dfrac{\sin t + \tan t}{\cot t + \csc t} = \tan t \sin t$

37. $\dfrac{\sec^2 t + \csc^2 t}{\sec t \csc t} = \tan t + \cot t$

38. $\dfrac{\cos t}{1 - \cos t \tan t} = \dfrac{1 + \cos t \tan t}{\cos t}$

39. $\dfrac{\sin t - \cos t}{\sec t - \csc t} = \dfrac{\cos t}{\csc t}$

40. $\dfrac{\sin t}{\cos t} + \dfrac{1 + \cos t}{\sin t} = \dfrac{\cos t + 1}{\sin t \cos t}$

Show that given equations are *not* identities.

41. $\cos(t + u) = \cos t + \cos u$

42. $\cos(t - u) = \cos t - \cos u$

43. $(\sin t + \cos t)^2 = \sin^2 t + \cos^2 t$

44. $(\sin t - \cos t)^2 = \sin^2 t - \cos^2 t$

45. $\sqrt{\sin^2 t + \cos^2 t} = \sin t + \cos t$

46. $\sqrt{\sin^2 t - \cos^2 t} = \sin t - \cos t$

47. $\dfrac{\sin t + \tan t}{\cos t + \tan t} = \dfrac{\sin t}{\cos t}$

48. $\dfrac{\cos t + \cot t}{\sin t + \cot t} = \dfrac{\cos t}{\sin t}$

49. $(\tan t + 1)^2 = \sec^2 t$

50. $(\cot t + 1)^2 = \csc^2 t$

51. $\sin 2t = 2\sin t$

52. $\cos 2t = 2\cos t$

Sum and Difference Formulas
Double-Angle Formulas
Half-Angle Formulas

8 • 2 Special Formulas for the Cosine Function

In this section some special formulas for the cosine function are derived. These are the formulas for the cosine of the **sum and difference** of two angles and the **double-angle** and **half-angle formulas** for cosines.

The first identity to be developed is the identity for the cosine of the sum of two angles.

SECTION 8.2 SPECIAL FORMULAS FOR THE COSINE FUNCTION

Theorem 8.1: $\cos(u + v) = \cos u \cos v - \sin u \sin v$

Proof: The identity is established assuming that u, v, and $u + v$ are acute angles. Similar arguments can be made for angles that do not fit these conditions.

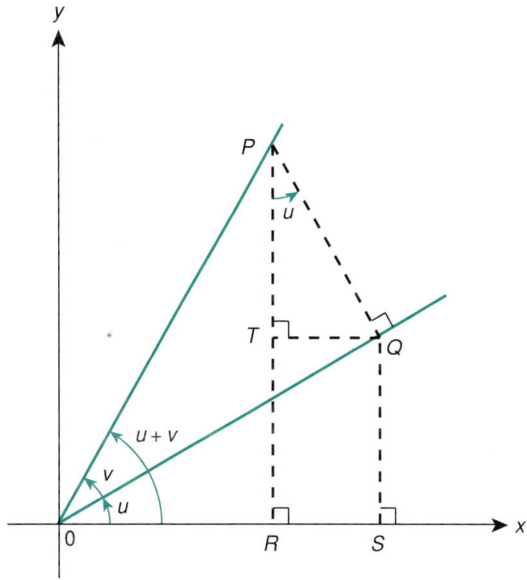

Figure 8.2

In figure 8.2, the following line segments have been added to angles u and v on a coordinate system.

1. $\overline{PR}$ is perpendicular to the x-axis at R.
2. $\overline{PQ}$ is perpendicular to the common side of angles u and v at Q. (P is an arbitrary point on the terminal side of angle v.)
3. $\overline{QT}$ is perpendicular to $\overline{PR}$ at T.
4. $\overline{QS}$ is perpendicular to the x-axis at S.

$$\cos(u + v) = \frac{OR}{OP} = \frac{OS - RS}{OP} \qquad \text{(Triangle } ORP\text{)}$$

$$= \frac{OS - TQ}{OP} = \frac{OS}{OP} - \frac{QT}{OP} \qquad \text{(Substitute } QT \text{ for } RS\text{)}$$

$$\cos u = \frac{OS}{OQ} \quad \text{or} \quad OS = OQ \cos u \quad \text{(In triangle } OQS\text{)}$$

$$\sin u = \frac{QT}{PQ} \quad \text{or} \quad QT = PQ \sin u. \quad \text{(In triangle } PTQ\text{)}$$

Substitute these expressions into the expression for $\cos(u + v)$:

$$\cos(u + v) = \cos u \frac{OQ}{OP} - \sin u \frac{PQ}{OP}.$$

However,

$$\frac{OQ}{OP} = \cos v \text{ and } \frac{PQ}{OP} = \sin v. \quad \text{(In triangle } OPQ\text{)}$$

Substitute these expressions into the expression for $\cos(u + v)$:

$$\cos(u + v) = \cos u \cos v - \sin u \sin v.$$

Theorem 8.2: $\cos(u - v) = \cos u \cos v + \sin u \sin v.$

Proof: Replace v with $-v$ in theorem 8.1

$$\cos(u - v) = \cos(u + (-v))$$
$$= \cos u \cos(-v) - \sin u \sin(-v).$$

Since $\cos(-v) = \cos v$ and $\sin(-v) = -\sin v$, we can make these substitutions into the expression for $\cos(u - v)$:
$$\cos(u - v) = \cos u \cos v + \sin u \sin v.$$

Theorem 8.1 can be used to obtain a double angle formula for cosines.

Theorem 8.3: $\cos 2t = \cos^2 t - \sin^2 t$
$$= 1 - 2 \sin^2 t$$
$$= 2 \cos^2 t - 1.$$

Proof: Using the sum formula for cosines with $u = t$ and $v = t$, we get

$$\cos(u + v) = \cos u \cos v - \sin u \sin v,$$
$$\cos(t + t) = \cos t \cos t - \sin t \sin t$$
$$\cos 2t = \cos^2 t - \sin^2 t. \tag{8.1}$$

SECTION 8.2 SPECIAL FORMULAS FOR THE COSINE FUNCTION

This is the first equation in the theorem.
Substituting $\cos^2 t = 1 - \sin^2 t$ into equation 8.1 gives

$$\cos 2t = \cos^2 t - \sin^2 t$$
$$= (1 - \sin^2 t) - \sin^2 t$$
$$\cos 2t = 1 - 2\sin^2 t.$$

This is the second equation in theorem 8.2.
Substituting $\sin^2 t = 1 - \cos^2 t$ into equation 8.1 gives

$$\cos 2t = \cos^2 t - \sin^2 t$$
$$= \cos^2 t - (1 - \cos^2 t)$$
$$\cos 2t = 2\cos^2 t - 1.$$

This is the last equation in theorem 8.3.

We can use the double-angle formula for cosines to obtain a half-angle formula.

Theorem 8.4: $\cos \dfrac{t}{2} = \pm \sqrt{\dfrac{1 + \cos t}{2}}.$

Proof: We restate one of the double-angle formulas in the following form:

$$\cos 2u = 2\cos^2 u - 1.$$

Substitute $u = \dfrac{t}{2}$ into this expression:

$$\cos 2\left(\frac{t}{2}\right) = 2\cos^2\left(\frac{t}{2}\right) - 1$$

$$\cos t = 2\cos^2 \frac{t}{2} - 1.$$

Solve this equation for $\cos \dfrac{t}{2}$:

$$2\cos^2 \frac{t}{2} = 1 + \cos t$$

$$\cos \frac{t}{2} = \pm \sqrt{\frac{1 + \cos t}{2}}.$$

Theorems 8.1 to 8.4 can be used to find exact expressions for cosines of angles that can be written as combinations of angles with known sines and cosines.

• *Example 8.7:* Obtain an exact expression for $\cos(105°)$.

Solution: Since $105° = 45° + 60°$, we can use the sum rule for cosines, using $u = 45°$ and $v = 60°$:

$$\cos 105° = \cos(45° + 60°)$$
$$= \cos(45°)\cos(60°) - \sin(45°)\sin(60°)$$
$$= \frac{\sqrt{2}}{2}\frac{1}{2} - \frac{\sqrt{2}}{2}\frac{\sqrt{3}}{2}$$
$$= \frac{\sqrt{2}(1 - \sqrt{3})}{4}.$$

• *Example 8.8:* Obtain an exact expression for $\cos\frac{\pi}{8}$.

Solution: Since $\frac{\pi}{8} = \frac{1}{2} \cdot \frac{\pi}{4}$, we can use the half-angle formula for cosines with $t = \frac{\pi}{4}$:

$$\cos\frac{\pi}{8} = \pm\sqrt{\frac{1 + \cos\frac{\pi}{4}}{2}}$$
$$= \sqrt{\frac{1 + \frac{\sqrt{2}}{2}}{2}}$$
$$= \sqrt{\frac{2 + \sqrt{2}}{4}}$$
$$= \frac{\sqrt{2 + \sqrt{2}}}{2}.$$

We use the positive square root since $t = \frac{\pi}{8}$ is in the first quadrant.

SECTION 8.2 SPECIAL FORMULAS FOR THE COSINE FUNCTION

• **Example 8.9:** Find $\cos \dfrac{t}{2}$ if $\cos t = \dfrac{12}{13}$, and t is in the fourth quadrant.

Solution: Since t is in the fourth quadrant,

$$\frac{3\pi}{2} < t < 2\pi$$

$$\frac{3\pi}{4} < \frac{t}{2} < \pi.$$

This places $\dfrac{t}{2}$ in the second quadrant and we use the negative square root in the half-angle formula:

$$\cos \frac{t}{2} = -\sqrt{\frac{1 + \cos t}{2}}$$

$$= -\sqrt{\frac{1 + \dfrac{12}{13}}{2}}$$

$$= -\sqrt{\frac{25}{26}} = -\frac{5}{\sqrt{26}}.$$

Many applied problems lead to equations of the form

$$y = a \cos kt + b \sin kt.$$

This function is periodic with a period of $\dfrac{2\pi}{k}$. However, the amplitude and the phase shift are not obvious. The following identity can be used to simplify graphing the function as well as to find the amplitude and the phase shift.

Theorem 8.5: $\quad a \cos kt + b \sin kt = m \cos(kt - c) \qquad (8.2)$

provided $m = \sqrt{a^2 + b^2}$ and $\tan c = \dfrac{b}{a}$.

Proof: Begin by using the difference formula for cosines on the right-hand side of equation 8.2:

$$m \cos(kt - c) = m \cos(kt) \cos c + m \sin(kt) \sin c. \qquad (8.3)$$

Then set the left-hand side of equation (8.2) equal to the right-hand side of equation (8.3):

$$a \cos kt + b \sin kt = m \cos(kt) \cos c + m \sin(kt) \sin c.$$

This equation is valid provided

$$a = m \cos c \quad \text{and} \quad b = m \sin c. \tag{8.4}$$

Square the expressions in equation (8.4) and add:

$$a^2 = m^2\cos^2 c \quad \text{and} \quad b^2 = m^2\sin^2 c$$
$$a^2 + b^2 = m^2(\sin^2 c + \cos^2 c) = m^2(1) = m^2.$$

Thus,

$$m^2 = a^2 + b^2 \quad \text{and} \quad m = \sqrt{a^2 + b^2}.$$

Also, divide the expressions in equation (8.4):

$$\frac{b}{a} = \frac{m \sin c}{m \cos c} = \tan c.$$

Note that the terminal side of c must contain the point (a,b).

Examples 8.10 and 8.11 illustrate the use of theorem 8.5 as an aid in graphing equations of the form $y = a \cos kt + b \sin kt$.

• **Example 8.10:** Graph one period of $y = 4 \sin 2t + 4 \cos 2t$ by converting the function to a function that contains only $\cos 2t$.

Solution: Use $a = 4$, $b = 4$, and $k = 2$ in the formula of theorem 8.5 (see figure 8.3):

$$m = \sqrt{4^2 + 4^2} = \sqrt{32} = 4\sqrt{2}$$
$$c = \tan^{-1}\frac{4}{4} = \tan^{-1}(1) = \frac{\pi}{4}.$$

The equation for y now becomes

$$y = 4\sqrt{2} \cos\left(2t - \frac{\pi}{4}\right).$$

The amplitude is $4\sqrt{2}$, the period is $\frac{2\pi}{2} = \pi$, and the phase shift is $\frac{\left(\frac{\pi}{4}\right)}{2} = \frac{\pi}{8}$. Table 8.1 shows the coordinates of several points on the graph.

SECTION 8.2 SPECIAL FORMULAS FOR THE COSINE FUNCTION

TABLE 8.1

	$\cos\left(2t - \dfrac{\pi}{4}\right)$	1	0	-1	0	1
	$2t - \dfrac{\pi}{4}$	0	$\dfrac{\pi}{2}$	π	$\dfrac{3\pi}{2}$	2π
	t	$\dfrac{\pi}{8}$	$\dfrac{3\pi}{8}$	$\dfrac{5\pi}{8}$	$\dfrac{7\pi}{8}$	$\dfrac{9\pi}{8}$
$4\sqrt{2}\cos\left(2t - \dfrac{\pi}{4}\right)$		$4\sqrt{2}$	0	$-4\sqrt{2}$	0	$4\sqrt{2}$

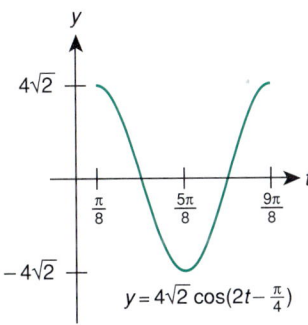

Figure 8.3

- **Example 8.11:** The current in a particular circuit can be represented by the equation $i = -1.1 \sin 400t + 3.3 \cos 400t$. Write the equation as a function of $\cos 400t$ and sketch one period of the graph.

Solution: For $i = -1.1 \sin 400t + 3.3 \cos 400t$ let $a = 3.3$, $b = -1.1$, and $k = 400$. Use the formula

$$a \cos kt + b \sin kt = m \cos(kt - c),$$

where $m = \sqrt{a^2 + b^2}$ and $c = \tan^{-1}\left(\dfrac{b}{a}\right)$:

$$m = \sqrt{(3.3)^2 + (-1.1)^2} = \sqrt{10.89 + 1.21} = \sqrt{12.1} = 3.5$$

$$c = \tan^{-1}\left(-\dfrac{1.1}{3.3}\right) = -0.32.$$

The equation now becomes $i = 3.5 \cos(400t + 0.32)$.

The amplitude is 3.5, the period is $\dfrac{2\pi}{400} = \dfrac{\pi}{200}$, and the phase shift is -0.0008. Table 8.2 shows the coordinates of several points on the graph.

TABLE 8.2

cos(400t + 0.32)	1	0	−1	0	1
400t + 0.32	0	$\frac{\pi}{2}$	π	$\frac{3\pi}{2}$	2π
t	−0.0008	0.0031	0.0071	0.0110	0.0149
3.5 cos(400t + 0.32)	3.5	0	−3.5	0	3.5

The graph of the function is shown in figure 8.4.

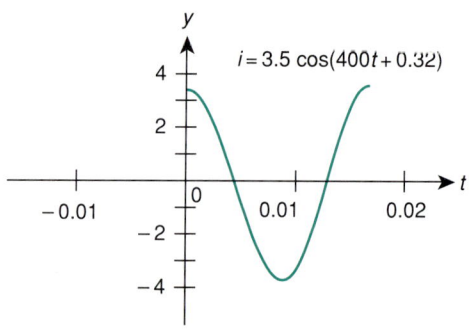

Figure 8.4

SUMMARY OF COSINE IDENTITIES

8.1 $\cos(u + v) = \cos u \cos v - \sin u \sin v.$
8.2 $\cos(u - v) = \cos u \cos v + \sin u \sin v.$
8.3 $\cos(2t) = \cos^2 t - \sin^2 t$
$\qquad = 1 - 2\sin^2 t$
$\qquad = 2\cos^2 t - 1.$
8.4 $\cos\left(\dfrac{t}{2}\right) = \pm\sqrt{\dfrac{1 + \cos t}{2}}.$
8.5 $a \cos kt + b \sin kt = m \cos(kt - c),$
provided $m = \sqrt{a^2 + b^2}$, $\tan c = \dfrac{b}{a}.$

Trial Problems 8.2

Before you begin the section exercises, warm up with these problems. Complete answers are included in the answer key.

Given $\cos u = \dfrac{\sqrt{2}}{2}$, $\sin u = \dfrac{\sqrt{2}}{2}$, $\cos v = \dfrac{2\sqrt{5}}{5}$, and $\sin v = \dfrac{\sqrt{5}}{5}$, calculate:

1. $\cos(u + v)$
2. $\cos(u - v)$
3. $\cos 2u$
4. $\cos \dfrac{u}{2}$
5. $\cos 2v$

Exercises 8.2

Use the exact values of the cosine function at $\frac{\pi}{6}$ (30°), $\frac{\pi}{4}$ (45°), $\frac{\pi}{3}$ (60°), and $\frac{\pi}{2}$ (90°), and their multiples to find exact values for the following cosines.

1. $\cos 75°$
2. $\cos 105°$
3. $\cos \frac{11\pi}{12}$
4. $\cos(-22.5°)$
5. $\cos \frac{3\pi}{8}$
6. $\cos \frac{9\pi}{8}$
7. $\cos \frac{5\pi}{8}$
8. $\cos 22.5°$

If $\cos u = \frac{3}{5}$, $\cos v = \frac{12}{13}$, and u and v are in the first quadrant, determine the value of each expression.

9. $\sin u$
10. $\sin v$
11. $\cos(u + v)$
12. $\cos(u - v)$
13. $\cos 2u$
14. $\cos 2v$
15. $\cos \frac{u}{2}$
16. $\cos \frac{v}{2}$

If $\cos 2t = \frac{1}{4}$ and t is in the first quadrant, determine:

17. $\cos t$
18. $\sin t$
19. $\tan t$
20. $\cot t$
21. $\sec t$
22. $\csc t$

23. If $\sin t = \frac{5}{13}$ and t is in the second quadrant, determine $\sin 2t$, $\cos 2t$, and $\tan 2t$, where $0 \leq t \leq 2\pi$.

24. If $\cos t = -\frac{12}{13}$ and t is in the second quadrant, determine $\sin 2t$, $\cos 2t$, and $\tan 2t$, where $0 \leq t \leq 2\pi$.

25. If $\cos t = -\frac{5}{13}$ and t is in the third quadrant, determine $\sin \frac{t}{2}$, $\cos \frac{t}{2}$, and $\tan \frac{t}{2}$, where $0 \leq t \leq 2\pi$.

26. If $\tan t = -\frac{5}{12}$ and t is in the fourth quadrant, determine $\sin \frac{t}{2}$, $\cos \frac{t}{2}$, and $\tan \frac{t}{2}$, where $0 \leq t \leq 2\pi$.

Verify each identity.

27. $\cos(t + 2\pi) = \cos t$
28. $\cos\left(t + \frac{\pi}{2}\right) = -\sin t$
29. $\cos\left(t - \frac{\pi}{2}\right) = \sin t$
30. $\cos(\pi - t) = -\cos t$
31. $\cos 2t = \frac{1 - \tan^2 t}{1 + \tan^2 t}$
32. $\cos 2t = \frac{\cot t - \tan t}{\cot t + \tan t}$

33. Use the sum formulas for cosines to find an expression for $\cos 6t$ in terms of $\cos 3t$.

34. Use the sum formulas for cosines to find an expression for $\cos 4t$ in terms of $\cos t$.

35. Prove that $\cos\dfrac{7\pi}{12} + \cos\dfrac{\pi}{12} = \dfrac{\sqrt{2}}{2}$.

36. Prove that $\cos\dfrac{7\pi}{12} - \cos\dfrac{\pi}{12} = -\dfrac{\sqrt{6}}{2}$.

37. Use the half-angle formula twice to obtain an exact value for $\cos\dfrac{\pi}{16}$. $\left(\text{Let } t = \dfrac{\pi}{4}, \dfrac{t}{2} = \dfrac{\pi}{8}, \dfrac{\frac{t}{2}}{2} = \dfrac{\pi}{16}.\right)$

38. Use the half-angle formula twice to obtain an exact value for  $\left(\text{Let } t = \dfrac{\pi}{6}, \dfrac{t}{2} = \dfrac{\pi}{12}, \dfrac{\frac{t}{2}}{2} = \dfrac{\pi}{24}.\right)$

39. Prove that $\cos(u+v) + \cos(u-v) = 2\cos u \cos v$.

40. Prove that $\cos(u+v) - \cos(u-v) = -2\sin u \sin v$.

41. Write the equation $y = 3.2\cos 20t + 1.7\sin 20t$ using only the cosine.

42. Write the equation $y = 1.8\cos 40t + 2.6\sin 20t$ using only the cosine.

43. Use theorem 8.4 as an aid in graphing the equation $y = 3\cos 2t + 4\sin 2t$.

44. Use theorem 8.4 as an aid in graphing the equation $y = 4\cos 2t + 3\sin 2t$.

45. The current in an electrical circuit is represented by the equation $i = 1.5\cos 20t + 2\sin 20t$. Graph one period of the function.

46. Solve exercise 45 given that $i = 2\cos 30t + 1.5\sin 30t$.

47. A vibrating weight moves vertically from an equilibrium point according to the equation

$$y = 0.25\sin 2t + 0.20\cos 2t.$$

Write an equivalent equation in the form $y = a\cos(bt + c)$.

8 • 3 Special Formulas for the Sine and Tangent Functions

In this section the sum, difference, half-angle, and double-angle formulas for the sine and tangent functions are developed.

Theorem 8.6 is included since the two properties developed in the theorem are needed to prove the sum rule for sines.

Theorem 8.6: a. $\sin\left(\dfrac{\pi}{2} - t\right) = -\cos t$ b. $\cos\left(\dfrac{\pi}{2} - t\right) = \sin t$.

Proof: These relationships can be verified by considering the graphs of $y = \sin t$ and $y = \cos t$ in figure 8.5. Since the cosine function is a sine function with a phase shift of $\dfrac{\pi}{2}$,

$$\sin\left(\dfrac{\pi}{2} + t\right) = \cos t.$$

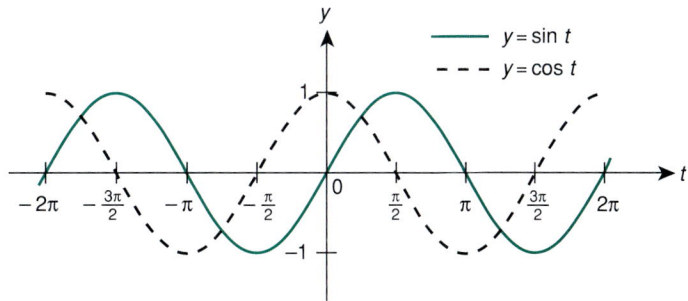

Figure 8.5

a. Use this identity and substitute $-t$ for t:

$$\sin\left(\frac{\pi}{2} - t\right) = \cos(-t).$$

However,

$$\cos(-t) = -\cos t.$$

Therefore,

$$\sin\left(\frac{\pi}{2} - t\right) = -\cos t.$$

b. Use the identity $\sin\left(\frac{\pi}{2} + t\right) = \cos t$ and substitute $\frac{\pi}{2} - t$ for t:

$$\sin\left(\frac{\pi}{2} + \left(\frac{\pi}{2} - t\right)\right) = \cos\left(\frac{\pi}{2} - t\right)$$

$$\sin(\pi - t) = \cos\left(\frac{\pi}{2} - t\right).$$

Substitute $\sin t$ for $\sin(\pi - t)$:

$$\sin t = \cos\left(\frac{\pi}{2} - t\right).$$

Theorem 8.7: **a.** $\sin(u + v) = \sin u \cos v + \cos u \sin v$
 b. $\sin(u - v) = \sin u \cos v - \cos u \sin v.$

Proof: Begin with theorem 8.6b:

$$\sin t = \cos\left(\frac{\pi}{2} - t\right) = \cos(90° - t).$$

Substitute $u + v$ for t in this expression:

$$\sin(u + v) = \cos(90° - (u + v)) = \cos((90° - u) - v).$$

Now use theorem 8.2, replacing $\cos(90° - u)$ with $\sin u$:

$$\sin(u + v) = (\cos(90° - u))\cos v - (\sin(90° - u))\sin v$$

replace with $\sin u$. (theorem 8.6a)
replace with $-\cos u$. (theorem 8.6b)

$$\sin(u + v) = \sin u \cos v - (-\cos u)\sin v$$
$$= \sin u \cos v + \cos u \sin v.$$

The b. part of the theorem can be established by substituting $-v$ for v in theorem 8.7a and simplifying the results.

Theorem 8.8: $\sin 2t = 2 \sin t \cos t.$

Proof: Substitute $u = t$ and $v = t$ into the sum formula of theorem 8.7a:

$$\sin 2t = \sin(t + t)$$
$$= \sin t \cos t + \cos t \sin t$$
$$= 2 \sin t \cos t.$$

The half-angle formula for sines can be established using the double-angle formula for cosines.

Theorem 8.9: $\sin \dfrac{t}{2} = \pm \sqrt{\dfrac{1 - \cos t}{2}}.$

 a. Use the positive root if $\dfrac{t}{2}$ is in quadrant I or II.

 b. Use the negative root if $\dfrac{t}{2}$ is in quadrant III or IV.

SECTION 8.3 SPECIAL FORMULAS FOR THE SINE AND TANGENT FUNCTIONS

Proof: Begin with the identity of theorem 8.3:

$$\cos 2t = 1 - 2\sin^2 t.$$

Replace t with $\dfrac{t}{2}$ in this identity:

$$\cos 2\left(\frac{t}{2}\right) = 1 - 2\sin^2 \frac{t}{2}$$

$$\cos t = 1 - 2\sin^2 \frac{t}{2}.$$

Solve this equation for $\sin \dfrac{t}{2}$:

$$2\sin^2 \frac{t}{2} = 1 - \cos t \qquad \text{(Add } 2\sin^2 \frac{t}{2} - \cos t \text{ to both sides)}$$

$$\sin^2 \frac{t}{2} = \frac{1 - \cos t}{2} \qquad \text{(Divide both sides by 2)}$$

$$\sin \frac{t}{2} = \pm\sqrt{\frac{1 - \cos t}{2}}. \qquad \text{(Take the square root of both sides)}$$

Examples 8.12, 8.13, and 8.14 illustrate some of the uses of these theorems.

• **Example 8.12:** Obtain an exact expression for $\sin 15°$.

Solution: Since $15° = 60° - 45°$, and we already have exact expressions for $\sin 60°$, $\cos 60°$, $\sin 45°$, and $\cos 45°$, we can use the difference formula for sines:

$$\sin 15° = \sin(60° - 45°)$$
$$= (\sin 60°)(\cos 45°) - (\cos 60°)(\sin 45°)$$
$$= \left(\frac{\sqrt{3}}{2}\right)\left(\frac{\sqrt{2}}{2}\right) - \left(\frac{1}{2}\right)\left(\frac{\sqrt{2}}{2}\right)$$
$$= \frac{\sqrt{2}(\sqrt{3} - 1)}{4}.$$

• • • • • • • • •

• **Example 8.13:** Obtain an exact expression for $\sin \dfrac{\pi}{12}$.

Solution: Since $\dfrac{\pi}{12} = \dfrac{\left(\dfrac{\pi}{6}\right)}{2}$, we can use the half-angle formula for sines:

$$\sin \frac{\pi}{12} = \sin \frac{\left(\dfrac{\pi}{6}\right)}{2} = \pm \sqrt{\frac{1 - \cos\left(\dfrac{\pi}{6}\right)}{2}}.$$

Since $\dfrac{\pi}{12}$ is in the first quadrant, we use the positive square root:

$$\sin \frac{\pi}{12} = +\sqrt{\frac{1 - \cos\left(\dfrac{\pi}{6}\right)}{2}}$$

$$= \sqrt{\frac{1 - \left(\dfrac{\sqrt{3}}{2}\right)}{2}}$$

$$= \sqrt{\frac{2 - \sqrt{3}}{4}} = \frac{\sqrt{2 - \sqrt{3}}}{2}.$$

• **Example 8.14:** Find $\sin 2\theta$ if $\sin \theta = \dfrac{5}{13}$ and θ is in quadrant I.

Solution: We use the formula $\sin 2\theta = 2 \sin \theta \cos \theta$. First we need to find $\cos \theta$. Using figure 8.6 and the Pythagorean theorem,

$$OA = \sqrt{13^2 - 5^2} = \sqrt{144} = 12$$

$$\cos \theta = \frac{OA}{OB} = \frac{12}{13}$$

$$\sin 2\theta = 2 \sin \theta \cos \theta$$

$$= (2)\left(\frac{5}{13}\right)\left(\frac{12}{13}\right) = \frac{120}{169}.$$

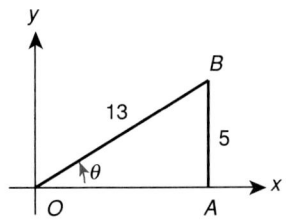

Figure 8.6

The tangent identities for sums, differences, half-angles, and double-angles can be derived using the identities for the sines and cosines.

SECTION 8.3 SPECIAL FORMULAS FOR THE SINE AND TANGENT FUNCTIONS

Theorem 8.10:
$$\tan(u + v) = \frac{\tan u + \tan v}{1 - \tan u \tan v}$$

$$\tan(u - v) = \frac{\tan u - \tan v}{1 + \tan u \tan v}.$$

Proof: We use the definition of the tangent function and the sum formulas for the sine and cosine:

$$\tan(u + v) = \frac{\sin(u + v)}{\cos(u + v)} = \frac{\sin u \cos v + \cos u \sin v}{\cos u \cos v - \sin u \sin v}.$$

We divide the numerator and denominator of the right-hand side of the expression by $\cos u \cos v$:

$$\tan(u + v) = \frac{\dfrac{\sin u \cos v}{\cos u \cos v} + \dfrac{\cos u \sin v}{\cos u \cos v}}{\dfrac{\cos u \cos v}{\cos u \cos v} - \dfrac{\sin u \sin v}{\cos u \cos v}}.$$

Simplifying the right-hand side of the equation we obtain

$$\tan(u + v) = \frac{\dfrac{\sin u}{\cos u} + \dfrac{\sin v}{\cos v}}{1 - \dfrac{\sin u \sin v}{\cos u \cos v}} = \frac{\tan u + \tan v}{1 - \tan u \tan v}.$$

Corollary: $\tan(u - v) = \dfrac{\tan u - \tan v}{1 + \tan u \tan v}.$

Proof: We substitute $-v$ for v in the formula of theorem 8.10.

$$\tan(u - v) = \tan(u + (-v)) = \frac{\tan u - \tan(-v)}{1 + \tan(u)\tan(-v)}.$$

Since $\tan(-v) = -\tan v$,

$$\tan(u - v) = \frac{\tan u - \tan v}{1 + \tan u \tan v}.$$

The double-angle formula for tangents can be developed from the sum rule for tangents.

Theorem 8.11: $\tan 2t = \dfrac{2 \tan t}{1 - \tan^2 t}$.

Proof: We substitute $u = t$ and $v = t$ into the formula for the sum of tangents:

$$\tan 2t = \tan(t + t) = \frac{\tan t + \tan t}{1 - (\tan t)(\tan t)} = \frac{2 \tan t}{1 - \tan^2 t}.$$

The half-angle formula for tangents can be developed from the ratio

$$\tan u = \frac{\sin u}{\cos u}$$

and the half-angle formulas for sines and cosines.

Theorem 8.12: $\tan \dfrac{t}{2} = \dfrac{1 - \cos t}{\sin t}$

Proof: $\tan \dfrac{t}{2} = \dfrac{\sin \dfrac{t}{2}}{\cos \dfrac{t}{2}}$

We multiply the numerator and denominator of the right-hand side of the equation by $2 \sin \dfrac{t}{2}$:

$$\tan \frac{t}{2} = \frac{2 \sin^2 \dfrac{t}{2}}{2 \sin \dfrac{t}{2} \cos \dfrac{t}{2}}.$$

Since

$$2 \sin^2 \frac{t}{2} = 1 - \cos t \quad \text{(Theorem 8.3, replacing } t \text{ with } \frac{t}{2}\text{)}$$

and

$$2 \sin \frac{t}{2} \cos \frac{t}{2} = \sin t \quad \text{(Theorem 8.8, replacing } t \text{ with } \frac{t}{2}\text{)}$$

then

$$\tan \frac{t}{2} = \frac{1 - \cos t}{\sin t}.$$

Examples 8.15, 8.16, and 8.17 illustrate some of the uses of the tangent formulas.

• **Example 8.15:** Find an expression for the exact value of tan 165°.

 Solution: Note that 165° = 45° + 120°. Use the sum rule for tangents and substitute $\tan 120° = -\sqrt{3}$ and $\tan 45° = 1$:

$$\tan 165° = \tan(45° + 120°)$$
$$= \frac{\tan 45° + \tan 120°}{1 - (\tan 45°)(\tan 120°)}$$
$$= \frac{1 + (-\sqrt{3})}{1 - (1)(-\sqrt{3})}$$
$$= \frac{1 - \sqrt{3}}{1 + \sqrt{3}}.$$

• **Example 8.16:** Find an expression for $\tan(45° - 2t)$ in terms of tan t.

 Solution: We attack the problem in two stages.

 I. Use the difference rule for tangents on $\tan(45° - 2t)$.

 II. Use the double-angle rule for tangents on the resulting expression in tan $2t$.

 I. $\tan(45° - 2t) = \dfrac{\tan 45° - \tan 2t}{1 + (\tan 45°)(\tan 2t)}$, $\tan 45° = 1$

 $= \dfrac{1 - \tan 2t}{1 + \tan 2t}.$

 II. Substitute $\dfrac{2 \tan t}{1 - \tan^2 t}$ for each occurrence of tan $2t$ in the result of step I:

$$\tan(45° - 2t) = \frac{1 - \left(\dfrac{2 \tan t}{1 - \tan^2 t}\right)}{1 + \left(\dfrac{2 \tan t}{1 - \tan^2 t}\right)}.$$

 Multiply the numerator and denominator by $1 - \tan^2 t$:

$$\tan(45° - 2t) = \frac{1 - \tan^2 t - 2 \tan t}{1 - \tan^2 t + 2 \tan t}.$$

• **Example 8.17:** Given $\cos t = -\dfrac{3}{5}$, for $\dfrac{\pi}{2} < t < \pi$.

a. Find $\tan 2t$. **b.** Find $\tan \dfrac{t}{2}$.

Solution: **a.** To use the double-angle formula, we first must find $\tan t$:

$$\sin t = \sqrt{1 - \cos^2 t} = \sqrt{1 - \left(-\dfrac{3}{5}\right)^2} = \sqrt{1 - \dfrac{9}{25}} = \dfrac{4}{5}$$

$$\tan t = \dfrac{\sin t}{\cos t} = \dfrac{\frac{4}{5}}{-\frac{3}{5}} = -\dfrac{4}{3}$$

$$\tan 2t = \dfrac{2 \tan t}{1 - \tan^2 t} = \dfrac{2\left(-\dfrac{4}{3}\right)}{1 - \left(-\dfrac{4}{3}\right)^2}$$

$$= \dfrac{-\dfrac{8}{3}}{1 - \dfrac{16}{9}} = \dfrac{-\dfrac{8}{3}}{-\dfrac{7}{9}} = \dfrac{24}{7}.$$

b. $\tan \dfrac{t}{2} = \dfrac{1 - \cos t}{\sin t} = \dfrac{1 - \left(-\dfrac{3}{5}\right)}{\dfrac{4}{5}} = \dfrac{\dfrac{8}{5}}{\dfrac{4}{5}} = 2.$

SUMMARY OF SINE AND TANGENT IDENTITIES

8.6 $\sin\left(\dfrac{\pi}{2} - t\right) = -\cos t$

$\cos\left(\dfrac{\pi}{2} + t\right) = \sin t$

8.7 $\sin(u + v) = \sin u \cos v + \cos u \sin v$
$\sin(u - v) = \sin u \cos v - \cos u \sin v$

8.8 $\sin 2t = 2 \sin t \cos t$

8.9 $\sin \dfrac{t}{2} = \pm\sqrt{\dfrac{1 - \cos t}{2}}$

8.10 $\tan(u + v) = \dfrac{\tan u + \tan v}{1 - \tan u \tan v}$

$\tan(u - v) = \dfrac{\tan u - \tan v}{1 + \tan u \tan v}$

SECTION 8.3 SPECIAL FORMULAS FOR THE SINE AND TANGENT FUNCTIONS

$$8.11 \quad \tan 2t = \frac{2 \tan t}{1 - \tan^2 t}$$

$$8.12 \quad \tan \frac{t}{2} = \frac{1 - \cos t}{\sin t}$$

Trial Problems 8.3

Before you begin the section exercises, warm up with these problems. Complete answers are included in the answer key.

If $\sin u = \frac{\sqrt{2}}{2}$, $\cos u = \frac{\sqrt{2}}{2}$, $\sin v = \frac{\sqrt{5}}{5}$, and $\cos v = \frac{2\sqrt{5}}{5}$, calculate:

1. $\sin(u + v)$
2. $\sin(u - v)$
3. $\sin 2v$
4. $\sin \frac{v}{2}$
5. $\tan \frac{v}{2}$

Exercises 8.3

Find an exact expression for each sine or tangent.

1. $\sin 75°$
2. $\tan 75°$
3. $\sin \frac{\pi}{12}$
4. $\tan \frac{\pi}{12}$
5. $\sin 165°$
6. $\tan 165°$
7. $\tan \frac{3\pi}{8}$
8. $\sin \frac{3\pi}{8}$

If $\sin u = \frac{4}{5}$ and $\sin v = \frac{5}{13}$ and u and v are in the first quadrant, determine the value of each expression.

9. $\cos u$
10. $\cos v$
11. $\sin(u + v)$
12. $\sin(u - v)$
13. $\sin 2u$
14. $\sin 2v$
15. $\sin \frac{u}{2}$
16. $\sin \frac{v}{2}$
17. $\tan(u + v)$
18. $\tan(u - v)$
19. $\tan 2u$
20. $\tan 2v$
21. $\tan \frac{u}{2}$
22. $\tan \frac{v}{2}$

If $\cos 2t = \frac{2}{3}$ and $2t$ is in the first quadrant, determine:

23. $\sin t$
24. $\cos t$
25. $\tan t$
26. $\cot t$
27. $\sec t$
28. $\csc t$

Verify each identity.

29. $\sin(t + 2\pi) = \sin t$

30. $\sin\left(t + \dfrac{\pi}{2}\right) = \sin\left(t - \dfrac{\pi}{2}\right)$

31. $\sin(t + \pi) = -\sin(t - \pi)$

32. $\tan(t + \pi) = -\tan(\pi - t)$

33. $\cot 2t = \dfrac{1 - \tan^2 t}{2 \tan t}$

34. $\cot 2t = \dfrac{\cot t - \tan t}{2}$

35. $\sin 3t = 3 \sin t - 4 \sin^3 t$

36. $\sin 4t = 4 \sin t \cos t - 8 \sin^3 t \cos t$

37. Express $\sin 8t$ in terms of functions of $4t$.

38. Express $\sin 6t$ in terms of functions of $3t$.

39. Use the half-angle formula to obtain an exact value for $\sin \dfrac{\pi}{16}$. $\left(\text{If } t = \dfrac{\pi}{4},\ \dfrac{t}{2} = \dfrac{\pi}{8},\ \dfrac{t}{4} = \dfrac{\pi}{16}.\right)$

40. Use the half-angle formula to obtain an exact value for $\sin\left(\dfrac{\pi}{24}\right)$. $\left(\text{If } t = \dfrac{\pi}{6},\ \dfrac{t}{2} = \dfrac{\pi}{12},\ \dfrac{t}{4} = \dfrac{\pi}{24}.\right)$

41. Prove that $\sin(u + v) + \sin(u - v) = 2 \sin u \cos v$.

42. Prove that $\sin(u + v) - \sin(u - v) = 2 \cos u \sin v$.

43. Obtain an expression for $\sec \dfrac{t}{2}$ in terms of $\cos t$ and $\sin t$.

44. Obtain an expression for $\csc \dfrac{t}{2}$ in terms of $\cos t$ and $\sin t$.

45. Show that $\sec(u + v) = \dfrac{\csc u \csc v}{\cot u \cot v - 1}$.

46. Show that $\sec(u - v) = \dfrac{\sec u \sec v}{1 + \tan u \tan v}$.

47. In a problem in mechanics it is necessary to show that
$$Wr \tan(t - \phi) = Wr \left(\dfrac{1 - \tan t}{1 + \tan t}\right), \quad \text{if } \phi = \dfrac{\pi}{4}.$$
Verify this identity.

48. In a problem in mechanics it is necessary to show that
$$\dfrac{W(\sin t + a \cos t)}{\cos t - a \sin t} = W \tan(t + \phi), \quad \text{if } a = \tan \phi.$$
Verify this identity.

49. A calculus problem requires the following identity:
$$\frac{\sin(x+h) - \sin x}{h} = (\cos x)\left(\frac{\sin h}{h}\right) - (\sin x)\left(\frac{1 - \cos h}{h}\right).$$

Verify this identity.

50. A calculus problem requires the following identity.
$$\frac{\cos(x+h) - \cos x}{h} = (\cos x)\left(\frac{\cos h - 1}{h}\right) - (\sin x)\left(\frac{\sin h}{h}\right).$$

Verify this identity.

8 • 4 Sums, Differences, and Products of Sines and Cosines

Sometimes it is convenient to write a sum of sines or cosines as a product or to write a product of sines or cosines as a sum of trigonometric functions. The following formulas are used for such cases.

Theorem 8.13: $\sin u \cos v = \dfrac{\sin(u+v) + \sin(u-v)}{2}.$

Proof: We prove this identity by using the sum and difference formulas for sines on the right-hand side of the equation:

$$\frac{\sin(u+v) + \sin(u-v)}{2}$$
$$= \frac{\sin u \cos v + \cos u \sin v + \sin u \cos v - \cos u \sin v}{2}$$
$$= \frac{2 \sin u \cos v}{2}$$
$$= \sin u \cos v.$$

Theorem 8.14: $\sin u \sin v = \dfrac{\cos(u-v) - \cos(u+v)}{2}.$

The proof is left as an exercise.

Theorem 8.15: $\cos u \cos v = \dfrac{\cos(u+v) + \cos(u-v)}{2}.$

The proof is left as an exercise.
Examples 8.18–8.20 illustrate some of the uses of these formulas.

• **Example 8.18:** Write $(\sin \pi t)\left(\cos \dfrac{\pi}{2}t\right)$ as an expression using sines.

Solution: Use theorem 8.13 with $u = \pi t$ and $v = \dfrac{\pi}{2}t$.

$$\sin \pi t \cos \dfrac{\pi}{2}t = \dfrac{\sin\left(\pi t + \dfrac{\pi}{2}t\right) + \sin\left(\pi t - \dfrac{\pi}{2}t\right)}{2}$$

$$= \dfrac{\sin \dfrac{3\pi}{2}t + \sin \dfrac{\pi}{2}t}{2}.$$

• **Example 8.19:** Write $\cos \dfrac{t}{2} \cos \dfrac{t}{4}$ as the sum of cosines.

Solution: Use theorem 8.15 with $u = \dfrac{t}{2}$ and $v = \dfrac{t}{4}$.

$$\cos u \cos v = \dfrac{\cos\left(\dfrac{t}{2} + \dfrac{t}{4}\right) + \cos\left(\dfrac{t}{2} - \dfrac{t}{4}\right)}{2}$$

$$= \dfrac{\cos \dfrac{3t}{4} + \cos \dfrac{t}{4}}{2}.$$

• **Example 8.20:** Write the given expression as a sum: $\sin(2t)\sin(6t)$.

Solution: Use theorem 8.14 with $u = 2t$ and $v = 6t$.

$$\sin 2t \sin 6t = \dfrac{\cos(2t - 6t) - \cos(2t + 6t)}{2}$$

$$= \dfrac{\cos(-4t) - \cos(8t)}{2}$$

$$= \dfrac{\cos 4t - \cos 8t}{2}.$$

Note: $(\cos(-w) = \cos w)$.

Theorems 8.13, 8.14, and 8.15 are usually called the *product formulas,* in which we replace a product with a sum or difference. Reversing the process, the following formulas are used to replace a sum or difference with a product.

Section 8.4 Sums, Differences, and Products of Sines and Cosines

Theorem 8.16: $\cos u + \cos v = 2 \cos\left(\dfrac{u+v}{2}\right) \cos\left(\dfrac{u-v}{2}\right).$

Proof: Write theorem 8.15 replacing u with x and v with y:

$$\cos x \cos y = \frac{\cos(x+y) + \cos(x-y)}{2}.$$

Now substitute $x = \dfrac{u+v}{2}$ and $y = \dfrac{u-v}{2}$ to get

$$\cos\left(\frac{u+v}{2}\right)\cos\left(\frac{u-v}{2}\right) = \frac{\cos\left[\left(\dfrac{u+v}{2}\right) + \left(\dfrac{u-v}{2}\right)\right] + \cos\left[\left(\dfrac{u+v}{2}\right) - \left(\dfrac{u-v}{2}\right)\right]}{2}$$

$$\cos\left(\frac{u+v}{2}\right)\cos\left(\frac{u-v}{2}\right) = \frac{\cos\left(\dfrac{2u}{2}\right) + \cos\left(\dfrac{2v}{2}\right)}{2} \qquad \text{(Combine the fractions)}$$

$$2\cos\left(\frac{u+v}{2}\right)\cos\left(\frac{u-v}{2}\right) = \cos u + \cos v. \qquad \text{(Multiply both sides by 2)}$$

Theorem 8.17: $\sin u + \sin v = 2 \sin\left(\dfrac{u+v}{2}\right) \cos\left(\dfrac{u-v}{2}\right).$

Theorem 8.18: $\cos u - \cos v = -2 \sin\left(\dfrac{u+v}{2}\right) \sin\left(\dfrac{u-v}{2}\right).$

Theorem 8.19: $\sin u - \sin v = 2 \cos\left(\dfrac{u+v}{2}\right) \sin\left(\dfrac{u-v}{2}\right).$

The proofs of these theorems are left as exercises.
Examples 8.21–8.24 illustrate some of the uses of these theorems.

- **Example 8.21:** Write cos 2*t* + cos 4*t* as a product.

  Solution: We use theorem 8.16 with $u = 2t$, $v = 4t$:

  $$\cos 2t + \cos 4t = 2 \cos\left(\frac{2t + 4t}{2}\right) \cos\left(\frac{2t - 4t}{2}\right)$$
  $$= 2 \cos 3t \cos(-t)$$
  $$= 2 \cos 3t \cos t.$$

- **Example 8.22:** Factor the expression sin 5*t* − sin 3*t*.

  Solution: Factoring an expression means writing the expression as a product of factors. Applying theorem 8.19 to the expression with $u = 5t$ and $v = 3t$ gives us such a pair of factors:

  $$\sin(5t) - \sin(3t) = 2 \cos\left(\frac{5t + 3t}{2}\right) \sin\left(\frac{5t - 3t}{2}\right)$$
  $$= 2 \cos 4t \sin t.$$

- **Example 8.23:** Graph the function $f(t) = \sin(50t + \pi) + \sin(50t - \pi)$.

  Solution: First simplify $\sin(50t + \pi) + \sin(50t - \pi)$ using theorem 8.17 with $u = 50t + \pi$ and $v = 50t - \pi$:

  $$\sin(50t + \pi) + \sin(50t - \pi)$$
  $$= 2 \sin\left[\frac{(50t + \pi) + (50t - \pi)}{2}\right] \cos\left[\frac{(50t + \pi) - (50t - \pi)}{2}\right]$$
  $$= 2(\sin 50t)(\cos \pi) \quad \text{(Substitute } -1 \text{ for cos } \pi\text{)}$$
  $$= 2(\sin 50t)(-1)$$
  $$= -2 \sin 50t.$$

  Table 8.3 contains data for the function $f(t) = -2 \sin 50t$. This function is graphed in figure 8.7.

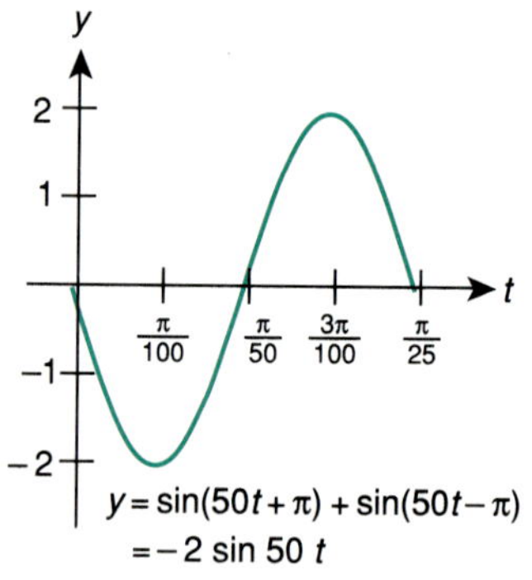

TABLE 8.3

sin 50*t*	0	1	0	−1	0
50*t*	0	$\frac{\pi}{2}$	π	$\frac{3\pi}{2}$	2π
t	0	$\frac{\pi}{100}$	$\frac{\pi}{50}$	$\frac{3\pi}{100}$	$\frac{\pi}{25}$
−2 sin 50*t*	0	−2	0	2	0

Figure 8.7

$y = \sin(50t + \pi) + \sin(50t - \pi)$
$= -2 \sin 50\ t$

SECTION 8.4 SUMS, DIFFERENCES, AND PRODUCTS OF SINES AND COSINES

• **Example 8.24:** A formula for calculating the instantaneous power to a load is $p = vi$, where v represents the voltage and i represents the current. Given $v = 50 \sin\left(40t + \dfrac{\pi}{6}\right)$ and $i = 10 \sin\left(40t + \dfrac{2\pi}{3}\right)$. Write an expression for the power as a sum or difference of cosines.

Solution:
$$p = vi = \left[50 \sin\left(40t + \frac{\pi}{6}\right)\right]\left[10 \sin\left(40t + \frac{2\pi}{3}\right)\right]$$
$$= 500 \sin\left(40t + \frac{\pi}{6}\right)\sin\left(40t + \frac{2\pi}{3}\right).$$

Use theorem 8.14 with $u = 40t + \dfrac{\pi}{6}$ and $v = 40t + \dfrac{2\pi}{3}$:
$$\sin u \sin v = \frac{\cos(u - v) - \cos(u + v)}{2},$$

$$p = \frac{500\left\{\cos\left[\left(40t + \dfrac{\pi}{6}\right) - \left(40t + \dfrac{2\pi}{3}\right)\right] - \cos\left[\left(40t + \dfrac{\pi}{6}\right) + \left(40t + \dfrac{2\pi}{3}\right)\right]\right\}}{2}$$

$$= 250\left[\cos\left(\frac{\pi}{6} - \frac{2\pi}{3}\right) - \cos\left(80t + \frac{\pi}{6} + \frac{2\pi}{3}\right)\right]$$

$$= 250\left[\cos\left(-\frac{\pi}{2}\right) - \cos\left(80t + \frac{5\pi}{6}\right)\right]$$

$$= 250\left[0 - \cos\left(80t + \frac{5\pi}{6}\right)\right]$$

$$= -250 \cos\left(80t + \frac{5\pi}{6}\right).$$

• • • • • • • • • •

Trial Problems 8.4

Before you begin the section exercises, warm up with these problems. Complete answers are included in the answer key.

Given $u = \dfrac{\pi}{8}$ and $v = \dfrac{\pi}{8}$, find an expression for problems 1–3.

1. $\sin u \cos v$
2. $\sin u \sin v$
3. $\cos u \cos v$

4. Use theorem 8.16 to find an expression for
$$2 \cos\left(\frac{t + \pi}{2}\right)\cos\left(\frac{t - \pi}{2}\right).$$

5. Use theorem 8.19 to find an expression for
$$2\cos\left(\frac{t+\pi}{2}\right)\sin\left(\frac{t-\pi}{2}\right).$$

Exercises 8.4

Express each product as a sum or difference.

1. $\sin(4t)\cos(6t)$
2. $\sin(3t)\cos(5t)$
3. $\cos\dfrac{t}{3}\cos\dfrac{t}{6}$
4. $\cos\dfrac{t}{\pi}\cos\dfrac{t}{2\pi}$
5. $\sin(6t-2)\sin(6t+2)$
6. $\sin(4t+7)\sin(4t-7)$
7. $\sin\left(4t+\dfrac{\pi}{2}\right)\cos\left(4t+\dfrac{3\pi}{2}\right)$
8. $\sin\left(2t-\dfrac{\pi}{2}\right)\cos 2t$
9. $\cos\left(2t-\dfrac{\pi}{3}\right)\cos\left(2t+\dfrac{\pi}{6}\right)$
10. $\cos\left(8t+\dfrac{\pi}{3}\right)\cos\left(8t+\dfrac{5\pi}{6}\right)$

Express each sum or difference as a product.

11. $\sin 5t - \sin 3t$
12. $\sin 4t - \sin t$
13. $\cos 4t - \cos t$
14. $\cos 5t - \cos 3t$
15. $\sin(3t-2) + \sin(2t-3)$
16. $\sin(2t-3) + \sin(5t-7)$
17. $\cos(4t-1) + \cos(4t+1)$
18. $\cos(2t-1) + \cos(2t+1)$
19. $\sin\left(50t+\dfrac{\pi}{3}\right) + \sin\left(50t+\dfrac{2\pi}{3}\right)$
20. $\cos\left(20t+\dfrac{\pi}{2}\right) + \cos(20t+\pi)$

Factor each expression.

21. $\sin 3t + \sin 5t$
22. $\sin 4t + \sin t$
23. $\cos 4t + \cos t$
24. $\cos 5t + \cos 3t$
25. $\sin(3t-2) - \sin(2t-3)$
26. $\sin(2t-3) - \sin(5t-7)$
27. $\cos(4t-1) - \cos(4t+1)$
28. $\cos(2t-1) - \cos(2t+1)$

Verify each identity.

29. $\dfrac{\sin u - \sin v}{\cos u + \cos v} = \tan\left(\dfrac{u-v}{2}\right)$
30. $\dfrac{\sin u + \sin v}{\cos u - \cos v} = \cot\left(\dfrac{v-u}{2}\right)$
31. $\dfrac{\sin t + \sin 3t}{\cos t - \cos 3t} = \cot t$
32. $\dfrac{\cos 2t - \cos 4t}{\sin 2t + \sin 4t} = \tan t$

33. Evaluate $\sin\dfrac{5\pi}{12}\cos\dfrac{\pi}{12}$ exactly using theorem 8.13.

34. Evaluate $\sin\dfrac{11\pi}{12}\sin\dfrac{7\pi}{12}$ exactly using theorem 8.14.

Graph each equation. Simplify the right-hand side of the equation using the formulas of this section before graphing.

35. $y = \sin\left(12t + \dfrac{\pi}{4}\right) + \sin\left(12t - \dfrac{\pi}{4}\right)$

36. $y = \cos\left(12t + \dfrac{\pi}{4}\right) + \cos\left(12t - \dfrac{\pi}{4}\right)$

37. $y = \sin\left(20t + \dfrac{3\pi}{4}\right) - \sin\left(20t + \dfrac{\pi}{4}\right)$

38. $y = \cos\left(20t + \dfrac{\pi}{6}\right) - \cos\left(20t + \dfrac{5\pi}{6}\right)$

Using the power formula of example 8.24 of this section, write an expression for the instantaneous power in terms of sums of differences of sines and cosines.

39. $v = 500 \sin\left(100t + \dfrac{\pi}{3}\right),\ i = 40 \sin\left(100t + \dfrac{5\pi}{6}\right)$

40. $v = 280 \sin\left(50t + \dfrac{\pi}{4}\right),\ i = 50 \sin\left(50t + \dfrac{3\pi}{4}\right)$

41. $v = 50 \cos\left(\omega t + \dfrac{8\pi}{9}\right),\ i = 5 \sin\left(\omega t + \dfrac{17\pi}{9}\right)$

42. $v = 36 \sin\left(\omega t + \dfrac{8\pi}{9}\right),\ i = 4 \sin\left(\omega t + \dfrac{25\pi}{18}\right)$

43. $v = 12 \sin(20t + b),\ i = 4 \sin(20t + b)$

44. $v = 35 \sin(400t + b),\ i = 7 \cos(400t + b)$

45. Prove theorem 8.14.

46. Prove theorem 8.15.

47. Prove theorem 8.17.

48. Prove theorem 8.19.

49. A formula from the field of broadcasting is
$i = a(\sin 2\pi k_1 t + b \sin 2\pi k_2 t \sin 2\pi k_1 t)$. Use theorem 8.14 on the last term to write the expression for i as the sum of three terms.

50. A general formula for instantaneous power (see example 8.24) is $p = vi$, where $v = V_m \sin(bt + c_1)$ and $i = I_m \sin(bt + c_2)$. Use theorem 8.14 to write a formula for p as a difference of cosines.

8•5 Trigonometric Equations

Trigonometric Equations

In this chapter we have considered trigonometric identities, equations that are valid for all admissible values of the variables involved in the equations. Some **trigonometric equations** are valid only for particular values of the variables. In previous chapters we studied such equations, which were called conditional equations. For example, $3^x = 81$ has a particular value of x, namely $x = 4$, for which the equation is valid.

The equation $x^2 - 5x + 6 = 0$ has two values, $x = 2$ and $x = 3$, for which the equation is valid. In this section we consider such conditional trigonometric equations.

We have already considered very simple trigonometric equations such as $\sin t = \frac{1}{2}$. We determined that since $\sin \frac{\pi}{6} = \frac{1}{2}$, then $t = \frac{\pi}{6}$ is a solution to this equation. However, $t = \frac{\pi}{6}$ is not the only solution to the equation. Consider the graph of $y = \sin t$.

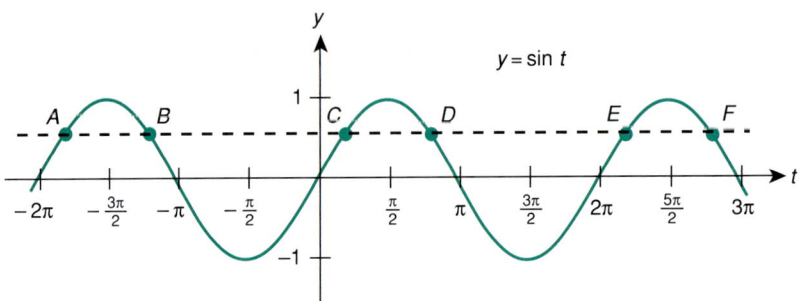

Figure 8.8

There are an infinite number of values of t for which $\sin t = \frac{1}{2}$. Six of these values are shown in figure 8.8. Using points A, C, and E we see that $t = -2\pi + \frac{\pi}{6}$, $t = \frac{\pi}{6}$, and $t = 2\pi + \frac{\pi}{6}$ are solutions. Using points B, D, and F, other solutions are $t = -2\pi + \frac{5\pi}{6}$, $t = \frac{5\pi}{6}$, and $2\pi + \frac{5\pi}{6}$. From the property $\sin t = \sin(t \pm 2\pi)$, it follows that any t of the form

$$t = 2\pi k + \frac{\pi}{6} \quad \text{or} \quad t = 2\pi k + \frac{5\pi}{6}$$

is a solution to the equation $\sin t = \frac{1}{2}$. The solution can be written as follows:

$$t \in \left\{ 2\pi k + \frac{\pi}{6} \text{ or } 2\pi k + \frac{5\pi}{6}, \text{ where } k \text{ is any integer} \right\}.$$

Examples 8.25–8.27 illustrate how to solve trigonometric equations that involve only one of the trigonometric functions.

SECTION 8.5 TRIGONOMETRIC EQUATIONS

• **Example 8.25:** Solve for t, $2 \sin 3t - 2 = 0$.

Solution: First solve the equation for $\sin 3t$:

$$2 \sin 3t = 2$$
$$\sin 3t = 1.$$

Now find the values of $3t$ for which $\sin 3t = 1$:

$$\sin 3t = 1 \quad \text{if } 3t \in \left\{ \ldots, \frac{-7\pi}{2}, \frac{-3\pi}{2}, \frac{\pi}{2}, \frac{5\pi}{2}, \frac{9\pi}{2}, \ldots \right\}$$

$$\text{or} \quad \text{if } t \in \left\{ \ldots, \frac{-7\pi}{6}, \frac{-3\pi}{6}, \frac{\pi}{6}, \frac{5\pi}{6}, \frac{9\pi}{6}, \ldots \right\}.$$

(Divide each solution in the first set by 3.)

• **Example 8.26:** Solve the equation $\sin t = \cos t$, $0 \leq t \leq 2\pi$.

Solution: If both sides of the equation are divided by $\cos t$, the following equation results:

$$\frac{\sin t}{\cos t} = 1 \quad \text{or} \quad \tan t = 1.$$

The restriction $0 \leq t \leq 2\pi$ gives two values of t for which $\tan t = 1$, namely $t = \dfrac{\pi}{4}$ and $t = \dfrac{5\pi}{4}$, so the solution set is

$$\left\{ \frac{\pi}{4}, \frac{5\pi}{4} \right\}.$$

• **Example 8.27:** Solve for t, $2 \cos^2 t - \cos t - 1 = 0$ for $0° \leq t \leq 360°$.

Solution: Since this is a quadratic equation in $\cos t$, we first try to factor the left-hand side of the equation. If it is not factorable, we use the quadratic formula:

$$2 \cos^2 t - \cos t - 1 = 0$$
$$(2 \cos t + 1)(\cos t - 1) = 0.$$

Indeed the equation is factorable, so we set each factor equal to zero and solve these equations subject to $0° \leq t \leq 360°$.

I. $\quad 2 \cos t + 1 = 0$
$$\cos t = -\frac{1}{2}$$
$$t = 120° \quad \text{or} \quad t = 240°.$$

II. $\quad \cos t - 1 = 0$
$$\cos t = 1$$
$$t = 0° \quad \text{or} \quad t = 360°.$$

The solution set is $\{0°, 120°, 240°, 360°\}$.

• **Example 8.28:** Solve $\cos t + \sin t \cot t = 1.5$ for $0 \leq t \leq 2\pi$.

Solution: This equation contains more than one trigonometric function. We attack the problem by attempting to write an equivalent equation containing only one of the trigonometric functions.

If we substitute $\dfrac{\cos t}{\sin t}$ for $\cot t$, the equation becomes

$$\cos t + \sin t \left(\frac{\cos t}{\sin t} \right) = 1.5$$
$$\cos t + \cos t = 1.5$$
$$2 \cos t = 1.5$$
$$\cos t = 0.75.$$

We can find one solution from the calculator.

$$t = \cos^{-1}(0.75) = 0.7227 \text{ (approx.)}.$$

0.75 │INV│ │COS│.

Display: 0.7227342.

From figure 8.9 and the fact that $\cos u = \cos(2\pi - u)$, we can obtain the second solution in the interval $0 \leq t \leq 2\pi$:

$$t = (2\pi - 0.7227) = 5.5605.$$

The solution set is $\{0.7227, 5.5605\}$.

Many times we can simplify a trigonometric equation by using one or more of the trigonometric identities that we have developed in this chapter. The following set of suggestions provides a summary of some rules for attacking trigonometric equations.

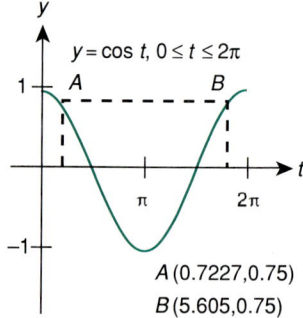

Figure 8.9

SOLVING TRIGONOMETRIC EQUATIONS

1. For equations containing one trigonometric function, solve the equation for that function.

2. For equations containing more than one function, try to factor the left-hand side after making the right-hand side zero.

3. If the equation is not factorable, try to change the form of the equation by using the trigonometric identities.

• *Example 8.29:* Solve $\sin 2t + \sin t = 0$ for $0 \leq t \leq 2\pi$.

Solution: Use the identity $\sin 2t = 2 \sin t \cos t$:

$$2 \sin t \cos t + \sin t = 0.$$

Factor the left-hand side of the equation:

$$(\sin t)(2 \cos t + 1) = 0.$$

Set each of the factors equal to zero and solve each of the resulting equations:

$$\sin t = 0 \qquad\qquad 2 \cos t + 1 = 0$$
$$t \in \{0, \pi, 2\pi\} \qquad\qquad \cos t = -\frac{1}{2}$$
$$\qquad\qquad\qquad t \in \left\{\frac{2\pi}{3}, \frac{4\pi}{3}\right\}.$$

The solution set is $\left\{0, \dfrac{2\pi}{3}, \pi, \dfrac{4\pi}{3}, 2\pi\right\}$.

• *Example 8.30:* The current in a circuit is represented by the equation $i = 50(\sin 20t + \sin 60t)$, $t \geq 0$. Determine the first three positive values of t for which $i = 0$.

Solution: We need to solve the equation

$$50(\sin 20t + \sin 60t) = 0$$

for the three smallest positive values of t.

Since the equation contains $\sin 20t$ and $\sin 60t$, one of the ways we can proceed is to factor the expression using theorem 8.17:

$$\sin u + \sin v = 2 \sin\left(\frac{u+v}{2}\right) \cos\left(\frac{u-v}{2}\right).$$

In the problem $u = 20t$ and $v = 60t$

$$50(\sin 20t + \sin 60t)$$
$$= 100 \sin\left(\frac{20t + 60t}{2}\right) \cos\left(\frac{20t - 60t}{2}\right)$$
$$= 100(\sin 40t)(\cos(-20t))$$
$$= 100(\sin 40t)(\cos 20t) = 0$$

We now set each of the factors equal to zero:

$\sin 40t = 0$	$\cos 20t = 0$
$40t \in \{0, \pi, 2\pi, 3\pi, \ldots\}$	$20t \in \left\{\dfrac{\pi}{2}, \dfrac{3\pi}{2}, \dfrac{5\pi}{2}, \dfrac{7\pi}{2}, \ldots\right\}$
$t \in \left\{0, \dfrac{\pi}{40}, \dfrac{2\pi}{40}, \dfrac{3\pi}{40}, \ldots\right\}$	$t \in \left\{\dfrac{\pi}{40}, \dfrac{3\pi}{40}, \dfrac{5\pi}{40}, \dfrac{7\pi}{40}, \ldots\right\}.$
(Divide the elements in the first set by 40)	(Divide the elements in the first set by 20)

These sets can be combined into the solution set

$$t \in \left\{\frac{k\pi}{40}, \text{ where } k \text{ is a nonnegative integer}\right\}.$$

The first three positive values of t, where the current is zero, are

$$t = \frac{\pi}{40}, \; t = \frac{\pi}{20}, \text{ and } t = \frac{3\pi}{40}.$$

• • • • • • • • • •

Trial Problems 8.5

Before you begin the section exercises, warm up with these problems. Complete answers are included in the answer key.

Find the smallest positive value of t that is a solution to the given equation.

1. $\sin 3t = 1$
2. $\cos \pi t = 0$
3. $2 \cos t + 1 = 0$
4. $\sin^2 t - 2 \sin t + 1 = 0$
5. $\sin 3t + \sin t = \cos t$

Exercises 8.5

Find all solutions for the following equations. Use a calculator for inverse functions when necessary.

1. $\sin t = \dfrac{1}{2}$
2. $\cos t = \dfrac{1}{2}$
3. $\tan t = 1$
4. $\tan t = -1$
5. $\cos 2t = 0$
6. $\sin 2t = 1$
7. $\cos \pi t = \dfrac{\sqrt{3}}{2}$
8. $\sin \pi t = \dfrac{\sqrt{3}}{2}$
9. $\sin 2t - 1 = \cos 2t$
10. $\cos 2t - 1 = \sin 2t$

Determine the solutions of the following equations for $0 \le t \le 2\pi$.

11. $\tan^2 t - 1 = 0$

12. $\cot^2 t - 1 = 0$

13. $2 \sin t + 1 = \sin t + \dfrac{1}{2}$

14. $2 \cos t - 1 = \cos t - \dfrac{1}{2}$

15. $(\sin t - 1)(\cos t + 1) = 0$

16. $(\cos t - 1)(\sin t + 1) = 0$

17. $\sin t (2 \cos t - 1) = 0$

18. $\cos t (2 \sin t - 1) = 0$

19. $(2 \sin 2t)(1 - \cos 2t) = 0$

20. $(2 \cos 2t)(1 - \sin 2t) = 0$

21. $\sin^2 t - \dfrac{1}{4} = 0$

22. $\cos^2 t - \dfrac{3}{4} = 0$

23. $2 \sin^2 t + 9 \sin t - 5 = 0$

24. $2 \cos^2 t - 7 \cos t + 3 = 0$

25. $\sec^2 t - 2 \tan t = 0$

26. $\csc^2 t - 2 \cot t = 0$

27. $2 \sin^2 t + 3 \cos t - 3 = 0$

28. $2 \cos^2 t - 3 \sin t - 3 = 0$

Solve each equation over the specified interval.

29. $\sin t + \sin 3t = \cos t, \quad 0 \le t \le 2\pi$

30. $\cos 5t - \cos 3t = \sin t, \quad 0 \le t \le 2\pi$

31. $\sec^2 t - 2 \sec^2 t \csc^2 t + \csc^2 t = 0, \quad 0 \le t \le 2\pi$

32. $\tan t + \cot t = \sec t \csc t, \quad 0 \le t \le 2\pi$

33. $\cot^2 t - 5 \cot t + 4 = 0, \quad 0 \le t \le 2\pi$

34. $\sec^2 t + 3 \tan t - 11 = 0, \quad 0 \le t \le 2\pi$

35. $\sin 4t - 2 \sin 2t = 0$, all real numbers

36. $\sin 3t + \sin 2t = 0$, all real numbers

37. $\dfrac{1}{2} \sin\left(2t + \dfrac{\pi}{4}\right) = 0, \quad -\dfrac{\pi}{2} \le t \le \pi$

38. $\sin\left(t - \dfrac{\pi}{3}\right) = 1, \quad 0 \le t \le 2\pi$

39. $-2 \sin(3t + 1) = 2, \quad \left(-\dfrac{\pi}{6} - \dfrac{1}{3}\right) \le t \le \left(\dfrac{11\pi}{6} - \dfrac{1}{3}\right)$

40. $3 \sin(2t - \pi) = 3, \quad \dfrac{\pi}{2} \le t \le \dfrac{3\pi}{2}$

41. The current through a coil is represented by the equation $i = 30 \sin 30t$. Determine the first three positive values of t for which $i = 0$.

42. Solve exercise 41 using the equation $i = 25 \sin 20t$.

43. The voltage across a capacitor is represented by the equation $v = -120 \sin\left(200t + \dfrac{\pi}{6}\right)$. Find the first three positive values of t for which $v = 120$.

44. Solve exercise 43 using the equation $v = 120 \sin\left(200t + \dfrac{\pi}{3}\right)$.

45. The voltage in a circuit is represented by the equation $v = 36 \sin\left(600t + \dfrac{\pi}{4}\right)$ and the current by the equation $i = 4\left(\sin 600t + \dfrac{3\pi}{4}\right)$. Determine the smallest positive value where $p = iv = 60$.

46. Solve exercise 45 given $v = 80 \sin\left(200t - \dfrac{\pi}{4}\right)$, $i = 2 \sin\left(200t + \dfrac{\pi}{4}\right)$.

47. The maximum height attained by a projectile fired at an angle θ with the horizontal plane with initial velocity v_0 is

$$h = \frac{v_0^2 \sin^2\theta}{2g},$$

where $g = 32$ ft/sec^2 and $0 < \theta < \dfrac{\pi}{2}$. Given that $v_0 = 160$ ft/sec and $h = 160$ ft, determine θ.

48. The range of a projectile fired at an angle θ with the horizontal plane having an initial velocity v_0 is

$$d = \frac{v_0^2 \sin 2\theta}{g},$$

where $g = 32$ ft/sec^2 and $0 < \theta < \dfrac{\pi}{2}$. Given that $v_0 = 160$ ft/sec and $d = 693$ ft, determine θ.

8·6 Summary of Terms, Formulas, Rules, and Procedures

Terms

Double-Angle Formulas (p. 336)
Half-Angle Formulas (p. 336)
Pythagorean Identities (p. 330)

Quotient Identities (p. 330)
Reciprocal Identities (p. 330)
Sign Identities (p. 330)

Sum and Difference Formulas (p. 336)
Trigonometric Equations (p. 363)

Formulas

Reciprocal Identities   (8.1)

$$\sin t = \frac{1}{\csc t} \qquad \cos t = \frac{1}{\sec t} \qquad \tan t = \frac{1}{\cot t}$$

$$\cot t = \frac{1}{\tan t} \qquad \sec t = \frac{1}{\cos t} \qquad \csc t = \frac{1}{\sin t}.$$

Quotient Identities   (8.1)

$$\tan t = \frac{\sin t}{\cos t} \qquad \cot t = \frac{\cos t}{\sin t}.$$

Sign Identities   (8.1)

$\sin(-t) = -\sin t \qquad \cos(-t) = \cos t \qquad \tan(-t) = -\tan t$
$\cot(-t) = -\cot t \qquad \sec(-t) = \sec t \qquad \csc(-t) = -\csc t.$

Pythagorean Identities   (8.1)

$$\sin^2 t + \cos^2 t = 1 \qquad \tan^2 t + 1 = \sec^2 t$$
$$\cot^2 t + 1 = \csc^2 t.$$

Cosine Identities   (8.2)

8.1 $\cos(u + v) = \cos u \cos v - \sin u \sin v$
8.2 $\cos(u - v) = \cos u \cos v + \sin u \sin v$
8.3 $\cos 2t = \cos^2 t - \sin^2 t = 1 - 2\sin^2 t = 2\cos^2 t - 1$

8.4 $\cos \dfrac{t}{2} = \pm\sqrt{\dfrac{1 + \cos t}{2}}$

8.5 $a \cos kt + b \sin kt = m \cos(kt - c)$, provided
$m = \sqrt{a^2 + b^2}$, $\tan c = \dfrac{b}{a}$

Sine Identities   (8.3)

8.6 $\sin\left(\dfrac{\pi}{2} - t\right) = -\cos t$

$\cos\left(\dfrac{\pi}{2} + t\right) = \sin t$

8.7 $\sin(u + v) = \sin u \cos v + \cos u \sin v$
$\sin(u - v) = \sin u \cos v - \cos u \sin v$

8.8 $\sin 2t = 2 \sin t \cos t$

8.9 $\sin \dfrac{t}{2} = \pm\sqrt{\dfrac{1 - \cos t}{2}}$

Tangent Identities   (8.3)

8.10 $\tan(u + v) = \dfrac{\tan u + \tan v}{1 - \tan u \tan v}$

$\tan(u - v) = \dfrac{\tan u - \tan v}{1 + \tan u \tan v}$

8.11 $\tan 2t = \dfrac{2 \tan t}{1 - \tan^2 t}$

8.12 $\tan \dfrac{t}{2} = \dfrac{1 - \cos t}{\sin t}$

Identities for Sums, Differences, and Products of Sines and Cosines   (8.4)

8.13 $\sin u \cos v = \dfrac{\sin(u + v) + \sin(u - v)}{2}$

8.14 $\sin u \sin v = \dfrac{\cos(u - v) - \cos(u + v)}{2}$

8.15 $\cos u \cos v = \dfrac{\cos(u + v) + \cos(u - v)}{2}$

8.16 $\cos u + \cos v = 2 \cos\left(\dfrac{u + v}{2}\right)\cos\left(\dfrac{u - v}{2}\right)$

8.17 $\sin u + \sin v = 2 \sin\left(\dfrac{u + v}{2}\right)\cos\left(\dfrac{u - v}{2}\right)$

8.18 $\cos u - \cos v = -2 \sin\left(\dfrac{u + v}{2}\right)\sin\left(\dfrac{u - v}{2}\right)$

8.19 $\sin u - \sin v = 2 \cos\left(\dfrac{u + v}{2}\right)\sin\left(\dfrac{u - v}{2}\right)$

Rules and Procedures

- Suggestions for Proving Identities   (8.1)
  1. If possible, (a) simplify one side of the equation until it is the same as the other side, or (b) simplify both sides of the equation until they are the same.
  2. If the identity contains fractions, (a) combine the sum or the difference of the two fractions into a single fraction, or (b) split a single fraction into the sum or difference of two fractions.
  3. Write all of the expressions in the equation in terms of sines and cosines, then simplify the equation.
  4. Usually it is best to work with the most complex side of the equation.

- Suggestions for Solving Trigonometric Equations   (8.5)
  1. For equations containing one trigonometric function, solve the equation for that function.
  2. For equations containing more than one trigonometric function, try to factor the left-hand side after making the right-hand side zero.
  3. If the equation is not factorable, try to change the form of the equation by using the trigonometric identities.

8·7 Chapter 8 Review Exercises

Simplify each expression.

1. $\sin t \csc t$
2. $\cos t \sec t$
3. $\sin^2 t(1 + \cot^2 t)$
4. $\dfrac{\sec^2 t - 1}{\sec^2 t}$
5. $\dfrac{\tan^2 t}{\sec^2 t} + \cos^2 t$
6. $\dfrac{\sin^2 t + \cos^2 t}{\csc^2 t - \cot^2 t}$
7. $\cos t(\tan t + \sec t)$
8. $\tan t(\cot t + \csc t)$
9. $\cos t \sin^3 t + \sin t \cos^3 t$
10. $\sin^4 t + 2\cos^2 t \sin^2 t + \cos^4 t$

Verify each identity.

11. $\dfrac{\sin^2 t}{\cos t} + \cos t = \sec t$
12. $\dfrac{\cos^2 t}{\sin t} + \dfrac{1}{\csc t} = \csc t$
13. $\dfrac{(1 - \sin t)(1 + \sin t)}{\cos t} = \cos t$
14. $\dfrac{(1 - \cos t)(1 + \cos t)}{\sin t} = \sin t$
15. $\sin\left(\dfrac{\pi}{2} + t\right) = \cos t$
16. $\sin\left(\dfrac{3\pi}{2} + t\right) = -\cos t$
17. $\sin 3t = 3\sin t - 4\sin^3 t$
18. $\cos 3t = -3\cos t + 4\cos^3 t$
19. $\sin(u + v) + \sin(u - v) = 2\sin u \cos v$
20. $\sin(u + v) - \sin(u - v) = 2\cos u \sin v$

If u and v are in the first quadrant, $\sin u = \dfrac{3\sqrt{13}}{13}$ and $\sin v = \dfrac{2\sqrt{5}}{5}$, find an expression for each of the following.

21. $\cos u$
22. $\sin(u + v)$
23. $\sin 2u$
24. $\sin \dfrac{u}{2}$
25. $\tan 2u$

Write each product as a sum or difference of trigonometric functions.

26. $2\sin 2t \sin 4t$
27. $4\cos 4y \cos 6y$
28. $\sin 5t \cos 3t$
29. $\sin 2t \cos 4t$

Write each sum or difference as a product of trigonometric functions.

30. $\sin(3t + 1) - \sin(3t - 1)$
31. $\cos(3t + 1) - \cos(3t - 1)$
32. $\sin 2t + \sin 4t$
33. $\cos 2t + \cos 4t$

Solve each trigonometric equation for the indicated interval.

34. $\sin t = \dfrac{\sqrt{2}}{2}, \quad 0 \leq t \leq 2\pi$
35. $\cos t = -\dfrac{\sqrt{2}}{2}, \quad 0 \leq t \leq 2\pi$
36. $\sin\left(3t + \dfrac{\pi}{2}\right) = 1, \quad 0 \leq t \leq 4\pi$
37. $\cos\left(3t + \dfrac{\pi}{2}\right) = 1, \quad 0 \leq t \leq 4\pi$
38. $\sin^2 t - \sin t = 0, \quad 0 \leq t \leq 2\pi$
39. $\cos^2 t + \cos t = 0, \quad 0 \leq t \leq 4\pi$
40. $\sin 2t + \sin t = 0, \quad 0 \leq t \leq 2\pi$
41. $\cos t + \cos 2t = 0, \quad 0 \leq t \leq 2\pi$
42. $\cos \dfrac{t}{2} = \cos t + 1, \quad 0 \leq t \leq 2\pi$
43. $\sin \dfrac{t}{2} = 1 - \cos t, \quad 0 \leq t \leq 2\pi$

44. $\cos\left(50t + \dfrac{\pi}{2}\right) = -1, \quad 0 \le t \le \dfrac{\pi}{50}$

45. $\sin\left(50t - \dfrac{\pi}{2}\right) = 1, \quad 0 \le t \le \dfrac{\pi}{50}$

46. A generator produces a current according to the equation $i = 30 \sin 100\pi t$. Find the smallest positive value of t for which $i = 20$ amps if t is in seconds.

47. Solve exercise 46 for $i = 15$ amps.

48. According to Snell's law, when light passes through a medium the relationship between the angle of incidence, θ_1, and the angle of refraction, θ_2, is stated as follows:

$$\dfrac{\sin \theta_1}{\sin \theta_2} = k,$$

a constant, where k is the index of refraction of the medium. Given that $k = 2.42$ and $\theta_1 = \dfrac{\pi}{6}$, determine θ_2. (See figure 8.10.)

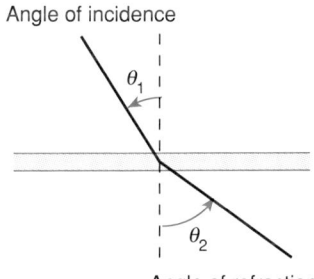

Figure 8.10

49. Solve exercise 48 given that $\theta_1 = \dfrac{\pi}{6}$ and $k = 1.42$.

50. A coil in a magnetic field induces a voltage given by the equation $v = 15 \sin\left(\dfrac{\pi t}{6} - \dfrac{\pi}{4}\right)$. Determine the smallest value of t for which $v = 6.72$.

51. Solve exercise 50 given that $v = 8.61$.

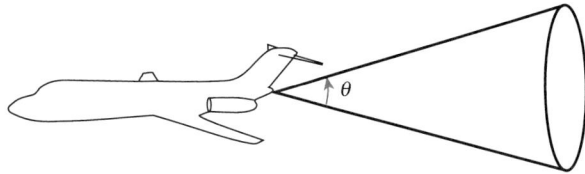

Figure 8.11

52. A plane (figure 8.11) flying at the speed of sound is said to be flying at Mach 1. The Mach number is the ratio of the speed of the plane to the speed of sound. When flying faster than speed of sound, a cone created by the sound waves produces a sonic boom with vertex angle θ. If $M > 1$ then $M = \dfrac{1}{\sin \dfrac{\theta}{2}}$. Given that $M = 1.23$, determine θ.

53. Solve exercise 52 given that $M = 2.42$.

54. The current in a circuit is represented by the equation $i = \dfrac{1}{4} \cos 8t - \dfrac{6}{5} \sin 8t$. Determine the first value of $t > 0$ for which the current is 0. *Hint:* Use theorem 8.5 to write an expression for i in terms of cosines.

55. Solve exercise 54 given that $i = 0.3 \cos 8t - 1.2 \sin 8t$.

56. The displacement of a particular pendulum is represented by the equation $y = \dfrac{7}{3} \cos \dfrac{\pi t}{20} + \dfrac{10}{3} \sin \dfrac{\pi t}{20}$. Rewrite the equation in the form $y = a \cos(bt + c)$.

57. Solve exercise 56 using the equation $y = \dfrac{5}{3} \cos \dfrac{\pi t}{20} + \dfrac{8}{3} \sin \dfrac{\pi t}{20}$.

8·8 Chapter 8 Test

Simplify each expression.

1. $(1 - \sin t)(\sec t + \tan t)$

2. $\dfrac{\tan^2 t - \sin^2 t}{\tan^2 t}$

3. $\left(\dfrac{\sec t + 1}{\sec t - 1}\right)\left(\dfrac{1 - \cos t}{1 + \cos t}\right)$

4. $\dfrac{\sec^2 t - \csc^2 t}{\sin^4 t - \cos^4 t}$

5. $\dfrac{\sec t - \cos t}{\tan t}$

Verify each identity.

6. $\dfrac{1 + \sin t}{\cos t} = \dfrac{\cos t}{1 - \sin t}$

7. $\sec^2 t + \csc^2 t = (\tan t + \cot t)^2$

8. $\dfrac{\sin t}{1 + \cos t} + \dfrac{1 + \cos t}{\sin t} = \dfrac{2}{\sin t}$

9. $(\sin t + \cos t)^2 + (\sin t - \cos t)^2 = 2$

10. $\dfrac{\tan^2 t}{1 + \tan^2 t} = \sin^2 t$

If u and v are in the first quadrant, $\sin u = \dfrac{\sqrt{15}}{5}$ and $\cos v = \dfrac{3\sqrt{10}}{10}$, calculate each of the following.

11. $\cos u$

12. $\sin(u + v)$

13. $\sin 2u$

14. $\cos \dfrac{v}{2}$

15. $\tan 2v$

Write each product as a sum or difference of trigonometric functions.

16. $2 \cos 2t \cos 4t$

17. $\sin 4t \cos 2t$

Write each sum or difference as a product of trigonometric functions.

18. $\cos(2t + 1) - \cos(2t - 1)$

19. $\cos 3t + \cos 6t$

Solve the equations for the indicated interval.

20. $\sin t = -\dfrac{\sqrt{2}}{2}, \quad 0 \leq t \leq 2\pi$

21. $\cos\left(2t + \dfrac{\pi}{4}\right) = 1, \quad 0 \leq t \leq 4\pi$

22. $\sin^2 t - 2 \sin t = -1, \quad 0 \leq t \leq 2\pi$

23. $4 \sin^2 t - 1 = 0, \quad 0 \leq t \leq 2\pi$

24. $\sin^2 2t - \sin 2t = 0, \quad 0 \leq t \leq \pi$

25. $2 \sin^2 t - 5 \sin t = 3, \quad 0 \leq t \leq 2\pi$

26. The instantaneous power in a circuit is represented by the equation $y = -2 \sin 50t$, where t is in seconds. Find the two smallest positive values of t for which $y = 2$.

27. The current (i) in amperes in a circuit is represented by the equation $i = 20 \sin 60(t - 0.087)$, where t is in seconds. Find the smallest positive value of t for which the current is 12.0 amperes.

28. The current in a circuit is represented by the equation $i = 0.23 \cos 8t - 1.35 \sin 8t$. Write an equation, in the form $i = m \cos(kt - c)$, that represents the current in the circuit.

8·9 Cumulative Review—Chapters 6, 7, and 8

1. Find the length of the hypotenuse of a right triangle that has legs that measure 8.75 cm and 6.25 cm.

2. A triangle has sides that measure 15.0 mm, 24.0 mm, and 36.0 mm. A second triangle, which is similar to the first triangle, has a side 10.5 mm long. This side corresponds to the 15-mm side of the first triangle. Find the lengths of the remaining sides of the second triangle.

3. If the sine of an acute angle is 0.7880, find the cosine and the tangent of the angle.

4. The sides of a right triangle measure 50.0 in., 120.0 in., and 130.0 in. Find the sine, cosine, and tangent of the angle opposite the shortest side of the triangle.

Use figure 8.12 as a guide in solving exercises 5–7.

5. If $a = 6.4$ and $b = 5.3$, find A.

6. If $A = 46°$ and $a = 19.8$, find c.

7. If $B = 23°35'$ and $b = 152.0$, find c.

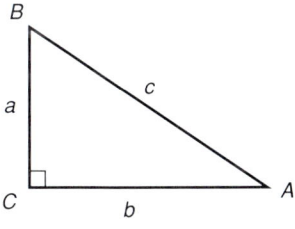

Figure 8.12

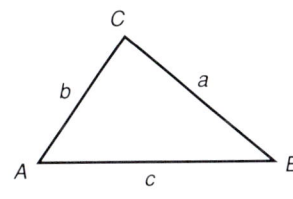

Figure 8.13

Use figure 8.13 as a guide in exercises 8–10.

8. If $A = 42°$, $C = 86°$, and $a = 9.3$, find b and c.

9. If $a = 6.42$, $b = 8.71$, and $A = 82.3°$, find C and c.

10. If $a = 12.0$, $b = 14.0$, and $c = 18.0$, find A and B.

11. $81°32' = $ __?__ radians

12. 1.42 radians = __?__ degrees

13. A linear velocity of 12.6 m/sec along a circle with radius 0.6 m represents what angular velocity?

14. Calculate $y = 0.42 \sin 10t$ when $t = \dfrac{\pi}{20}$.

15. Calculate $1.83\left(\cos \dfrac{\pi}{16} + \sin \dfrac{\pi}{16}\right)$.

16. Graph $y = 1.8 \sin t$ for $0 \le t \le 2\pi$.

17. Graph $y = \cos\left(t + \dfrac{\pi}{8}\right)$ for $0 \le t \le 2\pi$.

Graph each function for one period.

18. $y = 1.5 \cos\left(t + \dfrac{\pi}{4}\right)$

19. $y = 20 \sin\left(30t + \dfrac{\pi}{2}\right)$

20. $y = \cos t + \sin 2t$

Simplify each expression.

21. $\dfrac{\cos t \tan t}{\sin t}$

22. $\dfrac{(\tan t)(1 - \sin^2 t)}{\cos t}$

23. A wave traveling in a string is represented by the equation $y = 1.84 \sin(10t - 0.35)$. Graph the function for two periods.

If u and v are in the second quadrant, $\sin u = \dfrac{3}{5}$, and $\cos v = -\dfrac{1}{4}$, find each of the following values.

24. $\sin v$

25. $\cos(u + v)$

26. $\cos 2u$

27. $\sin \dfrac{v}{2}$

28. $\tan \dfrac{u}{2}$

Solve for t.

29. $\cos t = \dfrac{\sqrt{3}}{2}$, for $0 \le t \le 2\pi$

30. $\sin\left(3t - \dfrac{\pi}{4}\right) = 1$, for $0 \le t \le 2\pi$

31. $\cos^2 t - \cos t = 0$, for $0 \le t \le 2\pi$

32. $2 \sin^2 t + \sin t = 1$, for $0 \le t \le 2\pi$

33. The displacement (y) of an object from a position of equilibrium is represented by the equation

$y = 1.75 \cos\left(0.520t + \dfrac{\pi}{6}\right)$ at any time t. Find the two smallest values of t for which the displacement is 1.50.

Prove each identity.

34. $\dfrac{\cos t}{\tan t + \sec t} = 2 + \dfrac{\cos t}{\tan t - \sec t}$

35. $\dfrac{\tan t \csc t}{\tan^2 t + 1} = \cos t$

Chapter 9

Vectors and Complex Numbers

*U*p until the time of an ancient Greek mathematician named Pythagoras (about 569 B.C. to 501 B.C.), it was believed that all numbers were rational numbers. Pythagoras formed a secret society whom he taught to worship numbers, to remain anonymous, and to sign Pythagoras' name to any mathematical discovery. The Pythagoreans are credited with the first proof of the theorem of Pythagoras, although the theorem itself was known centuries earlier.

More importantly, the Pythagoreans discovered that $\sqrt{2}$ is an irrational number, a discovery that they regretted and threatened death to any member who revealed the secret.

The idea of considering the square root of a negative number as actually being a number (complex or imaginary number) may have begun with a sixteenth century mathematician named Girolamo Cardano in connection with his work in calculating roots of certain kinds of cubic equations. The first mathematician to accept the existence of complex numbers may have been Rafael Bombelli, later in the sixteenth century, also in connection with his work in solving cubic equations.

Number
Unit
Scalar Quantity
Direction
Vector Quantity

Magnitude of a Vector

Initial Point
Terminal Point

9 • 1 Vectors and Vector Operations

If you wanted to find the area of a rectangle, you would use the lengths of the sides. These quantities consist of a **number** and a **unit**. Quantities like length, width, temperature, and mass are called **scalar quantities**.

Other quantities that require not only a number and a unit but also a **direction** are called **vector quantities**. Examples of vector quantities are acceleration, force, and velocity.

Vector quantities can be represented geometrically as a line segment having a terminal arrow to indicate the direction. The length of the segment represents the **magnitude** of the vector.

Figure 9.1 shows a representation of several vectors. The vector on the right represents a force of 100 pounds applied at an angle of 62° with the horizontal line. Vector OA can be written as **OA**. The first letter (O) indicates the **initial point** of the vector, and the second letter (A) indicates the **terminal point** of the vector. The *magnitude* of **OA** is written as $|\mathbf{OA}|$. The magnitude of **B** is written as $|\mathbf{B}|$ and equals 100 lb.

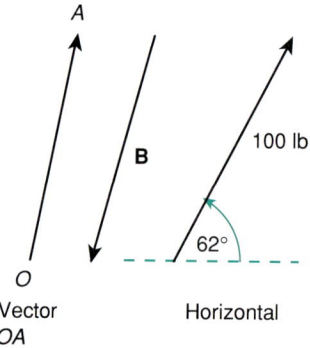

Figure 9.1

EQUAL VECTORS

Two vectors are equivalent or **equal vectors** if they have the same direction and the same magnitude (length).

The vectors in figure 9.2 satisfy the following properties

OA = **PB** and **C** = **D**

E ≠ **F,** since they have different directions

G ≠ **H,** since they have different magnitudes

MN ≠ **QR,** since they have different directions and different magnitudes

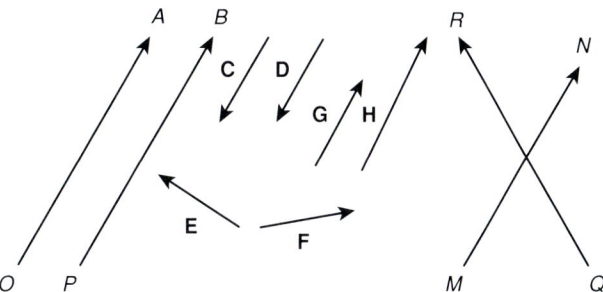

Figure 9.2

SCALAR MULTIPLES

If two vectors are parallel but not necessarily equal, each vector is a **scalar multiple** of the other vector.

Figure 9.3 shows some vectors that are scalar multiples of vector **V**. Of course, since all of the vectors are parallel, each vector is a scalar multiple of each of the other vectors.

A pair of vectors **A** and **B** can be added geometrically to obtain another vector **A + B** in two ways.

The first method is to place the initial point of **B** at the terminal point of **A**. The sum vector **A + B** is the vector with its initial point the same as the initial point of **A** and its terminal point the same as the terminal point of **B**. See figure 9.4.

The second method is to position **A** and **B** so that their initial points coincide. The sum vector **A + B** is represented by the diagonal of the parallelogram formed by using **A** and **B** as adjacent sides. See figure 9.5.

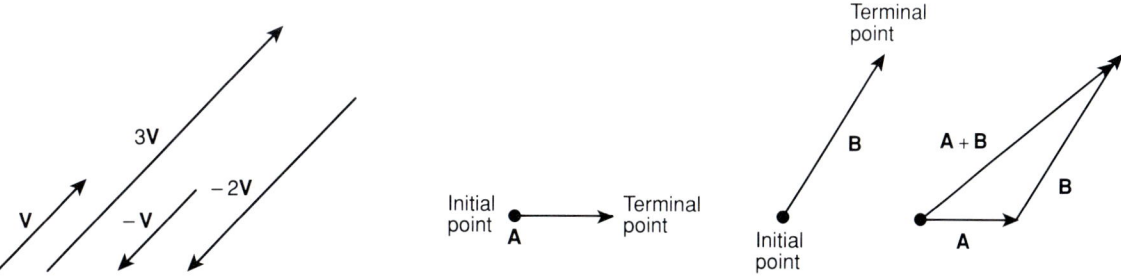

Figure 9.3

Figure 9.4

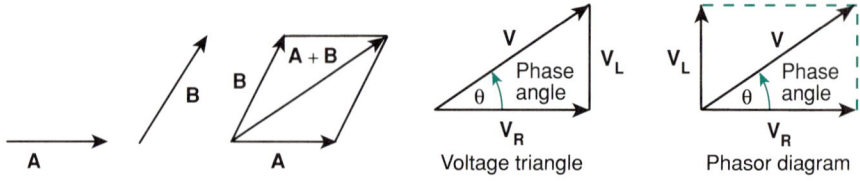

Figure 9.5 Figure 9.6

Note that a vector can be moved around provided the length and the direction are kept constant.

Resultant The sum $\mathbf{A} + \mathbf{B}$ is called the **resultant** of $\mathbf{A}$ and $\mathbf{B}$.

Many applications-oriented problems can be represented in terms of vectors. For example, the voltage triangle and the phasor diagram for a series circuit containing a resistance and an inductance are shown in figure 9.6.

Given a vector $\mathbf{A}$, the vector $-\mathbf{A}$ (the opposite of $\mathbf{A}$) is the vector that has the same magnitude as $\mathbf{A}$ but has the opposite direction from $\mathbf{A}$.

VECTOR SUBTRACTION

Vectors $\mathbf{A}$ and $\mathbf{B}$ are subtracted using the pattern $\mathbf{A} - \mathbf{B} = \mathbf{A} + (-\mathbf{B})$.

Figure 9.7 shows how to subtract vectors geometrically.

When a vector is placed on a coordinate system, the initial point is usually placed at the origin. The lengths of the vectors $\mathbf{V}_x$ and $\mathbf{V}_y$ in figure 9.8 are called the **horizontal** **Horizontal Components** **and vertical components** of the original vector $\mathbf{V}$. Using the sine and cosine ratios on **Vertical Components** right triangle OAB produces these results:

$$\sin \theta = \frac{AB}{|\mathbf{V}|} = \frac{OC}{|\mathbf{V}|} = \frac{|\mathbf{V}_y|}{|\mathbf{V}|}$$

$$\cos \theta = \frac{OA}{|\mathbf{V}|} = \frac{|\mathbf{V}_x|}{|\mathbf{V}|}.$$

Solving these equations for $|\mathbf{V}_x|$ and $|\mathbf{V}_y|$ yields these results.

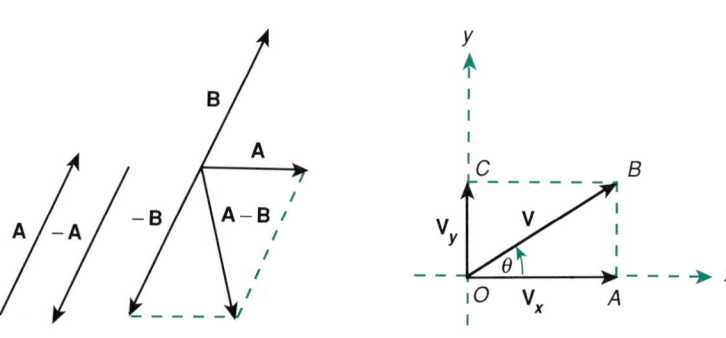

Figure 9.7 Figure 9.8

COMPONENTS OF A VECTOR (V)

Horizontal Component $= |V_x| = |V| \cos \theta$
Vertical Component $= |V_y| = |V| \sin \theta$.

Examples 9.1–9.4 illustrate some vector applications.

- **Example 9.1:** A force of 40 newtons makes an angle of 37° with the horizontal. Determine the magnitudes of the horizontal and vertical components of the force.

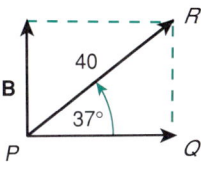

Figure 9.9

Solution: We need to determine the magnitudes (lengths) of vectors **A** and **B** in figure 9.9. Using triangle PQR, the magnitude of **B** is the length of QR. Use the sine ratio:

$$\sin 37° = \frac{QR}{PR}$$

$$QR = (PR)(\sin 37°)$$
$$= 40 \sin 37° = 24.1.$$

The vertical component is 24.1 newtons, since the magnitude of **B** is the same as the length of QR.

Using the same triangle,

$$\cos 37° = \frac{PQ}{PR}$$

$$PQ = (PR)(\cos 37°)$$
$$= 40 \cos 37° = 31.9.$$

Since the magnitude of **A** is the same as the length of PQ, the horizontal component is 31.9 newtons.

- **Example 9.2:** Two tugboats are pushing a ship as shown in figure 9.10a. Each tug exerts a force of 420 lb. Find the magnitude of the resultant force.

Solution: Use the force diagram in figure 9.10b. Since we know two sides of triangle ABC and with the two known angles, we can easily calculate the third angle, either the law of sines or the law of cosines can be used to calculate AC.

Suppose we use the law of cosines:

$$\text{angle } B = 180° - (20° + 20°) = 140°$$
$$AC = \sqrt{(AB)^2 + (BC)^2 - 2(AB)(BC) \cos 140°}$$
$$= \sqrt{(420)^2 + (420)^2 - 2(420)(420) \cos 140°}$$
$$= 789 \text{ lb}.$$

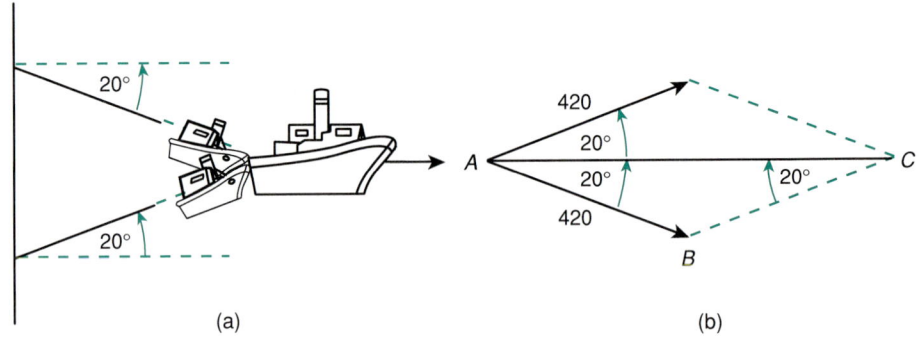

Figure 9.10

- **Example 9.3:** A plane flies 38° north of due east for 550 mi and then due east for another 410 mi. (a) How far is the plane from the starting point? (b) What angle does the resultant vector make with the "due east" line? See figure 9.11.

Solution: **a.** Find the magnitude of **OP** in figure 9.12. $QP = OR = 550$, and

$$\text{angle } Q = 180° - 38° \quad \text{(Consecutive angles of a parallelogram} \\ = 142°. \quad \text{are supplementary)}$$

Use the law of cosines on triangle OPQ:

$$OP = \sqrt{(OQ)^2 + (QP)^2 - 2(OQ)(QP)\cos\angle OQP}$$
$$= \sqrt{(410)^2 + (550)^2 - 2(410)(550)\cos 142°}$$
$$= 909 \text{ mi.}$$

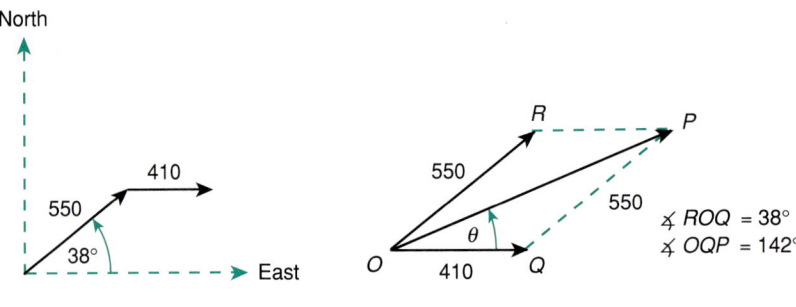

Figure 9.11   Figure 9.12

b. Find angle θ. Use the law of sines in triangle OQP:

$$\frac{OP}{\sin Q} = \frac{PQ}{\sin \theta}$$

$$\sin \theta = \frac{PQ \sin Q}{OP} \quad \text{(Solve for sin } \theta\text{)}$$

$$= \frac{550 \sin 142°}{909} \quad \text{(Substitute in the expression for sin } \theta\text{)}$$

$$\sin \theta = 0.3725$$

$$\theta = 22°. \quad \text{(Inverse sine of 0.3725)}$$

• • • • • • • • • •

Vectors are used to solve many electrical problems. One of the applications is the calculating of the impedance in a circuit. Figure 9.13 shows a series circuit containing a resistor, a capacitor and an inductor. The reactance of the circuit is calculated as follows:

$X = X_L - X_C$, where X is the reactance, X_L is the inductive reactance, and X_C is the capacitive reactance.

The magnitude of the impedance (retardance of the flow of current) in an AC circuit is defined as follows:

$Z(\text{impedance}) = \sqrt{R^2 + X^2}$, where X is the reactance and R is the resistance.

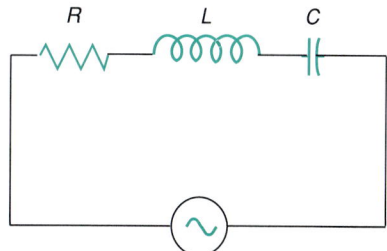

Figure 9.13

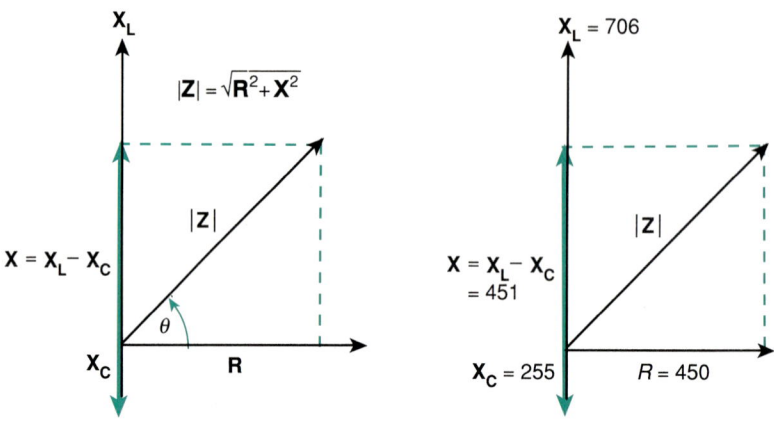

Figure 9.14   Figure 9.15

Figure 9.14 shows a vector diagram for the circuit. The phase angle is the angle between Z and R. θ is calculated using the following formula:

$$\text{Phase angle } (\theta) = \tan^{-1}\left(\frac{X}{R}\right).$$

• **Example 9.4:** An *RLC* circuit has an inductive reactance of 706 ohms, a capacitive reactance of 255 ohms and a resistance of 450 ohms. Calculate the impedance of the circuit and the phase angle in figure 9.15.

Solution:
$$X = X_L - X_C = 706 - 255 = 451 \text{ ohms}$$
$$Z = \sqrt{R^2 + X^2}$$
$$= \sqrt{(450)^2 + (451)^2} = 637 \text{ ohms}$$
$$\theta = \tan^{-1}\left(\frac{X}{R}\right) = \tan^{-1}\left(\frac{451}{450}\right) = 45°.$$

Trial Problems 9.1

Before you begin the section exercises, warm up with these problems. Complete answers are included in the answer key.

Find the horizontal and vertical components of the vectors.

1. $\theta = 48°, |V| = 8.6$ 2. $\theta = 32°, |V| = 6.0$

3. $\theta = 1.31$ radians, $|V| = 2.34$

Given two vectors, **V** and **W**, and the angle θ between the vectors, find the magnitude of the resultant vector.

4. $|V| = 110, |W| = 220, \theta = 41.3°$ 5. $|V| = 0.53, |W| = 0.76, \theta = 2.34°$

Exercises 9.1

Exercises 1–4 refer to the vectors in figure 9.16.

1. Which vectors appear to be equal?
2. Which vectors appear to be scalar multiples of other vectors?
3. Which pairs of vectors appear to be opposites, that is, one is the negative of the other?
4. Which pair of vectors appears to be of equal magnitude?

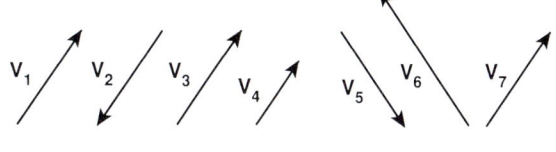

Figure 9.16

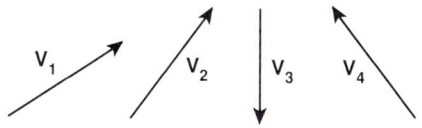

Figure 9.17

Exercises 5–18 refer to the vectors shown in figure 9.17. Make a sketch that illustrates the following vectors:

5. $-V_1$
6. $-V_2$
7. $2V_1$
8. $3V_2$
9. $-3V_1$
10. $-2V_2$
11. $V_1 + V_2$
12. $V_3 - V_4$
13. $V_1 - V_2$
14. $V_3 + V_4$
15. $V_1 + (V_2 + V_3)$
16. $V_4 + (V_2 + V_3)$
17. $2V_1 + 3V_2$
18. $3V_1 + 2V_2$

Given the magnitude $|V|$ and the angle θ, which the vector makes with the horizontal, find the horizontal and vertical components.

19. $\theta = 66°$, $|V| = 6.3$
20. $\theta = 58°$, $|V| = 7.2$
21. $\theta = 46.3°$, $|V| = 6.43$
22. $\theta = 48.5°$, $|V| = 8.42$
23. $\theta = 1.23$ rad, $|V| = 4.05$
24. $\theta = 0.87$ rad, $|V| = 6.45$

The magnitudes of vectors V and W and the angle θ between the vectors are given. Find the magnitude of the resultant vector.

25. $|V| = 250$, $|W| = 268$, $\theta = 36.5°$
26. $|V| = 270$, $|W| = 295$, $\theta = 43.2°$
27. $|V| = 3.82$, $|W| = 4.62$, $\theta = 61.5°$
28. $|V| = 1.87$, $|W| = 2.63$, $\theta = 66.5°$

29. Two forces of 30 newtons and 40 newtons act on a point. The angle between the forces is 30°. Find the magnitude of the resultant force.

30. Two forces of 30 newtons and 40 newtons act on a point. The angle between the forces is 45°. Find the magnitude of the resultant force.

31. A force of 25 newtons makes an angle of 62° with the horizontal. Determine the magnitude of the horizontal and vertical components of the force.

32. Solve exercise 31 if the angle is 58°.

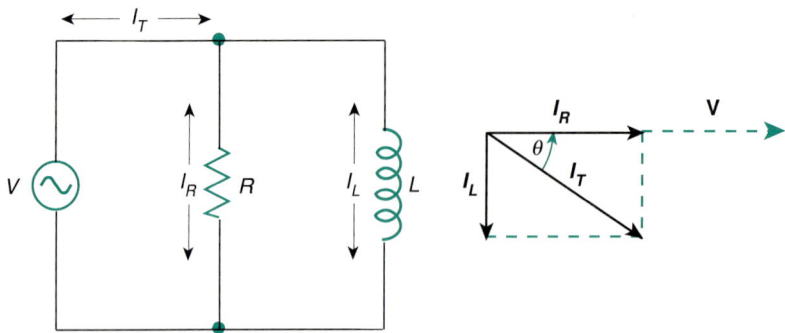

Figure 9.18

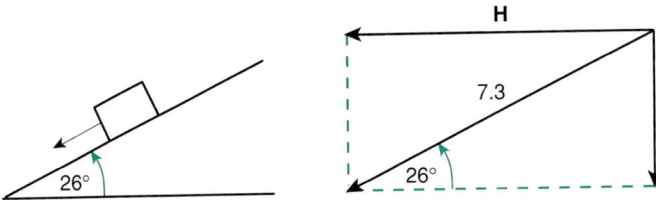

Figure 9.19

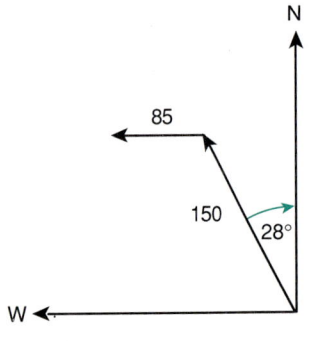

Figure 9.20

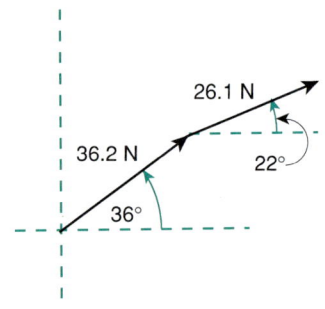

Figure 9.21

33. When a resistor and an inductor are connected in parallel across a sinusoidal voltage, the total current drawn is the vector sum of the branches of the current. See figure 9.18. Given that $I_R = 0.32$ amps and $I_L = 0.27$ amps, find I_T and the phase angle θ.

34. Solve exercise 33 given that $I_R = 0.63$ amps and $I_L = 0.55$ amps.

35. An object is sliding down a ramp that is inclined at an angle of 26° with the horizontal. If the object is sliding at 7.3 m/sec, find the horizontal and vertical components of the velocity. See figure 9.19.

36. Solve exercise 35 if the angle is changed to 28°.

37. A plane flies 150 mi in a direction that is 28° west of due north. The plane then turns due west and flies 85 mi. How far is the plane from the starting point? See figure 9.20.

38. Find the resultant of the vectors shown in figure 9.21.

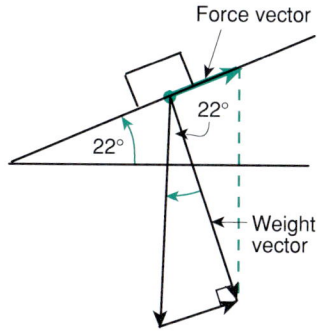

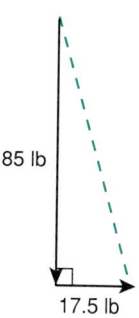

Figure 9.22   Figure 9.23

39. Find the force required to pull a 35 lb weight up a ramp that is inclined 22° with the horizontal. See figure 9.22. Note that the weight can be considered as a vector that is perpendicular to the horizontal plane.

40. An object weighing 85.0 lb is suspended by a rope. The object is being pulled by a horizontal force of 17.5 lb. Find the angle between the rope and the vertical. See figure 9.23. Note that the weight can be considered as a vector that is perpendicular to the horizontal plane.

Exercises 41–44 assume a series *RLC* circuit with an AC source.

41. A circuit has a reactance of 4.5 ohms and a resistance of 4.6 ohms. Find the magnitude of the impedance and the phase angle.

42. A circuit has a reactance of 6.5 ohms and a resistance of 4.8 ohms. Find the magnitude of the impedance and the phase angle.

43. A circuit has a capacitive reactance of 14.6 ohms, an inductive reactance of 24.3 ohms, and a resistance of 9.8 ohms. Find the magnitude of the reactance and the phase angle.

44. A circuit has a capacitive reactance of 9.8 ohms, an inductive reactance of 14.3 ohms, and a total impedance of 7.4 ohms. Find the phase angle and the resistance.

45. An object is dragged 25 ft across a floor using a force of 65 lb. If the work (W) done is calculated using $W = FD$, where F is the horizontal component of the applied force and D is the horizontal distance that the object is moved, find the work done if the force on the object is applied at an angle of 65° above the horizontal.

46. Solve exercise 45 given that an object is moved horizontally 15 ft using a force of 80 lb applied at an angle of 65° above the horizontal.

9 • 2 Operations with Complex Numbers

In section 5.4, we discussed solutions to quadratic equations, that is, equations that can be written as $ax^2 + bx + c = 0$, where a, b, and c are real constants. The quadratic formula was used to find the pair of solutions:

$$x = \frac{-b \pm \sqrt{b^2 - 4ac}}{2a}.$$

If the discriminant $b^2 - 4ac = 0$, there is one real solution. If $b^2 - 4ac > 0$, there are two real solutions. We did not consider the third case where $b^2 - 4ac < 0$. In this case, there are two solutions, neither of which is real, since the square root of a negative number is not a real number.

To handle this third case as well as a number of other problems, we introduce the idea of a complex number.

Complex Number A **complex number** is a number that can be written as $a + bj$, where a and b are real numbers and $j = \sqrt{-1}$ ($j^2 = -1$).

Any real number a can be written in complex number form as $a + 0j$. A complex number of the form $0 + bj$, where $b \neq 0$, is called an **imaginary number**. A complex number that is written in the form $a + bj$ or $a + jb$ is in **standard form**. The reason for using the latter form is that the complex number $3 + \sqrt{2}j$ can easily be mistaken for $3 + \sqrt{2j}$, so we write it as $3 + j\sqrt{2}$. Some examples follow:

Imaginary Number
Standard Form of a Complex Number

$3 + 4j$ complex, not imaginary, not real

$-2j = 0 - 2j$ complex, purely imaginary, not real

$-4 = -4 + 0j$ complex, not imaginary, real.

Equal Complex Numbers Equality for complex numbers is defined as follows.

$$a + bj = c + dj, \quad \text{if and only if} \quad a = c, b = d.$$

Thus, $\dfrac{3}{2} + \dfrac{1}{4}j = 1.5 + 0.25j$, since $\dfrac{3}{2} = 1.5$ and $\dfrac{1}{4} = 0.25$.

Addition of Complex Numbers Addition of complex numbers is defined as follows.

$$(a + bj) + (c + dj) = (a + c) + (b + d)j.$$

Thus, to add two complex numbers, add the corresponding real parts and the corresponding imaginary parts.

• *Example 9.5:*   a. $(3 + 2j) + (-1 + 4j)$
$\qquad\qquad\qquad\quad = (3 + (-1)) + (2 + 4)j$
$\qquad\qquad\qquad\quad = 2 + 6j.$

                     b. $(-7 - 6j) + (3 - 2j)$
$\qquad\qquad\qquad\quad = (-7 + 3) + ((-6) + (-2))j$
$\qquad\qquad\qquad\quad = -4 - 8j.$

Subtraction of Complex Numbers

Subtraction of complex numbers is defined in terms of addition.

$$(a + bj) - (c + dj) = (a + bj) + (-c - dj)$$
$$= (a - c) + (b - d)j.$$

Thus, to subtract two complex numbers, subtract the corresponding real parts and the corresponding imaginary parts.

• **Example 9.6:**
a. $(3 + 2j) - (5 - 7j) = (3 + 2j) + (-5 + 7j)$
$= (3 + (-5)) + (2 + 7)j = -2 + 9j.$
b. $(-4 - 6j) - (-3 - 2j) = (-4 - 6j) + (3 + 2j)$
$= -1 - 4j.$

••••••••••

The product of two complex numbers can be found by multiplying the numbers as if they were binomials and using the fact that $j^2 = -1$:

$$(a + bj)(c + dj) = a(c + dj) + bj(c + dj)$$
$$= ac + adj + bcj + bdj^2 \quad \text{(Substitute } -1 \text{ for } j^2\text{)}$$
$$= ac + adj + bcj + bd(-1)$$
$$= (ac - bd) + (ad + bc)j.$$

Multiplication of Complex Numbers

This leads to the following definition of **multiplication for complex numbers.**

$$(a + bj)(c + dj) = (ac - bd) + (ad + bc)j.$$

Rather than use the definition directly, it is usually more convenient to multiply complex numbers as we did with binomial expressions.

• **Example 9.7:**
a. $(2 + 3j)(4 + j) = 2(4 + j) + 3j(4 + j)$
$= 8 + 2j + 12j + 3j^2$
$= 8 + 2j + 12j + 3(-1)$
$= (8 - 3) + (2 + 12)j$
$= 5 + 14j.$
b. $(2 - 3j)(-4 + 2j) = 2(-4 + 2j) - 3j(-4 + 2j)$
$= -8 + 4j + 12j - 6j^2$
$= -8 + 4j + 12j - 6(-1)$
$= (-8 + 6) + (4 + 12)j$
$= -2 + 16j.$

••••••••••

Note the pattern for determining the various powers of j.

$j^1 = j$ $j^2 = -1$ $j^3 = (j)(j^2) = -j$ $j^4 = j^2 j^2 = 1$
$j^5 = j^4 j = j$ $j^6 = j^4 j^2 = -1$ $j^7 = j^4 j^3 = -j$ $j^8 = j^4 j^4 = 1$
$j^9 = j^8 j = j$ $j^{10} = j^8 j^2 = -1$ $j^{11} = j^8 j^3 = -j$ $j^{12} = j^8 j^4 = 1$
⋮   ⋮   ⋮   ⋮
$j^{4k+1} = j$ $j^{4k+2} = -1$ $j^{4k+3} = -j$ $j^{4k} = 1.$

This pattern allows us to calculate any power of j, that is,

$$j^n \in \{j, -1, -j, 1\}, \quad \text{for any positive integer } n.$$

- **Example 9.8:** Calculate (a) j^{33} and (b) j^{1602}.

  Solution: Divide the exponents by 4 and write as $4k + m$, where m is the remainder:

  a. $33 = 4(8) + 1$
  $j^{33} = j^{4(8)+1} = (j^4)^8 j^1 = (1)^8 j = j.$

  b. $1602 = 4(400) + 2$
  $j^{1602} = j^{4(400)+2} = (1)^{400} j^2 = j^2 = -1.$

  ....

Before defining division of complex numbers, the idea of the conjugate of a complex number is needed. The **conjugate** of the complex number $a + bj$ is the complex number $a - bj$. The conjugates of several complex numbers are listed below.

Number	Conjugate
$3 + 2j$	$3 - 2j$
$6 - 4j$	$6 + 4j$
$-3 + 2j$	$-3 - 2j$
$4j$	$-4j$
6	6

Theorem 9.1 $(a + bj)(a - bj) = a^2 + b^2.$

Proof: $(a + bj)(a - bj) = a(a - bj) + bj(a - bj)$
$= a^2 - abj + abj - b^2 j^2$
$= a^2 - b^2(-1)$
$= a^2 + b^2.$

Note that this theorem ensures that the product of a complex number and its conjugate is always a real number. We use the idea of conjugates to divide complex numbers.

DIVISION OF COMPLEX NUMBERS

To divide a pair of complex numbers, write the quotient as a fraction and multiply the numerator and denominator of the fraction by the conjugate of the denominator.

• *Example 9.9:* Divide $3 + 2j$ by $1 + 2j$.

Solution: Write the quotient as a fraction:

$$\frac{3 + 2j}{1 + 2j}$$

Multiply the numerator and denominator by the conjugate of $1 + 2j$, which is $1 - 2j$:

$$\frac{(3 + 2j)(1 - 2j)}{(1 + 2j)(1 - 2j)} = \frac{3 - 6j + 2j - 4j^2}{1 + 2j - 2j - 4j^2}$$

$$= \frac{3 - 4j - 4(-1)}{1 - 4(-1)}$$

$$= \frac{7 - 4j}{5}$$

$$= \frac{7}{5} - \frac{4}{5}j.$$

• *Example 9.10:* Simplify $\left(\frac{4 - 3j}{2 + 5j}\right)\left(\frac{2 + j}{1 - j}\right)$.

Solution: First we multiply the numerators and denominators to perform the multiplication:

$$\left(\frac{4 - 3j}{2 + 5j}\right)\left(\frac{2 + j}{1 - j}\right) = \frac{(4 - 3j)(2 + j)}{(2 + 5j)(1 - j)}$$

$$= \frac{8 + 4j - 6j - 3j^2}{2 - 2j + 5j - 5j^2}$$

$$= \frac{8 - 2j + 3}{2 + 3j + 5}$$

$$= \frac{11 - 2j}{7 + 3j}.$$

Now we multiply by the conjugate of $7 + 3j$ to perform the division:

$$\left(\frac{4 - 3j}{2 + 5j}\right)\left(\frac{2 + j}{1 - j}\right) = \frac{(11 - 2j)(7 - 3j)}{(7 + 3j)(7 - 3j)}$$

$$= \frac{77 - 33j - 14j + 6j^2}{49 - 21j + 21j - 9j^2}$$

$$= \frac{71 - 47j}{58}$$

$$= \frac{71}{58} - \frac{47}{58}j.$$

∙∙∙∙∙∙∙∙∙∙

When we solve equations, we can replace $\sqrt{-a}$ $(a > 0)$ with

$$\sqrt{a}\sqrt{-1} = \sqrt{a}\,j = j\sqrt{a}$$

since $(\sqrt{-1})(\sqrt{-1}) = -1 = j^2$. Thus,

$$\sqrt{-80} = \sqrt{80}\sqrt{-1} = \sqrt{16}\sqrt{5}\sqrt{-1} = 4\sqrt{5}j.$$

One of the applications of complex numbers occurs in the use of phasor diagrams to represent electrical quantities. For example, in a series circuit containing a resistor and a coil in series, the impedance can be represented as $a + bj$, where a is the resistance and b is the reactance of the coil. Thus, in a series *RL* circuit containing a 10-ohm resistor and a coil with a reactance of 30 ohms, the impedance Z is represented as $Z = 10 + 30j$.

• *Example 9.11:* If two impedances Z_1 and Z_2 are connected in parallel, their total impedance is

$$Z_t = \frac{Z_1 Z_2}{Z_1 + Z_2}.$$

Given two impedances $Z_1 = 3 + 6j$ and $Z_2 = 4 + 5j$ in parallel, find the total impedance.

Solution:
$$Z_t = \frac{Z_1 Z_2}{Z_1 + Z_2} = \frac{(3 + 6j)(4 + 5j)}{(3 + 6j) + (4 + 5j)}$$

$$= \frac{12 + 39j + 30j^2}{7 + 11j}$$

$$= \frac{-18 + 39j}{7 + 11j} = \frac{(-18 + 39j)(7 - 11j)}{(7 + 11j)(7 - 11j)}$$

$$= \frac{303 + 471j}{170} = 1.78 + 2.77j.$$

∙∙∙∙∙∙∙∙∙∙

We now return to the problem of solving quadratic equations that have complex solutions.

- **Example 9.12:** Solve $2x^2 + x + 1 = 0$.

  Solution: Use the quadratic formula

  $$x = \frac{-b \pm \sqrt{b^2 - 4ac}}{2a},$$

  where $a = 2$, $b = 1$, and $c = 1$.

  $$\begin{aligned}x &= \frac{-(1) \pm \sqrt{(1)^2 - 4(2)(1)}}{2(2)} \\ &= \frac{-1 \pm \sqrt{1 - 8}}{4} \\ &= \frac{-1 \pm \sqrt{-7}}{4} \\ &= \frac{-1 \pm \sqrt{-1}\sqrt{7}}{4} \\ &= \frac{-1 \pm j\sqrt{7}}{4} \\ &= -\frac{1}{4} \pm j\frac{\sqrt{7}}{4}.\end{aligned}$$

  The solutions are $-\frac{1}{4} + j\frac{\sqrt{7}}{4}$ and $-\frac{1}{4} - j\frac{\sqrt{7}}{4}$.

Trial Problems 9.2

Before you begin the section exercises, warm up with these problems. Complete answers are included in the answer key.

Perform the indicated operations.

1. $(2 + 3j) - (5 - 5j)$
2. $(2 + j)(3 - j)$
3. $\dfrac{2 + j}{3 - j}$
4. Solve the equation $x^2 + 2x + 6 = 0$.
5. Calculate j^{15}.

Exercises 9.2

Perform the indicated operations. Write the answer in the form $a + bj$.

1. $(7 - 3j) + (2 - 5j)$
2. $(7 + 3j) + (2 + 5j)$
3. $(4 + 3j) + (-1 + 7j)$
4. $(4 - 3j) + (1 - 7j)$
5. $(7 - 3j) - (2 + 5j)$
6. $(2 - j) - (4 - 2j)$
7. $(2 - j) - (4 - j)$
8. $(4 - j) - (4 + j)$
9. $(4 + 3j)(3 - 2j)$
10. $(4 - 3j)(3 + 2j)$
11. $(3 - 2j)(3 - 2j)$
12. $(3 + 2j)(6 - j)$
13. $(4 + j)(4 - j)$
14. $(7 + j)(7 - j)$
15. $(2 - 3j)(2 + 3j)$
16. $(4 + 2j)(4 - 2j)$
17. $(6 - 11j)(2 + 4j)$
18. $(11 - 6j)(4 - 2j)$
19. $\dfrac{1}{1 + j}$
20. $\dfrac{2}{2 - j}$
21. $\dfrac{3}{2 + 3j}$
22. $\dfrac{3}{2 - 3j}$
23. $\dfrac{3 + 2j}{3 - 2j}$
24. $\dfrac{3 - 2j}{3 + 2j}$
25. $\dfrac{j}{1 + j}$
26. $\dfrac{j}{1 - j}$
27. $\dfrac{1 + j}{1 - j}$
28. $\dfrac{1 - j}{1 + j}$
29. $\dfrac{j}{1 + j} - \dfrac{j}{1 - j}$
30. $\dfrac{1 + j}{1 - j} + \dfrac{1 - j}{1 + j}$

Simplify each expression using the properties of j.

31. $\sqrt{-4}$
32. $\sqrt{-9}$

Note in exercises 33–42 that for a and $b < 0$, $\sqrt{ab} \neq \sqrt{a}\sqrt{b}$.

33. $\sqrt{-3}\sqrt{-4}$
34. $\sqrt{-6}\sqrt{-12}$
35. $\dfrac{\sqrt{30}}{\sqrt{-5}}$
36. $\dfrac{\sqrt{50}}{\sqrt{-5}}$
37. j^6
38. j^8
39. j^{-2}
40. j^{-4}
41. j^{78}
42. j^{93}

Solve the quadratic equations.

43. $x^2 + x + 1 = 0$
44. $x^2 - x + 1 = 0$
45. $x^2 - 2x + 6 = 0$
46. $x^2 - 2x + 4 = 0$
47. $5x^2 + 6x + 4 = 0$
48. $5x^2 - 6x + 4 = 0$
49. $7x^2 + 6x + 3 = 0$
50. $7x^2 - 6x + 3 = 0$

51. Evaluate the polynomial $x^3 + x^2 - x + 1$, when $x = 1 + j$.

52. Evaluate the polynomial $x^3 + x^2 - x + 1$, when $x = 1 - j$.

53. a. Compute $(1 + j)^2$.
    b. Compute $(1 + j)^{18}$. (Use the result from part a.)

54. a. Compute $(1 - j)^2$.
    b. Compute $(1 - j)^{16}$. (Use the result of part a.)

In an AC circuit, complex numbers are used to describe current *I*, impedance *Z*, and voltage *E*. If $E = IZ$, find the missing quantities.

55. $Z = 6 - 3j, I = 2 - j$
56. $Z = 4 - j, I = 2 - 2j$
57. $I = 7.2 + 5.1j, E = 22.4 + 41.2j$
58. $I = 8.6 + 4.7j, E = 18.4 + 36.2j$
59. $Z = 6.4 + 4.1j, E = 32.2 + 48.6j$
60. $Z = 8.1 + 3.4j, E = 32.2 + 48.6j$
61. Two impedances $Z_1 = 20 + 8j$ and $Z_2 = 8 + 10j$ are connected in parallel. Find the total impedance Z_t. (See example 9.11.)
62. Two impedances $Z_1 = 8 + 10j$ and $Z_2 = 4 + j$ are connected in parallel. Find the total impedance Z_t. (See example 9.11.)

9•3 Geometric Representation of Complex Numbers

In this section we represent complex numbers as vectors in a plane. In order to graph a complex number, we must change the coordinate system. The standard way of doing this is to call the horizontal axis the **real axis** and the vertical axis the **imaginary axis** or the *j-axis*. See figure 9.24.

Real Axis
Imaginary Axis
j-axis

On this coordinate system we represent the complex number $a + bj$ as a position vector with terminal point (a,b). Thus, $-2 + 4j$ would be represented by a vector **V**, which has its initial point at the origin and its terminal point at a point $P(-2,4)$.

We know from section 9.2 that the sum of two complex numbers $a + bj$ and $c + dj$ is the complex number $(a + c) + (b + d)j$.

Graphically, the sum is represented by the position vector that is the resultant of the vectors that represent $a + bj$ and $c + dj$. This sum is illustrated in figure 9.25.

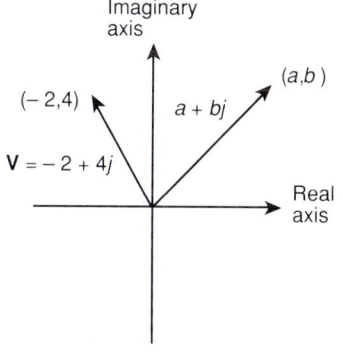

Figure 9.24

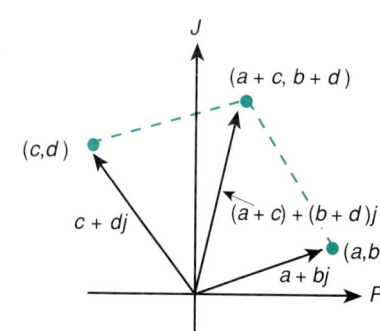

Figure 9.25

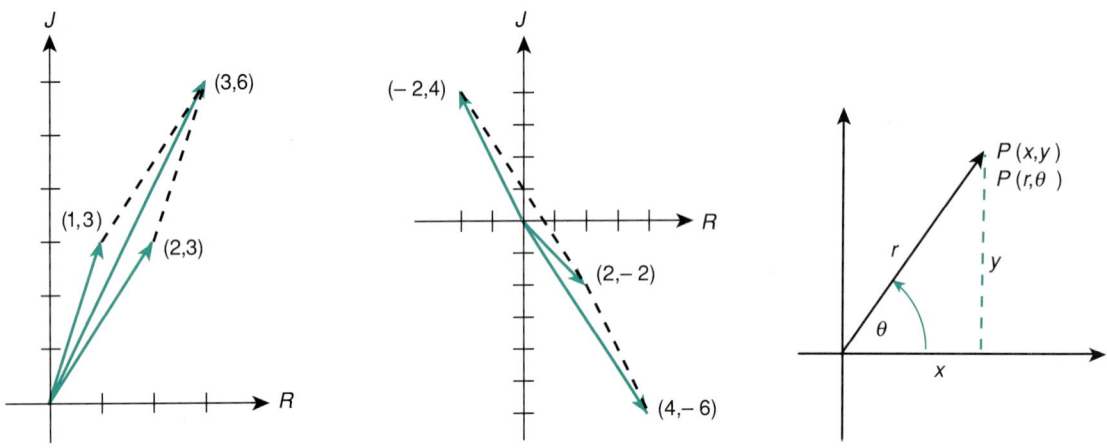

Figure 9.26   Figure 9.27

For example, the sums

$$(1 + 3j) + (2 + 3j) = 3 + 6j$$

and

$$(-2 + 4j) + (4 - 6j) = 2 - 2j$$

are shown in figure 9.26.

Figure 9.27 shows a complex number $x + yj$, which determines a vector **V** in the complex plane. Let θ be the angle that the vector makes with the positive x-axis, measured in a counterclockwise direction, and let $r = $ the length of the vector. If the pair (r,θ) is used to represent the complex number, the pair (r,θ) is called the **polar coordinates** of the point $P(x,y)$.

Polar Coordinates

Using right triangle OAP, the following relationships can be obtained.

$$r = \sqrt{a^2 + b^2} \quad \text{(Pythagorean theorem)}$$

$$\tan \theta = \frac{y}{x} \quad \text{(Definition of the tangent)}$$

$$\sin \theta = \frac{y}{r} \quad \text{(Definition of the sine)}$$

$$\cos \theta = \frac{x}{r} \quad \text{(Definition of the cosine)}$$

These relationships produce the following formulas for converting the coordinates of points between the two types of coordinate systems.

CONVERTING COORDINATES

Rectangular to Polar Coordinates:

$$(x,y) \to (r,\theta)$$
$$r = \sqrt{x^2 + y^2}, \tan \theta = \frac{y}{x}.$$

Polar to Rectangular Coordinates:

$$(r,\theta) \to (x,y)$$
$$x = r \cos \theta, \; y = r \sin \theta.$$

Similar arguments can be made for points in the other quadrants.

- **Example 9.13:** Find the polar coordinates of the complex number $3 - 3j$ (see figure 9.28).

    Solution: $r = \sqrt{a^2 + b^2} = \sqrt{3^2 + (-3)^2} = \sqrt{18} = 3\sqrt{2}$

    $\theta = \tan^{-1}\left(\frac{-3}{3}\right) = \tan^{-1}(-1).$

    Since $P(3,-3)$ is in the fourth quadrant,

    $$\theta = \frac{7\pi}{4}.$$

    Thus, $\left(3\sqrt{2}, \frac{7\pi}{4}\right)$ are the polar coordinates of $3 - 3j$.

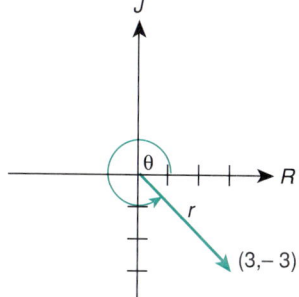

Figure 9.28

- **Example 9.14:** Write $2\left(\cos \frac{\pi}{6} + j \sin \frac{\pi}{6}\right)$ in the form $a + bj$.

    Solution: Since $\cos \frac{\pi}{6} = \frac{\sqrt{3}}{2}$ and $\sin \frac{\pi}{6} = \frac{1}{2}$,

    $$2\left(\cos \frac{\pi}{6} + j \sin \frac{\pi}{6}\right) = 2\left(\frac{\sqrt{3}}{2} + \frac{1}{2}j\right) = \sqrt{3} + j.$$

Polar Form
Modulus
Argument

In the expression $r(\cos\theta + j\sin\theta)$, the **trigonometric** or **polar** form of a complex number, we call r the **modulus** or **absolute value** of $a + bj$ and θ the **argument** of $a + bj$.

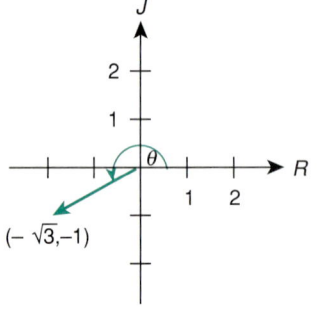

Figure 9.29

• *Example 9.15:* Determine the trigonometric form of the complex number $-\sqrt{3} - j$. Refer to figure 9.29.

Solution: Since $a = -\sqrt{3}$, $b = -1$, and $r = \sqrt{a^2 + b^2}$
$= \sqrt{(-\sqrt{3})^2 + (-1)^2} = \sqrt{4} = 2$,

$$\tan\theta = \frac{b}{a} = \frac{-1}{-\sqrt{3}} = \frac{\sqrt{3}}{3}.$$

Since $(a,b) = (-\sqrt{3},-1)$ is in quadrant III, $\theta = \frac{7\pi}{6}$. Thus,

$$-\sqrt{3} - j = 2\left(\cos\frac{7\pi}{6} - j\sin\frac{7\pi}{6}\right).$$

••••••••••

We can use the polar form of a pair of complex numbers to determine the product and quotient of the numbers. The following theorem shows the procedure:

Theorem 9.2: If $u = r(\cos\alpha + j\sin\alpha)$ and $v = s(\cos\beta + j\sin\beta)$ are complex numbers, then

a. $uv = rs[\cos(\alpha + \beta) + j\sin(\alpha + \beta)]$

b. $\dfrac{u}{v} = \dfrac{r}{s}[\cos(\alpha - \beta) + j\cos(\alpha - \beta)].$

Proof of a: $uv = r(\cos\alpha + j\sin\alpha)s(\cos\beta + j\sin\beta)$
$= rs[(\cos\alpha\cos\beta + j^2\sin\alpha\sin\beta)$
$\quad + j(\sin\alpha\cos\beta + \cos\alpha\sin\beta)]$
$= rs[(\cos\alpha\cos\beta - \sin\alpha\sin\beta)$
$\quad + j(\sin\alpha\cos\beta + \cos\alpha\sin\beta)].$

From sections 8.2 and 8.4 we know that

$$\cos(\alpha + \beta) = \cos\alpha\cos\beta - \sin\alpha\sin\beta$$

and

$$\sin(\alpha + \beta) = \sin\alpha\cos\beta + \cos\alpha\sin\beta.$$

We substitute these expressions into the formula for uv:

$$uv = rs[\cos(\alpha + \beta) + j\sin(\alpha + \beta)].$$

The proof of b is similar to this proof. It is left as an exercise for the student. The results of theorem 9.2 can be stated in the following form.

SECTION 9.3 GEOMETRIC REPRESENTATION OF COMPLEX NUMBERS

> The product of two complex numbers in polar form can be found by multiplying the moduli and adding the arguments of the numbers. The quotient is found by dividing the moduli and subtracting the arguments.

• **Example 9.16:** Given $u = 2 + 2j$, $v = \sqrt{3} + j$.

  a. Use theorem 9.2a to calculate uv.

  b. Use theorem 9.2b to calculate $\dfrac{u}{v}$.

Solution: 1. Change the numbers to polar form:

$$u = 2 + 2j,$$
$$r = \sqrt{2^2 + 2^2} = \sqrt{8} = 2\sqrt{2},$$
$$\alpha = \tan^{-1}\left(\frac{2}{2}\right) = \tan^{-1} 1 = \frac{\pi}{4}$$
$$u = 2\sqrt{2}\left(\cos\frac{\pi}{4} + j\sin\frac{\pi}{4}\right)$$
$$v = \sqrt{3} + j, \quad s = \sqrt{(\sqrt{3})^2 + 1^2} = 2,$$
$$\beta = \tan^{-1}\left(\frac{1}{\sqrt{3}}\right) = \frac{\pi}{6}$$
$$v = 2\left(\cos\frac{\pi}{6} + j\sin\frac{\pi}{6}\right).$$

2. Multiply by multiplying the moduli and adding the arguments:

$$uv = \left(2\sqrt{2}\left(\cos\frac{\pi}{4} + j\sin\frac{\pi}{4}\right)\right)\left(2\left(\cos\frac{\pi}{6} + j\sin\frac{\pi}{6}\right)\right)$$
$$= 4\sqrt{2}\left(\cos\left(\frac{\pi}{4} + \frac{\pi}{6}\right) + j\sin\left(\frac{\pi}{4} + \frac{\pi}{6}\right)\right)$$
$$= 4\sqrt{2}\left(\cos\frac{5\pi}{12} + j\sin\frac{5\pi}{12}\right).$$

3. Divide u and v by dividing the moduli and subtracting the arguments of u and v:

$$\frac{u}{v} = \frac{2\sqrt{2}}{2}\left(\cos\left(\frac{\pi}{4} - \frac{\pi}{6}\right) + j\sin\left(\frac{\pi}{4} - \frac{\pi}{6}\right)\right)$$
$$= \sqrt{2}\left(\cos\frac{\pi}{12} + j\sin\frac{\pi}{12}\right).$$

The following alternate notation is often used to represent complex numbers in polar form.

NOTATION

$r \operatorname{cjs} \theta$ is equivalent to $r(\cos \theta + j \sin \theta)$.

Theorem 9.2 written using the alternate notation is as follows.

Theorem 9.2 (alternate form):
If $u = r \operatorname{cjs} \alpha$ and $v = s \operatorname{cjs} \beta$, then

a. $uv = rs \operatorname{cjs}(\alpha + \beta)$

b. $\dfrac{u}{v} = \dfrac{r}{s} \operatorname{cjs}(\alpha - \beta)$.

Examples 9.17 and 9.18 illustrate the advantages of using the alternate notation for complex numbers in polar form.

• **Example 9.17:** Calculate $\dfrac{(1 + j)(-2 + 2j)}{(-\sqrt{3} + j)}$ using the polar form of the complex numbers. Write the answer in the form $a + bj$.

Solution: 1. Change the numbers to polar form:

$$1 + j = \sqrt{2}(\cos 45° + j \sin 45°) = \sqrt{2} \operatorname{cjs} 45°$$
$$-2 + 2j = 2\sqrt{2}(\cos 135° + j \sin 135°) = 2\sqrt{2} \operatorname{cjs} 135°$$
$$-\sqrt{3} + j = 2(\cos 150° + j \sin 150°) = 2 \operatorname{cjs} 150°.$$

2. Perform the indicated operations:

$$\dfrac{(1 + j)(-2 + 2j)}{-\sqrt{3} + j} = \dfrac{(\sqrt{2} \operatorname{cjs} 45°)(2\sqrt{2} \operatorname{cjs} 135°)}{2 \operatorname{cjs} 150°}$$

$$= \dfrac{4 \operatorname{cjs}(45° + 135°)}{2 \operatorname{cjs} 150°} \quad \text{(Multiply the moduli and add the arguments in the numerator)}$$

$$= \dfrac{4 \operatorname{cjs} 180°}{2 \operatorname{cjs} 150°}$$

$$= 2 \operatorname{cjs}(180° - 150°) \quad \text{(Divide the moduli and subtract the arguments)}$$

$$= 2 \operatorname{cjs} 30°$$
$$= 2(\cos 30° + j \sin 30°)$$
$$= 2\left(\dfrac{\sqrt{3}}{2} + \dfrac{1}{2}\right)$$
$$= \sqrt{3} + 1.$$

SECTION 9.3 GEOMETRIC REPRESENTATION OF COMPLEX NUMBERS

• **Example 9.18:** Calculate $\dfrac{(3+j)(-3+2j)}{(1-2j)(4-j)}$ using theorem 9.2. Write the answer in the form $a+bi$.

Solution:

1. Change the numbers to polar form. Use radians to express the numbers in polar form (an arbitrary choice):

$$3 + j = \sqrt{10}\ cjs(0.3218)$$
$$-3 + 2j = \sqrt{13}\ cjs(2.5536)$$
$$1 - 2j = \sqrt{5}\ cjs(-1.1071)$$
$$4 - j = \sqrt{17}\ cjs(-0.2450).$$

2. Use theorem 9.2 to perform the calculations. Multiply and divide to find the modulus. Add and subtract to find the argument.

$$\frac{(3+j)(-3+2j)}{(1-2j)(4-j)}$$

$$= \frac{(\sqrt{10}\ cjs(0.3218))(\sqrt{13}\ cjs(2.5536))}{(\sqrt{5}\ cjs(-1.1071))(\sqrt{17}\ cjs(-0.2450))}$$

$$= \frac{\sqrt{10}\sqrt{13}}{\sqrt{5}\sqrt{17}}\ cjs(0.3218 + 2.5536$$
$$\quad - (-1.1071) - (-0.2450))$$

$$= \sqrt{\frac{26}{17}}\ cjs\ 4.2275 \qquad \text{(Simplify the expression)}$$

$$= \sqrt{\frac{26}{17}}(\cos 4.2275 + j\sin 4.2275) \qquad \text{(Write in trigonometric form)}$$

$$= -0.5764 - 1.0941j. \qquad \text{(Write in nontrigonometric form)}$$

• • • • • • • • •

In electrical problems the polar form of a complex number is often written in the following form:

$$M\underline{/\theta},$$

where M is the magnitude and θ is the phase angle. Suppose we have two phasor quantities, $M\underline{/\alpha}$ and $N\underline{/\beta}$, then,

a. $(M\underline{/\alpha})(N\underline{/\beta}) = MN\underline{/\alpha + \beta}$.

b. $\dfrac{M\underline{/\alpha}}{N\underline{/\beta}} = \dfrac{M}{N}\underline{/\alpha - \beta}$.

• **Example 9.19:** Given two phasor quantities $A = 3.2 \underline{/82°}$, $B = 4.3 \underline{/-20°}$, calculate AB and $\dfrac{A}{B}$.

Solution:
$$AB = (3.2\underline{/82°})(4.3\underline{/-20°})$$
$$= (3.2)(4.3)\underline{/82° + (-20°)} \quad \text{(Multiply the moduli and add the arguments)}$$
$$= 13.8\underline{/62°}$$

$$\frac{A}{B} = \frac{3.2\underline{/82°}}{4.3\underline{/-20°}} \quad \text{(Divide the moduli and subtract the arguments)}$$
$$= \frac{3.2}{4.3}\underline{/82° - (-20°)}$$
$$= 0.74\underline{/102°}.$$

Trial Problems 9.3

Before you begin the section exercises, warm up with these problems. Complete answers are included in the answer key.

1. Write the complex number $\sqrt{3} - j$ in polar form.

2. Write the complex number $2\left(\cos\dfrac{\pi}{2} + j\sin\dfrac{\pi}{2}\right)$ in rectangular form.

Given $u = 2\left(\cos\dfrac{\pi}{3} + j\sin\dfrac{\pi}{3}\right)$ and $v = 4\left(\cos\dfrac{\pi}{6} + j\sin\dfrac{\pi}{6}\right)$,

3. Find uv.   4. Find $\dfrac{u}{v}$.   5. Find u^2.

Exercises 9.3

Find the polar coordinates of each complex number and plot as a point in the complex plane.

1. $\dfrac{\sqrt{3}}{2} + \dfrac{1}{2}j$
2. $\dfrac{\sqrt{3}}{2} - \dfrac{1}{2}j$
3. $2 - j$
4. $-2 + j$
5. $2 + 2j$
6. $-2 + 2j$
7. $2 + 4j$
8. $3 + 6j$
9. $\sqrt{3} + j$
10. $1 - \sqrt{3}j$

Write each complex number in the form $a + bj$.

11. $\cos\dfrac{\pi}{4} + j\sin\dfrac{\pi}{4}$
12. $\cos\dfrac{3\pi}{4} - j\sin\dfrac{3\pi}{4}$
13. $2(\cos\pi + j\sin\pi)$
14. $3(\cos\pi - j\sin\pi)$
15. $7(\cos 120° + j\sin 120°)$
16. $5(\cos 150° + j\sin 150°)$
17. $\sqrt{2}\left(\cos\dfrac{5\pi}{4} - j\sin\dfrac{5\pi}{4}\right)$
18. $\sqrt{2}\left(\cos\dfrac{7\pi}{4} + j\sin\dfrac{7\pi}{4}\right)$
19. $\sqrt{10}(\cos(-1.249) + j\sin(-1.249))$
20. $\sqrt[3]{5}(\cos 1.785 + j\sin(1.785))$

Determine the product, in polar form, of each pair of complex numbers.

21. $u = 6\left(\cos \dfrac{\pi}{3} + j \sin \dfrac{\pi}{3}\right)$, $v = 4\left(\cos \dfrac{\pi}{4} + j \sin \dfrac{\pi}{4}\right)$ 22. $u = 3\left(\cos \dfrac{\pi}{4} + j \sin \dfrac{\pi}{4}\right)$, $v = 4\left(\cos \dfrac{\pi}{3} + j \sin \dfrac{\pi}{3}\right)$

23. $u = 0.5\ cjs\ 20°$, $v = 0.3\ cjs\ 40°$ 24. $u = 0.4\ cjs\ 25°$, $v = 0.6\ cjs\ 35°$

25. $u = 0.36\ cjs\ 1.23$, $v = 0.24\ cjs\ 0.85$ 26. $u = 0.72\ cjs\ 0.46$, $v = 0.28\ cjs\ 1.72$

Determine the quotient, in polar form, for each pair of complex numbers.

27. $u = 6\left(\cos \dfrac{\pi}{3} + j \sin \dfrac{\pi}{3}\right)$, $v = 4\left(\cos \dfrac{\pi}{4} + j \sin \dfrac{\pi}{4}\right)$ 28. $u = 3\left(\cos \dfrac{\pi}{4} + j \sin \dfrac{\pi}{4}\right)$, $v = 2\left(\cos \dfrac{\pi}{3} + j \sin \dfrac{\pi}{3}\right)$

29. $u = 0.6\ cjs\ 68°$, $v = 0.3\ cjs\ 38°$ 30. $u = 0.8\ cjs\ 65°$, $v = 0.4\ cjs\ 35°$

31. $u = 0.82\ cjs\ 1.78$, $v = 0.46\ cjs\ 1.28$ 32. $u = 0.96\ cjs\ 2.65$, $v = 0.48\ cjs\ 1.15$

Perform the following operations both algebraically and by first changing the numbers to trigonometric form.

33. $(1 + j)(\sqrt{3} + j)$ 34. $(1 - j)(\sqrt{3} + j)$

35. $\dfrac{1 + j}{\sqrt{3} - j}$ 36. $\dfrac{\sqrt{3} + j}{1 + j}$

37. $(1 + j)(1 + j\sqrt{3})(-1 + j)\left(\dfrac{1}{2} + \dfrac{\sqrt{3}}{2}j\right)$ 38. $(1 - j)(1 - j\sqrt{3})(-1 + j)\left(\dfrac{1}{2} - \dfrac{\sqrt{3}}{2}j\right)$

39. $\dfrac{(\sqrt{3} + j)(-1 - j)}{(1 + j\sqrt{3})(2\sqrt{3} - 2j)}$ 40. $\dfrac{(1 + j\sqrt{3})(2\sqrt{3} - 2j)}{(\sqrt{3} + j)(-1 - j)}$

41. Given $I = 6.4\underline{/-40°}$ amps and $X_L = 4.2\underline{/90°}$ ohms, calculate $V_L = IX_L$.

42. Given $I = 2.5\underline{/-40°}$ amps and $Z = 8.0\underline{/45°}$ ohms, calculate $E = IZ$.

43. Given $I = 0.88\underline{/68°}$ amps and $R = 40\underline{/0°}$ ohms, calculate $V_R = IR$.

44. Given $I = 0.88\underline{/68°}$ amps and $X_C = 95\underline{/-90°}$ ohms, calculate $V_C = IX_C$.

45. Given $V = 35\underline{/0°}$ volts and $I = 5.2\underline{/-40°}$ amps, calculate $Z = \dfrac{V}{I}$.

46. Given $V = 110\underline{/0°}$ volts and $Z = 140\underline{/-47.5°}$ ohms, calculate $I = \dfrac{V}{Z}$.

9•4 Powers and Roots of Complex Numbers

The results of section 9.3 can be used to determine the square of a complex number in polar form. Thus,

$$[r(\cos \theta + j \sin \theta)]^2 = r(\cos \theta + j \sin \theta)\, r(\cos \theta + j \sin \theta)$$
$$= r^2[\cos(\theta + \theta) + j \sin(\theta + \theta)]$$
$$= r^2(\cos 2\theta + j \sin 2\theta).$$

Similarly, we can show that

$$[r(\cos\theta + j\sin\theta)]^3 = r^3(\cos 3\theta + j\sin 3\theta)$$
$$[r(\cos\theta + j\sin\theta)]^4 = r^4(\cos 4\theta + j\sin 4\theta)$$
$$\vdots$$
$$[r(\cos\theta + j\sin\theta)]^n = r^n(\cos n\theta + j\sin n\theta).$$

De Moivre's Theorem This result is known as **De Moivre's theorem.** We can use this theorem to find powers of complex numbers that are in polar form.

• **Example 9.20:** Calculate $(1 + j)^{20}$.

Solution: We first convert $1 + j$ to polar form:

$$1 + j = \sqrt{2}\left(\cos\frac{\pi}{4} + j\sin\frac{\pi}{4}\right).$$

We now apply De Moivre's theorem:

$$(1 + j)^{20} = \left[\sqrt{2}\left(\cos\frac{\pi}{4} + j\sin\frac{\pi}{4}\right)\right]^{20}$$
$$= (\sqrt{2})^{20}\cos\left(20 \cdot \frac{\pi}{4}\right) + j\sin\left(20 \cdot \frac{\pi}{4}\right)$$
$$= 1024\,(\cos 5\pi + j\sin 5\pi)$$
$$= 1024\,(\cos \pi + j\sin \pi)$$
$$= 1024\,(-1 - 0j) = -1024 + 0j.$$

• • • • • • • • • •

• **Example 9.21:** Calculate $(\sqrt{3} + j)^8$.

Solution:
$$\sqrt{3} + j = 2\left(\cos\frac{\pi}{6} + j\sin\frac{\pi}{6}\right)$$
$$(\sqrt{3} + j)^8 = 2^8\left[\cos\left(8 \cdot \frac{\pi}{6}\right) + j\sin\left(8 \cdot \frac{\pi}{6}\right)\right]$$
$$= 256\left(\cos\frac{4\pi}{3} + j\sin\frac{4\pi}{3}\right)$$
$$= 256\left(-\frac{1}{2} - \frac{\sqrt{3}}{2}j\right)$$
$$= -128 - 128\sqrt{3}\,j.$$

• • • • • • • • • •

Not all problems work out as conveniently as examples 9.20 and 9.21. Example 9.22 illustrates this.

SECTION 9.4 POWERS AND ROOTS OF COMPLEX NUMBERS

• **Example 9.22:** Calculate $(1 - 3j)^{10}$.

Solution:
$$1 - 3j = \sqrt{10}[\cos(-1.2490) + j\sin(-1.2490)]$$
$$(1 - 3j)^{10} = (\sqrt{10})^{10}[\cos(10)(-1.2490) + j\sin(10)(-1.2490)]$$
$$= 100{,}000[\cos(-12.490) + j\sin(-12.490)]$$
$$= 100{,}000[0.997085 + j(0.0762964)]$$
$$= 99708.5 + 7629.64j.$$

⋯⋯⋯⋯⋯

Theorem 9.3 illustrates the form of the roots of a complex number.

Theorem 9.3:
$$\left[\sqrt[n]{r}\left(\cos\frac{\theta + 2\pi k}{n} + j\sin\frac{\theta + 2\pi k}{n}\right)\right]^n = r(\cos\theta + j\sin\theta), \text{ for}$$
k a nonnegative integer.

Proof: Use De Moivre's theorem on the left-hand side of the equation.

$$\left[\sqrt[n]{r}\left(\cos\frac{\theta + 2\pi k}{n} + j\sin\frac{\theta + 2\pi k}{n}\right)\right]^n$$
$$= (\sqrt[n]{r})^n\left(\cos n\left(\frac{\theta + 2\pi k}{n}\right) + j\cos n\left(\frac{\theta + 2\pi k}{n}\right)\right)$$
$$= r(\cos(\theta + 2\pi k) + j\sin(\theta + 2\pi k))$$
$$= r(\cos\theta + j\sin\theta)$$

$(\cos(\theta + 2\pi k) = \cos\theta$ and $\sin(\theta + 2\pi k) = \sin\theta)$.

Theorem 9.3 yields the following rule for finding the roots of a complex number.

FINDING THE nth ROOTS OF A COMPLEX NUMBER

The nth roots of the complex number $r[\cos\theta + j\sin\theta]$ are:

$$Z_1 = r^{1/n}\left[\cos\frac{\theta}{n} + j\sin\frac{\theta}{n}\right]$$

$$Z_2 = r^{1/n}\left[\cos\frac{\theta + 2\pi}{n} + j\sin\frac{\theta + 2\pi}{n}\right]$$

$$Z_3 = r^{1/n}\left[\cos\frac{\theta + 4\pi}{n} = j\sin\frac{\theta + 4\pi}{n}\right]$$

$$\vdots$$

$$Z_n = r^{1/n}\left[\cos\frac{\theta + (n-1)2\pi}{n} + \sin\frac{\theta + (n-1)(2\pi)}{n}\right].$$

There are precisely n different roots. Continuing the above sequence would repeat these same roots.

The following formulas are alternate forms of this rule.

ALTERNATE METHODS FOR FINDING THE nth ROOTS OF A COMPLEX NUMBER

I. The nth roots of the complex number $r(\cos \theta + j \sin \theta)$ are

$$Z = \sqrt[n]{r}\left(\cos \frac{\theta + 2\pi k}{n} + j \sin \frac{\theta + 2\pi k}{n}\right),$$

for $k = 0, 1, 2, \ldots (n-1)$.

II. The nth roots of the complex number $r \, cjs \, \theta$ are

$$Z = \sqrt[n]{r}\left(cjs \, \frac{\theta + 2\pi k}{n}\right), \quad \text{for } k = 0, 1, 2, \ldots, (n-1).$$

• **Example 9.23:** Calculate the three distinct roots of -1.

Solution:
$$-1 + 0j = \cos \pi + j \sin \pi$$
$$Z_1 = [\cos \pi + j \sin \pi]^{1/3}$$
$$= \cos \frac{\pi}{3} + j \sin \frac{\pi}{3} = \frac{1}{2} + \frac{\sqrt{3}j}{2}$$
$$Z_2 = \left[\cos \frac{(\pi + 2\pi)}{3} + j \sin \frac{(\pi + 2\pi)}{3}\right]$$
$$= \cos \pi + j \sin \pi = -1 + 0j$$
$$Z_3 = \left[\cos \frac{(\pi + 4\pi)}{3} + j \sin \frac{(\pi + 4\pi)}{3}\right]$$
$$= \cos \frac{5\pi}{3} + j \sin \frac{5\pi}{3} = \frac{1}{2} - \frac{\sqrt{3}j}{2}.$$

Note that if we attempt to calculate an additional root Z_4, we get

$$Z_4 = \cos \frac{\pi + 6\pi}{3} + j \sin \frac{\pi + 6\pi}{3}$$
$$= \cos \frac{7\pi}{3} + j \sin \frac{7\pi}{3},$$

which is a repeat of Z_1.

• **Example 9.24:** Calculate the four fourth roots of $1 + j$.

Solution: $1 + j = \sqrt{2}\left(\cos \frac{\pi}{4} + j \sin \frac{\pi}{4}\right) = \sqrt{2} \, cjs \, \frac{\pi}{4}$.

For $k = 0$,

$$Z_1 = (\sqrt{2})^{1/4} \, cjs \, \frac{\frac{\pi}{4}}{4} = \sqrt[8]{2} \, cjs \, \frac{\pi}{16}.$$

For $k = 1$,
$$Z_2 = \sqrt[8]{2} \; cjs \; \frac{\frac{\pi}{4} + 2\pi}{4} = \sqrt[8]{2} \; cjs \; \frac{9\pi}{16}.$$

For $k = 2$,
$$Z_3 = \sqrt[8]{2} \; cjs \; \frac{\frac{\pi}{4} + 4\pi}{4} = \sqrt[8]{2} \; cjs \; \frac{17\pi}{16}.$$

For $k = 3$,
$$Z_4 = \sqrt[8]{2} \; cjs \; \frac{\frac{\pi}{4} + 6\pi}{4} = \sqrt[8]{2} \; cjs \; \frac{25\pi}{16}.$$

Note that if we plot the four roots on a coordinate system, they are equally spaced on a circle with a radius of $\sqrt[8]{2}$. See figure 9.30.

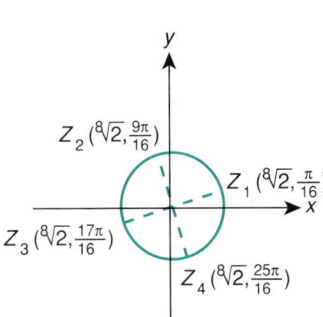

Figure 9.30

The complex roots of a complex number are always equally spaced along the circumference of a circle with the center at the origin. Example 9.25 makes use of this fact in finding the roots of a complex number without the necessity of calculating them individually.

• **Example 9.25:** Calculate the four fourth roots of $-8 + 8\sqrt{3} \, j$.

Solution: $-8 + 8\sqrt{3} \, j = 16 \; cjs \; \frac{2\pi}{3}$ (see figure 9.31). Use theorem 9.3 to find one root:

$$Z_1 = (16)^{1/4} \; cjs \; \frac{\frac{2\pi}{3}}{4} = 2 \; cjs \; \frac{\pi}{6}$$

$$= 2\left(\cos \frac{\pi}{6} + j \sin \frac{\pi}{6}\right)$$

$$= \sqrt{3} + j.$$

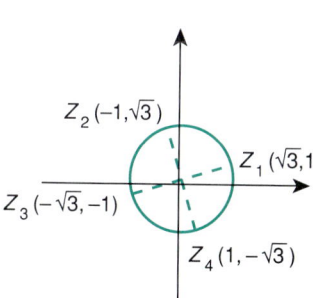

Figure 9.31

Plot the point $(\sqrt{3}, 1)$ and draw a circle with the center at the origin that contains the point on the circumference. Since the other roots are on the same circle and the roots are equally spaced, the other roots can be plotted by dividing the circle into four equal parts starting at $(\sqrt{3}, 1)$. Thus, the other roots are

$$Z_2 = -1 + \sqrt{3} \, j, \qquad Z_3 = -\sqrt{3} - j, \qquad Z_4 = 1 - \sqrt{3} \, j.$$

408 CHAPTER 9 VECTORS AND COMPLEX NUMBERS

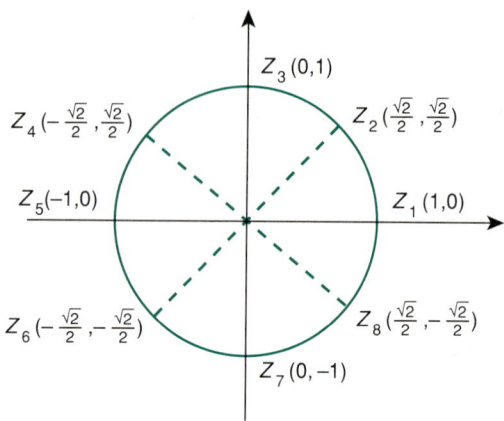

Figure 9.32

- **Example 9.26:** Calculate the eight eighth roots of 1.

    Solution: Since one of the roots is 1, we can draw a circle of radius 1 and divide it into 8 equal parts to find the other roots (see figure 9.32). The roots are $1, \frac{\sqrt{2}}{2} + j\frac{\sqrt{2}}{2}, j, -\frac{\sqrt{2}}{2} + j\frac{\sqrt{2}}{2}, -1, -\frac{\sqrt{2}}{2} - j\frac{\sqrt{2}}{2}, -j,$ and $\frac{\sqrt{2}}{2} - j\frac{\sqrt{2}}{2}$.

....

Trial Problems 9.4

Before you begin the section exercises, warm up with these problems. Complete answers are included in the answer key.

Calculate.

1. $[2(\cos(0.6) + j \sin(0.6))]^5$
2. $\left[\sqrt{2}\left(\cos\frac{\pi}{6} + j \sin\frac{\pi}{6}\right)\right]^4$
3. $(1 + j)^4$
4. Find one square root of $1 + j$.
5. Find one square root of j.

Exercises 9.4

Use De Moivre's theorem to calculate each power of a complex number.

1. $\left(\cos\frac{\pi}{4} + j \sin\frac{\pi}{4}\right)^8$
2. $\left(\cos\frac{\pi}{4} + j \sin\frac{\pi}{4}\right)^6$
3. $\left[2\left(\cos\frac{\pi}{2} + j \sin\frac{\pi}{2}\right)\right]^{12}$
4. $\left[2\left(\cos\frac{\pi}{2} + j \sin\frac{\pi}{2}\right)\right]^{10}$
5. $(\cos 1.32 + j \sin 1.32)^4$
6. $(\cos 0.85 + j \sin 0.85)^4$
7. $[3(\cos 1.87 + j \sin 1.87)]^3$
8. $[2(\cos 2.32 + j \sin 2.32)]^3$
9. $(1 + j)^6$

10. $(1-j)^6$
11. $(\sqrt{3}+j)^8$
12. $(\sqrt{3}-j)^6$
13. $(0.372+0.516j)^5$
14. $(0.436+0.172j)^4$
15. $(1+j)^{-6}$
16. $(\sqrt{3}-j)^{-5}$

Calculate the indicated roots of each complex number using De Moivre's theorem. Write the solutions in the form $a + bj$.

17. The fourth roots of $-\dfrac{9}{2} - \dfrac{9\sqrt{3}}{2}j$
18. The square roots of $\dfrac{9}{2} - \dfrac{9\sqrt{3}}{2}j$
19. The sixth root of $-64j$
20. The cube roots of $8j$
21. The fifth roots of $-16\sqrt{2} - 16j\sqrt{2}$
22. The square roots of $2 + 2j\sqrt{3}$
23. The fourth roots of $2 + 2j\sqrt{3}$
24. The cube roots of $2 + 2j\sqrt{3}$
25. The cube roots of $1.62 + 4.78j$
26. The cube roots of $1.87 + 2.63j$

Find and graph the following roots.

27. The fourth roots of 1
28. The eighth roots of 1
29. The fourth roots of j
30. The sixth roots of j

Solve each equation for Z.

31. $Z^3 + j = 0$
32. $Z^3 - j = 0$
33. $Z^2 = 2 + 2j$
34. $Z^2 = 2 - 2j$

9·5 Summary of Terms, Formulas, Rules, and Procedures

Terms

Argument (p. 398)
Complex Number (p. 388)
Conjugate (p. 390)
De Moivre's Theorem (p. 404)
Direction (p. 378)
Equal Vectors (p. 378)
Horizontal Components (p. 380)
Imaginary Axis (p. 395)
Imaginary Number (p. 388)

Initial Point (p. 378)
j-axis (p. 395)
Magnitude of a Vector (p. 378)
Modulus (p. 398)
Number (p. 378)
Polar Coordinates (p. 396)
Polar Form of a Complex Number (p. 398)
Real Axis (p. 395)
Resultant (p. 380)

Root of a Complex Number (pp. 405–408)
Scalar Multiple (p. 379)
Scalar Quantity (p. 378)
Standard Form of a Complex Number (p. 388)
Terminal Point (p. 378)
Unit (p. 378)
Vector Quantity (p. 378)
Vertical Components (p. 380)

Formulas

Vector Subtraction: (9.1)
$$\mathbf{A} - \mathbf{B} = \mathbf{A} + (-\mathbf{B})$$

Series RLC AC Circuits: (9.1)
$$X = X_L - X_C,$$

where X is the total reactance, X_C is the capacitive reactance, and X_L is the inductive reactance.
$$Z = R^2 + X^2,$$

where Z is the impedance, X is the total reactance, and R is the resistance. Phase angle $(\theta) = \tan^{-1}\left(\dfrac{X}{R}\right)$.

Complex Numbers:

Equality: (9.2)

$$a + bj = c + dj, \quad \text{if } a = c \text{ and } b = d.$$

Addition: (9.2)

$$(a + bj) + (c + dj) = (a + c) + (b + d)j.$$

Subtraction: (9.2)

$$(a + bj) - (c + dj) = (a - c) + (b - d)j.$$

Multiplication: (9.2)

$$(a + bj)(c + dj) = (ac - bd) + (ad + bc)j.$$

Conjugates: (9.2)

$a + bi$ and $a - bj$ are conjugate complex numbers.

Division: (9.2)
To divide complex numbers, write the quotient as a fraction and multiply the numerator and the denominator by the conjugate of the denominator.

Polar Form: (9.3)

$$a + bj = r(\cos \theta + j \sin \theta) = r \text{ cjs } \theta$$

$(a,b) \to (r,\theta)$ where $r = \sqrt{a^2 + b^2}$, $\tan \theta = \dfrac{b}{a}$

$(r,\theta) \to (a,b)$ where $a = r \cos \theta$ and $b = r \sin \theta$.

Multiplication and Division in Polar Form: If $u = r \text{ cjs } \alpha$, $v = s \text{ cjs } \beta$, (9.3)

a. $uv = rs \text{ cjs}(\alpha + \beta)$

b. $\dfrac{u}{v} = \dfrac{r}{s} \text{ cjs}(\alpha - \beta).$

Powers: (9.4)

$$(r \text{ cjs } \theta)^n = r^n \text{cjs}(n\theta).$$

Roots: (9.4)

$$(r \text{ cjs } \theta)^{1/n} = \sqrt[n]{r}\left(\text{cjs}\left(\frac{\theta + 2\pi k}{n} \right) \right),$$

for $k = 0, 1, 2, \ldots (n - 1)$.

9·6 Chapter 9 Review Exercises

Draw a vector V_1 that makes an angle of 30° with the horizontal. Draw a second vector V_2 that is horizontal. Use these vectors to draw the following vectors.

1. $-V_1$
2. $-V_2$
3. $2V_2$
4. $2V_1$
5. $V_1 + V_2$
6. $V_1 - V_2$
7. $2V_1 - V_2$
8. $2V_1 + V_2$

Given the magnitude $|V|$ and the angle θ, which the vector makes with the horizontal, find the horizontal and vertical components of the vector.

9. $|V| = 3.8, \theta = 46°$
10. $|V| = 6.7, \theta = 62°$
11. $|V| = 625, \theta = 32.3°$
12. $|V| = 480, \theta = 18.3°$

13. Given two vectors **V** and **W**, with $|V| = 130$, $|W| = 120$, and the angle between **V** and **W**, $\theta = 63°$, find the magnitude of the resultant vector.

14. Given two vectors **V** and **W**, with $|V| = 220$, $|W| = 150$, and the angle between the vectors, $\theta = 40°$, find the magnitude of the resultant vector.

15. Forces of 50 newtons and 120 newtons act on a point. The angle between the directions of the forces is 60°. Find the magnitude of the resultant force.

16. Solve exercise 15 given that the forces are 60 and 24 newtons and the angle is 120°.

17. Find the force necessary to pull an object weighing 120 lb up a ramp that is inclined at 27.5° with the horizontal.

18. A 60-lb weight is suspended from a rope. If a horizontal force of 15 lb is applied to the object, find the angle between the rope and the vertical.

19. Given the circuit of figure 9.18 with $I_R = 0.35$ amps and $I_L = 0.18$ amps, determine I_T and the phase angle.

20. Given the circuit of figure 9.18 with $I_R = 0.65$ amps and $I_L = 0.48$ amps, find I_T and the phase angle.

Perform the indicated operations.

21. $(4 - 2j) + (3 - j)$
22. $(6 + 3j) + (2 - j)$
23. $(3 - j) - (2 - 6j)$
24. $(3 + 2j) - (6 + j)$
25. $(2 - 3j)(4 + j)$
26. $(2 + j)(4 - 3j)$
27. $\dfrac{1}{3 + 2j}$
28. $\dfrac{1}{2 - 3j}$
29. $\dfrac{2 + j}{3 - 2j}$
30. $\dfrac{2 - j}{3 + 2j}$
31. $\left(\dfrac{\sqrt{2}}{2} + \dfrac{\sqrt{2}}{2}j\right)^{10}$ *Hint:* Square the number and use the result.
32. $\left(\dfrac{\sqrt{2}}{2} - \dfrac{\sqrt{2}}{2}j\right)^{8}$ *Hint:* Square the number and use the result.

In an AC circuit, $E = IZ$, where E is the voltage, I is the current, and Z is the impedance. Find the missing quantity.

33. $I = 6.7 + 5.3j$, $E = 21.3 + 18.6j$
34. $I = 3.8 + 7.6j$, $E = 18.1 + 12.2j$
35. $Z = 5 - 4j$, $E = 7 + 2j$
36. $Z = 8 - 3j$, $E = 6 + 3j$

Write each complex number in polar form.

37. $3 - 3j$
38. $1 + j\sqrt{3}$
39. $\dfrac{\sqrt{2}}{2} + \dfrac{\sqrt{2}}{2}j$
40. $\dfrac{1}{2} - \dfrac{\sqrt{3}}{2}j$

Write each complex number in the form $a + bj$.

41. $\cos\dfrac{5\pi}{4} + j\sin\dfrac{5\pi}{4}$
42. $\cos\dfrac{\pi}{6} + j\sin\dfrac{\pi}{6}$
43. $\sqrt{2}(\cos 45° + j\sin 45°)$
44. $\sqrt{3}(\cos 150° + j\sin 150°)$

Find the product of each pair of complex numbers.

45. $u = \text{cjs}\,\dfrac{5\pi}{4}$, $v = 2\,\text{cjs}\,\dfrac{\pi}{4}$
46. $u = \text{cjs}\,30°$, $v = \sqrt{3}\,\text{cjs}\,150°$

Find $\dfrac{u}{v}$ for the given numbers.

47. $u = \sqrt{3}\,\text{cjs}\,\dfrac{5\pi}{6}$, $v = \text{cjs}\,\dfrac{\pi}{4}$
48. $u = \text{cjs}\,225°$, $v = \sqrt{3}\,\text{cjs}\,30°$

49. Use the polar form to calculate $(2 - 2\sqrt{3}j)^6$
50. Use the polar form to calculate $(4 - 4j)^5$
51. Calculate the fourth roots of $1 + j$
52. Calculate the cube roots of $2 - 2j$

9·7 Chapter 9 Test

Draw a vertical vector V_1 that is 3 cm long. Draw a vector V_2 that is 2 cm long and that makes an angle of 60° with the horizontal. Use these vectors for problems 1–4. Draw these vectors.

1. $-V_2$
2. $2V_1$
3. $V_1 + V_2$
4. $V_2 - V_1$

Find the horizontal and vertical components of vectors with the given magnitudes and directions.

5. $|V| = 17.3, \theta = 56°$
6. $|V| = 560, \theta = 25.3°$

7. Given two vectors with magnitudes 180 and 240. If the angle between the vectors is 56°, find the magnitude of the resultant vector.

8. Forces of 60 and 80 newtons act on a body. If the angle between the forces is 41°, find the magnitude of the resultant force.

Perform the indicated operations.

9. $(3 + j) + (2 - 3j)$
10. $(4 + 2j) - (3 - j)$
11. $(1 + j)(2 - 3j)$
12. $(4.3 + 1.2j)(5.6 - 3.2j)$
13. $\dfrac{1}{1 - 3j}$
14. $\dfrac{2 - j}{4 + j}$

Write each complex number in polar form.

15. $3 + 3j$
16. $1 - j\sqrt{3}$

Write each complex number in standard form.

17. $\cos \dfrac{5\pi}{6} + j \sin \dfrac{5\pi}{6}$
18. $\cos 120° + j \sin 120°$

Given $u = \text{cjs}\,\dfrac{2\pi}{3}$ and $v = 4\,\text{cjs}\,\dfrac{\pi}{6}$, perform the indicated calculations.

19. uv
20. $\dfrac{u}{v}$
21. u^3

22. Calculate $(1 + j\sqrt{3})^4$.
23. Calculate $(1 + j)^8$.
24. Calculate the fourth roots of $16j$.

CHAPTER 10
Exponential and Logarithmic Functions

*J*ohn Napier (1550–1617) lived most of his life in the family castle near Edinburgh, Scotland. He expended much of his energy as a political and religious activist. As a relaxation from his political and religious activity, Napier amused himself with the study of mathematics and science. One of the many results of his "relaxation" was the invention of logarithms. The significance of logarithms as a computing device lies in the fact that by them, multiplication and division are reduced to the simpler operations of addition and subtraction.

Napier published his discussion of logarithms in 1614 in a book whose title roughly translates to *A Description of the Wonderful Law of Logarithms*. The logarithms of Napier are often called natural logarithms.

In the year following the publication by Napier, Henry Briggs (1561–1630), a professor in London, traveled to Edinburgh to visit the great inventor of logarithms. As a result of that visit, Napier and Briggs agreed that logarithms would be more useful if they were altered to fit the fact that our number system is based on 10. Thus were born the so-called Briggsian, or common logarithms. Common logarithms have a base 10, while natural logarithms, those invented by Napier, have a base *e*.

It is interesting that although natural logarithms were developed first, and common logarithms later, natural logarithms are the most useful in engineering and scientific work today.

Logarithms were invented before exponents were widely in use. However, today the concept of exponential functions is often developed first, then logarithmic functions are introduced. This chapter begins with the study of exponential functions, then develops logarithmic functions using the concept of inverse functions.

10 • 1 Exponential Functions

In section 1.3 we introduced exponents and some of the properties related to them. We considered only those exponents that were integers. In section 1.5 we extended the definition so that rational numbers could be used as exponents. Now we extend the concept of exponents once more so that any real number can be an exponent. Later we develop logarithms that are very closely related to exponents.

There are many applications of exponential functions and logarithmic functions in science and technology. The voltage in a given circuit can be expressed using exponents. The value of money in an investment can be determined through the use of exponents. The intensity of earthquakes is measured by a logarithmic scale. The intensity of light related to the thickness of the material through which it passes can be expressed using exponents. The distinction between acids and bases in chemistry is measured in terms of logarithms.

If b is a real number greater than zero, then for each real exponent x we assume b^x is a unique real number. Since for each real x there is one and only one b^x, the equation

$$y = b^x, \quad (b > 0) \quad (10.1)$$

Exponential Function defines a function. We call such an equation an **exponential function**.
Constant Function If $b = 1$, we say that equation (10.1) is a **constant function**, since $y = 1^x$ becomes $y = 1$ for all real values of x. We graphed equations of the form $y = k$ in section 3.2.

Exponential functions can be visualized more clearly by considering their graphs. The graphs also indicate that the domain of an exponential function is the set of real numbers and the range of the exponential function is the set of positive real numbers. Consider example 10.1.

• **Example 10.1:** Graph $y = 2^x$.

   Solution: By letting $x = -3, -2, -1, 0, 1, 2, 3$, we find corresponding values for y. These ordered pairs are tabulated and graphed in figure 10.1.

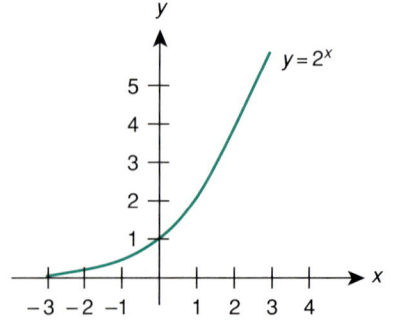

x	y
-3	$\frac{1}{8}$
-2	$\frac{1}{4}$
-1	$\frac{1}{2}$
0	1
1	2
2	4
3	8

Figure 10.1

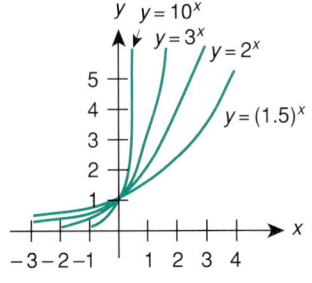

Figure 10.2

In figure 10.2 we have graphed many exponential functions on the same rectangular coordinate system. Note how the curve changes as b changes when $b > 1$. The larger the value of the base b, the steeper is the curve.

In figure 10.3 we have graphed $y = b^x$ for different values of b when $0 < b < 1$. Some of the pairs for $y = \left(\dfrac{1}{2}\right)^x$ have been tabulated with the figure. You should check these values. Note the nature of these curves compared with those in figure 10.2. The smaller the value of the base b, the steeper is the curve.

For functions with b between 0 and 1, the graphs have the same shape as the graphs in figure 10.3. For functions with b greater than 1, the graphs have the same shape as those in figure 10.2.

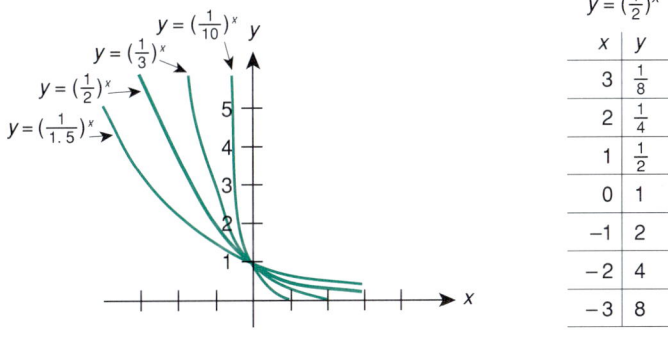

Figure 10.3

Note also that each of the graphs in figure 10.2 and figure 10.3 passes through the point (1,0). This is true since the equation $y = b^x$ becomes $1 = b^0$ when $x = 0$. This means the y-intercept is 1. Since the graphs approach the x-axis, but never touch it, we say the x-axis is an **asymptote** for each of the exponential functions.

Asymptote

Notice how the graph of $y = b^{-x}$ compares to the graph of $y = b^x$. Example 10.2 illustrates this concept.

• **Example 10.2:** Graph $y = 2^{-x}$.

    Solution: By letting $x = 3, 2, 1, 0, -1, -2,$ and -3, the value of $-x$ is $-3, -2, -1, 0, 1, 2,$ and 3. The corresponding values of y are tabulated and graphed in figure 10.4.

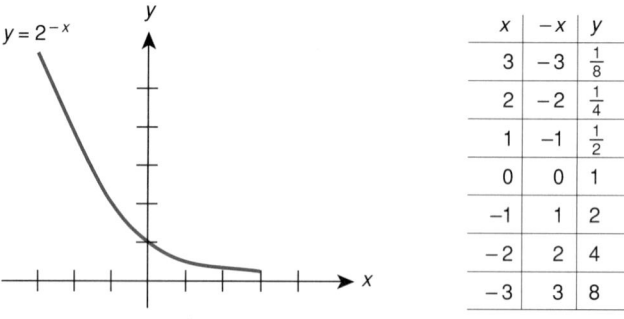

Figure 10.4

Since $y = 2^{-x}$ can be written as $y = \dfrac{1}{2^x}$ or as $y = \left(\dfrac{1}{2}\right)^x$, the graph of $y = 2^{-x}$ and the graph of $y = \left(\dfrac{1}{2}\right)^x$ are the same.

Decreasing Function
Increasing Function

If the value of y decreases as values of x increase, the function is called a **decreasing function**. If the value of y increases as values of x increase, the function is called an **increasing function**. The graphs in figure 10.2 are increasing functions while the graphs in figures 10.3 and 10.4 are decreasing functions.

Two exponential functions are of special interest. One is defined by

$$y = 10^x,$$

and the other by

$$y = e^x,$$

where e is an irrational number whose decimal approximation to eight digits is 2.7182818. We obtain approximations for powers 10^x and e^x for all real numbers x by methods that are introduced later. However, scientific calculators can be used to find approximations for these powers directly. Consider examples 10.3 and 10.4.

• *Example 10.3:* Approximate $10^{1.31}$.

  Solution: Enter

$$10 \; \boxed{y^x} \; 1.31 \; \boxed{=} \; .$$

The display indicates 20.4173794, so $10^{1.31} = 20.42$ (approximately).

• • • • • • • • • •

• *Example 10.4:* Approximate $e^{0.83}$.

Solution: Enter

$$0.83 \boxed{\text{2nd}} \boxed{e^x}.$$

The display indicates 2.29331874, so $e^{0.83} = 2.29$ (approximately).

When graphing functions, we often use values for x to determine the corresponding values for y. However, once we have a graph of the function, we can use values for y to find corresponding values for x, or vice versa. Examples 10.5 and 10.6 illustrate this concept.

• *Example 10.5:* Use figure 10.1 to find y when $x = 1.5$.

Solution: To find y when $x = 1.5$, begin on the x-axis at 1.5 and move vertically until you reach the curve. Then move horizontally until you reach the y-axis. The point on the y-axis (approximately 2.8) should be the value of y when $x = 1.5$.

• *Example 10.6:* Use figure 10.1 to find x when $y = 3.5$.

Solution: To find x when $y = 3.5$, begin on the y-axis at 3.5 and move horizontally until you reach the curve. Then move vertically until you reach the x-axis. This point on the x-axis (approximately 1.8) should be the value of x when $y = 3.5$.

The graphs and examples of this section demonstrate the following properties of exponential functions and their graphs.

PROPERTIES OF EXPONENTIAL FUNCTIONS AND THEIR GRAPHS

1. If $b > 1$, $y = b^x$ is an increasing function.
2. If $0 < b < 1$, $y = b^x$ is a decreasing function.
3. The domain of $y = b^x$ consists of all real numbers.
4. The range of $y = b^x$ consists of the positive real numbers.
5. The graph of $y = b^x$ approaches the x-axis as an asymptote.
6. All functions of the form $y = b^x$ intersect the y-axis at the point (0,1).

CHAPTER 10 EXPONENTIAL AND LOGARITHMIC FUNCTIONS

Trial Problems 10.1

Before you begin the section exercises, warm up with these problems. Complete answers are included in the answer key.

1. For the exponential function $y = 2^x$, find y when $x = -2, -1, 0, 1,$ and 2.
2. For the exponential function $y = \left(\frac{1}{3}\right)^x$, find y when $x = -2, -1, 0, 1,$ and 2.
3. For the exponential function $y = 10^x$, find x when $y = 1, 10, 100,$ and 1000.
4. For the exponential function $y = \left(\frac{1}{2}\right)^x$, find x when $y = \frac{1}{4}, \frac{1}{2}, 2,$ and 4.
5. Is the exponential function $y = \left(\frac{1}{3}\right)^x$ an increasing or a decreasing function?

Exercises 10.1

Using figure 10.1 and the values given, find approximations for the required values.

1. $x = 2$, find y
2. $x = 1$, find y
3. $x = \frac{1}{2}$, find y
4. $x = \frac{1}{3}$, find y
5. $x = -1$, find y
6. $x = -2$, find y
7. $y = \frac{1}{2}$, find x
8. $y = \frac{1}{3}$, find x
9. $y = \frac{3}{2}$, find x
10. $y = \frac{5}{2}$, find x

Graph the following for the given values of x.

11. $y = 3^x, -3 \le x \le 0$
12. $y = 2^x, -3 \le x \le 0$
13. $y = 3^{-x}, 0 \le x \le 3$
14. $y = 2^{-x}, 0 \le x \le 3$
15. $y = \left(\frac{1}{2}\right)^x, -3 \le x \le 3$
16. $y = \left(\frac{1}{3}\right)^x, -2 \le x \le 2$

Solve.

17. The number N that a certain culture of bacteria will have after t hours can be expressed by the equation

$$N = 5000 \times 2^{t/7}.$$

How many bacteria will there be in 21 hours?

18. How many bacteria will there be in exercise 17 after 2.1 hours?

19. A certain radioactive substance decays according to the formula

where A is the amount left after t years and I is the initial amount. If the initial amount is 40 grams, how long will it take for half of the initial amount to decay?

20. The amount A of money that results when interest is compounded annually can be represented by the formula

$$A = P(1 + r)^n,$$

where P is the principal invested,
  r is the rate of interest,
and n is the number of years.

How many years will it take to double the principal if the money is invested at 8%?

10 • 2 Logarithmic Functions

The exponential function $y = b^x$ with $b > 1$ has an inverse that is also a function. This inverse function is called the **logarithmic function with base b**. We write the logarithmic function with base b as

Logarithmic Function with Base b

Logarithmic Function

$$y = \log_b x. \tag{10.2}$$

One of the properties of a function and its inverse function is that their graphs are symmetric with respect to the line $y = x$. Figure 10.5 illustrates this concept.

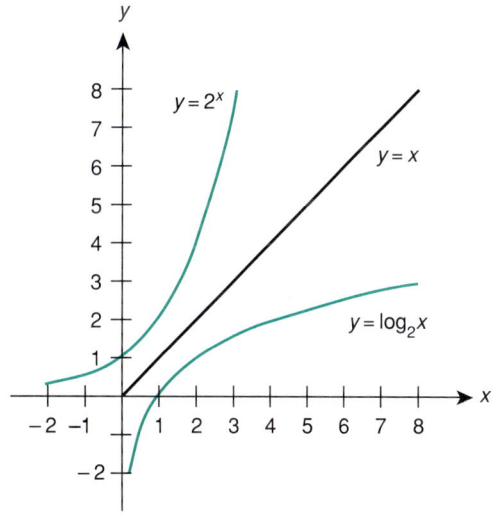

Figure 10.5

One way to create the inverse of a function is to interchange the role of x and y. That is, if we replace each x with y and each y with x in a given equation, we have created the inverse. Thus we see that $y = b^x$ and $x = b^y$ are inverses. But we have previously stated that $y = \log_b x$ is the inverse of $y = b^x$. Therefore,

$$x = b^y \quad \text{and} \quad y = \log_b x$$

are equivalent expressions.

With this fact in mind, we see that we can take an expression in exponential form and write it in logarithmic form, and we can take an expression in logarithmic form and write it in exponential form.

The following is a list of some equivalent exponential and logarithmic expressions:

Exponential Relation	**Logarithmic Relation**
$b^x = y$	$\log_b y = x$
$3^3 = 27$	$\log_3 27 = 3$
$3^2 = 9$	$\log_3 9 = 2$
$3^1 = 3$	$\log_3 3 = 1$
$3^0 = 1$	$\log_3 1 = 0$
$3^{-1} = \dfrac{1}{3}$	$\log_3 \dfrac{1}{3} = -1$
$3^{-2} = \dfrac{1}{9}$	$\log_3 \dfrac{1}{9} = -2$
$3^{1/2} = \sqrt{3}$	$\log_3 \sqrt{3} = \dfrac{1}{2}$

Examples 10.7 and 10.8 illustrate how to convert from exponential form to logarithmic form.

• **Example 10.7:** Express $y = 2^x$ in logarithmic form.

   Solution: $x = \log_2 y$.

   Notice that the base 2 in the exponential expression is also the base in the logarithmic expression.

• **Example 10.8:** Express $y = \log_3 x$ in exponential form.

   Solution: $x = 3^y$.

Examples 10.9 and 10.10 illustrate how to evaluate certain logarithmic expressions by first converting them to exponential expressions.

SECTION 10.2 LOGARITHMIC FUNCTIONS

• **Example 10.9:** Find $\log_2 32$.

Solution: Set $x = \log_2 32$, then convert the equation to its equivalent exponential expression:

$$2^x = 32.$$

But

$$2^x = 2^5,$$

therefore,

$$x = 5.$$

• **Example 10.10:** Find $\log_5 \frac{1}{25}$.

Solution: Set $x = \log_5 \frac{1}{25}$, then write this equation in its equivalent exponential form:

$$5^x = \frac{1}{25}.$$

But

$$5^x = \left(\frac{1}{5}\right)^2,$$

or

$$5^x = 5^{-2}.$$

Therefore,

$$x = -2.$$

Trial Problems 10.2

Before you begin the section exercises, warm up with these problems. Complete answers are included in the answer key.

1. Express the equation $5^2 = 25$ in logarithmic notation.
2. Express the equation $r^s = t$ in logarithmic notation.
3. Express $\log_2 8 = 3$ in exponential notation.
4. Express $\log_{10} 100 = 2$ in exponential notation.
5. Determine the value of x in the equation $\log_6 36 = x$.

Exercises 10.2

Express each equation in logarithmic notation.

1. $3^2 = 9$
2. $4^2 = 16$
3. $16^{1/4} = 2$
4. $64^{1/3} = 4$
5. $8^{-1/3} = \dfrac{1}{2}$
6. $64^{-1/6} = \dfrac{1}{2}$
7. $10^0 = 1$
8. $2^0 = 1$
9. $10^{-1} = 0.1$
10. $10^{-2} = 0.01$
11. $a^{-3} = \dfrac{1}{8}$
12. $b^{-4} = \dfrac{1}{16}$
13. $x^2 = 100$
14. $y^4 = 81$
15. $10^x = 100$
16. $10^y = 1000$
17. $2^3 = x$
18. $3^2 = y$
19. $3^y = 9$
20. $5^x = 125$
21. $a^b = 8$
22. $x^y = 16$
23. $2^x = y$
24. $3^y = x$
25. $b^x = y$
26. $a^y = x$
27. $r^s = \dfrac{1}{2}$
28. $t^r = \dfrac{1}{3}$
29. $a^{x+y} = s$
30. $b^{x+y} = t$

Express each equation in exponential notation.

31. $\log_6 36 = 2$
32. $\log_2 64 = 6$
33. $\log_{10} 1000 = 3$
34. $\log_{10} 10{,}000 = 4$
35. $\log_{10} 1 = 0$
36. $\log_5 1 = 0$
37. $\log_{10} 0.01 = -2$
38. $\log_{10} 0.0001 = -4$
39. $\log_4 x = 3$
40. $\log_3 y = 5$
41. $\log_a \dfrac{1}{8} = -3$
42. $\log_b \dfrac{1}{16} = -4$
43. $\log_x 100 = 2$
44. $\log_y 81 = 4$
45. $\log_{10} 100 = x$
46. $\log_{10} 1000 = y$
47. $\log_2 x = 3$
48. $\log_3 y = 2$
49. $\log_3 9 = y$
50. $\log_5 125 = x$
51. $\log_a 8 = b$
52. $\log_x 16 = y$
53. $\log_2 y = x$
54. $\log_3 x = y$
55. $\log_b y = x$
56. $\log_a x = y$
57. $\log_r \dfrac{1}{2} = s$
58. $\log_t \dfrac{1}{3} = r$
59. $\log_a s = x + y$
60. $\log_b t = x + y$

Find the value of each logarithm.

61. $\log_4 16$
62. $\log_3 81$
63. $\log_7 49$
64. $\log_2 32$
65. $\log_5 5$
66. $\log_3 3$
67. $\log_{10} 10$
68. $\log_8 8$
69. $\log_5 1$
70. $\log_7 1$

Solve for the unknown value.

71. $\log_2 x = 3$
72. $\log_3 x = 2$
73. $\log_3 9 = y$
74. $\log_5 125 = y$
75. $\log_2 \dfrac{1}{8} = y$
76. $\log_5 \dfrac{1}{5} = y$

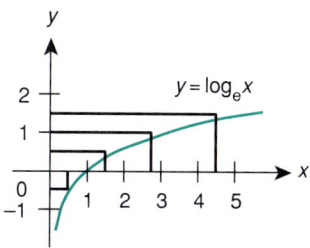

Figure 10.6

Figure 10.6 illustrates the graph of the logarithmic function $y = \log_e x$. Use the graph to approximate the values for the unknowns.

77. $y = \log_e 4.5$ **78.** $y = \log_e 2.7$ **79.** $-0.7 = \log_e x$ **80.** $0.4 = \log_e x$

81. If B_1 is the apparent brightness of a star of magnitude H and B_2 is the apparent brightness of a star of magnitude J, then

$$\log_{10} \frac{B_1}{B_2} = 0.4(J - H).$$

Express this formula in exponential form.

82. For a certain electric circuit, the voltage v is expressed in terms of the time t by the formula

$$v = e^{-0.2t}.$$

Express this formula in logarithmic form.

10 • 3 Properties of Logarithms

From the properties of exponents and from the relation between exponential expressions and logarithmic expressions, we can develop many properties of logarithms. Of course, we must keep in mind that we have placed certain restrictions on the logarithmic base. In section 10.1 we restricted the base to be any real number greater than zero. We also commented that a base of 1 gave us a constant function, so we eliminate 1 as an acceptable base in logarithms.

With these comments in mind, we are ready to develop some properties of logarithms.

We know that $b^1 = b$. If we write this exponential expression in logarithmic form, we get

$$\log_b b = 1. \tag{10.3}$$

We also know that $b^0 = 1$, so

$$\log_b 1 = 0. \tag{10.4}$$

If we take the identity

$$\log_b x = \log_b x$$

and treat it as a logarithmic expression, the equivalent exponential expression becomes

$$b^{\log_b x} = x. \tag{10.5}$$

424   CHAPTER 10 EXPONENTIAL AND LOGARITHMIC FUNCTIONS

Three more properties of logarithms are

Logarithm of a Product
$$\log_b xy = \log_b x + \log_b y \qquad (10.6)$$

Logarithm of a Quotient
$$\log_b \frac{x}{y} = \log_b x - \log_b y \qquad (10.7)$$

Logarithm of a Power
$$\log_b x^p = p \log_b x. \qquad (10.8)$$

These properties can be proved by using the laws of exponents that were developed in section 1.3 of chapter 1. The laws of exponents that will be necessary for the proofs of these logarithmic properties are listed here for convenience.

Laws of Exponents
$$a^m a^n = a^{m+n}$$
$$\frac{a^m}{a^n} = a^{m-n}$$
$$(a^m)^n = a^{mn}.$$

To develop property (10.6), the logarithm of a product, it will be necessary to use the following substitutions.

Let $m = \log_b x$ and $n = \log_b y$. In exponential form these become
$$x = b^m, \quad y = b^n.$$

Using the first law of exponents and substituting these values gives
$$xy = b^m b^n$$

or
$$xy = b^{m+n}.$$

Converting this last equation to logarithmic form gives
$$\log_b xy = m + n.$$

Substituting for m and n gives
$$\log_b xy = \log_b x + \log_b y.$$

Logarithm of a Product
Stated in words, the formula for the logarithm of a product becomes: **The logarithm of the product of two numbers is equal to the sum of the logarithms of the numbers.**

Example 10.11 illustrates the use of this property.

• **Example 10.11:** Express $\log_b 30$ as the sum of two logarithms.

Solution: $\log_b 30 = \log_b(5)(6) = \log_b 5 + \log_b 6$ or
$$= \log_b(2)(15) = \log_b 2 + \log_b 15.$$

• • • • • • • • • •

You should also note from the formula and from this example that the sum of two logarithms can be written as the logarithm of a product. Hence,
$$\log_b 5 + \log_b 6 = \log_b(5)(6) = \log_b 30.$$

The property dealing with the logarithm of a quotient equation (10.7) can be developed by using some of the same substitutions that were used to prove the property dealing with the logarithm of a product. Hence,

$$\frac{x}{y} = \frac{b^m}{b^n}$$

or

$$\frac{x}{y} = b^{m-n}.$$

Expressed in logarithmic form, this equation becomes

$$\log_b \frac{x}{y} = m - n.$$

Substituting the values for m and n gives

$$\log_b \frac{x}{y} = \log_b x - \log_b y.$$

Logarithm of a Quotient

This property, when expressed in words states: **The logarithm of a quotient of two numbers is equal to the logarithm of the numerator minus the logarithm of the denominator.**

Example 10.12 illustrates how expressions can be altered by using this property.

• *Example 10.12:* **a.** $\log_b \dfrac{5}{8} = \log_b 5 - \log_b 8$

b. $\log_b 6 = \log_b \dfrac{30}{5} = \log_b 30 - \log_b 5$

c. $\log_b 15 - \log_b 2 = \log_b \dfrac{15}{2}.$

The third property of logarithms dealing with the logarithm of a power, equation (10.8), can be proved as follows:

If

$$x = b^m,$$

then

$$x^p = (b^m)^p$$

or

$$x^p = b^{mp}.$$

Expressing this last equation in logarithmic form gives

$$\log_b x^p = mp$$

or

$$\log_b x^p = pm.$$

Substituting $m = \log_b x$ we get

$$\log_b x^p = p \log_b x.$$

What appears to be happening in the case of the logarithm of a power (property 10.8) is that the exponent p is being "moved to the front" of the expression as a factor. We say that **the logarithm of the pth power of a number is p times the logarithm of the number.**

Logarithm of a Power

Example 10.13 illustrates the use of this property.

• **Example 10.13:** a. $\log_b 10^2 = 2 \log_b 10$

  b. $3 \log_b x = \log_b x^3$

• **Example 10.14:** Express $\log_b \sqrt{\dfrac{xy}{z}}$ as the sum or difference of simpler logarithms. Assume that all variables represent positive real numbers.

Solution:
$$\log_b \sqrt{\frac{xy}{z}} = \log_b \left(\frac{xy}{z}\right)^{1/2} \quad \text{(Convert radical to exponent)}$$
$$= \frac{1}{2} \log_b \frac{xy}{z} \quad \text{(Property 10.8)}$$
$$= \frac{1}{2}[\log_b xy - \log_b z] \quad \text{(Property 10.7)}$$
$$= \frac{1}{2}[\log_b x + \log_b y - \log_b z]. \quad \text{(Property 10.6)}$$

• **Example 10.15:** Express $\log_b 12 - \log_b 4 + 2 \log_b 5$ in terms of a single logarithm.

Solution: $\log_b 12 - \log_b 4 + 2 \log_b 5$
$$= \log_b \frac{12}{4} + 2 \log_b 5 \quad \text{(Property 10.7)}$$
$$= \log_b 3 + 2 \log_b 5 \quad \text{(Divide)}$$
$$= \log_b 3 + \log_b 5^2 \quad \text{(Property 10.8)}$$
$$= \log_b (3)(5^2) \quad \text{(Property 10.6)}$$
$$= \log_b (3)(25) \quad \text{(Square 5)}$$
$$= \log_b 75. \quad \text{(Multiply)}$$

SECTION 10.3 PROPERTIES OF LOGARITHMS

By using properties of logarithms, it may be possible to express a rather complicated expression in a much more convenient form. This is illustrated in example 10.16.

• **Example 10.16:** Simplify $3 \log_6 36 - 2 \log_2 \frac{1}{32}$.

Solution:
$$3 \log_6 36 - 2 \log_2 \frac{1}{32}$$
$$= 3 \log_6 6^2 - 2 \log_2 \frac{1}{2^5} \quad \text{(Substitute } 6^2 \text{ for 36 and } 2^5 \text{ for 32)}$$
$$= 3 \log_6 6^2 - 2 \log_2 2^{-5} \quad \left(\text{Substitute } 2^{-5} \text{ for } \frac{1}{2^5}\right)$$
$$= 3(2) \log_6 6 - 2(-5) \log_2 2 \quad \text{(Property 10.8)}$$
$$= 3(2)(1) - 2(-5)(1) \quad \text{(Property 10.3)}$$
$$= 6 + 10 \quad \text{(Multiply)}$$
$$= 16. \quad \text{(Add)}$$

Trial Problems 10.3

Before you begin the section exercises, warm up with these problems. Complete answers are included in the answer key.

1. Express $\log_a 3x$ as a sum of two logarithms.
2. Express $\log_b \frac{x}{a}$ as a difference of two logarithms.
3. Express $\log_x b^3$ as a logarithm of a value without an exponent.
4. Express $\log_b r + \log_b s$ as a single logarithm.
5. Express $\log_b u - \log_b v$ as a single logarithm.

Exercises 10.3

Express as the sum or difference of simpler logarithmic quantities or as a single log without an exponent. Assume that all variables denote positive real numbers.

1. $\log_b 2x$
2. $\log_b 3x$
3. $\log_b 3xy$
4. $\log_b 2xy$
5. $\log_b \frac{y}{z}$
6. $\log_b \frac{z}{x}$
7. $\log_b x^3$
8. $\log_b y^3$
9. $\log_b x^{1/3}$
10. $\log_b y^{1/3}$
11. $\log_b \sqrt{x^3}$
12. $\log_b \sqrt[3]{x^2}$
13. $\log_{10} 2\pi \sqrt{\frac{1}{g}}$
14. $\log_{10} \frac{2L}{R^2}$

Express as a single logarithm with a coefficient of 1.

15. $\log_b x + \log_b y$

16. $\log_b y + \log_b z$

17. $3 \log_b x - 2 \log_b y$

18. $5 \log_b x - 3 \log_b y$

19. $\dfrac{1}{3}(\log_b x + \log_b y - 2 \log_b z)$

20. $\dfrac{1}{2}(\log_b x - 3 \log_b y + \log_b z)$

Solve.

21. If $a = \log_8 225$ and $b = \log_2 15$, express a in terms of b.

22. Find the value of $\log_5 \dfrac{(125)(625)}{25}$.

23. Find the value of $10^{\log_{10} 7}$.

24. Find the value of $\log_3(27 \sqrt[4]{9} \sqrt[3]{9})$.

25. Find the value of $\log_2 0.0625$.

10 • 4 Natural Logarithms and Common Logarithms

In section 10.1 we stated that there are two exponential functions that are of special interest, namely $y = 10^x$ and $y = e^x$. Likewise, there are two logarithmic functions that are of special interest. They are

$$y = \log_{10} x$$

and

$$y = \log_e x.$$

Common Logarithm

The logarithm of x to the base 10 is called the **common logarithm**. We often write log x (without the base indicated) instead of $\log_{10} x$.

Natural Logarithms

Logarithms of x to the base e are called **natural logarithms**. The expression ln x is used instead of $\log_e x$. On a scientific calculator there is $\boxed{\text{log}}$ key for logarithms to the base 10 and an $\boxed{\text{ln}}$ key for logarithms to the base e.

For practical computations the base 10 is most useful, because it works so well with decimal notation. However, for theoretical work and for work in engineering, the most useful base is the irrational number e. To ten places,

$$e = 2.7182818285.$$

Until recently, computations with logarithms were done using tables of common logarithms. Now hand-held calculators and other electronic devices have made calculations with log tables obsolete. Examples 10.17 and 10.18 illustrate how a calculator can be used to find the logarithms of numbers.

• **Example 10.17:** Calculate log 256.

Solution: Key in

$$256 \boxed{\text{log}}.$$

The answer on the display should be 2.408239965. This means $256 = 10^{2.408\cdots}$.

Section 10.4 Natural Logarithms and Common Logarithms

• *Example 10.18:* Calculate ln 256.

Solution: Key in

256 **ln** .

The display should read 5.545177445. This means $256 = e^{5.545\cdots}$.

Antilog   If we have the logarithm of a number and wish to find the number, we must find what is called the **antilog**. This suggests an inverse operation and is accomplished with the **INV** key or its equivalent. Remember that the e^x function and the ln function are inverses, and the 10^x function and the log x function are also inverses. We use this idea to find a number when its logarithm is known. Consider examples 10.19 and 10.20.

• *Example 10.19:* Find x if log $x = 3$.

Solution: Key in

3 **INV** **log** .

The display should indicate 1000. This means log 1000 = 3 or $10^3 = 1000$.

• *Example 10.20:* Find x if ln $x = 3.86$.

Solution: Key in

3.86 **INV** **ln** .

The display should indicate 47.46535137.
This means ln 47.465 . . . = 3.86 or $e^{3.86} = 47.465\ldots$.

Occasionally it becomes necessary to find the logarithm of a number to a base other than 10 or e. For example if we want to find $\log_b x$, where b is not 10 or e, we write $y = \log_b x$ and express $y = \log_b x$ in exponential notation to get

$$b^y = x.$$

When we take the log of both sides of this equation, we get

$$\log b^y = \log x$$

or

$$y \log b = \log x.$$

Solving for y, we get

$$y = \frac{\log x}{\log b}.$$

If we substitute this value for y in $y = \log_b x$, we get the property

Change of Base Property

$$\log_b x = \frac{\log x}{\log b}. \qquad (10.9)$$

This property is illustrated in example 10.21.

• *Example 10.21:* Find $\log_7 54$.

Solution: Using property (10.9) we have

$$\log_7 54 = \frac{\log 54}{\log 7}.$$

On a calculator, key in

54 $\boxed{\log}$ ÷ 7 $\boxed{\log}$ $\boxed{=}$.

The display should show 2.049932289.
This means $\log_7 54 = 2.04\ldots$, or

$$7^{2.04\ldots} = 54.$$

Trial Problems 10.4

Before you begin the section exercises, warm up with these problems. Complete answers are included in the answer key.

1. Calculate log 5.28 to four significant digits.
2. Calculate log 300 to five significant digits.
3. Calculate log 0.004 to five significant digits.

Calculate x to three significant digits.

4. $\ln x = 0.9163$
5. $\ln x = 3.127$

SECTION 10.5 EXPONENTIAL AND LOGARITHMIC EQUATIONS

Exercises 10.4

Calculate to six places.

1. log 3.285
2. log 2.385
3. log 0.0123
4. log 0.0321
5. ln 3.14159
6. ln 1.41421
7. ln 0.2367
8. ln 0.3458

Find x if

9. $\log x = 3$
10. $\log x = 2$
11. $\log x = 1.816$
12. $\log x = 0.49715$
13. $\ln x = 1.14473$
14. $\ln x = 2.718282$
15. $\ln x = 0.003214$
16. $\ln x = 0.004321$

Given that $\log_a 2 = 0.69$, $\log_a 3 = 1.10$, $\log_a 5 = 1.61$, and $\log_a 7 = 1.95$, find:

17. $\log_a \dfrac{2}{3}$
18. $\log_a \dfrac{3}{2}$
19. $\log_a 3^2$
20. $\log_a 2^3$
21. $\log_a 35$
22. $\log_a 14$
23. $\log_a \dfrac{27}{25}$
24. $\log_a \dfrac{35}{27}$
25. $\log_a \sqrt{\dfrac{2}{3}}$
26. $\log_a \sqrt[3]{\dfrac{2}{3}}$

10 • 5 Exponential and Logarithmic Equations

An equation of the type $10^x = 2$ is of the form $b^x = a$, where a and b are positive numbers. If $b^x = a$ is expressed in logarithmic form, it becomes $x = \log_b a$, which is a solution for the equation $b^x = a$. However, this is a theoretical solution and is not very practical, since the base b may not be either 10 or e. We could use equation (10.9) from section 10.4 to write $x = \log_b a = \dfrac{\log a}{\log b}$. However, there is another way to solve $b^x = a$ for x. Simply take the log of both sides of the equation to get

$$\log b^x = \log a.$$

Now use property (10.8) from section 10.3 to get

$$x \log b = \log a,$$

then

$$x = \dfrac{\log a}{\log b}.$$

Consider examples 10.22 and 10.23.

• **Example 10.22:** Solve the equation $2^x = 7$ for x.

Solution: By taking the logarithm of both sides of the equation, we obtain

$$\log 2^x = \log 7$$

or

$$x \log 2 = \log 7, \quad \text{(Property 10.8)}$$

so

$$x = \frac{\log 7}{\log 2}. \quad \text{(Divide by log 2)}$$

With a calculator we can get an approximate solution by keying in

$$7 \; \boxed{\log} \; \boxed{\div} \; 2 \; \boxed{\log} \; \boxed{=}.$$

The display should read approximately 2.80735. Hence, $2^{2.80\cdots} = 7$.

• • • • • • • • • •

• **Example 10.23:** Solve $2^x = 3^{x+1}$ for x.

Solution: Take the log of both sides of this equation to get

$$\log 2^x = \log 3^{x+1},$$

then

$$x \log 2 = (x + 1)\log 3,$$

or

$$x \log 2 = x \log 3 + \log 3$$
$$x \log 2 - x \log 3 = \log 3$$
$$x(\log 2 - \log 3) = \log 3$$
$$x = \frac{\log 3}{\log 2 - \log 3}.$$

An approximate solution can be obtained from a calculator by keying in

$$2 \; \boxed{\log} \; \boxed{-} \; 3 \; \boxed{\log} \; \boxed{=} \; \boxed{1/x} \; \boxed{\times} \; 3 \; \boxed{\log} \; \boxed{=}.$$

The display should show -2.7095113.

• • • • • • • • • •

SECTION 10.5 EXPONENTIAL AND LOGARITHMIC EQUATIONS

Now let us look at a logarithmic equation. Again we use the inverse relationship between exponentials and logarithms to solve this equation.

• **Example 10.24:** Solve the equation $2 \log x - \log 10x = 0$.

Solution: We can use the properties of logarithms and exponents to simplify the left-hand side of the given equation:

$$
\begin{aligned}
2 \log x - \log 10x &= 0 \\
2 \log x - (\log 10 + \log x) &= 0 \quad &\text{(Property 10.6)} \\
2 \log x - \log 10 - \log x &= 0 \quad &\text{(Distributive property)} \\
\log x - \log 10 &= 0 \quad &\text{(Subtract)} \\
\log x - 1 &= 0 \quad &\text{(Substitute 1 for log 10)} \\
\log x &= 1 \quad &\text{(Transpose 1)} \\
x &= 10^1 \quad &\text{(Express in exponential form)} \\
x &= 10.
\end{aligned}
$$

Example 10.25 illustrates how a formula from astronomy can be solved for one of its variables by using the properties of logarithmic and exponential functions. Two slightly different approaches to the solution of this problem are given.

• **Example 10.25:** The apparent magnitude m of a star is given by the formula

$$m = M + 5 \log \frac{r}{10},$$

where M is the absolute magnitude of the star and r is the distance of the star from the earth. Solve this formula for r.

Solution 1:

$$
\begin{aligned}
m &= M + 5 \log\left(\frac{r}{10}\right) \\
m - M &= 5 \log\left(\frac{r}{10}\right) \quad &\text{(Transpose M)} \\
m - M &= 5(\log r - \log 10) \quad &\text{(Property 10.7)} \\
m - M &= 5 \log r - 5 \log 10 \quad &\text{(Distribute 5)} \\
m - M &= 5 \log r - 5(1) \quad &\text{(Property 10.3)} \\
m - M &= 5 \log r - 5 \quad &\text{(Multiply)} \\
m - M + 5 &= 5 \log r \quad &\text{(Transpose } -5\text{)} \\
\frac{m - M + 5}{5} &= \log r \quad &\text{(Divide by 5)} \\
r &= 10^{(m - M + 5)/5}. \quad &\text{(Express in exponential form)}
\end{aligned}
$$

Solution 2:
$$m = M + 5 \log \frac{r}{10}$$

$$m - M = 5 \log \frac{r}{10} \quad \text{(Transpose } M\text{)}$$

$$\frac{m - M}{5} = \log \frac{r}{10} \quad \text{(Divide by 5)}$$

$$\frac{r}{10} = 10^{(m - M)/5} \quad \text{(Express in exponential form)}$$

$$r = 10(10^{(m - M)/5}). \quad \text{(Multiply by 10)}$$

It can be shown that these two solutions yield equivalent results:

$$r = 10(10^{(m - M)/5}) = 10^{5/5}(10^{(m - M)/5})$$
$$= 10^{5/5 + (m - M)/5} = 10^{(m - M + 5)/5}.$$

Example 10.26 illustrates that there may be two solutions to a logarithmic equation.

• **Example 10.26:** Solve for x in the logarithmic equation

$$\log(x^2 - 9x + 120) = 2.$$

Solution: Express the given equation in exponential form to get

$$x^2 - 9x + 120 = 10^2.$$
$$x^2 - 9x + 120 = 100 \quad \text{(Square 10)}$$
$$x^2 - 9x + 120 - 100 = 0 \quad \text{(Transpose)}$$
$$x^2 - 9x + 20 = 0 \quad \text{(Subtract)}$$
$$(x - 5)(x - 4) = 0. \quad \text{(Factor)}$$

Setting each factor equal to zero we get

$$x - 5 = 0 \quad \text{or} \quad x - 4 = 0$$
$$x = 5 \qquad\qquad x = 4.$$

The student should check these values in the original equation.

Although the computer and the hand-held calculator have made calculations with logarithms almost obsolete, example 10.27 illustrates how a problem can be solved by using logarithms and a calculator, and how the same problem can be solved by using the calculator alone without logarithms.

SECTION 10.5 EXPONENTIAL AND LOGARITHMIC EQUATIONS

• **Example 10.27:** The formula for the period T of a simple ideal pendulum is expressed as

$$T = 2\pi\sqrt{\frac{L}{g}},$$

where L is the length of the pendulum in feet and g is 32 feet per second^2. Use logarithms to calculate T when L is 3 feet. Then use a calculator to find T without using logarithms.

Solution: Using logarithms,

$$T = 2\pi\sqrt{\frac{L}{g}}$$

$$\log T = \log\left(2\pi\sqrt{\frac{L}{g}}\right) \quad \text{(Take the log of both sides)}$$

$$= \log 2 + \log \pi + \log\sqrt{\frac{L}{g}}$$

$$= \log 2 + \log \pi + \log\left(\frac{L}{g}\right)^{1/2}$$

$$= \log 2 + \log \pi + \frac{1}{2}(\log L - \log g)$$

$$= \log 2 + \log 3.142 + \frac{1}{2}(\log 3 - \log 32)$$

$$= 0.3010 + 0.4971 + 0.5(0.4771 - 1.5052)$$

$$\log T = 0.2841$$
$$T = 1.924.$$

Using a calculator,

$$T = 2\pi\sqrt{\frac{L}{g}}$$

$$= 2(3.142)\sqrt{\frac{3}{32}}.$$

Now enter

$$3 \;\boxed{\div}\; 32 \;\boxed{=}\; \boxed{\sqrt{}} \;\boxed{\times}\; 3.142 \;\boxed{\times}\; 2 \;\boxed{=}$$

The display should now read 1.924. Note that this is approximately the same value for T as was calculated by logs.

Trial Problems 10.5

Before you begin the section exercises, warm up with these problems. Complete answers are included in the answer key.

Solve for x. Round to five significant digits where appropriate.

1. $5^x = 3$
2. $6^{-x} = 10$
3. $\log(x - 4) = 4$
4. $\log x - \log 4 = \log(x - 1)$
5. $\log(x^2 - 5x - 4) = 1$

Exercises 10.5

Solve for x.

1. $3^x = 2$
2. $2^x = 3$
3. $4^x = \dfrac{5}{3}$
4. $3^x = \dfrac{5}{4}$
5. $7^{-x} = 8$
6. $8^{-x} = 7$
7. $5^{1/x} = 6$
8. $6^{1/x} = 5$
9. $\log x = -b - \log n$
10. $\log 125 = x - 3 \log 2$
11. $\log(x^2 - 15x) = 2$
12. $\log(x^2 - 3x + 6) = 1$
13. $\log_6 x = 2.5$
14. $\log_2 x = -4$
15. $\log x - 5 \log 3 = -2$
16. $\log_{2x} 216 = 3$
17. $\log(x + 9) + \log x = 1$
18. $\log x + \log(x + 21) = 2$
19. $\log(x + 2) + \log(x - 1) = 1$
20. $\log(x + 3) - \log(x - 1) = 1$

Solve.

21. The number of milligrams of radium present at the end of t years is given by the formula $A = A_0 10^{-0.000174t}$. Solve for t.

22. The intensity I of an X-ray beam after passing through x centimeters of a certain substance is given by the formula $I = I_0 e^{-kx}$. Solve for x.

23. The surface area of a sphere is given by the formula $s = 4\pi r^2$. Use logs to find the surface area of the earth if $r = 3959$ miles and $\pi = 3.142$. Now use a calculator to find s without using logs.

24. The effective value of the current I in an alternating current circuit is given by the formula $I = \dfrac{I_m}{\sqrt{2}}$. Use logs to determine I if $I_m = 2.542$ amp. Now, use a calculator to find I without using logs.

10·6 Summary of Terms, Formulas, Rules, and Procedures

Terms

Antilog (p. 429)
Asymptote (p. 415)
Common Logarithm (p. 428)
Constant Function (p. 414)
Decreasing Function (p. 416)
Exponential Function (p. 414)
Increasing Function (p. 416)
Logarithmic Function with Base b (p. 419)
Logarithm of a Power (p. 424)
Logarithm of a Product (p. 424)
Logarithm of a Quotient (p. 424)
Natural Logarithms (p. 428)

Formulas

Exponential Function: $y = b^x$ $(b > 0)$. (10.1)

Logarithmic Function: $y = \log_b x$. (10.2)

Rules and Procedures

- **PROPERTIES OF EXPONENTIAL FUNCTIONS AND THEIR GRAPHS.** (10.1)
  1. If $b > 1$, $y = b^x$ is an increasing function.
  2. If $0 < b < 1$, $y = b^x$ is a decreasing function.
  3. The domain of $y = b^x$ consists of all real numbers.
  4. The range of $y = b^x$ consists of the positive real numbers.
  5. The graph of $y = b^x$ approaches the x-axis as an asymptote.
  6. All functions of the form $y = b^x$ intersect the y-axis at the point $(0,1)$.

- **PROPERTIES OF LOGARITHMS.** (10.3)
$$\log_b b = 1.$$
$$\log_b 1 = 0.$$
$$b^{\log_b x} = x.$$

- **LOGARITHM OF A PRODUCT.**
$$\log_b xy = \log_b x + \log_b y. \quad (10.3)$$

- **LOGARITHM OF A QUOTIENT.**
$$\log_b \frac{x}{y} = \log_b x - \log_b y. \quad (10.3)$$

- **LOGARITHM OF A POWER.**
$$\log_b x^p = p \log_b x. \quad (10.3)$$

- **LAWS OF EXPONENTS.** (10.3)
$$a^m a^n = a^{m+n}$$
$$\frac{a^m}{a^n} = a^{m-n}$$
$$(a^m)^n = a^{mn}.$$

- **CHANGE OF BASE PROPERTY.** (10.4)
$$\log_b x = \frac{\log x}{\log b}.$$

10·7 Chapter 10 Review Exercises

Graph the following for the given values of x.

1. $y = 3^x$, $-2 \leq x \leq 2$
2. $y = 2^x$, $-3 \leq x \leq 3$

Express each in logarithmic notation.

3. $2^3 = 8$
4. $3^2 = 9$
5. $49^{1/2} = 7$
6. $64^{1/6} = 2$
7. $64^{-1/3} = \frac{1}{4}$
8. $8^{-1/3} = \frac{1}{2}$

Express in exponential notation.

9. $\log_5 25 = 2$
10. $\log_7 49 = 2$
11. $\log 0.01 = -2$
12. $\log 0.001 = -3$
13. $\ln x = 3$
14. $\ln y = 5$

Evaluate.

15. $\log_4 16$
16. $\log_5 25$
17. $\log_5 \sqrt{5}$
18. $\log_3 \sqrt{3}$
19. $\log 100$
20. $\log 1000$

Solve for the variable.

21. $\log_3 x = 2$
22. $\log_2 x = 5$
23. $\log_3 9 = y$
24. $\log_2 8 = y$
25. $\log_x 8 = 3$
26. $\log_y 9 = 2$
27. $\log_a 25 = 2$
28. $\log_b 81 = 4$

Express each of the following as a single logarithm with a coefficient of 1.

29. $\log_b 2 + \log_b x$
30. $\log_b 3 + \log_b y$
31. $\log y - \log z$
32. $\log 3 - \log x$
33. $3 \log_b x - 5 \log_b y$
34. $5 \log_b x + 2 \log_b y$

Find the exact values of each.

35. $\log_5 \dfrac{(125)(625)}{25}$
36. $\log_4 \dfrac{(64)(1024)}{256}$
37. $10^{\log 7}$
38. $10^{\log 8}$

Calculate to 6 places.

39. $\log 32.58$
40. $\log 23.85$
41. $\ln 0.632$
42. $\ln 0.853$

Calculate the value of the variable to six places.

43. $\log x = 4$
44. $\log x = 5$
45. $\ln x = 3.75$
46. $\ln x = 5.37$

Given that $\log_a 2 = 0.69$, $\log_a 3 = 1.10$, $\log_a 5 = 1.61$, and $\log_a 7 = 1.95$, find:

47. $\log_a \dfrac{15}{21}$
48. $\log_a \dfrac{35}{6}$
49. $\log_a \sqrt{\dfrac{5}{7}}$
50. $\log_a \sqrt[3]{\dfrac{7}{2}}$

Solve for x.

51. $4^x = 5$
52. $5^x = 4$
53. $3^x = \dfrac{5}{2}$
54. $2^x = \dfrac{5}{3}$
55. $3^{1/x} = 5$
56. $4^{1/x} = 6$
57. $\log x = -5 - \log n$
58. $\log x = 5 \log 3 - 2$

Solve the following.

59. The amount of light passing through a certain medium is expressed by $L = 100(10^{-0.3010})$. Find L using logarithms.

60. In a certain circuit the voltage is expressed as
$$v = (1.35 \times 10^{-2})(56.8).$$
Use logarithms to find v.

61. The maximum bending moment of a beam supported at both ends is given by the formula $M = \dfrac{wl^2}{8}$, where w is the weight per unit length and l is the distance between supports. Use logarithms to find M if $l = 25.3$ ft and $w = 4.6$ lbs.

62. The intensity level of a sound is given by the formula

$$B = 10 \log \frac{I}{I_0},$$

where B is expressed in decibels. Find B if $I_0 = 10^{-15}$ watt/cm² and $I = 10^{-9}$ watt/cm².

63. The number of *E. coli* bacteria present in an experimental colony is given by the equation

$$N(t) = B_0 2^t,$$

where $N(t)$ is the number of bacteria at time t and B_0 is the number of bacteria initially. Find the number of bacteria present in this colony after five hours if there were 1,400,000 million initially. Use logarithms.

64. The process of radioactive decay can be expressed by the formula

$$I = I_0 e^{-Ct},$$

where I is the intensity at time t, I_0 is the initial intensity, and C is the decay constant. Use logarithms to find I if I_0 is 100 grams, the decay constant C is 0.000431/sec and t is 2320 seconds. Solve this problem using common logarithms, then solve the problem using natural logarithms.

65. Experimentation has shown that an approximate rule for atmospheric pressure at altitudes less than 80 km is as follows. Standard atmospheric pressure, 1035 grams per square centimeter, is halved for each 5.8 km of vertical ascent. Letting P denote atmospheric pressure at altitudes less than 80 km and h the altitude in km, we have

$$P = 1035 \left(\frac{1}{2}\right)^{h/5.8} \text{ g/cm}^2.$$

Use logarithms to compute the atmospheric pressure at an altitude of 40 km.

66. Use the formula from exercise 65 to find the altitude at which the pressure is 20% of standard atmospheric pressure. Use logarithms.

•••• 10·8 Chapter 10 Test

1. Using figure 10.1, if $x = 1$, find y.

2. Graph the equation $y = 3^x$ for $0 \le x \le 2$.

3. The number N that a certain culture of bacteria will have after t hours can be expressed by the equation $N = 3000(2^{t/5})$. How many bacteria will there be in 15 hours?

4. Express $3^y = x$ in logarithmic notation.

5. Express $\log_x 16 = y$ in exponential notation.

6. Find the value of $\log_2 32$.

7. Use the graph in figure 10.5 to approximate the value of x in the equation $1 = \log_e x$.

8. Express $\log 5x$ as the sum of two logarithms.

9. Express $5 \log_b x - 3 \log_b y$ as a single logarithm with a coefficient of 1.

10. Find the value of $\log_3(81) \frac{\sqrt[4]{243}}{9}$.

11. Use your calculator to find log 3.852. Round your answer to six significant digits.

12. Use your calculator to solve ln $x = 1.447$. Round your answer to five significant digits.

13. Given that log 2 = 0.30, log 3 = 0.48, and log 5 = 0.70, find $\log \sqrt[5]{\frac{2}{3}}$ to two significant digits.

14. Solve $4^x = \frac{3}{5}$ for x.

15. Solve log $x - 4$ log $3 = -1$ for x.

Chapter 11

Some Nonlinear Equations

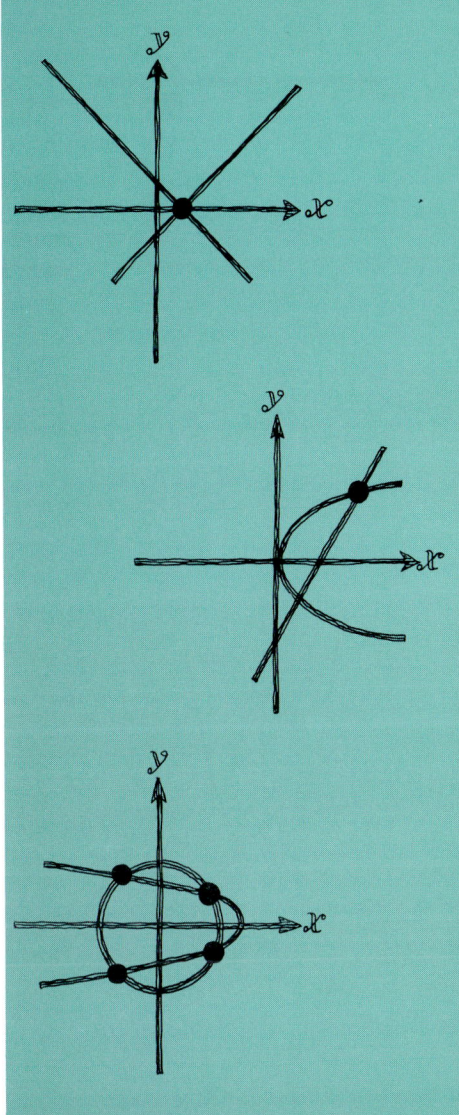

*I*n chapter 3, linear equations were introduced. These equations were called linear since their graphs were straight lines. Any equation whose graph is other than a straight line might be considered nonlinear. Our first encounter with nonlinear equations was in chapter 5 when quadratic equations were discussed. Later in chapter 10 exponential equations and logarithmic equations were developed. Both were nonlinear.

In this chapter we consider methods of solving certain systems of nonlinear equations. In some cases the system might involve one linear equation and one nonlinear equation. In other cases both of the equations in the system are nonlinear equations.

Just as the graphs of two noncoincident linear equations could meet in at most one point, the graphs of a linear equation and that of a parabola can meet in at most two points. A parabola and a circle could have as many as four points in common.

The chapter begins with the graphic solution of systems of nonlinear equations since the graph helps us to visualize the common solutions. Later the technique of eliminating a variable by substitution, by addition, or by subtraction is developed much like it was in chapter 4. In these processes the graphs are not drawn, however having an idea what the graphs might look like helps us to determine that we have found the proper solutions to the systems in question.

442 CHAPTER 11 SOME NONLINEAR EQUATIONS

11 • 1 Graphic Solutions of a System of Equations

In chapter 3 we graphed a few linear, quadratic, and other nonlinear equations. In chapter 4 we solved systems of linear equations by various means.

In this chapter we consider additional nonlinear equations as well as systems of nonlinear equations. We consider only the relatively simple case of two equations in two unknowns. The solution of such a system can be found (sometimes only approximately) by sketching the graphs of the two equations on the same coordinate system, and determining the points where the two graphs intersect. Examples 11.1 and 11.2 illustrate the solution of systems that involve some nonlinear equations by graphing.

• *Example 11.1:* Solve the following system by graphing:

$$x^2 - y = 4$$
$$x + y = -2.$$

Solution: The graph of $x^2 - y = 4$ or $y = x^2 - 4$ is a parabola opening upward. If we let $x = 0$, we find $y = -4$. Thus, the parabola passes through the point $(0, -4)$. The point $(0, -4)$ is also an extreme (minimum) point. Let $y = 0$, then $x = \pm 2$ and the curve crosses the x-axis at $(2,0)$ and $(-2,0)$. The graph of $x + y = -2$ is a straight line. The points $(0, -2)$ and $(-2, 0)$ are the intercepts. Hence, we see in figure 11.1 that the straight line intersects the parabola at two points. From figure 11.1 we can estimate these points of intersection to be approximately $(-2, 0)$ and $(1, -3)$.

We find that both of these solutions check when we substitute them into the given system of equations.

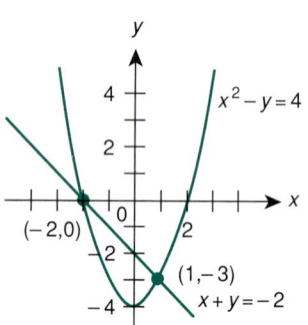

Figure 11.1

• *Example 11.2:* Solve the following system by graphing:

$$x^2 - y = 3$$
$$x^2 + y = 5.$$

Solution: The graph of $x^2 - y = 3$ is a parabola opening upward. It intersects the y-axis at $(0, -3)$ and the x-axis at $(\sqrt{3}, 0)$ and $(-\sqrt{3}, 0)$. The point $(0, -3)$ is the extreme (minimum) point. The graph of $x^2 + y = 5$ is also a parabola, but this one opens downward. It intersects the axes at the points $(0, 5)$, $(\sqrt{5}, 0)$, and $(-\sqrt{5}, 0)$. The point $(0, 5)$ is an extreme (maximum) point. The graph of this system is shown in figure 11.2.

We can estimate the two points of intersection to be $(2, 1)$ and $(-2, 1)$. Both of these solutions check when substituted into the given system.

SECTION 11.1 GRAPHIC SOLUTIONS OF A SYSTEM OF EQUATIONS 443

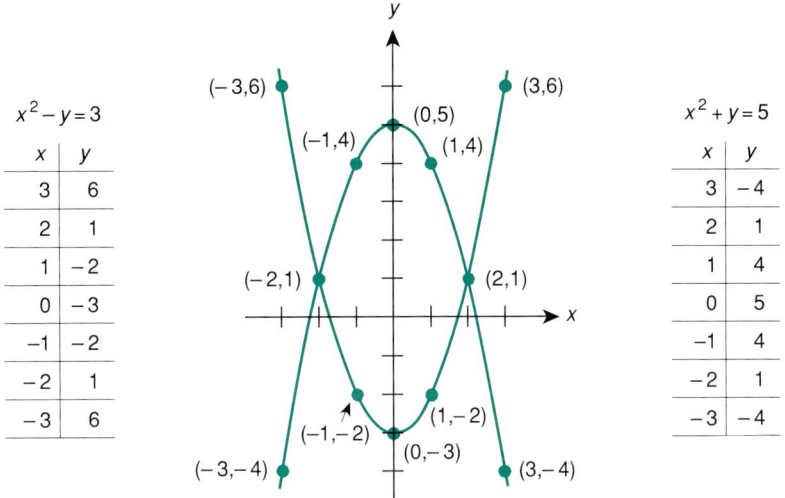

Figure 11.2

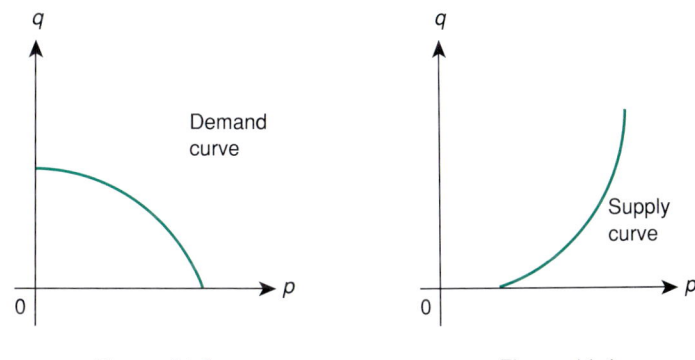

Figure 11.3   Figure 11.4

Demand Curve   If p denotes the price per unit of a commodity, and q denotes the quantity demanded in the marketplace at price p, a graph of q as a function of p produces a **demand curve**. Figure 11.3 illustrates a demand curve.

On the other hand, if p represents the price per unit of a commodity in the marketplace, and q denotes the quantity that the manufacturers are willing to supply at **Supply Curve**   that price, the graph of q as a function of p produces a **supply curve**. See figure 11.4.

If the demand curve and the supply curve are graphed on the same coordinate **Market Equilibrium Point**   system, the point of intersection of the graphs is called the **market equilibrium point**. At the equilibrium price p the quantity supplied equals the quantity demanded. This is illustrated in figure 11.5.

In example 11.3, the equations that produce the supply curve and the demand curve represent parabolas. Under the usual interpretations of price, supply, and demand, only the portions of the supply and demand curves that fall in the first quadrant are economically meaningful.

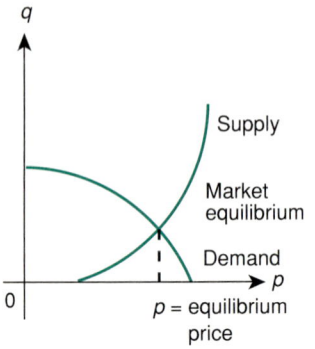

Figure 11.5

• **Example 11.3:** Suppose that the quantity q (in million of gallons) of diesel fuel demanded at a price p in dollars per gallon produces the demand equation $8p^2 + 5q = 100$. Suppose also that an oil company is willing to supply q million gallons of diesel fuel per week at the market price of p dollars per gallon according to the supply equation $6p^2 - p - 3q = 5$. Find the market equilibrium point and the equilibrium price.

Solution: The market equilibrium point is found by solving the system

$$8p^2 + 5q = 100 \quad \text{(Demand)}$$
$$6p^2 - p - 3q = 5 \quad \text{(Supply)}$$

The demand equation intersects the q-axis at $(0,20)$ and the p-axis at $(+\sqrt{12.5},0)$ and $(-\sqrt{12.5},0)$. By letting $p = 1, 2, 3,$ and 4 we find q to be $\dfrac{92}{5}, \dfrac{68}{5}, \dfrac{28}{5}$, and $-\dfrac{28}{5}$, respectively. The supply equation intersects the q-axis at $\left(0, -\dfrac{5}{3}\right)$ and the p-axis at $\left(-\dfrac{5}{6}, 0\right)$ and $(1,0)$. By letting $p = 1, 2, 3,$ and 4 we find q to be $0, \dfrac{17}{3}, \dfrac{46}{3}$, and 29, respectively.

Figure 11.6 illustrates the demand and supply curves on the same coordinate system.

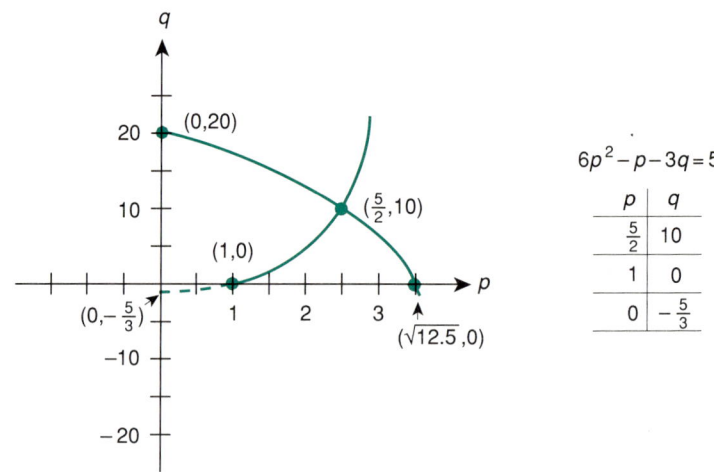

Figure 11.6

We can now estimate the market equilibrium point to be $\left(\dfrac{5}{2}, 10\right)$. Thus the equilibrium price is $\dfrac{5}{2}$ dollars or $2.50 per gallon. At that price the oil company is willing to supply 10 million gallons.

Trial Problems 11.1

Before you begin the section exercises, warm up with these problems. Complete answers are included in the answer key.

1. The graphs of a parabola and a straight line might have how many points in common?
2. The graphs of two parabolas might have how many points in common?
3. The graphs of a supply curve and a demand curve usually have how many points in common?
4. What is the name of the point where the graph of a supply curve crosses the graph of a demand curve?
5. The graph of $y = \log x$ and the graph of $y = x^2$ have how many points in common?

Exercises 11.1

Sketch graphs of each system of equations and estimate the solutions.

1. $x^2 - 2y = 0$
   $x + 2y = 6$

2. $2x^2 - y = 0$
   $x + y = 1$

3. $y^2 - 3x = 3$
   $3x + 3y = 1$

4. $y^2 - 6x = -7$
   $3x - 2y = 2$

5. $2x^2 + y = 4$
   $-3x - y = 1$

6. $x^2 + y = 9$
   $y - 2x = 1$

Find the market equilibrium point by graphing the supply and demand equations. Interpret your results.

7. $p^2 - 2q = 0$ (Supply equation)
   $3p + 2q = 10$ (Demand equation)

8. $p^2 - 3q = 0$ (Supply equation)
   $5p + 3q = 24$ (Demand equation)

9. If the supermarket price for a certain cut of beef is p dollars per pound, then q million pounds will be sold according to the demand equation $p^2 + q = 5$. When the supermarket price is p dollars per pound, a packing company agrees to supply q million pounds of the meat according to the supply equation $p - q - 1 = 0$. Solve graphically to find the equilibrium price and the number of pounds supplied at that price.

10. If a contractor's price for a missile component is p dollars per unit, then the government will purchase q hundred units according to the demand equation $p^2 + q = 402$. When the government price is p dollars per unit, a contractor is willing to supply q hundred units according to the supply equation $5p - q = 98$. Solve graphically to find the equilibrium price and the number of units supplied at that price.

11. A rectangular region has an area of 20 square meters and a perimeter of 18 meters. Find two equations in two unknowns and solve the system graphically to find the length and width of the region.

12. The variables x and y are related by the logarithmic equation $y = \log_2 x$ and the equation of a parabola $y = x^2 - 1$. Solve the system of equations graphically.

13. The kinetic energy of an object is determined with the formula

$$\text{K.E.} = \frac{1}{2}mv^2,$$

where K.E. is measured in J (joules), m is the mass in kilograms, and v is the velocity in meters per second. If the kinetic energy of two objects whose velocities are v_1 and v_2, is expressed by the equation $3v_1^2 + 4v_2^2 = 3991$, and $v_1 - v_2 = 2$, solve the system graphically.

11 • 2 Algebraic Solutions of a System of Equations

In section 11.1 we solved systems of nonlinear equations by graphing. Graphing was not only a tedious process but our solutions were often only approximations. In chapter 4 we solved systems of linear equations by various methods including substitution, addition, and subtraction. We now use these same techniques to solve some nonlinear systems. We refer to these techniques as the **algebraic solution** of systems of equations.

Algebraic Solution

The system of equations in example 11.4 is first graphed to illustrate the nature of the solutions, then the system is solved by the method of addition.

• **Example 11.4:** Solve algebraically the system of equations

$$x^2 - 2y = 0$$
$$x + 2y = 6.$$

Solution: The graph of $x^2 - 2y = 0$ is a parabola opening upward with vertex at the origin. The vertex is also the extremum (minimum) point. The graph of $x + 2y = 6$ is a straight line that intersects the parabola at two points. This is illustrated in figure 11.7.

To find the two solutions algebraically, we use the method of addition since the coefficients of the y terms are additive inverses (-2 and $+2$) and their sum produces $0y$. If the second equation is added to the first, we obtain

$$x^2 - 2y = 0$$
$$\underline{x + 2y = 6}$$
$$x^2 + x + 0y = 6.$$

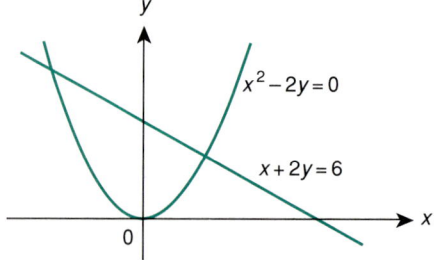

Figure 11.7

Using $x^2 + x = 6$ as our new first equation we have the equivalent system

$$x^2 + x = 6$$
$$x + 2y = 6.$$

Note that we have eliminated y from the first equation. The first equation is now a quadratic in x, which we solve by factoring. Thus,

$$x^2 + x = 6$$
$$x^2 + x - 6 = 0 \quad \text{(Transpose 6)}$$
$$(x + 3)(x - 2) = 0. \quad \text{(Factor)}$$

Therefore,

$$x = -3 \text{ or } x = 2. \quad \text{(Solve for } x\text{)}$$

Substituting these values of x into the second equation, we get

$$x + 2y = 6$$
$$(-3) + 2y = 6 \quad \text{(Substitute } -3 \text{ for } x\text{)}$$
$$2y = 6 + 3 \quad \text{(Transpose } -3\text{)}$$
$$2y = 9 \quad \text{(Add)}$$
$$y = \frac{9}{2}, \quad \text{(Divide by 2)}$$

and

$$x + 2y = 6$$
$$(2) + 2y = 6 \quad \text{(Substitute 2 for } x\text{)}$$
$$2y = 6 - 2 \quad \text{(Transpose 2)}$$
$$2y = 4 \quad \text{(Subtract)}$$
$$y = 2. \quad \text{(Divide by 2)}$$

Therefore,

$$y = \frac{9}{2} \quad \text{when } x = -3,$$

and

$$y = 2 \quad \text{when } x = 2.$$

Hence, our two solutions are $\left(-3, \dfrac{9}{2}\right)$ and $(2,2)$. Check these solutions in both of the original equations of the system.

448 CHAPTER 11 SOME NONLINEAR EQUATIONS

• *Example 11.5:* Solve algebraically the system

$$x^2 + y^2 = 25$$
$$x^2 + y = 13.$$

Solution: Subtracting the second equation from the first (to eliminate x^2), we obtain the equivalent system

$$y^2 - y = 12$$
$$x^2 + y = 13.$$

The first equation is now a quadratic in y and can be solved by factoring. Thus,

$$y^2 - y = 12$$
$$y^2 - y - 12 = 0 \quad \text{(Transpose 12)}$$
$$(y - 4)(y + 3) = 0. \quad \text{(Factor)}$$

Hence,

$$y = 4, \quad \text{or} \quad y = -3. \quad \text{(Solve for } y\text{)}$$

Substituting $y = 4$ into the second equation, we get

$$x^2 + y = 13$$
$$x^2 + (4) = 13 \quad \text{(Substitute 4 for } y\text{)}$$
$$x^2 = 13 - 4 \quad \text{(Transpose 4)}$$
$$x^2 = 9 \quad \text{(Subtract)}$$
$$x = \pm 3. \quad \text{(Extract roots)}$$

Substituting $y = -3$ into the second equation, we get

$$x^2 + y = 13$$
$$x^2 + (-3) = 13 \quad \text{(Substitute } -3 \text{ for } y\text{)}$$
$$x^2 = 13 + 3 \quad \text{(Transpose)}$$
$$x^2 = 16 \quad \text{(Add)}$$
$$x = \pm 4. \quad \text{(Extract roots)}$$

Hence, our solutions are

$$(3,4)(-3,4)(4,-3)(-4,-3).$$

Had we graphed the equations, the graphs would have looked much like those in figure 11.8.

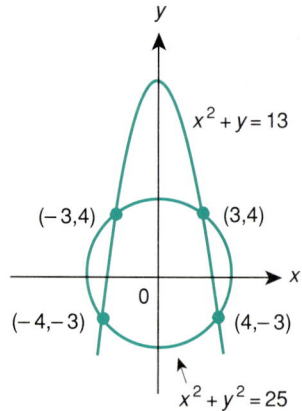

Figure 11.8

• *Example 11.6:* Solve algebraically the system of equations

$$y - \log_4(6x + 10) = 1$$
$$y + \log_4 x = 2.$$

Solution: If we subtract the first equation from the second, we eliminate the y term and obtain
$$\log_4 x + \log_4(6x + 10) = 1.$$
Using the properties of logarithms and exponentials from chapter 10, we rewrite the last equation as
$$\log_4[x(6x + 10)] = 1 \quad \text{(Property 10.6)}$$
or
$$x(6x + 10) = 4^1. \quad \text{(Exponential form)}$$
Solving this equation for x, we get
$$\begin{aligned} x(6x + 10) &= 4 \\ 6x^2 + 10x &= 4 & \text{(Distribute } x\text{)} \\ 6x^2 + 10x - 4 &= 0 & \text{(Transpose 4)} \\ 3x^2 + 5x - 2 &= 0 & \text{(Divide by 2)} \\ (3x - 1)(x + 2) &= 0. & \text{(Factor)} \end{aligned}$$
Hence,
$$x = \frac{1}{3}, \quad \text{or} \quad x = -2. \quad \text{(Solve for } x\text{)}$$

Substituting these values of x into either of the two given equations produces the corresponding values of y. We choose to substitute the values of x into the second equation. Thus, when $x = \frac{1}{3}$ we get
$$\begin{aligned} y + \log_4 x &= 2 \\ y + \log_4\left(\frac{1}{3}\right) &= 2 & \left(\text{Substitute } \frac{1}{3} \text{ for } x\right) \\ y &= 2 - \log_4 \frac{1}{3} & \text{(Transpose)} \\ y &= 2 - \log_4 3^{-1} & \left(\text{Express } \frac{1}{3} \text{ as } 3^{-1}\right) \\ y &= 2 + \log_4 3. & \text{(Property 10.8)} \end{aligned}$$

Hence, when $x = \frac{1}{3}$, $y = 2 + \log_4 3$. This solution could be expressed as the ordered pair $\left(\frac{1}{3}, 2 + \log_4 3\right)$.

Since $\log_4 x$ is not defined for negative values of x, there is no corresponding value for y when $x = -2$. Hence the only solution is $\left(\frac{1}{3}, 2 + \log_4 3\right)$.

Example 11.7 uses the method of substitution to solve the given system algebraically.

- **Example 11.7:** Solve algebraically the system

$$y = 10^x - 15$$
$$y = 10^{2-x}.$$

Solution: If we substitute the value of y from the first equation into the second equation we obtain

$$10^x - 15 = 10^{2-x}.$$

Using the properties of exponents from chapter 10, we rewrite this last equation and solve for x:

$$10^x - 10^{2-x} - 15 = 0 \quad \text{(Transpose)}$$

$$10^x - \frac{10^2}{10^x} - 15 = 0 \quad \left(\text{Express } 10^{2-x} \text{ as } \frac{10^2}{10^x}\right)$$

$$(10^x)^2 - 10^2 - 15(10^x) = 0 \quad \text{(Multiply by } 10^x \text{ to clear fractions)}$$

$$(10^x)^2 - 15(10^x) - 100 = 0 \quad \text{(Substitute 100 for } 10^2\text{)}$$

$$(10^x - 20)(10^x + 5) = 0. \quad \text{(Factor)}$$

Thus,

$$10^x - 20 = 0 \quad \text{or} \quad 10^x + 5 = 0$$
$$10^x = 20 \quad \text{or} \quad 10^x = -5.$$

Hence,

$$x = \log 20$$

or,

$$x = \log(-5). \quad \text{(Not a solution)}$$

Substitute this value of x into either of the given equations to obtain the corresponding value of y. If we substitute into the first equation, we get

$$y = 10^x - 15$$
$$y = 10^{\log 20} - 15 \quad \text{(Substitute log 20 for } x\text{)}$$
$$y = 20 - 15 \quad \text{(Simplify)}$$
$$y = 5. \quad \text{(Subtract)}$$

Thus, our solution is $(\log 20, 5)$. Check this solution in the original system of equations.

Trial Problems 11.2

Before you begin the section exercises, warm up with these problems. Complete answers are included in the answer key.

1. What is the maximum number of points that the graph of a circle and the graph of a parabola might have in common?

2. What is the minimum number of points that the graph of a parabola and the graph of a straight line might have in common?

3. What is the maximum number of points that the graphs of two straight lines might have in common?

4. Solve algebraically the system

$$y = x^2 + 1$$
$$2x + y = 4.$$

5. Solve algebraically the system

$$y = \log x$$
$$y = \log 4(x - 1).$$

Exercises 11.2

Solve algebraically the following systems of equations.

1. $x^2 - 2y = 0$
   $3x + 2y = 10$

2. $x^2 - 3y = 0$
   $5x + 3y = 6$

3. $x^2 + y^2 = 18$
   $x - y = 0$

4. $x^2 + y^2 = 8$
   $x - y = 0$

5. $y - \log_6(x - 1) = 1$
   $y + \log_6 x = 2$

6. $y - \log(x + 3) = 1$
   $y + \log(x + 4) = 2$

7. The ratio of two positive numbers is 5 to 8, and their product is 160. Find the numbers.

8. The ratio of two positive numbers is 5 to 6, and their product is 270. Find the numbers.

9. Find two numbers whose sum is 4 and whose product is -96.

10. What are the dimensions of a rectangle whose perimeter is 36 meters and whose area is 80 square meters?

11. The product of two numbers is 5. The sum of their cubes is 30. Find the numbers.

12. The difference of two numbers is 2. The sum of their squares is 100. Find the numbers.

13. A rectangular piece of fiberglass has an area of 14.0 cm² and a perimeter of 15.0 cm. Set up two equations in two variables and solve the system algebraically to find the dimensions of the piece.

14. The height h_a (in meters) of an aircraft at a given time t (in seconds) is given by the formula

$$h_a = 18{,}000 - 300t.$$

When the aircraft is at 18,000 m a missile is fired from the ground. The height of the missile is given by the formula

$$h_m = 3000t - 150t^2.$$

Find the time t when the aircraft and the missile will be at the same height. What will be the height of the aircraft and the missile at this time? Interpret your answer.

15. The kinetic energy of two objects with velocities v_1 and v_2 is expressed in the following equation

$$4v_1^2 + 3v_2^2 = 4503.$$

The difference in the velocities of the two objects is expressed by the formula

$$v_1 - v_2 = 4.$$

Solve this system algebraically.

11•3 Radical Equations

The concept of a radical was introduced in section 1.4. Now we are ready to consider the solutions of equations that contain radicals. An equation in which the unknown appears in a radicand is called a **radical equation**. For example,

Radical Equation

$$\sqrt[3]{x+3} = 5 \quad \text{and} \quad \sqrt{x+5} + \sqrt{x+12} = 7$$

are radical equations.

The procedure for solving a radical equation might be summarized as follows.

SOLVING RADICAL EQUATIONS

1. Identify one of the radical expressions.
2. Isolate this radical expression by transposing all other terms to the other side of the equation.
3. Eliminate the radical that has been isolated by raising both sides of the equation to a power equal to the index of the isolated radical.
4. If any radicals remain, repeat steps 1, 2, and 3.
5. Check all possible answers.

Examples 11.8–11.11 illustrate this procedure.

• **Example 11.8:** Solve the equation

$$\sqrt{x-3} - 5 = 0.$$

Solution: First we must isolate the radical by transposing the -5. This yields

$$\sqrt{x-3} = 5.$$

Now we raise both sides to the power of 2 (that is, we square both sides of the equation) and obtain

$$(\sqrt{x-3})^2 = (5)^2$$

or

$$x - 3 = 25$$
$$x = 25 + 3.$$

Hence,

$$x = 28.$$

To check the answer we substitute 28 for x in the original equation, and obtain

$$\sqrt{x-3} - 5 = 0$$
$$\sqrt{(28)-3} - 5 = 0$$
$$\sqrt{25} - 5 = 0$$
$$0 = 0.$$

Hence, the solution checks.

• *Example 11.9:* Solve the equation

$$\sqrt[3]{x-4} - 2 = 0.$$

Solution: First we isolate the radical by transposing the -2. This yields

$$\sqrt[3]{x-4} = 2.$$

Now we raise both sides to the power of 3 and obtain

$$(\sqrt[3]{x-4})^3 = (2)^3$$
$$x - 4 = 8$$
$$x = 8 + 4.$$

Hence,

$$x = 12.$$

We find that this solution checks. Hence, $x = 12$ is an acceptable solution.

CHAPTER 11 SOME NONLINEAR EQUATIONS

• **Example 11.10:** Solve the equation
$$\sqrt{3x+7} + \sqrt{x+2} = 1.$$

Solution: Leave the radical expression $\sqrt{3x+7}$ on one side of the equation. Isolate it by transposing all other terms to the other side of the equation. This gives us the equation
$$\sqrt{3x+7} = 1 - \sqrt{x+2}.$$

Next square both sides of this equation and simplify as follows:

$(\sqrt{3x+7})^2 = (1 - \sqrt{x+2})^2$ (Square both sides)
$3x + 7 = 1 - 2\sqrt{x+2} + x + 2$ (Expand)
$3x + 7 = x - 2\sqrt{x+2} + 3$ (Commute and add)
$2x + 4 = -2\sqrt{x+2}$ (Transpose and add)
$x + 2 = -\sqrt{x+2}.$ (Divide by 2)

Since there is still one radical expression remaining, square both sides of the equation again. This gives

$(x+2)^2 = (-\sqrt{x+2})^2$ (Square both sides)
$x^2 + 4x + 4 = x + 2$ (Expand)
$x^2 + 3x + 2 = 0$ (Transpose and add)
$(x+2)(x+1) = 0$ (Factor)

Hence,
$$x + 2 = 0 \quad \text{or} \quad x + 1 = 0, \quad \text{(Solve for } x\text{)}$$

so
$$x = -2 \quad \text{or} \quad x = -1.$$

Checking these roots by substituting them into the original equation, we see that $x = -2$ checks, but $x = -1$ does not. Hence, the only solution is $x = -2$.

• **Example 11.11:** Solve the equation
$$\sqrt{y-5} - \sqrt{y} = 1.$$

Solution: Leave the radical expression $\sqrt{y-5}$ on the left-hand side of the equation. Isolate it by transposing all other terms to the right-hand side of the equation. This gives us the equation
$$\sqrt{y-5} = \sqrt{y} + 1.$$

Next square both sides of the equation and simplify as follows:

$$(\sqrt{y-5})^2 = (\sqrt{y}+1)^2 \quad \text{(Square both sides)}$$
$$y - 5 = y + 2\sqrt{y} + 1 \quad \text{(Expand)}$$
$$-6 = 2\sqrt{y} \quad \text{(Transpose and add)}$$
$$-3 = \sqrt{y}. \quad \text{(Divide by 2)}$$

Since there is still one radical expression remaining, square both sides of the equation again. This yields

$$(-3)^2 = (\sqrt{y})^2$$
$$9 = y.$$

Replacing y with 9 in the original equation gives

$$\sqrt{y-5} - \sqrt{y} = 1$$
$$\sqrt{9-5} - \sqrt{9} = 1$$
$$\sqrt{4} - \sqrt{9} = 1$$
$$2 - 3 = 1$$
$$-1 = 1.$$

This is obviously not true, so the equation has no solution. Careful observation of the equation

$$-3 = \sqrt{y}$$

should have indicated that there would be no solution, since $\sqrt{y}$ cannot be negative. Had we noticed this earlier we could have saved some work.

• • • • • • • • • •

Trial Problems 11.3

Before you begin the section exercises, warm up with these problems. Complete answers are included in the answer key.

Solve for x.

1. $\sqrt{x} - 6 = 2$
2. $\sqrt{x} - 7 = 9$
3. $\sqrt[3]{x} = 5$
4. $\sqrt{x-2} = 5$
5. $\sqrt{x+6} = x + 4$

Exercises 11.3

Solve and check.

1. $\sqrt{x} - 5 = 3$
2. $\sqrt{x} - 4 = 1$
3. $\sqrt{x+6} = 2$
4. $\sqrt{x-3} = 5$
5. $3x + 4 = \sqrt{3x+10}$
6. $2x - 3 = \sqrt{7x-3}$
7. $\sqrt[3]{y} = -3$
8. $\sqrt[3]{y} = -2$
9. $\sqrt[4]{t-1} = 2$

10. $\sqrt[4]{t-1} = 3$
11. $2s + 1 = \sqrt{10s + 5}$
12. $4s + 5 = \sqrt{3s + 4}$
13. $\sqrt{x - 3} + \sqrt{x} = 2$
14. $\sqrt{x - 3} + \sqrt{x} = 6$
15. $\sqrt{x + 4} = \sqrt{x + 20} - 2$
16. $\sqrt{2x - 5} = \sqrt{x - 2} + 2$
17. $\sqrt{5x + 6} - \sqrt{x + 1} = 1$
18. $\sqrt{4x + 17} + \sqrt{x + 1} = 4$
19. $4\sqrt{y} + \sqrt{1 + 16y} = 5$
20. $6\sqrt{y} - \sqrt{10y - 1} = 3$

21. Find the dimensions of a rectangle that has an area of 60 square centimeters and a diagonal that is 13 centimeters long.

22. In meteorology the heat loss (wind chill) is given by
$$H = (A + B\sqrt{V} - CV)(S - T).$$
Solve for the wind velocity V in terms of H, the neutral skin temperature S, the air temperature T, and the constants A, B, and C.

23. A spacecraft is at an altitude h above the earth. The distance from the spacecraft to the horizon is represented by d and the radius of the earth is r. The formula relating d, h, and r is
$$d = \sqrt{2rh + h^2}.$$
Solve this formula for r.

24. Solve the formula in exercise 23 for h.

25. The circular orbital velocity v is determined by the formula
$$v = \sqrt{\frac{GM}{r}},$$
where G is the constant of universal gravitation, M is the mass of the body, and r is the radius of the orbit measured from the center of mass of the body. Solve this formula for r.

26. The perimeter of a rectangular piece of metal is 42 cm. If the length is x and the width is $\sqrt{x + 9}$, find the length and the width of the piece.

11·4 Summary of Terms, Rules, and Procedures

Terms

Algebraic Solution (p. 446)   Market Equilibrium Point (p. 443)   Supply Curve (p. 443)
Demand Curve (p. 443)   Radical Equation (p. 452)

Rules and Procedures

- PROCEDURE FOR SOLVING EQUATIONS
  INVOLVING RADICALS   (11.3)
  1. Identify one of the radical expressions.
  2. Isolate this radical expression by transposing all other terms to the other side of the equation.
  3. Eliminate the radical that has been isolated by raising both sides of the equation to a power equal to the index of the isolated radical.
  4. If any radicals remain, repeat steps 1, 2, and 3.
  5. Check all possible answers.

11·5 Chapter 11 Review Exercises

Estimate the solutions by graphing.

1. $x^2 - y = 4$
   $x + y = -2$

2. $x^2 + y = 2$
   $x - y = 0$

Find the market equilibrium point by solving (by any method) these supply and demand equations.

3. $p^2 - 3q = 0$ (Supply equation)
   $3p + 3q = 10$ (Demand equation)

4. $p^2 - 4q = 0$ (Supply equation)
   $5p + 4q = 24$ (Demand equation)

Solve for the required variable.

5. The formula for the lateral surface area A of a right circular cone with height h and base radius r is

   $$A = \pi r \sqrt{r^2 + h^2}.$$

   Solve for r in terms of A and h.

6. Solve the equation for lateral surface area A from exercise 5 for h in terms of r and A.

Solve the systems.

7. The sum of two numbers is 13. The difference of their square roots is 1. Find the numbers.

8. The sum of two numbers is 16 and their product is 63. Find the numbers.

9. The sides of a certain triangle are $\sqrt{2x-1}$, $\sqrt{x+4}$, and 11. If the perimeter of the triangle is 17, find the value of x and the lengths of the sides of the triangle.

10. The impedance in a series circuit is given by the formula

    $$Z = \sqrt{R_t^2 + X_t^2},$$

    where R_t is the total resistance and X_t is the total reactance. Solve this formula for X_t.

11. The area of a properly designed chimney is determined by the formula

    $$A = \frac{0.03Q}{\sqrt{h}},$$

    where Q is the quantity of coal, in pounds, burned per hour and h is the height of the chimney in feet. Solve this formula for h.

12. The resonance frequency, f_r, of a parallel circuit can be calculated by the formula

    $$f_r = \frac{1}{2\pi\sqrt{LC}},$$

    where L is the inductance and C is the capacitance. Solve the formula for C.

13. The deflection D, in inches, of a simple beam is calculated by the formula

    $$D = \frac{PL^3}{34.9SA},$$

    where P is the total load in pounds, L is the length of the beam in feet, S is the safe load per square inch, and A is the cross section area of the beam. Solve this formula for L.

11·6 Chapter 11 Test

1. Find the point of market equilibrium by graphing the supply equation $p^2 - 2q = 5$ and the demand equation $3p + 2q = 13$.

2. Solve the system by graphing.
$$x^2 + y = 15$$
$$x^2 - y = 17$$

3. Solve algebraically the system
$$x^2 - 2y = 10$$
$$3x + 2y = 18.$$

4. Solve algebraically the system
$$x^2 + y^2 = 6$$
$$x^2 - y^2 = 2.$$

5. The ratio of two positive numbers is 4 to 3 and their product is 108. Find the numbers.

6. Solve $3x - 5 = \sqrt{3x + 7}$. Check your answers.

7. The difference of two numbers is 16. The difference of their square roots is 2. Find the numbers.

11·7 Cumulative Review—Chapters 9, 10, and 11

1. Given the magnitude $|\mathbf{V}| = 7.2$ and the angle $\theta = 35°$, which the vector makes with the horizontal, find the horizontal and vertical components.

2. If the magnitude of a vector $\mathbf{V}$ is 250, that of a vector $\mathbf{W}$ is 350, and the angle between the vectors is 25°, find the magnitude of the resultant vector.

3. Find the force required to pull a 65 pound weight up a ramp that is inclined 18° with the horizontal.

4. Find the sum of $(3 - 7j)$ and $(2 + 9j)$.

5. Find the product of $(11 - 6j)$ and $(4 + 2j)$.

6. Divide $(3 - 2j)$ by $(3 + 2j)$.

7. Solve the quadratic equation $x^2 - 3x + 4 = 0$.

8. Find the polar coordinates of $\dfrac{\sqrt{3}}{2} + \dfrac{1}{2}j$.

9. Write $\sin \dfrac{\pi}{4} + j \sin \dfrac{\pi}{4}$ in the form $a + bj$.

10. Find the product in polar form of $3\left(\cos \dfrac{\pi}{4} + j \sin \dfrac{\pi}{4}\right)$ and $5\left(\cos \dfrac{\pi}{3} + j \sin \dfrac{\pi}{3}\right)$.

11. Use De Moivre's theorem to calculate $\left(\cos \dfrac{\pi}{3} + j \sin \dfrac{\pi}{3}\right)^6$.

12. Use De Moivre's theorem to calculate $(1 + j)^5$.

13. Use De Moivre's theorem to find the fifth roots of $-16\sqrt{2} + 16\sqrt{2}\,j$.

14. Use the figure below to find an approximation for y when $x = 2$.

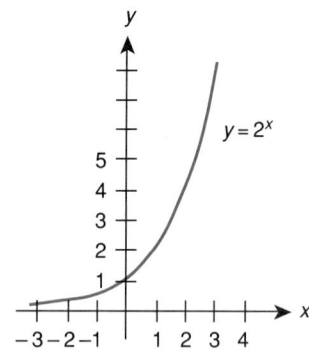

15. Use the figure at the right to find an approximation for x when $y = \dfrac{3}{2}$.

16. Express $2^x = y$ in logarithmic notation.

17. Express $\log_a 8 = b$ in exponential notation.

18. Find the value of $\log_5 25$.

19. Solve $\log_3 x = 2$ for x.

20. Express $\log_b \sqrt[3]{x^2}$ as a single logarithm without an exponent.

21. Express $14\log_b x - 6\log_b x$ as a single logarithm with a coefficient of 1.

22. Find the value of $\log_4 \dfrac{(64)(256)}{16}$.

23. Calculate log 2.358 to five significant digits.

24. Find x to five significant digits if $\log x = 1.618$.

25. If $\log_a 2 = 0.69$ and $\log_a 3 = 1.10$, find $\log_a \dfrac{9}{2}$.

26. If $4^x = \dfrac{5}{3}$, solve for x.

27. If $\log 8 = x - 3\log 5$, solve for x.

28. If $\log(x^2 - 8x + 25) = 1$, solve for x.

29. Sketch the graphs of the system of equations
$$2x^2 + y = 4$$
$$2x - y = 0$$
and estimate the solutions.

30. Solve algebraically the system
$$x^2 - y = 13$$
$$x^2 + y^2 = 25.$$

31. Solve algebraically the system
$$xy = 3$$
$$x - 3y = 0.$$

32. Solve the equation
$$\sqrt{x-1} = x - 3.$$
Check the results.

33. Solve the equation
$$\sqrt{3x+4} + \sqrt{x+2} = 8.$$
Check the results.

Chapter 12

Inequalities

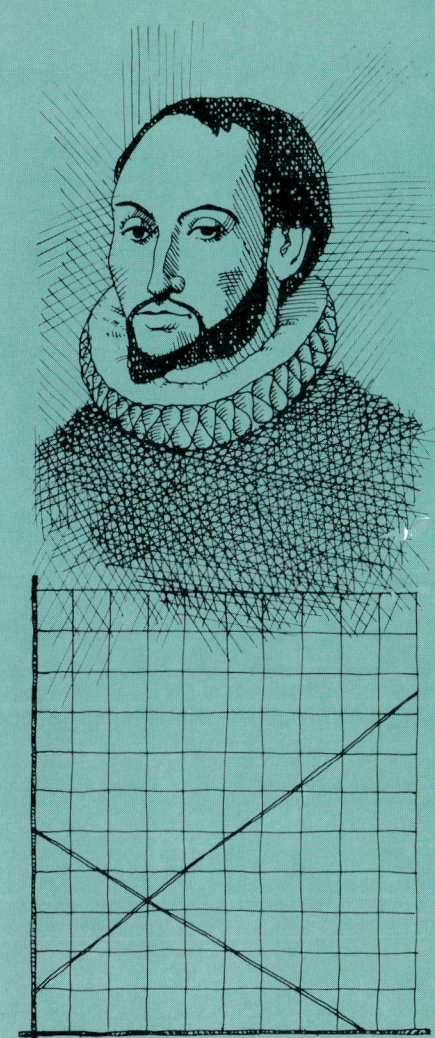

One of the greatest achievements in the history of mathematics has been the development of mathematical symbols. It is difficult to envision solving mathematical problems without writing equations and inequalities using symbols such as $=$, $\sqrt{}$, $\leq$, $\geq$, $+$, $-$, and superscripts for exponents.

The ancient Egyptians, for example, did not have notation that permitted them to write fractions unless the numerator of each fraction was one. Their notation for a fraction was an elliptical symbol over an integer. Thus, meant $\frac{1}{3}$ and meant $\frac{1}{4}$.

As late as the sixteenth century, equations were written without the benefit of exponents. It was common practice to use different symbols or letters to represent different powers of a variable. A fifteenth century mathematician, François Viète, was the first to use the same letter to represent different powers of a variable. He used x quad for x^2 and x cubum for x^3.

The $=$ symbol for equality was first used by Robert Recorde in 1557. Other symbols that improved mathematical notation were $<$ and $>$ for less or more, first used by Thomas Harriot in 1631. The $+$ and $-$ symbols first appeared in print in 1489, where they referred to surpluses and deficits in business problems. A Dutch mathematician, Vander Hoecke, first used $+$ and $-$ for plus and minus in algebraic expressions in 1514.

12•1 Linear Inequalities

In chapter 1 we defined the symbols $<$, $\leq$, $>$, and $\geq$, and we learned how to solve linear equations. In this section we consider the solutions to linear inequalities. An equation is a statement that two expressions are equal, while an inequality is a statement that one expression is less than ($<$), less than or equal to ($\leq$), greater than ($>$), or greater than or equal to ($\geq$) another expression.

LINEAR INEQUALITY

A **linear inequality** is defined as a statement that can be put in the form $ax + b < c$, where x is a variable and a, b, and c are constants.

Note that the symbol $<$ in the definition can be replaced by $\leq$, $>$, or $\geq$. We use the symbol $<$ in the rules for solving inequalities with the understanding that the same rules apply for all of the symbols.

As with equations, the set of values for which the inequality is a true statement is called the **solution set** of the inequality. Two inequalities that have the same solution set are **equivalent inequalities**.

Solution Set
Equivalent Inequalities

Inequalities are solved using the following rules.

PROPERTIES OF INEQUALITIES

Given that x, y, and b are real numbers,

1. If $x < y$, then $x + b < y + b$ and $x - b < y - b$.

2. If $x < y$ and $b > 0$, then $bx < by$ and $\dfrac{x}{b} < \dfrac{y}{b}$.

3. If $x < y$ and $b < 0$, then $bx > by$ and $\dfrac{x}{b} > \dfrac{y}{b}$.

Properties 1 and 2 correspond to the rules for solving linear equations. Property 3 states that multiplying or dividing both sides of an inequality by a negative number reverses the inequality symbol.

Property 3 can be illustrated as follows:

$$-5 < -3 \text{ is a valid statement.}$$

If we multiply both sides by -4,

$$(-4)(-5) > (-4)(-3) \qquad \text{or} \qquad 20 > 12$$

is a valid statement.

We solve linear inequalities using properties 1–3 in much the same manner as we solve linear equations. Examples 12.1 and 12.2 illustrate the use of these properties:

• **Example 12.1:** Solve $2x - 7 \leq -4$.

Solution: Add 7 to both sides of the inequality (Property 1):
$$(2x - 7) + 7 \leq -4 + 7$$
$$2x \leq 3.$$

Divide both sides of the inequality by 2 (Property 2):
$$\frac{2x}{2} \leq \frac{3}{2}$$
$$x \leq \frac{3}{2}.$$

Thus, any real number less than or equal to $\frac{3}{2}$ satisfies the inequality.

• **Example 12.2:** Solve $-3x + 17 < -13$.

Solution: Add -17 to both sides of the inequality (Property 1):
$$-3x + 17 + (-17) < -13 + (-17)$$
$$-3x < -30.$$

Divide both sides of the inequality by -3 (Property 3):
$$\frac{-3x}{-3} > \frac{-30}{-3} \quad \text{(Note that the inequality symbol is reversed since we divided by a negative number)}$$
$$x > 10.$$

Thus, any real number greater than 10 satisfies the inequality.

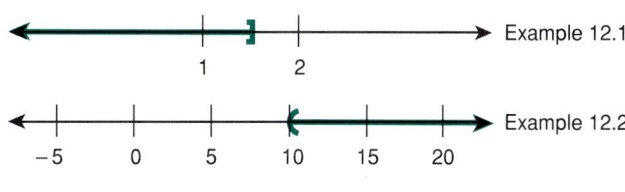

Figure 12.1

The results of examples 12.1 and 12.2 are shown graphically in figure 12.1. By custom we use a bracket] or [to indicate that the endpoint is included in the set and a parenthesis (or) to indicate that the endpoint is not included in the set. We can also

Interval Notation use the following notation to designate these solutions. We call this the **interval notation:**

$$x \le \frac{3}{2} \qquad \left(-\infty, \frac{3}{2}\right]$$

$$x > 10 \qquad (10, \infty).$$

The $-\infty$ symbol translates as "unbounded below," and the ∞ symbol translates as "unbounded above."

The following table illustrates various types of intervals, their inequality notation, and the corresponding interval notation. Figure 12.2 indicates how these intervals might be graphed on a real number line.

Type of Interval	Set	Interval Notation
open	$a < x$	(a, ∞)
open	$x < b$	$(-\infty, b)$
open	$a < x < b$	(a, b)
half-open	$x \ge a$	$[a, \infty)$
half-open	$x \le b$	$(-\infty, b]$
half-open	$a \le x < b$	$[a, b)$
half-open	$a < x \le b$	$(a, b]$
closed	$a \le x \le b$	$[a, b]$

Figure 12.2

• **Example 12.3:** Illustrate the given sets of numbers graphically and by using the interval notation.

    a. $x \ge 3$ **b.** $x < 7$

    c. $3 < x \le 7$ **d.** $-3 < x < -1$

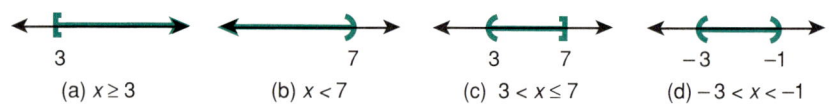

Figure 12.3

Solution: The graphs are illustrated in figure 12.3.

    a. $[3, \infty)$ **b.** $(-\infty, 7)$

    c. $(3, 7]$ **d.** $(-3, -1)$

SECTION 12.1 LINEAR INEQUALITIES

Double Inequality

An expression of the form $a \leq x \leq b$ is called a **double inequality**. Linear inequalities of this type can be solved in a manner similar to the method used in examples 12.1 and 12.2.

The inequality $5 < 2x - 3 < 7$ means that the quantity $2x - 3$ must be greater than 5 and also less than 7. That is,

$$5 < 2x - 3 \quad \text{and} \quad 2x - 3 < 7.$$

If we solve these inequalities separately, we obtain

$$\begin{aligned} 5 &< 2x - 3 \\ 8 &< 2x \\ 4 &< x \end{aligned} \quad \text{and} \quad \begin{aligned} 2x - 3 &< 7 \\ 2x &< 10 \\ x &< 5. \end{aligned}$$

We can write this result as

$$4 < x < 5 \quad \text{or} \quad x \in (4,5). \quad \text{(Read as } x \text{ is an element of the interval (4,5))}$$

We could have solved the inequality by using the rules for inequalities on the original statement:

$$\begin{aligned} 5 < 2x - 3 &< 7 \\ 8 < 2x &< 10 \quad \text{(Add 3)} \\ 4 < x &< 5. \quad \text{(Divide by 2)} \end{aligned}$$

However, it is not always possible to use this approach. Example 12.4 illustrates this.

• Example 12.4: Solve $5x + 6 \leq 3x + 3 \leq 2x - 7$.

Solution: The appearance of the $5x$, $3x$, and $2x$ in the different parts of the inequality makes it necessary to solve the pair of inequalities separately and to combine the results:

$$\begin{aligned} 5x + 6 &\leq 3x + 3 \\ 2x + 6 &\leq 3 \\ 2x &\leq -3 \\ x &\leq -\frac{3}{2} \end{aligned} \quad \text{and} \quad \begin{aligned} 3x + 3 &\leq 2x - 7 \\ x + 3 &\leq -7 \\ x &\leq -10. \end{aligned}$$

We now ask which numbers satisfy both statements $x \leq -\dfrac{3}{2}$ and $x \leq -10$.

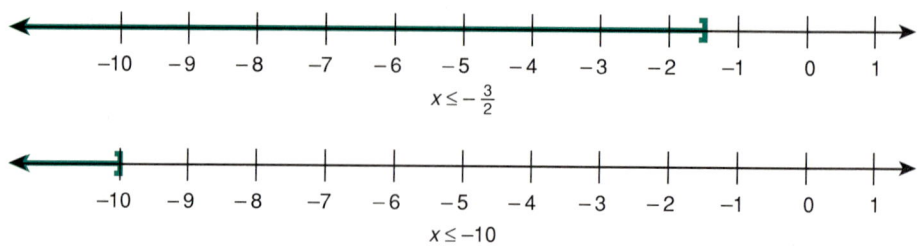

Figure 12.4

We can see from figure 12.4 that the set of numbers that fits both of the conditions is $x \leq -10$. Thus the solution set is $x \leq -10$.

• **Example 12.5:** A heat treatment for steel sheets requires that the annealing furnace be kept between 420 and 460 degrees Fahrenheit. What temperature range is this on the Celsius scale? ($F = \frac{9}{5}C + 32$)

Solution: This is equivalent to solving the following inequality:

$$420 \leq \frac{9}{5}C + 32 \leq 460$$

$$420 - 32 \leq \frac{9}{5}C \leq 460 - 32 \quad \text{(Subtract 32)}$$

$$388 \leq \frac{9}{5}C \leq 428$$

$$\frac{5}{9}(388) \leq \left(\frac{5}{9}\right)\left(\frac{9}{5}\right)C \leq \frac{5}{9}(428) \quad \left(\text{Multiply by } \frac{5}{9}\right)$$

$$216 \leq C \leq 238. \quad \text{(Approximately)}$$

• **Example 12.6:** The demand for power in a 110-volt circuit ranges from 300 to 3600 watts. What is the range of current in the circuit? ($P = EI$, where P is the power in watts, E is the voltage in volts, and I is the current in amperes.)

Solution: The power demands are given by the inequality

$$300 \leq P \leq 3600.$$

We substitute EI for P:

$$300 \leq EI \leq 3600.$$

SECTION 12.1 LINEAR INEQUALITIES

However, $E = 110$:

$$300 \leq 110I \leq 3600$$

$$\frac{300}{110} \leq I \leq \frac{3600}{110} \quad \text{(Divide by 110)}$$

$$2.8 \leq I \leq 32.7 \quad \text{(Approximately)}$$

Trial Problems 12.1

Before you begin the section exercises, warm up with these problems. Complete answers are included in the answer key.

Write each inequality in interval notation.

1. $-6 \leq x < 5$
2. $x \geq -5$

Write each interval as an inequality.

3. $(-\infty, 4)$
4. $(-4, 6]$

Solve for x.

5. $3x + 5 < 20$
6. $7 \leq 2x - 3 \leq 11$

Exercises 12.1

Write each inequality in interval notation.

1. $-2 < x < 8$
2. $-7 < x < -3$
3. $-5 \leq x \leq -3$
4. $3 \leq x \leq 6$
5. $-2 < x$
6. $3 < x$
7. $4 \leq x$
8. $-4 \leq x$
9. $x < -7$
10. $x < -2$
11. $x \leq 3$
12. $x \leq -1$
13. $3 \leq x < 7$
14. $5 < x \leq 10$

Write each interval as an equality and graph on a number line.

15. $(-1, 3)$
16. $(7, 11)$
17. $[-5, 2]$
18. $[-3, -1]$
19. $(-2, 3]$
20. $[-2, 1)$
21. $(-\infty, 5]$
22. $(-\infty, -3]$
23. $(-\infty, 3)$
24. $(-\infty, 7)$
25. $(3, \infty)$
26. $(-2, \infty)$
27. $[2, \infty)$
28. $[-1, \infty)$

Solve each inequality.

29. $x + 3 \leq 7$
30. $x - 2 \leq 5$
31. $2x - 7 \leq 8$
32. $3x - 4 \geq 4$
33. $\dfrac{x + 1}{3} \leq 4$
34. $\dfrac{x - 2}{5} \leq 6$
35. $\dfrac{2x - 3}{5} \leq 6$
36. $\dfrac{5x + 7}{2} \leq 17$
37. $4x - 3 < 2x + 7$
38. $5x + 6 \leq x - 4$
39. $3 - 2x \leq 2x + 1$
40. $2 - x \leq 3x + 4$

41. $6 \leq 2x - 5 \leq 18$ **42.** $-3 \leq 2x + 1 \leq 3$ **43.** $5 \leq 3x + 7 \leq 25$

44. $35 \leq 2x - 2 \leq 45$ **45.** $x - 2 \leq 2x + 3 \leq x + 8$ **46.** $x + 2 \leq 2x + 5 \leq x + 8$

47. $4x + 3 < 2x - 3 < -17$ **48.** $18 < 2x + 12 < 4x$

49. The EZ Rental Agency rents cars at $29.95 per day plus $.12 per mile. How many miles can be driven per day if the cost-per-day cannot exceed $85?

50. Solve exercise 49 given that the cost-per-day cannot exceed $105.

51. A rectangle must have an area between 15 cm² and 60 cm². If the width of the rectangle is 3 cm, what is the range of values for the length?

52. A parallelogram must have an area between 30 cm² and 60 cm². If the height is 5 cm, what is the range of values for the base?

53. The power in a 220 volt electrical circuit must range between 600 and 2400 watts. What is the range of values for the current? (Use $P = EI$).

54. Solve exercise 53 given that the power must range between 500 and 3000 watts.

55. The current in a 110 volt circuit ranges between 5 amps and 15 amps. What is the range of the power in the circuit?

56. The current in a 110 volt circuit ranges between 0 amps and 30 amps. What is the range of the power in the circuit?

57. The temperature range for storage of a product is $0° < t < 5°$ (C). What is the corresponding range in Fahrenheit degrees?

58. A chemical process requires the temperature in Celsius degrees to be between 80° and 95°. What is the corresponding range in Fahrenheit degrees?

59. If the maximum speed on a highway is 55 mi/hr and the minimum speed is 40 mi/hr, how long will it take to travel 360 miles?

60. Solve exercise 59 for a trip of 220 miles.

61. A bridge has "expansion joints," which are small gaps between the sections in the pavement. Given that the gap width g is related to the Fahrenheit temperature t by the formula $g = -0.00625t + 0.75$. What is the range of g when the temperature varies between 62° and 86°?

62. Solve exercise 61 for the temperature range 20° and 100°.

63. Boyle's law for a particular gas states that $pv = 180$, where p is the pressure in pounds per square inch and v is the volume in cubic inches. If $3 \leq p \leq 6$, what is the range for v?

64. In exercise 63, if $30 \leq v \leq 60$, what is the corresponding range for p?

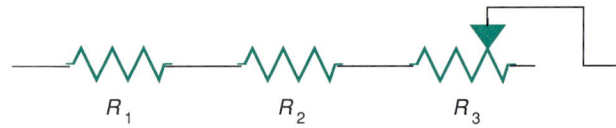

Figure 12.5

65. For the circuit shown in figure 12.5 the total resistance R_T is given by the formula $R_T = R_1 + R_2 + R_3$. Given that $R_1 = 7.2$ ohms and $R_2 = 8.3$ ohms, if R_3 is a variable resistor and $2.3 \leq R_3 \leq 5.8$, what is the range of R_T?

66. In exercise 65, given that $R_1 = 7.2$ ohms and $R_2 = 8.3$ ohms, determine the range of R_3 if $17.8 \leq R_T \leq 21.3$.

Nonlinear Inequalities

12 • 2 Nonlinear Inequalities

In chapter 5 we learned to solve quadratic inequalities by graphing and by factoring. In this section, we extend this method to solve some other types of **nonlinear inequalities**.

Let us review the factoring method for solving quadratic inequalities.

SECTION 12.2 NONLINEAR INEQUALITIES

• **Example 12.7:** Solve $x^2 - 5x + 6 < 0$.

Solution: We factor the left side of the inequality:
$$(x - 2)(x - 3) < 0.$$
In order for the product to be negative, one of the factors must be positive and the other factor must be negative.

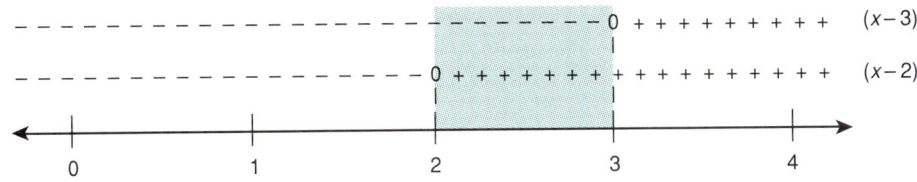

Figure 12.6

We can see from figure 12.6 that the factor $(x - 3)$ is negative to the left of $x = 3$ and positive to the right of $x = 3$. Similarly, the factor $x - 2$ is negative to the left of $x = 2$ and positive to the right of $x = 2$. The interval $2 < x < 3$ has one positive and one negative factor. The solution to the inequality is $2 < x < 3$.

The same technique can be used on any expression that has linear factors. Examples 12.8 and 12.9 illustrate how to extend the technique.

• **Example 12.8:** Solve $\dfrac{(x - 2)(x - 3)}{x + 2} \leq 0$.

Solution: We now have three factors. In order for the expression to be negative, exactly one or exactly three of the factors $(x - 2)$, $(x - 3)$, and $(x + 2)$ must be negative.

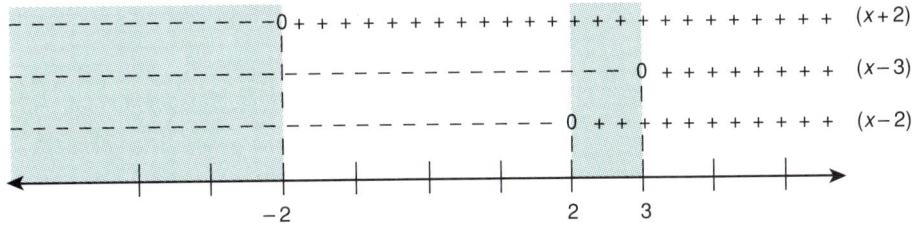

Figure 12.7

Figure 12.7 shows the regions that fit the solution to the inequality

$$\frac{(x - 2)(x - 3)}{x + 2} \leq 0.$$

All three of the factors are negative if $x < -2$. One of the factors is negative for $2 < x < 3$. The expression is equal to zero if $x = 2$ or $x = 3$. The solution to the original inequality is

$$x < -2 \quad \text{or} \quad 2 \leq x \leq 3.$$

Do you see why $x = -2$ is not a solution? (Substituting $x = -2$ results in a zero denominator in the inequality.)

••••••••••

If an algebraic expression is not factorable into linear factors, we can sometimes obtain the solution of an inequality using the quadratic formula. Example 12.9 illustrates this technique.

• **Example 12.9:** Solve $\dfrac{(x + 3)(x^2 + 4x + 2)}{(x - 2)} > 0$.

Solution: Since the factor $x^2 + 4x + 2$ is not easily factorable, we consider this factor separately. Using the quadratic formula on the equation

$$x^2 + 4x + 2 = 0,$$

we obtain

$$x = \frac{-4 \pm \sqrt{16 - 4(2)(1)}}{2(1)}$$

$$x = \frac{-4 \pm \sqrt{8}}{2}$$

$$= \frac{-4 \pm 2\sqrt{2}}{2}$$

$$= -2 \pm \sqrt{2}.$$

$$x = -3.414 \quad \text{or} \quad x = -0.586. \quad \text{(Approximately)}$$

We sketch a graph of the function $f(x) = x^2 + 4x + 2$ and determine the interval for which the expression is positive and negative. These results are obtained from the graph in figure 12.8:

$$x^2 + 4x + 2 > 0 \quad \text{for} \quad x < -2 - \sqrt{2} \quad \text{or} \quad x > -2 + \sqrt{2}$$
$$x^2 + 4x + 2 < 0 \quad \text{for} \quad -2 - \sqrt{2} < x < -2 + \sqrt{2}.$$

We can now proceed to solve the original inequality using the procedure of example 12.8. This is illustrated in figure 12.9.

In order for the expression to be positive, either none or exactly two of the factors must be negative. This occurs in three intervals:

$$(-\infty, -2 - \sqrt{2}), \quad (-3, -2 + \sqrt{2}), \quad (2, \infty).$$

The solution is $x < -2 - \sqrt{2}$ or $-3 < x < -2 + \sqrt{2}$ or $x > 2$.

••••••••••

SECTION 12.2 NONLINEAR INEQUALITIES   471

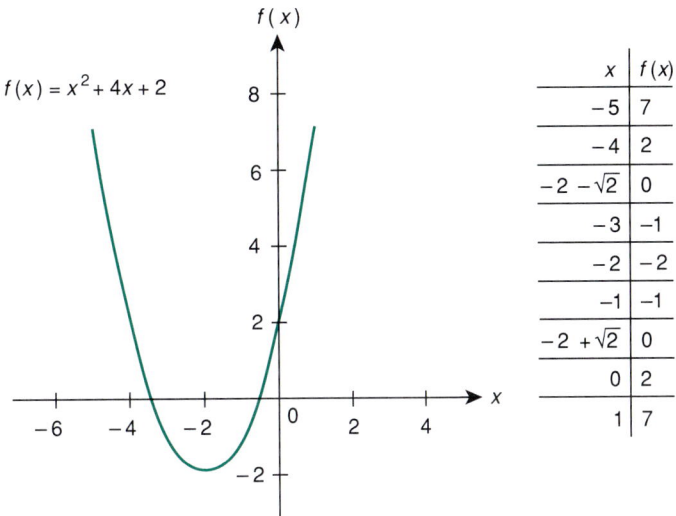

Figure 12.8

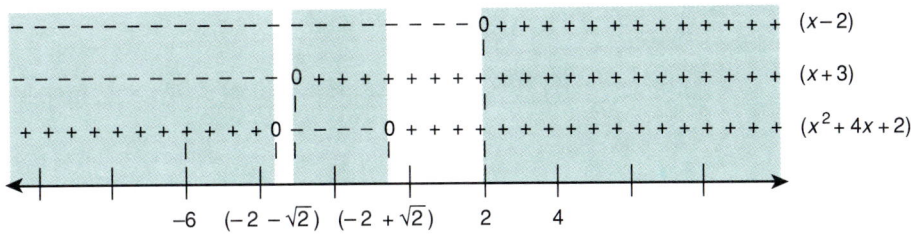

Figure 12.9

Sometimes it is necessary to simplify an inequality before using this procedure for solving it. Be careful when multiplying or dividing both sides of an inequality by an algebraic expression. You need to know whether the expression is positive or negative before multiplying or dividing. Recall that multiplying or dividing both sides of an inequality by a negative number reverses the direction of the inequality symbol.

• **Example 12.10:** Solve $\dfrac{3}{x-2} < \dfrac{4}{2x+1}$.

Solution: One might be tempted to multiply both sides of the inequality by $(x-2)(2x+1)$ to clear fractions. However, we do not know whether $(x+2)(2x+1)$ is positive or negative. We would not know which of the symbols $\leq$ or $\geq$ to use in the resulting inequality. Our approach to the problem is to subtract $\dfrac{4}{2x+1}$

from both sides of the inequality. The result is

$$\frac{3}{x-2} - \frac{4}{2x+1} \leq 0.$$

We now combine the fractions on the left-hand side:

$$\frac{3(2x+1)}{(x-2)(2x+1)} - \frac{4(x-2)}{(2x+1)(x-2)} \leq 0$$

$$\frac{(6x+3) - (4x-8)}{(x-2)(2x+1)} \leq 0$$

$$\frac{2x+11}{(x-2)(2x+1)} \leq 0.$$

We can now use the techniques of the previous examples. See figure 12.10.

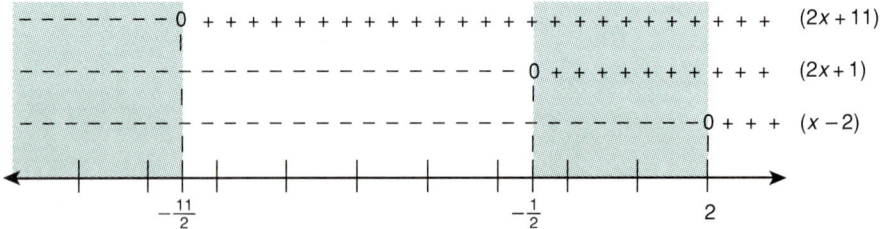

Figure 12.10

This expression is negative, provided exactly one or exactly three of the factors are negative. This occurs for the intervals $\left(-\infty, -\frac{11}{2}\right)$ and $\left(-\frac{1}{2}, 2\right)$. The value $x = -\frac{11}{2}$ also satisfies the inequality, since the numerator is zero for this value. The solution is

$$x \leq -\frac{11}{2} \quad \text{or} \quad -\frac{1}{2} < x < 2.$$

• **Example 12.11:** Under certain conditions the power P in a circuit is related to the current I and the resistance R according to the equation $P = EI - RI^2$. Given that $E = 120$ and $R = 15$, for what values of I does P fit the conditions $0 \leq P \leq 225$?

Solution: We must solve the inequality

$$0 \leq EI - RI^2 \leq 225.$$

We first substitute $E = 120$ and $R = 15$:

$$0 \leq 120I - 15I^2 \leq 225.$$

We solve the two inequalities separately.

	$0 \leq 120I - 15I^2$	$120I - 15I^2 \leq 225$	
(Divide by -15)	$0 \geq -8I + I^2$	$-15I^2 + 120I - 225 \leq 0$	
	$0 \geq I(I - 8)$	$I^2 - 8I + 15 \geq 0$	(Divide by -15)
		$(I - 5)(I - 3) \geq 0.$	

Figures 12.11 and 12.12 illustrate the solutions of each of these inequalities:

$$0 \leq I \leq 8 \quad \text{and} \quad (I \leq 3 \quad \text{or} \quad I \geq 5).$$

Since I must fit both of the inequalities, the solution set is $0 \leq I \leq 3$ or $5 \leq I \leq 8$. This is illustrated in figure 12.13.

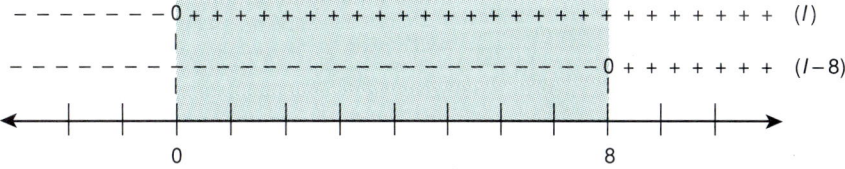

Figure 12.11

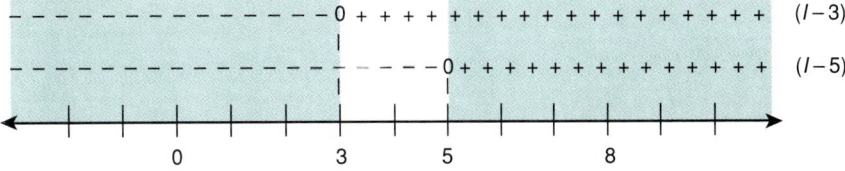

Figure 12.12

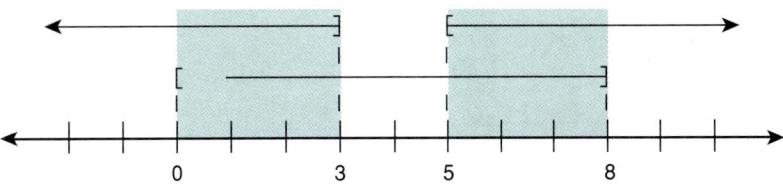

Figure 12.13

Trial Problems 12.2

Before you begin the section exercises, warm up with these problems. Complete solutions are included in the answer key.

Solve for x.

1. $(x + 3)(x + 2) > 0$
2. $x^2 - 7x + 6 \leq 0$
3. $\dfrac{x - 1}{x - 6} \geq 0$
4. $x(2x - 1)(x - 3) > 0$
5. $\dfrac{(x - 1)(x + 2)}{x - 3} < 0$

Exercises 12.2

Solve each inequality.

1. $(x - 3)(x + 2) > 0$
2. $(x - 2)(x + 3) > 0$
3. $(x - 4)(x + 2) < 0$
4. $(x + 4)(x - 2) < 0$
5. $(2x - 1)(3x - 2) \geq 0$
6. $(5x + 2)(3x - 1) \geq 0$
7. $(2x - 3)(5x + 4) \leq 0$
8. $(3x - 6)(x - 4) \leq 0$
9. $x^2 - x - 12 \geq 0$
10. $x^2 - x + 20 \geq 0$
11. $x^2 + 5x + 6 < 0$
12. $x^2 - 7x - 8 \leq 0$
13. $x^2 + 4x + 4 < 0$
14. $x^2 - 4x + 4 > 0$
15. $x^2 - 5x + 5 \geq 0$
16. $x^2 - 5x - 6 \leq 0$
17. $x^2 + x + 1 \geq 0$
18. $x^2 - x + 1 \leq 0$
19. $(x - 2)(x - 3)(x - 4) < 0$
20. $(x - 1)(x - 3)(x + 2) < 0$
21. $(x - 2)(x - 4)(x + 3) \geq 0$
22. $(x - 3)(x + 2)(x + 4) \geq 0$
23. $\dfrac{2x - 1}{3x + 1} < 0$
24. $\dfrac{3x - 2}{x + 1} < 0$
25. $\dfrac{3x + 6}{2x - 4} > 0$
26. $\dfrac{2x + 3}{2x - 3} > 0$
27. $\dfrac{x(x - 1)}{x + 2} < 0$
28. $\dfrac{x(3x - 7)}{x - 2} < 0$
29. $\dfrac{x(x + 1)(x - 2)}{x + 4} < 0$
30. $\dfrac{x(x - 3)(2x + 4)}{x + 3} < 0$
31. $\dfrac{(x + 1)(x - 1)(x + 3)}{(x - 2)(x - 4)} > 0$
32. $\dfrac{x(x + 1)(x - 1)}{(x^2 - x - 2)} > 0$
33. $\dfrac{3x - 2}{3x - 9} < \dfrac{1}{x - 3}$
34. $\dfrac{5x - 7}{2(x - 2)} < \dfrac{1}{x - 2}$
35. $\dfrac{(x^2 - 2x - 4)(x - 1)}{x + 3} < 0$
36. $\dfrac{(x^2 - 2x - 2)(x - 3)}{x + 1} \geq 0$

37. The equation $h = 304t - 16t^2$ represents the height h of an object above the ground after t seconds. Find the range of values of t for which the object is higher than 1408 feet.

38. Using $h = 640t - 16t^2$ solve exercise 37 for $h \geq 4800$.

39. For the engineering formula $I = \dfrac{nE}{nR_1 + R_2}$ given that $E = 120$, $R_1 = 10$, and $R_2 = 15$. For what values of n does I fit the inequality $0 \leq I \leq 8$?

40. Solve exercise 39 using the data $E = 120$, $R_1 = 10$, $R_2 = 15$, and $0 \leq I \leq 9$.

41. Given the electrical formula $P = EI - RI^2$ (see example 12.11) with $E = 12$ and $R = 3$. Determine the values of I for which $0 \leq P \leq 9$.

42. Solve exercises 41 given $E = 20$, $R = 4$, and $0 \leq P \leq 16$.

43. For what values of x does this expression produce real numbers?

$$\frac{\sqrt{(x+1)^3(x-2)}}{x}$$

44. For what values of x does this expression produce real numbers?

$$\frac{\sqrt{(x-2)^3(x+1)}}{(x-1)}$$

45. An open-topped box is to be constructed by cutting squares from the corners of a 10 inch by 20 inch rectangular piece of material. The sides are then folded up to form the box. If the base of the box must have an area of at least 100 square inches, what are the possible heights of the box?

46. Solve exercise 45 given the length and width of the material to be 20 inches and 40 inches.

12 • 3 Absolute Value Equations and Inequalities

Absolute Value

In this section we consider methods of solving equations and inequalities that contain absolute values. By the **absolute value** of a number we mean the distance on the number line from that number to zero.

If a is a real number

$$|a| = a \quad \text{if } a \geq 0$$
$$ = -a \quad \text{if } a < 0.$$

Thus, $|3| = 3$ and $|-3| = 3$.

Figure 12.14 illustrates the distance from 3 to 0, and the distance from -3 to 0.

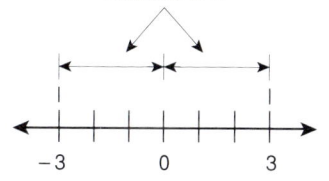

Figure 12.14

• **Example 12.12:** Solve $|x - 3| = 4$.

Solution: The expression $|x - 3|$ represents the distance between x and 3. To solve the equation $|x - 3| = 4$, we must determine all real numbers that are 4 units from 3.

We can see from figure 12.15 that there are two real numbers, -1 and 7, that fit this condition. The solution set is $\{-1, 7\}$.

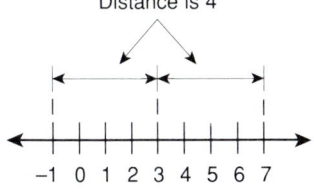

Figure 12.15

This procedure would be tedious in solving equations containing absolute values. Theorem 12.1 simplifies the procedure.

Theorem 12.1: If b is a positive number, then $|y| = b$ has the solutions $y = b$ or $y = -b$. The proof of the theorem follows from the definition of absolute value.

• **Example 12.13:** Solve $|2x - 5| = 7$.

Solution: Use theorem 12.1 with $y = 2x - 5$ and $b = 7$:

$$2x - 5 = 7 \quad \text{or} \quad 2x - 5 = -7$$
$$2x = 12 \quad\quad\quad\quad 2x = -2$$
$$x = 6 \quad\quad\quad\quad x = -1.$$

The solution set is $\{-1, 6\}$.

• **Example 12.14:** Solve $|2x - 3| = |5x + 6|$.

Solution: Use theorem 12.1 with $y = 2x - 3$ and $b = 5x + 6$:

$$2x - 3 = 5x + 6 \quad \text{or} \quad 2x - 3 = -(5x + 6)$$
$$-3 = 3x + 6 \quad\quad\quad\quad 2x - 3 = -5x - 6$$
$$-9 = 3x \quad\quad\quad\quad\quad 7x - 3 = -6$$
$$-3 = x \quad\quad\quad\quad\quad\quad 7x = -3$$
$$x = -\frac{3}{7}.$$

The solution set is $\left\{-3, -\dfrac{3}{7}\right\}$.

We now consider inequalities that contain absolute values. Example 12.15 illustrates the meaning of an expression like $|y| < b$ and $|y| > b$.

• **Example 12.15:** Solve (a) $|x| < 3$ and (b) $|x| > 3$.

Solution: **a.** Since absolute value means the distance of a particular number from the number zero, $|x| < 3$ is satisfied by all numbers whose distance from 0 is less than 3.

From figure 12.16, we can see that any number in the interval $(-3, 3)$ fits this condition. The solution set is

$$(-3, 3) \quad \text{or} \quad \{-3 < x < 3\}.$$

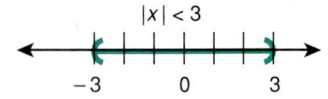

Figure 12.16

b. Similarly, $|x| > 3$ can be solved by determining the numbers whose distance from 0 is greater than 3. The solution set is

$$(-\infty, -3) \cup (3, \infty) = \{x < -3 \text{ or } x > 3\}.$$

This solution is graphed in figure 12.17.

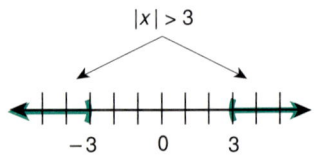

Figure 12.17

This procedure would be tedious in solving more complicated problems. Theorem 12.2 simplifies the solution process.

> **Theorem 12.2:** **a.** $|y| < b$ is equivalent to $-b < y < b$ $(b > 0)$.
>
> **b.** $|y| > b$ is equivalent to $y < -b$ or $y > b$ $(b > 0)$.

These statements can be proved using an argument similar to the one used in solving example 12.14. We use theorem 12.2 as an aid in solving examples 12.16 and 12.17.

- **Example 12.16:** Solve $|x - 3| < 7$.

    Solution: Using theorem 12.2a with $y = x - 3$ and $b = 7$, we obtain this equivalent statement:

    $$-7 < x - 3 < 7$$
    $$3 - 7 < x < 3 + 7 \quad \text{(Add 3 to each expression)}$$
    $$-4 < x < 10.$$

- **Example 12.17:** Solve $|3x - 2| \geq 5$.

    Solution: Use theorem 12.2b with $y = 3x - 2$ and $b = 5$:

    $$3x - 2 \leq -5 \quad \text{or} \quad 3x - 2 \geq 5$$
    $$\text{(Add 2)} \quad 3x \leq -3 \qquad\qquad 3x \geq 7 \quad \text{(Add 2)}$$
    $$\text{(Divide by 3)} \quad x \leq -1 \qquad\qquad x \geq \frac{7}{3} \quad \text{(Divide by 3)}$$

    The solution set is $\left\{ x \leq -1 \text{ or } x \geq \frac{7}{3} \right\}$.

Note that not every inequality has a real solution and that every real number might be a solution to an inequality. For example, $|x - 2| < -2$ has the empty set as its solution set since absolute values are nonnegative, and $|x - 2| > -2$ has $(-\infty, \infty)$ as its solution set since any real number x satisfies the inequality.

- **Example 12.18:** Solve $\left| \frac{3x + 2}{x - 1} \right| \leq 12$.

    Solution: Use theorem 12.2a with $y = \frac{3x + 2}{x - 1}$ and $b = 12$:

    $$-12 \leq \frac{3x + 2}{x - 1} \leq 12.$$

In order to be a solution to this inequality, *both* of the following inequalities must be satisfied:

and

$$\frac{3x+2}{x-1} \geq -12 \qquad\qquad \frac{3x+2}{x-1} \leq 12$$

$$\frac{3x+2}{x-1} + 12 \geq 0 \qquad\qquad \frac{3x+2}{x-1} - 12 \leq 0$$

$$\frac{3x+2}{x-1} + \frac{12(x-1)}{x-1} \geq 0 \qquad\qquad \frac{3x+2}{x-1} - \frac{12(x-1)}{x-1} \leq 0$$

$$\frac{15x-10}{x-1} \geq 0 \qquad\qquad \frac{-9x+14}{x-1} \leq 0$$

$$\frac{5(3x-2)}{x-1} \geq 0 \qquad\qquad \frac{9x-14}{x-1} \geq 0 \quad \text{(Multiply by } -1\text{)}$$

$$x \leq \frac{2}{3} \quad \text{or} \quad x > 1 \qquad\qquad x < 1 \quad \text{or} \quad x \geq \frac{14}{9}.$$

Since $x < \frac{2}{3}$ and $x \geq \frac{14}{9}$ fits both of the inequalities, the solution set is

$$\left\{ x \leq \frac{2}{3} \text{ or } x \geq \frac{14}{9} \right\}.$$

Figure 12.18, 12.19, and 12.20 help to illustrate the solutions.

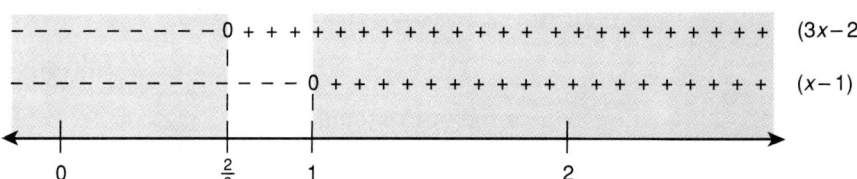

Figure 12.18

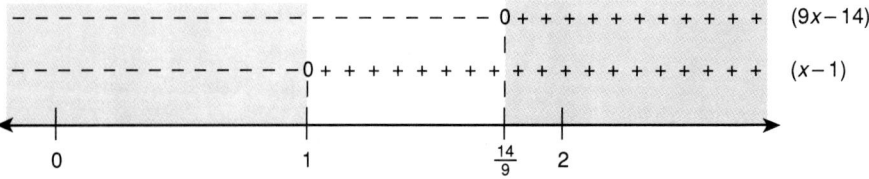

Figure 12.19

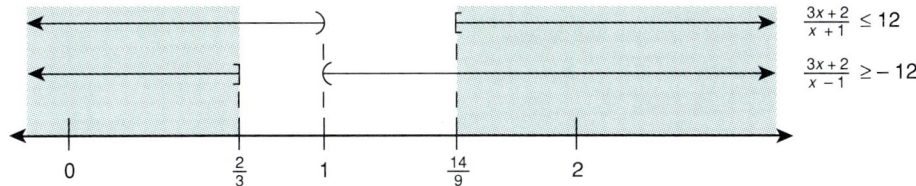

Figure 12.20

• **Example 12.19:** Solve $\left|\dfrac{3x-5}{x-1}\right| \geq 2$.

Solution: Use theorem 12.2b with $y = \dfrac{3x-5}{x-1}$ and $b = 2$:

$$\dfrac{3x-5}{x-1} \leq -2 \quad \text{or} \quad \dfrac{3x-5}{x-1} \geq 2$$

$$\dfrac{3x-5}{x-1} + 2 \leq 0 \qquad \dfrac{3x-5}{x-1} - 2 \geq 0$$

$$\dfrac{3x-5}{x-1} + \dfrac{2(x-1)}{(x-1)} \leq 0 \qquad \dfrac{3x-5}{x-1} - \dfrac{2(x-1)}{(x-1)} \geq 0$$

$$\dfrac{5x-7}{x-1} \leq 0 \qquad \dfrac{x-3}{x-1} \geq 0$$

$$1 < x \leq \dfrac{7}{5} \qquad x < 1 \quad \text{or} \quad x \geq 3.$$

Figures 12.21 and 12.22 indicate that the solution set consists of all real numbers that fit at least one of the conditions $1 < x \leq \dfrac{7}{5}$, $x < 1$, or $x \geq 3$.

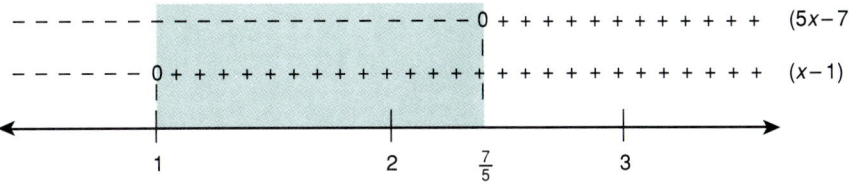

Figure 12.21

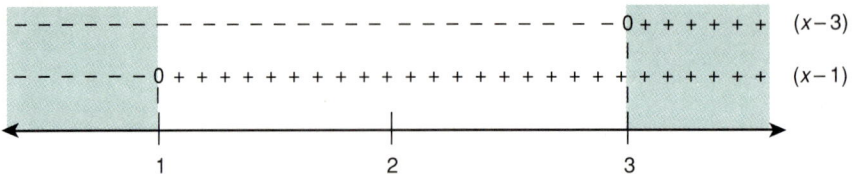

Figure 12.22

Figure 12.23

Figure 12.23 shows that we could also write the solution set as

$$x < \frac{7}{5}, \, x \neq 1 \quad \text{or} \quad x \geq 3.$$

• **Example 12.20:** Statisticians often obtain a "confidence interval" by solving for y in an inequality of the form

$$|y - \bar{x}| \leq \frac{Zs}{\sqrt{n}},$$

where $\bar{x}$ is the average of some data, s is the standard deviation of the data, n represents the size of the sample, and Z (standard score) is a number obtained from a table. Given $n = 103$, $\bar{x} = 174.8$, $s = 19.7$, and $Z = 2.43$, find a confidence interval for y.

Solution: We substitute the given values into the inequality:

$$|y - 174.8| \leq \frac{(2.43)(19.7)}{\sqrt{103}}$$

$$|y - 174.8| \leq 4.7. \text{ (Approximately)}$$

We now use theorem 12.2a:

$$-4.7 \leq y - 174.8 \leq 4.7.$$

SECTION 12.3 ABSOLUTE VALUE EQUATIONS AND INEQUALITIES

Add 174.8 to each member of the inequalities:
$$174.8 - 4.7 \leq y \leq 174.8 + 4.7$$
$$170.1 \leq y \leq 179.5$$

The confidence interval is [170.1, 179.5].

Trial Problems 12.3

Before you begin the section exercises, warm up with these problems. Complete answers are included in the answer key.

Solve for x.

1. $|x - 3| = 5$
2. $|2x - 3| = |x + 4|$
3. $|2x - 3| \leq 4$
4. $|5x - 4| > 17$
5. $\dfrac{|x - 6|}{|x - 3|} < 2$

Exercises 12.3

Solve for x.

1. $|x| = 8$
2. $|x| = 4$
3. $|x| = 0$
4. $|x| = -1$
5. $|x - 3| = |x + 4|$
6. $|x - 1| = |x + 3|$
7. $|4x - 4| = |x - 2|$
8. $|3x - 6| = |x - 1|$
9. $|2x - 2| = |x - 1|$
10. $|3x - 6| = |x - 3|$
11. $|x - 4| = 7$
12. $|x + 4| = 6$
13. $|2x - 5| = 3$
14. $|3x + 7| = 10$

Solve for x.

15. $|x| < 4$
16. $|x| < 5$
17. $|x| > 5$
18. $|x| > 3$
19. $|x| \geq -2$
20. $|x| \leq -2$
21. $|x - 4| < 7$
22. $|x + 3| < 2$
23. $|x + 3| > 2$
24. $|x - 4| > 2$
25. $|2x - 3| \leq 6$
26. $|3x - 5| \leq 8$
27. $|4x - 5| > 8$
28. $|2x + 3| > 14$
29. $\left|\dfrac{x - 3}{x - 2}\right| < 10$ Note: $\left|\dfrac{a}{b}\right| = \dfrac{|a|}{|b|}$
30. $\left|\dfrac{x - 1}{x - 2}\right| < 4$ Note: $\left|\dfrac{a}{b}\right| = \dfrac{|a|}{|b|}$
31. $\left|\dfrac{x - 1}{x - 2}\right| \geq 4$
32. $\left|\dfrac{x - 3}{x - 2}\right| \geq 10$
33. $\left|\dfrac{2x - 3}{x - 5}\right| < 2$
34. $\left|\dfrac{x - 5}{2x - 3}\right| < 3$
35. $\left|\dfrac{2x - 1}{3x + 5}\right| \geq 1$

36.  $\left|\dfrac{3x+5}{2x-1}\right| \geq 1$

37. $\dfrac{2x-8}{3x+1} < \dfrac{1}{2}$

38. $\dfrac{3x+1}{2x-8} > \dfrac{1}{2}$

39. Solve example 12.20 given that $\bar{x} = 40.8$, $s = 4.3$, $Z = 2.05$, and $n = 64$.

40. Solve example 12.20 given that $\bar{x} = 218.3$, $s = 22.9$, $Z = 2.24$, and $n = 132$.

41. A cruise control on a car changes the speed of the vehicle if the speed varies from the set speed by 1.2 mi/hr or more. Write an inequality that shows this relationship using m as the set speed and x as the actual speed of the vehicle.

42. A pressure-monitoring device on a gas line will adjust the pressure in the line if it varies from a predetermined pressure by more than 4.5 lb/in.^{2}. Write an inequality that shows the relationship using m as the predetermined set pressure and x as the actual pressure in the line.

12 • 4 Linear Inequalities in Two Variables

In previous chapters linear equations of the form $ax + by = c$ were graphed. The graph of such an equation is a line in the coordinate plane. If we replace the equality symbol by one of the symbols $<$, $\leq$, $>$, or $\geq$, the result is called a **linear inequality in two variables**.

Linear Inequality in Two Variables

The following steps yield the graph of a linear inequality of the form $ax + by < c$.

GRAPHING A LINEAR INEQUALITY

1. Graph the corresponding linear equation
$$ax + by = c.$$

2. The graph of $ax + by < c$ is one of the half-planes (A or B in figure 12.24) determined by this line.

3. Select a point that is obviously in one of the half-planes and check whether the coordinates of the point satisfy the inequality.

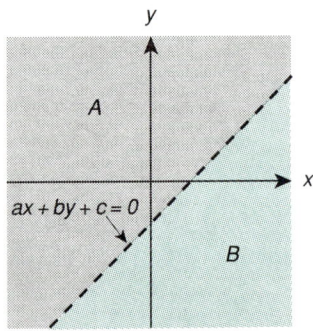

Figure 12.24

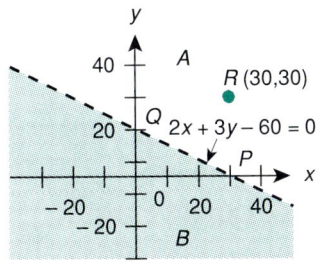

Figure 12.25

- **Example 12.21:** Graph the inequality $2x + 3y < 60$.

  Solution: 1. Graph the line $2x + 3y = 60$. Two points on the line are $P(30,0)$ and $Q(0,20)$ (see figure 12.25).

  2. Point $R(30,30)$ is obviously in half-plane A.

  3. Substitute $x = 30$, $y = 30$ into the expression
  $$2x + 3y$$
  $$2(30) + 3(20) = 120.$$

  Since 120 is not less than 60, half-plane A is not the solution set. Therefore, half-plane B is the solution set to the inequality. Line $2x + 3y = 60$ is dashed to indicate that it is not part of the solution set.

- **Example 12.22:** Graph the solution set to the following inequalities:
  $$4x + 3y - 12 \geq 0$$
  $$2x - 3y - 6 \geq 0.$$

  Solution: 1. Graph the lines $4x + 3y - 12 = 0$ and $2x - 3y - 6 = 0$.

  2. The graph of $4x + 3y - 12 \geq 0$ is the line $4x + 3y - 12 = 0$ and the half-plane to the right of this line.

  3. The graph of $2x - 3y - 6 \geq 0$ is the line $2x - 3y - 6 = 0$ and the half-plane below this line.

  4. The region that represents the solution set is the part of the graph indicated by the shaded part of figure 12.26. Note that the solid lines indicate that the edges are included in the region.

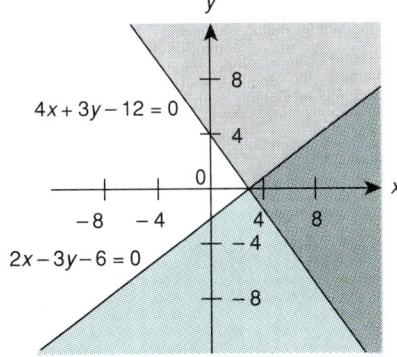

Figure 12.26

Example 12.23 illustrates how to graph a solution to a larger set of inequalities.

• **Example 12.23:** Graph the solution set to the inequalities

$$x - y \geq 0$$
$$x \geq 3$$
$$y \geq -2$$
$$2x + 3y - 20 \leq 0.$$

Solution:
1. Graph the line $x - y = 0$. All points to the right of and on this line satisfy the inequality $x - y \geq 0$.

2. Graph the line $x = 3$. All points on and to the right of this line satisfy the inequality $x \geq 3$.

3. Graph the line $y = -2$. All points on and above this line satisfy the inequality $y \geq -2$.

4. Graph the line $2x + 3y - 20 = 0$. All points on and below this line satisfy the inequality $2x + 3y - 20 \leq 0$.

5. The solution set is indicated by the shaded region in figure 12.27.

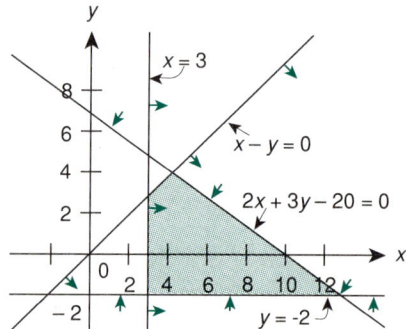

Figure 12.27

Polygonal Set
Feasible Solution

A set of points like the one obtained as the solution set in example 12.23 is called a **polygonal set**. In general, a polygonal set is one that has a polygon as its boundary. Any point in the region that satisfies the defining set of inequalities is called a **feasible solution** to the set of inequalities.

SECTION 12.4 LINEAR INEQUALITIES IN TWO VARIABLES   **485**

- *Example 12.24:* The EZ Corporation makes two types of containers, types A and B. Experience has shown that the costs are limited by the restriction $2x + 3y < 900$, where x and y are the number of types A and B containers to produce per hour. Experience has also shown that production should be limited by the sales restrictions $x \leq 300$, $y \leq 200$. Graph the region of feasible solutions.

Solution: Implied in the problem are the restrictions $x \geq 0$ and $y \geq 0$, since it makes no sense to produce a negative number of containers. We need, therefore, to graph the region that satisfies these inequalities:

$$x \geq 0$$
$$y \geq 0$$
$$x \leq 300$$
$$y \leq 200$$
$$2x + 3y \leq 900.$$

We graph the lines $x = 0$ (y-axis), $y = 0$ (x-axis), $x = 300$, $y = 200$, and $2x + 3y = 900$ and determine which values of x and y fit all of the restrictions.

The region of feasible solutions is

1. to the right of and on the y-axis ($x \geq 0$),
2. above and on the x-axis ($y \geq 0$),
3. to the left of and on the line $x = 300$ ($x \leq 300$),
4. on and below the line $y = 200$ ($y \leq 200$), and
5. on and below the line $2x + 3y = 900$ ($2x + 3y \leq 900$).

The shaded region in figure 12.28 is the solution.

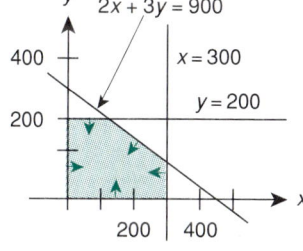

Figure 12.28

Trial Problems 12.4

Before you begin the section exercises, warm up with these problems. Complete answers are included in the answer key.

Graph the region that satisfies the given set of inequalities.

1. $4x + 5y \leq 20$
2. $3x - 4y \geq 12$
3. $x \geq 3$, $y \geq 4$
4. $x \geq 0$, $y \geq 0$, $x + y \leq 800$
5. $x \geq 0$, $y \geq 0$, $2x + 3y \leq 600$, $x \geq 200$

Exercises 12.4

Graph the region that satisfies the given system of inequalities.

1. $2x - 3y \leq 6$
2. $3x + 2y \leq 6$
3. $4x - 5y \geq 20$
4. $5x - 4y \geq 20$
5. $2x + 3y \leq 600$
6. $3x + 2y \leq 1200$
7. $0.03x + 0.04y \leq 12$
8. $0.04x + 0.03y \leq 6$
9. $x > 3, y \leq 6$
10. $x < 6, y \geq 1$
11. $2x - 3y + 7 \geq 0, \quad x - y - 4 \leq 0$
12. $x - 5y + 10 \leq 0, \quad 3x - 2y + 6 \leq 0$
13. $x - y + 4 \geq 0, \quad 2x + y - 3 \geq 0$
14. $2x - y + 3 \geq 0, \quad x - y + 1 \leq 0$
15. $x \geq 0, \quad y \geq 0, \quad x + y - 2 \leq 0$
16. $x \geq 0, \quad y \geq 0, \quad 2x + 3y \leq 6$
17. $x \geq 0, \quad y \geq 0, \quad 2x + 3y \leq 1800$
18. $x \geq 0, \quad y \geq 0, \quad 2x + 3y \leq 2400$
19. $x \geq 3, \quad x \leq 5, \quad x - y \leq 0$
20. $y \leq 1, \quad y \leq 5, \quad x - y \leq 0$
21. $x + y - 5 \leq 0, \quad 5x - 3y + 15 \geq 0,$ $x - 3y + 3 \leq 0$
22. $x + y - 1 \geq 0, \quad 2x - y + 2 \geq 0,$ $2x + y - 6 \leq 0$
23. $y \leq 3, \quad 4x + y + 5 \geq 0, \quad x - y \leq 0$
24. $x \leq 10, \quad x - y \geq 0, \quad x + 2y \geq 20$
25. $x \geq 0, \quad y \geq 0, \quad x \leq 6, \quad y \leq 7, \quad 5x + 3y \leq 36$
26. $x \geq 0, \quad y \geq 0, \quad x \leq 10, \quad y \leq 8, \quad x + y \leq 12$
27. $x \geq 0, \quad y \geq 0, \quad x \leq 6, \quad y \leq 7, \quad 5x + 3y \leq 36, \quad x + y \geq 2$
28. $x \geq 0, \quad y \geq 0, \quad x \leq 10, \quad y \leq 8, \quad x + y \leq 12, \quad 2x + 5 \geq 10$

29. A company manufactures two types of bicycles. Sales restrictions are given by the inequality $20x + 30y \leq 9000$, where x and y are the number of type A and B bicycles that can be sold per time period. A manufacturing restriction is given by the inequality $20x + 30y \geq 6000$. Graph the region of feasible solutions assuming that $x \geq 0$ and $y \geq 0$.

30. Solve exercise 29 given these restrictions:

    | Sales restriction | $5x + 9y \leq 4500$ |
    | Manufacturing restriction | $3x + 7y \geq 2100$ |

31. In exercise 29, add the restrictions $x \leq 400$ and $y \leq 250$ and graph the new region.

32. In exercise 30, add restrictions $x \leq 500$ and $y \leq 250$ and graph the new region.

33. A company produces two products, A and B. The major inputs for the products are energy, raw materials and labor. Table 12.1 shows the input requirements and the daily total available.

TABLE 12.1

	Product		Daily Availability
	A	B	
Energy	300 kwh	600 kwh	30,000 kwh
Raw Materials	500 lb	700 lb	42,000 lb
Labor	5 hr	4 hr	400 hr

Write a set of constraints for the table and graph the region of feasible solutions. (Assume $x \geq 0$ and $y \geq 0$, where x and y are the number of product A and product B to produce per hour.)

34. Solve exercise 33 using table 12.2.

TABLE 12.2

	Product		Daily Availability
	A	B	
Energy	200 kwh	300 kwh	24,000 kwh
Raw Materials	100 kwh	100 lb	9000 lb
Labor	26 hr	15 hr	195 hr

35. In exercise 33 add the constraints $x \geq 30$ and $y \leq 30$.

36. In exercise 34 add the constraints $x \geq 20$ and $y \geq 20$.

12•5 Graphical Linear Programming

Many problems in business and industry are concerned with finding the maximum or the minimum value of a function subject to constraints on the variable. In section 12.4 we graphed a region determined by a set of inequalities. The coordinates of any point in the region constituted a feasible solution to the set of inequalities. We now consider the problem of maximizing or minimizing a linear function defined on a polygonal region.

Suppose we have a system of inequalities that has as its solution set the polygon **ABCDE** and its interior (figure 12.29). Let $f(x,y) = ax + by + c$ be a function defined for each point in the polygonal region. Thus, for any point (r,s) in the region, $f(r,s) = ar + bs + c$ is a unique real number. The largest such number is called the **maximum of** f over the region, and the smallest such number is called the **minimum of** f over the region. If the region is bounded, the maximum and minimum always exist.

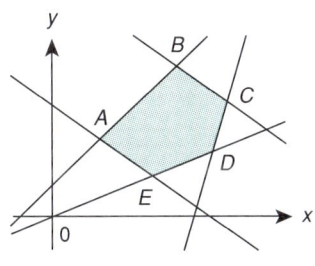

Figure 12.29

Maximum of f
Minimum of f

• *Example 12.25:*

Solution: Let T be the region defined by the set of inequalities of example 12.23 of section 12.4 (see figure 12.30):

$$x - y \geq 0, \qquad x \geq 3, \qquad y \geq -2, \qquad 2x + 3y - 20 \leq 0.$$

Let $f(x,y) = 2x - 3y + 6$ be defined over the region T. Points $A(3,-2)$ and $P(4,1)$ are obviously in T.

At A, $f(3,-2) = 2(3) - 3(-2) + 6 = 18$.

At P, $f(4,1) = 2(4) - 3(1) + 6 = 11$.

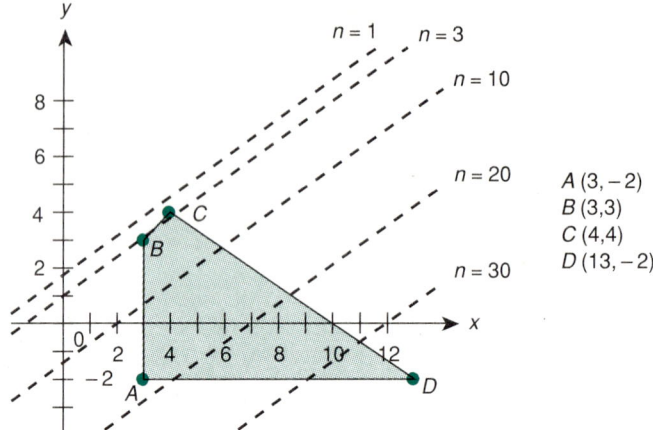

Figure 12.30

We want to determine the maximum and minimum values of the function $f(x,y)$ over the region T. Let $n = 2x - 3y + 6$.

If $n = 1$, then $2x - 3y + 6 = 1$ or $2x - 3y + 5 = 0$.

If $n = 3$, then $2x - 3y + 6 = 3$ or $2x - 3y + 3 = 0$.

If $n = 10$, then $2x - 3y + 6 = 10$ or $2x + 3y - 4 = 0$.

If we superimpose these three lines and the lines for $n = 20$ and $n = 30$ on the graph of the region T, we get a series of parallel lines for which the value of the function $f(x,y)$ increases as the line moves from upper left to lower right. Note that for each of the choices of n, the resulting value of point (x,y) on the line is constant. Thus, $f(x,y) = 1$ at each point on the line $2x - 3y + 5 = 0$. Similarly, $f(x,y) = 30$ on the line $2x - 3y + 6 = 30$ or $2x - 3y - 24 = 0$.

The maximum value of $f(x,y)$ occurs at point D and the minimum value of $f(x,y)$ occurs at either B or C (or both if one of the lines contains the segment BC).

At B, $f(3,3) = 2(3) - 3(3) + 6 = 3$.

At C, $f(4,4) = 2(4) - 3(4) + 6 = 2$.

The minimum value of $f(x,y)$ over T is 2 at point $C(4,4)$. The maximum value of $f(x,y)$ over T is at point D.

At D, $f(13,-2) = 2(13) - 3(-2) + 6 = 38$. (Maximum)

SECTION 12.5 GRAPHICAL LINEAR PROGRAMMING

The general procedure for maximizing or minimizing a linear function over a polygonal region is outlined in the following steps.

MAXIMIZING-MINIMIZING A LINEAR FUNCTION

1. Graph the polygonal region and find the coordinates of the vertices of the polygon.
2. Evaluate the function at each of the vertices.
3. The maximum value of the function over the region is the largest number obtained in step 2.
4. The minimum value of the function over the region is the smallest number obtained in step 2.

We use this procedure to solve examples 12.26 and 12.27.

• **Example 12.26:** Calculate the maximum and minimum values of the function $f(x,y) = -5x - 3y + 17$ over the region defined by the following inequalities: $x \geq 0$, $y \geq 0$, $x + y - 1 \geq 0$, $2x - y + 2 \geq 0$, and $2x + y - 6 \leq 0$.

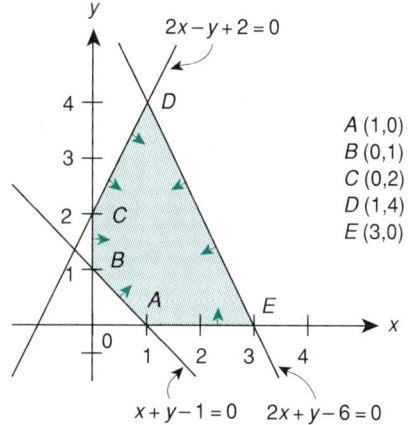

Figure 12.31

Solution: 1. Graph the region and find the coordinates of the vertices. The coordinates of A, B, C, and E are obvious from the graph in figure 12.31. To find the coordinates of D, we solve the pair of equations of the lines that intersect at point D.

$$\begin{cases} 2x - y + 2 = 0 \\ 2x + y - 6 = 0 \end{cases} \rightarrow \begin{cases} 2x - y + 2 = 0 \\ x = 1 \end{cases} \rightarrow \begin{cases} y = 4 \\ x = 1. \end{cases}$$

2. Evaluate the function at each vertex.

At A, $f(1,0) = (-5)(1) - 3(0) + 17 = 12$.
At B, $f(0,1) = (-5)(0) - 3(1) + 17 = 14$.
At C, $f(0,2) = (-5)(0) - 3(2) + 17 = 11$.
At D, $f(1,4) = (-5)(1) - 3(4) + 17 = 0$.
At E, $f(3,0) = (-5)(3) - 3(0) + 17 = 2$.

3. The maximum of f over the region is the maximum of the set $\{12, 14, 11, 0, 2\}$. The maximum is 14.

4. The minimum of f over the region is the minimum of the set $\{12, 14, 11, 0, 2\}$. The minimum is 0.

Linear Programming
Objective Function
Region of Feasible Solutions

This procedure is called **linear programming**. The function that is defined over the region is called the **objective function** and the region is called the **region of feasible solutions**.

• *Example 12.27:* The EZ Corporation makes two types of containers, type A and type B. Each type A container requires 2 minutes of chemical-operations time and 3 minutes of finishing-room time. Each type B container uses 3 minutes of chemical-operations time and 2 minutes of finishing-room time. Each week the company has available 30 hours of operating time in its chemical-operations department and 40 hours of operating time in the finishing department. The profit function is $P(x,y) = 0.40x + 0.50y$, where x and y represent the numbers of type A and B containers produced. Determine the weekly production schedule that would maximize the profits.

Solution: Implied restrictions are

$$x \geq 0 \quad \text{and} \quad y \geq 0.$$

TABLE 12.3

Container Type	A	B
Number of Containers	x	y
Chemical Operations Time	$2x$	$3y \leq 1800$
Finishing Time	$3x$	$2y \leq 2400$

Since there are 30 hours (1800 minutes) of chemical-operations time available,

$$2x + 3y \leq 1800.$$

Since there are 40 hours (2400 minutes) of finishing time available,

$$3x + 2y \leq 2400.$$

1. We graph the region of feasible solutions and determine the corner points (figure 12.32).

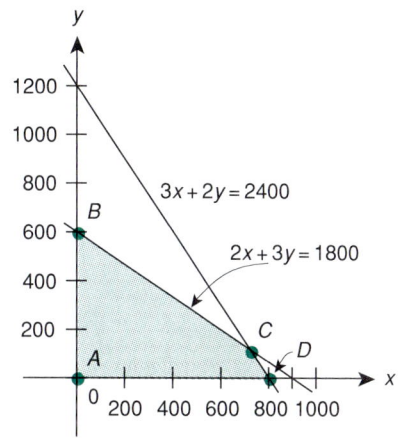

Figure 12.32

2. Determine the value of the objective function at each of the corner points.

   At A, $P(0,0) = 0.4(0) + 0.5(0) = 0$.

   At B, $P(0,600) = 0.4(0) + 0.5(600) = 300$.

   At C, $P(720,120) = 0.4(720) + 0.5(120) = 348$.

   At D, $P(800,0) = 0.4(800) + 0.5(0) = 320$.

3. The maximum of P over the region is the largest number in the set $\{0, 300, 348, 320\}$. The maximum profit is obtained by producing 720 of the type A containers and 120 of the type B containers per week.

••••••••••

When solving problems in industry, it is often necessary to change production schedules to meet changing conditions in the market. Example 12.28 illustrates such a situation.

- *Example 12.28:* The sales department for the company in example 12.27 has determined that the maximum number of type A and B containers that can be sold per week is 450 and 400, respectively. Assuming these added restrictions, determine a production schedule that maximizes the profit.

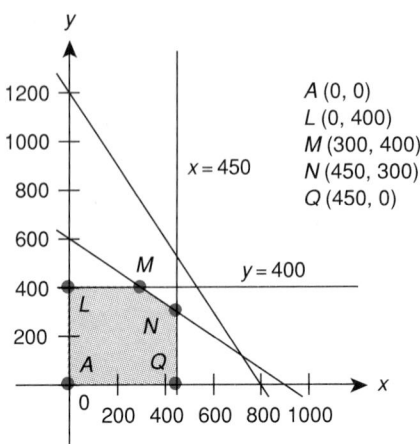

Figure 12.33

Solution: When we add the restrictions $x \leq 450$ and $y \leq 400$ to the graph for example 12.27, we obtain the polygon $AQNML$ in figure 12.33 as the region of feasible solutions.

We now determine the values of $P(x,y)$ at the corner points.

At A, $P(0,0) = 0$.

At L, $P(0,400) = 0.4(0) + 0.5(400) = 200$.

At M, $P(300,400) = 0.4(300) + 0.5(400) = 320$.

At N, $P(450,300) = 0.4(450) + 0.5(300) = 330$.

At Q, $P(450,0) = 0.4(450) + 0.5(0) = 180$.

The maximum profit of 330 is obtained by producing 450 type A and 300 type B containers per week.

Trial Problems 12.5

Before you begin the section exercises, warm up with these problems. Complete answers are included in the answer key.

For the region shown on page 493:

1. Maximize $f(x,y) = x - y$
2. Maximize $f(x,y) = 2x + 3y$
3. Maximize $f(x,y) = 0.3x + 0.5y$
4. Minimize $f(x,y) = 2x - 3y$
5. Minimize $f(x,y) = 0.5x + 0.6y$

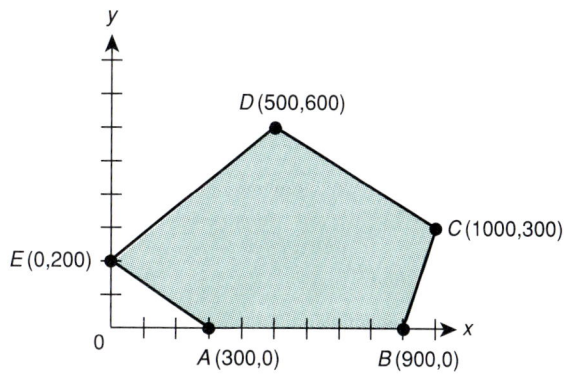

Exercises 12.5

Determine the maximum and minimum values of the given objective function $f(x,y)$ over the region defined by the given set of inequalities.

1. $f(x,y) = 2x + 3y$
   $x \geq 0, y \geq 0$
   $x - y - 2 \leq 0$

2. $f(x,y) = x - 2y$
   $x \geq 0, y \geq 0$
   $2x + 3y \leq 6$

3. $f(x,y) = x - y + 2$
   $x \geq 0, y \geq 0$
   $2x + 3y \leq 1800$

4. $f(x,y) = x - y + 2$
   $x \geq 0, y \geq 0$
   $2x + 3y \leq 2400$

5. $f(x,y) = 2x + 3y + 5$
   $x \geq 3, y \leq 5$
   $x - y \leq 30$

6. $f(x,y) = 7x - 8y + 3$
   $y \geq 1, y \leq 5, x \geq 0$
   $x - y \leq 0$

7. $f(x,y) = 30x + 45y + 100$
   $x - y - 5 \leq 0$
   $5x - 3y + 15 \geq 0$
   $x - 3y + 3 \geq 0$

8. $f(x,y) = 10x + 20y + 14$
   $x - y \leq 5$
   $5x - 3y \geq -3$
   $x - 3y \geq -3$

9. $f(x,y) = 2x - 3y$
   $x - y + 1 \leq 0$
   $2x - y + 2 \geq 0$
   $2x + y - 6 \leq 0$

10. $f(x,y) = 3x - 4y$
    $x - y + 1 \leq 0$
    $2x - y + 2 \geq 0$
    $2x + y - 6 \leq 0$

11. $f(x,y) = 7x - 8y$
    $y \leq 3$
    $4x + y + 5 \geq 0$
    $x - y \geq 0$

12. $f(x,y) = 5x - 3y$
    $y \leq 3$
    $4x + y + 5 \geq 0$
    $x - y \geq 0$

13. $f(x,y) = 0.3x + 0.4y$
    $x \leq 10$
    $x - y \geq 0$
    $x + 2y \leq 20$

14. $f(x,y) = 0.5x - 0.6y$
    $x \leq 10$
    $x - y \geq 0$
    $x + 2y \leq 20$

15. $f(x,y) = 14x + 17y - 11$
    $x - y \geq 0$
    $x \geq 3$
    $y \geq -2$
    $2x + 3y - 20 \leq 0$

16. $f(x) = 12x + 15y - 10$
    $x - y \geq 0$
    $x \geq 3$
    $y \geq -2$
    $2x + 3y - 20 \leq 0$

17. $f(x,y) = 1.5x - 1.6y + 3$
    $x \geq 0, y \geq 0$
    $2x - y + 2 \geq 0$
    $x + y - 1 \geq 0$
    $2x + y - 6 \leq 0$

18. $f(x,y) = 3.2x - 2.1y + 4$
    $x \geq 0, y \geq 0$
    $2x - y + 2 \geq 0$
    $x + y - 1 \geq 0$
    $2x + y - 6 \leq 0$

19. $f(x,y) = 4.5x + 2.3y - 10.5$
    $x \geq 0, y \geq 0$
    $x \leq 75, y \leq 60$
    $x + y \leq 100$
    $2x + 3y \geq 60$

20. $f(x,y) = 3.2x + 1.7y + 11.2$
    $x \geq 0, y \geq 0$
    $x \leq 75, y \leq 60$
    $x + y \leq 100$
    $2x + 3y \geq 60$

21. $f(x,y) = 3x - 4y + 2$
    $x \geq 0, y \geq 0$
    $y \leq 5$
    $4x - 3y + 12 \geq 0$
    $x - y - 1 \leq 0$

22. $f(x,y) = 3x - 4y + 2$
    $x \geq 0, y \geq 0$
    $x \leq 10, y \leq 8$
    $x + y \leq 12$

23. $f(x,y) = 100x + 200y$
    $x \geq 0, y \geq 0$
    $x \leq 6, y \leq 7$
    $5x + 3y \leq 36$
    $x - 6y + 6 \leq 0$

24. $f(x) = 100x + 200y$
    $x \geq 0, y \geq 0$
    $x \leq 10, y \leq 8$
    $x + y \leq 12$
    $x + 6y \geq 12$

25. Suppose we add the restriction $x \leq 750$ and $y \leq 500$ to the conditions in example 12.27. Find the values of x and y that yield the maximum profit.

26. Suppose we add the restrictions $x \leq 600$ and $y \leq 400$ to the conditions in example 12.27. Find the values of x and y that yield the maximum profit.

27. The EZ Company manufactures two models of electronic components. The labor costs are indicated in the following table:

Labor Type	Model 201	Model 301
I	3 hours	2 hours
II	2 hours	4 hours

A total of 40 hours of each type of labor is available each week. If $70 profit is realized on each model 201 and $60 profit is realized on each model 301, determine a production schedule to maximize profits.

28. Solve exercise 27 given that the profit function is $f(x,y) = 80x + 70y$, where x and y are the numbers of model 201 and 301 produced.

29. A rental agency has a maximum of $194,000 to invest in the purchase of automobiles. They normally purchase two types of cars. The cost per car and expected profit per car are shown in the following table:

Type	Cost	Profit
I	$10,000	$3000
II	$12,000	$4000

Experience has shown that the company should buy no more than 11 cars of type I and no more than 12 cars of type II. Calculate the number of each type to buy to realize the maximum profit.

30. Solve exercise 29 given that the profit function is $P(x,y) = 3500x + 4000y$ instead of $P(x,y) = 3000x + 4000y$.

31. A production unit in a power plant uses a mixture of two grades of coal. The kilowatt-hour (kwh) production per ton, the emissions of hydrocarbons per ton, and the emissions of solids per ton are listed in the following table.

Grade	kwh/ton	Hydrocarbons/ton	Solids/ton
I	2500	1.5 lb	10 lb
II	3500	1 lb	12 lb

The maximum emissions allowable per hour are 100 pounds of solids and 10 pounds of hydrocarbons. The plant can get at most 5 tons of grade I and 7 tons of grade II coal per hour for each of the production units. What mixture of coal should be used to gain the maximum kwh output from the production units?

32. Given that the production units in exercise 31 must have an output of at least 25,000 kwh and type I coal costs $20 per ton and type II coal costs $15 per ton, determine the mixture of coal to use in order to operate at minimum cost. Assume that all of the other restrictions in exercise 31 must still be met.

33. The EZ Corporation has two mines that each produce two grades of iron ore. Production capabilities are listed in the following table:

Hourly Maximum Production

Mine	Grade A	Grade B	Hours of Operation Per Day
x	24 tons	17 tons	x
y	15 tons	18 tons	y

The company's mills cannot use more than 270 tons of Grade A ore nor more than 265 tons of Grade B ore per day. The mining operation yields a profit of $1200 per hour at mine x and $1300 per hour at mine y. Determine a production schedule that maximizes the profit.

34. The EZ Manufacturing Corporation has a machine that produces type 201 and type 301 tapered roller bearings. The profits per bearing are $20 and $10, respectively. The following table shows the material and labor costs.

Type	Labor Costs	Material Costs
201	$1	$3
301	$4	$1

The machine cannot produce more than 600 bearings per day, the labor costs cannot exceed $1200 per day, and material costs cannot exceed $1000 per day. Determine a daily production schedule that maximizes profits.

12 • 6 Summary of Terms, Rules, and Procedures

Terms

Absolute Value (p. 475)
Double Inequality (p. 465)
Equivalent Inequalities (p. 462)
Feasible Solution (p. 484)
Interval Notation (p. 464)

Linear Inequality (p. 462)
Linear Inequality in Two Variables (p. 482)
Linear Programming (p. 490)
Maximum of f (p. 487)
Minimum of f (p. 487)

Nonlinear Inequalities (p. 468)
Objective Function (p. 490)
Polygonal Set (p. 484)
Region of Feasible Solutions (p. 490)
Solution Set (p. 462)

Rules and Procedures

- SOLVING LINEAR INEQUALITIES (12.1)
  If x, y, and b are real numbers:
  1. If $x < y$, then $x + b < y + b$ and $x - b < y - b$.
  2. If $x < y$ and $b > 0$, then $bx < by$ and $\dfrac{x}{b} < \dfrac{y}{b}$.
  3. If $x < y$ and $b < 0$, then $bx > by$ and $\dfrac{x}{b} > \dfrac{y}{b}$.

- GRAPHING LINEAR INEQUALITIES (12.1)

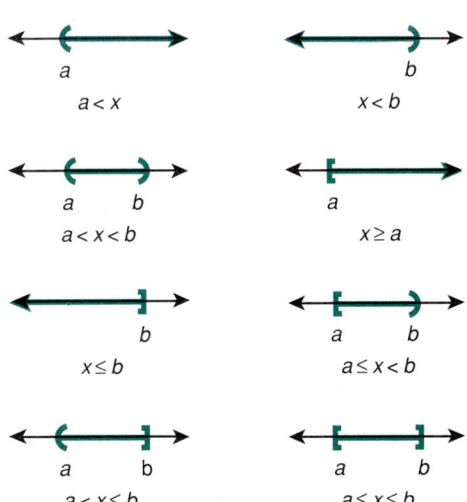

Figure 12.34

- **ABSOLUTE VALUE** (12.3)

$$|a| = a \quad \text{if } a \geq 0$$
$$ = -a \quad \text{if } a < 0.$$

- **THEOREM 12.1** (12.3)

  If $b > 0$ then $|y| = b$ has the solutions $y = b$ or $y = -b$.

- **THEOREM 12.2** (12.3)
  a. $|y| < b$ is equivalent to $-b < y < b$ $(b > 0)$.
  b. $|y| > b$ is equivalent to $y < -b$ or $y > b$ $(b > 0)$.

- **GRAPHING A LINEAR INEQUALITY OF THE FORM**
  $ax + by < c$ (12.4)
  1. Graph the equation $ax + by = c$.
  2. The graph of $ax + by < c$ is one of the half-planes determined by the line $ab + by = c$.
  3. Select a point that is obviously in one of the half-planes and check whether the coordinates of the point satisfy the inequality.

- **SOLVING LINEAR PROGRAMMING PROBLEMS** (12.5)
  1. Graph the polygonal region determined by the inequalities and find the coordinates of the vertices of the polygon.
  2. Evaluate the objective function at each of the corner points.
  3. The maximum value of the objective function over the region is the largest number obtained in step 2.
  4. The minimum value of the objective function over the region is the smallest number obtained in step 2.

12·7 Chapter 12 Review Exercises

Write each inequality in interval notation.

1. $2 \leq x < 3$
2. $-3 < x \leq 4$
3. $x < 4$
4. $x \geq -3$

Write each interval as an inequality and graph on a number line.

5. $(-6, -2)$
6. $(-2, 3)$
7. $[3, 7)$
8. $(-4, 6]$
9. $(2, \infty)$
10. $(-\infty, 5]$

Solve for x.

11. $2x + 4 \leq 5$
12. $3x - 6 \geq 10$
13. $\dfrac{x+4}{3} \leq 6$
14. $\dfrac{x-3}{2} \geq 4$
15. $5 \leq 3x - 2 \leq 8$
16. $-4 \leq 2x - 1 \leq 9$
17. $2 - x < 3x + 2 < x + 4$
18. $4x - 8 \leq x + 1 \leq 6x + 2$
19. $x^2 - 6x + 8 \leq 0$
20. $x^2 - 7x + 12 \geq 0$
21. $x^2 + 5x + 5 > 0$
22. $x^2 + 3x + 4 \geq 0$
23. $(x + 1)(x - 2)(x + 3) \leq 0$
24. $(x - 1)(x + 2)(x + 3) < 0$
25. $\dfrac{2x(x - 1)}{x + 2} \leq 0$
26. $\dfrac{x(2x - 3)}{x + 1} > 0$
27. $|2x - 4| = 3$
28. $|3x - 7| = 8$

29. $\left|\dfrac{x}{3}\right| \le 3$

30. $\left|\dfrac{x}{2}\right| \le 1$

31. $|2x - 6| < 7$

32. $|3x + 4| \le 11$

33. $|x + 3| > 5$

34. $|x - 3| > 4$

35. $\left|\dfrac{2x - 1}{3x - 2}\right| > 4$

36. $\left|\dfrac{2x + 3}{2x - 3}\right| \le 7$

Graph the region that satisfies the given system of inequalities.

37. $3x + 4y \le 12$

38. $4x + 3y \ge 12$

39. $3x + 4y \le 12, \quad x + y \ge 6$

40. $3x + 4y \ge 12, \quad x + y \le 6$

41. $x \ge 0, \quad 4x + 2y \le 100, \quad y \ge 0$

42. $x \ge 0, \quad y \ge 0, \quad 2x + 3y \le 120$

43. $x \ge 0, \quad y \ge 0, \quad 5x + 4y \le 2000,$
    $5x + 12y \le 3000$

44. $x \ge 0, \quad y \ge 0, \quad 2x + y \le 50, \quad 3x - y \le 25$

45. Maximize the function $f(x,y) = 10x + 15y$ over the region in exercise 43.

46. Maximize the function $f(x,y) = 3x + 5y$ over the region in exercise 44.

12·8 Chapter 12 Test

Write each inequality in interval notation.

1. $5 \le x \le 6$

2. $x \ge -6$

Write each interval as an inequality.

3. $(-\infty, -2]$

4. $[3, 7]$

5. $(2, \infty)$

6. $(-2, 3)$

Solve for x.

7. $3x - 5 \le 7$

8. $\dfrac{x - 2}{3} > 6$

9. $4 \le 2x - 4 \le 8$

10. $x^2 - 7x + 6 \ge 0$

11. $(x - 2)(x + 4)(x + 1) < 0$

12. $\dfrac{3x(2x - 1)}{x - 3} > 0$

13. $|2x + 3| < 5$

14. $|2x + 1| \ge 9$

15. $\left|\dfrac{x}{2}\right| < 6$

16. $|2x + 1| = 9$

Graph the region that satisfies the given set of inequalities.

17. $5x + 6y \le 30$

18. $5x + 6y \le 30, \quad x + 2y \ge 4$

19. $x \ge 0, y \ge 0, 5x + 6y \le 60$

20. $x \ge 0, y \ge 0, 4x + 5y \le 2000$

21. $y \ge 0, 3x - 4y \ge 0, 3x + 4y \le 240$

22. Maximize $f(x,y) = 0.3x + 0.5y$ subject to these constraints: $x \geq 50$, $y \geq 60$, $x + y \leq 200$, $4x + 5y \leq 900$

23. The shop boss of the state prison is interested in holding down the costs in the shop. They produce tables and chairs for the prison system. He must schedule at least 60 hours/day of carpentry time and at least 40 hours/day of common labor to keep the prisoners busy. The labor requirements and the material costs are shown in the table.

	Table	Chair
Carpentry Time	10 hr	5 hr
Common Labor	8 hr	2 hr
Material Costs	$80	$17

The prison system cannot use a production of more than 10 tables and 8 chairs per day. Find a daily production schedule that minimizes the material costs.

CHAPTER

13 *Matrices and Determinants*

Matrices were invented in the nineteenth century by Arthur Cayley (1821–1895), a British mathematician. He developed matrices as a tool that he used in his work in analytic geometry. Cayley's discovery of the rules for matrix operations helped to establish a new branch of mathematics, abstract algebra.

Matrices were originally not expected to have any practical value, but in 1925 Walter Heisenberg found that matrix theory was an excellent tool when he formed a new mathematical theory of quantum mechanics based on the frequencies and amplitudes of the emitted radiation and on the energy levels of the atomic system.

Many other applications of matrices have been found, not the least of which is the development of the subject of input-output analysis by Wassily Leontief, who won the Nobel Prize in economics in 1973 for his input-output model of the economy.

In this chapter we introduce methods of solving larger systems of linear equations using matrix methods. A matrix is simply a rectangular array of numbers. We study the various operations on matrices and the properties of matrices. We also consider determinants of square matrices along with their properties.

13 • 1 Addition and Scalar Multiplication of Matrices

Consider the following system of equations:

$$x + y + z = 6$$
$$2x - y + z = 3$$
$$3x - 2y + 2z = 5.$$

Recall that in the solution process for this system of equations we had to write and rewrite the variables. If we dispense with the writing of the variables and write only the coefficients of the variables and the constants, we obtain the following configuration:

$$\begin{bmatrix} 1 & 1 & 1 & 6 \\ 2 & -1 & 1 & 3 \\ 3 & -2 & 2 & 5 \end{bmatrix}.$$

Matrix   A rectangular array of numbers like this is called a **matrix.** We refer to this matrix as a "3 by 4" matrix, since it has three rows and four columns.

m by n Matrix   An ***m* by *n* matrix** is a rectangular array of numbers having m rows and n columns.
Dimensions of a Matrix   The numbers m and n are called the **dimensions** of the matrix.

$$\begin{bmatrix} a_{11} & a_{12} & \cdots & a_{1n} \\ a_{21} & a_{22} & \cdots & a_{2n} \\ \vdots & \vdots & & \vdots \\ a_{m1} & a_{m2} & \cdots & a_{mn} \end{bmatrix}$$

This matrix illustrates how to represent the general m by n matrix using subscripts. The entry a_{ij} means that this entry appears in the ith row and the jth column of the matrix. For example, a general 3 by 4 matrix is represented as

$$\begin{bmatrix} a_{11} & a_{12} & a_{13} & a_{14} \\ a_{21} & a_{22} & a_{23} & a_{24} \\ a_{31} & a_{32} & a_{33} & a_{34} \end{bmatrix}.$$

Some particular examples of matrices are:

1. 2 by 3 matrix $\begin{bmatrix} 1 & 0 & 4 \\ -1 & 3 & 2 \end{bmatrix}$

2. 1 by 5 matrix $[0.46 \quad 1.5 \quad 7.2 \quad 0 \quad 3.6]$

3. 4 by 1 matrix $\begin{bmatrix} 0.03 \\ 0.21 \\ 0.37 \\ 0.46 \end{bmatrix}$

SECTION 13.1 ADDITION AND SCALAR MULTIPLICATION OF MATRICES

4. 4 by 4 matrix
$$\begin{bmatrix} 1 & 0 & 0 & 4 \\ -1 & 3 & 2 & 7 \\ 0 & 0 & 1 & 2 \\ 4 & -3 & 7 & 5 \end{bmatrix}.$$

EQUAL MATRICES

Two matrices are equal, provided:

1. they have the same dimensions, and
2. each element in the first matrix is equal to the element in the corresponding position in the second matrix.

• **Example 13.1:**
$$\begin{bmatrix} \frac{1}{2} & 4 & \frac{3}{2} \\ 2 & \frac{1}{2} & 3 \end{bmatrix} = \begin{bmatrix} 0.5 & 4 & 1.5 \\ 2 & 0.5 & 3 \end{bmatrix}.$$

The matrices in example 13.1 are equal, since both matrices are 2 by 3 and the elements in the corresponding positions in the matrices are equal.

Note that the matrices

$$\mathbf{A} = \begin{bmatrix} 4 & 2 & 3 \\ 7 & 0 & 1 \end{bmatrix} \quad \text{and} \quad \mathbf{B} = \begin{bmatrix} 4 & 7 \\ 2 & 0 \\ 3 & 1 \end{bmatrix}$$

are not equal, since **A** is a 2 by 3 matrix and **B** is a 3 by 2 matrix.

• **Example 13.2:**
$$\begin{bmatrix} (x-1) & (z-1) \\ (y-2) & (w-4) \end{bmatrix} \quad \text{and} \quad \begin{bmatrix} 6 & 5 \\ 2 & -1 \end{bmatrix}$$

could be equal matrices, since they are both 2 by 2 matrices. However, in order for the matrices to be equal, these conditions must be met:

$$\begin{array}{lll} x - 1 = 6 & \text{or} & x = 7 \\ y - 2 = 2 & \text{or} & y = 4 \\ z - 1 = 5 & \text{or} & z = 6 \\ w - 4 = -1 & \text{or} & w = 3. \end{array}$$

An expression such as

$$\begin{bmatrix} (x-1) & (z-1) \\ (y-2) & (w-4) \end{bmatrix} = \begin{bmatrix} 6 & 5 \\ 2 & -1 \end{bmatrix}$$

Matrix Equation is called a **matrix equation**. The solution can be written as $x = 7$, $y = 4$, $z = 6$, $w = 3$, or as

$$\begin{bmatrix} x & z \\ y & w \end{bmatrix} = \begin{bmatrix} 7 & 6 \\ 4 & 3 \end{bmatrix}.$$

A pair of matrices having the same dimensions can be added, using the pattern in the following example.

• **Example 13.3:**
$$\begin{bmatrix} 1 & 0 & 4 & 3 \\ 6 & -1 & 7 & 2 \end{bmatrix} + \begin{bmatrix} 1 & -6 & 3 & 5 \\ 8 & 2 & -1 & 3 \end{bmatrix} =$$
$$\begin{bmatrix} (1+1) & (0-6) & (4+3) & (3+5) \\ (6+8) & (-1+2) & (7-1) & (2+3) \end{bmatrix} = \begin{bmatrix} 2 & -6 & 7 & 8 \\ 14 & 1 & 6 & 5 \end{bmatrix}.$$

In general, we add matrices according to this pattern:

Addition of Matrices

$$\begin{bmatrix} a_{11} & a_{12} & \cdots & a_{1n} \\ a_{21} & a_{22} & \cdots & a_{2n} \\ \vdots & \vdots & & \vdots \\ a_{m1} & a_{m2} & \cdots & a_{mn} \end{bmatrix} + \begin{bmatrix} b_{11} & b_{12} & \cdots & b_{1n} \\ b_{21} & b_{22} & \cdots & b_{2n} \\ \vdots & \vdots & & \vdots \\ b_{m1} & b_{m2} & \cdots & b_{mn} \end{bmatrix}$$

$$= \begin{bmatrix} (a_{11}+b_{11}) & (a_{12}+b_{12}) & \cdots & (a_{1n}+b_{1n}) \\ (a_{21}+b_{21}) & (a_{22}+b_{22}) & \cdots & (a_{2n}+b_{2n}) \\ \vdots & \vdots & & \vdots \\ (a_{m1}+b_{m1}) & (a_{m2}+b_{m2}) & \cdots & (a_{mn}+b_{mn}) \end{bmatrix}.$$

We add the elements in the corresponding positions in the two matrices. Note that the pair of matrices being added must have the same dimensions.

• **Example 13.4:** The EZ Manufacturing Company has three warehouses that stock three types of automobile generators. An inventory reveals the following stocks:

Warehouse	Generator Type		
	X_1	X_2	X_3
A	500	600	800
B	200	100	350
C	400	1200	550

SECTION 13.1 ADDITION AND SCALAR MULTIPLICATION OF MATRICES

The company has ordered the following supplies for the three warehouses:

$$\begin{array}{c} \text{Warehouse} \\ A \\ B \\ C \end{array} \begin{array}{c} \text{Generator Type} \\ \begin{array}{ccc} X_1 & X_2 & X_3 \end{array} \\ \begin{bmatrix} 450 & 300 & 150 \\ 250 & 150 & 150 \\ 300 & 0 & 100 \end{bmatrix} \end{array}$$

Write a matrix that represents the total inventory, assuming that the orders are delivered.

Solution: We obtain the desired matrix by adding the "present inventory" and "on order" matrices:

$$\begin{bmatrix} 500 & 600 & 800 \\ 200 & 100 & 350 \\ 400 & 1200 & 550 \end{bmatrix} + \begin{bmatrix} 450 & 300 & 150 \\ 250 & 150 & 150 \\ 300 & 0 & 100 \end{bmatrix}$$

$$= \begin{bmatrix} 950 & 900 & 950 \\ 450 & 250 & 500 \\ 700 & 1200 & 650 \end{bmatrix}.$$

The total inventory matrix is

$$\begin{array}{c} \text{Warehouse} \\ A \\ B \\ C \end{array} \begin{array}{c} \text{Generator Type} \\ \begin{array}{ccc} X_1 & X_2 & X_3 \end{array} \\ \begin{bmatrix} 950 & 900 & 950 \\ 450 & 250 & 500 \\ 700 & 1200 & 650 \end{bmatrix} \end{array}.$$

Can you see any significance in the sums of the columns of the matrix? (The sum of a column represents the total inventory of that particular type of generator.)

Scalar Multiplication

Scalar   One type of operation on matrices that is useful in many problems is **scalar multiplication** When a matrix is multiplied by a real number, called a **scalar,** the result is a matrix having the same dimensions as the original matrix. Example 13.5 shows the pattern for scalar multiplication.

• *Example 13.5:* $5 \begin{bmatrix} 1 & 0 & 3 \\ 6 & 8 & -10 \end{bmatrix} = \begin{bmatrix} (5)(1) & (5)(0) & (5)(3) \\ (5)(6) & (5)(8) & (5)(-10) \end{bmatrix}$

$$= \begin{bmatrix} 5 & 0 & 15 \\ 30 & 40 & -50 \end{bmatrix}.$$

We multiply each entry in the matrix by the scalar. The general pattern for scalar multiplication is illustrated as follows:

$$k \begin{bmatrix} a_{11} & a_{12} & \cdots & a_{1n} \\ a_{21} & a_{22} & \cdots & a_{2n} \\ \vdots & \vdots & & \vdots \\ a_{m1} & a_{m2} & \cdots & a_{mn} \end{bmatrix}$$

$$= \begin{bmatrix} (ka_{11}) & (ka_{12}) & \cdots & (ka_{1n}) \\ (ka_{21}) & (ka_{22}) & \cdots & (ka_{2n}) \\ \vdots & \vdots & & \vdots \\ (ka_{m1}) & (ka_{m2}) & \cdots & (ka_{mn}) \end{bmatrix}.$$

• **Example 13.6:** The EZ Corporation is assembling four modules for electronic equipment. The modules use the following parts as components:

		Part Number			
Module	I	II	III	IV	V
A	4	2	0	2	3
B	2	0	0	1	4
C	0	2	3	6	2
D	1	4	6	4	1

The company produces 600 of each of the modules per day. We represent the parts matrix for daily production as follows:

$$600 \begin{bmatrix} 4 & 2 & 0 & 2 & 3 \\ 2 & 0 & 0 & 1 & 4 \\ 0 & 2 & 3 & 6 & 2 \\ 1 & 4 & 6 & 4 & 1 \end{bmatrix}$$

$$= \begin{bmatrix} & I & II & III & IV & V \\ & 2400 & 1200 & 0 & 1200 & 1800 \\ & 1200 & 0 & 0 & 600 & 2400 \\ & 0 & 1200 & 1800 & 3600 & 1200 \\ & 600 & 2400 & 3600 & 2400 & 600 \end{bmatrix}.$$

• **Example 13.7:** Perform the following matrix operations:

$$4 \begin{bmatrix} 1 & 0 & 2 & 3 \\ -3 & 6 & 0 & -7 \\ 3 & -4 & 2 & 3 \end{bmatrix} - 5 \begin{bmatrix} 1 & 6 & 2 & 1 \\ 0 & 0 & 5 & 3 \\ 6 & -5 & 2 & -1 \end{bmatrix}.$$

Solution: We first find the scalar products:

$$\begin{bmatrix} 4 & 0 & 8 & 12 \\ -12 & 24 & 0 & -28 \\ 12 & -16 & 8 & 12 \end{bmatrix} - \begin{bmatrix} 5 & 30 & 10 & 5 \\ 0 & 0 & 25 & 15 \\ 30 & -25 & 10 & -5 \end{bmatrix}.$$

Finally we perform the subtractions. Note that in a subtraction problem we simply subtract the numbers in the corresponding positions in the two matrices.

$$\begin{bmatrix} (4-5) & (0-30) & (8-10) & (12-5) \\ (-12-0) & (24-0) & (0-25) & (-28-15) \\ (12-30) & (-16-(-25)) & (8-10) & (12-(-5)) \end{bmatrix}$$

$$= \begin{bmatrix} -1 & -30 & -2 & 7 \\ -12 & 24 & -25 & -43 \\ -18 & 9 & -2 & 17 \end{bmatrix}.$$

Of particular interest in the addition and subtraction of matrices are matrices that have zeros for all of the entries.

Zero Matrix An *m* by *n* matrix in which all of the entries are zero is called a **zero matrix**. The following matrices are zero matrices:

$$\begin{bmatrix} 0 & 0 & 0 \\ 0 & 0 & 0 \\ 0 & 0 & 0 \end{bmatrix} \quad \begin{bmatrix} 0 \\ 0 \\ 0 \end{bmatrix} \quad [0 \ 0 \ 0 \ 0].$$

3 by 3   3 by 1   1 by 4

From the definition of matrix addition and the definition of zero matrices, the following properties are evident:

PROPERTIES OF ZERO MATRICES

If **A**, **B**, and **C** are *m* by *n* matrices and θ represents the *m* by *n* zero matrix, then these properties hold.

1. $\mathbf{A} + \theta = \theta + \mathbf{A} = \mathbf{A}$.
2. Addition of matrices is commutative: $\mathbf{A} + \mathbf{B} = \mathbf{B} + \mathbf{A}$.
3. Addition of matrices is associative: $(\mathbf{A} + \mathbf{B}) + \mathbf{C} = \mathbf{A} + (\mathbf{B} + \mathbf{C})$.

Check these properties using particular examples of matrices. See exercises 19 and 20 in the following exercise set.

CHAPTER 13 MATRICES AND DETERMINANTS

Trial Problems 13.1

Before you begin the section exercises, warm up with these problems. Complete answers are included in the answer key.

Perform the indicated operations.

1. $[1 \quad 5 \quad 3] + [1 \quad 6 \quad 2]$

2. $5 \begin{bmatrix} 1 \\ 3 \end{bmatrix} - 2 \begin{bmatrix} -6 \\ -2 \end{bmatrix}$

3. $4 \begin{bmatrix} 6 \\ -2 \\ 2 \end{bmatrix} + 7 \begin{bmatrix} 5 \\ 6 \\ -3 \end{bmatrix}$

4. $\begin{bmatrix} 1 & 2 & 5 \\ 6 & -3 & 1 \\ 5 & 1 & -3 \end{bmatrix} + \begin{bmatrix} 1 & 4 & 6 \\ -3 & 0 & 1 \\ 0 & 2 & 3 \end{bmatrix}$

Solve.

5. $[x \quad y] + 2[3 \quad 6] = [5 \quad 2]$

Exercises 13.1

Perform the indicated matrix operations in exercises 1–10.

1. $\begin{bmatrix} 1 & 0 & 3 \\ 6 & 5 & -1 \\ 3 & 0 & 1 \end{bmatrix} + \begin{bmatrix} 6 & 5 & -3 \\ 0 & 4 & -2 \\ 1 & 1 & 1 \end{bmatrix}$

2. $\begin{bmatrix} 5 & 0 & 3 \\ -1 & -1 & -1 \end{bmatrix} + \begin{bmatrix} 3 & 2 & 4 \\ 1 & -1 & 1 \end{bmatrix}$

3. $2 \begin{bmatrix} 1 \\ 4 \\ 3 \end{bmatrix} + 3 \begin{bmatrix} 2 \\ 0 \\ 1 \end{bmatrix}$

4. $-3 \begin{bmatrix} -1 \\ 0 \\ 4 \end{bmatrix} + 2 \begin{bmatrix} 4 \\ 6 \\ -3 \end{bmatrix}$

5. $[3 \quad 2 \quad 0] - 3[1 \quad 4 \quad -3]$

6. $14[2 \quad 0 \quad 1] + \dfrac{1}{2}[6 \quad 4 \quad -8]$

7. $\dfrac{1}{2} \begin{bmatrix} 8 & -4 & 2 \\ 16 & 0 & 14 \end{bmatrix} - \dfrac{2}{3} \begin{bmatrix} 9 & 0 & 6 \\ 12 & 3 & 3 \end{bmatrix}$

8. $0.5 \begin{bmatrix} 1.2 & 3.4 \\ 6.8 & 5.2 \end{bmatrix} - 0.4 \begin{bmatrix} 10 & -20 \\ 4.5 & 1.6 \end{bmatrix}$

9. $5 \begin{bmatrix} 1 & 0 & 3 & 6 \\ 4 & 6 & 2 & 3 \\ 8 & 0 & 0 & -2 \end{bmatrix} + \begin{bmatrix} 4 & 1 & -3 & 3 \\ 1 & 0 & 4 & 6 \\ 7 & 3 & -4 & 2 \end{bmatrix} - 2 \begin{bmatrix} 1 & 0 & 4 & 7 \\ 5 & -6 & 3 & 3 \\ 0 & 0 & 4 & 5 \end{bmatrix}$

10. $-4 \begin{bmatrix} 3 & 6 & 3 \\ 1 & 0 & 3 \\ 1 & 6 & 1 \end{bmatrix} + 2 \begin{bmatrix} 7 & 8 & 9 \\ 4 & -3 & 2 \\ 11 & -3 & 3 \end{bmatrix} + \begin{bmatrix} 4 & 0 & -1 \\ 3 & 2 & 2 \\ 7 & -5 & 1 \end{bmatrix}$

Determine the values of the variables in each of the matrix equations in exercises 11–16.

11. $[5x \quad 4y \quad 3z] = [10 \quad 16 \quad 18]$

12. $\begin{bmatrix} 4x \\ 3y \\ 2z \end{bmatrix} = \begin{bmatrix} 7 \\ 3 \\ 5 \end{bmatrix}$

13. $\begin{bmatrix} x - y \\ x + 2y \end{bmatrix} = \begin{bmatrix} 3 \\ 9 \end{bmatrix}$

14. $[x - y \quad 3x - 2y] = [3 \quad 11]$

15. $\begin{bmatrix} x \\ y \\ z \end{bmatrix} + 2 \begin{bmatrix} 3 \\ 1 \\ 4 \end{bmatrix} = \begin{bmatrix} 6 \\ 5 \\ 6 \end{bmatrix}$

16. $[x \quad y \quad z] + 2[1 \quad 0 \quad 4] = [6 \quad 4 \quad -3]$

17. The EZ Corporation has two warehouses, each of which stocks five types of sparkplugs. The following matrix represents the present inventories:

$$\begin{array}{c} \text{Warehouse} \\ A \\ B \end{array} \begin{array}{ccccc} & & \text{Type} & & \\ 017 & 018 & 019 & 020 & 021 \\ \left[\begin{array}{ccccc} 4000 & 2600 & 3000 & 150 & 3200 \\ 6000 & 3000 & 2100 & 4000 & 8200 \end{array}\right]. \end{array}$$

The projected sales for the month are represented by this matrix:

$$\begin{array}{c} \text{Warehouse} \\ A \\ B \end{array} \begin{array}{ccccc} & & \text{Type} & & \\ 017 & 018 & 019 & 020 & 021 \\ \left[\begin{array}{ccccc} 2500 & 1000 & 1200 & 100 & 1000 \\ 2000 & 1200 & 900 & 850 & 2600 \end{array}\right]. \end{array}$$

a. Write a matrix that represents the projected inventory at the end of the month, assuming that no new supplies are ordered.
b. A 15% increase in sales is projected for next month. Write a matrix that represents next month's sales.
c. If next month's sales increase materializes, write a matrix that represents the projected inventory at the end of that month. (Note that the negative entries are permissible, since they represent the shortages that will occur if no new supplies are received.)

18. At the beginning of the month, an electronics supply house has an inventory of four types of transistors in two warehouses:

$$\begin{array}{c} \text{Warehouse} \\ A \\ B \end{array} \begin{array}{cccc} & \text{Transistor Type} & & \\ 030 & 040 & 050 & 060 \\ \left[\begin{array}{cccc} 4000 & 5000 & 6000 & 3500 \\ 2000 & 6500 & 5400 & 3200 \end{array}\right]. \end{array}$$

The predicted sales for the month are

$$\begin{array}{c} \text{Warehouse} \\ A \\ B \end{array} \begin{array}{cccc} & \text{Transistor Type} & & \\ 030 & 040 & 050 & 060 \\ \left[\begin{array}{cccc} 1100 & 1200 & 1300 & 1250 \\ 800 & 2000 & 1200 & 800 \end{array}\right]. \end{array}$$

a. Write a matrix that represents the projected inventory for the end of the month.
b. A 15% increase in sales is predicted for next month. Write a matrix that represents next month's sales.
c. Assuming that the 15% increase in sales materializes, write a matrix for the projected inventory at the end of next month.

19. Use the following matrices to verify the indicated properties of matrix operations:

$$A = \begin{bmatrix} 1 & 0 & 4 \\ 6 & 2 & -1 \end{bmatrix} \quad B = \begin{bmatrix} 2 & 1 & 7 \\ -3 & 4 & 2 \end{bmatrix}$$

$$C = \begin{bmatrix} 1 & 4 & 3 \\ -2 & -3 & 6 \end{bmatrix} \quad \theta = \begin{bmatrix} 0 & 0 & 0 \\ 0 & 0 & 0 \end{bmatrix}.$$

a. $A + \theta = \theta + A = A$
b. $A + B = B + A$
c. $(A + B) + C = A + (B + C)$

20. Use the following matrices to verify the indicated properties of matrix operations:

$$A = \begin{bmatrix} 1 & 4 \\ -1 & 3 \\ 2 & 7 \end{bmatrix} \quad B = \begin{bmatrix} 0 & 0 \\ 1 & 4 \\ 2 & 0 \end{bmatrix}$$

$$C = \begin{bmatrix} 1 & -2 \\ 4 & 3 \\ 3 & 6 \end{bmatrix} \quad \theta = \begin{bmatrix} 0 & 0 \\ 0 & 0 \\ 0 & 0 \end{bmatrix}.$$

a. $A + \theta = \theta + A = A$
b. $A + B = B + A$
c. $(A + B) + C = A + (B + C)$

13 • 2 Solving Systems of Equations Using Matrices

In this section we solve n by n systems of equations using operations on the rows of the matrices formed from the systems of equations. Recall that in chapter 4 we solved such systems by addition and subtraction.

The following set of rules for solving a system of equations can be rewritten in terms of operations on the rows of the matrix for the system. Given a system of equations that has a unique solution, we can

1. interchange any pair of equations,
2. multiply (or divide) any equation by a nonzero constant, and
3. add to any equation, a multiple of another equation.

This set of rules can be adapted to form a set of row operations on matrices. Consider this set of linear equations.

$$a_{11}x + a_{12}y + a_{13}z = k_1$$
$$a_{21}x + a_{22}y + a_{23}z = k_2$$
$$a_{31}x + a_{32}y + a_{33}z = k_3.$$

NOTE: For purposes of writing matrices for a system of equations, the equations must be in this form.

The matrix

$$\begin{bmatrix} a_{11} & a_{12} & a_{13} \\ a_{21} & a_{22} & a_{23} \\ a_{31} & a_{32} & a_{33} \end{bmatrix}$$

Coefficient Matrix

Augmented Coefficient Matrix

is called the **coefficient matrix** of the system of equations. If we annex an additional column to the matrix, namely the set of constants on the right-hand side of the equations, we obtain the **augmented coefficient matrix**

$$\begin{bmatrix} a_{11} & a_{12} & a_{13} & \vdots & k_1 \\ a_{21} & a_{22} & a_{23} & \vdots & k_2 \\ a_{31} & a_{32} & a_{33} & \vdots & k_3 \end{bmatrix}.$$

The vertical dashed line is merely a convenience to separate the coefficients of the variables from the constants.

We can rewrite the set of rules for performing operations on systems of equations as a set of rules for operating on the augmented coefficient matrix of the system.

When the following row operations are performed on the augmented coefficient matrix of a system of equations, a matrix of an equivalent system results.

1. Interchange any pair of rows.
2. Multiply (or divide) any row by a nonzero constant.
3. Add to any row a multiple of another row.

The three row operations mentioned here lead to an equivalent system whether the original system has a unique solution or not. Example 13.8 illustrates how to use these rules to solve a 3 by 3 system of equations.

Section 13.2 Solving Systems of Equations Using Matrices

• *Example 13.8:* Solve the system of equations

$$2x - 3y + 2z = 14$$
$$x + 2y - z = -6$$
$$3x - 2y + z = 10.$$

Solution: First, write the augmented coefficient matrix for the system of equations:

$$\begin{bmatrix} 2 & -3 & 2 & | & 14 \\ 1 & 2 & -1 & | & -6 \\ 3 & -2 & 1 & | & 10 \end{bmatrix}.$$

We solve the system by using the row operations to obtain a matrix of an equivalent system that has the following form:

$$\mathbf{A} = \begin{bmatrix} 1 & 0 & 0 & | & m \\ 0 & 1 & 0 & | & n \\ 0 & 0 & 1 & | & p \end{bmatrix}.$$

This gives the solution $x = m$, $y = n$, $z = p$, since the system of equations for this matrix can be written as

$$\begin{array}{lll} 1x + 0y + 0z = m & \text{or} & x = m \\ 0x + 1y + 0z = n & \text{or} & y = n \\ 0x + 0y + 1z = p & \text{or} & z = p. \end{array}$$

•••••••••

Pivot Positions
Pivot Elements
Pivot Operation

The positions a_{11}, a_{22}, and a_{33}, which contain the "ones" in matrix **A** are called the **pivot positions**. The numbers in these positions are called **pivot elements**. The basic idea is to select these pivot elements, one at a time, and perform the following **pivot operation** on each of these elements.

1. Obtain the "1" in the pivot position.

2. Obtain zeros in all the other positions in the column containing the pivot element.

Since the matrix already has a "1" in one of the pivot positions, (the a_{33} position), select this element as the pivot element. This saves step 1 in the pivot position and avoids using fractions, at least temporarily:

$$\begin{bmatrix} 2 & -3 & \boxed{2} & | & 14 \\ 1 & 2 & \boxed{-1} & | & -6 \\ 3 & -2 & \boxed{1} & | & 10 \end{bmatrix}.$$

Step 2 in the pivot operation requires that we obtain zeros in the other positions in column 3. This is accomplished in two stages.

1. Add to row 1, -2 times row 3. The -2 (the negative of the element in the a_{13} position was chosen so that we can obtain a zero in this position $((-2)(1) + 2 = 0)$:

$$\begin{bmatrix} 2 & -3 & 2 & | & 14 \\ 1 & 2 & -1 & | & -6 \\ 3 & -2 & 1 & | & 10 \end{bmatrix}$$ (Add to row 1, -2, times row 3.)

The result is the following matrix, which has the zero in the a_{13} position,

$$\begin{bmatrix} -4 & 1 & 0 & | & -6 \\ 1 & 2 & -1 & | & -6 \\ 3 & -2 & 1 & | & 10 \end{bmatrix}.$$

2. Add to row 2, one times row 3. This operation obtains a zero in the a_{23} position

$$\begin{bmatrix} -4 & 1 & 0 & | & -6 \\ 1 & 2 & -1 & | & -6 \\ 3 & -2 & 1 & | & 10 \end{bmatrix}.$$ (Add to row 2, 1 times row 3.)

The result of this operation is a zero in the desired position:

$$\begin{bmatrix} -4 & 1 & 0 & | & -6 \\ 4 & 0 & 0 & | & 4 \\ 3 & -2 & 1 & | & 10 \end{bmatrix}.$$

The matrix now has column 3 in the desired form. Now select a new pivot element. The choices are the -4 in the a_{11} position and the 0 in the a_{22} position. We can interchange rows 1 and 2 to obtain a "1" in the a_{22} position:

$$\begin{bmatrix} 4 & 0 & 0 & | & 4 \\ -4 & 1 & 0 & | & -6 \\ 3 & -2 & 1 & | & 10 \end{bmatrix}.$$ (Interchange row 1 and row 2.)

Since the pivot position in column 2 is already a 1 and since there is already a 0 in one of the positions in column 2, we need only obtain a 0 in the a_{32} position. We do this by adding to row 3, 2 times row 2:

$$\begin{bmatrix} 4 & 0 & 0 & | & 4 \\ -4 & 1 & 0 & | & -6 \\ -5 & 0 & 1 & | & -2 \end{bmatrix}.$$ (Add to row 3, 2 times row 2.)

We now have columns 2 and 3 in the desired form. The pivot position in column 1 contains a 4. To obtain a 1 in this position, we multiply row 1 by $\dfrac{1}{4}$:

$$\begin{bmatrix} 1 & 0 & 0 & | & 1 \\ -4 & 1 & 0 & | & -6 \\ -5 & 0 & 1 & | & -2 \end{bmatrix}.$$ (Multiply $\dfrac{1}{4}$ times row 1.)

To complete the operations on column 1 we first obtain a zero in the a_{21} position. Add to row 2, 4 times row 1:

$$\begin{bmatrix} 1 & 0 & 0 & | & 1 \\ 0 & 1 & 0 & | & -2 \\ -5 & 0 & 1 & | & -2 \end{bmatrix}. \quad \text{(Add to row 2, 4 times row 1.)}$$

We complete the operation by adding to row 3, 5 times row 1:

$$\begin{bmatrix} 1 & 0 & 0 & | & 1 \\ 0 & 1 & 0 & | & -2 \\ 0 & 0 & 1 & | & 3 \end{bmatrix}. \quad \text{(Add to row 3, 5 times row 1.)}$$

We can now read the solution to the system as $x = 1, y = -2, z = 3$. Substitution of these values in the original three equations verifies that we have the correct solution to the system.

While this method of solution seems complicated at first, a little practice should convince you that, for a 3 by 3 system, it is no more trouble to use this system than it is to solve the system algebraically. For systems larger than 3 by 3, the matrix method proves to be less trouble. Let us try the procedure on a 4 by 4 system.

• *Example 13.9:* Solve the following system of equations:

$$x + y + z + w = 10$$
$$x - y + 2z - w = 1$$
$$2x - 3y + 2z + 3w = 14$$
$$3x + 2y - 2w = -1.$$

Solution: The matrix for the system is

$$\begin{bmatrix} 1 & 1 & 1 & 1 & | & 10 \\ 1 & -1 & 2 & -1 & | & 1 \\ 2 & -3 & 2 & 3 & | & 14 \\ 3 & 2 & 0 & -2 & | & -1 \end{bmatrix}.$$

We need to obtain a matrix for an equivalent system of equations that has the following form:

$$\begin{bmatrix} 1 & 0 & 0 & 0 & | & m \\ 0 & 1 & 0 & 0 & | & n \\ 0 & 0 & 1 & 0 & | & p \\ 0 & 0 & 0 & 1 & | & q \end{bmatrix}.$$

Since column 1 has a 1 in the pivot position, we work on column 1 first. We obtain the zeros in the remainder of the first column with the following three operations.

1. Add to row 2, -1 times row 1.
2. Add to row 3, -2 times row 1.
3. Add to row 4, -3 times row 1.

The result of these actions is

$$\begin{bmatrix} 1 & 1 & 1 & 1 & \vdots & 10 \\ 0 & -2 & \boxed{1} & -2 & \vdots & -9 \\ 0 & -5 & \boxed{0} & 1 & \vdots & -6 \\ 0 & -1 & -3 & -5 & \vdots & -31 \end{bmatrix}.$$

We can obtain a 1 in the row 3 pivot position by interchanging row 2 and row 3. (We cannot get a "1" in the position by multiplying the row by a constant.)

$$\begin{bmatrix} 1 & 1 & \boxed{1} & 1 & \vdots & 10 \\ 0 & -5 & 0 & 1 & \vdots & -6 \\ 0 & -2 & \boxed{1} & -2 & \vdots & -9 \\ 0 & -1 & \boxed{-3} & -5 & \vdots & -31 \end{bmatrix}.$$

We can obtain zeros in the remaining positions in column 3 by the following operations.

1. Add to row 1, -1 times row 3.
2. Add to row 4, 3 times row 3.

The result is the following matrix:

$$\begin{bmatrix} 1 & \boxed{3} & 0 & 3 & \vdots & 19 \\ 0 & \boxed{-5} & 0 & 1 & \vdots & -6 \\ 0 & \boxed{-2} & 1 & -2 & \vdots & -9 \\ 0 & \boxed{-7} & 0 & \boxed{-11} & \vdots & -58 \end{bmatrix}.$$

We must now decide whether to work on column 2 or column 4 next. Column 4 looks like the more promising of the two columns, since it contains a 1 in row 2, the only 1 in that row. We interchange rows 2 and 4:

$$\begin{bmatrix} 1 & 3 & 0 & 3 & \vdots & 19 \\ 0 & -7 & 0 & -11 & \vdots & -58 \\ 0 & -2 & 1 & -2 & \vdots & -9 \\ 0 & -5 & 0 & \boxed{1} & \vdots & -6 \end{bmatrix}.$$

We now obtain the zeros in the remaining positions in column 4 by the following operations:

1. Add to row 1, -3 times row 4.
2. Add to row 2, 11 times row 4.
3. Add to row 3, 2 times row 4.

The result of these actions is the following matrix:

$$\begin{bmatrix} 1 & 18 & 0 & 0 & | & 37 \\ 0 & -62 & 0 & 0 & | & -124 \\ 0 & -12 & 1 & 0 & | & -21 \\ 0 & -5 & 0 & 1 & | & -6 \end{bmatrix}.$$

We now must obtain a 1 in the a_{22} position. We can accomplish this by dividing row 2 by -62:

$$\begin{bmatrix} 1 & 18 & 0 & 0 & | & 37 \\ 0 & 1 & 0 & 0 & | & 2 \\ 0 & -12 & 1 & 0 & | & -21 \\ 0 & -5 & 0 & 1 & | & -6 \end{bmatrix}.$$

We complete the solution of the problem by obtaining zeros in the remaining positions in column 2.

1. Add to row 1, -18 times row 2.
2. Add to row 3, 12 times row 2.
3. Add to row 4, 5 times row 2.

$$\begin{bmatrix} 1 & 0 & 0 & 0 & | & 1 \\ 0 & 1 & 0 & 0 & | & 2 \\ 0 & 0 & 1 & 0 & | & 3 \\ 0 & 0 & 0 & 1 & | & 4 \end{bmatrix} \quad \begin{array}{l} x = 1 \\ y = 2 \\ x = 3 \\ x = 4 \end{array}.$$

In the examples so far we have managed to avoid using fractions and decimals in the matrices by choosing the pivot elements carefully. However, it is not always possible to do this. Example 13.10 illustrates this. A calculator is helpful in solving such problems.

• **Example 13.10:** Analysis of the resistor circuit in figure 13.1 leads to the following set of equations for I_1, I_2, and I_3, the currents in the loops of the circuit:

$$(5 + 4)I_1 - 5I_3 - 4I_2 = 15$$
$$(4 + 6 + 2)I_2 - 4I_1 - 6I_3 = 0$$
$$(5 + 8 + 6)I_3 - 5I_1 - 6I_2 = 0.$$

Solve the system for I_1, I_2, and I_3.

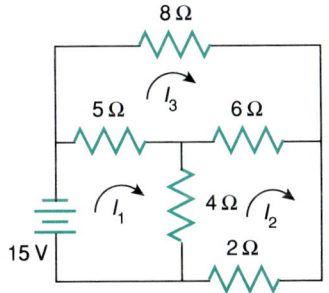

Figure 13.1

Solution: We rearrange the equations to the usual form:

$$9I_1 - 4I_2 - 5I_3 = 15$$
$$-4I_1 + 12I_2 - 6I_3 = 0$$
$$-5I_1 - 6I_2 + 19I_3 = 0.$$

The matrix for the system is

$$\begin{bmatrix} 9 & -4 & -5 & | & 15 \\ -4 & 12 & -6 & | & 0 \\ -5 & -6 & 19 & | & 0 \end{bmatrix}.$$

Observing the first column of the matrix, it appears that -4 is the easiest element to work with, so we interchange rows 1 and 2 to get the -4 into the a_{11} position. (Fourths are simpler than ninths.)

$$\begin{bmatrix} -4 & 12 & -6 & | & 0 \\ 9 & -4 & -5 & | & 15 \\ -5 & -6 & 19 & | & 0 \end{bmatrix}.$$

Divide the first row by -4 to obtain a 1 in the pivot position:

$$\begin{bmatrix} 1 & -3 & 1.5 & | & 0 \\ 9 & -4 & -5 & | & 15 \\ -5 & -6 & 19 & | & 0 \end{bmatrix}.$$

Obtain zeros in the a_{21} and a_{31} positions by adding, to row 2, -9 times row 1, and by adding to row 3, 5 times row 1:

$$\begin{bmatrix} 1 & -3 & 1.5 & | & 0 \\ 0 & \boxed{23} & -18.5 & | & 15 \\ 0 & \boxed{-21} & 26.5 & | & 0 \end{bmatrix}.$$

We could now use any of the four elements -23, -18.5, -21, and 26.5 as our next pivot element. The only advantage seems to be the presence of the zero in the last row and column. (The choice is arbitrary.)

Divide row 3 by 26.5 to obtain a 1 in the a_{33} position. The decimals are rounded to three places:

$$\begin{bmatrix} 1 & -3 & 1.5 & | & 0 \\ 0 & 23 & -18.5 & | & 15 \\ 0 & -0.792 & \boxed{1} & | & 0 \end{bmatrix}.$$

Obtain zeros in the a_{31} and a_{32} positions by adding, to row 2, 18.5 times row 3, and adding to row 1, -1.5 times row 3:

$$\begin{bmatrix} 1 & -1.812 & 0 & | & 0 \\ 0 & 8.348 & 0 & | & 15 \\ 0 & -0.792 & 1 & | & 0 \end{bmatrix}.$$

Section 13.2 Solving Systems of Equations Using Matrices

Obtain a 1 in the a_{22} position by dividing row 2 by 8.348:

$$\begin{bmatrix} 1 & -1.812 & 0 & | & 0 \\ 0 & \boxed{1} & 0 & | & 1.797 \\ 0 & -0.792 & 1 & | & 0 \end{bmatrix}.$$

Complete the solution by obtaining zeros in the a_{21} and a_{31} positions.

$$\begin{bmatrix} 1 & 0 & 0 & | & 3.256 \\ 0 & \boxed{1} & 0 & | & 1.797 \\ 0 & 0 & 1 & | & 1.423 \end{bmatrix} \qquad \begin{array}{l} I_1 = 3.256 \text{ amps} \\ I_2 = 1.797 \text{ amps} \\ I_3 = 1.423 \text{ amps}. \end{array}$$

Trial Problems 13.2

Before you begin the section exercises, warm up with these problems. Complete answers are included in the answer key.

Solve using matrices.

1. $x - 2y = -1$
   $2x + y = 3$

2. $x + 2y = 8$
   $2x - y = 1$

3. $x + y + z = 2$
   $2x - y + 2z = 7$
   $3x - 2y + z = 7$

4. $x + 2y - 2z = 10$
   $3x - y + z = 2$
   $5x + 2y + 3z = 13$

Exercises 13.2

Solve the following systems of equations using matrices.

1. $x + y + z = 10$
   $x - y + z = 2$
   $2x + 2y + z = 15$

2. $x + y + z = 9$
   $x - y + z = 5$
   $2x + 2y + z = 14$

3. $x + y + z = 2$
   $3x - 2y + z = 7$
   $2x - 3y + 4z = 13$

4. $x + y + z = -2$
   $3x - 2y + z = 10$
   $2x - 3y + 4z = 0$

5. $2x + y - z = 0$
   $x - 3y - 2z = 13$
   $-2x + 3y - 3z = 8$

6. $2x + y - z = 5$
   $x - 3y - 2z = 10$
   $-2x + 3y - 2z = -23$

7. $x + y + z = 90$
   $2x - 3y + 2z = 30$
   $3x + 4y + 3z = 300$

8. $x + y + z = 85$
   $2x - 3y + 2z = 45$
   $3x + 4y + 3z = 280$

9. $x + y + z + w = 6$
   $x - y - z + 2w = -1$
   $2x + 3y + 2z - w = 10$
   $3x + y - z + 2w = 3$

10. $x + y + z + w = 5$
    $x - y - z + 2w = 6$
    $2x + 3y + 2z - w = 3$
    $3x + y - z + 2w = 12$

11. $2x + 2y - z + w = 70$
    $x + 2y + 2z - 3w = -10$
    $4x + 5y - z + 2w = 190$
    $x + 2y + z - w = 40$

12. $2x + 2y - z + w = 70$
    $x + 2y + 2z - 3w = -10$
    $4x + 5y - z + 2w = 220$
    $x + 2y + z - w = 30$

13. $0.01x + 0.01y + 0.01z = 1$
    $0.2x + 0.2y + 0.3z = 24$
    $0.04x - 0.04y + 0.05z = 2$

14. $0.01x + 0.01y + 0.01z = 18$
    $0.2x + 0.2y + 0.3z = 44$
    $0.04x + 0.04y + 0.05z = 8$

15. Analysis of an electrical circuit yields the following set of equations for the loop currents I_1, I_2, and I_3:

$$8I_1 - 2I_2 + 3I_3 = 12$$
$$-I_1 + 8I_2 + 5I_3 = 0$$
$$9I_1 + 15I_2 + 36I_3 = 0.$$

Determine the loop currents to two decimal places.

16. Solve exercise 15 given the following set of equations:

$$9I_1 + 6I_3 = 30$$
$$6I_1 - 12I_2 + 27I_3 = 0$$
$$27I_2 - 12I_3 = -45.$$

17. The circuit in figure 13.2 yields the following set of equations for the voltages V_1, V_2, and V_3. Determine these voltages.

$$\left(\frac{1}{3} + \frac{1}{3} + \frac{1}{12}\right)V_1 - \frac{1}{3}V_2 - \frac{1}{12}V_3 = 0$$

$$\left(\frac{1}{3} + \frac{1}{2}\right)V_2 - \frac{1}{3}V_1 - \frac{1}{2}V_3 = 5$$

$$\left(\frac{1}{12} + \frac{1}{2} + \frac{1}{4}\right)V_3 - \frac{1}{2}V_2 - \frac{1}{12}V_1 = 0.$$

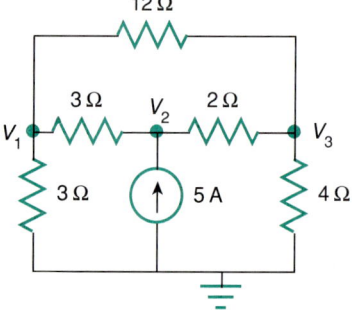

Figure 13.2

18. The circuit in figure 13.3 yields the following set of equations for the voltages V_1, V_2, and V_3. Find these voltages.

$$\left(\frac{1}{4} + \frac{1}{4} - \frac{1}{2}\right)V_1 + \frac{1}{4}V_2 - \frac{1}{2}V_3 = 12.5$$

$$\left(\frac{1}{4} + \frac{1}{4} + \frac{1}{1}\right)V_2 - \frac{1}{4}V_1 - \frac{1}{4}V_3 = 0$$

$$\left(\frac{1}{4} + \frac{1}{2} + \frac{1}{2}\right)V_3 - \frac{1}{2}V_1 - \frac{1}{4}V_2 = 0.$$

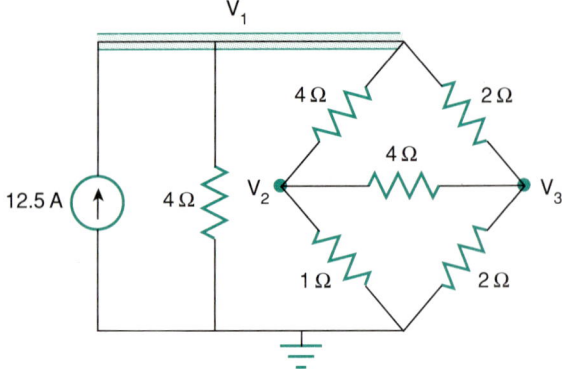

Figure 13.3

19. The EZ Manufacturing Company manufactures three types of modules for electronic equipment. Each type of module requires machine time in three different departments. These time requirements are shown in the following table.

Department	Module Type M	N	P
A	3 min	2 min	4 min
B	5 min	4 min	6 min
C	12 min	8 min	10 min

The company has available 52,400 min, 89,600 min, and 180,800 min of machine time available in departments A, B, and C, respectively. How many of each type of module should the company schedule for production in order to use all of the available machine time?

20. Solve exercise 19 given that the following amounts of machine time are available.

Department	
A	71,900 min
B	121,300 min
C	245,000 min

21. Three types of fertilizer/weed killer/insecticide are available. They contain the percentages of nitrogen, herbicide and insecticide shown in the following table.

Type	Nitrogen	Herbicide	Insecticide
I	20	10	3
II	10	4	8
III	10	8	4

How much of each type should be used in order to obtain a mixture of 800 pounds that contains 15%, 7.5%, and 5%, respectively, of nitrogen, herbicide, and insecticide?

22. Solve exercise 21 if the mixture of 1000 pounds must contain 15%, 8%, and 4.5%, respectively, of nitrogen, herbicide, and insecticide.

23. The EZ Manufacturing Company produces three items. The production costs and selling prices of the items are shown in the following table.

Item	Selling Price	Production Costs
A	$6	$2
B	$8	$4
C	$10	$8

Given that the total number of items was 900, the production costs totaled $4800 and the sales totaled $7600, how many items were produced and sold?

24. Solve exercise 23 using the same selling price and production costs, if the total number of items sold was 1500, the total production cost was $7000 and the sales totaled $12,000.

25. The equation of a circle can be written in the form $x^2 + y^2 + ax + by + c = 0$. Find the equation of the circle that contains the points $P(0,3)$, $Q(6,3)$, and $R(7,-4)$.

26. Solve exercise 25 using the points $D(3,-1)$, $E(5,3)$, and $F(6,2)$.

27. Find the equation of the curve of the form $y = ax^3 + bx^2 + cx + d$ that contains the points $P(1,-2)$, $Q(-1,10)$, $R(3,14)$, and $S(-2,2)$.

28. Solve exercise 27 using the points $E(1,-8)$, $F(2,-2)$, $G(-1,4)$, and $H(3,7)$.

13·3 Multiplication of Matrices

Column Matrix
Column Vector

An n by 1 matrix, consisting of a single column of numbers, is called a **column matrix** or a **column vector**. Some examples are

$$\begin{bmatrix} 1 \\ 2 \\ -1 \\ 3 \end{bmatrix}, \quad \begin{bmatrix} -3 \\ 4 \\ 0.5 \end{bmatrix}, \quad \begin{bmatrix} a_1 \\ a_2 \\ a_3 \\ a_4 \end{bmatrix}, \quad \begin{bmatrix} a_1 \\ a_2 \\ \cdot \\ \cdot \\ \cdot \\ a_n \end{bmatrix}$$

Row Matrix
Row Vector

A 1 by n matrix, consisting of a single row, is called a **row matrix** or a **row vector**. The following matrices are row matrices:

$$[1 \quad 4 \quad -1 \quad 3], \quad [x_1 \quad x_2 \quad x_3], \quad [a_1 \quad a_2 \quad \cdots \quad a_n].$$

We can multiply a row matrix times a column matrix (in that order) provided the row and column matrices have the same number of entries. We multiply according to the pattern indicated in Example 13.11.

- **Example 13.11:** $[1 \quad 2 \quad 3 \quad 4] \begin{bmatrix} 5 \\ 6 \\ 7 \\ 8 \end{bmatrix}$

$$= [(1)(5) + (2)(6) + (3)(7) + (4)(8)]$$
$$= [(5 + 12 + 21 + 32)]$$
$$= [70].$$

Note that the product is always a 1 by 1 matrix. In general, the pattern for multiplying a 1 by n row matrix times an n by 1 column matrix is

$$[a_1 \quad a_2 \quad a_3 \quad \cdots \quad a_n] \begin{bmatrix} b_1 \\ b_2 \\ b_3 \\ \vdots \\ b_n \end{bmatrix}$$

$$= [a_1 b_1 + a_2 b_2 + a_3 b_3 + \cdots + a_n b_n].$$

Note that the product matrix, a 1 by 1 matrix, is defined only if the row matrix has the same number of entries as the column matrix. Note also that the row matrix must be on the left.

- **Example 13.12:** $[1 \quad 4 \quad 3] \begin{bmatrix} -1 \\ -3 \\ 2 \end{bmatrix} = [(1)(-1) + (4)(-3) + (3)(2)]$

$$= [(-1 - 12 + 6)] = [-7].$$

- **Example 13.13:** An electrical supply house buys respectively, 100, 150, 200, 400, 600, and 140 of six different transformers from a supplier at costs per item of 3.10, 6.50, 3.00, 4.25, 1.85, and 5.85, respectively. Use matrix multiplication to determine the total cost.

    Solution: We represent the numbers of the items purchased by the row matrix:

    $$\mathbf{A} = [100 \quad 150 \quad 200 \quad 400 \quad 600 \quad 140].$$

We represent the respective costs of the items by the column matrix:

$$\mathbf{B} = \begin{bmatrix} 3.10 \\ 6.50 \\ 3.00 \\ 4.25 \\ 1.85 \\ 5.85 \end{bmatrix}.$$

The total cost is represented by the product **AB**:

$$\mathbf{AB} = \begin{bmatrix} 100 & 150 & 200 & 400 & 600 & 140 \end{bmatrix} \begin{bmatrix} 3.10 \\ 6.50 \\ 3.00 \\ 4.25 \\ 1.85 \\ 5.85 \end{bmatrix}$$

$$= [(100)(3.10) + (150)(6.50) + (200)(3) \\ + (400)(4.25) + (600)(1.85) + (140)(5.85)]$$
$$= [(310 + 975 + 600 + 1700 + 1110 + 819)]$$
$$= [5514].$$

In general, we can multiply a pair of matrices, provided the first matrix has the same number of columns as the second matrix has rows. Thus, if **A** and **B** are matrices, **AB** is defined, provided that the number of columns in matrix **A** equals the number of rows in matrix **B**.

Multiplication of Matrices

$$\mathbf{A} = \begin{bmatrix} a_{11} & a_{12} & \cdots & a_{1k} \\ a_{21} & a_{22} & \cdots & a_{2k} \\ \vdots & \vdots & & \vdots \\ a_{m1} & a_{m2} & \cdots & a_{mk} \end{bmatrix} \qquad \mathbf{B} = \begin{bmatrix} b_{11} & b_{12} & \cdots & b_{1n} \\ b_{21} & b_{22} & \cdots & b_{2n} \\ \vdots & \vdots & & \vdots \\ b_{k1} & b_{k2} & \cdots & b_{kn} \end{bmatrix}.$$

m rows, k columns $\qquad\qquad\qquad\qquad$ k rows, n columns

$$m \text{ by } \underbrace{k \leftarrow \text{ must be } \rightarrow k}_{\text{the same}} \text{ by } n$$

The dimensions of the product matrix is (the number of rows in the first matrix) by (the number of columns in the second matrix).

DIMENSIONS OF THE PRODUCT MATRIX

$$\underbrace{\mathbf{A}}_{m \text{ by } k} \quad \underbrace{\mathbf{B}}_{k \text{ by } n}$$

The product matrix is an *m* by *n* matrix.

NOTE: If these dimension conditions are not met, the product is undefined.

Example 13.14 illustrates the pattern for multiplying matrices.

• **Example 13.14:** If

$$\mathbf{A} = \begin{bmatrix} 1 & 2 & 3 \\ 4 & 5 & 6 \end{bmatrix} \quad \text{and} \quad \mathbf{B} = \begin{bmatrix} 7 & 10 & 13 \\ 8 & 11 & 14 \\ 9 & 12 & 15 \end{bmatrix},$$

determine the product matrix **AB**.

Solution: $\underbrace{\mathbf{A}}_{2 \text{ by } 3} \quad \underbrace{\mathbf{B}}_{3 \text{ by } 3}$

The product matrix is defined, since the number of columns in matrix **A** equals the number of rows in matrix **B**.

$$\underbrace{\mathbf{A}}_{2 \text{ by } 3} \quad \underbrace{\mathbf{B}}_{3 \text{ by } 3}$$

The product matrix is a 2 by 3 matrix.

The entries in the product matrix are found by using a procedure similar to the one that we used in multiplying row and column matrices:

$$\begin{bmatrix} 1 & 2 & 3 \\ 4 & 5 & 6 \end{bmatrix} \begin{bmatrix} 7 & 10 & 13 \\ 8 & 11 & 14 \\ 9 & 12 & 15 \end{bmatrix} = \begin{bmatrix} c_{11} & c_{12} & c_{13} \\ c_{21} & c_{22} & c_{23} \end{bmatrix}$$

$$c_{11} = (1 \ 2 \ 3) \begin{pmatrix} 7 \\ 8 \\ 9 \end{pmatrix} = \quad \begin{array}{l}(1)(7) + (2)(8) + (3)(9) = 50 \\ \text{(Row 1 times column 1)}\end{array}$$

$$c_{12} = (1 \ 2 \ 3) \begin{pmatrix} 10 \\ 11 \\ 12 \end{pmatrix} = \quad \begin{array}{l}(1)(10) + (2)(11) + (3)(12) = 68 \\ \text{(Row 1 times column 2)}\end{array}$$

$$c_{13} = (1 \ 2 \ 3) \begin{pmatrix} 13 \\ 14 \\ 15 \end{pmatrix} = \quad \begin{array}{l}(1)(13) + (2)(14) + (3)(15) = 86 \\ \text{(Row 1 times column 3)}\end{array}$$

$$c_{21} = \begin{pmatrix} 4 & 5 & 6 \end{pmatrix} \begin{pmatrix} 7 \\ 8 \\ 9 \end{pmatrix} = \begin{array}{l} (4)(7) + (5)(8) + (6)(9) = 122 \\ \text{(Row 2 times column 1)} \end{array}$$

$$c_{22} = \begin{pmatrix} 4 & 5 & 6 \end{pmatrix} \begin{pmatrix} 10 \\ 11 \\ 12 \end{pmatrix} = \begin{array}{l} (4)(10) + (5)(11) + (6)(12) = 167 \\ \text{(Row 2 times column 2)} \end{array}$$

$$c_{23} = \begin{pmatrix} 4 & 5 & 6 \end{pmatrix} \begin{pmatrix} 13 \\ 14 \\ 15 \end{pmatrix} = \begin{array}{l} (4)(13) + (5)(14) + (6)(15) = 212. \\ \text{(Row 2 times column 3)} \end{array}$$

The product matrix is

$$\begin{bmatrix} 50 & 68 & 86 \\ 122 & 167 & 212 \end{bmatrix}.$$

In general, for the product

$$\begin{bmatrix} a_{11} & a_{12} & \cdots & a_{1k} \\ a_{21} & a_{22} & \cdots & a_{2k} \\ \vdots & \vdots & & \vdots \\ a_{m1} & a_{m2} & \cdots & a_{mk} \end{bmatrix} \begin{bmatrix} b_{11} & b_{12} & \cdots & b_{1n} \\ b_{21} & b_{22} & \cdots & b_{2n} \\ \vdots & \vdots & & \vdots \\ b_{k1} & b_{k2} & \cdots & b_{kn} \end{bmatrix}$$

$$= \begin{bmatrix} c_{11} & c_{12} & \cdots & c_{1n} \\ c_{21} & c_{22} & \cdots & c_{2n} \\ \vdots & \vdots & & \vdots \\ c_{m1} & c_{m2} & \cdots & c_{mn} \end{bmatrix}.$$

The c_{ij} entry in the product matrix is found by multiplying the ith row in the first matrix by the jth column in the second matrix. That is,

$$c_{ij} = \begin{pmatrix} a_{i1} & a_{i2} & \cdots & a_{ik} \end{pmatrix} \begin{pmatrix} b_{1j} \\ b_{2j} \\ \vdots \\ b_{kj} \end{pmatrix}$$

$$= a_{i1}b_{1j} + a_{i2}b_{2j} + \cdots + a_{ik}b_{kj}.$$

• *Example 13.15:* Calculate **AB** given that

$$\mathbf{A} = \begin{bmatrix} 1 & 4 & -1 \\ 3 & -2 & 6 \end{bmatrix} \quad \text{and} \quad \mathbf{B} = \begin{bmatrix} 3 & 6 \\ -1 & 4 \\ 2 & -3 \end{bmatrix}.$$

Solution: Since **A** is a 2 by 3 matrix and **B** is a 3 by 2 matrix, **AB** is defined and is a 2 by 2 matrix.

$$\mathbf{AB} = \begin{bmatrix} 1 & 4 & -1 \\ 3 & -2 & 6 \end{bmatrix} \begin{bmatrix} 3 & 6 \\ -1 & 4 \\ 2 & -3 \end{bmatrix} = \begin{bmatrix} c_{11} & c_{12} \\ c_{21} & c_{22} \end{bmatrix},$$

where

$c_{11} = (1)(3) + (4)(-1) + (-1)(2) = -3$ (Row 1 times column 1)
$c_{12} = (1)(6) + (4)(4) + (-1)(-3) = 25$ (Row 1 times column 2)
$c_{21} = (3)(3) + (-2)(-1) + (6)(2) = 23$ (Row 2 times column 1)
$c_{22} = (3)(6) + (-2)(4) + (6)(-3) = -8$ (Row 2 times column 2),

$$\mathbf{AB} = \begin{bmatrix} -3 & 25 \\ 23 & -8 \end{bmatrix}.$$

••••••••••

In section 13.1 we defined a zero matrix as a matrix that has zero for all of its entries. We now define a special matrix called an *identity matrix,* which is useful in the multiplication of matrices.

Identity Matrix

An **identity matrix** is a square matrix that has ones on the diagonal from upper left to lower right and zeros for all of the other entries. Here are some examples of identity matrices:

$$\begin{bmatrix} 1 & 0 & 0 \\ 0 & 1 & 0 \\ 0 & 0 & 1 \end{bmatrix} \quad \begin{bmatrix} 1 & 0 \\ 0 & 1 \end{bmatrix} \quad \begin{bmatrix} 1 & 0 & 0 & 0 \\ 0 & 1 & 0 & 0 \\ 0 & 0 & 1 & 0 \\ 0 & 0 & 0 & 1 \end{bmatrix}.$$

3 by 3,   2 by 2,   4 by 4

> If **A** is an n by n matrix and **I** is the n by n identity matrix, then $\mathbf{AI} = \mathbf{IA} = \mathbf{A}$.

The following example verifies this property for a 2 by 2 case:

$$\mathbf{AI} = \begin{bmatrix} a_{11} & a_{12} \\ a_{21} & a_{22} \end{bmatrix} \begin{bmatrix} 1 & 0 \\ 0 & 1 \end{bmatrix}$$

$$= \begin{bmatrix} a_{11}(1) + a_{12}(0) & a_{11}(0) + a_{12}(1) \\ a_{21}(1) + a_{22}(0) & a_{21}(0) + a_{22}(1) \end{bmatrix}$$

$$= \begin{bmatrix} a_{11} & a_{12} \\ a_{21} & a_{22} \end{bmatrix} = \mathbf{A}.$$

Multiplication of matrices is not, in general, commutative. $(\mathbf{A} \cdot \mathbf{B}) \neq (\mathbf{B} \cdot \mathbf{A})$, since both products are not necessarily defined.

For example,

$$\begin{bmatrix} 1 & 2 & 3 \\ 4 & 5 & 6 \end{bmatrix} \begin{bmatrix} 7 & 10 & 13 \\ 8 & 11 & 14 \\ 9 & 12 & 15 \end{bmatrix}$$

2 by ③   ③ by 3

is defined, while,

$$\begin{bmatrix} 7 & 10 & 13 \\ 8 & 11 & 14 \\ 9 & 12 & 15 \end{bmatrix} \begin{bmatrix} 1 & 2 & 3 \\ 4 & 5 & 6 \end{bmatrix}$$

3 by ③   ② by 3

is not defined. Do you see why the second product is not defined? (The first matrix has 3 as its second dimension, while the second matrix has 2 as its first dimension.)

Example 13.16 shows that even if the products **AB** and **BA** for a pair of matrices are defined and the product matrices have the same dimensions, the product matrices may not be equal.

• *Example 13.16:* Given the matrices **A** and **B**, calculate **AB** and **BA**

$$\mathbf{A} = \begin{bmatrix} 1 & -1 \\ 2 & -3 \end{bmatrix} \quad \mathbf{B} = \begin{bmatrix} 2 & 1 \\ 0 & 0 \end{bmatrix}$$

Solution: $\mathbf{AB} = \begin{bmatrix} 1 & -1 \\ 2 & -3 \end{bmatrix} \begin{bmatrix} 2 & 1 \\ 0 & 0 \end{bmatrix} = \begin{bmatrix} 2 & 1 \\ 4 & 2 \end{bmatrix}$

$\mathbf{BA} = \begin{bmatrix} 2 & 1 \\ 0 & 0 \end{bmatrix} \begin{bmatrix} 1 & -1 \\ 2 & -3 \end{bmatrix} = \begin{bmatrix} 4 & -5 \\ 0 & 0 \end{bmatrix}.$

Indeed, $\mathbf{AB} \neq \mathbf{BA}$.

524 CHAPTER 13 MATRICES AND DETERMINANTS

The following properties of matrix multiplication are valid, assuming that the size restrictions of the matrices are met. Let **A**, **B**, and **C** be matrices.

> **PROPERTIES OF MATRIX MULTIPLICATION**
>
> 1. Multiplication is *associative*:
> $$(AB)C = A(BC).$$
> 2. Multiplication *distributes over addition*:
> $$A(B + C) = AB + AC.$$

Example 13.17 illustrates the second of these properties.

- **Example 13.17:** Given the matrices **A**, **B**, and **C**, show that $A(B + C) = AB + AC$:

$$A = \begin{bmatrix} 1 & 4 \\ 3 & -1 \end{bmatrix} \quad B = \begin{bmatrix} 1 & 6 \\ -1 & 3 \end{bmatrix} \quad C = \begin{bmatrix} 2 & 2 \\ 1 & 1 \end{bmatrix}.$$

Solution:
$$A(B + C) = \begin{bmatrix} 1 & 4 \\ 3 & -1 \end{bmatrix} \left(\begin{bmatrix} 1 & 6 \\ -1 & 3 \end{bmatrix} + \begin{bmatrix} 2 & 2 \\ 1 & 1 \end{bmatrix} \right)$$

$$= \begin{bmatrix} 1 & 4 \\ 3 & -1 \end{bmatrix} \begin{bmatrix} 3 & 8 \\ 0 & 4 \end{bmatrix} = \begin{bmatrix} 3 & 24 \\ 9 & 20 \end{bmatrix}$$

$$AB + AC = \begin{bmatrix} 1 & 4 \\ 3 & -1 \end{bmatrix} \begin{bmatrix} 1 & 6 \\ -1 & 3 \end{bmatrix} + \begin{bmatrix} 1 & 4 \\ 3 & -1 \end{bmatrix} \begin{bmatrix} 2 & 2 \\ 1 & 1 \end{bmatrix}$$

$$= \begin{bmatrix} -3 & 18 \\ 4 & 15 \end{bmatrix} + \begin{bmatrix} 6 & 6 \\ 5 & 5 \end{bmatrix} = \begin{bmatrix} 3 & 24 \\ 9 & 20 \end{bmatrix}.$$

Thus, $A(B + C) = AB + AC$.

Example 13.18 illustrates an industrial production scheduling problem involving matrix operations. The problem has been simplified to keep the matrices involved reasonable in size. Since computers are available in many industries, they can handle matrices with much greater dimensions.

- **Example 13.18:** The EZ Instruments Company makes three different models of portable calculators. The production schedule for three months is shown in matrix **P**:

$$\begin{array}{r} \\ \text{Model } M_1 \\ P = \text{Model } M_2 \\ \text{Model } M_3 \end{array} \begin{array}{ccc} \text{Oct.} & \text{Nov.} & \text{Dec.} \\ \begin{bmatrix} 1000 & 2000 & 3000 \\ 1500 & 1200 & 4000 \\ 1000 & 1500 & 1000 \end{bmatrix} \end{array}.$$

The calculators contain three different types of modules. The number of modules per model is shown in matrix **M**:

$$\mathbf{M} = \begin{array}{c} \text{Module} \\ x \\ y \\ z \end{array} \begin{array}{c} \text{Model} \\ \begin{array}{ccc} M_1 & M_2 & M_3 \end{array} \\ \begin{bmatrix} 2 & 4 & 2 \\ 3 & 1 & 1 \\ 2 & 4 & 3 \end{bmatrix} \end{array}.$$

The modules contain two different types of silicon chips. The number of chips per module is shown in matrix **C**:

$$\mathbf{C} = \begin{array}{c} \text{chip} \\ c_1 \\ c_2 \end{array} \begin{array}{c} \text{Module} \\ \begin{array}{ccc} x & y & z \end{array} \\ \begin{bmatrix} 3 & 6 & 2 \\ 1 & 4 & 2 \end{bmatrix} \end{array}$$

a. Determine the chips-per-model matrix.

b. Determine the production matrix for the two chip-types for the months of October, November, and December.

Solution a: The chips per model matrix is obtained as follows:

$$\frac{\text{Chips}}{\text{Model}} = \left(\frac{\text{Chips}}{\text{Module}}\right)\left(\frac{\text{Module}}{\text{Model}}\right).$$

This is equivalent to **CM**

$$\mathbf{CM} = \begin{bmatrix} 3 & 6 & 2 \\ 1 & 4 & 2 \end{bmatrix} \begin{bmatrix} 2 & 4 & 2 \\ 3 & 1 & 1 \\ 2 & 4 & 3 \end{bmatrix} = \begin{bmatrix} 28 & 26 & 18 \\ 18 & 16 & 12 \end{bmatrix}.$$

Solution b: The chips-per-month matrix is determined as follows:

$$\frac{\text{Chips}}{\text{Month}} = \left(\frac{\text{Chips}}{\text{Model}}\right)\left(\frac{\text{Model}}{\text{Month}}\right).$$

This is equivalent to **(CM)P**:

$$\mathbf{(CM)P} = \begin{bmatrix} 28 & 26 & 18 \\ 18 & 16 & 12 \end{bmatrix} \begin{bmatrix} 1000 & 2000 & 3000 \\ 1500 & 1200 & 4000 \\ 1000 & 1500 & 1000 \end{bmatrix}$$

$$= \begin{bmatrix} 85{,}000 & 114{,}200 & 206{,}000 \\ 54{,}000 & 73{,}200 & 130{,}000 \end{bmatrix}.$$

Thus the production matrix for the chips is:

$$\begin{array}{c} \text{chip} \\ c_1 \\ c_2 \end{array} \begin{array}{ccc} \text{October} & \text{November} & \text{December} \\ \left[\begin{array}{ccc} 85{,}000 & 114{,}200 & 206{,}000 \\ 54{,}000 & 73{,}200 & 130{,}000 \end{array}\right] \end{array}.$$

Note that since multiplication is associative, we could have obtained the same results with the product **C(MP)**. This product is

$$\mathbf{C(MP)} = \begin{bmatrix} 3 & 6 & 2 \\ 1 & 4 & 2 \end{bmatrix} \left(\begin{bmatrix} 2 & 4 & 2 \\ 3 & 1 & 1 \\ 2 & 4 & 3 \end{bmatrix} \begin{bmatrix} 1000 & 2000 & 3000 \\ 1500 & 1200 & 4000 \\ 1000 & 1500 & 1000 \end{bmatrix} \right)$$
$$\phantom{\mathbf{C(MP)} =} \text{Chips/Module} \quad \text{Module/Model} \quad \text{Model/Month}$$

$$= \begin{bmatrix} 3 & 6 & 2 \\ 1 & 4 & 2 \end{bmatrix} \begin{bmatrix} 10{,}000 & 11{,}800 & 24{,}000 \\ 5500 & 8700 & 14{,}000 \\ 11{,}000 & 13{,}300 & 25{,}000 \end{bmatrix}$$
$$ \text{Chips/Module} \quad\quad \text{Module/Month}$$

$$= \begin{bmatrix} 85{,}000 & 114{,}200 & 206{,}000 \\ 54{,}000 & 73{,}200 & 130{,}000 \end{bmatrix}.$$
$$\text{Chips/Month}$$

Trial Problems 13.3

Before you begin the section exercises, warm up with these problems. Complete answers are included in the answer key.

Find the products.

1. $[-1 \ \ 0 \ \ 1] \begin{bmatrix} 3 \\ 2 \\ -3 \end{bmatrix}$

2. $\begin{bmatrix} 3 \\ 2 \\ 3 \end{bmatrix} [-1 \ \ 0 \ \ 1]$

3. $[1 \ \ 4] \begin{bmatrix} 1 & 2 \\ 3 & -4 \end{bmatrix}$

4. $\begin{bmatrix} 1 & 2 \\ 3 & -4 \end{bmatrix} [1 \ \ 4]$

5. $\begin{bmatrix} 1 & 4 \\ 3 & 2 \end{bmatrix} \begin{bmatrix} 0 & 1 \\ 1 & 0 \end{bmatrix}$

Exercises 13.3

Find the product of the following pairs of matrices.

1. $[1 \ \ 4 \ \ 3] \begin{bmatrix} 3 \\ 6 \\ -1 \end{bmatrix}$

2. $[3 \ \ 6 \ \ -1] \begin{bmatrix} 1 \\ 4 \\ 3 \end{bmatrix}$

3. $[1 \ \ 0 \ \ 0 \ \ 4] \begin{bmatrix} 0 \\ 1 \\ 0 \\ 1 \end{bmatrix}$

4. $[4 \ \ 5 \ \ 0 \ \ 6] \begin{bmatrix} 1 \\ 0 \\ 1 \\ 0 \end{bmatrix}$

5. $\begin{bmatrix} 3 \\ 6 \\ -1 \end{bmatrix} [1 \ \ 4 \ \ 3]$

6. $\begin{bmatrix} 1 \\ 4 \\ 3 \end{bmatrix} [3 \ \ 6 \ \ -1]$

7. $\begin{bmatrix} 1 & 4 \\ -2 & 1 \end{bmatrix} \begin{bmatrix} 6 & 3 \\ -2 & 2 \end{bmatrix}$

8. $\begin{bmatrix} 6 & 3 \\ -2 & 1 \end{bmatrix} \begin{bmatrix} 1 & 4 \\ -2 & 1 \end{bmatrix}$

9. $\begin{bmatrix} 1 & 4 \\ -2 & 1 \end{bmatrix} \begin{bmatrix} 2 & 1 \\ 0 & 0 \end{bmatrix}$

10. $\begin{bmatrix} 2 & 1 \\ 0 & 0 \end{bmatrix} \begin{bmatrix} 1 & 4 \\ -2 & 1 \end{bmatrix}$

11. $\begin{bmatrix} 1 & 4 & 1 & 1 \\ 2 & 3 & 0 & 1 \end{bmatrix} \begin{bmatrix} 4 \\ 3 \\ -1 \\ 2 \end{bmatrix}$

12. $\begin{bmatrix} 2 & 3 & -1 & 2 \\ 1 & 0 & 0 & 3 \end{bmatrix} \begin{bmatrix} 6 \\ 5 \\ 0 \\ 1 \end{bmatrix}$

13. $\begin{bmatrix} 1 & 4 & 3 \\ 6 & 5 & 2 \\ 0 & -1 & -3 \end{bmatrix} \begin{bmatrix} 1 & 1 \\ 4 & 2 \end{bmatrix}$

14. $\begin{bmatrix} 1 & 1 \\ 4 & 2 \end{bmatrix} \begin{bmatrix} 1 & 4 & 3 \\ 6 & 5 & 2 \\ 0 & -1 & -3 \end{bmatrix}$

15. $\begin{bmatrix} 1 & 5 & -1 \\ 2 & -3 & -1 \\ 1 & 4 & -3 \end{bmatrix} \begin{bmatrix} 0 & 0 & 0 \\ 0 & 0 & 0 \\ 0 & 0 & 0 \end{bmatrix}$

16. $\begin{bmatrix} 0 & 0 & 0 \\ 0 & 0 & 0 \\ 0 & 0 & 0 \end{bmatrix} \begin{bmatrix} 1 & 4 & -1 \\ 3 & 2 & 7 \\ 1 & 0 & 3 \end{bmatrix}$

17. $\begin{bmatrix} 1 & 5 & 2 \\ 6 & 5 & 1 \\ 0 & 4 & 7 \end{bmatrix} \begin{bmatrix} 1 & 0 & 0 \\ 0 & 1 & 0 \\ 0 & 0 & 1 \end{bmatrix}$

18. $\begin{bmatrix} 1 & 0 & 0 \\ 0 & 1 & 0 \\ 0 & 0 & 1 \end{bmatrix} \begin{bmatrix} -6 & 4 & 3 \\ 1 & 5 & 2 \\ -1 & 6 & 5 \end{bmatrix}$

19. $\begin{bmatrix} 1.755 & 0.851 & 0.745 \\ 1.330 & 2.766 & 1.170 \\ 0.691 & 0.638 & 1.809 \end{bmatrix} \begin{bmatrix} 1000 \\ 5000 \\ 3000 \end{bmatrix}$

20. $\begin{bmatrix} 1.755 & 1.330 & 0.690 \\ 0.851 & 2.766 & 0.638 \\ 0.745 & 1.170 & 1.809 \end{bmatrix} \begin{bmatrix} 680 \\ 950 \\ 1100 \end{bmatrix}$

21. $\begin{bmatrix} 1 & -2 & 1 & 0 \\ 1 & -2 & 2 & -3 \\ 0 & 1 & -1 & 1 \\ -2 & 3 & -2 & 3 \end{bmatrix} \begin{bmatrix} 1 & 2 & 3 & 1 \\ 1 & 3 & 3 & 2 \\ 2 & 4 & 3 & 3 \\ 1 & 1 & 1 & 1 \end{bmatrix}$

22. $\begin{bmatrix} 1 & 2 & 3 & 1 \\ 1 & 3 & 3 & 2 \\ 2 & 4 & 3 & 3 \\ 1 & 1 & 1 & 1 \end{bmatrix} \begin{bmatrix} 1 & -2 & 1 & 0 \\ 1 & -2 & 2 & -3 \\ 0 & 1 & -1 & 1 \\ -2 & 3 & -2 & 3 \end{bmatrix}$

23. Given that

$$\mathbf{A} = \begin{bmatrix} 0.2 & 0 & 0.1 \\ 0.3 & 0.1 & 0.2 \\ 0.2 & 0.4 & 0 \end{bmatrix} \quad \mathbf{P} = \begin{bmatrix} 150 \\ 200 \\ 200 \end{bmatrix} \quad \mathbf{Q} = \begin{bmatrix} 300 \\ 400 \\ 450 \end{bmatrix},$$

calculate $\mathbf{A}(\mathbf{P} + \mathbf{Q})$.

24. Given that

$$\mathbf{A} = \begin{bmatrix} 0.2 & 0.3 & 0.2 \\ 0 & 0.1 & 0.4 \\ 0.1 & 0.2 & 0 \end{bmatrix} \quad \mathbf{P} = \begin{bmatrix} 150 \\ 200 \\ 200 \end{bmatrix} \quad \mathbf{Q} = \begin{bmatrix} 300 \\ 400 \\ 450 \end{bmatrix},$$

calculate $\mathbf{A}(\mathbf{P} + \mathbf{Q})$.

25. The EZ Corporation produces modules for electronic equipment. The modules are made up of parts according to the following parts-per-model matrix, **P**:

$$\mathbf{P} = \begin{array}{c} \\ \text{Parts} \\ \begin{array}{c} 1 \\ 2 \\ 3 \\ 4 \end{array} \end{array} \overset{\text{Module}}{\begin{array}{c} A \quad B \quad C \end{array}} \begin{bmatrix} 3 & 4 & 6 \\ 5 & 1 & 7 \\ 6 & 0 & 4 \\ 5 & 5 & 6 \end{bmatrix}.$$

The module-per-model matrix is

$$\begin{array}{c} \text{Module} \end{array} \begin{array}{c} \text{Model} \\ \begin{array}{ccc} 201 & 202 & 203 \end{array} \end{array}$$

$$\mathbf{M} = \begin{array}{c} A \\ B \\ C \end{array} \begin{bmatrix} 7 & 8 & 3 \\ 5 & 0 & 4 \\ 4 & 6 & 3 \end{bmatrix}.$$

The cost-per-part matrix is

$$\begin{array}{c} \text{Parts} \\ \begin{array}{cccc} 1 & 2 & 3 & 4 \end{array} \end{array}$$

$$\mathbf{C} = \text{cost} \begin{bmatrix} 0.83 & 1.05 & 4.53 & 6.85 \end{bmatrix}.$$

a. Calculate the parts-per-model matrix.

b. Calculate the cost-per-module matrix.

c. Calculate the cost-per-model matrix.

26. Solve exercise 25 given the following matrices:

$$\mathbf{M} = \begin{bmatrix} 8 & 6 & 11 \\ 0 & 7 & 10 \\ 5 & 3 & 9 \end{bmatrix} \quad \mathbf{C} = \begin{bmatrix} 1.05 & 1.83 & 6.23 & 5.99 \end{bmatrix}.$$

27. A survey of owners of three major brands of home computers yields the following data about future purchases:

$$\begin{array}{c} \text{Next Purchase} \\ \begin{array}{ccc} A & B & C \end{array} \end{array}$$

$$\begin{array}{c} \text{Present} \\ \text{Brand} \end{array} \begin{array}{c} A \\ B \\ C \end{array} \begin{bmatrix} 0.5 & 0.3 & 0.2 \\ 0.2 & 0.6 & 0.2 \\ 0.3 & 0.2 & 0.5 \end{bmatrix} = \mathbf{N}.$$

Line one [0.5 0.3 0.2], for example, indicates that of the people who presently have brand A, 0.5 will purchase another brand A, 0.3 will purchase brand B, and 0.2 will purchase brand C. Suppose the three companies presently have these shares of that part of the market:

$$\begin{array}{c} \phantom{\text{Part of Market }} \begin{array}{ccc} A & B & C \end{array} \\ \text{Part of Market } \begin{bmatrix} 0.35 & 0.30 & 0.35 \end{bmatrix} = \mathbf{P}. \end{array}$$

For each of the three companies, determine their projected share of the "next purchase" market.

28. Solve exercise 27 for the given matrices:

$$\mathbf{N} = \begin{bmatrix} 0.50 & 0.25 & 0.25 \\ 0.30 & 0.45 & 0.25 \\ 0.30 & 0.10 & 0.60 \end{bmatrix} \quad \mathbf{P} = \begin{bmatrix} 0.25 & 0.40 & 0.35 \end{bmatrix}.$$

13•4 The Inverse of a Square Matrix

Multiplicative Inverse of a Square (n by n) Matrix

Some applications of matrices require the calculation of the **multiplicative inverse of a square (n by n) matrix**. By this we mean the matrix, if it exists, by which we multiply a given square matrix to obtain an identity matrix.

In the real number system $\dfrac{1}{2}$ is the multiplicative inverse of 2 since $2\left(\dfrac{1}{2}\right) = 1$.

For a 3 by 3 matrix,

$$\mathbf{A} = \begin{bmatrix} a_{11} & a_{12} & a_{13} \\ a_{21} & a_{22} & a_{23} \\ a_{31} & a_{32} & a_{33} \end{bmatrix}.$$

The inverse matrix $\mathbf{A}^{-1}$ is the 3 by 3 matrix, if it exists,

$$\mathbf{A}^{-1} = \begin{bmatrix} x & y & z \\ u & v & w \\ r & s & t \end{bmatrix},$$

which satisfies the following matrix equation:

$$\mathbf{A}\mathbf{A}^{-1} = \mathbf{I} \text{ (the 3 by 3 identity matrix)},$$

or

$$\mathbf{A}\mathbf{A}^{-1} = \begin{bmatrix} a_{11} & a_{12} & a_{13} \\ a_{21} & a_{22} & a_{23} \\ a_{31} & a_{32} & a_{33} \end{bmatrix} \begin{bmatrix} x & y & z \\ u & v & w \\ r & s & t \end{bmatrix} = \begin{bmatrix} 1 & 0 & 0 \\ 0 & 1 & 0 \\ 0 & 0 & 1 \end{bmatrix} = \mathbf{I}.$$

For the 2 by 2 case, if

$$\mathbf{B} = \begin{bmatrix} b_{11} & b_{12} \\ b_{21} & b_{22} \end{bmatrix}, \quad \mathbf{C} = \begin{bmatrix} c_{11} & c_{12} \\ c_{21} & c_{22} \end{bmatrix}, \quad \text{and } \mathbf{BC} = \begin{bmatrix} 1 & 0 \\ 0 & 1 \end{bmatrix},$$

then $\mathbf{C}$ is the inverse of $\mathbf{B}$, or $\mathbf{C} = \mathbf{B}^{-1}$.

Calculation of the inverse of a square matrix makes use of a set of row operations similar to the rules that we used in section 13.2 to solve systems of equations. Example 13.19 illustrates this procedure for a 3 by 3 matrix.

• **Example 13.19:** Determine the inverse, if it exists, of the given matrix

$$\mathbf{A} = \begin{bmatrix} 1 & 4 & 3 \\ 2 & 6 & 1 \\ 1 & 0 & 3 \end{bmatrix}.$$

Augmented Matrix

Solution: We first write what we refer to as an **augmented matrix,** a 3 by 6 matrix formed by using the three columns of **A** along with the three columns of the 3 by 3 identity matrix:

$$\begin{bmatrix} 1 & 4 & 3 & | & 1 & 0 & 0 \\ 2 & 6 & 1 & | & 0 & 1 & 0 \\ 1 & 0 & 3 & | & 0 & 0 & 1 \end{bmatrix}.$$

The inverse of **A** can be determined by using row operations on this matrix to transform it into the following form:

$$\begin{bmatrix} 1 & 0 & 0 & | & x & y & z \\ 0 & 1 & 0 & | & u & v & w \\ 0 & 0 & 1 & | & r & s & t \end{bmatrix}.$$

When we have accomplished this,

$$\mathbf{A}^{-1} = \begin{bmatrix} x & y & z \\ u & v & w \\ r & s & t \end{bmatrix}.$$

Thus $\mathbf{AI} \rightarrow \mathbf{IA}^{-1}$ solves the problem.

$$\begin{bmatrix} 1 & 4 & 3 & | & 1 & 0 & 0 \\ 2 & 6 & 1 & | & 0 & 1 & 0 \\ 1 & 0 & 3 & | & 0 & 0 & 1 \end{bmatrix}$$

$$\begin{bmatrix} 1 & 4 & 3 & | & 1 & 0 & 0 \\ 0 & -2 & -5 & | & -2 & 1 & 0 \\ 0 & -4 & 0 & | & -1 & 0 & 1 \end{bmatrix}$$ (We have a 1 in the a_{11} position. We obtain zeros in the a_{21} and a_{31} position by adding, to row 2, -2 times row 1 and adding, to row 3, -1 times row 1.)

$$\begin{bmatrix} 1 & 4 & 3 & | & 1 & 0 & 0 \\ 0 & \boxed{1} & \frac{5}{2} & | & 1 & -\frac{1}{2} & 0 \\ 0 & -4 & 0 & | & -1 & 0 & 1 \end{bmatrix}$$ (We now obtain a 1 in the a_{22} position by multiplying row 2 by $-\frac{1}{2}$.)

$$\begin{bmatrix} 1 & 0 & -7 & | & -3 & 2 & 0 \\ 0 & 1 & \frac{5}{2} & | & 1 & -\frac{1}{2} & 0 \\ 0 & 0 & \boxed{10} & | & 3 & -2 & 1 \end{bmatrix}$$ (Obtain zeros in the a_{12} and a_{32} positions. (Add to row 1, -4 times row 2 and add, to row 3, 4 times row 2.))

$$\begin{bmatrix} 1 & 0 & -7 & | & -3 & 2 & 0 \\ 0 & 1 & \frac{5}{2} & | & 1 & -\frac{1}{2} & 0 \\ 0 & 0 & \boxed{1} & | & \frac{3}{10} & -\frac{2}{10} & \frac{1}{10} \end{bmatrix}$$ (Obtain a 1 in the a_{33} position by multiplying row 3 by $\frac{1}{10}$.)

$$\begin{bmatrix} 1 & 0 & 0 & \vdots & -\dfrac{9}{10} & \dfrac{3}{5} & \dfrac{7}{10} \\ 0 & 1 & 0 & \vdots & \dfrac{1}{4} & 0 & -\dfrac{1}{4} \\ 0 & 0 & 1 & \vdots & \dfrac{3}{10} & -\dfrac{2}{10} & \dfrac{1}{10} \end{bmatrix}$$ (Complete the solution by obtaining zeros in the a_{13} and a_{23} positions.)

$$\mathbf{A}^{-1} = \begin{bmatrix} -\dfrac{9}{10} & \dfrac{3}{5} & \dfrac{7}{10} \\ \dfrac{1}{4} & 0 & -\dfrac{1}{4} \\ \dfrac{3}{10} & -\dfrac{2}{10} & \dfrac{1}{10} \end{bmatrix}.$$

Check that this is indeed the inverse of matrix $\mathbf{A}$ by multiplying $\mathbf{AA}^{-1}$.

• *Example 13.20:* Find the inverse of the matrix

$$\mathbf{B} = \begin{bmatrix} 1 & 2 \\ 3 & 1 \end{bmatrix}.$$

Solution: $\begin{bmatrix} \boxed{1} & 2 & \vdots & 1 & 0 \\ 3 & 1 & \vdots & 0 & 1 \end{bmatrix}$ (Write the augmented matrix.)

$\begin{bmatrix} 1 & 2 & \vdots & 1 & 0 \\ 0 & -5 & \vdots & -3 & 1 \end{bmatrix}$ (Add, to row 2, -3 times row 1.)

$\begin{bmatrix} 1 & 2 & \vdots & 1 & 0 \\ 0 & \boxed{1} & \vdots & \dfrac{3}{5} & -\dfrac{1}{5} \end{bmatrix}$ (Multiply row 2 by $-\dfrac{1}{5}$.)

$\begin{bmatrix} 1 & 0 & \vdots & -\dfrac{1}{5} & \dfrac{2}{5} \\ 0 & 1 & \vdots & \dfrac{3}{5} & -\dfrac{1}{5} \end{bmatrix}$ (Add to row 1, -2 times row 2.)

$$\mathbf{A}^{-1} = \begin{bmatrix} -\dfrac{1}{5} & \dfrac{2}{5} \\ \dfrac{3}{5} & -\dfrac{1}{5} \end{bmatrix}.$$

Note that we used the following row operations in the solution of examples 13.19 and 13.20.

---

1. Any row may be multiplied by a nonzero constant.

2. Any row can be replaced by adding to it a constant times another row.

The other row operation that we used in solving systems of equations (interchanging rows) *should not* be used in calculating inverses, since it unnecessarily complicates the solution process.

Some square matrices do not have inverses. Example 13.21 illustrates such a matrix.

• *Example 13.21:* Calculate the inverse of the matrix

$$\mathbf{A} = \begin{bmatrix} 3 & 6 \\ 2 & 4 \end{bmatrix}.$$

Solution: $\begin{bmatrix} 3 & 6 & | & 1 & 0 \\ 2 & 4 & | & 0 & 1 \end{bmatrix}$ (Write the augmented matrix.)

$\begin{bmatrix} 1 & 2 & | & \frac{1}{3} & 0 \\ 2 & 4 & | & 0 & 1 \end{bmatrix}$ (Multiply row 1 by $\frac{1}{3}$.)

$\begin{bmatrix} 1 & 2 & | & \frac{1}{3} & 0 \\ 0 & 0 & | & -\frac{2}{3} & 1 \end{bmatrix}.$ (Add, to row 2, -2 times row 1.)

At this point it is not possible to obtain a 1 in the a_{22} position. We conclude that this matrix does not have an inverse.

• • • • • • • • • •

We note in passing that although multiplication of matrices is not, in general, commutative, the following property does indeed hold. For a square matrix **A**, if $\mathbf{A}^{-1}$ exists, then

$$\mathbf{A}\mathbf{A}^{-1} = \mathbf{A}^{-1}\mathbf{A} = \mathbf{I}.$$

The inverse of a matrix can be used to solve a system of linear equations. When using the "paper and pencil" approach, the inverse-matrix method is not the most efficient method of solving a system of equations. However, if a computer has built-in matrix operations, as many computers do, the following method is an excellent choice for such problems. Consider this 3 by 3 system of equations:

$$a_{11}x + a_{12}y + a_{13}z = c_1$$
$$a_{21}x + a_{22}y + a_{23}z = c_2$$
$$a_{31}x + a_{32}y + a_{33}z = c_3.$$

We can write this system of equations in the following matrix form:

$$\begin{bmatrix} a_{11} & a_{12} & a_{13} \\ a_{21} & a_{22} & a_{23} \\ a_{31} & a_{32} & a_{33} \end{bmatrix} \begin{bmatrix} x \\ y \\ z \end{bmatrix} = \begin{bmatrix} c_1 \\ c_2 \\ c_3 \end{bmatrix} \quad \text{or} \quad \mathbf{AZ} = \mathbf{C},$$

where

$$\mathbf{A} = \begin{bmatrix} a_{11} & a_{12} & a_{13} \\ a_{21} & a_{22} & a_{23} \\ a_{31} & a_{32} & a_{33} \end{bmatrix} \quad \mathbf{Z} = \begin{bmatrix} x \\ y \\ z \end{bmatrix} \quad \mathbf{C} = \begin{bmatrix} c_1 \\ c_2 \\ c_3 \end{bmatrix}.$$

$\mathbf{A}^{-1}(\mathbf{AZ}) = \mathbf{A}^{-1}\mathbf{C}$ (Multiply both sides of the equation by $\mathbf{A}^{-1}$)
$(\mathbf{A}^{-1}\mathbf{A})\mathbf{Z} = \mathbf{A}^{-1}\mathbf{C}$ (Multiplication is associative)
$\mathbf{IZ} = \mathbf{A}^{-1}\mathbf{C}$ ($\mathbf{A}^{-1}\mathbf{A} = \mathbf{I}$ (3 by 3 identity matrix))
$\mathbf{Z} = \mathbf{A}^{-1}\mathbf{C}.$ ($\mathbf{IZ} = \mathbf{Z}$)

If we substitute for **A**, **Z**, and **C**, the following matrix equation results:

$$\begin{bmatrix} x \\ y \\ z \end{bmatrix} = \begin{bmatrix} a_{11} & a_{12} & a_{13} \\ a_{21} & a_{22} & a_{23} \\ a_{31} & a_{32} & a_{33} \end{bmatrix}^{-1} \begin{bmatrix} c_1 \\ c_2 \\ c_3 \end{bmatrix}.$$

Example 13.22 illustrates how to solve a system of equations using inverse matrices.

• **Example 13.22:** Solve the given system of equations.

$$\begin{aligned} x + 4y + 3z &= 7 \\ 2x + 6y + z &= 18 \\ x + 3z &= -5. \end{aligned}$$

Solution: In matrix form, this system is

$$\begin{bmatrix} 1 & 4 & 3 \\ 2 & 6 & 1 \\ 1 & 0 & 3 \end{bmatrix} \begin{bmatrix} x \\ y \\ z \end{bmatrix} = \begin{bmatrix} 7 \\ 18 \\ -5 \end{bmatrix},$$

or

$$\begin{bmatrix} x \\ y \\ z \end{bmatrix} = \begin{bmatrix} 1 & 4 & 3 \\ 2 & 6 & 1 \\ 1 & 0 & 3 \end{bmatrix}^{-1} \begin{bmatrix} 7 \\ 18 \\ -5 \end{bmatrix}.$$

We have already calculated the inverse of the 3 by 3 matrix in example 13.19.

$$\begin{bmatrix} 1 & 4 & 3 \\ 2 & 6 & 1 \\ 1 & 0 & 3 \end{bmatrix}^{-1} = \begin{bmatrix} -\frac{9}{10} & \frac{3}{5} & \frac{7}{10} \\ \frac{1}{4} & 0 & -\frac{1}{4} \\ \frac{3}{10} & -\frac{2}{10} & \frac{1}{10} \end{bmatrix}.$$

If we substitute this matrix into the matrix equation, we get

$$\begin{bmatrix} x \\ y \\ z \end{bmatrix} = \begin{bmatrix} -\frac{9}{10} & \frac{3}{5} & \frac{7}{10} \\ \frac{1}{4} & 0 & -\frac{1}{4} \\ \frac{3}{10} & -\frac{2}{10} & \frac{1}{10} \end{bmatrix} \begin{bmatrix} 7 \\ 18 \\ -5 \end{bmatrix}$$

$$[x] = \begin{bmatrix} -\frac{9}{10} & \frac{3}{5} & \frac{7}{10} \end{bmatrix} \begin{bmatrix} 7 \\ 18 \\ -5 \end{bmatrix} = [1]$$

$$[y] = \begin{bmatrix} \frac{1}{4} & 0 & -\frac{1}{4} \end{bmatrix} \begin{bmatrix} 7 \\ 18 \\ -5 \end{bmatrix} = [3]$$

$$[z] = \begin{bmatrix} \frac{3}{10} & -\frac{2}{10} & \frac{1}{10} \end{bmatrix} \begin{bmatrix} 7 \\ 18 \\ -5 \end{bmatrix} = [-2].$$

Hence,

$$\begin{bmatrix} x \\ y \\ z \end{bmatrix} = \begin{bmatrix} 1 \\ 3 \\ -2 \end{bmatrix}.$$

• **Example 13.23:** One of the most useful applications of matrices comes from the field of economics. Suppose we have a subsystem of the economy, consisting of two related industries, the coal and steel industries. The following matrix describes a **consumption matrix**.

Consumption Matrix

$$\text{Producer} \begin{array}{c} \\ \text{Coal} \\ \text{Steel} \end{array} \overset{\begin{array}{cc} \text{User} \\ \text{Coal} \quad \text{Steel} \end{array}}{\begin{bmatrix} 0.10 & 0.50 \\ 0.20 & 0.30 \end{bmatrix}} = \mathbf{A}.$$

SECTION 13.4 THE INVERSE OF A SQUARE MATRIX

Internal Demands

This matrix describes the **internal demands** of the system. Column 1 shows that it requires 0.10 units of coal and 0.20 units of steel to produce 1 unit of coal, and column 2 shows that it requires 0.50 units of coal and 0.30 units of steel to produce 1 unit of steel. A second matrix,

$$\begin{matrix} \text{Coal} \\ \text{Steel} \end{matrix} \begin{bmatrix} 800 \\ 1000 \end{bmatrix} = \mathbf{D}$$

External Demand Matrix

called an **external demand matrix**, gives the number of units, 800 and 1000, of coal and steel that are needed by other industries. We wish to find a third matrix,

$$\begin{matrix} \text{Coal} \\ \text{Steel} \end{matrix} \begin{bmatrix} p_1 \\ p_2 \end{bmatrix} = \mathbf{P},$$

Total Production

which describes the **total production** of coal and steel to meet both the internal and external demands of the system.

Solution: The total production matrix (**P**) can be found by solving the following matrix equation:

$$\mathbf{P} - \mathbf{AP} = \mathbf{D}$$
$$(\mathbf{I} - \mathbf{A})\mathbf{P} = \mathbf{D} \quad \text{(Factor the } \mathbf{P} \text{ on the left-hand side (I is the 2 by 2 identity matrix))}$$
$$(\mathbf{I} - \mathbf{A})^{-1}(\mathbf{I} - \mathbf{A})\,\mathbf{P} = (\mathbf{I} - \mathbf{A})^{-1}\mathbf{D} \quad \text{(Multiply both sides of the equation by } (\mathbf{I} - \mathbf{A})^{-1}\text{)}$$
$$\mathbf{I} \cdot \mathbf{P} = (\mathbf{I} - \mathbf{A})^{-1}\mathbf{D}$$
$$\mathbf{P} = (\mathbf{I} - \mathbf{A})^{-1}\mathbf{D}.$$

Substitution of the appropriate matrices yields the following matrix equation:

$$\begin{bmatrix} p_1 \\ p_2 \end{bmatrix} = \left(\begin{bmatrix} 1 & 0 \\ 0 & 1 \end{bmatrix} - \begin{bmatrix} 0.10 & 0.50 \\ 0.20 & 0.30 \end{bmatrix} \right)^{-1} \begin{bmatrix} 800 \\ 1000 \end{bmatrix}$$
$$= \begin{bmatrix} 0.90 & -0.50 \\ -0.20 & 0.70 \end{bmatrix}^{-1} \begin{bmatrix} 800 \\ 1000 \end{bmatrix}.$$

We use the method of this section to find the inverse:

$$\left[\begin{array}{cc|cc} \dfrac{9}{10} & -\dfrac{1}{2} & 1 & 0 \\ -\dfrac{1}{5} & \dfrac{7}{10} & 0 & 1 \end{array}\right] \rightarrow \left[\begin{array}{cc|cc} 1 & -\dfrac{5}{9} & \dfrac{10}{9} & 0 \\ -\dfrac{1}{5} & \dfrac{7}{10} & 0 & 1 \end{array}\right]$$

$$\rightarrow \left[\begin{array}{cc|cc} 1 & -\dfrac{5}{9} & \dfrac{10}{9} & 0 \\ 0 & \dfrac{53}{90} & \dfrac{2}{9} & 1 \end{array}\right]$$

$$\rightarrow \left[\begin{array}{cc|cc} 1 & -\dfrac{5}{9} & \dfrac{10}{9} & 0 \\ 0 & 1 & \dfrac{20}{53} & \dfrac{90}{53} \end{array}\right]$$

$$\rightarrow \left[\begin{array}{cc|cc} 1 & 0 & \dfrac{70}{53} & \dfrac{50}{53} \\ 0 & 1 & \dfrac{20}{53} & \dfrac{90}{53} \end{array}\right]$$

$$\left[\begin{array}{cc} \dfrac{9}{10} & -\dfrac{1}{2} \\ -\dfrac{1}{5} & \dfrac{7}{10} \end{array}\right]^{-1} = \left[\begin{array}{cc} \dfrac{70}{53} & \dfrac{50}{53} \\ \dfrac{20}{53} & \dfrac{90}{53} \end{array}\right] = \dfrac{1}{53}\left[\begin{array}{cc} 70 & 50 \\ 20 & 90 \end{array}\right].$$

The solution to the matrix equation is

$$\begin{bmatrix} p_1 \\ p_2 \end{bmatrix} = \dfrac{1}{53}\begin{bmatrix} 70 & 50 \\ 20 & 90 \end{bmatrix} \cdot \begin{bmatrix} 800 \\ 1000 \end{bmatrix} = \begin{bmatrix} 2000 \\ 2000 \end{bmatrix}.$$

We need to produce 2000 units each of coal and steel to satisfy both the internal and external demands of the system.

● ● ● ● ● ● ● ● ● ●

Trial Problems 13.4

Before you begin the section exercises, warm up with these problems. Complete answers are included in the answer key.

Find an inverse matrix, if it exists, for each matrix.

1. $\begin{bmatrix} 1 & -1 \\ 2 & 2 \end{bmatrix}$

2. $\begin{bmatrix} 3 & 0 \\ 0 & 0 \end{bmatrix}$

3. $\begin{bmatrix} 1 & -2 & 2 \\ 0 & 1 & 0 \\ -2 & -2 & 4 \end{bmatrix}$

Exercises 13.4

Find an inverse, if it exists, for each matrix.

1. $\begin{bmatrix} -2 & 1 \\ \frac{3}{2} & \frac{1}{2} \end{bmatrix}$

2. $\begin{bmatrix} 1 & 2 \\ 3 & 4 \end{bmatrix}$

3. $\begin{bmatrix} 1 & 5 \\ -2 & 4 \end{bmatrix}$

4. $\begin{bmatrix} 1 & 4 \\ -2 & 3 \end{bmatrix}$

5. $\begin{bmatrix} 0.7 & -0.4 \\ -0.5 & 0.8 \end{bmatrix}$

6. $\begin{bmatrix} 0.6 & -0.4 \\ -0.2 & 0.6 \end{bmatrix}$

7. $\begin{bmatrix} 2 & 2 & 3 \\ 0 & 1 & 1 \\ 1 & 1 & 1 \end{bmatrix}$

8. $\begin{bmatrix} 0 & -1 & 1 \\ -1 & 1 & 2 \\ 1 & 0 & -2 \end{bmatrix}$

9. $\begin{bmatrix} 2 & 1 & -1 \\ 3 & 2 & 5 \\ 1 & 1 & 5 \end{bmatrix}$

10. $\begin{bmatrix} 1 & 1 & 1 \\ 0 & 1 & 0 \\ 2 & -1 & 3 \end{bmatrix}$

11. $\begin{bmatrix} 1 & 0 & 0 \\ 0 & -1 & 0 \\ 1 & 0 & 1 \end{bmatrix}$

12. $\begin{bmatrix} 1 & 4 & 3 \\ 2 & 6 & 1 \\ 1 & 0 & 3 \end{bmatrix}$

13. $\begin{bmatrix} 15 & 4 & -5 \\ -12 & -3 & 4 \\ -4 & -1 & 1 \end{bmatrix}$

14. $\begin{bmatrix} -5 & 6 & 7 \\ 10 & -11 & 13 \\ -1 & 1 & -1 \end{bmatrix}$

15. $\begin{bmatrix} 1 & -2 & 1 & 0 \\ 1 & -2 & 2 & -3 \\ 0 & 1 & -1 & 1 \\ -2 & 3 & -2 & 3 \end{bmatrix}$

16. $\begin{bmatrix} 1 & 2 & 3 & 1 \\ 1 & 3 & 3 & 2 \\ 2 & 4 & 3 & 3 \\ 1 & 1 & 1 & 1 \end{bmatrix}$

17. $\begin{bmatrix} 2 & -1 & 4 \\ 3 & -2 & 2 \\ 1 & -3 & 1 \end{bmatrix}$

18. $\begin{bmatrix} 2 & -1 & 1 \\ 1 & 2 & -1 \\ 3 & -4 & 2 \end{bmatrix}$

19. $\begin{bmatrix} 2 & -1 & 3 & -1 \\ 1 & -4 & 1 & 0 \\ 3 & 0 & -5 & 2 \\ 8 & 1 & 4 & -2 \end{bmatrix}$

20. $\begin{bmatrix} 1 & 0 & -2 & 3 \\ 1 & 1 & -3 & 2 \\ 2 & 1 & 0 & 3 \\ 0 & 1 & 0 & -3 \end{bmatrix}$

Use the method of this section to solve the following systems of equations.

21. $2x + y - z = 4$
$3x + 2y + 5z = 21$
$x + y + 5z = 15$
(Use the inverse from exercise 9)

22. $-y + z = -3$
$-x + y + 2z = 6$
$x - 2z = -1$
(Use the inverse from exercise 8)

23. $2x - y + 4z = -1$
$3x - 2y + 2z = 11$
$x - 3y + z = 5$
(Use the inverse from exercise 17)

24. $2x - y + z = -5$
$x + 2y - z = 3$
$3x - 4y + 2z = 9$
(Use the inverse from exercise 18)

25. $x - 2y + z = 2$
$x - 2y + 2z - 3w = -2$
$y - z + w = 0$
$-2x + 3y - 2z + 3w = 2$
(Use the inverse from exercise 15)

26. $x + 2y + 3z + w = -5$
$x + 3y + 3z + 2w = -4$
$x + y + z + w = 0$
$2x + 4y + 3z + 3w = -4$
(Use the inverse from exercise 16)

27. Given the internal demand matrix **A** and the external demand matrix **D**, calculate the production matrix for the system. (See example 13.23.)

$$\mathbf{A} = \begin{bmatrix} 0.3 & 0.4 \\ 0.5 & 0.2 \end{bmatrix} \quad \mathbf{D} = \begin{bmatrix} 14{,}000 \\ 15{,}000 \end{bmatrix}.$$

28. Given the internal demand matrix **A** and the external demand matrix **D**, calculate the production matrix for the system. (See example 13.23).

$$\mathbf{A} = \begin{bmatrix} 0.2 & 0.1 \\ 0.3 & 0.4 \end{bmatrix} \quad \mathbf{D} = \begin{bmatrix} 40{,}000 \\ 50{,}000 \end{bmatrix}.$$

13 • 5 The Determinant of a Square Matrix

In section 4.5 we used determinants along with Cramer's rule to evaluate 2 by 2 and 3 by 3 determinants. In this section we introduce methods of evaluating higher-ordered determinants using the method of **expansion by minors**. Let us first illustrate the method with a 3 by 3 determinant. Given the general 3 by 3 matrix,

Expansion by Minors

$$\mathbf{A} = \begin{bmatrix} a_{11} & a_{12} & a_{13} \\ a_{21} & a_{22} & a_{23} \\ a_{31} & a_{32} & a_{33} \end{bmatrix}.$$

The determinant of **A** ($\det(A)$) is a real number that is obtained by the method of section 4.5.

$$\det(A) = \begin{vmatrix} a_{11} & a_{12} & a_{13} \\ a_{21} & a_{22} & a_{23} \\ a_{31} & a_{32} & a_{33} \end{vmatrix}$$ (Note the use of bars instead of brackets to denote the determinant of a square matrix)

$$= a_{11}a_{22}a_{33} + a_{13}a_{21}a_{32} + a_{12}a_{23}a_{31} - a_{11}a_{23}a_{32} - a_{13}a_{22}a_{31} - a_{12}a_{21}a_{33}.$$

We can, however, obtain the same results using a method called expansion by minors. This process consists of first selecting a row or column. Suppose we use the first column:

$$\begin{vmatrix} \boxed{a_{11}} & a_{12} & a_{13} \\ \boxed{a_{21}} & a_{22} & a_{23} \\ \boxed{a_{31}} & a_{32} & a_{33} \end{vmatrix}.$$

Minor of an Element of a Determinant

The **minor of an element** in a 3 by 3 determinant is the 2 by 2 determinant obtained by eliminating the column and row in which the element appears. Thus,

in $\begin{vmatrix} \boxed{a_{11}} & a_{12} & a_{13} \\ \boxed{a_{21}} & a_{22} & a_{23} \\ \boxed{a_{31}} & a_{32} & a_{33} \end{vmatrix}$, the minor of a_{11} is $\begin{vmatrix} a_{22} & a_{23} \\ a_{32} & a_{33} \end{vmatrix}$.

In $\begin{vmatrix} a_{11} & a_{12} & a_{13} \\ \boxed{a_{21}} & a_{22} & a_{23} \\ a_{31} & a_{32} & a_{33} \end{vmatrix}$, the minor of a_{21} is $\begin{vmatrix} a_{12} & a_{13} \\ a_{32} & a_{33} \end{vmatrix}$.

In $\begin{vmatrix} a_{11} & a_{12} & a_{13} \\ a_{21} & a_{22} & a_{23} \\ \boxed{a_{31}} & a_{32} & a_{33} \end{vmatrix}$, the minor of a_{31} is $\begin{vmatrix} a_{12} & a_{13} \\ a_{22} & a_{23} \end{vmatrix}$.

Now consider the following sum

$$a_{11}\begin{vmatrix}a_{22} & a_{23}\\ a_{32} & a_{33}\end{vmatrix} - a_{21}\begin{vmatrix}a_{12} & a_{13}\\ a_{32} & a_{33}\end{vmatrix} + a_{31}\begin{vmatrix}a_{12} & a_{13}\\ a_{22} & a_{23}\end{vmatrix}$$
$$= a_{11}a_{22}a_{33} - a_{11}a_{23}a_{32} - a_{21}a_{12}a_{33} + a_{21}a_{13}a_{32} + a_{31}a_{12}a_{23} - a_{31}a_{13}a_{22}.$$

Except for the ordering of terms and factors, this is the same expression that we obtained using the method from chapter 4.

$$\det\begin{vmatrix}a_{11} & a_{12} & a_{13}\\ a_{21} & a_{22} & a_{23}\\ a_{31} & a_{32} & a_{33}\end{vmatrix} = a_{11}\begin{vmatrix}a_{22} & a_{23}\\ a_{32} & a_{33}\end{vmatrix} - a_{21}\begin{vmatrix}a_{12} & a_{13}\\ a_{32} & a_{33}\end{vmatrix} + a_{31}\begin{vmatrix}a_{12} & a_{13}\\ a_{22} & a_{23}\end{vmatrix}.$$

Example 13.24 illustrates how to use this procedure.

• *Example 13.24:* Evaluate the determinant of the matrix

$$\mathbf{A} = \begin{bmatrix}1 & 4 & 7\\ 2 & 5 & 8\\ 3 & 6 & 9\end{bmatrix}.$$

Solution: Suppose we expand using the first column:

$$\begin{vmatrix}1 & 4 & 7\\ 2 & 5 & 8\\ 3 & 6 & 9\end{vmatrix} = 1\begin{vmatrix}5 & 8\\ 6 & 9\end{vmatrix} - 2\begin{vmatrix}4 & 7\\ 6 & 9\end{vmatrix} + 3\begin{vmatrix}4 & 7\\ 5 & 8\end{vmatrix}$$
$$= (45 - 48) - 2(36 - 42) + 3(32 - 35)$$
$$= (-3) + 12 - 9 = 0.$$

Note that we could have used any row or column. However, the following sign restrictions must be observed:

$$\begin{vmatrix}+ & - & +\\ - & + & -\\ + & - & +\end{vmatrix}.$$

Thus, if we use the second row, the first and third products are negative. Using the second row, we get

$$\begin{vmatrix}1 & 4 & 7\\ 2 & 5 & 8\\ 3 & 6 & 9\end{vmatrix} = -2\begin{vmatrix}4 & 7\\ 6 & 9\end{vmatrix} + 5\begin{vmatrix}1 & 7\\ 3 & 9\end{vmatrix} - 8\begin{vmatrix}1 & 4\\ 3 & 6\end{vmatrix}$$
$$= (-2)(36 - 42) + 5(9 - 21) + (-8)(6 - 12)$$
$$= 12 - 60 + 48 = 0.$$

For a 4 by 4 determinant, the pattern of signs for expansion by minors is illustrated by the following configuration:

$$\begin{vmatrix}+ & - & + & -\\ - & + & - & +\\ + & - & + & -\\ - & + & - & +\end{vmatrix}.$$

• • • • • • • • •

In general, if the determinant is expanded by using an odd-numbered row or column, the first term is preceded by a plus sign. If it is expanded using an even-numbered row or column, the first term is preceded by a negative sign. In either case, the signs of the terms alternate thereafter.

• *Example 13.25:* Evaluate the following determinant by the method of expansion by minors:

$$D = \begin{vmatrix} 1 & 1 & 1 & 1 \\ 3 & 0 & 4 & 0 \\ 2 & -1 & 3 & -2 \\ -1 & -2 & 5 & 1 \end{vmatrix}.$$

Solution: Since row 2 of the determinant has two zeros, we use this row for expansion. The zeros make things easier, since we have to evaluate only two 3 by 3 determinants instead of four.

$$\begin{vmatrix} 1 & 1 & 1 & 1 \\ 3 & 0 & 4 & 0 \\ 2 & -1 & 3 & -2 \\ -1 & -2 & 5 & 1 \end{vmatrix} = (-3) \begin{vmatrix} 1 & 1 & 1 \\ -1 & 3 & -2 \\ -2 & 5 & 1 \end{vmatrix}$$

$$+ 0 \begin{vmatrix} \cdot & \cdot & \cdot \\ \cdot & \cdot & \cdot \\ \cdot & \cdot & \cdot \end{vmatrix} + (-4) \begin{vmatrix} 1 & 1 & 1 \\ 2 & -1 & 2 \\ -1 & -2 & 1 \end{vmatrix} + 0 \begin{vmatrix} \cdot & \cdot & \cdot \\ \cdot & \cdot & \cdot \\ \cdot & \cdot & \cdot \end{vmatrix}.$$

It is not necessary to fill in the second and fourth minors, since they are multiplied by zeros. We now calculate the other two minors. (Note that the method of section 4.5 could be used to evaluate the determinants.)

$$\begin{vmatrix} 1 & 1 & 1 \\ -1 & 3 & -2 \\ -2 & 5 & 1 \end{vmatrix} = (1) \begin{vmatrix} 3 & -2 \\ 5 & 1 \end{vmatrix} - (1) \begin{vmatrix} -1 & -2 \\ -2 & 1 \end{vmatrix} + (1) \begin{vmatrix} -1 & 3 \\ -2 & 5 \end{vmatrix}$$

$$= (3 + 10) - (-1 - 4) + (-5 + 6) = 19$$

$$\begin{vmatrix} 1 & 1 & 1 \\ 2 & -1 & -2 \\ -1 & -2 & 1 \end{vmatrix} = (1) \begin{vmatrix} -1 & -2 \\ -2 & 1 \end{vmatrix} - (1) \begin{vmatrix} 2 & -2 \\ -1 & 1 \end{vmatrix} + (1) \begin{vmatrix} 2 & -1 \\ -1 & -2 \end{vmatrix}$$

$$= (-1 - 4) - (2 - 2) + (-4 - 1) = -10.$$

We now substitute these values into the original expansion:

$$\begin{vmatrix} 1 & 1 & 1 & 1 \\ 3 & 0 & 4 & 0 \\ 2 & -1 & 3 & -2 \\ -1 & -2 & 5 & 1 \end{vmatrix} = (-3)(19) + (-4)(-10) = -17.$$

This appears to be an undue amount of work for the results obtained. However, we use some of the properties of determinants in section 13.6 to simplify the procedure.

Section 13.5 The Determinant of a Square Matrix

Trial Problems 13.5

Before you begin the section exercises, warm up with these problems. Complete answers are included in the answer key.

Calculate each determinant.

1. $\begin{vmatrix} 2 & 4 & 3 \\ 0 & -6 & 2 \\ 0 & -5 & 1 \end{vmatrix}$

2. $\begin{vmatrix} 4 & 0 & 0 \\ 1 & -5 & 2 \\ 3 & 1 & 6 \end{vmatrix}$

3. $\begin{vmatrix} 1 & 1 & 1 \\ 4 & 1 & 2 \\ -3 & 0 & 2 \end{vmatrix}$

4. $\begin{vmatrix} 2 & -3 & 1 \\ 1 & 4 & 2 \\ 0 & 3 & -1 \end{vmatrix}$

Exercises 13.5

Evaluate each determinant by the method of this section.

1. $\begin{vmatrix} 1 & 2 & 3 \\ 0 & 4 & 5 \\ 0 & 6 & 7 \end{vmatrix}$

2. $\begin{vmatrix} 1 & 0 & 0 \\ 2 & 3 & 4 \\ 5 & 6 & 7 \end{vmatrix}$

3. $\begin{vmatrix} 1 & 0 & 1 \\ 2 & 4 & 3 \\ -1 & 3 & 2 \end{vmatrix}$

4. $\begin{vmatrix} 1 & 4 & 2 \\ 0 & 5 & 1 \\ 1 & -1 & 6 \end{vmatrix}$

5. $\begin{vmatrix} 1 & 5 & -1 \\ 3 & 4 & -2 \\ 4 & 6 & -3 \end{vmatrix}$

6. $\begin{vmatrix} 6 & 3 & -2 \\ 5 & 4 & -1 \\ -1 & 2 & 3 \end{vmatrix}$

7. $\begin{vmatrix} 5 & 6 & -2 \\ 4 & 3 & -1 \\ 5 & 6 & -2 \end{vmatrix}$

8. $\begin{vmatrix} 4 & 6 & 3 \\ 1 & 0 & 4 \\ 4 & -1 & 3 \end{vmatrix}$

9. $\begin{vmatrix} 0.3 & 0.4 & 1.2 \\ 1.7 & 0.3 & 1.1 \\ -0.5 & -0.6 & 1.3 \end{vmatrix}$

10. $\begin{vmatrix} 0.5 & 0.4 & 0.3 \\ 1.1 & 1.2 & 1.4 \\ -0.6 & -0.4 & 0.7 \end{vmatrix}$

11. $\begin{vmatrix} 3 & 1 & 4 & 2 \\ 0 & 0 & 6 & 3 \\ 0 & 1 & 5 & 1 \\ 0 & -1 & 2 & 3 \end{vmatrix}$

12. $\begin{vmatrix} 2 & 0 & 0 & 0 \\ 1 & 4 & 2 & 3 \\ 0 & 1 & 6 & 7 \\ -3 & 2 & -2 & 1 \end{vmatrix}$

13. $\begin{vmatrix} 1 & 4 & 3 & 2 \\ 0 & 3 & 0 & 1 \\ -5 & 3 & -2 & 1 \\ 4 & 6 & -1 & 2 \end{vmatrix}$

14. $\begin{vmatrix} 3 & 1 & -5 & 6 \\ -1 & 0 & 2 & 4 \\ 3 & 0 & -7 & 5 \\ 4 & 3 & 2 & 1 \end{vmatrix}$

15. $\begin{vmatrix} 1 & 5 & 6 & -1 \\ -3 & 2 & 4 & -1 \\ 5 & 6 & -2 & 3 \\ 4 & 4 & -1 & 7 \end{vmatrix}$

16. $\begin{vmatrix} 4 & -5 & 2 & 3 \\ -5 & 1 & 1 & 3 \\ 1 & 0 & 4 & -2 \\ 3 & -1 & -1 & 6 \end{vmatrix}$

542 CHAPTER 13 MATRICES AND DETERMINANTS

Given three points A, B, and C on a coordinate system (see figure 13.4), the area of the triangle having A, B, and C as vertices can be found using the formula

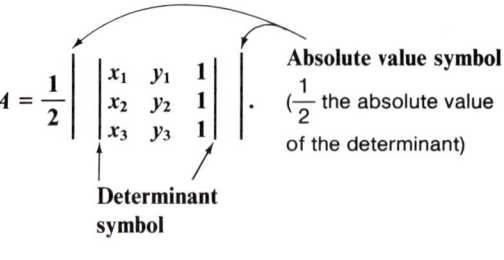
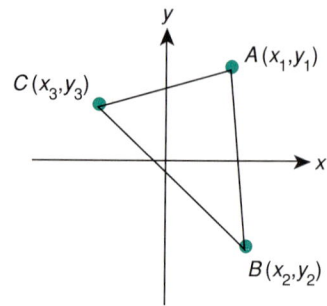

Figure 13.4

For each of the following exercises, determine the areas of the triangle having the given vertices.

17. $A(2, -3)$, $B(1, 4)$, $C(-4, -3)$
18. $A(4, 4)$, $B(6, 8)$, $C(-4, -3)$
19. $A(1, 1)$, $B(4, 4)$, $C(8, -3)$
20. $A(5, -10)$, $B(3, 10)$, $C(-3, 0)$

Three points on a coordinate plane, $A(x_1, y_1)$, $B(x_2, y_2)$, and $C(x_3, y_3)$ are collinear, provided the determinant

$$D = \begin{vmatrix} x_1 & y_2 & 1 \\ x_2 & y_2 & 1 \\ x_3 & y_3 & 1 \end{vmatrix} = 0.$$

Determine whether the given sets of points are collinear.

21. $A(0, -4)$, $B(2, 2)$, $C(5, 11)$
22. $A(0, 6)$, $B(2, 2)$, $C(-2, 10)$
23. $A(10, 150)$, $B(25, 300)$, $C(50, 550)$
24. $A(10, 40)$, $B(-10, 120)$, $C(0, 80)$

13•6 Properties of Determinants

We mentioned at the end of section 13.5 that determinants have some properties that can shorten the calculations involved in evaluation of determinants. This section is devoted to a study of these properties.

PROPERTY 1: If two rows or columns of a determinant are interchanged, the sign of the determinant changes.

For example,

$$\begin{vmatrix} 1 & 2 \\ 3 & 4 \end{vmatrix} = -\begin{vmatrix} 3 & 4 \\ 1 & 2 \end{vmatrix} \quad \text{(Rows interchanged)}$$

$$\begin{vmatrix} 5 & 6 \\ 7 & 8 \end{vmatrix} = -\begin{vmatrix} 6 & 5 \\ 8 & 7 \end{vmatrix}. \quad \text{(Columns interchanged)}$$

You can check this very easily.

PROPERTY 2: If you multiply every entry in one row or column by a nonzero constant, the determinant is multiplied by that constant. That is,

$$k \begin{vmatrix} a_{11} & a_{12} & a_{13} \\ a_{21} & a_{22} & a_{23} \\ a_{31} & a_{32} & a_{33} \end{vmatrix} = \begin{vmatrix} a_{11} & (ka_{12}) & a_{13} \\ a_{21} & (ka_{22}) & a_{23} \\ a_{31} & (ka_{32}) & a_{33} \end{vmatrix}$$

$$= k \det(A)$$

Note the difference from scalar multiplication of matrices, where all of the entries of the matrix were multiplied by k.

$$k \begin{vmatrix} a_{11} & a_{12} & a_{13} \\ a_{21} & a_{22} & a_{23} \\ a_{31} & a_{32} & a_{33} \end{vmatrix} = \begin{vmatrix} (ka_{11}) & (ka_{12}) & (ka_{13}) \\ a_{21} & a_{22} & a_{23} \\ a_{31} & a_{32} & a_{33} \end{vmatrix}.$$

$$= k \det(A)$$

This property allows you to remove a common factor from a row or column of a determinant. For example,

$$\begin{vmatrix} 10 & -1 & 4 \\ 20 & 0 & 3 \\ 30 & 2 & 1 \end{vmatrix} = 10 \begin{vmatrix} 1 & -1 & 4 \\ 2 & 0 & 3 \\ 3 & 2 & 1 \end{vmatrix}$$

and

$$\begin{vmatrix} 1 & 0 & 7 \\ 8 & 16 & 24 \\ 2 & -3 & 3 \end{vmatrix} = 8 \begin{vmatrix} 1 & 0 & 7 \\ 1 & 2 & 3 \\ 2 & -3 & 3 \end{vmatrix}.$$

PROPERTY 3: If two rows or two columns of a determinant are identical, the value of the determinant is zero.

For example,

$$\begin{vmatrix} 1 & 5 & 1 \\ 4 & 8 & 4 \\ 2 & -3 & 2 \end{vmatrix} = 0 \quad \text{and} \quad \begin{vmatrix} 3 & 5 & 8 & 6 \\ 1 & 0 & 4 & 6 \\ 3 & 5 & 8 & 6 \\ -3 & 4 & 2 & 1 \end{vmatrix} = 0.$$

While properties 1–3 are useful, in some instances, for simplifying the procedure in evaluating a determinant, they do not shorten the procedure. Properties 4 and 5, which follow, are useful in developing a shorter procedure for evaluating determinants.

PROPERTY 4:

If we add to one row a multiple of another row, the value of the determinant is not changed. The rule also applies to columns.

For example,

$$\text{let } D = \begin{vmatrix} 1 & -1 & 2 \\ 4 & 3 & 2 \\ 5 & -1 & 3 \end{vmatrix}.$$

If we want to evaluate D by expanding by minors using the first column, it would be advantageous to have two zeros in the column. Suppose we use property 4 to obtain zeros in the a_{21} and a_{31} positions.

1. Add, to row 2, -4 times row 1:

$$\begin{vmatrix} 1 & -1 & 2 \\ 0 & 7 & -6 \\ 5 & -1 & 3 \end{vmatrix}.$$

2. Add, to row 3, -5 times row 1:

$$\begin{vmatrix} 1 & -1 & 2 \\ 0 & 7 & -6 \\ 0 & 4 & -7 \end{vmatrix}.$$

Expanding by minors using the first column now gives

$$D = 1 \begin{vmatrix} 7 & -6 \\ 4 & -7 \end{vmatrix} = -49 + 24 = -25.$$

The next property makes the procedure even simpler. Before taking up this property, we need two additional definitions.

Main Diagonal of a Determinant The **main diagonal of a determinant** is the diagonal going from upper left to lower right:

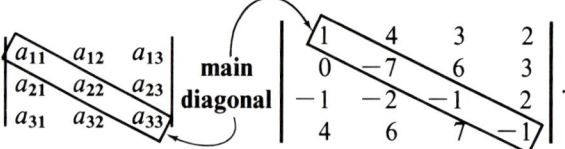

Triangular Form of a Determinant A determinant is in **triangular form** if all of the entries above or below the main diagonal are zeros:

$$D_1 = \begin{vmatrix} 1 & 4 & 3 \\ 0 & 2 & 5 \\ 0 & 0 & -2 \end{vmatrix} \quad D_2 = \begin{vmatrix} 1 & 5 & 6 & 1 \\ 0 & 4 & 2 & -1 \\ 0 & 0 & -3 & -2 \\ 0 & 0 & 0 & 4 \end{vmatrix}.$$

Determinants D_1 and D_2 are in triangular form.

SECTION 13.6 PROPERTIES OF DETERMINANTS

PROPERTY 5: The value of a determinant in triangular form is the product of the entries on the main diagonal.

For example, the value of D_1 and D_2 are

$$D_1 = (1)(2)(-2) = -4$$
$$D_2 = (1)(4)(-3)(4) = -48.$$

We can now use properties 1–4 to get a determinant in triangular form and use property 5 to evaluate the determinant.

• *Example 13.26:* Evaluate the determinant

$$D = \begin{vmatrix} 1 & 1 & 1 \\ -1 & 3 & -2 \\ -2 & 4 & 1 \end{vmatrix}.$$

Solution:
1. We obtain zeros in the a_{21} and a_{31} positions by adding row 1 to row 2 and by adding, to row 3, 2 times row 1:

$$D = \begin{vmatrix} 1 & 1 & 1 \\ 0 & 4 & -1 \\ 0 & 6 & 3 \end{vmatrix}.$$

2. We now obtain a zero in the a_{32} position by adding, to column 2, -2 times column 3.

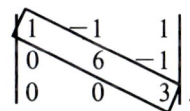

3. We can now use property 5 to evaluate D:

$$D = (1)(6)(3) = 18.$$

Let us try this procedure on the 4 by 4 determinant of example 13.25 of the last section.

• *Example 13.27:* Evaluate

$$D = \begin{vmatrix} 1 & 1 & 1 & 1 \\ 3 & 0 & 4 & 0 \\ 2 & -1 & 3 & -2 \\ -1 & -2 & 5 & 1 \end{vmatrix}.$$

Solution 1:

$$\begin{vmatrix} 1 & 1 & 1 & 1 \\ 3 & 0 & 4 & 0 \\ 2 & -1 & 3 & -2 \\ -1 & -2 & 5 & 1 \end{vmatrix}$$

$$= \begin{vmatrix} 1 & 1 & 1 & 1 \\ 0 & -3 & 1 & -3 \\ 0 & -3 & 1 & -4 \\ 0 & -1 & 6 & 2 \end{vmatrix}$$ (Add, to row 2, -3 times row 1)
(Add, to row 3, -2 times row 1)
(Add, to row 4, 1 times row 1)

$$= -\begin{vmatrix} 1 & 1 & 1 & 1 \\ 0 & 1 & -3 & -3 \\ 0 & 1 & -3 & -4 \\ 0 & 6 & -1 & 2 \end{vmatrix}$$ (Interchange columns 2 and 3 to get a 1 in the a_{22} position. Remember to change the sign of the determinant.)

$$= -\begin{vmatrix} 1 & 1 & 1 & 1 \\ 0 & 1 & -3 & -3 \\ 0 & 0 & 0 & -1 \\ 0 & 0 & 17 & 20 \end{vmatrix}$$ (Add, to row 3, -1 times row 2)
(Add, to row 4, -6 times row 2)

The determinant is in triangular form when we obtain a zero in the a_{43} position. We can accomplish this by interchanging rows 3 and 4. Again, we must change the sign of the determinant:

$$D = +\begin{vmatrix} 1 & 1 & 1 & 1 \\ 0 & 1 & -3 & -3 \\ 0 & 0 & 17 & 20 \\ 0 & 0 & 0 & -1 \end{vmatrix} = (1)(1)(17)(-1) = -17.$$

Solution 2: There are many ways to get a determinant into triangular form. Suppose we take a different approach. Since there is a zero available in the a_{22} position of the original determinant, we can begin by interchanging columns 1 and 2 to put the zero below the diagonal. Note that this procedure changes the sign of the determinant:

$$D = -\begin{vmatrix} 1 & 1 & 1 & 1 \\ 0 & 3 & 4 & 0 \\ -1 & 2 & 3 & -2 \\ -2 & -1 & 5 & 1 \end{vmatrix}.$$

We can now obtain zeros in the a_{31} and a_{41} positions.

$$D = -\begin{vmatrix} 1 & 1 & 1 & 1 \\ 0 & 3 & 4 & 0 \\ 0 & 3 & 4 & -1 \\ 0 & 1 & 7 & 3 \end{vmatrix}$$ (Add row 1 to row 3)
(Add, to row 4, 2 times row 1)

We can now obtain zeros in the a_{32} and a_{42} positions:

$$D = - \begin{vmatrix} 1 & 1 & 1 & 1 \\ 0 & 3 & 4 & 0 \\ 0 & 0 & 0 & -1 \\ 0 & 0 & \frac{17}{3} & 3 \end{vmatrix} \quad \begin{array}{l} \text{(Add, to row 3, } -1 \text{ times row 2)} \\ \\ \text{(Add, to row 4, } -\frac{1}{3} \text{ times row 2)} \end{array}$$

The determinant will be in triangular form if we interchange rows 3 and 4 (remember to change the sign of the determinant):

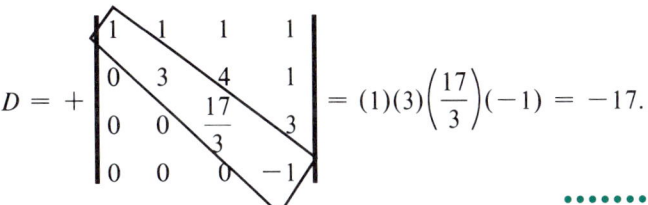

$$D = + \begin{vmatrix} 1 & 1 & 1 & 1 \\ 0 & 3 & 4 & 1 \\ 0 & 0 & \frac{17}{3} & 3 \\ 0 & 0 & 0 & -1 \end{vmatrix} = (1)(3)\left(\frac{17}{3}\right)(-1) = -17.$$

Trial Problems 13.6

Before you begin the section exercises, warm up with these problems. Complete answers are included in the answer key.

Evaluate each determinant.

1. $\begin{vmatrix} 1 & 1 & 1 \\ 0 & 1 & -2 \\ 0 & 0 & 4 \end{vmatrix}$

2. $\begin{vmatrix} 1 & -4 & 3 \\ 0 & 1 & 3 \\ 0 & 2 & 4 \end{vmatrix}$

3. $\begin{vmatrix} 4 & -4 & 0 \\ 2 & 2 & 0 \\ 1 & 3 & -6 \end{vmatrix}$

4. $\begin{vmatrix} 1 & 4 & -3 \\ 2 & 6 & -1 \\ -3 & 2 & 4 \end{vmatrix}$

Exercises 13.6

Evaluate each determinant by the method of section 13.6.

1. $\begin{vmatrix} 1 & 1 & 1 \\ 2 & -3 & 1 \\ -1 & 2 & -2 \end{vmatrix}$

2. $\begin{vmatrix} 1 & -1 & 1 \\ 2 & -3 & 1 \\ -1 & 2 & -2 \end{vmatrix}$

3. $\begin{vmatrix} 2 & 1 & 1 \\ 7 & -3 & 1 \\ -7 & 2 & -2 \end{vmatrix}$

4. $\begin{vmatrix} 6 & -1 & 1 \\ 13 & -3 & 1 \\ -11 & 2 & -2 \end{vmatrix}$

5. $\begin{vmatrix} 1 & 2 & 1 \\ 2 & 7 & 1 \\ -1 & -7 & -2 \end{vmatrix}$

6. $\begin{vmatrix} 1 & 6 & 1 \\ 2 & 13 & 1 \\ -1 & -11 & -2 \end{vmatrix}$

Solve each system of equations using Cramer's rule (see section 4.5).

7. $x + y + z = 2$
$2x - 3y + z = 7$
$-x + 2y - 2z = -7$
(Use the results of exercises 1, 3, and 5.)

8. $x - y + z = 6$
$2x - 3y + z = 13$
$-x + 2y - 2z = -11$
(Use the results of exercises 2, 4, and 6.)

Evaluate each determinant.

9. $\begin{vmatrix} 1 & 1 & 1 & 1 \\ 2 & -1 & -2 & 1 \\ -1 & 2 & 1 & -2 \\ 3 & 4 & -2 & -1 \end{vmatrix}$

10. $\begin{vmatrix} 1 & 1 & 1 & -1 \\ 2 & -1 & -2 & 1 \\ -1 & 2 & 1 & -2 \\ 3 & 4 & -2 & -1 \end{vmatrix}$

11. $\begin{vmatrix} 8 & 1 & 1 & 1 \\ -2 & -1 & -2 & 1 \\ 4 & 2 & 1 & -2 \\ 15 & 4 & -2 & -1 \end{vmatrix}$

12. $\begin{vmatrix} -1 & 1 & 1 & -1 \\ 3 & -1 & -2 & 1 \\ -6 & 2 & 1 & -2 \\ 5 & 4 & -2 & -1 \end{vmatrix}$

13. $\begin{vmatrix} 1 & 8 & 1 & 1 \\ 2 & -2 & -2 & 1 \\ -1 & 4 & 1 & -2 \\ 3 & 15 & -2 & -1 \end{vmatrix}$

14. $\begin{vmatrix} 1 & -1 & 1 & -1 \\ 2 & 3 & -2 & 1 \\ -1 & -6 & 1 & -2 \\ 3 & -5 & -2 & -1 \end{vmatrix}$

15. $\begin{vmatrix} 1 & 1 & 8 & 1 \\ 2 & -1 & -2 & 1 \\ -1 & 2 & 4 & -2 \\ 3 & 4 & 15 & -1 \end{vmatrix}$

16. $\begin{vmatrix} 1 & 1 & -1 & -1 \\ 2 & -1 & 3 & 1 \\ -1 & 2 & -6 & -2 \\ 3 & 4 & -5 & -1 \end{vmatrix}$

Solve each system of equations using Cramer's rule.

17. $x + y + z + w = 8$
$2x - y - 2z + w = -2$
$-x + 2y + z - 2w = 4$
$3x + 4y - 2z - w = 15$
(Use the results of exercises 9, 11, 13, and 15.)

18. $x + y + z - w = -1$
$2x - y - 2z + w = 3$
$-x + 2y + z - 2w = -6$
$3x + 4y - 2z - w = -5$
(Use the results of exercises 10, 12, 14, and 16.)

19. Given matrices
$$\mathbf{A} = \begin{bmatrix} 1 & 4 & 3 \\ 6 & -1 & 2 \\ -1 & 3 & 1 \end{bmatrix},$$
$$\mathbf{B} = \begin{bmatrix} -1 & 2 & 1 \\ 4 & 3 & -1 \\ 1 & 0 & 2 \end{bmatrix},$$
show that $\det(\mathbf{AB}) = (\det \mathbf{A})(\det \mathbf{B})$.

20. Given matrices
$$\mathbf{C} = \begin{bmatrix} 1 & 3 & -1 \\ 2 & 0 & 3 \\ 1 & 6 & 1 \end{bmatrix} \quad \mathbf{D} = \begin{bmatrix} 1 & 2 & -1 \\ 5 & 6 & 2 \\ -1 & 0 & -1 \end{bmatrix},$$
show that $\det(\mathbf{CD}) = (\det \mathbf{C})(\det \mathbf{D})$.

21. In a course in differential equations, to show that the functions $f_1(x) = e^x$, $f_2(x) = e^{2x}$, and $f_3(x) = e^{3x}$ fit a property called linear independence, it is necessary to evaluate the following determinant:
$$W = \begin{vmatrix} e^x & e^{2x} & e^{3x} \\ e^x & 2e^{2x} & 3e^{3x} \\ e^x & 4e^{2x} & 9e^{3x} \end{vmatrix}.$$
Find an expression for W.

22. In a course in differential equations, to show that the functions $f_1(x) = x$, $f_2(x) = \cos x$, and $f_3(x) = \sin x$ fit a property called linear independence, it is necessary to evaluate the following determinant:
$$W = \begin{vmatrix} x & \cos x & \sin x \\ 1 & -\sin x & \cos x \\ 0 & -\cos x & -\sin x \end{vmatrix}.$$
Find an expression for W.

13·7 Summary of Terms, Rules, and Procedures

Terms

Addition of Matrices (p. 502)
Augmented Coefficient Matrix (p. 508)
Augmented Matrix (p. 530)
Coefficient Matrix (p. 508)
Column Matrix (p. 517)
Column Vector (p. 517)
Consumption Matrix (p. 534)
Dimensions of a Matrix (p. 500)
Expansion by Minors (p. 538)
External Demand Matrix (p. 535)
Identity Matrix (p. 522)

Internal Demands (p. 535)
Main Diagonal of a Determinant (p. 544)
Matrix (p. 500)
Matrix Equation (p. 502)
m by n Matrix (p. 500)
Minor of an Element of a Determinant (p. 538)
Multiplication of Matrices (p. 519)
Multiplicative Inverse of a Square (n by n) Matrix (p. 529)
Pivot Elements (p. 509)

Pivot Operation (p. 509)
Pivot Positions (p. 509)
Row Matrix (p. 517)
Row Vector (p. 517)
Scalar (p. 503)
Scalar Multiplication (p. 503)
Triangular Form of a Determinant (p. 544)
Total Production (p. 535)
Zero Matrix (p. 505)

Rules and Procedures

- **ADDITION OF MATRICES** (13.1)
  Two matrices having the same dimensions can be added by adding the elements in the corresponding positions in the original matrices.

- **ADDITION PROPERTIES OF MATRICES** (13.1)
  1. $\mathbf{A} + \theta = \theta + \mathbf{A} = \mathbf{A}$
  2. $\mathbf{A} + \mathbf{B} = \mathbf{B} + \mathbf{A}$
  3. $(\mathbf{A} + \mathbf{B}) + \mathbf{C} = \mathbf{A} + (\mathbf{B} + \mathbf{C})$

- **SCALAR MULTIPLICATION** (13.1)
  To multiply a matrix by a scalar, multiply each element in the matrix by the scalar.

- **ROW OPERATIONS FOR MATRICES** (13.2)
  For an augmented coefficient matrix of a system of equations, the following operations result in an equivalent matrix.
  1. Interchange any pair of rows.
  2. Multiply or divide any row by a nonzero constant.
  3. Add any row to a multiple of another row.

- **PIVOT OPERATION** (13.2)
  1. Obtain a "1" in the pivot position.
  2. Use the row operations to obtain zeros in all of the other positions in the column containing the pivot element.

- **MATRIX MULTIPLICATION** (13.3)
  The c_{ij} entry in a product matrix is obtained by multiplying the ith row in the first matrix by the jth column in the second matrix.

- **ROW OPERATIONS TO FIND AN INVERSE MATRIX** (13.4)
  1. Any row can be multiplied by a nonzero constant.
  2. Any row can be replaced by the sum of that row and a constant times another row.

- **MINOR OF AN ELEMENT IN AN n BY n DETERMINANT** (13.5)
  The $n - 1$ by $n - 1$ determinant obtained by eliminating the column and the row in which the element appears.

- **PROPERTIES OF DETERMINANTS** (13.6)
  1. If two rows or columns of a determinant are interchanged, the sign of the determinant changes.
  2. If you multiply every entry in one row or column by a nonzero constant, the determinant is multiplied by that constant.
  3. If two rows or two columns of a determinant are identical, the value of the determinant is zero.
  4. If you add to one row a multiple of another row, the value of the determinant is not changed. The same rule applies to columns.
  5. The value of a determinant in triangular form is the product of the entries on the main diagonal.

13·8 Chapter 13 Review Exercises

Perform the indicated matrix operations.

1. $3\begin{bmatrix} 1 & 4 \\ 6 & 3 \\ -1 & 2 \end{bmatrix} + 4\begin{bmatrix} -3 & 2 \\ -1 & 4 \\ 3 & 5 \end{bmatrix} - 5\begin{bmatrix} 1 & 1 \\ 0 & 1 \\ -2 & 3 \end{bmatrix}$

2. $2\begin{bmatrix} 1 \\ 3 \\ 1 \end{bmatrix} - 5\begin{bmatrix} 6 \\ -1 \\ 3 \end{bmatrix} + 4\begin{bmatrix} -1 \\ 2 \\ 3 \end{bmatrix}$

Solve the following systems of equations using matrix row operations.

3. $x + y + z = -6$
   $2x - 3y + z = 0$
   $-x + y - z = 2$

4. $x + y + z = -5$
   $2x - 3y + z = -3$
   $-x + y - z = 3$

5. $x + y + z + w = 4$
   $3x - 2y + z - w = 5$
   $2x - y + 2z - 2w = 3$
   $4x - 3y - z + 2w = 9$

6. $x + y + z + w = 2$
   $3x - 2y + z - w = 5$
   $2x - y + 2z - 2w = 9$
   $4x - 3y - z + 2w = -5$

7. $x + y + z = 60$
   $-3x + 2y - z = -20$
   $4x + 5y - 2z = 80$

8. $x + y + z = 60$
   $-3x + 2y - z = -70$
   $4x + 5y - 2z = 70$

Given the following matrices:

$A = \begin{bmatrix} 2 & 3 & -1 \\ 1 & 2 & 1 \\ -1 & -1 & 3 \end{bmatrix}$ $B = \begin{bmatrix} 1 & -1 & 1 \\ 0 & 2 & -1 \\ 2 & 3 & 0 \end{bmatrix}$ $C = \begin{bmatrix} -1 & 2 & 3 \\ -1 & 2 & 4 \\ 1 & -1 & -2 \end{bmatrix}$ $D = \begin{bmatrix} 0 & -1 & 1 \\ 1 & 5 & -4 \\ 1 & 2 & -2 \end{bmatrix}$,

calculate the following matrices.

9. AB
10. BC
11. CD
12. AD
13. A(B + C)
14. D(B + C)
15. A^{-1}
16. B^{-1}
17. C^{-1}
18. D^{-1}
19. $A^{-1}B$
20. $B^{-1}A$

Use inverse matrices to solve the systems of equations.

21. $2x + 3y - z = 10$
    $x + 2y + z = 7$
    $-x - y + 3z = -2$
    (Use the results of exercise 15.)

22. $x - y + z = -2$
    $y - z = 3$
    $2x + 3y = 14$
    (Use the results of exercise 16.)

23. $-x + 2y + 3z = 140$
    $-x + 2y + 4z = 170$
    $x - y - 2z = -90$
    (Use the results of exercise 17.)

24. $-y + z = -30$
    $x + 5y - 4z = 160$
    $x + 2y - 2z = 80$
    (Use the results of exercise 18.)

Evaluate the following determinants.

25. $\begin{vmatrix} 1 & 4 & 3 \\ 2 & -1 & 3 \\ 4 & 6 & 2 \end{vmatrix}$

26. $\begin{vmatrix} 1 & 5 & -6 \\ 2 & 3 & 4 \\ 0 & 7 & 1 \end{vmatrix}$

27. $\begin{vmatrix} 1 & 1 & 1 & 1 \\ -1 & 4 & 2 & 3 \\ 0 & 6 & 9 & 1 \\ 0 & 2 & 4 & 1 \end{vmatrix}$

28. $\begin{vmatrix} 1 & 1 & 1 & 1 \\ 4 & -1 & 2 & 3 \\ 0 & 5 & -6 & 2 \\ 0 & 1 & -1 & 3 \end{vmatrix}$

Solve the following systems of equations using determinants.

29. $x + 4y + 3z = 20.3$
    $2x - y + 3z = 10.6$
    $4x + 6y + 2z = 24$
    (Use the results of exercise 25.)

30. $x + 5y - 6z = -1.9$
    $2x + 3y + 4z = 5.1$
    $7y + z = 1.3$
    (Use the results of exercise 26.)

31. $x + y + z + w = 3$
    $-x + 4y + 2z + 3w = 5$
    $6y + 9z + w = 2$
    $2y + 4z + w = 2$
    (Use the results of exercise 27.)

32. $x + y + z + w = 3$
    $4x - y + 2z + 3w = 11$
    $5y - 6z + 2w = 2$
    $y - z + 3w = 3$
    (Use the results of exercise 28.)

33. The following matrix represents the January production totals for three items X, Y, and Z, which are produced at two factories A and B.

    $$\begin{array}{c} \text{Factory} \\ \begin{array}{c} M \\ N \end{array} \end{array} \begin{array}{c} \text{Item} \\ \begin{bmatrix} 2000 & 4000 & 6000 \\ 3500 & 8000 & 2000 \end{bmatrix} \end{array}$$

    a. If February production increased by 20% for each item at each factory write a production matrix for February.
    b. If March production was 15% below the January production, write a production matrix for March.

    The following matrix represents the selling prices of the items produced at the factories.

    $$\begin{array}{c} \text{Item} \\ \begin{array}{c} X \\ Y \\ Z \end{array} \end{array} \begin{array}{c} \text{Selling Price} \\ \begin{bmatrix} \$8.72 \\ \$4.65 \\ \$7.62 \end{bmatrix} \end{array}$$

    c. Find the income-per-factory matrix for January.
    d. Find the income-per-factory matrix for February.
    e. Find the income-per-factory matrix for March.
    f. Find the income-per-factory matrix for the year's first quarter.

34. Solve exercise 33 for the following matrices.

    $$\begin{array}{c} \text{Factory} \\ \begin{array}{c} M \\ N \\ P \end{array} \end{array} \begin{array}{c} \text{Item} \\ \begin{array}{ccc} X & Y & Z \end{array} \\ \begin{bmatrix} 6000 & 4500 & 1500 \\ 8000 & 7200 & 0 \\ 9000 & 8500 & 3000 \end{bmatrix} \end{array} \quad \begin{array}{c} \text{Item} \\ \begin{array}{c} X \\ Y \\ Z \end{array} \end{array} \begin{array}{c} \text{Price} \\ \begin{bmatrix} \$3.50 \\ \$5.25 \\ \$6.75 \end{bmatrix} \end{array}$$

13·9 Chapter 13 Test

Perform the indicated operations.

1. $2\begin{bmatrix} 3 \\ 5 \end{bmatrix} - 2\begin{bmatrix} -1 \\ 3 \end{bmatrix}$

2. $3\begin{bmatrix} 8 & 6 \\ 2 & 3 \end{bmatrix} - 4\begin{bmatrix} 6 & 3 \\ 1 & 2 \end{bmatrix}$

Use matrix row operations to solve the systems of equations.

3. $x - y + 2z = -3$
   $2x + y - z = 5$
   $3x + 2y + 4z = 3$

4. $x - 3y + z = 1$
   $4x + 5y - z = -8$
   $3x - 2y - 2z = 1$

Use the following matrices for exercises 5–10.

$$A = \begin{bmatrix} 0 & 1 & 2 \\ -1 & 1 & 2 \\ 1 & -2 & -5 \end{bmatrix} \quad B = \begin{bmatrix} -1 & 22 & -8 \\ 0 & -5 & 2 \\ 1 & -19 & 7 \end{bmatrix} \quad C = \begin{bmatrix} 1 & -2 & 3 & 1 \\ 0 & 1 & 2 & -4 \\ 0 & 0 & 1 & 0 \\ 0 & 0 & 0 & 1 \end{bmatrix}$$

5. $A + B$
6. AB
7. B^{-1}
8. A^{-1}
9. C^{-1}
10. $(A + B)C$

11. Use the results of exercise 7 to solve the following system of equations.

    $-x + 22y - 8z = 35$
    $-5y + 2z = -8$
    $x - 19y + 7z = -30$

12. Use the results of exercise 8 to solve the following system of equations.

    $y + 2z = 8$
    $-x + y + 2z = 2$
    $x - 2y - 5z = -11$

Evaluate the following determinants.

13. $\begin{vmatrix} 1 & 3 & 5 \\ 7 & 9 & 11 \\ 13 & 15 & 17 \end{vmatrix}$

14. $\begin{vmatrix} 1 & -1 & 2 & 3 \\ 3 & 1 & 5 & 2 \\ 0 & 3 & -1 & 2 \\ 0 & 4 & 2 & -6 \end{vmatrix}$

Solve the following systems of equations using determinants.

15. $13x + 15y + 17z = 45$
    $7x + 9y + 11z = 27$
    $x + 3y + 5z = 9$

16. $3x + 2y + 3z = 320$
    $x + 2y + 4z = 310$
    $2x + 3y + z = 230$

13·10 Cumulative Review—Chapters 12 and 13

Write each inequality in interval notation.

1. $-2 < x \leq 4$
2. $x \leq 3.5$

Write each interval as an inequality.

3. $(-3, \infty)$
4. $[-2, 0)$

Solve for x.

5. $2x + 4 > 7$
6. $\dfrac{2x + 1}{3} \leq 5$
7. $(x - 2)(x + 4) \geq 0$
8. $\dfrac{x(x - 1)}{3x - 2} < 0$
9. $|2x - 1| \leq 7$
10. $|3x - 5| > 7$

Graph the region that satisfies the inequalities.

11. $4x + 6y \leq 12$
12. $x \geq 0, y \geq 0, 7x + 10y \leq 280$
13. $x \geq 0, y \geq 0, -x + 2y \leq 20, x + y \leq 40$
14. $x \geq 40, y \geq 60, 15x + 8y \geq 1200, 5x + 4y \leq 600$
15. Maximize $f(x, y) = 2.5x + 1.5y$ over the region in exercise 13.
16. Maximize $f(x, y) = 3.50x + 2.40y$ over the region in exercise 14.
17. Perform the indicated matrix operations.

$$4\begin{bmatrix} 6 & -1 \\ 2 & 3 \\ 4 & 0 \end{bmatrix} + 5\begin{bmatrix} 0 & 1 \\ 4 & -3 \\ 2 & 2 \end{bmatrix}$$

18. Use matrix row operations to solve the system of equations.

$$2x + 4y - 5z = 6$$
$$3x - 8y + z = 3$$
$$-x - y + z = 1$$

Use the following matrices for exercises 19–27.

$$A = \begin{bmatrix} -2 & -1 & 1 \\ 0 & 1 & 2 \\ -1 & 1 & 3 \end{bmatrix} \quad B = \begin{bmatrix} 1 & 4 & -3 \\ -2 & -5 & 4 \\ 1 & 3 & -2 \end{bmatrix} \quad C = \begin{bmatrix} 1 & 4 & 2 \\ 0 & 1 & 0 \\ -3 & 1 & 4 \end{bmatrix}$$

19. Calculate $(A + B) - C$
20. Calculate A^{-1}
21. Calculate B^{-1}
22. Calculate AC
23. Calculate CA
24. Calculate $(AB)^{-1}$
25. Calculate $\det A$
26. Calculate $\det B$
27. Use determinants to solve the system of equations.

$$x + 4y - 3z = -4$$
$$-2x - 5y + 4z = -9$$
$$x + 3y - 2z = 5$$

Chapter 14
Theory of Equations

The history of attempts to solve polynomial equations of degree higher than two can be traced to the eleventh century. Omar Khayyam, a Persian poet, astronomer and mathematician, used geometric methods to solve equations of the form $x^3 + b^2x + a^3 = cx^2$.

One of the most bizarre stories in the history of mathematics centers on two sixteenth century Italian mathematicians, Girolamo Cardano and Nicola Fontana (commonly known as Tartaglia). In one version of the story, Tartaglia discovered a method for solving cubic equations that have no quadratic term. He used the method to win mathematical contests. Cardano, pledging secrecy, managed to get the method from Tartaglia, but broke his pledge and published the method in *Ars Marna* in 1545. Thus began one of the most bitter feuds in the history of mathematics. Other versions of the story exist, since neither participant seemed to care much about telling the truth.

14•1 Introduction

In this chapter, we consider techniques for solving polynomial equations that have degree greater than 2. Some equations of this type can be solved by techniques that we have already considered.

For example, the equation $x^3 - 5x^2 + 6x = 0$ can be solved by factoring the expression $x^3 - 5x^2 + 6x$ and setting the individual factors equal to 0.

Thus,

$$x^3 - 5x^2 + 6x = 0$$
$$x(x^2 - 5x + 6) = 0$$
$$x(x - 3)(x - 2) = 0$$

$x = 0$	$x - 3 = 0$	$x - 2 = 0$
	$x = 3$	$x = 2.$

The solution set is $\{0, 3, 2\}$.

The equation $x^4 + 5x^2 + 6 = 0$ can be also solved by factoring the expression $x^4 + 5x^2 + 6$:

$$x^4 + 5x^2 + 6 = (x^2)^2 + 5(x^2) + 6$$
$$= (x^2 + 3)(x^2 + 2).$$

Thus,

$$x^4 + 5x^2 + 6 = 0$$

can be replaced by

$$(x^2 + 3)(x^2 + 2) = 0.$$

We set the individual factors equal to zero, and solve the resulting equations.

$x^2 + 3 = 0$	$x^2 + 2 = 0$
$x^2 = -3$	$x^2 = -2$
$x = j\sqrt{3}$ or $x = -j\sqrt{3}$	$x = j\sqrt{2}$ or $x = -j\sqrt{2}$.

The solution set is $\{j\sqrt{3}, -j\sqrt{3}, j\sqrt{2}, -j\sqrt{2}\}$.

Both of the previous equations are special cases. The first equation was solvable because the left-hand side had an x that could be factored out, leaving a quadratic factor that we could set equal to zero and factor or use the quadratic formula to solve. The second equation was factorable as a quadratic equation in x^2.

Let us consider a third equation that illustrates a difficulty that we encounter when solving polynomial equations.

The equation $x^3 - x^2 - x - 2 = 0$ is a polynomial equation, but its roots are not obvious. If we knew that $x - 2$ was a factor of $x^3 - x^2 - x - 2$, then the remaining quadratic factor could be found by long division:

$$\begin{array}{r} x^2 + x + 1 \\ x - 2 \overline{) x^3 - x^2 - x - 2} \\ \underline{x^3 - 2x^2} \\ x^2 - x \\ \underline{x^2 - 2x} \\ x - 2 \\ \underline{x - 2} \end{array}$$

We could then solve the equation as follows:

$$x^3 - x^2 - x - 2 = 0$$
$$(x - 2)(x^2 + x + 1) = 0.$$

Set the individual factors equal to zero:

$$\begin{array}{l|l} x - 2 = 0 & x^2 + x + 1 = 0 \\ x = 2 & x = \dfrac{-(1) \pm \sqrt{(1) - 4(1)(1)}}{2(1)} \\ & x = \dfrac{-1 \pm \sqrt{-3}}{2} = \dfrac{-1 \pm j\sqrt{3}}{2} \end{array}$$

The solution set is $\left\{ 2, \dfrac{-1 + j\sqrt{3}}{2}, \dfrac{-1 - j\sqrt{3}}{2} \right\}$.

The solution of this equation, however, depended on the fact that we knew one of the factors in advance. Unfortunately, most of the time we do not have this kind of information. The techniques developed in this chapter are devoted to the problem of solving polynomial equations for which the special approaches like factoring a common expression, special cases of quadratic equations in x^2, or knowing one of the factors in advance are inappropriate.

14 • 2 Synthetic Division

Synthetic Division

To solve polynomial equations with degree greater than 2, it is frequently necessary to divide polynomials by expressions of the form $x - a$. To facilitate this process, we introduce a shortened algorithm for the long division process called **synthetic division**.

Consider the following long division problem:

$$\begin{array}{r} 3x^2 + 2x + 11 \\ x - 2 \overline{) 3x^3 - 4x^2 + 7x - 5} \\ \underline{3x^3 - 6x^2} \\ 2x^2 + 7x \\ \underline{2x^2 - 4x} \\ 11x - 5 \\ \underline{11x - 22} \\ 17. \end{array}$$

If we rewrite this algorithm omitting the variables, we get the following arrangement:

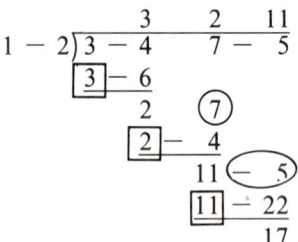

Observe that the numbers in the squares are repetitions of the coefficients of the variables in the quotient, and that the circled numbers are repetitions of some of the coefficients of the dividend. If we delete these repetitions, the problem takes the following form. Note that we can also delete the "1" in the divisor:

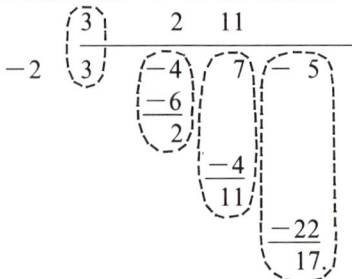

We can rearrange these numbers to the following form:

$$\begin{array}{c|cccc} -2 & 3 & -4 & 7 & -5 \\ & & -6 & -4 & -22 \\ \hline & 3 & 2 & 11 & 17 \end{array}$$

Note that the coefficients of the quotient and the remainder appear in order in the last row of this representation. Finally, rather than performing the subtractions, we can change the signs on the remaining number in the divisor and on the numbers in the second row and perform additions instead. The algorithm then has the following configuration:

$$\text{Negative of the divisor} \rightarrow \boxed{2} \begin{array}{|cccc} 3 & -4 & 7 & -5 \\ & 6 & 4 & 22 \\ \hline 3 & 2 & 11 & \boxed{17} \end{array} \begin{array}{l} \leftarrow \text{Coefficients} \\ \text{of dividend} \\ \leftarrow \text{Remainder } (R = 17) \end{array}$$

The synthetic division procedure is done according to the following pattern.

$$\begin{array}{c|cccc} 2 & 3 & -4 & 7 & -5 \\ \hline & & & & \end{array}$$

Step 1: In the first row list the negative of the constant term of the divisor (namely 2) and the coefficients of the dividend in descending powers of x.

$$\begin{array}{c|cccc} 2 & 3 & -4 & 7 & -5 \\ & \downarrow & & & \\ \hline & 3 & & & \end{array}$$

Step 2: Bring down the first dividend coefficient to the third row.

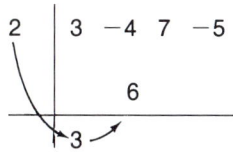

Step 3: Multiply (2)(3) and insert the answer in the second row.

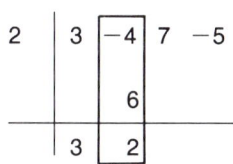

Step 4: Add (−4) and 6 and write the answer in the third row.

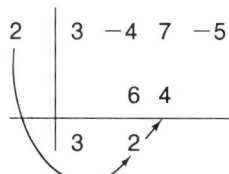

Step 5: Multiply (2)(2) and write the answer in the second row.

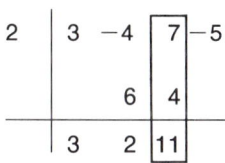

Step 6: Add the third column.

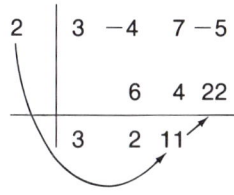

Step 7: Multiply (2)(11) and write the answer in the second row.

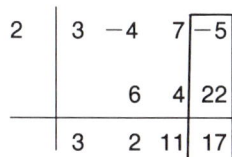

Step 8: Add the fourth column.

We obtained the same three-line summary that we obtained by long division. The divisor and dividend are listed in the first row. The quotient and remainder can be read directly from the third row.

NOTE: This method can be used only when the divisor is of the form $x - a$ or $x + a$ (since $x + a$ can be written as $x - (-a)$).

Example 14.1 carefully illustrates each step of this method.

• Example 14.1: Find the quotient and remainder when $x^3 - x^2 - x - 2$ is divided by $x - 2$.

Solution: This example was solved by long division in section 14.1. However, we use synthetic division to produce the same result.

Step 1: The first row contains the negative of (-2) followed by the coefficients of the dividend.

$$\begin{array}{c|cccc} 2 & 1 & -1 & -1 & -2 \\ \hline & & & & \end{array}$$

Step 2: Bring down the 1.

$$\begin{array}{c|cccc} 2 & 1 & -1 & -1 & -2 \\ & \downarrow & & & \\ \hline & 1 & & & \end{array}$$

Step 3: Multiply $(2)(1)$ and insert the product in the second column.

Step 4: Add the second column.

Step 5: Multiply $(2)(1)$ and insert the product in the third column.

Step 6: Add the third column.

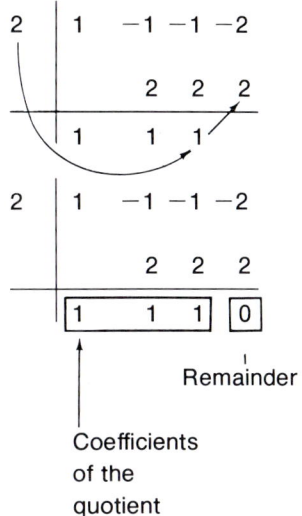

Coefficients
of the
quotient

Step 7: Multiply (2)(1) and insert the product in the fourth column.

Step 8: Add the last column.

Since the dividend has degree 3 and the divisor has degree 1, the quotient has degree 2. The quotient is $1x^2 + 1x + 1$ or $x^2 + x + 1$, and the remainder is zero. Thus, $x - 2$ is a factor of $x^3 - x^2 - x - 2$.

If there are any terms missing when the dividend is arranged in descending powers of the variable, we must insert zeros for the coefficients of the missing terms. Example 14.2 illustrates how to solve this type of problem.

• **Example 14.2:** Use synthetic division to divide

$$16x^4 + 8x^2 + 1 \quad \text{by } x - \frac{1}{2}.$$

Solution: When we write the coefficients of the dividend in row 1 of the algorithm, we must insert zeros for the missing powers. This is allowable, since the dividend can be written as

$$16x^4 + 8x^2 + 1 = 16x^4 + 0x^3 + 8x^2 + 0x + 1.$$

Hence the algorithm looks like this.

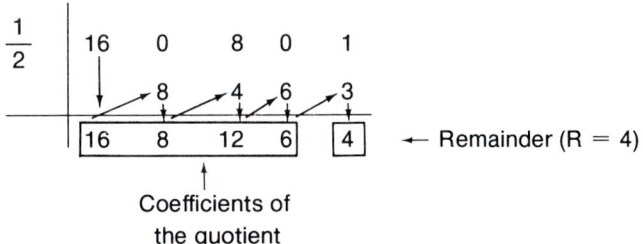

Since the dividend has degree 4 and the divisor has degree 1, the quotient has degree 3. The quotient is $16x^3 + 8x^2 + 12x + 6$ and the remainder is 4.

In example 14.3 synthetic division is used to help find the factors of the polynomial.

- **Example 14.3:** Determine whether or not $x + 4$ is a factor of
$$x^3 - x^2 - 14x + 24.$$

Solution: We divide $x^3 - x^2 - 14x + 24$ by $x + 4$ to see if the remainder is zero. If so, then $x + 4$ is a factor. If the remainder is not zero, then $x + 4$ is not a factor:

$$\begin{array}{r|rrrr} -4 & 1 & -1 & -14 & 24 \\ & & -4 & 20 & -24 \\ \hline & 1 & -5 & 6 & \boxed{0} \end{array} \leftarrow \text{Remainder } (R = 0)$$

Since the remainder is zero, $x + 4$ is a factor of $x^3 - x^2 - 14x + 24$.

In fact, $x^3 - x^2 - 14x + 24 = (x + 4)(x^2 - 5x + 6)$, since the division problem can be stated as

$$\frac{x^3 - x^2 - 14x + 24}{x + 4} = x^2 - 5x + 6.$$

• • • • • • • • • •

Earlier in this section we noted that the synthetic division process works only for divisors of the form $x - a$. However, we can revise the procedure slightly and use it for divisors of the form $ax + b$. Example 14.4 illustrates this.

- **Example 14.4:** Given that $2x - 3$ is a factor of $2x^4 + 3x^3 - 23x^2 + 25x - 6$, find the other factor.

Solution: If $2x - 3$ is a factor of $2x^4 + 3x^3 - 23x^2 + 25x - 6$ then $x - \frac{3}{2}$ is also a factor, since $2x - 3$ can be written as $2\left(x - \frac{3}{2}\right)$. We first divide by $x - \frac{3}{2}$ using synthetic division:

$$\begin{array}{r|rrrrr} \frac{3}{2} & 2 & 3 & -23 & 25 & -6 \\ & & 3 & 9 & -21 & 6 \\ \hline & 2 & 6 & -14 & 4 & \boxed{0} \end{array}$$

Note $R = 0$, therefore $(2x - 3)$ is a factor. Thus,

$$2x^4 + 3x^3 - 23x^2 + 25x - 6$$
$$= \left(x - \frac{3}{2}\right)(2x^3 + 6x^2 - 14x + 4)$$
$$= (2)\left(x - \frac{3}{2}\right)\left(\frac{2x^3 + 6x^2 - 14x + 4}{2}\right) \quad \text{(Multiply by 2 and divide by 2)}$$
$$= (2x - 3)(x^3 + 3x^2 - 7x + 2)$$

The other factor is $x^3 + 3x^2 - 7x + 2$.

Trial Problems 14.2

Before you begin the section exercises, warm up with these problems. Complete answers are included in the answer key.

Use synthetic division to find the quotient and the remainder.

1. $\dfrac{x^3 - 2x - 5}{x - 2}$

2. $\dfrac{x^4 - 16}{x - 2}$

3. $\dfrac{x^4 - 3x^2 - x - 6}{x + 3}$

4. $\dfrac{2x^4 - x^3 - 6x^2 + 4x - 8}{x - 2}$

Exercises 14.2

Use synthetic division to find the quotient and remainder.

1. $(x^3 + 2x^2 - 2x + 3) \div (x + 3)$

2. $(5x^3 - 2x^2 + 3x - 4) \div (x - 3)$

3. $(-3x^4 - 3x^3 + 3x^2 + 2x - 4) \div (x + 2)$

4. $\dfrac{x^3 - 6x^2 - 4x - 21}{x - 7}$

5. $\dfrac{x^3 - 3x^2 - 9x + 2}{x + 2}$

6. $\dfrac{x^3 - 3x^2 + x - 4}{x - 1}$

7. $\dfrac{6x^3 - x^2 + 2x + 1}{x + \frac{1}{3}}$

8. $\dfrac{2x^3 + 3x^2 + 5}{x - \frac{1}{2}}$

9. $(x^4 + x^2 + 2x - 1) \div (x + 3)$

10. $\dfrac{2x^3 + x - 5}{x + 1}$ (*Hint:* remember that the coefficient of x^2 is zero.)

11. $\dfrac{x^4 - 5x^2 + 4}{x - 3}$

12. $\dfrac{-2x^3 + 13x^2 - 3x - 5}{2x + 1}$

13. $\dfrac{-3x^3 - 25x^2 + 30x - 8}{3x - 2}$

14. $\dfrac{2x^5 - 7x^4 + 15x^3 - 6x^2 - 10x + 5}{2x - 1}$

15. $\dfrac{x^6 - 1}{x - 1}$

16. $\dfrac{x^5 + 243}{x + 3}$

Given that $A(x)$ is a factor of $B(x)$, write $B(x)$ in the form $B(x) = A(x)C(x)$.

17. $A(x) = x - 1$
    $B(x) = 2x^4 - 6x^3 + 3x^2 + 4x - 3$

18. $A(x) = x + 1$
    $B(x) = 3x^5 + x^4 + x^3 - 8x^2 - 5x + 6$

19. $A(x) = 2x - 1$
    $B(x) = 4x^4 - 14x^3 + 10x^2 - 6x + 2$

20. $A(x) = 2x - 3$
    $B(x) = 4x^5 - 12x^4 + 5x^3 + 20x^2 - 33x + 18$

Find the quotients. *Hint:* Use synthetic division twice.

21. $\dfrac{x^3 + x^2 - 33x + 63}{(x - 3)(x + 7)}$

22. $\dfrac{x^3 - 7x^2 + 14x - 8}{(x - 4)(x - 1)}$

Use a calculator and synthetic division to find the quotient and remainder.

23. $\dfrac{3.23x^3 - 13.009x^2 - 14.423x - 39.008}{x - 5.3}$

24. $\dfrac{6.84x^4 - 28.35x^3 - 98.6x^2 + 61.11x - 100}{x - 6.25}$

14 • 3 Rational Roots of Polynomial Equations

In section 14.2 we used synthetic division to solve the problem
$$(3x^3 - 4x^2 + 7x - 5) \div (x - 2).$$
The result is

```
2 | 3   -4    7   -5
  |      6    4   22
  |_____
    3    2   11   17
```

which translates into
$$\frac{3x^3 - 4x^2 + 7x - 5}{x - 2} = 3x^2 + 2x + 11 + \frac{17}{x - 2}.$$
Multiplying both sides of the equation by $x - 2$ gives the form
$$3x^3 - 4x^2 + 7x - 5 = (x - 2)(3x^2 + 2x + 11) + 17$$
Dividend = Divisor × Quotient + Remainder.

Let us now consider the function $f(x) = 3x^3 - 4x^2 + 7x - 5$. We calculate $f(2)$:
$$\begin{aligned} f(2) &= 3(2)^3 - 4(2)^2 + 7(2) - 5 \\ &= 24 - 16 + 14 - 5 \\ &= 17. \end{aligned}$$

This is precisely the remainder that we obtained in the division problem. We would expect this to happen if we wrote $f(x)$ as $(x - 2)(3x^2 + 2x + 11) + 17$, since substituting 2 in the factor $x - 2$ gives

$$f(2) = 0(3x^2 + 2x + 11) + 17$$
$$= 0 + 17 = 17.$$

This example suggests the following theorems.

THE REMAINDER THEOREM

If a polynomial $P(x)$ is divided by $(x - a)$, then the remainder is precisely $P(a)$.

Example 14.5 illustrates the remainder theorem.

• **Example 14.5:** What is the remainder if $x^3 - 2x^2 + x - 7$ is divided by $x - 2$?

Solution: Using the remainder theorem, we can calculate $f(2)$ instead of dividing $x^3 - 2x^2 + x - 7$ by $x - 2$:

$$f(2) = (2)^3 - 2(2)^2 + 2 - 7$$
$$= 8 - 8 + 2 - 7 = -5.$$

Checking the answer by synthetic division gives

```
2 | 1   -2   1   -7
  |      2   0    2
  |_____
    1    0   1   -5
```

and, indeed, the remainder is -5.

THE FACTOR THEOREM

$(x - a)$ is a factor of a polynomial $P(x)$ if and only if $P(a) = 0$.

The factor theorem is illustrated in example 14.6.

• **Example 14.6:** Determine whether

a. $x - 2$ is a factor of $P(x) = x^4 - 3x^3 + 2x^2 + x - 2$, or
b. $x + 3$ is a factor of $S(x) = 2x^4 - 3x^3 + 7x + 27$.

Solution: **a.** $P(2) = (2)^4 - 3(2)^3 + 2(2)^2 + (2) - 2$
$= 16 - 24 + 8 + 2 - 2$
$= 0.$

Since $P(2) = 0$, then, by the factor theorem, $x - 2$ is a factor of $x^4 - 3x^3 + 2x^2 + x - 2$

b. $S(-3) = 2(-3)^4 - 3(-3)^3 + 7(-3) + 27$
$= 162 + 81 - 21 + 27$
$= 249$

Since $S(-3) \neq 0$, then $(x - (-3))$ or $(x + 3)$ is not a factor of $S(x)$.

We can use these results to find the *rational roots* of polynomial equations. (Roots of the form $\frac{p}{q}$, $q \neq 0$, where p and q are integers.)

Zero of the Polynomial

Given a polynomial $P(x)$, a number c is called a **zero of the polynomial**, provided $P(c) = 0$.

We attack the problem of finding the zeros of a polynomial by examining the polynomial for possible *rational zeros*, that is, zeros of the polynomial that are rational numbers. The following theorem is useful in the search for rational zeros.

RATIONAL ZERO THEOREM

Given the polynomial $P(x) = a_n x^n + a_{n-1} x^{n-1} + \cdots + a_1 x + a_0$ with $a_n \neq 0$. If the coefficients are integers and $\frac{p}{q}$ is a rational zero in lowest terms, then p must be a divisor of a_0 and q must be a divisor of a_n.

Examples 14.7 and 14.8 illustrate how to use this theorem.

• *Example 14.7:* Use the rational zero theorem and the factor theorem to find all of the rational zeros of the polynomial $P(x) = 2x^3 - 3x^2 - 2x + 3$.

Solution: Using the rational zero theorem, any rational zero must have the form $\frac{p}{q}$, where p is a divisor of 3 and q is a divisor of 2. The possibilities for p are $\pm 1, \pm 2, \pm 3$. The possibilities for q are $\pm 1, \pm 2$. The *different* possibilities for $\frac{p}{q}$ are these:

$$\pm \frac{1}{2}, \pm \frac{3}{2}, \pm 1, \pm 2, \pm 3.$$

Using the factor theorem, a search for rational zeros can be made by substituting the possible zeros into $P(x)$:

$$P\left(-\frac{1}{2}\right) = 2\left(-\frac{1}{2}\right)^3 - 3\left(-\frac{1}{2}\right)^2 - 2\left(-\frac{1}{2}\right) + 3$$

$$= -\frac{1}{4} - \frac{3}{4} + 1 + 3 = 3 \neq 0,$$

$$P\left(\frac{1}{2}\right) = 2\left(\frac{1}{2}\right)^3 - 3\left(\frac{1}{2}\right)^2 - 2\left(\frac{1}{2}\right) + 3$$

$$= \frac{1}{4} - \frac{3}{4} - 1 + 3 = \frac{3}{2} \neq 0,$$

$$P\left(\frac{3}{2}\right) = 2\left(\frac{3}{2}\right)^3 - 3\left(\frac{3}{2}\right)^2 - 2\left(\frac{3}{2}\right) + 3$$

$$= \frac{27}{4} - \frac{27}{4} - 3 + 3 = 0.$$

Since $P\left(\frac{3}{2}\right) = 0$, $\left(x - \frac{3}{2}\right)$ is a factor of $P(x)$ and $x = \frac{3}{2}$ is a zero of $P(x)$.

Use synthetic division to factor $P(x)$:

$$\begin{array}{c|cccc} \frac{3}{2} & 2 & -3 & -2 & 3 \\ & & 3 & 0 & -3 \\ \hline & 2 & 0 & -2 & 0 \end{array}$$

$P(x)$ factors as follows:

$$2x^3 - 3x^2 - 2x + 3 = \left(x - \frac{3}{2}\right)(2x^2 - 2)$$

$$= \left(x - \frac{3}{2}\right)(2)(x^2 - 1)$$

$$= \left(x - \frac{3}{2}\right)(2)(x + 1)(x - 1).$$

$P(x) = 0$ if $x = \frac{3}{2}$, $x = 1$, or $x = -1$. The set of zeros is $\left\{\frac{3}{2}, 1, -1\right\}$.

• • • • • • • • • •

• *Example 14.8:* Find all of the rational zeros of $P(x) = 12x^3 - 16x^2 - 7x + 6$.

Solution: This is equivalent to solving the equation

$$12x^3 - 16x^2 - 7x + 6 = 0.$$

Using the rational zero theorem, if $\frac{p}{q}$ is a rational zero of $P(x)$, then q must be a divisor of 12 and p must be a divisor of 6. p is in the set $\{\pm 1, \pm 2, \pm 3, \pm 6\}$. q is in the set $\{\pm 1, \pm 2, \pm 3, \pm 4, \pm 6, \pm 12\}$.

The set of fractions that can be obtained using these numerators and denominators is:

$\pm 1, \pm 2, \pm 3, \pm 6$ \hfill $(q = 1)$

$\pm \frac{1}{2}, \pm \frac{2}{2}, \pm \frac{3}{2}, \pm \frac{6}{2}$ \hfill $(q = 2)$

$\pm \frac{1}{3}, \pm \frac{2}{3}, \pm \frac{3}{3}, \pm \frac{6}{3}$ \hfill $(q = 3)$

$\pm \frac{1}{4}, \pm \frac{2}{4}, \pm \frac{3}{4}, \pm \frac{6}{4}$ \hfill $(q = 4)$

$\pm \frac{1}{6}, \pm \frac{2}{6}, \pm \frac{3}{6}, \pm \frac{6}{6}$ \hfill $(q = 6)$

$\pm \frac{1}{12}, \pm \frac{2}{12}, \pm \frac{3}{12}, \pm \frac{6}{12}$ \hfill $(q = 12)$

The list of *different* fractional zeros in lowest terms is somewhat smaller:

$$\left\{ \pm 1, \pm 2, \pm 3, \pm 6, \pm \frac{1}{2}, \pm \frac{3}{2}, \pm \frac{1}{3}, \pm \frac{2}{3}, \pm \frac{1}{4}, \pm \frac{3}{4}, \pm \frac{1}{6}, \pm \frac{1}{12} \right\}.$$

We need to find a fraction, a, in this set so that $P(a) = 0$. Starting from left to right

$$P(-1) = -12 - 16 + 7 + 6 \neq 0$$
$$P(1) = 12 - 16 - 7 + 6 \neq 0$$
$$P(-2) = -96 - 64 + 14 + 6 \neq 0$$
$$P(2) = 96 - 64 - 14 + 6 \neq 0$$
$$P(-3) = -324 - 144 + 21 + 6 \neq 0$$
$$P(3) = 324 - 144 - 21 + 6 \neq 0$$
$$P(-6) = -2592 - 576 + 42 + 6 \neq 0$$
$$P(6) = 2592 - 576 - 42 + 6 \neq 0$$
$$P\left(-\frac{1}{2}\right) = -\frac{3}{2} - 4 + \frac{7}{2} + 6 \neq 0$$
$$P\left(\frac{1}{2}\right) = \frac{3}{2} - 4 - \frac{7}{2} + 6 = 0.$$

Since $P\left(\dfrac{1}{2}\right) = 0$, then $x = \dfrac{1}{2}$ is a zero of $P(x)$ and $x - \dfrac{1}{2}$ is a factor of $P(x)$. We use synthetic division to factor $P(x)$:

$$\begin{array}{c|cccc} \dfrac{1}{2} & 12 & -16 & -7 & +6 \\ & & 6 & -5 & -6 \\ \hline & 12 & -10 & -12 & 0. \end{array}$$

$P(x)$ factors as

$$12x^3 - 16x^2 - 7x + 6 = \left(x - \dfrac{1}{2}\right)(12x^2 - 10x - 12)$$

$$= \left(x - \dfrac{1}{2}\right)(2)(6x^2 - 5x - 6).$$

We could continue with the list of possible rational zeros of $P(x)$. However, we know how to factor quadratic expressions.

$$P(x) = \left(x - \dfrac{1}{2}\right)(2)(6x^2 - 5x - 6)$$

$$= \left(x - \dfrac{1}{2}\right)(2)(3x + 2)(2x - 3).$$

The solution set to $P(x) = 0$ can now be obtained:

1. If $x - \dfrac{1}{2} = 0$ then $x = \dfrac{1}{2}$.

2. If $3x + 2 = 0$ then $x = -\dfrac{2}{3}$.

3. If $2x - 3 = 0$ then $x = \dfrac{3}{2}$.

Sometimes the context of the problem allows us to reduce the number of possibilities of rational zeros of a polynomial. Example 14.9 illustrates such a case.

• **Example 14.9:** A manufacturer has determined that the profit $P(x)$ (in thousands of dollars) is related to the number of units produced (x) by the polynomial $P(x) = x^3 - 16x^2 + 16x - 15$. Determine the break-even point, that is, the value of x for which $P(x) = 0$.

Solution: The break-even point will occur when
$$P(x) = x^3 - 16x^2 + 16x - 15 = 0.$$

We need to consider only positive real values of x, since it makes no sense to manufacture a negative number of units of a product. The possible rational solutions to the problem must also be divisors of -15. These are 1, 3, 5, 15. Hence,

$$P(1) = 1 - 16 + 16 - 15 \neq 0$$
$$P(3) = 27 - 144 + 48 - 15 \neq 0$$
$$P(5) = 125 - 400 + 80 - 15 \neq 0$$
$$P(15) = 3375 - 3600 + 240 - 15 = 0.$$

Since $P(15) = 0$, then $x - 15$ is a factor of $P(x)$. We use synthetic division to factor $P(x)$:

$$\begin{array}{r|rrrr} 15 & 1 & -16 & 16 & -15 \\ & & 15 & -15 & 15 \\ \hline & 1 & -1 & 1 & 0 \end{array}$$

$$x^3 - 16x^2 + 16x - 15 = (x - 15)(x^2 - x + 1).$$

Since $x^2 - x + 1 = 0$ gives complex numbers for x, then $x = 15$ is the only usable solution to the problem. Hence the break-even point occurs when 15 units are produced.

∙∙∙∙∙∙∙∙∙∙∙

Descartes's Rule of Signs

A procedure that often reduces the number of possibilities for real roots of a polynomial equation is called **Descartes's rule of signs**. This rule helps us to determine the numbers of positive and negative real roots of a polynomial equation $P(x) = 0$ by counting the number of *variations in sign* that occur in $P(x)$ and $P(-x)$.

Variation in Sign

A **variation in sign** is counted when the coefficients of two successive terms in a polynomial differ in sign. For example the polynomial

$$+5x^4 \quad -3x^3 \quad +2x^2 \quad +x \quad -7$$
$$123$$
(pos. to neg.) (neg. to pos.) (pos. to neg.)

has three variations in sign. The polynomial

$$x^4 \quad +3x^2 \quad +7x \quad -4$$
$$1$$
(pos. to neg.)

has one variation in sign.

SECTION 14.3 RATIONAL ROOTS OF POLYNOMIAL EQUATIONS

DESCARTES'S RULE OF SIGNS

Let $P(x) = 0$ be a polynomial equation with real coefficients.

1. The number of positive real roots is either equal to the number of variations in sign occurring in $P(x)$ or is less than the number of variations by a positive even integer.

2. The number of negative real roots is either equal to the number of variations in sign occurring in $P(-x)$ or is less than the number of variations by a positive even integer.

Applied to the polynomial $5x^4 - 3x^3 + 2x^2 + x - 7$, which has three variations in sign, Descartes's rule yields the following results:

1. Since $P(x)$ has three variations, the number of positive real roots is either 3 or $3 - 2 = 1$. We can expect $P(x) = 0$ to have either three or one positive real root.

2. Calculate $P(-x)$. Recall that this is accomplished by substituting $-x$ for x in $P(x)$:

$$P(-x) = 5(-x)^4 - 3(-x)^3 + 2(-x)^2 + (-x) - 7$$
$$= 5x^4 + 3x^3 + 2x^2 - x - 7.$$

Since there is one variation in sign in $P(-x)$, we can expect one negative real root for $P(x) = 0$.

Note that there are four roots for a fourth degree polynomial equation. Thus, there are three possibilities:
  a. one negative and three positive real roots
  b. one negative and one positive real root and a pair of complex roots.

Example 14.10 illustrates how to use Descartes's rule of signs.

• *Example 14.10:* Find the real roots of the equation

$$12x^5 + 92x^4 + 257x^3 + 323x^2 + 180x + 36$$

by examining the equation for rational roots.

Solution: Since $P(x) = 12x^5 + 92x^4 + 257x^3 + 323x^2 + 180x + 36 = 0$ has no variations in sign, there are no positive roots.

$$P(-x) = 12(-x)^5 + 92(-x)^4 + 257(-x)^3 + 323(-x)^2 + 180(-x) + 36$$
$$= -12x^5 + 92x^4 - 257x^3 + 323x^2 - 180x + 36.$$
$$\quad\ \ \underbrace{}_{1}\ \underbrace{}_{2}\ \underbrace{}_{3}\ \underbrace{}_{4}\ \underbrace{}_{5}$$

Since $P(-x)$ has five variations in sign, the number of negative real roots of $P(x) = 0$ is 5, 3, or 1. We examine the equation for rational roots. The positive factors of 12 are $\{1,2,3,4,6,12\}$. The positive factors of 36 are $\{1,2,3,4,6,9,12,18,36\}$. Since a rational root $\dfrac{p}{q}$ must have p as a divisor of 36 and q as a divisor of 12, the possible roots are

$$\left\{-1, -2, -3, -4, -6, -9, -12, -18, -36, -\dfrac{1}{2}, -\dfrac{3}{2}, -\dfrac{9}{2}, -\dfrac{1}{3}, -\dfrac{2}{3}, -\dfrac{4}{3}, -\dfrac{1}{4}, -\dfrac{3}{4}, -\dfrac{9}{4}, -\dfrac{1}{6}, -\dfrac{1}{12}\right\}.$$

Suppose we use synthetic division to search for rational roots, beginning with -1:

$$\begin{array}{r|rrrrrr}
-1 & 12 & 92 & 257 & 323 & 180 & 36 \\
 & & -12 & -80 & -177 & -146 & -34 \\
\hline
 & 12 & 80 & 177 & 146 & 34 & \boxed{2.} \quad \text{Remainder}
\end{array}$$

Since the remainder is not zero, -1 is not a root. Check -2 as a possible root:

$$\begin{array}{r|rrrrrr}
-2 & 12 & 92 & 257 & 323 & 180 & 36 \\
 & & -24 & -136 & -242 & -162 & -36 \\
\hline
 & 12 & 68 & 121 & 81 & 18 & \boxed{0.} \quad \text{Remainder}
\end{array}$$

Since the remainder is zero, -2 is a root and $P(x)$ factors as $(x + 2)(12x^4 + 68x^3 + 121x^2 + 81x + 18)$. Check -3 as a possible root:

$$\begin{array}{r|rrrrr}
-3 & 12 & 68 & 121 & 81 & 18 \\
 & & -36 & -96 & -75 & -18 \\
\hline
 & 12 & 32 & 25 & 6 & \boxed{0.} \quad \text{Remainder}
\end{array}$$

Since the remainder is zero, -3 is a root and $P(x)$ factors as $(x + 2)(x + 3)(12x^3 + 32x^2 + 25x + 6)$. Observe that in attempting to obtain factors of $A(x) = 12x^3 + 32x^2 + 25x + 6$ that in order for $\left(x - \dfrac{p}{q}\right)$ to be a factor of $A(x)$, p must be a divisor of 6. This eliminates $-4, -9, -12, -18, -36, -\dfrac{9}{2}, -\dfrac{4}{3}, -\dfrac{9}{4}$ from consideration as roots. The remaining possible roots are

$$\left\{-6, -\dfrac{1}{2}, -\dfrac{3}{2}, -\dfrac{1}{3}, -\dfrac{2}{3}, -\dfrac{1}{4}, -\dfrac{3}{4}, -\dfrac{1}{6}, -\dfrac{1}{12}\right\}.$$

Check -6 as a possible root:

$$\begin{array}{r|rrrr} -6 & 12 & 32 & 25 & 6 \\ & & -72 & 240 & -1590 \\ \hline & 12 & -40 & 265 & \boxed{-1584.} \end{array} \quad \text{Remainder}$$

Since the remainder is not zero, -6 is not a root. Check $-\dfrac{1}{2}$ as a possible root:

$$\begin{array}{r|rrrr} -\dfrac{1}{2} & 12 & 32 & 25 & 6 \\ & & -6 & -13 & -6 \\ \hline & 12 & 26 & 12 & \boxed{0.} \end{array} \quad \text{Remainder}$$

Since the remainder is zero, $-\dfrac{1}{2}$ is a root and $x + \dfrac{1}{2}$ is a factor of $A(x)$ and of $P(x)$:

$$P(x) = (x + 2)(x + 3)\left(x + \frac{1}{2}\right)(12x^2 + 26x + 12)$$
$$= (x + 2)(x + 3)\left(x + \frac{1}{2}\right)(2)(6x^2 + 13x + 6)$$
$$= (x + 2)(x + 3)(2x + 1)(6x^3 + 13x + 6).$$

We can now find the remaining two roots by employing our usual methods to solve $6x^2 + 13x + 6 = 0$. In this particular problem the left-hand side is factorable, and $P(x) = (x + 2)(x + 3)(2x + 1)(2x + 3)(3x + 2)$. The roots of $P(x) = 0$ are $\left\{-2, -3, -\dfrac{1}{2}, -\dfrac{3}{2}, -\dfrac{2}{3}\right\}$.

•••••••••

It should be pointed out that while this procedure seems to be tedious, alternative methods of finding roots of polynomial equations of degree three or greater are either more tedious or nonexistent. If the real roots are not rational, for example, except in special cases it is not possible to determine the roots exactly. We pursue some methods of approximating the nonrational real roots of polynomial equations in section 14.4.

Trial Problems 14.3

Before you begin the section exercises, warm up with these problems. Complete answers are included in the answer key.

Find the integral zeros of each polynomial.

1. $x^3 + 8x^2 + 13x + 6$
2. $x^4 - 4x^3 + 5x^2 - 2x - 2$

Find the rational zeros of each polynomial.

3. $2x^3 - x^2 - 4x + 2$
4. $96x^3 - 16x^2 - 6x + 1$

Exercises 14.3

Determine the rational zeros for each polynomial.

1. $P(x) = x^3 - 6x^2 + 11x - 6$
2. $P(x) = x^3 - 3x - 2$
3. $P(x) = x^3 - x^2 - 14x + 24$
4. $P(x) = x^3 + 5x^2 + 2x - 8$
5. $P(x) = 2x^3 - 13x^2 + 24x - 9$
6. $P(x) = 2x^3 + 5x^2 - 4x - 12$
7. $P(x) = 12x^3 - 65x^2 + 74x - 24$
8. $P(x) = 6x^3 + 29x^2 - 40x + 12$
9. $P(x) = 8x^3 - 12x^2 + 6x - 1$
10. $P(x) = 27x^3 - 18x^2 + 12x - 8$
11. $P(x) = 12x^3 - 20x^2 + 11x - 2$
12. $P(x) = 18x^3 + 3x^2 - 4x - 1$

Determine all of the zeros for each polynomial.

13. $3x^3 + x^2 + x - 2$
14. $3x^3 - x^2 + x + 2$
15. $2x^3 - 9x^2 + 16x - 12$
16. $2x^3 - 9x^2 + 13x - 12$
17. $x^4 - 4x^3 + 2x^2 + x + 6$
18. $x^4 - 7x^3 + 8x^2 + 7x + 15$
19. $6x^4 - 25x^3 + 24x^2 + 2x - 3$
20. $6x^4 + 11x^3 - 18x^2 - 16x - 3$
21. $6x^5 - 31x^4 + 49x^3 - 22x^2 - 5x + 3$
22. $6x^5 - x^4 - 40x^3 + 20x^2 + 29x + 6$

23. If $P(t)$ represents the position of a point moving on a straight line at any given time t, determine the value of t for which $P(t) = 0$. $P(t) = t^3 + t^2 - 3t + 1$.

24. Solve exercise 23 given that $P(t) = t^3 + t^2 - 7t + 2$.

25. An insecticide reduces an initial population of 800,000 mosquitoes (in an idealized experiment) according to the formula $P(t) = -t^5 + 300t^3 + 800{,}000$, where t represents the time in hours and $P(t)$ represents the mosquito population remaining after t hours. Determine how long it would take before the population theoretically reaches zero.

26. Solve exercise 25 using the formula
$$P(t) = -t^4 + 200t^2 + 630{,}000.$$

27. The number of people in a given population who have contracted a particular disease is estimated by the equation $P(t) = -2t^3 + 50t^2 + 9000$, where $P(t)$ represents the number of people who have the disease after t days. Determine the number of days that it should take for the disease to die out in the population.

28. Solve exercise 27 given that
$$P(t) = -2t^3 + 70t^2 + 16{,}000.$$

29. In an electrical circuit, three resistors having resistances R_1, R_2, and R_3 are connected in parallel. The total resistance R_T is given by the formula

$$R_T = \frac{R_1 R_2 R_3}{R_1 R_2 + R_2 R_3 + R_1 R_3}.$$

Suppose that it is known that $R_1 = x$, $R_2 = x + 3$, $R_3 = x + 2$, and $R_T = \frac{10}{7}$. Determine the values of R_1, R_2, and R_3.

30. Solve exercise 29 given that $R_1 = x$, $R_2 = x$, $R_3 = x + 3$, and $R_T = 1.2$.

31. The total cost of producing x items is represented by the equation $C(x) = \frac{x^3}{3} - 2x^2 + 3x + 7$ for $x \geq 3$. Determine the value of x for which $C(x) = 25$.

32. In exercise 31, determine the value of x for which $C(x) = 115$.

33. The concentration of a compound in solution in parts/million as a function of time t in seconds is represented by the following function:

$$f(t) = -3t^4 + 70t^3 - 26t^2 + 90t.$$

Determine the value of t for which $f(t) = 1683$.

34. In exercise 33, determine the value of t for which $f(t) = 3656$.

35. Under certain conditions the deflection d (in inches) of a loaded beam at a point x feet from one end of the beam is $d = 0.002(-\frac{1}{8}x^3 + 200x)$, for $0 \leq x \leq 30$. Find the value of x for which $d = 3.072$ in.

36. In exercise 35, find the value of x for which $d = 1.584$ in.

14•4 Approximate Roots of Polynomial Equations

So far we have solved problems in which we found the roots of polynomial equations that had at least one rational root. When there are no rational roots, the problems can be very difficult. Consequently, it is often necessary to settle for an approximation of the roots to some stated degree of accuracy. The following theorem provides some help in locating roots of polynomial equations.

LOCATION THEOREM
For a polynomial equation, $P(x) = 0$, if $P(a) < 0$ and $P(b) > 0$ for real numbers a and b, then $P(x)$ has a zero between a and b.

Example 14.11 shows how to use this theorem.

• **Example 14.11:** Approximate the zeros of $P(x) = x^3 + 3x^2 - 6x + 1$ to the nearest tenth.

Solution: First sketch a graph of the function $P(x)$ to get a general idea of the number and location of the roots.

$$P(-5) = -125 + 75 + 30 + 1 = -19$$
$$P(-4) = -64 + 48 + 24 + 1 = 9$$
$$P(-3) = -27 + 27 + 18 + 1 = 19$$
$$P(-2) = -8 + 12 + 12 + 1 = 17$$
$$P(-1) = -1 + 3 + 6 + 1 = 9$$
$$P(0) = 0 + 0 + 0 + 1 = 1$$
$$P(1) = 1 + 3 - 6 + 1 = -1$$
$$P(2) = 8 + 12 - 12 + 1 = 9$$
$$P(3) = 27 + 27 - 18 + 1 = 37$$

x	-5	-4	-3	-2	-1	0	1	2	3
$P(x)$	-19	9	19	17	9	1	-1	9	37

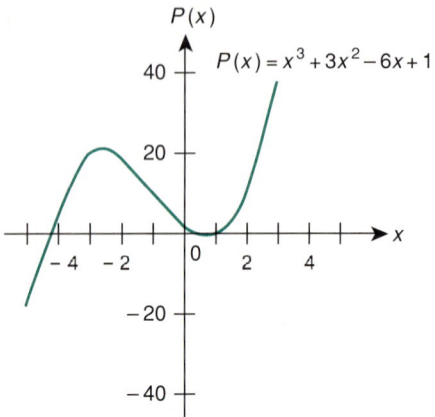

Figure 14.1

We can see from the graph in figure 14.1 that there are zeros of the polynomial between -5 and -4, between 0 and 1, and between 1 and 2.

1. Let us first approximate the zero that is between 0 and 1. Using either synthetic division or direct substitution, we calculate the following values for $P(x)$:

x	0.1	0.2	0.3
$P(x)$	0.431	-0.072	-0.503

Since $P(0.1)$ is positive and $P(0.2)$ is negative, the zero of the polynomial is between 0.1 and 0.2. We now subdivide this interval and calculate the following values for $P(x)$ (rounded to 4 places).

x	0.11	0.12	0.13	0.14	0.15	0.16	0.17	0.18	0.19
$P(x)$	0.3776	0.3249	0.2729	0.2215	0.1709	0.1209	0.0716	0.0230	-0.0248

Using the location theorem, we see that the zero of $P(x)$ is between 0.18 and 0.19.

Thus $x = 0.2$ is an approximate zero of $P(x)$.

2. We now locate the zero that is between 1 and 2

x	1.1	1.2	1.3
$P(x)$	-0.639	-0.152	0.467

A zero of $P(x)$ is between 1.2 and 1.3. Looking at the values of $P(1.2)$ and $P(1.3)$ we can probably be safe in assuming that the zero is closer to $x = 1.2$. Let us check this assumption.

x	1.21	1.22
P(x)	−0.0961	0.0390

Since $P(1.21)$ is negative and $P(1.22)$ is positive, the assumption is correct, and $x = 1.2$ is an approximate zero of $P(x)$.

3. We now approximate the zero of $P(x)$ that is between -5 and -4. From the graph, it appears that the root will be closer to $x = -4$, so we start with -4.1 on calculating $P(x)$.

x	−4.1	−4.2	−4.3	−4.4	−4.5
P(x)	7.109	5.032	2.763	0.296	−2.375

The zero of $P(x)$ is between $x = -4.4$ and $x = -4.5$.

We can safely assume that the zero is closer to $x = -4.4$, however, to check this assumption, let us calculate $P(-4.44) = -0.7476$. Since $P(-4.44)$ is negative and $P(-4.4)$ is positive, the zero is between -4.4 and -4.44. Thus, to the nearest tenth, $x = -4.4$ is the desired zero of $P(x)$. The set of zeros of $P(x) = x^3 + 3x^2 - 6x + 1$ is $\{0.2, 1.2, -4.4\}$ (to the nearest tenth).

• • • • • • • • • • •

Sometimes the context of the problem makes it unnecessary to find all of the zeros of a polynomial function. Example 14.12 illustrates this idea.

• **Example 14.12:** An open container is constructed by cutting squares from the corners of a 16 inch square piece of material (see figure 14.2). The volume of the box must be approximately 280 cubic inches. To the nearest tenth of an inch, what should be the dimension of the box?

Solution: The volume of the box will be $V(x) = (16 - 2x)(16 - 2x)(x) = 4x^3 - 64x^2 + 256x$. Since we want the volume to be 280 cubic inches, the volume can be written as

$$4x^3 - 64x^2 + 256x = 280$$

or

$$4x^3 - 64x^2 + 256x - 280 = 0.$$

Since the equation can be divided by 4, we solve the equation in the form

$$x^3 - 16x^2 + 64x - 70 = 0.$$

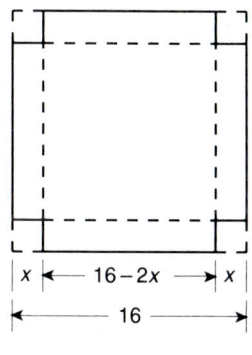

Figure 14.2

To determine the practical solutions for the equation, the ones where $0 < x < 8$, we sketch an approximate graph of $P(x) = x^3 - 16x^2 + 64x - 70$. Note that any number less than zero or greater than 8 would make one or more of the dimensions of the box negative.

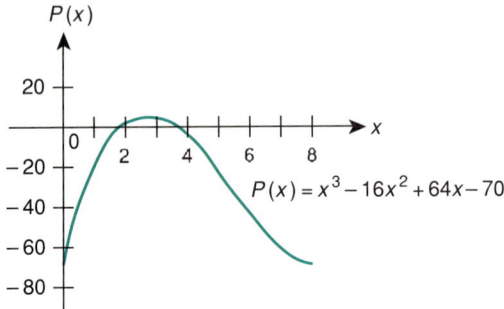

Figure 14.3

x	0	1	2	3	4	5	6	7	8
$P(x)$	-70	-21	2	5	-6	-25	-46	-63	-70

From the graph in figure 14.3, we can see that there are two practical solutions, x between 1 and 2 and x between 3 and 4.

1. For x between 1 and 2: From the graph, it appears that the desired value of x is closer to $x = 2$ than it is to $x = 1$. We will start with $x = 1.9$ and decrease x by 0.1 each step.

x	1.9	1.8
$P(x)$	0.699	-0.808

Thus, a zero of $P(x)$ is between 1.8 and 1.9. We could proceed with $x = 1.81, 1.82, \ldots$, however, we have the function $V(x) = x(16 - 2x)^2$. Let us substitute $x = 1.8$ and $x = 1.9$ in $V(x)$ to see which value $V(1.8)$ or $V(1.9)$ gives a volume closer to 280 cubic inches:

$$V(1.8) = (1.8)(16 - 2(1.8))^2 = (1.8)(12.4)^2 = 276.8$$
$$V(1.9) = (1.9)(16 - 2(1.9))^2 = (1.9)(12.2)^2 = 282.8.$$

The dimensions that give the closer value are

$$x = 1.9, \quad (16 - 2x) = 12.2$$

2. For x between 3 and 4: From the graph, it appears that x should be somewhere around $x = 3.5$.

x	3.5	3.6
$P(x)$	0.875	-0.304

The zero of $P(x)$ is between 3.5 and 3.6:

$$V(3.5) = (3.5)(16 - 2(3.5))^2 = (3.5)(9)^2 = 283.5$$
$$V(3.6) = (3.6)(16 - 2(3.6))^2 = (3.6)(8.8)^2 = 278.784.$$

$V(3.6)$ is closer to 280 cubic inches than $V(3.5)$. The dimensions of the box are

$$x = 3.6, \quad (16 - 2x) = 8.8.$$

Hence, there are two sets of dimensions that produce a box whose volume is approximately 280 cubic inches. One box is 12.2 inches square and 1.9 inches high. The other is 8.8 inches square and 3.6 inches high.

Trial Problems 14.4

Before you begin the section exercises, warm up with these problems. Complete answers are included in the answer key.

Find the zero of the polynomial that is in the given interval. Calculate to the nearest tenth.

1. $x^3 - 2x - 5$, between 2 and 3
2. $x^3 + 6x^2 + 8x + 8$, between -5 and -4
3. $x^3 + 3x^2 - 2x - 5$, between 1 and 2
4. Find the real zero of $P(x) = x^3 + 3x^2 - 2x - 5$ that is between -2 and -1. Calculate to the nearest hundredth.

Exercises 14.4

Determine the real zero of each polynomial to the nearest tenth that is in the given interval.

1. $P(x) = x^3 - x^2 - 7x + 8$ between 2 and 3
2. $P(x) = 3x^3 - 2x^2 - 9x + 6$ between -2 and -1
3. $P(x) = 2x^3 - 5x^2 - 8x + 5$ between 3 and 4
4. $P(x) = 2x^3 - 13x^2 + 25x - 15$ between 3 and 4
5. $P(x) = 5x^3 + 23x^2 + 5x - 6$ between -5 and -4
6. $P(x) = x^5 - 2x^2 - 7$ between 1 and 2
7. $P(x) = x^4 + 2x^3 + 3x^2 + 7x - 4$ between 0 and 1
8. $P(x) = x^4 + 3x^3 - 2x^2 + 7x - 5$ between 0 and 1
9. $P(x) = x^4 - 2x^3 + 2x^2 + 7x - 4$ between -2 and -1
10. $P(x) = x^4 - 2x^3 - 2x^2 + 7x - 4$ between -2 and -1

Determine all of the real zeros of each polynomial to the nearest tenth. First make a rough sketch of the graph of the polynomial function.

11. $P(x) = x^3 - 3x + 1$
12. $P(x) = x^3 + 2x^2 - 7x + 1$
13. $P(x) = x^3 + 2x^2 - 3x - 5$
14. $P(x) = x^3 - 5x - 1$
15. $P(x) = x^4 - 2x^3 + x^2 - 1$
16. $P(x) = x^4 - 3x^3 - x^2 + 3x + 3$
17. $P(x) = x^4 + x - 5$
18. $P(x) = x^4 - 2x^3 - 5x^2 + 6x + 2$

19. If the length of the side of a cube is increased by 1 cm, the volume of the cube is doubled. Determine the length of the side of the original cube to the nearest hundredth of a cm.

20. A sphere floating in water will sink to a depth x given by the smaller root of the equation

$$x^3 - 3rx^2 + 4r^3s = 0$$

where r = radius of the sphere and $s < 1$ is the specific gravity of the sphere. Determine x if $r = 1$ and $s = 0.25$.

21. The resistance of a particular resistor varies by temperature according to the formula

$$R^2 = 0.001t^4 - 5t + 80$$

where R is the resistance in ohms and t is the temperature in degrees celsius. Determine the temperature at which the resistor will have $R = 12.3$ ohms.

22. Solve exercise 21 given that $R = 10.7$ ohms.

23. The number of bacteria (in millions) present in a colony as a function of time t (in minutes) is described by the equation $f(t) = -0.0075t^5 + t^3 + 2.5$. Determine (to the nearest tenth) the value of t for which $f(t) = 15.7$.

24. Solve exercise 23 for $f(t) = 44.7$.

14·5 Summary of Terms, Rules, and Procedures

Terms

Descartes's Rule of Signs (p. 570)
Rational Zero of a Polynomial (p. 566)
Synthetic Division (p. 557)
Variation in Sign (p. 570)
Zero of a Polynomial (p. 566)

Rules and Procedures

- **THE REMAINDER THEOREM** (14.3)
  If a polynomial $P(x)$ is divided by $(x - a)$, then the remainder is precisely $P(a)$.

- **THE FACTOR THEOREM** (14.3)
  $(x - a)$ is a factor of a polynomial $P(x)$ if and only if $P(a) = 0$.

- **RATIONAL ZERO OF A POLYNOMIAL** (14.3)
  A rational number $\dfrac{p}{q}$ is a zero of a polynomial $P(x)$ if $P\left(\dfrac{p}{q}\right) = 0$.

- **RATIONAL ZERO THEOREM** (14.3)
  Given the polynomial
  $$a_n x^n + a_{n-1} x^{n-1} + \cdots + a_1 x + a_0.$$
  If the coefficients are integers and $\dfrac{p}{q}$ is a rational zero in lowest terms, then p must be a divisor of a_0 and q must be a divisor of a_n.

- **DESCARTES'S RULE OF SIGNS** (14.3)
  Let $P(x) = 0$ be a polynomial equation with real coefficients.

1. The number of positive real roots is either equal to the number of variations in sign occurring in $P(x)$ or is less than the number of variations by a positive even integer.
2. The number of negative real roots is either equal to the number of variations in sign occurring in $P(-x)$ or less than the number of variations by a positive even integer.

- LOCATION THEOREM (14.4)
  For a polynomial equation, $P(x) = 0$, if $P(a) < 0$ and $P(b) > 0$ for real numbers a and b, then $P(x)$ has a zero between a and b.

14·6 Chapter 14 Review Exercises

Use synthetic division to find the quotient and remainder.

1. $\dfrac{x^3 + x^2 + x + 1}{x - 1}$
2. $\dfrac{x^4 + 3x^2 - 3x + 2}{x - 2}$
3. $\dfrac{x^3 + 2x^2 - 5x + 7}{x - 1}$
4. $\dfrac{x^3 + 2x^2 - 5x + 2}{x - 1}$
5. $\dfrac{2x^3 + 4x^2 - 10x + 12}{x - 3}$
6. $\dfrac{2x^3 + 4x^2 - 10x + 12}{x + 3}$
7. $\dfrac{3x^4 - 6x^3 + 7x^2 - 4x + 11}{x + 2}$
8. $\dfrac{5x^4 - 8x^3 + 7x^2 - 3x - 38}{x - 2}$
9. $\dfrac{2.7x^3 - 4.3x^2 + 7.8x - 2.0259}{x - 0.3}$
10. $\dfrac{3.4x^3 + 3.02x^2 + 2.6x + 0.6}{x + 0.3}$

Determine the rational zeros for each polynomial.

11. $x^3 - 4x^2 + x + 6$
12. $x^3 + 4x^2 + x - 6$
13. $3x^3 + x^2 + x - 2$
14. $3x^3 + 2x^2 + 2x - 1$
15. $3x^3 - 10x^2 + 9x - 2$
16. $3x^3 - 11x^2 + 12x - 4$
17. $x^4 - 2x^3 - x + 2$
18. $2x^4 - x^3 - 2x + 1$

Determine all of the zeros for each polynomial.

19. $2x^3 - 3x^2 - x + 1$
20. $3x^3 - 8x^2 - 2x + 4$
21. $6x^3 - 19x^2 + 18x - 5$
22. $24x^3 - 34x^2 + 15x - 2$

Determine the real zeros of each polynomial to the nearest tenth.

23. $3x^3 - 14x^2 + 18x - 5$
24. $x^3 + 2x^2 - 3x - 2$
25. $x^4 - 3x^3 + x^2 + x + 1$
26. $x^4 - 2x^3 - 4x^2 - 5x - 2$

27. A 20 inch by 40 inch rectangular piece of cardboard is made into a topless box by cutting squares from the corners and folding the sides up. The box must have a volume of 1500 cu in. Find the dimensions of the box. Note that there is more than one possibility.

28. Solve exercise 27 if the piece of cardboard is a 16 by 16 inch square and the volume must be 288 cubic inches.

29. A three inch thick layer is cut from a cube. The remaining volume is 320 cubic inches. What is the volume of the original cube?

30. A three inch thick layer is added to one face of a cube. The new volume is 704 cu in. What is the volume of the original cube?

31. Find 3 consecutive positive integers whose product is 4080.

32. Find 3 consecutive positive integers whose product is 17,550.

33. The resistance R in ohms of a resistor in a particular circuit is related to the temperature in degrees Celsius by the formula $R^2 = 0.001t^4 - 4t + 430$. At what temperature would the resistor have a resistance of 20 ohms?

34. Solve exercise 33 using the formula

$$R^2 = 0.002t^3 - 2t + 249$$

and a resistance of 15 ohms.

35. The viscosity v of a particular type of motor oil is related to the temperature t, in hundreds of degrees Fahrenheit, according to the following formula:

$$v = -t^3 + 9t^2 - 24t + 70, \quad \text{for } 0 \le t \le 10.$$

Determine the value(s) of t for which $v = 54$.

36. Solve exercise 35 for $v = 34$.

14·7 Chapter 14 Test

Use synthetic division to find the quotient and the remainder.

1. $\dfrac{3x^3 + 2x^2 - 5x + 7}{x - 4}$

2. $\dfrac{2x^3 - 3x^2 + 4x - 6}{x + 1}$

3. $\dfrac{4x^4 + 2x^2 - 5}{x - 1}$

4. $\dfrac{5x^4 + 3x^3 - 4x^2 + x - 1}{x + 2}$

5. $\dfrac{2.4x^3 + 3.5x^2 + 0.6x - 1.6}{x - 1.2}$

Find the rational zeros for each polynomial.

6. $2x^3 + 5x^2 + 4x + 1$

7. $x^3 - x^2 - 10x - 8$

8. $8x^3 - 12x^2 + 6x - 1$

9. $4x^3 - 31x + 15$

10. $x^5 + 4x^4 - 4x^3 - 34x^2 - 45x - 18$

Find all of the zeros for each polynomial.

11. $x^3 - 2x^2 - 3x + 6$

12. $x^3 + 6x^2 - 3x - 4$

13. $8x^4 + 6x^3 - 13x^2 - x + 3$

14. $x^3 - 6x^2 + 5x + 6$

15. $2x^3 + x^2 - 2x - 6$

16. To the nearest tenth, approximate the zero of the polynomial $x^3 - 2x^2 + x + 3$ that is between -1 and 0.

17. To the nearest hundredth, approximate the zero of the polynomial $x^4 + x^3 - 3x^2 - x - 4$ that is between 1 and 2.

18. An open box is made by cutting equal squares from the corners of a 40 in. by 30 in. rectangle and turning up the sides. If the volume of the box is 875 cu in., find the size of the squares that are cut from the corners of the rectangle.

Chapter 15

Sequences and Series

Population, when unchecked, increases in a geometric ratio. Subsistence increases only in an arithmetical ratio. This means that if we start with one unit of population and one unit of subsistence, after one unit of time we have two units of each. After another unit of time we have four units of population and three units of subsistence, and so on. This can be illustrated with the following table.

> Population: 1, 2, 4, 8, 16, 32, 64, 128, . . .
> Subsistence: 1, 2, 3, 4, 5, 6, 7,   8, . . .

Each of these sets of numbers represents a sequence. The first is a geometric sequence and the second is an arithmetic sequence. As you can see by the two sequences, if population goes unchecked, and if the production of the earth remains constant, the population eventually outstrips the earth's ability to feed it.

In this chapter we are concerned with sequences like those above and with many others. Different techniques are used to find various terms of a sequence as well as the sum of any number of terms of a particular sequence. On the basis of the information that we might have about a sequence we are able to predict the behavior of that sequence.

15•1 Sequences

Throughout this book we have discussed many types of functions. We have studied linear functions, quadratic functions, logarithmic functions, and others. In this section we study another function called a sequence. A **sequence** is a function in which the first components, or domain elements, are natural numbers. The second components, or elements of the range, can be any numbers. For example,

$$\{(1,1), (2,4), (3,9), (4,16) \ldots \}$$

is a sequence, since the domain elements are 1, 2, 3, 4, ... In this example, the second components are 1, 4, 9, 16, ...

In practice we often do not bother to write the domain elements, since they will always be 1, 2, 3, 4, ... We simply express a sequence by writing the range elements, or second components.

For example, instead of writing the sequence as

$$\{(1,1) \ (2,4) \ (3,9) \ (4,16) \ldots \},$$

we would write

$$1, \quad 4, \quad 9, \quad 16, \ldots$$

Sequence

Infinite Sequence

If we have a sequence such as 1, 4, 9, 16, . . . , it is called an **infinite sequence** since it has an infinite number of elements. The three dots suggest that the sequence goes on indefinitely. A sequence like 1, 4, 9, 16 (without the three dots) has a finite number of elements and is called a **finite sequence.**

Finite Sequence

When dealing with sequences we are often concerned with finding additional elements of the sequence once we know the first few elements of the sequence. Examples 15.1, 15.2, and 15.3 illustrate how the next few elements of certain sequences may be found.

• **Example 15.1:** Find the next three elements of the sequence 2, 4, 6, 8, ...

Solution: It is obvious that the next three elements are 10, 12, and 14. However, we might also note that $(2)(1) = 2$, $(2)(2) = 4$, $(2)(3) = 6$, and $(2)(4) = 8$. So the next three elements should be $(2)(5)$ or 10, $(2)(6)$ or 12, and $(2)(7)$ or 14.

• **Example 15.2:** Find the next three elements of the sequence 2, 4, 8, 16, ...

Solution: We might note that $2^1 = 2$, $2^2 = 4$, $2^3 = 8$, and $2^4 = 16$. Therefore, the next three numbers should be 2^5, 2^6, and 2^7 or 32, 64, and 128.

• *Example 15.3:* Find the next three terms of the sequence 2, 5, 10, 17, 26, 37, ...

Solution 1: If we subtract each number from the succeeding number, we see a pattern develop. That is,

$$5 - 2 = 3$$
$$10 - 5 = 5$$
$$17 - 10 = 7$$
$$26 - 17 = 9$$
$$37 - 26 = 11.$$

If we place these differences above the sequence as follows

$$\begin{array}{ccccccc} 3 & & 5 & & 7 & & 9 & & 11 \\ 2, & 5, & & 10, & & 17, & & 26, & & 37, \ldots, \end{array}$$

the next three terms should be $37 + 13$ or 50, $50 + 15$ or 65, and $65 + 17$ or 82.

Solution 2: We might also note that

$$2 = 1^2 + 1$$
$$5 = 2^2 + 1$$
$$10 = 3^2 + 1$$
$$17 = 4^2 + 1$$
$$26 = 5^2 + 1$$
$$37 = 6^2 + 1.$$

So the next three terms of the sequence should be

$$7^2 + 1 \text{ or } 50$$
$$8^2 + 1 \text{ or } 65$$
$$9^2 + 1 \text{ or } 82.$$

General Term
nth Term

When working with sequences we are often concerned with developing an expression for the **general term** of a sequence. This is also called the **nth term,** as $n = 1, 2, 3, 4, \ldots$.

In example 15.1 the first term is $(2)(1)$, the second term is $(2)(2)$, the third term is $(2)(3)$, the fourth term is $(2)(4)$, and so forth. This suggests that the nth term should be $2n$.

In example 15.2, the first, second, third, and fourth terms are, respectively, 2^1, 2^2, 2^3, and 2^4. Therefore, the nth term should be 2^n.

In example 15.3, the second solution suggests a pattern that indicates that the nth term should be $n^2 + 1$.

If the general, or nth term, of a sequence is known, then any term of the sequence can be found. Using functional notation, we let $S(n)$ (read "S of n") represent the nth term. Then $S(1)$ represents the first term, $S(2)$ the second, $S(3)$ the third, and so on. This notation is illustrated in example 15.4.

• **Example 15.4:** Given a sequence where $S(n) = \dfrac{3}{2n-1}$, find $S(5)$, $S(7)$, and $S(25)$.

Solution: If
$$S(n) = \dfrac{3}{2n-1},$$
then
$$S(5) = \dfrac{3}{2(5)-1} = \dfrac{3}{10-1} = \dfrac{3}{9},$$
$$S(7) = \dfrac{3}{2(7)-1} = \dfrac{3}{14-1} = \dfrac{3}{13},$$
and
$$S(25) = \dfrac{3}{2(25)-1} = \dfrac{3}{50-1} = \dfrac{3}{49}.$$

Recursion Formula

A formula that uses one or more numbers to find succeeding numbers is called a *recursion formula*. A sequence can be defined with a recursion formula. When using a recursion formula to designate a sequence, the first term of the sequence is usually given. Then the formula determines the a_n term by using the previous, or a_{n-1}, term. Example 15.5 illustrates the use of a recursion formula to designate a sequence.

• **Example 15.5:** If the first term of a sequence is $a_1 = 1$, and if $a_n = 3a_{n-1}$ for $n > 1$, find a_2, a_3, and a_4.

Solution: Since $a_n = 3a_{n-1}$, for $n > 1$, then
$$a_2 = 3a_{2-1} = 3a_1 = (3)(1) = 3,$$
$$a_3 = 3a_{3-1} = 3a_2 = (3)(3) = 9,$$
and
$$a_4 = 3a_{4-1} = 3a_3 = (3)(9) = 27.$$

Hence, the first four terms of the sequence are 1, 3, 9, and 27.

Example 15.6 illustrates how a recursion formula uses the previous *two* terms to find the next term of a sequence.

- *Example 15.6:* If $a_1 = 1$ and $a_2 = 2$, and if $a_n = a_{n-2} + a_{n-1}$, for $n > 2$, find a_3, a_4, and a_5.

    Solution: Since $a_n = a_{n-2} + a_{n-1}$, for $n > 2$, then
    $$a_3 = a_{3-2} + a_{3-1},$$
    so
    $$a_3 = a_1 + a_2$$
    or
    $$a_3 = (1) + (2) = 3,$$
    and
    $$a_4 = a_3 + a_2$$
    so
    $$a_4 = (3) + (2) = 5,$$
    and
    $$a_5 = a_4 + a_3$$
    so
    $$a_5 = (5) + (3) = 8.$$
    Hence, the first five terms of the sequence are 1, 2, 3, 5, 8.

Trial Problems 15.1

Before you begin the section exercises, warm up with these problems. Complete answers are included in the answer key.

1. Find the next two terms of the sequence 2, 4, 6, 8, ...
2. Find the next two terms of the sequence 2, 4, 8, 16, ...
3. Find the next three terms of the sequence 1, -2, 4, -8, ...
4. Find an expression for the nth term of the sequence 1, 5, 9, 13, 17, ...
5. If $a_1 = 1$ and $a_n = 3a_{n-1}$, find a_2, a_3, and a_4.

Exercises 15.1

Find the next three terms of the given sequences.

1. 3, 5, 7, 9, . . .
2. 1, 3, 5, 7, . . .
3. 1, 4, 7, 10, . . .
4. 2, 5, 8, 11, . . .
5. 3, 6, 9, 12, . . .
6. 0, 3, 6, 9, . . .
7. −1, 2, 5, 8, . . .
8. −1, 3, 7, 11, . . .
9. 2, 5, 10, 17, . . .
10. 0, 3, 8, 15, . . .
11. $\dfrac{3}{1}, \dfrac{3}{3}, \dfrac{3}{5}, \dfrac{3}{7}, \ldots$
12. $\dfrac{2}{3}, \dfrac{2}{5}, \dfrac{2}{7}, \dfrac{2}{9}, \ldots$
13. $x, x^3, x^5, x^7, \ldots$
14. $x^3, x^5, x^7, x^9, \ldots$
15. $\dfrac{1}{2}, \dfrac{2}{3}, \dfrac{3}{4}, \dfrac{4}{5}, \ldots$
16. $\dfrac{2}{1}, \dfrac{3}{2}, \dfrac{4}{3}, \dfrac{5}{4}, \ldots$

Find an expression for the nth term in the following sequences.

17. 3, 5, 7, 9, . . .
18. 1, 3, 5, 7, . . .
19. 1, 4, 7, 10, . . .
20. 2, 5, 8, 11, . . .
21. 3, 6, 9, 12, . . .
22. 0, 3, 6, 9, . . .
23. 0, 2, 6, 12, . . .
24. 2, 6, 12, 20, . . .
25. 2, 5, 10, 17, . . .
26. 0, 3, 8, 15, . . .
27. $\dfrac{3}{1}, \dfrac{3}{3}, \dfrac{3}{5}, \dfrac{3}{7}, \ldots$
28. $\dfrac{2}{3}, \dfrac{2}{5}, \dfrac{2}{7}, \dfrac{2}{9}, \ldots$
29. $x, x^3, x^5, x^7, \ldots$
30. $x^3, x^5, x^7, x^9, \ldots$
31. $\dfrac{1}{2}, \dfrac{2}{3}, \dfrac{3}{4}, \dfrac{4}{5}, \ldots$
32. $\dfrac{2}{1}, \dfrac{3}{2}, \dfrac{4}{3}, \dfrac{5}{4}, \ldots$

Use the given recursion formula to find a_2, a_3, and a_4 in each of the following.

33. $a_1 = 1, a_n = 2a_{n-1}$
34. $a_1 = 1, a_n = 4a_{n-1}$
35. $a_1 = 2, a_n = 3 + a_{n-1}$
36. $a_1 = 3, a_n = 2 + a_{n-1}$
37. $a_1 = -1, a_n = -1a_{n-1}$
38. $a_1 = -2, a_n = -1a_{n-1}$

39. An object dropped from an aircraft falls in such a manner that it drops 24 feet the first second. During each succeeding second it falls 36 feet more than it did during the preceding second. Write a sequence representing the distance the object falls during each of the first five seconds.

40. Suppose that you save $20.00 one week and that each week thereafter you save $1.50 more. Write a sequence to represent the money saved each week for the first six weeks.

41. A pyramid of boxes has 28 boxes in the bottom row and three fewer boxes in each row thereafter. Write a sequence that represents the number of boxes in each of the first four rows.

42. A certain investment pays 9% interest compounded annually. Write a sequence to represent the value of $3000 investment at the end of each of the first five years.

15·2 Series

Series

Associated with any sequence is a series. A **series** is defined as the sum of the terms of the sequence and is denoted by S_n. For example, associated with the sequence

$$4, 7, 10, \ldots, 3n + 1 \tag{15.1}$$

is the series

$$S_n = 4 + 7 + 10 + \cdots + (3n + 1). \tag{15.2}$$

Associated with the sequence

$$y, y^2, y^3, y^4, \ldots, y^n$$

is the series

$$S_n = y + y^2 + y^3 + y^4 + \cdots + y^n.$$

If we know the general term of a sequence, we can express the series in a convenient form by using the symbol Σ (sigma or summation) together with the general term. For example, the series in equation (15.2) can be written

$$S_n = \sum_{k=1}^{n} (3k + 1).$$

Expanded Form

If we successively replace k with the numbers 1, 2, 3, ..., we obtain the terms of the series in **expanded form.** Thus,

$$S_5 = \sum_{k=1}^{5} (3k + 1)$$

in expanded form becomes

$$[3(1) + 1] + [3(2) + 1] + [3(3) + 1] + [3(4) + 1] + [3(5) + 1]$$

or $\quad S_5 = \quad 4 \quad + \quad 7 \quad + \quad 10 \quad + \quad 13 \quad + \quad 16.$

Index of Summation

The variable used in conjunction with summation notation is called the **index of summation.** In the example, k was the index of summation. The expression $k = 1$, below the summation symbol, means that the first value to be used for k is 1. The number above the summation symbol, in this case 5, means the last value to be used for k is 5. Other letters, such as $i, j, l, m, n,$ and the like may also be used as the index of summation. Consider examples 15.7 and 15.8.

• **Example 15.7:** Find the expanded form for

$$S_4 = \sum_{i=1}^{4} (2i + 1).$$

Solution: $\quad S_4 = [2(1) + 1] + [2(2) + 1] + [2(3) + 1] + [2(4) + 1]$
$\qquad\qquad = \quad 3 \quad + \quad 5 \quad + \quad 7 \quad + \quad 9.$

• **Example 15.8:** Find the expanded form for

$$\sum_{m=3}^{7} (2^m - 1).$$

Solution: The index of summation does not have to begin with 1. In this example the index, m, goes from 3 to 7. Therefore,

$$\sum_{m=3}^{7} (2^m - 1) = (2^3 - 1) + (2^4 - 1) + (2^5 - 1) + (2^6 - 1) + (2^7 - 1)$$
$$= 7 + 15 + 31 + 63 + 127.$$

In examples 15.7 and 15.8 we were given a series in summation notation and were asked to find the expanded form of the series. If we are given the expanded form of the series, we can also express the series in summation notation. Consider examples 15.9 and 15.10.

• **Example 15.9:** Write the series

$$5 + 8 + 11 + 14$$

in summation notation.

Solution: First we must find the *n*th term or general term. The following analysis should help us to find the general term.

The first term is 5 or $3(1) + 2$,

the second term is 8 or $3(2) + 2$,

the third term is 11 or $3(3) + 2$,

the fourth term is 14 or $3(4) + 2$.

Therefore, the *n*th term, or general term, is $3n + 2$. Thus, the series, in summation notation is

$$\sum_{n=1}^{4} (3n + 2).$$

• **Example 15.10:** Write the series

$$3 + 5 + 7 + 9$$

in summation notation.

Solution 1: First we must find an expression for the *n*th, or general, term.

The first term is 3 or $2(1) + 1$,

the second term is 5 or $2(2) + 1$,

the third term is 7 or $2(3) + 1$, and

the fourth term is 9 or $2(4) + 1$.

Therefore the *n*th term, or general term can be expressed as $2n + 1$.

Thus, the series, in summation notation, may be

$$\sum_{n=1}^{4} (2n + 1).$$

Solution 2: You might also note that

the first term is 3 or $2(0) + 3$,

the second term is 5 or $2(1) + 3$,

the third term is 7 or $2(2) + 3$, and

the fourth term is 9 or $2(3) + 3$.

Thus, the series, in summation notation, may be

$$\sum_{n=0}^{3} (2n + 3).$$

Trial Problems 15.2

Before you begin the section exercises, warm up with these problems. Complete answers are included in the answer key.

Write in expanded form.

1. $\sum_{k=1}^{3} (2k - 1)$
2. $\sum_{n=0}^{3} (n^2 + n)$
3. $\sum_{n=3}^{5} (n + 4)$

Express each series in summation notation. There may be alternate forms for some answers.

4. $2 + 4 + 6 + 8$
5. $1 + 5 + 9 + 13$

Exercises 15.2

Write the following in expanded form.

1. $\sum_{k=1}^{4} (3k - 1)$
2. $\sum_{k=1}^{4} (2k + 1)$
3. $\sum_{n=0}^{3} (2^n + 1)$
4. $\sum_{n=0}^{3} (3^n - 1)$
5. $\sum_{n=3}^{7} (n^2 - 3)$
6. $\sum_{n=2}^{6} (n^2 - 2)$
7. $\sum_{n=1}^{5} \frac{(-1)^n}{3n}$
8. $\sum_{n=1}^{5} \frac{(-1)^n}{4^n}$

Express each series in summation notation. There may be alternate forms for some answers.

9. $3 + 5 + 7 + 9$
10. $1 + 3 + 5 + 7$
11. $1 + 4 + 7 + 10 + 13$
12. $2 + 5 + 8 + 11 + 14$
13. $-1 + 2 + 5 + 8$
14. $-1 + 3 + 7 + 11$
15. $\frac{2}{1} + \frac{2}{3} + \frac{2}{5} + \frac{2}{7}$
16. $\frac{3}{3} + \frac{3}{5} + \frac{3}{7} + \frac{3}{9}$
17. $x^3 + x^5 + x^7 + x^9$
18. $x + x^2 + x^3 + x^4 + x^5$
19. $1 + y^2 + y^4 + y^6 + y^8$
20. $1 + \frac{1}{x} + \frac{1}{x^2} + \frac{1}{x^3}$

21. An auditorium has a total of 15 rows with 60 seats in the back row. Each row thereafter has two fewer seats than the row behind it. Express as a sequence, the number of seats in each of the back 5 rows.

22. A taxi company charges $1.00 for the first $\frac{1}{4}$ mile and 50¢ for each $\frac{1}{4}$ mile thereafter. Express as a sequence the fare for each of the first six quarter miles.

23. A carpenter is going to shingle a small roof in the shape of an isosceles triangle. The roof requires eight rows of shingles. The bottom row uses 15 shingles, and each row thereafter uses two fewer shingles. Express as a sequence the number of shingles in the eight rows.

24. A professional scuba diver charges $3.00 per minute to work at a depth of 20 feet. For each additional 10 feet of depth, the fee is increased by $1.50 per minute. What is the fee per minute to work at a depth of 80 feet?

15 • 3 Arithmetic Progressions

Arithmetic Progression

An **arithmetic progression** (abbreviated A.P.) is a sequence that can be determined by the equation

$$a_n = a_1 + (n - 1)d, \qquad (15.3)$$

Common Difference

Arithmetic Sequence

where a_1 is the first term of the sequence, d is the **common difference** between any term and the preceding term, and n is the number of terms in the sequence. Simply stated, this means that we get from one term to the succeeding term by adding the value of d. An arithmetic progression is sometimes called an **arithmetic sequence.** For example, the following sequences are arithmetic progressions.

SECTION 15.3 ARITHMETIC PROGRESSIONS

1, 3, 5, 7, 9, 11, . . .	(Add 2 to each term)
2, 4, 6, 8, 10, 12, . . .	(Add 2 to each term)
0, 3, 6, 9, 12, 15, . . .	(Add 3 to each term)
1, 5, 9, 13, 17, 21, . . .	(Add 4 to each term)
$-3, -1, 1, 3, 5, 7, \ldots$	(Add 2 to each term)
$-4, -2, 0, 2, 4, 6, \ldots$	(Add 2 to each term)

Since equation (15.3) determines an arithmetic progression, the equation can be manipulated to find any one of the variables in the equation, provided that the other variables are known. The examples 15.11, 15.12, 15.13, and 15.14 illustrate how to find a specific term, the common difference, the number of terms, or the first term when the other parts of the formula are known.

• **Example 15.11:** Find the twelfth term of the A.P. if the first five terms are 5, 8, 11, 14, and 17.

Solution: We know that $a_1 = 5$, $d = 3$, and $n = 12$. Therefore, the general equation

$$a_n = a_1 + (n - 1)d$$

becomes

$$a_{12} = 5 + (12 - 1)3$$
$$a_{12} = 5 + (11)3$$
$$a_{12} = 5 + 33.$$

Hence,

$$a_{12} = 38.$$

• **Example 15.12:** Find the common difference d of an A.P. if the first term is 3, and the fifth term is 19.

Solution: We know that $a_1 = 3$, and $a_5 = 19$ when $n = 5$. Therefore, the general equation

$$a_n = a_1 + (n - 1)d$$

becomes

$$19 = 3 + (5 - 1)d$$
$$19 = 3 + (4)d$$
$$16 = 4d.$$

Hence,

$$4 = d.$$

• **Example 15.13:** How many terms are there in an A.P. if the first term is -2, the nth term is 23, and the common difference is 5?

Solution: We know that $a_1 = -2$, $a_n = 23$, and $d = 5$. Therefore the general equation

$$a_n = a_1 + (n-1)d$$

becomes

$$23 = -2 + (n-1)5$$
$$23 = -2 + 5n - 5$$
$$23 = -7 + 5n$$
$$30 = 5n.$$

Hence,

$$6 = n.$$

• **Example 15.14:** Find the first term in an A.P. if $a_5 = -3$ and $d = -2$.

Solution: Substituting these values into the general equation

$$a_n = a_1 + (n-1)d$$

we get

$$-3 = a_1 + (5-1)(-2)$$
$$-3 = a_1 + 4(-2)$$
$$-3 = a_1 - 8.$$

Hence,

$$5 = a_1.$$

Frequently it is necessary to find the sum of a specified number of terms of an A.P. The formula for the sum of the first n terms of an A.P. is

$$S_n = n\frac{(a_1 + a_n)}{2}. \qquad (15.4)$$

Formula (15.4) states that the sum of consecutive terms of an A.P. is obtained by multiplying the average of the first and last terms considered by the number of terms being added. Consider examples 15.15 and 15.16.

• **Example 15.15:** Find the sum of the first 50 positive even integers.

Solution: Since $a_1 = 2$, $a_{50} = 100$, and $n = 50$, equation (15.4) becomes

$$S_{50} = \frac{50(2 + 100)}{2}$$

$$S_{50} = \frac{50(102)}{2}$$

$$S_{50} = 50(51)$$

$$S_{50} = 2550.$$

• *Example 15.16:* Two hundred oil well pipes are stacked so that there are 20 pipes in the bottom row, 19 pipes in the next row, and so on. How many rows of pipes are there?

Solution: The number of pipes in each row of the pile forms an A.P. whose first few terms are 20, 19, 18, 17, ... Thus, $a_1 = 20$ and $d = -1$. Let n be the number of rows of pipes in the pile. Then

$$S_n = 200,$$

and

$$a_n = 20 + (n - 1)(-1)$$

or

$$a_n = 21 - n.$$

Hence

$$S_n = \frac{n(a_1 + a_n)}{2}$$

becomes

$$200 = \frac{n(20 + 21 - n)}{2}$$

$$200 = \frac{n(41 - n)}{2}$$

$$400 = n(41 - n)$$

$$400 = 41n - n^2$$

$$n^2 - 41n + 400 = 0$$

$$(n - 16)(n - 25) = 0.$$

Therefore, $n = 16$ or $n = 25$. But 25 does not make sense, since there are not more than 20 rows of pipes. Hence, $n = 16$ is the answer.

Trial Problems 15.3

Before you begin the section exercises, warm up with these problems. Complete answers are included in the answer key.

Find the indicated value in each A.P.

1. If $a_1 = 3$ and $d = 2$, find a_5.
2. If $a_1 = 2$ and $d = -3$, find a_4.
3. If $a_1 = 3$ and $a_5 = 11$, find d.
4. If $a_6 = 11$ and $d = 2$, find a_1.
5. If $a_1 = 2$ and $a_5 = 13$, find S_5.

Exercises 15.3

Find the indicated value in each A.P.

1. If $a_1 = 2$ and $d = -3$, find a_7.
2. If $a_1 = 2$ and $d = 3$, find a_7.
3. If $a_1 = 3$ and $a_6 = 13$, find d.
4. If $a_1 = 3$ and $a_5 = 11$, find d.
5. If $a_5 = 18$ and $d = 5$, find a_1.
6. If $a_7 = 28$ and $d = 5$, find a_1.
7. If $a_1 = 5$ and $a_5 = 19$, find S_5.
8. If $a_1 = 7$ and $a_5 = 23$, find S_5.

Solve.

9. A carpenter was hired to build 192 window frames. The first day he made five frames, and each day thereafter he made two more frames than he had made the day before. How many days did it take him to finish the job?

10. How many feet of material are needed to make the rungs of a ladder if the ladder has 12 equally spaced rungs and the bottom rung is 18 inches long when the top rung is 12 inches long?

11. A body falling in a vacuum near the surface of the earth will fall 16 feet the first second. Thereafter, each second that the body falls, it falls 32 feet farther than it did the second before. How far does a body fall in ten seconds?

12. The number of bacteria in a culture on the first day was 24,000. The number of bacteria in the same culture on the fifth day was 50,000. What was the daily rate of increase if the rate is assumed to be constant?

13. The population of a city was 39,000 five years ago. Each year the city planners project a population increase of 540. What is the population in five years?

14. An employee earns a salary of $15,000 per year. If the salary is increased $1000 per year, what is the salary during the tenth year?

15·4 Geometric Progressions

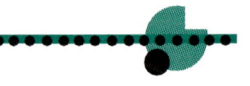

Geometric Progression

A **geometric progression** (abbreviated G.P.) is a sequence a_n that can be determined by the equation

$$a_n = a_1 r^{n-1}, \tag{15.5}$$

where a_1 is the first term of the sequence, r is any nonzero number and n is the number of terms in the sequence.

From equation (15.5) we see that $a_{n+1} = a_1 r^n$, hence

$$\frac{a_{n+1}}{a_n} = \frac{a_1 r^n}{a_1 r^{n-1}} = r.$$

SECTION 15.4 GEOMETRIC PROGRESSIONS

Common Ratio

Thus, the *quotient* of two consecutive terms of a G.P. is the constant value r. This value is often called the **common ratio**. Remember, the difference between two consecutive terms of an A.P. was the constant value d, called the common difference. The following sequences are examples of geometric progressions.

$$1, 2, 4, 8, 16, 32, \ldots \quad \text{(Multiply each term by 2)}$$
$$1, 3, 9, 27, 81, 243, \ldots \quad \text{(Multiply each term by 3)}$$
$$1, \frac{1}{2}, \frac{1}{4}, \frac{1}{8}, \frac{1}{16}, \frac{1}{32}, \ldots \quad \text{(Multiply each term by } \frac{1}{2}\text{)}$$
$$2, \frac{4}{3}, \frac{8}{9}, \frac{16}{27}, \frac{32}{81}, \frac{64}{243}, \ldots \quad \text{(Multiply each term by } \frac{2}{3}\text{)}$$
$$3, -3, 3, -3, 3, -3, \ldots \quad \text{(Multiply each term by } -1\text{)}$$
$$3, 6, 12, 24, 48, 96, \ldots \quad \text{(Multiply each term by 2)}$$

Formula (15.5) has four variables: a_n, a_1, r, and n. If any three of these values are known, the remaining value can be determined. Examples 15.17 through 15.21 illustrate how this might be done.

• **Example 15.17:** Find the fifth term of a G.P. if the first term is $\frac{1}{3}$ and the common ratio is $\frac{1}{2}$.

Solution: We know that $a_1 = \frac{1}{3}$, $r = \frac{1}{2}$, and $n = 5$. Therefore, the general equation

$$a_n = a_1 r^{n-1}$$

becomes

$$a_5 = \left(\frac{1}{3}\right)\left(\frac{1}{2}\right)^{5-1}$$

$$a_5 = \left(\frac{1}{3}\right)\left(\frac{1}{2}\right)^4$$

$$a_5 = \left(\frac{1}{3}\right)\left(\frac{1}{16}\right).$$

Hence,

$$a_5 = \frac{1}{48}.$$

- **Example 15.18:** The population of a country is 5.1 million today and the population is growing at 3% per year. What is the population of the country in 24 years?

  Solution: The population figures represent a G.P. A growth of 3% per year, when added to the previous year, and expressed as a decimal, makes the value of $r = 1.03$. The value of $a_1 = 5.1$ million, and $n = 24$. Substituting these values in formula (15.5), gives

  $$a_n = a_1 r^{(n-1)}$$
  $$a_{24} = (5.1)(1.03)^{24-1}$$
  $$a_{24} = (5.1)(1.03)^{23} \quad \text{(Use a calculator)}$$
  $$a_{24} = (5.1)(1.9735865)$$
  $$a_{24} = 10.07 \text{ million.} \quad \text{(Approximately)}$$

- **Example 15.19:** Find the common ratio of a G.P. if the first term is 3 and the fifth term is 48.

  Solution: We know that $a_1 = 3$, and $a_5 = 48$ when $n = 5$. Therefore, the general equation

  $$a_n = a_1 r^{n-1}$$

  becomes

  $$48 = (3)r^{5-1}$$
  $$48 = 3r^4$$
  $$16 = r^4.$$

  Hence,

  $$r = \pm 2.$$

  If $r = 2$, the sequence is 3, 6, 12, 24, 48. If $r = -2$, the sequence is 3, -6, 12, -24, 48.

- **Example 15.20:** How many terms are there in a G.P. if the first term is 4, the nth term is 64, and the common ratio is 2?

  Solution: We know that $a_1 = 4$, $a_n = 64$, and $r = 2$. Therefore, the general equation

  $$a_n = a_1 r^{n-1}$$

  becomes

  $$64 = (4)(2)^{n-1}$$
  $$16 = 2^{n-1}$$
  $$2^4 = 2^{n-1}.$$

Therefore,
$$n - 1 = 4.$$
Hence,
$$n = 5.$$

• *Example 15.21:* Find the first term of a G.P. if $a_5 = -8$, and $r = -2$.

Solution: Substituting these values into the general equation
$$a_n = a_1 r^{n-1}$$
we get
$$-8 = a_1(-2)^{5-1}$$
$$-8 = a_1(-2)^4$$
$$-8 = 16a_1.$$
Therefore,
$$-\frac{1}{2} = a_1.$$

Frequently it is necessary to find the sum of a specified number of terms of a G.P. The formula for the sum of the first n terms of a G.P. is
$$S_n = a_1 \frac{(1-r^n)}{1-r}. \tag{15.6}$$

• *Example 15.22:* Find the sum of the first four terms of a G.P. if $a_1 = 3$ and $r = 5$.

Solution: Substituting the given values in equation (15.6), we get
$$S_4 = \frac{3(1-5^4)}{1-5}$$
$$S_4 = \frac{3(1-625)}{-4}$$
$$S_4 = \frac{3(-624)}{-4}$$
$$S_4 = 3(156)$$
$$S_4 = 468.$$

• **Example 15.23:** You have two parents, four grandparents, eight great-grandparents, and so on. How many ancestors do you have if you go back ten generations?

Solution: We need to find the sum of the terms of a G.P. in which $a_1 = 2$, $r = 2$, and $n = 10$. Substituting these values in equation (15.6) we get

$$S_{10} = \frac{2(1 - 2^{10})}{1 - 2}$$

$$S_{10} = \frac{2(1 - 1024)}{-1}$$

$$S_{10} = \frac{2(-1023)}{-1}$$

$$S_{10} = 2046.$$

Trial Problems 15.4

Before you begin the section exercises, warm up with these problems. Complete answers are included in the answer key.

Find the indicated value in each G.P.

1. If $a_1 = 3$ and $r = 2$, find a_4.

2. If $a_1 = 2$ and $r = -\frac{1}{2}$, find a_4.

3. If $a_1 = \frac{2}{3}$ and $a_5 = \frac{27}{8}$, find r.

4. If $a_5 = 16$ and $r = 2$, find a_1.

5. If $a_1 = 4$ and $a_4 = 32$, find S_4.

Exercises 15.4

Find the indicated value in each G.P.

1. If $a_1 = 2$ and $r = 3$, find a_5.
2. If $a_1 = 3$ and $r = 2$, find a_5.
3. If $a_1 = 1$ and $a_6 = 243$, find r.
4. If $a_1 = 1$ and $a_6 = 32$, find r.
5. If $a_5 = 16$ and $r = 2$, find a_1.
6. If $a_5 = 48$ and $r = 2$, find a_1.
7. If $a_1 = 1$ and $a_5 = 16$, find S_5.
8. If $a_1 = 1$ and $a_5 = 81$, find S_5.

Solve.

9. A certain type of bacteria doubles its population every 20 minutes. Assuming that no bacteria die, how many bacteria will there be after 2 hours if there are 1 million bacteria now?

10. A certain type of insect doubles its population every 8 hours. Assuming that no insects die, how many insects will there be after 2 days if there are 1 million now?

11. The **half-life** of a radioactive substance is defined to be the time required for half of any given amount to change its form. Given 20 grams of a substance having a half-life of 2 days, how much is unchanged after 30 days?

12. Given 30 grams of a substance having a half-life of 3 days, how much of the substance is unchanged after 30 days? See exercise 11 for the definition of half-life.

13. The final step in processing a photographic print is to immerse the print in a chemical called a "fixer." The print is then washed in running water. Under normal conditions 95% of the fixer on the print is removed with 15 minutes of washing. How much of the original fixer remains after one hour of washing?

14. A farmer has 10,000 pounds of potatoes in storage. If he sells them today he receives 10 cents per pound. He anticipates a price increase of 0.1 cents per pound for each of the next six months. However, the potatoes lose 2% of their weight per month. If he sells the potatoes in six months, how much does he get for them?

15·5 Infinite Series

The indicated sum of all the terms of an infinite sequence, such as

$$a_1 + a_2 + a_3 + \cdots + a_n + \cdots$$

Infinite Series is called an **infinite series**. Using the sigma notation, we can write such a series more compactly as

$$\sum_{k=1}^{\infty} a_k.$$

Although we cannot literally add an infinite number of terms, it is sometimes convenient to assign a numerical value as the "sum" of an infinite series. This is accomplished by using the concept of a **limit**, which is studied in calculus.

Limit

We attempt to get a feeling for the concept of the sum of an infinite series by considering an infinite geometric series.

Consider the infinite geometric series of the form

$$\sum_{k=1}^{\infty} a_1 r^{k-1} = a_1 + a_1 r + a_1 r^2 + \cdots + a_1 r^{n-1} + \cdots,$$

where $-1 < r < 1$.

Equation (15.6) in section 15.4 indicates that the sum of the first n terms of this series is

$$\sum_{k=1}^{n} a_1 r^{k-1} = a_1 + a_1 r + a_1 r^2 + \cdots + a_1 r^{n-1}$$

$$= a_1 \frac{(1-r^n)}{1-r}.$$

Since $-1 < r < 1$, it follows that r^n gets smaller and smaller as n gets larger and larger. Therefore, as n gets larger and larger, r^n gets closer and closer to zero and $a_1 \dfrac{(1 - r^n)}{1 - r}$ gets closer and closer to $a_1 \dfrac{(1 - 0)}{1 - r} = \dfrac{a_1}{1 - r}$. That is, as we add more and more terms of such a geometric series, the sum gets closer to $\dfrac{a_1}{1 - r}$.

Therefore, we define the sum of such an infinite geometric series to be

$$S = \frac{a_1}{1 - r}. \tag{15.7}$$

The examples 15.24, 15.25, and 15.26 illustrate this concept.

• **Example 15.24:** A force is applied to a particle that moves in a straight line. Each second the particle moves only one-half of the distance it moved during the preceding second. If the particle moves 5 centimeters the first second, approximately how far does it move before coming to rest?

Solution: The particle moves 5 cm the first second, $\dfrac{5}{2}$ cm the second second, $\dfrac{5}{4}$ cm the third second, and so on. These values are the first three terms of the infinite geometric series

$$5 + \frac{5}{2} + \frac{5}{4} + \cdots$$

in which $a_1 = 5$ and $r = \dfrac{1}{2}$. Using equation (15.7) we have

$$S = \frac{a}{1 - r}$$

$$= \frac{5}{1 - \dfrac{1}{2}}$$

$$= \frac{5}{\dfrac{1}{2}}$$

$$S = 10.$$

Hence, the particle moves 10 cm.

• **Example 15.25:** The arc length through which a pendulum bob moves is nine-tenths of its preceding arc length. How far does the pendulum bob move before coming to rest if the first arc length is 25 cm?

Solution: The arc lengths form the terms of an infinite geometric series with $a_1 = 25$ and $r = 0.9$. Thus, using equation (15.7) we have

$$S = \frac{a_1}{1-r}$$

$$= \frac{25}{1-0.9}$$

$$= \frac{25}{0.1}$$

$$= 250.$$

Hence, the bob moves 250 cm.

• **Example 15.26:** Find a rational number that is equal to the repeating decimal 0.454545. . . .

Solution: The repeating decimal can be rewritten as a series as

$$\frac{45}{100} + \frac{45}{10,000} + \frac{45}{1,000,000} + \cdots.$$

This is an infinite geometric series with $a_1 = \frac{45}{100}$ and $r = \frac{1}{100}$. Using equation (15.7) we have

$$S = \frac{a_1}{1-r}$$

$$= \frac{\frac{45}{100}}{1 - \frac{1}{100}}$$

$$= \frac{\frac{45}{100}}{\frac{99}{100}}$$

$$= \frac{45}{99}.$$

Hence $\frac{45}{99} = 0.454545\ldots$ or $\frac{5}{11} = 0.454545\ldots$

Trial Problems 15.5

Before you begin the section exercises, warm up with these problems. Complete answers are included in the answer key.

Find the sum of the infinite geometric series.

1. $1 + \dfrac{2}{3} + \dfrac{4}{9} + \cdots$

2. $3 + \dfrac{3}{2} + \dfrac{3}{4} + \cdots$

3. $\dfrac{4}{3} + \dfrac{2}{3} + \dfrac{1}{3} + \dfrac{1}{6} + \cdots$

4. $0.2 + 0.02 + 0.002 + \cdots$

5. Find a rational number that is equal to the repeating decimal $0.353535\ldots$

Exercises 15.5

Find the sum of each infinite geometric series.

1. $\dfrac{9}{2} + 3 + 2 + \cdots$

2. $3 + 2 + \dfrac{4}{3} + \cdots$

3. $12 + 6 + 3 + \cdots$

4. $8 + 4 + 2 + \cdots$

5. $1 + \dfrac{2}{3} + \dfrac{4}{9} + \cdots$

6. $1 + \dfrac{3}{4} + \dfrac{9}{16} + \cdots$

7. $\dfrac{3}{4} - \dfrac{1}{2} + \dfrac{1}{3} + \cdots$

8. $\dfrac{2}{3} - \dfrac{1}{2} + \dfrac{3}{8} + \cdots$

9. $\dfrac{1}{81} - \dfrac{2}{243} + \dfrac{4}{729} - \cdots$

10. $\dfrac{1}{49} - \dfrac{1}{56} + \dfrac{1}{64} + \cdots$

Find a rational number equal to each repeating decimal.

11. $0.333\ldots$

12. $0.666\ldots$

13. $0.313131\ldots$

14. $0.272727\ldots$

15. $2.410410\ldots$

16. $3.027027\ldots$

17. $0.12888\ldots$

18. $0.81333\ldots$

Solve.

19. A golf ball is dropped from a height of 6 feet. It returns two-thirds of its previous height on each bounce. What is the total distance the ball travels before coming to rest?

20. A ball is dropped from a height of 250 centimeters. On each rebound it rises $\dfrac{1}{2}$ the height from which it fell. Find the total distance the ball travels before coming to rest.

21. A substance initially weighing 64 grams is decaying at a rate such that after 8 hours there are only 32 grams remaining. In another 4 hours there are only 16 grams left, and after 2 more hours only 8 grams remain. How much time passes before there is none of the substance remaining?

22. A pendulum is swinging such that each swing is 60% as long as the previous swing. If the first swing of the pendulum is 30 cm long, how far does the pendulum travel before coming to rest?

23. A moving object is decelerating in such a manner that it travels 90.0 meters during the first second, 81.0 meters during the next second, 72.9 during the third, and so on. How far does the object travel before coming to rest?

24. An object oscillating up and down at the end of a spring has an initial oscillation of 40.0 mm. Each successive oscillation is 30% of the previous one. What is the total distance the object travels before coming to rest?

15•6 The Binomial Theorem

Occasionally it is necessary to represent the product of consecutive positive integers. To do this, we introduce the symbol, $n!$ (read "n factorial" or "factorial n"), which is defined by

$$n! = n(n-1)(n-2)(n-3)\ldots(2)(1).$$

Thus,

$$5! = (5)(4)(3)(2)(1),$$

and

$$7! = (7)(6)(5)(4)(3)(2)(1).$$

Factorial In words, the **factorial** of a positive integer n is the product of n and all smaller positive integers.

Factorial notation can also be used to represent products of any collection of consecutive positive integers. This is illustrated in example 15.27.

• *Example 15.27:* Represent $(8)(7)(6)(5)$ in factorial notation.

Solution: $(8)(7)(6)(5) = \dfrac{(8)(7)(6)(5)(4)(3)(2)(1)}{(4)(3)(2)(1)} = \dfrac{8!}{4!}$

Since

$$n! = n(n-1)(n-2)(n-3)\ldots(5)(4)(3)(2)(1)$$

and since

$$(n-1)! = (n-1)(n-2)(n-3)\ldots(5)(4)(3)(2)(1)$$

we see that $n!$ can be expressed by the recursive formula

$$n! = n(n-1)! \qquad \text{where } n > 1. \tag{15.8}$$

Thus

$$7! = (7)6!$$

and

$$35! = (35)34!.$$

If we let $n = 1$, equation (15.8) becomes

$$n! = n(n-1)!$$

or

$$1! = 1(1-1)!$$
$$1! = (1)0!.$$

Therefore, for consistency, we must define

$$0! = 1.$$

Binomial Coefficients

Using factorial notation, we introduce special symbols that we ultimately use as **binomial coefficients.** If n and j are integers with $0 \le j \le n$, the symbol $\binom{n}{j}$ is defined as follows:

$$\binom{n}{j} = \frac{n!}{j!(n-j)!}. \tag{15.9}$$

• **Example 15.28:** Evaluate a. $\binom{5}{2}$, b. $\binom{7}{6}$, c. $\binom{n}{0}$, and d. $\binom{n}{n}$, assuming that n is a nonnegative integer.

Solution:

a. $\binom{5}{2} = \dfrac{5!}{2!(5-2)!} = \dfrac{5!}{2!3!} = \dfrac{(5)(4)(3)(2)(1)}{(2)(1)(3)(2)(1)} = \dfrac{(5)4}{2} = 10.$

b. $\binom{7}{6} = \dfrac{7!}{6!(7-6)!} = \dfrac{7!}{6!1!} = \dfrac{7(6!)}{6!} = 7.$

c. $\binom{n}{0} = \dfrac{n!}{0!(n-0)!} = \dfrac{n!}{0!n!} = \dfrac{1}{0!} = \dfrac{1}{1} = 1.$

d. $\binom{n}{n} = \dfrac{n!}{n!(n-n)!} = \dfrac{n!}{n!0!} = \dfrac{1}{0!} = \dfrac{1}{1} = 1.$

● ● ● ● ● ● ● ● ● ●

• **Example 15.29:** If k and j are integers with $0 \le j \le k$, show that

$$\binom{k}{j-1} + \binom{k}{j} = \binom{k+1}{j}.$$

Solution: $\binom{k}{j-1} + \binom{k}{j}$

$= \dfrac{k!}{(j-1)![k-(j-1)]!} + \dfrac{k!}{j!(k-j)!}$

$= \dfrac{k!}{(j-1)!(k-j+1)!} + \dfrac{k!}{j!(k-j)!}$

$= \dfrac{jk!}{j(j-1)!(k-j+1)!} + \dfrac{(k-j+1)k!}{(k-j+1)j!(k-j)!}$

$= \dfrac{jk!}{j!(k-j+1)!} + \dfrac{(k+1)k! - jk!}{(k-j+1)!j!}$

$= \dfrac{(k+1)k!}{j!(k-j+1)!}$

$= \dfrac{(k+1)!}{j![(k+1)-j]!}$

$= \binom{k+1}{j}.$

● ● ● ● ● ● ● ● ● ●

With the ideas that we have just developed, we are now ready to establish a general formula for the expansion of $(a + b)^n$, where n is an arbitrary positive integer. The expression $a + b$ is a binomial, so the expansion of $(a + b)^n$ is called the **binomial expansion** or **binomial theorem**.

Binomial Expansion
Binomial Theorem

Direct calculation gives us the expansion of $(a + b)^n$ for small values of n. For instance,

$$(a + b)^0 = 1$$
$$(a + b)^1 = a + b$$
$$(a + b)^2 = a^2 + 2ab + b^2$$
$$(a + b)^3 = a^3 + 3a^2b + 3ab^2 + b^3$$
$$(a + b)^4 = a^4 + 4a^3b + 6a^2b^2 + 4ab^3 + b^4$$
$$(a + b)^5 = a^5 + 5a^4b + 10a^3b^2 + 10a^2b^3 + 5ab^4 + b^5$$

By arranging the binomial coefficients in a triangle as shown in figure 15.1,

```
              1
            1   1
          1   2   1
        1   3   3   1
      1   4   6   4   1
    1   5  10  10   5   1
    :   :   :   :   :   :
    :   :   :   :   :   :
```

Figure 15.1

Pascal Triangle

we obtain the **Pascal triangle**, named in honor of the French mathematician, Blaise Pascal (1623–1662). Notice that each number in the Pascal triangle (other than the first and last in each row, which is equal to 1) can be obtained by adding the two numbers diagonally above it. Thus, we could continue to write rows of the Pascal triangle as is illustrated in example 15.30.

• *Example 15.30:* Find the entries in the seventh horizontal row of the Pascal triangle.

Solution: The seventh row is obtained from the sixth row by the method just described, that is,

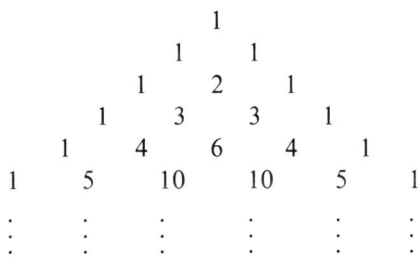

It can be shown that the Pascal triangle represented in figure 15.1 can be expressed using the notation for binomial coefficients developed in formula (15.9). Figure 15.2 illustrates the Pascal triangle in this notation.

$$\binom{0}{0}$$

$$\binom{1}{0} \quad \binom{1}{1}$$

$$\binom{2}{0} \quad \binom{2}{1} \quad \binom{2}{2}$$

$$\binom{3}{0} \quad \binom{3}{1} \quad \binom{3}{2} \quad \binom{3}{3}$$

$$\binom{4}{0} \quad \binom{4}{1} \quad \binom{4}{2} \quad \binom{4}{3} \quad \binom{4}{4}$$

$$\binom{5}{0} \quad \binom{5}{1} \quad \binom{5}{2} \quad \binom{5}{3} \quad \binom{5}{4} \quad \binom{5}{5}$$

$$\vdots \quad \vdots \quad \vdots \quad \vdots \quad \vdots \quad \vdots$$

Figure 15.2

Where there are a sufficient number of terms in a binomial expansion, we can observe that:

1. There are $n + 1$ terms, beginning with a^n and ending with b^n.

2. As we move from each term to the next, the powers of a decrease by 1 and the powers of b increase by 1. Thus, the sum of the exponents of a and b is n.

3. Successive terms of the binomial expansion can be written in the form

$$\binom{n}{j} a^{n-j} b^j,$$

for $j = 0, 1, 2, \ldots, n$.

We can generalize these observations to obtain an expression for the expansion of a binomial where a and b are any two numbers and n is a positive integer.

$$(a + b)^n = \binom{n}{0} a^n b^0 + \binom{n}{1} a^{n-1} b^1 + \binom{n}{2} a^{n-2} b^2 + \cdots$$
$$+ \binom{n}{j} a^{n-j} b^j + \cdots + \binom{n}{n} a^0 b^n.$$

This binomial expansion is often called the *binomial theorem*. Notice that b^0 in the first term of the expansion and a^0 in the last term can be omitted, since they are both equal to 1. They are placed there to emphasize the pattern. Consider examples 15.31 through 15.35.

SECTION 15.6 THE BINOMIAL THEOREM

• *Example 15.31:* Use the binomial theorem and Pascal's triangle to expand $(2x + y)^5$.

Solution:
$$(2x + y)^5 = \binom{5}{0}(2x)^5(y)^0 + \binom{5}{1}(2x)^4(y)^1 + \binom{5}{2}(2x)^3(y)^2$$
$$+ \binom{5}{3}(2x)^2(y)^3 + \binom{5}{4}(2x)^1(y)^4 + \binom{5}{5}(2x)^0(y)^5$$
$$= 1(32x^5) + 5(16x^4)y + 10(8x^3)y^2$$
$$+ 10(4x^2)y^3 + 5(2x)y^4 + 1(1)y^5$$
$$= 32x^5 + 80x^4y + 80x^3y^2 + 40x^2y^3 + 10xy^4 + y^5.$$

• *Example 15.32:* Use the binomial theorem to expand $(a + 3b)^4$.

Solution:
$$(a + 3b)^4 = \binom{4}{0}(a)^4(3b)^0 + \binom{4}{1}(a)^3(3b)^1 + \binom{4}{2}(a)^2(3b)^2$$
$$+ \binom{4}{3}(a)^1(3b)^3 + \binom{4}{4}(a)^0(3b)^4$$
$$= 1(a^4) + 4(a^3)(3b) + 6(a^2)(9b^2)$$
$$+ 6(a)(27b^3) + 1(a^0)(81b^4)$$
$$= a^4 + 12a^3b + 54a^2b^2 + 162ab^3 + 81b^4.$$

• *Example 15.33:* Find the fourth term of $(s^2 - t)^7$.

Solution: The fourth term of the general binomial expansion is

$$\binom{n}{3}a^{n-3}b^3.$$

Thus, the fourth term of the binomial $(s^2 - t)^7$ is

$$\binom{7}{3}(s^2)^{7-3}(-t)^3$$

or

$$\frac{7!}{3!(7-3)!}(s^2)^4(-t)^3$$

or

$$\frac{(7)(6)(5)4!}{3!4!}s^8(-t)^3$$

or

$$-35s^8t^3.$$

CHAPTER 15 SEQUENCES AND SERIES

• **Example 15.34:** Find the third term of $(x^2 - 5y^3)^8$.

Solution: The third term of the general binomial expression is

$$\binom{n}{2} a^{n-2} b^2.$$

Thus, the third term of $(x^2 + (-5y^3))^8$ is

$$\binom{8}{2}(x^2)^{8-2}(-5y^3)^2$$

or

$$\binom{8}{2}(x^2)^6(-5y^3)^2$$

or

$$(28)(x^{12})(25y^6)$$

or

$$700 x^{12} y^6.$$

• **Example 15.35:** Find the term involving x^3 in $(\sqrt{x} + \sqrt{y})^8$.

Solution: The term involving x^3 must come from $(\sqrt{x})^6$. Thus, the term must be of the form

$$\binom{8}{2}(\sqrt{x})^6(\sqrt{y})^2.$$

Hence, the term is $28x^3 y$.

Trial Problems 15.6

Before you begin the section exercises, warm up with these problems. Complete answers are included in the answer key.

Find the numerical value of each expression.

1. $5!$
2. $\dfrac{9!}{7!}$
3. $\dfrac{6!\,4!}{3!\,2!}$

Write the first three factors in the expansion of each of the following.

4. $m!$
5. $(3k)!$

Exercises 15.6

Find the numerical value of each expression.

1. $7!$
2. $8!$
3. $\dfrac{7!}{5!3!}$
4. $\dfrac{8!}{6!4!}$
5. $\dfrac{8!7!}{5!4!}$
6. $\dfrac{5!4!}{3!2!}$
7. $\binom{7}{3}$
8. $\binom{7}{4}$
9. $\binom{12}{2}$
10. $\binom{12}{10}$
11. $\binom{5}{5}$
12. $\binom{6}{6}$

Write the first three factors of the expansion of each of the following.

13. $n!$
14. $k!$
15. $(2n)!$
16. $(3k)!$
17. $2(n)!$
18. $3(k)!$
19. $(n+1)!$
20. $(k+1)!$
21. $(n-2)!$
22. $(k-2)!$

23. Find the eighth horizontal row of Pascal's triangle.
24. Find the ninth horizontal row of Pascal's triangle.

Use the binomial theorem and Pascal's triangle to expand each binomial expression.

25. $(2a+b)^4$
26. $(a+3b)^4$
27. $(3a-2b)^5$
28. $(2a-3b)^5$
29. $(c+2)^6$
30. $(d+2)^5$
31. $\left(\dfrac{x}{2}-3y\right)^4$
32. $\left(3x-\dfrac{2}{y}\right)^4$

Find and simplify the specified term in the binomial expansion of the indicated expression.

33. The third term of $(r-s)^5$
34. The fourth term of $(s-r)^5$
35. The fifth term of $\left(2x^2 - \dfrac{y^3}{4}\right)^8$
36. The fifth term of $\left(\dfrac{x^3}{4} - 2y^2\right)^8$
37. The term involving y^3 in $(x+y)^5$
38. The term involving x^3 in $(x+y)^5$
39. The term involving a^6 in $(a+2b)^{10}$
40. The term involving b^4 in $(a+2b)^{10}$
41. The term involving x^2 in $(\sqrt{x}+\sqrt{y})^7$
42. The term involving y^2 in $(\sqrt{x}+\sqrt{y})^5$

43. Show that Pascal's triangle is symmetric about a vertical line through its apex by showing that for integers n and j, where $0 \leq j \leq n$,

$$\binom{n}{j} = \binom{n}{n-j}.$$

44. Let S be the probability of success for a particular event and F be the probability of failure for the same event and assume that $S = F$. If the event is performed seven times, the expression

$$(S + F)^7$$

can be used to determine the probability of a particular number of successes. The various coefficients of the terms of the binomial expansion give the number of chances out of 128 that an event will be a success. For example, the coefficient of the S^3F^4 term gives the chances in 128 of succeeding three times and failing four times. Expand the binomial and find the coefficient of the S^3F^4 term.

45. Approximate $(0.99)^8$ using the first four terms of the binomial expansion of $(1 - 0.01)^8$. Then compute $(0.99)^8$ on your calculator and compare the results.

15·7 Summary of Terms, Formulas, Rules, and Procedures

Terms

Arithmetic Progression (p. 592)
Arithmetic Sequence (p. 592)
Binomial Coefficients (p. 606)
Binomial Expansion (p. 607)
Binomial Theorem (p. 607)
Common Difference (p. 592)
Common Ratio (p. 597)
Expanded Form (p. 589)

Factorial (p. 605)
Finite Sequence (p. 584)
General Term (p. 585)
Geometric Progression (p. 596)
Half-Life (p. 601)
Index of Summation (p. 589)
Infinite Sequence (p. 584)

Infinite Series (p. 601)
Limit (p. 601)
nth Term (p. 585)
Pascal Triangle (p. 607)
Recursion Formula (p. 586)
Sequence (p. 584)
Series (p. 589)

Formulas

Formula for the parts of an arithmetic progression: An *arithmetic progression* (abbreviated A.P.) is a sequence that can be determined by the equation

$$a_n = a_1 + (n - 1)d,$$

where a_1 is the first term of the sequence, d is the *common difference* between any term and the preceding term, and n is the number of terms in the sequence. (15.3)

Formula for the sum of the first n terms of an arithmetic progression: $S_n = n\dfrac{(a_1 + a_n)}{2}$. (15.3)

Formula for the parts of a geometric progression: A *geometric progression* (abbreviated G.P.) is a sequence a_n that can be determined by the equation

$$a_n = a_1 r^{n-1},$$

where a_1 is the first term of the sequence, r is any number and n is the number of terms in the sequence. (15.4)

Formula for the sum of the first n terms of a geometric progression: $S_n = a_1 \dfrac{(1 - r^n)}{1 - r}$. (15.4)

Formula for the sum of the terms of an infinite geometric series: $S = \dfrac{a_1}{1 - r}$. (15.5)

Formula defining $n!$:
$n! = n(n - 1)(n - 2)(n - 3) \ldots (2)(1)$. (15.6)

Recursive formula defining $n!$: $n! = n(n - 1)!$ where $n > 1$. (15.6)

Formula defining binomial coefficients: $\binom{n}{j} = \dfrac{n!}{j!(n - j)!}$. (15.6)

Binomial theorem: $(a + b)^n = \binom{n}{0}a^n b^0 + \binom{n}{1}a^{n-1}b^1 + \binom{n}{2}a^{n-2}b^2 + \cdots + \binom{n}{j}a^{n-j}b^j + \cdots + \binom{n}{n}a^0 b^n$. (15.6)

Rules and Procedures

- An infinite series can be represented by the expression

$$\sum_{k=1}^{\infty} a_k = a_1 + a_2 + a_3 + \cdots . \qquad (15.5)$$

- Zero factorial is defined as

$$0! = 1 \qquad (15.6)$$

15·8 Chapter 15 Review Exercises

Find the next three terms of the given sequences.

1. $5, 7, 9, \ldots$
2. $4, 6, 8, \ldots$
3. $-1, 2, -4, \ldots$
4. $-1, 3, -9, \ldots$

Find an expression for the nth term in each of the following sequences.

5. $5, 7, 9, \ldots$
6. $4, 6, 8, \ldots$
7. $-1, 2, -4, \ldots$
8. $-1, 3, -9, \ldots$

Use the given recursion formula to find a_2, a_3, and a_4 in each of the following.

9. $a_1 = 1, a_n = 2a_{n-1}$
10. $a_1 = 1, a_n = 3a_{n-1}$
11. $a_1 = 2, a_n = 4 + a_{n-1}$
12. $a_1 = 3, a_n = 2 + a_{n-1}$

Write the following in expanded form.

13. $\sum_{j=1}^{4} (3j - 1)$
14. $\sum_{j=1}^{4} (2j + 1)$
15. $\sum_{n=0}^{3} (2^n - 1)$
16. $\sum_{n=0}^{3} (4^n - 1)$

Express each series in summation notation. Answers may vary.

17. $5 + 7 + 9 + 11$
18. $4 + 6 + 8 + 10$
19. $x + x^3 + x^5 + x^7$
20. $x + x^4 + x^7 + x^{10}$

Find the indicated value in each A.P.

21. If $a_1 = 2$ and $d = 3$, find a_6.
22. If $a_1 = 3$ and $d = 2$, find a_5.
23. If $a_5 = 18$ and $d = 5$, find a_1.
24. If $a_7 = 28$ and $d = 5$, find a_1.
25. If $a_1 = 5$ and $a_5 = 19$, find S_5.
26. If $a_1 = 7$ and $a_5 = 23$, find S_5.

Find the indicated value in each G.P.

27. If $a_1 = 2$ and $r = 3$, find a_5.
28. If $a_1 = 3$ and $r = 2$, find a_5.
29. If $a_1 = 1$ and $a_6 = 32$, find r.
30. If $a_1 = 1$ and $a_6 = 243$, find r.
31. If $a_5 = 16$ and $r = 2$, find a_1.
32. If $a_5 = 48$ and $r = 2$, find a_1.
33. If $a_1 = 1$ and $a_5 = 16$, find S_5.
34. If $a_1 = 1$ and $a_5 = 81$, find S_5.

Find the sum of each infinite geometric series.

35. $3 + 2 + \dfrac{4}{3} + \cdots$

36. $4 + 3 + \dfrac{9}{4} + \cdots$

37. $18 + 9 + \dfrac{9}{2} + \cdots$

38. $27 + 9 + 3 + \cdots$

Find the numerical value of each expression.

39. $9!$

40. $6!$

41. $\dfrac{9!}{6!}$

42. $\dfrac{6!}{9!}$

43. $\binom{10}{4}$

44. $\binom{10}{6}$

45. $\binom{8}{2}$

46. $\binom{8}{6}$

Write the first three factors of the expansion of each of the following.

47. $(k - 1)!$

48. $(n - 1)!$

49. $(2k + 1)!$

50. $(2n + 1)!$

Expand the following.

51. $(a + 2b)^4$

52. $(3a + b)^5$

53. $\left(\dfrac{1}{a} - \dfrac{1}{b}\right)^5$

54. $\left(\dfrac{1}{x} - \dfrac{1}{y}\right)^4$

55. Find the third term of $(2x - y)^5$.

56. Find the third term of $(x - 2y)^4$.

57. Find the term involving x^2 in $(\sqrt{x} + y)^5$.

58. Find the term involving y^2 in $(x - \sqrt{y})^5$.

59. An insect population is growing in such a way that each generation is 1.5 times as numerous as the previous generation. If there are 1000 insects in the first generation, how many are there in the fifth generation?

60. A sky diver falls 10 meters during the first second, 20 meters during the next second, 30 meters during the third second, and so on. How many meters does the diver fall during the eighth second?

15·9 Chapter 15 Test

1. Find the next three terms of the sequence $1, 5, 9, 13, \ldots$.

2. Find the next three terms of the sequence $\dfrac{1}{3}, \dfrac{2}{3}, \dfrac{4}{3}, \dfrac{8}{3}, \ldots$.

3. Find an expression for the nth term of the sequence $1, 4, 7, 10, \ldots$.

4. If the first term of a sequence is $a_1 = 3$, and if $a_n = 4a_{n-1}$, find $a_2, a_3,$ and a_4.

5. Find the expanded form for $\sum\limits_{i=1}^{4} (2i + 1)$.

6. Find the expanded form for $\sum\limits_{k=2}^{5} (k^2 - k)$.

7. Express $2 + 5 + 8 + 11 + 14$ in summation notation.

8. Express $1 + \dfrac{1}{y} + \dfrac{1}{y^2} + \dfrac{1}{y^3}$ in summation notation.

9. Find the eighth term of an A.P. if the first five terms are 4, 9, 14, 19, and 24.

10. How many terms are there in an A.P. if the first term is -3, the nth term is 21, and the common difference is 4?

11. Find the common difference d in an A.P. if the first term is -3 and the fifth term is 17.

12. How many terms are there in an A.P. if the first term is 7, the nth term is -18, and the common difference is -5?

13. Find the sum of the first 100 positive integers.

14. Find the fourth term of a G.P. if the first term is $\frac{1}{2}$ and the common ratio is $\frac{1}{3}$.

15. Find the common ratio in a G.P. if the first term is 4 and the fifth term is 64.

16. How many terms are there in a G.P. if the first term is 2, the nth term is 162, and the common ratio is 3?

17. Find the sum of the first four terms of a G.P. if $a_1 = 5$ and $r = 3$.

18. If the fifth term in a G.P. is 256 and r is 4, find a_1.

19. Find the sum of the infinite geometric series
$$9 + \frac{9}{2} + \frac{9}{4} + \cdots.$$

20. Find the sum of the infinite geometric series $.2 + 0.02 + 0.002 + \cdots$.

21. Evaluate $7!$.

22. Evaluate $\binom{7}{2}$.

23. Use the binomial theorem to expand $(x + 2y)^5$.

24. Find the third term of $(2x - 5y^2)^7$.

25. Find the term involving y^3 in $(\sqrt{x} + 2\sqrt{y})^8$.

15·10 Cumulative Review—Chapters 14 and 15

Use synthetic division to find the quotient and the remainder.

1. $\dfrac{x^4 - 3x^2 + 4x - 6}{x - 2}$

2. $\dfrac{3x^3 - 5x^2 + 6x + 4}{x + 1}$

Find the rational zeros of the polynomials.

3. $x^3 + 3x^2 - 4$

4. $4x^3 - 13x + 6$

Find all of the zeros for each polynomial.

5. $x^3 + 2x^2 - 9x + 6$

6. $8x^3 - 12x^2 + 6x - 1$

7. $2x^3 + x^2 - 9x + 4$

8. $6x^4 + 5x^3 + 16x^2 - 4x - 3$

9. To the nearest tenth, approximate the zero of the polynomial $x^3 - 4x^2 - 5x + 20$ that is between 2 and 3.

10. To the nearest hundredth, approximate the zero of the polynomial $x^4 + x^3 - 6x^2 - 7x - 7$ that is between -3 and -2.

Find an expression for the nth term in each sequence.

11. $3, 6, 9, \ldots$

12. $-2, 0, 2, \ldots$

Use the recursion formula to find a_2, a_3, and a_4.

13. $a_1 = 2$, $a_n = 3a_{n-1}$

14. $a_1 = 1$, $a_n = 3 + 2a_{n-1}$

Express each series in summation notation. Answers may vary.

15. $3 + 7 + 11 + 15$

16. $x + x^5 + x^9 + x^{13}$

Write in expanded form.

17. $\sum_{j=1}^{4} (j - 2)$

18. $\sum_{n=0}^{3} (3^n + 1)$

Find the indicated value in each arithmetic progression.

19. If $a_1 = 4$ and $d = 3$, find a_5.

20. If $a_5 = 10$ and $d = 3$, find a_1.

Find the indicated value in each geometric progression.

21. If $a_1 = 1$ and $r = 4$, find a_4.

22. If $a_1 = 1$ and $a_4 = 27$, find S_4.

Find the sum of each infinite series.

23. $36 + 9 + \dfrac{9}{4} + \cdots$

24. $6 + 3 + \dfrac{3}{2} + \cdots$

Find the numerical value of each expression.

25. $\dfrac{8!}{4!2!}$

26. $\dbinom{14}{3}$

Write the first three factors of the expansion.

27. $(2k - 3)!$

28. $(n + 3)!$

Expand each expression.

29. $(2a - b)^5$

30. $\left(\dfrac{2}{x} + y\right)^4$

31. Find the third term of $(3x - 2y)^5$.

32. Find the term involving x^2 in $(x - 2y)^6$.

Chapter 16

Analytic Geometry

Before the seventeenth century algebra and geometry were commonly treated as being independent of one another. A French mathematician, René Descartes, published a work called *La Geometrie* in 1637 in which he introduced the idea of a coordinate system as a device for unifying algebra and geometry.

Coordinate systems make it possible to represent algebraic equations as graphs and to apply algebraic methods to the study of geometry. Descartes provided an extraordinary breakthrough in mathematical thought, since analytic methods today enter into a wide variety of theoretical and applied mathematics.

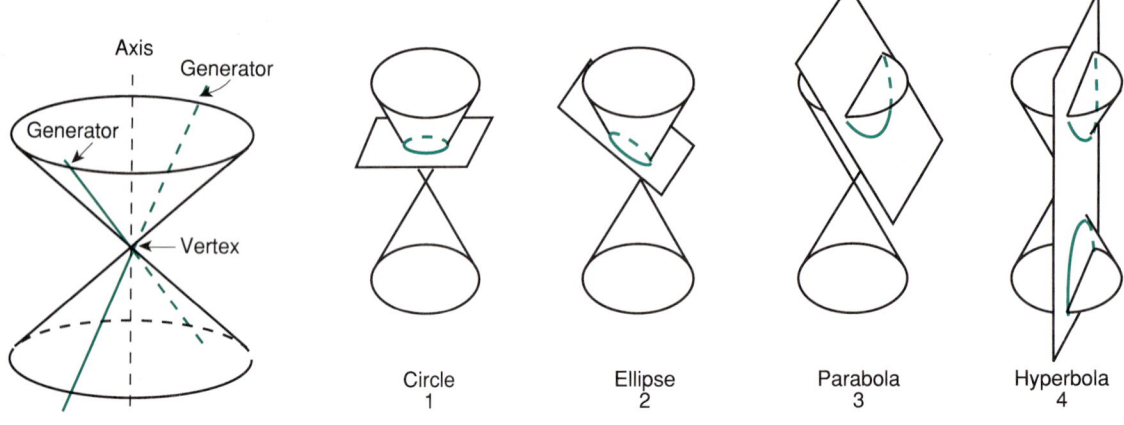

Figure 16.1

Figure 16.2

16 • 1 Introduction

In previous chapters, we considered the graphs of various types of functions. In this chapter, we consider the graphs of linear functions and the graphs of the general second degree equation; that is, an equation of the form

$$Ax^2 + Bxy + Cy^2 + Dx + Ey + F = 0,$$

where A, B, C, D, E, and F are real numbers and not all of A, B, and C are zeros. For different values of A, B, and C, the graph of the equation can be any of four curves called a *circle*, an *ellipse*, a *parabola* or a *hyperbola*. (In the event $A = B = C = 0$, the equation becomes a line.) These four curves are called **conic sections**, since they can be obtained by passing a plane through a cone. A cone can be defined as the surface generated by rotating a line about an axis. This is illustrated in figure 16.1.

The following conditions show how each of the four conic sections is obtained from a cone. They are illustrated in figure 16.2.

Conic Sections

CONIC SECTIONS

1. If the plane is perpendicular to the axis and does not contain the vertex, the intersection is a *circle*.

2. If the plane is not parallel to the line that generated the cone, cuts only the top or bottom section of the cone, and does not contain the vertex, the intersection is an *ellipse*.

3. If the plane is parallel to the line that generated the cone and does not contain the vertex, the intersection is a *parabola*.

4. If the plane cuts both the top and bottom sections of the cone, is parallel to the axis, and does not contain the vertex, the intersection is a *hyperbola*.

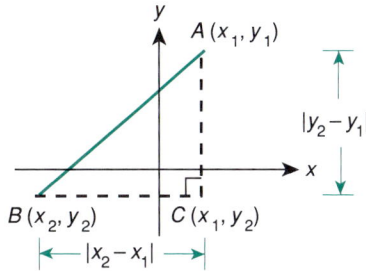

Figure 16.3

Note that if the plane contains the vertex or the axis, the intersection could be a point, a line, or a pair of lines. These cases are usually called *degenerate cases* and do not qualify to be conic sections.

16 • 2 Linear Equations

In previous chapters it was shown that the graph of an equation of the form $ax + by + c = 0$ is a line in the coordinate plane. In this section we review the use of the distance formula and introduce some alternate forms of linear equations. The distance formula is derived using figure 16.3.

Let $A(x_1, y_1)$ and $B(x_2, y_2)$ be arbitrary points in a coordinate plane. Point $C(x_1, y_2)$ is a vertex that makes triangle ABC a right triangle with the right angle at C. Use the Pythagorean relationship on the triangle:

$$(AB)^2 = (AC)^2 + (BC)^2$$
$$= |y_2 - y_1|^2 + |x_2 - x_1|^2$$
$$= (y_2 - y_1)^2 + (x_2 - x_1)^2.$$

Taking the square root of both sides of the equation yields the following formula.

DISTANCE FORMULA

If $A(x_1, y_1)$ and $B(x_2, y_2)$ are points in a coordinate plane, then the distance between the points is

$$AB = \sqrt{(y_2 - y_1)^2 + (x_2 - x_1)^2}.$$

For example, if $A(-4, -2)$ and $B(2, 6)$ are the given points, let $x_1 = -4$, $y_1 = -2$, $x_2 = 2$, and $y_2 = 6$. Substitute these values into the distance formula:

$$AB = \sqrt{(6 - (-2))^2 + (2 - (-4))^2}$$
$$= \sqrt{(8)^2 + (6)^2}$$
$$= \sqrt{64 + 36}$$
$$= \sqrt{100}$$
$$= 10.$$

This formula becomes very important in the analytic formulation of the equations for the conic sections.

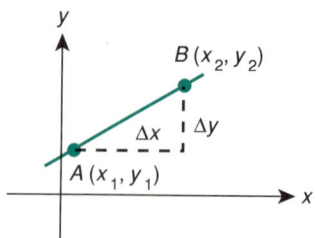

Figure 16.4

Given a line in the coordinate plane containing the points $A(x_1,y_1)$ and $B(x_2,y_2)$, as shown in figure 16.4, the **slope** of the line is defined as follows.

Slope

SLOPE OF A LINE

m (slope) = $\dfrac{\text{change in } y}{\text{change in } x}$, from point A to point B. Let Δy and Δx represent the changes in y and x. Then

$$m = \frac{\Delta y}{\Delta x} = \frac{y_2 - y_1}{x_2 - x_1}, \qquad x_2 \neq x_1.$$

Note that $\dfrac{y_2 - y_1}{x_2 - x_1} = \dfrac{-(y_2 - y_1)}{-(x_2 - x_1)} = \dfrac{y_1 - y_2}{x_1 - x_2}$, so it makes no difference which point is called (x_1,y_1) and which point is called (x_2,y_2).

Figure 16.5 shows several lines with different slopes. Note that for a line with a negative slope, the value of y decreases as x increases. For a line with a positive slope, the value of y increases as x increases. A horizontal line has slope zero, and the slope of a vertical line is undefined.

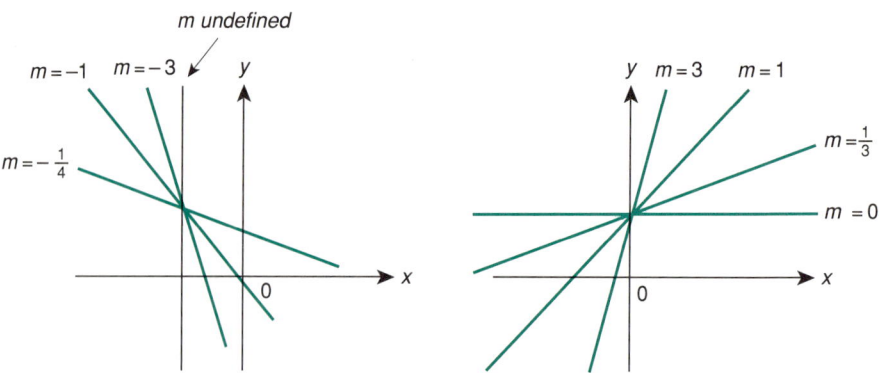

Figure 16.5

- **Example 16.1:** Given that the points $P(800,0)$, $Q(720,120)$, $R(0,1200)$ and $S(1200,-600)$ are on the line $3x + 2y = 2400$, use the pairs P and Q, Q and R, and R and S to show that the slope of the line is constant.

  Solution: Using P and Q:
  $$m = \frac{120 - 0}{720 - 800} = \frac{120}{-80} = -\frac{3}{2}.$$

  Using Q and R:
  $$m = \frac{1200 - 120}{0 - 720} = \frac{1080}{-720} = -\frac{3}{2}.$$

  Using R and S:
  $$m = \frac{-600 - 1200}{1200 - 0} = \frac{-1800}{1200} = -\frac{3}{2}.$$

  Since the slope $\frac{\Delta y}{\Delta x} = -\frac{3}{2}$ can be written as $\frac{-3}{+2}$ or as $\frac{+3}{-2}$, y changes by -3 units for a $+2$ unit change in x, or y changes by $+3$ units for a -2 unit change in x.

  If the equation of the line in example 16.1 is solved for y, the following equation is obtained:
  $$3x + 2y = 2400$$
  $$2y = -3x + 2400$$
  $$y = -\frac{3}{2}x + 1200.$$

  Note that the slope, $-\frac{3}{2}$ is the coefficient of x and that the line crosses the y-axis at $x = 0$, $y = 1200$. The number $b = 1200$ is called the y-intercept of the line.

SLOPE-INTERCEPT FORM OF A LINE

In general, a line written in the form $y = mx + b$ is called the slope-intercept form of the line, where m is the slope and b is the y-intercept of the line.

- **Example 16.2:** A car with a purchase price of $10,500 *depreciates* at the rate of $1200 per year for 8 years. Write an equation for the value of the car y in terms of the time t for $0 \leq t \leq 8$ (see figure 16.6).

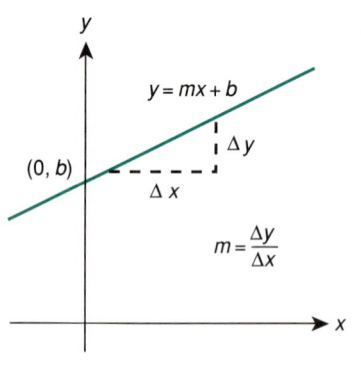

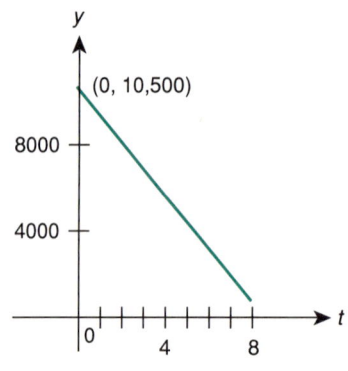

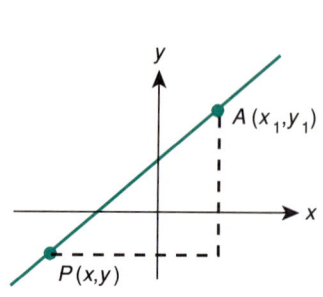

Figure 16.6

Figure 16.7

Solution: The change in y for each unit increase in the age of the car is -1200. The initial cost of the car ($t = 0$) is $10,500$. Thus, the point $(0, 10500)$ is on the line and the slope is -1200. Thus, $m = -1200$, $b = 10,500$ and the equation of the line is

$$y = -1200t + 10,500, \qquad 0 \leq t \leq 8.$$

Suppose the slope (m) and a point $A(x_1, y_1)$ on a line are given. Let $P(x, y)$ represent any other point on the line in figure 16.7. The formula for the slope must hold for any choice of P.

$$m = \frac{y - y_1}{x - x_1}.$$

This equation can be arranged into the following form.

POINT-SLOPE FORM OF A LINE

$y - y_1 = m(x - x_1)$, where m is the slope of the line and $P(x_1, y_1)$ is any point on the line.

• **Example 16.3:** Write an equation, in the form $ax + by = c$, for the line that contains the point $Q(-3, 2)$ and that has slope $-\frac{1}{2}$.

Solution: Use the equation
$$y - y_1 = m(x - x_1)$$
with $m = -\dfrac{1}{2}$, $x_1 = -3$, and $y_1 = 2$:
$$y - 2 = -\dfrac{1}{2}(x - (-3)).$$

Rewrite the equation in the form $ax + by = c$

$2y - 4 = -1(x + 3)$ (Multiply both sides by 2)
$x + 2y = 1.$ (Add 4 and add x to both sides)

If $m = \dfrac{y_2 - y_1}{x_2 - x_1}$ is substituted for the slope, m, in the point-slope form of the equation for a line, the following two-point equation is obtained.

TWO-POINT FORM OF A LINE

$$y - y_1 = \left(\dfrac{y_2 - y_1}{x_2 - x_1}\right)(x - x_1),$$

where $A(x_1, y_1)$ and $B(x_2, y_2)$ are points on the line.

• *Example 16.4:* The relationship between the load (x) in pounds and the length (y) of a spring is a linear relationship. If the length of the spring is 11 inches with a load of 1 pound and the length is 13.4 inches with a load of 3 pounds, write an equation, in the form $y = mx + b$, that describes the relationship.

Solution: Write an equation for the line that contains the points $(1,11)$ and $(3,13.4)$. Let $x_1 = 1$, $y_1 = 11$, $x_2 = 3$, $y_2 = 13.4$ and use the two-point form of a line:

$$y - 11 = \left(\dfrac{13.4 - 11}{3 - 1}\right)(x - 1)$$
$$y - 11 = 1.2(x - 1).$$

Solve this equation for y:
$$y - 11 = 1.2x - 1.2$$
$$y = 1.2x + 9.8.$$

- **Example 16.5:** The pressure p in newtons per square meter at a point x meters below the surface of a body of water is a linear function of x. If the pressure at 10 meters is 10.913 newtons per square meter and the pressure at 20 meters is 20.813 newtons per square meter, write an equation for p in terms of x.

    Solution: Let $(x_1, p_1) = (10, 10.913)$, $(x_2, p_2) = (20, 20.813)$ be two points on the line that represents the function. Use the two-point formula:

    $$p - p_1 = \left(\frac{p_2 - p_1}{x_2 - x_1}\right)(x - x_1)$$

    $$p - 10.913 = \left(\frac{20.813 - 10.913}{20 - 10}\right)(x - 10)$$

    $$p - 10.913 = 0.99(x - 10)$$
    $$p - 10.913 = 0.99x - 9.9$$
    $$p = 0.99x + 1.013.$$

There are two special cases of lines to consider.

1. If the line is a horizontal line, all of the points on the line have the same y-coordinate. The slope $\frac{y_2 - y_1}{x_2 - x_1}$ has a numerator of zero. Thus the slope is zero, and the line can be represented as $y = k$, for some constant k.

2. If the line is a vertical line, all of the points on the line have the same x-coordinate. The slope $\frac{y_2 - y_1}{x_2 - x_1}$ is undefined since the denominator is zero. The equation for the line can be represented as $x = c$, for some constant c.

SUMMARY OF LINEAR EQUATIONS

Form of the Equation	Explanation
1. $ax + by + c = 0$	Standard form of a line
2. $y = mx + b$	Slope $= m$, y-intercept $= b$
3. $y - y_1 = m(x - x_1)$	Slope $= m$, $P(x_1, y_1)$ is a point on the line
4. $y - y_1 = \left(\frac{y_2 - y_1}{x_2 - x_1}\right)(x - x_1)$	$A(x_1, y_1)$ and $B(x_2, y_2)$ are points on the line
5. $y = k$	Line parallel to the x-axis
6. $x = c$	Line parallel to the y-axis

Trial Problems 16.2

Before you begin the section exercises, warm up with these problems. Complete answers are included in the answer key.

1. Find the distance between the points $(3,-1)$ and $(2,4)$.

Write an equation for the line that fits the given conditions.

2. Slope $= -\dfrac{3}{4}$, intercept $= -3$

3. Slope $= -2$, contains the point $(-3,2)$

4. Contains the points $(3,-6)$ and $(4,-2)$

5. Is parallel to the y-axis, contains the point $(4,3)$

Exercises 16.2

Find the distance between each pair of points.

1. $(5,5)$, $(2,6)$
2. $(4,2)$, $(8,1)$
3. $(5,12)$, $(-1,-3)$
4. $(6,2)$, $(-5,-3)$
5. $(5,2)$, (h,k)
6. $(6,-2)$, (h,k)

Write the equation of the line that has slope m and y-intercept b.

7. $m = \dfrac{1}{2}$, $b = 4$
8. $m = \dfrac{1}{3}$, $b = 2$
9. $m = -2$, $b = -6$
10. $m = -3$, $b = 5$
11. $m = -\dfrac{1}{2}$, $b = -3$
12. $m = -\dfrac{2}{3}$, $b = -2$

Write the equation, in the form $ax + by + c = 0$, for the line that has slope, m, and that contains the point with the given coordinates.

13. $m = 2$, $(3,5)$
14. $m = 3$, $(4,6)$
15. $m = \dfrac{1}{2}$, $(-3,2)$
16. $m = \dfrac{1}{3}$, $(-4,5)$
17. $m = -3$, $(-3,-2)$
18. $m = -2$, $(-5,-2)$
19. $m = -\dfrac{1}{3}$, $(4,-3)$
20. $m = -\dfrac{2}{3}$, $(3,-1)$

Write the equation of the line, in the form $ax + by + c = 0$, that contains the given pair of points.

21. $A(5,5)$, $B(2,6)$
22. $C(4,2)$, $D(8,1)$
23. $M(5,12)$, $N(-1,-3)$
24. $P(6,2)$, $Q(-5,-3)$
25. $E(0,600)$, $F(720,120)$
26. $G(800,0)$, $H(720,120)$

27. A machine that costs $40,000 depreciates linearly for 7 years to a value of $5000. Write an equation for the value y of the machine as a function of the time t.

28. A machine that costs $50,000 depreciates linearly for 8 years to a value of $18,000. Write an equation for the value y of the machine as a function of the time t.

29. The volume of a particular gas varies linearly with the temperature in degrees Celsius. If 530 ml of the gas at 45° is heated to 60°, the volume increases to 555 ml. Write an equation for the volume v in terms of the temperature t.

30. Solve exercise 29 if the gas occupies 680 ml at 75° and 480 ml at 15°.

31. An object is falling at a constant velocity, that is, the height h is a linear function of the time t. If the height is 3300 feet after 20 seconds and 2700 feet after 30 seconds, write an equation for the relationship.

32. Solve exercise 31 if the height is 3000 feet after 25 seconds and 2100 after 40 seconds.

33. If a frictional force is a linear function of the weight of an object, find an equation that relates a frictional force to the weight if an 8-lb block produces a frictional force of 2.8 lb and a 12-lb block produces a force of 4.2 lb.

34. Solve exercise 33 given that blocks weighing 6 lb and 10 lb produce frictional forces of 2.8 lb and 4.5 lb, respectively.

35. The R-value of insulation is a measure of its ability to resist heat transfer. The R-value is a linear function of the thickness of the insulation in inches. If 3.0 in. of fiberglass insulation yields an R-value of 9.4 and 8.0 in. produces an R-value of 25.4, find the R-value produced by 6.5 in. of the insulation.

36. Solve exercise 35 given that 3.0 in. and 8.0 in. of cellulose insulation produce R-values of 11.2 and 29.2, respectively.

37. The resistance in ohms for a 10 km length of wire used in transmission lines at different temperatures in degrees Celsius is approximately a linear relationship between resistance and temperature. At 0°, the resistance is 5.1 ohms, and at 60° the resistance is 5.9 ohms. Find the approximate resistance at 40°.

38. The solubility in grams per 100 grams of water for potassium chloride is an approximate linear function of the temperature in degrees Celsius for temperatures between 0° and 100°. If the solubility at 20° is 34.0 g and the solubility at 60° is 45.5 g, what is the approximate solubility at 35°?

39. Within limits the length of a copper rod is linearly related to the temperature in degrees Celsius. If a rod is 15.0 cm long at 10° and 15.04 cm long at 20°, write an equation relating the two variables.

40. Solve exercise 39 given that a steel pipe is 861.22 cm long at 40° and 861.95 cm long at 80°.

41. The current in a resistor is a linear function of the applied voltage. Find an equation that relates the current to the voltage if 5.0 volts produces 0.42 amps and 8.0 volts produces 0.66 amps.

42. Solve exercise 41 given that 6.0 volts produces 0.15 amps and 8.4 volts produces 0.21 amps.

43. The boiling point of water, in degrees Fahrenheit, is a linear function of the altitude (in thousands of feet above sea level). The line that shows this relationship has a slope of -1.8 and a y-intercept of 212. Find the boiling point at an altitude of 50,000 ft.

44. The boiling point of water, in degrees Celsius, is a linear function of the altitude (in thousands of meters above sea level). If the line that shows this relationship has a slope of -3.3 and a y-intercept of 100, find the boiling point at 15,000 meters.

45. A testing agency determined that, for a particular model of car, the mileage (in mi/gal) is approximately a linear function of the speed (in mi/hr) for speeds between 30 and 70 mi/hr. If a speed of 30 mi/hr produces 32.5 mi/gal and a speed of 60 mi/hr produces 25.0 mi/gal, what mileage could you expect at 50 mi/hr?

46. A testing agency has found that the expected mileage (in mi/gal) is linearly related to the engine displacement (in liters). If a 1.5 liter engine gets 28.0 mi/gal and a 3.7 liter engine gets 24.4 mi/gal, what mileage would you expect from a 2.7 liter engine?

A pair of lines $y = m_1x + b_1$ and $y = m_2x + b_2$ are parallel if $m_1 = m_2$. The lines are perpendicular if $m_1m_2 = -1$.

47. Write the equation of the line that contains the point $A(1,2)$ and that is parallel to the line $2x - 3y = 4$.

48. Write the equation of the line that contains the point $B(-2,3)$ and that is parallel to the line $4x + 5y = 20$.

49. Write the equation of the line that contains the point $C(-2,3)$ and that is perpendicular to the line $4x + 5y = 20$.

50. Write the equation of the line that contains the point $D(1,2)$ and that is perpendicular to the line $2x - 3y = 4$.

16·3 The Circle

In this section we consider the equation for a circle in the coordinate plane.

Circle   A **circle** (figure 16.8) can be defined as the set of points that are equidistant from a fixed point called the *center of the circle*. Let $C(h,k)$ be the center of the circle. Let $P(x,y)$ be any point on the circle, and let r be the radius of the circle. The distance formula can be used to obtain an equation for the circle:

$$r = CP = \sqrt{(x-h)^2 + (y-k)^2}.$$

Squaring both sides of the equation yields the following equation.

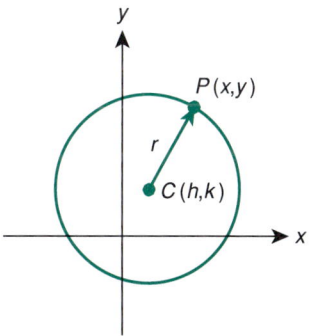

Figure 16.8

EQUATION OF A CIRCLE

$$(x - h)^2 + (y - k)^2 = r^2,$$

where (h,k) is the center of the circle, and r is the radius of the circle.

• **Example 16.6:** Find an equation for the circle that has center at the origin and radius 6.

Solution: Use the formula $(x - h)^2 + (y - k)^2 = r^2$, with $r = 6$, $h = 0$, and $k = 0$:

$$(x - 0)^2 + (y - 0)^2 = (6)^2$$
$$x^2 + y^2 = 36.$$

Suppose we wish to find an equation for a circle for which the coordinates of the center are known and the coordinates of a point on the circle are known. Example 16.7 illustrates the procedure.

• **Example 16.7:** Find an equation for a circle that has its center at $(3,-5)$ and that contains the point $(7,6)$.

Solution: First use the distance formula to find the radius and then use the standard formula for a circle.

1. The radius is the distance between the points $(3,-5)$ and $(7,6)$:

$$r = \sqrt{(7 - 3)^2 + (6 - (-5))^2} = \sqrt{16 + 121} = \sqrt{137}.$$

2. Substitute $h = 3$, $k = -5$, and $r = \sqrt{137}$ into the equation

$$(x - h)^2 + (y - k)^2 = r^2$$
$$(x - 3)^2 + (y - (-5))^2 = (\sqrt{137})^2$$
$$(x - 3)^2 + (y + 5)^2 = 137.$$

The coordinates of the center and the radius of a circle can be found by putting an equation for a circle into the standard form.

• **Example 16.8:** Find the center and radius of a circle that has the equation $x^2 + 8x + y^2 - 14y + 64 = 0$.

Solution: Put the equation in the standard form:

$$(x - h)^2 + (y - k)^2 = r^2.$$

To accomplish this it is necessary to complete the square for each of the expressions

$$x^2 + 2x \quad \text{and} \quad y^2 - 14y.$$

Recall from section 5.3 that an expression like

$$z^2 + bz + \frac{b^2}{4} = \left(z + \frac{b}{2}\right)^2$$

is a square trinomial. Thus, to complete the square on $z^2 + bz$, the quantity $\frac{b^2}{4}$ or $\left(\frac{b}{2}\right)^2$ must be added. Rewrite the equation of the circle as follows:

$$(x^2 + 8x + \underline{}) + (y^2 - 14y + \underline{}) = -64.$$

$\left(\frac{8}{2}\right)^2 = 16$ will complete the square for $x^2 + 8x$, and

$\left(\frac{-14}{2}\right)^2 = 49$ will complete the square for $y^2 - 14y$. We add 16 and 49 to both sides of the equation of the circle:

$$(x^2 + 8x + 16) + (y^2 - 14y + 49) = -64 + 16 + 49$$
$$(x + 4)^2 + (y - 7)^2 = 1.$$

Replacing 4 with $-(-4)$ in the first term yields the standard form for a circle:

$$(x - (-4))^2 + (y - 7)^2 = 1.$$

The center of the circle is $(-4, 7)$ and the radius is 1.

• **Example 16.9:** Find an equation for the circle that contains the points $A(0,3)$, $B(6,3)$, and $D(7,-4)$, as shown in figure 16.9.

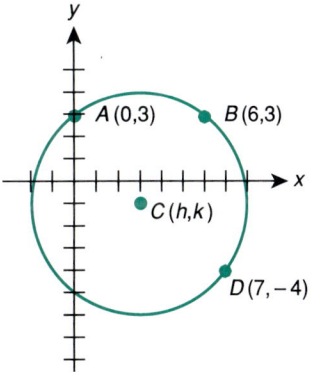

Figure 16.9

Solution: Solve the problem in three stages.

1. Use the fact that the distances of the points A, B, and D from the center $C(h,k)$ are equal to find the center.

2. Use any of the points and the center to find the radius using the distance formula.

3. Substitute the results of the first step into the formula for the standard form of a circle:

1. $(AC)^2 = (0 - h)^2 + (3 - k)^2$
   $(BC)^2 = (6 - h)^2 + (3 - k)^2$
   $(DC)^2 = (7 - h)^2 + (-4 - k)^2$.

   Since $(AC)^2 = (BC)^2$ and $(AC)^2 = (DC)^2$, we can substitute into these equations to obtain a pair of equations that contain h and k. Solving these equations gives us the coordinates of the center of the circle.
   Using $(AC)^2 = (BC)^2$,
   $$(0 - h)^2 + (3 - k)^2 = (6 - h)^2 + (3 - k)^2$$
   $$h^2 = 36 - 12h + h^2$$
   $$12h = 36$$
   $$h = 3.$$
   Using $(AC)^2 = (DC)^2$,
   $$(0 - h)^2 + (3 - k)^2 = (7 - h)^2 + (-4 - k)^2$$
   $$h^2 + 9 - 6k + k^2 = 49 - 14h + h^2 + 16 + 8k + k^2$$
   $$-6k + 9 = -14h + 8k + 65$$
   $$14h - 14k = 56$$
   $$h - k = 4.$$
   Substitute $h = 3$ into this equation to obtain $k = -1$. The center of the circle is $(3, -1)$.

2. Use one of the points A, B, or D along with the center to find the radius using the distance formula. Suppose we use $A(0,3)$ and $C(3,-1)$:
   $$r = \sqrt{(0 - 3)^2 + (3 - (-1))^2}$$
   $$= \sqrt{9 + 16} = \sqrt{25} = 5.$$
   The radius of the circle is 5.

3. Use the formula $(x - h)^2 + (y - k)^2 = r^2$ with the substitutions $h = 3$, $k = -1$, and $r = 5$:
   $$(x - 3)^2 + (y - (-1))^2 = 5^2$$
   $$(x - 3)^2 + (y + 1)^2 = 25.$$

• • • • • • • • • •

Trial Problems 16.3

Before you begin the section exercises, warm up with these problems. Complete answers are included in the answer key.

Find the center and the radius of each circle.

1. $(x + 2)^2 + (y + 3)^2 = 25$
2. $x^2 + y^2 - 4x - 10y + 4 = 0$
3. $x^2 + y^2 + 12x - 16y = 0$

Write an equation for each circle.

4. Center $(0, -5)$, radius $= 4$
5. Center $(0,0)$, contains the point $(4,2)$

Exercises 16.3

Determine the center and radius of each circle.

1. $x^2 + y^2 = 49$
2. $x^2 + y^2 = 64$
3. $x^2 + y^2 = 18$
4. $x^2 + y^2 = 30$
5. $x^2 + (y - 2)^2 = 25$
6. $x^2 + (y - 3)^2 = 36$
7. $(x + 3)^2 + y^2 = 18$
8. $(x + 2)^2 + y^2 = 27$
9. $(x - 2)^2 + (y - 3)^2 = 81$
10. $(x - 3)^2 + (y - 4)^2 = 100$
11. $(x + 2)^2 + (y + 4)^2 = 16$
12. $(x + 1)^2 + (y + 3)^2 = 25$
13. $\left(x + \dfrac{3}{2}\right)^2 + \left(y - \dfrac{1}{2}\right)^2 = \dfrac{25}{9}$
14. $\left(x - \dfrac{1}{4}\right)^2 + \left(y + \dfrac{2}{3}\right)^2 = \dfrac{36}{25}$

Write the equation of the circle with the given center C and radius r.

15. $C(0,0)$, $r = 3$
16. $C(0,0)$, $r = 5$
17. $C(0,0)$, $r = \sqrt{5}$
18. $C(0,0)$, $r = \sqrt{7}$
19. $C(3,2)$, $r = 4$
20. $C(3,4)$, $r = 5$
21. $C(-2,0)$, $r = 3$
22. $C(-3,0)$, $r = 4$
23. $C(2,-3)$, $r = 7$
24. $C(-3,2)$, $r = 6$
25. $C(-3,-5)$, $r = \dfrac{1}{2}$
26. $C(-2,-3)$, $r = \dfrac{1}{3}$
27. $C(-2,-4)$, $r = \sqrt{7}$
28. $C(-5,-4)$, $r = \sqrt{13}$

Determine the center and radius of each circle.

29. $x^2 + y^2 - 2x - 8y + 7 = 0$
30. $x^2 + y^2 - 6x + 10y + 25 = 0$
31. $x^2 + y^2 + 6x + 8y + 9 = 0$
32. $x^2 + y^2 - 12x + 10y + 25 = 0$
33. $x^2 + y^2 - 4x + 2y - 4 = 0$
34. $x^2 + y^2 + 4x - 10y + 20 = 0$
35. $x^2 + y^2 + 10x - 6y + 30 = 0$
36. $2x^2 + 2y^2 + 4x + 4y + 3 = 0$
37. $x^2 + y^2 + 6x - \dfrac{2y}{3} + \dfrac{64}{9} = 0$
38. $x^2 + y^2 + 3x - 5y - \dfrac{1}{2} = 0$

39. Write the equation, in standard form, of the circle that contains the point $(3,-2)$ and that has its center at the origin.

40. Write the equation, in standard form, of the circle that contains the point $(-3,-4)$ and that has its center at the origin.

41. Write the equation, in standard form, of the circle that contains the point $(6,1)$ and that has its center at the point $(3,-2)$.

42. Write the equation, in standard form, of the circle that contains the point $(5,2)$ and that has its center at the point $(1,4)$.

43. Write the equation of the circle that contains the points $(2,3)$, $(4,-1)$, and $(5,2)$.

44. Write the equation of the circle that contains the points $(4,3)$, $(2,7)$, and $(-3,-8)$.

45. Write the equation of the circle that contains the points $(4,9)$, $(6,5)$, and $(2,-3)$.

46. Write the equation of the circle that contains the points $(3,-1)$, $(5,3)$, and $(6,2)$.

47. A plan for a circular swimming pool with a 24 ft diameter calls for a drain with a 3 inch diameter to be placed 1 ft from the outer edge of the pool. Write equations for the perimeter of the pool and the circular drain if the center of the pool is placed at the origin and the center of the drain is placed on the positive x-axis.

48. Solve exercise 47 placing the center of the pool at $(0,-24)$ and the center of the drain on the negative y-axis.

49. A company is looking for a warehouse site that must be within 3 miles of the main plant. When a coordinate system using $\frac{1}{4}$ mile units is imposed on a map of the city, the main plant has $(14,-3)$ as its coordinates. Write an equation for the curve that encloses the potential warehouse sites.

50. Solve exercise 49 given that the scale units on the coordinate system represent $\frac{1}{2}$ mile, the main plant has coordinates $(5,11)$ and the warehouse must be located within 8 miles of the main plant.

51. The outside perimeter of a racetrack has the dimensions shown in figure 16.10. Placing the origin at point A, write equations for the semicircles at the ends of the track.

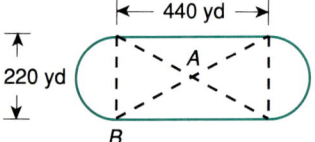

Figure 16.10

52. Solve exercise 51 using point B as the origin.

53. A plan calls for four equally-spaced circles along the circumference of the circle $x^2 + y^2 = 36$. If each circle has a radius of 1 unit, has a center on the given circle, and one hole has center $(6,0)$, write equations for the four circles.

54. Solve exercise 53 using 6 equally-spaced circles with one of the circles centered at $(6,0)$.

16 • 4 The Parabola

Parabola
Axis of Symmetry

In section 3.3, we graphed equations of the form $y = ax^2 + bx + c$. The result was a U-shaped curve called a **parabola.** In this section, we present a more general definition of a parabola. This allows us to consider cases where the **axis of symmetry** of the parabola is not necessarily a vertical line.

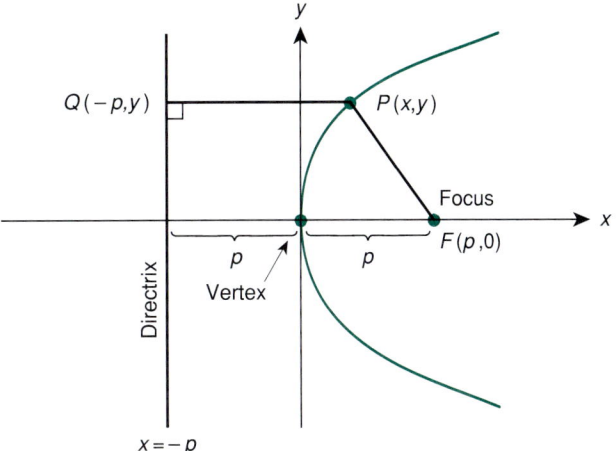

Figure 16.11

A *parabola* is the set of points that are equidistant from a fixed line and a fixed point that is not on the line. The set of points, the fixed line, and the fixed point must all lie in the same plane. The fixed point is called the **focus,** and the fixed line is called the **directrix.**

Focus
Directrix

To show why such a set of points describes a parabola, we set up a coordinate system with the focus at the point $(p,0)$ and the directrix as the line $x = -p$, as shown in figure 16.11. Let $P(x,y)$ be any point such that the distance between F and P is equal to the distance between P and the line $x = -p$.

Thus, $PF = PQ$ and $(PF)^2 = (PQ)^2$. Use the distance formula to obtain expressions for $(PF)^2$ and $(PQ)^2$:

$$(PF)^2 = (x - p)^2 + (y - 0)^2 = x^2 - 2px + p^2 + y^2$$
$$(PQ)^2 = (x + p)^2 = x^2 + 2px + p^2.$$

Set these two expressions equal, and solve the resulting equation for y^2:

$$x^2 - 2px + p^2 + y^2 = x^2 + 2px + p^2$$
$$y^2 = 4px.$$

This is the equation of the parabola with vertex $(0,0)$, focus $(p,0)$ and directrix $x = -p$.

A similar approach can be used to derive equations for the parabolas that have vertices at the origin, foci at points $(-p,0)$, $(0,p)$, and $(0,-p)$ with the corresponding directrices $x = p$, $y = -p$, and $y = p$. These are called the *standard forms* of parabolas with vertices at the origin. The standard forms are illustrated in figure 16.12.

634 CHAPTER 16 ANALYTIC GEOMETRY

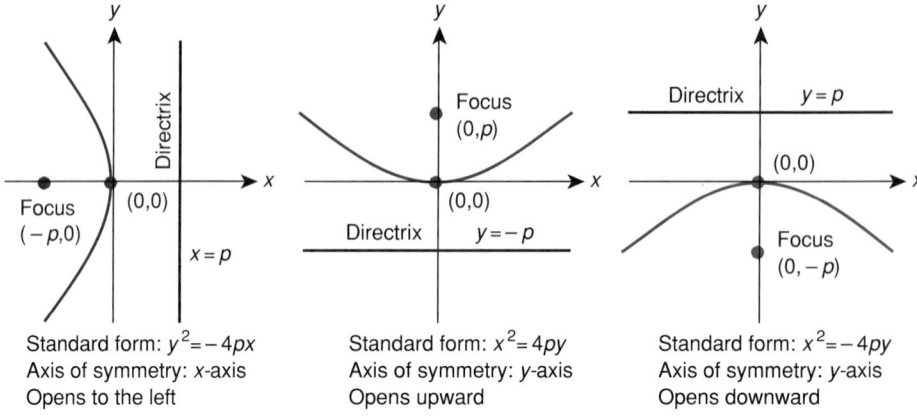

Figure 16.12

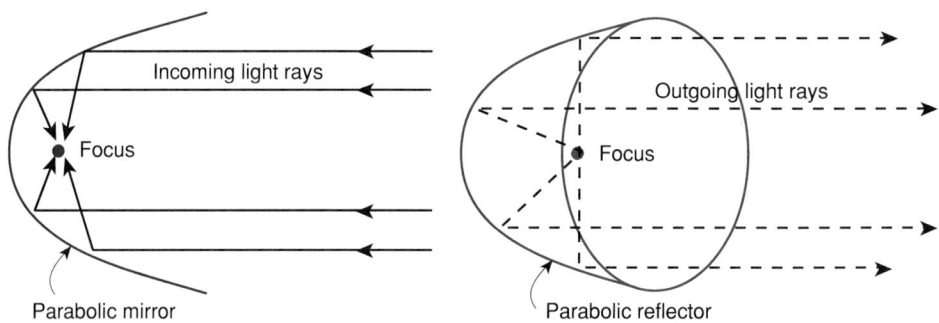

Figure 16.13

Practically speaking, the focus of a parabola is important. For a parabolic reflector on a telescope, the light entering the reflector is reflected into the focus. For a light with a parabolic reflector, a light source at the focus reflects the rays parallel to the axis of the parabola. These ideas are illustrated in figure 16.13.

- **Example 16.10:** Graph the parabola $x^2 = 12y$, showing the focus and the directrix.

    Solution: This parabola opens upward. The vertex is (0,0). We write the equation in the form $x^2 = 4py$:

    $$x^2 = 4(3)y$$

    Thus, $p = 3$, the focus is at (0,3), and the directrix is the line $y = -3$, as in figure 16.14.

• • • • • • • • • •

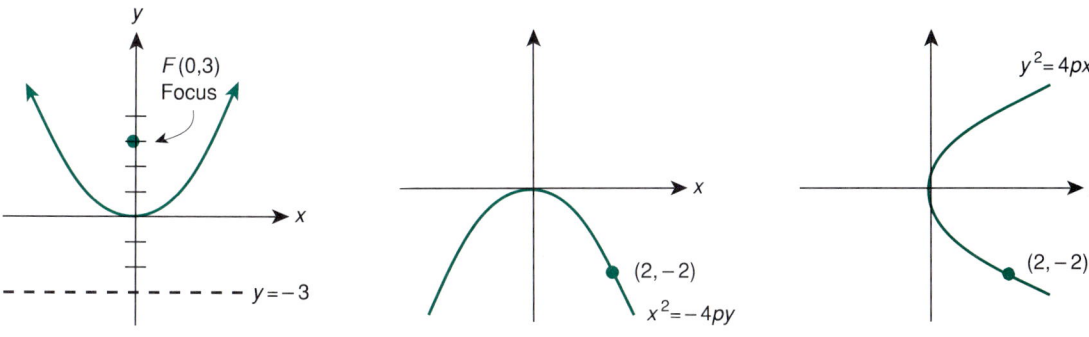

Figure 16.14

Figure 16.15

- **Example 16.11:** Find an equation for a parabola that has its vertex at the origin and that contains the point (2,2).

  Solution: An examination of a rough sketch (figure 16.15) of the graph of the parabola shows that two different parabolas fit the conditions of the problem, one with the focus on the y-axis opening downward and one with the focus on the x-axis opening to the right.

  1. For the parabola with the focus on the y-axis, the form of the equation for the parabola is $x^2 = -4py$. Since the point $(2,-2)$ is on the curve, substitute $x = 2$ and $y = -2$ into $x^2 = -4py$ and solve for p:

  $$2^2 = -4p(-2)$$
  $$4 = 8p$$
  $$p = \frac{1}{2}.$$

  The equation of the parabola is $x^2 = -4(\frac{1}{2})y$ or $x^2 = -2y$.

  The focus is $(-\frac{1}{2}, 0)$ and the directrix is $y = \frac{1}{2}$.

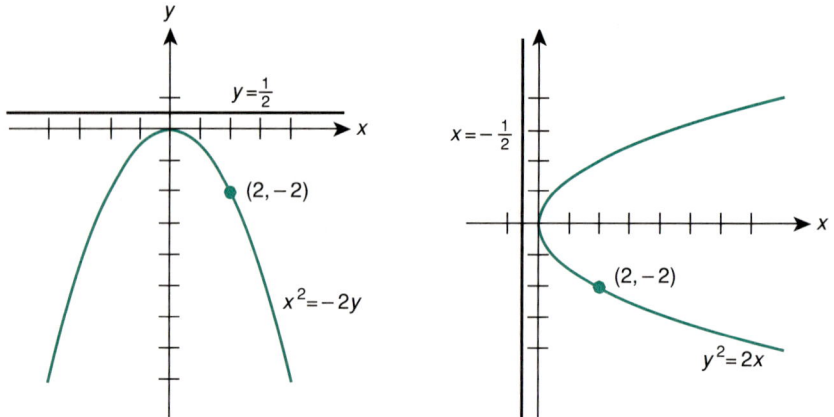

Figure 16.16

2. For the parabola with the focus on the x-axis, the form of the equation for the parabola is $y^2 = 4px$. Since the point $(2,-2)$ is on the parabola, substitute $x = 2$ and $y = -2$ into $y^2 = 4px$ and solve for p:

$$(-2)^2 = 4p(2)$$
$$4 = 8p$$
$$p = \frac{1}{2}.$$

The equation of the parabola is $y^2 = 4(\frac{1}{2})x$ or $y^2 = 2x$. The focus is $(\frac{1}{2}, 0)$ and the directrix is $x = -\frac{1}{2}$. Figure 16.16 shows the two solutions to the problem.

• **Example 16.12:** A bridge cable between two supporting towers sags to form an arc of a parabola. The supporting towers are 240 ft apart and 35 ft high and the cable sags 25 ft between the towers.

1. Place the origin of a coordinate system at the lowest point of the cable and write an equation for the position of the cable.

2. If the roadway is at the level of the bottoms of the towers, how far is the cable above the roadway at a point 50 ft from one of the towers?

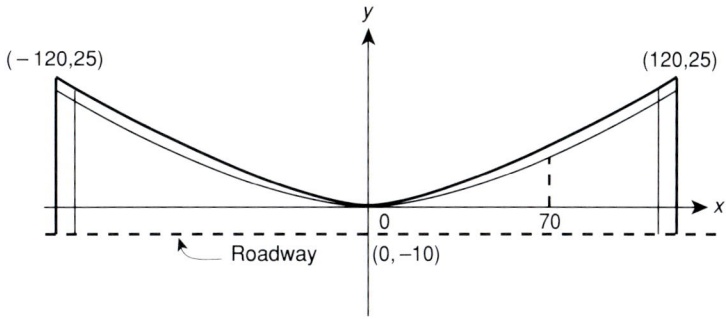

Figure 16.17

Solution: Figure 16.17 provides a description of the facts in the problem.

1. Since the parabola opens upward, the equation for the cable is of the form $x^2 = 4py$. Substitute $x = 120$ and $y = 25$. Solve for p:

$$(120)^2 = 4p(25)$$

$$p = \frac{(120)^2}{(4)(25)} = 144.$$

The equation for the parabola is

$$x^2 = 144(4)y \quad \text{or} \quad x^2 = 576y.$$

2. A distance of 50 ft from the tower on the right yields

$$x = 120 - 50 = 70.$$

Substitute $x = 70$ into $x^2 = 576y$ and solve for y:

$$(70)^2 = 576y$$

$$y = \frac{(70)^2}{576} = 8.5 \text{ ft.}$$

Since the roadway is at $y = -10$, the height of the cable above the roadway is $10 + 8.5 = 18.5$ ft.

Sometimes it is more meaningful in a problem to have the vertex of a parabola at a point other than the origin. If the vertex is at (h,k) and the directrix is parallel to one of the axes, the standard forms for the equation of the parabola are illustrated in figure 16.18.

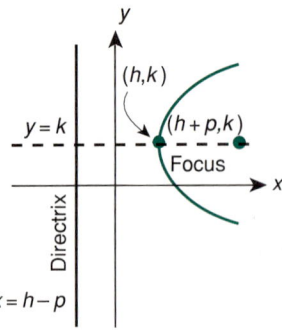
Standard form: $(y-k)^2 = 4p(x-h)$
Axis of symmetry: $y = k$
Opens to the right

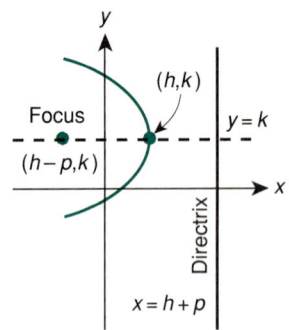
Standard form: $(y-k)^2 = -4p(x-h)$
Axis of symmetry: $y = k$
Opens to the left

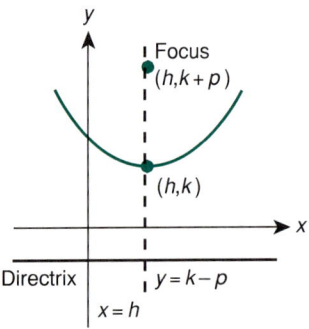
Standard form: $(x-h)^2 = 4p(y-k)$
Axis of symmetry: $x = h$
Opens upward

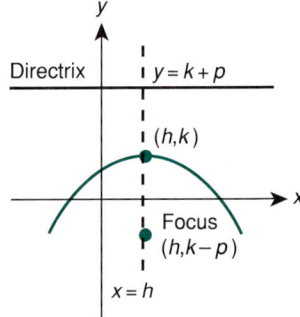
Standard form: $(x-h)^2 = -4p(y-k)$
Axis of symmetry: $x = h$
Opens downward

Figure 16.18

Example 16.13 illustrates one of these forms.

• **Example 16.13:** Find the vertex, the focus and the equation of the directrix for the parabola $x^2 + 2x + 8y - 15 = 0$. Graph the parabola.

Solution: Since the equation contains an x^2 term but no y^2 term, the equation could be either of the standard forms

$$(x-h)^2 = 4p(y-k) \quad \text{or} \quad (x-h)^2 = -4p(y-k).$$

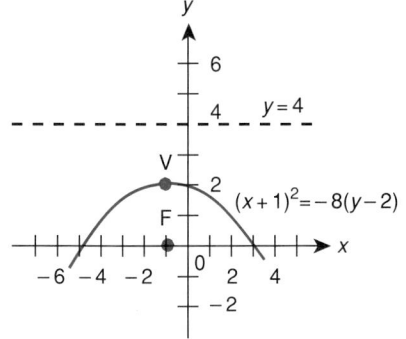
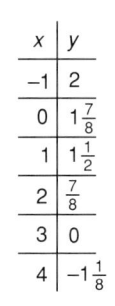
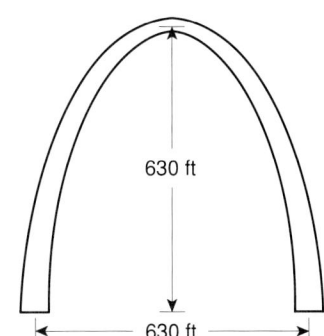

Figure 16.19

Figure 16.20

Complete the square on the expression $x^2 + 2x$ and move the remaining terms to the right-hand side of the equation:

$$x^2 + 2x + 8y - 15 = 0$$
$$(x^2 + 2x + \underline{}) = -8y + 15 \quad \text{(Add } -8y + 15 \text{ to both sides)}$$
$$(x^2 + 2x + 1) = -8y + 16 \quad \text{(Add 1 to both sides)}$$
$$(x + 1)^2 = -4(2)(y - 2) \quad \text{(Put the equation into one of}$$
$$(x - (-1))^2 = -4(2)(y - 2). \quad \text{the standard forms)}$$

Thus, $h = -1$, $k = 2$, and $p = 2$.

Vertex: $(h,k) = (-1,2)$

Focus: $(h, k - p) = (-1, 0)$

Directrix: $y = k + p$ or $y = 4$.

The graph is shown in figure 16.19.

• **Example 16.14:** The Gateway Arch in St. Louis (figure 16.20) has an approximate parabolic shape with a maximum height of 630 ft. The distance between the legs of the arch is about 630 ft. Place the origin directly under the highest point on the arch and write an equation for the arch. Find the focus, an equation for the directrix, and graph the equation (see figure 16.21).

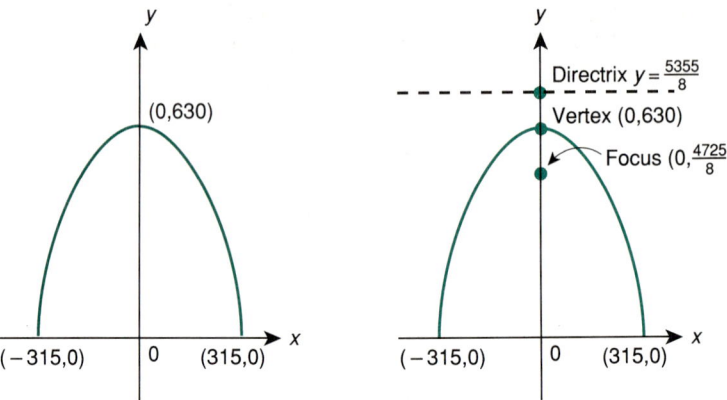

Figure 16.21   Figure 16.22

Solution: Since the parabola opens downward, the standard form is $(x - h)^2 = -4p(y - k)$. Since the vertex is (0,630), $h = 0$ and $k = 630$. Substitution of these values into the standard equation yields the following equation:

$$x^2 = -4p(y - 630).$$

Use the fact that (315,0) is a point on the curve to substitute $x = 315$ and $y = 0$ into the equation and solve for p:

$$(315)^2 = -4p(0 - 630)$$

$$p = \frac{(315)^2}{(-4)(-630)} = \frac{315}{8}.$$

Substitute $p = \dfrac{315}{8}$ into the previously obtained equation:

$$x^2 = -4\left(\frac{315}{8}\right)(y - 630).$$

Focus: $(h, k - p) = \left(0, 630 - \dfrac{315}{8}\right) = \left(0, \dfrac{4725}{8}\right).$

Directrix: $y = k + p$ or $y = 630 + \dfrac{315}{8}$ or $y = \dfrac{5355}{8}.$

The graph is shown in figure 16.22.

Trial Problems 16.4

Before you begin the section exercises, warm up with these problems. Complete answers are included in the answer key.

Find the vertex, the focus, and the equation of the directrix for each parabola.

1. $x^2 = 8y$
2. $(y - 2)^2 = x + 1$
3. $(y + 2)^2 = -\dfrac{1}{3}(x - 12)$

Find an equation for the parabola that fits the given conditions.

4. Vertex (0,0), directrix: $x = 2$
5. Vertex (1,2), opens upward, and contains the point $(-2,5)$

Exercises 16.4

Graph each of the following parabolas. Find the coordinates of the focus and the equation of the directrix.

1. $y^2 = 12x$
2. $y^2 = 8x$
3. $x^2 = 8y$
4. $x^2 = 12y$
5. $x^2 = -4y$
6. $x^2 = -8y$
7. $y^2 = -8x$
8. $y^2 = -4x$
9. $x^2 = -2y$
10. $x^2 = -y$
11. $y^2 = -3x$
12. $y^2 = -5x$
13. $5y^2 + 11x = 0$
14. $3x^2 + 5y = 0$

Find the equation of the parabola having its vertex at the origin given the conditions listed.

15. Focus is (4,0)
16. Focus is (8,0)
17. Focus is $(-3,0)$
18. Focus is $(-4,0)$
19. Focus is $(0,-2)$
20. Focus is $(0,-5)$
21. Directrix is $x + 5 = 0$
22. Directrix is $x + 4 = 0$
23. Directrix is $y - 3 = 0$
24. Directrix is $y - 2 = 0$
25. X-axis is the axis of symmetry and contains the point (3,2)
26. X-axis is the axis of symmetry and contains the point (5,3)
27. Y-axis is the axis of symmetry and contains the point (6,3)
28. Y-axis is the axis of symmetry and contains the point (4,2)

Write an equation, in standard form, for each parabola.

29. Vertex (3,2), opens to the right, and contains the point (5,8)
30. Vertex $(-3,-2)$, opens to the left, and contains the point $(-5,0)$
31. Vertex (1,5), axis of symmetry $y = 5$, and contains the point (6,8)
32. Vertex (2,4), axis of symmetry $x = 2$, and contains the point (4,6)

Write an equation, in standard form for each parabola. Find the vertex, the focus, and the equation of the directrix. Sketch the graph.

33. $x + y^2 - 2y = 0$
34. $x^2 + 4x + 4y - 12 = 0$
35. $x^2 + 2x + 4y - 7 = 0$
36. $8x - y^2 + 2y + 15 = 0$
37. $4x - y^2 + 6y - 17 = 0$
38. $4x + y^2 + 2y - 15 = 0$

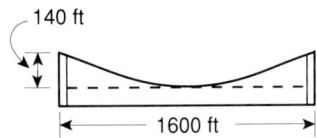

Figure 16.23

39. The Brooklyn Bridge is approximately 1600 feet long. The lowest point of the cables is 140 feet below the points of support, as indicated in figure 16.23. If the cables form a parabola, find the equation for the cables by taking the lowest point of a cable as the origin and the y-axis as the axis of symmetry.

40. Solve exercise 39 for a bridge that is 800 feet long and for which the lowest point of the cables is 60 feet below the points of support.

41. An 8-inch parabolic mirror has a maximum depth of 2.5 inches. Write an equation for the parabola indicated in figure 16.24, placing the origin at point A. Find the coordinates of the focus.

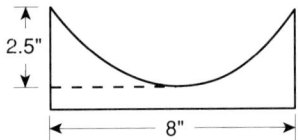

Figure 16.24

42. Solve exercise 41 for a 10-inch mirror with a maximum depth of $3\frac{1}{8}$ inches.

43. A parabolic arch has a maximum height of 85 feet and the distance between the feet of the arch is 120 feet. Write an equation for the arch, placing the origin directly under the highest point on the arch.

44. Solve exercise 43 given a height of 60 feet and a distance between the feet of the arch of 100 feet.

45. The height y in feet of a projectile fired at an angle of 30° with the horizontal after t seconds is described by the equation $y = -16t^2 + \frac{1}{2}v_0 t$, where v_0 represents the initial velocity. If $v_0 = 768$ ft/sec, find the maximum height of the projectile.

46. Solve exercise 45 given an initial velocity of 896 ft/sec.

47. An underpass is parabolic in shape with a width at its base of 26 ft and a maximum height of 18 ft. If a "wide-load" vehicle with a width of 12 ft is permitted to travel the center line of the underpass, find the maximum height for the vehicle to fit through the underpass.

48. Solve exercise 47 if the width of the underpass is 28 ft and the maximum height of the underpass is 20 ft.

49. The power P (in watts) dissipated by a resistor in an electrical circuit is described by the equation $P = 110I - 100I^2$, where I is in amps and $0 \leq I \leq 1.1$. Sketch a graph of the equation and find the value of I that maximizes P.

50. The power P (in watts) supplied to a circuit varies according to the equation $P = 12I - 0.40I^2$, where I is in amps and $0 \leq I \leq 30$. Sketch the graph of the equation and find the value of I that maximizes P.

51. The bending moment (y) of an 8-ft beam is described by the equation $y = 2000x - 250x^2$ for $0 \leq x \leq 8$, where x is the distance from one end of the beam. Graph the equation and find the value of x for which the bending moment is the greatest.

52. Solve exercise 51 using the equation $y = 8000x - 500x^2$ for x between 0 and 16 ft.

53. A cable on a suspension bridge forms an arc of a parabola. The supporting towers are 30 ft high and 200 ft apart. The lowest point on the cable is 8 ft above the roadway. Find the length of a supporting rod that is 50 ft from the center of the bridge.

54. In exercise 53, how far is the cable above the roadway at a point that is 40 ft from the center of the bridge?

16·5 The Ellipse

In this section we consider the equations for the oval-shaped curve called the ellipse.

Ellipse An **ellipse** is the set of points in the plane for which the sum of the distances from two fixed points is constant. Each of the fixed points is called a *focus* (plural, foci).

An ellipse can be constructed by placing a pair of tacks at the foci (F_1 and F_2 in figure 16.25). Then take a piece of string (longer than F_1F_2) and tie the ends of the string to the tacks. Pull the string taut with a pencil and move the pencil, keeping the string taut. The resulting figure is an ellipse. The sum of the distances from the foci is constant.

If we place an ellipse on a coordinate system with the foci (F_1 and F_2) on the x-axis, with the origin as the midpoint of $\overline{F_1F_2}$, the ellipse is said to be in a *standard position*. Points V_1 and V_2 in figure 16.26 are called the **vertices** of the ellipse, segment V_1V_2 is called the **major axis**, segment M_1M_2 is called the **minor axis**, and 0 is called the **center** of the ellipse.

Vertices
Major Axis
Minor Axis
Center

Using $(-c,0)$ and $(c,0)$ as the coordinates of the foci, we can now derive the equation for an ellipse in one of the standard positions. Refer to figure 16.27. Let $P(x,y)$ be an arbitrary point on the ellipse. By the definition of an ellipse, the sum of the distances PF_1 and PF_2 must be constant.

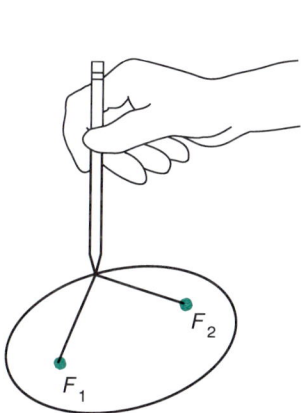

Figure 16.25

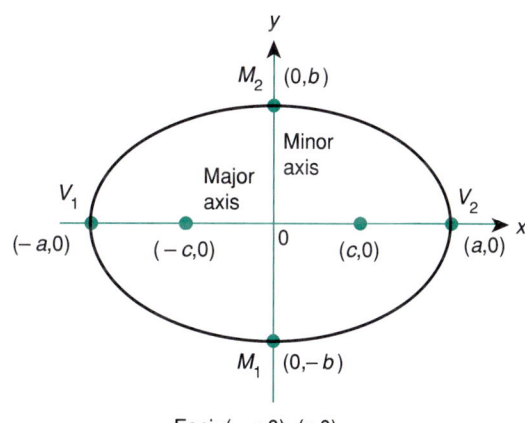

Foci: $(-c,0)$, $(c,0)$
Vertices: $(-a,0)$, $(a,0)$

Figure 16.26

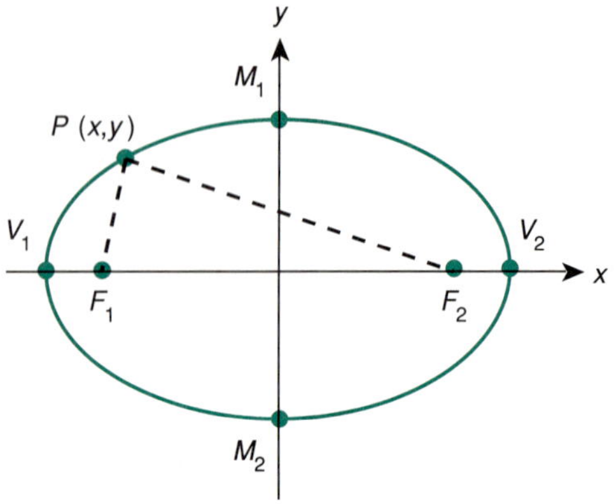

Figure 16.27

Since $V_2(a,0)$ is a point on the ellipse, and
$$V_2F_2 + V_2F_1 = (a - c) + (a + c) = 2a$$
the constant must be $2a$.

Thus, $PF_1 + PF_2 = 2a$, and
$$[(PF_1) + (PF_2)]^2 = 4a^2$$
$$(PF_1)^2 + 2(PF_1)(PF_2) + (PF_2)^2 = 4a^2.$$
$$[(x + c)^2 + (y - 0)^2] + 2\sqrt{(x + c)^2 + (y - 0)^2}\sqrt{(x - c)^2 + (y - 0)^2}$$
$$+ [(x - c)^2 + (y - 0)^2] = 4a^2. \quad \text{(Use the distance formula)}$$

We can rearrange this expression in the following form:
$$a\sqrt{(x + c)^2 + y^2} = a^2 + xc.$$

We square both sides of this equation:
$$a^2(x^2 + 2xc + c^2 + y^2) = a^4 + 2a^2xc + x^2c^2$$
$$a^2x^2 + 2a^2xc + a^2c^2 + a^2y^2 = a^4 + 2a^2xc + x^2c^2.$$

We can rewrite this equation in the following form:
$$a^2x^2 - x^2c^2 + a^2y^2 = a^4 - a^2c^2$$
$$(a^2 - c^2)x^2 + a^2y^2 = a^2(a^2 - c^2). \qquad (16.1)$$

From figure 16.28 we can see that since $M_1(0,b)$ is on the ellipse, $F_1M_1 + F_2M_1 = 2a$.

$$F_1M_1 = F_2M_1 = \sqrt{b^2 + c^2} \quad \text{(Use the Pythagorean relationship)}$$
$$2\sqrt{b^2 + c^2} = 2a$$
$$b^2 + c^2 = a^2$$
$$a^2 - c^2 = b^2.$$

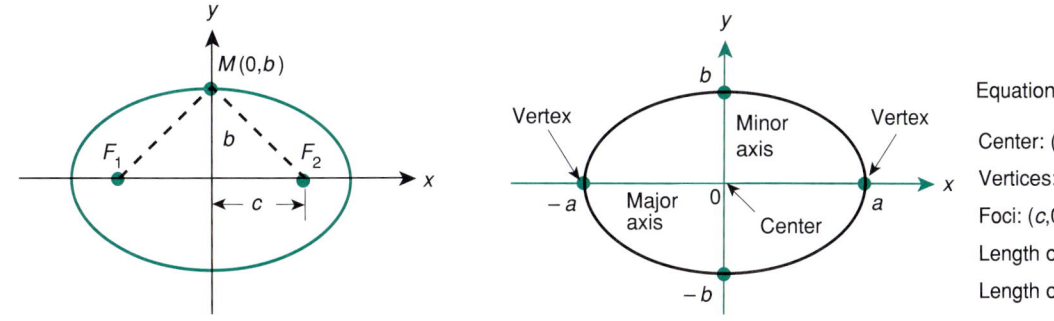

y
M(0,b)
b
F1
F2
x
c

Figure 16.28

y
b
Vertex
Minor
axis
Vertex
x
−a
Major
axis
0
Center
a
−b
Equation: $\frac{x^2}{a^2}+\frac{y^2}{b^2}=1,\quad a>b$
Center: (0,0)
Vertices: (a,0) and (−a,0)
Foci: (c,0) and (−c,0)
Length of major axis: 2a
Length of minor axis: 2b

Figure 16.29

y
a
Vertex
Major
axis
Minor
axis
x
−b
0
b
Center
−a
Vertex
Equation: $\frac{x^2}{b^2}+\frac{y^2}{a^2}=1,\quad a>b$
Center: (0,0)
Vertices: (0,a) and (0,−a)
Foci: (0,c) and (0,−c)
Length of major axis: 2a
Length of minor axis: 2b

Figure 16.30

We substitute b^2 for $a^2 - c^2$ in equation (16.1).

$$b^2x^2 + a^2y^2 = a^2b^2$$

$$\frac{x^2}{a^2} + \frac{y^2}{b^2} = 1. \quad \text{(Divide both sides of the equation by } a^2b^2\text{)}$$

This is the general equation for an ellipse, in standard position, on a coordinate system.

We summarize these results in figure 16.29.

If we plot an ellipse with the major axis on the y-axis of a coordinate system and the minor axis along the x-axis of the coordinate system, the results are similar. The results are summarized in figure 16.30.

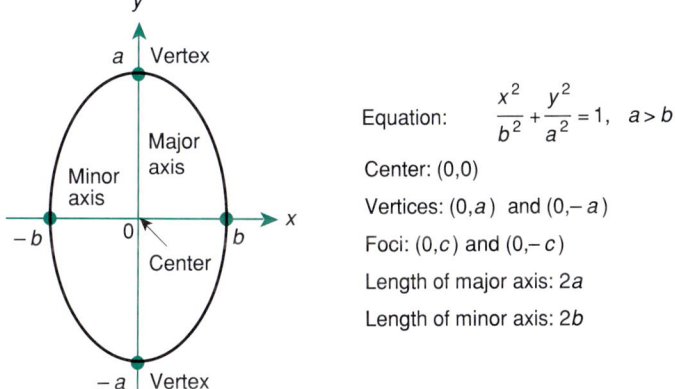

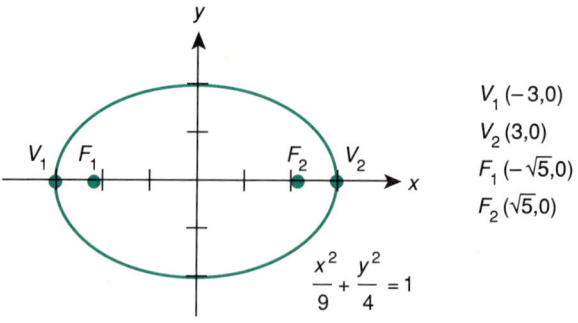

Figure 16.31

- **Example 16.15:** Graph the ellipse $4x^2 + 9y^2 = 36$. Determine the lengths of the major and minor axes and the coordinates of the vertices and the foci.

    Solution: First, put the equation into one of the standard forms by dividing both sides of the equation by 36, to obtain a "1" on the right-hand side of the equation:

    $$4x^2 + 9y^2 = 36$$

    $$\frac{4x^2}{36} + \frac{9y^2}{36} = \frac{36}{36}$$

    $$\frac{x^2}{9} + \frac{y^2}{4} = 1$$

    $$\frac{x^2}{3^2} + \frac{y^2}{2^2} = 1.$$

    Thus $a = 3$, $b = 2$, and $c = \sqrt{a^2 - b^2} = \sqrt{3^2 - 2^2} = \sqrt{9 - 4} = \sqrt{5}$. Length of major axis $= 2a = 2(3) = 6$. Length of minor axis $= 2b = 2(2) = 4$. Vertices at $(a,0)$ and $(-a,0)$ or at $(3,0)$ and $(-3,0)$. Foci at $(c,0)$ and $(-c,0)$ or at $(\sqrt{5},0)$ and $(-\sqrt{5},0)$. The graph of the ellipse is shown in figure 16.31.

- **Example 16.16:** Determine the equation, in standard form, for an ellipse that has center $(0,0)$, vertices $(0,-4)$ and $(0,4)$, and foci $(0,-3)$ and $(0,3)$.

    Solution: Figure 16.32 shows a sketch of the graph of the ellipse. The standard form for this ellipse is

    $$\frac{x^2}{b^2} + \frac{y^2}{a^2} = 1.$$

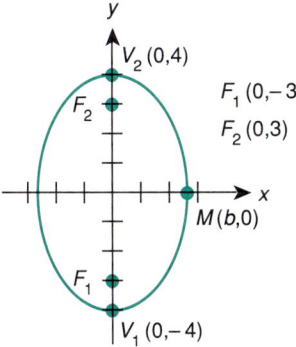

Figure 16.32

Since (0,4) is a vertex, $a = 4$. Since (0,3) is a focus, $c = 3$. Substitute $a = 4$ and $c = 3$ into the equation $a^2 = b^2 + c^2$ and solve for b^2:

$$4^2 = b^2 + 3^2$$
$$b^2 = 7.$$

Substitute $a^2 = 16$ and $b^2 = 7$ into the standard form

$$\frac{x^2}{7} + \frac{y^2}{16} = 1.$$

• • • • • • • • • •

• **Example 16.17:** The point (2,0.9) lies on an ellipse that has vertices (−2.5,0) and (2.5,0). Write an equation for the ellipse.

Solution: Since (2.5,0) is one of the vertices, $a = 2.5$, and the equation of the ellipse is of the form

$$\frac{x^2}{a^2} + \frac{y^2}{b^2} = 1.$$

Substitute 2.5 for a in the equation:

$$\frac{x^2}{(2.5)^2} + \frac{y^2}{b^2} = 1.$$

Since (2,0.9) is a point on the ellipse, we can substitute $x = 2$ and $y = 0.9$ into the last equation and solve for b:

$$\frac{2^2}{(2.5)^2} + \frac{(0.9)^2}{b^2} = 1$$

$$\frac{4}{6.25} + \frac{0.81}{b^2} = 1.$$

Multiply both sides of the equation by $6.25b^2$:

$$4b^2 + 5.0625 = 6.25b^2$$

$$b^2 = \frac{5.0625}{6.25 - 4} = 2.25.$$

Substitute $a^2 = 6.25$ and $b^2 = 2.25$ into the standard form

$$\frac{x^2}{a^2} + \frac{y^2}{b^2} = 1.$$

The equation for the ellipse is

$$\frac{x^2}{6.25} + \frac{y^2}{2.25} = 1.$$

• • • • • • • • • •

• **Example 16.18:** An ellipse on a drawing that is on a coordinate system has its center at the origin and passes through the points (4,2) and (1,3). Find an equation for the ellipse.

Solution: To solve the problem, we use the form $\dfrac{x^2}{a^2} + \dfrac{y^2}{b^2} = 1$ and substitute the coordinates of the points into this equation to obtain a pair of equations in a and b. Then we solve these equations for a and b and substitute these values into the standard equation.

1. Obtain a pair of equations in a and b.

   Use $x = 4, y = 2$ $\qquad$ Use $x = 1, y = 3$

   $\dfrac{16}{a^2} + \dfrac{4}{b^2} = 1$ $\qquad$ $\dfrac{1}{a^2} + \dfrac{9}{b^2} = 1.$

2. Solve the pair of equations for a and b. Obtain a $\dfrac{16}{a^2}$ term in both equations by multiplying the equation on the right by 16:

   $$\dfrac{16}{a^2} + \dfrac{4}{b^2} = 1 \qquad \dfrac{16}{a^2} + \dfrac{144}{b^2} = 16.$$

   Subtract the left equation from the right equation:

   $$\dfrac{140}{b^2} = 15, \qquad b^2 = \dfrac{140}{15} = \dfrac{28}{3}.$$

   Substitute $b^2 = \dfrac{28}{3}$ into the equation $\dfrac{1}{a^2} + \dfrac{9}{b^2} = 1$ and solve for a^2:

   $$\dfrac{1}{a^2} + \dfrac{27}{28} = 1$$

   $$\dfrac{1}{a^2} = 1 - \dfrac{27}{28} = \dfrac{1}{28}$$

   $$a^2 = 28.$$

3. Substitute $a^2 = 28$ and $b^2 = \dfrac{28}{3}$ into the equation:

$$\frac{x^2}{a^2} + \frac{y^2}{b^2} = 1.$$

An equation for the ellipse is

$$\frac{x^2}{28} + \frac{y^2}{\frac{28}{3}} = 1.$$

• • • • • • • • • •

Just as we did with circles and parabolas, we can write standard forms for ellipses that have their centers at points other than the origin. We restrict ourselves to the standard forms of the equations for ellipses that have their axes parallel to the axes of the coordinate system. The graphs along with some useful information are shown in figure 16.33.

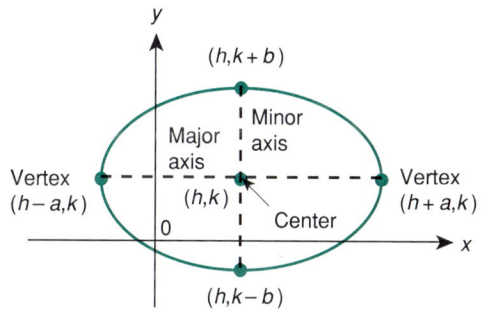

Equation: $\dfrac{(x-h)^2}{a^2} + \dfrac{(y-k)^2}{b^2} = 1, \quad a > b$

Center: (h, k)

Vertices: $(h-a, k)$ and $(h+a, k)$

Foci: $(h-c, k)$ and $(h+c, k)$, where $c^2 = a^2 - b^2$

Ends of minor axis: $(h, k+b)$ and $(h, k-b)$

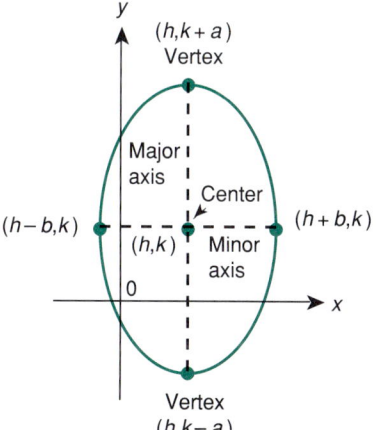

Equation: $\dfrac{(x-h)^2}{b^2} + \dfrac{(y-k)^2}{a^2} = 1, \quad a > b$

Center: (h, k)

Vertices: $(h, k-a)$ and $(h, k+a)$

Foci: $(h, k-c)$ and $(h, k+c)$, where $c^2 = a^2 - b^2$

Ends of minor axis: $(h-b, k)$ and $(h+b, k)$

Figure 16.33

• **Example 16.19:** Graph the equation $16x^2 - 128x + 25y^2 + 150y = -81$.

Solution: First, complete the square on each of the expressions $16x^2 - 128x$ and $25y^2 + 150y$:

$$(16x^2 - 128x + \underline{}) + (25y^2 + 150y + \underline{}) = -81.$$

Factor 16 from the expression $16x^2 - 128x$ and 25 from the expression $25y^2 + 150y$:

$$16(x^2 - 8x + \underline{}) + 25(y^2 + 6y + \underline{}) = -81$$

$$\underset{\text{(Add } 16 \times 16 = 256 \text{ to both sides)}}{\overset{16}{\uparrow}} \qquad \underset{\text{(Add } 25 \times 9 = 225 \text{ to both sides)}}{\overset{9}{\uparrow}}$$

$$16(x^2 - 8x + 16) + 25(y^2 + 6y + 9) = -81 + 256 + 225$$
$$16(x - 4)^2 + 25(y + 3)^2 = 400.$$

Divide both sides of the equation by 400:

$$\frac{(x - 4)^2}{25} + \frac{(y + 3)^2}{16} = 1.$$

Put the equation in standard form:

$$\frac{(x - 4)^2}{5^2} + \frac{(y - (-3))^2}{4^2} = 1.$$

Thus $h = 4$, $k = -3$, $a = 5$, and $b = 4$.

The center of the ellipse is the point $(4, -3)$, the major axis is parallel to the x-axis, the major axis is $2a = 10$ units long, and the minor axis is $2b = 8$ units long. Plot the center and the axes of the ellipse and sketch the curve. The results are shown in figure 16.34.

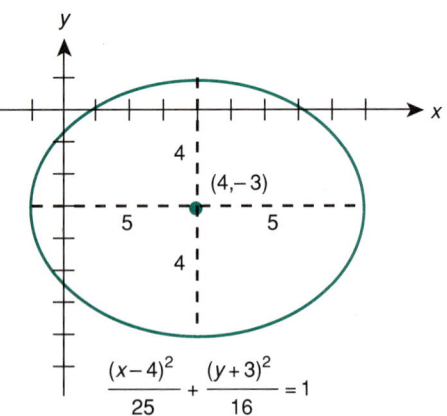

Figure 16.34

SECTION 16.5 THE ELLIPSE

Trial Problems 16.5

Before you begin the section exercises, warm up with these problems. Complete answers are included in the answer key.

For each ellipse, find the center, the vertices, and the foci.

1. $4x^2 + 36y^2 = 144$
2. $\dfrac{(x-3)^2}{4} + \dfrac{(y+2)^2}{9} = 1$

Find an equation of the ellipse that fits the given conditions.

3. Center $(0,0)$, x-intercepts ± 4, and y-intercepts ± 2
4. Vertices $(-2,-3)$ and $(-2,9)$, contains the point $(2,3)$, and center $(-2,3)$

Exercises 16.5

For each ellipse, find the vertices, the foci and sketch the graph.

1. $\dfrac{x^2}{25} + \dfrac{y^2}{16} = 1$
2. $\dfrac{x^2}{16} + \dfrac{y^2}{25} = 1$
3. $\dfrac{x^2}{144} + \dfrac{y^2}{121} = 1$
4. $\dfrac{x^2}{121} + \dfrac{y^2}{144} = 1$
5. $\dfrac{x^2}{49} + \dfrac{y^2}{25} = 1$
6. $\dfrac{x^2}{25} + \dfrac{y^2}{49} = 1$
7. $\dfrac{x^2}{4} + \dfrac{y^2}{5} = 1$
8. $\dfrac{x^2}{5} + \dfrac{y^2}{4} = 1$
9. $\dfrac{x^2}{10} + \dfrac{y^2}{12} = 1$
10. $\dfrac{x^2}{12} + \dfrac{y^2}{10} = 1$
11. $\dfrac{x^2}{\frac{1}{4}} + \dfrac{y^2}{\frac{1}{9}} = 1$
12. $\dfrac{x^2}{\frac{1}{9}} + \dfrac{y^2}{\frac{1}{4}} = 1$
13. $\dfrac{x^2}{\frac{1}{9}} + \dfrac{y^2}{4} = 1$
14. $\dfrac{x^2}{4} + \dfrac{y^2}{\frac{1}{9}} = 1$

Write the given equation in standard form. Sketch the graphs.

15. $9x^2 + 4y^2 = 36$
16. $4x^2 + 9y^2 = 36$
17. $9x^2 + 25y^2 = 225$
18. $25x^2 + 9y^2 = 225$
19. $16x^2 + 9y^2 = 144$
20. $9x^2 + 16y^2 = 144$
21. $16x^2 + 144y^2 = 2304$
22. $144x^2 + 16y^2 = 2304$
23. $3x^2 + 9y^2 = 27$
24. $9x^2 + 3y^2 = 27$
25. $3x^2 + 4y^2 = 12$
26. $4x^2 + 3y^2 = 12$
27. $2x^2 + y^2 = 1$
28. $x^2 + 2y^2 = 1$

Write the equation, in standard form, of the ellipse that fits the given conditions.

29. Center at $(0,0)$, length of major axis $= 9$, and length of minor axis $= 4$
30. Center at $(0,0)$, length of major axis $= 6$, and length of minor axis $= 2$
31. $F_2(0,3)$, $V_2(0,4)$, and center at $(0,0)$
32. $F_2(3,0)$, $V_2(4,0)$, and center at $(0,0)$
33. Center at $(0,0)$, $V_2(0,8)$, and contains the point $(6,0)$
34. Center at $(0,0)$, $V_2(8,0)$, and contains the point $(0,6)$
35. Center at $(0,0)$, contains the points $(4,0)$ and $(3,2)$
36. Center at $(0,0)$, contains the points $(0,4)$ and $(2,3)$

Write the equation for each ellipse in standard form by completing the squares and sketch the graph.

37. $4x^2 - 32x + 25y^2 - 150y = -189$

38. $25x^2 - 150x + 9y^2 - 90y = -225$

39. $16x^2 - 64x + 9y^2 + 18y = -37$

40. $16x^2 + 64x + 9y^2 - 54y = -1$

41. $9x^2 - 18x + 4y^2 + 32y = -37$

42. $25x^2 - 200x + 9y^2 + 18y = -184$

43. A race track is constructed using a pair of concentric ellipses. The larger ellipse has a major axis 2500 feet long and a minor axis of 600 feet long. The smaller ellipse has a major axis 2400 feet long and a minor axis 500 feet long. Write the equations for the ellipses using the origin as the center. Find the area of the track. (Area $= \dfrac{\pi}{4}ab$, where a and b are the lengths of the axes.)

44. Solve exercise 43 given that the axes for the larger ellipse are 3600 feet and 700 feet and the axes for the smaller ellipse are 3500 feet and 600 feet.

45. An arch is constructed in a semielliptical form. If the inside dimensions of the arch are 24 feet between the ends of the arch and the maximum height is 10 feet, write an equation for the arch using the origin as the center of the ellipse.

46. Solve exercise 45, given that there are 30 feet between the ends of the arch and the maximum height is 12 feet.

47. The orbit of Mars is an ellipse with the sun at one of the foci. If we place the ellipse on a coordinate system with the sun at the focus $F_1(-6.5,0)$, the length of the major axis is 140.8 where the numbers represent millions of miles. Write the equation, in standard form, for the orbit.

48. Solve exercise 47 for the orbit of Pluto given the sun is at $F_1(-0.9,0)$ and the length of the major axis is 7.4 (the numbers are in millions of miles).

49. A semielliptical arch is 40 ft across at the base with a maximum height of 14 ft. Find the height of the arch at a point that is 12 ft from the center of the arch.

50. Solve exercise 49 for an arch that is 20 ft across, 12 ft high, and the reference point is 4 ft from the center.

51. Find the area of the ellipse described by the equation $9x^2 + 16y^2 = 144$.

52. Find the area of the ellipse described by the equation $16x^2 + 9y^2 = 144$.

53. An underpass is a semiellipse with a major axis 50 ft long and a minor axis 20 ft long (height). Find the clearance at a point that is 8 ft from a vertex.

54. An underpass is a semiellipse with a base 60 ft long. The maximum height is 20 ft. Find the clearance at a point that is 15 ft from the center.

A "whispering gallery" has a dome with the shape of an ellipse that has been revolved about its major axis. The cross sections of the room are semiellipses. A whisper at one focus can be heard distinctly at the other focus.

55. Find the foci of a whispering gallery with a cross section that is a semiellipse with a major axis 40 ft long and a height of 12 ft.

56. Solve exercise 55 given a major axis 50 ft long and a height of 15 ft.

A ripple tank is a tank in the shape of an ellipse. A disturbance at one of the foci produces a spout at the other focus.

57. Find the foci for a ripple tank that is described by the equation $10.24x^2 + 6.25y^2 = 64.00$.

58. Solve exercise 57 using the equation $3.24x^2 + 6.25y^2 = 20.25$.

16•6 The Hyperbola

In section 16.5 we defined an ellipse as the set of points such that the sum of the distances of the points on the curve from two fixed foci is constant. If we consider the differences of the distances from the foci as constant instead of the sum of the distances, the resulting curve is called a **hyperbola** (see figure 16.35).

Hyperbola
Foci

Let F_1 and F_2 be the fixed points (**foci**). Let $2c$ be the distance between the foci. Let $a > 0$ and $a < c$. A hyperbola is the set of points, P, such that

$$|F_1P - F_2P| = 2a.$$

Center
Transverse Axis
Vertices

We consider hyperbolas that are in *standard positions*. The first such hyperbola has the center at the origin and the foci on the x-axis. The **center** is the midpoint of the line segment connecting the foci and the **transverse axis** is the line segment connecting the **vertices**.

Referring to figure 16.36, let $F_1(-c,0)$ and $F_2(c,0)$ be the foci and $(0,0)$ be the center of a hyperbola. Let $V_1(-a,0)$ and $V_2(a,0)$ be the vertices. Suppose $P(x,y)$ is any point on the hyperbola.

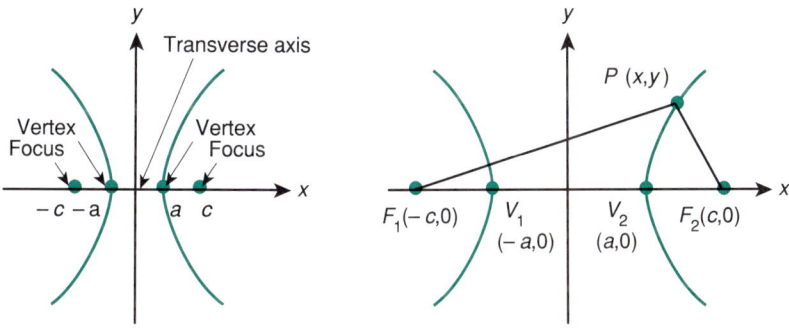

Figure 16.35   Figure 16.36

Since $V_2(a,0)$ is a point on the hyperbola, $|V_2F_1 - V_2F_2|$ is the constant required in the definition.

$$V_2F_1 = c - a \quad \text{and} \quad V_2F_2 = c + a$$
$$|V_2F_1 - V_2F_2| = |(c + a) - (c - a)| = |2a| = 2a.$$

Thus, the constant required in the definition is $2a$.
Let $P(x,y)$ be a general point on the curve:

$$|PF_2 - PF_1| = 2a$$
$$(PF_2 - PF_1)^2 = 4a^2 \quad \text{(Square both sides of the equation)}$$

$$(PF_2)^2 - 2(PF_1)(PF_2) + (PF_1)^2 = 4a^2.$$

$$[(x - c)^2 + y^2] - 2\sqrt{(x + c)^2 + y^2}\sqrt{(x - c)^2 + y^2}$$
$$+ [(x + c)^2 + y^2] = 4a^2. \quad \text{(Use the distance formula)}$$

This expression simplifies to the following form:
$$(c^2 - a^2)x^2 - a^2y^2 = a^2(c^2 - a^2).$$

If we make this substitution
$$b^2 = c^2 - a^2 \qquad (c^2 - a^2 > 0 \text{ since } c > a).$$

The equation for the hyperbola becomes
$$b^2x^2 - a^2y^2 = a^2b^2$$

$$\frac{x^2}{a^2} - \frac{y^2}{b^2} = 1. \quad \text{(Divide both sides by } a^2b^2\text{)}$$

This is the equation for a hyperbola with the vertices on the x-axis and with center at the origin.

A hyperbola becomes easier to graph if we note that the lines $y = \frac{b}{a}x$ and

Asymptotes $y = -\frac{b}{a}x$ are **asymptotes** to the hyperbola. In general, a line is an asymptote for a curve if the distance between the line and the curve approaches zero as we move out farther on the line. To graph a hyperbola, first plot the points $(-a,b)$, (a,b), $(-a,-b)$, and $(a,-b)$ on the coordinate system (see figure 16.37). Draw the rectangle having

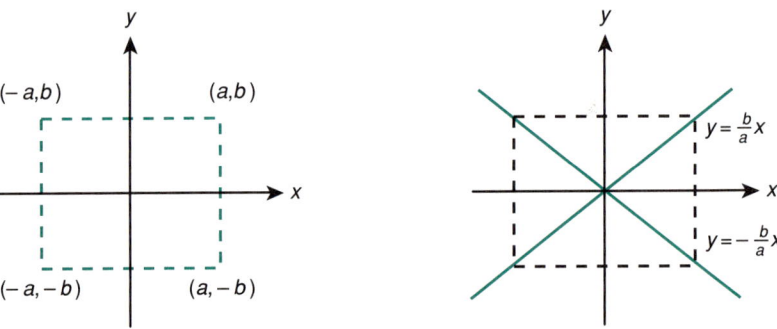

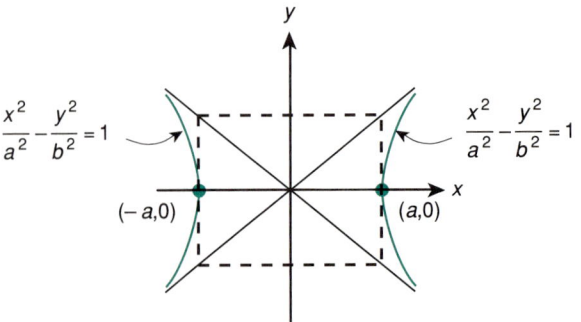

Figure 16.37

these four points as vertices. The lines that contain the diagonals of the rectangle, $y = \dfrac{b}{a}x$ and $y = -\dfrac{b}{a}x$, are the asymptotes of the hyperbola. Plot the vertices $(a,0)$ and $(-a,0)$ and enough additional points to obtain a smooth curve.

The standard form of a hyperbola with the foci on the y-axis and the center at the origin could be found in a similar manner. The equation of this curve is $\dfrac{y^2}{a^2} - \dfrac{x^2}{b^2} = 1$ (see figure 16.38).

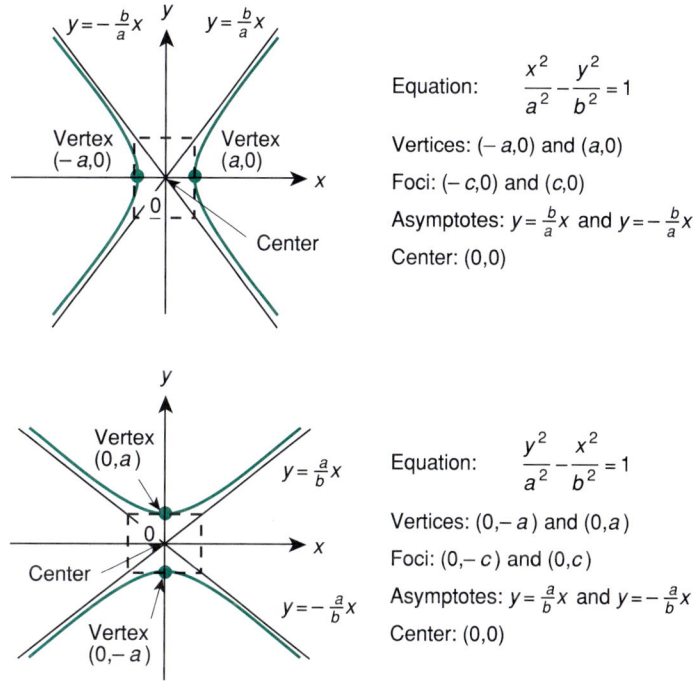

Equation: $\dfrac{x^2}{a^2} - \dfrac{y^2}{b^2} = 1$

Vertices: $(-a,0)$ and $(a,0)$

Foci: $(-c,0)$ and $(c,0)$

Asymptotes: $y = \dfrac{b}{a}x$ and $y = -\dfrac{b}{a}x$

Center: $(0,0)$

Equation: $\dfrac{y^2}{a^2} - \dfrac{x^2}{b^2} = 1$

Vertices: $(0,-a)$ and $(0,a)$

Foci: $(0,-c)$ and $(0,c)$

Asymptotes: $y = \dfrac{a}{b}x$ and $y = -\dfrac{a}{b}x$

Center: $(0,0)$

Figure 16.38

• **Example 16.20:** Determine the vertices, foci, and the equations of the asymptotes of the hyperbola $16x^2 - 9y^2 = 144$. Sketch the graph.

Solution: We first obtain the equation in standard form by dividing both sides of the equation by 144:

$$\frac{16x^2}{144} - \frac{9y^2}{144} = 1$$

$$\frac{x^2}{9} - \frac{y^2}{16} = 1$$

$$\frac{x^2}{3^2} - \frac{y^2}{4^2} = 1.$$

$a = 3$, $b = 4$, and $c = \sqrt{a^2 + b^2} = \sqrt{3^2 + 4^2} = \sqrt{25} = 5$. The vertices are $V_1(-a,0)$ and $V_2(a,0)$ or $V_1(-3,0)$ and $V_2(3,0)$. The foci are $F_1(-c,0)$ and $F_2(c,0)$ or $F_1(-5,0)$ and $F_2(5,0)$. The asymptotes are $y = \frac{b}{a}x$ and $y = -\frac{b}{a}x$ or $y = \frac{4}{3}x$ and $y = -\frac{4}{3}x$.

We now plot the lines $y = \frac{4}{3}x$ and $y = -\frac{4}{3}x$ and the points $V_1(-3,0)$ and $V_2(3,0)$. Points $A(3,4)$, $B(-3,4)$, $C(-3,-4)$, and $D(3,-4)$ allow us to draw the asymptotes. We can now determine several points on the hyperbola and connect them with a smooth curve, as illustrated in figure 16.39.

x	±3.5	±4	±5	±6
y	±2.40	±3.53	±5.33	±6.93

• **Example 16.21:** Determine the vertices, foci, and the equations of the asymptotes of the hyperbola $25y^2 - 9x^2 = 1$. Sketch the graph.

Solution: We first find the standard form of the equation:

$$\frac{y^2}{\left(\frac{1}{5}\right)^2} - \frac{x^2}{\left(\frac{1}{3}\right)^2} = 1.$$

$a = \frac{1}{5}$, $b = \frac{1}{3}$, and $c = \sqrt{\left(\frac{1}{5}\right)^2 + \left(\frac{1}{3}\right)^2}$

$$= \sqrt{\frac{34}{225}} = \frac{\sqrt{34}}{15}.$$

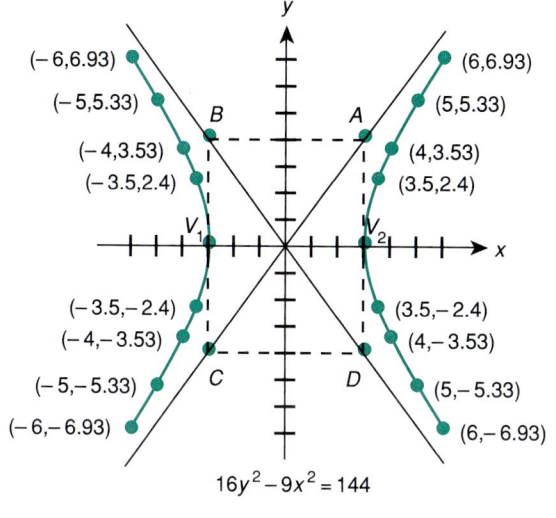

Figure 16.39

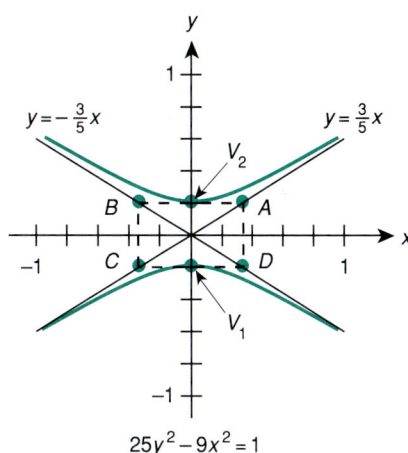

Figure 16.40

The vertices are $V_1(0,-a)$ and $V_2(0,a)$ or $V_1\left(0,-\dfrac{1}{5}\right)$ and $V_2\left(0,\dfrac{1}{5}\right)$. The foci are $F_1(0,-c)$ and $F_2(0,c)$ or $F_1\left(0,-\dfrac{\sqrt{34}}{15}\right)$ and $F_2\left(0,\dfrac{\sqrt{34}}{15}\right)$. The asymptotes are $y=-\dfrac{a}{b}x$ and $y=\dfrac{a}{b}x$ or $y=-\dfrac{3}{5}x$ and $y=\dfrac{3}{5}x$.

We graph the lines $y=-\dfrac{3}{5}x$ and $y=\dfrac{3}{5}x$ and the points $V_1\left(0,-\dfrac{1}{5}\right)$, $V_1\left(0,\dfrac{1}{5}\right)$. Points $A\left(\dfrac{1}{3},\dfrac{1}{5}\right)$, $B\left(-\dfrac{1}{3},\dfrac{1}{5}\right)$, $C\left(-\dfrac{1}{3},-\dfrac{1}{5}\right)$, and $D\left(\dfrac{1}{3},-\dfrac{1}{5}\right)$ helps us to draw the asymptotes. We now plot several points on the hyperbola and connect them with a smooth curve, as illustrated in figure 16.40.

x	0.2	0.5	1
y	0.23	0.36	0.63

• **Example 16.22:** Find the equation of the hyperbola that has vertices $V_1(-4,0)$ and $V_2(4,0)$ and that passes through the point $(8,2)$.

Solution: The equation for the parabola is of this form.

$$\frac{x^2}{a^2} - \frac{y^2}{b^2} = 1. \quad \text{(The vertices are on the x-axis)}$$

Since $V_2(4,0)$ is on the curve, $a = 4$:

$$\frac{x^2}{16} - \frac{y^2}{b^2} = 1.$$

We can substitute $x = 8$ and $y = 2$ in this equation and solve for b^2:

$$\frac{64}{16} - \frac{2^2}{b^2} = 1$$

$$4 - \frac{4}{b^2} = 1$$

$$4 - 1 = \frac{4}{b^2}$$

$$3 = \frac{4}{b^2}$$

$$b^2 = \frac{4}{3}.$$

The equation for the hyperbola is

$$\frac{x^2}{16} - \frac{y^2}{\frac{4}{3}} = 1.$$

•••••••••

• **Example 16.23:** Find the equation of the hyperbola that has vertices $V_1(0,-6)$ and $V_2(0,6)$ and asymptotes with equations $y = \frac{1}{2}x$ and

$$y = -\frac{1}{2}x.$$

Solution: The equation for the hyperbola is of this form:

$$\frac{y^2}{a^2} - \frac{x^2}{b^2} = 1. \quad \text{(The vertices are on the y-axis)}$$

Since $V_2(0,6)$ is a vertex $a = 6$. The asymptote $y = \frac{1}{2}x$ can be written in the form:

$$y = \frac{a}{b}x = \frac{1}{2}x.$$

Since $a = 6$,

$$\frac{6x}{b} = \frac{1x}{2}$$

$$\frac{6}{b} = \frac{1}{2}$$

$$b = 12.$$

The equation is $\frac{y^2}{6^2} - \frac{x^2}{(12)^2} = 1$ or $\frac{y^2}{36} - \frac{x^2}{144} = 1$.

• • • • • • • • • • •

As with the other conics, equations for hyperbolas that have center (h,k) and that have their axes parallel to the coordinate axes can be developed. The details are shown in figure 16.41.

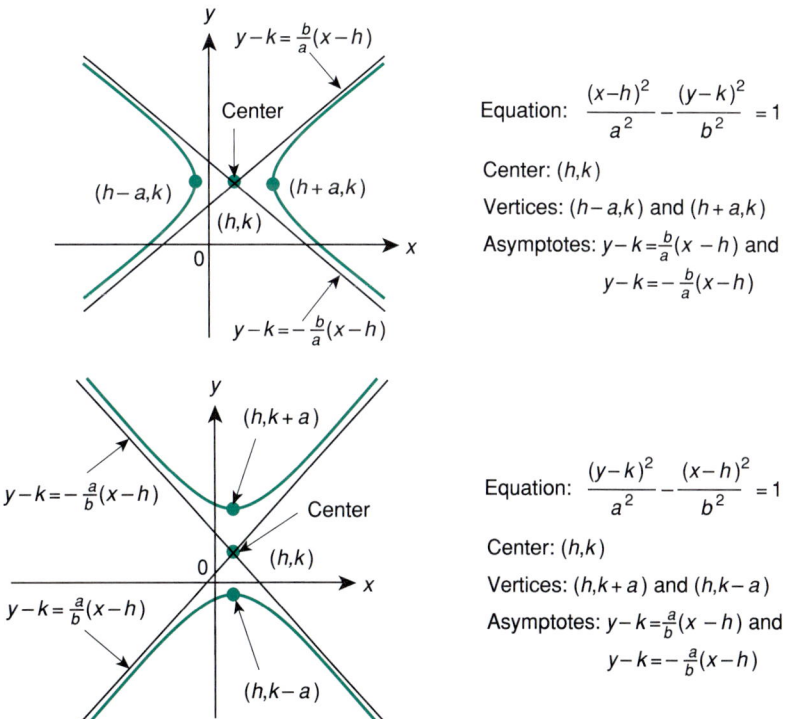

Figure 16.41

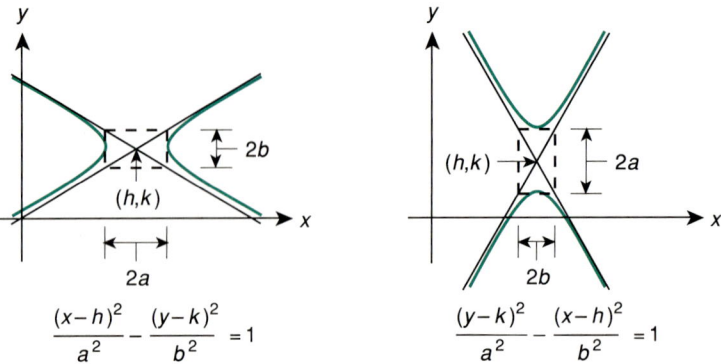

Figure 16.42

Practically speaking, a graph of such a hyperbola can be drawn by first drawing a rectangle similar to the ones we used for hyperbolas with center at the origin. Then draw the asymptotes and plot some additional points on the hyperbola. See figure 16.42.

• **Example 16.24:** Graph the conic represented by the following equation:

$$9x^2 - 18x - 4y^2 - 16y - 11 = 0.$$

Solution: Since the equation contains both x^2 and y^2 terms, we need to complete the square on both of these expressions.

$$9x^2 - 18x + \underline{} \quad \text{and} \quad -4y^2 - 16y + \underline{}.$$

Since the y^2 term is negative, the equation takes this form:

$$\frac{(x-h)^2}{a^2} - \frac{(y-k)^2}{b^2} = 1$$

$$(9x^2 - 18x) - (4y^2 + 16y) = 11$$
$$9(x^2 - 2x + \underline{}) - 4(y^2 + 4y + \underline{}) = 11$$
$$9(x^2 - 2x + 1) - 4(y^2 + 4y + 4) = 11 + 9 - 16$$
$$9(x - 1)^2 - 4(y + 2)^2 = 4$$
$$\frac{9(x-1)^2}{4} - \frac{4(y+2)^2}{4} = 1$$
$$\frac{(x-1)^2}{\frac{4}{9}} - \frac{(y+2)^2}{1} = 1$$
$$\frac{(x-1)^2}{\left(\frac{2}{3}\right)^2} - \frac{(y-(-2))^2}{(1)^2} = 1.$$

Thus, $a = \frac{2}{3}$, $b = 1$, $h = 1$, and $k = -2$. The curve is a hyperbola with center at $(1,-2)$. The vertices are $(h - a,k)$ and $(h + a,k)$ or $\left(\frac{1}{3},-2\right)$ and $\left(\frac{5}{3},-2\right)$. Draw the required rectangle and the asymptotes. Plot some additional points and sketch the curve. The results are shown in figure 16.43.

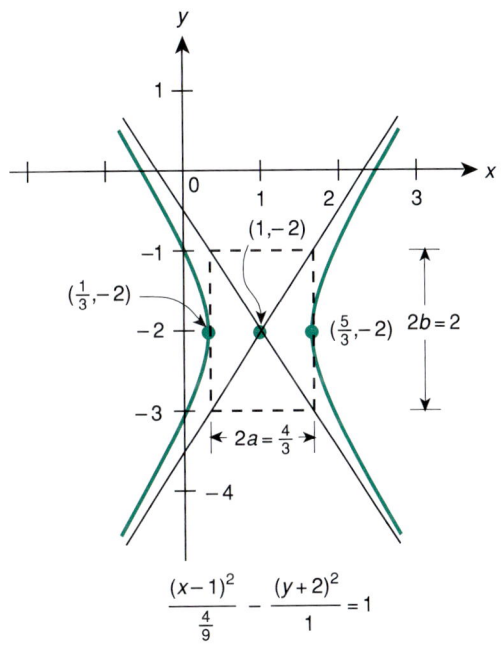

Figure 16.43

Trial Problems 16.6

Before you begin the section exercises, warm up with these problems. Complete answers are included in the answer key.

Find the vertices, the foci, and the asymptotes for each hyperbola.

1. $x^2 - y^2 = 9$

2. $\dfrac{(x + 2)^2}{9} - \dfrac{(y + 2)^2}{4} = 1$

Find an equation for the hyperbola that fits the given conditions.

3. Vertices $(0,-2)$ and $(0,2)$, foci $(0,-4)$ and $(0,4)$

4. Vertices $(-4,0)$ and $(4,0)$, contains the point $(8,3)$

Exercises 16.6

Find the vertices, foci, and asymptotes for each hyperbola. Sketch a graph.

1. $\dfrac{x^2}{9} - \dfrac{y^2}{16} = 1$
2. $\dfrac{y^2}{9} - \dfrac{x^2}{16} = 1$
3. $\dfrac{x^2}{16} - \dfrac{y^2}{9} = 1$
4. $\dfrac{y^2}{16} - \dfrac{x^2}{9} = 1$
5. $\dfrac{y^2}{64} - \dfrac{x^2}{81} = 1$
6. $\dfrac{x^2}{64} - \dfrac{y^2}{81} = 1$
7. $\dfrac{y^2}{81} - \dfrac{x^2}{64} = 1$
8. $\dfrac{x^2}{81} - \dfrac{y^2}{64} = 1$
9. $\dfrac{x^2}{10} - \dfrac{y^2}{5} = 1$
10. $\dfrac{y^2}{10} - \dfrac{x^2}{5} = 1$
11. $\dfrac{x^2}{\frac{1}{4}} - \dfrac{y^2}{2} = 1$
12. $\dfrac{y^2}{\frac{1}{4}} - \dfrac{x^2}{2} = 1$
13. $\dfrac{y^2}{\frac{1}{4}} - \dfrac{x^2}{\frac{1}{9}} = 1$
14. $\dfrac{x^2}{\frac{1}{4}} - \dfrac{y^2}{\frac{1}{9}} = 1$

Write each equation in standard form. Sketch the graph.

15. $4x^2 - 9y^2 = 36$
16. $4y^2 - 9x^2 = 36$
17. $9y^2 - 4x^2 = 36$
18. $9x^2 - 4y^2 = 36$
19. $x^2 - 4y^2 = 4$
20. $y^2 - 4x^2 = 4$
21. $4y^2 - 4x^2 = 16$
22. $4x^2 - 4y^2 = 16$
23. $25x^2 - 9y^2 = 225$
24. $25y^2 - 9x^2 = 225$
25. $25x^2 - 4y^2 = 100$
26. $25y^2 - 4x^2 = 100$
27. $9x^2 - 16y^2 = 1$
28. $9y^2 - 16x^2 = 1$
29. $225y^2 - 196x^2 = 36$
30. $225x^2 - 196y^2 = 36$

Write the equation, in standard form, for an hyperbola that fits the given conditions. Assume that the center is the origin.

31. $F_1(-4,0)$, $F_2(4,0)$, $V_1(-3,0)$, and $V_2(3,0)$
32. $F_1(0,-4)$, $F_2(0,4)$, $V_1(0,-3)$, and $V_2(0,3)$
33. $V_1(0,-3)$ and $V_2(0,3)$, asymptotes $y = \dfrac{5}{3}x$ and $y = -\dfrac{5}{3}x$
34. $V_1(-3,0)$ and $V_2(3,0)$, asymptotes $y = \dfrac{3}{5}x$ and $y = -\dfrac{3}{5}x$
35. $F_1(-5,0)$ and $F_2(5,0)$, asymptotes $y = 2x$ and $y = -2x$
36. $F_1(0,-5)$ and $F_2(0,5)$, asymptotes $y = \dfrac{x}{2}$ and $y = -\dfrac{x}{2}$
37. Contains the points $\left(\sqrt{5}, \dfrac{1}{2}\right)$ and $(2\sqrt{10}, 3)$
38. Contains the points $\left(\dfrac{1}{2}, \sqrt{5}\right)$ and $(3, 2\sqrt{10})$
39. Contains the points $(\sqrt{3}, 2)$ and $(3, 4)$
40. Contains the points $(2, \sqrt{3})$ and $(4, 3)$

Sketch the graph for each equation.

41. $4x^2 + 8x - 9y^2 + 90y - 257 = 0$

42. $9x^2 - 16y^2 - 160y - 544 = 0$

43. $16x^2 - 96x - 9y^2 + 36y - 36 = 0$

44. $4x^2 - 24x - 9y^2 - 90y - 225 = 0$

Write an equation for each hyperbola.

45. Hyperbola, center at $(7, -2)$, opens horizontally, $a = 8$, $b = 6$

46. Hyperbola, center at $(4, -6)$, opens vertically, $a = 8$, $b = 4$

> The equation $xy = k$, where k is a constant, represents a hyperbola that is not in one of the standard positions.

47. Graph the hyperbola $xy = 16$ and determine the asymptotes.

48. Graph the hyperbola $xy = 25$ and find the asymptotes.

49. For a gas at constant temperature, the product of the temperature and the volume is constant ($pv = k$). If the constant for a gas is 450, plot a graph of p (in KPa) against v (in cu m).

50. Solve exercise 49 given that the constant is 4500.

51. In a particular circuit the product of the current (i) and the resistance (r) in a resistor is constant. Graph the relationship between the variables if the constant is 300.

52. Solve exercise 51 given that the constant is 600.

53. Verify the fact that the conditions of the hyperbola $y^2 - x^2 = 1$ are satisfied by $x = \sinh t$, $y = \cosh t$, where the *hyperbolic sine* (sinh t) and the *hyperbolic cosine* (cosh t) are defined as follows:

$$\sinh t = \frac{e^t - e^{-t}}{2} \qquad \cosh t = \frac{e^t + e^{-t}}{2}.$$

54. Verify that the point $\left(\dfrac{br}{s^2 - r^2}, \dfrac{as}{s^2 - r^2}\right)$ is a point on the hyperbola $\dfrac{y^2}{a^2} - \dfrac{x^2}{b^2}$ given that $s^2 - r^2 > 0$.

55. Two receiving stations 12,000 feet apart record an explosion. One of the stations records the explosion 6 seconds after the other station. Given that sound travels at about 1100 ft/sec, write an equation for the possible locations of the explosion using the stations as the foci for the conic. *Hint:* Put the foci on the x-axis.

56. Solve exercise 55 given that the receiving stations are 9000 ft apart and the difference in times is 3 seconds.

16·7 Polar Equations

Polar Coordinates

In chapter 9 we used the polar form of a complex number to determine powers and roots of complex numbers. In this section, we introduce the **polar coordinate** system and the techniques for graphing equations on the polar coordinate plane.

Given a point $P(x,y)$ on a rectangular coordinate system, we can determine a pair (r,θ) in polar coordinates using the following transformation equations. Refer to figure 16.44.

$$r^2 = x^2 + y^2, \quad \theta = \tan^{-1}\left(\frac{y}{x}\right) \quad \left(\tan\theta = \frac{y}{x}\right)$$

$$x = r\cos\theta, \quad y = r\sin\theta. \quad \left(\cos\theta = \frac{x}{r}, \sin\theta = \frac{y}{r}\right)$$

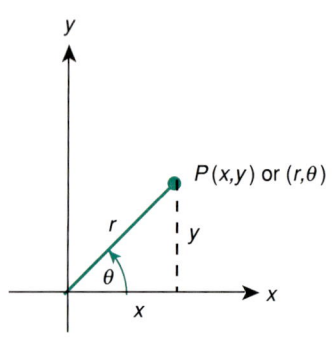

Figure 16.44

For example, the point $(x,y) = (2, 2\sqrt{3})$ in rectangular coordinates can be changed to polar coordinates as follows:

$$r^2 = x^2 + y^2 = (2)^2 + (2\sqrt{3})^2 = 4 + 12 = 16$$
$$r = \sqrt{16} = 4$$

$$\theta = \tan^{-1}\left(\frac{y}{x}\right) = \tan^{-1}\left(\frac{2\sqrt{3}}{2}\right) = \tan^{-1}\sqrt{3} = \frac{\pi}{3}.$$

Hence, $(r,\theta) = \left(4, \frac{\pi}{3}\right)$.

Given the point $\left(8, \frac{\pi}{3}\right)$ in polar coordinates, we can change to rectangular coordinates as follows:

$$x = r\cos\theta = 8\left(\cos\frac{\pi}{3}\right) = 8\left(\frac{1}{2}\right) = 4$$

$$y = r\sin\theta = 8\left(\sin\frac{\pi}{3}\right) = 8\left(\frac{\sqrt{3}}{2}\right) = 4\sqrt{3}.$$

Hence, $(x,y) = (4, 4\sqrt{3})$.

We can also use these equations to transform an equation from one system to the other system. Examples 16.25 through 16.28 illustrate these transformations.

Polar Equation

• *Example 16.25:* Transform the equation of the circle $x^2 + y^2 = 25$ into a **polar equation**.

Solution: Substitute $x = r\cos\theta$ and $y = r\sin\theta$ into the equation $x^2 + y^2 = 25$:

$$(r\cos\theta)^2 + (r\sin\theta)^2 = 25$$
$$r^2\cos^2\theta + r^2\sin^2\theta = 25$$
$$r^2(\cos^2\theta + \sin^2\theta) = 25.$$

Since $\cos^2 \theta + \sin^2 \theta = 1$,

$$r^2 = 25$$
$$r = 5.$$

We can see that when $r = 5$ is plotted on a coordinate system, we are plotting points of the form $(5,\theta)$, where θ represents any number such that $0 \leq \theta \leq 2\pi$, and we get a circle with center at the origin with a radius of 5. This circle is graphed on both coordinate systems in figure 16.45.

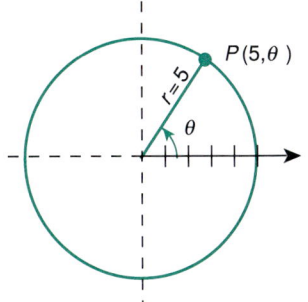

Figure 16.45

• **Example 16.26:** Write the equation $\dfrac{x^2}{9} + \dfrac{y^2}{16} = 1$ in polar form.

Solution: Substitute $x = r \cos \theta$ and $y = r \sin \theta$ into the equation:

$$\frac{(r \cos \theta)^2}{9} + \frac{(r \sin \theta)^2}{16} = 1$$

$$\frac{r^2 \cos^2 \theta}{9} + \frac{r^2 \sin^2 \theta}{16} = 1$$

$$16 r^2 \cos^2 \theta + 9 r^2 \sin^2 \theta = 144.$$

Replace $16 r^2 \cos^2 \theta$ with $7 r^2 \cos^2 \theta + 9 r^2 \cos^2 \theta$, so that we obtain the expression $9 r^2 \cos^2 \theta + 9 r^2 \sin^2 \theta$ in the equation:

$$7 r^2 \cos^2 \theta + 9 r^2 (\cos^2 \theta + \sin^2 \theta) = 144.$$

Use the relation $\cos^2 \theta + \sin^2 \theta = 1$:

$$7 r^2 \cos^2 \theta + 9 r^2 = 144$$
$$r^2 (7 \cos^2 \theta + 9) = 144$$

$$r^2 = \frac{144}{7 \cos^2 \theta + 9}.$$

• **Example 16.27:** Change the equation $r = \dfrac{6}{2 \cos \theta + 3 \sin \theta}$ to rectangular form.

Solution: Multiply both sides of the equation by $2 \cos \theta + 3 \sin \theta$:

$$r(2 \cos \theta + 3 \sin \theta) = 6$$
$$2r \cos \theta + 3r \sin \theta = 6.$$

Substitute x for $r \cos \theta$ and y for $r \sin \theta$:

$$2x + 3y = 6,$$

the equation of a line.

• **Example 16.28:** Change $r = \dfrac{4}{1 + \sin \theta}$ to rectangular coordinates.

Solution: Multiply both sides of the equation by $1 + \sin \theta$:

$$r(1 + \sin \theta) = 4$$
$$r + r \sin \theta = 4.$$

Substitute $r = \sqrt{x^2 + y^2}$ and $y = r \sin \theta$:

$$\sqrt{x^2 + y^2} + y = 4$$
$$\sqrt{x^2 + y^2} = -y + 4.$$

Square both sides:

$$x^2 + y^2 = (-y + 4)^2$$
$$x^2 + y^2 = y^2 - 8y + 16$$
$$x^2 + 8y - 16 = 0,$$

the equation of a parabola.

• • • • • • • • • • •

We now consider the techniques for plotting curves in polar form. Recall the method we used in plotting complex numbers in polar form. We use a polar coordinate system as illustrated in figure 16.46.

Pole
Polar Axis

We call the origin the **pole** and the horizontal axis the **polar axis**. We plot a point (r,θ) by determining the angle θ, measured counterclockwise from the polar axis, and move r units on the terminal side of the angle. If r is negative, we move r units on the

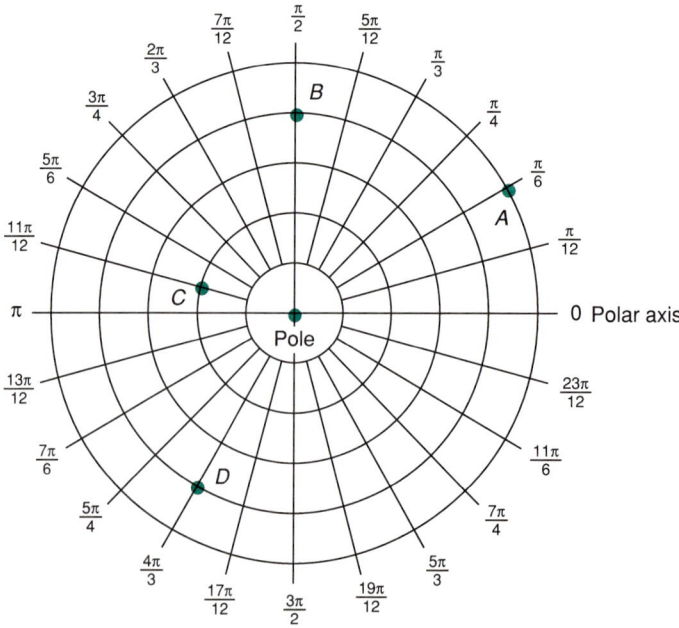

Figure 16.46

opposite ray from the terminal side of the angle. Figure 16.46 shows the appropriate locations of the points $A\left(5,\dfrac{\pi}{6}\right)$ or $\left(5,\dfrac{7\pi}{6}\right)$, $B\left(4,\dfrac{\pi}{2}\right)$ or $\left(-4,\dfrac{3\pi}{2}\right)$, $C\left(2,\dfrac{11\pi}{12}\right)$ or $\left(-2,\dfrac{23\pi}{12}\right)$ and $D\left(3,\dfrac{4\pi}{3}\right)$ or $\left(-3,\dfrac{\pi}{3}\right)$.

Note that we can use more than one set of coordinates for the same point. In general, $P(r,\theta)$ and $P(-r,\theta + \pi)$ represent the same points.

We now consider the graphs of equations in polar coordinates. Just as we did for equations to be graphed on a rectangular coordinate system, we find a set of ordered pairs that are solutions to the polar equation and graph these ordered pairs. If a pattern develops, we join the plotted points with a smooth curve.

• **Example 16.29:** Graph the equation $r = \dfrac{4}{\cos \theta}$.

Solution: We first construct a table of values that satisfy the equation.

θ	0	$\dfrac{\pi}{6}$	$\dfrac{\pi}{4}$	$\dfrac{\pi}{3}$	$\dfrac{\pi}{2}$	$\dfrac{2\pi}{3}$	$\dfrac{3\pi}{4}$	$\dfrac{5\pi}{6}$	π
r	4	4.62	5.66	8	undefined	-8	-5.66	-4.62	-4

After plotting these points on figure 16.47, it becomes obvious that the graph is a line.

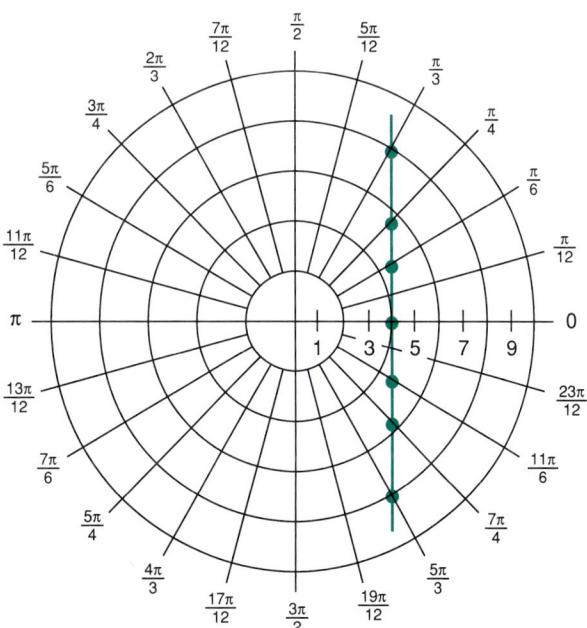

Figure 16.47

Example 16.30: Graph the equation $r = 1 - \cos \theta$.

Solution: Construct a table of values that satisfy the equation.

θ	0	$\frac{\pi}{6}$	$\frac{\pi}{4}$	$\frac{\pi}{3}$	$\frac{\pi}{2}$	$\frac{2\pi}{3}$	$\frac{3\pi}{4}$	$\frac{5\pi}{6}$	π	$\frac{7\pi}{6}$	$\frac{5\pi}{4}$	$\frac{4\pi}{3}$	$\frac{3\pi}{2}$	$\frac{5\pi}{3}$	$\frac{7\pi}{4}$	$\frac{11\pi}{6}$	2π
r	0.0	0.13	0.29	0.5	1.0	1.5	1.71	1.87	2.0	1.87	1.71	1.5	1.0	0.5	0.29	0.13	0.0

Connect the points in order with a smooth curve as shown in figure 16.48. The figure is called a cardioid because of its heartlike shape.

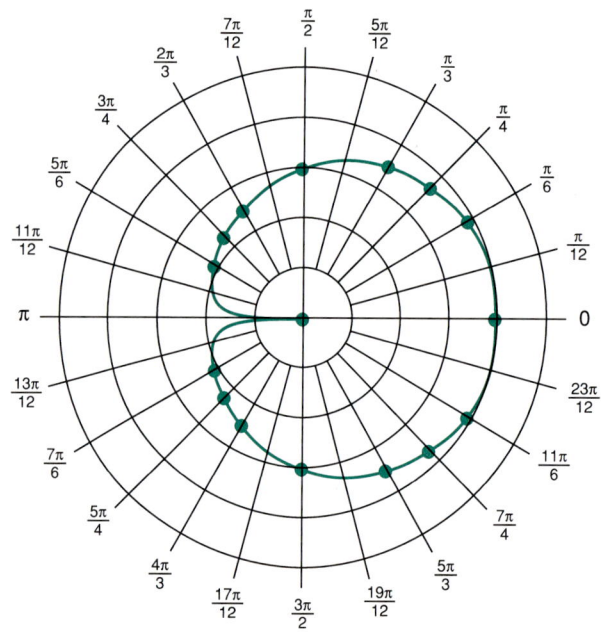

Figure 16.48

Example 16.31: Graph the equation $r = 2 \sin 2\theta$.

Solution: Construct a table of values that satisfy the equation.

θ	0	$\frac{\pi}{6}$	$\frac{\pi}{4}$	$\frac{\pi}{3}$	$\frac{\pi}{2}$	$\frac{2\pi}{3}$	$\frac{3\pi}{4}$	$\frac{5\pi}{6}$	π	$\frac{7\pi}{6}$	$\frac{5\pi}{4}$	$\frac{4\pi}{3}$	$\frac{3\pi}{2}$	$\frac{5\pi}{3}$	$\frac{7\pi}{4}$	$\frac{11\pi}{6}$	2π
r	0	1.73	2	1.73	0	-1.73	-2	-1.73	0	1.73	2	1.73	0	-1.73	-2	-1.73	0

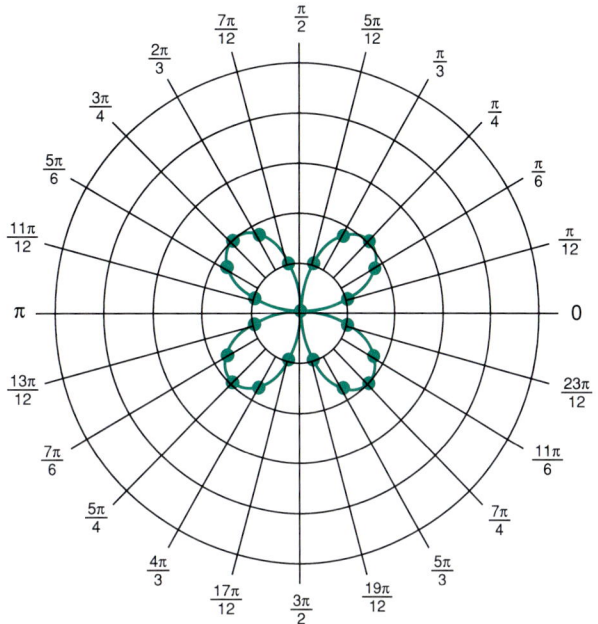

Figure 16.49

These points provide a pattern as indicated in figure 16.49. However, we can get a more accurate picture if we plot more points.

θ	$\dfrac{\pi}{12}$	$\dfrac{5\pi}{12}$	$\dfrac{7\pi}{12}$	$\dfrac{11\pi}{12}$	$\dfrac{13\pi}{12}$	$\dfrac{17\pi}{12}$	$\dfrac{19\pi}{12}$	$\dfrac{23\pi}{12}$
r	1	1	-1	-1	1	1	-1	-1

The resulting figure is called a four-leaved rose.

• **Example 16.32:** Graph the equation $r = \theta$ for $\theta \geq 0$.

Solution: When $\theta = k\pi$, for k a positive integer, the graph intersects the polar axis or the line $\theta = \dfrac{\pi}{2}$. When $\theta = \dfrac{k\pi}{2}$, where k is an odd positive integer, the graph intersects one of the lines $\theta = \dfrac{\pi}{2}$ or $\theta = \dfrac{3\pi}{2}$. Use these facts to plot a number of points on the graph.

θ	0	$\frac{\pi}{2}$	$\frac{3\pi}{2}$	2π	$\frac{5\pi}{2}$	3π	$\frac{7\pi}{2}$	4π
r	0	$\frac{\pi}{2}$	$\frac{3\pi}{2}$	2π	$\frac{5\pi}{2}$	3π	$\frac{7\pi}{2}$	4π

The resulting *spiral*, obtained by connecting these points with a smooth curve, is shown in figure 16.50.

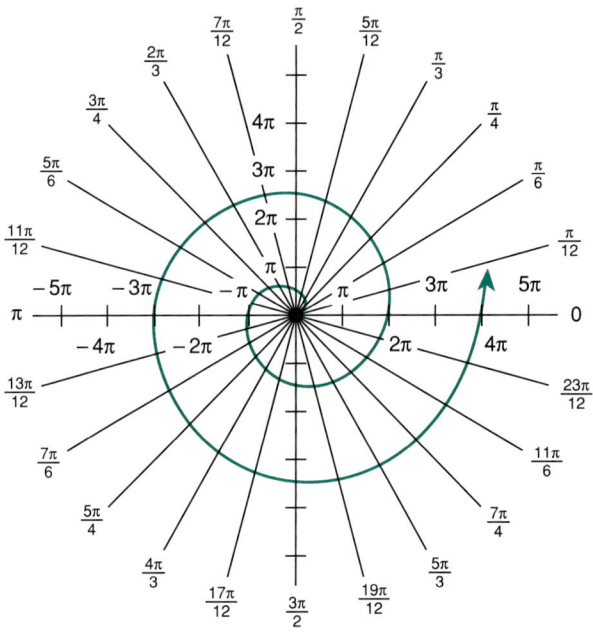

Figure 16.50

Trial Problems 16.7

Before you begin the section exercises, warm up with these problems. Complete answers are included in the answer key.

1. Change the point $\left(3, \frac{7\pi}{6}\right)$ to rectangular coordinates.

2. Change the point $(4, -4)$ to polar coordinates.

3. Write the equation $x^2 - y^2 = 9$ in polar form.

4. Write the equation $r = \dfrac{4}{2 - \cos\theta}$ in rectangular form.

5. Graph $r = 1 - \sin\theta$ (cardioid).

Exercises 16.7

Change each point (r,θ) from polar coordinates to rectangular coordinates.

1. $(2,0)$
2. $(-2,\pi)$
3. $\left(-2,\dfrac{3\pi}{2}\right)$
4. $\left(2,\dfrac{\pi}{2}\right)$
5. $\left(1,\dfrac{\pi}{4}\right)$
6. $\left(-1,\dfrac{5\pi}{4}\right)$
7. $\left(1,\dfrac{5\pi}{4}\right)$
8. $\left(-1,\dfrac{\pi}{4}\right)$
9. $\left(2\sqrt{2},\dfrac{\pi}{4}\right)$
10. $\left(2\sqrt{2},\dfrac{7\pi}{4}\right)$
11. $\left(\dfrac{1}{2},\dfrac{\pi}{6}\right)$
12. $\left(\dfrac{1}{2},\dfrac{5\pi}{6}\right)$
13. $\left(-5,\dfrac{7\pi}{4}\right)$
14. $\left(5,\dfrac{5\pi}{4}\right)$
15. $\left(2,\dfrac{\pi}{12}\right)$
16. $\left(2,\dfrac{11\pi}{12}\right)$

Change each point from rectangular coordinates to polar coordinates.

17. $(2,0)$
18. $(-2,0)$
19. $(0,-4)$
20. $(4,0)$
21. $(3,3)$
22. $(2,-2)$
23. $(\sqrt{3},1)$
24. $(-\sqrt{3},-1)$
25. $(\sqrt{2},-\sqrt{2})$
26. $(-\sqrt{2},\sqrt{2})$

Change each equation to an equation in polar coordinates.

27. $x + y = 5$
28. $x - y = 6$
29. $x^2 + y^2 = 16$
30. $x^2 + y^2 = 9$
31. $x^2 - y^2 = 9$
32. $x^2 - y^2 = 16$
33. $x^2 - 2x + y^2 = 10$
34. $x^2 + y^2 - 2y = 0$
35. $y = 2x^2$
36. $y = 4x^2$
37. $x^2 + 4y^2 = 25$
38. $4x^2 + y^2 = 25$

Change each equation to an equation in rectangular coordinates.

39. $r = 2$
40. $r = 3$
41. $r \cos \theta = 2$
42. $r \sin \theta = 3$
43. $\theta = \dfrac{\pi}{4}$
44. $\theta = \dfrac{3\pi}{4}$
45. $r = 4 \cos \theta$
46. $r = 6 \cos \theta$
47. $3r \cos \theta - 4r \sin \theta = 7$
48. $2r \cos \theta - 3r \sin \theta = 6$
49. $r = \dfrac{15}{4 - 4 \cos \theta}$
50. $r = \dfrac{3}{2 - 2 \cos \theta}$
51. $r = \dfrac{4}{\cos \theta - 2}$
52. $r = \dfrac{2}{\cos \theta - 1}$

Graph each equation in polar coordinates.

53. $r = 4$
54. $r = 2$
55. $r = 4 \cos \theta$
56. $r = 4 \sin \theta$
57. $r = -4 \sin \theta$
58. $r = -4 \cos \theta$
59. $r = 4 + 4 \sin \theta$ (cardioid)
60. $r = 1 + \cos \theta$ (cardioid)
61. $r^2 = 4 \cos 2\theta$ (lemniscate)
62. $r^2 = 4 \sin 2\theta$ (lemniscate)
63. $r = 4 \cos 2\theta$ (four-leaved rose)
64. $r = 4 \sin 2\theta$ (four-leaved rose)

65. $r = \dfrac{\theta}{\pi}$ for $0 \leq \theta \leq 4\pi$ (spiral)

66. $r = \dfrac{2\theta}{\pi}$ for $0 \leq \theta \leq 4\pi$ (spiral)

67. $r = \sin 3\theta$ (three-leaved rose)

68. $r = \cos 3\theta$ (three-leaved rose)

69. Halley's comet has an orbit with respect to the sun that can be approximated by the equation $r = \dfrac{1.07}{1 + 0.97 \sin \theta}$, where the sun is at the pole. r is measured in astronomical units (AU) and 1 AU = 93 million miles. What is the shape of the orbit of the comet?

70. A comet has an orbit with respect to the sun that can be approximated by the equation $r = \dfrac{40{,}000{,}000}{1 - \cos \theta}$, where r is measured in miles, and the sun is at the pole. What is the shape of the orbit?

16·8 Summary of Terms and Formulas

Terms

Circle (p. 627)
Conic Sections (p. 618)
Ellipse (p. 643)
    Center (p. 643)
    Major Axis (p. 643)
    Minor Axis (p. 643)
    Vertices (p. 643)

Hyperbola (p. 653)
    Asymptotes (p. 654)
    Center (p. 653)
    Foci (p. 653)
    Transverse Axis (p. 653)
    Vertices (p. 653)

Parabola (p. 632)
    Axis of Symmetry (p. 632)
    Directrix (p. 633)
    Focus (p. 633)
Point-Slope Form of a Line (p. 622)
Polar Axis (p. 666)
Polar Coordinates (p. 664)
Polar Equation (p. 664)
Pole (p. 666)
Slope of a Line (p. 620)

Formulas

Equations for a line: (16.2)

Form of the Equation	Explanation
$ax + by + c = 0$	Standard form of a line
$y = mx + b$	Slope $= m$, y-intercept $= b$
$y - y_1 = m(x - x_1)$	Slope $= m$, $P(x_1, y_1)$ is a point on the line
$y - y_1 = \left(\dfrac{y_2 - y_1}{x_2 - x_1}\right)(x - x_1)$	$A(x_1, y_1)$ and $B(x_2, y_2)$ are points on the line
$y = k$	Line parallel to the x-axis
$x = c$	Line parallel to the y-axis

Equation of a Circle: (16.3)

$$(x - h)^2 + (y - k)^2 = r^2$$

Equations for Parabolas: (16.4)

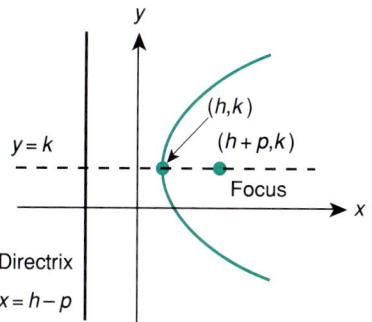

Standard form: $(y-k)^2 = 4p(x-h)$
Axis of symmetry: $y = k$
Opens to the right

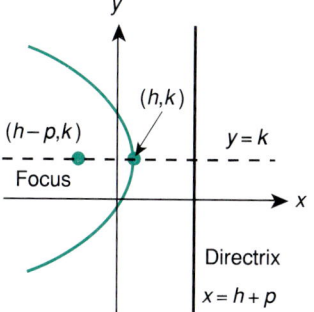

Standard form: $(y-k)^2 = -4p(x-h)$
Axis of symmetry: $y = k$
Opens to the left

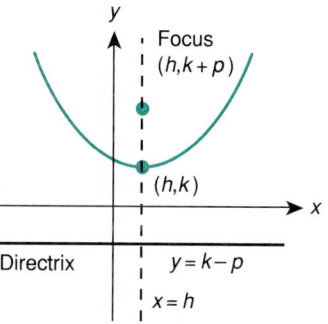

Standard form: $(x-h)^2 = 4p(y-k)$
Axis of symmetry: $x = h$
Opens upward

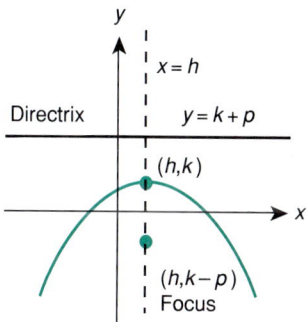

Standard form: $(x-h)^2 = -4p(y-k)$
Axis of symmetry: $x = h$
Opens downward

Equations for Ellipses: (16.5)

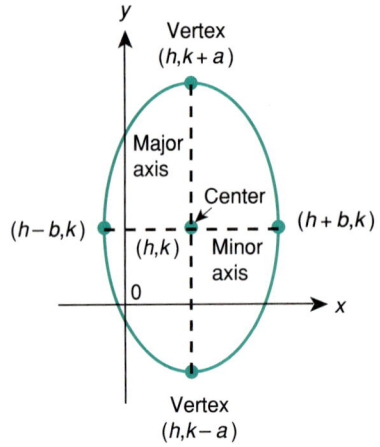

Equation: $\dfrac{(x-h)^2}{b^2} + \dfrac{(y-k)^2}{a^2} = 1, \quad a > b$

Center: (h,k)

Vertices: $(h, k-a)$ and $(h, k+a)$

Foci: $(h, k-c)$ and $(h, k+c)$, where $c^2 = a^2 - b^2$

Ends of minor axis: $(h-b, k)$ and $(h+b, k)$

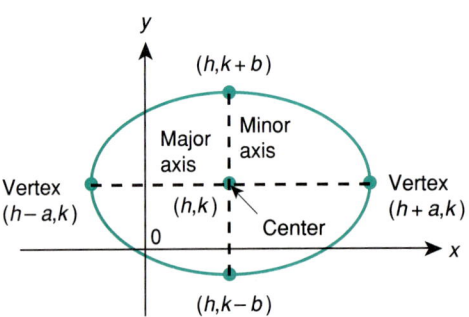

Equation: $\dfrac{(x-h)^2}{a^2} + \dfrac{(y-k)^2}{b^2} = 1, \quad a > b$

Center: (h,k)

Vertices: $(h-a, k)$ and $(h+a, k)$

Foci: $(h-c, k)$ and $(h+c, k)$, where $c^2 = a^2 - b^2$

Ends of minor axis: $(h, k+b)$ and $(h, k-b)$

Equations for Hyperbolas: (16.6)

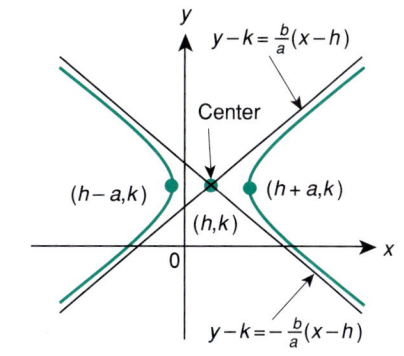

Equation: $\dfrac{(x-h)^2}{a^2} - \dfrac{(y-k)^2}{b^2} = 1$

Center: (h,k)

Vertices: $(h-a,k)$ and $(h+a,k)$

Asymptotes: $y-k = \dfrac{b}{a}(x-h)$ and
$y-k = -\dfrac{b}{a}(x-h)$

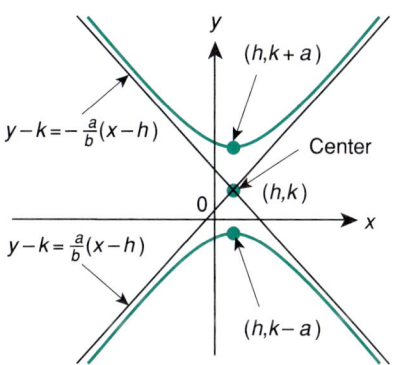

Equation: $\dfrac{(y-k)^2}{a^2} - \dfrac{(x-h)^2}{b^2} = 1$

Center: (h,k)

Vertices: $(h,k+a)$ and $(h,k-a)$

Asymptotes: $y-k = \dfrac{a}{b}(x-h)$ and
$y-k = -\dfrac{a}{b}(x-h)$

Changing between Rectangular and Polar Coordinates: (16.7)

Rectangular to Polar

$r^2 = x^2 + y^2$

$\theta = \tan^{-1}\left(\dfrac{y}{x}\right)$

Polar to Rectangular

$x = r \cos \theta$

$y = r \sin \theta$

16·9 Chapter 16 Review Exercises

Find the distance between the given pairs of points.

1. $A(-3,4)$, $B(2,-3)$
2. $C(-4,3)$, $D(6,-2)$
3. $E(2,6)$, $F(-3,-4)$
4. $G(-5,-3)$, $H(5,3)$

Find the slope of the line containing the given pair of points.

5. $C(4,-3)$, $D(6,-2)$
6. $A(-3,4)$, $B(2,-3)$
7. $G(-5,-3)$, $H(5,3)$
8. $E(2,6)$, $F(-3,-4)$

Write the equation, in the form $ax + by + c = 0$, for the line that fits the given conditions.

9. Slope $= -2$, y-intercept $= 5$
10. Slope $= -3$, y-intercept $= -2$
11. Slope $= -2$, contains the point $A(-3,4)$
12. Slope $= -3$, contains the point $B(2,-3)$
13. Contains the points $A(-3,4)$ and $B(2,-3)$
14. Contains the points $C(-4,3)$ and $D(6,-2)$
15. Contains the point $E(2,6)$, is parallel to the line $2x + 3y = 6$
17. Contains the point $G(-5,-3)$, is perpendicular to the line $2x + 3y = 6$
16. Contains the point $F(-3,-4)$, is parallel to the line $2x + 3y = 6$
18. Contains the point $H(5,3)$, is perpendicular to the line $2x + 3y = 6$

Write the equation of the circle that fits the given conditions.

19. Center at the origin, radius $= 3.5$
22. Center at $(2,4)$, radius $= 3$
20. Center at the origin, radius $= 1.5$
23. Center at $(3,0)$, contains the point $(5,6)$
21. Center at $(3,2)$, radius $= 4$
24. Center at $(0,3)$, contains the point $(4,-2)$

Write the equation of the ellipse that fits the given conditions.

25. Foci $(-3,0)$, $(3,0)$, vertices $(-4,0)$, $(4,0)$
28. Center at the origin, contains the points $(0,4)$, $(2,3)$
26. Foci $(0,-3)$, $(0,3)$, vertices $(0,-4)$, $(0,4)$
29. Center at $(2,-1)$, vertices at $(2,3)$, $(2,-5)$, contains the point $(5,-1)$
27. Center at the origin, contains the points $(4,0)$, $(3,2)$
30. Center at $(5,7)$, vertices at $(8,7)$, $(2,7)$, length of minor axis $= 2$

Write the equation of the parabola that fits the given conditions.

31. Vertex $(3,0)$, focus $(5,0)$
34. Vertex $(0,-2)$, contains the point $(1,-3)$
32. Vertex $(0,3)$, focus $(0,5)$
35. Vertex $(3,2)$, vertical axis, contains the point $(6,1)$
33. Vertex $(2,0)$, contains the point $(3,1)$
36. Vertex $(2,3)$, horizontal axis, contains the point $(1,6)$

Write the equation of the hyperbola that fits the given conditions.

37. Vertices $(-1,0)$, $(1,0)$, asymptotes $y = 2x$, $y = -2x$

38. Vertices $(0,-1)$, $(0,1)$, asymptotes $y = \dfrac{x}{2}$, $y = -\dfrac{x}{2}$

39. Vertices $(-3,0)$, $(3,0)$, foci $(-4,0)$, $(4,0)$

40. Vertices $(0,-3)$, $(0,3)$, foci $(0,-4)$, $(0,4)$

41. Center $(-1,2)$, vertices $(-1,-1)$, $(-1,5)$, asymptotes $3x - 2y + 7 = 0$, $3x + 2y - 1 = 0$

42. Center $(-1,3)$, vertices $(-4,3)$, $(2,3)$, asymptotes $2x - 3y + 11 = 0$, $2x + 3y - 7 = 0$

Graph the following equations.

43. $x^2 - 6x + y^2 + 8y + 17 = 0$

44. $x^2 + 8x + y^2 - 6y + 17 = 0$

45. $16x^2 + 7y^2 - 112 = 0$

46. $7x^2 + 16y^2 - 112 = 0$

47. $16x^2 - 96x + 25y^2 - 100y - 156 = 0$

48. $25x^2 - 100x + 16y^2 - 96y - 156 = 0$

49. $x - y^2 + 2y + 2 = 0$

50. $x^2 - 2x - y - 2 = 0$

51. $5x^2 - 3y^2 - 15 = 0$

52. $3x^2 - 5y^2 + 15 = 0$

53. $9x^2 + 90x - 16y^2 + 64y + 17 = 0$

54. $9x^2 - 36x - 16y^2 + 96y + 36 = 0$

Graph the given equations on a polar coordinate system.

55. $r = 3 \cos \theta$

56. $r = 3 \sin \theta$

57. $r = 2 - 2 \cos \theta$

58. $r = 2 + 2 \cos \theta$

59. $r = \dfrac{3}{2 - 2 \cos \theta}$

60. $r = \dfrac{2}{1 - \cos \theta}$

16·10 Chapter 16 Test

Find the distance between the pairs of points.

1. $A(1,-1)$, $B(2,-3)$

2. $C(-3,-2)$, $D(5,6)$

Find the slope of the line that contains the pair of points.

3. $E(6,-1)$, $F(7,-3)$

4. $G(2,2)$, $H(-3,0)$

Write an equation, in the form $ax + by + c = 0$, for the line that fits the given conditions.

5. Slope $= -\dfrac{1}{2}$, contains the point $A(4,2)$

6. Contains the point $B(-2,3)$, is parallel to the line $3x + 2y = 12$

Write an equation for a circle that fits the given conditions.

7. Center at $(2,2)$, radius $= 3$

8. Center at $(0,3)$, contains the point $(5,3)$

Write an equation for an ellipse that fits the given conditions.

9. Foci: $(-1,0)$ and $(1,0)$, vertices: $(-3,0)$ and $(3,0)$

10. Center at $(4,2)$, vertices: $(1,2)$ and $(7,2)$, length of minor axis $= 2$

Write an equation for a parabola that fits the given conditions.

11. Vertex at $(2,0)$, contains the point $(4,2)$

12. Vertex at $(3,3)$, contains the point $(4,6)$

Write an equation for an hyperbola that fits the given conditions.

13. Vertices: $(-2,0)$ and $(2,0)$, foci: $(-4,0)$ and $(4,0)$

14. Center at $(3,2)$, vertices: $(1,2)$ and $(5,2)$, asymptotes: $3x - 2y = 5$ and $3x + 2y = 13$

Graph the following equations.

15. $x^2 - 4x + y^2 - 4y + 4 = 0$

16. $25x^2 + 16y^2 = 400$

17. $9x^2 - 36x + 4y^2 - 24y + 36 = 0$

18. $y^2 = 12x$

19. $x^2 - 4x + 4y + 16 = 0$

20. $9x^2 - 18x - 16y^2 + 96y - 279 = 0$

21. $r = 2 - 2 \sin \theta$ (cardioid)

22. $r = 3 \sin(5\theta)$ (five-leaved rose)

Chapter 17: Statistics

The beginnings of statistics can be traced to the gathering of data on populations, wealth, trade, and manufacturing in ancient times. As a discipline, statistics began in London in the 1600s with John Graunt (1620–1674). In 1662, he published *Natural and Political Observations Made upon the Bills of Mortality* in which he used sets of data to make statistical inferences from data taken from the *Bills of Mortality* in London. He concluded such things as "women live longer than men" and "the number of male births is greater than the number of female births."

The concept of a "normal frequency curve," discussed in section 17.3, was devised by Abraham de Moivre (1667–1754) in the early 1700s. De Moivre is known as one of the founders of modern probability theory. He published *The Doctrine of Chances,* a classic publication in probability.

The statistical methods used in linear regression (section 17.4) were pioneered by Sir Frances Galton (1822–1911). He was interested in the similar characteristics of parents and their children, and he developed a mathematical method of measuring the strength of a relationship between two variables.

You probably know some things about statistics. Newspapers, magazines, and television provide a wide variety of statistics in many forms. The business section of the newspaper and the sports pages provide the results of much effort in compiling and analyzing data. For example, the batting champion in baseball might have a batting average of 0.351. If you wanted to determine the batting champion without any help, you would need the official score sheets of several hundred games and a considerable amount of time. The study of statistics involves the organization and presentation of data in an understandable form and the use of that data to make intelligent decisions.

17•1 Organizing Data

The amount of data available in real-world situations is often difficult to interpret unless it is grouped in some usable form. For example, if a group of inspectors in a roller bearing plant collects data on the diameters of the bearings being produced, the data collected might involve the results of inspecting 500 bearings each hour. The resulting hourly quality report is more readable if the numbers representing the diameters are grouped in a table.

TABLE 17.1

Bearing Diameter	Frequency
0.027″	22
0.028″	47
0.029″	95
0.030″	201
0.031″	85
0.032″	33
0.033″	17
	Total 500

The table provides a much better look at the results of the inspections than merely listing the 500 measurements.

When there is a wider range of data available, it is often advantageous to group the data into categories or **intervals**. Example 17.1 illustrates the procedure.

Intervals

• **Example 17.1:** In a recent year, 27 states produced electricity using nuclear energy. The numbers of millions of megawatt-hours (Mwh) produced by these states are shown in the list.

25.2 7.7 3.9 0.75 11.6 14.8 7.8 28.0 2.3 5.7 11.7 6.1

16.4 11.8 6.1 6.3   16.4 12.4 4.9 3.7 14.7 25.6 14.1 2.9

25.6   3.5 9.3

This data can best be displayed by grouping the data into intervals. An arbitrary choice for the intervals is shown in the following table.

TABLE 17.2

Interval	Tally	Frequency
0–4.9	𝍲 II	7
5–9.9	𝍲 II	7
10–14.9	𝍲 II	7
15–19.9	II	2
20–24.9		0
25–29.9	IIII	4
		Total 27

Frequency

The data in Example 17.1 was grouped into six *intervals* or *classes*. The number of items of data in each class is called the **frequency** of that class. While there are no strict rules on selecting the number and size of the classes when grouping data, some commonsense guidelines are

GUIDELINES FOR GROUPING DATA

1. The number of classes should not be too large or too small. For most problems somewhere between 5 and 12 classes is adequate.

2. Each bit of data must belong to exactly one class. Make certain that the classes do not overlap.

3. In most problems, each class should have the same width as the other classes.

4. The classes in the distribution must include all of the data.

Example 17.2 illustrates the procedure to use in representing a set of data using a frequency table.

• *Example 17.2:* An electrician made and recorded 40 measurements of resistance (in ohms). The measurements are shown below. Arrange the data into a frequency table.

525 652 700 755 805 871 620 667 715 761
814 935 632 661 712 777 825 652 722 785
837 690 726 790 841 684 740 766 820 673
735 765 816 667 735 787 737 792 720 788

Range

Solution: The largest number in the table is 935 and the smallest number is 525. The difference between these numbers is the **range** of the data. In this example,

$$935 - 525 = 410 \text{ is the range.}$$

A reasonable choice for the number of classes is 9, and 50 is a reasonable choice for the size of each of the classes.

$$9 \times 50 = 450.$$

This will cover a range of 410.

Suppose we make our lowest class 500–549 and the highest class 900–949. This will accommodate the smallest and largest bits of data, 525 and 935, respectively.

TABLE 17.3

Class	Tally	Frequency (f)
500–549	/	1
550–599		0
600–649	//	2
650–699	⊮ |||	8
700–749	⊮ ⊮	10
750–799	⊮ ⊮	10
800–849	⊮ ||	7
850–899	/	1
900–950	/	1
		Total N = 40

Note that the choice of nine classes of length 50 was arbitrary, but reasonable. The choice was not the only one that could have been made. Table 17.4 illustrates an alternate choice.

TABLE 17.4

Class	Tally	Frequency (f)
500–599	/	1
600–699	⊮ ⊮	10
700–799	⊮ ⊮ ⊮ ⊮	20
800–899	⊮ |||	8
900–999	/	1
		Total N = 40

Table 17.3 appears to provide the better representation of the data in the problem.

Relative Frequencies of the Classes

A very useful column that can be added to a frequency distribution table is called the **relative frequencies of the classes**. To obtain the relative frequency of a class, we divide the frequency of the class by the total number of bits of data being represented in the table:

RELATIVE FREQUENCY

$$\text{Relative frequency} = \frac{f}{N}.$$

Table 17.5 shows the relative frequencies for the data in example 17.2.

TABLE 17.5

Class	f	Relative Frequency
500–549	1	$\frac{1}{40} = 0.025$
550–599	0	$\frac{0}{40} = 0.000$
600–649	2	$\frac{2}{40} = 0.050$
650–699	8	$\frac{8}{40} = 0.200$
700–749	10	$\frac{10}{40} = 0.250$
750–799	10	$\frac{10}{40} = 0.250$
800–849	7	$\frac{7}{40} = 0.175$
850–899	1	$\frac{1}{40} = 0.025$
900–950	1	$\frac{1}{40} = 0.025$
	N = 40	Total = 1.000

Note that, in general, the sum of the relative frequency column in the table is 1. However, in some problems the sum may be slightly different from 1 due to rounding of the relative frequencies.

684 CHAPTER 17 STATISTICS

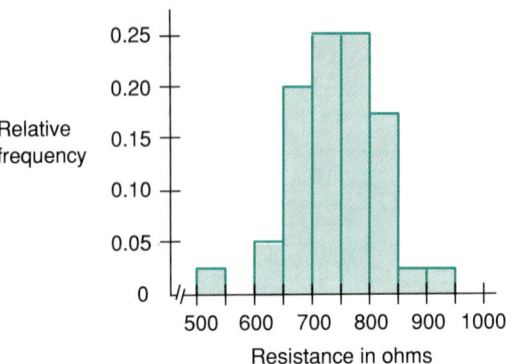

Figure 17.1

The relative frequency distribution in table 17.5 can also be represented pictorially by a graph, as shown in figure 17.1. The height of the bars are the relative frequencies for the intervals, and the widths of the bars are the lengths of the intervals. The resulting bar graph is called a **relative frequency histogram** for the data.

Relative Frequency Histogram

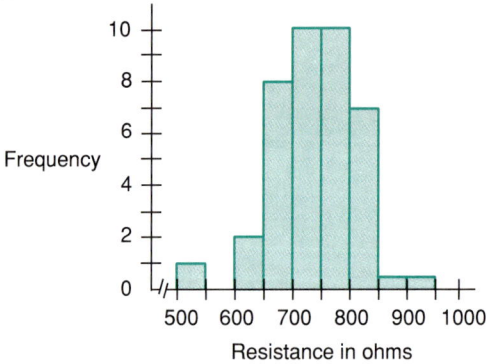

Figure 17.2

Figure 17.2 shows a histogram constructed using the frequencies as the heights of the bars and the lengths of the intervals as the widths of the bars. This graph is called a **frequency histogram**. These histograms highlight some of the characteristics of the data.

Frequency Histogram

1. Most of the data fall between 600 ohms and 850 ohms.
2. The average of the data is somewhere near 750 ohms.

Frequency tables and histograms are common ways of summarizing sets of data. The following steps describe a procedure for constructing a relative frequency distribution, a frequency histogram, and a relative frequency histogram for a set of data.

Constructing Histograms

1. Estimate the approximate number k of classes or intervals to use.

2. Determine the class width w.
   a. $w = \dfrac{\text{largest number} - \text{smallest number}}{k}$.
   b. Adjust k upward to obtain a convenient number for w.

3. Find the interval endpoints.
   a. Select a convenient number less than or equal to the smallest data number as the first endpoint.
   b. Obtain the remaining endpoints by adding multiples of the class length to the first endpoint.

4. Obtain the frequencies by tallying the data in the classes.

5. Compute the relative frequencies, and list the results in the table.

6. Construct a frequency histogram using the class lengths as the widths of the bars and the frequencies as the heights of the bars.

7. Construct a relative frequency histogram using the class lengths as the widths of the bars and the relative frequencies as the heights of the bars.

• *Example 17.3:* The normal annual precipitation for reporting stations in each of the 50 states is shown below (in inches).

67.0 54.7   7.1 48.5 19.5 15.5 43.4 40.3 54.5 48.3

22.9 11.5 34.4 38.7 30.9 30.6 43.1 56.8 40.8 40.5

42.5 31.7 25.9 49.2 35.9 11.4 30.2   8.5 36.2 45.5

 7.8 33.4 42.7 16.2 40.0 31.4 37.6 36.5 39.9 52.1

19.4 17.1 20.3 42.2 32.5 44.7 38.8 14.7 38.4 30.3

Arrange the data into classes, construct the frequency column and the relative frequency column, and construct frequency and relative frequency histograms.

Solution:
1. The numbers range from 7.1 to 67.0. Seven appears to be a reasonable choice for the number of classes.

2. $w = \dfrac{67.0 - 7.1}{7} = 8.6$. A convenient number to use for the class length is 10.

3a. Zero is a convenient number less than 7.1 to use for the first endpoint.

3b. The other endpoints are $0 + 10, 0 + 2(10), 0 + 3(10), 0 + 4(10), 0 + 5(10)$, and $0 + 6(10)$ or 10, 20, 30, 40, 50, 60.

4–5. Tally the data and compute the relative frequencies.

TABLE 17.6

Class	Tally	f	Relative Frequency
0–9.9	∕∕∕	3	$\frac{3}{50} = 0.06$
10–19.9	₩₩ ∕∕∕	8	$\frac{8}{50} = 0.16$
20–29.9	∕∕∕	3	$\frac{3}{50} = 0.06$
30–39.9	₩₩ ₩₩ ₩₩ ∕∕	17	$\frac{17}{50} = 0.34$
40–49.9	₩₩ ₩₩ ∕∕∕∕	14	$\frac{14}{50} = 0.28$
50–59.9	∕∕∕∕	4	$\frac{4}{50} = 0.08$
60–69.9	∕	1	$\frac{1}{50} = 0.02$
		$N = 50$	

6. The width of the bars is 10 and the heights of the bars are the frequencies on the frequency histogram. See figure 17.3.

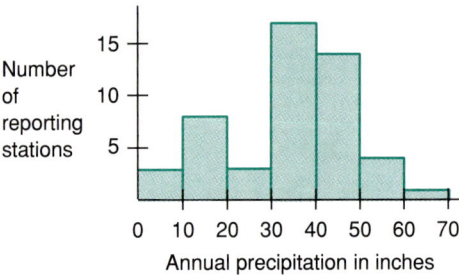

Figure 17.3

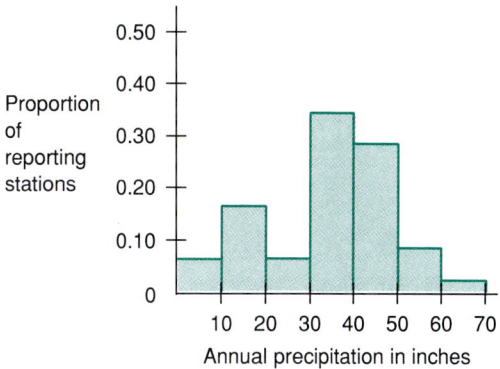

Figure 17.4

7. The width of the bars is 10 and the heights of the bars are the relative frequencies on the relative frequency histogram. See figure 17.4.

Trial Problems 17.1

Before you begin the section exercises, warm up with these problems. Complete answers are included in the answer key.

Class	Frequency
10–19.9	4
20–29.9	15
30–39.9	18
40–49.9	10
50–59.9	3

1. For the data shown above, construct a frequency histogram.
2. For the data shown above, construct a relative frequency histogram.

41.3	50.0	63.7	80.3	60.1	79.9	55.4	77.0	78.1	56.8
65.4	64.9	66.7	51.2	73.5	82.4	52.3	76.2	67.3	69.8
57.3	72.4	86.7	51.2	95.3	45.6	97.2	74.0	75.2	62.6
65.4	71.3	57.2	56.1	70.0	68.3	56.1	61.3	62.4	47.3

3. For the data shown above, construct a frequency histogram.
4. For the data shown above, construct a relative frequency histogram.

Exercises 17.1

1. The following table shows the results of a quality control survey that recorded the weights in grams of containers that were filled by a packaging machine.

Class	Frequency
390–392	2
393–395	8
396–398	25
399–401	68
402–404	31
405–407	9
408–410	1

    Construct a frequency histogram and a relative frequency histogram for the data.

2. The following table shows the results of a time study conducted on the time in minutes required to perform a specific operation in a manufacturing plant.

Class	Frequency
7.0–7.9	3
8.0–8.9	4
9.0–9.9	15
10.0–10.9	18
11.0–11.9	25
12.0–12.9	16
13.0–13.9	12
14.0–14.9	0
15.0–15.9	2

    Construct a frequency histogram and a relative frequency histogram for the data.

3.
```
35  52  12  61  55   6  80  30  10  55   7  84  17  28  63   3
62  24  25   9  81  20  65  18  23  90  17  12  64  50   1  34
10  33   8   5  76   2  54  15   7  70   1  35  20   7  42  41  56
```

Using the data given above, construct:
  a. A frequency chart that includes a relative frequency column using the intervals 0–5, 6–10, and so forth.
  b. A frequency histogram.
  c. A relative frequency histogram.

4. Use the data in exercise 3 to construct:
  a. A frequency chart that includes a relative frequency column using the intervals 0–10, 11–20, and so forth.
  b. A frequency histogram.
  c. A relative frequency histogram.

5. A sample of 50 households yields the following data for the number of millions of BTUs of energy consumed by the household.

```
 52   80   90   61   73   65   82  100   69   91
 91   65   85   43   90  127   50   84  102  100
 99   86   63   51   77  135   93  157   57   93
108  112   98   96  107   50   98   89  113  117
122   53  110  136   62  155   58   78   63  138
```

  a. Construct a frequency table that includes a relative frequency column using 40–49 as the first interval.
  b. Construct a frequency histogram.
  c. Construct a relative frequency histogram.

6. Solve exercise 5 using 40–59 as the first interval.

7. The following data represent the measured diameters in mm of 40 bearings after 1000 hours of use.

```
15.0  15.3  15.2  13.8  12.5  11.5  14.0  13.7  14.9  14.5
14.4  10.5  14.9  10.1  14.1  12.9  14.1  10.8  15.0  13.8
15.4  13.1  15.0  10.2  10.7  10.3  12.7  15.3  15.0  10.6
13.6  10.7  13.9  14.9  10.9  15.2  11.7  10.3  12.1  10.4
```

  a. Construct a frequency table that includes a relative frequency column using 10.1–10.5 as the first class.
  b. Construct a frequency histogram.
  c. Construct a relative frequency histogram.

8. Solve exercise 7 using intervals 10.1–11.0, 11.1–12.0, and so forth.

9. A survey after a car-pooling campaign yields the following data on the number of people in cars crossing a toll bridge. Construct (a) a frequency histogram and (b) a relative frequency histogram for the data.

1	2	2	3	4	2	2	5	3	5	3	4	2	2	2
3	2	2	3	2	3	2	4	2	6	2	4	2	3	1
1	2	3	1	4	1	4	1	5	1	2	1	1	2	2
2	1	3	1	3	3	3	3	3	3	3	3			

10. Use the data given below to construct a relative frequency histogram. Use ten intervals.

8.6	4.5	4.1	3.1	1.7	6.5	1.2	2.0	2.1	6.4
4.3	1.1	5.0	3.3	7.3	9.6	3.4	7.9	9.2	9.4
3.1	6.5	6.3	8.3	7.3	3.7	7.6	1.3	3.3	8.9
1.2	4.0	6.6	7.7	6.2	6.5	9.4	8.1	7.2	9.7
3.9	1.9	5.9	3.9	6.7	3.8	1.6	4.8	8.4	6.2
0.2	3.5	8.3	5.1	1.4	7.7	2.6	1.6	7.4	1.6
8.4	9.4	8.9	6.8	5.3	5.5	8.3	4.4	4.8	9.4
1.1	9.1	2.7	8.8	7.7	2.8	9.4	8.2	7.3	9.8

17·2 Measures of Central Tendency

In section 17.1 we learned how to make frequency distributions and how to present sets of data in the form of frequency and relative frequency histograms.

Often it is more meaningful to represent a set of data with a single number or a set of numbers that indicate the "location" of the data. Such a number or set of numbers are called **measures of central tendency**. The three most commonly used measures of central tendency are the mode, the median, and the mean.

Measures of Central Tendency

MODE

The mode of a set of numbers is the number that appears in the set most frequently.

For the set 1 1 ⎡2⎤ 5 4 6 8 ⎡2⎤ ⎡2⎤ 4 3 ⎡2⎤ 7, the mode is 2 since 2 appears more frequently than any of the other numbers.

The mode of a set of numbers does not have to be unique. For the set 1 1 ⎡2 2 2⎤ ⎡3 3 3⎤ ⎡4 4 4⎤ 5 5 6 7 there are three modes, namely 2, 3, and 4.

If each of the numbers in a set occurs only once, the set is described as having no mode.

MEDIAN

The median of a set of numbers is the number that occurs in the middle of the set when the numbers are arranged in either increasing or decreasing order.

The following rule of thumb gives the method of finding the median of a set of numbers.

FINDING THE MEDIAN

1. If the set contains an odd numbers of entries, the median is the middle number. That is, if there are n entries in the set, the median is the $\frac{n+1}{2}$ entry.

2. If a set contains an even number of entries, the median is the average of the two middle numbers. That is, if there are k numbers in the set, the median is the average of the $\frac{k}{2}$ and the $\frac{k}{2} + 1$ entries.

• *Example 17.4:* Find the median for each set of numbers.

    a. 3.1 3.3 3.7 3.7 3.8 3.9 4.0 4.3 4.3 5.1 5.3

    b. 161 170 175 183 187 191 205 210 214 217

Solution: **a.** Since there are 11 numbers in the set, the median is the $\frac{11+1}{2}$ or the sixth entry. Counting from either left to right or from right to left, the sixth entry is 3.9. Thus, the median is 3.9.

b. Since there are 10 numbers in the set, the median is the average of the middle two numbers or the average of the $\frac{10}{2}$ and the $\frac{10}{2} + 1$ entries. To find the median, we average the fifth and sixth entries.
161 170 175 183 |187 191| 205 210 214 217
The median is $\frac{187+191}{2} = 189$.

QUARTILES

The quartiles, Q_1, Q_2, and Q_3 of a set of numbers are the numbers that divide the set into four equal parts when the numbers are arranged in either increasing order or in decreasing order.

A rule of thumb for finding the quartiles is as follows.

FINDING QUARTILES

1. Q_2 is the median of the set of numbers.
2. Q_1 is the median of one of the subsets of the original set that were created by Q_2.
3. Q_3 is the median of the other subset created by Q_2.

Example 17.5 illustrates how to find quartiles.

• **Example 17.5:** An electrical component of an electrical appliance is guaranteed for 48 months. The following data represent the lifetimes of the component in a test sample. Find the quartiles for the data in the sample.

18.0   28.1   36.7   37.4 $\boxed{41.2\ \ 50.5}$ 51.3   52.0   53.1 $\boxed{54.4}$
$\boxed{54.5}$ 55.6   56.2   67.1 $\boxed{67.1\ \ 74.3}$ 80.3   80.4   81.1   83.7

Solution: Since there are 20 numbers in the set, Q_2 is the average of the tenth and eleventh entries:

$$Q_2 = \frac{54.4 + 54.5}{2} = 54.45.$$

Q_1 is the median of the first ten numbers in the set, or the average of the fifth and sixth entries:

$$Q_1 = \frac{41.2 + 50.5}{2} = 45.85.$$

Q_3 is the median of the last ten numbers in the set, or the average of the fifteenth and sixteenth entries:

$$Q_3 = \frac{67.1 + 74.3}{2} = 70.7.$$

The quartiles are useful in displaying a set of data in a diagram called a *box-and-whisker diagram*. The following steps outline a procedure for constructing such a diagram for a set of numbers.

CONSTRUCTING BOX-AND-WHISKER DIAGRAMS

1. Arrange the data in increasing order and find the median, Q_2.
2. Find the lowest and highest numbers in the set.
3. Find Q_1 and Q_3 for the data.
4. Draw a scale that includes the highest and lowest numbers. Mark the low, high, Q_1, Q_2, and Q_3 above the scale.
5. Complete the diagram as shown in figure 17.5.

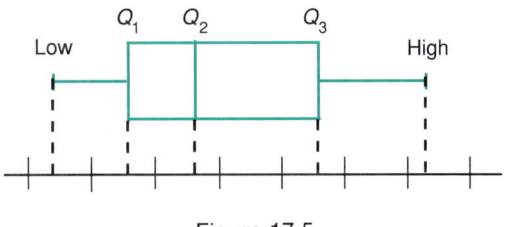

Figure 17.5

Note that the box shows the upper and lower limits of the middle half of the data, the left whisker shows the limits of the lowest quarter of the data, and the right whisker shows the limits of the highest quarter of the data.

• *Example 17.6:* Draw a box-and-whisker diagram for the data in example 17.5.

Solution: 1. The median Q_2 is 54.45.

2. 18.0 is the low and 83.7 is the high.

3. Q_1 is 45.85 and Q_3 is 70.7.

4–5. Figure 17.6 shows the complete diagram.

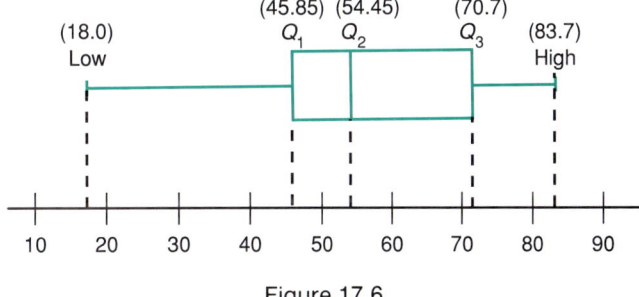

Figure 17.6

The diagram shows that half of the data is represented by the box; one quarter of the data by the left whisker and one quarter of the data by the right whisker.

The arithmetic mean of a set of data is the arithmetic average of the numbers in the set. If we represent n numbers in a set of data by $x_1, x_2, x_3, \ldots, x_n$, then the mean is calculated as follows.

ARITHMETIC MEAN

$$\bar{x} = \frac{x_1 + x_2 + x_3 + \cdots + x_n}{n} = \frac{\Sigma x_i}{n}.$$

For grouped data where we have $x_1, x_2, x_3, \ldots, x_n$, a set of numbers with the corresponding frequencies $f_1, f_2, f_3, \ldots, f_n$, the formula for the mean is as follows.

ARITHMETIC MEAN GROUPED DATA

$$\bar{x} = \frac{x_1 f_1 + x_2 f_2 + x_3 f_3 + \cdots + x_n f_n}{f_1 + f_2 + f_3 + \cdots + f_n} = \frac{\Sigma x_i f_i}{\Sigma f_i}$$

Examples 17.7 and 17.8 illustrate the use of these formulas.

• **Example 17.7:** Calculate the mean of the given set of numbers:

1.6   3.8   7.2   9.1   11.2   2.3   4.7   9.4   8.6.

Solution: There are nine numbers in the set, so $n = 9$:

$$\bar{x} = \frac{1.6 + 3.8 + 7.2 + 9.1 + 11.2 + 2.3 + 4.7 + 9.4 + 8.6}{9}$$

$$= \frac{57.9}{9} = 6.4. \quad \text{(To two significant digits)}$$

• **Example 17.8:** Calculate the mean for the data represented in the given frequency table.

x	f
0.027	22
0.028	47
0.029	95
0.030	201
0.031	85
0.032	33
0.033	17

Solution: To calculate Σf find the sum of the f-column. Form an xf-column by multiplying the entries in the x-column by the corresponding entries in the f-column. Add the xf-column to obtain Σxf.

x	f	xf
0.027	22	0.594
0.028	47	1.316
0.029	95	2.755
0.030	201	6.030
0.031	85	2.635
0.032	33	1.056
0.033	17	0.561
	$\Sigma f = 500$	$\Sigma xf = 14.947$

$$\bar{x} = \frac{\Sigma xf}{\Sigma f} = \frac{14.947}{500} = 0.030.$$

MEAN USING GROUPED DATA

To find the mean when the data is given in the form of a frequency table using intervals, and when we do not have the original data, an approximation to the mean can be calculated using the grouped-data formula where the x's in the formula are the midpoints of the intervals. Example 17.9 illustrates such a problem.

• *Example 17.9:* The following table shows the results of the time-study survey in exercise 2 of section 17.1.

Class	f
7.0–7.9	3
8.0–8.9	4
9.0–9.9	15
10.0–10.9	18
11.0–11.9	25
12.0–12.9	16
13.0–13.9	12
14.0–14.9	0
15.0–15.9	2

Find an approximate mean for the data.

Solution: First, find the *x*-column that contains the midpoints of the class intervals. Then proceed as in example 17.8. For example, the *x*-value for the 7.0–7.9 class is $\dfrac{7.0 + 7.9}{2} = 7.45$.

Class	f	x	xf
7.0–7.9	3	7.45	22.35
8.0–8.9	4	8.45	33.80
9.0–9.9	15	9.45	141.75
10.0–10.9	18	10.45	188.10
11.0–11.9	25	11.45	286.25
12.0–12.9	16	12.45	199.20
13.0–13.9	12	13.45	161.40
14.0–14.9	0	14.45	0.00
15.0–15.9	2	15.45	30.90
	$\Sigma f = 95$		$\Sigma xf = 1063.75$

$$\bar{x} = \frac{\Sigma xf}{\Sigma f} = \frac{1063.75}{95} = 11.20.$$

Trial Problems 17.2

Before you begin the section exercises, warm up with these problems. Complete answers are included in the answer key.

1.3 1.8 2.6 2.7 2.7 3.0 3.6 3.6 3.8 3.9
4.0 4.2 4.3 4.7 4.7 5.6 5.9 6.0 6.3 6.8

1. For the data shown above, find the median and the mean.
2. Find the quartiles of the data shown above.
3. Draw a box-and-whisker diagram for the data shown above.
4. Find the mean of the given data.

x	f
30	6
35	9
40	10
45	12
50	11
55	8
60	7

5. Approximate the mean of the given data.

Class	f
10–19.9	4
20–29.9	15
30–39.9	18
40–49.9	10
50–59.9	3

Exercises 17.2

In exercises 1–4, find the mode, median, and mean for the given set of numbers.

1. 17   18   20   20   22   22   24   27   30
2. 4.3   4.6   4.6   5.0   5.1   5.2   5.2   5.4   5.6   5.8   6.1
3. 15   12   17   12   15   18   26   41   16   14
   35   19   22   27   33   36   5   8   19   39
4. 16.8   9.4   18.3   15.1   16.3   15.2   14.0   12.1
   9.0   7.3   14.1   14.6   18.6   11.1   12.4   17.2

In exercises 5–8, find the quartiles and construct a box-and-whisker diagram for the given set of numbers.

5. The numbers of millions of Mwh of electricity produced by states using nuclear power plants.

   25.2   7.7   3.9   0.75   11.6   14.8   7.8   28.0   2.3
   5.7   11.7   6.1   16.4   11.8   6.1   6.3   16.4   12.4
   4.9   3.7   14.7   25.6   14.1   2.9   25.6   3.5   9.3

6. Forty measurements of resistance in ohms made by an electrician.

   525   652   700   755   805   871   620   667   715   761
   814   935   632   661   712   777   825   652   722   785
   837   690   726   790   841   684   740   766   820   673
   735   765   816   667   735   787   737   792   720   788

7. The normal annual precipitation in inches for reporting stations in each of the 50 states.

67.0	54.7	7.1	48.5	19.5	15.5	43.4	40.3	54.5	48.3
22.9	11.5	34.4	38.7	30.9	30.6	43.1	56.8	40.8	40.5
42.5	31.7	25.9	49.2	35.9	11.4	30.2	8.5	36.2	45.5
7.8	33.4	42.7	16.2	40.0	31.4	37.6	36.5	39.9	52.1
19.4	17.1	20.3	42.2	32.5	44.7	38.8	14.7	38.4	30.3

8. The daily emission levels of a pollutant from a refinery.

18.3	12.1	24.3	6.9	17.2	9.1	4.3	12.3
26.2	3.2	5.7	11.6	9.8	14.8	19.1	1.3
14.2	19.1	4.6	5.7	18.3	16.1	15.2	16.1
1.8	4.7	16.1	14.3	8.1	29.3	26.1	10.5

In exercises 9–14, find the mean of the given data.

9.
x	f
10	1
11	16
12	19
13	15
14	13
15	4

10.
x	f
260	17
270	40
280	52
290	51
300	38
310	16

11.
Class	f
10.0–10.9	1
11.0–11.9	19
12.0–12.9	35
13.0–13.9	33
14.0–14.9	17
15.0–15.9	5

12.
Class	f
0–9	14
10–19	73
20–29	85
30–39	99
40–49	91
50–59	72
60–69	51
70–79	11

13. The results of a time study on the time in minutes required for a particular industrial operation.

Class	f
7.0–7.9	2
8.0–8.9	8
9.0–9.9	16
10.0–10.9	27
11.0–11.9	28
12.0–12.9	19
13.0–13.9	9
14.0–14.9	1
15.0–15.9	1

14. The number of defective parts in each of 100 samples that were tested before shipping.

x	f
0	57
1	20
2	14
3	7
4	1
5	0
6	1

17 • 3 Measures of Dispersion

The measures of central tendency studied in the last section provided several measures that are useful in analyzing a set of data. However, these are not the only ones available. To get an accurate and meaningful summary of a set of measures, measures of dispersion are very useful.

The following sets of data have the same mean, 50, and the same median, 50. The sets of data are quite different.

Set A	Set B
100	52
70	51
50	50
30	49
0	48

The variability in the numbers in set A is much larger than the variability in the numbers in set B. For example, the range of set A is $100 - 0 = 100$, while the range of set B is $52 - 48 = 4$.

Another measure of dispersion for a set of data is the variance of the data. The variance is defined as follows.

VARIANCE (s^2)

$$s^2 = \frac{\Sigma(x - \bar{x})^2}{n - 1},$$

where $x - \bar{x}$ represents the deviation of the individual x's from the mean $\bar{x}$, and n is the number of bits of data in the sample.

For the data in set A, we calculate the variance as follows.

x	$x - \bar{x}$	$(x - \bar{x})^2$
100	100 − 50 = 50	2500
70	70 − 50 = 20	400
50	50 − 50 = 0	0
30	30 − 50 = −20	400
0	0 − 50 = −50	2500
		5800 = $\Sigma(x - \bar{x})^2$

$$s^2 = \frac{\Sigma(x - \bar{x})^2}{n - 1} = \frac{5800}{5 - 1} = \frac{5800}{4} = 1450.$$

For the data in set B, we calculate the variance as follows.

x	$x - \bar{x}$	$(x - \bar{x})^2$
52	52 − 50 = 2	4
51	51 − 50 = 1	1
50	50 − 50 = 0	0
49	49 − 50 = −1	1
48	48 − 50 = −2	4
		10 = $\Sigma(x - \bar{x})^2$

$$s^2 = \frac{10}{5 - 1} = \frac{10}{4} = 2.5.$$

Related to the variance is a measure called the standard deviation for a set of data.

STANDARD DEVIATION (s)

$$s = \sqrt{s^2} = \sqrt{\frac{\Sigma(x - \bar{x})^2}{n - 1}}$$

For sets A and B, the standard deviations are as follows:

Set A: $s = \sqrt{1450} = 38.08$
Set B: $s = \sqrt{2.5} = 1.58$.

There are alternate formulas for calculating the variance and the standard deviation of a set of numbers. In most cases, the calculations are easier using the alternate formulas. They are usually called the "short-cut" formulas.

VARIANCE AND STANDARD DEVIATION ALTERNATE FORMULAS

$$s^2 = \frac{n\Sigma x^2 - (\Sigma x)^2}{n(n-1)}$$

$$s = \sqrt{s^2} = \sqrt{\frac{n\Sigma x^2 - (\Sigma x)^2}{n(n-1)}}$$

Note that

$$\Sigma x^2 = x_1^2 + x_2^2 + \cdots + x_n^2,$$

while

$$(\Sigma x)^2 = (x_1 + x_2 + \cdots + x_n)^2.$$

Using the alternate formulas on the data in sets A and B yields the following results.

Set A:

x	x^2
100	10,000
70	4900
50	2500
30	900
0	0
$\Sigma x = 250$	$\Sigma x^2 = 18,300$

$$s^2 = \frac{5(18,300) - (250)^2}{(5)(4)} = 1450$$

The keystrokes to determine s^2 are

 5 $\times$ 18300 $-$ 250 x^2 $=$ $\div$ 5 $\div$ 4 $=$

$$s = \sqrt{1480} = 38.08.$$

Set B:

x	x^2
52	2704
51	2601
50	2500
49	2401
48	2304
$\Sigma x = 250$	$\Sigma x^2 = 12,510$

$$s^2 = \frac{n\Sigma x^2 - (\Sigma x)^2}{n(n-1)}$$

$$= \frac{5(12,510) - (250)^2}{(5)(4)}$$

$$= 2.5$$

$$s = \sqrt{2.5} = 1.58.$$

When a problem involves grouped data, the form of the formulas must be altered slightly.

VARIANCE AND STANDARD DEVIATION GROUPED DATA

$$s^2 = \frac{n\Sigma x^2 f - (\Sigma xf)^2}{n(n-1)}, \quad \text{where } n = \Sigma f$$

$$s = \sqrt{s^2} = \sqrt{\frac{n\Sigma x^2 f - (\Sigma xf)^2}{n(n-1)}}, \quad \text{where } n = \Sigma f.$$

Example 17.10 illustrates how to use these formulas.

• **Example 17.10:** Use the data in table 17.1 of section 17.1, and calculate the variance and the standard deviation.

Bearing diameter	Frequency
0.027	22
0.028	47
0.029	95
0.030	201
0.031	85
0.032	33
0.033	17

Solution: We attack the problem in two stages. Since we need $\Sigma x^2 f$ and Σxf to substitute in the formula for the variance, we first form an xf column and an $x^2 f$ column in the table and sum the columns. Then we substitute these results into the formula.

x	f	xf	$x^2 f$
0.027	22	0.594	0.016038
0.028	47	1.316	0.036848
0.029	95	2.755	0.079895
0.030	201	6.030	0.180900
0.031	85	2.635	0.081685
0.032	33	1.056	0.033792
0.033	17	0.561	0.018513
$\Sigma f = n = 500$		$\Sigma xf = 14.947$	$\Sigma x^2 f = 0.447671$

Note that $x^2f = x(xf)$. When calculating an "xf" entry, multiply the result by x to obtain the corresponding "x^2f" entry.

$$s^2 = \frac{n\Sigma x^2 f - (\Sigma xf)^2}{n(n-1)}$$

$$= \frac{500(0.447671) - (14.947)^2}{(500)(499)}$$

$$= 0.0000017.$$

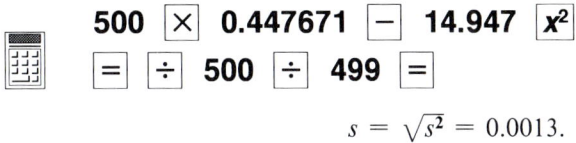

$$s = \sqrt{s^2} = 0.0013.$$

The standard deviation is like a geometric mean of the deviations of the individual numbers from the mean. If we merely took the arithmetic mean of the deviations, the fact that some of the deviations are positive and some are negative would negate any usefulness of the measure.

• **Example 17.11:** To estimate the number of trees of "cutting size" in a tract, a lumber company selects 40 sub-tracts of equal size and counts the number of harvest-ready trees in each sub-tract. The results are shown in the table.

Number of Trees	Frequency
12	3
11	5
10	6
9	12
8	6
7	5
6	2
5	0
4	1

Calculate the mean and standard deviation for the data.

Solution: For the formulas to calculate the mean and standard deviation, we need to calculate $n = \Sigma f$, Σxf, and $\Sigma x^2 f$, so we add xf and $x^2 f$ columns to the table.

x	f	xf	$x^2 f$
12	3	36	432
11	5	55	605
10	6	60	600
9	12	108	972
8	6	48	384
7	5	35	245
6	2	12	72
5	0	0	0
4	1	4	16
	$n = \Sigma f = 40$	$\Sigma xf = 358$	$\Sigma x^2 f = 3326$

$$\bar{x} = \frac{\Sigma xf}{n} = \frac{358}{40} = 8.95$$

$$s = \sqrt{\frac{n\Sigma x^2 f - (\Sigma xf)^2}{n(n-1)}}$$

$$= \sqrt{\frac{40(3326) - (358)^2}{(40)(39)}} = 1.77.$$

40 × 3326 − 358 x^2 = ÷ 40 ÷ 39
= $\sqrt{x}$

One of the uses of standard deviations is the result of a theorem called Chebyshev's Theorem.

CHEBYSHEV'S THEOREM

For any set of data and any $k > 1$, the proportion of the data that lies within k standard deviations on either side of the mean is at least $1 - \frac{1}{k^2}$.

Thus, if $k = 2$, the theorem states that at least $1 - \frac{1}{2^2}$ or 75% of the data must lie within two standard deviations of the mean, that is, in the interval $(\bar{x} - 2s, \bar{x} + 2s)$.

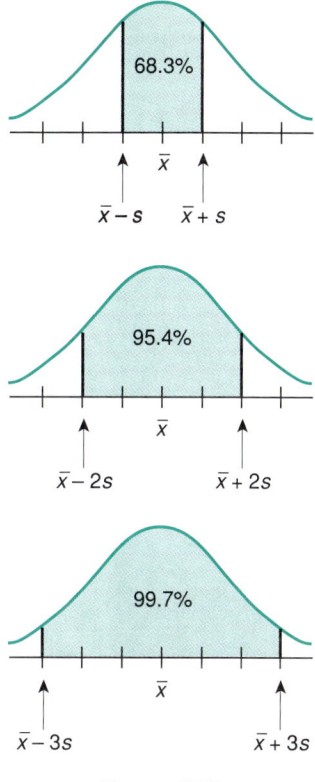

Figure 17.7

For the data in example 17.11, the mean is 8.95 and the standard deviation is 1.77. The interval $(\bar{x} - 2s, \bar{x} + 2s)$ is the interval

$$(8.95 - 2(1.77), \quad 8.95 + 2(1.77)) = (5.41, 12.49).$$

At least 75% of the data must lie within this interval.

If the data in a problem consists of a relatively large number of measurements, we could expect most of the measurements to be close to the mean. If the set of numbers has the general shape of the cross section of a bell (a **normal distribution**), then a stronger statement can be made about the distribution of the data. (See figure 17.7.)

Normal Distribution

EMPIRICAL RULE

For an approximately normal distribution, about 68.3% of the data lies within one standard deviation of the mean, about 95.4% of the data lies within two standard deviations of the mean, and about 99.7% of the data lies within three standard deviations of the mean.

Figure 17.8 shows some examples of normal curves with different means and standard deviations.

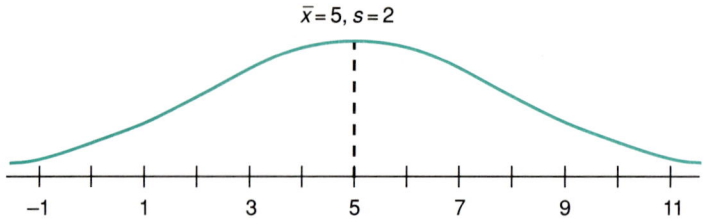

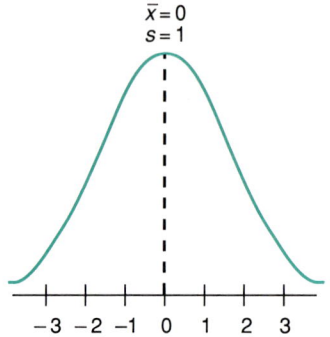

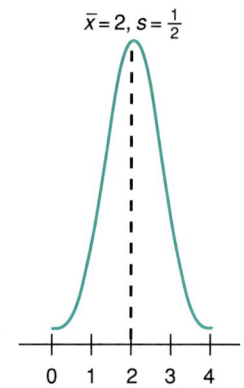

Figure 17.8

EQUATION OF A NORMAL CURVE

The equation of a general normal curve is

$$y = \left(\frac{1}{s\sqrt{2\pi}}\right) e^{-(1/2)[(x - \bar{x})/s]^2}.$$

If $\bar{x} = 0$ and $s = 1$, the curve is called the **standard normal curve,** and the equation becomes

$$y = \left(\frac{1}{\sqrt{2\pi}}\right) e^{-(x^2/2)}$$

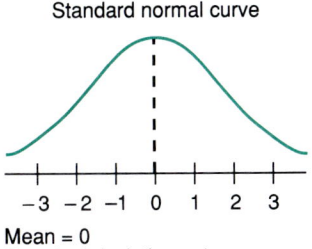

Standard normal curve

Mean = 0
Standard deviation = 1

Figure 17.9

A standard normal curve is shown in figure 17.9.

For the data in example 17.10, a curve for the numbers can be obtained by drawing a frequency histogram and connecting the midpoints of the bars with a smooth curve. The results are shown in figure 17.10. For the data in this example, the mean is 0.030 and the standard deviation is 0.0013. Since the graph is approximately bell-shaped, using the emperical rule, about 68.3% of the data should lie in the interval

$$(\bar{x} - s, \bar{x} + s) = (0.030 - 0.0013, 0.030 + 0.0013)$$
$$= (0.0287, 0.0313),$$

about 95.4% of the data should lie in the interval

$$(\bar{x} - 2s, \bar{x} + 2s) = (0.030 - 2(0.0013), 0.030 + 2(0.0013))$$
$$= (0.0274, 0.0326),$$

and about 99.7% of the data should lie in the interval

$$(\bar{x} - 3s, \bar{x} + 3s) = (0.030 - 3(0.0013), 0.030 + 3(0.0013))$$
$$= (0.0261, 0.0339).$$

Example 17.12 illustrates the use of the emperical rule for data that is grouped into classes.

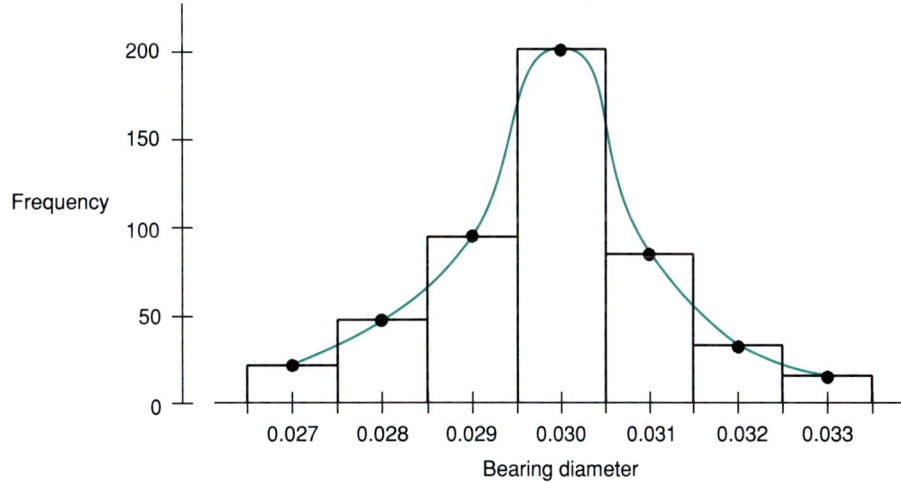

Figure 17.10

• **Example 17.12:** The results of a time study on the number of minutes required to perform a particular industrial operation are shown in the following table.

Class	Frequency
7.0–7.9	2
8.0–8.9	8
9.0–9.9	18
10.0–10.9	27
11.0–11.9	28
12.0–12.9	19
13.0–13.9	9
14.0–14.9	2
15.0–15.9	1

Given that $\bar{x} = 10.03$ and $s = 1.55$, find an interval that would contain

a. About 68.3% of the data,

b. About 95.4% of the data, and

c. About 99.7% of the data.

Solution: **1.** Check the data curve for normality.
The graph shown in figure 17.11 is approximately bell-shaped, so the data is approximately normal.

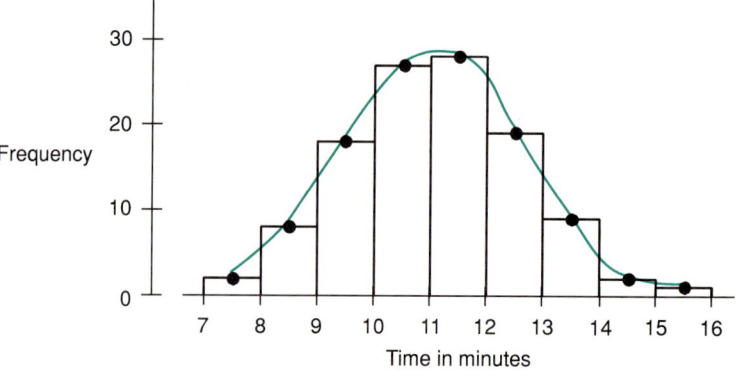

Figure 17.11

2. Find the intervals using the emperical rule.

   a. $(\bar{x} - s, \bar{x} + s) = (10.03 - 1.55, 10.03 + 1.55)$
   $= (8.48, 11.58)$.

   b. $(\bar{x} - 2s, \bar{x} + 2s) = (10.03 - 2(1.55), 10.03 + 2(1.55))$
   $= (6.93, 13.13)$.

   c. $(\bar{x} - 3s, \bar{x} + 3s) = (10.03 - 3(1.55), 10.03 + 3(1.55))$
   $= (5.38, 14.68)$.

• • • • • • • • • •

Trial Problems 17.3

Before you begin the section exercises, warm up with these problems. Complete answers are included in the answer key.

Find the standard deviation using the given data.

1. $n = 7$, $\Sigma x = 42$, $\Sigma(x - \bar{x})^2 = 84$

2. $n = 14$, $\Sigma x = 84$, $\Sigma x^2 = 672$

3. $n = 38$, $\Sigma x^2 f = 565{,}700$, $\Sigma xf = 4590$

4. 17.1 20.2 35.3 18.6 19.4
   35.2 26.1 19.0 20.4 26.3

5.
x	f
15	5
16	12
17	18
18	20
19	12
20	9
21	4

Exercises 17.3

For exercises 1–10, find the variance and the standard deviation.

1. $\Sigma(x - \bar{x})^2 = 463$, $n = 32$

2. $\Sigma(x - \bar{x})^2 = 19.43$, $n = 18$

3. $\Sigma x^2 = 1156.2$, $\Sigma x = 84$, $n = 35$

4. $\Sigma x^2 = 5643.85$, $\Sigma x = 125$, $n = 41$

5. A set of temperature measurements taken during a chemical reaction.

   163 165 265 262 265 266 266 268 263 269
   266 270 267 267 264 268 265 262 261 269
   271 263 272 261 262 262 265 273 262 264

6. A set of 24 diameters of parts produced by a milling machine.

19.4   19.4   19.3   19.1   19.5   19.7   19.6   19.5   19.9   19.5   19.8   19.7
19.6   19.7   19.6   20.0   19.6   19.6   19.7   19.9   19.6   19.0   19.6   19.5

7. A set of electrical measurements in ohms.

x	f
40.3	6
40.2	22
40.1	38
40.0	46
39.9	44
39.8	36
39.7	20
39.6	5

8. The results of testing a sample of pieces of 10-pound test line manufactured by a company.

x(lbs)	f	x(lbs)	f
9.6	1	10.1	18
9.7	4	10.2	10
9.8	10	10.3	3
9.9	19	10.4	2
10.0	26	10.6	1

9. The number of items produced per hour during a 40-hour work week.

x (items/hr)	f
21	1
22	1
23	2
24	8
25	18
26	7
27	3

10. The number of defective circuit boards detected in 60 samples that were tested.

x (number of defectives)	f
0	42
1	8
2	4
3	2
4	2
5	1
6	1

In exercises 11–14, use Chebyshev's theorem to find an interval that fits the given conditions.

11. $\bar{x} = 37.85$, $s = 3.62$, $k = 2$

12. $\bar{x} = 115.62$, $s = 9.41$, $k = 1.5$

13. $\bar{x} = 75.12$, $s = 4.32$,   75% of data in the interval

14. $\bar{x} = 0.25$, $s = 0.015$,   84% of data in the interval (*Hint:* Find k.)

15. A sample yields a mean of 136.2 and a standard deviation of 3.8. Between which two numbers will approximately 68.3% of the data fall, assuming a normal distribution?

16. A sample yields a mean of 18.35 and a standard deviation of 0.26. Into what interval will approximately 95.4% of the data fall, assuming a normal distribution?

17. A sample of bottles filled by a bottling machine shows a mean contents of 12.0 oz with a standard deviation of 0.25 oz. What percent of the bottles filled by the machine should contain between 11.5 and 12.5 oz? (Assume a normal distribution.)

18. A machine produces parts that are acceptable provided their lengths are between 21.90 cm and 22.10 cm. If the mean length of the parts produced by the machine is 22.00 cm and the standard deviation is 0.05 cm, what percent of the parts produced by the machine are acceptable? (Assume a normal distribution.)

In exercises 19–20, find intervals in which approximately the following percentages of data are located. a. 68.3 % b. 95.4% c. 99.7%

19. The force in hundreds of pounds per square inch required to break samples of an alloy.

Force required	f
23–23.9	3
24–24.9	7
25–25.9	10
26–26.9	24
27–27.9	10
28–28.9	6
29–29.9	2

$\Sigma xf = 1634.9$

$\Sigma x^2 f = 43{,}227.855$

20. The amount of silicon in grams present in 49 samples of a steel alloy.

Amount of silicon	f
8.0–8.9	2
9.0–9.9	3
10.0–10.9	9
11.0–11.9	21
12.0–12.9	9
13.0–13.9	4
14.0–14.9	1

$\Sigma xf = 560.05$

$\Sigma x^2 f = 6474.1225$

17 • 4 The Least Squares Line

Bivariate Data

In many statistical studies, information is available on more than one variable. Data in which information is available on two variables is called **bivariate data**. For example, the table shown below shows the data from an experiment relating the temperature in degrees Celsius and the solubility of potassium chloride in grams per 100 ml of water.

Temperature	0	10	20	30	40	50	60	70	80	90	100
Solubility	27.6	31.0	34.0	37.0	40.0	42.6	45.5	48.3	51.1	54.0	56.7

Scatter Diagram

If the data is plotted on a coordinate system as a collection of points, the resulting graph is called a **scatter diagram**. Figure 17.12 shows the scatter diagram for the data in the table.

You might suspect from looking at the scatter diagram that there may be a linear relationship between the temperature and the solubility, at least for the temperature range from 0 to 100 degrees.

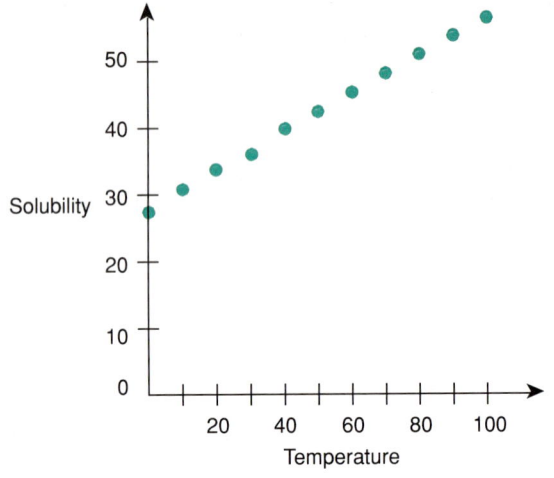

Figure 17.12

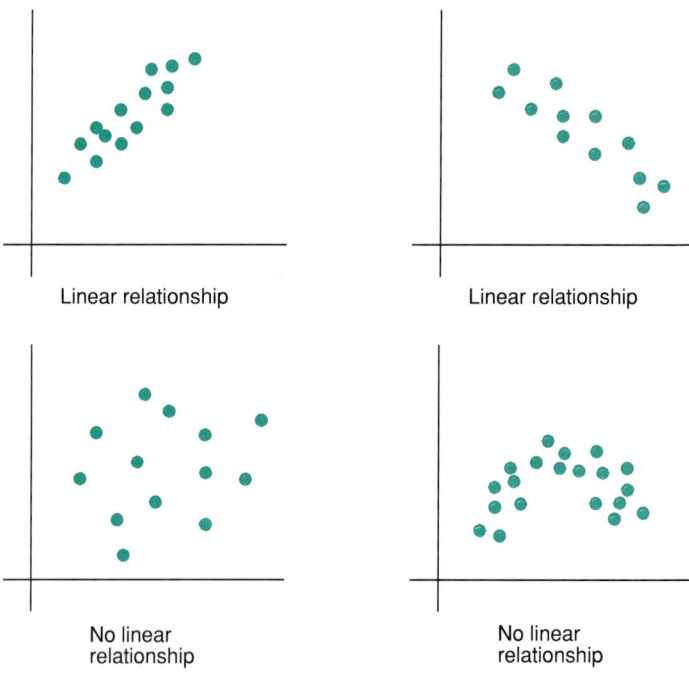

Figure 17.13

The main focus of this section is to find a linear equation that best represents a set of bivariate data for which an approximately linear relationship exists. Since not all sets of bivariate data are linearly related, the first step in the process is to construct a scatter diagram. Figure 17.13 shows scatter diagrams for some linear and some non-linear sets of data.

Before attempting to write an equation of a line that approximates a set of bivariate data, it is useful to calculate a measure of the strength of the linear relationship between the variables. One of the most common such measures is the **linear coefficient of correlation**, r. The coefficient indicates the degree of linearity between the variables. The following formula can be used to calculate this coefficient.

Linear Coefficient of Correlation

LINEAR CORRELATION COEFFICIENT

$$r = \frac{n(\Sigma xy) - (\Sigma x)(\Sigma y)}{\sqrt{n(\Sigma x^2) - (\Sigma x)^2} \sqrt{n(\Sigma y^2) - (\Sigma y)^2}},$$

where n represents the number of data points.

The linear correlation coefficient is always a number between -1 and $+1$. Figure 17.14 shows some examples of scatter diagrams along with the corresponding correlation coefficients.

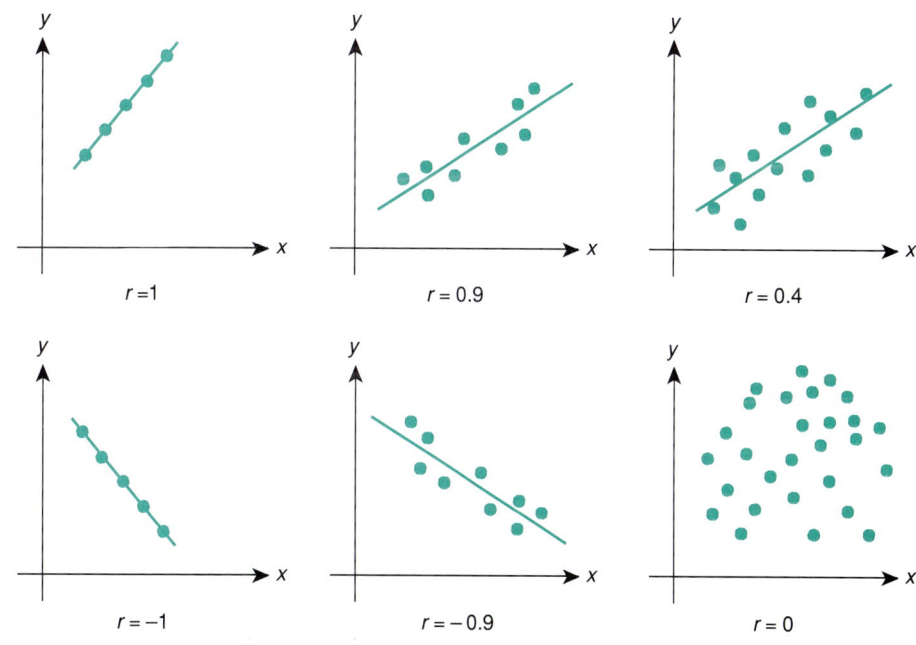

Figure 17.14

If $r > 0$, then the variables are positively correlated, and an increase in one of the variables yields an increase in the other variable. If $r < 0$, then the variables are negatively correlated, and an increase in one of the variables yields a decrease in the other variable.

Examples 17.13 and 17.14 illustrate how to calculate the linear correlation coefficient for a set of bivariate data.

• **Example 17.13:** Given that the following data is linearly related, calculate the correlation coefficient.

x	0.2	0.4	0.6	0.8	1.0
y	3	10	13	15	20

Solution: 1. The formula calls for Σx, Σy, Σx^2, Σy^2, and Σxy. Make a table that includes these columns and sum the columns.

x	y	x^2	y^2	xy
0.2	3	0.04	9	0.6
0.4	10	0.16	100	4.0
0.6	13	0.36	169	7.8
0.8	15	0.64	225	12.0
1.0	20	1.00	400	20.0
$\Sigma x = 3.0$	$\Sigma y = 61$	$\Sigma x^2 = 2.20$	$\Sigma y^2 = 903$	$\Sigma xy = 44.4$

2. Substitute these values into the formula. Note that $n = 5$, since there are 5 pairs of data.

$$r = \frac{n(\Sigma xy) - (\Sigma x)(\Sigma y)}{\sqrt{n(\Sigma x^2) - (\Sigma x)^2} \sqrt{n(\Sigma y^2) - (\Sigma y)^2}}$$

$$= \frac{5(44.4) - (3.0)(61)}{\sqrt{5(2.20) - (3.0)^2} \sqrt{5(903) - (61)^2}}$$

$$= \frac{39}{(1.4142136)(28.178006)} = 0.979.$$

An r of 0.979 shows a strong positive correlation between the variables.

• • • • • • • • • •

SECTION 17.4 THE LEAST SQUARES LINE

• **Example 17.14:** Find the linear correlation coefficient for the solubility data given at the beginning of the section.

Solution: Let x represent the temperature and y represent the solubility.

x	y	x^2	y^2	xy
0	27.6	0	761.76	0
10	31.0	100	961.00	310
20	34.0	400	1156.00	680
30	37.0	900	1369.00	1110
40	40.0	1600	1600.00	1600
50	42.6	2500	1814.76	2130
60	45.5	3600	2070.25	2730
70	48.3	4900	2332.89	3381
80	51.1	6400	2611.21	4088
90	54.0	8100	2916.00	4860
100	56.7	10,000	3214.89	5670
550	467.8	38,500	20,807.76	26,559
Σx	Σy	Σx^2	Σy^2	Σxy

Note that $n = 11$.

$$r = \frac{11(26{,}559) - 550(467.8)}{\sqrt{11(38{,}500) - (550)^2}\sqrt{11(20{,}807.76) - (467.8)^2}}$$

$$= \frac{34{,}859}{(347.8505)(100.2423)} = 0.9997.$$

The result, $r = 0.9997$, indicates virtually perfect positive linear correlation between the variables.

••••••••••

Assuming that two variables are approximately linearly related, our next task is to find an equation for the line that "best fits" a set of given data. Two methods are introduced, one that uses the correlation coefficient, and one that does not use it. By a **Least Squares Line** **least squares line** we mean the line that minimizes the sum of the squares of the deviations of the y-values in the table from the line.

CHAPTER 17 STATISTICS

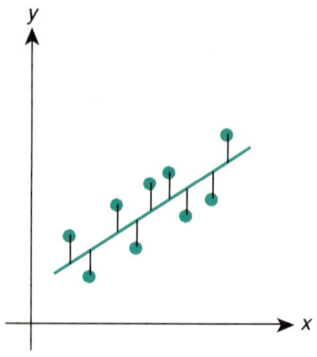

Figure 17.15

Figure 17.15 shows the deviations of a set of points from a line as the lengths of a set of vertical segments. The idea is to find the equation of a line that minimizes the sum of the squares of these deviations.

Merely adding the deviations and minimizing this sum would not provide the "best fit" since some of the deviations are positive and some are negative.

One of the methods of finding the equation for a least squares line for a set of bivariate data uses the point-slope form of the equation of a line. Recall from section 16.2 that a line that contains a particular point (x_1, y_1) and that has slope m can be written as follows:

$$y - y_1 = m(x - x_1).$$

Two properties of the least square line that make this form useful are

1. The point $(\bar{x}, \bar{y})$ is on the line, where $\bar{x}$ and $\bar{y}$ are the means of the two sets of data.

2. The slope of the line is $m = r\left(\dfrac{s_y}{s_x}\right)$, where r is the correlation coefficient and s_x and s_y are the standard deviations of the two sets of data.

Substituting these values in the point-slope form of the equation for a line yields the following equation.

LEAST SQUARES LINE

$$y - \bar{y} = r\left(\dfrac{s_y}{s_x}\right)(x - \bar{x}),$$

where $\bar{x}$ is the mean of the x's, $\bar{y}$ is the mean of the y's, s_x is the standard deviation of the x's, s_y is the standard deviation of the y's, and r is the coefficient of correlation.

Example 17.15 illustrates how to use this formula.

• **Example 17.15:** Find an equation for the least squares line using the data and the results of example 17.13.

Solution: From example 17.13, $r = 0.979$, $\Sigma x = 3.0$, $\Sigma y = 61$, $\Sigma x^2 = 2.20$, and $\Sigma y^2 = 903$.

1. $\bar{x} = \dfrac{\Sigma x}{n} = \dfrac{3.0}{5} = 0.60$.

2. $\bar{y} = \dfrac{\Sigma y}{n} = \dfrac{61}{5} = 12.2$.

3. $s_x = \sqrt{\dfrac{n\Sigma x^2 - (\Sigma x)^2}{n(n-1)}} = \sqrt{\dfrac{5(2.20) - (3.0)^2}{5(4)}} = 0.316$.

4. $s_y = \sqrt{\dfrac{n\Sigma y^2 - (\Sigma y)^2}{n(n-1)}} = \sqrt{\dfrac{5(903) - (61)^2}{5(4)}} = 6.30.$

5. Substitute these values into the equation

$$y - \bar{y} = r\left(\dfrac{s_y}{s_x}\right)(x - \bar{x})$$

$$y - 12.2 = (0.979)\dfrac{(6.30)}{(.316)}(x - 0.60).$$

If we solve this equation for y, this equation results:

$$y = 19.5x + 0.5.$$

This equation can be used to predict a y-value for any given x-value. For example, if $x = 0.65$,

$$y = 19.5(0.65) + 0.5 = 13.18.$$

Care should be exercised in using the least squares equation to make predictions if the x-value used is outside the range of the given data. Many linear relationships are in fact linear for only a restricted set of values of x and y.

If the coefficient of correlation is not needed, for example in instances where you are convinced of the linearity of the relationship from the scatter diagram, an alternate formula for the least squares line can be used.

LEAST SQUARES LINE ALTERNATE FORMULA

where

$$y = mx + b,$$

$$m = \dfrac{n(\Sigma xy) - (\Sigma x)(\Sigma y)}{n\Sigma x^2 - (\Sigma x)^2}$$

$$b = \dfrac{\Sigma y - m(\Sigma x)}{n}.$$

Using the data from example 17.13, $\Sigma x = 3.0$, $\Sigma y = 61$, $\Sigma x^2 = 2.20$, $\Sigma y^2 = 903$, and $\Sigma xy = 44.4$, the formula works like this:

$$m = \dfrac{5(44.4) - 3.0(61)}{5(2.20) - (3.0)^2} = 19.5$$

$$b = \dfrac{61 - 19.5(3.0)}{5} = 0.5.$$

The equation is, $y = 19.5x + 0.05$, which is identical to the equation that we obtained in example 17.15.

CHAPTER 17 STATISTICS

• **Example 17.16:** Use the data from example 17.14 and the alternate formula to obtain a least squares line for the temperature/solubility data.

Solution: $n = 11$, $\Sigma x = 550$, $\Sigma y = 467.8$, $\Sigma x^2 = 38,500$, $\Sigma y^2 = 20,807.76$, $\Sigma xy = 26,559$,

$$m = \frac{n(\Sigma xy) - (\Sigma x)(\Sigma y)}{n(\Sigma x^2) - (\Sigma x)^2}$$

$$m = \frac{11(26,559) - 550(467.8)}{11(38,500) - (550)^2} = \frac{34,859}{121,000} = 0.288$$

$$b = \frac{\Sigma y - m(\Sigma x)}{n} = \frac{(467.8) - 0.288(550)}{11} = 28.13.$$

The least squares equation is

$$y = mx + b$$

or

$$y = 0.288x + 28.13.$$

In most problems, the alternate formula is easier to use. However, if the means and standard deviations of the variables have already been calculated in a problem, then the original formula may be easier to use.

Trial Problems 17.4

Before you begin the section exercises, warm up with these problems. Complete answers are included in the answer key.

1. Find the coefficient of correlation using the following data:
   $n = 30$, $\Sigma x = 33$, $\Sigma y = 95$, $\Sigma x^2 = 21,485$, $\Sigma y^2 = 16,820$, $\Sigma xy = 6810$.

2. Find an equation for the least squares line using the following data:
   $\bar{x} = 60.4$, $\bar{y} = 68.3$, $s_x = 21.4$, $s_y = 16.8$, $r = 0.83$.

3. Find an equation for the least squares line using the following data:
   $n = 10$, $\Sigma x = 180$, $\Sigma y = 335$, $\Sigma x^2 = 5805$, $\Sigma y^2 = 12,725$, $\Sigma xy = 4325$.

x	1	2	3	4	5	6	7
y	19	30	48	49	62	71	83

4. For the data in the table above, find the coefficient of correlation.

5. Find an equation for the least squares line using the data in the table shown above.

Exercises 17.4

In exercises 1–8, draw a scattergram and find the coefficient of correlation.

1.
x	0.2	0.4	0.6	0.8	1.0
y	2	10	13	15	20

2.
x	6	2	7	4	6	8	7	1	5	2	8	3
y	8	5	11	8	5	10	8	5	6	3	11	4

3.
x	34	42	37	55	47	43	52	39
y	6.3	8.1	7.9	9.8	8.6	8.4	9.1	8.6

4.
x	5	2	6	4	5	3	6
y	4	12	1	8	6	10	9

5. The number y of parts/million of a drug in the bloodstream of a patient after x hours.

x	2	3	4	5	6	7	8
y	4.3	4.0	3.3	3.1	2.9	2.5	2.1

6. The number y of miles/gallon of consumption at speed x in mi/hr.

x	30	40	50	60	70
y	32.2	30.2	29.6	25.1	23.4

7. The reduction y of pollutants in exhaust gases for the addition of x grams/liter of gasoline additive.

x	2.0	2.3	2.5	2.7	3.0	3.2	3.5
y	3.54	3.68	3.78	3.87	4.10	4.10	4.24

8. The length y of a spring in cm when a force of x kg is applied.

x	50	100	150	200	250	300
y	39.0	42.2	44.6	46.8	49.3	51.5

9. **a.** Write an equation for the least squares line in exercise 5.
   b. Use the equation to estimate the number of parts/million after 9 hours.

10. **a.** Write an equation for the least squares line in exercise 6.
    b. Use the equation to predict the consumption at 55 mi/hr.

11. **a.** Write an equation for the least squares line in exercise 7.
    b. Use the equation to estimate the reduction in pollutants for $x = 3.7$ grams/liter.

12. **a.** Write an equation for the least squares line in exercise 8.
    b. Use the equation to predict the length of the spring when a force of 180 kg is applied.

13. An experiment relating the pressure y in newtons/square meter on an object that is x meters below the surface of water yields the following results.

x	20	30	40	50	60
y	20.82	30.73	40.61	50.50	60.36

   a. Find an equation for the least squares line.
   b. Use the equation to predict the pressure at 45 m.

14. An experiment on the volume y in ml of a gas at constant pressure at a temperature of x degrees Celsius yielded these results.

x	30	40	50	60	70	80
y	50.4	56.9	63.2	69.7	76.3	82.7

a. Find an equation for the least squares line.
b. Use the equation to estimate the volume at 63°.

15. An experiment to determine the relationship between the resistance y in ohms of a resistor at various temperatures x in degrees Celsius yields the results shown in the table.

x	40	80	120	160	200
y	102.5	103.7	104.8	106.0	107.1

a. Find an equation for the least squares line.
b. Use the equation to predict the resistance at 100°.

16. Automobile engines with displacements y in liters were tested for mileage x in mi/gal. The results are shown in the table.

x	1.5	1.7	2.0	2.3	3.5	3.7	4.3	4.7
y	34.4	33.8	33.0	32.0	28.1	28.0	26.4	24.1

a. Find an equation for the least squares line.
b. Use the equation to predict the mileage of a 2.7 liter engine.

17. An electric company lists the rates shown in the table below.

kwh x	200	500	800	1000	1500	2000	5000
Cost y	$27.42	$48.34	$65.75	$77.35	$102.17	$126.98	$275.89

a. Find an equation for the least squares line.
b. Estimate the cost of 3500 kwh.

17 • 5 Nonlinear Curve Fitting

When the relationship between two variables is not linear, a set of data can often be fitted to a curve other than a line. We will consider several of these models in this section.

For models that fit the general form $y = mf(x) + b$, the procedure for fitting a least squares curve to the data is not very different from the procedure for finding a least squares line. The procedure involves performing the substitution $x^* = f(x)$ on the data. This substitution yields the equation $y = mx^* + b$, and the linear least squares procedure yields the values for m and b that are needed for the original equation.

Transformation
Transformed Equation

A substitution such as this one is called a **transformation**, and the resulting equation $y = mx^* + b$ is the **transformed equation**.

SECTION 17.5 NONLINEAR CURVE FITTING

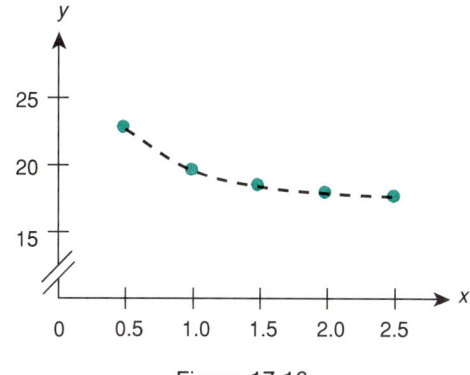

Figure 17.16

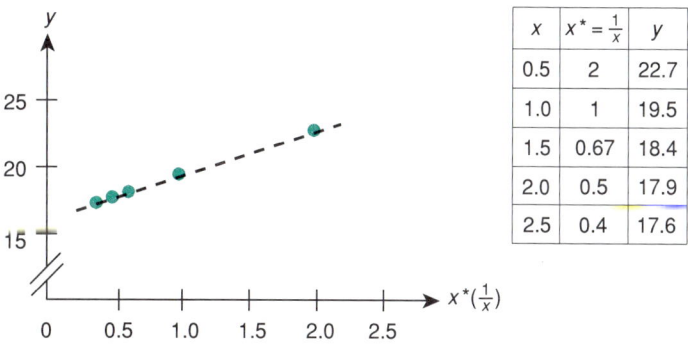

Figure 17.17

For example, consider the following table on x and y values.

x	0.5	1.0	1.5	2.0	2.5
y	22.7	19.5	18.4	17.9	17.6

A scatter diagram for the data, shown in figure 17.16, shows that the data is nonlinear. However, if we plot a scatter diagram using $x^* = \dfrac{1}{x}$ instead of x, the scatter diagram in figure 17.17 shows an approximately linear relationship between x^* and y.

To obtain a least squares equation for x^* and y, it is necessary to obtain these values: Σx^*, Σy, $\Sigma (x^*)^2$, and $\Sigma x^* y$. Construct a table containing the appropriate columns.

x	$x* = \dfrac{1}{x}$	y	$(x*)^2$	$x*y$
0.5	2.00	22.7	4.0000	45.4
1.0	1.00	19.5	1.0000	19.5
1.5	0.67	18.4	0.4489	12.328
2.0	0.50	17.9	0.2500	8.95
2.5	0.40	17.6	0.1600	7.04
	4.57	96.1	5.8589	93.218
	$\Sigma x*$	Σy	$\Sigma(x*)^2$	$\Sigma x*y$

$$m = \frac{n\Sigma(x*y) - (\Sigma x*)(\Sigma y)}{n\Sigma(x*)^2 - (\Sigma x*)^2}$$

$$= \frac{5(93.218) - (4.57)(96.1)}{5(5.8589) - (4.57)^2} = \frac{26.913}{8.4096} = 3.20$$

$$b = \frac{\Sigma y - m\Sigma x*}{n}$$

$$= \frac{(96.1) - (3.2)(4.57)}{(5)} = 16.3.$$

The equation for the least squares line is

$$y = mx* + b,$$

where $m = 3.20$ and $b = 16.3$. The least squares equation for the original data is

$$y = \left(\frac{1}{x}\right)m + b,$$

where $m = 3.20$ and $b = 16.3$.

$$y = \frac{3.20}{x} + 16.3$$

is the equation that approximates the data in the original table.

Example 17.17 illustrates another model of the form $y = mf(x) + b$.

• **Example 17.17:** The solubility y in grams of a sugar in 100 grams of water at x degrees Celsius is shown in the table.

x	10	20	30	40	50	60	70
y	190	204	220	238	260	287	321

Find an equation of the form $y = mx^2 + b$ that approximates the data.

Solution: The transformation $x^* = x^2$ yields the linear equation $y = mx^* + b$. We find a least squares line using Σx^*, Σy, $\Sigma (x^*)^2$, and $\Sigma x^* y$.

x	$x^* = x^2$	y	$(x^*)^2$	x^*y
10	100	190	10,000	19,000
20	400	204	160,000	81,600
30	900	220	810,000	198,000
40	1600	238	2,560,000	380,800
50	2500	260	6,250,000	650,000
60	3600	287	12,960,000	1,033,200
70	4900	321	24,010,000	1,572,900
	14,000	1720	46,760,000	3,935,500
	Σx^*	Σy	$\Sigma (x^*)^2$	$\Sigma (x^*y)$

$$m = \frac{n\Sigma(x^*y) - (\Sigma x^*)(\Sigma y)}{n\Sigma(x^*)^2 - (\Sigma x^*)^2}$$

$$= \frac{7(3,935,500) - (14,000)(1720)}{7(46,760,000) - (14,000)^2} = \frac{3.4685 \times 10^6}{1.3132 \times 10^8} = 0.0264$$

$$b = \frac{\Sigma y - m\Sigma x^*}{n}$$

$$= \frac{1720 - (0.0264)(14,000)}{7} = 193.$$

The least squares line is $y = 0.0264x^* + 193$. Using the transformation $x^* = x^2$, the least squares curve for the original data is $y = 0.0264x^2 + 193$.

⋯⋯⋯⋯⋯

There are other transformations that can be made on a nonlinear equation to obtain a transformed equation that is linear. One of these transformations uses the following property of natural logarithms:

$$\ln(e^{f(x)}) = f(x)$$

For example, given an equation of the form

$$y = e^{mx + b},$$

taking the natural logarithm of both sides yields this equation.

$$\ln y = \ln(e^{mx + b}) = mx + b$$
$$\ln y = mx + b.$$

The transformation $y^* = \ln y$ yields the following linear equation:

$$y^* = mx + b.$$

Example 17.18 illustrates this transformation.

724 CHAPTER 17 STATISTICS

• **Example 17.18:** The following table shows the wind chill index y at 35° F for a wind velocity of x mi/hr.

x	5	10	15	20	25	30	35	40	45
y	33	22	16	12	8	6	4	3	2

Find an equation for the least squares curve of the form $y = e^{mx + b}$ that approximates the data.

Solution: Use the transformation $y^* = \ln x$.

x	y	$y^* = \ln y$	x^2	xy^*
5	33	3.4965	25	17.4825
10	22	3.0910	100	30.9100
15	16	2.7726	225	41.589
20	12	2.4849	400	49.698
25	8	2.0794	625	51.985
30	6	1.7918	900	53.754
35	4	1.3863	1225	48.5205
40	3	1.0986	1600	43.944
45	2	0.6931	2025	31.1895
225		18.8942	7125	369.0725
Σx		Σy^*	Σx^2	Σxy^*

$$m = \frac{n\Sigma(xy^*) - (\Sigma x)(\Sigma y^*)}{n\Sigma(x^2) - (\Sigma x)^2}$$

$$= \frac{9(369.0725) - (225)(18.8942)}{9(7125) - (225)^2}$$

$$= \frac{-929.5425}{13,500} = -0.06686$$

$$b = \frac{\Sigma y^* - m\Sigma x}{n}$$

$$= \frac{18.8942 - (-0.06886)(225)}{9} = 3.821.$$

The least squares equation is $y = e^{-0.06886x + 3.821}$.

We can use this equation to estimate the wind chill index for other wind velocities. For example for $x = 13$ mi/hr,

$$y = e^{-0.06886(13) + 3.821} = 18.6° \text{ F}.$$

0.06886   13 3.821

 = INV ln x .

SECTION 17.5 NONLINEAR CURVE FITTING

The following table shows some of the common transformations that result in linear equations.

Nonlinear Equation	Transformation	Transformed Equation
1. $y = m(f(x)) + b$	$x^* = f(x)$	$y = mx^* + b$
2. $y = e^{mx+b}$	$y^* = \ln y$	$y^* = mx + b$
3. $y = \dfrac{1}{mx+b}$	$y^* = \dfrac{1}{y}$	$y^* = mx + b$
4. $y = be^{mx}$	$y^* = \ln y$ $b^* = \ln b$	$y^* = mx + b^*$
5. $y = bx^m$	$y^* = \ln y$ $b^* = \ln b$ $x^* = \ln x$	$y^* = mx^* + b^*$

Example 17.19 illustrates fitting a curve of the form $y = bx^m$ to a set of bivariate data.

• **Example 17.19:** The stopping distance y in feet for a particular model of automobile traveling at x mi/hr is shown in the following table.

x	20	30	40	50	60	70
y	47	91	146	210	283	364

Fit a curve of the form $y = bx^m$ to the data.

Solution: If we take the natural logarithm of both sides of the equation $y = bx^m$ we get these results:

$$\ln y = \ln(bx^m) = \ln b + m(\ln x).$$

The transformations $y^* = \ln y$, $b^* = \ln b$, and $x^* = \ln x$ yield the linear equation $y^* = mx^* + b^*$.

x	y	$x^* = \ln x$	$y^* = \ln y$	$(x^*)^2$	$x^* y^*$
20	47	2.9957	3.8501	8.9742	11.5337
30	91	3.4012	4.5109	11.5682	15.3425
40	146	3.6889	4.9836	13.6080	18.3840
50	210	3.9120	5.3471	15.3037	20.9179
60	283	4.0943	5.6454	16.7633	23.1140
70	364	4.2485	5.8972	18.0498	25.0543
		22.3406	30.2343	84.2672	114.3464
		Σx^*	Σy^*	$\Sigma (x^*)^2$	$\Sigma x^* y^*$

$$m* = \frac{n(\Sigma x*y*) - (\Sigma x*)(\Sigma y*)}{n(\Sigma(x*)^2) - (\Sigma x*)^2}$$

$$= \frac{6(114.3464) - (22.3406)(30.2343)}{6(84.2672) - (22.3406)^2} = 1.635$$

$$b* = \frac{\Sigma y* - m\Sigma x*}{n} = \frac{30.2343 - 1.635(22.3406)}{6} = -1.0488.$$

Since $b* = \ln b$ and $\ln b = -1.0488$,

$$b = 0.350.$$

The least squares curve is

$$y = bx^m$$

where $b = 0.350$ and $m = 1.635$, or

$$y = 0.350 x^{1.635}.$$

Using the equation to estimate the stopping distance at $x = 55$ mi/hr yields these results:

$$y = 0.350(55)^{1.635} = 245 \text{ ft.}$$

 0.350 × 55 y^x 1.635 =

Display: 245.22291.

Trial Problems 17.5

Before you begin the section exercises, warm up with these problems. Complete answers are included in the answer key.

1.
x	10	20	30	40	50	60	70
y	23.1	28.8	33.2	36.9	40.1	43.1	45.8

Fit the data to a curve of the form $y = m\sqrt{x} + b$.

2.
x	2	4	8	10	20	40
y	125	75	50	45	35	30

Fit the data to a curve of the form $y = \dfrac{m}{x} + b$.

3.
x	0.5	1.0	1.5	2.0	3.0	3.5
y	12.8	16.5	21.2	27.2	44.8	57.5

Fit the data to a curve of the form $y = me^{bx}$.

Exercises 17.5

1. Fit the data in the table to a curve of the form $y = m\left(\dfrac{1}{x}\right) + b$.

x	10	20	30	40	50
y	35	24	21	19	18

2. Fit the data in the table to a curve of the type $y = m\left(\dfrac{1}{x}\right) + b$.

x	5	10	15	20	25	30
y	41	24	19	16	14	13

3. Fit the data in the table to a curve of the form $y = m\sqrt{x} + b$.

x	5	10	15	20	25	30
y	14.5	17.7	20.1	22.1	24.0	25.6

4. Fit the data in the table to a curve of the form $y = m\sqrt{x} + b$.

x	10	20	30	40	50	60
y	6.9	9.1	10.8	12.2	13.4	14.5

5. Fit the data in the table to a curve of the form $y = mx^2 + b$.

x	0.5	1.0	1.5	2.0	2.5	3.0	3.5
y	30.2	79.2	160.1	275.1	422.0	601.6	813.8

6. Fit the data in the table to a curve of the form $y = mx^2 + b$.

x	1.5	2.0	2.5	3.0	3.5	4.0	4.5
y	11.4	12.8	14.5	16.7	19.2	22.2	25.5

7. An experiment yields the following data for the resistance y in ohms/cm³ and the number of grams x of silver nitrate per 100 grams of solution.

x	5	10	15	20	25	30	40	50
y	39.0	21.0	14.6	11.5	9.5	8.1	6.4	5.3

   a. Find a least squares curve of the type $y = m\left(\dfrac{1}{x}\right) + b$ that approximates the data.

   b. Find an estimate for the resistance for $x = 35$ grams/cm³.

8. An experiment yields the following data on the intensity y of a light from above the surface of the water at a depth of x meters.

x	2	4	6	8	10	12	14
y	64.0	41.0	26.2	16.8	10.7	6.9	4.4

a. Find a least squares curve of the type $y = m\left(\dfrac{1}{x}\right) + b$ that approximates the data.
b. Find an estimate for the intensity at a depth of 9.0 meters.

9. An experiment relating the temperature y in degrees F of an object after x seconds yields the following data.

x	2	4	6	8	10	12	14
y	50.8	57.7	62.1	65.0	66.8	67.9	68.7

a. Find a least squares equation of the form $y = m(\ln x) + b$ that approximates the data.
b. Find an estimate for the temperature after 9 seconds.

10. The velocity y in ft/sec of a stream of liquid through an orifice that is at a depth x in feet below the surface is shown in the table.

x	1	3	5	7	10	12
y	8.1	14.0	18.1	21.4	25.6	28.1

a. Find an equation of the form $y = m\sqrt{x} + b$ that approximates the data.
b. Estimate the velocity at a depth of 8.5 ft.

11. Find an equation of the form $y = e^{mx+b}$ that approximates the following data.

x	20	30	40	50	60	70
y	38.9	16.9	7.4	3.2	1.4	0.6

12. Find an equation of the form $y = e^{mx+b}$ that approximates the following data.

x	10	20	30	40	50	60
y	58.6	27.7	13.1	6.2	2.9	1.4

13. At 105° F, the discomfort index in degrees F is related to the humidity x in percent according to the following table.

x	10	20	30	40	50	60
y	100	105	113	123	135	149

a. Fit the data to an equation of the form $y = be^{mx}$.
b. Estimate the discomfort index for a humidity of 45%.

14. After an injection of penicillin, the amount of the drug y in mg in the bloodstream after x hours is shown in the following table.

x	1	2	5	7	10
y	102.3	52.4	7.0	1.8	0.25

a. Fit the data to an equation of the form $y = be^{mx}$.
b. Estimate the value for y at $x = 4$ hours.

15. Fit the data in the table to a curve of the form $y = bx^m$.

x	4.0	8.0	16.0	20.0
y	0.50	0.11	0.08	0.06

16. Fit the data in the table to an equation of the form $y = bx^m$.

x	3.1	3.6	4.1	4.6	5.1	5.6	6.1
y	11.1	10.5	10.0	9.6	9.2	8.9	8.6

17·6 Summary of Terms, Formulas, Rules, and Procedures

Terms

Arithmetic Mean (p. 694)
Bivariate Data (p. 711)
Coefficient of Correlation (p. 713)
Frequency (p. 681)
Frequency Histogram (p. 684)
Intervals (p. 680)

Least Squares Line (p. 715)
Measures of Central Tendency (p. 690)
Median (p. 691)
Normal Distribution (p. 705)
Range (p. 682)
Relative Frequencies of the Classes (p. 683)

Relative Frequency Histogram (p. 684)
Scatter Diagram (p. 711)
Standard Normal Curve (p. 706)
Transformation (p. 720)
Transformed Equation (p. 720)

Formulas

Arithmetic mean of $x_1, x_2, x_3, \ldots, x_n$: $\bar{x} = \dfrac{\Sigma x_i}{n}$ (17.2)

Arithmetic Mean for Grouped Data: $\bar{x} = \dfrac{\Sigma xf}{\Sigma f} = \dfrac{\Sigma xf}{n}$ (17.2)

Variance of a Set of Data: $s^2 = \dfrac{\Sigma(x - \bar{x})^2}{n - 1}$ (17.3)

Standard Deviation of a Set of Data: $s = \sqrt{\dfrac{\Sigma(x - \bar{x})^2}{n - 1}}$ (17.3)

Variance of a Set of Data (alternate form): $s^2 = \dfrac{n\Sigma x^2 - (\Sigma x)^2}{n(n - 1)}$ (17.3)

Standard Deviation of a Set of Data (alternate form): $s = \sqrt{\dfrac{n\Sigma x^2 - (\Sigma x)^2}{n(n - 1)}}$ (17.3)

Variance (grouped data): $s^2 = \dfrac{n\Sigma x^2 f - (\Sigma xf)^2}{n(n - 1)}$, where $n = \Sigma f$ (17.3)

Standard Deviation (grouped data): $s = \sqrt{\dfrac{n\Sigma x^2 f - (\Sigma xf)^2}{n(n - 1)}}$, where $n = \Sigma f$ (17.3)

Linear Correlation Coefficient: $r = \dfrac{n(\Sigma xy) - (\Sigma x)(\Sigma y)}{\sqrt{n(\Sigma x^2) - (\Sigma x)^2} \sqrt{n(\Sigma y^2) - (\Sigma y)^2}}$ (17.4)

Least Squares Line: (17.4)

$$y - \bar{y} = r\left(\dfrac{s_y}{s_x}\right)(x - \bar{x}),$$

where $\bar{x}$ = x-mean, $\bar{y}$ = y-mean,
   s_x = standard deviation of x's,
   s_y = standard deviation of y's,
and r = linear correlation coefficient.

Least Squares Line (alternate form):   (17.4)

$$y = mx + b, \text{ where } m = \dfrac{n\Sigma xy - (\Sigma x)(\Sigma y)}{n\Sigma x^2 - (\Sigma x)^2}, \; b = \dfrac{\Sigma y - m(\Sigma x)}{n}$$

Some Common Transformations Resulting in Linear Equations: (17.5)

Nonlinear Equation	Transformation	Transformed Equation
1. $y = m(f(x)) + b$	$x^* = f(x)$	$y = mx^* + b$
2. $y = e^{mx + b}$	$y^* = \ln y$	$y^* = mx + b$
3. $y = \dfrac{1}{mx + b}$	$y^* = \dfrac{1}{y}$	$y^* = mx + b$
4. $y = be^{mx}$	$y^* = \ln y$ $b^* = \ln b$	$y^* = mx + b^*$
5. $y = bx^m$	$y^* = \ln y$ $b^* = \ln b$ $x^* = \ln x$	$y^* = mx^* + b^*$

Rules and Procedures

- Guidelines for constructing a frequency table   (17.1)
  1. The number of classes should be between 5 and 12 for most problems.
  2. Each bit of data must belong to exactly one class.
  3. In most problems, all classes should have equal widths.
  4. The classes in the distribution must include all of the data.

- Constructing frequency distributions, frequency histograms, and relative frequency histograms   (17.1)
  1. Estimate the number of classes to use.
  2. Determine the class width.
  3. Find the interval endpoints.
  4. Obtain the frequencies.
  5. Compute the relative frequencies.
  6. Construct a frequency histogram using the frequencies as the heights of the bars.
  7. Construct a relative frequency histogram using the relative frequencies as the heights of the bars.

Box-and-whisker diagram for a set of data (17.2)

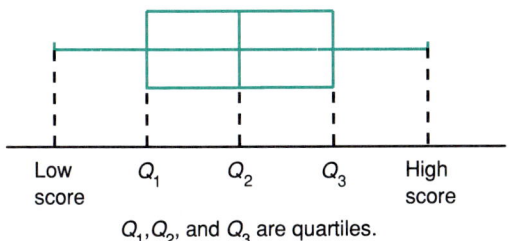

Q_1, Q_2, and Q_3 are quartiles.

- Chebyshev's Theorem (17.3)

   For any set of data and any $k > 1$, the proportion of the data that is within k standard deviations on either side of the mean is at least $1 - \dfrac{1}{k^2}$.

- Emperical Rule (17.3)

   For a set of data that is approximately normally distributed, about 68.3% of the data lies within one standard deviation of the mean, about 95.4% within two standard deviations of the mean, and about 99.7% within three standard deviations of the mean.

 17·7 Chapter 17 Review Exercises

1. The following data represent the results of testing the breaking strength in pounds for a material.

115.8	119.3	118.1	119.8	117.1	116.6	114.7	117.4	117.1	117.5
120.7	120.1	116.0	123.3	117.6	115.8	117.9	113.7	116.1	113.3
117.2	116.4	122.6	114.5	116.7	120.2	115.9	118.3	119.3	120.3
117.7	120.5	120.6	121.6	117.0	115.9	120.5	116.8	116.3	122.8

   a. Construct a frequency table for the data that includes a relative frequency column using 113.0–114.9 as the first class.
   b. Construct a frequency histogram for the data.
   c. Construct a relative frequency histogram for the data.

2. The following data shows the results of testing the amount of radon in picocuries/liter in 30 homes.

3.9	6.3	4.5	6.0	4.8	0.5	5.5	2.8	5.7	2.7
5.3	15.2	2.8	5.4	9.3	6.2	5.0	8.6	5.5	7.9
2.5	20.3	10.1	5.8	5.2	5.4	1.6	5.3	2.5	3.7

   a. Find the mean.
   b. Find the median.
   c. Find the quartile scores.
   d. Construct a box-and-whisker diagram for the data.
   e. Calculate the variance for the data.
   f. Calculate the standard deviation for the data.

In exercises 3–6, find the mean and standard deviation for the data.

3. $\Sigma x = 67$, $n = 28$, $\Sigma x^2 = 1143.2$

4. $\Sigma x = 116$, $n = 39$, $\Sigma x^2 = 5814.7$

5. | x | f |
    |---|---|
    | 16 | 3 |
    | 17 | 9 |
    | 18 | 15 |
    | 19 | 22 |
    | 20 | 16 |
    | 21 | 8 |
    | 22 | 3 |

6. | x | f |
    |---|---|
    | 2.5 | 4 |
    | 3.5 | 8 |
    | 4.5 | 18 |
    | 5.5 | 25 |
    | 6.5 | 17 |
    | 7.5 | 9 |
    | 8.5 | 3 |

In exercises 7–8, use the data and Chebyshev's theorem to find an interval that fits the given conditions.

7. $\bar{x} = 19.6$, $s = 2.7$, $k = 2$

8. $\bar{x} = 28.3$, $s = 3.5$, 75% of the data in the interval

9. An approximately normally distributed sample yields a mean of 85.2 and a standard deviation of 6.3. Find an interval that would contain:
   a. About 68.3% of the data
   b. About 94.5% of the data
   c. About 99.7% of the data

10. Given $\Sigma xf = 540.7$, $\Sigma x^2 f = 6584.3$, and $n = 48$, find an interval that contains approximately the following percents of the data, assuming an approximately normal distribution.
    a. 68.3%   b. 95.4%   c. 99.7%

11. Given a set of data with $n = 10$, $\Sigma x = 351$, $\Sigma y = 380$, $\Sigma xy = 14{,}257$, $\Sigma x^2 = 13{,}717$, and $\Sigma y^2 = 15{,}298$, calculate:
    a. The coefficient of correlation
    b. The equation of the least squares line

12. Given a set of data with $n = 10$, $\Sigma x = 517$, $\Sigma y = 516$, $\Sigma xy = 25{,}990$, $\Sigma x^2 = 28{,}165$, and $\Sigma y^2 = 26{,}996$, calculate:
    a. The coefficient of correlation
    b. The equation of the least squares line

13. The length y of a steel pipe in cm is measured at temperatures x in degree Celsius. The results are shown in the table.

x	20	40	60	80	100
y	860.86	861.22	861.59	861.95	862.31

    a. Find an equation for the least squares line.
    b. Estimate the length of the pipe at 50° C.

14. The resistance y in ohms for a 10 km length of wire used in overhead transmission lines at temperature x in degrees Celsius is shown in the table.

x	−20	0	20	40	60	80
y	4.30	4.71	5.11	5.52	5.92	6.32

    a. Find an equation for the least squares line.
    b. Estimate the resistance at 45° C.

15. Six measurements of the altitude y in km above sea level and the atmospheric pressure x in g/cm² yielded the following data.

x	0.5	0.75	1.25	2.0	4.0	8.0
y	63.6	60.2	56.0	52.1	46.3	40.3

  a. Fit the data to an equation of the form $y = m(\ln x) + b$.
  b. Estimate the altitude for a pressure reading of 54 g/cm².

16. The number of grams y of unconverted substance in a chemical reaction after x seconds is shown in the table.

x	2	5	10	20	30	40
y	42.0	37.6	31.3	21.7	15.1	10.5

  a. Fit the data to an equation of the form $y = be^{mx}$.
  b. Estimate y for $x = 50$.

17. At 90° F the discomfort index y in degrees is related to the humidity x in percent according to the following table.

x	10	30	50	70	90
y	85	90	96	106	122

  a. Fit the data to an equation of the form $y = be^{mx}$.
  b. Estimate the discomfort index for a humidity of 60%.

18. The cost in dollars per item, y, of producing x items (in 1000's) is shown in the table.

x	10	20	30	40	50	60
y	36.60	19.10	13.30	10.10	8.60	7.40

  a. Fit the data to an equation of the form $y = m\left(\dfrac{1}{x}\right) + b$.
  b. Estimate the per item cost for producing 80,000 items.

17·8 Chapter 17 Test

1. Given the following data:

15.4   20.5   22.6   25.0   28.1   30.0   30.1   30.6   31.2   33.4
35.6   35.9   37.8   39.3   40.3   41.2   41.7   44.2   46.8   46.8
47.3   48.9   52.3   53.7   55.6   58.1   58.1   60.3   61.5   66.4

  a. Construct a frequency table for the data that includes a relative frequency column. Use 10–19.9 as the first interval.
  b. Construct a frequency histogram.
  c. Construct a relative frequency histogram.
  d. Find the median.
  e. Find the quartile scores.
  f. Construct a box-and-whisker diagram.
  g. Calculate the variance using the frequency table.
  h. Calculate the standard deviation using the frequency table.

Calculate the mean and standard deviation for the data in problems 2 and 3.

2. | x | f |
    |---|---|
    | 0 | 3 |
    | 1 | 18 |
    | 2 | 25 |
    | 3 | 32 |
    | 4 | 33 |
    | 5 | 25 |
    | 6 | 16 |
    | 7 | 14 |
    | 8 | 5 |

3. | x | f |
    |---|---|
    | 1.5 | 3 |
    | 2.5 | 9 |
    | 3.5 | 18 |
    | 4.5 | 25 |
    | 5.5 | 16 |
    | 6.5 | 8 |
    | 7.5 | 1 |

4. Use the following data and Chebyshev's theorem to find an interval that fits the conditions: $\bar{x} = 146.2$, $s = 19.7$, and $k = 2$.

5. An approximately normally distributed sample yields a mean of 72.4 and a standard deviation of 9.6. Find an interval that would contain about 68.3% of the data in the sample.

6. Given a set of data with $n = 9$, $\Sigma x = 516$, $\Sigma y = 950$, $\Sigma xy = 73{,}750$, $\Sigma x^2 = 37{,}720$, and $\Sigma y^2 = 149{,}050$, calculate:
   a. The coefficient of correlation.
   b. An equation for the least squares line.

7. | x | 3 | 8 | 10 | 12 | 15 | 18 | 20 | 25 |
    |---|---|---|---|---|---|---|---|---|
    | y | 12 | 16 | 19 | 19 | 22 | 23 | 26 | 31 |

   For the data in the table:
   a. Calculate the coefficient of correlation.
   b. Find an equation for the least squares line.

8. | x | 5 | 10 | 15 | 20 | 25 | 30 | 35 | 40 |
    |---|---|---|---|---|---|---|---|---|
    | y | 36 | 60 | 80 | 100 | 125 | 145 | 165 | 180 |

   For the data in the table:
   a. Calculate the coefficient of correlation.
   b. Find an equation for the least squares line.

9. | x | 10 | 20 | 30 | 40 | 50 | 60 |
    |---|---|---|---|---|---|---|
    | y | 22.8 | 26.2 | 28.9 | 31.1 | 33.0 | 34.8 |

   a. Fit the data in the table to a curve of the form $y = m\sqrt{x} + b$.
   b. Predict a y-value for $x = 35$.

10.

x	5	10	15	20	25	30
y	2.66	2.00	1.52	1.15	0.87	0.66

    a. Fit the data in the table to a curve of the form $y = be^{mx}$.
    b. Predict a y-value for $x = 35$.

17·9 Cumulative Review—Chapters 16 and 17

1. Find the distance between the points $A(-3,1)$ and $B(-4,-4)$.

2. Find the slope of the line containing the points $A(2,2)$ and $B(5,-1)$.

3. Find an equation for the line that contains the point $C(-3,2)$ and that has a slope of $\frac{1}{3}$.

Write the equation, in standard form, for the conic that fits the given conditions.

4. Circle with center at $(1,-3)$ with a radius of 4

5. Ellipse with foci at $(0,-2)$ and $(0,2)$, and with vertices at $(0,-4)$ and $(0,4)$

6. Ellipse with center at $(2,3)$ that contains the points $(5,3)$ and $(2,5)$

7. Parabola with vertex at $(4,0)$ that contains the point $(6,2)$

8. Hyperbola with vertices $(-3,0)$ and $(3,0)$, and with foci at $(-5,0)$ and $(5,0)$

Graph the following equations.

9. $x^2 + y^2 - 6x - 8y = 39$

10. $x^2 + 4y^2 - 2x - 24y = -21$

11. $4x^2 + 16x - y^2 - 6y = 9$

12. $x^2 - 6x - 4y = -1$

13. $r = \sin(3\theta)$

14. Given the following data:

    41.3   45.3   50.3   51.0   52.6   55.4   56.0   59.8   60.0   61.1
    62.3   62.3   65.6   65.8   66.3   68.3   69.0   70.1   70.3   72.4
    72.4   75.6   75.8   76.2   79.1   85.3   88.2   89.1   91.3   93.4

    a. Construct a frequency table for the data using 40–49.9 as the first interval.
    b. Construct a frequency histogram for the data.
    c. Find the mean.
    d. Calculate the standard deviation.

15. Calculate the mean and standard deviation for the data.

x	f
10	4
20	9
30	30
40	40
50	35
60	10
70	2

16.

x	20	21	22	23	24	25	26	27	28
y	16.8	17.2	17.5	17.9	18.3	18.6	18.8	19.3	19.7

For the data in the table:
a. Calculate the coefficient of correlation.
b. Find an equation for the least squares line.

17.

x	0.1	0.2	0.3	0.4	0.5	0.6	0.7
y	7.33	4.87	3.64	3.07	2.30	1.72	1.29

The data in the table represents the charge (y) on a capacitor after x seconds. Fit the data to an equation of the form $y = be^{mx}$.

Chapter 18
Differential Calculus

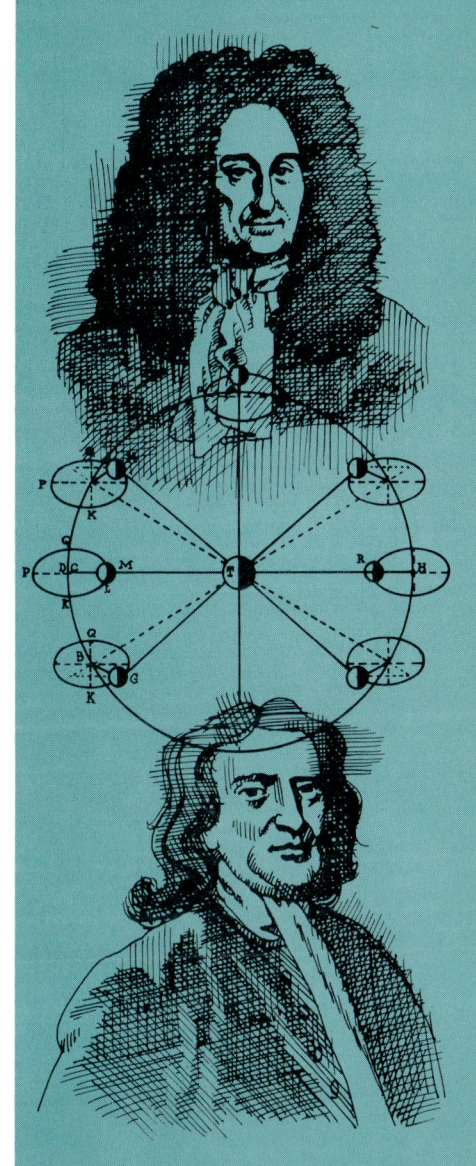

*I*n the remaining six chapters of this book a different branch of mathematics is presented: the calculus. The development of calculus is one of the most outstanding achievements, not only in the field of mathematics but also in the history of human thought. The impact of calculus on the mathematical community and on society in general has been astonishing. In the chapters that follow, fundamental ideas of calculus are presented, and several applications of these results are investigated.

The ideas of calculus can be traced back to Greek mathematicians who lived several hundred years before the birth of Christ. Many mathematicians and scientists since then provided useful contributions; but, the major credit for the introduction of calculus as an organized body of knowledge goes to two geniuses of the seventeenth century, Gottfried Wilhelm Leibniz (1646–1716) and Sir Isaac Newton (1642–1727). Working independently, these two men developed an understandable calculational system—the calculus—that was immediately put to use solving previously explored problems relating to the physical world. Today, this same system has proved to be an indispensable tool in such areas as chemistry, physics, astronomy, engineering, as well as economics, sociology, ecology, and other social and life sciences.

The calculus we study today is presented in two parts. The first is called *differential calculus.* Its historical basis stems from the study of the problem of finding the tangent line to a curve in the plane. The second is called *integral calculus* and stems from the search for a solution to the problem of finding the area of a region in a plane.

Fundamental to the study of calculus is the concept of limit. At first, this is a difficult concept to grasp, so an intuitive approach to the understanding of this concept is presented. Several theorems are stated following the intuitive approach. Proofs of theorems are kept to a minimum throughout the text, with the intention that just enough proofs be given so that rules are appreciated and a deeper understanding of the mathematics is offered. A delicate balance between memorizing the rules of calculus and appreciating how these rules were developed is attempted.

CHAPTER 18 DIFFERENTIAL CALCULUS

Limit

18 • 1 The Limit of a Function

The concept of the **limit** of a function is fundamental for the study of both differential and integral calculus. For some functions, $y = f(x)$, as x changes according to a certain pattern, $f(x)$ may change in a certain pattern that suggests that it may have a limiting value. The simple function $y = 2x + 3$ attains values closer and closer to 5 as x is given successive values close to 1.

$$x = 0, \quad f(0) = 3 \qquad\qquad x = 2, \quad f(2) = 7$$

$$x = \frac{1}{2}, \quad f\left(\frac{1}{2}\right) = 4 \qquad\qquad x = 1\frac{1}{2}, \quad f\left(1\frac{1}{2}\right) = 6$$

$$x = \frac{3}{4}, \quad f\left(\frac{3}{4}\right) = 4\frac{1}{2} \qquad\qquad x = 1\frac{1}{4}, \quad f\left(1\frac{1}{4}\right) = 5\frac{1}{2}$$

$$x = \frac{7}{8}, \quad f\left(\frac{7}{8}\right) = 4\frac{3}{4} \qquad\qquad x = 1\frac{1}{8}, \quad f\left(1\frac{1}{8}\right) = 5\frac{1}{4}$$

$$x = \frac{9}{10}, \quad f\left(\frac{9}{10}\right) = 4\frac{4}{5} \qquad\qquad x = 1\frac{1}{10}, \quad f\left(1\frac{1}{10}\right) = 5\frac{1}{5}$$

$$\cdot \qquad\qquad \cdot$$
$$\cdot \qquad\qquad \cdot$$
$$\cdot \qquad\qquad \cdot$$

$$x = \frac{99}{100}, \quad f\left(\frac{99}{100}\right) = 4\frac{49}{50} \qquad\qquad x = 1\frac{1}{100}, \quad f\left(1\frac{1}{100}\right) = 5\frac{1}{50}$$

As x approaches 1 from the left, $f(x)$ approaches 5. As x approaches 1 from the right, $f(x)$ approaches 5.

As we choose x-values closer and closer to 1, from both sides, we see that the values for $f(x)$ get closer and closer to 5. Using symbols, $f(x) \to 5$ means $f(x)$ approaches 5. Similarly, $x \to 1$ means x approaches 1. We say that the limit of the function $f(x)$ is 5 as x approaches 1, and we write

$$\lim_{x \to 1} f(x) = 5 \quad \text{or} \quad \lim_{x \to 1} (2x + 3) = 5.$$

It is not necessary to assign numerous values to x in order to find the limit of a function. We make use of several theorems on limits, and where appropriate, we use algebra to simplify functions.

Theorem 18.1 $\lim_{x \to a} c = c$, where a is a real number, and c is any constant.

• **Example 18.1:** $\lim_{x \to 1} 5 = 5.$

Theorem 18.2 $\lim_{x \to a} x^n = a^n$.

• **Example 18.2:** $\lim_{x \to 2} x^3 = 2^3 = 8$.

Theorem 18.3 $\lim_{x \to a} [f(x) + g(x)] = \lim_{x \to a} f(x) + \lim_{x \to a} g(x)$.

This theorem states that the limit of the sum of two functions is the sum of the limit of each function.

• **Example 18.3:**
$$\lim_{x \to 2} (x^2 + 3x) = \lim_{x \to 2} x^2 + \lim_{x \to 2} 3x$$
$$= 2^2 + 3 \cdot 2$$
$$= 4 + 6 = 10.$$

Theorem 18.4 $\lim_{x \to a} f(x)g(x) = \lim_{x \to a} f(x) \lim_{x \to a} g(x)$.

The limit of the product of two functions is the product of the limit of each function.

• **Example 18.4:**
$$\lim_{x \to 1} x^2(x + 2) = \lim_{x \to 1} x^2 \lim_{x \to 1} (x + 2)$$
$$= 1 \cdot 3 = 3.$$

Theorem 18.5 $\lim_{x \to a} \frac{f(x)}{g(x)} = \frac{\lim_{x \to a} f(x)}{\lim_{x \to a} g(x)}$, provided $\lim_{x \to a} g(x) \neq 0$.

The limit of a quotient of two functions is the quotient of the limit of each function.

• **Example 18.5:**
$$\lim_{x \to 4} \frac{x^2}{x + 1} = \frac{\lim_{x \to 4} x^2}{\lim_{x \to 4} (x + 1)}$$
$$= \frac{16}{5}.$$

NOTE: $\lim_{x \to a} f(x)$ may exist even though f is not defined at a. Example 18.6 illustrates this fact.

• **Example 18.6:** Find $\lim_{x \to 0} \dfrac{(x+1)^2 - 1}{x}$ (note: $x \neq 0$, x approaches 0).

Solution: In this case, $\lim_{x \to 0} x = 0$, which leads to the indeterminant form $\dfrac{0}{0}$.

Employing some algebraic manipulations, we obtain the following:

$$\lim_{x \to 0} \frac{(x+1)^2 - 1}{x} = \lim_{x \to 0} \frac{x^2 + 2x + 1 - 1}{x}$$
$$= \lim_{x \to 0} \frac{x^2 + 2x}{x}$$
$$= \lim_{x \to 0} \frac{x(x+2)}{x}$$
$$= \lim_{x \to 0} (x+2) = 2.$$

• **Example 18.7:** The current in a circuit is given by $i = \dfrac{t^3 - 8}{t - 2}$. Find $\lim_{t \to 2} i$.

Solution:
$$\lim_{t \to 2} i = \lim_{t \to 2} \frac{t^3 - 8}{t - 2}$$
$$= \lim_{t \to 2} \frac{(t-2)(t^2 + 2t + 4)}{t - 2}$$
$$= \lim_{t \to 2} (t^2 + 2t + 4)$$
$$= 4 + 4 + 4 = 12.$$

• **Example 18.8:** Evaluate $\lim_{x \to 2} \dfrac{1}{x - 2}$.

Solution:
$$\lim_{x \to 2} \frac{1}{x - 2} = \frac{1}{2 - 2} = \frac{1}{0},$$

which is undefined. We say that the limit does not exist.

This example illustrates that neither the function exists at $x = 2$ nor the limit exists as x approaches 2.

SECTION 18.1 THE LIMIT OF A FUNCTION

Understanding the concept of the limit of a function is essential for further study of functions. The limit concept allows us to define the notions of continuity and derivative of a function, and thus leads us to the applications of the calculus to scientific and technical problems.

Trial Problems 18.1

Before you begin the section exercises, warm up with these problems. Complete answers are included in the answer key.

Evaluate the following limits.

1. $\lim\limits_{x \to 2} x$
2. $\lim\limits_{x \to 2} x^2$
3. $\lim\limits_{x \to 0} \dfrac{x}{2}$
4. $\lim\limits_{x \to 1} \dfrac{3}{x}$
5. $\lim\limits_{x \to -1} 125$

Exercises 18.1

Use the limit theorems to evaluate the following limits.

1. $\lim\limits_{x \to 2} 0$
2. $\lim\limits_{x \to -2} (3x^3 + 2x + 7)$
3. $\lim\limits_{x \to 2} \dfrac{4}{x^2 + 1}$
4. $\lim\limits_{x \to 0} \dfrac{x^2 + x}{x}$
5. $\lim\limits_{x \to \frac{1}{2}} (x^2 + x + 1)$
6. $\lim\limits_{x \to -\frac{1}{2}} \dfrac{x + \frac{1}{2}}{x - \frac{1}{2}}$
7. $\lim\limits_{x \to 5} 3$
8. $\lim\limits_{x \to 3} \sqrt{15}$
9. $\lim\limits_{x \to 4} (x^2 + 3)(x - 4)$
10. $\lim\limits_{x \to 2} \dfrac{x - 2}{x^3 - 8}$
11. $\lim\limits_{x \to 4} \dfrac{6x - 1}{2x - 9}$
12. $\lim\limits_{x \to 1} \left(\dfrac{x^2}{x - 1} - \dfrac{x}{x - 1} \right)$
13. $\lim\limits_{x \to 4} (x^2 - 5x + 4)$
14. $\lim\limits_{x \to -1} \dfrac{x^3 + 1}{x + 1}$
15. $\lim\limits_{h \to 0} \dfrac{4 - \sqrt{16 + h}}{h}$
16. $\lim\limits_{x \to 5} 4\sqrt{x - 1}$
17. $\lim\limits_{x \to 6} \dfrac{x^2 - 4x - 12}{x - 6}$
18. $\lim\limits_{t \to 0} 2t$
19. $\lim\limits_{x \to 2} \dfrac{x^2 - x - 6}{x^2 - 4}$
20. $\lim\limits_{x \to 0} \dfrac{x + 5}{x - 2}$
21. $\lim\limits_{x \to 2} \dfrac{x + 5}{x - 2}$
22. $\lim\limits_{x \to \pi} (x - 3.14)$
23. $\lim\limits_{x \to 3} \dfrac{x^2 - 2x - 3}{3 - x}$
24. $\lim\limits_{x \to \frac{1}{3}} \dfrac{3x - 1}{3x^2 + 5x - 2}$

25. In the formula for the velocity of sound, V, in air, $V = 331.5 + 0.607t$, find the limit of V as t approaches $10°$ C.

26. The length of a pendulum is given by $L = \dfrac{g}{4}\pi^2 T^2$, where T denotes time and g denotes the force of gravity. Find $\lim\limits_{T \to 2} L$.

27. The distance s above the ground of a rocket fired vertically is given by $s = v_0 t - \dfrac{1}{2}gt^2$. Find $\lim\limits_{t \to 1} s$ if $v_0 = 196$ m/s and $g = 9.8$ m/s².

28. In an AC circuit, the resistance R, the impedance Z, and the reactance X are related as follows: $Z^2 = R^2 + X^2$. Find $\lim\limits_{R \to 0} Z$.

18·2 Continuity

When evaluating the $\lim\limits_{x \to a} f(x)$, note that x approaches a, but, this does not mean that x necessarily equals a. In fact, $f(x)$ may not even be defined at $x = a$. For example, if $f(x) = \dfrac{1}{x}$ and we wish to find $\lim\limits_{x \to 0} \dfrac{1}{x}$, it is clear that $f(x)$ is not defined at $x = 0$ $\left(\text{since } \dfrac{1}{0} \text{ is impossible to evaluate}\right)$ and that $\lim\limits_{x \to 0} \dfrac{1}{x}$ does not exist. Similarly, if $f(x) = \dfrac{x^2 - x - 6}{x - 3}$, note that $f(x)$ is not defined at $x = 3$. In this case,

$$\lim_{x \to 3} \dfrac{x^2 - x - 6}{x - 3} = \lim_{x \to 3} \dfrac{(x - 3)(x + 2)}{(x - 3)} \quad \text{(Factoring)}$$

$$= \lim_{x \to 3} (x + 2) \quad \text{(Reducing)}$$

$$= 5.$$

Functions are given a special classification if

1. $\lim\limits_{x \to a} f(x)$ exists,

2. $f(x)$ is defined at $x = a$, and

3. $\lim\limits_{x \to a} f(x) = f(a)$.

Continuous Under these three conditions, we say that the function is **continuous** at $x = a$.

• **Example 18.9:** $f(x) = \sqrt{2x - 5}$ is continuous at $x = 4$ since

1. $\lim\limits_{x \to 4} f(x) = \sqrt{3}$, so limit exists,

2. $f(x)$ is defined for $x = 4$, and

3. $\lim\limits_{x \to 4} f(x) = f(4)$.

• **Example 18.10:** $f(x) = \dfrac{x}{x-2}$ is not continuous at $x = 2$, since $f(x)$ is not defined at $x = 2$. $f(x)$ is continuous at any other values of x, such as $x = a$, since

$$f(a) = \dfrac{a}{a-2} \quad \text{and}$$

$$\lim_{x \to a} \dfrac{x}{x-2} = \dfrac{a}{a-2} = f(a), \quad a \neq 2.$$

Discontinuous If a function is not continuous at $x = a$, we say that the function is **discontinuous** at $x = a$. Graphically, this means that there is a "hole" in the graph of the function. See figure 18.1.

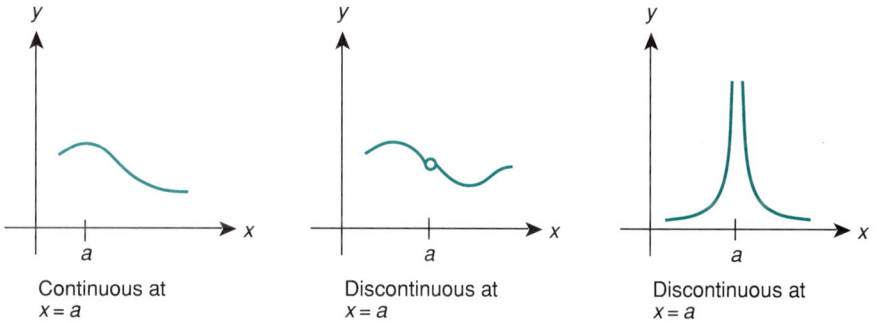

Continuous at $x = a$ Discontinuous at $x = a$ Discontinuous at $x = a$

Figure 18.1

• **Example 18.11:** Find all values for which $f(x) = \dfrac{3x - 5}{2x^2 - x - 3}$ is discontinuous.

Solution: $f(x) = \dfrac{3x - 5}{2x^2 - x - 3}$

$= \dfrac{3x - 5}{(2x - 3)(x + 1)}$

$x = -1 \quad \text{and} \quad x = \dfrac{3}{2}$

are not in the domain of $f(x)$. Therefore, $f(x)$ is discontinuous at $x = -1$ and $x = \dfrac{3}{2}$, but continuous at all other values of x.

Discontinuous functions can occur in applications, as example 18.12 illustrates.

• **Example 18.12:** It cost 25 cents to mail a letter that weighs up to 1 ounce. Each additional ounce requires 20 cents more in postage. Graph the postage $P(x)$ for mailing x ounces.

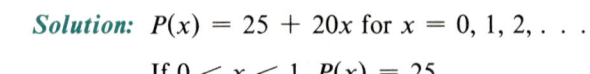

Solution: $P(x) = 25 + 20x$ for $x = 0, 1, 2, \ldots$

If $0 < x < 1$, $P(x) = 25$.

If $1 \leq x < 2$, $P(x) = 25 + 20(1) = 45$.

If $2 \leq x < 3$, $P(x) = 25 + 20(2) = 65$.

This is sufficient to illustrate the graph of the discontinuous function. See figure 18.2. Note that the line segments in the figure are not connected; there are "holes" at $x = 1, 2, 3, \ldots$. This is an example of a "step" function.

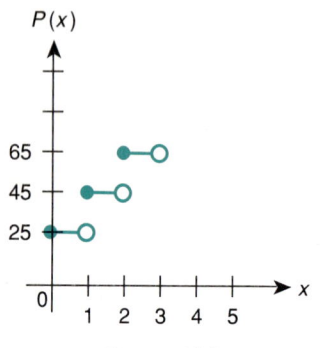

Figure 18.2

Trial Problems 18.2

Before you begin the section exercises, warm up with these problems. Complete answers are included in the answer key.

Sketch the graph of each of the following and determine where f is continuous or discontinuous.

1. $f(x) = x$
2. $f(x) = x + 2$
3. $f(x) = x - 2$
4. $f(x) = 4$
5. $f(x) = \sqrt{x}$

Exercises 18.2

Determine where f is continuous or discontinuous for each of exercises 1–10.

1. $f(x) = \dfrac{x}{x^2 - 4}$
2. $f(x) = 3x^2 + 7$
3. $f(x) = \dfrac{1}{x}$
4. $f(x) = \sqrt{x}$
5. $f(x) = \sqrt{x - 4}$
6. $f(x) = \dfrac{1}{x - 1}$
7. $f(x) = \dfrac{x}{x^2 + 1}$
8. $f(x) = \dfrac{5}{x^3 - x^2}$
9. $f(x) = 5$
10. $f(x) = \dfrac{\sqrt{9 - x}}{\sqrt{x - 6}}$

11. Given a circuit with battery voltage E, internal resistance r, and circuit resistance R, $I = \dfrac{E}{(R + r)}$ is the resulting current. For what values of E is I continuous, provided R and r are constant? When is there no current?

12. In a circuit, the current I is given as a function of time t, $I = \dfrac{t}{t + 1}$. Is I continuous for all nonnegative t?

13. A trucking company charges 50 cents per mile to deliver a package up to 200 miles and 40 cents per mile for each mile exceeding 200. Graph the cost $c(x)$ of sending a package x miles. Is $c(x)$ a continuous or discontinuous function at $x = 200$?

18·3 Limits and Infinity

The graph of $f(x) = \dfrac{1}{x}$ is given in figure 18.3. Note that as the x-values are selected closer and closer to zero, from the right, the y-values, or $f(x)$, become larger in the positive direction. In limit form we write

$$\lim_{x \to 0^+} \frac{1}{x} = +\infty.$$

Similarly, as the x-values approach zero from the left, $f(x)$ goes to negative infinity, and we write: $\lim\limits_{x \to 0^-} \dfrac{1}{x} = -\infty$. In other words, "the left limit is not equal to the right limit." In either case, we say that the limit does not exist, and simply write: $\lim\limits_{x \to 0} \dfrac{1}{x}$ does not exist.

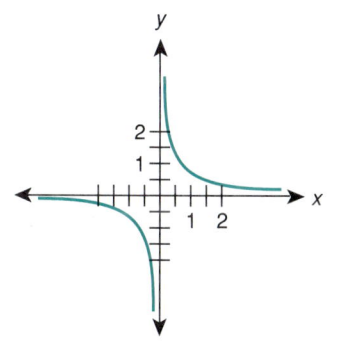

Figure 18.3

Investigating the graph further, we see that as the x-values become larger and larger in the positive direction, $f(x)$ gets closer and closer to zero. In limit form we write

$$\lim_{x \to +\infty} \frac{1}{x} = 0.$$

Similarly, as $x \to -\infty$, $f(x) = 0$, or

$$\lim_{x \to -\infty} \frac{1}{x} = 0.$$

We can combine both limit statements above and write

$$\lim_{x \to \pm\infty} \frac{1}{x} = 0.$$

To help us evaluate limits where infinity is involved, we use theorems 18.6 and 18.7 and show how algebra can make our work easier.

> **Theorem 18.6** $\lim_{x \to +\infty} \dfrac{c}{x^k} = 0$ and $\lim_{x \to -\infty} \dfrac{c}{x^k} = 0$, where c is a constant and k is a positive exponent.

Proof:
$$\lim_{x \to +\infty} \frac{c}{x^k} = c \lim_{x \to +\infty} \frac{1}{x^k} = c \cdot 0 = 0$$
$$\lim_{x \to -\infty} \frac{c}{x^k} = c \lim_{x \to -\infty} \frac{1}{x^k} = c \cdot 0 = 0$$

• *Example 18.13:* $\lim_{x \to \infty} \dfrac{3}{x^2} = 0.$

• *Example 18.14:* Find $\lim_{x \to \infty} \dfrac{4x}{x^2 + 9}$.

Solution: As $x \to \infty$, the fraction takes on the form $\dfrac{\infty}{\infty}$. Since ∞ is not a real number, $\dfrac{\infty}{\infty} \neq 1$. Using algebra gets us around this problem. Divide every term by x^2 to obtain an expression easier to evaluate:

$$\lim_{x \to \infty} \frac{4x}{x^2 + 9} = \lim_{x \to \infty} \frac{\frac{4x}{x^2}}{\frac{x^2}{x^2} + \frac{9}{x^2}}$$

$$= \lim_{x \to \infty} \frac{\frac{4}{x}}{1 + \frac{9}{x^2}}$$

$$= \frac{\lim_{x \to \infty} \frac{4}{x}}{\lim_{x \to \infty} 1 + \lim_{x \to \infty} \frac{9}{x^2}}$$

$$= \frac{0}{1 + 0}$$

$$= 0.$$

NOTE: Generally, it is best to divide both the numerator and the denominator by x to the highest power appearing in the expression.

• **Example 18.15:** Evaluate $\lim\limits_{x \to -\infty} \dfrac{2x^2 - 5}{3x^2 + x + 2}$.

Solution: Divide every term by x^2:

$$\lim_{x \to -\infty} \frac{2x^2 - 5}{3x^2 + x + 2} = \lim_{x \to -\infty} \frac{2 - \dfrac{5}{x^2}}{3 + \dfrac{1}{x} + \dfrac{2}{x^2}}$$

$$= \frac{2 - 0}{3 + 0 + 0} = \frac{2}{3}.$$

• • • • • • • • • •

Theorem 18.6 tells us that $\dfrac{1}{x^n}$ approaches 0 as x assumes larger and larger values. Next, we investigate the form $\dfrac{1}{(x-a)^n}$, where a is any fixed number. Theorem 18.7 indicates what happens as x approaches the number a.

Theorem 18.7 a. $\lim\limits_{x \to a} \dfrac{1}{(x-a)^n} = \infty$ if n is an even positive integer.

b. $\lim\limits_{x \to a^+} \dfrac{1}{(x-a)^n} = \infty$ if n is an odd positive integer.

c. $\lim\limits_{x \to a^-} \dfrac{1}{(x-a)^n} = -\infty$ if n is an odd positive integer.

It is sufficient to show the proofs for (a) and (b) only.

Proof of (a): Suppose $a = 1$ and n is any even positive number. Sketch the graph of $f(x) = \dfrac{1}{(x-1)^n}$.

Arbitrarily choose values for x close to 1, say $x = \dfrac{1}{2}$ and $x = \dfrac{3}{4}$, and see what happens to $f(x)$.

If we let $x = \dfrac{1}{2}$, then,

$$f(x) = \dfrac{1}{\left(\dfrac{1}{2} - 1\right)^n}$$

$$= \dfrac{1}{\left(-\dfrac{1}{2}\right)^n}$$

$$= \dfrac{1}{\left(\dfrac{1}{2}\right)^n} \quad \text{(Since } n \text{ is even)}$$

$$= 2^n.$$

If we let $x = \dfrac{3}{4}$, then

$$f(x) = \dfrac{1}{\left(\dfrac{3}{4} - 1\right)^n}$$

$$= \dfrac{1}{\left(-\dfrac{1}{4}\right)^n}$$

$$= \dfrac{1}{\left(\dfrac{1}{4}\right)^n}$$

$$= 4^n.$$

As $x \to 1$, $f(x)$ becomes a larger and larger number raised to the nth power. Hence,

$$\lim_{x \to 1} f(x) = \infty \quad \text{or} \quad \lim_{x \to 1} \dfrac{1}{(x-1)^n} = \infty.$$

SECTION 18.3 LIMITS AND INFINITY 749

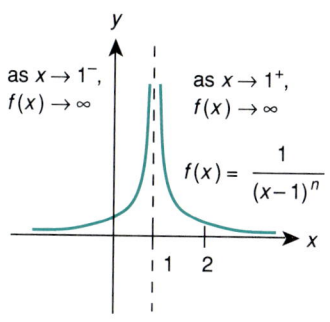

as $x \to 1^-$, $f(x) \to \infty$
as $x \to 1^+$, $f(x) \to \infty$
$f(x) = \dfrac{1}{(x-1)^n}$

Figure 18.4

Proof of (b): Similarly, if we choose $x > 1$ and let x approach 1, we obtain the right-hand side of the graph. In general,

$$\lim_{x \to a} \frac{1}{(x - a)^n} = \infty.$$

The unbounded nature of the function near $x = 1$ is represented by a vertical asymptote, as shown in figure 18.4.

• *Example 18.16:* $\displaystyle\lim_{x \to 2} \frac{1}{(x - 2)^4} = \infty.$

• *Example 18.17:* $\displaystyle\lim_{x \to 2^+} \frac{1}{(x - 2)^3} = \infty.$

• *Example 18.17:* $\displaystyle\lim_{x \to 2^+} \frac{1}{(x - 2)^3} = \infty.$

• *Example 18.18:* $\displaystyle\lim_{x \to 2^-} \frac{1}{(x - 2)^3} = -\infty.$

Solution: Using theorem 18.7, $a = -4$, we can write

$$\lim_{x \to -4} \frac{1}{(x + 4)^3} = \lim_{x \to -4} \frac{1}{[x - (-4)]^3}$$
$$= \infty.$$

• *Example 18.20:* Sketch the graph of $f(x) = \dfrac{1}{(x - 2)^2}$ to verify that

a. $\displaystyle\lim_{x \to \infty} \frac{1}{(x - 2)^2} = 0,$

b. $\displaystyle\lim_{x \to -\infty} \frac{1}{(x - 2)^2} = 0,$

c. $\displaystyle\lim_{x \to 2^+} \frac{1}{(x - 2)^2} = \infty,$ and

d. $\displaystyle\lim_{x \to 2^-} \frac{1}{(x - 2)^2} = \infty.$

Solution: The graph in figure 18.5 illustrates the value of each limit.

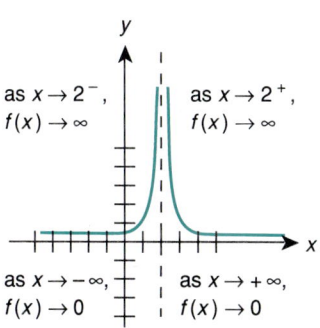

as $x \to 2^-$, $f(x) \to \infty$
as $x \to 2^+$, $f(x) \to \infty$
as $x \to -\infty$, $f(x) \to 0$
as $x \to +\infty$, $f(x) \to 0$

Figure 18.5

• **Example 18.21:** Find $\lim_{x \to \infty} \dfrac{\frac{1}{x}}{1 + 2x}$.

Solution: As $x \to \infty$, $\dfrac{1}{x} \to 0$ and $1 - 2x \to \infty$. Thus,

$$\lim_{x \to \infty} \frac{\frac{1}{x}}{1 + 2x} = \frac{0}{\infty},$$

which is not a real number. If we divide both numerator and denominator by x, we obtain

$$\lim_{x \to \infty} \frac{\frac{1}{x}}{1 + 2x} = \lim_{x \to \infty} \frac{\frac{1}{x^2}}{\frac{1}{x} + 2}$$

$$= \frac{0}{0 + 2} = \frac{0}{2} = 0.$$

Therefore,

$$\lim_{x \to \infty} \frac{\frac{1}{x}}{1 + 2x} = 0.$$

Trial Problems 18.3

Before you begin the section exercises, warm up with these problems. Complete answers are included in the answer key.

Evaluate the following limits.

1. $\lim_{x \to \infty} \dfrac{1}{x}$
2. $\lim_{x \to \infty} \dfrac{1}{x^2}$
3. $\lim_{x \to -\infty} \dfrac{1}{x^2}$
4. $\lim_{x \to \infty} \dfrac{1}{1 + \frac{1}{x}}$
5. $\lim_{x \to 3} \dfrac{1}{x - 3}$

Exercises 18.3

Evaluate the following limits.

1. $\lim\limits_{x \to \infty} \dfrac{x}{1 + 2x}$

2. $\lim\limits_{x \to \infty} \dfrac{5}{1 + \dfrac{2}{x}}$

3. $\lim\limits_{x \to -2} \sqrt{x}(x + 2)$

4. $\lim\limits_{x \to 1} \dfrac{1}{x - 1}$

5. $\lim\limits_{x \to \infty} \dfrac{2x^2 - 1}{x^2}$

6. $\lim\limits_{x \to \infty} \dfrac{x + 2}{x^2 + 3}$

7. $\lim\limits_{x \to 4} \dfrac{5}{x - 4}$

8. $\lim\limits_{x \to -8} \dfrac{3x}{(x + 8)^2}$

9. $\lim\limits_{x \to -\infty} \dfrac{4 - 7x}{2 + 3x}$

10. $\lim\limits_{x \to \frac{3}{7}} \dfrac{-4}{7x - 3}$

11. $\lim\limits_{x \to 2} \dfrac{2x^2}{x^2 - x - 2}$

12. Find the following limits and use this information to sketch a graph of $y = \dfrac{2x^2}{9 - x^2}$.

   a. $\lim\limits_{x \to \infty} \dfrac{2x^2}{9 - x^2}$

   b. $\lim\limits_{x \to 3^+} \dfrac{2x^2}{9 - x^2}$

   c. $\lim\limits_{x \to 3^-} \dfrac{2x^2}{9 - x^2}$

   d. $\lim\limits_{x \to -3^+} \dfrac{2x^2}{9 - x^2}$

   e. $\lim\limits_{x \to -3^-} \dfrac{2x^2}{9 - x^2}$

   f. $\lim\limits_{x \to -\infty} \dfrac{2x^2}{9 - x^2}$

13. Find $\lim\limits_{x \to \infty} \dfrac{\dfrac{3}{x}}{1 - 3x}$

14. Find $\lim\limits_{x \to \infty} \dfrac{5}{1 + \dfrac{3}{x^2}}$

15. Find $\lim\limits_{x \to \infty} \dfrac{2x^3 + 5}{x^3 - 7}$

16. Find $\lim\limits_{x \to \infty} \dfrac{x - 2}{5x + 4}$

17. Find $\lim\limits_{x \to \infty} \dfrac{5x - 10}{x^2 - 4}$

18. Find $\lim\limits_{x \to \infty} \dfrac{4 - x^2}{x^3 - 2}$

19. Find $\lim\limits_{h \to 0} \dfrac{(x + h)^2 - x^2}{h}$

20. Let $f(x) = 2x + 3$. Find $\lim\limits_{h \to 0} \dfrac{f(x + h) - f(x)}{h}$

Note that exercises 19 and 20 involve limit expressions that are very useful in our future work.

21. Find **a.** $\lim\limits_{x \to 0^+} 3^{1/x}$ and **b.** $\lim\limits_{x \to 0^-} 3^{1/x}$ (*Hint:* Sketch the graph of $f(x) = 3^{1/x}$.)

22. The current I in a circuit is given as a function of time:
$I = \dfrac{t}{t + 1}$. Find $\lim\limits_{t \to \infty} I$.

23. The formula for the adiabatic expansion of air is $pv^{1.4} = c$, where p denotes pressure, v denotes volume, and c is a constant. Find $\lim\limits_{v \to \infty} p$.

24. The volume V of water in a conical tank is given by $V = \dfrac{1}{3}\pi r^2 h$. Find $\lim\limits_{h \to 2^+} V$ when $r = 1$, and $\lim\limits_{h \to 2^-} V$ when $r = 1$. What one limit would give the same result?

18•4 The Slope of a Tangent Line to a Curve

Recall that the slope, m, of a line that passes through two points, $P(x_1,y_1)$ and $Q(x_2,y_2)$ is given by

$$m_{PQ} = \frac{y_2 - y_1}{x_2 - x_1} = \frac{\text{change in } y}{\text{change in } x} = \frac{\Delta y}{\Delta x}.$$

Figure 18.6 illustrates this concept.

Tangent Line
Secant Line
Recall that a line that touches a curve at one and only one point is called a **tangent line**. If a line crosses a curve at two points, it is called a **secant line**. See figure 18.7.

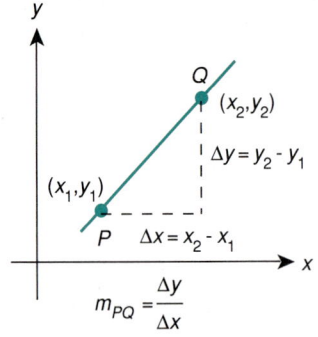

Figure 18.6

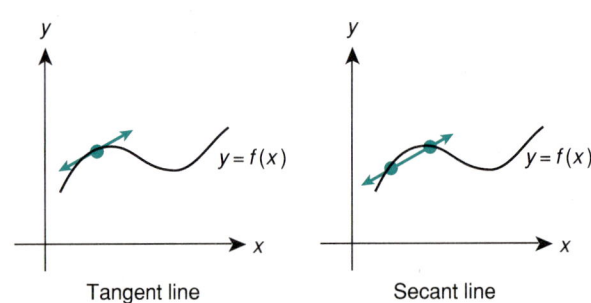

Tangent line   Secant line

Figure 18.7

Slope of a Tangent Line
We wish to determine the **slope of a tangent line** to a curve. In order to do this, we make use of slopes of secant lines and limits.

Suppose line l is a secant line to the curve $y = f(x)$, crossing the curve at (x_1,y_1) and (x_2,y_2) which we will label P and Q, respectively. Points P and Q may also be labeled $P(x_1, f(x_1))$ and $Q(x_2, f(x_2))$ with $x = x_2 - x_1$ and $y = f(x_2) - f(x_1)$. The slope of line l is given by

$$m_{PQ} = \frac{\Delta y}{\Delta x} = \frac{f(x_2) - f(x_1)}{\Delta x} = \frac{f(x_1 + \Delta x) - f(x_1)}{\Delta x},$$

since $x_2 = x_1 + \Delta x$. The secant line is illustrated in figure 18.8.

Suppose point P is fixed, but that Q may move along the curve $y = f(x)$. Let Q move toward P. As Q gets closer to P, the line through P and Q remains a secant line. The closer Q gets to P, the closer the secant line through P and Q gets to the tangent line at P. As Q approaches P, Δx gets smaller and smaller; written symbolically, as $Q \to P$ then $\Delta x \to 0$. This is illustrated in figure 18.9. Furthermore, the slope of each new secant line, as $Q \to P$, becomes closer and closer to the slope of the tangent line at P.

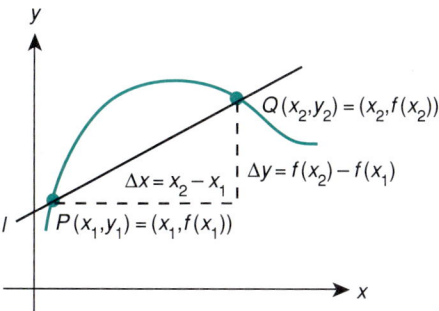

Figure 18.8

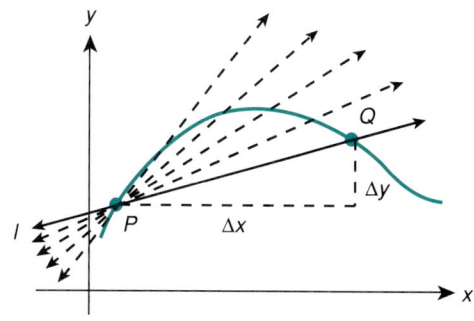

Figure 18.9

LIMIT OF THE SLOPE OF SECANT LINE

The limit of the slopes of the secant lines are defined as the *slope of the tangent line to the curve at point P*.

As $\Delta x \to 0$, $\dfrac{f(x_1 + \Delta x) - f(x_1)}{\Delta x}$ becomes the slope of the tangent line at P.

Using m as the slope of the tangent line, we write

$$m = \lim_{\Delta x \to 0} \frac{f(x_1 + \Delta x) - f(x_1)}{\Delta x},$$

the slope of tangent line at $P(x_1, y_1)$.

• **Example 18.22:** Find the slope of the tangent line to the curve $y = x^2$ at the point $(2, 4)$. The tangent line is shown in figure 18.10.

Solution: Since $m_{(x_1, y_1)} = \lim\limits_{\Delta x \to 0} \dfrac{f(x_1 + \Delta x) - f(x_1)}{\Delta x}$,

$$\begin{aligned}
m_{(2,4)} &= \lim_{\Delta x \to 0} \frac{f(2 + \Delta x) - f(2)}{\Delta x} \\
&= \lim_{\Delta x \to 0} \frac{(2 + \Delta x)^2 - (2)^2}{\Delta x} \\
&= \lim_{\Delta x \to 0} \frac{4 + 4\Delta x + (\Delta x)^2 - 4}{\Delta x} \\
&= \lim_{\Delta x \to 0} \frac{\Delta x(4 + \Delta x)}{\Delta x} \\
&= \lim_{\Delta x \to 0} (4 + \Delta x) \\
&= 4 + 0 = 4.
\end{aligned}$$

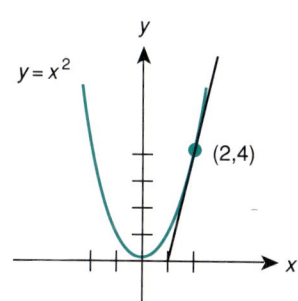

Figure 18.10

• **Example 18.23:** Find the slope of the tangent line to the curve $y = 2x^2 - 3$ at the point $(1, -1)$.

Solution:
$$m = \lim_{\Delta x \to 0} \frac{f(1 + \Delta x) - f(1)}{\Delta x}$$
$$= \lim_{\Delta x \to 0} \frac{[2(1 + \Delta x)^2 - 3] - [2(1)^2 - 3]}{\Delta x}$$
$$= \lim_{\Delta x \to 0} \frac{2 + 4\Delta x + 2(\Delta x)^2 - 3 - 2 + 3}{\Delta x}$$
$$= \lim_{\Delta x \to 0} \frac{\Delta x(4 + 2\Delta x)}{\Delta x}$$
$$= \lim_{\Delta x \to 0} (4 + 2\Delta x)$$
$$= 4 + 2(0) = 4.$$

• • • • • • • • • •

• **Example 18.24:** Find the point on the curve $y = x^2 - 2x + 1$ at which the tangent line has slope zero.

Solution:
$$m = \lim_{\Delta x \to 0} \frac{f(x + \Delta x) - f(x)}{\Delta x}$$
$$= \lim_{\Delta x \to 0} \frac{[(x + \Delta x)^2 - 2(x + \Delta x) + 1] - [x^2 - 2x + 1]}{\Delta x}$$
$$= \lim_{\Delta x \to 0} \frac{x^2 + 2x\Delta x + (\Delta x)^2 - 2x - 2\Delta x + 1 - x^2 + 2x - 1}{\Delta x}$$
$$= \lim_{\Delta x \to 0} \frac{2x\Delta x - 2\Delta x + (\Delta x)^2}{\Delta x}$$
$$= \lim_{\Delta x \to 0} \frac{\Delta x(2x - 2 + \Delta x)}{\Delta x} =$$
$$\lim_{\Delta x \to 0} (2x - 2 + \Delta x) = 2x - 2$$

$m = 0$ when $2x - 2 = 0$, $2x = 2$, or $x = 1$
$m = 0$ at $(1, 0)$.

• • • • • • • • • •

• **Example 18.25:** Find the slope of the tangent line to the curve $f(x) = \frac{1}{x}$ at the point $(1,1)$.

$$\text{Solution: } m = \lim_{\Delta x \to 0} \frac{f(x + \Delta x) - f(x)}{\Delta x}$$

$$= \lim_{\Delta x \to 0} \frac{\dfrac{1}{x + \Delta x} - \dfrac{1}{x}}{\Delta x}$$

$$= \lim_{\Delta x \to 0} \frac{\dfrac{x - x - \Delta x}{x(x + \Delta x)}}{\Delta x} \quad \text{(Find common denominator and combine fractions)}$$

$$= \lim_{\Delta x \to 0} \frac{-\Delta x}{x(x + \Delta x)} \cdot \frac{1}{\Delta x} \quad \text{(Simplify, invert, and multiply)}$$

$$= \lim_{\Delta x \to 0} \frac{-1}{x(x + \Delta x)}$$

$$= -\frac{1}{x^2}$$

$$= -\frac{1}{1^2} \quad (x = 1)$$

$$= -1 \quad \text{at } (1,1).$$

Trial Problems 18.4

Before you begin the section exercises, warm up with these problems. Complete answers are included in the answer key.

Find $\dfrac{f(x + \Delta x) - f(x)}{\Delta x}$ for each of the following.

1. $f(x) = 2x$
2. $f(x) = x^2$
3. $f(x) = 4$
4. $y = x + 1$
5. $y = 1 - x$

Exercises 18.4

Find the slope of the function at the given point.

1. $y = x + 2$, $(1,3)$
2. $y = 2 - x$, $(1,1)$
3. $y = x^2 + x$, $(2,6)$
4. $y = x^2 - 3x$, $(0,0)$
5. $y = x^2 - x + 1$, $(2,3)$
6. $y = \dfrac{1}{x}$, $(1,1)$
7. $y = x^3$, $(1,1)$
8. $y = \dfrac{1}{x^2}$, $\left(2, \dfrac{1}{4}\right)$
9. $y = x - \dfrac{1}{x}$, $(1,0)$
10. $y = \dfrac{x^2}{4}$, $(x = a)$
11. $y = x^2$, $(x = a)$
12. $y = x^2 + 1$, $(x = a)$

13. $y = x^3 - 1$, $(x = a)$
14. $y = -x^2 + x$, $(x = a)$
15. $f(x) = 3$, $(x = a)$
16. $f(x) = \dfrac{1}{x}$, $(x = a)$
17. $f(x) = x + \dfrac{1}{x}$, $(-1, -2)$
18. $f(x) = 1 - \dfrac{1}{x}$, $(1, 0)$
19. $f(x) = 1 + \dfrac{1}{x}$, $(-1, 0)$

18·5 Definition of Derivative

The limit used in finding the slope of a tangent line to a curve is fundamental in the study of the calculus. It has many applications other than slope of tangents to curves. For slopes of tangent lines we write

$$m = \lim_{\Delta x \to 0} \frac{f(x + \Delta x) - f(x)}{\Delta x}.$$

If we replace Δx with h and replace m with $f'(x)$, we have

$$f'(x) = \lim_{h \to 0} \frac{f(x + h) - f(x)}{h}.$$

Derivative
Differentiation

Replacing Δx with h is simply a matter of convenience. $f'(x)$ is read, "f prime of x," and the limit expression is defined as the **derivative** of the function $y = f(x)$. The process of finding the derivative of a function is called **differentiation.** There are other standard symbols used for expressing derivatives. In addition to $f'(x)$, we may also use y', read "y prime," $\dfrac{dy}{dx}$, which is read, "the derivative of y with respect to x," and $D_x y$, which is read the same as $\dfrac{dy}{dx}$.

• **Example 18.26:** Find the derivative of $y = 2x - 3$.

Solution: $f'(x) = \lim\limits_{h \to 0} \dfrac{f(x + h) - f(x)}{h}$

$= \lim\limits_{h \to 0} \dfrac{[2(x + h) - 3] - [2x - 3]}{h}$

$= \lim\limits_{h \to 0} \dfrac{2x + 2h - 3 - 2x + 3}{h}$

$= \lim\limits_{h \to 0} \dfrac{2h}{h}$

$= \lim\limits_{h \to 0} 2$

$= 2,$

which is the slope of the line $y = 2x - 3$.

SECTION 18.5 DEFINITION OF DERIVATIVE

• **Example 18.27:** Find $\dfrac{dy}{dx}$ if $y = 3x^2$.

Solution:
$$\begin{aligned}
\frac{dy}{dx} &= \lim_{h \to 0} \frac{f(x+h) - f(x)}{h} \\
&= \lim_{h \to 0} \frac{3(x+h)^2 - 3x^2}{h} \\
&= \lim_{h \to 0} \frac{3x^2 + 6xh + 3h^2 - 3x^2}{h} \\
&= \lim_{h \to 0} \frac{h(6x + 3h)}{h} \\
&= \lim_{h \to 0} (6x + 3h) \\
&= 6x.
\end{aligned}$$

In example 18.27, we see that the derivative of $y = 3x^2$ is $6x$. The slope of the tangent line to $y = 3x^2$ at the point $(1,3)$ can be evaluated by substituting $x = 1$ in the derivative. Thus, the slope is $m = 6(1) = 6$. If $f'(x) = 6x$, then the value of $f'(x)$ at $x = 1$ is written $f'(1)$, and $f'(1) = 6(1) = 6$.

• **Example 18.28:** Find $f'(2)$ for $y = \dfrac{1}{x}$.

Solution:
$$\begin{aligned}
f'(x) &= \lim_{h \to 0} \frac{f(x+h) - f(x)}{h} \\
&= \lim_{h \to 0} \frac{\dfrac{1}{x+h} - \dfrac{1}{x}}{h} \\
&= \lim_{h \to 0} \frac{\dfrac{x - x - h}{x(x+h)}}{h} \\
&= \lim_{h \to 0} \frac{-h}{x(x+h)} \cdot \frac{1}{h} \\
&= \lim_{h \to 0} \frac{-1}{x(x+h)} \\
&= \left(-\frac{1}{x^2}\right)
\end{aligned}$$

$$f'(2) = -\frac{1}{2^2} = -\frac{1}{4}.$$

• **Example 18.29:** The velocity, v, of a particle moving along a straight path is determined by finding $\dfrac{ds}{dt}$, where s is the distance traveled in time t. If $s(t) = t^2 + 3t$, find the velocity of the particle at $t = 2.1$ sec.

Solution:
$$v = \frac{ds}{dt}$$
$$= \lim_{h \to 0} \frac{s(t+h) - s(t)}{h}$$
$$= \lim_{h \to 0} \frac{(t+h)^2 + 3(t+h) - (t^2 + 3t)}{h}$$
$$= \lim_{h \to 0} \frac{t^2 + 2th + h^2 + 3t + 3h - t^2 - 3t}{h}$$
$$= \lim_{h \to 0} \frac{2th + h^2 + 3h}{h}$$
$$= \lim_{h \to 0} (2t + h + 3)$$
$$= 2t + 3$$
$$= 2(2.1) + 3 \quad \text{at } t = 2.1$$
$$= 7.2.$$

Trial Problems 18.5

Before you begin the section exercises, warm up with these problems. Complete answers are included in the answer key.

1. Find $f'(x)$ if $f(x) = x$.
2. Find $\dfrac{dy}{dx}$ if $y = x + 1$.
3. Find y' if $y = 1 - x$.
4. Find $f'(1)$ if $f(x) = x^2$.
5. Find $g'(0)$ if $g(x) = x^2 + x + 2$.

Exercises 18.5

1. Find $f'(x)$ if $f(x) = 6$.
2. Find $f'(x)$ if $f(x) = 6x + 5$.
3. Find $f'(x)$ if $f(x) = 5 - x^2$.
4. Find y' if $y = 2x$.
5. Find y' if $y = 2x - 1$.
6. Find $\dfrac{dy}{dx}$ if $y = x^2 - 3x + 2$.
7. Find $f'(2)$ if $y = 6x + 5$.
8. Find $f'(1)$ if $y = 5 - x^2$.
9. Find $f'(0)$ if $y = x^2 - 3x + 2$.
10. Find the slope of $y = x^3$ at $x = 1$.

11. Find $\dfrac{dy}{dx}$ if $y = \dfrac{1}{x^2}$.

12. Find y' if $y = \dfrac{x^2}{4}$.

13. Find $f'(a)$ if $y = \dfrac{4}{x^2}$.

14. Find $\dfrac{dy}{dx}$ if $y = x - \dfrac{1}{x}$.

15. Find the slope of the tangent line to the curve $y = 5 - x^2$ at the point $(1,4)$.

16. Find the slope of the tangent line to the curve $y = x^2 - 3x + 2$ at the point $(2,0)$.

Using the symbolization of your choice, find the derivative of each of the following.

17. $y = \dfrac{1}{x + 1}$

18. $f(x) = \dfrac{x}{x - 1}$

19. $f(x) = \dfrac{1}{x^2 + 1}$

20. $y = \dfrac{10}{x^2}$

21. $y = \sqrt{x}$ $\left(\text{Hint: } \dfrac{\sqrt{a} - \sqrt{b}}{h} = \dfrac{\sqrt{a} - \sqrt{b}}{h} \cdot \dfrac{\sqrt{a} + \sqrt{b}}{\sqrt{a} + \sqrt{b}}\right)$

22. $f(x) = \sqrt{x + 1}$

23. If velocity, v, is given by $v = \dfrac{ds}{dt}$, find the velocity of a particle moving along a straight path when the distance s traveled in time t is given by $s(t) = t^2 - 2t + 3$. Find the velocity at $t = 0.1$ sec.

24. The current i in a circuit can be determined by finding $\dfrac{dq}{dt}$, where q is the charge transferred in a circuit in time t sec. Find the current if $q = \dfrac{2}{t}$. Find i when $t = 0.5$ sec.

25. If a function is differentiable at $x = a$, discuss the possibility of f being continuous at $x = a$.

18 • 6 Rules for Finding Derivatives

As you worked the exercises in the previous sections, you may have reached the conclusion that finding derivatives is not an easy task. The task is easier if certain theorems are used.

Theorem 18.8 If $f(x) = c$, where c is any constant, then $f'(x) = 0$.

NOTE: This theorem states that the derivative of a constant function is zero.

Proof: Since $f(x) = c$ for every value of x, it follows that $f(x + h) = c$. Hence,

$$f'(x) = \lim_{h \to 0} \frac{f(x + h) - f(x)}{h}$$
$$= \lim_{h \to 0} \frac{c - c}{h}$$
$$= \lim_{h \to 0} \frac{0}{h}$$
$$= 0.$$

• **Example 18.30:** $y = 6$
$y' = 0$.

Theorem 18.9 If $f(x) = x$, then $f'(x) = 1$.

Proof: Since $f(x) = x$, $f(x + h) = x + h$. Hence,

$$f'(x) = \lim_{h \to 0} \frac{f(x + h) - f(x)}{h}$$
$$= \lim_{h \to 0} \frac{x + h - x}{h}$$
$$= \lim_{h \to 0} \frac{h}{h}$$
$$= \lim_{h \to 0} 1$$
$$= 1.$$

Theorem 18.10 If $f(x) = x^n$, then $f'(x) = nx^{n-1}$, n any integer. The derivative of a power of x is the power times x raised to the power decreased by 1.

Proof (for n positive):

$$f'(x) = \lim_{h \to 0} \frac{f(x+h) - f(x)}{h}$$

$$= \lim_{h \to 0} \frac{(x+h)^n - x^n}{h}$$

$$= \lim_{h \to 0} \frac{x^n + nx^{n-1}h + n(n-1)x^{n-2}h^2 + \cdots + h^n - x^n}{h}$$

(Using the binomial theorem)

$$= \lim_{h \to 0} \frac{nx^{n-1}h + n(n-1)x^{n-2}h^2 + \cdots + h^n}{h}$$

$$= \lim_{h \to 0} \frac{h(nx^{n-1} + n(n-1)x^{n-2}h + \cdots + h^{n-1})}{h}$$

$$= \lim_{h \to 0} (nx^{n-1} + n(n-1)x^{n-2}h + \cdots + h^{n-1})$$

$$= nx^{n-1}.$$

- **Example 18.31:** $y = x^3$
  $$\frac{dy}{dx} = 3x^{3-1} = 3x^2.$$

- **Example 18.32:** $y = x^{-5}$
  $$y' = -5x^{-5-1} = -5x^{-6}.$$

Theorem 18.11 If $y = cf(x)$, then $y' = cf'(x)$.

- **Example 18.33:** $y = 3x$
  $$y' = 3f'(x) = 3(1) = 3.$$

- **Example 18.34:** $y = 3x^5$
  $$y' = 3(5x^4) = 15x^4.$$

Theorem 18.12 If $y = f(x) + g(x)$, then $y' = f'(x) + g'(x)$. The derivative of a sum is the sum of the derivatives.

- **Example 18.35:** $y = 2x^4 + 5x^3 - x^2 + 4x - 1$
  $$y' = 8x^3 + 15x^2 - 2x + 4.$$

- **Example 18.36:** $y = 2x^{-3} + x^2 - 4x^{-1} + 5$
$y' = -6x^{-4} + 2x + 4x^{-2}$.

Theorem 18.13 The Product Rule

If $y = f(x)g(x)$, then $y' = f(x)g'(x) + g(x)f'(x)$.

Proof: We want to show that $y' = f(x)g'(x) + g(x)f'(x)$:

$$y' = \lim_{h \to 0} \frac{f(x+h)g(x+h) - f(x)g(x)}{h}$$

$$= \lim_{h \to 0} \frac{f(x+h)g(x+h) - f(x+h)g(x) + f(x+h)g(x) - f(x)g(x)}{h} \quad \text{(Adding and subtracting } f(x+h)g(x)\text{)}$$

$$= \lim_{h \to 0} f(x+h) \frac{g(x+h) - g(x)}{h} + g(x) \frac{f(x+h) - f(x)}{h}$$

$$= \lim_{h \to 0} f(x+h) \lim_{h \to 0} \frac{g(x+h) - g(x)}{h} + \lim_{h \to 0} g(x) \lim_{h \to 0} \frac{f(x+h) - f(x)}{h}$$

$$= f(x)g'(x) + g(x)f'(x).$$

The product rule states that the derivative of a product of two functions is the first function times the derivative of the second function plus the second function times the derivative of the first function.

- **Example 18.37:** $y = (x - 1)(x^2 + 3)$.

Solution: To find the derivative, apply theorem 18.13 with $f(x) = x - 1$ and $g(x) = x^2 + 3$:

$$y' = (x - 1)(2x) + (x^2 + 3)(1)$$
$$= 2x^2 - 2x + x^2 + 3$$
$$= 3x^2 - 2x + 3.$$

• **Example 18.38:** $y = \dfrac{3x^2 + 5}{x^4}$.

Solution: Rewrite the expression as

$$y = (3x^2 + 5)(x^{-4}).$$

Then

$$\begin{aligned} y' &= (3x^2 + 5)(-4x^{-5}) + x^{-4}(6x) \\ &= \dfrac{-4(3x^2 + 5)}{x^5} + \dfrac{6x}{x^4} \\ &= \dfrac{-4(3x^2 + 5) + 6x^2}{x^5} = \dfrac{-12x^2 - 20 + 6x^2}{x^5} \\ &= \dfrac{-6x^2 - 20}{x^5}. \end{aligned}$$

・・・・・・・・・・

There is another way of finding the derivative of the expression in example 18.38. The expression indicates that $3x^2 + 5$ is divided by x^4; thus, it is called a quotient. Theorem 18.14 shows us how to find the derivative of a quotient.

Theorem 18.14 The Quotient Rule

If $y = \dfrac{f(x)}{g(x)}$, then $y' = \dfrac{g(x)f'(x) - f(x)g'(x)}{[g(x)]^2}$.

The quotient rule states that the derivative of the quotient of two functions is the denominator times the derivative of the numerator, minus the numerator times the derivative of the denominator, all divided by the square of the denominator.

• **Example 18.39:** $y = \dfrac{2x - 3}{4x + 5}$.

Solution: $y' = \dfrac{(4x + 5)(2) - (2x - 3)(4)}{(4x + 5)^2}$

$= \dfrac{8x + 10 - 8x + 12}{(4x + 5)^2}$

$= \dfrac{22}{(4x + 5)^2}.$

・・・・・・・・・・

- **Example 18.40:** $y = \dfrac{3x^2 + 5}{x^4}$.

    Solution: $y' = \dfrac{x^4(6x) - (3x^2 + 5)(4x^3)}{(x^4)^2}$

    $= \dfrac{6x^5 - 12x^5 - 20x^3}{x^8}$

    $= \dfrac{-6x^5 - 20x^3}{x^8}$

    $= \dfrac{x^3(-6x^2 - 20)}{x^8}$

    $= \dfrac{-6x^2 - 20}{x^5}$.

- **Example 18.41:** Find $f'(5)$ if $f(x) = \dfrac{25}{x + 5} + 50$.

    Solution: First, find $f'(x)$. Then substitute 5 for x to find $f'(5)$:

    $f'(x) = \dfrac{(x + 5)(0) - 25(1)}{(x + 5)^2} + 0$

    $= \dfrac{-25}{(x + 5)^2}$

    $f'(5) = \dfrac{-25}{(5 + 5)^2}$

    $= \dfrac{-25}{(10)^2}$

    $= \dfrac{-25}{100} = -\dfrac{1}{4}$.

- **Example 18.42:** Find the slope of the tangent line to the curve $y = \dfrac{x^2 + 1}{x - 1}$ at the point $(0, -1)$. Give the equation of the tangent line.

Solution: $m = f'(x) = \dfrac{(x-1)(2x) - (x^2+1)(1)}{(x-1)^2}$

$= \dfrac{2x^2 - 2x - x^2 - 1}{(x-1)^2}$

$= \dfrac{x^2 - 2x - 1}{(x-1)^2}.$

At $x = 0$, $m = f'(0) = \dfrac{0 - 0 - 1}{(0-1)^2} = -1.$

Using the point-slope formula, we have:

$y - y_1 = m(x - x_1)$
$y - (-1) = -1(x - 0)$
$y + 1 = -x$
$x + y + 1 = 0$

is the equation of the tangent line at $(0, -1)$.

• *Example 18.43:* The acceleration, a, of an object moving along a straight path is determined by finding $\dfrac{dv}{dt}$, where v is the velocity of the particle at any time t. If the velocity of the particle is given by $v = t^2 - 2t + 3$, find the acceleration.

Solution: The acceleration of the particle is $a = \dfrac{dv}{dt}$.

Since $\dfrac{dv}{dt} = 2t - 2$, $a = 2t - 2$.

Trial Problems 18.6

Before you begin the section exercises, warm up with these problems. Complete answers are included in the answer key.

Differentiation may be less difficult if the expression is simplified first. Simplify the following expressions.

1. $y = (x + 1)(x - 1)$
2. $y = (2x^2)^3$
3. $y = (x^2 + 1)^2$
4. $y = 3x(x^2 + 2)$
5. $y = \dfrac{x^2 + 3x - 10}{x - 2}$

Exercises 18.6

Differentiate the following functions.

1. $y = 2x^3 + 5$
2. $y = -6x$
3. $y = 6x^3 - 5x^2 + x + 9$
4. $y = 10$
5. $y = (x + 1)(x - 2)$
6. $y = \dfrac{-x^3}{3} + \dfrac{x^2}{2} + x - 6$
7. $y = \dfrac{4}{x^2} + \dfrac{1}{x}$
8. $y = (x + 1)^2$
9. $f(x) = (3x)^4$
10. $f(x) = (3x)^{-4}$
11. $f(x) = (x + 1)^{-2}$
12. $f(x) = \dfrac{3x^2 - 5x + 8}{7}$
13. $f(x) = \dfrac{2x^3 - 7x^2 + 4x + 3}{x}$
14. $y = 10x^2 + x - 4$
15. $y = x^7 + x^6$
16. $f(x) = 6x^2 - 6x + 5$
17. $f(x) = \dfrac{1}{3}x^3 + \dfrac{1}{2}x^2$
18. $f(x) = \dfrac{1}{4}x^8 - \dfrac{1}{2}x^4 + 2$
19. $f(x) = 13x^4 - 6x^3 + x + 1$
20. $y = x^2(4x - 5)$
21. $y = 3x(x^3 - 2)$
22. $y = (x + 2)(3x - 5)$
23. $y = 2x^3(x^3 - 2)$
24. $f(x) = (x^6 - 4)(2x^6 + 5)$
25. $f(x) = (x^5 - 6x^3 + x)(2 - 3x^3)$
26. $y = \dfrac{x}{2x - 3}$
27. $y = \dfrac{1}{x^2 - 1}$
28. $y = \dfrac{x + 2}{3x - 2}$
29. $y = \dfrac{x + 7}{x^3 + x + 1}$
30. $y = \dfrac{x^2}{2 - 3x}$
31. $y = \dfrac{3x}{4x^2 - 3x + 5}$
32. $f(x) = \dfrac{x}{x + 1}$
33. $f(s) = 15 - s + 4s^2 - 5s^4$
34. $g(x) = (x^3 - 7)(2x^2 + 3)$
35. $h(r) = r^2(3r^4 - 7r + 2)$
36. $f(t) = 12 - 3t^4 + 4t^6$
37. $S(x) = 2x + \dfrac{1}{2x}$
38. $G(v) = \dfrac{v^3 - 1}{v^2 + 1}$
39. $f(x) = \dfrac{1}{1 + x + x^2 + x^3}$
40. $p(x) = 1 + \dfrac{1}{x} + \dfrac{1}{x^2} + \dfrac{1}{x^3}$
41. $T(s) = (3s)^{-4}$
42. $Q(w) = (2w + 1)^3$
43. $P(v) = 4v(v - 1)(2v - 3)$
44. $f(t) = \dfrac{\frac{3}{5}t - 1}{2t^{-2} + 7}$
45. $f(x) = \sqrt[3]{x}$ (Hint: $\sqrt[3]{x} = x^{1/3}$)
46. $r(x) = x^2\sqrt[3]{x^2}$

47. Find $\dfrac{dy}{dx}$ by means of: (1) the quotient rule and (2) the product rule.
  a. $y = \dfrac{3x - 1}{x^2}$
  b. $y = \dfrac{x^2 + 1}{x^4}$

48. Find an equation of the tangent line to the graph of $y = 2x^3 + 4x^2 - 5x - 3$ at the point $P(0, -3)$.

49. Find an equation of the tangent line to the graph of $y = \dfrac{5}{1 + x^2}$ at the point $P(-2, 1)$.

Compute $f'(a)$ in exercises 50–57.

50. $f(x) = 10$, $a = 1$

51. $f(x) = 2 - x^2$, $a = 5$

52. $f(x) = x^3 - x^2$, $a = 1$

53. $f(x) = 3x + 5$, $a = 2$

54. $f(x) = x^2 - 3x + 2$, $a = 2$

55. $f(x) = \dfrac{4}{x^2} + \dfrac{1}{x}$, $a = 1$

56. $f(x) = x(x + 2)$, $a = -1$

57. $f(x) = \dfrac{5}{x^2 - 1}$, $a = 2$

58. The derivative of $y = f(x)$ is y' or $f'(x)$ or $\dfrac{dy}{dx}$. The second derivative of $y = f(x)$ is written y'' or $f''(x)$ or $\dfrac{d^2y}{dx^2}$. For example, if $y = 5x^3 + x^2 - 3$, then $y' = 15x^2 + 2x$ and $y'' = 30x + 2$. Find the second derivative of the following problems:

a. $y = 7x^4 - 3x^3 + 2x - 1$
b. $y = 3x^{1/3}$
c. $y = 25$
d. $y = 5x^{1/2}$
e. $y = -3x^6 + 2x^4 - \dfrac{1}{2}x^2 + 1$
f. $f(x) = \dfrac{2}{x}$
g. $f(x) = \dfrac{2x + 1}{x}$
h. $f(x) = (x^2 + 1)(x - 1)$

59. For *Charles's law*, $V = kt$, where k is a constant, find $\dfrac{dV}{dt}$.

60. In the formula for the length of a pendulum, $I = kT^2$, find $\dfrac{dI}{dT}$.

61. In a formula for distance, $D = k\sqrt{h}$, where k is a constant, find $\dfrac{dD}{dh}$.

62. In the formula, $s = v_0 t - \dfrac{1}{2}gt^2$, find s' if v_0 and g are constants.

63. For *Kelvin's law*, $C = k_1 a + \dfrac{k_2}{a}$, where k_1 and k_2 are constants, find $\dfrac{dC}{da}$.

64. In the formula $E = kn + \dfrac{b}{n + a}$, where k, a, and b are constants, find $\dfrac{dE}{dn}$.

65. The study of the formation of producer gas in chemical engineering leads to the formula $(b - a)v = x^2 - x$, where a and b are constants. Find v'.

66. For what value(s) of x is the slope of the tangent to the curve $y = \dfrac{x}{x^2 + 1}$ equal to zero?

67. The formula $P = \dfrac{E^2 r}{R^2 + 2Rr + r^2}$ gives the electric power produced by a certain source. E is the voltage of the source, R is the resistance of the source, and r is the resistance in the circuit. Find $\dfrac{dP}{dr}$, assuming that the other quantities remain constant.

18·7 The Chain Rule

Many functions are composed of other functions, but are written in the form $y = f(x)$. For example, in the expression $y = (x^2 + 1)^3$, y is called a composite function since it can be expressed as $y = u^3$, where u is the function $u = x^2 + 1$. Thus, y is composed of the cubic function u^3 and u is the function $x^2 + 1$.

Chain Rule To find the derivative of composite functions, we will use the **chain rule**.

> **Theorem 18.15 The Chain Rule:** If y is a function of u and u is a function of x, then $\dfrac{dy}{dx} = \dfrac{dy}{du} \cdot \dfrac{du}{dx}$.

Proof: Let $h = a - x$. Then as $h \to 0$, $a - x \to 0$ or $x \to a$. Also, if $y = f(u)$ and $u = g(x)$, then $y = f(g(x))$. Furthermore,

$$y' = f'(x) = \lim_{h \to 0} \frac{f(x + h) - f(x)}{h} \text{ becomes}$$

$$y' = f'(x) = \lim_{x \to a} \frac{f(x + a - x) - f(x)}{a - x}$$

$$= \lim_{x \to a} \frac{f(a) - f(x)}{a - x}$$

$$= \lim_{x \to a} \frac{f(x) - f(a)}{x - a}.$$

Now, for $y = f(u) = f(g(x))$, we have

$$y' = f'(u) = f'(g(x))$$

$$= \lim_{x \to a} \frac{f(g(x)) - f(g(a))}{x - a}$$

$$= \lim_{x \to a} \frac{f(g(x)) - f(g(a))}{x - a} \cdot \frac{g(x) - g(a)}{g(x) - g(a)}$$

$$= \lim_{x \to a} \frac{f(g(x)) - f(g(a))}{g(x) - g(a)} \cdot \frac{g(x) - g(a)}{x - a}$$

$$= \lim_{x \to a} \frac{f(g(x)) - f(g(a))}{g(x) - g(a)} \lim_{x \to a} \frac{g(x) - g(a)}{x - a}$$

$$= f'(g(x)) \cdot g'(x),$$

or

$$y' = \frac{dy}{dx} = f'(u) \frac{du}{dx} = \frac{dy}{du} \cdot \frac{du}{dx}.$$

• **Example 18.44:** If $y = (x^2 + 1)^3$, find $\dfrac{dy}{dx}$.

    Solution: Let $u = x^2 + 1$ then $y = u^3$, and $\dfrac{dy}{du} = 3u^2$. Since $u = x^2 + 1$, $\dfrac{du}{dx} = 2x$,

$$\frac{dy}{dx} = \frac{dy}{du} \cdot \frac{du}{dx}$$
$$= 3u^2 \cdot 2x = 6xu^2$$
$$= 6x(x^2 + 1)^2,$$

since $u = x^2 + 1$.

• **Example 18.45:** $y = (x^5 - 4x + 8)^7$.

    Solution: In this example, $y = u^7$ where $u = x^5 - 4x + 8$. Therefore, $y' = 7(x^5 - 4x + 8)^6 (5x^4 - 4)$.

• **Example 18.46:** $y = \left(\dfrac{5}{x^2 - 2}\right)^3$.

    Solution: In this case, $y = u^3$ where $u = \dfrac{5}{x^2 - 2}$. Thus, to find $\dfrac{du}{dx}$ we use the quotient rule.

$$y' = 3\underbrace{\left(\frac{5}{x^2 - 2}\right)^2}_{\frac{dy}{du}} \underbrace{\frac{(x^2 - 2)(0) - 5(2x)}{(x^2 - 2)^2}}_{\frac{du}{dx}}$$

$$= 3\left(\frac{5}{x^2 - 2}\right)^2 \cdot \frac{-10x}{(x^2 - 2)^2}$$

$$= \frac{3(5^2)(-10x)}{(x^2 - 2)^2(x^2 - 2)^2}$$

$$= \frac{-750x}{(x^2 - 2)^4}.$$

• **Example 18.47:** $f(x) = \sqrt[3]{3x^2 + 4x - 5}$.

Solution: Write $f(x) = (3x^2 + 4x - 5)^{1/3}$
then,

$$f'(x) = \frac{1}{3}(3x^2 + 4x - 5)^{1/3 - 1} \frac{d}{dx}(3x^2 + 4x - 5)$$

$$= \frac{1}{3}(3x^2 + 4x - 5)^{-2/3}(6x + 4)$$

$$= \frac{6x + 4}{3(3x^2 + 4x - 5)^{2/3}}.$$

••••••••••

• **Example 18.48:** As the number of automobiles in a certain city increases (at a specific rate), the pollution index P is given by

$$P = (2t + 3)^3 + \frac{1}{2t + 3}, \text{ where } t \text{ is time in months. Find } \frac{dP}{dt}.$$

Solution: Let $u = 2t + 3$. Then $\frac{du}{dt} = 2$. Also, $P = u^3 + \frac{1}{u} = u^3 + u^{-1}$:

$$\frac{dP}{dt} = 3u^2 \frac{du}{dt} + (-1)u^{-2} \frac{du}{dt}$$

$$= 3u^2(2) - u^{-2}(2)$$

$$= 6u^2 - \frac{2}{u^2}$$

$$= \frac{6u^4 - 2}{u^2}$$

$$= \frac{6(2t + 3)^4 - 2}{(2t + 3)^2}.$$

••••••••••

Trial Problems 18.7

Before you begin the section exercises, warm up with these problems. Complete answers are included in the answer key.

Use the chain rule to find the derivative of each of the following.

1. $y = (x + 1)^2$
2. $f(x) = (x + 1)^{-1}$
3. $y = (x + 1)^{1/2}$
4. $g(x) = (x + 1)^{-1/2}$
5. $y = \sqrt[3]{2x + 3}$

Exercises 18.7

Use the chain rule to find the derivative of each of the following.

1. $y = (2x + 5)^3$
2. $y = (3 - 2x)^4$
3. $f(x) = (8x - 7)^{-5}$
4. $f(x) = 2x(3x - 1)^3$
5. $y = \dfrac{3x}{(x^2 + 1)^2}$
6. $f(x) = \sqrt{3x - 2}$
7. $y = (4x^5 - 3x^3 + 2x)^{-2}$
8. $f(x) = \left(x^2 - \dfrac{1}{x^2}\right)^6$
9. $f(x) = \sqrt{(x + 1)(x + 2)}$
10. $y = \dfrac{x^4 - 3x^2 + 1}{(2x + 3)^4}$
11. $k(x) = \sqrt[3]{8x^3 + 27}$
12. $h(t) = (2t^2 - 9t + 8)^{-2/3}$
13. $r(s) = \dfrac{1}{\sqrt{3s - 5}}$
14. $F(x) = \left(\dfrac{x}{x^2 + 1}\right)^{3/2}$
15. $f(x) = \sqrt[4]{x^2 + 9}\,(4x + 7)^4$
16. $y = (7x + \sqrt{x^2 + 5})^6$
17. $f(t) = \sqrt{t^2(5t + 1)^5}$
18. $p(s) = \dfrac{1}{\sqrt{s^2 + 1}}$
19. $y = \sqrt{1 - x}$
20. $y = \sqrt[3]{1 - 4x^2}$
21. $y = x\sqrt{1 - x}$
22. $y = x^2(1 - 2x)^4$
23. $y = \dfrac{\sqrt{4x + 1}}{2x + 1}$
24. $y = \dfrac{x\sqrt{x - 2}}{2x - 1}$

25. Find an equation of the tangent line to the graph of $y = (4x^2 - 8x + 3)^4$ at the point $P(2,81)$.

26. Find the slope of the tangent line to the curve $y = \dfrac{x^2}{\sqrt{x^2 + 1}}$ at $x = 0$.

27. Find the value(s) of x for which the derivative of $y = \dfrac{x^2}{\sqrt{x^2 + 1}}$ is zero.

28. Find the second derivative of each of the following.
    a. $y = (x^2 - 1)^7$
    b. $f(x) = \dfrac{1}{x^2 + 1}$

29. Find the slope of the tangent line to the curve $y = \sqrt{3x}$ at $(3,3)$.

30. In the formula for electrical power, $P = I^2R$, the current I is given by $I = \dfrac{t}{t + 1}$, and the resistance R is given by $R = 0.5t^2 + 1.0t + 0.5$. Use the chain rule to find $\dfrac{dP}{dt}$.

31. The curved surface S of a right circular cone having altitude h and base radius r is given by $S = \pi r \sqrt{r^2 + h^2}$. Find $\dfrac{ds}{dr}$.

32. The formula $V = \dfrac{kq}{\sqrt{x^2 + b^2}}$ gives the electrical potential along a line due to the presence of a charge, q, under certain conditions. Find $\dfrac{dV}{dx}$ assuming that k, q, and b are constants.

33. The current in a circuit containing a resistance, R, and an inductance, L, is given by $I = \dfrac{R}{\sqrt{k^2 + (kL)^2}}$. Find $\dfrac{dI}{dL}$ assuming that the other quantities remain constant.

18•8 The Derivative as a Rate of Change

Suppose an object moves along a straight path according to the function

$$s = f(t) = t^2 + 2t + 1,$$

where s denotes position on the path at any time $t \geq 0$. If t is measured in seconds and s in feet, then the object moves 9 feet in 2 seconds. We see that s changes as t changes. The change in s over the time period of 2 seconds can be computed by: $f(2) - f(0) = 9 - 1 = 8$. The average rate of change of s over the time period of 2 seconds is found by computing: $\dfrac{f(2) - f(0)}{2} = \dfrac{8}{2} = 4$ feet per second. We can write this as

$$\frac{\Delta s}{\Delta t} = \frac{f(2) - f(0)}{2} = \frac{\text{change in } s \text{ from 1 ft to 9 ft}}{\text{change in } t \text{ from 0 sec to 2 sec}}.$$

$\dfrac{\Delta s}{\Delta t}$ AVERAGE RATE OF CHANGE

$\dfrac{\Delta s}{\Delta t}$ is called the **average rate of change** of s with respect to t over the time interval $[0,2]$.

In general, we can write $\dfrac{\Delta s}{\Delta t} = \dfrac{f(t + \Delta t) - f(t)}{\Delta t}$, which is a familiar form used in limits. If we take the limit of both sides as Δt goes to 0, we have

$$\lim_{\Delta t \to 0} \frac{\Delta s}{\Delta t} = \lim_{\Delta t \to 0} \frac{f(t + \Delta t) - f(t)}{\Delta t}$$

$$= \frac{ds}{dt} \quad \text{(The derivative of } s \text{ with respect to } t\text{)}$$

$$= s'(t).$$

INSTANTANEOUS RATE OF CHANGE

The limit of the average rate of change of s with respect to t is called the **instantaneous rate of change** of s with respect to t.

Thus, we have the derivative as a rate of change.

• **Example 18.49:** Suppose $y = 2x^2 + 1$.

1. Find the instantaneous rate of change of y with respect to x, using limits.

2. Find the instantaneous rate of change of y with respect to x when $x = 1$, using limits.

3. Find the instantaneous rate of change of y with respect to x when $x = 1$, using the derivative.

Solution: **1.** The instantaneous rate of change of y with respect to x is

$$\lim_{\Delta x \to 0} y = \lim_{\Delta x \to 0} \frac{[2(x + \Delta x)^2 + 1] - [2x^2 + 1]}{\Delta x}$$

$$= \lim_{\Delta x \to 0} \frac{2x^2 + 4x\Delta x + 2(\Delta x)^2 + 1 - 2x^2 - 1}{\Delta x}$$

$$= \lim_{\Delta x \to 0} \frac{\Delta x(4x + 2\Delta x)}{\Delta x}$$

$$= \lim_{\Delta x \to 0} (4x + 2\Delta x)$$

$$= 4x.$$

2. $\lim_{\Delta x \to 0} \dfrac{\Delta y}{\Delta x} = 4(1), \qquad$ when $x = 1$

$\qquad\qquad = 4.$

3. $y = 2x^2 + 1 \rightarrow \dfrac{dy}{dx} = 4x$

$\qquad\qquad\qquad\qquad = 4(1), \qquad$ when $x = 1$
$\qquad\qquad\qquad\qquad = 4.$

• *Example 18.50:* When a certain substance is heated, the Celsius temperature after t minutes, where $0 \leq t \leq 5$, can be found by using the formula $C(t) = 30t + 6\sqrt{t} + 6$.

1. Find the average rate of change of C during the time interval [3, 3.21].

2. Find the rate of change of C at $t = 3$.

Solution: **1.** Let $t = 3$ and $\Delta t = 0.21$. The average rate of change of C in [3, 3.21] is

$$\frac{\Delta C}{\Delta t} = \frac{C(t + \Delta t) - C(t)}{\Delta t} = \frac{C(3 + 0.21) - C(3)}{0.21}$$

$$= \frac{C(3.21) - C(3)}{0.21}$$

$$= \frac{[30(3.21) + 6\sqrt{3.21} + 6] - [30(3) + 6\sqrt{3} + 6]}{0.21}$$

$$= \frac{113.05 - 106.39}{0.21} = \frac{6.66}{0.21}$$

$$= 31.71° \ C/\text{min}.$$

2. The instantaneous rate of change of C at $t = 3$ is found by

$$C'(t) = 30 + 6\left(\frac{1}{2}t^{-1/2}\right) + 0$$

$$= 30 + \frac{3}{\sqrt{t}}$$

and

$$C'(3) = 30 + \frac{3}{\sqrt{3}}$$

$$= 31.73° \text{ C/min}.$$

••••••••••

• *Example 18.51:* A certain liquid is being poured into a large preparation vat. After t hours there are $6t - t^{1/2}$ gallons in the vat. At what rate is the liquid pouring into the vat (in gallons per hour) when $t = 4$?

Solution: Let G be the number of gallons of liquid in the vat after t hours. Then $G = 6t - t^{1/2}$, and the rate of change of G with respect to t is given by $\frac{dG}{dt}$:

$$\frac{dG}{dt} = 6 - \frac{1}{2}t^{-1/2}$$

$$= 6 - \frac{1}{2t^{1/2}}.$$

When $t = 4$,

$$\frac{dG}{dt} = 6 - \frac{1}{2(4)^{1/2}}$$

$$= 6 - \frac{1}{2(2)}$$

$$= 6 - \frac{1}{4}$$

$$= 5\frac{3}{4}.$$

The liquid is pouring into the vat at the rate of $5\frac{3}{4}$ gallons per hour at time $t = 4$ hours.

••••••••••

Trial Problems 18.8

Before you begin the section exercises, warm up with these problems. Complete answers are included in the answer key.

1. Find $\dfrac{\Delta y}{\Delta x}$ for $y = x^2$.
2. Find $\lim\limits_{\Delta x \to 0} \dfrac{\Delta y}{\Delta x}$ for $y = x^2$.
3. Find y' for $y = x^2$.
4. Find $f'(3)$ for $y = x^2$.
5. Find the instantaneous rate of change of y with respect to x if $y = \dfrac{1}{x}$ when $x = 2$.

Exercises 18.8

Find the instantaneous rate of change of y with respect to x, using limits.

1. $y = x + 3$
2. $y = 2 - 3x$
3. $y = x^2 - 4$
4. $y = 2 - x^2$
5. $y = x + \dfrac{1}{x}$
6. $y = \dfrac{1}{2x + 3}$

Find the instantaneous rate of change of y with respect to x for the given value of x.

7. $y = 4x + 10$, $x = 3$ (use limits)
8. $y = 20 - 8x$, $x = 2$ (use limits)
9. $y = 1 - 2x^2$, $x = 4$ (use the derivative)
10. $y = 32x - 16x^2$, $x = \dfrac{1}{2}$ (use the derivative)
11. $y = \sqrt{x}$, $x = 1$ (use the derivative)
12. **a.** For $r = 30x - 0.2x^2$, use the derivative to find the instantaneous rate of change of r with respect to x.
    b. How fast does r change with respect to x when $x = 4$?
13. The current I (in amperes) in a certain electrical circuit is given by $I = \dfrac{100}{R}$, where R denotes resistance (in ohms). Find the rate of change of I with respect to R when the resistance is 20 ohms. Is the current increasing or decreasing?
14. As a spherical balloon is being inflated, its radius (in centimeters) after t minutes is given by $r(t) = 3\sqrt[3]{t + 8}$, where $0 \le t \le 10$. (a) What is the rate of change of r with respect to t at $t = 8$? (b) What is the rate of change of V (volume of the balloon) with respect to t at $t = 8$?
15. Suppose that t seconds after a jogger begins to run, his pulse rate (in beats/min) is found by $P(t) = 58 + 2t^2 - t$, where $0 \le t \le 7$. Find the rate of change of $P(t)$ with respect to t at (a) $t = 1$, (b) $t = 3$, (c) $t = 5$, and (d) $t = 7$.
16. *Boyle's law* for gasses states that $pv = c$, where p denotes pressure, v the volume, and c is a constant. Find the rate of change of p with respect to v.
17. Suppose $s(t) = 3t^2 - 12t + 1$ gives the position of an object, moving in a straight line, at any time t. The velocity v of the object is found by $v = \dfrac{ds}{dt}$. Find the velocity of the object at $t = 4$.
18. Do the same as in exercise 17 for $s(t) = t + \dfrac{4}{t}$ at $t = 1$.
19. Do the same as in exercise 17 for $s(t) = t^3 + 1$ at $t = 2$.
20. Do the same as in exercise 17 for $s(t) = 2\sqrt{t} + \dfrac{1}{t}$ at $t = 4$.

21. Temperature conversions from Fahrenheit F to Celsius C are given by $C = \frac{5}{9}(F - 32)$. Find the rate of change of F with respect to C.

22. The formula $\frac{1}{f} = \frac{1}{p} + \frac{1}{q}$ is used in optics, where f is the focal length of a convex lens, and p and q are the distances from the lens to the object and image, respectively. If f is fixed, find the rate of change of q with respect to p.

23. A 5-ohm resistor and a variable resistor of resistance R are placed in parallel. The resulting resistance, R_T, is given by $R_T = \frac{5R}{(5 + R)}$. Find the instantaneous rate of change of R_T with respect to R when $R = 8$ ohms.

24. Find the instantaneous rate of change of the area of a circle with respect to its radius. Find this change when the radius is 2.

18 • 9 Differentials

Increment In the equation $y = f(x)$, Δx was used to represent a *small change*, or **increment** of x. If x changes from x_1 to x_2, we write $\Delta x = x_2 - x_1$. Similarly, Δy is used to denote the change in y that corresponds to the change Δx. We write $\Delta y = f(x_2) - f(x_1) = f(x_1 + \Delta x) - f(x_1)$. (See figure 18.11.)

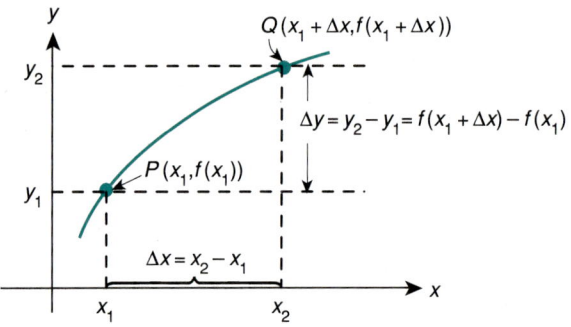

Figure 18.11

In the expression for the derivative of y with respect to x, we have

$$f'(x) = \lim_{\Delta x \to 0} \frac{f(x + \Delta x) - f(x)}{\Delta x}$$

$$= \lim_{\Delta x \to 0} \frac{\Delta y}{\Delta x}.$$

From this we conclude that $\frac{\Delta y}{\Delta x}$ is approximately equal to $f'(x)$ for Δx close to 0. We write $\frac{\Delta y}{\Delta x} \approx f'(x)$ if $\Delta x \approx 0$. Since Δy and Δx are real numbers, we can write $\Delta y \approx f'(x)\Delta x$ if $\Delta x \approx 0$. (Multiply both sides of the equation by Δx.) The expression $f'(x)\Delta x$ is given a special name.

DIFFERENTIALS

Differential of y
Differential of x

1. $dy = f'(x)\Delta x$ is called the **differential of y,** where f is differentiable and Δx is an increment of x.

2. dx is the **differential of x** and $dx = \Delta x$.

From the above definition, we conclude that

$$\Delta y \approx dy = f'(x)dx, \qquad \text{when } \Delta x \approx 0.$$

The increment of y is approximately the same as the *differential of y*, provided $dx = \Delta x$ is "small."

In figure 18.12, we see that Δy is the vertical distance from P to Q on the curve, whereas dy is the vertical distance from P on the curve to R on the tangent line. As $\Delta x \to 0$, these distances are "almost" the same. That is, $\Delta y \approx dy$ as $\Delta x \to 0$.

Differentials are useful in finding approximations when using formulas.

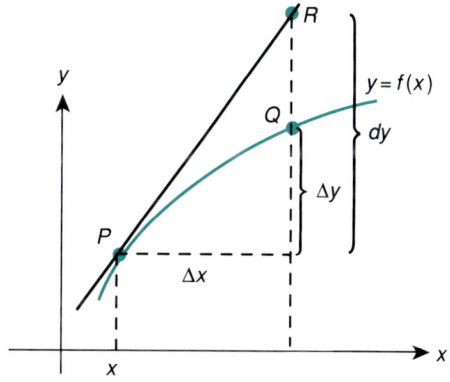

Figure 18.12

• **Example 18.52:** Find the differential of y in terms of x and dx for the function $y = x^2 + \sqrt{x}$.

Solution:
$$dy = f'(x)dx$$
$$= \left(2x + \frac{1}{2}x^{-1/2}\right)dx$$
$$= \left(2x + \frac{1}{2\sqrt{x}}\right)dx.$$

• **Example 18.53:** Find dy for $y = 8 - 5x^2$, where $x = 2$ and $dx = 0.01$.

Solution: $dy = f'(x)dx$
$= (-10x)dx$
$= [-10(2)](0.01)$
$= (-20)(0.01)$
$= -0.2$.   (y decreases approximately 0.2 units when x increases 0.01 units)

• **Example 18.54:** The radius of a circular disk is estimated to be 6 inches with a maximum error in measurement of 0.06 inches. Use differentials to estimate the maximum error in the calculated area of one side of the disk.

Solution: $A = \pi r^2$
$dA = 2\pi r\, dr$
$= 2\pi(6)(0.06)$
$= 0.72\pi$ in.^{2}.

Note: $A_1 = \pi(6)^2 = 36\pi$ in.^{2} when $r = 6$ inches.
$A_2 = \pi(6.06)^2 = 36.7236\pi$ in.^{2} when $r = 6.06$ inches
($dr = 0.06$).

$A = A_2 - A_1 = 0.7236\pi$ in.^{2}
$\approx dA$.

(The exact change is 0.7236π in.^{2}, whereas the differential is 0.72π in.^{2}. Hence, the differential is a good approximation to the exact change when the change in r is relatively small.)

• **Example 18.55:** The current in a circuit changes according to the formula $i = 0.2t^3 + 6$, where i is the current in amperes (A) and t is time in seconds. What is the approximate change in current from $t = 2.00$ sec to $t = 2.02$ sec?

Solution: $i = 0.2t^3 + 6$
$di = 3(0.2)t^2\, dt$
Let $t = 2$, then $dt = 0.02$
$di = 3(0.2)(2)^2(0.02)$
$= 0.048$.

The current changes approximately $0.05\, A$ from $t = 2.00$ to $t = 2.02$ sec.

Trial Problems 18.9

Before you begin the section exercises, warm up with these problems. Complete answers are included in the answer key.

1. Find dy for $y = x$.
2. Find dy for $y = x^2$.
3. Find dy for $y = x^2$, where $x = 1$ and $dx = \dfrac{1}{2}$.
4. Find dy for $y = x + 1$.
5. Find dy for $y = x^2 + 1$, where $x = 0$ and $dx = 0.1$.

Exercises 18.9

Find the differential of the following functions.

1. $y = 2x - 6$
2. $y = 6$
3. $f(x) = \sqrt{3x - 2}$
4. $f(x) = (3x^2 - 5x + 2)^3$
5. $g(x) = \dfrac{3x + 4}{x^2 + 1}$
6. $g(x) = (2 + x)(x^2 - 1)^4$
7. $y = \dfrac{1}{x}$
8. $y = x^{-2}$

Evaluate dy for the indicated values of x and dx in exercises 9–12.

9. $f(x) = 5 - 3x$, $x = 2$, and $dx = 0.02$
10. $g(x) = 4x^2 - x + 5$, $x = -1$, and $dx = 0.3$
11. $y = \sqrt{25 - x^2}$, $x = 3$, and $dx = -0.1$
12. $h(x) = -3x^3 + 8x - 7$, $x = 4$, $dx = 0.04$

13. If $f(x) = x^3 - 3x^2 + 2x - 7$ and $w = f(x)$, find dw. Use dw to approximate the change in w if x changes from 4 to 3.95.

14. If $s = g(t) = \dfrac{1}{2 - t^2}$, find ds and use it to approximate the change in s if t changes from 1 to 1.2.

15. Use differentials to estimate the maximum error in the calculated area of a square whose side is estimated to be 1 foot with a maximum error in measurement of $\dfrac{1}{8}$ inch.

16. An ice cube has edges that change from 10 inches to 9.9 inches in a certain period of time. Use differentials to approximate the decrease in volume of the ice cube.

17. Newton's law of gravitation is given by $F = \dfrac{gm_1 m_2}{s^2}$. Suppose g, m_1, and m_2 are constants. Use differentials to find the change in force, F, as s changes from 10 cm to 12 cm.

18. Boyle's law can be given by the formula $pv = c$, where c is a constant. Use differentials to show that $p\, dv + v\, dp = 0$.

19. For a certain thermocouple, $E = 5.9T + 0.0002T^3$ gives the voltage as a function of temperature. Find the approximate change in voltage if temperature changes from 100° C to 101° C.

20. The formula $r = \dfrac{10R}{10 + R}$ gives the combined resistance of a 10.0-ohm resistor and a variable resistor placed in parallel. Find the relative error of the combined resistance if R has a measurement of 40.0 ohms with a possible measurement error of 1.5 ohms.

18 • 10 Implicit Differentiation and Higher-Order Derivatives

Very frequently, it is not feasible to express y *explicitly* as a function of x in the usual way, $y = f(x)$. Consider the equation $x^2 + y^2 = 1$, which is not given in the form $y = f(x)$. We could solve for y, but in doing so we obtain two equations: $y = \sqrt{1 - x^2}$ and $y = -\sqrt{1 - x^2}$. In both cases, y is a function of x. In the form $x^2 + y^2 = 1$, we may *imply* that y is a function of x. In this case, we say that y is an **Implicit Function** implicit function of x.

If we want to differentiate $x^2 + y^2 = 1$, we could solve for y and obtain

$$y' = \frac{1}{2}(1 - x^2)^{-1/2}(-2x) = \frac{-x}{\sqrt{1 - x^2}}$$

and

$$y' = -\frac{1}{2}(1 - x^2)^{-1/2}(-2x) = \frac{x}{\sqrt{1 - x^2}}.$$

Implicit Differentiation Another possibility is to use **implicit differentiation.** This method is illustrated in the steps below.

1. $x^2 + y^2 = 1$

2. $\dfrac{d}{dx}(x^2 + y^2) = \dfrac{d}{dx}(1)$ (Take the derivative with respect to x of both sides of the equation)

3. $\dfrac{d(x^2)}{dx} + \dfrac{d(y^2)}{dx} = 0$

4. $2x + 2y\dfrac{dy}{dx} = 0$ (Since y is implied to be a function of x, we must use both the power rule and the chain rule for y)

5. $2y\dfrac{dy}{dx} = -2x$

6. $\dfrac{dy}{dx} = -\dfrac{x}{y}$ (With implicit differentiation the answer usually is given in terms of both variables)

Many times it is not convenient to express y as a function of x. Then, implicit differentiation gives us the desired result.

• **Example 18.56:** Find the slope of the tangent line to the graph
$y^4 - 3y - 4x^3 = 5x + 4$ at the point $P(0, -1)$.

SECTION 18.10 IMPLICIT DIFFERENTIATION AND HIGHER-ORDER DERIVATIVES

Solution: It would not be feasible to solve for y. Differentiate implicitly:

$$\frac{d}{dx}(y^4 - 3y - 4x^3) = \frac{d}{dx}(5x + 4)$$

$$4y^3 \frac{dy}{dx} - 3\frac{dy}{dx} - 12x^2 = 5$$

or

$$4y^3 y' - 3y' - 12x^2 = 5.$$

Now solve for y':

$$y'(4y^3 - 3) = 12x^2 + 5$$

$$y' = \frac{12x^2 + 5}{4y^3 - 3}.$$

To find the slope at $P(0, -1)$, substitute $x = 0$ and $y = -1$ in the above equation:

$$m = y' = \frac{12(0)^2 + 5}{4(-1)^3 - 3} = \frac{5}{-7} = -\frac{5}{7}.$$

• **Example 18.57:** Find y' if $2xy^3 - x^2y + x^4 - 5x + 2 = 0$.

Solution: $\frac{d}{dx}(2xy^3) - \frac{d}{dx}(x^2y) + \frac{d}{dx}(x^4) - \frac{d}{dx}(5x) + \frac{d}{dx}(2) = \frac{d}{dx}(0).$

Apply the product rule to the first two differentiations:

$$\left[2x\frac{d}{dx}(y^3) + y^3\frac{d}{dx}(2x)\right] - \left[x^2\frac{d}{dx}(y) + y\frac{d}{dx}(x^2)\right]$$
$$+ 4x^3 - 5 + 0 = 0$$
$$2x(3y^2)y' + y^3(2) - x^2y' - y(2x) + 4x^3 - 5 = 0$$
$$6xy^2y' + 2y^3 - x^2y' - 2xy + 4x^3 - 5 = 0$$
$$(6xy^2 - x^2)y' + 2y^3 - 2xy + 4x^3 - 5 = 0$$
$$y' = \frac{5 - 2y^3 + 2xy - 4x^3}{6xy^2 - x^2}.$$

• **Example 18.58:** The relationship between the distance traveled s and time t of a particle is found to be $s^3 - t^2 = 23$. Find the rate of change of the distance traveled in feet with respect to time at $t = 2$.

Solution: We wish to find $\dfrac{ds}{dt}$ at $t = 2$. Differentiating implicitly, we obtain

$$3s^2 \frac{ds}{dt} - 2t = 0$$

$$3s^2 \frac{ds}{dt} = 2t$$

$$\frac{ds}{dt} = \frac{2t}{3s^2}$$

When $t = 2$,

$$s^3 - (2)^2 = 23,$$

or

$$s^3 = 27$$
$$s = 3,$$

therefore,

$$\frac{ds}{dt} = \frac{2(2)}{3(3)^2} = \frac{4}{27} \approx 0.15.$$

The rate of change of the particle is 0.15 feet at $t = 2$ sec.

Whenever we differentiate, explicitly or implicitly, the result is another function. This new function can be differentiated also. For example, if

$$f(x) = x^3 - 5x^2 + 3x + 5,$$

then

$$f'(x) = 3x^2 - 10x + 3 = g(x).$$

Now $g'(x) = 6x - 10$.

First and Second Derivative

The **first derivative** of $g(x)$ is the **second derivative** of $f(x)$. We write:

$$g'(x) = f''(x).$$

We can continue this process. Since

$$g'(x) = 6x - 10 = h(x),$$
$$h'(x) = 6 \quad \text{and} \quad h'(x) = f'''(x) = 6.$$

Furthermore, $f^{iv}(x) = 0$. Again, finding successive derivatives, we simply can write

$$f(x) = x^3 - 5x^2 + 3x + 5$$
$$f'(x) = 3x^2 - 10x + 3 \qquad \text{(First-order derivative)}$$
$$f''(x) = 6x - 10 \qquad \text{(Second-order derivative)}$$
$$f'''(x) = 6 \qquad \text{(Third-order derivative)}$$
$$f^{iv}(x) = 0. \qquad \text{(Fourth-order derivative)}$$

Higher-Order Derivatives Later, we make use of **higher-order derivatives.**

Section 18.10 Implicit Differentiation and Higher-Order Derivatives

• **Example 18.59:** If $y = 2x^4 + 4x^3 - 10x^2 + 5x - \dfrac{2}{x}$, find $f'''(x)$.

Solution: Write $f(x) = 2x^4 + 4x^3 - 10x^2 + 5x - 2x^{-1}$.
Then

$$f'(x) = 8x^3 + 12x^2 - 20x + 5 + 2x^{-2}$$
$$f''(x) = 24x^2 + 24x - 20 - 4x^{-3}$$
$$f'''(x) = 48x + 24 + 12x^{-4}$$

or

$$f'''(x) = 48x + 24 + \dfrac{12}{x^4}.$$

• **Example 18.60:** Find y'' if $x^2 + 4y^2 = 4$.

Solution: Using implicit differentiation, $2x + 8yy' = 0$. Then

$$y' = \dfrac{-2x}{8y} = \dfrac{-x}{4y}.$$

Differentiating again, using the quotient rule, we have

$$y'' = \dfrac{4y(-1) - (-x)(4y')}{(4y)^2}$$

$$= \dfrac{-4y + 4xy'}{16y^2}$$

$$= \dfrac{-4y + 4x\left(\dfrac{-x}{4y}\right)}{16y^2} \qquad \left(\text{Substituting } y' = -\dfrac{x}{4y}\right)$$

$$= \dfrac{-4y - \dfrac{4x^2}{4y}}{16y^2}$$

$$= \dfrac{-16y^2 - 4x^2}{4y} \cdot \dfrac{1}{16y^2}$$

$$= \dfrac{-16y^2 - 4x^2}{64y^3}$$

$$= \dfrac{-4y^2 - x^2}{16y^3}, \qquad \text{(Dividing by 4)}$$

or

$$y'' = -\dfrac{x^2 + 4y^2}{16y^3}.$$

Note that for higher-order derivatives as well as first derivatives, we can use various symbolizations:

$$y' = f'(x) = \frac{dy}{dx} = D_x(y),$$

$$y'' = f''(x) = \frac{d^2y}{dx^2} = D_x^2(y),$$

$$y''' = f'''(x) = \frac{d^3y}{dx^3} = D_x^3(y),$$

and so forth.

• **Example 18.61:** An electrical potential is given by $v(t) = \dfrac{6}{t}$. Find the first and second derivatives of this potential.

Solution:
$$v(t) = \frac{6}{t} = 6t^{-1}$$
$$v'(t) = (-1)(6)t^{-2}$$
$$= -\frac{6}{t^2}, \quad \text{or } -6t^{-2}$$
$$v''(t) = (-2)(-6)t^{-3}$$
$$= 12t^{-3}$$
$$= \frac{12}{t^3}.$$

Trial Problems 18.10

Before you begin the section exercises, warm up with these problems. Complete answers are included in the answer key.

1. Find y'' if $y = 2x^2$.
2. Find y''' if $y = 2x^2$.
3. Find y'' if $y = 2x^{-2}$.
4. Find y' if $x + y = 0$.
5. Find y' if $2x + 3y = 0$.

Exercises 18.10

Find y'' for each of the following.

1. $y = 4x^3 - 10x^2 + 4x + 1$
2. $y = -2x - x^2$
3. $y = \dfrac{1}{2x^3}$
4. $y = 2 - \dfrac{1}{x}$
5. $y = x$
6. $y = \sqrt{1-x}$

7. $y = \dfrac{1}{2x+3}$

8. $y = \dfrac{x}{x^2+1}$

9. $y = (3x+2)^4$

10. $y = (x+1)^2(x-1)$

Find the first and second derivatives of the following functions.

11. $f(x) = 3x^8$

12. $h(x) = 6$

13. $k(s) = \sqrt[3]{s} + \dfrac{2}{s^2}$

14. $s(t) = (t^2+4)^{2/3}$

15. $g(x) = \sqrt[3]{2-9x}$

16. $r(t) = \dfrac{t}{t^2-1}$

Assuming $y = f(x)$, find y' for each of the following.

17. $2x + 3y = 5$

18. $5x - 2y = 7$

19. $4x^2 - 3y = x$

20. $x^4 - 3y = 4 - x$

21. $x^2 - y^2 = 9$

22. $2x^2 + 3y^2 = 18$

23. $y^3 = x^2 - 1$

24. $y^4 = x^3 - 2x^2$

25. $x^2 + y^2 + y = 4$

26. $x^2 + 2x + y^2 = 7$

27. $x + xy + y = 2$

28. $xy^2 + 3y + x^2 = 9$

29. $x^3 - y^3 = 1$

30. $\sqrt{x} + 4\sqrt{y} = 4$

31. $x^2 - 3xy + y^2 = 4$

32. $xy + y - x = 4$

33. Find all the nonzero derivatives of
$$f(x) = x^6 - 3x^4 + 2x^3 - x + 2.$$

34. If $f(x) = \dfrac{1}{x^n}$, how many derivatives can you find? How could you find the nth derivative, n any positive integer, without finding the previous derivatives?

35. If $f(x) = x^4 - x^3 - 6x^2 + 7x$, find the slope of the tangent line to the graph of $f(x)$ at the point $P(2,-2)$.

36. If $f(x) = x^4 - 10x^2 + x + 2$, use differentials to approximate the change in $f'(x)$ if x changes from 2 to 2.01.

37. In a right triangle of fixed hypotenuse a and sides of x and y, find the rate of change of y with respect to x.

38. The relationship between the distance s traveled in feet and time t of a particle is found to be $s^3 + t^2 = 31$. Find the rate of change of the distance traveled with respect to time t at $t = 2$ sec.

39. The displacement of an object is given by $s = 12 + 5t^3$. Find the first and second derivatives of this displacement.

40. For certain functions the expression $t^2y'' + ty' - y$ is equal to zero. Show that this is true for $y = 3t + \dfrac{2}{t}$.

18·11 Summary of Terms and Theorems

Terms

Average Rate of Change (p. 772)
Chain Rule (p. 768)
Continuous (p. 742)
Derivative (p. 756)
Differentials (p. 777)
Differentiation (p. 756)
Discontinuous (p. 743)

First Derivative (p. 782)
Higher-Order Derivatives (p. 782)
Implicit Differentiation (p. 780)
Implicit Function (p. 780)
Increment (p. 776)
Instantaneous Rate of Change (p. 772)
Limit (p. 738)

Product Rule (p. 762)
Quotient Rule (p. 763)
Rate of Change (p. 772)
Secant Line (p. 752)
Second Derivative (p. 782)
Slope of Tangent Line (p. 752)
Tangent Line (p. 752)

Theorems

$\lim_{x \to a} c = c$, where a is a real number, and c is any constant.   (18.1)

$\lim_{x \to a} x^n = a^n$ (18.1)

$\lim_{x \to a} [f(x) + g(x)] = \lim_{x \to a} f(x) + \lim_{x \to a} g(x)$ (18.1)

$\lim_{x \to a} f(x)g(x) = \lim_{x \to a} f(x) \lim_{x \to a} g(x)$ (18.1)

$\lim_{x \to a} \dfrac{f(x)}{g(x)} = \dfrac{\lim_{x \to a} f(x)}{\lim_{x \to a} g(x)}$, provided $\lim_{x \to a} g(x) \neq 0$.   (18.1)

$\lim_{x \to \infty} \dfrac{c}{x^k} = 0$ and $\lim_{x \to -\infty} \dfrac{c}{x^k} = 0$, c a constant and k a positive number.   (18.3)

$\lim_{x \to a} \dfrac{1}{(x-a)^n} = \infty$ if n is an even positive integer.   (18.3)

$\lim_{x \to a^+} \dfrac{1}{(x-a)^n} = \infty$ if n is an odd positive integer.   (18.3)

$\lim_{x \to a^-} \dfrac{1}{(x-a)^n} = -\infty$ if n is an odd positive integer.   (18.3)

If $f(x) = c$, where c is any constant, then $f'(x) = 0$.   (18.6)

If $f(x) = x$, then $f'(x) = 1$.   (18.6)

If $f(x) = x^n$, then $f'(x) = nx^{n-1}$, n any integer.   (18.6)

If $y = cf(x)$, then $y' = cf'(x)$.   (18.6)

If $y = f(x) + g(x)$, then $y' = f'(x) + g'(x)$.   (18.6)

If $y = f(x)g(x)$, then $y' = f(x)g'(x) + g(x)f'(x)$.   (18.6)

If $y = \dfrac{f(x)}{g(x)}$, then $y' = \dfrac{g(x)f'(x) - f(x)g'(x)}{[g(x)]^2}$. (18.6)

If y is a function of u and u is a function of x, then $\dfrac{dy}{dx} = \dfrac{dy}{du} \cdot \dfrac{du}{dx}$. (18.7)

18·12 Chapter 18 Review Exercises

Find the limit of each of the following as indicated.

1. $\lim\limits_{x \to 0} (x^2 + 2x + 1)$

2. $\lim\limits_{x \to \infty} \left(\dfrac{1}{x^2} + \dfrac{1}{x} + 1 \right)$

3. $\lim\limits_{x \to 1} \left(x + \dfrac{1}{x} \right)$

4. $\lim\limits_{x \to -1} \left(x - \dfrac{1}{x} \right)$

5. $\lim\limits_{x \to 2} \dfrac{3x + 10}{\sqrt{x - 1}}$

6. $\lim\limits_{x \to 0} \dfrac{1}{x - 2}$

7. $\lim\limits_{h \to 0} \dfrac{(x + h)^3 - x^3}{h}$

8. $\lim\limits_{x \to \infty} \left(1 + \dfrac{1}{x} \right)$

9. $\lim\limits_{x \to \infty} \left(x + \dfrac{1}{x} \right)$

10. $\lim\limits_{x \to \infty} \dfrac{3x^2 - 3x + 1}{3 - x^2}$

11. Find the discontinuities of $f(x) = \dfrac{x^2 - x - 2}{x^2 - 2x}$.

12. Use the *definition* of derivative to find $f'(x)$ if $f(x) = \dfrac{2}{5x^2 + 1}$.

Find the first derivative of each of the following.

13. $y = 7x^5 - x - 2$

14. $y = 2\sqrt{x} - \dfrac{1}{x}$

15. $y = (3x^2 - 1)^5$

16. $y = 3x^4 - 2x^2 + 7x - 1$

17. $g(t) = \sqrt{2t + 3}$

18. $g(x) = \sqrt[3]{6x^3 - x + 3}$

19. $f(w) = \sqrt[5]{3w^2}$

20. $y = \dfrac{5}{(2x^3 - 1)^2}$

21. $G(x) = (x^2 - x^{-2})^{-2}$

22. $r(s) = \left(\dfrac{3s^2 - 1}{1 - s^3} \right)^3$

23. $F(x) = (x^5 + 1)^4 (x + 2)^3$

24. $k(s) = \dfrac{1}{\sqrt{6s + 1}}$

25. $q(x) = \dfrac{(x - 1)(x - 3)}{(x + 1)(x + 3)}$

26. Find y' if $3x^2 - x^2y^2 + 4y^2 - 5 = 0$

27. Find y' if $y^2 + xy + x^2 = 2$

28. The combined capacities for two capacitors connected in series is given by $C_T = \dfrac{C_1 C_2}{C_1 + C_2}$.
  a. Find $\lim\limits_{C_1 \to C_2} C_T$
  b. Find $\lim\limits_{C_1 \to 0} C_T$

29. Find the slope of the tangent line to the curve $y = 3x^4 - x^3$ at the point $(-1, 4)$.

30. Find the equation of the tangent line to the graph of $y = 2x - \dfrac{1}{\sqrt{x}}$ at the point $P(1, 1)$.

31. Find y' and y'' if $x^2 + 2xy - y^2 = 4$.

32. If $y = \dfrac{x}{x^2 + 1}$, use differentials to approximate the change in y if x changes from 2 to 1.98.

33. Find the rate of change of the area A of a circle, with respect to its radius r when $r = 2$.

34. The resistance of a certain resistor changes according to the change in temperature. If R is the resistance and t is temperature and $R = 30 + 0.160t + 0.0003t^2$, find $\dfrac{dR}{dt}$ at $t = 30°$ C.

35. The relationship between two resistors placed in parallel is given by $\tau = \dfrac{5R}{5 + R}$. Find $\dfrac{d\tau}{dR}$.

36. Velocity is the rate of change of distance with respect to time. If the distance s traveled by an object in time t is given by $s = \dfrac{16}{t^2}$, find the velocity when $t = 4$.

37. The period P of a pendulum is directly proportional to the square root of its length L. Find $\dfrac{dP}{dL}$ if $L = 2$ ft and $P = \dfrac{\pi}{2}$ seconds.

38. By the Pythagorean theorem, the legs x and y of a right triangle and the hypotenuse c are related by the equation $x^2 + y^2 = c^2$. If c is a constant, find an expression for $\dfrac{dy}{dx}$.

39. Find $\dfrac{d^2E}{dr^2}$, where E is the electric field at a distance r from a point charge, and $E = \dfrac{k}{r}$, k a constant.

40. The formula $s = 10t - 2t^2$ gives the distance that an object travels in t seconds. If velocity is the rate of change of s with respect to t, determine how far the object travels before coming to rest.

41. If the price of a product is related to quantity x of the product by the equation $p = \dfrac{x + 14}{x + 4}$, find the rate of change of price with respect to quantity, x.

42. In an experiment, molecular change takes place according to $m_c = \dfrac{4t}{t + 1}$, where t is measured in seconds. Find the rate of change of m_c with respect to t.

43. The curved surface of a cone is given by $S = \pi r \sqrt{r^2 + h^2}$, where r is the radius and h the height. If S is kept constant and r is varied, h will change. Find $\dfrac{dh}{dr}$.

44. If a force has constant vertical component 5 and variable horizontal component h, then the magnitude of the force is given by $F = \sqrt{h^2 + 25}$. Find the rate at which F changes with respect to h when $h = 6$.

45. If the distance D in feet required for a certain driver to stop an automobile is given by $D = 0.5v + 0.02v^2$, where v is velocity and $0 \leq v \leq 100$, find the rate at which D increased with respect to v when v was 30 ft per sec.

46. The pulse rate u of a string is the function of the tension T on the string. If $u = \sqrt{\dfrac{T}{p}}$, p a constant, find the rate at which u increases with T when $T = 16$.

47. The relationship between the workload L of a certain machine and time t in hours is given by $3\sqrt{L} - 0.5t^2 = 10$. Find $\dfrac{dL}{dt}$.

48. An object is thrown straight up into the air. Its distance s from the ground t seconds after being thrown is $s = t^2 + t$. Find s' and s''.

18·13 Chapter 18 Test

Evaluate the following.

1. $\lim_{x \to 3} (x^2 - 3x + 2)$

2. $\lim_{x \to 2} \dfrac{x^2 + x - 6}{x^2 - 4}$

Determine where f is continuous or discontinuous.

3. $f(x) = \sqrt{x - 2}$

4. $f(x) = \dfrac{1}{x - 3}$

Evaluate the following limits.

5. $\lim_{x \to 3} \dfrac{6}{x - 3}$

6. $\lim_{x \to \infty} \dfrac{4x^2 + 9}{x^2 - 5}$

7. Use the limit concept to find the slope of $y = x + \dfrac{1}{x}$ at $\left(2, \dfrac{5}{2}\right)$.

8. Find $f'(2)$ if $y = x^2 - 3x + 2$.

Differentiate the following.

9. $y = (x - 3)^2$

10. $f(x) = 2x(x^2 - 5)$

11. $f(x) = (x - 2)(3x + 5)$

12. $y = \dfrac{x - 2}{4x^2 + 3}$

13. $y = (4x - 6)^3$

14. $f(x) = \sqrt{1 - 4x^2}$

15. The current I in a certain electrical circuit is given by $I = \dfrac{50}{R - 10}$. Find the rate of change of I with respect to R when $R = 35$ ohms.

16. The position of an object moving in a straight line, at any time t, is given by $s = 3t^3 - 8t + 7$. Find ds and use it to approximate the change in s if t changes from 1 to 1.3.

Find the first and second derivative of the following.

17. $f(x) = 4x^2$

18. $g(x) = \sqrt{4 - x}$

Assuming $y = f(x)$, find y' for each of the following.

19. $x^2 + y^2 = 4$

20. $x + xy - y = 10$

CHAPTER 19
Derivatives of Transcendental Functions

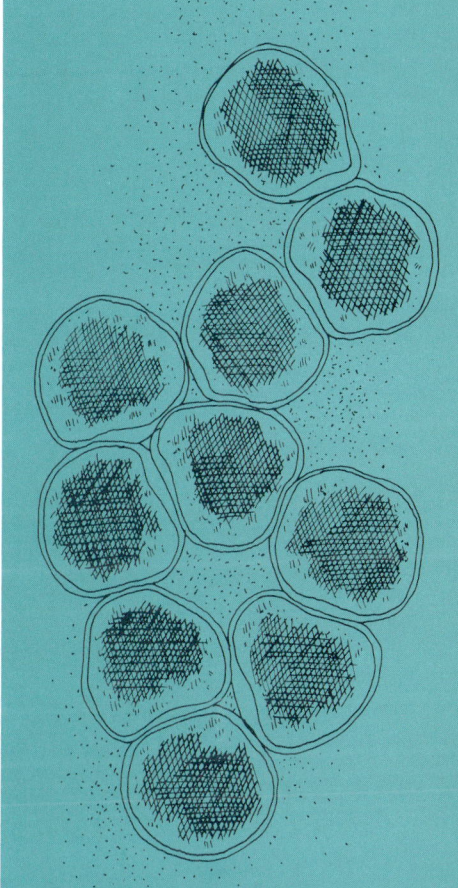

*I*n earlier chapters logarithmic, exponential, and trigonometric functions were investigated extensively. These functions are known as transcendental functions. Knowledge of these functions is essential for the study of several areas of applied mathematics. As the use of calculus is applied to the study of technical mathematics, we need to learn how to apply the calculus to transcendental functions. In this chapter formulas, rules, and procedures for differentiating transcendental functions are carefully developed. Some technical applications of these processes are presented, but extensive applications of the derivative, as applied to transcendental functions, are presented in chapter 20.

19 • 1 Derivatives of Logarithmic Functions

The general logarithmic function $y = \log_a x$ and the natural logarithmic function $y = \ln x$ were investigated in chapter 10. In the calculus, the natural logarithmic function is used in all of our work. First, we find the derivative of $y = \ln x$. Then, we use the chain rule to find the derivative of $y = \ln u$, where u is a function of x, $u = f(x)$.

In order to find the derivative of $y = \ln x$, we use the definition of y'.

$$y' = \lim_{h \to 0} \frac{f(x+h) - f(x)}{h}$$

$$= \lim_{h \to 0} \frac{\ln(x+h) - \ln x}{h}$$

$$= \lim_{h \to 0} \frac{\ln \frac{x+h}{x}}{h} \qquad \left(\ln a - \ln b = \ln \frac{a}{b}\right)$$

$$= \lim_{h \to 0} \frac{1}{h} \ln\left(1 + \frac{h}{x}\right) \qquad \left(\text{Dividing by } h \text{ is equivalent to multiplying by } \frac{1}{h}\right)$$

$$= \lim_{h \to 0} \frac{1}{x} \frac{x}{h} \ln\left(1 + \frac{h}{x}\right) \qquad \left(\frac{1}{x} \cdot \frac{x}{h} = \frac{x}{x} \cdot \frac{1}{h} = 1 \cdot \frac{1}{h} = \frac{1}{h}\right)$$

$$= \lim_{h \to 0} \frac{1}{x} \ln\left(1 + \frac{h}{x}\right)^{x/h} \qquad (p \ln b = \ln b^p)$$

$$\left(\text{Let } t = \frac{h}{x}. \text{ Then } \frac{1}{t} = \frac{x}{h}\right)$$

$$= \lim_{t \to 0} \frac{1}{x} \ln(1+t)^{1/t} \qquad (\text{As } h \to 0,\ t \to 0)$$

$$= \frac{1}{x} \lim_{t \to 0} \ln(1+t)^{1/t}$$

$$= \frac{1}{x} \ln e \qquad \left(\lim_{t \to 0} \ln(1+t)^{1/t} \text{ is a special limit whose value is } e\right)$$

$$= \frac{1}{x}. \qquad (\text{Since } \ln e = 1)$$

Therefore, $y' = \dfrac{1}{x}$.

SECTION 19.1 DERIVATIVES OF LOGARITHMIC FUNCTIONS

Theorem 19.1 If $y = \ln x$, then $y' = \dfrac{1}{x}$.

• **Example 19.1:** $y = 3 \ln x$.

  Solution: $y' = 3\left(\dfrac{1}{x}\right) = \dfrac{3}{x}$.

• **Example 19.2:** $y = x \ln x$.

  Solution: $y' = x\left(\dfrac{1}{x}\right) + \ln x (1)$ (Product rule)

  $= 1 + \ln x$.

• **Example 19.3:** $y = (\ln x)^2$

  Solution: $y' = 2(\ln x) D_x(\ln x)$ (Power rule)

  $= 2(\ln x)\left(\dfrac{1}{x}\right)$

  $= \dfrac{2}{x} \ln x$.

Theorem 19.2 If $y = \ln u$, where $u = f(x)$, then $y' = \dfrac{1}{u} D_x u$. ($D_x u$ is "the derivative of u with respect to x.")

Proof: Let $y = \ln u = \ln(f(x))$

By the chain rule, $y' = [D_x \ln(f(x))] f'(x)$.

Therefore, $y' = \dfrac{1}{f(x)} f'(x)$

$= \dfrac{1}{u} D_x u$.

• **Example 19.4:** $y = \ln(x^2 + 2)$.

Solution: In this example, $y = \ln u$, where $u = x^2 + 2$. Therefore, $\dfrac{du}{dx} = 2x$. Thus,

$$y' = \underbrace{\dfrac{1}{x^2 + 2}}_{\dfrac{dy}{du}} \underbrace{(2x)}_{\dfrac{du}{dx}}$$

$$= \dfrac{2x}{x^2 + 2}.$$

• **Example 19.5:** $y = 3 \ln x^2$.

Solution: $u = x^2$ and $\dfrac{du}{dx} = 2x$. Therefore,

$$y' = 3\left(\dfrac{1}{x^2}\right)(2x)$$

$$= \dfrac{6x}{x^2}$$

$$= \dfrac{6}{x}.$$

Also, the derivative can be found by noticing that

$$\ln x^2 = 2 \ln x.$$

Then

$$y = 3 \ln x^2 = 6 \ln x$$

and

$$y' = 6 \cdot \dfrac{1}{x} = \dfrac{6}{x}.$$

• **Example 19.6:** $y = x^3 \ln(x^2 + 2)$.

Solution: We need to use the product rule with $f(x) = x^3$ and $g(x) = \ln(x^2 + 2)$. Therefore,

$$y' = \underbrace{x^3}_{f(x)} \underbrace{\left[\frac{1}{x^2 + 2}(2x)\right]}_{g'(x)} + \underbrace{[\ln(x^2 + 2)]}_{g(x)} \underbrace{(3x^2)}_{f'(x)}$$

$$= \frac{2x^4}{x^2 + 2} + 3x^2 \ln(x^2 + 2).$$

••••••••••

• **Example 19.7:** The distance s a car travels varies with time t according to $s = \ln(t^3 + 1)$. Find $\dfrac{ds}{dt}$, the rate of change of the distance traveled with respect to time.

Solution: We need to use the chain rule. Let $u = t^3 + 1$, then $\dfrac{du}{dt} = 3t^2$:

$$s = \ln(t^3 + 1) = \ln u$$

$$\frac{ds}{dt} = \frac{1}{u} \cdot \frac{du}{dt}$$

$$= \frac{1}{t^3 + 1} \cdot 3t^2$$

$$= \frac{3t^2}{t^3 + 1}.$$

••••••••••

• **Example 19.8:** The energy dissipated by a resistor during an experiment is given by $E = 6 + \dfrac{1}{2}t - \ln(t + 1)$ joules. Find E' and E''.

Solution:
$$E = 6 + \frac{1}{2}t - \ln(t + 1)$$

$$E' = \frac{1}{2} - \frac{1}{t + 1}, \quad \text{or} \quad \frac{1}{2} - (t + 1)^{-1}$$

$$E'' = -(-1)(t + 1)^{-2}$$

$$= \frac{1}{(t + 1)^2}.$$

••••••••••

Trial Problems 19.1

Before you begin the section exercises, warm up with these problems. Complete answers are included in the answer key.

Find the derivative of each of the following.

1. $y = \ln x$
2. $y = \ln 2x$
3. $y = \ln\left(\dfrac{1}{2}x\right)$
4. $y = \ln(x + 1)$
5. $y = \ln(1 - x)$

Exercises 19.1

Find the derivative of each of the following.

1. $y = \ln 10x$
2. $y = \ln(2 + x^2)$
3. $f(x) = \ln(9x + 4)$
4. $y = \ln(x^3 + 1)$
5. $f(x) = \ln(1 - 3x)$
6. $f(x) = x^2 \ln x$
7. $y = \ln(2 - 3x)^4$
8. $f(x) = \ln(3x^2 + 1)^5$
9. $r(x) = \ln(3x^2 - 2x + 1)$
10. $s(t) = \ln(4t^3 - t^2 + 2)$
11. $f(x) = x \ln x$
12. $f(x) = \dfrac{x^2}{\ln x}$
13. $y = \ln(\ln x)$
14. $y = (\ln x)^3$
15. $g(x) = \ln(x^2 - 1)$
16. $g(x) = \ln\sqrt{x}$
17. $y = x^2 + \ln x$
18. $y = \dfrac{\ln x}{x}$
19. $y = \ln\left(\dfrac{1 + x}{1 - x}\right)$
20. $y = \ln\sqrt{\dfrac{x^2 - 1}{x^2 + 1}}$
21. $y = \sqrt{2 + \ln x}$
22. $y = \ln(x + \sqrt{1 + x^2})$

Find the second derivative of each of the following.

23. $y = \ln x$
24. $y = \ln 3x$
25. $y = \ln x^2$
26. $y = (\ln x)^2$
27. $y = \ln\left(\dfrac{1}{x}\right)$
28. $y = \dfrac{\ln x}{x}$
29. $y = \ln(1 - 3x)$
30. $y = x(\ln x)$
31. $y = x + \ln x$

Find the equation of the tangent line to each curve at the indicated point. Sketch the graph of the function and draw the tangent line at the indicated point.

32. $y = \ln x$, (1,0)
33. $y = x \ln x$, (e,e)

Differentiate implicitly. Show all steps used.

34. $\ln(xy) + x = 4$
35. $y \ln x = x \ln y$

36. At an inflation rate of r percent a year, prices double in approximately n years, where $n = \dfrac{\ln 2}{\ln(1+r)}$. Find $\dfrac{dn}{dr}$.

37. In a chemical reaction, the concentration x of a substance at time t, is related by:

$$t = \dfrac{1}{c_1(c_2 - c_3)} \ln \dfrac{c_3(c_2 - x)}{c_2(c_3 - x)}.$$

If c_1, c_2, and c_3 are constant, show that

$$\dfrac{dx}{dt} = c_1(c_2 - x)(c_3 - x). \quad \left(\text{Hint: } \dfrac{dx}{dt} = \dfrac{1}{\dfrac{dt}{dx}}.\right)$$

38. The tensile strength s of a certain material is a function of temperature T. What is the rate of change of tensile strength with respect to temperature if

$$s = \ln\left(\dfrac{3}{1 + 3T}\right) \text{ psi?}$$

39. The current i in a certain inductor varies as $i = \ln(2t - 1)$ A, where t is time in seconds. The expression for induced voltage in the coil is given by $\dfrac{di}{dt}$. Find $\dfrac{di}{dt}$ when $t = 3$ sec.

40. The distance s an object travels in time t is given by the expression $s = \ln(t^3 - 1)$. Find s' and s''. (*Note:* It will be shown later the s' is the velocity of the object and s'' is the acceleration of the object, at any time t.)

41. In an experiment with bacteria, it is observed that the relative activeness of the bacteria colony is

$$A = 6 \ln\left(\dfrac{T}{a - T} - a\right),$$

where a is a constant and T is the surrounding temperature. $\dfrac{dA}{dT}$ gives an expression that indicates how A changes with respect to time T. Find $\dfrac{dA}{dT}$.

19 • 2 Derivatives of Exponential Functions

Common and Natural Exponential Functions

The **common exponential function** $y = a^x$ and the **natural exponential function** $y = e^x$ are inverse functions of $y = \log_a x$ and $y = \ln x$, respectively. Recall that:

> $y = a^x$ if and only if $x = \log_a y$, and $y = e^x$ if and only if $x = \ln y$.

Furthermore, we found that $y = a^x$ can be expressed in the form $y = e^u$. If $a^x = e^u$, then

$$x \ln a = u \ln e. \quad \text{(Taking ln of both sides of the equation)}$$

But $u \ln e = u$, so $x \ln a = u$. Therefore,

$$a^x = e^{x \ln a}.$$

Thus, as with natural logarithmic functions, in the calculus we use the natural exponential function $y = e^x$ in most of our work.

The derivative of $y = e^x$ can be found by taking the natural logarithm of both sides of the equation:

$$y = e^x$$
$$\ln y = \ln e^x = x \ln e = x$$
$$D_x(\ln y) = D_x(x)$$
$$\frac{1}{y} y' = 1. \quad \text{(Using the derivative of ln and using the chain rule)}$$

Therefore, $y' = y$. Since $y = e^x$, we have $y' = e^x$.

Theorem 19.3 If $y = e^x$, then $y' = e^x$.

Theorem 19.4 If $y = e^u$, where $u = f(x)$, then $y' = e^u \dfrac{du}{dx}$.

Proof: Let $y = e^u = e^{f(x)}$. By the chain rule, $y' = D_x[e^{f(x)}] f'(x)$.
Therefore,
$$y' = e^{f(x)} f'(x)$$
$$= e^u \frac{du}{dx}.$$

• **Example 19.9:** $y = 2e^x$.

  Solution: $y' = 2e^x$.

• **Example 19.10:** $f(x) = e^{3x}$.

  Solution: In this example, $f(x) = e^u$, where $u = 3x$ and $\dfrac{du}{dx} = 3$.
  Therefore,
  $$f'(x) = \underbrace{(e^{3x})}_{D_x f(x)} \underbrace{(3)}_{D_x u}$$
  $$= 3e^{3x}.$$

• **Example 19.11:** $y = e^{x^2}$.

  Solution: $y' = (e^{x^2})(2x)$
  $= 2xe^{x^2}$.

• **Example 19.12:** $g(x) = xe^x$.

Solution: Using the product rule, we have
$$g'(x) = x(e^x) + (e^x)(1)$$
$$= xe^x + e^x.$$

••••••••••

• **Example 19.13:** $y = e^{-5x}$.

Solution: $y' = (e^{-5x})(-5)$
$$= -5e^{-5x}.$$

••••••••••

• **Example 19.14:** $y = (e^{3x})^4$.

Solution: This example is in the form $y = t^4$, where $t = e^u$, $u = 3x$, and $\dfrac{du}{dx} = 3$. Hence,

$$y' = 4t^3 \frac{dt}{dx} \text{ and } \frac{dt}{dx} = \frac{dt}{du} \cdot \frac{du}{dx} = (e^u)(3) = (e^{3x})(3).$$

Therefore,
$$y' = 4(e^{3x})^3(e^{3x})(3)$$
$$= 12(e^{3x})^3(e^{3x})$$
$$= 12e^{12x}.$$

••••••••••

• **Example 19.15:** The shear s of a simple beam is equal to $D_x M$, where M is the bending moment and x is the distance measured from one end of the beam. Find the shear equation of a beam whose bending moment is $M = xe^x$.

Solution: We have $s = D_x M$ and $M = xe^x$. We want to find s as a function of x, $s = f(x)$.

$$s = D_x M = D_x(xe^x)$$
$$= x D_x(e^x) + e^x D_x x \quad \text{(Chain rule)}$$
$$= xe^x + e^x(1).$$

Therefore,
$$s = xe^x + e^x,$$

or
$$s = e^x(x + 1).$$

••••••••••

Trial Problems 19.2

Before you begin the section exercises, warm up with these problems. Complete answers are included in the answer key.

Find the derivative of each of the following.

1. $y = e^x$
2. $y = e^{-x}$
3. $y = -e^x$
4. $y = e^{2x}$
5. $y = 5^x$

Exercises 19.2

Find the derivative of each of the following.

1. $y = e^{6x}$
2. $y = e^{-6x}$
3. $y = 3e^{-1/3x}$
4. $y = 2 - 3e^{-2x}$
5. $f(x) = e^x + e^{-x}$
6. $g(x) = e^x - e^{-x}$
7. $h(x) = 2^x - 5xe^{-2x}$
8. $y = 2^x + 3^x$
9. $y = xe^x$
10. $r(x) = \dfrac{e^x}{x}$
11. $y = e^{\sqrt{x}}$
12. $y = e^{x^2}$
13. $f(x) = e^{x^2 + 1}$
14. $f(x) = e^{(3 - 5x)}$
15. $y = x^2 e^{-x^2}$
16. $y = e^{(1 + \sqrt{x})}$
17. $y = \dfrac{e^x - 1}{e^x + 1}$
18. $y = e^{2x}(x + 1)$
19. $y = e^{\ln x}$
20. $y = e^{\ln(x^2 + 1)}$
21. $f(x) = e^{x \ln x}$
22. $y = e^{-x} \ln x$

Find the second derivative of each of the following.

23. $y = e^x$
24. $y = e^{-x}$
25. $y = xe^x$
26. $y = e^x \ln x$

Find the equation of the tangent line to each curve at the indicated point.

27. $y = e^x$ at $x = 0$
28. $y = e^{-x}$ at $x = 1$

29. Certain warm objects lose temperature according to the formula $T = 50(e)^{-0.067t}$. Find $\dfrac{dT}{dt}$.

30. In a formula for mass, $m = 200 - 2^{t/2}$. Find $\dfrac{dm}{dt}$. (Hint: $2^x = e^{x \ln 2}$).

31. $V = 100(1 - e^{-0.1t})$ gives the voltage in a particular electric circuit at time t. Find $\dfrac{dV}{dt}$.

32. Use differentials to approximate the change in $f(x) = xe^{x^2}$ if x changes from 1.00 to 1.01.

33. If P_0 is the population of a country at $t = 0$, and $P = P_0 e^{kt}$ is the population after t years, k a constant, show that $\dfrac{dP}{dt} = kP$.

34. In an experiment, the scientist Rutherford showed that the number of alpha particles N scattered through an angle greater than or equal to α is given by

$$N = N_0 e^{-(\alpha/a)^2},$$

where N_0 and a are constants. $\dfrac{dN}{d\alpha}$ gives an expression that indicates how N changes with respect to α. Find $\dfrac{dN}{d\alpha}$.

19 • 3 Derivatives of Trigonometric Functions

In order to find derivatives of trigonometric functions, we use two limit theorems and the basic definition of the derivative.

Theorem 19.5 $\quad \lim\limits_{x \to 0} \dfrac{\sin x}{x} = 1.$

Theorem 19.6 $\quad \lim\limits_{x \to 0} \dfrac{\cos x - 1}{x} = 0.$

The proof of these theorems is not given. It is suggested that a calculator be used to test each statement by selecting smaller and smaller values of x, $x \to 0$, and observe the resulting values.

The form of the definition of the derivative that we use is

$$f'(x) = \lim_{h \to 0} \dfrac{f(x + h) - f(x)}{h}.$$

For the sine function, $y = \sin x$, we can write

$$\begin{aligned}
f'(x) &= \lim_{h \to 0} \dfrac{\sin(x + h) - \sin x}{h} \\
&= \lim_{h \to 0} \dfrac{\sin x \cos h + \cos x \sin h - \sin x}{h} \\
&= \lim_{h \to 0} \dfrac{\sin x(\cos h - 1) + \cos x \sin h}{h} \quad \text{(Using the identity } \sin(a + b) \\
&\qquad\qquad\qquad\qquad\qquad\qquad\qquad\qquad = \sin a \cos b + \cos a \sin b\text{)} \\
&= \lim_{h \to 0} \left[\sin x \left(\dfrac{\cos h - 1}{h} \right) + \cos x \left(\dfrac{\sin h}{h} \right) \right] \\
&= \sin x(0) + \cos x(1) \quad \text{(Theorem 19.5 and theorem 19.6)} \\
&= \cos x.
\end{aligned}$$

Thus, the derivative of the sine function is the cosine function. The derivatives of the other trigonometric functions can be derived in a similar manner. We use the following theorems to determine the derivatives of trigonometric functions.

Theorem 19.7 If $y = \sin u$, where $u = f(x)$, then $y' = \cos u \, D_x u$.

Proof: Let $y = \sin u = \sin(f(x))$. By the chain rule,
$$y' = [D_x \sin(f(x))] f'(x).$$
Therefore,
$$y' = \cos(f(x)) f'(x)$$
$$= \cos u \, D_x u.$$

• **Example 19.16:** $y = \sin 2x$.

Solution: $y' = \cos 2x (2)$ $\quad \left(\text{Using the chain rule with } u = 2x, \dfrac{du}{dx} = 2\right)$
$= 2 \cos 2x.$

• **Example 19.17:** $y = x \sin x^2$.

Solution: $y' = x(2x \cos x^2) + \sin x^2 (1)$ $\quad$ (Product rule)
$= 2x^2 \cos x^2 + \sin x^2.$

Theorem 19.8 If $y = \cos u$, where $u = f(x)$, then $y' = -\sin u \, D_x u$.

Proof: First, consider $y = \cos x$. Then,
$$y' = \lim_{h \to 0} \frac{\cos(x+h) - \cos x}{h}$$
$$= \lim_{h \to 0} \frac{\cos x \cos h - \sin x \sin h - \cos x}{h}$$
$$= \lim_{h \to 0} \left[\cos x \left(\frac{\cos h - 1}{h}\right) - \sin x \left(\frac{\sin h}{h}\right)\right]$$
$$= \cos x \lim_{h \to 0} \frac{\cos h - 1}{h} - \sin x \lim_{h \to 0} \frac{\sin h}{h}$$
$$= \cos x \cdot (0) - \sin x \cdot (1)$$
$$= -\sin x.$$

It follows that if $y = \cos u$, then $y' = -\sin u \, D_x u$.

SECTION 19.3 DERIVATIVES OF TRIGONOMETRIC FUNCTIONS

• **Example 19.18:** $y = 2 \cos 3x$.

Solution: $y = 2(-\sin 3x) D_x(3x)$ (Chain rule)
$= 2(-\sin 3x)(3)$
$= -6 \sin 3x$.

• **Example 19.19:** $y = \dfrac{\cos x}{x}$.

Solution: $y' = \dfrac{x(-\sin x) - \cos x (1)}{x^2}$ (Quotient rule)

$= \dfrac{-x \sin x - \cos x}{x^2}$.

Theorem 19.9   If $y = \tan u$, where $u = f(x)$, then $y' = \sec^2 u \, D_x u$.

Proof: First consider $y = \tan x = \dfrac{\sin x}{\cos x}$. Then,

$y' = \dfrac{\cos x (D_x \sin x) - \sin x (D_x \cos x)}{\cos^2 x}$ (Quotient rule and chain rule)

$= \dfrac{\cos x \cos x - \sin x(-\sin x)}{\cos^2 x}$

$= \dfrac{\cos^2 x + \sin^2 x}{\cos^2 x}$

$= \dfrac{1}{\cos^2 x}$

$= \sec^2 x$.

It follows that if $y = \tan u$, then $y' = \sec^2 u \, D_x u$.

• **Example 19.20:** $y = 3 \tan x^2$.

Solution: $y' = 3(\sec^2 x^2) D_x(x^2)$ (Chain rule)
$= 3(\sec^2 x^2)(2x)$
$= 6x \sec^2 x^2$.

• **Example 19.21:** $y = \sin x \tan x$.

Solution: $y' = (\sin x)(\sec^2 x) + (\tan x)(\cos x)$ (Product rule)
$= \sin x \sec^2 x + \tan x \cos x$
$= \sin x \sec^2 x + \dfrac{\sin x}{\cos x} \cos x$
$= \sin x \sec^2 x + \sin x$
$= \sin x(\sec^2 x + 1)$.

Theorem 19.10 If $y = \sec u$, where $u = f(x)$, then $y' = \sec u \tan u \, D_x u$.

Proof: First consider $y = \sec x = \dfrac{1}{\cos x}$. Then,

$$y' = \dfrac{\cos x [D_x 1] - 1[D_x \cos x]}{\cos^2 x} \quad \text{(Quotient rule)}$$

$$= \dfrac{\cos x \cdot 0 - (-\sin x)}{\cos^2 x}$$

$$= \dfrac{\sin x}{\cos^2 x} = \dfrac{1}{\cos x} \cdot \dfrac{\sin x}{\cos x}$$

$$= \sec x \tan x.$$

It follows that if $y = \sec u$, then $y' = \sec u \tan u \, D_x u$.

• **Example 19.22:** $y = 3 \sec x^2$.

Solution: $y' = 3(\sec x^2 \tan x^2)(2x)$ (Chain rule)
$= 6x \sec x^2 \tan x^2$.

Theorem 19.11 If $y = \cot u$, where $u = f(x)$, then $y' = -\csc^2 u \, D_x u$.

• **Example 19.23:** $y = x^2 \cot x$.

Solution: $y' = x^2(-\csc^2 x) + \cot x \, (2x)$ (Product rule)
$= -x^2 \csc^2 x + 2x \cot x$.

SECTION 19.3 DERIVATIVES OF TRIGONOMETRIC FUNCTIONS

Theorem 19.12 If $y = \csc u$, where $u = f(x)$, then $y' = -\csc u \cot u\, D_x u$.

• **Example 19.24:** $y = \csc(3x - 1)$.

Solution:
$$\begin{aligned} y' &= -\csc(3x - 1)\cot(3x - 1)\, D_x(3x - 1) \\ &= -\csc(3x - 1)\cot(3x - 1)(3) \\ &= -3\csc(3x - 1)\cot(3x - 1). \end{aligned}$$

The following table contains a summary of the rules for differentiating logarithmic, exponential, and trigonometric functions.

Function	Derivative
1. $y = \ln u$	$y' = \dfrac{1}{u} D_x u$
2. $y = e^u$	$y' = e^u D_x u$
3. $y = \sin u$	$y' = \cos u\, D_x u$
4. $y = \cos u$	$y' = -\sin u\, D_x u$
5. $y = \tan u$	$y' = \sec^2 u\, D_x u$
6. $y = \cot u$	$y' = -\csc^2 u\, D_x u$
7. $y = \sec u$	$y' = \sec u \tan u\, D_x u$
8. $y = \csc u$	$y' = -\csc u \cot u\, D_x u$

• **Example 19.25:** The force F needed to move an object up an incline is given by $F = W \cos\theta$, where W is the weight of the object and θ is the angle of inclination of the inclined plane. Find the rate of change of F with respect to θ.

Solution: We assume that W is a constant. The rate of change of F with respect to θ is given by $\dfrac{dF}{d\theta}$:

$$\begin{aligned} \dfrac{dF}{d\theta} &= W\, D_\theta(\cos\theta) \\ &= W(-\sin\theta) \\ &= -W\sin\theta. \end{aligned}$$

CHAPTER 19 DERIVATIVES OF TRANSCENDENTAL FUNCTIONS

• **Example 19.26:** Find the slope m of the tangent line to the curve $y = x^2 \sin x$ at $x = \dfrac{\pi}{2}$.

Solution: The slope is given by $m = \dfrac{dy}{dx}$.

$$m = \frac{dy}{dx} = x^2(\cos x) + \sin x(2x) \quad \text{(Product rule)}$$
$$= x^2 \cos x + 2x \sin x.$$

At $x = \dfrac{\pi}{2}$,

$$m = \left(\frac{\pi}{2}\right)^2 \cos \frac{\pi}{2} + 2\left(\frac{\pi}{2}\right) \sin \frac{\pi}{2}$$
$$= \left(\frac{\pi}{2}\right)^2 (1) + \pi(0)$$
$$= \left(\frac{\pi}{2}\right)^2 \approx 2.5.$$

Trial Problems 19.3

Before you begin the section exercises, warm up with these problems. Complete answers are included in the answer key.

Find the derivative of each of the following.

1. $y = \sin 2x$
2. $y = \cos 2x$
3. $y = \tan 2x$
4. $y = \cot 2x$
5. $y = \sec 2x$

Exercises 19.3

Find the derivative of each of the following.

1. $y = \sin(8x - 3)$
2. $y = 2 \sin 3x$
3. $y = \cos \dfrac{x}{2}$
4. $y = 4 \tan 5x$
5. $y = \dfrac{1}{2} \cos x^2$
6. $y = \cot(x^3 - 2x)$
7. $y = \cos 3x^2$
8. $y = \cos^2 3x$
9. $y = \sin 3^{-2x}$
10. $y = x \csc \dfrac{1}{x}$
11. $y = x^2 \sec^3 4x$
12. $y = \sin(2x + 3)^4$

13. $y = e^x \tan x$

14. $y = \ln \sin x$

15. $y = 3 \csc(3x - 2)^2$

16. $y = 3 \cot^{2/3}(x - 5)$

17. $y = 2 \sin^4 \sqrt{3x + 2}$

18. $y = \sec \sqrt{1 - x}$

19. $y = \sin x \cos x$

20. $y = (2x + 5)^2 \sin^2(3x)$

21. $y = 2 \sin(2x - 1)$

22. $f(x) = 6 \cos(3 - x)$

23. $g(x) = \sin x^2 \cos 2x$

24. $g(x) = 6 \sin 3x \cos 2x^2$

25. $y = \dfrac{\sin 3x}{x}$

26. $y = \dfrac{2x}{\sin 3x}$

27. $y = \sqrt{\sec 3x}$

28. $y = \sec^3 x$

29. $f(x) = \dfrac{\cos 4x}{1 + \cot 3x}$

30. $f(x) = \dfrac{\tan^2 2x}{2 + \sin x}$

31. $r(x) = \tan 2x - \sec 2x$

32. $s(x) = \csc 2x - 2 \cot 2x$

Find the second derivative of each of the following.

33. $y = \sin x$

34. $y = \cos x$

35. $y = \tan x$

36. $y = \sec x$

37. $y = \sin x - x \cos x$

38. $f(x) = x \sin x$

39. Find the slope of the tangent line to the curve $y = 3 \sin 2x$ at $x = \dfrac{\pi}{8}$.

40. Find the slope of the tangent line to the curve $y = 2 \cot 3x$ at $x = \dfrac{\pi}{12}$.

41. Find the slope of the tangent line to the curve $y = e^{-x} \sin x$ at $x = 0$.

42. A weight hanging from a spring vibrates in simple harmonic motion, given by $y = 5.5 \sin(1.5t)$. Find $\dfrac{dy}{dt}$.

43. A linear speed of a point on the tread of an automobile is given by the formula $v = 32\pi \sin 8\pi t$. Find $\dfrac{dv}{dt}$.

44. Use differentials to approximate the change in $\cot x$ if x changes from $45°$ to $46°$.

45. The current in a certain electric circuit at any time t is given by $i = 10 \cos\left(120t + \dfrac{\pi}{6}\right)$. Find $\dfrac{di}{dt}$.

46. In a nuclear physics experiment, the impact parameter p of a moving particle is a function of the scattering angle θ and is given by

$$p = \dfrac{b}{2} \cot \dfrac{\theta}{2},$$

where b is a constant. $\dfrac{dp}{d\theta}$ gives an expression that indicates how p changes as θ changes. Find $\dfrac{dp}{d\theta}$.

19 • 4 Derivatives of Inverse Trigonometric Functions

Inverse Trigonometric Functions

In our earlier study of trigonometry we identified. The functions $f(x) = \sin x$ and $g(x) = \sin^{-1} x$ are inverses of each other with the restrictions that in $f(x) = \sin x$, x is an angle between $-\dfrac{\pi}{2}$ and $\dfrac{\pi}{2}$, and in $g(x) = \sin^{-1} x$, x is a number between -1 and 1.

In the form $y = \sin^{-1}x$, we can write the inverse function as $\sin y = x$. That is, $y = \sin^{-1}x$ implies $x = \sin y$. To find the derivative of $y = \sin^{-1}x$, we use the form $x = \sin y$ and differentiate implicitly.

With $x = \sin y$, we have

$$D_x x = D_x \sin y$$
$$1 = \cos y (y')$$
$$y' = \frac{1}{\cos y}$$
$$= \frac{1}{\sqrt{1 - \sin^2 y}} \quad \text{(Since } \sin^2 y + \cos^2 y = 1\text{)}$$
$$= \frac{1}{\sqrt{1 - x^2}}. \quad \text{(Since } x = \sin y\text{)}$$

Theorem 19.13 If $y = \sin^{-1}x$, $y' = \dfrac{1}{\sqrt{1 - x^2}}$, for $-1 < x < 1$.

• **Example 19.27:** $y = 3 \sin^{-1}x$.

Solution: $y' = \dfrac{3}{\sqrt{1 - x^2}}$.

Theorem 19.14 If $y = \sin^{-1}u$, where $u = f(x)$, then $y' = \dfrac{1}{\sqrt{1 - u^2}} D_x u$.

• **Example 19.28:** $f(x) = \sin^{-1} x^2$

Solution: $f'(x) = \dfrac{1}{\sqrt{1 - (x^2)^2}}(2x) \quad \text{(Chain rule)}$
$$= \frac{2x}{\sqrt{1 - x^4}}.$$

In a similar manner, we can obtain the derivatives of the other inverse trigonometric functions. We state these as theorems and give examples of each.

Theorem 19.15 If $y = \cos^{-1}u$, where $u = f(x)$, then $y' = -\dfrac{1}{\sqrt{1 - u^2}} \cdot D_x u$, provided $-1 < x < 1$.

Section 19.4 Derivatives of Inverse Trigonometric Functions

- **Example 19.29:** $y = 3 \cos^{-1} \sqrt{x}$.
$$\left(u = x \text{ and } D_x u = \frac{1}{2} x^{-1/2}. \right)$$

Solution: $y' = (3) \dfrac{-1}{\sqrt{1 - \sqrt{x}^2}} \left(\dfrac{1}{2} x^{-1/2} \right)$ (Chain rule)

$$= -\frac{3}{2\sqrt{x}\sqrt{1-x}}.$$

- **Example 19.30:** Find $\dfrac{dy}{dx}$ if $y = \sin^{-1} 3x - \cos^{-1} 3x$.

Solution: $\dfrac{dy}{dx} = \dfrac{1(3)}{\sqrt{1 - 9x^2}} - \dfrac{-1(3)}{\sqrt{1 - 9x^2}}$

$$= \frac{3}{\sqrt{1 - 9x^2}} + \frac{3}{\sqrt{1 - 9x^2}}$$

$$= \frac{6}{\sqrt{1 - 9x^2}}.$$

Theorem 19.16 $D_x \tan^{-1} u = \dfrac{1}{1 + u^2} D_x u$, where $u = f(x)$.

- **Example 19.31:** $f(x) = \tan^{-1} e^{2x}$.
$(u = e^{2x}$ and $D_x u = 2e^{2x}.)$

Solution: $f'(x) = \dfrac{1}{1 + (e^{2x})^2}(2e^{2x})$ (Chain rule)

$$= \frac{2e^{2x}}{1 + e^{4x}}.$$

Theorem 19.17 If $y = \sec^{-1} u$, where $u = f(x)$, then $y' = \dfrac{1}{u\sqrt{u^2 - 1}} D_x u$, provided $|u| > 1$.

• **Example 19.32:** $y = \sec^{-1} x^2$.
($u = x^2$.)

Solution: $y' = \dfrac{1}{x^2\sqrt{(x^2)^2 - 1}}(2x)$ (Chain rule)

$= \dfrac{2}{x\sqrt{x^4 - 1}}.$

• **Example 19.33:** Find the slope of the curve $y = 3x \tan^{-1}(2x)$, at $x = \dfrac{1}{2}$.

Solution: Using the product rule, we obtain:

$m = y' = 3x \cdot \dfrac{1}{1 + (2x)^2} D_x(2x) + \tan^{-1}(2x)\, D_x(3x)$

$= \dfrac{3x}{1 + 4x^2}(2) + \tan^{-1}(2x)(3)$

$= \dfrac{6x}{1 + 4x^2} + 3\tan^{-1}(2x).$

At $x = \dfrac{1}{2}$,

$m = \dfrac{6\left(\dfrac{1}{2}\right)}{1 + 4\left(\dfrac{1}{2}\right)^2} + 3\tan^{-1}\left(2 \cdot \dfrac{1}{2}\right)$

$= \dfrac{3}{2} + 3\tan^{-1} 1$

$= \dfrac{3}{2} + 3\left(\dfrac{\pi}{4}\right)$

$\approx 3.86.$

• **Example 19.34:** A balloon rises vertically from a point 1000 ft from an observer. If the balloon is rising at the rate of 15 ft/sec, at what rate is the angle of elevation changing when the balloon is 200 ft high?

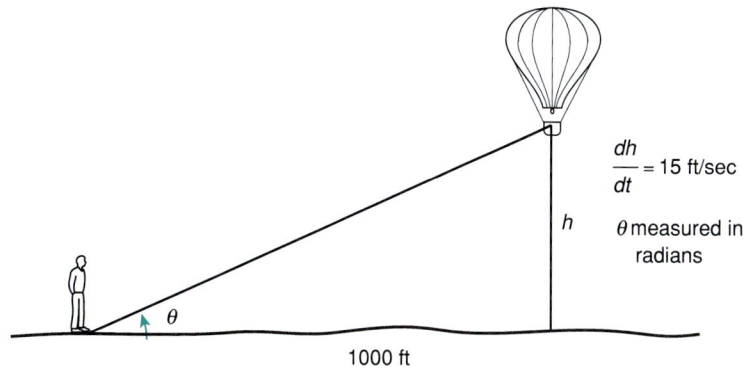

Figure 19.1

Solution: Referring to figure 19.1, θ can be expressed as

$$\theta = \tan^{-1}\left(\frac{h}{1000}\right). \quad (\theta \text{ measured in radians})$$

We need to find $\dfrac{d\theta}{dt}$ when $h = 200$. Taking the derivative of θ, and using the chain rule, we obtain

$$\frac{d\theta}{dt} = \frac{d}{dh}\left[\tan^{-1}\left(\frac{h}{1000}\right)\right]\frac{dh}{dt}$$

$$= \frac{1}{1 + \left(\dfrac{h}{1000}\right)^2} \cdot \frac{1}{1000} \cdot \frac{dh}{dt}.$$

When $h = 200$,

$$\frac{d\theta}{dt} = \frac{1}{1 + \left(\dfrac{200}{1000}\right)^2} \cdot \frac{1}{1000} \cdot 15 \quad \left(\frac{dh}{dt} = 15\right)$$

$$= \frac{1}{1 + \dfrac{1}{25}} \cdot \frac{1}{1000} \cdot 15$$

$$\approx 0.014 \text{ rad/sec.}$$

CHAPTER 19 DERIVATIVES OF TRANSCENDENTAL FUNCTIONS

Trial Problems 19.4

Before you begin the section exercises, warm up with these problems. Complete answers are included in the answer key.

Find the derivative of each of the following.

1. $y = \sin^{-1} 2x$
2. $y = \cos^{-1} 2x$
3. $y = \tan^{-1} 2x$
4. $y = \sec^{-1} 2x$
5. $y = \sin^{-1} x + \cos^{-1} x$

Exercises 19.4

Find the first derivative of each of the following.

1. $y = \sin^{-1} \sqrt{x}$
2. $y = \sin^{-1}\left(\dfrac{x}{3}\right)$
3. $f(x) = \tan^{-1}(2x + 1)$
4. $f(x) = \tan^{-1}(x^2)$
5. $g(x) = 2x \sin^{-1}(3x - 1)$
6. $g(x) = \dfrac{1}{x} \cos^{-1}(x^2)$
7. $y = \sec^{-1}\sqrt{x^2 + 1}$
8. $y = x^2 \sec^{-1} 3x$
9. $h(x) = x^2 \tan^{-1}(x^2)$
10. $h(x) = e^x \sec^{-1}(3^x)$
11. $y = (\cos^{-1}(3x) + 1)^2$
12. $y = \cos^{-1} \dfrac{2x - 3}{3x + 1}$
13. $f(x) = \dfrac{1}{\cos^{-1} x}$
14. $f(x) = \dfrac{e^x}{\sin^{-1}(5x)}$
15. $g(x) = \cos^{-1}(3x + 1)^2$
16. $g(x) = (\tan^{-1} 3x)^2$
17. $y = e^{\sin^{-1}(x^3)}$
18. $y = \sin^{-1}(\ln x)$
19. $y = \cos^{-1}(\cos e^x)$
20. $y = \sqrt{\sec^{-1} 2x}$
21. $x^2 + x \sin^{-1} y = ye^x$
22. $\ln(x + y) = \tan^{-1} xy$

Find the second derivative of each of the following.

23. $y = \sin^{-1} x$
24. $y = \cos^{-1} x$
25. $y = \tan^{-1} x$
26. $y = \sec^{-1} x$

27. Use differentials to approximate the change in $y = \sin^{-1} x$ if x changes from 0.20 to 0.21.

28. Find the equation of the tangent line to the graph of $y = \sin^{-1}(x - 1)$ at the point $P\left(\dfrac{3}{2}, \dfrac{\pi}{6}\right)$.

29. When an arc generator is applied to a series RLC circuit, the voltage and current are out of phase by some angle ϕ. The magnitude of ϕ is described by

$$\phi = \tan^{-1} \dfrac{X}{R},$$

where X is the reactance of the circuit and R is the resistance.

a. Find $\dfrac{d\phi}{dR}$.

b. Find $\dfrac{d\phi}{dX}$.

30. The sides opposite and adjacent to an angle θ of a right triangle are measured as 10 feet and 7 feet, respectively, with a possible error of $\frac{1}{2}$ inch in the 10-foot measurement. Use the differential of an inverse trigonometric function to approximate the error in the calculated value of θ.

31. An airplane 5 miles high and traveling at a speed of 500 miles per hour is flying directly away from an observer on the ground. Use inverse trigonometric functions to find the rate at which the angle of elevation is changing when the airplane is over a point 2 miles from the observer.

19·5 Summary of Terms, Rules, and Theorems

Terms

Common Exponential Function (p. 797)   Inverse Trigonometric Functions (p. 807)   Natural Exponential Function (p. 797)

Rules

$y = a^x$ if and only if $x = \log_a y$. (19.2)

$y = e^x$ if and only if $x = \ln y$. (19.2)

$a^x = e^{x \ln a}$. (19.2)

Theorems

If $y = \ln x$, then $y' = \dfrac{1}{x}$. (19.1)

If $y = \ln u$, where $u = f(x)$, then $y' = \dfrac{1}{u} D_x u$. (19.1)

If $y = e^x$, then $y' = e^x$. (19.2)

If $y = e^u$, where $u = f(x)$, then $y' = e^u \dfrac{du}{dx}$. (19.2)

$\lim\limits_{x \to 0} \dfrac{\sin x}{x} = 1$. (19.3)

$\lim\limits_{x \to 0} \dfrac{\cos x - 1}{x} = 0$. (19.3)

If $y = \sin u$, where $u = f(x)$, then $y' = \cos u \, D_x u$. (19.3)

If $y = \cos u$, where $u = f(x)$, then $y' = -\sin u \, D_x u$. (19.3)

If $y = \tan u$, where $u = f(x)$, then $y' = \sec^2 u \, D_x u$. (19.3)

If $y = \sec u$, where $u = f(x)$, then $y' = \sec u \tan u \, D_x u$.   (19.3)

If $y = \cot u$, where $u = f(x)$, then $y' = -\csc^2 u \, D_x u$.   (19.3)

If $y = \csc u$, where $u = f(x)$, then $y' = -\csc u \cot u \, D_x u$.   (19.3)

If $y = \sin^{-1} x$, $y' = \dfrac{1}{\sqrt{1-x^2}}$,   for $-1 < x < 1$.   (19.4)

If $y = \sin^{-1} u$, where $u = f(x)$, then $y' = \dfrac{1}{\sqrt{1-u^2}} D_x u$.   (19.4)

If $y = \cos^{-1} u$, where $u = f(x)$, then $y' = -\dfrac{1}{\sqrt{1-u^2}} D_x u$, provided $-1 < u < 1$.   (19.4)

If $y = \tan^{-1} u$, then $y' = \dfrac{1}{1+u^2} D_x u$, where $u = f(x)$.   (19.4)

If $y = \sec^{-1} u$, where $u = f(x)$, then $y' = \dfrac{1}{u\sqrt{u^2-1}} D_x u$, provided $|u| > 1$.   (19.4)

19·6 Chapter 19 Review Exercises

Find the derivative of each of the following.

1. $y = 2 \ln(x^2 - 1)$
2. $y = \ln(1 - 3x)^3$
3. $f(x) = (e^{x-2})^3$
4. $g(x) = \ln(2 + \sin x^2)$
5. $y = [\ln(2 + \sin x)]^2$
6. $y = e^{\cos x}$
7. $g(x) = \dfrac{e^{2x}}{x^2 + 1}$
8. $h(x) = \dfrac{\ln x}{3x - 1}$
9. $y = \tan\sqrt{2 - x}$
10. $y = 2\sin^3\sqrt{x}$
11. $y = \sin^{-1}(\cos x)$
12. $y = \sin(\tan^{-1} x)$
13. $y = 2x \ln(2x)$
14. $y = \dfrac{1}{\ln(3x^2 + 1)}$
15. $f(x) = \dfrac{\ln x}{e^{2x} + 1}$
16. $g(x) = e^{\ln(x^2 + 3)}$
17. $h(x) = x^2 e^{-x^2}$
18. $r(x) = (1 + \sqrt{x})^e$
19. $F(x) = \ln e^{\sqrt{x}}$
20. $y = \cos\sqrt{3x}$
21. $y = (\sec x + \tan x)^3$
22. $G(x) = x^2 \tan 2x$
23. $y = x^2 \sec^{-1} x^2$
24. $f(x) = \tan^{-1}(\ln x)$
25. $r(x) = e^x \sin^{-1} 5x$
26. $y = \dfrac{\sin(2x)}{2x}$
27. $y = e^{\sin 3x}$
28. $y = \ln(\cos 2x)$
29. $f(x) = \dfrac{1}{2x + \sec^2 x}$
30. $g(x) = \sin\left(\dfrac{1}{x}\right) + \dfrac{1}{\sin x}$

31. $h(x) = (\cos x)^e$

32. $y = \ln \sin^2(2x)$

33. $G(x) = e^{\tan^{-1} x}$

34. $w(x) = \sec(2x) \tan(2x)$

35. $y = \dfrac{1 - x^2}{\cos^{-1} x}$

36. $y = \sin^{-1} \sqrt{1 - x^2}$

37. $y = (\tan x + \tan^{-1} x)^2$

38. $f(x) = \sin^3(e^{-2x})$

Differentiate implicitly.

39. $e^{-y} + \ln(xy) + x = 5$

40. $\sin x + \cos y = e^{xy}$

41. $1 + xy = e^{xy}$

42. $\ln(x + y) + x^2 - y^3 = 1$

Find the second derivative of each of the following.

43. $y = x \cos x$

44. $y = e^{-x} \ln x$

45. $y = e^{-x} \ln x$

46. $y = e^x \sin^{-1} x$

47. $f(x) = \ln(\cos x)$

48. $f(x) = e^x \tan^{-1} x$

49. Find the slope of the tangent line to the graph of $y = xe^{1/x^3} + \ln|2 - x^2|$ at the point $P(1, e)$.

50. Find the points on the graph of $y = \sin^{-1} 3x$ at which the tangent line is parallel to the line through $A(2, -3)$ and $B(4, 7)$.

51. The position of a moving point on a coordinate line is given by $s(t) = a \sin kt + b \cos kt$, where a, b, and k are constants. Find $\dfrac{ds}{dt}$.

52. Use differentials to approximate $\tan^{-1}(0.98)$. (*Hint:* Let $y = \tan^{-1} x$. Then $x = 1$ and $dx = -0.02$.)

53. According to *Newton's law of cooling*, the temperature $f(t)$ of a certain substance at any time t is given by $f(t) = T + (1 - t)e^{-kt}$, where T is the temperature of the surrounding medium and k is a positive constant. Find $f'(t)$.

54. For simple harmonic motion, the formula $x = b \sin 2kt$ describes the distance traveled after a certain period of time. Find $\dfrac{dx}{dt}$ and $\dfrac{d^2x}{dt^2}$ (first and second derivatives) (b and k are constants).

55. The work exerted in an electric circuit is given by $W = 10 \cos 2t$, where t denotes time. If power P is defined as the rate of change of work with respect to time, find P as a function of t.

56. Find the equation of the tangent line to the curve $y = 4 \cos^2 x^2$ at $x = 1$.

57. A right triangle with vertical leg of 10 cm, horizontal leg of x cm, and hypotenuse of y cm has angle θ between sides x and y. x, y, and θ are changing. Express x in terms of θ and find the instantaneous rate of change of x with respect to θ when $\theta = \dfrac{\pi}{4}$.

58. Current in a given circuit at any time t is given by $i = 10e^{-t/100}$. Find the rate of change of the current with respect to time.

59. If inflation makes the dollar worth 5% less each year, then the value of \$100 in t years is $V = 100e^{-0.05t}$. What is the approximate change in the value during the fourth year?

19·7 Chapter 19 Test

Find the derivative of each of the following.

1. $y = \dfrac{1}{3}\ln(2 + 3x)$
2. $f(x) = x\ln(3x)$
3. $g(x) = e^x - e^{-x}$
4. $y = \dfrac{x}{e^x}$
5. $f(x) = \dfrac{1}{2}\sin(2x - 3)$
6. $h(x) = \ln\cos x$
7. $y = \tan x^2$
8. $f(x) = e^x \csc x$
9. $y = \sin^{-1}(x^2)$
10. $g(x) = \tan^{-1}(2x - 1)$
11. $r(x) = \cos^{-1}(\ln x)$
12. $y = \sec^{-1}\sqrt{x^2 - 1}$

Find the second derivative of each of the following.

13. $f(x) = \ln(\sin x)$
14. $y = e^x \cos x$

Differentiate implicitly.

15. $\sin x + \cos y = e^x$
16. $\tan y = e^{xy}$

17. Find the slope of the tangent line to the graph of $y = xe^x + \ln x$ at the point $(1, e)$.

18. The position of a moving point on a coordinate line is given by $s(t) = 2\sin 3t + 3\cos 2t$. Find the velocity $\dfrac{ds}{dt}$, at any time t.

19. Annual output from the Shiney Silver Mine is given by $f(t) = 200e^{-0.05t}$ tons per year. What is the approximate change in output during the fourth year?

20. If work exerted in an electric circuit is given by $W = \tan^{-1} t$, where t denotes time, use differentials to approximate the change in W from $t = 1$ to $t = 2$.

CHAPTER 20
Applications of Derivatives

*T*he applications of differential calculus to mathematics, science, business, the social sciences, and many other areas of technical mathematics, are seemingly endless. Now that we have investigated derivatives of a variety of functions, we study specific applications of the derivative. We see that derivatives assist us in finding equations of tangent and normal lines, in sketching the essential points for the graphs of various functions, in finding maximum and minimum values for functions occurring in many areas of technical mathematics, in finding velocity and acceleration of moving objects, and in applying these concepts and procedures to solving many practical problems.

20·1 Tangent Lines and Normal Lines

Point-Slope Form

In earlier chapters we learned that the equation for a straight line can be written in the **point-slope form** $y - y_1 = m(x - x_1)$, where m is the slope and (x_1, y_1) is a given point on the line. Also, we found that slope can be determined by the first derivative: $m = \dfrac{dy}{dx}$. Thus, the formula for the equation of a line having a given slope and passing through a given point is

POINT-SLOPE EQUATION OF A LINE

$$y - y_1 = \frac{dy_1}{dx_1}(x - x_1).$$

In figure 20.1, C is a curve and T is a line tangent to the curve at point $P(x_1, y_1)$. The slope of T can be found by substituting the values (x_1, y_1) into the first derivative. The equation then can be found by substituting the slope of the line and coordinates of P into the point-slope form.

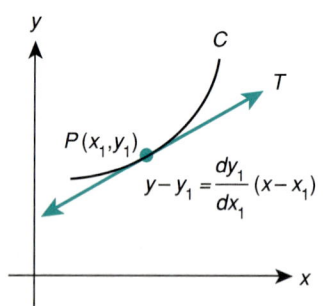

Figure 20.1

• **Example 20.1:** Find the equation of the tangent line to the curve $y = x^3$ at the point $(1,1)$.

Solution: For $y = x^3$, $\dfrac{dy}{dx} = 3x^2$. At the point $(1,1)$ $x = 1$, therefore, $\dfrac{dy}{dx} = 3(1)^2 = 3$, and the slope of the tangent line is 3.

Substituting into the point-slope form, $y - y_1 = \dfrac{dy_1}{dx_1}(x - x_1)$ we obtain $y - 1 = 3(x - 1)$ or $y = 3x - 2$ as the equation of the tangent line. See figure 20.2.

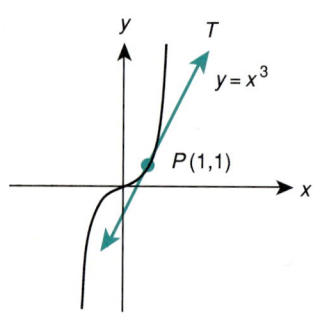

Figure 20.2

• **Example 20.2:** Find a point on the graph of $y = x^2 + 2$ where the tangent line is horizontal.

Solution: If the tangent line is horizontal, then the slope is 0: $\dfrac{dy}{dx} = 0$.

For $y = x^2 + 2$, $\dfrac{dy}{dx} = 2x$. $\dfrac{dy}{dx} = 2x = 0$, when $x = 0$.

Therefore, a horizontal tangent line to the curve $y = x^2 + 2$ occurs at $(0, 2)$. See figure 20.3.

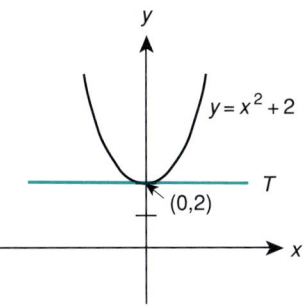
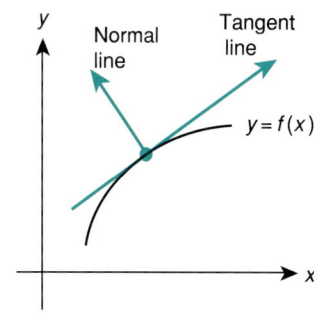

Figure 20.3   Figure 20.4

Normal Line A line perpendicular to the tangent line at the point of tangency is called the **normal line**. See figure 20.4. From precalculus work it was found that perpendicular lines have negative reciprocal slopes. If m_1 is the slope of the tangent line and m_2 is the slope of the normal line, then $m_2 = -\dfrac{1}{m_1}$. Using derivatives, we write

$$\text{slope of normal line} = -\dfrac{1}{f'(x_1)},$$

where $f'(x_1)$ gives the slope of the tangent line.

The equation of the normal line is given by

$$y - y_1 = -\dfrac{1}{f'(x_1)}(x - x_1),$$

where (x_1, y_1) is the point of tangency.

• **Example 20.3:** Find the equation of the normal line to the curve $4x^2 + 9y^2 = 25$ at the point $(2,1)$.

Solution: Differentiating implicitly gives

$$8x + 18yf'(x) = 0 \quad \text{or} \quad f'(x) = -\dfrac{4x}{9y}.$$

Therefore,

$$-\dfrac{1}{f'(x)} = \dfrac{9y}{4x}$$

$$= \dfrac{9(1)}{4(2)} = \dfrac{9}{8} \text{ at the point } (2,1).$$

The equation of the normal line is

$$y - 1 = \dfrac{9}{8}(x - 2) \quad \text{or} \quad 9x - 8y = 10.$$

Trial Problems 20.1

Before you begin the section exercises, warm up with these problems. Complete answers are included in the answer key.

Here are examples of the types of expressions you will need to differentiate in this exercise set. Find the derivative of each of the following.

1. $y = x^4 + 2x^3 - 5x^2 + x - 10$
2. $x^2 + 2x + y^2 = 12$
3. $y = e^{3x}$
4. $y = \ln\sqrt{x}$
5. $y = \tan\dfrac{x}{3}$

Exercises 20.1

1. Find the equation of the line tangent to the curve $y^2 - 2x - 4y - 1 = 0$ at $(-2,1)$.

2. Find the equation of the line normal to the curve $xy + 2x - 5y - 2 = 0$ at $(3,2)$.

3. Find the equation of the tangent line and of the normal line to the graph of $f(x) = x^4 + x^2 + 1$ at $P(-1,3)$.

4. Find the equation of the tangent line and of the normal line to the curve $x^3 + y + xy = 0$ at the point where $x = -2$.

5. Find the equation of the tangent line and of the normal line to the graph of $f(x) = 8x^2 - 5x + 7$ at $P(-2,49)$.

6. Find the equation of the normal line to the graph of $y = 3x^2 + 4x - 6$ if the normal line is parallel to the line $2x + 5y = 1$. (Recall that the slopes of parallel lines are equal.)

7. Find an equation of the line through the point $P(5,9)$ that is tangent to the graph of $y = x^2$.

8. How many tangent lines can be drawn from the point $(0,0)$ to the curve $xy = 1$?

9. Find the x-value of the point on the graph of $x^2 + xy + y^2 = 12$ where the tangent line is horizontal.

10. How many horizontal tangent lines does the circle $x^2 + y^2 = r^2$ have? Vertical tangent lines?

11. Sketch the graph of $y = |x|$ and determine where the graph has the following. (Be careful!)
    a. A horizontal tangent line
    b. A vertical tangent line
    c. No tangent line

12. Find the equation of the tangent line to the curve at the given point for each of the following.
    a. $y = e^x$, $(0,1)$
    b. $y = e^{2x}$, $(0,1)$
    c. $y = e^{-x}$, $(0,1)$
    d. $y = e^{-2x}$, $(0,1)$

13. Find the equation of the tangent line to the curve at the given point for each of the following.
    a. $y = \ln x^2$, $(1,0)$
    b. $y = \ln x^{3/2}$, $(1,0)$
    c. $y = \ln x$, $(1,0)$

14. Find the equation of the tangent line to the curve at the given point for each of the following.
    a. $y = \sin x$, $(0,0)$
    b. $y = \sin 3x$, $(0,0)$
    c. $y = \sin\dfrac{x}{2}$, $(0,0)$
    d. $y = \tan x$, $(0,0)$
    e. $y = \cos x$, $(0,1)$

15. In an electric field, the lines of force are perpendicular to the curves of equal electric potential. The equal electric potential of a certain curve is given by $y = 6x^{2/3}$. Find the equation of the line along which the force acts on a particle located on this curve where $x = 8$.

16. An object moves along a path given by $s = 2t^2 + 4t^4$. At $t = 1$ sec the object encounters a second object moving along a path that is normal to the path of the first object. Find the equation for the path of the second object.

20 • 2 The First Derivative Test

Increasing-Decreasing Curves

Future study of the calculus and its applications requires us to know if a curve is increasing or decreasing on an interval. By sketching the graph of a curve we can see the intervals where it is increasing or decreasing. The first derivative will be very useful for determining whether a **curve** is **increasing** or **decreasing** *at particular values of x.*

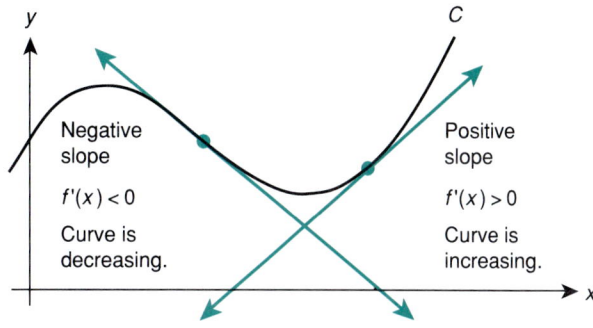

Figure 20.5

Look at figure 20.5. Tangent lines have been drawn to the curve C at particular points. Visualize tangent lines being drawn at several points on the curve. Two observations can be made.

1. The curve is increasing when $f'(x)$ is positive.

2. The curve is decreasing when $f'(x)$ is negative.

• **Example 20.4:** If $f(x) = x^2 + 2x - 3$, find the interval on which f is increasing and the interval on which f is decreasing.

Solution: Sketch the graph of $y = x^2 + 2x - 3$.

$$f'(x) = 2x + 2 = 2(x + 1) = 0 \text{ when } x = -1.$$

$f'(x) = 0$ means that the slope of the tangent line is 0. Lines with slope 0 are horizontal lines. Therefore, there is a horizontal tangent line at $x = -1$. This is illustrated in figure 20.6.

For the interval $(-\infty, -1)$ $f'(x) = 2x + 2$ is negative. So, the slopes of all tangent lines are negative. Therefore, the curve is decreasing on this interval.

For the interval $(-1, \infty)$ $f'(x) = 2x + 2$ is positive. So, the slopes of all tangent lines are positive. Therefore, the curve is increasing on this interval.

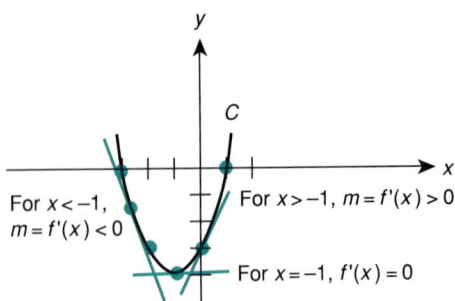

Figure 20.6

We can put this information in tabular form.

Interval	$2(x+1)$	$f'(x)$	f
$(-\infty, -1)$	−	−	Decreasing
$x = -1$	0	0	Level
$(-1, \infty)$	+	+	Increasing

Critical Number

Minimum Value

Maximum Value

In example 20.4, $f'(x) = 0$ at the point $x = -1$. A value of x where $f'(x) = 0$ or $f'(x)$ is not defined is called a **critical number**. Since the curve is decreasing ($f'(x) < 0$) on the left side of $x = -1$ and increasing ($f'(x) > 0$) on the right side of $x = -1$, we conclude that the curve has a **minimum value** at $x = -1$.

Knowing whether a curve increases or decreases near a critical number tells us whether a function has a *minimum value* or a *maximum value* at the critical number. The procedure we use to determine the maximum and minimum values of a function is called the **first derivative test**. We state it as follows.

THE FIRST DERIVATIVE TEST

Suppose $x = c$ is a critical number.

1. $f(x)$ has a maximum value at $x = c$ if $f'(c) = 0$, and $f'(x)$ changes from + to − near $x = c$.

2. $f(x)$ has a minimum value at $x = c$ if $f'(c) = 0$, and $f'(x)$ changes from − to + near $x = c$.

• **Example 20.5:** Find the maximum and minimum value of the function $y = x^3 - 27x + 2$.

Solution: a. $f'(x) = 3x^2 - 27$
$= 3(x^2 - 9)$
$= 3(x - 3)(x + 3)$.

Since $f'(x) = 0$ when $x = +3$ and $x = -3$, $+3$ and -3 are critical numbers, and horizontal tangent lines exist at these values of x.

b. Check $f'(x)$ near $x = -3$ for a change in signs (check a number to the left of -3 first, and a number to the right of -3).
At $x = -4$,
$$f'(x) = 3(-4)^2 - 27 = 3(16) - 27$$
$$= 48 - 27 = 21$$
$$f'(x) > 0 \ (+). \quad \text{(Curve is increasing)}$$

At $x = -2$,
$$f'(x) = 3(-2)^2 - 27 = 3(4) - 27$$
$$= 12 - 27 = -15$$
$$f'(x) < 0 \ (-). \quad \text{(Curve is decreasing)}$$

Since $f'(x)$ changes from $+$ to $-$ near $x = -3$, $f(x)$ has a maximum value at $x = -3$.

c. Check $f'(x)$ near $x = +3$.
At $x = 2$,
$$f'(x) = 3(2)^2 - 27 = 3(4) - 27$$
$$= 12 - 27 = -15.$$

Therefore,
$$f'(x) < 0. \quad \text{(Curve is decreasing)}$$

At $x = 4$,
$$f'(x) = 3(4)^2 - 27 = 3(16) - 27$$
$$= 48 - 27 = 21.$$

Therefore,
$$f'(x) > 0. \quad \text{(Curve is increasing)}$$

Since $f'(x)$ changes from $-$ to $+$ near $x = +3$, $f(x)$ has a minimum value at $x = +3$.

d. Sketch the graph of $y = x^3 - 27x + 2$. The graph is shown in figure 20.7.

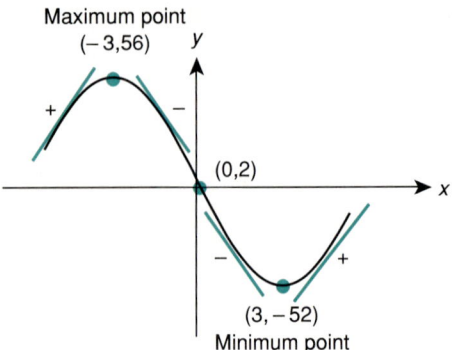

Figure 20.7

NOTE: If $f'(x)$ does not change signs near the critical number $x = c$, then we cannot conclude that $f(x)$ has a maximum or minimum value at $x = c$. More information is needed.

There are many technical and industrial applications of finding minimum and maximum values. In section 20.5 we investigate several of these applications.

• **Example 20.6:** The profit function, $P(x)$, for selling computer memory chips is given by

$$P(x) = 1.45x - \frac{x^2}{20,000} - 500,$$

where x is the number of memory chips. Find the number of chips that will produce a maximum profit.

Solution: **a.** $P'(x) = 1.45 - \dfrac{2x}{20,000} - 0$

$ = 1.45 - \dfrac{x}{10,000}.$

$P'(x) = 0$

when

$$1.45 - \frac{x}{10,000} = 0,$$

or when

$$\frac{x}{10,000} = 1.45$$

$$x = 14,500.$$

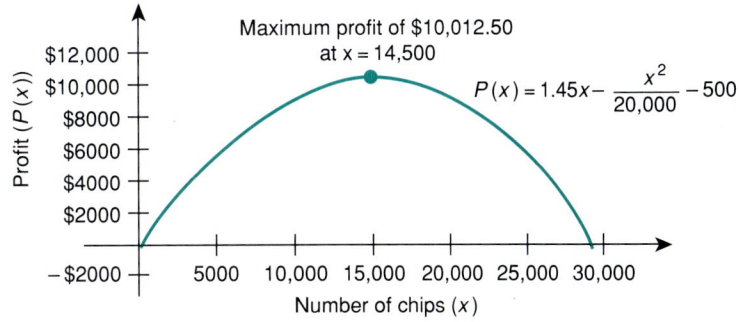

Figure 20.8

b. For

$$x < 14{,}500, \qquad P'(x) > 0. \quad \text{(Check any } x < 14{,}500 \text{ but close to 14,500)}$$

For

$$x = 0, \qquad P'(x) = 0.$$

For

$$x > 14{,}500, \qquad P'(x) < 0. \quad \text{(Check any } x > 14{,}500 \text{ but close to 14,500)}$$

Since $P'(x)$ changes from $+$ to 0 to $-$ near $x = 14{,}500$, $P(x)$ has a maximum value at $x = 14{,}500$.

c. Figure 20.8 is a sketch of the profit curve. For 14,500 chips, the profit is $10,012. (Substituting $x = 14{,}500$ in $P(x)$.)

Trial Problems 20.2

Before you begin the section exercises, warm up with these problems. Complete answers are included in the answer key.

Here are examples of the types of expressions you will need to differentiate in this exercise set. Find the derivative of each of the following.

1. $f(x) = x^3 - \dfrac{5}{2}x^2 + 2$

2. $y = \dfrac{x}{x^2 + 2}$

3. $g(x) = \sin x$

4. $h(t) = t\sqrt{3 - t}$

5. $y = e^{1/2x}$

Exercises 20.2

Find critical numbers, if they exist, and state the intervals on which the function is increasing or decreasing. Test for maximum and minimum values. Sketch the graph of each function.

1. $f(x) = 3x + 4$
2. $f(x) = x^2 - 2x$
3. $f(x) = 4 - 3x$
4. $f(x) = 2x - x^2$
5. $f(x) = \dfrac{x}{x^2 + 4}$
6. $y = \dfrac{1}{x^2 + 1}$
7. $f(x) = \sqrt{x}$
8. $y = \sqrt{4 - x^2}$
9. $f(x) = \sqrt[3]{x - 1}$
10. $f(x) = x\sqrt{3 - x}$
11. $y = x^2 - 4x$
12. $y = x^2 + 6x + 8$
13. $y = x^3 - 9x$
14. $y = x^3 - 3x^2 + 6x - 2$

Test each of the following for maximum and minimum values.

15. $f(x) = \sin x,\ 0 \le x \le 2\pi$
16. $f(x) = \cos x,\ -\dfrac{\pi}{2} \le x \le \dfrac{3\pi}{2}$
17. $y^2 = \sin x,\ y \ge 0,\ 0 \le x \le \pi$
18. $y = xe^x$
19. $f(x) = \dfrac{\ln x}{x}$
20. $f(x) = x \ln x$
21. $y = e^{(x^2 - x)}$

22. Find the intervals on which $f(x) = x^3 - \dfrac{3}{2}x^2$ is increasing or decreasing. Give maximum and minimum points.

23. Show that $y = x^3 - 3x^2 + 3x$ is always increasing.

24. The profit, P, in selling certain transistors is given by
$$P = 2.50x - \dfrac{x^2}{20{,}000} - 5000,\ 0 \le x \le 35{,}000.$$
a. Find intervals on which P is increasing or decreasing.
b. Find the number of transistors that will produce a maximum profit.

25. The level of oxygen in a pond is given by
$$L(t) = \dfrac{t^2 - t + 1}{t^2 + 1},$$
where t is time measured in weeks and $L(t) = 1$ is the normal level of oxygen. When $t = 0$, some organic waste is dumped into the pond and begins to oxidize, changing the oxygen level of the pond.
a. When is the oxygen level lowest?
b. When is the oxygen level highest?

26. A baseball is hit into the air and its height at any time t is given by $h(t) = 96t - 16t^2$. How high will the ball go?

27. The electric power P in watts in a direct current circuit with two resistors of resistance R_1 and R_2 connected in series is
$$P = \dfrac{v\,R_1 R_2}{(R_1 + R_2)^2},$$
where v is the voltage. If v and R_1 are held constant, what resistance R_2 produces the maximum power?

28. The deflection D of a particular beam of length L is given by $D = 2x^4 - 5Lx^3 + 3L^2 x^2$, where x is the distance in feet from one end of the beam. Find the value of x that yields the maximum deflection.

29. The profit equation for advertising by a company is given by $P(x) = 230 + 20x - \dfrac{1}{2}x^2$, where x is the amount in dollars spent by the company. What dollar amount of advertising gives the maximum profit?

30. The equation $E = \dfrac{T}{(x^2 + a^2)^{3/2}}$ gives the electric field intensity on the axis of a uniformly charged ring, where T is the total charge on the ring and a is the radius of the ring. At what value of x is E maximum?

20 • 3 The Second Derivative Test

If we observe the first derivative closely, we can find a pattern that is useful for quickly determining maximum and minimum values of $f(x)$. In figure 20.9 a maximum value of $f(x)$ is given at R. Notice that:

1. At point R, $f'(x) = 0$,
2. To the left of R, $f'(x) > 0$, and
3. To the right of R, $f'(x) < 0$.

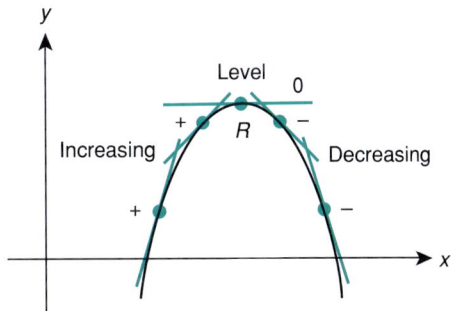

Figure 20.9

Decreasing Function

The slope of $f(x)$ changes from $+$ to 0 to $-$ (moving from left to right). Therefore $f'(x)$ is a **decreasing function**. From our work in section 20.2, it must be true that the derivative of $f'(x)$ is negative or zero. Hence $f''(x) < 0$ when $f'(x) = 0$ tells us that $f(x)$ has a maximum value at R.

In a similar manner, looking at figure 20.10, it can be seen that $f(x)$ has a minimum at S, and that $f'(x)$ changes from $-$ to 0 to $+$ near S. Hence, $f'(x)$ is an **increasing function**; and $f''(x)$ must be positive or $f''(x) > 0$.

Increasing Function

The above information is summarized and stated as *the second derivative test*.

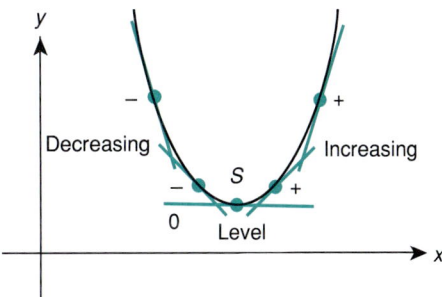

Figure 20.10

THE SECOND DERIVATIVE TEST

Suppose $x = c$ is a critical number, and $f'(c) = 0$.

1. $f(x)$ has a maximum value at $x = c$ if $f''(c) < 0$.
2. $f(x)$ has a minimum value at $x = c$ if $f''(c) > 0$.

NOTE: If $f''(c) = 0$, the second derivative test is not applicable. Use the first derivative test.

• **Example 20.7:** Find maximum and minimum values for $y = x^3 - 48x + 8$.

Solution: First, determine the critical values by setting

$$f'(x) = 0.$$
$$f'(x) = 3x^2 - 48$$
$$= 3(x^2 - 16)$$
$$= 3(x + 4)(x - 4)$$
$$= 0 \text{ at } x = -4 \text{ and } x = 4.$$

Hence, $x = -4$ and $x = 4$ are critical numbers.

Now check the second derivative for maximum and minimum values:

$$f''(x) = 6x$$
$$f''(-4) = 6(-4) = -24.$$

Since $f''(-4) < 0$, $f(x)$ has a *maximum value at* $x = -4$.
Therefore, $f(-4) = 136$ is a relative maximum.

$$f''(4) = 6(4) = 24.$$

Since $f''(4) > 0$, $f(x)$ has a *minimum value at* $x = 4$.
Therefore, $f(4) = -120$ is a relative minimum.

• **Example 20.8:** Find maximum and minimum values for $y = x^3$.

Solution: $f'(x) = 3x^2$
$\quad\quad\quad = 0$

at $x = 0$, the only critical number.

$$f''(0) = 0.$$

Since the second derivative is zero, we cannot determine maximum and minimum values for $f(x) = x^3$ using the second derivative test.

Using the first derivative test, we obtain

Interval	$3x^2$	$f'(x)$	f
$(-\infty, 0)$	+	+	increasing
$(0, \infty)$	+	+	increasing

Therefore, we conclude that $f(x) = x^3$ has no maximum or minimum at $x = 0$. (See figure 20.11.)

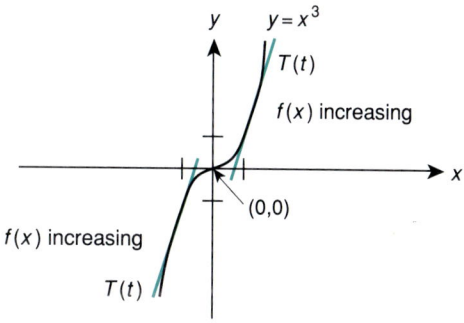

Figure 20.11

In section 20.5, we can use the second derivative test to help solve problems involving applications of the derivatives.

•••••••••

• **Example 20.9:** Find maximum and minimum values for $f(x) = e^x - x$.

Solution: $f'(x) = e^x - 1$
$= 0$ when $x = 0$, the only critical number.

$f''(x) = e^x$, which is always positive. ($e^x > 0$ for all x.)

Therefore, $f''(0) > 0$, and $f(x)$ has a minimum at $x = 0$. (See figure 20.12.)

•••••••••

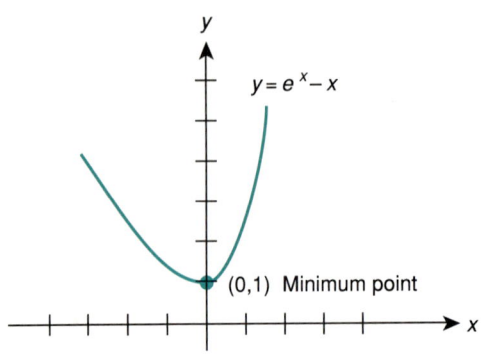

Figure 20.12

• **Example 20.10:** The work-time relation for a solenoid is given by
$$w(t) = -t^3 + t^2 + 5.$$
The power p transferred by the solenoid is determined by $p = \dfrac{dw}{dt}$. Find the time ($t > 0$) of maximum power transfer.

Solution: $w(t) = -t^3 + t^2 + 5$

$$p = \frac{dw}{dt} = -3t^2 + 2t$$

$$p' = -6t + 2$$

$$= 0 \text{ when } t = \frac{1}{3}$$

$p'' = -6$ which is less than zero for all t. Therefore, the maximum power transfer occurs at $t = \dfrac{1}{3}$.

Trial Problems 20.3

Before you begin the section exercises, warm up with these problems. Complete answers are included in the answer key.

In this exercise set you will need to find the second derivative of certain functions. Find the second derivative of each of the following.

1. $y = (x^2 + 3)^2$
2. $y = \sin 2x$
3. $f(x) = x^2 + \dfrac{1}{x}$

4. $g(x) = 10xe^{x/10}$
5. $w = \ln(x - 1)$

Exercises 20.3

Use the second derivative test to find maximum and minimum points for each of the following.

1. $y = x^3 - 4x + 3$
2. $y = 4 + 3x - x^3$
3. $f(x) = \dfrac{x}{x+1}$
4. $f(x) = x^3 - 3x^2 + 2$
5. $y = x^4 - 32x + 48$
6. $f(x) = (x^2 - 1)^2$
7. $f(x) = x^2 - x - 2$
8. $y = -x^3 + 3x^2 - 2$
9. $f(x) = \dfrac{24}{x^2 + 12}$
10. $f(x) = x + \dfrac{4}{x}$
11. $y = x^3 + 1$
12. $f(x) = x^2\sqrt{25 - 4x^2}$
13. $y = \sin x, \ 0 \le x \le 2\pi$
14. $y = \cos x, \ \dfrac{\pi}{2} \le x \le \dfrac{3\pi}{2}$
15. $y = xe^x$
16. $y = x \ln x$
17. $y = \sin x + \cos x, \ 0 \le x \le 2\pi$

18. Does the second derivative indicate a maximum or minimum value for $y = -3x^5 + 5x^3$ at $x = 0$?

19. Why doesn't the graph of $f(x) = x^4 + x^3 - 3x^2 + 1$ have neither a maximum nor a minimum at $x = \dfrac{1}{2}$?

20. Power is defined as the rate of change of work with respect to time. If work is given as
$$w(t) = 3t^2 - t^4 \text{ ft/lb},$$
find the time when the power p is a maximum.

21. The expected income E received from the sale of c computer circuit boards per month is
$$E = 200 \, ce^{-c/100}.$$
How many items should be sold per month so that E is a maximum?

22. The energy dissipated by a resistor at time t is given by $E = \ln(t + 2) - 0.13t + 6$ joules. When is the energy output a maximum?

23. A large highway concrete machine moves along a given path according to
$$s = \sqrt{t} + \cos\sqrt{t},$$
where t is time in minutes. As the concrete load is being used, the machine moves faster. At what time is the distance moved a maximum before a new load of concrete is needed?

24. What is the maximum strength of a new polymer film if the strength varies with temperature according to $s = 4.9 + 6.5T - 0.05\,T^2$ lb?

25. The work done by a machine varies with time in seconds according to
$$w = 12t^2 + 2t^3 - t^4 \text{ ft-lb}.$$
The power output for the machine is determined by
$$p = \dfrac{dw}{dt}.$$
What is the maximum power output, assuming $t \ge 0$?

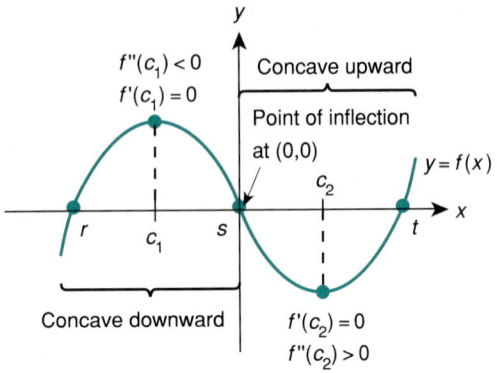

Figure 20.13

20 • 4 Curve Sketching

The first and second derivatives are very useful for obtaining an accurate sketch of a curve without plotting a large number of points. We found that if $f'(c) = 0$, then $x = c$ is a critical number. If we let $f''(x) = 0$ and solve for x, we will determine additional information about the curve for these values. The sketch in figure 20.13 will help in putting together the information we need for sketching curves.

Concave Downward
Concave Upward
Point of Inflection

Notice that the curve is open downward on the interval $(-\infty,0)$ and open upward on the interval $(0,\infty)$. We say that the curve is **concave downward** on $(-\infty,0)$ and **concave upward** on $(0,\infty)$. At the point $x = 0$, the curve changes concavity. We call this point a **point of inflection**. Points of inflection sometimes occur at the value of x for which $f''(x) = 0$. If a curve changes concavity at the point where $f''(x) = 0$, then a point of inflection occurs at that point. Consequently, we include these points on the list of critical points. Example 20.11 helps us in using all of the concepts involving the first and second derivatives of a function.

• **Example 20.11:** Discuss and sketch the curve $y = 2x^2 - x^4$.

Solution: **a.** Determine critical numbers by setting $f'(x) = 0$:

$$f'(x) = 4x - 4x^3$$
$$= 4x(1 - x^2)$$
$$= 4x(1 + x)(1 - x).$$

$x = 0, -1, 1$ are critical numbers.

b. Use the second derivative test:

$$f''(x) = 4 - 12x^2$$
$$f''(0) = 4 - 12(0)^2$$
$$= 4, \text{ which is positive.}$$

f has a minimum at $x = 0$.
$$f''(-1) = 4 - 12(-1)^2$$
$$= -8, \text{ which is negative.}$$
f has a maximum at $x = -1$.
$$f''(1) = 4 - 12(1)^2$$
$$= -8, \text{ which is negative.}$$
f has a maximum at $x = 1$.

c. Check for possible points of inflection by setting $f''(x) = 0$:
$$f''(x) = 4 - 12x^2$$
$$= 4(1 - 3x^2)$$
$$= 0 \text{ when } 1 - 3x^2 = 0.$$

$3x^2 = 1$, $x^2 = \dfrac{1}{3}$, or $x = \pm\sqrt{\dfrac{1}{3}} = \pm\dfrac{\sqrt{3}}{3}$. Possible points of inflection occur at $x = \pm\dfrac{\sqrt{3}}{3}$.

1. For $x < -\dfrac{\sqrt{3}}{3}$, $f''(x) = 4(1 - 3x^2) < 0$ and $f(x)$ is concave downward. For $-\dfrac{\sqrt{3}}{3} < x < \dfrac{\sqrt{3}}{3}$, $f''(x) = 4(1 - 3x^2) > 0$, and $f(x)$ is concave upward. Therefore, at $x = -\dfrac{\sqrt{3}}{3}$ there is a point of inflection.

2. For $-\dfrac{\sqrt{3}}{3} < x < \dfrac{\sqrt{3}}{3}$, $f''(x) = 4(1 - 3x^2) > 0$ and $f(x)$ is concave upward. For $x > \dfrac{\sqrt{3}}{3}$, $f''(x) = 4(1 - 3x^2) < 0$ and $f(x)$ is concave downward. Therefore, at $x = \dfrac{\sqrt{3}}{3}$ there is a point of inflection.

d.

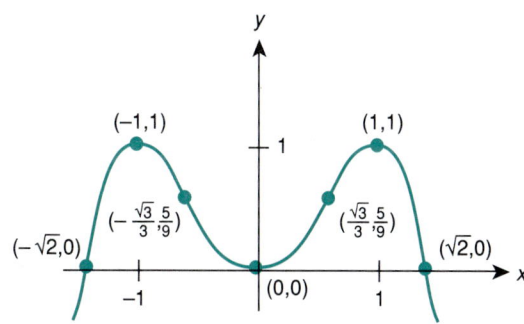

Figure 20.14

e. The intercepts are determined from the original equation $y = 2x^2 - x^4$:

$$y = 2x^2 - x^4$$
$$= x^2(2 - x^2) = x^2(\sqrt{2} + x)(\sqrt{2} - x)$$
$$= 0 \text{ when } x = 0, -\sqrt{2}, \sqrt{2}.$$

The intercepts are $(0,0)$, $(-\sqrt{2},0)$, and $(\sqrt{2},0)$.

The y-coordinates of maximum and minimum points and points of inflection are found by substituting the appropriate x value in the *original equation*.

Maximum: At $x = -1$,

$$y = 2(-1)^2 - (-1)^4$$
$$= 1,$$

and $(-1,1)$ is a maximum point.
At $x = 1$,

$$y = 2(1)^2 - (1)^4$$
$$= 1,$$

and $(1,1)$ is a maximum point.

Minimum: At $x = 0$,

$$y = 2(0)^2 - (0)^4$$
$$= 0,$$

and $(0,0)$ is a minimum point.

Points of inflection: At $x = -\dfrac{\sqrt{3}}{3}$,

$$y = 2\left(-\dfrac{\sqrt{3}}{3}\right)^2 - \left(-\dfrac{\sqrt{3}}{3}\right)^4$$
$$= \dfrac{5}{9},$$

and $\left(-\dfrac{\sqrt{3}}{3}, \dfrac{5}{9}\right)$ is a point of inflection.

At $x = \dfrac{\sqrt{3}}{3}$,

$$y = 2\left(\dfrac{\sqrt{3}}{3}\right)^2 - \left(\dfrac{\sqrt{3}}{3}\right)^4$$
$$= \dfrac{5}{9},$$

and $\left(\dfrac{\sqrt{3}}{3}, \dfrac{5}{9}\right)$ is a point of inflection. All of the critical points are illustrated in figure 20.14.

f. From the sketch we see that
1. $f(x)$ is increasing on $(-\infty,-1)$ and $(0,1)$.
2. $f(x)$ is decreasing on $(-1,0)$ and $(1,\infty)$.
3. $f(x)$ is concave downward on $\left(-\infty, -\dfrac{\sqrt{3}}{3}\right)$ and $\left(\dfrac{\sqrt{3}}{3}, \infty\right)$.
4. $f(x)$ is concave upward on $\left(-\dfrac{\sqrt{3}}{3}, \dfrac{\sqrt{3}}{3}\right)$.

••••••••••

• **Example 20.12:** Discuss and sketch the curve $y = \dfrac{x^3}{6} - x^2 + \dfrac{3x}{2} + 1$.

Solution: **a.** $f'(x) = \dfrac{x^2}{2} - 2x + \dfrac{3}{2}$

$= \dfrac{1}{2}(x^2 - 4x + 3)$ $\quad$ (Factoring out $\dfrac{1}{2}$)

$= \dfrac{1}{2}(x-1)(x-3)$.

$x = 1, 3$ are critical numbers.

b. $f''(x) = x - 2$ $\quad$ (Using the second derivative test)
$f''(1) = 1 - 2$
$\quad\quad = -1$, which is negative.

Therefore, f has a maximum at $x = 1$.

$f''(3) = 3 - 2$
$\quad\quad = 1$, which is positive.

Therefore, f has a minimum at $x = 3$.

c. $f''(x) = x - 2$
$\quad\quad\quad = 0 \quad$ at $x = 2$.

For $x < 2, f''(x) < 0$ and $f(x)$ is concave downward.
For $x > 2, f''(x) > 0$ and $f(x)$ is concave upward and at $x = 2$, there is a point of inflection.

d. The above results are illustrated in figure 20.15.

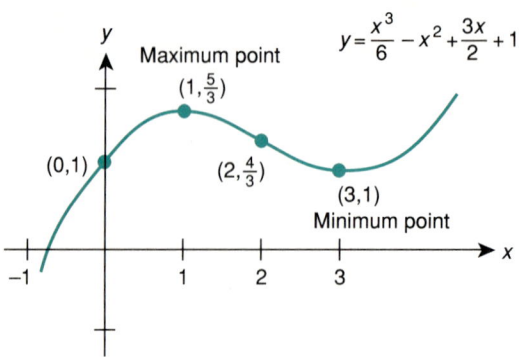

Figure 20.15

NOTE: The y-intercept is found by setting $x = 0$. Thus, the point $(0,1)$ is convenient for sketching the curve.

e. 1. $f(x)$ is increasing on $(-\infty,1)$ and $(3,\infty)$.
   2. $f(x)$ is decreasing on $(1,3)$.
   3. $f(x)$ is concave downward on $(-\infty,2)$.
   4. $f(x)$ is concave upward on $(2,\infty)$.

• **Example 20.13:** Discuss and sketch the curve $y = x^3$.

Solution: a. $f'(x) = 3x^2$
$= 0$ when $x = 0$.

b. $f''(x) = 6x$
$= 0$ when $x = 0$.

The only critical number is $x = 0$. $f''(0) = 0$ tells us nothing about maximum or minimum points.

NOTE: The first derivative test could be used, but it proves to be more efficient to check for concavity in this example.

c. For $x < 0, f''(x) < 0$ and for $x > 0, f''(x) > 0$. $f(x)$ changes concavity and at $x = 0$ there is a point of inflection.

d. Sketch the curve. (See figure 20.16.)

e. $f(x)$ is increasing on $(-\infty,\infty)$, $f(x)$ is concave downward on $(-\infty,0)$, $f(x)$ is concave upward on $(0,\infty)$, and there are no maximum points and no minimum points.

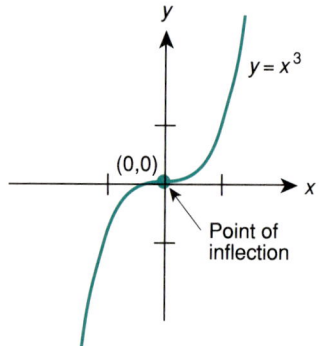

Figure 20.16

Trial Problems 20.4

Before you begin the section exercises, warm up with these problems. Complete answers are included in the answer key.

In this exercise set you will need to find the first and second derivatives of various functions. Find the first and second derivatives of each of the following.

1. $h(t) = t^3 - \dfrac{5}{2}t^2$
2. $f(x) = x^4 + 1$
3. $y = x^{1/3} + 1$
4. $r(t) = 8t - t^3$
5. $g(x) = \dfrac{x}{1 + x^2}$

Exercises 20.4

Sketch the graph of each function and label maximum and minimum points and points of inflection. Indicate intervals of concavity.

1. $f(x) = x^2 + 3$
2. $f(x) = x^3 - x^2 + 2$
3. $y = 3x - x^3$
4. $y = x^3 - 3x^2 + 3x - 3$
5. $f(x) = x^4 + 4x^3$
6. $f(x) = x^4 - 3x^3$
7. $y = x^5 + 1$
8. $y = x^5 - 5x$
9. $y = \dfrac{x}{x^2 + 1}$
10. $y = \dfrac{x^2}{x^2 + 3}$
11. $y = x^4 - 4x^3 + 16x$
12. $y = x^4 - 4x^3 + 16x - 16$
13. $f(x) = x^{1/3}$
14. $f(x) = x^{2/3}$
15. $f(x) = x^3 + 3x^2$
16. $f(x) = \dfrac{1}{3}x^3 - \dfrac{3}{2}x^2 + 2x + 1$
17. $y = x^{1/3} + 2x$
18. $y = x^{2/3} + 1$

19. Consider any quadratic function.
    a. Does the graph have any points of inflection?
    b. Under what condition is the graph always concave upward?
    c. Under what condition is the graph always concave downward?
    d. Does the graph always cross the x-axis?

20. Current in a particular circuit is given by $i = t^2 - t^3$. Sketch the graph for values of t in the interval $[0,3]$.

21. The stress S in a cantilever beam is given by $S = 12d - d^3$, where d is the distance from the fixed end. Sketch the graph of S for distances between 0 and 4 feet inclusive.

22. The production rate of a small milling machine varies with time according to $r(t) = 6t^2 - t^3$. Sketch the rate-time curve for $t \geq 0$.

23. The power output for a machine is given by $p(t) = 12t + 3t^2 - 2t^3$. Sketch the curve for the power out assuming $t \geq 0$.

24. The pressure in a hydraulic hose is found to vary in time in accordance with

$$p = t^3 - 6t^2 \text{ psi.}$$

Sketch the pressure-time curve for $t \geq 0$.

20·5 Maxima and Minima Problems

Relative Maxima
Relative Minima

When we find the *x*-values that give us maximum and minimum points, the *y*-values of these points give us **maxima** and/or **minima** for the function. Since a function can have two or more maxima and/or minima, they are referred to as **relative maxima** and **relative minima**.

In our work we focus our attention on the many practical situations in which the determination of maximum or minimum values is useful.

• **Example 20.14:** A rectangular picture has a perimeter of 60 in. Find the dimensions that will give the maximum area.

Solution: 1. We wish to maximize the area A, which is given by $A = \ell w$. The perimeter P is given by $P = 2\ell + 2w$.

$$P = 2\ell + 2w = 60$$
$$\ell = \frac{60 - 2w}{2}$$

or

$$\ell = 30 - w$$
$$A = \ell w = (30 - w)w$$
$$= 30w - w^2$$

2. $A' = 30 - 2w \quad \left(\text{or } \dfrac{dA}{dw} = 30 - 2w\right)$

$ = 0$ when $w = 15$.

$A'' = -2$, which is always negative. Therefore, $w = 15$ gives a maximum value for A.

3. Since $\ell = 30 - w$ and $w = 15$, $\ell = 15$ also. The dimensions of $\ell = 15$ and $w = 15$ (a square) give the maximum area.

• **Example 20.15:** Find two positive numbers whose sum is 20 and whose product is as large as possible.

Solution: 1. Let one number be x. Then the second number is $20 - x$.

2. $P = x(20 - x) \quad$ (Product)
$ = 20x - x^2$
$P' = 20 - 2x$
$ = 0$ when $x = 10$.

$P'' = -2$, which is always negative. Therefore, P will be maximum when $x = 10$.

3. The two numbers are
$$x = 10, \quad \text{and} \quad 20 - x = 10.$$
Therefore, the sum $= 20$ and product $= 100$.

• **Example 20.16:** The power lost in a resistor, R, is given by the equation
$$P = \frac{100R}{(R + 0.2)^2}.$$
Find the value of R for which the power dissipated in R is a maximum.

Solution: 1. $\dfrac{dP}{dR} = \dfrac{(R + 0.2)^2(100) - 100R(2)(R + 0.2)}{(R + 0.2)^4}$ (Using the quotient rule)

$= \dfrac{100(R + 0.2)[R + 0.2 - 2R]}{(R + 0.2)^4}$

$= \dfrac{100(R + 0.2)(0.2 - R)}{(R + 0.2)^4}$

$= \dfrac{100(0.2 - R)}{(R + 0.2)^3}$

$= 0$ when $R = 0.2$.

2. $\dfrac{d^2P}{dR^2} = \dfrac{(R + 0.2)^3(-100) - 100(0.2 - R)(3)(R + 0.2)^2}{(R + 0.2)^6}$

(Second derivative)

$= \dfrac{-100(R + 0.2)^2[R + 0.2 + 3(0.2 - R)]}{(R + 0.2)^6}$

$= \dfrac{-100[R + 0.2 + 0.6 - 3R]}{(R + 0.2)^4}$

$= \dfrac{-100[0.8 - 2R]}{(R + 0.2)^4}.$

At $R = 0.2$,
$$\frac{d^2P}{dR^2} = -\frac{40}{(0.2 + 0.2)^4} < 0.$$

Therefore, the power dissipated in R is a maximum when $R = 0.2$.

CHAPTER 20 APPLICATIONS OF DERIVATIVES

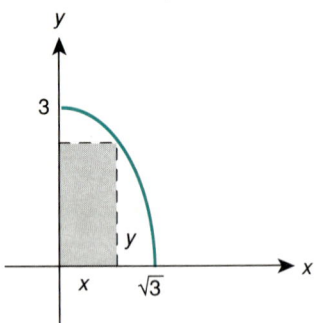

Figure 20.17

• **Example 20.17:** Find the dimensions of the rectangle of maximum area that can be inscribed in the region bounded by $y = 3 - x^2$, the y-axis, and the x-axis. See figure 20.17.

Solution: $A = xy$ (Area of region).
Since $y = 3 - x^2$,
$A = x(3 - x^2) = 3x - x^3$.
For A to be maximum, set $A' = 0$:

$$A' = 3 - 3x^2 = 0, \text{ when } x = 1.$$

Since $A'' = -6x$ and $A''(1) = -6 < 0$, $x = 1$ gives a maximum A.

When $x = 1$, $y = 3 - 1^2 = 2$. Therefore, the dimensions of the rectangle are $x = 1$ and $y = 2$ and $A = 1 \cdot 2 = 2$ sq. units.
Note: Also $A' = 0$ when $x = -1$; however, this would give a negative dimension for the rectangle.

Trial Problems 20.5

Before you begin the section exercises, warm up with these problems. Complete answers are included in the answer key.

In this exercise set the first, and maybe the second, derivative may be required. Find the first and second derivatives of each of the following.

1. $y = 3x^2 - 5x + 6$
2. $g(t) = 500t - 16t^2$
3. $P = 32 \ln x + 5$
4. $r(x) = 3 \sin 2x$
5. $y = x^2 + \sqrt{x}$

Exercises 20.5

1. Find the greatest product that can be obtained by multiplying two numbers whose sum is 32.

2. Find two numbers whose difference is 4 and whose product is a minimum.

3. Find the minimum power developed by a motor given that $p(t) = 3t^2 - 3t + 4$ represents the power as a function of time.

4. Show that the rectangle of largest area having a perimeter of 12 inches is a square.

5. The current in a particular circuit is given by
$i(t) = \frac{1}{3}t^3 - \frac{7}{2}t^2 + 10t, t > 0$. Find the maximum value of the current for the interval $t = 0$ to $t = 5$.

6. The deflection D of a particular type of beam is given by $D = x^4 - 5x^3 + 7x^2$, where x is the distance (in feet) from one end of the beam. Find the value of x that yields the minimum deflection.

7. A ball is thrown straight up into the air. Its height in feet at any time t is given by $s(t) = 1000t - 16t^2$. Find the maximum height that the ball rises above the ground.

8. The electric field intensity on the axis of a uniformly charged ring is given by $E = \dfrac{C}{(x^2 + a^2)^{3/2}}$, where C is the total charge and a is the radius of the ring. For what value of x is E maximum?

9. The electric power (in watts) produced by a certain source is given by $P = \dfrac{144r}{(r + 0.6)^2}$. For what value of r (resistance) is the power a maximum?

10. The velocity of a rocket is given by $v = 4t - t^2$. At what time t does the rocket reach its maximum velocity? What is the maximum velocity?

11. The power produced by a solenoid is given by $p = 12t - 4t^3$, $t > 0$. Find the greatest power produced by the solenoid.

12. The current in a particular circuit is given by $i = 4 + 6t - t^2$. At what time t does the current reach its maximum value? What is the maximum current?

13. An open box of greatest possible capacity (volume) is to be made from a square piece of cardboard whose sides are each 36 inches long, by cutting equal small squares out of the corners and folding up the remaining pieces. What should be the length of each side of the small squares?

14. A rectangular animal pen is to be constructed against one side of a barn. What is the maximum area that can be enclosed if 200 feet of fence are used?

15. The speed of signaling in a marine cable is given by $S = kx^2 \ln x$, where k is a constant and x is the ratio of the radius of the core of the cable to that of the entire cable. Which x gives the maximum speed?

16. A riverboat can carry 600 passengers. A total of x passengers will ride the boat if the cost is y dollars per ticket. If $x = 1400 - 400y$, what price should be charged to maximize the income?

17. The energy w in an inductance coil varies with time (in seconds) according to $w = 3 + 4t - 3t^2$. Find the maximum energy of the coil.

18. The profit P in selling certain transistors is given by $P = 2.50x - \dfrac{x^2}{20{,}000} - 5000$, $0 \le x \le 35{,}000$. How many transistors must be sold for the profit to be a maximum?

19. Find the dimensions of the rectangle of greatest area that can be inscribed in a semicircle of radius r.

20. A UPS regulation says that no package can be accepted if the sum of its length and cross-sectional perimeter exceeds 60 inches. What are the dimensions of the package of largest volume that can be mailed if the shape of the package is a circular cylinder?

21. An open trough is to be made from a rectangular sheet of metal of length L by bending up the long edges so as to give the trough a rectangular cross section. If the width of the sheet is 8 inches, how deep should the trough be made so that it has a maximum carrying capacity?

22. Find the coordinates of the point on the curve $y = \sqrt{x}$ closest to the point $(4, 0)$.

23. A segment of the path of an amusement park ride is described by $p(x) = 2 \sin x - \cos 2x$ for $0 \le x \le \pi$. Find the relative maxima and/or minima and sketch the graph of $p(x)$.

24. The normal average daily temperature, in degrees Fahrenheit, for a certain city is approximated by

$$T = 45 - 23 \cos \dfrac{2\pi(t - 32)}{365},$$

where t is the time in days with $t = 1$ corresponding to January 1. Find the expected date of
a. The warmest day.
b. The coldest day.

20 • 6 Velocity and Acceleration

Average Velocity

For an object moving in a straight path, we define its **average velocity** as the distance traveled per unit of time. For example, if a car travels 100 miles in 2 hours, we say that its average velocity is 50 miles per hour. Using s for distance traveled and t for the time required to travel the distance, we have

$$\text{average velocity} = \frac{s}{t}.$$

INSTANTANEOUS VELOCITY

The **instantaneous velocity** at any particular time t is given by

$$v = \lim_{t \to 0} \frac{s}{t} = \frac{ds}{dt}.$$

The first derivative $\dfrac{ds}{dt}$ is interpreted as "the rate of change in distance s with respect to time t." Thus, velocity (instantaneous) v is given as the first derivative of distance s, where s is a function of time t. Since $s = f(t)$, $v = \dfrac{ds}{dt}$ or $v = s'$.

• **Example 20.18:** An object moves in a straight line according to the rule $s = 120t - 16t^2$, $t \geq 0$, where s is in feet and t is in seconds. Find the velocity of the object at any time t. Find the velocity at the end of the third second.

Solution: Since $s = 120t - 16t^2$,

$$v = \frac{ds}{dt} = 120 - 32t.$$

At the end of the third second, $t = 3$ and

$$\begin{aligned} v &= 120 - 32(3) \\ &= 120 - 96 \\ &= 24 \text{ ft/sec.} \end{aligned}$$

Acceleration

Just as velocity is the rate of change of distance with respect to time, we can consider the rate of change of velocity with respect to time. We define the rate of change of velocity with respect to time as **acceleration** and denote it as a. Thus,

ACCELERATION

$$a = \frac{dv}{dt} = \frac{d^2s}{dt^2}, \quad \text{or } a = v' = s''.$$

• **Example 20.19:** A projectile is fired into the air. Its distance traveled at any time t is given by $s = 30t^3 - 50t$. Find the velocity and acceleration after 1 second. (s is given in feet and t in seconds.)

Solution: $s = 30t^3 - 50t$

$$v = \frac{ds}{dt} = 90t^2 - 50.$$

At the end of 1 second,
$$v = 90(1)^2 - 50$$
$$= 40 \text{ ft/sec}.$$

$$a = \frac{dv}{dt} = 180t$$

At the end of 1 second,
$$a = 180 \text{ ft/sec/sec}. \quad \text{or}$$
$$a = 180 \text{ ft/sec}^2.$$

• **Example 20.20:** A rocket is moving in such a way that its distance after t seconds is given by $s = 192t - 16t^2$. Find the velocity at the end of 5 sec, at the end of 8 sec, and at the end of 6 sec. How high does the rocket travel?

Solution: $s = 192t - 16t^2$
$v = s' = 192 - 32t.$

The velocity at the end of 5 sec is
$$v = 192 - 32(5) = 192 - 160$$
$$= 32 \text{ ft/sec}.$$

The velocity at the end of 8 sec is
$$v = 192 - 32(8) = 192 - 256$$
$$= -64 \text{ ft/sec}.$$

(v negative means the rocket is coming down.)
The rocket reaches its highest point when $v = 0$:
$$v = 192 - 32t$$
$$0 = 192 - 32t$$
$$t = 6.$$

At $t = 6$, $s = 192(6) - 16(6)^2 = 1152 - 576 = 576$ ft.
The rocket reaches a height of 576 ft.

Trial Problems 20.6

Before you begin the section exercises, warm up with these problems. Complete answers are included in the answer key.

In order to find velocity and acceleration you will need to find first and second derivatives of functions. Find the first and second derivatives of each of the following.

1. $s = 7t^2 + 2t$
2. $s = 6t^{2/3}$
3. $s(t) = 2 \cos \pi t$
4. $s(t) = t^2 e^{-t}$
5. $s(t) = t^2 + \ln t$

Exercises 20.6

The following equations give the distance traveled (in a straight line) by an object in t seconds. Find the velocity and acceleration at the indicated times.

1. $s = 3t + t^3$, $t = 2$
2. $s = 10t - 5t^2$, $t = 1$
3. $s = 2t^3 - 6t^2$, $t = 1$
4. $s = t^2 + 3t$, $t = 5$
5. $s = \frac{1}{3}t + 3$, $t = 1$
6. $s = 2t^3 - 4$, $t = 2$
7. $s = -3t^2 + 2t + 1$, $t = 0$
8. $s = 4 - 2t^2$, $t = 2$
9. $s = 4t^3 + 3t + 2$, $t = 1$
10. $s = t^2 - 6t - 3$, $t = 3$

11. Find the equation for the velocity of an object whose distance at any time is given by $s = t^2 + 2t$.

12. If $s = v_0 t + \frac{1}{2}at^2$, where v_0 is an initial velocity (a constant) and a is any constant acceleration, show that
    (a) $v = v_0 + at$, and (b) $as = \frac{1}{2}(v^2 - v_0^2)$.

13. An object dropped from the top of a tower travels according to the equation $s = 16t^2$. Find the velocity of the object after 3 seconds and the acceleration after 5 seconds.

14. The formula $s = v_0 t^3 + 2t^2$ describes the distance in feet traveled by an object that is given a push (initial velocity) of v_0 ft/sec. If acceleration at $t = 1$ sec is 10 ft/sec², find the initial velocity, v_0.

15. The position at any time t of an accelerating car is given by $s(t) = 10t^{3/2}$, $0 \leq t \leq 10$. Find the velocity of the car when $t = 0$, $t = 1$, $t = 4$, and $t = 9$.

16. If a projectile travels according to the rule $s = 16t^2 - 4t + 5$, show that the acceleration of the projectile is constant.

17. If a rocket is fired straight up from the ground and travels according to the law $s = 500t - 16t^2$, what is its initial velocity? (*Hint:* $t = 0$.)

18. The plunger of a solenoid has a velocity $v = 4\sqrt{t}$ cm/sec. Find the acceleration of the plunger when $t = 0.01$ sec.

19. Find the maximum height of an arrow whose height above ground level is given by $s = 64t - 16t^2$.

20. If an object is dropped from a balloon 256 feet above the ground, then its distance above ground after t seconds is given by $s(t) = 256 - 16t^2$. Find the velocity at $t = 1$, $t = 2$, and $t = 3$. What is the velocity when the object strikes the ground?

21. The distance-time equation for a charged particle in an electrical field is given by $s = 0.42t^{1.2} + 2.56t$, with s measured in cm and t in sec. What is the velocity and acceleration of the particle when $t = \frac{5}{2}$ sec?

22. An object rolls down an inclined plane according to the rule $s = 5t^2 + 2$, where s is in inches and t is in seconds. What is the velocity after 1 second? 2 seconds? When is the velocity 28 inches/second?

23. An object moves in a straight line according to the rule $s = t^2 - 4\ln(t+1)$, $0 \le t \le 4$. Find the velocity and acceleration at $t = 1$, $t = 2$, and $t = 4$.

24. The position of a particle moving in a straight line is given by $s(t) = (t^2 + 2t)e^{-t}$.
   a. When is the velocity 0?
   b. When is acceleration 0?

25. If a particle moves in a straight line according to the rule $s(t) = a \cos wt$ or $s(t) = a \sin wt$, where a and w are constants, then the motion of the particle is called *simple harmonic*. Find the velocity and acceleration for the simple harmonic motion.
   a. $s(t) = 3 \cos 2t$
   b. $s(t) = 2 \sin \pi t$

20 • 7 Related Rates

In many applications, two variables may be related to each other according to some function or equation; yet, each may change according to a third variable, usually time t. The variables x and y may be related by the equation

$$x^2 - y^3 + 4x - 3y^2 + 5 = 0,$$

and each variable may change at a different rate with respect to time. Differentiating implicitly with respect to time t results in the following equation:

$$2x\frac{dx}{dt} - 3y^2\frac{dy}{dt} + 4\frac{dx}{dt} - 6y\frac{dy}{dt} = 0,$$

where $\dfrac{dx}{dt}$ and $\dfrac{dy}{dt}$ are the respective rates of change of x and y with respect to time t.

Related Rates The derivatives $\dfrac{dx}{dt}$ and $\dfrac{dy}{dt}$ are called **related rates**, since they are related by the given equation.

• **Example 20.21:** A 13-ft ladder leans against a building. If the bottom of the ladder is moved away from the building at the rate of 2 ft/sec, how fast is the top of the ladder moving down the building when the ladder is 5 ft from the building?

Solution: Make a sketch as in figure 20.18. Since x is changing at the rate of 2 ft/sec, $\dfrac{dx}{dt} = 2$. We want to find $\dfrac{dy}{dt}$. x and y are related by the equation $x^2 + y^2 = 169$, by the Pythagorean theorem. Differentiating with respect to t gives

$$2x\frac{dx}{dt} + 2y\frac{dy}{dt} = 0$$

or

$$\frac{dy}{dt} = -\frac{x}{y} \cdot \frac{dx}{dt}.$$

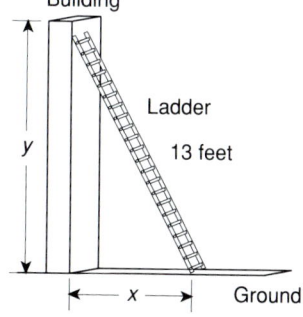

Figure 20.18

In the equation $x^2 + y^2 = 169$, it is given that x is 5 ft. To find y, we have $25 + y^2 = 169$ or $y^2 = 144$ or $y = 12$. Therefore,

$$\frac{dy}{dt} = -\frac{x}{y} \cdot \frac{dx}{dt}.$$

$$= -\frac{5}{12} \cdot 2$$

$$= -\frac{5}{6} \text{ ft/sec.}$$

Thus, the ladder is moving down the building at a rate of $\frac{5}{6}$ ft/sec when the ladder is 5 ft from the building.

• **Example 20.22:** Sand is being poured into a pile that is in the shape of a cone. As the conical pile increases in height, the radius always remains the same as the height. If the height of the pile is increasing at the rate of 6 ft/min, find the rate at which the sand is increasing in volume when the height is 10 ft.

Solution: The volume of a cone is given by $V = \frac{1}{3}\pi r^2 h$. Since $r = h$ and $\frac{dh}{dt} = 6$, we can write

$$V = \frac{1}{3}\pi h^3 \quad \text{and} \quad \frac{dV}{dt} = \pi h^2 \frac{dh}{dt}.$$

Therefore,

$$\frac{dV}{dt} = \pi(10)^2(6) = 600\pi.$$

Thus, the sand pile is increasing in volume at the rate of 600π ft³/min.

• **Example 20.23:** The electrical resistance of a wire varies with temperature according to the equation $R = 45 - 0.2T + 0.001T^2$. What is the rate at which the resistance is changing when the temperature is 200° if the temperature is changing at the rate of 3°/min?

Solution: $\dfrac{dR}{dt} = -0.2\dfrac{dT}{dt} + 0.002T\dfrac{dT}{dt}$

$= -0.2(3) + 0.002(200)(3)$

$= -0.6 + 1.2$

$= 0.6$ ohms/min.

• **Example 20.24:** The radius of a heated circular plate increases at the rate of 0.02 in./min. What is the rate at which the area of the plate increases when the radius is 6 in.?

Solution: The area of the circular plate is given by $A = \pi r^2$. Therefore,

$\dfrac{dA}{dt} = 2\pi r \dfrac{dr}{dt}$

$= 2\pi(6)(0.02)$

$= 0.24\pi$.

Thus, the area of the plate is increasing at the rate of 0.24π in.²/min.

Trial Problems 20.7

Before you begin the section exercises, warm up with these problems. Complete answers are included in the answer key.

In this exercise set you will need to find the derivative of certain functions. Find the derivative of each of the following.

1. $y = \dfrac{4 + x}{4 - x}$
2. $y = x + \sqrt{x}$
3. $y = \dfrac{6}{x^2}$
4. $y = \sqrt{25 - x^2}$
5. $y = \pi x^2$

Exercises 20.7

1. The sides of a square are increasing at the constant rate of 2 in./min. At what rate is the area of the square increasing when the sides are 4 in.?

2. *Boyle's law for gases* is given by $pv = c$, where p denotes pressure, v is the volume, and c is a constant. At a certain instant $v = 75$ in.³, $p = 30$ lbs/in.², and the pressure is decreasing at the rate of 2 lbs/in.² every minute. At what rate is the volume changing at this instant?

3. A spherical balloon is filled with gas at the rate of 5 ft³/sec. Find the rate at which the radius of the balloon is increasing when the radius is 2 ft.

4. A machine produces two products and the number of each product is related by $y = \dfrac{10(20 + x)}{50 - x}$, where x and y denote the number of each product produced. How fast is y changing if x is changing at the rate of 2 per hour when $x = 40$?

5. The length of a rectangular metal plate is three times its width and is increasing at the rate of 2 cm/min. How fast is the area of the plate changing?

6. An object moves along the graph given by $y = \sqrt{x}$ with its x-coordinate increasing at the rate of 4 cm²/sec. How fast is the y-coordinate of the object changing when the object is at the point (16,4)?

7. A boat is being pulled by rope into a dock that is 8 ft above water level. If the rope is being pulled in at the rate of 2 ft/sec, how fast is the boat moving toward the dock when it is 25 ft from the dock?

8. When two electrical resistances R_1 and R_2 are connected in parallel, the total resistance R is given by $\frac{1}{R} = \frac{1}{R_1} + \frac{1}{R_2}$. If R_1 and R_2 are increasing at rates of 0.01 ohms/sec and 0.02 ohms/sec, respectively, at what rate is R changing when $R_1 = 30$ ohms and $R_2 = 90$ ohms?

9. Water is flowing into an 8-ft diameter conical tank that is 12 ft deep. If the rate of water flow is 20 cu ft/min, what is the rate of water level rise when the depth of the water is 4 ft?

10. If $y = 2x + 3x^3$, and x is increasing at the rate of 0.5 units/sec, how fast is the slope of the graph changing when $x = 2$?

11. An ice cube is melting in such a manner that each edge is decreasing in length by 0.10 cm/min. How fast is the volume changing when an edge measures 2.5 cm?

12. The voltage V and current I in a certain wire are related by the equation $V = 0.02I/r^2$, where r is the radius of the wire. If the current increases at the rate of 0.02 amps/sec in a wire of 0.03 in radius, find the rate at which the voltage is increasing.

13. Find the rate of change of the area of a circle with respect to its radius. Does this have any relationship to the circumference of the circle?

14. A certain bacteria grows in a circular colony. As it grows the surface area it covers is directly proportional to its population and contains 10^6 members when the area is 1 cm². How fast is the population increasing when the radius of the circle is 10 cm and if the radius of the circle is increasing at the rate of 2 cm/hr?

15. A metal rod has the shape of a right circular cylinder. As it is being heated its length increases at the rate of 0.005 cm/min and its diameter increases at 0.002 cm/min. How fast is the volume changing when the rod is 20 cm and the diameter is 3 cm?

16. The power in a resistor is given by $p = Ri^2$, where R is measured in ohms and i in amps. If the current in a 50 ohm resistor is changing at a rate of 0.1 amps/sec, at what rate is the power changing when the current is 2 amps?

17. Oil is being pumped into a cylinder that raises a piston. If the cylinder has a radius of 8.128 cm and oil is being pumped in at the rate of 5.900 cm³/sec, find the rate at which the piston is rising.

18. The period T of a pendulum of length L is given by $T = \sqrt{\frac{L}{96}}$. If the length of the pendulum is decreasing at the rate of 0.254 cm/sec, find the rate of change of the period when the length is 40.64 cm.

19. A hot air balloon is moving at an altitude of 300 meters and is being pushed horizontally by the wind at the rate of 2 meters/sec. At what rate is the distance between the balloon and its ground post changing at the time this distance is 500 meters?

20. A 15-ft ladder leans against a building. If the bottom of the ladder is pulled away from the building at the rate of 10 ft/sec, how fast is the angle between the ladder and the ground changing when the top of the ladder is 5 ft from the ground?

20·8 Summary of Terms and Rules

Terms

Acceleration (p. 842)
Average velocity (p. 842)
Concave downward (p. 832)
Concave upward (p. 832)
Critical number (p. 822)
Decreasing function (p. 827)

First derivative test (p. 822)
Increasing/decreasing curves (p. 821)
Increasing function (p. 827)
Instantaneous velocity (p. 842)
Maxima/minima (p. 838)
Maximum value (p. 822)

Minimum value (p. 822)
Normal line (p. 819)
Point of inflection (p. 832)
Point-slope form (p. 818)
Related rates (p. 845)
Relative maxima/minima (p. 838)

Rules

- **THE FIRST DERIVATIVE TEST.** Suppose $x = c$ is a critical number. (20.2)

  1. $f(x)$ has a maximum value at $x = c$ if $f'(c) = 0$, and $f'(x)$ changes from $+$ to $-$ near $x = c$.
  2. $f(x)$ has a minimum value at $x = c$, if $f'(c) = 0$, and $f'(x)$ changes from $-$ to $+$ near $x = c$.

- **THE SECOND DERIVATIVE TEST.** Suppose $x = c$ is a critical number, and $f'(c) = 0$. (20.3)

  1. $f(x)$ has a maximum value at $x = c$ if $f''(c) < 0$.
  2. $f(x)$ has a minimum value at $x = c$ if $f''(c) > 0$.

  NOTE: If $f''(c) = 0$, the second derivative test is not applicable. Use the first derivative test.

20·9 Chapter 20 Review Exercises

1. Find equations of the tangent and normal lines to the graph of the equation $y = -x^3 + 4x^2 - 3x$ at the point $(0,0)$.

2. Find the equations of the tangent lines to the graph of the equation $x^2 + y^2 = 1$ at the point $\left(\dfrac{\sqrt{2}}{2}, \dfrac{\sqrt{2}}{2}\right)$.

Give the intervals where the graph is increasing and decreasing, and sketch each curve.

3. $y = 2x^3 - 9x^2 + 12x$
4. $y = x^3 - x^2 - x - 1$
5. $y = 2x^2 - 6x + 5$
6. $y = 3x^4 - x^3 + 2$
7. $y = e^{-2x}$
8. $y = \sqrt{x}$
9. $y = 2 \sin x$
10. $y = 3 \ln x$

11. Use the first derivative test to find maximum and minimum points for $f(x) = -x^3 + 4x^2 + 3x$. Sketch the graph.

12. Use the first derivative test to find maximum and minimum points for $f(x) = \dfrac{1}{1 + x^2}$. Sketch the graph.

Determine maximum and minimum points, points of inflection, concavity, and sketch the graph for each of the following:

13. $y = 2x^3 - 9x^2 + 12x$
14. $y = x^3 + x^2 - x - 1$
15. $y = 3x^4 - x^3 + 2$
16. $y = e^{-2x}$
17. $y = -x^3 + 4x^2 + 3x$
18. $f(x) = 40x^3 - x^6$
19. $f(x) = \dfrac{1}{1+x^2}$
20. $f(x) = 4x + x^{-1}$
21. $f(x) = \dfrac{x-1}{x+1}$
22. $f(x) = x^2 - \dfrac{1}{6}x^3$
23. $y = \sqrt{x}$
24. $y = 2\sin x$
25. $y = 3\ln x$

26. The slope of a curve at any point (x,y) is given by $f'(x) = 6(x-1)(x-2)^2(x-3)^2(x-4)^4$.
   a. For what value(s) of x is y a maximum?
   b. For what value(s) of x is y a minimum?

27. Find the equation of the line that is tangent to $y = x^3 + 6x^2 - 2x + 8$ at its point of inflection.

28. A wire 5 feet long is cut into two pieces. One of the pieces is bent into a circular shape and the other into a square shape. Where should the wire be cut so that the sum of the areas of the circle and square is (a) a maximum and (b) a minimum?

29. The sum of two numbers is 20. What are the two numbers if the product of one with the square of the other is to be a maximum?

30. The perimeter p and area A of a circular sector of radius r and arc length s are given by $p = 2r + s$ and $A = \dfrac{1}{2}rs$, respectively. If the perimeter is 100 feet, what value of r produces a maximum area?

31. A soup can in the form of a right circular cylinder has a volume of 16π in.³. What dimensions require the least amount of material to make the can?

32. Find the velocity of an object after 2 seconds if its distance at any time t is given by
$$s = \dfrac{t}{2t+1}.$$

33. Electrical current i is defined as $i = \dfrac{dq}{dt}$, where q is the quantity of electrical charge in coulombs and t is time in seconds. What is the current in a circuit at $t = 0.5$ sec if $q = t^3 + 2t + 1$?

34. Power is defined as the rate of change of work with respect to time. If work is given as $w = 3t^2 + t^4$ ft/lb, find the power p after $t = 2$ seconds.

35. Work being done by a machine is given by $w = 18t^2 + 4t^3 - t^4$ ft/lb, $t > 0$. What is the maximum power output? (See exercise 34.)

36. The distance in feet traveled by an elevated train after the brakes are applied is given by $s = 36t - 4.5t^2$. After the brakes are applied, how long does it take for the train to come to a stop?

37. The velocity of an object is given by $v = 3t^2 - t^3$ ft/sec. At what time does the maximum acceleration occur?

38. The distance traveled by an object is given by $s = 2t \sec \sqrt{t}$. Find the velocity when $t = 0.05$.

39. Find the equation of the normal line to the curve $y = \tan^{-1}\left(\dfrac{x}{2}\right)$ at $x = 3$.

40. Find the equation of the normal line to the curve $y = \ln|\cos x|$ at $x = \dfrac{\pi}{6}$.

41. Find the equation of the normal line to the curve $y = e^{x^2}$ at $x = \dfrac{1}{2}$.

42. The charge q on a certain capacitor is given by $q = e^{-0.1t}(0.2 \sin 100t + \cos 100t)$. Find the current i at $t = 0$. (See exercise 33.)

43. The gross income I received from the sale of r items per month was $I = 100re^{-r/10}$. How many items should be sold per month so that I is a maximum?

44. The energy dissipated by a resistor at a particular time is given by $E = \ln(t + 1) - 0.5t + 10$ joules. When is the energy output a maximum?

45. If the velocity of an object is given by $v(t) = 2e^{-(t-1)}$, find the acceleration at $t = 1$.

46. A point $P(x,y)$ moves on the graph given by the equation $y = x^3 + x^2 + x + 1$. The x-value changes at the rate of 2 units per second. How fast is the y-value changing at the point $(1,4)$?

47. Airplane A is 100 miles east of an airport and flying east at 300 mph. Airplane B is 50 miles south of the airport and flying north at 200 mph. Assuming the planes are at the same altitude, how fast is the distance between them changing?

48. The dimensions of a box are 15 cm, 18 cm, and 24 cm. If the shorter sides are decreasing at the rate of 0.2 in./min, and the longest side is increasing at the rate of 0.3 in./min, how fast is the volume changing?

49. Water is flowing into a conical tank 12 ft in diameter and 12 ft deep. If the water is rising at the rate of 1 in./min, what is the rate of flow of the water when the water is 6 ft deep?

50. The expansion of a gas under certain conditions is given by $PV^{1.5} = c$, where P is pressure, V is volume, and c is a constant. At what rate is the pressure of the gas increasing if $P = 15$ psi, $V = 5.2$ ft³, and the volume is decreasing at the rate of 0.06 ft³/sec?

20·10 Chapter 20 Test

1. Find the equation of the line tangent to the curve $y = x^2 + 3$ at the point $(1,3)$.

2. Find the equation of the normal line to the graph of $y = 3 - x^2$ at the point $(1,2)$.

3. Find the equation of the line tangent to $y = \sin x$ at the point $\left(\dfrac{\pi}{2}, 1\right)$.

4. Use the first derivative test to find maximum and minimum values of $y = x^3 - 4x$. Sketch the graph.

5. Use the second derivative test to find maximum and minimum points on the graph of $y = x^3 - 2x^2 - x + 2$. Sketch the graph.

6. Current in a particular circuit is given by $i = 12t - t^3$. Sketch the graph for value of t from 0 to 4. Label maximum and minimum points and points of inflection. Indicate the intervals of concavity.

7. The velocity of a particle is given by $v = t + \cos t$, $0 \le t \le \pi$ radians. For what value of t does the particle reach its maximum velocity?

8. An object moves in a straight line according to the rule $s = t^2 - 12 \ln(t + 1)$, $0 \le t \le 4$. Find the velocity and acceleration at $t = 3$.

9. The pollutants in the atmosphere near a power plant will be p parts per million t years from now, where $y = 30 + 125e^{-0.06t}$. Find the rate of change of the pollutants now, and also 4 and 12 years from now.

10. Find the coordinates of the maximum and minimum values for $x^2 + y^2 = 4$. Sketch the graph.

20 · 11 Cumulative Review—Chapters 18, 19, and 20

Evaluate the following limits.

1. $\lim_{x \to -2} \dfrac{x+2}{x-2}$
2. $\lim_{t \to 0} \dfrac{3 - \sqrt{9+t}}{t}$
3. $\lim_{x \to \infty} \dfrac{x}{4+3x}$
4. $\lim_{x \to 4^+} \dfrac{2x^2}{16-x^2}$

Determine where f is continuous or discontinuous for each of the following.

5. $f(x) = \dfrac{1}{\sqrt{x-4}}$
6. $f(x) = \dfrac{6}{x^2 - x}$

Find the slope of the function at the given point.

7. $y = x^2 + x - 1$, $(2,5)$
8. $f(x) = x + \dfrac{1}{x}$, $(1,2)$
9. $h(x) = \ln 5x$, $(1, \ln 5)$
10. $y = 2 - 3e^{-2x}$, $(0, -1)$
11. $y = 2 \cos 2x$, $\left(\dfrac{\pi}{8}, \sqrt{2}\right)$
12. $f(x) = \sin^{-1}(x-1)$, $\left(\dfrac{3}{2}, \dfrac{\pi}{6}\right)$

Differentiate the following functions.

13. $f(x) = 4x(x^3 - 1)$
14. $y = (2x)^{-5}$
15. $g(x) = (x-2)(3x+4)$
16. $r(x) = \dfrac{x}{3x-2}$
17. $f(x) = \dfrac{x^2}{4-3x}$
18. $y = \sqrt{2x}$
19. $y = \ln(x^2 - 5)$
20. $f(x) = \ln\left(\dfrac{x+1}{x-1}\right)$
21. $f(x) = x^2 \ln x$
22. $y = e^{2x} + e^{-2x}$
23. $y = 2xe^x$
24. $h(x) = \dfrac{e^{x^2}}{x}$
25. $f(x) = \sin 3x^2 + \cos 2x^3$
26. $y = e^x \sin(x^2 - 1)$
27. $y = \dfrac{3x}{\tan 3x}$
28. $f(x) = \sin^{-1}(2x+1)$
29. $h(x) = \dfrac{1}{x} \tan^{-1} x$
30. $y = e^x \cos^{-1} x$

Evaluate dy for the indicated values of x and dx.

31. $y = 3x^2 + x - 6$, $x = 1$, and $dx = 0.1$
32. $y = \dfrac{1}{2 - x^2}$, $x = 1$, and $dx = 1.2$

Differentiate the following implicitly.

33. $x^2 - y^2 = 25$
34. $x^2 - 2xy + y^2 = 6$
35. $e^x + y + x = xy$
36. $\sin x \cos y + y^2 = 6$

Find the second derivative of each of the following.

37. $y = 3x^4 + 2x^2 - x + 5$
38. $f(x) = \dfrac{1}{3x - 2}$
39. $f(x) = \sqrt{3 - 2x}$
40. $y = x - \ln x$
41. $g(x) = 1 - e^{-x}$
42. $y = \sin x + \cos x$

43. The current I (in amperes) in a certain electrical circuit is given by $I = 75R^{-2}$, where R denotes resistance (in ohms). Find the rate of change of I with respect to R when the resistance is 30 ohms. Is the current increasing or decreasing?

44. Find the equations of the lines tangent and normal to the curve $y = 3x^2 - 5x + 6$ at the point $P(2,8)$.

45. Use the first derivative test to find the maximum and minimum points for $f(x) = \frac{2}{3}x^3 + x^2 - 12x + 1$.

46. Use the second derivative test to find maximum and minimum points for $f(x) = x^3 - \frac{3}{2}x^2$.

47. Sketch the graph of $y = \frac{15}{2}t^2 - \frac{5}{3}t^3$ and label maximum and minimum points and points of inflection. Indicate intervals of concavity.

48. Political advisors for a certain candidate estimate that if x thousand people vote in the next election, their candidate will receive y thousand notes, where

$$y = 10\sqrt{x} - 0.5x - 5.$$

Find the intervals on which the candidate's vote is increasing and those on which it is decreasing. Find the candidates maximum number of votes.

49. A lake that is being deacidified has its pH y described by $y = 2\ln(3t + 6)$, where t is in years and $t = 0$ represents the start of the deacidification process. Find the rate at which the deacidification process is proceeding 5 and 10 years after its start. Use curve-sketching to graph the function and analyze the deacidification process as a function of time.

50. In t minutes a rocket travels $s(t) = 250t$ kilometers in a straight line. Find its velocity and acceleration t minutes after it started.

51. A ball is thrown straight up into the air. The altitude of the ball t seconds after it is released is given by $s(t) = -2t^2 + 20t$ feet. What is the altitude of the ball 2 seconds after it is released? What is its velocity at this time? Is it going up or coming down? What is its acceleration?

52. The amount of energy radiated by a body per unit of surface area is given by *Stefan's Law:*

$$y = kT^4$$

where k is the constant of proportionality and T is the temperature of the body in degrees kelvin (K). The surface temperature of the sun is approximately 6000 K. Find the rate of change of the energy radiated by the sun per unit of surface area.

53. The number of fish remaining in a lake t months after it was stocked is $y = 20(4000 - 10t)^{1/2}$. Use differentials to find the approximate number of fish in the lake 42 months after it was stocked.

54. The second derivative test is not always easier to apply than the first derivative test. This becomes evident when working with a function such as

$$f(x) = \frac{x + 1}{x^2 + 3}.$$

Use the first derivative test and then the second derivative test to find the relative maxima and relative minima of $f(x)$.

55. It has been determined that if a company spends x thousand dollars on advertising, its net profit will be $P(x)$ dollars, where

$$P(x) = x^3 - 240x^2 + 18{,}900x - 100{,}000,\ 0 \le x \le 100.$$

The firm's maximum profit occurs at a relative maximum of P. Use the second derivative test to find the maximum profit.

56. The number of worker-hours needed to produce the xth unit of a product are y, where $y = 4000 + 1000e^{-0.1x}$, $x \ge 1$. Find the rate at which production time is changing when x equals 1, 5, and 25. Sketch the graph and analyze production time as a function of the number of units produced.

57. An electronics company is introducing a new large-screen television. The firm estimates it will take $T(x)$ worker-hours to build the xth unit of the new computer, where

$$T(x) = \frac{300x^3 + 50x + 100}{x^3}.$$

Find and interpret $\lim\limits_{x \to +\infty} T(x)$.

58. The number of bacteria present t hours after the start of an experiment is y thousand, where

$$y = \begin{cases} 2t^2 + 100, & 0 \le t < 2 \\ 160 - 27t, & 2 \le t \le 6 \end{cases}.$$

Graph this function. Where is it continuous? Where is it discontinuous? Where is it not differentiable?

59. Over the past 5 years, the average productivity of a company's plant has been $y = 0.4t^2 - 30t + 1400$ units per day in month t, where $t = 0$ represents the beginning of the 5-year period.
   a. Graph this function.
   b. The plant's minimum productivity occurred at the point on the graph where the tangent line is horizontal. Find the plant's minimum average productivity over the 5-year period and the time when it occurred.

Chapter 21
Integral Calculus

*I*ntegral calculus has its origin in the study of the ancient problem of finding the area enclosed by a given curve. Nearly 4000 years ago the Egyptians knew how to find the areas of circles and rectangles. It was not until in ancient Greece that Archimedes of Syracuse (287–212 B.C.) devised a method for calculating the areas bounded by curves other than circles and rectangles. Again, it was not until Newton and Leibniz that a complete system was developed so that areas enclosed by any curve could be found in an efficient manner. The symbolic notation invented by these two men was further refined by subsequent mathematicians. These refinements led to many other applications besides finding the area bounded by a curve. In general, this new symbolic notation allows us to do just the opposite of what we did in differential calculus. In differential calculus we started with a function and found a derivative function; integral calculus involves processes to find a function when the derivative function is given. The beginning of the study of these processes involve finding "antiderivatives" for certain derivative functions. As we complete the study of several of these integration processes, we investigate several applications of the integral calculus, including finding the area bounded by a curve.

21 • 1 Antiderivatives

In our work with derivatives, we started with a function $y = f(x)$ and derived a new function $F(x)$, which we labeled in several different ways:

$$y' = F(x), \quad \frac{dy}{dx} = F(x), \quad f'(x) = F(x), \quad \text{and } D_x f(x) = F(x),$$

as well as other notations. In general, we started with one function, and through the process of differentiation, we derived another function.

Now, we start with the second function and see if we can obtain the first function. That is, if we know $f'(x)$, how can we find $f(x)$? For the function $y = x^3$, $y' = 3x^2$ is called the derivative of $y = x^3$. Likewise, $y = x^3$ is called an **antiderivative** of $y = 3x^2$.

Antiderivative

In order to obtain a process for finding the antiderivative of a function, we use two of the notations above:

$$\frac{dy}{dx} \text{ and } f'(x), \text{ where } \frac{dy}{dx} = f'(x).$$

Then, we can express this equation with differentials:

$$dy = f'(x)dx.$$

Our goal is to find a function $f(x)$ such that $dy = f'(x)dx$. The process of finding $f(x)$ is called **integration**, and $f(x)$ is said to be the **integral** of the differential $f'(x)dx$.

Integration
Integral

The notation for integration is given by $\int f'(x)dx$, where the symbol $\int$ is called the **integral sign** and $f'(x)$ is called the **integrand**. The symbol dx indicates that x is the variable of integration. In the above example, $\int 3x^2 \, dx = x^3$ since $\frac{d(x^3)}{dx} = 3x^2$, or $d(x^3) = 3x^2 \, dx$. Therefore, if $f'(x)dx = 3x^2 \, dx$, then $f(x) = x^3$.

Integral Sign
Integrand

In this view, we see integration as a process of finding an antiderivative. Examples 21.1–21.4 help us establish this process.

• **Example 21.1:** Find $\int 10x \, dx$.

Solution: $\int 10x \, dx = 5x^2$ since

$$\frac{d(5x^2)}{dx} = 10x \quad \text{or} \quad d(5x^2) = 10x \, dx,$$

$$f(x) = 5x^2,$$

where $f'(x) = 10x$.

• **Example 21.2:** Find $\int (2x + 5)dx$.

Solution: $\int (2x + 5)dx = x^2 + 5x$ since $\frac{d}{dx}(x^2 + 5x) = 2x + 5$.

• **Example 21.3:** Find $\int \sin x \, dx$.

Solution: $\int \sin x \, dx = -\cos x$ since

$$\frac{d}{dx}(-\cos x) = -\frac{d}{dx}(\cos x) = -(-\sin x)$$
$$= \sin x.$$

• **Example 21.4:** Find $\int \frac{1}{x} dx$.

Solution: $\int \frac{1}{x} dx = \ln x$ since

$$\frac{d}{dx}(\ln x) = \frac{1}{x}.$$

We need to refine the process of integration to help us with the following problem. What is the derivative of $f(x) = x^2 + 5$? Clearly, $f'(x) = 2x$. What is the derivative of $f(x) = x^2 - 6$? Clearly $f'(x) = 2x$. Therefore, does $\int 2x \, dx = x^2 + 5$ or does $\int 2x \, dx = x^2 - 6$? Notice that $x^2 + 5$ and $x^2 - 6$ differ only in the constant terms. We could have x^2 with any constant term and the derivative would be $2x$; that is, if $f(x) = x^2 + c$, c any constant, then $f'(x) = 2x$. Hence, we may write

$$\int 2x \, dx = x^2 + c,$$

Constant of Integration where c stands for any arbitrary constant and is called the **constant of integration**

• **Example 21.5:** Find $\int 3x^2 \, dx$.

Solution: $\int 3x^2 \, dx = x^3 + c$ since

$$\frac{d}{dx}(x^3 + c) = 3x^2.$$

• **Example 21.6:** Find $\int e^x \, dx$.

Solution: $\int e^x \, dx = e^x + c$ since

$$\frac{d}{dx}(e^x + c) = e^x.$$

Given the appropriate information, it is often possible to find a particular value of the constant of integration, c. We discuss this further in section 21.5.

In our work with differentiation, we defined velocity as the rate of change of distance with respect to time. Using $s = f(t)$ as a distance function in terms of time t we have $v = s' = f'(t) = \frac{ds}{dt}$ as velocity. If the velocity function is known, we can find the distance function by finding the antiderivative of the velocity function. Example 21.7 illustrates this.

• **Example 21.7:** If the velocity of an object moving in a straight line is given by $v = 2t$, find a general expression for the distance traveled by the object.

Solution: Since $v = \frac{ds}{dt} = 2t$, s is a function whose derivative is $2t$. That is, s is the antiderivative of $2t$. Therefore, $s = \int 2t \, dt = t^2 + c$

since $\frac{d}{dt}(t^2 + c) = 2t$. Again, the value of c can be found when more information is given.

In a similar manner, velocity can be considered as the antiderivative of acceleration. If $a = \frac{dv}{dt}$, then $v = \int a \, dt$.

- **Example 21.8:** An object is falling to earth, and thus, is under the influence of gravity. Acceleration due to gravity is essentially constant and is given as $a = -32$ ft/sec². Find a general expression for velocity v and distance s.

    Solution: Since $a = -32 = \dfrac{dv}{dt}$,

    $$v = \int -32 \, dt$$
    $$= -32t + c_1.$$

    Since $v = \dfrac{ds}{dt}$,

    $$s = \int v \, dt = \int (-32t + c_1) \, dt$$
    $$= -16t^2 + c_1 t + c_2.$$

- **Example 21.9:** The rate of growth, in thousands per year, of the population of a newly incorporated city is given by $P'(t) = 10 + 6t^2$, where t is the number of decades since the city was incorporated. Find an expression for the population of the city at any time $t > 0$.

    Solution:
    $$P'(t) = 10 + 6t^2$$
    $$\int P'(t) \, dt = \int (10 + 6t^2) \, dt \quad \text{(Integrating both sides)}$$
    $$P(t) = 10t + 6 \cdot \left(\frac{1}{3} t^3\right) + c.$$

    Since $\dfrac{d}{dt}(10t) = t$ and $\dfrac{d}{dt}\left(\dfrac{1}{3} t^3\right) = t^2$,

    $$P(t) = 10t + 2t^3 + c.$$

Trial Problems 21.1

Before you begin the section exercises, warm up with these problems. Complete answers are included in the answer key.

Find the antiderivative of each of the following.

1. $f'(x) = 2$
2. $f'(x) = x$
3. $f'(x) = 2x$
4. $f'(x) = x^7$
5. $f'(x) = 6x^2 + 2$

Exercises 21.1

Verify, by differentiating, the following integrals.

1. $\int 2x^3 \, dx = \dfrac{1}{2}x^4 + c$
2. $\int 4x^{3/2} \, dx = \dfrac{8}{5}x^{5/2} + c$
3. $\int -5x^{-2} \, dx = 5x^{-1} + c$
4. $\int 7\sqrt[3]{x^4} \, dx = 3\sqrt[3]{x^7} + c$
5. $\int -5 \, dx = -5x + c$

Derivative functions are given below. Find the antiderivative of each and check the result by differentiating.

6. $f'(x) = 5x^4$, $f(x) =$ _____
7. $f'(x) = 6x^5$, $f(x) =$ _____
8. $f'(x) = 3x^2 + 5$, $f(x) =$ _____
9. $f'(x) = 2x - 7$, $f(x) =$ _____
10. $f'(x) = 8x^3$, $f(x) =$ _____
11. $f'(x) = x^{3/4}$, $f(x) =$ _____
12. $f'(x) = 2\sqrt{x}$, $f(x) =$ _____
13. $f'(x) = \dfrac{3}{4}x^{3/4}$, $f(x) =$ _____
14. $f'(x) = 5x^{2/5}$, $f(x) =$ _____
15. $f'(x) = -\dfrac{3}{x^4}$, $f(x) =$ _____

Integrate each of the following and check the result by differentiating. Don't forget the constant of integration.

16. $\int x^6 \, dx$
17. $\int 2x \, dx$
18. $\int -\dfrac{2}{x} \, dx$
19. $\int 2x^{-2} \, dx$
20. $\int x^3 \, dx$
21. $\int (3x^2 + 2x) \, dx$
22. $\int (3x^2 + 2x + 1) \, dx$
23. $\int (x + 3)^2 \, dx$
24. $\int (3 - x^2)^2 \, dx$
25. $\int \cos x \, dx$
26. $\int x \, dx$
27. $\int dx$
28. $\int (x^2 - 7) \, dx$
29. $\int 5 \, dx$
30. $\int e^2 \, dx$
31. $\int \dfrac{3}{2}(6x + 1)^{1/2}(6) \, dx$
32. $\int 8(1 - x^2)^7(-2x) \, dx$

33. Find the equation of the graph having slope $\dfrac{dy}{dx} = 3x$ and passing through the point (1,2).

34. A missile accelerates from rest according to $a = 24t + 100$ ft/sec².
    a. Find an expression for velocity at any time t.
    b. Find an expression for distance s at any time t.

35. If acceleration of an object starting at rest is given by $a = \sqrt{t}$ ft/sec², find an expression for (a) its velocity, and (b) its distance.

36. A newly developed natural-gas well is producing gas at the instantaneous rate given by

$$R'(t) = 20 + \dfrac{5}{2}t - \dfrac{1}{4}t^2,$$

where $R(t)$ is the number of millions of cubic feet of gas produced after t years of operation. Find the expression for $R(t)$.

37. The work done, dw, in moving a large piece of machinery through a distance, ds, is

$$dw = F \cos \theta \, ds,$$

where F is the applied force and θ is the angle the force makes with the plane. Find the expression for work done in moving the machinery if θ is a constant and $F = s^2$ lb.

38. In the study of fluid resistance, the rate of change of the velocity of the fluid with respect to the distance from the boundary layer is

$$\frac{dv}{dy} = \frac{2.5s}{y},$$

where s is the shear velocity of the given fluid. Assuming s to be a constant, find an expression for the velocity of the fluid as a function of the distance y.

39. A manufacturer of skis finds that producing x pairs of skis yields a marginal profit, in dollars per pair of skis, that is given by

$$P'(x) = 100 - 0.04x.$$

Find an expression for the profit $P(x)$.

40. A radioactive substance's rate of decay is given by

$$Q'(t) = -100e^{-0.2t} \text{ grams per week,}$$

where t is in weeks. Find an expression for the quantity $Q(t)$ present at any time t.

21·2 Basic Rules

In order to solve the problems in the previous exercises, one may have employed the "trial-and-error" method. After selecting $f(x)$ as a solution to $\int f'(x)dx$, we tested it by differentiating.

As for differentiation, there are specific rules for integration. The first of these involves $\int dx$. Notice that $\int dx = \int 1 \ dx = x + c$ since $\frac{d}{dx}(x + c) = 1$, or $d(x + c) = 1 \ dx = dx$.

Since 1 is a constant, we should be able to replace it with any constant. Letting k be a constant, we have $\int k \ dx = kx + c$ since $\frac{d}{dx}(kx + c) = k$. Furthermore, we can write

$$\int k \ dx = k \int dx = kx + c.$$

[More precisely, $\int k \ dx = k \int dx = k(x + c_1) = kx + kc_1 = kx + c$, where kc_1 is simply another constant, say c.]

Putting the constant k in front of the integral sign suggests that we do the same for the following form:

$$\int kf'(x)dx = k \int f'(x)dx = kf(x) + c.$$

• **Example 21.10:** Find $\int 3 \ dx$.

Solution: $\int 3 \ dx = 3 \int dx = 3x + c.$

- **Example 21.11:** Find $\int -4x\,dx$.

   Solution:
   $$\int -4x\,dx = -4\int x\,dx$$
   $$= -4\left(\frac{x^2}{2}\right) + c$$
   $$= -2x^2 + c.$$

From our work in differentiation, we know that

$$\frac{d}{dx}[f(x) \pm g(x)] = \frac{d}{dx}[f(x)] \pm \frac{d}{dx}[g(x)].$$

Similarly, for integration we have

$$\int [f'(x) \pm g'(x)]dx = \int f'(x)dx \pm \int g'(x)dx.$$

- **Example 21.12:** Find $\int (3x^2 + x)dx$.

   Solution:
   $$\int (3x^2 + x)dx = \int 3x^2\,dx + \int x\,dx$$
   $$= 3\int x^2\,dx + \int x\,dx \qquad \left(\text{since}\right.$$
   $$= 3\left(\frac{1}{3}x^3 + c_1\right) + \frac{1}{2}x^2 + c_2 \qquad \frac{d}{dx}\left(\frac{1}{3}x^3\right) = x^2$$
   $$\text{and}$$
   $$= x^3 + 3c_1 + \frac{1}{2}x^2 + c_2 \qquad \left.\frac{d}{dx}\left(\frac{1}{2}x^2\right) = x\right)$$
   $$= x^3 + \frac{1}{2}x^2 + c, \text{ where } c = 3c_1 + c_2.$$

Another basic rule involves a power of a variable. In the previous example we found that $\int x^2\,dx = \frac{1}{3}x^3 + c$. We can write this as follows: $\int x^2\,dx = \frac{x^3}{3} + c$
$= \frac{x^{2+1}}{2+1} + c$. This suggests the rule

Power Rule
$$\int x^n\,dx = \frac{x^{n+1}}{n+1} + c, \text{ provided that } n \neq -1.$$

We can verify this by differentiating.
$$\frac{d}{dx}\frac{(x^{n+1}+c)}{(n+1)} = \frac{(n+1)x^{(n+1)-1}}{(n+1)} = x^n.$$

NOTE: $n \neq -1$ or else the denominator will equal zero, an impossible situation.

• **Example 21.13:** Evaluate $\int x^6 \, dx$.

Solution: $\int x^6 \, dx = \frac{x^{6+1}}{6+1} + c = \frac{x^7}{7} + c.$

••••••••••

• **Example 21.14:** Evaluate $\int (4x^3 + 6x^2 + x + 2) \, dx$.

Solution: $\int (4x^3 + 6x^2 + x + 2) \, dx = \frac{4x^4}{4} + \frac{6x^3}{3} + \frac{x^2}{2} + 2x + c$

$$= x^4 + 2x^3 + \frac{1}{2}x^2 + 2x + c.$$

••••••••••

• **Example 21.15:** Evaluate $\int (x^{1/2} + x^{-1/4}) \, dx$.

Solution: $\int (x^{1/2} + x^{-1/4}) \, dx = \frac{x^{1/2+1}}{\frac{1}{2}+1} + \frac{x^{-1/4+1}}{-\frac{1}{4}+1} + c$

$$= \frac{x^{3/2}}{\frac{3}{2}} + \frac{x^{3/4}}{\frac{3}{4}} + c$$

$$= \frac{2}{3}x^{3/2} + \frac{4}{3}x^{3/4} + c.$$

••••••••••

• **Example 21.16:** A missile accelerates from rest according to
$$a = 20t + 100 \text{ ft/sec}^2.$$

Find an expression for the velocity of the missile as a function of time.

Solution: Since acceleration is the rate of change of velocity, velocity is the antiderivative of acceleration.
Thus,

$$v = \int a \, dt$$

and

$$v = \int (20t + 100) \, dt,$$

or

$$v = 10t^2 + 100t + c \text{ ft/sec.}$$

We summarize the rules for integration as follows.

BASIC INTEGRATION RULES

1. $\int dx = x + c$ (Constant rule)

2. $\int kf(x) \, dx = k \int f(x) \, dx$ (Constant multiple rule)

3. $\int [f(x) \pm g(x)] \, dx = \int f(x) \, dx \pm \int g(x) \, dx$ (Sum/difference rule)

4. $\int x^n \, dx = \dfrac{x^{n+1}}{n+1} + c$, if $n \neq -1$ (Power rule)

Trial Problems 21.2

Before you begin the section exercises, warm up with these problems. Complete answers are included in the answer key.

Use the basic rules of integration to evaluate each integral.

1. $\int x \, dx$ 2. $\int (x + 2) \, dx$ 3. $\int x^{-2} \, dx$

4. $\int x^2 \cdot x^{1/2} \, dx$ 5. $\int \dfrac{1}{x^2} \, dx$

Exercises 21.2

Evaluate the following integrals. Check the answers by differentiating.

1. $\int (x^3 - 7) \, dx$ 2. $\int (x^2 - 3x + 4) \, dx$ 3. $\int (x^{3/2} + 2x + 1) \, dx$

4. $\int (\sqrt{x} + 5) \, dx$ (Hint: $\sqrt{x} = x^{1/2}$.)   5. $\int \sqrt[3]{x^2} \, dx$ 6. $\int \dfrac{1}{x^2} \, dx$ $\left(\text{Hint: } \dfrac{1}{x^2} = x^{-2}. \right)$

7. $\int \dfrac{1}{x^3} \, dx$ 8. $\int \dfrac{1}{6x^2} \, dx$ 9. $\int (2x + x^{-2}) \, dx$

10. $\displaystyle\int \frac{x^2 + 1}{x^2}\, dx$ (Hint: divide first.)

11. $\displaystyle\int (x + 2)(x - 1)\, dx$ (Hint: multiply first.)

12. $\displaystyle\int (x - 2)^2\, dx$ (Hint: expand first.)

13. $\displaystyle\int (2r^2 - 1)^2\, dr$

14. $\displaystyle\int (1 + 2t)t\, dt$

15. $\displaystyle\int 2s(8 - s^{3/2})\, ds$

16. $\displaystyle\int u\sqrt{u}\, du$

17. $\displaystyle\int x^2\left(x - \frac{2}{x}\right)^2 dx$

18. $\displaystyle\int 5(x + 2)^3\, dx$

19. $\displaystyle\int \frac{-2x^{-6/5}}{3}\, dx$

20. $\displaystyle\int \frac{dx}{\sqrt[3]{x}}$

21. Can you suggest a method of integrating $\dfrac{dx}{1 - x^2}$?

22. Is it possible to evaluate $\displaystyle\int \frac{dx}{x + 1}$ by the power rule? Why?

23. The velocity of an object moving in a straight line is given by $v = 3t^2 + 2t$. Find an expression for the distance, s.

24. A spaceship moving through a force field experiences an acceleration $a = 300 - 2t$ m/sec². Find an expression for the velocity of the spaceship as a function of time t.

25. The rate of change of the vertical deflection y with respect to the horizontal distance x from one end of a beam is given by

$$y' = k(x^5 + 1350x^3 - 7000x^2),$$

where k is a constant. Find y as a function of x.

26. The equation for the voltage across the plates of a capacitor is given by

$$v = \frac{1}{c}\int i\, dt,$$

where c is a constant and i is the current. Find the voltage as a function of time t if

$$i = 0.04t^5 + 0.5t^4 + 40t^3 + 2t.$$

27. The slope of the tangent line to the path of a particle moving in a coordinate system is

$$y' = 6x - 5.$$

Find a general expression for the path of the particle.

28. At an automotive test site in Detroit, the velocity of an experimental car was found to be

$$v = 180\sqrt{t},$$

where v is in ft/sec when t is in seconds. Find a general expression for the displacement of the car as a function of time.

21•3 Variations of the Power Rule

The integral $\int (2x + 1)^2 \, dx$ can be evaluated by expanding $(2x + 1)^2$ and using the basic rules for integration:

$$\int (2x + 1)^2 \, dx = \int (4x^2 + 4x + 1) \, dx = \frac{4}{3}x^3 + 2x^2 + x + c.$$

In this example, $(2x + 1)$ is raised to a positive integral power. Suppose, instead, that we take the square root of $(2x + 1)$ and then integrate:

$$\int \sqrt{2x + 1} \, dx = \int (2x + 1)^{1/2} \, dx.$$

Changing to the one-half power creates a much more complex situation than having a positive integral power. Expanding such an expression leads to an infinite expression.

In order to work with the expression $(2x + 1)^{1/2}$, and many others similar to it, we extend the power rule:

$$\int x^n \, dx = \frac{x^{n+1}}{n+1} + c, \qquad n \neq -1,$$

to the form

$$\int u^n \, du = \frac{u^{n+1}}{n+1} + c, \qquad \text{where } u \text{ is a function of } x, \text{ and } n \neq -1.$$

For $\int (2x + 1)^{1/2} \, dx$, $u = 2x + 1$ and $n = \frac{1}{2}$. Since $u = 2x + 1$,

$$du = 2 \, dx.$$

Notice the factor of 2 in front of dx. This 2 is not in the original integral, although we place it in the original integral if we multiply by $\frac{1}{2}$ at the same time:

$$\int (2x + 1)^{1/2} \, dx = \frac{1}{2} \int \underbrace{(2x + 1)^{1/2}}_{u^n} \underbrace{(2 \, dx)}_{du}.$$

SECTION 21.3 VARIATIONS OF THE POWER RULE

Now we have the form $\int u^n\, du$, where $u = 2x + 1$, $n = \dfrac{1}{2}$, and $du = 2\, dx$. Completing the integration, we have

$$\int (2x + 1)^{1/2}\, dx = \frac{1}{2} \int (2x + 1)^{1/2} (2\, dx)$$

$$= \frac{1}{2} \left[\frac{(2x + 1)^{1/2 + 1}}{\frac{1}{2} + 1} \right] + c$$

$$= \frac{1}{2} \left[\frac{(2x + 1)^{3/2}}{\frac{3}{2}} \right] + c$$

$$= \frac{1}{2} \cdot \frac{2}{3} \cdot (2x + 1)^{3/2} + c$$

$$= \frac{1}{3}(2x + 1)^{3/2} + c.$$

The form $\int u^n\, du$ has many variations. Examples 21.17, 21.18 and 21.19 include some of these variations.

NOTE: It is not always possible to obtain the form $\int u^n\, du$. Sometimes an appropriate du can not be determined. In this case, there are other integration techniques available. Some of these will be discussed in chapter 22.

• **Example 21.17:** Evaluate $\int 2x(x^2 + 1)^3\, dx$.

Solution: The expression raised to a power is $(x^2 + 1)$. Therefore, $u = x^2 + 1$ and $n = 3$. Since $u = x^2 + 1$, $du = 2x\, dx$. Hence,

$\int 2x(x^2 + 1)^3\, dx = \int (x^2 + 1)^3 (2x\, dx)$, which is in the form $\int u^n\, du$. Completing the work,

$$\int 2x(x^2 + 1)^3\, dx = \int \underbrace{(x^2 + 1)^3}_{u} \underbrace{(2x\, dx)}_{du}$$

$$= \frac{(x^2 + 1)^4}{4} + c$$

$$= \frac{1}{4}(x^2 + 1)^4 + c.$$

Checking the work by differentiating,

$$\frac{d}{dx}\left[\frac{1}{4}(x^2 + 1)^4 + c\right] = \frac{1}{4}(4)(x^2 + 1)^3(2x)$$
$$= (x^2 + 1)^3(2x)$$

and

$$d\left[\frac{1}{4}(x^2 + 1)^4 + c\right] = (x^2 + 1)^3(2\dot{x}\ dx)$$
$$= 2x(x^3 + 1)^3\ dx;$$

which is the original integrand.

• **Example 21.18:** Evaluate $\int (2x + 1)(x^2 + x)dx$.

Solution: Let $u = x^2 + x$. Then $n = 1$ and $du = (2x + 1)dx$. Hence,

$$\int (2x + 1)(x^2 + x)dx = \int (x^2 + x)^1 (2x + 1)dx$$
$$= \frac{(x^2 + x)^2}{2} + c$$
$$= \frac{1}{2}(x^2 + x)^2 + c.$$

• **Example 21.19:** Evaluate $\int \frac{x\ dx}{\sqrt{1 - x^2}}$.

Solution: Rewriting, we have

$$\int \frac{x\ dx}{\sqrt{1 - x^2}} = \int (1 - x^2)^{-1/2}(x\ dx).$$

Hence, $u = 1 - x^2$, $n = -\dfrac{1}{2}$, and $du = -2x\,dx$. Now,

$$\int \frac{x\,dx}{\sqrt{1-x^2}} = \int (1-x^2)^{-1/2}(x\,dx)$$

$$= -\frac{1}{2}\int (1-x^2)^{-1/2}(-2x\,dx)$$

$$= -\frac{1}{2}\left[\frac{(1-x^2)^{-1/2+1}}{-\dfrac{1}{2}+1}\right] + c$$

$$= -\frac{1}{2}\left[\frac{(1-x^2)^{1/2}}{\dfrac{1}{2}}\right] + c$$

$$= -\frac{1}{2}\cdot\frac{2}{1}(1-x^2)^{1/2} + c$$

$$= -\sqrt{1-x^2} + c.$$

● ● ● ● ● ● ● ● ● ●

NOTE: Only constant multipliers, or factors, can be placed inside and outside the integral sign. We cannot do the same with variable factors. See example 21.20.

• **Example 21.20:** Evaluate $\displaystyle\int (x^2+1)^3\,dx$.

Solution: In this example, if we let $u = x^2 + 1$ and $n = 3$, then $du = 2x\,dx$. Since there is no factor of x in the integrand, we cannot put this integral in the form $\displaystyle\int u^n\,du$. That is, it would be wrong to write

$$\int (x^2+1)^3\,dx = \frac{1}{2x}\int (x^2+1)^3(2x\,dx).$$

(Can you give reasons why this would not work?) At present, we have to evaluate this integral by expanding $(x^2+1)^3$:

$$\int (x^2+1)^3\,dx = \int (x^6 + 3x^4 + 3x^2 + 1)\,dx$$

$$= \frac{1}{7}x^7 + \frac{3}{5}x^5 + x^3 + x + c.$$

● ● ● ● ● ● ● ● ● ●

• **Example 21.21:** If the velocity of an object is given by $v = (5.2t + 1)^{0.6}$ ft/sec, find an expression for the distance s traveled by the object in t sec.

Solution: The distance traveled by the object is found by evaluating

$$s(t) = \int v \, dt.$$

Thus,

$$s(t) = \int (5.2t + 1)^{0.6} \, dt.$$

Let $u = 5.2t + 1$. Then $du = 5.2 \, dt$ and $n = 0.6$:

$$s(t) = \frac{1}{5.2} \int (5.2t + 1)^{0.6} (5.2 \, dt)$$

$$= 0.19 \left[\frac{(5.2t + 1)^{1.6}}{1.6} \right] + c$$

$$= 0.12(5.2t + 1)^{1.6} + c.$$

In the listing of the basic integration rules, rule number 5 can be included:

5. $\int u^n \, du = \dfrac{u^{n+1}}{n+1} + c$, where $u = f(x)$ and $n \neq -1$. (General power rule.)

Trial Problems 21.3

Before you begin the section exercises, warm up with these problems. Complete answers are included in the answer key.

Use the general power rule to evaluate the following integrals.

1. $\int (x + 1)^2 \, dx$
2. $\int 2x(x^2 + 1) \, dx$
3. $\int (x + 1)^{1/2} \, dx$
4. $\int 2x(x^2 + 1)^{1/2} \, dx$
5. $\int \dfrac{1}{(x + 1)^2} \, dx$

Exercises 21.3

Evaluate the following integrals using the general power rule. Check the answers by differentiating.

1. $\int (2 + 3x)^4 \, dx$
2. $\int x\sqrt{9 - x^2} \, dx$
3. $\int x^2(x^3 - 1)^4 \, dx$
4. $\int \dfrac{dx}{\sqrt{1 - x}}$
5. $\int \dfrac{4x}{1 + x^2} \, dx$
6. $\int \dfrac{3 \, dx}{(x + 1)^2}$

7. $\displaystyle\int \frac{x^2}{(1+x^3)^2}\,dx$

8. $\displaystyle\int \frac{dx}{2(x-5)^3}$

9. $\displaystyle\int 5x\sqrt[3]{1+x^2}\,dx$

10. $\displaystyle\int \frac{6r}{(1+r^2)^3}\,dr$

11. $\displaystyle\int 7(x-3)^{5/2}\,dx$

12. $\displaystyle\int \frac{4t+6}{(t^2+3t+7)^3}\,dt$

13. $\displaystyle\int \frac{x\,dx}{\sqrt[3]{1-x^2}}$

14. $\displaystyle\int (2+x)(2+x)\,dx$

15. $\displaystyle\int \frac{z+1}{(z^2+2z-3)^2}\,dz$

16. $\displaystyle\int x^3\sqrt{x^4+5}\,dx$

17. $\displaystyle\int \frac{1}{\sqrt{x}(1+\sqrt{x})^2}\,dx$

18. $\displaystyle\int \left(1+\frac{1}{x}\right)^3\left(\frac{1}{x^2}\right)dx$

19. $\displaystyle\int \frac{1}{(3s)^2}\,ds$

20. $\displaystyle\int \frac{1}{\sqrt{2r}}\,dr$

21. Integrate $\displaystyle\int (x^4-1)^2 x^3\,dx$ in two ways. Are the answers identical? Are they both correct?

22. Find the equation of the curve whose slope is $-x\sqrt{1-4x^2}$ and passes through the point $(0,7)$.

23. In the study of variable inductance electrical circuits, integrals of the form $\displaystyle\int \frac{dt}{(7-t)^2}$ must be evaluated. Evaluate this integral.

24. When finding distances traveled by particles, integrals of the form $\displaystyle\int \sqrt{6t-5}\,dt$ are involved. Evaluate this integral.

25. A manufacturing plant dumps harmless waste into a nearby stream at a rate, in tons per year, given by

$$A'(t) = \frac{\ln(t+1)}{t+1}, \quad t \geq 0,$$

where t is the number of years that the plant has been in operation. Find the amount of waste that the plant has dumped into the stream during the first t years of operation.

26. The rate of change of resistance of a certain resistor with respect to temperature is given by

$$R'(T) = \frac{0.002T}{\sqrt[3]{3T^2+1}}.$$

Find the resistance for any time T.

27. In a mechanical system the power p, which is the rate of change of work, is given by

$$p = 5(2t+1)^{0.2} \text{ ft-lb/sec},$$

where t is time in seconds. Find the work equation for any time t.

28. The velocity of a particle is given by

$$v(t) = 3t^2\sqrt{t^3+9} \text{ ft/sec}.$$

Find an expression for the distance traveled by the particle at any time t.

21 • 4 Indefinite and Definite Integrals

When we integrate a function, we must remember to add the constant of integration c. This tells us that there are many possible functions, each of whose derivative is equal to the integrand. When we write

$$\int f'(x)dx = f(x) + c$$

we are not finding a specific answer to the integration problem. Rather, we are finding one of many functions that differ only by a constant. For example,

$$\int 3x^2 \, dx = x^3 + 3,$$
$$= x^3 - 7, \quad \text{or}$$
$$= x^3 + c,$$

where c is any constant. Because we have not specified the value of c in the last form, we say that

$$\int 3x^2 \, dx = x^3 + c$$

Indefinite Integral is an **indefinite integral**.

Additional information is needed to determine the value of c in an indefinite integral. Once we have determined the value of c, we say that the integral has a **particular solution**.

Particular Solution

$$\int f'(x)dx = F(x),$$

where $F(x) = f(x) + c$ and c is known, is a particular solution of the indefinite integral.

We can determine c if we know a specific value of $f(x)$ for some x. For example, suppose we wish to evaluate $\int 3x^2 \, dx$ if we know that $f(0) = 1$. Then we have

$$\int 3x^2 \, dx = x^3 + c.$$

Since $f(0) = 1$, $(0)^3 + c = 1$ or

$$c = 1.$$

Therefore, $\int 3x^2 \, dx = x^3 + 1$ when $f(0) = 1$.

SECTION 21.4 INDEFINITE AND DEFINITE INTEGRALS

• **Example 21.22:** Evaluate $\int 2x(x^2 + 1)^3 \, dx$ if $f(1) = 6$.

Solution: $\int 2x(x^2 + 1)^3 \, dx = \dfrac{(x^2 + 1)^4}{4} + c.$

$(u = x^2 + 1, n = 3, du = 2x \, dx)$

Since

$$f(1) = 6, \quad \dfrac{(1^2 + 1)^4}{4} + c = 6,$$

$$\dfrac{2^4}{4} + c = 6$$

or,

$$4 + c = 6$$
$$c = 2.$$

Therefore, $\int 2x(x^2 + 1)^3 \, dx = \dfrac{1}{4}(x^2 + 1)^4 + 2.$

••••••••••

• **Example 21.23:** The marginal cost for producing x cameras is given by

$$\dfrac{dc}{dx} = 225 - 1.10x.$$

If it costs $250 to make 1 camera, find the cost of making 200 cameras.

Solution: We can write $dc = (225 - 1.10x)dx,$

or (cost): $c = \int (225 - 1.10x)dx$

$= 225x - \dfrac{1.10}{2}x^2 + h$ (*h* is the constant of integration)

$= 225x - 0.55x^2 + h.$

Since 1 camera costs $250, $f(1) = 250$.
Therefore, $225(1) - 0.55(1)^2 + h = 250$
or $224.45 + h = 250$
$h = 25.55.$

Hence, $c = 225x - 0.55x^2 + 25.55.$
The cost of making 200 cameras is

$$c = 225(200) - 0.55(200)^2 + 25.55$$
$$= 45{,}000 - 22{,}000 + 25.55$$
$$= \$23{,}025.55.$$

••••••••••

Initial Conditions

We can find the constant of integration if we know the **initial conditions** of $f(x)$. In other words, for

$$\int f'(x)\,dx = f(x) + c$$

we can determine the value of c if we know that $f(a) = b$, where a and b are specified values.

We investigated many applications of derivatives of functions. Similarly, we want to apply our knowledge of integration to solving problems in many technical fields. In order to do this, we need to know how to evaluate an integral and obtain an answer that is a number instead of an indefinite integral.

The process of obtaining a number for $\int f'(x)\,dx$ involves evaluating $f(x)$ at two different values and finding their difference. We can write

$$\int_a^b f'(x)\,dx = [f(x) + c]\Big|_a^b$$
$$= [f(b) + c] - [f(a) + c]$$
$$= f(b) + c - f(a) - c$$
$$= f(b) - f(a).$$

Limits of Integration
Lower Limit
Upper Limit

The numbers a and b are called the **limits of integration**, and the result $f(b) - f(a)$ is a number also. The number a is called the **lower limit** and b is called the **upper limit**

THE FUNDAMENTAL THEOREM OF CALCULUS

Definite Integral

The integral $\int_a^b f'(x)\,dx$ is called a **definite integral**, and

$$\int_a^b f'(x)\,dx = f(b) - f(a).$$ This is known as the **fundamental theorem of calculus.**

• *Example 21.24:* Evaluate $\int_1^4 x^3\,dx$.

Solution: $\int_1^4 x^3\,dx = \dfrac{x^4}{4}\Big|_1^4 = \dfrac{(4)^4}{4} - \dfrac{(1)^4}{4}$

$$= \dfrac{256}{4} - \dfrac{1}{4}$$

$$= \dfrac{255}{4}.$$

Notice that the constant of integration is not needed when evaluating definite integrals.

• **Example 21.25:** Evaluate $\int_0^1 x(x^2 + 1)^3 \, dx$.

Solution:
$$\int_0^1 x(x^2 + 1)^3 \, dx = \frac{1}{2} \int_0^1 (x^2 + 1)^3 (2x \, dx)$$
$$= \frac{1}{2} \cdot \frac{1}{4}(x^2 + 1)^4 \Big|_0^1$$
$$= \frac{1}{8}((1)^2 + 1)^4 - \frac{1}{8}((0)^2 + 1)^4$$
$$= \frac{1}{8} \cdot 16 - \frac{1}{8} = \frac{15}{8}.$$

• **Example 21.26:** Evaluate $\int_0^{2\pi} \cos x \, dx$.

Solution:
$$\int_0^{2\pi} \cos x \, dx = \sin x \Big|_0^{2\pi} \quad \left(\int \cos x \, dx = \sin x \text{ since } \frac{d}{dx}(\sin x) = \cos x. \right)$$
$$= \sin 2\pi - \sin 0$$
$$= 0 - 0$$
$$= 0.$$

• **Example 21.27:** A stone is dropped from the top of a building 50 feet tall. Find expressions for velocity and distance.

Solution: Neglecting air resistance, acceleration a is given by $a = -32$ ft/sec² (downward pull of gravity).
Therefore, $v = \int a \, dt = \int -32 \, dt = -32t + c_1$.
When $t = 0$, $v = 0$, since the stone has no velocity the instant it is released.
Hence, $0 = -32(0) + c_1$ or $c_1 = 0$.
Thus, $v = -32t$ (stone is traveling *down* at $32t$ ft/sec.)
Now, $s = \int v \, dt = \int -32t \, dt = -16t^2 + c_2$.
When $t = 0$, $s = 50$, since the stone is 50 ft. above the ground.
Therefore, $50 = -16(0)^2 + c_2$ or $c_2 = 50$.
Hence, $s = -16t^2 + 50$.

Trial Problems 21.4

Before you begin the section exercises, warm up with these problems. Complete answers are included in the answer key.

Evaluate the following integrals.

1. $\int_0^1 2 \, dx$
2. $\int_0^1 2x \, dx$
3. $\int_0^1 2x^2 \, dx$
4. $\int_0^4 \sqrt{x} \, dx$
5. $\int_1^2 2x(x^2 + 1) \, dx$

Exercises 21.4

Evaluate the following integrals given the initial conditions.

1. $\int 10x \, dx, \quad f(2) = 35$
2. $\int (6x + 2) \, dx, \quad f(-10) = 0$
3. $\int x^{1/3} \, dx, \quad f(1) = 1$
4. $\int x^{-1/3} \, dx, \quad f(1) = 1$
5. $\int 3(4x^2 + 3x + 1)^2 (8x + 3) \, dx, \quad f(0) = 1$
6. $\int (2 + 3x)^4 \, dx, \quad f(0) = 0$
7. $\int \frac{dx}{\sqrt{3-x}}, \quad f(1) = 2$
8. $\int 5x \sqrt[3]{x^2 + 1} \, dx, \quad f(0) = 5$
9. $\int \frac{dx}{4(x-5)^3}, \quad f(5) = 4$
10. $\int 2(x - 4)^{3/2} \, dx, \quad f(8) = 32$

Evaluate the following definite integrals.

11. $\int_0^1 (x + 1)^4 \, dx$
12. $\int_0^3 \sqrt{4 - t} \, dt$
13. $\int_0^1 \frac{r + 1}{r^2 + 2r} \, dr$
14. $\int_{1/2}^1 4z(4z^2 - 1)^{3/2} \, dz$
15. $\int_{-1}^{1/3} (2 - 3x)^{-3} \, dx$
16. $\int_0^1 \sqrt{x}(1 - x) \, dx$
17. $\int_0^2 \frac{t}{1 + 2t^2} \, dt$
18. $\int_1^2 \left(\frac{3}{w^2} - 1\right) dw$
19. $\int_1^4 \frac{x - 2}{\sqrt{x}} \, dx$
20. $\int_{-3}^3 t^{1/3} \, dt$
21. $\int_{-2}^{-1} \left(x - \frac{1}{x^2}\right) dx$
22. $\int_0^1 x \sqrt{1 - x^2} \, dx$

23. Find the total cost function for a product if the marginal cost of producing x units is $\frac{dc}{dx} = 2x - 12$ and the cost for 1 unit is $125. What would be the total cost of producing 50 units?

24. An evergreen nursery sells yews after 6 years of growth and shaping. The growth rate after n years is given by $\frac{dh}{dn} = \frac{1}{2}n + 2$, where $n = 0$ represents the yews as seedlings 5 inches tall ($h = 5$ when $n = 0$).
   a. What is the height of the yews after n years?
   b. How tall are the yews when they are sold?

25. Find the expression for velocity if $a = 8t$, given that $v = 8$ when $t = 1$.

26. A projectile is fired vertically upward from a tower 75 feet high. The initial velocity is 500 ft/sec.
   a. What is the velocity at any time t?
   b. What is the expression for distance at any time t?
   c. What is the distance traveled during the first 5 sec?

27. An epidemic of Asian flu has hit a certain city. The rate of growth of the disease, in new cases per day, is given by

$$G'(t) = 3t^2 \sqrt{t^3 + 9},$$

where t is the number of days after the start of the epidemic. It was determined that 30 people had the disease when the epidemic started. How many people will have been affected during the first 3 days after the beginning of the epidemic?

28. Power is the rate of change of work. The power generated by a mechanical system is

$$P = 5(2t + 1)^{0.2} \text{ ft-lb/sec.}$$

What is the work equation of the system if $w = 0$ when $t = 0$?

29. Let $P(t)$ be the total number of circuit boards assembled by a high tech assembly line after t hours of work. The rate of production at time t is $60 + 2t - \dfrac{3}{8}t^2$ boards per hour. Find the formula for $P(t)$. (*Hint:* $P(0) = 0$.)

30. The rate of change of a certain electric resistor with respect to temperature is given by $\dfrac{0.002T}{\sqrt[3]{3T^2 + 1}}$. Find the resistance as a function of temperature if $R = 0.5$ mΩ when $T = 0°$ C.

21•5 Integrals Involving Exponential and Logarithmic Functions

From our work in differentiation, we know that if $f(x) = e^x$, then $f'(x) = e^x$ also. Thus, $\int e^x \, dx = e^x + c$. In the case where the power of e is a more complex function of x, we must determine if the integral is in the form $\int e^u \, du$, where u is a function of x. Example 21.28 shows that

$$\int e^u \, du = e^u + c, \text{ where } u = f(x).$$

• **Example 21.28:** Evaluate $\int xe^{x^2} \, dx$.

Solution: Let $u = x^2$. Then $du = 2x \, dx$.

$$\int xe^{x^2} \, dx = \frac{1}{2} \int e^{x^2}(2x \, dx) \quad \left(\text{In the form } \int e^u \, du\right)$$

$$= \frac{1}{2} e^{x^2} + c. \quad \left(\int e^u \, du = e^u + c\right)$$

• **Example 21.29:** Evaluate $\int e^{x^2} \, dx$.

Solution: Let $u = x^2$, then $du = 2x \, dx$. We see that $\int e^{x^2} \, dx$ cannot be put into the form $e^u \, du$. Therefore, the integral cannot be evaluated at this time. More sophisticated procedures are needed. They are discussed later.

The general exponential function a^u, where u is a function of x, can be integrated if we can put it into the form $a^u \, du$. If $f(x) = a^x$, recall that $f'(x) = a^x \ln a$. This says that $a^x = \dfrac{1}{\ln a} f'(x)$, or $a^x = \dfrac{1}{\ln a} \dfrac{dy}{dx}$, and $a^x \, dx = \dfrac{1}{\ln a} dy$. Therefore,

$$\int a^x \, dx = \dfrac{1}{\ln a} a^x + c.$$

Similarly,

$$\int a^u \, du = \dfrac{1}{\ln a} a^u + c, \text{ where } u \text{ is a function of } x.$$

• **Example 21.30:** Evaluate $\int 3^{x^2} x \, dx$.

Solution: Let $u = x^2$. Then $du = 2x \, dx$.

$$\int 3^{x^2} x \, dx = \dfrac{1}{2} \int 3^{x^2} (2x \, dx)$$

$$= \dfrac{1}{2} \dfrac{1}{\ln 3} 3^{x^2} + c$$

$$= \dfrac{3^{x^2}}{2 \ln 3} + c.$$

· · · · · · · · · ·

We cannot evaluate $\int \log x \, dx$ and $\int \ln x \, dx$ at this time, but we can evaluate integrals of functions that contain $\log x$ or $\ln x$ *in the answer*. If $y = \ln x$, we found that $y' = \dfrac{1}{x}$. Changing symbolization, we have $\dfrac{dy}{dx} = \dfrac{1}{x}$ or $dy = \dfrac{1}{x} dx$. Therefore,

$$\int \dfrac{1}{x} dx = y + c$$

$$= \ln x + c. \quad \text{(Since } y = \ln x\text{)}$$

Notice that $\dfrac{1}{x} = x^{-1}$ and $\int x^{-1} dx$ is of the form $\int x^n \, dx$. Now we can evaluate the integral for any value of n.

In general, we can evaluate the form $\int \dfrac{1}{u} du$ for u a function of x.

$\int \dfrac{1}{u} du = \ln |u| + c$. We must use the absolute value sign since $\dfrac{1}{u}$ is defined for all values of u, except $u = 0$, but $\ln u$ is defined only for $u > 0$.

SECTION 21.5 INTEGRALS INVOLVING EXPONENTIAL AND LOGARITHMIC FUNCTIONS

• **Example 21.31:** Evaluate $\int \dfrac{x}{x^2 + 1} \, dx$.

Solution: Let $u = x^2 + 1$. Then $du = 2x \, dx$. Therefore,

$$\int \frac{x}{x^2 + 1} \, dx = \frac{1}{2} \int \frac{2x \, dx}{x^2 + 1}$$

$$= \frac{1}{2} \ln(x^2 + 1) + c.$$

(No absolute value signs needed since $x^2 + 1$ is always positive.)

• **Example 21.32:** Evaluate $\int \dfrac{e^x}{e^x + 1} \, dx$.

Solution: Let $u = e^x + 1$. Then $du = e^x \, dx$. Thus, the problem is already in the form $\int \dfrac{1}{u} \, du$:

$$\int \frac{e^x}{e^x + 1} \, dx = \int \frac{e^x \, dx}{e^x + 1} = \ln(e^x + 1) + c.$$

• **Example 21.33:** Evaluate $\int \dfrac{x + 1}{x - 1} \, dx$.

Solution: Before integrating, divide $x - 1$ into $x + 1$:

$$\begin{array}{r} 1 \\ x-1\overline{\smash{)}x+1} \\ \underline{x-1} \\ 2 \end{array}$$

Therefore,

$$\frac{x + 1}{x - 1} = 1 + \frac{2}{x - 1}.$$

Now,

$$\int \frac{x + 1}{x - 1} \, dx = \int \left(1 + \frac{2}{x - 1}\right) dx$$

$$= \int dx + \int \frac{2}{x - 1} \, dx. \quad \text{(Let } u = x - 1, \text{ then } du = dx\text{)}$$

$$= x + 2 \ln |x - 1| + c.$$

Therefore, $\int \dfrac{x+1}{x-1} dx = x + 2 \ln |x-1| + c.$

In the examples we expressed answers in terms of the natural logarithm, ln, instead of log. If necessary, we can always express ln as log by the conversion given in chapter 10.

• **Example 21.34:** The velocity of a charged particle in a magnetic field is found to be

$$v = \dfrac{10}{12t+1} \text{ cm/sec.}$$

How far does the particle move from $t = 0$ to $t = 1$ sec?

Solution: The distance s moved by the particle is found by integrating $v\, dt$, from $t = 0$ to $t = 1$. Thus,

$$s = \int_0^1 v\, dt$$

$$= \int_0^1 \dfrac{10}{12t+1} dt.$$

Let $u = 12t + 1$. Then $du = 12\, dt$. Then

$$s = \dfrac{1}{12} \int_0^1 \dfrac{10}{12t+1} (12\, dt)$$

$$= \dfrac{10}{12} \int_0^1 \dfrac{1}{12t+1} (12\, dt)$$

$$= \dfrac{5}{6} \ln (12t+1) \Big|_0^1$$

$$= \dfrac{5}{6} \ln (12(1)+1) - \dfrac{5}{6} \ln (12(0)+1)$$

$$= \dfrac{5}{6} \ln 13 - \dfrac{5}{6} \ln 1$$

$$= \dfrac{5}{6} \ln 13 - 0$$

$$\approx 2.14 \text{ cm/sec.}$$

SECTION 21.5 INTEGRALS INVOLVING EXPONENTIAL AND LOGARITHMIC FUNCTIONS

Trial Problems 21.5

Before you begin the section exercises, warm up with these problems. Complete answers are included in the answer key.

Evaluate the following integrals.

1. $\int e^x \, dx$
2. $\int e^{-x} \, dx$
3. $\int 2^x \, dx$
4. $\int \dfrac{2}{x} \, dx$
5. $\int \dfrac{2x}{x^2 - 1} \, dx$

Exercises 21.5

Evaluate the following integrals. Check the work by differentiation.

1. $\int \dfrac{3}{5} e^{3x} \, dx$
2. $\int e^{-3x} \, dx$
3. $\int \dfrac{4 \, dx}{5x}$
4. $\int \dfrac{dx}{2x + 3}$
5. $\int 5^{x/3} \, dx$
6. $\int 4^{2x-1} \, dx$
7. $\int \dfrac{dr}{2 - 3r}$
8. $\int \dfrac{2t \, dt}{t^2 + 2}$
9. $\int e^{2x^3} x^2 \, dx$
10. $\int e^x(e^x + 1) \, dx$
11. $\int \dfrac{v \, dv}{v^2 - 4}$
12. $\int \dfrac{x^2 \, dx}{x^3 - 1}$
13. $\int \dfrac{6 \, dx}{3^{2x}}$
14. $\int (e^{x/2} + e^{-x/2}) \, dx$
15. $\int \dfrac{x^2 - 1}{x^3} \, dx$
16. $\int \dfrac{2x^2 - 2}{x} \, dx$
17. $\int \dfrac{\ln x}{x} \, dx$
18. $\int \dfrac{e^{\sqrt{x}}}{\sqrt{x}} \, dx$
19. $\int \dfrac{x^3 \, dx}{x^2 - 1}$
20. $\int \dfrac{3x^2 - 2}{x^3 - 2x + 1} \, dx$
21. $\int \dfrac{e^x - e^{-x}}{e^x + e^{-x}} \, dx$
22. $\int \dfrac{e^x}{e^x + 1} \, dx$
23. $\int_1^2 \dfrac{3x}{x^2 + 4} \, dx$
24. $\int_{-1}^{0} \dfrac{1}{4 - 5x} \, dx$
25. $\int_0^1 e^{2x+3} \, dx$
26. $\int_1^2 5^{-2x} \, dx$
27. $\int_{-1}^{1} 2^{3x-1} \, dx$
28. $\int_0^1 x^2 2^{x^3} \, dx$
29. $\int e^{4x} \, dx$
30. $\int e^{2x/3} \, dx$
31. $\int_0^1 e^{-5x} \, dx$
32. $\int_0^4 e^{-3x/4} \, dx$
33. $\int_0^1 2^{-x} \, dx$
34. $\int_{-1}^{0} 10^{2x} \, dx$

35. The slope of the tangent line to a curve is given by $\dfrac{dy}{dx} = e^{x+2}$. Find the equation of the curve if it passes through the point $(1, 0)$.

36. The impulse I of the force acting on an object is given by $I = \int_{t_1}^{t_2} F \, dt$. If $F = \dfrac{1}{1 + 3t}$ lb, find the impulse from $t = 0$ to $t = \dfrac{1}{2}$ sec.

37. The velocity of a charged particle in a magnetic field is given by $v = \dfrac{12}{6t + 1}$ cm/sec. Find the distance that the particle moves from $t = 0$ to $t = 1$ sec.

38. The velocity of a charged particle is given by $v = e^{t/3}$ cm/sec. Find the distance traveled by the particle from $t = 0$ to $t = 5$ sec.

39. The temperature of a heated metal plate increases at the rate $\dfrac{dT}{dt} = 3e^{t/100}$ deg/min. Find the temperature of the plate 3 minutes after being heated from its initial temperature of 75°.

40. During the combustion of a certain flammable material, heat is radiated according to the formula
$$\frac{dH}{dt} = \frac{3t}{t^2 + 4} \text{ btu/sec.}$$
Find the amount of heat generated from $t = 0$ to $t = 2$ sec.

21•6 Integrals of the Trigonometric Functions

In this section we investigate integrals of trigonometric functions as well as integrals that lead to trigonometric solutions.

In chapter 19 we learned how to differentiate trigonometric functions. Using the rules for differentiation gives us corresponding rules for integration.

Differentiation

$D_x(\sin x) = \cos x$

$D_x(\cos x) = -\sin x$

$D_x(\tan x) = \sec^2 x$

$D_x(\cot x) = -\csc^2 x$

$D_x(\sec x) = \sec x \tan x$

$D_x(\csc x) = -\csc x \cot x$

Integration

$\int \cos x \, dx = \sin x + c$

$\int \sin x \, dx = -\cos x + c$

$\int \sec^2 x \, dx = \tan x + c$

$\int \csc^2 x \, dx = -\cot x + c$

$\int \sec x \tan x \, dx = \sec x + c$

$\int \csc x \cot x \, dx = -\csc x + c$

Our list should include four more basic integrals: $\int \tan x \, dx$, $\int \cot x \, dx$, $\int \sec x \, dx$, and $\int \csc x \, dx$.

Let us consider $\int \tan x \, dx$ first. We can write

$$\int \tan x \, dx = \int \frac{\sin x}{\cos x} \, dx \qquad \text{(Let } u = \cos x, \text{ then } du = -\sin x \, dx\text{)}$$

$$= -\int \frac{-\sin x \, dx}{\cos x} \qquad \left(\frac{1}{u} \, du \text{ form, where } u = \cos x\right)$$

$$= -\ln |\cos x| + c$$

$$= \ln [|\cos x|]^{-1} + c$$

$$= \ln \left|\frac{1}{\cos x}\right| + c$$

$$= \ln |\sec x| + c.$$

SECTION 21.6 INTEGRALS OF THE TRIGONOMETRIC FUNCTIONS

Therefore, $\int \tan x \, dx = \ln |\sec x| + c$. In a similar manner,

$\int \cot x \, dx = \ln |\sin x| + c$.

Finding the integral of sec x is not so straight-forward.

$$\int \sec x \, dx = \int \frac{\sec x(\sec x + \tan x)}{(\sec x + \tan x)} \, dx$$

$$= \int \frac{\sec^2 x + \sec x \tan x}{\sec x + \tan x} \, dx. \quad \text{(Multiplying and dividing by sec } x + \tan x\text{)}$$

Let $u = \sec x + \tan x$.

Then $du = (\sec x \tan x + \sec^2 x) dx$.

$\left(\text{The form is } \int \frac{du}{u} = \ln |u| + c.\right)$

Therefore,

$$\int \sec x \, dx = \ln |\sec x + \tan| + c.$$

In a similar manner, $\int \csc x \, dx = \ln |\csc x - \cot x| + c$.

All of the previous basic trigonometric integrals are in simplest form. For example, $\int \sin x \, dx$ in general form is $\int \sin u \, du$, where u is a function of x. Examples 21.35 through 21.40 help us be aware of the general form.

• **Example 21.35:** Evaluate $\int 3x^2 \sin x^3 \, dx$.

Solution: Let $u = x^3$. Then $du = 3x^2 \, dx$.

$$\int 3x^2 \sin x^3 \, dx = \int \sin x^3 \, (3x^2 \, dx) \qquad \text{(This is in the general form } \int \sin u \, du = -\cos u + c\text{)}$$
$$= -\cos x^3 + c.$$

• **Example 21.36:** Evaluate $\int \cos 5x \, dx$.

Solution: Let $u = 5x$. Then $du = 5 \, dx$.

$$\int \cos 5x \, dx = \frac{1}{5} \int \cos 5x (5 \, dx)$$
$$= \frac{1}{5} \sin 5x + c.$$

• **Example 21.37:** Evaluate $\int \dfrac{dx}{\sin^2 x}$.

Solution: Since $\csc^2 x = \dfrac{1}{\sin^2 x}$,

$$\int \dfrac{dx}{\sin^2 x} = \int \csc^2 x \, dx = -\cot x + c.$$

• **Example 21.38:** Evaluate $\int \sec^3 x \tan x \, dx$.

Solution: $\int \sec^3 x \tan x \, dx = \int \sec^2 x (\sec x \tan x) dx$

Let $u = \sec x$, $n = 2$. Then $du = \sec x \tan x \, dx$. Therefore,

$$\int \sec^3 x \tan x \, dx = \dfrac{\sec^3 x}{3} + c. \quad \left(\text{Using } \int u^n \, du = \dfrac{u^{n+1}}{n+1} + c, \text{ where } u = \sec x\right)$$

• **Example 21.39:** Evaluate $\int x \csc^2(x^2 + 1) dx$.

Solution: Let $u = x^2 + 1$. Then $du = 2x \, dx$,

$$\int x \csc^2(x^2 + 1) dx = \dfrac{1}{2} \int \csc^2(x^2 + 1)(2x \, dx)$$

$$= -\dfrac{1}{2} \cot(x^2 + 1) + c.$$

• **Example 21.40:** The total radiation through a hemispherical surface is given by

$$R = 2\pi h \int_0^{\pi/2} \cos\theta \sin\theta \, d\theta,$$

where h is a constant and θ is the angle of radiation. Find the total radiation.

Solution: $R = 2\pi h \int_0^{\pi/2} \cos\theta \sin\theta \, d\theta$.

Let $u = \sin\theta$. Then $du = \cos\theta \, d\theta$ and $n = 1$. Thus,

$$R = 2\pi h \int_0^{\pi/2} \cos\theta \sin\theta \, d\theta$$

$$= 2\pi h \int_0^{\pi/2} \underbrace{\sin\theta}_{u} \underbrace{(\cos\theta \, d\theta)}_{du}$$

$$= 2\pi h \left(\frac{\sin^2\theta}{2}\right)\Big|_0^{\pi/2} \quad \left(\int u^n \, du = \frac{u^{n+1}}{n+1}\right)$$

$$= \pi h \sin^2\theta \Big|_0^{\pi/2}$$

$$= \pi h \sin^2\frac{\pi}{2} - \pi h \sin^2 0$$

$$= \pi h(1)^2 - \pi h(0)$$

$$= \pi h.$$

A more complete table for integrating trigonometric forms is given below.

INTEGRATION OF THE TRIGONOMETRIC FUNCTIONS

$$\int \sin u \, du = -\cos u + c$$

$$\int \cos u \, du = \sin u + c$$

$$\int \tan u \, du = \ln|\sec u| + c$$

$$\int \cot u \, du = \ln|\sin u| + c$$

$$\int \sec u \, du = \ln|\sec u + \tan u| + c$$

$$\int \csc u \, du = \ln|\csc u - \cot u| + c$$

$$\int \sec^2 u \, du = \tan u + c$$

$$\int \csc^2 u \, du = -\cot u + c$$

$$\int \sec u \tan u \, du = \sec u + c$$

$$\int \csc u \cot u \, du = -\csc u + c$$

In chapter 19, we found derivatives of inverse trigonometric functions. Rules for finding derivatives enable us to evaluate integrals of certain algebraic functions. Again, in simplest form, we can tabulate the rules.

Differentiation	Integration
$D_x(\sin^{-1} x) = \dfrac{1}{\sqrt{1-x^2}}$ for $-1 < x < 1$	$\displaystyle\int \dfrac{1}{\sqrt{1-x^2}}\, dx = \sin^{-1} x + c$
$D_x(\cos^{-1} x) = -\dfrac{1}{\sqrt{1-x^2}}$ for $-1 < x < 1$	$\displaystyle\int \dfrac{1}{\sqrt{1-x^2}}\, dx = -\cos^{-1} x + c$
$D_x(\tan^{-1} x) = \dfrac{1}{1+x^2}$	$\displaystyle\int \dfrac{1}{1+x^2}\, dx = \tan^{-1} x + c$
$D_x(\sec^{-1} x) = \dfrac{1}{x\sqrt{x^2-1}},\ |x| > 1$	$\displaystyle\int \dfrac{1}{x\sqrt{x^2-1}}\, dx = \sec^{-1} x + c$

To help us with these more complicated forms, we use the more general rules, using a as a constant and u as a function of x.

RULES FOR INTEGRATION

$$\int \frac{1}{\sqrt{a^2 - u^2}}\, du = \sin^{-1}\left(\frac{u}{a}\right) + c, \quad \text{or} \quad -\cos^{-1}\left(\frac{u}{a}\right) + c$$

$$\int \frac{1}{a^2 + u^2}\, du = \frac{1}{a}\tan^{-1}\left(\frac{u}{a}\right) + c$$

$$\int \frac{1}{u\sqrt{u^2 - a^2}}\, du = \frac{1}{a}\sec^{-1}\left(\frac{u}{a}\right) + c$$

• **Example 21.41:** Evaluate $\displaystyle\int \frac{3}{4 + x^2}\, dx$.

Solution: $\displaystyle\int \frac{3}{4 + x^2}\, dx = 3\int \frac{1}{4 + x^2}\, dx$.

Let $a = 2$, $u = x$, and $du = dx$.
Therefore,

$$\int \frac{3}{4 + x^2}\, dx = \frac{3}{2}\tan^{-1}\left(\frac{x}{2}\right) + c.$$

SECTION 21.6 INTEGRALS OF THE TRIGONOMETRIC FUNCTIONS

- **Example 21.42:** Evaluate $\dfrac{dx}{x\sqrt{4x^2 - 25}}$.

  Solution: $\displaystyle\int \dfrac{dx}{x\sqrt{4x^2 - 25}} = \int \dfrac{2\,dx}{2x\sqrt{(2x)^2 - (5)^2}}$ (Multiply both numerator and denominator by 2 with $u = 2x$, $a = 5$, and $du = 2\,dx$)

  $= \dfrac{1}{5} \sec^{-1}\left(\dfrac{2x}{5}\right) + c.$

  Therefore,

  $$\int \dfrac{dx}{x\sqrt{4x^2 - 25}} = \dfrac{1}{5}\sec^{-1}\left(\dfrac{2x}{5}\right) + c.$$

  ∙∙∙∙∙∙∙∙∙∙

- **Example 21.43:** Evaluate $\displaystyle\int \dfrac{dx}{x^2 + 4x + 20}$.

  Solution: $\displaystyle\int \dfrac{dx}{x^2 + 4x + 20} = \int \dfrac{dx}{x^2 + 4x + 4 + 20 - 4}$ (Completing the square)

  $= \displaystyle\int \dfrac{dx}{(x + 2)^2 + 16}$ (Inverse tan u form)

  $= \dfrac{1}{4} \tan^{-1}\left(\dfrac{x + 2}{4}\right) + c.$

  Therefore,

  $$\int \dfrac{dx}{x^2 + 4x + 20} = \dfrac{1}{4}\tan^{-1}\left(\dfrac{x + 2}{4}\right) + c.$$

  ∙∙∙∙∙∙∙∙∙∙

- **Example 21.44:** Evaluate $\displaystyle\int \dfrac{4}{\sqrt{9 - 4x^2}}\,dx$.

  Solution: $\displaystyle\int \dfrac{4}{\sqrt{9 - 4x^2}}\,dx = 2\int \dfrac{2\,dx}{\sqrt{(3)^2 - (2x)^2}}$ ($a = 3$, $u = 2x$, $du = 2\,dx$)

  $= 2 \sin^{-1}\left(\dfrac{2x}{3}\right) + c.$

  ∙∙∙∙∙∙∙∙∙∙

• **Example 21.45:** The acceleration of an object in a resisting medium is

$$a = \frac{1}{1 + 9t^2} \text{ m/sec}^2.$$

Find an expression for the velocity of the object, given that $v = 100$ when $t = 0$.

Solution: Since velocity is the antiderivative of acceleration, we have

$$v = \int a\, dt = \int \frac{1}{1 + 9t^2}\, dt.$$

Let $u = 3t$. Then $du = 3\, dt$. Thus,

$$v = \frac{1}{3} \int \frac{1}{1 + (3t)^2} (3\, dt) \qquad (\tan^{-1} u\, du \text{ form})$$

$$= \frac{1}{3} \tan^{-1}(3t) + c.$$

When $t = 0$, $v = 0$, and

$$0 = \frac{1}{3} \tan^{-1}(3 \cdot 0) + c$$

$$0 = \frac{1}{3} \cdot 0 + c, \qquad (\tan^{-1} 0 = 0)$$

therefore, $c = 0$.
The velocity is given by

$$v = \frac{1}{3} \tan^{-1}(3t).$$

• • • • • • • • • •

Trial Problems 21.6

Before you begin the section exercises, warm up with these problems. Complete answers are included in the answer key.

Evaluate the following integrals.

1. $\int 2 \sin 2x\, dx$

2. $\int \tan 2x\, dx$

3. $\int \sec x \tan x\, dx$

4. $\int \sin x \cos x\, dx$

5. $\int \frac{1}{\sqrt{1 - t^2}}\, dt$

Exercises 21.6

Evaluate the following integrals and check by differentiation.

1. $\displaystyle\int \sin 3x \, dx$

2. $\displaystyle\int \cos \frac{2}{3}x \, dx$

3. $\displaystyle\int 3 \cos 4x \, dx$

4. $\displaystyle\int 2 \sin 2x \, dx$

5. $\displaystyle\int \tan \frac{\theta}{2} \, d\theta$

6. $\displaystyle\int \sec^2 3\theta \, d\theta$

7. $\displaystyle\int -\csc^2 5x \, dx$

8. $\displaystyle\int \cot \frac{x}{3} \, dx$

9. $\displaystyle\int \tan 2\theta \, d\theta$

10. $\displaystyle\int x \cot \frac{x^2}{3} \, dx$

11. $\displaystyle\int \frac{1}{3} \csc^2 3r \, dr$

12. $\displaystyle\int e^s \cot e^s \, ds$

13. $\displaystyle\int \frac{1}{x} \sin(\ln x) \, dx$

14. $\displaystyle\int x^3 \sec^2(x^4) \, dx$

15. $\displaystyle\int \sec \frac{\theta}{2} \, d\theta$

16. $\displaystyle\int \csc 2\theta \, d\theta$

17. $\displaystyle\int \sin^3 2x \cos 2x \, dx$

18. $\displaystyle\int (1 + \sin x)^2 \cos x \, dx$

19. $\displaystyle\int \frac{\cos 2x}{\sin^3 2x} \, dx$

20. $\displaystyle\int \sin x \cos^8 x \, dx$

21. $\displaystyle\int \frac{\cos \theta}{1 - \sin \theta} \, d\theta$

22. $\displaystyle\int e^{\sin 2\theta} \cos 2\theta \, d\theta$

23. $\displaystyle\int \sec 7x \tan 7x \, dx$

24. $\displaystyle\int \tan 3x \sec 3x \, dx$

25. $\displaystyle\int \frac{1}{\sqrt{9 - x^2}} \, dx$

26. $\displaystyle\int \frac{1}{9 + x^2} \, dx$

27. $\displaystyle\int \frac{x^2}{36 + x^6} \, dx$ (*Hint:* Let $u = x^3$.)

28. $\displaystyle\int \frac{x}{\sqrt{25 - x^4}} \, dx$

29. $\displaystyle\int \frac{\sec^2 \theta}{e^{\tan \theta}} \, d\theta$

30. $\displaystyle\int \frac{3x - 5}{4x^2 + 25} \, dx$ (*Hint:* separate into two fractions.)

31. $\displaystyle\int \sin 4x \sin 2x \, dx$ (*Hint:* $\sin 4x = \sin(2x + 2x)$)

32. $\displaystyle\int \sqrt[3]{\tan x} \sec^2 x \, dx$

33. $\displaystyle\int \frac{dx}{4x^2 + 9}$

34. $\displaystyle\int \frac{dx}{\sqrt{16 - 9x^2}}$

35. $\displaystyle\int \frac{dx}{x\sqrt{9x^2 - 25}}$

36. $\displaystyle\int \frac{dx}{x\sqrt{3x^2 - 4}}$

37. $\displaystyle\int \frac{dx}{3x^2 + 7}$

38. $\displaystyle\int \frac{x \, dx}{16 + x^4}$

Evaluate the following definite integrals.

39. $\int_0^{\pi/4} \cos\theta \sin^2\theta \, d\theta$

40. $\int_{\pi/4}^{\pi/3} \sec\theta \tan\theta \, d\theta$

41. $\int_0^1 x \cos x^2 \, dx$

42. $\int_{\pi/2}^{\pi} (\cos x + \sin x) dx$

43. The velocity of particle is given by $v = 2 \cos 3t$. Find the distance traveled from $t = 0$ to $t = 2$ sec.

44. The rate of radiation by an accelerated charge is given by $R = \int \sin^3\theta \, d\theta$. Find R.
    (*Hint:* $\sin^3\theta = \sin^2\theta \sin = (1 - \cos^2\theta) \sin\theta$.)

45. Determine the formula for current i in a certain inductor if $i = 10\int \dfrac{dt}{30 + t^2}$.

46. If velocity is given by $v = \dfrac{1}{10 + t^2}$, find the formula for distance s if $s = 0$ when $t = 0$.

47. The velocity of a particle is given by $v = \sin t \cos^3 t$ ft/sec. Find the formula for distance s if $s = 0$ when $t = 0$.

48. The acceleration of an object is found to be
    $$a = \dfrac{1}{1 + 4t^2} \text{ m/sec}^2.$$
    Find the formula for the velocity of the object if $v = 100$ when $t = 0$.

49. The study of the lateral displacement x of a swinging pendulum is an example of *simple harmonic motion*. To determine x it may be necessary to solve the following equation:
    $$\dfrac{dx}{\sqrt{A^2 - x^2}} = \sqrt{\dfrac{k}{m}} \, dt,$$
    where k, m, and A are constants. If $x = x_0$ when $t = 0$, find an equation for x as a function of t. (*Hint:* integrate both sides of the equation independently.)

50. A vertical column begins to buckle at the critical or *Euler load*. The slope of the column at this load is given by $\left(\dfrac{dy}{dx}\right)^2 = -k^2 y^2 + c^2$, where k and c are constants. Find an equation for x as a function of y.

21·7 Summary of Term and Rules

Terms

Antiderivative (p. 856)
Constant of integration (p. 857)
Fundamental theorem of calculus (p. 874)
Definite integral (p. 874)
Indefinite integral (p. 872)

Initial conditions (p. 874)
Integral (p. 856)
Integral sign (p. 856)
Integrand (p. 856)
Integration (p. 856)

Limits of integration (p. 874)
Lower limit (p. 874)
Particular solution (p. 872)
Power rule (p. 862)
Upper limit (p. 874)

Rules

$\int k \, dx = kx + c.$ (21.3)

$\int k f'(x) dx = k \int f'(x) dx = kf(x) + c.$ (21.3)

$\int [f'(x) + g'(x)] dx = \int f'(x) dx + \int g'(x) dx.$ (21.3)

$\int x^n \, dx = \dfrac{x^{n+1}}{n+1} + c,$ if $n \neq -1.$ (21.3)

$$\int u^n \, du = \frac{u^{n+1}}{n+1} + c, \text{ where } u = f(x), \text{ and } n \neq -1. \quad (21.4)$$

$$\int e^u \, du = e^u + c, \text{ where } u = f(x). \quad (21.6)$$

$$\int a^u \, du = \frac{1}{\ln a} a^u + c, \text{ where } u = f(x). \quad (21.6)$$

$$\int \frac{1}{u} \, du = \ln |u| + c, \text{ where } u = f(x), \, u \neq 0. \quad (21.6)$$

$$\int \sin u \, du = -\cos u + c, \text{ where } u = f(x). \quad (21.7)$$

$$\int \cos u \, du = \sin u + c, \text{ where } u = f(x). \quad (21.7)$$

$$\int \tan u \, du = \ln |\sec u| + c, \text{ where } u = f(x). \quad (21.7)$$

$$\int \cot u \, du = \ln |\sin u| + c, \text{ where } u = f(x). \quad (21.7)$$

$$\int \sec u \, du = \ln |\sec u + \tan u| + c, \text{ where } u = f(x). \quad (21.7)$$

$$\int \csc u \, du = \ln |\csc u - \cot u| + c, \text{ where } u = f(x). \quad (21.7)$$

$$\int \sec^2 u \, du = \tan u + c, \text{ where } u = f(x). \quad (21.7)$$

$$\int \csc^2 u \, du = -\cot u + c, \text{ where } u = f(x). \quad (21.7)$$

$$\int \sec u \tan u \, du = \sec u + c, \text{ where } u = f(x). \quad (21.7)$$

$$\int \csc u \cot u \, du = -\csc u + c, \text{ where } u = f(x). \quad (21.7)$$

$$\int \frac{1}{\sqrt{a^2 - u^2}} \, du = \sin^{-1}\left(\frac{u}{a}\right) + c, \text{ or } -\cos^{-1}\left(\frac{u}{a}\right) + c, \text{ where } u = f(x). \quad (21.7)$$

$$\int \frac{1}{a^2 + u^2} \, du = \frac{1}{a} \tan^{-1}\left(\frac{u}{a}\right) + c, \text{ where } u = f(x). \quad (21.7)$$

$$\int \frac{1}{u\sqrt{u^2 - a^2}} \, du = \frac{1}{a} \sec^{-1}\left(\frac{u}{a}\right) + c, \text{ where } u = f(x). \quad (21.7)$$

21·8 Chapter 21 Review Exercises

Evaluate the following integrals.

1. $\int 2x^4 \, dx$

2. $\int (x^3 - 4x^2 + 3x + 1) \, dx$

3. $\int (x + 1)^2 \, dx$

4. $\int (x + 1)(x - 1) \, dx$

5. $\int \frac{r^3 + 1}{r^2} \, dr$

6. $\int (y^2 - \sqrt{y}) \, dy$

7. $\int (1 + 2z)^3 \, dz$

8. $\int \frac{t^2}{(t^3 - 1)^2} \, dt$

9. $\int \frac{v^3}{\sqrt{1 + v^4}} \, dv$

10. $\int x^3 \sqrt{x^4 + 1} \, dx$

11. $\int \frac{e^{2x} + 2e^x + 1}{e^x} \, dx$

12. $\int e^x \sqrt{1 - e^x} \, dx$

13. $\int \frac{e^{-t}}{1 - e^{-t}} \, dt$

14. $\int \frac{t^2 + 2t + 3}{t^3 + 3t^2 + 9t + 1} \, dt$

15. $\int e^{10x} \, dx$

16. $\int 3e^{-x/2} \, dx$

17. $\int x3^{x^2} \, dx$

18. $\int 4^{x/2} \, dx$

19. $\int \frac{1}{1 + 9r^2} \, dr$

20. $\int \frac{dt}{\sqrt{1 - 2t^2}} \, dt$

21. $\int v(v^2 + 2)^{-1} \, dv$

22. $\int \sqrt[3]{x^5} \, dx$

23. $\int \sin^5 \theta \cos \theta \, d\theta$

24. $\int \cos x (1 + \sin x)^3 \, dx$

25. $\int \frac{(\ln x)^4}{x} \, dx$

26. $\int \sin \theta \cos^3 \theta \, d\theta$

27. $\int \frac{e^{\sqrt{x}}}{\sqrt{x}} \, dx$

28. $\int e^{\tan \theta} \sec^2 \theta \, d\theta$

29. $\int \frac{t^2}{\sqrt{t^3 + 8}} \, dt$

30. $\int \frac{4}{1 + r} \, dr$

31. $\int e^{-ax} \, dx$

32. $\int 3x^2 e^{x^3 + 1} \, dx$

33. $\int 2^{2x} \, dx$

34. $\int 2^{-2x} \, dx$

35. $\int \frac{4r}{1 + r^2} \, dr$

36. $\int \frac{\cos \theta}{1 + \sin \theta} \, d\theta$

37. $\int e^{\cos 2x} \sin 2x \, dx$

38. $\int \frac{dw}{\sqrt{4 - w^2}}$

39. $\int \sec 3\theta \tan 3\theta \, d\theta$

40. $\int e^x \csc e^x \, dx$

41. $\int v^2 \sec^2 v^3 \, dv$

42. $\int \frac{\sec^2 \theta}{e^{\tan \theta}} \, d\theta$

43. $\int \frac{x - 2}{4x^2 + 25} \, dx$

44. $\int \frac{x - 4}{x + 4} \, dx$

45. $\int (1 + \sin^3 2\theta) \cos 2\theta \, d\theta$

46. $\int \sqrt{\frac{2}{t}} \, dt$

Evaluate the following definite integrals.

47. $\int_0^2 (t+4)\,dt$

48. $\int_3^4 (r^2 - 9)\,dr$

49. $\int_0^1 (v - v^3)\,dv$

50. $\int_0^1 x(1-x)\,dx$

51. $\int_2^{\sqrt{29}} \dfrac{x}{\sqrt{x^2 - 4}}\,dx$

52. $\int_0^3 \dfrac{dt}{2t+1}$

53. $\int_{\pi/2}^{\pi/3} \sin\dfrac{\theta}{2}\,d\theta$

54. $\int_{\pi/3}^{\pi/6} \sec x \tan x \,dx$

55. $\int_0^1 \dfrac{dw}{w^2 + 4}$

56. $\int_0^{-1/2} e^{3x}\,dx$

57. $\int_1^e \dfrac{x+1}{x}\,dx$

58. $\int_1^{-2} (t^2 + 4)^2\,dt$

59. $\int_1^2 \left(\sqrt{x} - \dfrac{1}{\sqrt{x}}\right)dx$

60. $\int_0^{\sqrt{3}} \dfrac{x}{1 + x^4}\,dx$

61. $\int_0^{\pi/4} \cos x \sin^2 x\,dx$

62. $\int_{\pi/2}^{\pi} (\sin\theta + \cos\theta)\,d\theta$

63. $\int_0^{1/2}\left(x + \dfrac{1}{\sqrt{1-x^2}}\right)dx$

64. $\int_{-3}^{-2} 2^x\,dx$

65. $\int_{-1}^1 2^x\,dx$

66. $\int_{-1}^1 e^{-2x}\,dx$

67. $\int_0^1 (2t+1)(t^2 + t)^3\,dt$

68. $\int_{\ln 2}^{\ln 3} \dfrac{e^{\sqrt{x}}}{\sqrt{x}}\,dx$

69. $\int_0^2 (x^2 - 1)e^{x^3 - 3x + 1}\,dx$

70. $\int_1^2 e^{1-x}\,dx$

71. Find $\displaystyle\int_0^{\pi} r'\,d\theta$ if $r = \theta^3 + \sin\theta$.

72. Show that $\displaystyle\int_0^{\pi/2}\cos\theta\,d\theta + \int_{\pi/2}^{\pi}\cos\theta\,d\theta = 0$.

73. Show that $\displaystyle\int_a^c f(x)\,dx = \int_a^b f(x)\,dx + \int_b^c f(x)\,dx$.

74. The remaining mass m of a radioactive substance can be found if the rate of decay is known to be $\dfrac{dm}{dt} = -1.5e^{-0.3t}$. Find m after $t = 5$ years if the initial mass was 50 g.

75. Find the distance traveled by an object if the velocity is given $v = 100e^{1-t}$ ft from $t = 0$ to $t = 4$ sec.

76. Find the equation of the function whose graph passes through the point $(1,0)$ and whose derivative is
$$y' = \dfrac{3}{2 - 3x}.$$

77. If $F = \dfrac{\sin x}{1 + \cos x}$ represents a certain force and work done is given by $w = \displaystyle\int F\,dx$, find the amount of work done if $w = 0$ when $x = 0$.

78. Recall that in an electrical circuit, charge is given by $q = \displaystyle\int i\,dt$, where i is the current. Find q if $i = \dfrac{1}{t^2 + 25}$ and $q = 0$ when $t = 0$.

79. Find the expression for velocity if $a = t^2 - 2t$ and $v = 0$ when $t = 0$.

80. The slope m of the tangent line to a curve is given by $m = \sec^2 x$. Find the equation of the curve if it passes through the point $\left(\dfrac{\pi}{4}, 2\right)$.

81. Find the distance traveled in meters by an object from $t = 0$ to $t = \dfrac{1}{2}$ sec if the velocity is given by
$$v = \dfrac{1}{\sqrt{1 - t^2}}.$$

82. The current in a resistor is given by $i = \sqrt{3t + 5}$. Find the equation for the charge q if $q = 0$ when $t = 0$. (See problem 78.)

83. The current in a circuit is given by

$$i = \frac{1}{4(t^2 + 1)} \text{ A.}$$

Find the charge q for $t = 0$ to $t = 1$ sec. (See problem 78.)

84. A manufacturer's marginal cost $c'(x)$, in millions of dollars per thousand units, is given by

$$c'(x) = 2 + 3x - \sin\left(\frac{\pi x}{20}\right),$$

where x is the number of thousands of units made. The cost is $4 million when the level of production is zero. Find the cost c when the level of production is 10,000 units.

21·9 Chapter 21 Test

Integrate each of the following and check the result by differentiating.

1. $\int x^3 \, dx$

2. $\int (x + 1)^2 \, dx$

3. $\int \sin x \, dx$

4. $\int \frac{1}{x^2} \, dx$

5. $\int (1 + 2x)x \, dx$

6. $\int \frac{dx}{\sqrt{x}}$

7. $\int (1 + 3x)^4 \, dx$

8. $\int \frac{2 \, dx}{(x - 1)^2}$

9. $\int x^2 \sqrt{x^3 + 2} \, dx$

10. $\int \frac{2 \, dx}{3x}$

11. $\int e^{-2x} \, dx$

12. $\int \frac{t}{t^2 - 4} \, dt$

13. $\int 3 \sin 3x \, dx$

14. $\int \frac{\cos t}{1 - \sin t} \, dt$

15. $\int \frac{1}{16 + x^2} \, dx$

Evaluate the following definite integrals.

16. $\int_0^1 (1 - x)\sqrt{x} \, dx$

17. $\int_{\pi/2}^{\pi} (\sin x + \cos x) \, dx$

18. Find the equation of the graph having slope $\dfrac{dy}{dx} = 2x$ and passing through the point (4,0).

19. The velocity of an object moving in a straight line is given by $v = e^{1/2t} + \cos t$. Find an expression for the distance s.

20. During the testing of an experimental light bulb, light was emitted according to the expression

$$\frac{dL}{dt} = \frac{t}{t^2 + 4} \text{ lumens/sec.}$$

Find the amount of light emitted between $t = 0$ and $t = 2$ sec.

Chapter 22
Techniques of Integration

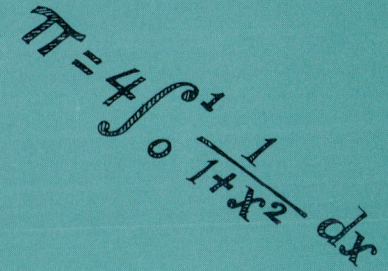

*A*lthough we have learned to integrate many basic functions, there is a need to investigate specific procedures that apply to functions that do not fall in the same category as those functions presented in chapter 21. In many areas of applied mathematics, functions are encountered that require different formulas, rules, and procedures in order to find their integrals. Basically, three new integration procedures are presented in this chapter. Some technical applications of integration are presented, but extensive applications are investigated in chapter 23.

22 • 1 Integration by Parts

There are many functions that cannot be integrated by using the basic forms given in chapter 21. For example, the function $y = \ln x$ in integral form $\int \ln x \, dx$ does not fit into any of the basic forms. In order to evaluate this integral, a special method must be used. In this chapter we investigate several special methods.

Integration by Parts

One of the most common of the special methods is known as **integration by parts**. Suppose y is a composite function of x in the form $y = h(x)g(x)$. Let $u = h(x)$ and $v = g(x)$. Then,

$$y = uv,$$

and

$$\frac{dy}{dx} = u\frac{dv}{dx} + v\frac{du}{dx}, \quad \text{(Product rule for derivatives)}$$

or

$$dy = u\frac{dv}{dx}dx + v\frac{du}{dx}dx \quad \text{(Using differentials)}$$
$$= u \, dv + v \, du.$$

Since $y = uv$, we have

$$d(uv) = u \, dv + v \, du.$$

Solving for $v \, du$, we obtain

$$v \, du = d(uv) - u \, dv.$$

Now, integrating both sides,

$$\int v \, du = \int d(uv) - \int u \, dv,$$
$$\int v \, du = uv - \int u \, dv,$$

or

$$\int u \, dv = uv - \int v \, du.$$

This form indicates that if an integral can be placed in the form $\int u \, dv$, it is possible that it can be separated into the product uv and an *easier* integral, $\int v \, du$. Example 22.1 illustrates this.

• **Example 22.1:** Evaluate $\int x \sin x \, dx$.

Solution: First, it should be clear that this is not an integral in one of our basic forms. Can we express $\int x \sin x \, dx$ in the form $\int u \, dv$?

Let $u = x$ and $dv = \sin x \, dx$.

Then $du = dx$ and $v = \int \sin x \, dx = -\cos x$.

Therefore,

$$\underbrace{\int x}_{u} \underbrace{\sin x \, dx}_{dv} = \underbrace{x}_{u}\underbrace{(-\cos x)}_{v} - \int \underbrace{(-\cos x)}_{v}\underbrace{dx}_{du}$$

$$= -x \cos x + \int \cos x \, dx$$

$$= -x \cos x + \sin x + c.$$

NOTE: There is no set rule for determining whether we let $u = x$ or $u = \sin x$. Until enough experience is obtained, a trial-and-error method prevails. Try to choose u so that du is in simpler form than u; dv must be something you can integrate.

• **Example 22.2:** Evaluate $\int xe^x \, dx$.

Solution: Let $u = e^x$ and $dv = x \, dx$.

Then $du = e^x \, dx$ and $v = \int x \, dx = \dfrac{x^2}{2}$.

Therefore,

$$\int xe^x \, dx = \underbrace{\int e^x}_{u}\underbrace{x \, dx}_{dv} = \underbrace{e^x}_{u \cdot v}\underbrace{\dfrac{x^2}{2}}_{} - \int \underbrace{\dfrac{x^2}{2}}_{v}\underbrace{e^x \, dx}_{du}.$$

The second integral, $\int \dfrac{x^2}{2} e^x \, dx$, is more difficult to evaluate than the original problem. In this case, we try the following.

Let $u = x$ and $dv = e^x \, dx$.

Then $du = dx$ and $v = \int e^x \, dx = e^x$.

Now we have,

$$\underbrace{\int x}_{u}\underbrace{e^x \, dx}_{dv} = \underbrace{x \, e^x}_{u \cdot v} - \int \underbrace{e^x}_{v}\underbrace{dx}_{du}$$

$$= xe^x - e^x + c.$$

• **Example 22.3:** Evaluate $\int \ln x \, dx$.

Solution: Let $u = \ln x$ and $dv = dx$. Then

$$du = \frac{1}{x} dx \text{ and } v = \int dx = x$$

$$\int \ln x \, dx = x \ln x - \int 1 \, dx$$
$$= x \ln x - x + c.$$

••••••••••

• **Example 22.4:** Evaluate $\int \sin^{-1} x \, dx$.

Soluton: Let $u = \sin^{-1} x$ and $dv = dx$.
Then $du = \dfrac{1}{\sqrt{1 - x^2}} dx$ and $v = x$.

$$\int \sin^{-1} x \, dx = \underbrace{(\sin^{-1} x)}_{u}\underbrace{(x)}_{dv} - \int \underbrace{x}_{v} \underbrace{\frac{1}{\sqrt{1 - x^2}} dx}_{du}$$

$$= x \sin^{-1} x - \int (1 - x^2)^{-1/2} x \, dx \quad (\int u^n \, du \text{ form})$$

$$= x \sin^{-1} x - \left(-\frac{1}{2}\right) \int (1 - x^2)^{-1/2}(-2x) dx$$

$$= x \sin^{-1} x + \frac{1}{2} \cdot \frac{(1 - x^2)^{1/2}}{\frac{1}{2}} + c$$

$$= x \sin^{-1} x + \sqrt{1 - x^2} + c.$$

••••••••••

• **Example 22.5:** Evaluate $\int x^2 \sin x \, dx$.

Solution: Let $u = x^2$ and $dv = \sin x \, dx$.
Then $du = 2x \, dx$ and $v = \int \sin x \, dx = -\cos x$.

$$\int x^2 \sin x \, dx = x^2(-\cos x) - \int (-\cos x)(2x \, dx)$$
$$= -x^2 \cos x + \int 2x \cos x \, dx.$$

In order to obtain the final answer, we must use integration by parts to evaluate $\int 2x \cos x \, dx$.

Let $u = 2x$ and $dv = \cos x \, dx$.
Then $du = 2 \, dx$ and $v = \sin x$.

$$\int x^2 \sin x \, dx = -x^2 \cos x + 2x \sin x - \int (\sin x)(2 \, dx)$$
$$= -x^2 \cos x + 2x \sin x + 2 \cos x + c.$$

Consider integration by parts if the integrand is

1. a product,
2. an inverse function, or
3. a logarithm, provided other integration formulas are not appropriate.

• **Example 22.6:** The current in a given circuit is given by $i = e^{-t} \cos t$. Find an expression for the amount of charge q that passes a given point in the circuit if $q = 0$ when $t = 0$.

Solution: The amount of charge q is found by

$$q = \int i \, dt = \int e^{-t} \cos t \, dt.$$

Integrating, we obtain

1. $\int e^{-t} \cos t \, dt = e^{-t} \sin t - \int (\sin t)(-e^{-t} \, dt) \quad \begin{array}{l}(u = e^{-t}, \\ dv = \cos t \, dt, \\ du = -e^{-t} \, dt, \\ \text{and } v = \sin t)\end{array}$

$= e^{-t} \sin t + \int e^{-t} \sin t \, dt.$

The integral $\int e^{-t} \sin t \, dt$ needs to be evaluated by parts, also.

2. $\int e^{-t} \sin t \, dt = (e^{-t})(-\cos t) - \int (-\cos t)(-e^{-t} \, dt)$

$= -e^{-t} \cos t - \int e^{-t} \cos t \, dt. \quad \begin{array}{l}(u = e^{-t}, \\ dv = \sin t \, dt, \\ du = -e^{-t} \, dt, \\ \text{and } v = -\cos t)\end{array}$

Putting 1 and 2 together, we have

$$\int e^{-t} \cos t \, dt = e^{-t} \sin t + (-e^{-t} \cos t - \int e^{-t} \cos t \, dt)$$

or,

$$\int e^{-t} \cos t \, dt = e^{-t} \sin t - e^{-t} \cos t - \int e^{-t} \cos t \, dt.$$

Adding $\int e^{-t}\cos t\, dt$ to both sides of the equation we have,

$$2\int e^{-t}\cos t\, dt = e^{-t}\sin t - e^{-t}\cos t.$$

Dividing by 2 and adding the constant of integration, we have

$$q = \int e^{-t}\cos t\, dt = \frac{1}{2} e^{-t}\sin t - \frac{1}{2} e^{-t}\cos t + c,$$

when $t = 0$, $q = 0$, or $0 = \frac{1}{2} e^{-0}\sin 0 - \frac{1}{2} e^{-0}\cos 0 + c.$

Therefore, $c = \frac{1}{2}$. Hence,

$$q = e^{-t}\sin t - e^{-t}\cos t + \frac{1}{2}.$$

Trial Problems 22.1

Before you begin the section exercises, warm up with these problems. Complete answers are included in the answer key.

When using the method of integration by parts, you must remember how to integrate some basic forms. Evaluate the following integrals.

1. $\int x^2\, dx$

2. $\int \frac{1}{x}\, dx$

3. $\int \sin 2x\, dx$

4. $\int e^{-2x}\, dx$

5. $\int \frac{x}{x^2 + 1}\, dx$

Exercises 22.1

Use the method of integration by parts to evaluate the following.

1. $\int x \cos x\, dx$

2. $\int x \ln x\, dx$

3. $\int x\, 2^x\, dx$

4. $\int x^2 e^x\, dx$

5. $\int \tan^{-1} x\, dx$

6. $\int x \sin 2x\, dx$

7. $\int x^3 e^{-x^2}\, dx$

8. $\int x^2 \sin 2x\, dx$

9. $\int x\sqrt{x + 1}\, dx$

10. $\int_1^2 \frac{x\, dx}{\sqrt{4 - x}}$

11. $\int \frac{\ln x}{x}\, dx$

12. $\int (\ln x)^2\, dx$

13. $\int \cos(\ln x)\, dx$

14. $\int e^x \sin x\, dx$

15. $\int x e^{-x}\, dx$

16. $\int (x^2 - 1)e^x \, dx$

17. $\int \dfrac{\ln x}{x^2} \, dx$

18. $\int \tan^{-1} x \, dx$

19. $\int_0^{\pi/2} \sin x \cos 2x \, dx$

20. $\int \dfrac{xe^x}{(x+1)^2} \, dx$

21. $\int \cos \sqrt{x} \, dx$ (*Hint:* let $w = \sqrt{x}$.)

22. $\int \sin x \ln(\cos x) \, dx$

23. The slope of the tangent line to a curve is given by $\dfrac{dy}{dx} = \dfrac{x}{\sqrt{1-x}}$. Find the equation of the curve if it passes through the origin.

24. The velocity of a particle is given by $v = te^{2t}$. Find the distance s in terms of t if $s = 0$ when $t = 0$.

25. The current in a particular circuit is given by $i = t \cos 2t$. Find the equation for the amount of charge q if $q = 0$ when $t = 0$.

26. A particle moves such that its velocity v is given by $v = t\sqrt{t+1}$. Find an expression for the distance s traveled by the particle if $s = 0$ when $t = 0$.

27. A model for the ability M, measured on a scale from 0 to 10, of a man to memorize the steps for operating a computerized lathe is given by

$$M = 10 - 0.05t \ln t, \quad 20 \leq t \leq 50,$$

where t is the man's age in years. Find the average value of M for a man
  a. between 20 and 30 years old.
  b. between 30 and 40 years old.

Hint: average value is found by computing $\dfrac{\int_a^b M \, dt}{b - a}$.

28. The inventory of a newly found computer memory chip manufacturer is growing at the rate, in chips per month, given by

$$c'(t) = te^{t/10}, \quad 0 \leq t \leq 20,$$

where t is the number of months after manufacturing startup. Find the inventory $c(t)$ after t months assuming $c(0) = 0$.

22 • 2 Trigonometric Substitution

The integral $\int \sin x \, dx$ is one of the basic integration forms, whereas $\int \sin^2 x \, dx$ is not. Many integrands containing trigonometric functions can be transformed into one of the basic forms by using an appropriate trigonometric substitution. Examples 22.7–22.10 illustrate some common substitutions.

• **Example 22.7:** Evaluate $\int \sin^2 x \, dx$.

Solution: From our work in trigonometry, recall that $\cos 2x = 1 - 2\sin^2 x$, or $\sin^2 x = \dfrac{1}{2}(1 - \cos 2x)$.

Therefore,

$$\int \sin^2 x \, dx = \int \frac{1}{2}(1 - \cos 2x) dx$$

$$= \frac{1}{2} \int (1 - \cos 2x) dx$$

$$= \frac{1}{2} \int dx - \frac{1}{2} \int \cos 2x \, dx \quad \left(\int \cos u \, du \text{ form} \right)$$

$$= \frac{1}{2} x - \frac{1}{2} \cdot \frac{1}{2} \int \cos 2x (2 \, dx) \quad \left(\int \cos u \, du \text{ form} \right)$$

$$= \frac{1}{2} x - \frac{1}{4} \sin 2x + c.$$

• **Example 22.8:** Evaluate $\int \sin^3 x \, dx$.

Solution:
$$\int \sin^3 x \, dx = \int \sin x \sin^2 x \, dx$$

$$= \int \sin x (1 - \cos^2 x) dx \quad (\sin^2 x = 1 - \cos^2 x)$$

$$= \int (\sin x - \cos^2 x \sin x) dx$$

$$= \int \sin x \, dx - \int \cos^2 x \sin x \, dx \quad \left(\int u^n \, du \text{ form} \right)$$

$$= -\cos x - (-) \int \cos^2 x (-\sin x \, dx). \quad (u = \cos x,$$
$$du = -\sin x \, dx,$$
$$\text{power rule})$$

Therefore,

$$\int \sin^3 x \, dx = -\cos x + \frac{\cos^3 x}{3} + c.$$

• **Example 22.9:** Evaluate $\int \sin^2 x \cos^4 x \, dx$.

Solution:
$$\int \sin^2 x \cos^4 x \, dx = \int \sin^2 x \cos^2 x \cos^2 x \, dx$$
$$= \int (\sin x \cos x)^2 \cos^2 x \, dx$$
$$\left(\sin x \cos x = \frac{1}{2}\sin 2x, \text{ and } \cos^2 x = \frac{1}{2}(1 + \cos 2x)\right)$$
$$= \int \left(\frac{1}{2}\sin 2x\right)^2 \frac{1}{2}(1 + \cos 2x) \, dx$$
$$= \int \frac{1}{4}\sin^2 2x \cdot \frac{1}{2}(1 + \cos 2x) \, dx$$
$$= \frac{1}{8}\int (\sin^2 2x + \sin^2 2x \cos 2x) \, dx$$
$$= \frac{1}{8}\left[\int \frac{1}{2}(1 - \cos 4x) \, dx + \int \sin^2 2x \cos 2x \, dx\right]$$
$$= \frac{1}{8}\left[\frac{1}{2}\int dx - \frac{1}{2}\cdot\frac{1}{4}\int \cos 4x (4 \, dx)\right.$$
$$\left. + \frac{1}{2}\int \sin^2 2x (2 \cos 2x \, dx)\right]$$
$$= \frac{1}{16}x - \frac{1}{64}\sin 4x + \frac{1}{16}\frac{\sin^3 2x}{3} + c$$
$$= \frac{1}{16}x - \frac{1}{64}\sin 4x + \frac{1}{48}\sin^3 2x + c.$$

• **Example 22.10:** Evaluate $\int \tan^4 x \, dx$.

Solution:
$$\int \tan^4 x = \int \tan^2 x \tan^2 x \, dx$$
$$= \int (\sec^2 x - 1)\tan^2 \, dx \quad (\tan^2 x = \sec^2 x - 1)$$
$$= \int (\sec^2 x \tan^2 x - \tan^2 x) \, dx$$
$$= \int \tan^2 x (\sec^2 x \, dx) - \int \tan^2 x \, dx$$
$$= \frac{\tan^3 x}{3} - \int (\sec^2 x - 1) \, dx$$
$$= \frac{\tan^3 x}{3} - \tan x + x + c.$$

CHAPTER 22 TECHNIQUES OF INTEGRATION

Trigonometric substitutions are frequently used when the integrand is algebraic and of the form $a^2 - u^2$ or $u^2 + a^2$, where u is a function of x and a is any constant. Using a right triangle aids us in selecting an appropriate trigonometric substitution. Again, examples 22.11–22.14 illustrate these forms.

• **Example 22.11:** Evaluate $\int \sqrt{4 - x^2}\, dx$.

NOTE: This is not of the form $\int u^n\, du$ since du would be $-2x\, dx$ if we were to choose $u = 4 - x^2$.

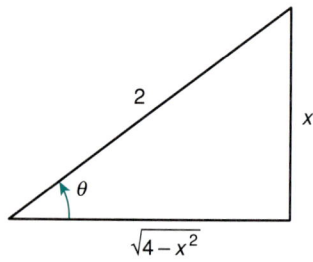

Figure 22.1

Solution: Let $a = 2$ and $u = x$. Draw a right triangle and label it as shown in figure 22.1. Notice that the sides of the triangle satisfy the Pythagorean theorem. Now, $\cos \theta = \dfrac{\sqrt{4 - x^2}}{2}$, or

$\sqrt{4 - x^2} = 2 \cos \theta$.

Also, $\sin \theta = \dfrac{x}{2}$, or $x = 2 \sin \theta$. Therefore, $\underline{dx = 2 \cos \theta\, d\theta}$.

Substituting the underlined quantities gives us

$$\int \sqrt{4 - x^2}\, dx = \int (2 \cos \theta)(2 \cos \theta\, d\theta)$$

$$= 4 \int \cos^2 \theta\, d\theta$$

$$= 4 \int \frac{1}{2}(1 + \cos 2\theta)\, d\theta$$

$$= 2 \int (1 + \cos 2\theta)\, d\theta$$

$$= 2\left[\theta + \frac{1}{2} \sin 2\theta\right] + c.$$

Since $\sin \theta = \dfrac{x}{2}$, $\theta = \sin^{-1}\left(\dfrac{x}{2}\right)$.

Also, since $\sin 2\theta = 2 \sin \theta \cos \theta$, $\dfrac{1}{2} \sin 2\theta = \sin \theta \cdot \cos \theta$.

Thus,

$$\int \sqrt{4 - x^2}\, dx = 2\left[\theta + \frac{1}{2} \sin 2\theta\right] + c$$

$$= 2\left[\sin^{-1}\left(\frac{x}{2}\right) + \sin \theta \cos \theta\right] + c$$

$$= 2\left[\sin^{-1}\left(\frac{x}{2}\right) + \frac{x}{2} \cdot \frac{\sqrt{4 - x^2}}{2}\right] + c$$

$$= 2 \sin^{-1}\left(\frac{x}{2}\right) + \frac{x\sqrt{4 - x^2}}{2} + c.$$

• **Example 22.12:** Evaluate $\int \dfrac{dx}{\sqrt{x^2 + 9}}$.

Solution: Let $u = x$ and $a = 3$.
Label a right triangle so that the Pythagorean theorem holds. See figure 22.2.

Now, $\tan \theta = \dfrac{x}{3}$, or $x = 3 \tan \theta$.

Thus, $dx = 3 \sec^2\theta\, d\theta$.

Also, $\sec \theta = \dfrac{\sqrt{x^2 + 9}}{3}$, or $\underline{\sqrt{x^2 + 9} = 3 \sec \theta}$.

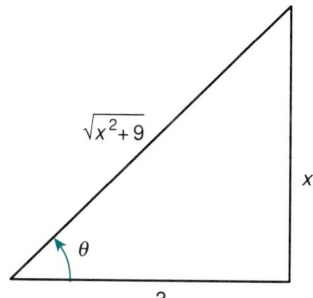

Figure 22.2

Making the substitutions for the underlined expressions, we have

$$\int \frac{dx}{\sqrt{x^2 + 9}} = \int \frac{3 \sec^2\theta\, d\theta}{3 \sec \theta}$$

$$= \int \sec \theta\, d\theta$$

$$= \ln |\sec \theta + \tan \theta| + c$$

$$= \ln \left|\frac{\sqrt{x^2 + 9}}{3} + \frac{x}{3}\right| + c$$

$$= \ln \left|\frac{\sqrt{x^2 + 9} + x}{3}\right| + c,$$

which reduces to

$$\int \frac{dx}{\sqrt{x^2 + 9}} = \ln \left|\sqrt{x^2 + 9} + x\right| + c. \quad \text{(Why?)}$$

• **Example 22.13:** Evaluate $\int \frac{\sqrt{x^2 - 16}}{x} dx$.

Solution: Let $u = x$ and $a = 4$.
Label a right triangle as shown in figure 22.3.
$\sec \theta = \frac{x}{4}$, or $x = 4 \sec \theta$.
Therefore, $dx = 4 \sec \theta \tan \theta \, d\theta$.
Also, $\tan \theta = \frac{\sqrt{x^2 - 16}}{4}$, or $\sqrt{x^2 - 16} = 4 \tan \theta$.

Making the substitutions for the underlined expressions, we have

$$\int \frac{\sqrt{x^2 - 16}}{x} dx = \int \frac{4 \tan \theta}{4 \sec \theta} (4 \sec \theta \tan \theta) d\theta$$

$$= \int 4 \tan^2 \theta \, d\theta$$

$$= 4 \int (\sec^2 \theta - 1) d\theta$$

$$= 4(\tan \theta - \theta) + c$$

$$= 4 \frac{\sqrt{x^2 - 16}}{4} - 4 \sec^{-1}\left(\frac{x}{4}\right) + c$$

$$\left(\text{since } \sec \theta = \frac{x}{4}, \theta = \sec^{-1}\left(\frac{x}{4}\right)\right)$$

$$= \sqrt{x^2 - 16} - 4 \sec^{-1}\left(\frac{x}{4}\right) + c.$$

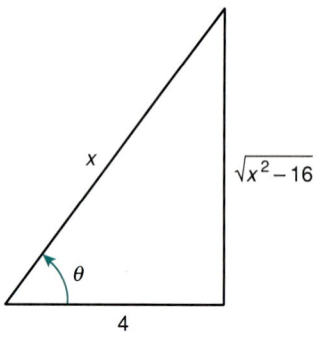

Figure 22.3

NOTE: There is no one way to label the right triangle. Label it to satisfy the Pythagorean theorem. If the results are not easy to work with upon the initial labeling, you may relabel the triangle, as long as the Pythagorean theorem is satisfied.

• **Example 22.14:** The current in a resistor varies as

$$i = \frac{t^2}{\sqrt{9 - t^2}} \text{ A}.$$

Find the amount of charge q transferred through the resistor at any time t if $q = 0$ when $t = 0$.

SECTION 22.2 TRIGONOMETRIC SUBSTITUTION

Solution: The charge q is given by

$$q = \int i\, dt = \int \frac{t^2}{\sqrt{9-t^2}}\, dt.$$

If we label the triangle as in figure 22.4, then $t = 3\sin\theta$ and $\frac{dt}{d\theta} = 3\cos\theta$.

$$q = \int \frac{t^2}{\sqrt{9-t^2}}\, dt = \int \frac{9\sin^2\theta}{\sqrt{9-9\sin^2\theta}}(3\cos\theta)d\theta$$

$$= \int \frac{9\sin^2\theta}{\sqrt{9(1-\sin^2\theta)}}(3\cos\theta)d\theta$$

$$= \int \frac{9\sin^2\theta}{\sqrt{9\cos^2\theta}}(3\cos\theta)d\theta$$

$$= \int \frac{9\sin^2\theta}{3\cos\theta}(3\cos\theta)d\theta$$

$$= 9\int \sin^2\theta\, d\theta$$

$$= 9\left(\frac{1}{2}\theta - \frac{1}{4}\sin 2\theta\right) + c \quad \text{(see example 21.6)}$$

$$= \frac{9}{2}\theta - \frac{9}{4}\sin 2\theta + c.$$

Since $\sin\theta = \frac{t}{3}$, we have $\theta = \sin^{-1}\left(\frac{t}{3}\right)$ (see triangle in figure 22.4). Also, $\sin 2\theta = 2\sin\theta\cos\theta$, and $\cos\theta = \frac{\sqrt{9-t^2}}{3}$.

Therefore,

$$\int \frac{t^2}{\sqrt{9-t^2}}\, dt = \frac{9}{2}\theta - \frac{9}{4}\sin 2\theta + c$$

$$= \frac{9}{2}\theta - \frac{9}{4}(2\sin\theta\cos\theta) + c$$

$$= \frac{9}{2}\sin^{-1}\left(\frac{t}{3}\right) - \frac{9}{2}\cdot\frac{t}{3}\cdot\frac{\sqrt{9-t^2}}{3} + c$$

$$= \frac{9}{2}\sin^{-1}\left(\frac{t}{3}\right) - \frac{1}{2}t\sqrt{9-t^2} + c,$$

when $t = 0$, $c = 0$, and $q = 0$. Therefore, the charge q is given by

$$q = \frac{9}{2}\sin^{-1}\left(\frac{t}{3}\right) - \frac{1}{2}t\sqrt{9-t^2}.$$

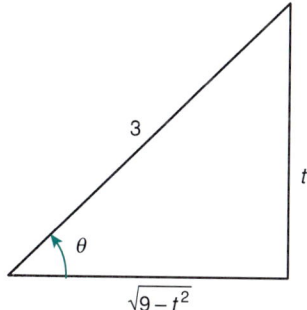

Figure 22.4

Trial Problems 22.2

Before you begin the section exercises, warm up with these problems. Complete answers are included in the answer key.

Integration of trigonometric expressions are needed in this exercise set. Evaluate the following integrals.

1. $\int \cos \dfrac{x}{2} \, dx$
2. $\int \csc^2 2x \, dx$
3. $\int \tan 3x \, dx$
4. $\int \dfrac{5}{1 + x^2} \, dx$
5. $\int \dfrac{1}{\sqrt{4 - x^2}} \, dx$

Exercises 22.2

Evaluate the following integrals.

1. $\int \cos^2 x \, dx$
2. $\int \sin^4 \theta \, d\theta$
3. $\int \cos^3 \theta \, d\theta$
4. $\int \sin^5 x \, dx$
5. $\int \cos^2 x \sin^2 x \, dx$
6. $\int \sin^2 \theta \cos \theta \, d\theta$
7. $\int \sin^3 \theta \cos \theta \, d\theta$
8. $\int \cos^3 x \sin x \, dx$
9. $\int \cot^2 x \, dx$
10. $\int \tan^2 2\theta \, d\theta$
11. $\int \sin \theta \cos^3 \theta \, d\theta$
12. $\int \dfrac{\cos \theta}{\sqrt{\sin \theta}} \, d\theta$
13. $\int (\tan \theta + \cot \theta)^2 \, d\theta$
14. $\int_0^1 \tan^2 \left(\dfrac{\pi x}{4}\right) dx$
15. $\int \dfrac{\sin 2x}{\cos^2 2x} \, dx$
16. $\int \dfrac{\sin^2 x}{\cos x} \, dx$
17. $\int \dfrac{\sec^2 x \, dx}{2 + \tan x}$
18. $\int \dfrac{\cos x}{(1 + \sin x)^2} \, dx$
19. $\int \dfrac{dx}{\sqrt{x^2 - 16}}$
20. $\int \sqrt{25 - x^2} \, dx$
21. $\int \dfrac{dx}{x^2 \sqrt{9 + x^2}}$
22. $\int \dfrac{1}{x \sqrt{x^2 + 1}} \, dx$
23. $\int x \sqrt{x^2 - 36} \, dx$ (Careful!)
24. $\int \dfrac{x^2 \, dx}{\sqrt{1 - x^2}}$
25. $\int \dfrac{dx}{(4 + x^2)^{3/2}}$
26. $\int \dfrac{dx}{x^2(x^2 - 1)}$
27. $\int_0^2 \dfrac{dx}{4 + x^2}$
28. $\int_0^1 \dfrac{dx}{4 - x^2}$
29. $\int \dfrac{x + 1}{\sqrt{4 - x^2}} \, dx$
30. $\int \dfrac{1}{\sqrt{e^{2x} + 1}} \, dx$

31. The slope of the tangent line to a curve at the point (x,y) on the curve is $2 - \sin^2 x$. If the point $(0,2)$ lies on the curve, find the equation for the curve.

32. The rate of radiation given off by an accelerated charge is given by

$$R = \int_a^b \sin^3 \theta \, d\theta.$$

Find R if $a = 0$ and $b = \pi$.

33. Show that $\int \sec^2 x \tan x \, dx$ can be integrated in two ways.

34. The acceleration of a certain power tool is given by $a = \sqrt{4 - t^2}$. Find the equation for velocity if $v = 0$ when $t = 0$.

35. The force acting on an object is given by $F = \dfrac{s^2}{\sqrt{16 - s^2}}$ lb. Find work done in moving the object from $s = 0$ to $s = 2$ ft.

36. The current in a resistor is given by $i = \dfrac{1}{t^2\sqrt{1 - t^2}}$. Find the amount of charge transferred through the resistor from $t = \dfrac{1}{2}$ to $t = 1$ sec.

22 • 3 Partial Fractions

Many integrands are in the form of a fraction where both numerator and denominator are polynomials in x. At first glance, it may appear that the integral is not one of the standard forms. After some study, and algebraic manipulation, it may be possible to transform the integrand into one that is easily integrable. For example, $\int \dfrac{x + 1}{x^2 - 3x + 2} \, dx$ does not appear to be in any of the standard forms studied thus far. The integrand can be transformed into a more familiar form by using an algebraic method called **partial fractions**. This method is illustrated below. It may be possible that $\dfrac{x + 1}{x^2 - 3x + 2}$ is the result of adding two fractions. To determine this, we factor the denominator:

Partial Fractions

$$\frac{x + 1}{x^2 - 3x + 2} = \frac{x + 1}{(x - 1)(x - 2)}.$$

Let $\dfrac{x + 1}{(x - 1)(x - 2)} = \dfrac{A}{x - 1} + \dfrac{B}{x - 2}$, where A and B are constants, to be determined, that hold for any x. Multiply both sides of the equation by $(x - 1)(x - 2)$. We obtain

$$x + 1 = A(x - 2) + B(x - 1).$$

Next, let $x = 1$. This eliminates B and gives an equation with one unknown. Then,

$$1 + 1 = A(1 - 2) + B(1 - 1),$$
$$2 = -A,$$

or

$$\boxed{-2 = A.}$$

Now, let $x = 2$. Then, this eliminates A and gives an equation that can be solved for B.

$$2 + 1 = A(2 - 2) + B(2 - 1),$$

or

$$\boxed{3 = B.}$$

Thus,

$$\frac{x + 1}{x^2 - 3x + 2} = \frac{-2}{x - 1} + \frac{3}{x - 2}.$$

Now we see that it is not difficult to evaluate $\int \frac{x + 1}{x^2 - 3x + 2} dx$:

$$\int \frac{x + 1}{x^2 - 3x + 2} dx = \int \left(\frac{-2}{x - 1} + \frac{3}{x - 2} \right) dx$$

$$= -2 \int \frac{dx}{x - 1} + 3 \int \frac{dx}{x - 2}$$

$$= -2 \ln |x - 1| + 3 \ln |x - 2| + c.$$

NOTE: Before using the method of partial fractions, be sure that the degree of the expression in the numerator is less than the degree of the expression in the denominator. Otherwise, it will be necessary to divide the numerator by the denominator before proceeding with the integration.

Examples 22.15–22.18 illustrate various approaches for employing the method of partial fractions.

• **Example 22.15:** Evaluate $\int \frac{5x^2 - 3x - 2}{x^2(x - 2)} dx$.

Solution: $\frac{5x^2 - 3x - 2}{x^2(x - 2)} = \frac{A}{x} + \frac{B}{x^2} + \frac{C}{x - 2}$. (All possible fractions whose LCD is $x^2(x - 2)$)

Multiply by $x^2(x - 2)$:

$$5x^2 - 3x - 2 = Ax(x - 2) + B(x - 2) + Cx^2.$$

Let $x = 0$ to eliminate A and C terms. Then B can be determined:

$$-2 = B(0 - 2),$$

or

$$\boxed{1 = B.}$$

Substitute $B = 1$ and let $x = 2$. This eliminates A so that C can be determined.

$$5 \cdot 2^2 - 3 \cdot 2 - 2 = C \cdot 2^2$$
$$12 = 4C,$$

or

$$\boxed{3 = C.}$$

Substitute 3 for C. The equation now contains only the unknown A. Expanding the right-hand side and simplifying gives a polynomial in A.

The coefficients of like terms on the left- and right-hand sides of the equation can be set equal as follows.

Let $B = 1$ and $C = 3$:

$$5x^2 - 3x - 2 = Ax(x - 2) + 1(x - 2) + 3x^2$$
$$5x^2 - 3x - 2 = Ax^2 - 2Ax + x - 2 + 3x^2$$
$$5x^2 - 3x - 2 = (A + 3)x^2 + (-2A + 1)x - 2.$$

Equating the coefficients of the x^2 terms:

$$5 = A + 3,$$

or

$$\boxed{A = 2.}$$

Note that equating the coefficients of the x terms will yield the same value for A.

Integrating, we have

$$\int \frac{5x^2 - 3x - 2}{x^2(x - 2)} \, dx = \int \left(\frac{2}{x} + \frac{1}{x^2} + \frac{3}{x - 2} \right) dx$$
$$= 2 \int \frac{1}{x} \, dx + \int x^{-2} \, dx + 3 \int \frac{1}{x - 2} \, dx$$
$$= 2 \ln |x| - \frac{1}{x} + 3 \ln |x - 2| + c.$$

• **Example 22.16:** Evaluate $\int \dfrac{x+1}{(x+2)(x-1)^2} \, dx$.

Solution: $\dfrac{x+1}{(x+2)(x-1)^2} = \dfrac{A}{x+2} + \dfrac{B}{x-1} + \dfrac{C}{(x-1)^2}$.

(All possible fractions whose LCD is $(x+2)(x-1)^2$)

Multiply by $(x+2)(x-1)^2$:

$$x + 1 = A(x-1)^2 + B(x+2)(x-1) + C(x+2).$$

Let $x = -2$ (this eliminates B and C):

$$-2 + 1 = A(-2-1)^2$$
$$-1 = 9A$$
$$\boxed{-\dfrac{1}{9} = A.}$$

Let $x = 1$:

$$1 + 1 = 3C$$
$$\boxed{\dfrac{2}{3} = C.}$$

Let $x = 0$, $A = -\dfrac{1}{9}$, and $C = \dfrac{2}{3}$:

$$1 = -\dfrac{1}{9}(0-1)^2 + B(0+2)(0-1) + \dfrac{2}{3}(0+2)$$

$$1 = -\dfrac{1}{9} - 2B + \dfrac{4}{3}$$

$$9 = -1 - 18B + 12$$

$$-2 = -18B$$

$$\boxed{\dfrac{1}{9} = B.}$$

Therefore,

$$\int \frac{x+1}{(x+2)(x-2)^2}\,dx = \int \left[\frac{-\frac{1}{9}}{x+2} + \frac{\frac{1}{9}}{x-1} + \frac{\frac{2}{3}}{(x-1)^2}\right]dx$$

$$= -\frac{1}{9}\int \frac{dx}{x+2} + \frac{1}{9}\int \frac{dx}{x-1} + \frac{2}{3}\int (x-1)^{-2}\,dx$$

$$= -\frac{1}{9}\ln|x+2| + \frac{1}{9}\ln|x-1| + \frac{2}{3}\frac{(x-1)^{-1}}{-1} + c$$

$$= -\frac{1}{9}\ln|x+2| + \frac{1}{9}\ln|x-1| - \frac{2}{3(x-1)} + c.$$

••••••••

• **Example 22.17:** Evaluate $\int \frac{x+4}{x(x^2+4)}\,dx$.

Solution: $\dfrac{x+4}{x(x^2+4)} = \dfrac{A}{x} + \dfrac{Bx+C}{x^2+4}$.

We use $Bx + C$ in the last fraction, since any numerator of this form ensures that the fraction is a proper fraction; that is, the degree of the numerator is less than the degree of the denominator.

Multiply by $x(x^2 + 4)$:

$$x + 4 = A(x^2 + 4) + (Bx + C)x$$
$$x + 4 = Ax^2 + 4A + Bx^2 + Cx,$$

or

$$0 \cdot x^2 + 1 \cdot x + 4 = (A + B)x^2 + Cx + 4A.$$

Now, $A + B = 0$, since $A + B$ is the coefficient of x^2 on the right-hand side, and 0 is the coefficient of x^2 on the left-hand side.
Similarly, $\boxed{C = 1}$ and $4 = 4A$, or $\boxed{A = 1}$. Since $A + B = 0$ and $A = 1$, we must have $B = -1$. Then

$$\int \frac{x+4}{x(x^2+4)}\,dx = \int \left(\frac{1}{x} + \frac{-x+1}{x^2+4}\right)dx$$

$$= \int \frac{1}{x}\,dx - \int \frac{x}{x^2+4}\,dx + \int \frac{1}{x^2+4}\,dx$$

$$= \ln|x| - \frac{1}{2}\ln|x^2+4| + \frac{1}{2}\tan^{-1}\left(\frac{x}{2}\right) + c.$$

••••••••

• **Example 22.18:** Evaluate $\int \dfrac{x^3}{(x-1)^2}\, dx$.

Solution: Since the degree of the numerator is greater than the degree of the denominator, we divide before attempting to integrate:

$$\frac{x^3}{(x-1)^2} = \frac{x^3}{x^2 - 2x + 1}$$

and

$$\begin{array}{r}
x + 2 + \dfrac{3x-2}{x^2-2x+1} \\
x^2 - 2x + 1 \overline{\smash{\big)} x^3 } \\
\underline{x^3 - 2x^2 + x} \\
2x^2 - x \\
\underline{2x^2 - 4x + 2} \\
3x - 2.
\end{array}$$

Therefore,

$$\int \frac{x^3}{(x-1)^2}\, dx = \int \left[x + 2 + \frac{3x-2}{(x-1)^2} \right] dx$$

$$= \int x\, dx + \int 2\, dx + \int \frac{3x-2}{(x-1)^2}\, dx$$

$$= \frac{x^2}{2} + 2x + \int \frac{3x-2}{(x-1)^2}\, dx.$$

Finding partial fractions, we have

$$\frac{3x-2}{(x-1)^2} = \frac{A}{x-1} + \frac{B}{(x-1)^2},$$

and

$$3x - 2 = Ax - A + B.$$

Therefore,

$$\boxed{A = 3}$$

and

$$-A + B = -2$$

or

$$-3 + B = -2$$

$$\boxed{B = 1.}$$

Then,

$$\frac{3x-2}{(x-1)^2} = \frac{3}{x-1} + \frac{1}{(x-1)^2}.$$

Now,

$$\int \frac{x^3}{(x-1)^2} dx = \frac{x^2}{2} + 2x + \int \frac{3}{x-1} dx + \int \frac{1}{(x-1)^2} dx$$

$$= \frac{x^2}{2} + 2x + 3 \ln |x - 1| - \frac{1}{x-1} + c.$$

(See example 22.16.)

Trial Problems 22.3

Before you begin the section exercises, warm up with these problems. Complete answers are included in the answer key.

When using the method of partial fractions, the following integral forms are often used. Evaluate each integral.

1. $\int \frac{3}{x-2} dx$
2. $\int x^{-2} dx$
3. $\int \frac{6}{(x-1)^2} dx$
4. $\int \frac{x}{x^2+4} dx$
5. $\int \frac{1}{x^2+4} dx$

Exercises 22.3

Evaluate the following.

1. $\int \frac{x+16}{(x+4)(x-2)} dx$
2. $\int \frac{5x^2 - 10x - 8}{x(x^2 - 4)} dx$
3. $\int \frac{2x^2 - 25x - 33}{(x+1)^2(x-5)} dx$
4. $\int \frac{2x^3 + 10x}{(x^2+1)^2} dx$
5. $\int \frac{dx}{x^2 - 5x + 6}$
6. $\int \frac{(x+2)dx}{(x+1)^2(x-3)}$
7. $\int \frac{(x+1)}{x^3 + 6x} dx$
8. $\int \frac{2 \, dx}{x^4 - 1}$
9. $\int \frac{(x-2)}{x^3 + x^2} dx$
10. $\int \frac{(x+35)}{x^2 - 25} dx$
11. $\int \frac{7x^2 + x - 3}{x^2(5x+3)} dx$
12. $\int \frac{3x^2 + 11x + 8}{(x+2)^3} dx$
13. $\int \frac{5x^2 + 13x + 2}{(x^2 - 4)(x+2)} dx$
14. $\int \frac{(2x+4)dx}{(1+x^2)(1-2x)}$
15. $\int \frac{7x^3 + 2x^2 + 5x - 2}{x^4 - 1} dx$
16. $\int \frac{9x^2 - 5x + 3}{x + x^3} dx$
17. $\int \frac{5x}{x^2 - 3x - 4} dx$
18. $\int \frac{x^2 + 6}{x^2 - 3x} dx$ (*Hint:* divide first.)
19. $\int \frac{x}{x^2 - 4} dx$
20. $\int \frac{x+1}{x^2(x+1)} dx$
21. $\int \frac{x^2 - x}{x^2 + x + 1} dx$

22. The acceleration of an object is given by $a = \dfrac{1}{4 - t^2}$. Find the expression for velocity if $v = 0$ when $t = 0$.

23. The current in a circuit is given by $i = \dfrac{t}{t^2 + 2t - 3}$. Find the equation for the charge if $q = 0$ when $t = 2$. (Recall that $q = \int i\, dt$.)

24. The velocity of an object moving along a straight line is given by
$$v = \dfrac{t + 3}{t^3 + t}\text{ ft/sec.}$$
How far does the object travel from $t = 0$ to $t = 1$ sec.

25. A medical technician determines that the logistics growth y of a certain bacteria culture is
$$y = \dfrac{10}{1 + 9e^{-kt}},$$
where t is time in seconds and k is the constant of proportionality for the growth. Show that $y = \dfrac{10}{1 + 9e^{-kt}}$ is a solution to the equation
$$\int \dfrac{1}{y(10 - y)}\, dy = \int k\, dt.$$
(To find the constant of integration, let $y = 1$ when $t = 0$.)

22 • 4 Integral Tables

Although we have investigated many integral forms, there are many more forms that are required for solving certain problems. As long as we have a good working knowledge of the basic forms, we can use a table of integrals to help us evaluate other forms. Sometimes we can find the exact integral form in such a table. At other times it may be necessary to use algebra and trigonometry to transform an integral into a form that we recognize in a table.

The following is a partial table of integrals. The table does not contain the basic forms nor any of those that can be evaluated by partial fractions. More extensive tables can be found in books of mathematical tables. Examples 22.19 through 22.24 illustrate how we can use the tables to evaluate particular integrals.

Table of Integrals

1. $\displaystyle\int \dfrac{du}{a + bu^2} = \dfrac{1}{\sqrt{ab}} \tan^{-1}\sqrt{\dfrac{b}{a}}\, u + c$, where a and b are constants.

2. $\displaystyle\int \dfrac{du}{a^2 - b^2 u^2} = \dfrac{1}{2ab} \ln\left|\dfrac{a + bu}{a - bu}\right| + c$

3. $\displaystyle\int \dfrac{u\, du}{a + bu^2} = \dfrac{1}{2b} \ln\left|u^2 + \dfrac{a}{b}\right| + c$

4. $\displaystyle\int \dfrac{u^2\, du}{a + bu^2} = \dfrac{u}{b} - \dfrac{a}{b}\int \dfrac{du}{a + bu^2}$

5. $\displaystyle\int \dfrac{du}{u(a + bu^2)} = \dfrac{1}{2a} \ln\left|\dfrac{u^2}{a + bu^2}\right| + c$

6. $\displaystyle\int \frac{du}{u^2(a+bu^2)} = -\frac{1}{au} + \frac{b}{a^2}\ln\left|\frac{a+bu}{u}\right| + c$

7. $\displaystyle\int \frac{du}{(a+bu^2)^2} = \frac{u}{2a(a+bu^2)} + \frac{1}{2a}\int \frac{du}{a+bu^2}$

8. $\displaystyle\int \sqrt{u^2 \pm a^2}\, du = \frac{1}{2}\left[u\sqrt{u^2 \pm a^2} \pm a^2 \ln\left|u + \sqrt{u^2 \pm a^2}\right|\right] + c$

9. $\displaystyle\int u^2\sqrt{u^2 \pm a^2}\, du = \frac{1}{8}\left[u(2u^2 \pm a^2)\sqrt{u^2 \pm a^2} - a^2 \ln\left|u + \sqrt{u^2 \pm a^2}\right|\right] + c$

10. $\displaystyle\int \frac{\sqrt{u^2+a^2}}{u}\, du = \sqrt{u^2+a^2} - a\ln\left|\frac{a+\sqrt{u^2+a^2}}{u}\right| + c$

11. $\displaystyle\int \frac{\sqrt{u^2+a^2}}{u^2}\, du = \frac{-\sqrt{u^2 \pm a^2}}{u} + \ln\left|u + \sqrt{u^2 \pm a^2}\right| + c$

12. $\displaystyle\int \frac{du}{\sqrt{u^2 \pm a^2}} = \ln\left|u + \sqrt{u^2 \pm a^2}\right| + c$

13. $\displaystyle\int \frac{du}{u\sqrt{u^2+a^2}} = -\frac{1}{a}\ln\left|\frac{a+\sqrt{u^2+a^2}}{u}\right| + c$

14. $\displaystyle\int \frac{u^2\, du}{\sqrt{u^2 \pm a^2}} = \frac{1}{2}\left[u\sqrt{u^2 \pm a^2} \mp a^2 \ln\left|u + \sqrt{u^2 \pm a^2}\right|\right] + c$

15. $\displaystyle\int \frac{du}{u^2\sqrt{u^2 \pm a^2}} = \mp \frac{\sqrt{u^2 \pm a^2}}{a^2 u} + c$

16. $\displaystyle\int \frac{du}{(u^2 \pm a^2)^{3/2}} = \frac{\pm u}{a^2\sqrt{u^2 \pm a^2}} + c$

17. $\displaystyle\int \sqrt{a^2 - u^2}\, du = \frac{u}{2}\sqrt{a^2-u^2} + \frac{a^2}{2}\sin^{-1}\left(\frac{u}{a}\right) + c$

18. $\displaystyle\int u^2\sqrt{a^2 - u^2}\, du = \frac{u}{8}(2u^2 - a^2)\sqrt{a^2-u^2} + \frac{a^4}{8}\sin^{-1}\left(\frac{u}{a}\right) + c$

19. $\displaystyle\int \frac{du}{(a^2-u^2)^{3/2}} = \frac{u}{a^2\sqrt{a^2-u^2}} + c$

20. $\displaystyle\int \frac{u^2\, du}{\sqrt{a^2-u^2}} = -\frac{u}{2}\sqrt{a^2-u^2} + \frac{a^2}{2}\sin^{-1}\left(\frac{u}{a}\right) + c$

21. $\displaystyle\int \frac{du}{\sqrt{ua^2-u^2}} = \frac{1}{a}\ln\left|\frac{u}{a+a^2-u^2}\right| + c$

22. $\displaystyle\int \frac{du}{u^2\sqrt{a^2-u^2}} = -\frac{\sqrt{a^2-u^2}}{a^2 u} + c$

23. $\displaystyle\int \frac{\sqrt{a^2-u^2}}{u}\,du = \sqrt{a^2-u^2} - a\ln\frac{a+\sqrt{a^2-u^2}}{u} + c$

24. $\displaystyle\int \frac{\sqrt{a^2-u^2}}{u^2}\,du = -\frac{\sqrt{a^2-u^2}}{u} = \sin^{-1}\left(\frac{u}{a}\right) + c$

25. $\displaystyle\int u\sqrt{a+bu}\,du = -\frac{2(2a-3bu)\sqrt{(a+bu)^3}}{15b^2} + c$

26. $\displaystyle\int u^2\sqrt{a+bu}\,du = \frac{2(8a^2-12abu+15b^2u^2)\sqrt{(a+bu)^3}}{105b^3} + c$

27. $\displaystyle\int \frac{u\,du}{a+bu} = -\frac{2(2a-bu)}{3b^2}\sqrt{a+bu} + c$

28. $\displaystyle\int \frac{du}{u\sqrt{a+bu}} = \frac{1}{\sqrt{a}}\ln\left|\frac{\sqrt{a+bu}-\sqrt{a}}{\sqrt{a+bu}+\sqrt{a}}\right| + c$, if $a > 0$

   $\displaystyle\qquad\qquad\qquad = \frac{2}{\sqrt{-a}}\tan^{-1}\sqrt{\frac{a+bu}{-a}} + c$, if $a < 0$

29. $\displaystyle\int \frac{du}{u^2\sqrt{a+bu}} = \frac{-\sqrt{a+bu}}{au} - \frac{b}{2a}\int \frac{du}{u\sqrt{a+bu}} + c$

30. $\displaystyle\int \frac{\sqrt{a+bu}}{u}\,du = 2\sqrt{a+bu} + a\int \frac{du}{u\sqrt{a+bu}} + c$

31. $\displaystyle\int \sin au\,\sin bu\,du = -\frac{\sin(a+b)u}{2(a+b)} + \frac{\sin(a-b)u}{2(a-b)} + c$

32. $\displaystyle\int \cos au\,\cos bu\,du = \frac{\sin(a+b)u}{2(a+b)} + \frac{\sin(a-b)u}{2(a-b)} + c$

33. $\displaystyle\int \sin au\,\cos bu\,du = -\frac{\cos(a+b)u}{2(a+b)} - \frac{\cos(a-b)u}{2(a-b)} + c$

34. $\displaystyle\int e^{au}\sin bu\,du = \frac{e^{au}(a\sin bu - b\cos bu)}{a^2+b^2} + c$

35. $\displaystyle\int e^{au}\cos bu\,du = \frac{e^{au}(b\sin bu + a\cos bu)}{a^2+b^2} + c$

36. $\displaystyle\int \frac{du}{u\ln u} = \ln(\ln u) + c$

37. $\displaystyle\int u\ln u\,du = \frac{1}{2}u^2\left(\ln|u| - \frac{1}{2}\right) + c$

38. $\int u^2 \ln u \, du = \frac{1}{3} u^3 \left(\ln |u| - \frac{1}{3} \right) + c$

39. $\int (\ln u)^2 \, du = u \left[2 - 2 \ln |u| + (\ln |u|)^2 \right] + c$

40. $\int \frac{du}{1 + e^u} = u - \ln(1 + e^u) + c$

41. $\int \frac{du}{1 + e^{2u}} = u - \frac{1}{2} \ln(1 + e^{2u}) + c$

• **Example 22.19:** Use the integral table to evaluate $\int \frac{4}{25 - x^2} \, dx$.

Solution: Using number 2 we have $a = 5, b = 1$:

$$\int \frac{4}{25 - x^2} \, dx = \frac{4}{2 \cdot 5 \cdot 1} \ln \left| \frac{5 + x}{5 - x} \right| + c$$

$$= \frac{2}{5} \ln \left| \frac{5 + x}{5 - x} \right| + c.$$

••••••••••

• **Example 22.20:** Use the integral table to evaluate $\int x \sqrt{x^4 - 9} \, dx$.

Solution: Using number 8 with $u = x^2$, $a = 3$, we have $du = 2x \, dx$, and

$$\int x \sqrt{x^4 - 9} \, dx = \frac{1}{2} \int \sqrt{(x^2)^2 - (3)^2} \, (2x \, dx)$$

$$= \frac{1}{2} \cdot \frac{1}{2} [x^2 \sqrt{x^4 - 9} - 9 \ln |x^2 + \sqrt{x^4 - 9}|] + c$$

$$= \frac{1}{4} [x^2 \sqrt{x^4 - 9} - 9 \ln |x^2 + \sqrt{x^4 - 9}|] + c.$$

••••••••••

• **Example 22.21:** Evaluate $\int \frac{dx}{x \sqrt{x^2 + 4}}$.

Solution: Using number 13 with $u = x$, $a = 2$, and $du = dx$, we have

$$\int \frac{dx}{x \sqrt{x^2 + 4}} = -\frac{1}{2} \ln \left| \frac{2 + \sqrt{x^2 + 4}}{x} \right| + c.$$

••••••••••

• **Example 22.22:** Evaluate $\int \sin 2x \cos 3x \, dx$.

Solution: Using number 33 with $a = 2$ and $b = 3$, we have

$$\int \sin 2x \cos 3x \, dx = -\frac{\cos(2+3)x}{2(2+3)} - \frac{\cos(2-3)x}{2(2-3)} + c$$

$$= -\frac{\cos 5x}{10} - \frac{\cos(-x)}{-2} + c$$

$$= -\frac{\cos 5x}{10} + \frac{\cos x}{2} + c.$$

• **Example 22.23:** Evaluate $\int \sqrt{9 - 16x^2} \, dx$.

Solution: Using number 17 with $a = 3$, $u = 4x$, and $du = 4 \, dx$, we have

$$\int \sqrt{9 - 16x^2} \, dx = \frac{1}{4} \int \sqrt{(3)^2 - (4x)^2} \, (4 \, dx)$$

$$= \frac{1}{4} \left[\frac{4x}{2} \sqrt{(3)^2 - (4x)^2} + \frac{(3)^2}{2} \sin^{-1}\left(\frac{4x}{3}\right) \right] + c$$

$$= \frac{x}{2} \sqrt{9 - 16x^2} + \frac{9}{8} \sin^{-1}\left(\frac{4x}{3}\right) + c.$$

• **Example 22.24:** Evaluate $\int \frac{dx}{1 + e^{4x}}$.

Solution: Using number 41 with $u = 2x$ and $du = 2 \, dx$, we have

$$\int \frac{dx}{1 + e^{4x}} = \frac{1}{2} \int \frac{2 \, dx}{1 + e^{4x}}$$

$$= \frac{1}{2} \left[2x - \frac{1}{2} \ln |1 + e^{4x}| \right] + c$$

$$= x - \frac{1}{4} \ln |1 + e^{4x}| + c.$$

Trial Problems 22.4

Before you begin the section exercises, warm up with these problems. Complete answers are included in the answer key.

The following can be integrated immediately by using a table of integrals. Evaluate each integral.

1. $\displaystyle\int \frac{dx}{x(1 + x^2)}$
2. $\displaystyle\int \frac{dx}{x\sqrt{x^2 + 1}}$
3. $\displaystyle\int \sin 3x \cos 2x \, dx$
4. $\displaystyle\int \frac{dx}{x \ln x}$
5. $\displaystyle\int \frac{dx}{1 + e^x}$

Exercises 22.4

Use the integral table to evaluate each of the following.

1. $\displaystyle\int \frac{dx}{x^2(2 + 3x)}$
2. $\displaystyle\int e^{3x}\sin 2x \, dx$
3. $\displaystyle\int \frac{dx}{5 - 2x^2}$
4. $\displaystyle\int \frac{x^2}{\sqrt{x^2 - 4}} dx$
5. $\displaystyle\int \frac{\sqrt{4x^2 + 9}}{x} dx$
6. $\displaystyle\int \frac{dx}{x^2\sqrt{3 - x^2}}$
7. $\displaystyle\int x^2 \ln x \, dx$
8. $\displaystyle\int [\ln 4x]^2 \, dx$
9. $\displaystyle\int \frac{x \, dx}{\sqrt{5 + x^2}}$
10. $\displaystyle\int \frac{\sqrt{9 - 4x^2}}{2x} dx$
11. $\displaystyle\int \cos 2x \sin 6x \, dx$
12. $\displaystyle\int \cos 6x \sin 2x \, dx$
13. $\displaystyle\int \sin 3x \sin x \, dx$
14. $\displaystyle\int_0^\pi \sin 2x \cos 2x \, dx$
15. $\displaystyle\int_1^2 x\sqrt{1 - 2x} \, dx$
16. $\displaystyle\int \frac{dx}{x\sqrt{5 + 3x}}$
17. $\displaystyle\int \frac{\sqrt{5 + 3x}}{x} dx$
18. $\displaystyle\int e^{4x}\cos 3x \, dx$
19. $\displaystyle\int \frac{dx}{x \ln x^2}$
20. $\displaystyle\int \frac{e^x}{1 + e^x} dx$
21. $\displaystyle\int \sqrt{4 + \frac{9}{x^2}} \, dx$
22. $\displaystyle\int_0^{1/2} \sqrt{1 - 4x^2} \, dx$

23. The electric current in a circuit is given by
$i = \dfrac{2}{\sqrt{t^2 + 100}}$. Find the formula for the amount of charge if $q = 0$ when $t = 0$.

24. The velocity of a particle is given by $v = \dfrac{t}{\sqrt{1 + t}}$. Find the formula for distance if $s = \dfrac{5}{3}$ when $t = 0$.

25. Find the work equation of a particle if its power equation is $p = \dfrac{1}{t \ln t}$, $t \geq 2$, and $w = 0$ when $t = 2$.

26. The growth in a bacteria culture is given by

$$N(t) = \dfrac{50}{1 + e^{4.8 - 1.9t}},$$

where t is the time in days. Find the average number in the culture between the third and fourth day.

$\left(\text{Hint: the average number is found by evaluating } \dfrac{1}{b-a} \int_a^b N(t)\, dt.\right)$

27. A chemical reaction leads to the equation

$$dt = \dfrac{kx\, dx}{2 + x^2},$$

where t is time, k is a constant, and x is the concentration of the chemical. Find t as a function of x.

28. An object moves along the curve $y = \dfrac{1}{2}x^2 + 3$. To find the distance s traveled by the object, it is necessary to evaluate

$$s = \int_a^b \sqrt{1 + (y')^2}\, dx.$$

Find the distance traveled by the object from $x = 0$ to $x = 5$.

22·5 Summary of Terms and Rules

Terms

Integration by parts (p. 896)   Partial fractions (p. 909)

Rules

- $\displaystyle\int u\, dv = uv - \int v\, du$, where $u = h(x)$ and $v = g(x)$.
  (22.1)

- TABLE OF INTEGRALS   (22.4)

  1. $\displaystyle\int \dfrac{du}{a + bu^2} = \dfrac{1}{\sqrt{ab}} \tan^{-1} \sqrt{\dfrac{b}{a}}\, u + c$, where a and b are constants.

  2. $\displaystyle\int \dfrac{du}{a^2 - b^2 u^2} = \dfrac{1}{2ab} \ln\left|\dfrac{a + bu}{a - bu}\right| + c$

  3. $\displaystyle\int \dfrac{u\, du}{a + bu^2} = \dfrac{1}{2b} \ln\left|u^2 + \dfrac{a}{b}\right| + c$

  4. $\displaystyle\int \dfrac{u^2\, du}{a + bu^2} = \dfrac{u}{b} - \dfrac{a}{b}\int \dfrac{du}{a + bu^2}$

  5. $\displaystyle\int \dfrac{du}{u(a + bu^2)} = \dfrac{1}{2a} \ln\left|\dfrac{u^2}{a + bu^2}\right| + c$

  6. $\displaystyle\int \dfrac{du}{u^2(a + bu^2)} = -\dfrac{1}{au} + \dfrac{b}{a^2} \ln\left|\dfrac{a + bu}{u}\right| + c$

7. $\displaystyle\int \frac{du}{(a + bu^2)^2} = \frac{u}{2a(a + bu^2)} + \frac{1}{2a}\int \frac{du}{a + bu^2}$

8. $\displaystyle\int \sqrt{u^2 \pm a^2}\, du = \frac{1}{2}\left[u\sqrt{u^2 \pm a^2} \pm a^2 \ln|u + \sqrt{u^2 \pm a^2}|\right] + c$

9. $\displaystyle\int u^2\sqrt{u^2 \pm a^2}\, du = \frac{1}{8}\left[u(2u^2 \pm a^2)\sqrt{u^2 \pm a^2} - a^4 \ln|u + \sqrt{u^2 \pm a^2}|\right] + c$

10. $\displaystyle\int \frac{\sqrt{u^2 + a^2}}{u}\, du = \sqrt{u^2 + a^2} - a \ln\left|\frac{a + \sqrt{u^2 + a^2}}{u}\right| + c$

11. $\displaystyle\int \frac{\sqrt{u^2 + a^2}}{u^2}\, du = \frac{-\sqrt{u^2 \pm a^2}}{u} + \ln|u + \sqrt{u^2 \pm a^2}| + c$

12. $\displaystyle\int \frac{du}{\sqrt{u^2 \pm a^2}} = \ln|u + \sqrt{u^2 \pm a^2}| + c$

13. $\displaystyle\int \frac{du}{u\sqrt{u^2 + a^2}} = -\frac{1}{a}\ln\left|\frac{a + \sqrt{u^2 + a^2}}{u}\right| + c$

14. $\displaystyle\int \frac{u^2\, du}{\sqrt{u^2 \pm a^2}} = \frac{1}{2}\left[u\sqrt{u^2 \pm a^2} \mp a^2 \ln|u + \sqrt{u^2 \pm a^2}|\right] + c$

15. $\displaystyle\int \frac{du}{u^2\sqrt{u^2 \pm a^2}} = \mp \frac{\sqrt{u^2 \pm a^2}}{a^2 u} + c$

16. $\displaystyle\int \frac{du}{(u^2 \pm a^2)^{3/2}} = \frac{\pm u}{a^2\sqrt{u^2 \pm a^2}} + c$

17. $\displaystyle\int \sqrt{a^2 - u^2}\, du = \frac{u}{2}\sqrt{a^2 - u^2} + \frac{a^2}{2}\sin^{-1}\left(\frac{u}{a}\right) + c$

18. $\displaystyle\int u^2\sqrt{a^2 - u^2}\, du = \frac{u}{8}(2u^2 - a^2)\sqrt{a^2 - u^2} + \frac{a^4}{8}\sin^{-1}\left(\frac{u}{a}\right) + c$

19. $\displaystyle\int \frac{du}{(a^2 - u^2)^{3/2}} = \frac{u}{a^2\sqrt{a^2 - u^2}} + c$

20. $\displaystyle\int \frac{u^2\, du}{\sqrt{a^2 - u^2}} = -\frac{u}{2}\sqrt{a^2 - u^2} + \frac{a^2}{2}\sin^{-1}\left(\frac{u}{a}\right) + c$

21. $\displaystyle\int \frac{du}{\sqrt{ua^2 - u^2}} = \frac{1}{a}\ln\left|\frac{u}{a + a^2 - u^2}\right| + c$

22. $\displaystyle\int \frac{du}{u^2\sqrt{a^2 - u^2}} = -\frac{\sqrt{a^2 - u^2}}{a^2 u} + c$

23. $\displaystyle\int \frac{\sqrt{a^2 - u^2}}{u}\, du = \sqrt{a^2 - u^2} - a \ln\frac{a + \sqrt{a^2 - u^2}}{u} + c$

24. $\displaystyle\int \frac{\sqrt{a^2 - u^2}}{u^2}\, du = -\frac{\sqrt{a^2 - u^2}}{u} - \sin^{-1}\left(\frac{u}{a}\right) + c$

25. $\displaystyle\int u\sqrt{a+bu}\,du = -\frac{2(2a-3bu)\sqrt{(a+bu)^3}}{15b^2} + c$

26. $\displaystyle\int u^2\sqrt{a+bu}\,du = \frac{2(8a^2-12abu+15b^2u^2)\sqrt{(a+bu)^3}}{105b^3} + c$

27. $\displaystyle\int \frac{u\,du}{a+bu} = -\frac{2(2a-bu)}{3b^2}\sqrt{a+bu} + c$

28. $\displaystyle\int \frac{du}{u\sqrt{a+bu}} = \frac{1}{\sqrt{a}}\ln\left|\frac{\sqrt{a+bu}-\sqrt{a}}{\sqrt{a+bu}+\sqrt{a}}\right| + c,$ if $a > 0$

$\qquad = \dfrac{2}{\sqrt{-a}}\tan^{-1}\sqrt{\dfrac{a+bu}{-a}} + c,$ if $a < 0$

29. $\displaystyle\int \frac{du}{u^2\sqrt{a+bu}} = \frac{-\sqrt{a+bu}}{au} - \frac{b}{2a}\int \frac{du}{u\sqrt{a+bu}}$

30. $\displaystyle\int \frac{\sqrt{a+bu}}{u}\,du = 2\sqrt{a+bu} + a\int \frac{du}{u\sqrt{a+bu}}$

31. $\displaystyle\int \sin au \sin bu\,du = -\frac{\sin(a+b)u}{2(a+b)} + \frac{\sin(a-b)u}{2(a-b)} + c$

32. $\displaystyle\int \cos au \cos bu\,du = \frac{\sin(a+b)u}{2(a+b)} + \frac{\sin(a-b)u}{2(a-b)} + c$

33. $\displaystyle\int \sin au \cos bu\,du = -\frac{\cos(a+b)u}{2(a+b)} - \frac{\cos(a-b)u}{2(a-b)} + c$

34. $\displaystyle\int e^{au}\sin bu\,du = \frac{e^{au}(a\sin bu - b\cos bu)}{a^2+b^2} + c$

35. $\displaystyle\int e^{au}\cos bu\,du = \frac{e^{au}(b\sin bu + a\cos bu)}{a^2+b^2} + c$

36. $\displaystyle\int \frac{du}{u\ln u} = \ln(\ln u) + c$

37. $\displaystyle\int u\ln u\,du = \frac{1}{2}u^2\left(\ln|u| - \frac{1}{2}\right) + c$

38. $\displaystyle\int u^2\ln u\,du = \frac{1}{3}u^3\left(\ln|u| - \frac{1}{3}\right) + c$

39. $\displaystyle\int (\ln u)^2\,du = u\left[2 - 2\ln|u| + (\ln|u|)^2\right] + c$

40. $\displaystyle\int \frac{du}{1+e^u} = u - \ln(1+e^u) + c$

41. $\displaystyle\int \frac{du}{1+e^{2u}} = u - \frac{1}{2}\ln(1+e^{2u}) + c$

22·6 Chapter 22 Review Exercises

Evaluate the following integrals using integration by parts, trigonometric substitutions, or partial fractions. If none of these three methods apply, use the integral tables.

1. $\displaystyle\int \frac{(3x-7)dx}{(x-1)(x-2)(x-3)}$

2. $\displaystyle\int x^2 e^x \, dx$

3. $\displaystyle\int \sqrt{x^2+1} \, dx$

4. $\displaystyle\int \frac{dt}{\sqrt{e^t+1}}$

5. $\displaystyle\int \frac{dx}{\sqrt{(25-x^2)^3}}$

6. $\displaystyle\int \frac{dx}{x(x^2+1)^2}$

7. $\displaystyle\int \frac{x+1}{x^2(x-1)} dx$

8. $\displaystyle\int \frac{x \, dx}{x^2+4x+3}$

9. $\displaystyle\int \frac{4 \, dx}{x^3+4x}$

10. $\displaystyle\int \frac{\sqrt{x^2-16}}{x} dx$

11. $\displaystyle\int e^x \cos 2x \, dx$

12. $\displaystyle\int x \sin^{-1} x \, dx$

13. $\displaystyle\int \frac{(x^3+x^2)dx}{x^2+x-2}$

14. $\displaystyle\int \frac{x \, dx}{(x-1)^2}$

15. $\displaystyle\int \frac{dx}{x^2\sqrt{36-x^2}}$

16. $\displaystyle\int x[\ln x^2]^2 \, dx$

17. $\displaystyle\int x^2 \sin x \, dx$

18. $\displaystyle\int (x+1)^2 e^x \, dx$

19. $\displaystyle\int x^3 e^{x^2} \, dx$

20. $\displaystyle\int \frac{x \, dx}{\sqrt{1-x}}$

21. $\displaystyle\int \ln(x^2+x) dx$

22. $\displaystyle\int \frac{dx}{e^{2x}+1}$

23. $\displaystyle\int_0^1 \ln(1+x) dx$

24. $\displaystyle\int \frac{dx}{(x^2+25)^{3/2}}$

25. $\displaystyle\int \frac{\sqrt{4-x^2}}{x} dx$

26. $\displaystyle\int \frac{x^3-20x^2-63x-198}{x^4-81} dx$

27. $\displaystyle\int \sin^3 x \cos^3 x \, dx$

28. $\displaystyle\int \frac{dx}{x^{3/2}+x^{1/2}}$

29. $\displaystyle\int x \sec x \tan x \, dx$

30. $\displaystyle\int \sin^3 x \sqrt{\cos x} \, dx$

31. $\displaystyle\int \frac{x^2 \, dx}{\sqrt{4x^2+25}}$

32. $\displaystyle\int \sin 2x \cos x \, dx$

33. $\displaystyle\int \sin 3x \cos 3x \, dx$

34. $\displaystyle\int \frac{dx}{x^2\sqrt{16-x^2}}$

35. The velocity of a particle is given by $v = \sin^{-1}(2t)$ cm/sec. Find the distance s traveled from $t = 0$ to $t = \frac{1}{2}$ sec.

36. The current in a circuit is given by $i = \frac{t}{t+1}$. Find the amount of charge transferred from $t = 0$ to $t = 3$ sec.

37. Find the equation of the curve for which $\frac{dy}{dx} = xe^x$ if the curve passes through the origin.

38. The average power delivered to an electric circuit is given by $P = \frac{4}{\pi}\int_0^{\pi/4} ei \, dt$. Find the average power if $e = 10 \cos 2t$ and $i = 2 \sin 2t$.

39. In a certain experiment in physics, the quantities t and x are related by $\dfrac{dt}{dx} = \dfrac{7 - 4x}{(1 - x)(2 - x)}$. Find t as a function of x.

40. If the velocity of an object is given by $v = \tan^{-1} t$ cm/sec, find the distance traveled from $t = 0$ to $t = 1$ sec.

41. The current in a resistor is given by $i = \dfrac{t^2}{\sqrt{9 - t^2}}$. Find the equation for the charge q if $q = 0$ when $t = 0$.

22·7 Chapter 22 Test

Use the method of integration by parts to evaluate the following.

1. $\displaystyle\int t e^t \, dt$

2. $\displaystyle\int x^2 \cos x \, dx$

3. $\displaystyle\int \dfrac{x \, dx}{e^x}$

4. $\displaystyle\int \dfrac{1}{x} \ln x \, dx$

Use trigonometric substitution to evaluate the following.

5. $\displaystyle\int \sin^2 x \cos^2 x \, dx$

6. $\displaystyle\int \dfrac{\cos^2 x}{\sin x} \, dx$

7. $\displaystyle\int \sqrt{9 - t^2} \, dt$

8. Use partial fractions to evaluate $\displaystyle\int \dfrac{3x - 2}{(x - 1)^2} \, dx$.

9. Use the integral tables to evaluate $\displaystyle\int \dfrac{dx}{x\sqrt{2 + 3x}}$.

10. The acceleration of an object is given by $a = \dfrac{5t}{t^2 - 3t - 4}$. Find the expression for velocity if $v = 0$ when $t = 0$.

CHAPTER 23

Applications of Integration

*I*n chapter 20 we investigated several applications of differential calculus in various areas of technical mathematics. In this chapter we investigate several applications of integral calculus. We begin with the classical example of finding the bounded area under a curve. Applications of this concept to business and probability theory, for example, are investigated. Formulas, rules, and procedures for finding volumes, work performed, centroids, moments of inertia, as well as others, are carefully developed. Several different types of applications help us appreciate the power of the calculus in providing solutions to problems in all areas of technical mathematics.

23 • 1 The Area under a Curve

One of the most important applications of integration is finding the area under a curve described by a function $y = f(x)$. We use a little geometry and consider a continuous curve as illustrated in figure 23.1. Our goal is to find the area under the curve $y = f(x)$ from $x = a$ to $x = b$, bounded below by the x-axis. In this case, $f(x) > 0$.

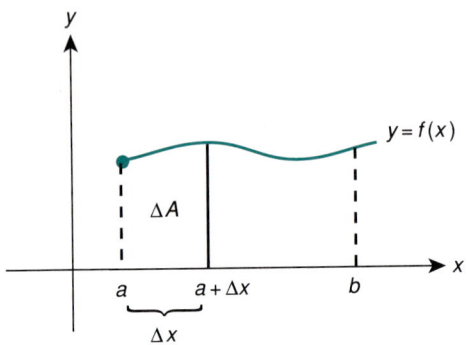

Figure 23.1

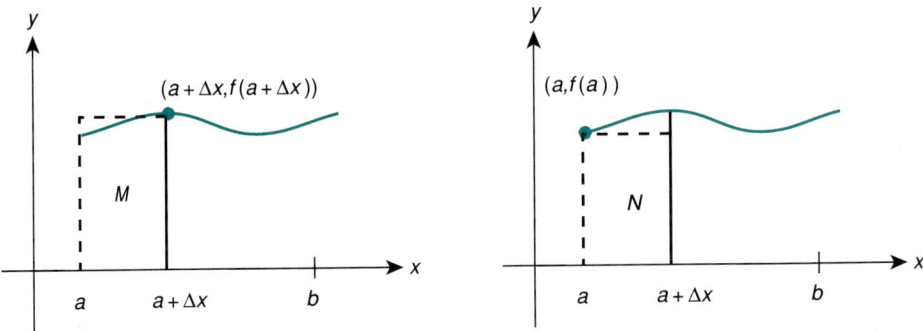

Figure 23.2

First, consider the region under the curve from a to $(a + \Delta x)$, where Δx is a small increment of x. The area of this region can be found by considering the rectangular regions above and below this region as shown in figure 23.2. Label the area of the region under the curve from a to $(a + \Delta x)$, ΔA. We call the other two regions M and N, respectively. It follows that:

$$\text{area } N \leq \Delta A \leq \text{area } M.$$

Now,
$$\text{area } N = f(a)\Delta x \quad \text{(Base } \Delta x \text{ times height } f(a)\text{)}$$

and
$$\text{area } M = f(a + \Delta x)\Delta x.$$

Therefore,
$$f(a)\Delta x \leq \Delta A \leq f(a + \Delta x)\Delta x.$$

Dividing by Δx,
$$f(a) \leq \frac{\Delta A}{\Delta x} \leq f(a + \Delta x).$$

If we let Δx get smaller and smaller, approaching 0, we have $a + \Delta x$ approaching a and $\lim_{\Delta x \to 0} \frac{\Delta A}{\Delta x} = A'$, from the definition of the first derivative. Now,
$$f(a) \leq A' \leq f(a),$$

which implies that
$$A' = f(a),$$
$$\frac{dA}{dx} = f(a),$$

or
$$dA = f(a)dx.$$

Since a is any particular x-value, we can write
$$dA = f(x)dx.$$

Integrating both sides,
$$\int dA = \int f(x)dx,$$

we obtain
$$A = \int f(x)dx$$
$$= F(x) + c,$$

where $D_x[F(x) + c] = f(x)$.

The value of the constant, c, can be determined as follows. If $\Delta x = 0$, then the area is zero (A = base × height = 0 · height = 0), and

$$A = F(a) + c = 0. \quad \text{(Substituting } a \text{ for } x \text{ in } A = F(x) + c\text{)}$$

This gives us

$$c = -F(a)$$
$$A = F(x) - F(a).$$

If we let Δx increase to the point where $x = b$, then we have

$$A = F(b) - F(a).$$

Since $A = \int f(x)dx$, we have

$$A = \int f(x)dx = F(b) - F(a).$$

Area under the Curve

Since $F(x)$ is the antiderivative of $f(x)$, and x goes from $x = a$ to $x = b$, we can use a definite integral to evaluate the **area under the curve** $y = f(x)$ from $x = a$ to $x = b$, by

$$A = \int_a^b f(x)dx.$$

• **Example 23.1:** Find the area under the curve $y = x^2$ from $x = 1$ to $x = 3$. (See figure 23.3.)

Solution: $A = \int_1^3 x^2\,dx = \dfrac{x^3}{3}\bigg|_1^3 = \dfrac{3^3}{3} - \dfrac{1^3}{3}$

$= \dfrac{27}{3} - \dfrac{1}{3}$

$= \dfrac{26}{3}$

or $8\dfrac{2}{3}$ square units.

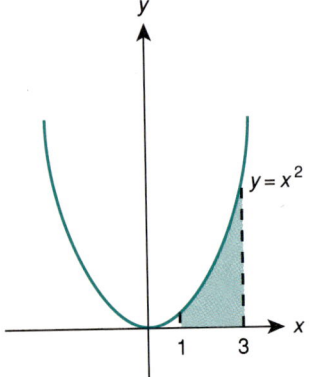

Figure 23.3

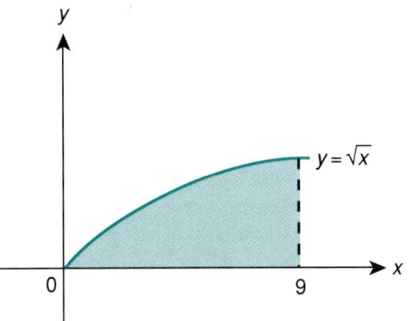

Figure 23.4

- **Example 23.2:** Find the area under the curve $y = \sqrt{x}$ from $x = 0$ to $x = 9$. (See figure 23.4.)

Solution:
$$A = \int_0^9 \sqrt{x}\, dx$$
$$= \int_0^9 x^{1/2}\, dx$$
$$= \left. \frac{x^{3/2}}{\frac{3}{2}} \right|_0^9$$
$$= \left. \frac{2}{3} x^{3/2} \right|_0^9$$
$$= \frac{2}{3}(9)^{3/2} - \frac{2}{3}(0)^{3/2}$$
$$= \frac{2}{3}(27) - 0$$
$$= 18 \text{ sq. units.}$$

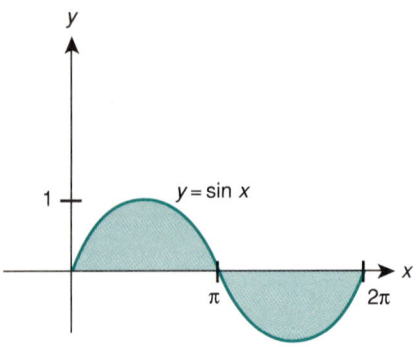

Figure 23.5

- **Example 23.3:** Find the area bounded by the curve $y = \sin x$ and $y = 0$ from $x = 0$ to $x = 2\pi$. (See figure 23.5.)

  Consider, first, the area from $x = 0$ to $x = \pi$.

  Solution: $A = \int_0^\pi \sin x \, dx$

  $= -\cos x \Big|_0^\pi$

  $= (-\cos \pi) - (-\cos 0)$

  $= -(-1) - (-1)$

  $= 1 + 1$

  $= 2$ sq. units.

  Similarly, $\int_\pi^{2\pi} \sin x \, dx = -2$ because $f(x)$ is negative for x between π and 2π. In this case the area is

  $\left| \int_\pi^{2\pi} \sin x \, dx \right| = \left| -2 \right| = 2.$

  Therefore, the total area is $2 + 2 = 4$ sq. units.

- **Example 23.4:** Find the area of the circle $x^2 + y^2 = 4$ in the first quadrant only. (See figure 23.6.)

  Solution: Since $x^2 + y^2 = 4$,

  $$y^2 = 4 - x^2$$

  and

  $$y = \pm \sqrt{4 - x^2}.$$

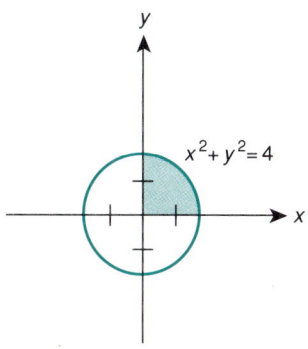

Figure 23.6

For $x = 0$ to $x = 2$,

$$y = \sqrt{4 - x^2}$$

$$A = \int_0^2 \sqrt{4 - x^2}\, dx$$

$$= \left[\frac{x}{2}\sqrt{4 - x^2} + \frac{4}{2}\sin^{-1}\frac{x}{2}\right]\Big|_0^2 \quad \text{(Number 17 in the integral tables)}$$

$$= \left[\frac{2}{2}\sqrt{4 - 2^2} + 2\sin^{-1}\frac{2}{2}\right] - \left[\frac{0}{2}\sqrt{4 - 0^2} + 2 \cdot \sin^{-1}\frac{0}{2}\right]$$

$$= \left(1 \cdot 0 + 2\frac{\pi}{2}\right) - (0 \cdot 2 + 2 \cdot 0)$$

$$= \pi \text{ sq. units.}$$

• **Example 23.5:** The velocity of a rocket t seconds after lift-off is given by $v = 0.3t^2 + 4t$ m/sec.
   a. Determine the distance the rocket travels during the time $t = 5$ to $t = 7$ seconds.
   b. Represent the answer to part a as an area.

Solution: a. Distance traveled by the rocket is given by $s(t) = \int v\, dt$:

$$s(t) = \int_5^7 (0.3t^2 + 4t)\,dt \quad \text{(Integrating from } t = 5 \text{ to } t = 7\text{)}$$

$$= [0.1t^3 + 2t^2]\Big|_5^7$$

$$= [0.1(7)^3 + 2(7)^2] - [0.1(5)^3 + 2(5)^2]$$

$$= 69.8 \text{ m.}$$

b. The distance traveled by the rocket can be represented by the area under the curve $v = 0.3t^2 + 4t$. The area found by integrating this function is shown in figure 23.7.

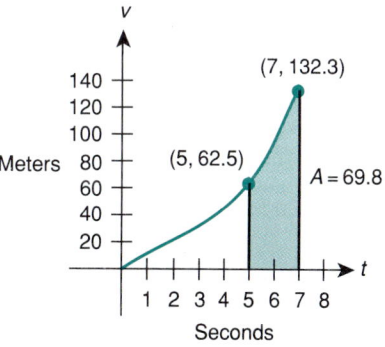

Figure 23.7

Trial Problems 23.1

Before you begin the section exercises, warm up with these problems. Complete answers are included in the answer key.

In order to find area under a curve, you will have to integrate expressions of various forms. Evaluate the following integrals.

1. $\int_0^1 (x^2 + 1)\,dx$
2. $\int_0^4 \sqrt{x}\,dx$
3. $\int_1^3 \frac{1}{x}\,dx$
4. $\int_0^{\pi/2} \cos x\,dx$
5. $\int_0^2 (x + 1)^2\,dx$

Exercises 23.1

Find the area bounded by the following curves and the x-axis. Sketch the curves and shade the area.

1. $y = x - x^2$, from $x = 0$ to $x = 1$
2. $y = -x^2 + 2x + 3$, from $x = -1$ to $x = 3$
3. $y = \frac{1}{x^2}$, from $x = 1$ to $x = 2$
4. $y = \frac{x^2}{4}$, from $x = 4$ to $x = 10$
5. $y = 6\sqrt{x}$, from $x = 4$ to $x = 9$
6. $y = \sqrt{x + 9}$, from $x = 0$ to $x = 16$
7. $y = x^3$, from $x = -1$ to $x = 1$ (*Hint:* find two separate areas and add.)
8. $y = x^2 - 4$, from $x = 2$ to $x = 6$
9. $y = x$, from $x = 0$ to $x = 4$
10. $y = \frac{4}{x}$, from $x = 1$ to $x = 4$
11. $y = \tan \theta$, from $\theta = 0$ to $\theta = \frac{\pi}{4}$
12. $y = 9x - x^3$, from $x = 0$ to $x = 3$
13. $y = e^x$, from $x = 0$ to $x = 1$
14. $y = \sin^2 x$, from $x = 0$ to $x = \pi$
15. $y = \frac{1}{\sqrt{2x + 1}}$, from $x = 0$ to $x = 4$
16. $y = (2x + 1)^2$, from $x = -1$ to $x = 3$
17. $y = x\sqrt{2x^2 + 1}$, from $x = 0$ to $x = 2$
18. $y = \frac{x}{(2x^2 + 1)^2}$, from $x = 0$ to $x = 2$
19. $y = \frac{4}{\sqrt{16 - x^2}}$, from $x = 0$ to $x = 2$
20. $y = \ln x$, from $x = 1$ to $x = 2$

Sketch the region whose area is given by the following integrals and evaluate the definite integral.

21. $\int_{-1/2}^{2}(2x+1)dx$

22. $\int_{3}^{4}(x^2-9)dx$

23. $\int_{0}^{1}(x-x^3)dx$

24. $\int_{0}^{1}(1-x^2)dx$

25. $\int_{3}^{8}\sqrt{1+x}\,dx$

26. $\int_{0}^{4}\dfrac{2\,dx}{(1+x)^2}$

27. $\int_{1}^{2}\dfrac{x^2+1}{x^2}dx$

28. Find the area bounded by the curves $y = x^2 - 5x$ and $y = 0$, from $x = 1$ to $x = 4$. (*Hint:* Since the region lies below the x-axis, $A = -\int_{a}^{b} f(x)dx$.)

29. Find the area bounded by the curves $y = x^2 - 4$ and $y = 0$, from $x = 0$ to $x = 5$. (*Hint:* Sketch the graph, find two separate areas and add.)

30. Find the area between the curves $y = x^2$ and $y = x + 2$. (*Hint:* Sketch the region and find points of intersection. Let $h(x) = x^2$ and $g(x) = x + 2$. Then $f(x) = g(x) - h(x)$.)

31. Using the hint from exercise 30, find the area of regions described below.
   a. The region between $y = x^2$ and $y = x^3$
   b. The region between $y = 8 - x^3$ and $y = 7x$ from $x = 0$ to $x = 1$
   c. The region between $y = x^2 + 2x - 8$ and $y = x + 4$
   d. The region between $y = 4 + x^2$ and $y = x - x^2$ from $x = -2$ to $x = 2$

32. Find the area of the region between the curves $y = x - 4$ and $y^2 = 2x$. (*Hint:* Integrate with respect to y.)

33. In designing circuit boards, a particular board has the shape of the region formed by the parabola $y = 4 - x^2$ from $x = 0$ to $x = 2$ cm. Find the area of this board.

34. If the velocity of an object is given by $v = t^3 + 2$, find the distance the object travels from $t = 0$ to $t = 2$ by finding the area under a curve.

35. After t hours of operation an assembly line is producing power lawn mowers at the rate of $21 - \dfrac{4}{5}t$ mowers per hour.
   a. How many mowers are produced during $t = 2$ to $t = 5$ hours?
   b. Sketch a curve and represent the answer to part a as an area.

36. Food is placed in a freezer and after t hours the temperature of the food is dropping at the rate given by
$$r(t) = 12 + \dfrac{4}{(t+3)^2},$$
where $r(t)$ is in degrees Fahrenheit.
   a. Compute the area under the graph of $y = r(t)$ from $t = 0$ to $t = 2$.
   b. What does the area in part **a** represent?

23 • 2 The Area Between Two Curves

The area of the region between two curves can be determined by finding the area under each curve and subtracting. Figure 23.8 illustrates this procedure.

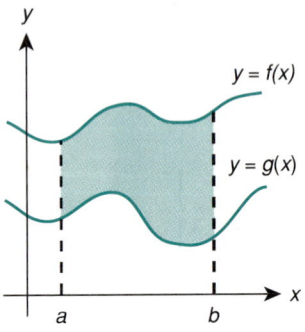

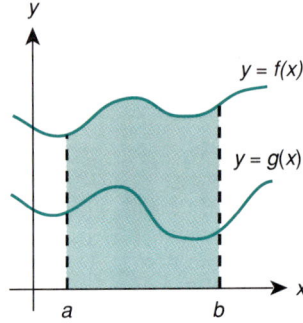

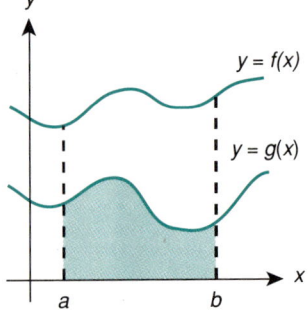

Figure 23.8

AREA OF THE REGION BETWEEN TWO CURVES

Area between $f(x)$ and $g(x)$ = area under $f(x)$ − area under $g(x)$.

$$A = \int_a^b f(x)\,dx - \int_a^b g(x)\,dx$$
$$= [F(b) - F(a)] - [G(b) - G(a)]$$
$$= F(b) - F(a) - G(b) + G(a)$$
$$= [F(b) - G(b)] - [F(a) - G(a)]$$
$$= \int_a^b [f(x) - g(x)]\,dx$$

In figure 23.8, both the curves for $f(x)$ and $g(x)$ lie above the x-axis. This need not be the case; the only requirement is that the curve for $g(x)$ lie below the curve for $f(x)$.

• **Example 23.6:** Find the area of the region between $y = x^2$ and $y = x - 2$ from $x = 0$ and $x = 2$. (See the graph in figure 23.9.)

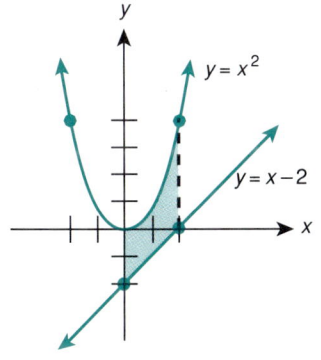

Figure 23.9

Solution:
$$A = \int_0^2 [x^2 - (x - 2)]\,dx$$
$$= \int_0^2 (x^2 - x + 2)\,dx$$
$$= \left[\frac{x^3}{3} - \frac{x^2}{2} + 2x\right]\Big|_0^2$$
$$= \left(\frac{8}{3} - \frac{4}{2} + 4\right) - \left(\frac{0}{3} - \frac{0}{2} + 0\right)$$
$$= 4\frac{2}{3} \text{ sq. units.}$$

• **Example 23.7:** Find the area of the region between the curves $y = 4x - x^2$ and $y = -x + 4$.

Solution: In this case, we need to find the points of intersection of the two curves. Since the y-values for the two curves are the same at the points of intersection, we can set the equations equal to each other and solve for x:

$$4x - x^2 = -x + 4,$$

or

$$x^2 - 5x + 4 = 0$$
$$(x - 1)(x - 4) = 0.$$

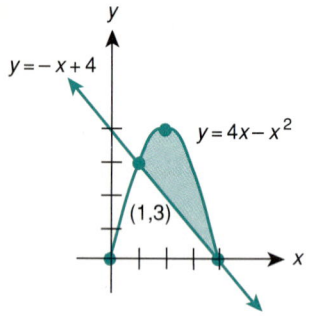

Figure 23.10

$x = 1$ and $x = 4$ are the x-coordinates of the points of intersection. (See the graph in figure 23.10.)

Now,

$$A = \int_1^4 [(4x - x^2) - (-x + 4)]dx$$

$$= \int_1^4 (4x - x^2 + x - 4)dx$$

$$= \int_1^4 (5x - x^2 - 4)dx$$

$$= \left[\frac{5x^2}{2} - \frac{x^3}{3} - 4x\right]\Big|_1^4$$

$$= \left(40 - \frac{64}{3} - 16\right) - 0$$

$$= 24 - 21\frac{1}{3}$$

$$= 2\frac{2}{3} \text{ sq. units.}$$

• **Example 23.8:** Find the area of the region between the curves $y = 4$ and $y = x^2$.

Solution: From the graph in figure 23.11, we see that the integral needed to find the area is

$$A = \int_{-2}^2 [4 - x^2]dx$$

$$= \left[4x - \frac{x^3}{3}\right]\Big|_{-2}^2$$

$$= \left(8 - \frac{8}{3}\right) - \left(-8 + \frac{8}{3}\right)$$

$$= 8 - \frac{8}{3} + 8 - \frac{8}{3}$$

$$= 16 - \frac{16}{3}$$

$$= 10\frac{2}{3} \text{ sq. units.}$$

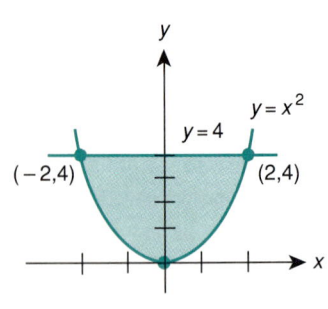

Figure 23.11

• **Example 23.9:** Find the area of the region between $y^2 = 2x$ and $y = x - 4$. See figure 23.12.

Solution: In this case, to find the points of intersection we have

$$x = \frac{y^2}{2} \quad \text{and} \quad x = y + 4$$

$$\frac{y^2}{2} = y + 4,$$

or

$$y^2 = 2y + 8,$$

and

$$y^2 - 2y - 8 = 0$$
$$(y - 4)(y + 2) = 0$$
$$y = 4 \quad \text{and} \quad y = -2.$$

Therefore, at $y = 4$, $x = 8$, and at $y = -2$, $x = 2$.

The area of the desired region can be found by splitting the region into two parts: from $x = 0$ to $x = 2$ and from $x = 2$ to $x = 8$. The reason for doing this is because the lower curve from $x = 0$ to $x = 2$ is different than the lower curve from $x = 2$ to $x = 8$.

$$A = \int_0^2 [\underset{\substack{\uparrow \\ \text{upper} \\ \text{curve}}}{\sqrt{2x}} - \underset{\substack{\uparrow \\ \text{lower} \\ \text{curve}}}{(-\sqrt{2x})}]\,dx + \int_2^8 [\sqrt{2x} - (x - 4)]\,dx \quad (y = \sqrt{2x} \text{ since } y^2 = 2x)$$

$$A = \int_0^2 2\sqrt{2}\, x^{1/2}\, dx + \int_2^8 [\sqrt{2} x^{1/2} - x + 4]\,dx$$

$$= \frac{4\sqrt{2}}{3} x^{3/2} \Big|_0^2 + \left[\frac{2\sqrt{2}}{3} x^{3/2} - \frac{x^2}{2} + 4x\right]\Big|_2^8$$

$$= \left(\frac{4\sqrt{2}}{3} \cdot 2\sqrt{2} - 0\right) + \left(\frac{2\sqrt{2}}{3} \cdot 16\sqrt{2} - 32 + 32\right)$$

$$\quad - \left(\frac{2\sqrt{2}}{3} \cdot 2\sqrt{2} - 2 + 8\right)$$

$$= \frac{16}{3} + \frac{64}{3} - \frac{8}{3} + 2 - 8$$

$$= \frac{72}{3} - 6$$

$$= 24 - 6$$

$$= 18 \text{ sq. units.}$$

(Investigate the possibility of solving this problem by integrating with respect to y.)

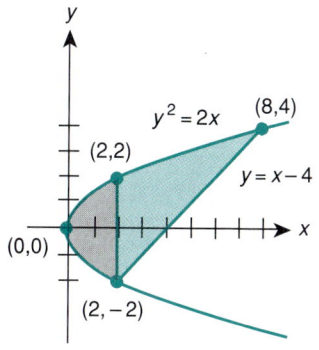

Figure 23.12

• • • • • • • • •

Trial Problems 23.2

Before you begin the section exercises, warm up with these problems. Complete answers are included in the answer key.

In order to find the area between two curves, you will have to integrate expressions of various forms. Evaluate the following integrals.

1. $\int_{-1}^{1} (1 - x^2) dx$
2. $\int_{0}^{1} (x^3 + x) dx$
3. $\int_{0}^{1} (e^x + 1) dx$
4. $\int_{1}^{2} \ln x \, dx$
5. $\int_{0}^{1} \frac{x}{x^2 + 1} dx$

Exercises 23.2

For exercises 1–16, find the area of the regions between the curves. Sketch the graphs first.

1. $y = x^2$ and $y = 2x$
2. $y = \sqrt{x}$ and $y = x$
3. $y = \sin x$ and $y = \cos x$, from $x = \frac{\pi}{4}$ to $x = \frac{\pi}{2}$
4. $y = 6 - x - x^2$ and the x-axis
5. $y = \sqrt{1 - x}$, $x = 0$ and $y = 0$
6. $y = x^2 + 3$ and $y = 9$
7. $y = x^2 + 2x + 1$ and $y = 3x + 3$
8. $y = x$, $y = 2 - x$, and the x-axis
9. $y = 3x^2 + 2x$ and $y = 8$
10. $y = x^{1/3}$ and $y = x$
11. $y = x^4 - 2x^2$ and $y = 2x^2$
12. $y = 18 - x^2$ and $y = x^2$
13. $y^2 = x^3$ and $x = 4$
14. $y = 1 - x^2$ and $y = x - 1$
15. $y = x^3 - x$ and $y = 0$
16. $y = e^{x/2}$ and $y = e$, $x = 0$

17. Find the area bounded by the curve $y = x^2 - 5$ and the straight line passing through the points $(-2, -1)$ and $(3, 4)$.

18. Find the area bounded by the curves $y = x^2 + 5$ and $y = x^2 + 2x$ from $x = -1$ to $x = 1$.

19. Find the area bounded by the curves $y = 8 - x^2$ and $y = x^2$ from $x = -1$ to $x = 1$.

20. Find the area bounded by the curves $y^2 = x$ and $y = 2x + 1$ from $y = 0$ to $y = 3$.

21. Find the area of the region bounded by the curves $y = e^x$ and $y = \sqrt{x}$ from $x = 0$ to $x = 1$.

22. Find the area of the region bounded by the curves $y = e^{-x}$ and $xy = 1$ from $x = 1$ to $x = 2$.

23. Find the area of the region bounded by the curves $y = \sec x$ and $y = x$ from $x = -\frac{\pi}{4}$ to $x = \frac{\pi}{4}$.

24. Find the area of the region bounded by the curves $y = \sin x$ and $y = \cos x$ from $x = -\frac{\pi}{2}$ to $x = \frac{\pi}{6}$.

25. Find the area of the region bounded by $y = \ln x$, $y = 0$, and $x = 4$.

26. Find the area of the region bounded by the curves $y = \ln x$ and $y = x - 1$ from $x = 1$ to $x = 5$.

27. Find the area of the region bounded by $y = \frac{x}{x^2 + 1}$, $y = 0$, and $x = 3$.

28. Find the area of the region bounded by $y = \frac{x^2 + 4}{x}$, $y = 0$, $x = 1$, and $x = 4$.

23•3 The Volume of a Solid of Revolution by Cylindrical Solids

If we draw the curve given by $y = f(x)$ from $x = a$ to $x = b$ and rotate this curve about the x-axis, we obtain a **solid of revolution**. Using integration, we are able to find the **volume of a solid of revolution**. Investigating figures 23.13 and 23.14 helps us develop a suitable process. As we did with area under a curve, we draw rectangles under the curve as in figure 23.13. Select any one of these rectangles and rotate it around the x-axis. By doing so, we create a cylindrical solid as illustrated in figure 23.14. This cylindrical solid is better viewed in figure 23.15. The volume of this cylindrical solid is given by $\Delta V = \pi r^2 h$, where r is the radius and h is the height (or thickness, in this case).

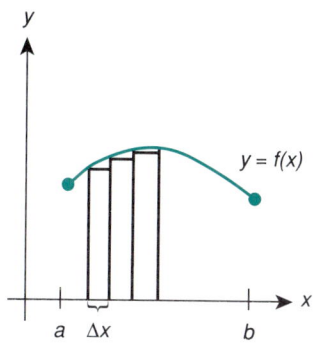

Figure 23.13

The radius r varies as we select different rectangles under $y = f(x)$, and since r is a y-value, we can substitute y in place of r. Furthermore, if we let $h = \Delta x$ the formula becomes

$$\Delta V = \pi r^2 h$$
$$= \pi y^2 \Delta x$$

In terms of differentials, we can write

$$dV = \pi y^2 \, dx.$$

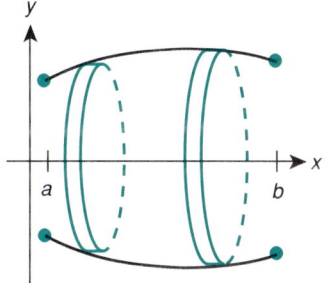

Figure 23.14

If we let Δx become smaller and smaller, we create more and more rectangles, thus, more and more cylindrical solids. Adding the volumes of all such cylindrical solids from $x = a$ to $x = b$ gives us the volume of the solid of revolution formed by rotating $y = f(x)$ about the x-axis. This can be expressed by integrating both sides of the equation

$$\int dV = \int \pi y^2 \, dx,$$

to obtain

$$V = \int \pi y^2 \, dx.$$

The definite integral from $x = a$ to $x = b$ gives the desired total volume:

$$V = \int_a^b \pi y^2 \, dx.$$

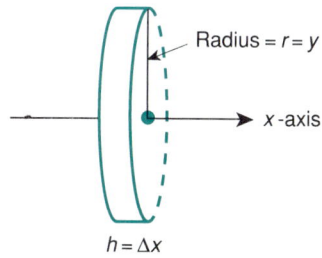

Figure 23.15

Since $y = f(x)$ and since π is constant, we can write:

VOLUME OF A SOLID OF REVOLUTION

$$V = \pi \int_a^b [f(x)]^2 \, dx$$

as the volume of the solid of revolution generated by rotating the curve of $y = f(x)$ about the x-axis.

• **Example 23.10:** Find the volume of the solid of revolution formed by revolving $y = x^2$ about the x-axis from $x = 0$ to $x = 2$. See figure 23.16.

Solution:
$$V = \pi \int_0^2 y^2 \, dx$$
$$= \pi \int_0^2 (x^2)^2 \, dx \quad \text{(Substituting } y = x^2\text{)}$$
$$= \pi \int_0^2 x^4 \, dx$$
$$= \pi \left[\frac{x^5}{5} \right]_0^2$$
$$= \pi \left[\frac{32}{5} - 0 \right] \quad \text{(Let } x = 2\text{; then } x = 0\text{)}$$
$$= \frac{32\pi}{5} \text{ cu units.}$$

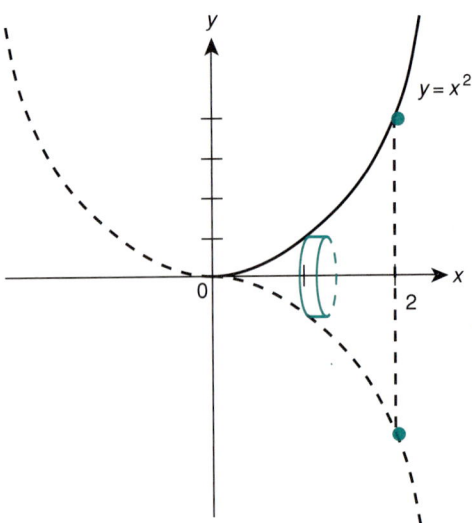

Figure 23.16

SECTION 23.3 THE VOLUME OF A SOLID OF REVOLUTION BY CYLINDRICAL SOLIDS

• **Example 23.11:** Find the volume of the sphere generated by revolving the circle $x^2 + y^2 = 4$ about the x-axis. The upper half of the circle is given by $y = \sqrt{4 - x^2}$. Revolving this about the x-axis generates the desired sphere. See figure 23.17.

Solution: Since the radius of the circle is 2, the curve intersects the x-axis at $x = -2$ and $x = 2$. These are the limits of integration.

$$V = \pi \int_{-2}^{2} y^2 \, dx$$

$$= \pi \int_{-2}^{2} [\sqrt{4 - x^2}]^2 \, dx \quad \text{(Substituting } y = \sqrt{4 - x^2}\text{)}$$

$$= \pi \int_{-2}^{2} (4 - x^2) \, dx$$

$$= \pi \left[4x - \frac{x^3}{3} \right]\Big|_{-2}^{2}$$

$$= \pi \left[\left(8 - \frac{8}{3} \right) - \left(-8 + \frac{8}{3} \right) \right]$$

$$= \pi \left[8 - \frac{8}{3} + 8 - \frac{8}{3} \right]$$

$$= \pi \left[16 - \frac{16}{3} \right]$$

$$= \pi \left(\frac{48 - 16}{3} \right)$$

$$= \frac{32}{3} \pi \text{ cu units.}$$

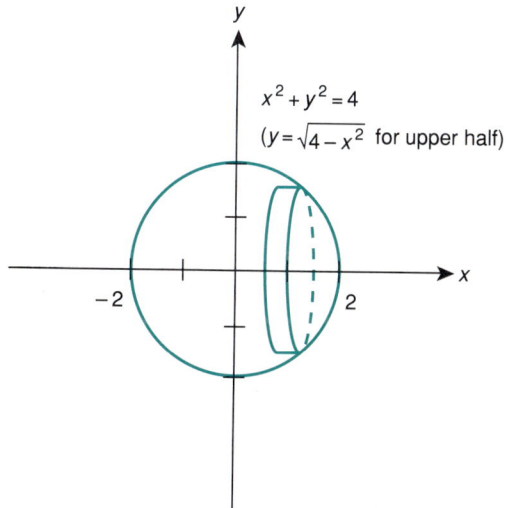

Figure 23.17

NOTE: In general, the volume of sphere is given by $V = \frac{4}{3}\pi r^3$. With $r = 2$ (radius of the circle on page 943),

$$V = \frac{4}{3}\pi(2)^3 = \frac{4}{3}\pi(8)$$

$$= \frac{32}{3}\pi \text{ cu units.}$$

• **Example 23.12:** Find the volume of the solid of revolution formed by revolving the curve of $y = \frac{1}{x}$ about the x-axis from $x = 1$ to $x = 3$. See figure 23.18.

Solution: $V = \pi \int_1^3 \left(\frac{1}{x}\right)^2 dx$

$\quad\quad\quad = \pi \int_1^3 x^{-2} dx \quad \left(\left(\frac{1}{x}\right)^2 = \frac{1}{x^2} = x^{-2}\right)$

$\quad\quad\quad = \pi \frac{[x^{-1}]}{-1}\Big|_1^3$

$\quad\quad\quad = \pi\left[\left(-\frac{1}{3}\right) - (-1)\right]$

$\quad\quad\quad = \pi\left[-\frac{1}{3} + 1\right]$

$\quad\quad\quad = \frac{2}{3}\pi \text{ cu units.}$

•••••••••

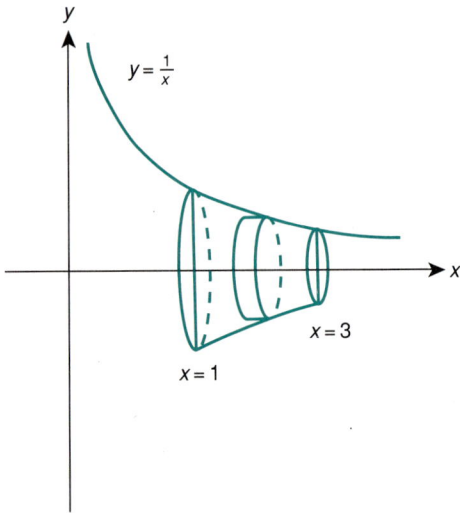

Figure 23.18

• **Example 23.13:** Find the volume of the solid of revolution formed by revolving the curve $y = x^{1/3}$ about the y-axis from $x = 0$ to $x = 8$.

Solution: Since the curve is being revolved about the y-axis, the representative rectangle is drawn as in figure 23.19. Note the height is $h = \Delta y$ and the radius is $f(y) = x^3$, solving $y = x^{1/3}$ for x. The limits of integration are y-values:

$$y = 0 \quad \text{when} \quad x = 0,$$

and

$$y = 2 \quad \text{when} \quad x = 8.$$

Thus,

$$V = \pi \int_0^2 [y^3]^2 \, dy$$

$$= \pi \int_0^2 y^6 \, dy$$

$$= \pi \left[\frac{y^7}{7} \right]_0^2$$

$$= \frac{128\pi}{7}$$

$$\approx 57.4 \text{ cu units.}$$

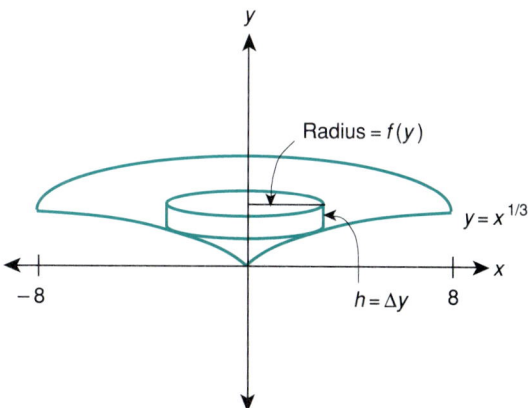

Figure 23.19

Trial Problems

Before you begin the section exercises, warm up with these problems. Complete answers are included in the answer key.

In order to find volumes of solids of revolution, you will have to integrate expressions of various forms. Evaluate the following integrals.

1. $\int_0^2 \pi(\sqrt{x})^2 \, dx$
2. $\int_0^\pi \sin^2 x \, dx$
3. $\int_0^1 e^{-2x} \, dx$
4. $\int_0^{\pi/4} \tan^2 x \, dx$
5. $\int_0^1 (1 - x^2)^2 \, dx$

Exercises 23.3

Find the volume of the solid of revolution formed by revolving the curve of the given equation about the x-axis.

1. $y = x$, from $x = 2$ to $x = 5$
2. $2y = x$, from $x = 0$ to $x = 4$
3. $y = \sqrt{x}$, from $x = 0$ to $x = 4$
4. $y = x^2 - 4x$, from $x = 0$ to $x = 4$
5. $y = x^3$, from $x = 0$ to $x = 2$
6. $y = 9 - x^2$, from $x = 0$ to $x = 3$
7. $y = \sin x$, from $x = 0$ to $x = \pi$
8. $x + y = 2$, from $x = 0$ to $x = 2$
9. $y = e^x$, from $x = 0$ to $x = 1$
10. $y = \sqrt{9 - x^2}$, from $x = 0$ to $x = 3$
11. $y = e^{-x}$, from $x = 1$ to $x = 2$
12. $y = xe^x$, from $x = 0$ to $x = 1$
13. $y = \cos x$, from $x = 0$ to $x = \dfrac{\pi}{2}$
14. $y = -x^2 + x$, from $x = 0$ to $x = 1$
15. $y = \dfrac{1}{\sqrt{x^2 + 4x + 13}}$, from $x = -2$ to $x = -1$
    (*Hint:* Complete the square and use trigonometric substitution or the tables.)
16. $y = \tan x$, from $x = 0$ to $x = \dfrac{\pi}{4}$
17. Find the volume of the solid of revolution formed by revolving the region bounded by $y = x^2$ and $y = 2$ about the y-axis. (*Hint:* Integrate with respect to y.)
18. Find the volume of the solid of revolution formed by revolving the region bounded by $y = e^{-x/2}$, $y = 0$, $x = 0$, and $x = 1$ about the y-axis.
19. Find the volume of the solid of revolution formed by revolving the region bounded by $y = \dfrac{1}{x}$, $y = 0$, $x = 1$, and $x = 4$ about the x-axis.
20. Find the volume of the solid of revolution formed by revolving the region bounded by $y = \sec x$, $y = 0$, $x = -\dfrac{\pi}{3}$, and $x = \dfrac{\pi}{3}$ about the x-axis.

Find the volume of the solid of revolution formed by revolving the indicated regions about the y-axis. (See exercise 17.)

21. $y = x$ and $x = 0$, from $y = 0$ to $y = 1$
22. $y = \dfrac{1}{x}$ and $x = 0$, from $y = 1$ to $y = 3$
23. $y = x^3$, from $y = 0$ to $y = 1$
24. $y^2 = x + 4$ and $x = 0$
25. $y = 3$, $y = 0$, $x = 0$, and $x = 2$
26. $y = \sqrt{x}$, $y = 4$, and $y = 0$

Find the volume of the solid of revolution formed by revolving the indicated regions about the x-axis. (*Hint:* **Find two different volumes and subtract.**)

27. $y = x^2$ and $y = 4 - x^2$

28. $y = x^2$ and $y = x$

29. Consider the region formed by the curves $y = x^2$, $x = 0$, and $y = 9$. Find the volume of the solid of revolution formed by revolving this region about the line $y = 9$. (*Hint:* radius of a representative disk will be $9 - x$.)

30. Do the same as in exercise 29 for the region $y = x^2$, $x = 2$, and $y = 0$ revolved about $x = 2$.

23 • 4 The Volume of a Solid of Revolution by Cylindrical Shells

Consider the region bounded by the curve of $y = 2x - x^2$ and the x-axis. In section 23.3 we determined the volume of the solid of revolution formed by revolving the region about the x-axis. Now, let us consider revolving this region about the y-axis. In so doing, we obtain a solid of revolution with a hole in it, as illustrated in figures 23.20 and 23.21. Again, drawing a representative rectangle under the curve, and revolving it about the y-axis, we obtain a solid of revolution called a **cylindrical shell**. In figure 23.22, we illustrate an enlarged cylindrical shell and proceed to find its volume. The shell has an

Cylindrical Shell

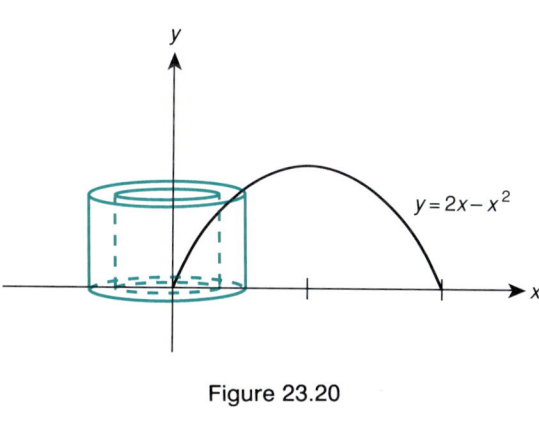

Figure 23.20

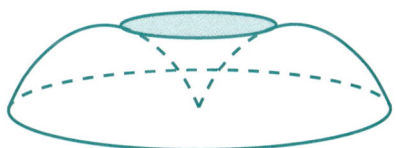

Figure 23.21

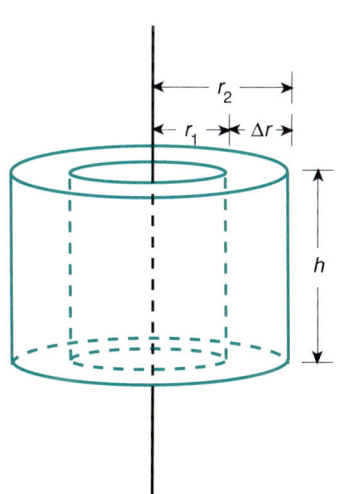

Figure 23.22

inside radius r_1 and an outside radius r_2. The height of the shell is h and the thickness is Δr. The volume is determined by finding the volume using the outside radius and subtracting the volume using the inside radius:

$$\begin{aligned}\Delta V &= \pi r_2^2 h - \pi r_1^2 h && \text{(Volume of entire shell minus} \\ & && \text{the volume of the ``hole'')} \\ &= \pi(r_2^2 - r_1^2)h \\ &= \pi(r_2 + r_1)(r_2 - r_1)h && \text{(Factoring the difference of two squares)} \\ &= 2\pi \frac{(r_2 + r_1)}{2}(r_2 - r_1)h. && \text{(Multiplying and dividing by 2)}\end{aligned}$$

Now, if we let $r = \dfrac{r_2 + r_1}{2}$ be the average radius and $r_2 - r_1 = \Delta r$, we have

$$\Delta V = 2\pi r \Delta r h,$$

or

$$\Delta V = 2\pi r h \Delta r.$$

Returning to the figure on the coordinate system, we see that r is any x-value; h is the y-value, or $f(x)$; and Δr can be written as dx. Hence, we can write

$$\begin{aligned}\Delta V &= 2\pi r h \Delta r \\ &= 2\pi x f(x) dx.\end{aligned}$$

Since this is the volume of one shell formed by revolving one rectangle about the y-axis, we can find the volume of all such shells by creating rectangles from $x = a$ to $x = b$. Thus, the total volume is

$$V = \int_a^b 2\pi x f(x) dx,$$

or

$$V = 2\pi \int_a^b x f(x) dx.$$

We complete the original problem as the first example in this section.

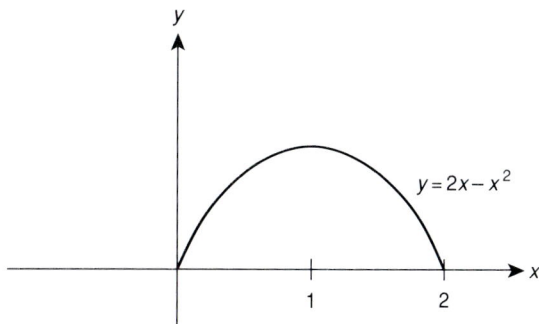

Figure 23.23

• **Example 23.14:** Find the volume of the solid of revolution formed by revolving about the y-axis the region bounded by the curve $y = 2x - x^2$ and the x-axis. See figure 23.23.

Solution: Since the curve intersects the x-axis at $x = 0$ and $x = 2$, these will be the limits of integration.

$$V = 2\pi \int_0^2 xf(x)dx$$
$$= 2\pi \int_0^2 x(2x - x^2)dx$$
$$= 2\pi \int_0^2 (2x^2 - x^3)dx$$
$$= 2\pi \left[\frac{2x^3}{3} - \frac{x^4}{4} \right]\Big|_0^2$$
$$= 2\pi \left[\left(\frac{16}{3} - 4\right) - (0) \right]$$
$$= 2\pi \left[\frac{4}{3} \right]$$
$$= \frac{8\pi}{3} \text{ cu units.}$$

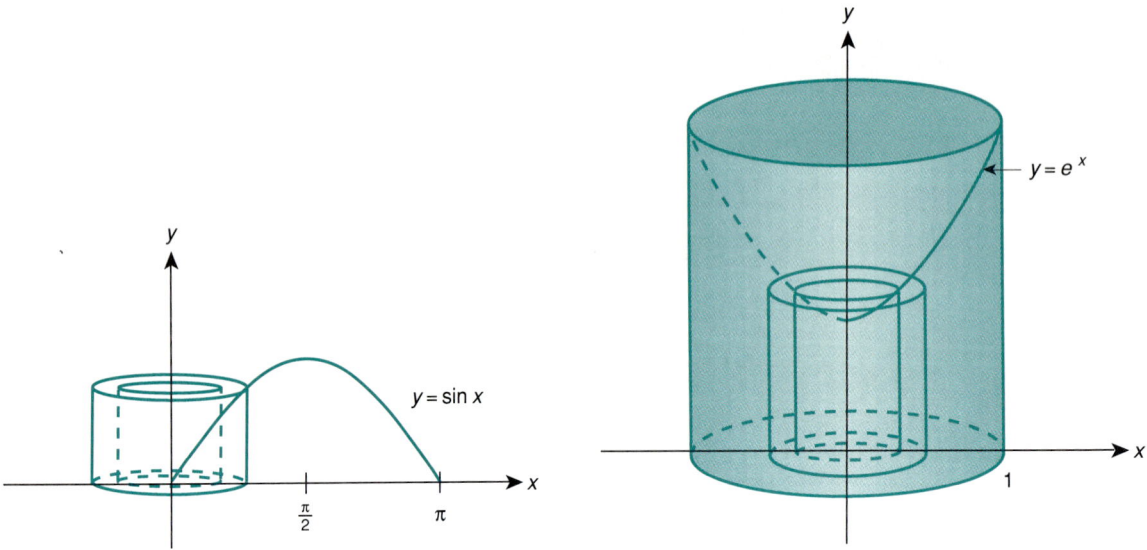

Figure 23.24

Figure 23.25

• **Example 23.15:** Find the volume of the solid of revolution formed by revolving about the y-axis the region bounded by $y = \sin x$ and the x-axis from $x = 0$ to $x = \dfrac{\pi}{2}$. See figure 23.24.

Solution:
$$V = 2\pi \int_0^{\pi/2} x \sin x \, dx$$
$$= 2\pi[-x \cos x + \sin x]\Big|_0^{\pi/2} \quad \text{(Integration by parts)}$$
$$= 2\pi\left[\left(-\frac{\pi}{2} \cdot 0 + 1\right) - (-0 \cdot 1 + 0)\right]$$
$$= 2\pi \text{ cu units.}$$

• **Example 23.16:** Find the volume of the solid of revolution formed by revolving about the y-axis the region bounded by $y = e^x$, the x-axis, and $x = 1$. See figure 23.25.

Solution:
$$V = 2\pi \int_0^1 x e^x \, dx$$
$$= 2\pi[x e^x - e^x]\Big|_0^1 \quad \text{(Integration by parts)}$$
$$= 2\pi[(e - e) - (0 - 1)]$$
$$= 2\pi \text{ cu units.}$$

Trial Problems 23.4

Before you begin the section exercises, warm up with these problems. Complete answers are included in the answer key.

When using the method of cylindrical shells to find volume, you will have to integrate expressions of various forms. Evaluate the following integrals.

1. $\int_0^3 x(5 - x)\,dx$
2. $\int_0^4 2x\sqrt{x}\,dx$
3. $\int_0^\pi x \sin x\,dx$
4. $\int_1^3 xe^x\,dx$
5. $\int_1^2 2x(1 + x^2)^{-1}\,dx$

Exercises 23.4

Use the method of cylindrical shells to find the volume of the indicated regions rotated about the y-axis.

1. $y = 3 - x$, the x-axis, from $x = 0$ to $x = 3$
2. $y = x^2 + 1$, the x-axis, from $x = 0$ to $x = 3$
3. $y = 2x + 1$, the x-axis, from $x = 1$ to $x = 2$
4. $y = 2\sqrt{x}$, the x-axis, from $x = 1$ to $x = 2$
5. $y = \dfrac{1}{x}$, the x-axis, from $x = 1$ to $x = 2$
6. $y = 2x - x^2$, the x-axis, from $x = 0$ to $x = 2$
7. $y = \sqrt{4 - x^2}$, the x-axis, from $x = 0$ to $x = 2$
8. $y = e^x$, the x-axis, from $x = 1$ to $x = 2$
9. $y = \dfrac{1}{1 + x^2}$, the x-axis, from $x = 1$ to $x = 2$
10. $y = 6 - x - x^2$, the x-axis, from $x = 0$ to $x = 2$
11. $y = 8 - x^3$, the x-axis, from $x = 0$ to $x = 2$
12. $y = \ln x$, the x-axis, from $x = 1$ to $x = 2$
13. $y = \cos x^2$, the x-axis, from $x = 0$ to $x = 1$
14. $y = \sin x$, the x-axis, from $x = 0$ to $x = \pi$
15. $y = \dfrac{\sqrt{x^2 - 16}}{x^2}$, the x-axis, from $x = 4$ to $x = 5$
16. $y = e^{-x}$, the x-axis, from $x = -2$ to $x = -1$
17. $2y = x$, the x-axis, from $x = 0$ to $x = 1$

Use the method of cylindrical shells to find the volume of the indicated regions rotated about the x-axis. (*Hint:* integrate with respect to y.)

18. $x = 2\sqrt{y}$, the y-axis, from $y = 0$ to $y = 4$
19. $2y = x$, $x = 4$, from $y = 0$ to $y = 2$
20. $y^3 = x^2$, the y-axis, from $y = 0$ to $y = 1$
21. $y = 3 - x$, the y-axis, from $y = 1$ to $y = 3$
22. $y = x^2$, the y-axis, from $y = 0$ to $y = 4$

23 • 5 Work

In physics, work is defined as the product of the force acting on an object by the distance the object is moved. If d is used for distance, F for force, and W for work, then

$$W = Fd.$$

This formula holds true if the acting force is a constant. If the acting force is a variable, then work is determined by evaluating a definite integral.

Suppose we wish to find the work done by applying a force that moves an object from $x = a$ to $x = b$. See figure 23.26. Assume that a constant force F is applied to an object in order to move it from point a to point $a + \Delta x$, where Δx is a "small" distance. Using W as the work done in this situation, we have

$$W = F\Delta x.$$

Using differentials, we can write

$$dW = F\,dx.$$

Integrating both sides, we have

$$W = \int F\,dx.$$

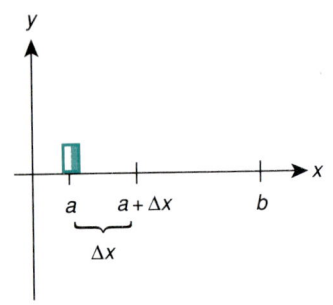

Figure 23.26

WORK

If we apply a constant force F for each Δx, although F may vary for each change in Δx, we find the total work completed in moving the object from $x = a$ to $x = b$ by

$$W = \int_a^b F\,dx.$$

If F varies, then F may be expressed as some function of x; that is, $F = f(x)$.

• **Example 23.17:** Find the work done in stretching a spring from its original length of 5 cm to a length of 10 cm, if it is known that a force of 1.5 newtons stretches it 1 cm.

Solution: By *Hooke's law*, $F = kx$, where x is the elongation. We first need to determine the value of k. Since $F = 1.5$ when $x = 1$, we have $1.5 = k \cdot 1$. $k = 1.5$.
Substituting in the general equation, $F = 1.5x$.

In stretching the spring from 5 cm (its original length) to 10 cm (a total stretching of 5 cm), we have $a = 0$ and $b = 5$. Therefore,

$$W = \int_0^5 1.5x \, dx$$

$$= \frac{1.5x^2}{2} \Big|_0^5$$

$$= \frac{1.5(5)^2}{2}$$

$$= 18.75 \text{ cm-kg.}$$

(Work is measured in cm per kilogram of force, or cm per gram of force)

or $\quad = 1875$ cm-g.

• **Example 23.18:** The force driving a piston varies with the piston displacement according to the law $F = \dfrac{3000}{x}$ lbs. Find the work done in moving a piston from $x = 2$ in. to $x = 6$ in.

Solution: $W = \int_2^6 \dfrac{3000}{x} dx = 3000 \ln x \Big|_2^6$

$= 3000(\ln 6 - \ln 2)$
$\approx 3000(1.792 - 0.693)$
≈ 3297 in.-lbs.

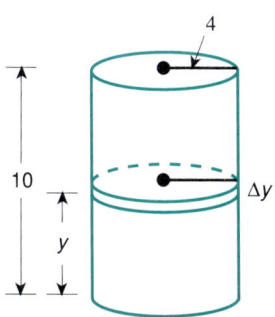

Figure 23.27

• **Example 23.19:** Find the work done in pumping the water over the top of a cylindrical tank filled with water and having a radius of 4 ft. and height of 10 ft. See figure 23.27.

Solution: Consider the work done in lifting a small disk of water over the top. The disk of water is y feet from the bottom. Water weighs approximately 62.5 lb/ft³. The volume of the disk is $\pi r^2 h$, and in this case the volume equals $\pi(4)^2 \Delta y$. Therefore, the weight of the disk of water is

$$(62.5)(16\pi)\Delta y.$$

(Weight of the disk of water is given by density $\times$ volume.)

The remaining distance this disk of water must travel to the top is $10 - y$ feet. With force as $(62.5)(16\pi)\Delta y$ and distance as $(10 - y)$, work is

$$W = F \cdot d$$
$$W = (62.5)(16\pi)\Delta y(10 - y).$$

The total work done in lifting all such disks from the bottom to the top of the cylinder is found by

$$W = \int_0^{10} (62.5)(16\pi)(10 - y)dy$$
$$= 1000\pi \int_0^{10} (10 - y)dy$$
$$= 1000\pi \left[10y - \frac{y^2}{2} \right]\Big|_0^{10}$$
$$= 1000\pi[100 - 50]$$
$$= 50{,}000\pi \text{ ft-lb.}$$

Trial Problems 23.5

Before you begin the section exercises, warm up with these problems. Complete answers are included in the answer key.

The following integrals are frequently used when finding work. Evaluate each integral.

1. $\int_0^5 \frac{3}{2}x \, dx$

2. $\int_0^5 1000\pi(10 - x) \, dx$

3. $\int_4^9 8x^{-1/2} \, dx$

4. $\int_0^1 \left(\frac{y}{2}\right)^2 (10 - y)dy$

5. $\int_3^7 \frac{1000}{x} \, dx$

Exercises 23.5

1. The force in pounds required to stretch a spring a distance of x inches is given by $F = 6x$. Find the work necessary to stretch the spring 4 inches from its original length if the force is given by $F = kx$.

2. A spring of length 5 inches requires a force of 4 lb to stretch it 3 inches. Find the work done in stretching it to a total of 12 inches if the force is given by $F = kx$.

3. How much work is done in moving directly against a force F from $x = 1$ to $x = 5$ if $F = -0.1x$?

4. A spring of natural length 10 inches stretches 1.5 inches under a weight of 8 pounds.
   a. Find the work done in stretching the spring from its natural length to a length of 14 inches.
   b. Find the work done in stretching the spring from a length of 11 inches to a length of 13 inches.

5. A force of 200 pounds is required to compress a spring of natural length 12 inches to a length of 10 inches. Find the work done in compressing the spring from its natural length to a length of 8 inches.

6. The force used in driving a certain piston is given as follows: $F = 10x^{-3/2}$. Find the work done in moving the piston between $x = 20$ and $x = 50$.

7. A cylindrical tank of radius 3 feet and height 12 feet is full of water. Find the work necessary to empty the tank.

8. A cylindrical tank of radius 4 feet and height 10 feet is one-half full of water. Find the work necessary to empty the tank.

9. A right circular conical tank of height 20 feet and radius of base 5 feet is full of water. Find the work done in pumping the water over the top of the tank.

10. A cylindrical tank of diameter 3 feet and height 6 feet is full of water.
   a. Find the work required to pump the water out over the top of the tank.
   b. Find the work required to pump the water out through a pipe that rises to a height of 3 feet above the top of the tank.

23 • 6 Centroids

The shaded region in figure 23.28 is formed by the curve of $y = e^x$ bounded by the x-axis, the y-axis, and the line $x = 1$. If we trace this region onto a rigid material of uniform density, we could find a point on the material such that it could be balanced on an object, such as a pencil tip, at this point. The coordinates of this point are given by $(\bar{x}, \bar{y})$, and the point is the center of gravity, or **centroid**. Simply stated, the centroid is the center of area of a geometrical shape. Conceptually, the centroid is the point at which all of the area of a given figure could be concentrated.

Centroid

Moment of Area

If we assume that the area of a plane, or flat, figure is concentrated at its centroid, then we can define the **moment of area**, M, as the product of the area times the perpendicular distance between the centroid and the axis of rotation. The moment of a given area A about the y-axis is denoted by $M_y = A\bar{x}$, and the moment of A about the x-axis is $M_x = A\bar{y}$.

We have seen that the area A under a curve is obtained by evaluating $A = \int_a^b f(x)\,dx$. From the definition of moment of area, we have

$$M_y = \int_a^b x f(x)\,dx.$$

Also, it can be shown that

$$M_x = \int_a^b \frac{1}{2}[f(x)]^2\,dx.$$

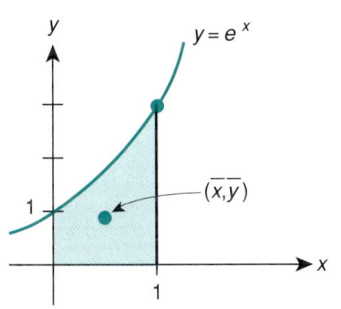

Figure 23.28

Since $M_y = A\bar{x}$ and $M_x = A\bar{y}$, we now have formulas for finding the centroid, $(\bar{x}, \bar{y})$:

$$\bar{x} = \frac{M_y}{A} = \frac{\int_a^b x f(x)\,dx}{\int_a^b f(x)\,dx}$$

$$\bar{y} = \frac{M_x}{A} = \frac{\int_a^b \frac{1}{2}[f(x)]^2\,dx}{\int_a^b f(x)\,dx}.$$

CENTROID FORMULAS

If we are to find the center of gravity, or centroid, of a solid of revolution of uniform density, we use the following formulas:

a. $\bar{x} = \dfrac{\int_a^b x[f(x)]^2\, dx}{\int_a^b [f(x)]^2\, dx}$, $\bar{y} = 0$ $\Bigg\}$ for solids of revolution about the x-axis.

b. $\bar{x} = 0$, $\bar{y} = \dfrac{\int_a^b y[g(y)]^2\, dy}{\int_a^b [g(y)]^2\, dy}$ $\Bigg\}$ for solids of revolution about the y-axis.

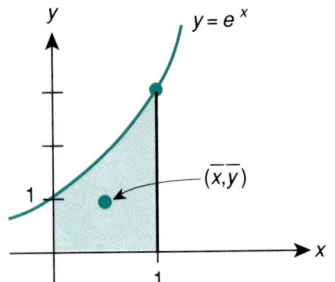

Figure 23.29

• **Example 23.20:** Find the centroid of the region bounded by $y = e^x$, the y-axis, the x-axis, and $x = 1$. See figure 23.29.

Solution:
$$\bar{x} = \frac{\int_0^1 xe^x\, dx}{\int_0^1 e^x\, dx}$$

$$= \frac{e^x(x-1)\big|_0^1}{e^x\big|_0^1} \qquad \text{(Integration by parts)}$$

$$= \frac{e^1(0) - [1(-1)]}{e^1 - e^0}$$

$$= \frac{1}{e-1} \approx 0.58$$

$$\bar{y} = \frac{\int_0^1 \tfrac{1}{2} e^{2x}\, dx}{\int_0^1 e^x\, dx}$$

$$= \frac{\tfrac{1}{2} \int_0^1 \tfrac{1}{2} e^{2x}(2\, dx)}{\int_0^1 e^x\, dx} \qquad \left(\int e^u\, du \text{ forms}\right)$$

$$= \frac{\frac{1}{4}e^{2x}\big|_0^1}{e-1}$$

$$= \frac{\frac{1}{4}(e^2-1)}{e-1} \qquad \text{(Evaluating at } x = 1 \text{ and } x = 0\text{)}$$

$$= \frac{1}{4}(e+1) \approx 0.93.$$

Therefore, $(\bar{x},\bar{y}) \approx (0.58, \; 0.93)$.

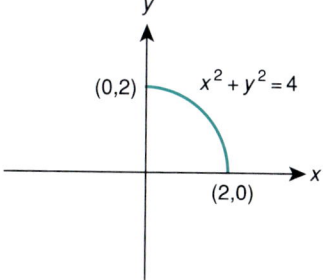

Figure 23.30

• **Example 23.21:** Find the centroid of the region bounded by $x^2 + y^2 = 4$, the y-axis, and the x-axis, $x \geq 0$, $y \geq 0$. See figure 23.30.

Solution:

$$\bar{x} = \frac{\int_0^2 x\sqrt{4-x^2}\,dx}{\int_0^2 \sqrt{4-x^2}\,dx}$$

$$= \frac{-\frac{1}{2}\int_0^2 -2x\sqrt{4-x^2}\,dx}{\int_0^2 \sqrt{4-x^2}\,dx}$$

$$= \frac{-\frac{1}{2}\cdot\frac{2}{3}(4-x^2)^{3/2}\big|_0^2}{\left[\frac{1}{2}x\sqrt{(4-x^2)} + 2\sin^{-1}\frac{x}{2}\right]\big|_0^2}$$

$$= \frac{-\frac{1}{3}[0-8]}{2\frac{\pi}{2}}$$

$$= \frac{8}{3}\cdot\frac{1}{\pi}$$

$$\approx 0.85.$$

Since the curve is symmetric about the axes, $\bar{y} = \bar{x}$. Therefore, $(\bar{x},\bar{y}) \approx (0.85, \; 0.85)$.

CHAPTER 23 APPLICATIONS OF INTEGRATION

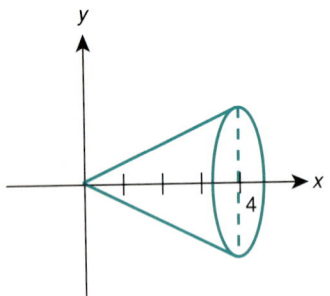

Figure 23.31

• **Example 23.22:** Find the centroid of the solid of revolution formed by revolving the line $y = x$ about the x-axis from $x = 0$ to $x = 4$. See figure 23.31.

Solution: Clearly $\bar{y} = 0$, since the formation is symmetric about the x-axis,

$$\bar{x} = \frac{\int_0^4 x[f(x)]^2 \, dx}{\int_0^4 [f(x)]^2 \, dx}$$

$$= \frac{\int_0^4 x \cdot x^2 \, dx}{\int_0^4 x^2 \, dx}$$

$$= \frac{\int_0^4 x^3 \, dx}{\int_0^4 x^2 \, dx}$$

$$= \frac{\left.\frac{x^4}{4}\right|_0^4}{\left.\frac{x^3}{3}\right|_0^4}$$

$$= \frac{64}{\frac{64}{3}} = \frac{64}{1} \cdot \frac{3}{64}$$

$$= 3.$$

Therefore, $(\bar{x}, \bar{y}) = (3, 0)$.

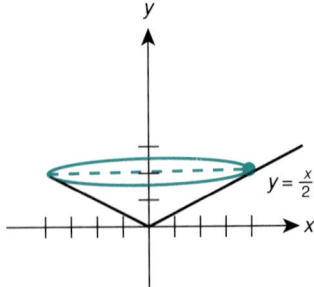

Figure 23.32

• **Example 23.23:** Find the centroid of the solid of revolution formed by revolving the line $y = \frac{x}{2}$ about the y-axis from $y = 0$ to $y = 2$. See figure 23.32.

Solution: In this case, $\bar{x} = 0$ and we use

$$\bar{y} = \frac{\int_0^2 y[g(y)]^2 \, dy}{\int_0^2 [g(y)]^2 \, dy}.$$

Next, $y = \dfrac{x}{2}$ means that $x = 2y = g(y)$.
Therefore,

$$\bar{y} = \dfrac{\int_0^2 y[2y]^2 \, dy}{\int_0^2 [2y]^2 \, dy}$$

$$= \dfrac{\int_0^2 4y^5 \, dy}{\int_0^2 4y^2 \, dy}$$

$$= \dfrac{\left.\dfrac{4y^6}{6}\right|_0^2}{\left.\dfrac{4y^3}{3}\right|_0^2} = \dfrac{\dfrac{2 \cdot 64}{3}}{\dfrac{4 \cdot 8}{3}} = \dfrac{128}{32} = \dfrac{4}{1} = 4.$$

The centroid is (0,4).

Trial Problems 23.6

Before you begin the section exercises, warm up with these problems. Complete answers are included in the answer key.

When finding centroids, you will have to integrate expressions of various forms. Evaluate the following integrals.

1. $\displaystyle\int_0^9 x\sqrt{x} \, dx$
2. $\displaystyle\int_0^1 (e^{2x})^2 \, dx$
3. $\displaystyle\int_0^1 x(2-x^2) \, dx$
4. $\displaystyle\int_{-1}^0 e^{-2x} \, dx$
5. $\displaystyle\int_0^1 \sqrt{1-x} \, dx$

Exercises 23.6

Sketch the region bounded by the graphs of the given equations and axes and find the centroid of the region.

1. $y = x^2$, the x-axis, $x = 2$
2. $y = \sin x$, the x-axis between $x = 0$ and $x = \pi$
3. $y = x$, the x-axis, $x = 4$
4. $y = \sqrt{x}$, the x-axis, $x = 9$
5. $xy = 1$, the x-axis, $x = 1$, $x = 2$
6. $y = e^{2x}$, the x-axis, $x = 0$, $x = 1$
7. $y = 4 - x^2$, the x-axis
8. $x = 0$, $x = 6$, $y = 0$, $y = 2$
9. $y = x^2$, the y-axis, $y = 9$
10. $y = x^3$, the y-axis, $y = 8$

11. $y^2 = 4x$, $y = x$

$\left(\text{Hint:} \int_a^b xf(x)dx = \int_a^b x[g(x) - h(x)]dx \right.$

$\left. \text{and } \int_a^b \frac{1}{2}[f(x)]^2 = \int_a^b \frac{1}{2}[(g(x))^2 - (h(x))^2]dx \right)$

12. $y = x^2$, $y = 4x$

13. $y = e^{-3x}$, the x-axis, $x = 0$, $x = 1$

14. $y = \cos x$, the x-axis between $x = 0$ and $x = \frac{\pi}{2}$

15. $y = \sqrt{1-x}$, the x-axis, $x = 0$, $x = 1$

Find the centroid of the solid of revolution generated by revolving about the x-axis the region bounded by the graphs of the given equations.

16. $y = 2x$ and the x-axis, from $x = 1$ to $x = 2$

17. $y = e^x$ and the x-axis, from $x = 0$ to $x = 1$

18. $y = x^2$ and the x-axis, from $x = 0$ to $x = 2$

19. $y = \sin x$ and the x-axis, from $x = 0$ to $x = \pi$

20. $y = \dfrac{1}{\sqrt{16 + x^2}}$ and the x-axis, from $x = 0$ to $x = 4$

Find the centroid of the solid of revolution generated by revolving about the y-axis the region bounded by the graphs of the given equations.

21. $y = x$, from $y = 0$ to $y = 3$, $x = 0$

22. $x = \sqrt{y - 1}$, from $y = 1$ to $y = 5$, $x = 0$

23. $x = \sqrt{y^2 - 4}$, from $y = 2$ to $y = 4$, $x = 0$

24. The x-coordinate of the centroid of a circular arc of radius 6 is given by $\bar{x} = \dfrac{4}{\pi} \int_0^{\pi/4} 6 \cos \theta \, d\theta$. Find $\bar{x}$.

23 • 7 Moments of Inertia

Moment of Inertia

If a particle of mass m is rotated about the y-axis, its **moment of inertia** is defined by

$$I_y = md^2,$$

where d is distance from the particle to the y-axis. If several particles of mass, $m_1, m_2, \ldots, m_n$, located at different distances from the y-axis, are rotated about the y-axis, then the moment of inertia of the group of particles is given by

$$I_y = m_1 d_1^2 + m_2 d_2^2 + \cdots + m_n d_n^2.$$

Now, suppose that an infinite number of such particles of mass are located in a particular region, and this region is revolved about the y-axis. Using integration as a means of summing up the moments of inertia of each individual particle:

MOMENT OF INERTIA FOR A REGION WITH RESPECT TO y-AXIS

$$I_y = k \int_a^b x^2 f(x) dx,$$

is the moment of inertia of the region with respect to the y-axis, where $kf(x)dx$ represents the mass of the region (k is mass per unit area or density) and x is the distance of each particle from the y-axis.

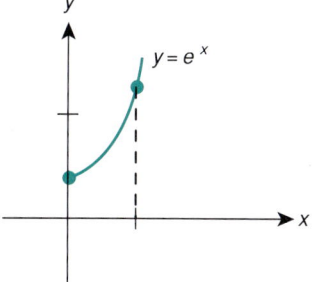

Figure 23.33

• **Example 23.24:** Find the moment of inertia with respect to the *y*-axis of the region bounded by $y = e^x$, the *x*-axis from $x = 0$ to $x = 1$ (assume $k = 1$). See figure 23.33.

Solution:
$$I_y = \int_0^1 x^2 e^x \, dx$$
$$= [x^2 e^x - 2xe^x + 2e^x]\Big|_0^1 \quad \text{(Integration by parts twice)}$$
$$= (e - 2e + 2e) - (0 - 0 + 2) \quad \text{(Substituting } x = 1, \text{ then } x = 0\text{)}$$
$$= e - 2.$$

MOMENT OF INERTIA FOR A REGION WITH RESPECT TO *x*-AXIS

To find the moment of inertia with respect to the *x*-axis, we use the formula

$$I_x = k \int_c^d y^2 f(y) \, dy.$$

• **Example 23.25:** Find I_x for the region bounded by $y = x^2$, $y = 4$, and the *y*-axis. Assume $k = 1$. See figure 23.34.

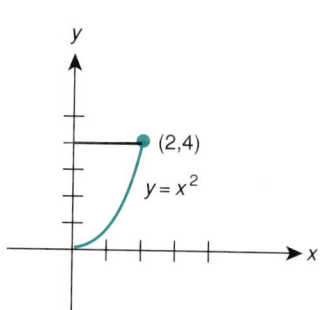

Figure 23.34

Solution: In this case, $x = \sqrt{y}$.

$$I_x = \int_0^4 y^2(4 - \sqrt{y}) \, dy \quad \text{(Since } y = x^2, x = \sqrt{y}\text{)}$$
$$= \int_0^4 (4y^2 - y^{5/2}) \, dy$$
$$= \left[\frac{4y^3}{3} - \frac{2}{7} y^{7/2}\right]\Big|_0^4$$
$$= \frac{256}{3} - \frac{256}{7} \quad \text{(Substituting } y = 4; \text{ then } y = 0\text{)}$$
$$= \frac{1024}{21}.$$

If a region is bounded by two curves $y_1 = f(x)$ and $y_2 = g(x)$, the general formulas for moments of inertia can be modified as follows:

$$I_y = k \int_a^b x^2 (y_2 - y_1) \, dx,$$

and similarly,

$$I_x = k \int_c^d y^2 (x_2 - x_1) \, dy.$$

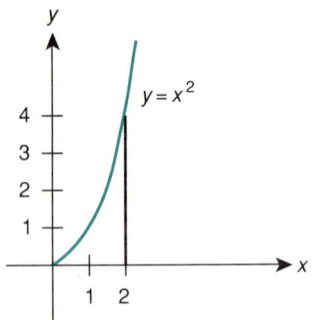

Figure 23.35

• **Example 23.26:** Find I_x for the region bounded by $y = x^2$ and $x = 2$, and the x-axis. Assume $k = 1$. See figure 23.35.

Solution:
$$I_x = \int_0^4 y^2(2 - \sqrt{y})dy \quad \text{(Since } y = x^2, x = \sqrt{y}\text{)}$$
$$= \int_0^4 (2y^2 - y^{5/2})dy$$
$$= \left[\frac{2y^3}{3} - \frac{2y^{7/2}}{7}\right]\Big|_0^4$$
$$= \frac{128}{3} - \frac{256}{7}$$
$$= \frac{128}{21}.$$

Moments of inertia of solids of revolution have many important applications. With respect to the axes, moments of inertia are defined to be mass (in terms of volume) multiplied by the square of the distance to the axes. To find I_y for a solid of revolution about the y-axis, we use the method of cylindrical shells. See figures 23.36 and 23.37.

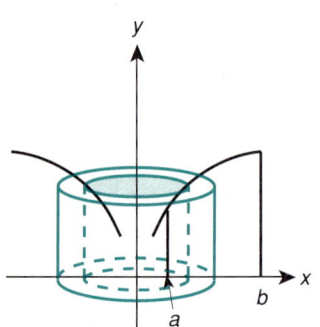

Figure 23.36

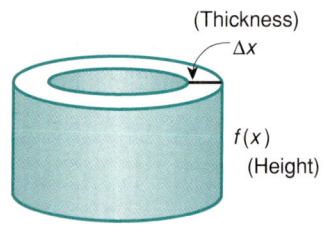

Figure 23.37

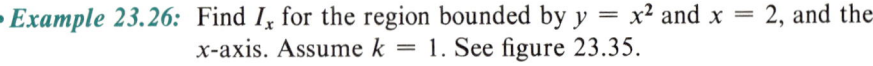

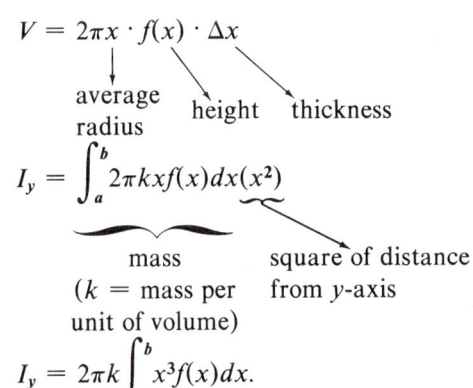

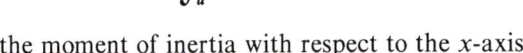

$$I_y = 2\pi k \int_a^b x^3 f(x)dx.$$

Similarly, for the moment of inertia with respect to the x-axis,

$$I_x = 2\pi k \int_c^d y^3 g(y)dy.$$

• **Example 23.27:** Find I_x and I_y for the solid of revolution generated by the curve of $y = \sqrt[3]{x}$ from $x = 0$ to $x = 8$. Assume $k = 1$. See figure 23.38.

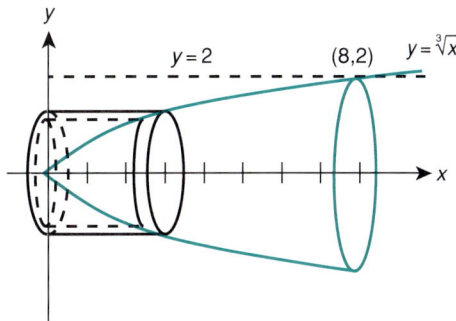

Figure 23.38

Solution: $I_x = 2\pi \int_0^2 y^3 \cdot g(y)\,dy$

$= 2\pi \int_0^2 y^3 \cdot y^3\,dy \quad (g(y) = x = y^3 \text{ since } y = \sqrt[3]{x})$

$= 2\pi \left[\dfrac{y^7}{7}\right]\Big|_0^2$

$= 2\pi \left(\dfrac{128}{7}\right)$

$= \dfrac{256}{7}\pi$

$I_y = 2\pi \int_0^8 x^3 f(x)\,dx$

$= 2\pi \int_0^8 x^3 \cdot x^{1/3}\,dx$

$= 2\pi \int_0^8 x^{10/3}\,dx$

$= 2\pi \dfrac{3x^{13/3}}{10}\Big|_0^8$

$= 2\pi \dfrac{[3 \cdot 2^{13}]}{10}$

$= \dfrac{3\pi}{5}(8192)$

$= \dfrac{24{,}576}{5}\pi.$

• • • • • • • • • •

The moment of inertia always turns out to be a number. This number represents a quantity of force needed for an object to resist a change in motion. Other quantities associated with centroids and moments of inertia include mass and radius of gyration. Integrals needed to compute these quantities are found in any general calculus text.

Trial Problems 23.7

Before you begin the section exercises, warm up with these problems. Complete answers are included in the answer key.

When finding moments of inertia, you will have to integrate expressions of various forms. Evaluate the following integrals.

1. $\int_0^4 x^2 \sqrt{x}\, dx$
2. $\int_0^1 x^2(4 - x^{1/3})\, dx$
3. $\int_0^1 y e^{y^2}\, dy$
4. $\int_0^1 x e^x\, dx$
5. $\int_1^8 y^2 \cdot y^{1/3}\, dy$

Exercises 23.7

Find I_x and I_y for each of the following regions. Assume $k = 1$ in all cases.

1. $y = \sqrt{x}$, the x-axis, and $x = 9$
2. $y = x^2$, $y = 4$
3. $y = x^3$, $y = 8$, and y-axis
4. $y = e^x$, $y = e$, and y-axis
5. Find I_y for the region bounded by $y = e^{x^3}$, the x-axis, and the y-axis, from $x = 0$ to $x = 1$.
6. Find I_y for the region bounded by $y = x$ and the x-axis, from $x = 1$ to $x = 3$.
7. Find I_y for the region bounded by $y = 4 - x^2$ and the x-axis, from $x = 0$ to $x = 2$.
8. Find I_y and I_x for the region bounded by $y = x^{3/2}$, the y-axis, and $y = 8$.

Find I_x or I_y for each of the following solids of revolution, as indicated. (Assume $k = 1$ in all cases.)

9. I_x: $y = \sqrt{x}$, $y = 0$, from $x = 0$ to $x = 4$, about the x-axis
10. I_y: $y = x$, $x = 0$, from $y = 0$ to $y = 5$, about the y-axis
11. I_y: $y = 4 - x^2$, $x = 0$, from $x = 0$ to $x = 2$, about the y-axis
12. I_y: $x = 0$, $x = 3$, $y = 0$, $y = 4$, about the y-axis
13. The moment of inertia of a circular area of radius r is given by $I_x = \dfrac{r^4}{4} \int_0^{2\pi} \sin^2 \theta\, d\theta$. Find I_x if $r = 2$.

23 • 8 Other Applications

There are many other applications of integration. Some involve methods usually studied in more advanced work in the calculus. Others do not require additional methods, but may require further depth of study than those we have already studied. In this section, we investigate examples of the latter.

Length of Arc The evaluation of a definite integral can be used to determine the length of a curve whose equation is known. This application is known as **length of arc**.

LENGTH OF ARC

If the curve of $y = f(x)$ is continuous from $x = a$ to $x = b$, we can find the length of the curve from $P_1(a, f(a))$ to $P_2(b, f(b))$ by using the formula

$$\text{length of arc} = S = \int_a^b \sqrt{1 + [f'(x)]^2}\, dx.$$

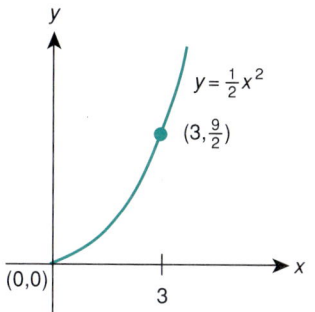

Figure 23.39

• **Example 23.28:** Find the length of arc of $y = \dfrac{1}{2}x^2$ from $x = 0$ to $x = 3$. See figure 23.39.

Solution: $f'(x) = 2 \cdot \dfrac{1}{2} \cdot x = x$

$[f'(x)]^2 = x^2$

$S = \displaystyle\int_0^3 \sqrt{1 + x^2}\, dx$

$= \dfrac{1}{2}[x\sqrt{1 + x^2} + \ln(x + \sqrt{1 + x^2})]\Big|_0^3$ (Number 8 in the table of integrals)

$= \dfrac{1}{2}[3\sqrt{10} + \ln|3 + \sqrt{10}|]$

$\approx \dfrac{1}{2}[9.49 + 1.82]$

≈ 5.66 units.

Another application of integration involves finding the force due to liquid pressure. In physics it is known that pressure is directly proportional to depth below the surface of a liquid. If W is the weight per unit volume of the liquid, and h is the depth of the liquid over an area, then the pressure or force F exerted by the liquid between depths $h = a$ and $h = b$ is given by

$$F = W \int_{h=a}^{h=b} lh\, dh,$$

where l is the length of the container or object.

• **Example 23.29:** Find the total pressure exerted on a vertical side of a fish tank 3 feet long and 18 inches high that is $\dfrac{2}{3}$ filled with water. See figure 23.40.

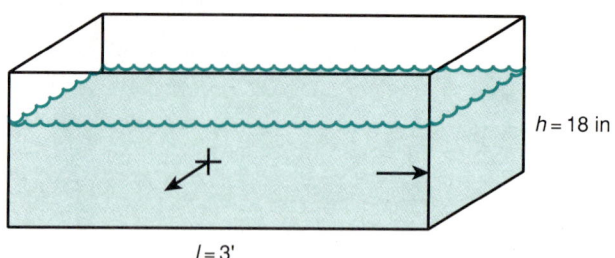

Figure 23.40

Solution: The height of the water h is $\frac{2}{3} \cdot 18 = 12$ inches, or $h = 1$ foot. The length l is 3 feet. $W =$ weight of water 62.5 lb/cu ft.

$$F = W \int_{h=a}^{h=b} lh \, dh$$

$$F = 62.5 \int_0^1 3h \, dh = 6.25 \left(\frac{3h^2}{2}\right)\Big|_0^1 = 62.5 \left[\frac{3-0}{2}\right]$$

$$= 93.75 \text{ lb.}$$

The integrals of certain functions play an important role in probability theory. Any function $f(x)$ with the following two properties is said to be a **probability density function**.

Probability Density Function

1. $f(x) \geq 0$ for $c \leq x \leq d$
2. $\int_c^d f(x) dx = 1$

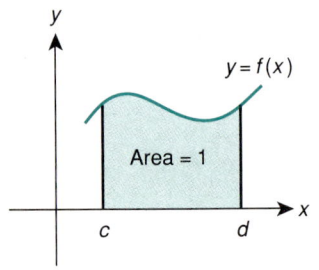

Figure 23.41

Graphically, this says that the curve of $f(x)$ lies above the x-axis from $x = c$ to $x = d$, and the area under the curve from $x = c$ to $x = d$ is 1. This is illustrated in figure 23.41.

• **Example 23.30:** Biologists have found that under ideal circumstances the proportion of cells whose ages are between c and d is given by the area under the curve of $f(x) = 2ke^{-kx}$ from $x = c$ to $x = d$, where $k = \frac{1}{3} \ln 2$. Show that $f(x)$ is a probability density function for the interval $x = 0$ to $x - 3$.

Solution: $2k = \dfrac{2}{3} \ln 2$ is a positive number. e^{-kx} is always positive for k and any x. Therefore, property 1 on page 966 is satisfied; that is, $f(x) \geq 0$ for $c \leq x \leq d$.

$$\int_0^3 2ke^{-kx}\,dx = -2\int_0^3 -ke^{-kx}\,dx \quad \left(\int e^u\,du \text{ form}\right)$$

Since $u = -kx$ and $du = -k\,dx$,

$$\int_0^3 2ke^{-kx}\,dx = -2[e^{-kx}]_0^3$$
$$= -2e^{-3k} - (-2)e^0$$
$$= 2 - 2e^{-3(1/3\,\ln\,2)}$$
$$= 2 - 2e^{-\ln 2}$$
$$= 2 - \dfrac{2}{e^{\ln 2}} \quad (e^{\ln 2} = 2)$$
$$= 2 - \dfrac{2}{2}$$
$$= 2 - 1$$
$$= 1.$$

Therefore, property 2 is satisfied since $\int_0^3 f(x)\,dx = 1$.

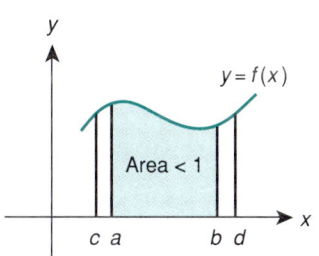

Figure 23.42

If we know that an event definitely will occur, we say that the probability is 1. If the event definitely will not occur, we say the probability is 0. Therefore, the probability that an event *may* occur has a value between 0 and 1. This value can be found by investigating the area under the curve of a probability density function for the interval $a \leq x \leq b$ where $c \leq a \leq b$. See figure 23.42.

PROBABILITY OF AN EVENT

Thus, if $f(x)$ is a probability density function for an event, where x varies randomly in some interval $a \leq x \leq b$, then the probability that the event will occur in that interval is found by

$$\int_a^b f(x)\,dx,$$

and will have value between 0 and 1.

968 CHAPTER 23 APPLICATIONS OF INTEGRATION

• **Example 23.31:** During a certain year a farming region produced x bushels of wheat on a given acre. Suppose the probability density function for the number of bushels of wheat produced on a given acre is

$$f(x) = \frac{1}{50}(x - 30) \qquad \text{for } 30 \leq x \leq 40.$$

What is the probability that an acre selected at random produced less than 35 bushels of wheat?

Solution: The probability can be determined by integrating $f(x)$ from $x = 30$ to $x = 35$.

$$\int_{30}^{35} \frac{1}{50}(x - 30)dx = \left[\frac{x^2}{100} - \frac{30x}{50}\right]\Big|_{30}^{35}$$

$$= \left[\frac{(35)^2}{100} - \frac{30(35)}{50}\right] - \left[\frac{(30)^2}{100} - \frac{30(30)}{50}\right]$$

$$= [12.25 - 21] - [9 - 18]$$

$$= -8.75 + 9$$

$$= 0.25.$$

The probability is 0.25 or $\dfrac{1}{4}$.

•••••••••••

The final discussion of integration may not seem to be directly applicable to physical situations, but the process can be related to areas under curves. The area under the curve $y = \dfrac{1}{x^2}$ from $x = 1$ to $x = b$ is given by

$$A = \int_{1}^{b} \frac{1}{x^2}\, dx.$$

See figure 23.43.

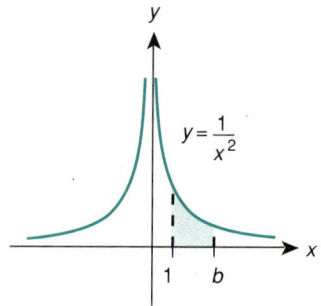

Figure 23.43

Suppose we increase b without limit. That is, suppose b goes to infinitely: $b \to \infty$. Then the integral becomes $A = \int_1^\infty \frac{1}{x^2}\, dx$. This integral can be evaluated in the following manner:

$$A = \int_1^\infty \frac{1}{x^2}\, dx = \lim_{b \to \infty} \int_1^b \frac{1}{x^2}\, dx = \lim_{b \to \infty} \int_1^b x^{-2}\, dx$$

$$= \lim_{b \to \infty} \left.\frac{x^{-1}}{-1}\right|_1^b = \lim_{b \to \infty} \left.-\frac{1}{x}\right|_1^b = \lim_{b \to \infty} \left(-\frac{1}{b} + 1\right)$$

$$= 0 + 1 \qquad \left(\text{As } b \to \infty,\ -\frac{1}{b} \to 0\right)$$

$$= 1.$$

Improper Integrals Integrals in which one or both of the limits of integration are infinite are called **improper integrals.** Examples of improper integrals include: $\int_a^\infty f(x)\, dx$, $\int_{-\infty}^b f(x)\, dx$, and $\int_{-\infty}^\infty f(x)\, dx$. Other types of improper integrals exist, but are not discussed in this text.

Not all improper integrals can be evaluated. Consider, for example, $\int_1^\infty \frac{1}{x}\, dx$:

$$\int_1^\infty \frac{1}{x}\, dx = \lim_{b \to \infty} \int_1^b \frac{1}{x}\, dx = \lim_{b \to \infty} \ln x \Big|_1^b$$

$$= \lim_{b \to \infty} [\ln b - \ln 1] = \lim_{b \to \infty} [\ln b - 0]$$

$$= \lim_{b \to \infty} \ln b.$$

Since $\ln b \to \infty$ as $b \to \infty$, $\lim_{b \to \infty} \ln b$ does not exist. Thus, the integral $\int_1^\infty \frac{1}{x}\, dx$ has no value since the limit involved does not exist.

• **Example 23.32:** Find the total area under the curve $y = \dfrac{2}{x^2 + 1}$. See figure 23.44.

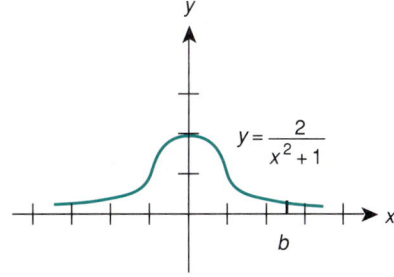

Figure 23.44

Solution: We find the total area by integrating from $x = 0$ to $x = \infty$ and multiplying by 2 (the y-axis is an axis of symmetry):

$$A = 2\int_0^\infty \frac{2}{x^2+1}\,dx = \lim_{b\to\infty}\left[2\int_0^b \frac{2}{x^2+1}\,dx\right]$$

$$= 2\lim_{b\to\infty}\int_0^b \frac{2}{x^2+1}\,dx = 4\lim_{b\to\infty}\int_0^b \frac{1}{x^2+1}\,dx$$

$$= 4\lim_{b\to\infty} \tan^{-1}x\Big|_0^b = 4\lim_{b\to\infty}[\tan^{-1}b - \tan^{-1}0]$$

$$= 4\lim_{b\to\infty}[\tan^{-1}b - 0]$$

$$= 4[\tan^{-1}\infty - 0] \quad \left(\begin{array}{l}\text{An angle whose tangent}\\ \text{is infinite is } \dfrac{\pi}{2}\end{array}\right)$$

$$= 4\left[\frac{\pi}{2}\right]$$

$$= \frac{4\pi}{2} = 2\pi \text{ sq. units.}$$

• **Example 23.33:** The function $f(x) = ke^{-kx}$, where k is a constant and $x \geq 0$, is an example of a probability density function. Show that

$$\int_0^\infty f(x)\,dx = 1.$$

Solution: $\displaystyle\int_0^\infty f(x)\,dx = \int_0^\infty ke^{-kx}\,dx$

$$= \lim_{b\to\infty}\left[\int_0^b ke^{-kx}\,dx\right] \quad \left(\int e^u\,du \text{ form}\right)$$

Since $u = -kx$, $du = -k\,dx$,

$$\int_0^\infty f(x)\,dx = \lim_{b\to\infty}\left[-\int_0^\infty -ke^{-kx}\,dx\right]$$

$$= \lim_{b\to\infty}[-e^{-kx}]_0^b$$

$$= \lim_{b\to\infty}[-e^{-bk} - (-e^0)]$$

$$= \lim_{b\to\infty}[1 - e^{1/bk}]$$

$$= 1 - 0$$

$$= 1.$$

Trial Problems 23.8

Before you begin the section exercises, warm up with these problems. Complete answers are included in the answer key.

Expressions of various forms need to be integrated in this exercise set. Evaluate the following integrals.

1. $\int_0^3 \sqrt{7}\, dx$
2. $\int_0^2 2x\sqrt{4-x^2}\, dx$
3. $\int \frac{\ln x}{x}\, dx$
4. $\int \frac{dx}{x^2+1}$
5. $\int \frac{k}{(1+t)^2}\, dt$

Exercises 23.8

Find the length of arc for each of the following curves.

1. $y = x$ from $x = 0$ to $x = 4$
   (Check the answer by using the Pythagorean theorem.)
2. $y = x$ from $x = 1$ to $x = 4$
   (Check the answer by using the Pythagorean theorem.)
3. $y = 2x$ from $x = 0$ to $x = 2$
4. $y = x + 5$ from $x = 0$ to $x = 2$
5. $y = \frac{2}{3}x^{3/2}$ from $x = 3$ to $x = 8$
6. $y = \frac{2}{3}(x-1)^{3/2}$ from $x = 1$ to $x = 4$
7. $y = \frac{2}{3}(x^2-1)^{3/2}$ from $x = 1$ to $x = 3$
8. $y^2 = 4x$ from $x = 0$ to $x = 2$
9. A steel producer finds that the amount x of steel sold daily varies randomly between 200 and 600 tons. Let x be measured in hundreds of tons so that $2 \le x \le 6$. The probability density function for x is given by
$$f(x) = \frac{3}{160}(x^2 - x).$$
   a. What is the probability that at least 300 tons will be sold on a certain day? (*Hint:* integrate from $x = 3$ to $x = 6$.)
   b. What is the probability that at most 400 tons will be sold on that day?
   c. What is the probability that at least 300 tons, but no more than 500 tons, will be sold on that day?

10. In a certain cell population cells divide every 10 days and the age x of a cell selected at random has the probability density function
$$f(x) = 2ke^{-kx} \quad \text{for } 0 \le x \le 10,$$
    where $k = (0.1) \ln 2$. Find the probability that a cell is at most 5 days old.

11. If $f(x) = kx^2$, determine the value of k that makes $f(x)$ a probability density function for $0 \le x \le 2$.

12. An automated machine produces an automobile part every 3 minutes. An inspector arrives at a random time and must wait t minutes for a part.
    a. Find the probability density function for t.
       (*Hint:* solve $\int_0^3 k\, dt = 1$ for k.)
    b. Find the probability that the inspector must wait at least 1 minute.
    c. Find the probability that the inspector must wait no more than 1 minute.

The following problems involve finding the force due to liquid pressure.

13. Find the force on one end of a rectangular tank 4 feet wide and 2 feet high if the tank is full of water.

14. Find the force on one side of a cubical tank each of whose edges are 3 feet long if the tank is half full of water.

15. A rectangular gate in a vertical dam is 10 feet wide and 6 feet deep. Find the total force exerted against this gate when the water level is 8 feet above the top of the gate. (*Hint:* the limits of integration are $h = 8$ to $h = 14$. Why?)

16. Find the force on the gate in exercise 15 if the water level drops to 6 feet above the top of the gate.

17. A horizontal cylindrical tank having a diameter of 8 feet is half full of oil weighing 60 lb/cu ft. Calculate the force exerted on one end. (*Hint:* the end of the tank is pictured in figure 23.45.)
   a. l is a variable length; let $l = 2x$.
   b. h is a variable height; let $h = y$.
   c. The radius of the tank is 4 feet and the edge of the tank end is a semicircle: $x^2 + y^2 = 16$, or $x = \sqrt{16 - y^2}$.
   d. The height of oil goes from 0 to 4 feet.
   e. Therefore,
   $$F = \int_0^4 (2x)y\, dy = \int_0^4 (2y)x\, dy$$
   $$= \int_0^4 2y\sqrt{16 - y^2}\, dy.$$
   Complete the problem.

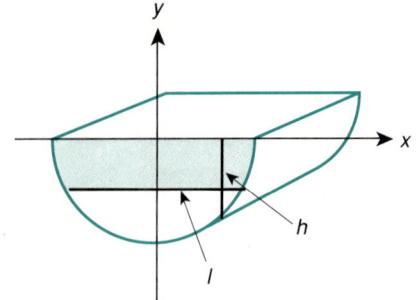

Figure 23.45

Evaluate the following improper integrals, if the integral exists.

18. $\int_2^\infty \dfrac{dx}{x^3}$

19. $\int_{-\infty}^1 e^x\, dx$ (*Hint:* $e^{-\infty} = \dfrac{1}{e^\infty} \to 0$)

20. $\int_0^\infty \cos x\, dx$

21. $\int_0^\infty e^{-x}\, dx$

22. $\int_0^\infty \dfrac{1}{(x-1)^2}\, dx$

23. $\int_1^\infty \ln x\, dx$

24. $\int_0^\infty e^{-x} \cos x\, dx$

25. $\int_1^\infty x\, dx$

26. $\int_{-\infty}^0 \dfrac{dx}{1 + x^2}$

27. $\int_{-\infty}^1 e^x\, dx$

28. $\int_{-\infty}^{-1} \dfrac{1}{x^3}\, dx$

29. $\int_{-\infty}^\infty xe^{-x^2}\, dx$
    (*Hint:* $\int_{-\infty}^\infty f(x)dx = \int_{-\infty}^c f(x)\, dx + \int_c^\infty f(x)dx$)

30. $\int_1^\infty \dfrac{\ln x}{x}\, dx$

31. $\int_0^\infty \dfrac{1}{x^2 - 1}\, dx$

32. A computer time-sharing service finds that the time t, in seconds, required by the average job has the probability density function
$$f(t) = \dfrac{2}{(2 + t)^2} \quad \text{if } t \geq 0.$$
What is the probability that a job will take at least 12 seconds?

23·9 Summary of Terms and Rules

Terms

Area under a curve (p. 930)
Centroid (p. 955)
Cylindrical shell (p. 947)
Improper integrals (p. 969)
Length of arc (p. 964)
Moment of area (p. 955)
Moment of inertia (p. 960)
Probability density function (p. 966)
Solid of revolution (p. 941)
Volume of a solid of revolution (p. 941)
Work (p. 952)

Rules

- Area under a curve $y = f(x)$ from $x = a$ to $x = b$ is given by

$$A = \int_a^b f(x)\,dx. \tag{23.1}$$

- Area between two curves $y = f(x)$ and $y = g(x)$ is given by

$$A = \int_a^b [f(x) - g(x)]\,dx, \tag{23.2}$$

where the $y = g(x)$ lies below $y = f(x)$.

- Volume of a solid of revolution generated by rotating the curve $y = f(x)$ about the x-axis is given by

$$V = \pi \int_a^b [f(x)]^2\,dx. \tag{23.3}$$

- Volume of a solid of revolution by cylindrical shells is given by

$$V = 2\pi \int_a^b x f(x)\,dx. \tag{23.4}$$

- Work completed in moving an object from $x = a$ to $x = b$ is given by

$$W = \int_a^b F\,dx, \tag{23.5}$$

where F is a force function.

- Coordinates of the centroid are given by

$$\bar{x} = \frac{M_y}{A} = \frac{\int_a^b x f(x)\,dx}{\int_a^b f(x)\,dx}, \quad \text{and}$$

$$\bar{y} = \frac{M_x}{A} = \frac{\int_a^b \frac{1}{2}[f(x)]^2\,dx}{\int_a^b f(x)\,dx}. \tag{23.6}$$

- Moment of inertia for a particle of mass with respect to the y-axis is given by

$$I_y = k\int_a^b x^2 f(x)\,dx. \tag{23.7}$$

- Moment of inertia for a particle of mass with respect to the x-axis is given by

$$I_x = k\int_c^d y^2 f(y)\,dy. \tag{23.7}$$

- Moment of inertia for a solid of revolution about the y-axis is given by

$$I_y = 2\pi k \int_a^b x^3 f(x)\,dx. \tag{23.7}$$

- Moment of inertia for a solid of revolution about the x-axis is given by

$$I_x = 2\pi k \int_c^d y^3 g(y)\,dy. \tag{23.7}$$

- Length of arc is given by

$$S = \int_a^b \sqrt{1 + [f'(x)]^2}\,dx. \tag{23.8}$$

- The pressure or force F exerted by a liquid between depths $h = a$ and $h = b$ is given by

$$F = W \int_{h=a}^{h=b} lh\,dh,$$

where W is the weight per unit volume of the liquid, and l is the length of the container or object. (23.8)

- The probability that an event will occur in some interval $a \leq x \leq b$, is given by

$$P = \int_a^b f(x)\,dx,$$

where $f(x)$ is the probability density function. (23.8)

- The improper integral $\int_a^\infty f(x)dx$ can be evaluated by

$$\int_a^\infty f(x)dx = \lim_{b \to \infty} \int_a^b f(x)dx,$$

provided the limit exists. (23.8)

- The improper integral $\int_{-\infty}^b f(x)dx$ can be evaluated by

$$\int_{-\infty}^b f(x)dx = \lim_{a \to -\infty} \int_a^b f(x)dx,$$

provided the limit exists. (23.8)

- The improper integral $\int_{-\infty}^\infty f(x)dx$ can be evaluated by

$$\int_{-\infty}^\infty f(x)dx = \int_{-\infty}^c f(x)dx + \int_c^\infty f(x)dx$$
$$= \lim_{a \to -\infty} \int_a^c f(x)\,dx + \lim_{b \to \infty} \int_c^b f(x)dx,$$

provided the limits exist. (23.8)

23·10 Chapter 23 Review Exercises

Find the area under each of the following curves.

1. $y = 3x^2$ from $x = 0$ to $x = 2$
2. $y = \dfrac{1}{x^3}$ from $x = 1$ to $x = 2$
3. $y = \cos x$ from $x = -\dfrac{\pi}{2}$ to $x = \dfrac{\pi}{2}$
4. $y = e^{2x}$ from $x = 0$ to $x = 1$
5. $y = x \ln x$ from $x = 1$ to $x = 2$
6. $y = -x^2 + x + 2$ from $x = 0$ to $x = 2$
7. $y = (x + 1)^2$ from $x = -1$ to $x = 1$
8. $y = \dfrac{1}{(x + 1)^2}$ from $x = 0$ to $x = 1$
9. $y = \dfrac{1}{x + 1}$ from $x = 0$ to $x = 2$
10. $y = \dfrac{1}{1 - x^2}$ from $x = -\dfrac{1}{2}$ to $x = \dfrac{1}{2}$
11. $y = \tan x$ from $x = \dfrac{\pi}{6}$ to $x = \dfrac{\pi}{3}$
12. $y = 1 - x^3$ from $x = 0$ to $x = 1$
13. $y = x^{1/3}$, bounded by the y-axis, from $y = 0$ to $y = 2$
14. $y = e^{-3x}$ from $x = 0$ to $x = 1$
15. $y = \cos \dfrac{\theta}{2}$ from $\theta = 0$ to $\theta = \pi$
16. $h = \dfrac{2t}{1 + t^2}$ from $t = 2$ to $t = 3$
17. $y = \sin^2 x \cos x$ from $x = \dfrac{\pi}{3}$ to $x = \dfrac{\pi}{2}$
18. $y = (x + 2)e^x$ from $x = 0$ to $x = 1$

Find the area of the region bounded by the following curves.

19. $y = x^2 + x$, the x-axis, from $x = 0$ to $x = 2$
20. $y^2 = x$, $x - y = 2$
21. $y = x\sqrt{x^2 + 1}$, the x-axis, from $x = 0$ to $x = 1$
22. $y = x - 4$, $y^2 = 2x$ (Hint: integrate with respect to y)
23. $y = x^3$, $y = \sqrt{x}$
24. $y = 3x + x^2$, $y = 4$

25. $y = \ln x$, $y = 0$, and $x = e$

26. $y = \tan x$, $y = 0$, $x = \dfrac{\pi}{6}$ to $x = \dfrac{\pi}{3}$

27. $y = \dfrac{1}{1+x^2}$, $x = 0$, $y = 0$, and $x = 2$

Find the volume of the solid of revolution generated by revolving the given region about the indicated axis.

28. $y = \sqrt{4x+1}$, $x = 0$, $x = 2$, $y = 0$, about the x-axis

29. $y = x^4$, $x = 1$, $y = 0$, about the y-axis

30. $y = x^3 + 1$, $x = 0$, $y = 0$, $x = 1$, about the y-axis

31. $y = x^2$, $x = 0$, $x = 2$
    (a) about the x-axis, $y = 0$ and (b) about the y-axis, $x = 0$

32. $y = \sin 2x$, $x = 0$ to $x = \dfrac{\pi}{2}$ and $y = 0$; about the x-axis

33. $y = \sqrt{x}$, $x = 1$, $x = 4$, $y = 0$; about the x-axis

34. Use the shell method to find the volume of the solid of revolution generated by revolving the region bounded by $y = 3x - x^2$ and the x-axis about the y-axis.

35. Use the shell method to find the volume of the solid of revolution generated by revolving the region bounded by $y = x^2$, the y-axis, $y = 1$, and $y = 4$ about the x-axis.

36. A cylindrical tank of radius 8 feet and height 30 feet is one-third full of water. How much work is required to empty the tank?

37. Find the work necessary to stretch a spring 6 inches from its original length if the force is given by $F = \dfrac{1}{2}x$.

38. A force of 100 pounds is required to compress a spring of natural length 10 inches to a length of 8 inches. Find the work done in compressing the spring from its natural length to a length of 6 inches.

39. A circular storage tank 8 feet in diameter and 20 feet high is half full of oil weighing 70 lb/ft³. Find the work done in emptying the tank through a pipe that extends 6 feet above the top of the tank.

40. Find the amount of work required to lift a 10 pound box out of a hold 16 feet deep using a rope that weighs 1 oz. per foot. (*Hint:* The force is 10 lbs. plus the weight of the rope at any depth, x.)

41. Find the centroid of the region bounded by $4y = x^2$, the x-axis, and the line $x = 4$.

42. Find the centroid of the region bounded by $y = e^x$, $x = 0$, the x-axis, and $x = 1$.

43. Find the centroid of the solid of revolution generated by revolving about the x-axis the region bounded by $y = \cos x$ and $x = 0$ from $x = -\dfrac{\pi}{2}$ to $x = \dfrac{\pi}{2}$.

44. Find the centroid of the solid of revolution generated by revolving about the x-axis the region bounded by $y = e^{-x}$ and $y = 0$ from $x = 0$ to $x = 1$.

45. Find I_x and I_y for the region $y = \sqrt[3]{x}$, the x-axis, and $x = 8$. Assume $k = 1$.

46. Find I_x and I_y for the solid of revolution generated by the curve $y = x^{2/3}$ from $x = 0$ to $x = 8$. Assume $k = 1$.

47. Find the length of arc of $y = \ln x$ from $x = 1$ to $x = 3$.

48. Find the length of arc of $y = \ln(\cos x)$ from $x = 0$ to $x = \dfrac{\pi}{3}$.

49. Find the force on one end of a rectangular tank 10 feet long and 3 feet high if the tank is half full of water.

50. A horizontal cylindrical tank has elliptical ends with the minor axis horizontal. The major axis is 6 feet and the minor axis is 4 feet. Find the force on one end if the tank is half full of water.

51. A gasoline dealer finds the amount x of unleaded gasoline sold weekly varies randomly between 2000 and 8000 gallons. Let x be measured in thousands of gallons so that $2 \le x \le 8$. The probability density function for x is given by

$$f(x) = \dfrac{20}{3} \cdot \dfrac{1}{(x+2)^2}.$$

If the owner orders 5000 gallons of unleaded gasoline for a certain week, what is the probability of running out of gasoline before the end of the week?

52. At a certain gas station, it takes an average of 4 minutes to get serviced. The probability density function for the service time t is for a car is given by

$$f(t) = \frac{1}{4}e^{-1/4t}, \quad t \geq 0.$$

What fraction of the cars are serviced within 2 minutes?

53. An experiment with outcomes between 0 and 1 has probability density function $f(x) = 4x^3$. What is the probability of an outcome larger than $\frac{1}{2}$?

Evaluate the following improper integrals, if the integral exists.

54. $\int_2^\infty \frac{1}{(x-1)^2} dx$

55. $\int_2^\infty \frac{1}{x-1} dx$

56. $\int_{-\infty}^\infty \frac{1}{1+x^2} dx$

57. $\int_0^\infty x^3 e^{-x^2} dx$

58. $\int_{-\infty}^{-1} \frac{dx}{x^2}$

59. $\int_0^\infty y = e^{-x}, \; x \geq 0$

60. $\int_{-\infty}^0 e^x \, dx$

61. $\int_1^\infty xe^{-x^2} dx$

62. $\int_0^\infty \frac{1}{\sqrt[3]{x+1}} dx$

63. Find the area under the curve $y = e^{-x}$ for $x \geq 0$.

64. A psychologist finds that the time t, in minutes, required by an average rat to go through a maze has the probability density function

$$f(t) = \frac{1}{(1+t)^2}, \quad \text{for } t \geq 0.$$

Find the probability that an average rat will take at least 3 minutes to go through the maze.

65. In exercise 52, what is the probability that a car will have to wait at least 4 minutes to be serviced?

23·11 Chapter 23 Test

1. Find the area bounded by $y = x^3 + 2$ and the x-axis from $x = 0$ to $x = 1$.

2. Find the area of the region between the curves $y = e^x$ and $y = \sqrt{x}$ from $x = 0$ to $x = 4$.

3. Find the volume of the solid of revolution formed by revolving the curve $y = \frac{1}{x^2}$ about the x-axis from $x = 1$ to $x = 3$.

4. Use the method of cylindrical shells to find the volume of the region formed by $y = \frac{1}{x^2 - 1}$ and the x-axis, from $x = 1$ to $x = 3$, rotated about the y-axis.

5. The force used in operating a hydraulic lifter is given by $F = 50\sqrt{x}$. Find the work done between $x = 25$ and $x = 36$.

6. Sketch the region by $y = \cos x$ and the x-axis from $x = -\frac{\pi}{2}$ to $x = \frac{\pi}{2}$, and find the centroid of the region.

7. Find I_x and I_y for the region bounded by $y = \dfrac{x^2}{2}$ and $y = 2$. Assume $k = 1$.

8. Evaluate $\displaystyle\int_{2}^{\infty} \dfrac{1}{x^2 - 1}\, dx$.

9. Annual output from a silver mine is given by
$$f(t) = 200 e^{-0.05t} \text{ tons/year.}$$
Find the area under the curve for the function from $t = 0$ to $t = 4$ and interpret your result.

10. Let x denote the time between contraction of a certain disease and death, measured in years. Suppose the probability density function for x is f, where
$$f(x) = \dfrac{1}{(x+1)\ln 11}, \quad \text{for } 0 \le x \le 10.$$
What is the probability that a person who contracts the disease will live for more than five years?

23·12 Cumulative Review—Chapters 21, 22, and 23

Find the antiderivative of each of the following derivative functions and check the result by differentiating.

1. $f'(x) = 6x^5$
2. $f'(x) = 3x^2 - 2$
3. $f'(x) = \dfrac{2}{3} x^{2/3}$
4. $f'(x) = 4\sqrt{x}$
5. $f'(x) = e^x + 2$
6. $f'(x) = \cos x$
7. $f'(x) = \dfrac{1}{x}$
8. $f'(x) = (x+2)^2$

Evaluate the following integrals.

9. $\displaystyle\int (x^{2/3} + x - 2)\, dx$
10. $\displaystyle\int \dfrac{1}{x^4}\, dx$
11. $\displaystyle\int 2x(8 - x^{3/2})\, dx$
12. $\displaystyle\int \dfrac{ds}{\sqrt[3]{s}}$
13. $\displaystyle\int (2 + 3t)^4\, dt$
14. $\displaystyle\int \dfrac{6x}{(1 + x^2)^3}\, dx$
15. $\displaystyle\int r^3 \sqrt{r^4 - 2}\, dr$
16. $\displaystyle\int \left(1 + \dfrac{1}{t}\right)^3 \left(\dfrac{1}{t^2}\right) dt$
17. $\displaystyle\int_{0}^{1} (x + 2)^4\, dx$
18. $\displaystyle\int_{1/2}^{1} x(3x^2 - 1)^3\, dx$
19. $\displaystyle\int_{-1}^{1/3} (2 - 3s)^{-3}\, ds$
20. $\displaystyle\int_{-2}^{-1} \left(t - \dfrac{1}{t^2}\right) dt$
21. $\displaystyle\int \dfrac{2x\, dx}{x^2 + 2}$
22. $\displaystyle\int \dfrac{\ln t}{t}\, dt$
23. $\displaystyle\int \dfrac{4x^3 - 3}{x^4 - 3x + 5}\, dx$
24. $\displaystyle\int \dfrac{e^x}{e^x + 1}\, dx$
25. $\displaystyle\int e^{3x}\, dx$
26. $\displaystyle\int_{-1}^{0} \dfrac{1}{2 - 5x}\, dx$
27. $\displaystyle\int_{1}^{2} 5^{-2t}\, dt$
28. $\displaystyle\int_{0}^{4} e^{-(3x)/4}\, dx$
29. $\displaystyle\int \sin 6x\, dx$
30. $\displaystyle\int \tan \dfrac{\theta}{3}\, d\theta$
31. $\displaystyle\int (1 + \cos x)^2 \sin x\, dx$
32. $\displaystyle\int \sec 3x \tan 3x\, dx$

33. $\displaystyle\int_0^1 t^2\cos t^3\, dt$ 34. $\displaystyle\int_0^{\pi/4} \cos t \sin t^2\, dt$ 35. $\displaystyle\int x \ln 2x\, dx$

36. $\displaystyle\int \frac{1}{r}\ln r\, dr$ 37. $\displaystyle\int \frac{x}{\sqrt{1-x}}\, dx$ 38. $\displaystyle\int e^x \cos x\, dx$

39. $\displaystyle\int \sqrt{4-x^2}\, dx$ 40. $\displaystyle\int \sin^4 t\, dt$ 41. $\displaystyle\int \frac{\cos^2 x}{\sin x}\, dx$

42. $\displaystyle\int \frac{1}{x\sqrt{1+x^2}}\, dx$ 43. $\displaystyle\int \frac{x+9}{(x+3)(x-2)}\, dx$ 44. $\displaystyle\int \frac{dx}{x^2+x-6}$

Find the area of the region bounded by the following curves and the x-axis.

45. $y = (2-x)^2$, from $x = 0$ to $x = 2$

46. $y = e^{2x}$, from $x = -1$ to $x = 1$

47. $y = 4x\sqrt{2x^2+1}$, from $x = 0$ to $x = 2$

Find the area of the regions between the curves. Sketch the graphs first.

48. $y = (2-x)^2$ and $y = x$

49. $y = \sin x$, $x = \dfrac{\pi}{2}$, and the x-axis

50. $y = \ln 2x$, $y = 0$, and $x = 2$

Find the volume of the solid of revolution formed by revolving the curve of the given equation about the x-axis.

51. $y = \sin x$, from $x = 0$ to $x = \dfrac{\pi}{2}$

52. $y = 2e^x$, from $x = -1$ to $x = 1$

53. $y = (2-x)^2$, from $x = 0$ to $x = 2$

Use the method of cylindrical shells to find the volume of indicated regions rotated about the y-axis.

54. $y = 3x - x^2$, the x-axis, from $x = 0$ to $x = 3$

55. $y = e^{2x}$, the x-axis, and $x = 1$

56. $y = \dfrac{1}{x}$, the x-axis, from $x = \dfrac{1}{2}$ to $x = 1$

57. The force driving a hydraulic plunger varies according to the law $F = \dfrac{2000}{x}$ lbs. Find the work done in moving the plunger from $x = 2$ ft to $x = 5$ ft.

58. A cylindrical tank of radius 6 feet and height 10 feet is one-half full of water. Find the work necessary to pump the water out over the top of the tank.

59. Find the centroid of the region bounded by $y = \sqrt{4-x}$, the x-axis, from $x = 0$ to $x = 4$.

60. Find the centroid of the solid of revolution generated by revolving about the x-axis the region bounded by the graph of $y = e^{-x}$, and $y = 0$ from $x = 0$ to $x = 1$.

61. Find I_x and I_y for the region bounded by $y = x^{3/2}$, the x-axis and $x = 8$. Assume $k = 1$.

62. Find the length of arc for the curve $y = \dfrac{1}{2}x^2$ from $x = 0$ to $x = 2$.

63. Find the force on one end of a rectangular tank 6 feet wide and 3 feet high if the tank is half full of water.

64. Evaluate the improper integral $\int_3^\infty \frac{1}{(x-2)^2}\,dx$.

65. The concentration of smog produced by automobiles in a large city changes with the temperature at a rate given by

$$f(t) = \frac{0.006}{(0.02t + 100)^{1/2}},$$

where t is temperature in degrees Celsius and the smog concentration is in parts per million (ppm). Find a function that describes the smog concentration if 6 ppm are produced at 0° C.

66. Maintenance costs for a building t years old increase at a rate of $y = 0.5t^{3/2}$ thousand dollars per year. Find the total maintenance cost for the building over its first 4 years of life.

67. The savings produced by new equipment accrue at a rate of $y = 10,000 - 400t$ dollars per year, where t is the number of years since the equipment was purchased. Find the area of the region bounded by the graph of this function and the coordinate axes, and interpret your result.

68. A firm's new milling machine is saving y thousand dollars per month in raw materials cost, where $y = 150 - x^2$. Also, the cost of labor needed to run the new machine is increasing at the rate of $c = 0.5x^2$ thousand dollars per month. Here x is months since the machine was installed. Find the net savings by finding the area bounded by the graphs of y and c.

69. Find the equation of the curve that passes through (1,2) if the slope of the tangent line is $3x^2 - 4x$.

70. A toy rocket is launched from the top of a mountain 200 km high. Its velocity v is described by

$$v(t) = 98 - 9.8t \text{ km/sec}.$$

a. Find $s(t)$, the distance of the rocket above ground after t seconds.
b. How high above the ground does the rocket go?

71. Let T denote the time, in weeks, between successive inventory reorders for a certain company. Suppose the probability density function for T is given by

$$f(T) = \frac{1}{2}(T + 1)^{-3/2}, \qquad t \geq 0.$$

Find the probability that the time between successive reorders will be at least one week.

72. Investment income is flowing in continuously to a blind trust at a rate of $25,000e^{-0.02t}$ dollars per year t. Find the value of this stream of income if it lasts forever.

Answer Key

Chapter 1 Answers

Trial Problems 1.1
1. <   2. >   3. =   4. −8   5. 4

Exercises 1.1
1. $3 < 5$ 3. $-5 < -2$ 5. $-2 < 9$ 7. $\frac{3}{4} < \frac{7}{4}$ 9. $35 > 32$ 11. $52 > -60$ 13. $-\frac{2}{3} < 4$ 15. $\frac{3}{4} > -1$
17. $\frac{9}{10} < 1$ 19. $\frac{3}{4} > \frac{5}{8}$ 21. $\frac{2}{3} > 0.66$ 23. $0.44 < 0.444$ 25. $\frac{1}{3} = 0.333\ldots$ 27. $2.0 = 1.999\ldots$
29. $4.0 = 3.999\ldots$ 31. $-\frac{2}{3} > -\frac{3}{4}$ 33. $-2.3 > -3.2$ 35. $-0.5 > -0.6$ 37. $0.66 < 0.\overline{6}$ 39. $0.\overline{2} = \frac{2}{9}$
41. $-0.67 < -\frac{2}{3}$ 43. $\sqrt{2} < \sqrt{3}$ 45. $\sqrt{7} > 2.64$ 47. $-2.33 > -2.333$ 49. $-\sqrt{5} > -\sqrt{6}$ 51. $+10$ 53. -8
55. $+14$ 57. -3 59. -56 61. $+24$ 63. $+48$ 65. $+144$ 67. -5 69. $+5$
71. $-12, -|-4|, -2, \frac{5}{8}, \sqrt{6}, |-7|, 8$ 73. $-\frac{18}{7}$ 75. 7   77. 4   79. -6 81. -2 83. 4   85. -4 87. -8
89. -8 91. $-14{,}388$ ft   93. $181°$ 95. 7 per 1000   97. $135°$ C   99. $4\sqrt{5} < 8.95$

Trial Problems 1.2
1. Commutative   2. Commutative   3. Associative   4. Distributive   5. Inverse for addition

Exercises 1.2
1. 5   3. 8   5. -8 7. 1   9. a 11. $3x, 3y$ 13. 4   15. -3 17. 3   19. 2   21. $12 + (-3)$ 23. $-3 + (-5)$
25. $-4 + (6)$ 27. $\frac{2}{3}\left(\frac{1}{4}\right)$ 29. $10(2)$ 31. -5 33. 5   35. 4   37. -2 39. -9 41. 6   43. 4   45. -9
47. 2   49. 37   51. 2340 volts   53. 265

Trial Problems 1.3
1. -8 2. $\frac{1}{81}$ 3. 64   4. $\frac{1}{18}$ 5. $x^5 y$

Exercises 1.3

1. 8 **3.** 9 **5.** $\frac{1}{9}$ **7.** $\frac{1}{25}$ **9.** 8 **11.** 1 **13.** $-\frac{27}{8}$ **15.** $\frac{1}{72}$ **17.** a^2b^5 **19.** $\frac{x^2}{y}$ **21.** $\frac{y^7}{x}$ **23.** 72 **25.** 2^4 **27.** 2^{-12}
29. 2^{-6} **31.** 2^{45} **33.** 2^0 **35.** 2^0 **37.** 5.32×10^1 **39.** 4.0×10^{-1} **41.** 2.48×10^0 **43.** 8.016×10^{-1}
45. 6.7×10^{-4} **47.** 2.9×10^1 **49.** 5.28×10^3 **51.** 3.0×10^5 **53.** 4.52×10^{11} **55.** 4.914×10^5 **57.** 1.08×10^{-4}
59. 2.43×10^{27} **61.** 1.5625×10^{-26} **63.** 6.336×10^4 in. **65.** 6.4×10^{-18} grams **67.** 7.88×10^4
69. 3.82×10^2 **71.** 1.38×10^{-58} **73.** 12,367.15 **75.** 1.892×10^{22}

Trial Problems 1.4

1. 7 **2.** -4 **3.** 0.3 **4.** $4a\sqrt[3]{a}$ **5.** $\frac{\sqrt{10}}{2}$

Exercises 1.4

1. 6 **3.** -12 **5.** 0.4 **7.** 2 **9.** -2 **11.** -5 **13.** 0.3 **15.** 3 **17.** -2 **19.** 2 **21.** 2 **23.** $5a^2$ **25.** $3xy^2$
27. $x^2y^3\sqrt{z}$ **29.** $2ab\sqrt{3bx}$ **31.** $\frac{4a}{5b}$ **33.** $\frac{6x}{5}$ **35.** $\frac{x}{a^4}\sqrt{\frac{2y}{5}}$ **37.** 22.36 **39.** 88.32 **41.** 0.2 **43.** 21.43 **45.** $2\sqrt[3]{2}$
47. $3a\sqrt[3]{a}$ **49.** $3x^2\sqrt[4]{x}$ **51.** $2\sqrt[5]{2x}$ **53.** $\frac{3b\sqrt[3]{2bc^2}}{16a}$ **55.** $\frac{\sqrt{6}}{3}$ **57.** $\frac{\sqrt[3]{128}}{2} = \frac{4\sqrt[3]{2}}{2} = 2\sqrt[3]{2}$ **59.** $\frac{5\sqrt{6}}{6}$ **61.** $\frac{\sqrt{15b}}{5b}$
63. $\sqrt{3} - \sqrt{2}$ **65.** $2\sqrt{3} + 2\sqrt{2} + \sqrt{10} + \sqrt{15}$ **67.** $\frac{a - 2\sqrt{ab}}{a - 4b}$ **69.** $\frac{-3x - 2\sqrt{xy}}{9x - 4y}$ **71.** $\frac{8}{7}$ **73.** 9 ft
75. $6\sqrt{15}$ in. **77.** $\frac{(\sqrt{d_1} - \sqrt{d_2})^2}{d_1 - d_2}, \frac{13 - 2\sqrt{3}}{7}$ **79.** $\frac{\sqrt{0.9375}}{0.75}$

Trial Problems 1.5

1. $\sqrt[5]{x^2}$ **2.** $6^{1/3}$ **3.** 5 **4.** $\frac{1}{2}$ **5.** -4

Exercises 1.5

1. $\sqrt{3}$ **3.** $\sqrt[4]{x}$ **5.** $\sqrt[5]{b^3}$ **7.** $\sqrt[3]{r^5} = r\sqrt[3]{r^2}$ **9.** $3^{1/2}$ **11.** $5^{1/3}$ **13.** $7^{1/2}$ **15.** $a^{2/5}b^{1/5}$ **17.** 3 **19.** 4 **21.** 3 **23.** -2
25. 16 **27.** 4 **29.** $\frac{1}{2}$ **31.** $\frac{1}{3375}$ **33.** $x = A^{1/2}$ **35.** $E = 100(1 - \sqrt[5]{R^{-2}})$

Trial Problems 1.6

1. 3 **2.** $5x^2 + 6x - 5$ **3.** $-x^3 - 3x^2 + 4x - 2$ **4.** $x^2 - 7x - 3$ **5.** $x^4 - 2x^3 - x^2 - x - 5$

Exercises 1.6

1. 3 **3.** 3 **5.** 3 **7.** 2 **9.** 5 **11.** 6 **13.** $5x^2 + 6x + 5$ **15.** $6x^3y - 10x^2y^2 - 2xy^3$ **17.** $8x^6$
19. $-x^3 - 3x^2 + 5x - 6$ **21.** $-3x^6 - 5$ **23.** $2x^2 + x + 3$ **25.** $3x^2y + 11xy^2$ **27.** $x^4 - 3x^3 - 4x^2 + 3x - 11$
29. $2x^2 - 2x + 14$ **31.** $0.6x^2 - 0.1y$ **33.** $4x^3 - 4x^2 + 4x + 8$

Answer Key—Chapter 1

Trial Problems 1.7

1. x^7 **2.** $-6x^5y^3$ **3.** $15x^2 + 10x$ **4.** $x^2 - 9$ **5.** $x^3 + 3x^2 + x - 2$

Exercises 1.7

1. x^9 **3.** $-12x^3y^3$ **5.** x^5 **7.** $2x^5y^3$ **9.** x^3y^5 **11.** $3x^2 + 6x + 9$ **13.** $10x^2 + 15x$ **15.** $4x^5 + 4x^4 + 8x^3$
17. $-2x^4 + 8x^3 - 10x^2 + 6x$ **19.** $x^2 - 4$ **21.** $4x^2 + 20xy + 25y^2$ **23.** $x^3 + 1$ **25.** $3x^3 - x^2y + 3xy^2 - y^3$
27. $6x^4 - 5x^2y^2 - 4y^4$ **29.** $\dfrac{n}{R_1} - \dfrac{n}{R_2} - \dfrac{1}{R_1} + \dfrac{1}{R_2}$ **31.** $-\dfrac{v_0^2 - v^2}{t^2}$ **33.** $(5)(7) = 27 + 8$

Trial Problems 1.8

1. $3x$ **2.** $-ax$ **3.** $\dfrac{y^3}{3z^4}$ **4.** $x + 5$ **5.** $t^3 - t^2 - 4t + 4$

Exercises 1.8

1. $4m^3$ **3.** $7s$ **5.** $-ax$ **7.** $-x^2y^2$ **9.** $\dfrac{2t^4}{c}$ **11.** $\dfrac{7t^2u}{6}$ **13.** $a - x^2 - ax$ **15.** $-x^2y^3 - 2xy^4$ **17.** $c^2 - ab^2c^5 + 1$
19. $-a^2b^2 + ab - 1 + ab^3$ **21.** $x - 3$ **23.** $y - 5$ **25.** $2s + 3$ **27.** $2t^2 - t + 3$ **29.** $x^3 + 2x^2 + x + 2$
31. $3x^3 + 7x^2 + 5x + 1 - \dfrac{3}{2x - 2}$ **33.** $\dfrac{1}{R_3} + \dfrac{1}{R_1} + \dfrac{1}{R_2}$ **35.** $ar^3 + ar^2 + ar + a$

Trial Problems 1.9

1. 6 **2.** 16 **3.** 6 **4.** 18 **5.** $\dfrac{17}{12}$

Exercises 1.9

1. $x = 6$ **3.** $y = 14$ **5.** $s = -2$ **7.** $x = \dfrac{9}{22}$ **9.** $x = -21$ **11.** $x = 5$
13. $x = 1 - \dfrac{2}{5}y$ **15.** $x = 18$ **17.** $x = 4$ **19.** $x = -\dfrac{5}{13}$ **21.** $x = \dfrac{d-b}{a-c}$ **23.** $t = 2$ **25.** $t = \dfrac{d}{55}$ **27.** $x + y = 30$
29. $V = 355$ **31.** $c = \dfrac{E}{T}$

Trial Problems 1.10

1. $\dfrac{20}{15}$ or $\dfrac{4}{3}$ **2.** $\dfrac{6}{3}$ or $\dfrac{2}{1}$ **3.** $\dfrac{55 \text{ mi}}{1 \text{ hr}}$ or 55 mph **4.** $\dfrac{\$200}{1 \text{ yr}}$ or $200/yr **5.** $x = 9$

Exercises 1.10

1. $\dfrac{3}{4}$ **3.** $\dfrac{5}{3}$ **5.** $\dfrac{9}{2}$ **7.** $\dfrac{20}{3}$ **9.** $\dfrac{49}{36}$ **11.** $\dfrac{1}{3}$ **13.** $\dfrac{15 \text{ mi}}{2 \text{ hr}}$ **15.** 1 lb/3 gal **17.** $125/1 yr **19.** 16 tiles/9 sq ft
21. 1 oz/50 lbs **23.** $x = 6$ **25.** $x = 3$ **27.** 416 tiles **29.** 100,000 people **31.** $e = 20'$, $d = 28'$ **33.** 38"
35. 72 threads

Trial Problems 1.11

1. $r = ks$ **2.** $u = kv$ **3.** $t = \dfrac{k}{r}$ **4.** $p = \dfrac{k}{q^2}$ **5.** $y = \dfrac{kx^2}{\sqrt{z}}$

Exercises 1.11

1. $s = kt$ **3.** $y = kz$ **5.** $p = \dfrac{k}{q}$ **7.** $s = \dfrac{k}{t}$ **9.** $r = kd^2$ **11.** $p = \dfrac{k}{q^2}$ **13.** $y = \dfrac{ks}{t}$ **15.** $v = \dfrac{ks^3}{t^2}$ **17.** $k = 4$
19. $k = 2$ **21.** $k = 4$ **23.** $s = 10$ **25.** $s = \dfrac{1}{2}$ **27.** $r = 6$ **29.** 200 lbs/sq ft **31.** $L = 1620$ lbs **33.** 76,800 joules
35. $f' = \dfrac{f}{1.28}$

Review Exercises 1.13

1. [1.1] $3 < 5$ **3.** [1.1] $4 = \dfrac{8}{2}$ **5.** [1.1] $\dfrac{2}{3} < \dfrac{5}{6}$ **7.** [1.2] -4 **9.** [1.2] 5 **11.** [1.2] -3 **13.** [1.2] 5 **15.** [1.2] 6
17. [1.2] 6 **19.** [1.2] -5 **21.** [1.2] 6 **23.** [1.3] 9 **25.** [1.3] $-\dfrac{2}{3}$ **27.** [1.3] 72 **29.** [1.3] 2^{-12} **31.** [1.3] 2^{45}
33. [1.3] 2^0 **35.** [1.3] 5.432×10^2 **37.** [1.3] 6.78×10^{-3} **39.** [1.3] 2.1152×10^{-28} **41.** [1.3] 1.2482×10^{-58}
43. [1.4] 13 **45.** [1.4] -3 **47.** [1.4] 3 **49.** [1.4] $4xy^2$ **51.** [1.4] $3a\sqrt[3]{a}$ **53.** [1.4] $2y\sqrt[4]{y}$ **55.** [1.4] 24.47
57. [1.4] 21.43 **59.** [1.4] 3 **61.** [1.4] $\sqrt{3} - \sqrt{2}$ **63.** [1.5] $\sqrt{3}$ **65.** [1.5] $\sqrt[5]{a^2}$ **67.** [1.5] $\sqrt[3]{(-8)^2}$ **69.** [1.5] $\dfrac{\sqrt{25^{-3}}}{\sqrt[4]{81^3}}$
71. [1.5] $3^{1/2}$ **73.** [1.5] $5^{1/3}$ **75.** [1.5] $a^{1/5}b^{1/5}$ **77.** [1.5] 3 **79.** [1.6] $5x^2 + 2x + 7$ **81.** [1.6] $3x^2y + 11xy^2$
83. [1.7] $6x^2 - x - 15$ **85.** [1.8] $x + 2$ **87.** [1.9] $x = -30 - \dfrac{3}{2}y$ **89.** [1.10] $x = 3$ **91.** [1.10] $\dfrac{\$40}{\text{yr}}$
93. [1.10] $\dfrac{3}{2}$ **95.** [1.11] $s = \dfrac{kt}{u^2}$ **97.** [1.11] $u = 64$ **99.** [1.11] $R = \dfrac{kl}{d^2}$, $l = \dfrac{Rd^2}{k}$ **101.** [1.11] $AR = \dfrac{kS^2}{W}$
103. [1.11] 12° F **105.** [1.11] $d = 50$ meters **107.** [1.11] $P = \dfrac{P_0 V_0}{V}$

Answer Key—Chapter 2

Test 1.14

1. [1.1] False **2.** [1.1] False **3.** [1.1] True **4.** [1.2] -2 **5.** [1.2] -4 **6.** [1.2] $-\dfrac{1}{5}$ **7.** [1.3] $\dfrac{x^2}{y}$ **8.** [1.3] 2
9. [1.3] 2.85×10^3 **10.** [1.3] 8.16×10^{-1} **11.** [1.3] 78,815.639 **12.** [1.3] 0.274538 **13.** [1.4] $6 - 3\sqrt{7}$
14. [1.4] $\dfrac{5\sqrt{3}}{12}$ **15.** [1.4] $\dfrac{4\sqrt{5} - 4\sqrt{2}}{3}$ **16.** [1.5] $\sqrt[3]{r^5} = r\sqrt[3]{r^2}$ **17.** [1.5] $a^{1/5}b^{2/5}$ **18.** [1.6] $2x^2 + x - 7$
19. [1.7] $x^3 + 1$ **20.** [1.8] $3s + 2$ **21.** [1.9] $x = -21$ **22.** [1.9] $x = 1$ **23.** [1.10] $\dfrac{3}{4}$ **24.** [1.10] $x = 6$
25. [1.11] $y = 20$ **26.** [1.11] $x = \dfrac{kyz}{d^2}$

Chapter 2 Answers

Trial Problems 2.1

1. $ar + as + br + bs$ **2.** $x^2 + 2xy + y^2$ **3.** $4x^2 - 12xy + 9y^2$ **4.** $a^2 - 4x^2$ **5.** 884

Exercises 2.1

1. $xr + xs + yr + ys$ **3.** $xu - xr + yu - yr$ **5.** $2x^2 - 3xy + y^2$ **7.** $ax + ay - bx - by$ **9.** $a^2 + 4ab + 4b^2$
11. $a^2 + 2ab + b^2$ **13.** $4x^2 + 4xy + y^2$ **15.** $4a^2 - 12ab + 9b^2$ **17.** $x^2 - y^2$ **19.** $a^2 - 4b^2$ **21.** $x^4 - 1$
23. $25 - 9t^2$ **25.** $4a^2 - 25$ **27.** 896 **29.** $m_1V_a^2 - m_1V_b^2$ **31.** $3R^2 - 4xR$ **33.** $z^2 - R^2$ **35.** $f_0^2 - 2f_0f_t + f_t^2$
37. $xy - 3x - 4y + 12$

Trial Problems 2.2

1. $(a + b + c)y$ **2.** $4x(x^2 + 2x + 3)$ **3.** $(x - 8)(x - 2)$ **4.** $(4x + 3y)(4x - 3y)$ **5.** $(x - 2)(x^2 + 2x + 4)$

Exercises 2.2

1. $(a + b + c)x$ **3.** $x(3x^2 + 6x + 1)$ **5.** $ab(1 + ab + a^2b^2)$ **7.** $4a(1 + 2a + 4a^2)$ **9.** $4(x^2yz + 4x - 3z)$
11. $7xy(x + 2y + 3)$ **13.** $a^2b^3c^2(a^2 + bc + c^2)$ **15.** $(a + 1)^2$ **17.** $(x - 2)(x - 7)$ **19.** $(x - 7)(x + 2)$
21. $5(x - 4)(x + 4)$ **23.** $(2x - 3y)(2x + 3y)$ **25.** $(x + 8)^2$ **27.** $(2x + 3)(x - 4)$ **29.** $(5x + 6)(2x - 3)$
31. $(3x - a)^2$ **33.** $z(3z - 1)(z - 1)$ **35.** $(a + b)(b + 1)$ **37.** $(x + y)(2x - 1)$ **39.** $(x + 2)(x^2 - 2x + 4)$
41. $(2x - y)(4x^2 + 2xy + y^2)$ **43.** $2(3a + 4b)(9a^2 - 12ab + 16b^2)$ **45.** $3(5x - 3y)(25x^2 + 15xy + 9y^2)$
47. $2x(18 - x)$ **49.** $K(t_2^2 + t_1^2)(t_2 + t_1)(t_2 - t_1)$ **51.** $(Z + R)(Z - R)$ **53.** $(x + 2y)(x - 2y)$ **55.** $r^2(4 - \pi)$

Trial Problems 2.3

1. $\dfrac{-2}{x - 2}$ **2.** $\dfrac{b}{a - b}$ **3.** $\dfrac{a^3}{3b^2}$ **4.** $x + 2$ **5.** $\dfrac{2b - 3}{b - 1}$

Exercises 2.3

1. $-\dfrac{2}{3}$ 3. $-\dfrac{3}{4}$ 5. $\dfrac{5}{7}$ 7. $-\dfrac{4}{7}$ 9. $\dfrac{2}{5}$ 11. $-\dfrac{3}{5}$ 13. $-\dfrac{3x}{y}, y \neq 0$ 15. $-\dfrac{2y}{x-3}, x \neq 3$ 17. $-\dfrac{2x}{3y^2}, y \neq 0$
19. $\dfrac{x^2}{y^2}, y \neq 0$ 21. 1 23. 4 25. $2 - x$ 27. x 29. x 31. $\dfrac{x^3}{3y^2}$ 33. $(x + 2)$ 35. $\dfrac{1}{(a+b)}$ 37. $\dfrac{1}{(x+4)}$
39. $\dfrac{-1}{b+a}$ 41. $a^2 - 3a + 2$ or $(a-2)(a-1)$ 43. $\dfrac{(2x+3y)}{(x+2y)}$ 45. $\dfrac{2a+3b}{a+3b}$ 47. $\dfrac{40+t}{50-t}$
49. $P = \dfrac{2h(w + am - b)}{s + k}$

Trial Problems 2.4

1. $\dfrac{6}{35}$ 2. $\dfrac{3}{2b}$ 3. $\dfrac{2}{9}$ 4. $\dfrac{10}{3}$ 5. $\dfrac{5}{12}$

Exercises 2.4

1. $\dfrac{8}{15}$ 3. $\dfrac{3}{2xy}$ 5. $\dfrac{2}{9}$ 7. $\dfrac{(x+4)}{x(x+1)}$ 9. $-\dfrac{y}{x}$ 11. $\dfrac{x+3}{x+1}$ 13. $\dfrac{x+3}{12-x}$ 15. $\dfrac{5}{6}$ 17. $\dfrac{a^2xy}{b^2}$ 19. $\dfrac{2ab^2x^2y^2}{9}$ 21. $\dfrac{5}{12}$
23. $\dfrac{(2x+1)(2x+2)}{6(x+1)} = \dfrac{2x+1}{3}$ 25. $\dfrac{a-1}{a-3}$ 27. $\dfrac{r}{3}$ 29. $\dfrac{6}{\pi}$

Trial Problems 2.5

1. 60 2. $18xy^2$ 3. $\dfrac{23}{54}$ 4. $\dfrac{1}{24}$ 5. $\dfrac{bx + ay}{ab}$

Exercises 2.5

1. 120 3. $12ab^2$ 5. $6ab^2$ 7. $2a - 2$ 9. $(x-1)^2(x-2)$ 11. $\dfrac{19}{54}$ 13. $\dfrac{11}{24}$ 15. $\dfrac{ay + bx}{xy}$ 17. $\dfrac{19}{2x}$ 19. $\dfrac{1}{2x}$
21. $\dfrac{7}{24}$ 23. $\dfrac{x+90}{45x}$ 25. $\dfrac{9x+2}{(x-2)(x+3)}$ 27. $\dfrac{2(x+23)}{(x-7)(x+3)}$ 29. $\dfrac{2a-5}{a-7}$ 31. $\dfrac{5y+12}{(y+2)(y+3)}$
33. $\dfrac{2}{(b-1)(b+1)^2}$ 35. $\dfrac{5a^2+19a}{(a+5)(a-2)(a+3)}$ 37. 0 39. $\dfrac{-1}{(r-4)(r-3)(r-2)}$ 41. $\dfrac{m(y+x)(y-x)}{(x^2+y^2)^2}$
43. $\dfrac{C_1 C_2 C_3}{C_2 C_3 + C_1 C_3 + C_1 C_2}$

Trial Problems 2.6

1. 10 2. 18 3. 15 4. 5 5. 2

Exercises 2.6

1. $x = 10$ 3. $x = \dfrac{24}{7}$ 5. $y = \dfrac{31}{14}$ 7. $r = \dfrac{19}{2}$ 9. $s = -2$ 11. $x = 1$ 13. $x = 1$ 15. $s = 3$ 17. $q = \dfrac{pf}{p-f}$
19. $N = \dfrac{2D_P}{D_0 - D_P}$ 21. $T_1 = \dfrac{Q_1 T_2}{Q_1 + W}$ 23. $D = \dfrac{h + 2d}{2}$ 25. $P = \dfrac{M - \dfrac{WL}{2}}{L}$ 27. $w = \dfrac{ps + p}{2h}$

Answer Key—Chapter 2

Review Exercises 2.8

1. [2.1] $ax + ay + bx + by$ **3.** [2.1] $ax - ay + bx - by$ **5.** [2.1] $2x^2 - 3xy + y^2$ **7.** [2.1] $4a^2 + 4ab + b^2$
9. [2.1] $x^2 + 2xy + y^2$ **11.** [2.1] $4a^2 + 12ab + 9b^2$ **13.** [2.1] $a^2 - 4ab + 4b^2$ **15.** [2.1] $a^2 - 25b^2$
17. [2.2] $a(x + y + z)$ **19.** [2.2] $(x + 1)^2$ **21.** [2.2] $(x - 7)(x + 2)$ **23.** [2.2] $(y + 4)(y - 4)$
25. [2.2] $(3r + 7s)(3r - 7s)$ **27.** [2.2] $(2x + 3)(x - 4)$ **29.** [2.2] $(y + 1)(x + y)$ **31.** [2.2] $3(x + 2)(x^2 - 2x + 4)$
33. [2.3] $\dfrac{x^3}{y^2}$ **35.** [2.3] $3(t + 2)$ **37.** [2.3] $(x - 2)(x - 1)$ **39.** [2.3] $\dfrac{2b - 3}{b - 1}$ **41.** [2.3] $\dfrac{3}{2ab}$ **43.** [2.4] $\dfrac{s(r + 4)}{r(r + 1)}$
45. [2.4] $\dfrac{(a + 3)}{(a + 1)}$ **47.** [2.4] $\dfrac{2ab^2xy^2}{9}$ **49.** [2.4] $\dfrac{5}{12}$ **51.** [2.5] $\dfrac{bx + ay}{ab}$ **53.** [2.5] $\dfrac{7}{6a}$ **55.** [2.5] $\dfrac{7(r - 1)}{(r - 3)(r + 4)}$
57. [2.5] $\dfrac{-t - 33}{(t - 3)(t + 6)}$ **59.** [2.5] $\dfrac{-2x + 5}{7 - x} = \dfrac{2x - 5}{x - 7}$ **61.** [2.5] $\dfrac{-2(3a + 2)}{(a + 4)(a - 4)(a - 1)}$ **63.** [2.6] $x = 10$
65. [2.6] $x = \dfrac{19}{2}$ **67.** [2.6] $x = 12$ **69.** [2.6] $d = \dfrac{(P - p)(2\pi^2 r^2)}{m^3}$ **71.** [2.6] $d = D - 2T$ **73.** [2.6] 2

Test 2.9

1. [2.1] $ax + ay + bx + by$ **2.** [2.1] $ur + us - vr - vs$ **3.** [2.1] $3x^2 - 4xy + y^2$ **4.** [2.1] $a^2 + 4ab + 4b^2$
5. [2.1] $c^2 + 2cd + d^2$ **6.** [2.1] $9a^2 - 4b^2$ **7.** [2.1] $(50)^2 - (1)^2 = 2499$ **8.** [2.2] $2a(x + 2y + 3z)$
9. [2.2] $(x - 9)(x + 2)$ **10.** [2.2] $5(x - 4)(x + 4)$ **11.** [2.2] $(x + 8)^2$ **12.** [2.2] $x(a + 1)^2$
13. [2.2] $2(2a - 5b)(4a^2 + 10ab + 25b^2)$ **14.** [2.3] $x + 2$ **15.** [2.3] $\dfrac{1}{x + 9}$ **16.** [2.3] $a^2 - 3a + 2$ or $(a - 2)(a - 1)$
17. [2.3] $\dfrac{2y - 3}{y - 1}$ **18.** [2.3] $\dfrac{r - 3s}{3r + s}$ **19.** [2.4] $\dfrac{3}{2x}$ **20.** [2.4] $\dfrac{3}{4}$ **21.** [2.4] $\dfrac{x + 3}{x + 1}$ **22.** [2.4] $\dfrac{a - 1}{a - 2}$ **23.** [2.5] $\dfrac{19}{2x}$
24. [2.5] $\dfrac{7}{24}$ **25.** [2.5] $\dfrac{-2(3y + 2)}{(y - 1)(y - 4)(y + 4)}$ **26.** [2.6] $x = 10$ **27.** [2.6] $s = -2$ **28.** [2.6] $t = 2$

Cumulative Review—Chapters 1 and 2—2.10

1. [1] False **2.** [1] False **3.** [1] True **4.** [1] True **5.** [1] False **6.** [1] 2 **7.** [1] -2 **8.** [1] 2 **9.** [1] $2^0 = 1$
10. [1] 7.6×10^{-4} **11.** [1] 2.594×10^3 **12.** [1] ≈ 382.32118 **13.** [1] $\approx 1.7739208 \times 10^{-58}$ **14.** [1] $\dfrac{y^2 \sqrt{6axy}}{4a^2}$
15. [1] $\dfrac{a - 3\sqrt{ab}}{a - 9b}$ **16.** [1] $\sqrt[5]{a^2}$ **17.** [1] $\sqrt[3]{s^7} = s^2 \sqrt[3]{s}$ **18.** [1] $6^{1/3}$ **19.** [1] $9^{1/2} = 3$ **20.** [1] -243
21. [1] $\dfrac{1}{3375}$ **22.** [1] $5x^2 + 6x + 5$ **23.** [1] $7x^2 - 4x + 10$ **24.** [1] $x^3 + 1$ **25.** [1] $8x^4 - 2x^2y^2 - 3y^4$
26. [1] $\dfrac{4c^2 p^3}{3y^5}$ **27.** [1] $2t^2 - t + 3$ **28.** [1] $x = \dfrac{9}{22}$ **29.** [1] $x = 4$ **30.** [1] $x = 12$ **31.** [1] $x = \pm 8$
32. [1] $s = 10$ **33.** [1] $v = \dfrac{9}{4}$ **34.** [2] $2x^2 - 3xy + y^2$ **35.** [2] $4a^2 - 12ab + 9b^2$ **36.** [2] $9 - y^4$
37. [2] $(2x + 3)(x - 4)$ **38.** [2] $(3x - a)^2$ **39.** [2] $2(2a - y)(4a^2 + 2ay + y^2)$ **40.** [2] $\dfrac{2x + 3y}{x + 2y}$ **41.** [2] $\dfrac{r - 3s}{3r + s}$
42. [2] $\dfrac{x + 5}{x + 4}$ **43.** [2] $\dfrac{4(2x - 1)}{3(2x + 1)}$ **44.** [2] $\dfrac{7}{24}$ **45.** [2] $\dfrac{-2(3y + 2)}{(y + 4)(y - 4)(y - 1)}$ **46.** [2] $x = 1$ **47.** [2] $t = 2$

Chapter 3 Answers

Trial Problems 3.1

1.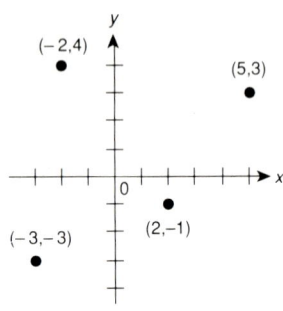

2.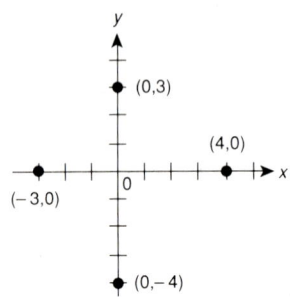

3. $A(5,1), B(0,2), C(-5,-2), D(-1,-2)$ 4. $E(3,0), F(0,0), G(-6,0), H(0,-5)$ 5. $(0,-5), (1,-2), (2,1), (3,4)$

Exercises 3.1

1.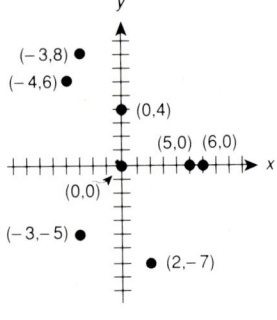

3. $A(2,3)$
$B(0,5)$
$C(-4,2)$
$D(-7,-5)$
$E(5,-4)$

5.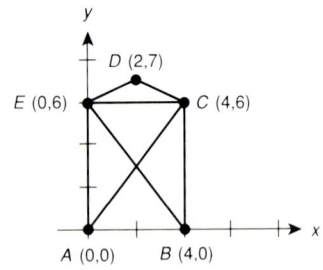

7.
x	C
10	54
20	58
30	62
40	66
50	70

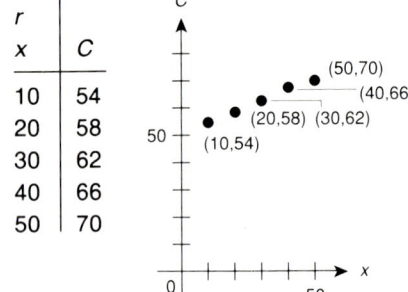

9. | N | D |
    |---|---|
    | 13 | 0.050 |
    | 18 | 0.036 |
    | 20 | 0.032 |
    | 24 | 0.027 |

11. | F | P |
    |---|---|
    | 50 | 10 |
    | 40 | 8 |
    | 30 | 6 |
    | 20 | 4 |
    | 10 | 2 |

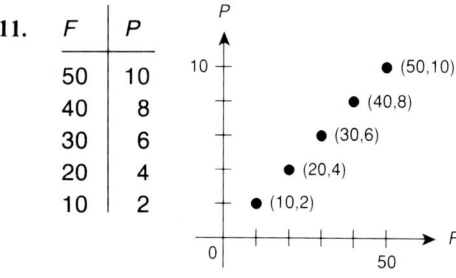

13. | W | P |
    |---|---|
    | 50 | 10 |
    | 45 | 9 |
    | 40 | 8 |
    | 35 | 7 |
    | 30 | 6 |

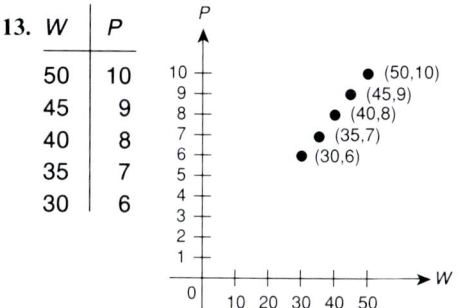

Trial Problems 3.2

1.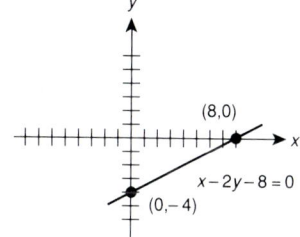

2.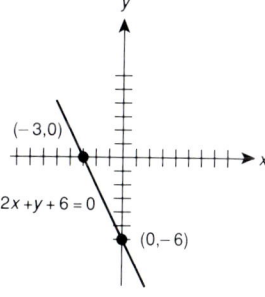

3.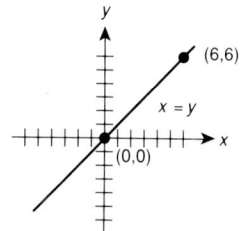

4.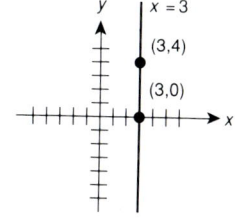

5.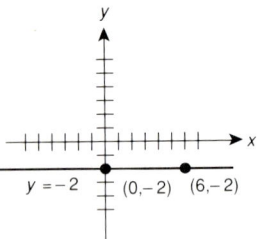

Exercises 3.2

1.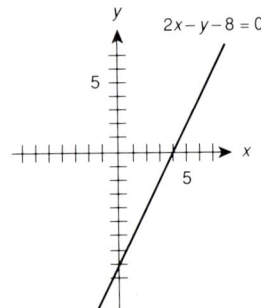

3.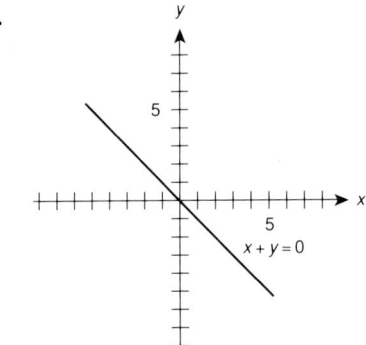

5.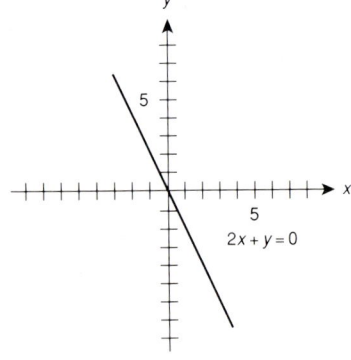

7.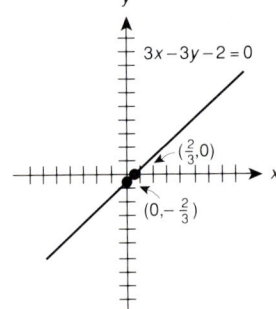

9.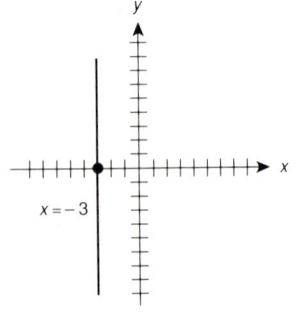

11.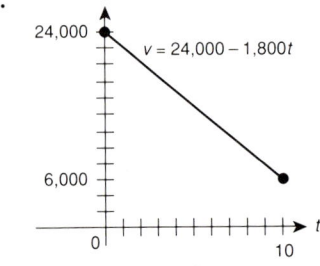

13.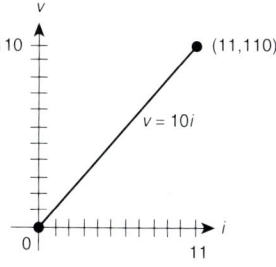

15.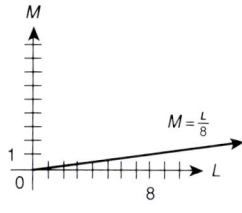

17.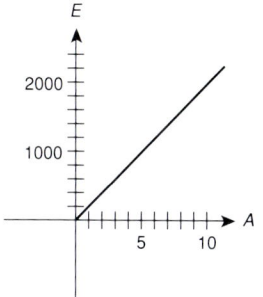

19.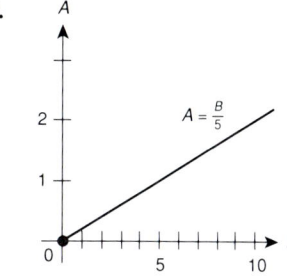

Trial Problems 3.3

1.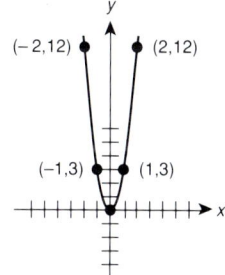

2.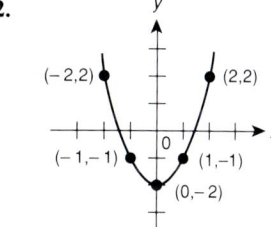

3.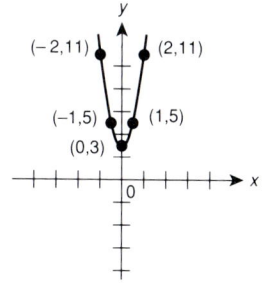

4.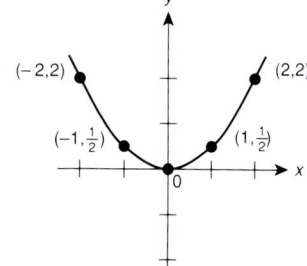

5.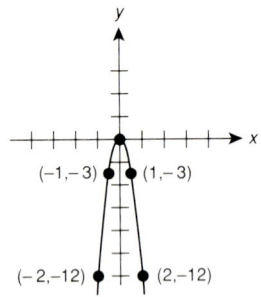

Exercises 3.3

1.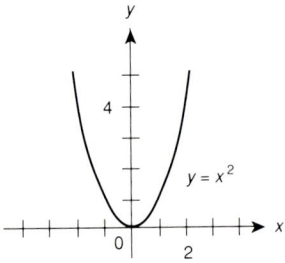

3.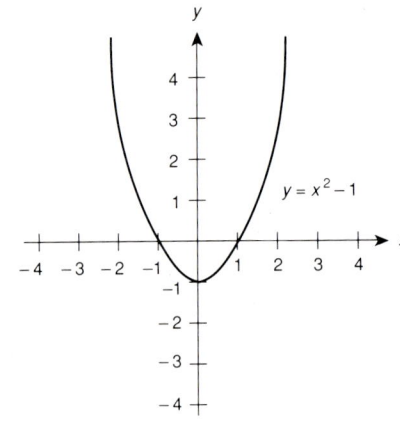

5.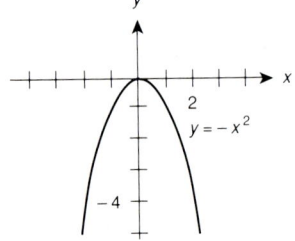

7.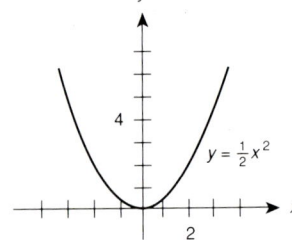

9.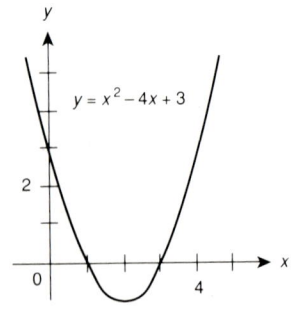

11.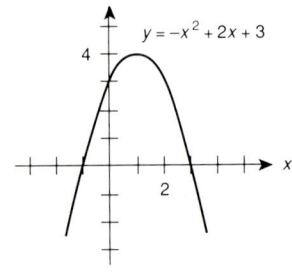

13. $A = -2x^2 + 28x$

15.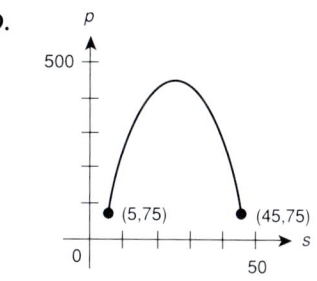

17. $p = \dfrac{v^2}{2g}$, $g = 32$ ft/sec²

p	v
0.390	5
1.563	10
3.516	15
6.25	20

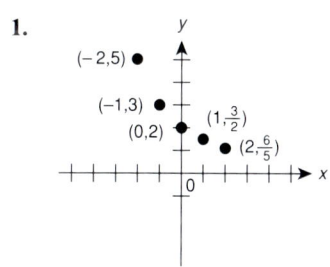

19.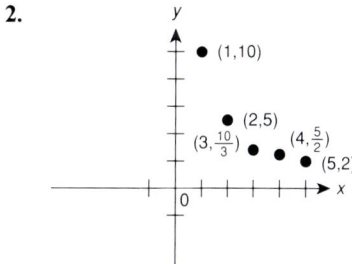

Trial Problems 3.4

1.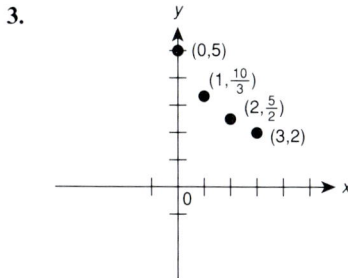

2.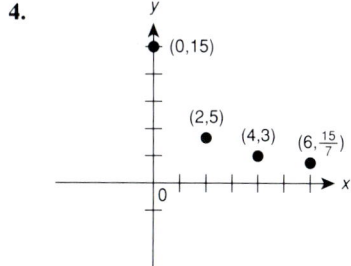

3.

(image not listed)

4.

(image not listed)

5.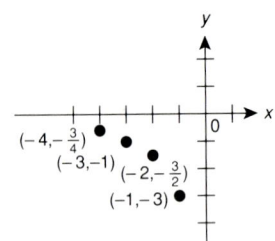

Exercises 3.4

1.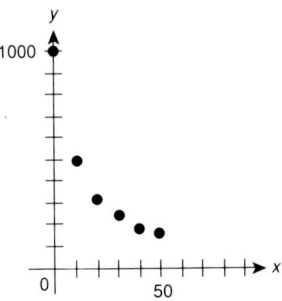

3.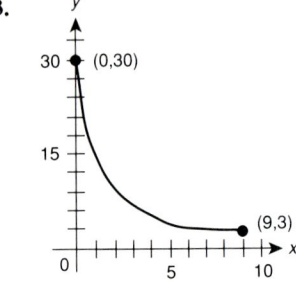

5.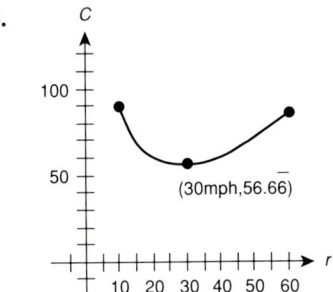

7.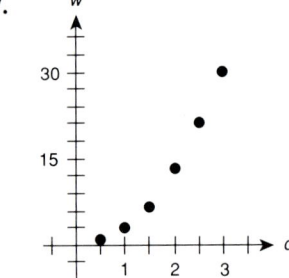

9.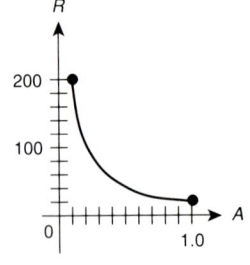

11.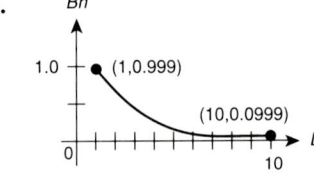

Answer Key—Chapter 3

Trial Problems 3.5

1. $f(x) = x - 3$ **2.** $f(x) = 5x^2$ **3.** 1   **4.** -5 **5.** $a^2 - 3a + 1$

Exercises 3.5

1. $f(x) = x + 3$ **3.** $f(t) = 16t^2$ **5.** $f(s) = s^2$ **7.** $f(w) = 2(5 + w)$ **9.** $f(x) = 3x^2 - 2$ **11.** $f(3) = 5$ **13.** $f(3) = 5$
15. $f(3) = 32$ **17.** $f(4) = 64$ **19.** $f(2) = 16$ **21.** $f(a) = a^3 + 2a - 1$
23. $f(x + h) = (x + h)^2 - 4(x + h) + 5 = x^2 + (2h - 4)x + h^2 - 4h + 5$ **25.** 1   **27.** 2   **29.** $8x + 4h - 3$
31. $C = 2\pi r$ **33.** $F = 0.50 + 0.10(4d - 1)$, d = miles   **35.** $C = 200 + 0.16x$ **37.** $R_1 = 1500 + R_2$

Review Exercises 3.7

1. [3.1]
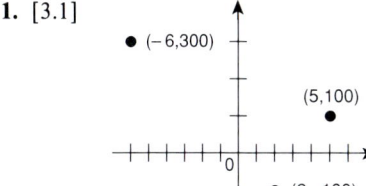

3. [3.1]
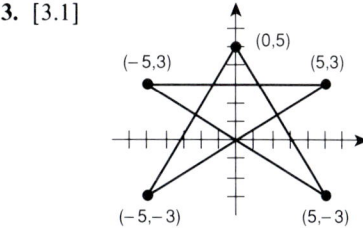

5. [3.1]
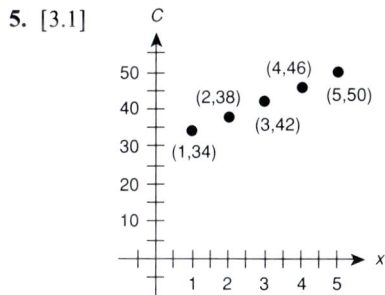

7. [3.2]
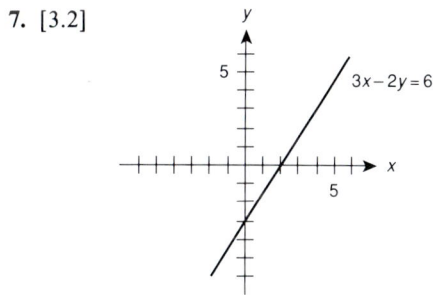

9. [3.3]
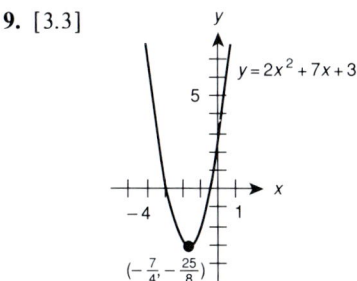

11. [3.3]
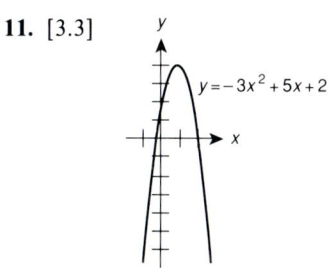

13. [3.4]

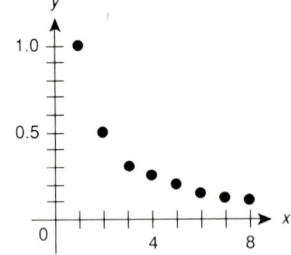

15. [3.5] $f(x) = x^2 + 3$ **17.** [3.5] $f(w) = 2(w + 5)$ **19.** [3.5] 1   **21.** [3.5] 24   **23.** [3.5] $3a^2 - 2a + 4$
25. [3.5] $p = 3s, f(s) = 3s$ **27.** [3.5] $C = \$35.00 + h(\$25.00)$ **29.** [3.5] $Q = 250U$ or $f(U) = 250U$

Test 3.8

1. [3.1]

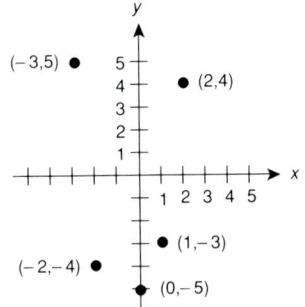

2. [3.1] (10,45), (20,50), (30,55), (40,60), (50,65)   **3.** [3.1] $A(4,0), B(3,2), C(0,3), D(-4,1), E(-2,-3), F(0,-4)$

4. [3.2]

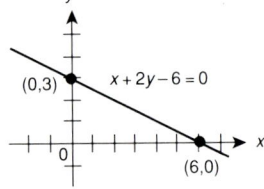

5. [3.2]

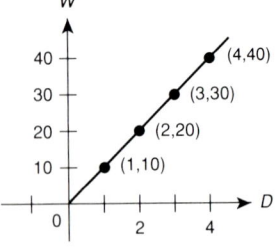

6. [3.3]

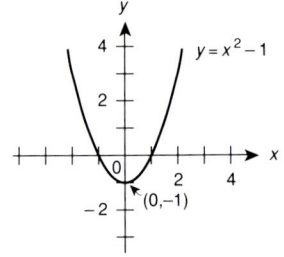

7. [3.3]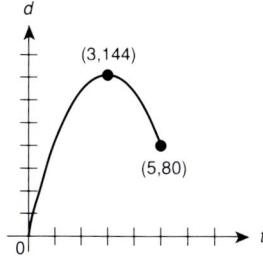

8. [3.4]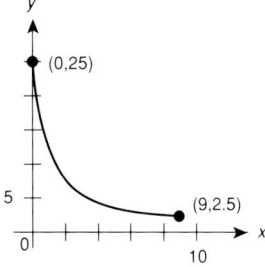

9. [3.4]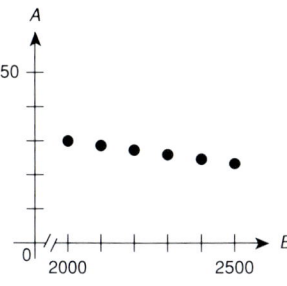

10. [3.5] $f(x) = x + 4$ **11.** [3.5] $f(w) = 2(5 + w)$ **12.** [3.5] $f(2) = 9$ **13.** [3.5] $C = 2\pi r$
14. [3.5] $f(x + h) = (x + h)^2 + 3(x + h) - 1$ **15.** [3.5] 16 **16.** [3.5] $8x + 4h - 3$

Chapter 4 Answers

Trial Problems 4.2

1.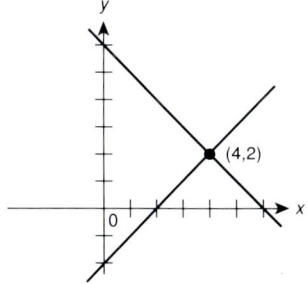

2.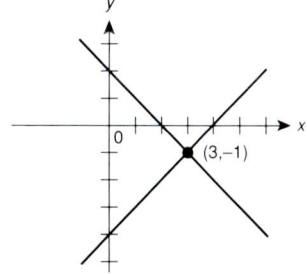

3.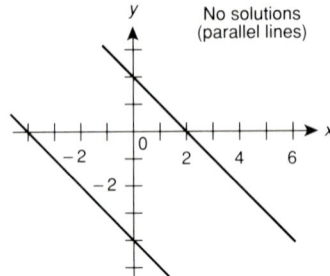

4.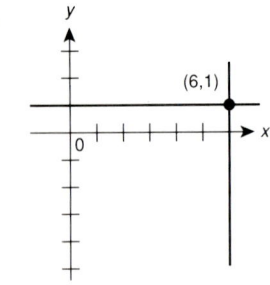

5.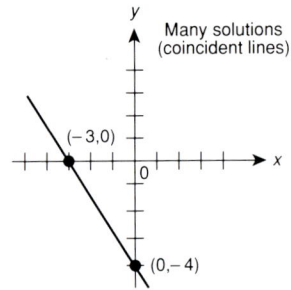

Exercises 4.2

1.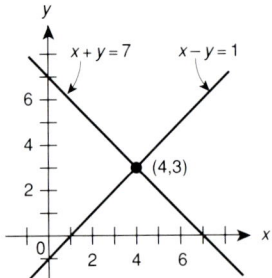

3.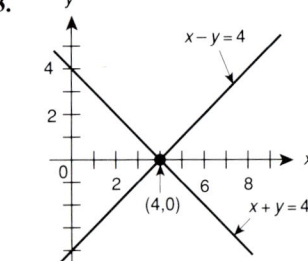

5.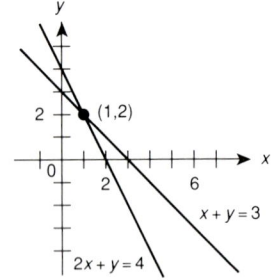

7.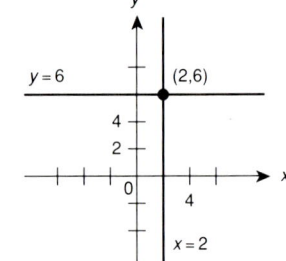

9.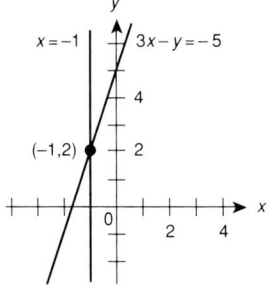

11.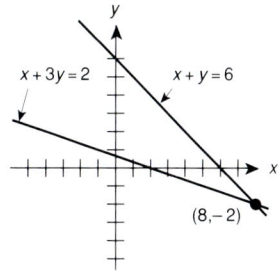

13.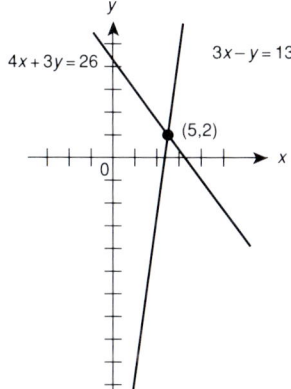

15.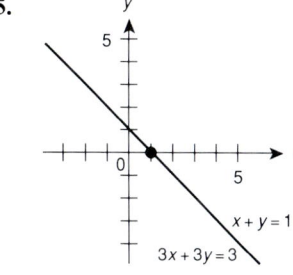

17.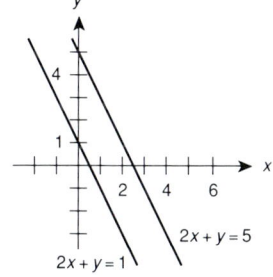

19.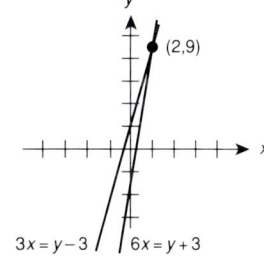

21.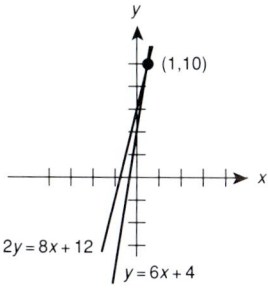

23. The break-even point is (100, 500).

25.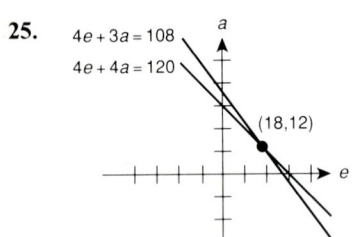

27.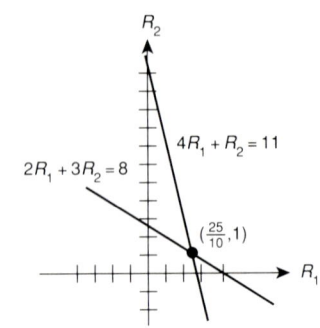

Trial Problems 4.3

1. $x = 2, y = -1$ **2.** $x = 6, y = -5$ **3.** $x = 1, y = 0$ **4.** $x = 7, y = -4$ **5.** $x = 3, y = 3$

Exercises 4.3

1. $x = \frac{1}{10}, y = \frac{9}{10}$ or $\left(\frac{1}{10}, \frac{9}{10}\right)$ **3.** $x = 6, y = 11$ (6,11)   **5.** $x = \frac{27}{5}, y = \frac{3}{5}$ $\left(\frac{27}{5}, \frac{3}{5}\right)$
7. $x = 5, y = -1$ (5,−1)   **9.** $x = -1, y = 2$ (−1,2)   **11.** $x = \frac{4}{5}, y = \frac{12}{25}$ $\left(\frac{4}{5}, \frac{12}{25}\right)$
13. $x = \frac{18}{5}, y = -\frac{2}{5}$ $\left(\frac{18}{5}, -\frac{2}{5}\right)$ **15.** No intersection   **17.** No intersection   **19.** $\left(\frac{54}{5}, 524\right)$
21. $r_1 = 120$ ohms, $r_2 = 200$ ohms   **23.** $\ell = 8$ meters, $w = 6$ meters   **25.** shorter rod = 5.575 cm, longer rod = 18.695 cm

Trial Problems 4.4

1. $x = 5, y = 2$ **2.** $x = 6, y = -1$ **3.** $x = -3, y = -4$ **4.** $x = -5, y = 1$ **5.** $x = 0, y = 1$

Exercises 4.4

1. (3,1)   **3.** $\left(2, \frac{9}{2}\right)$ **5.** (15,3)   **7.** $\left(\frac{4}{5}, \frac{12}{25}\right)$ **9.** (2,3)   **11.** (9,2)   **13.** (11,24)   **15.** $\left(\frac{7}{4}, 225\right)$ **17.** (24,0)
19. (210 ft, 150 ft)   **21.** (5 hrs, 4 hrs)   **23.** $\left(-\frac{1}{3}, 1\right)$ **25.** $y = 520, p = 12$

Trial Problems 4.5

1. 4   **2.** −7   **3.** 0   **4.** $x = 2, y = 1$ **5.** $x = -3, y = 2$

Exercises 4.5

1. (2,−1)   **3.** (2,3)   **5.** (3,−1)   **7.** (3,6)   **9.** $\left(2, \frac{2}{3}\right)$ **11.** Inconsistent   **13.** Dependent
15. $S_S = 16{,}000$ MPa, $S_A = -32{,}000$ MPa   **17.** $F_1 = 7, F_2 = 5$ **19.** $P = 6000, Q = 4000$

Trial Problems 4.6

1. $x = 3, y = 1, z = 2$ **2.** 5   **3.** 12   **4.** −24   **5.** $x = 2, y = 3, z = 1$

Exercises 4.6

1. -12 **3.** -45 **5.** 16 **7.** -18 **9.** 5 **11.** $\left(\dfrac{5}{2}, 2, \dfrac{3}{2}\right)$ **13.** $(1,0,0)$ **15.** $\left(-\dfrac{8}{15}, -\dfrac{13}{15}, \dfrac{22}{15}\right)$
17. $v_1 = 88, v_2 = 48, v_3 = 30$ **19.** $F_1 = 10, F_2 = 15, F_3 = 5$

Review Exercises 4.8

1. [4.2]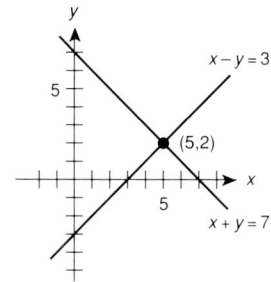

3. [4.2]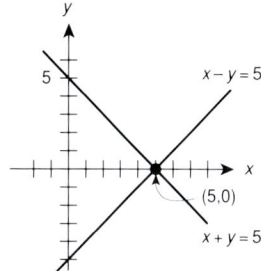

5. [4.2]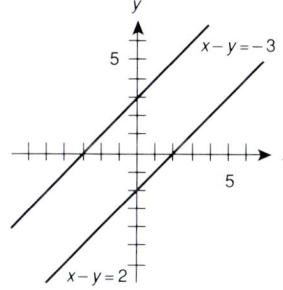

7. [4.2]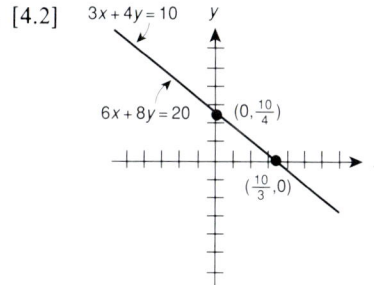

9. [4.3] $\left(\dfrac{1}{10}, \dfrac{9}{10}\right)$ **11.** [4.3] $(5,-1)$ **13.** [4.3] $(2,0)$ **15.** [4.4] $(3,1)$ **17.** [4.4] $\left(\dfrac{4}{5}, \dfrac{12}{25}\right)$ **19.** [4.4] $(11,24)$
21. [4.5] $(2,-1)$ **23.** [4.5] $(3,6)$ **25.** [4.6] -12 **27.** [4.6] -30 **29.** [4.5] $(5,2)$ **31.** [4.6] $\left(\dfrac{5}{2}, 2, \dfrac{3}{2}\right)$
33. [4.3] $(100,500)$ **35.** [4.3] $\left(\dfrac{54}{5}, 524\right)$ **37.** [4.4] Length $= 210$ ft, width $= 150$ ft   **39.** [4.6] $F_1 = 3, F_2 = 5, F_3 = 8$

Test 4.9

1. [4.2]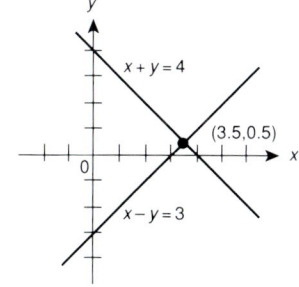

2. [4.3] $x = 1, y = 1$ **3.** [4.3] $x = \dfrac{12}{25}, y = \dfrac{4}{5}$ **4.** [4.4] $x = 5, y = 2$ **5.** [4.4] $x = \dfrac{4}{5}, y = \dfrac{12}{25}$

6. [4.4] $x = 11, y = 24$ **7.** [4.5] $x = 2, y = 1$ **8.** [4.5] $x = 2, y = \dfrac{2}{3}$

9. [4.5] $x = -\dfrac{5}{2}, y = \dfrac{11}{3}$, consistent and independent   **10.** [4.6] -2 **11.** [4.6] -18

12. [4.6] $x = \dfrac{5}{2}, y = 2, z = \dfrac{3}{2}$

Chapter 5 Answers

Trial Problems 5.1

1.
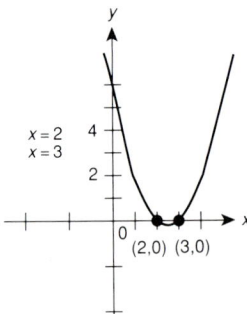

2.
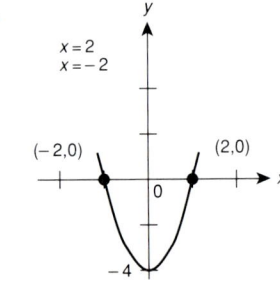

3.
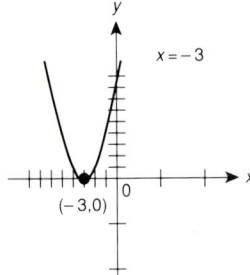

4.
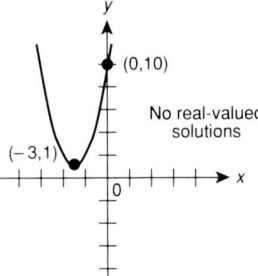

5.
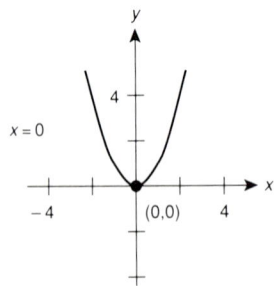

Exercises 5.1

1.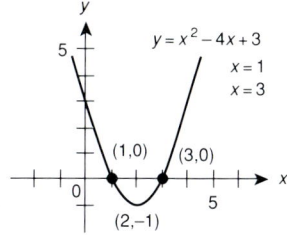

3.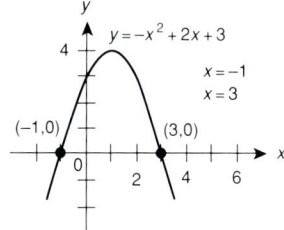

5.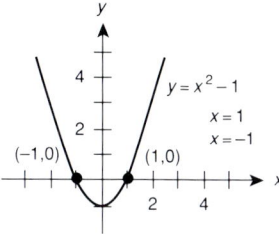

7.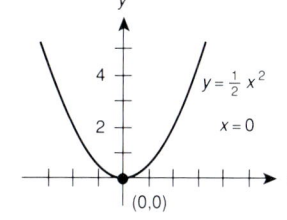

9.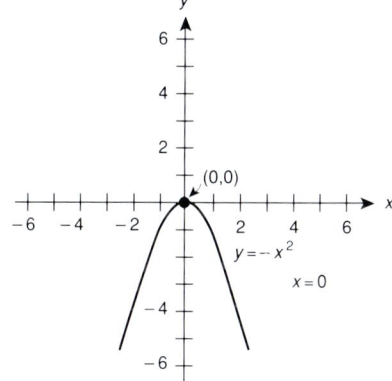

11.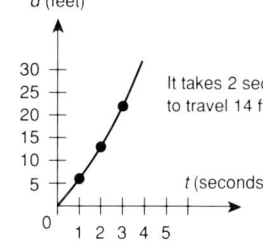

13.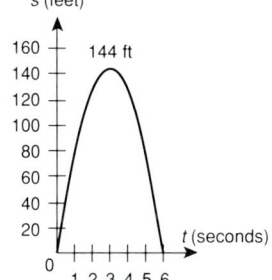

*projectile will strike the ground in 6 seconds

Answer Key—Chapter 5

Trial Problems 5.2

1. $(x-3)(x-2)$ 2. $(x+4)(x+4)$ 3. $(2x-3)(3x+5)$ 4. $x=-3, x=-y$ 5. $x=-\dfrac{3}{2}, x=\dfrac{5}{3}$

Exercises 5.2

1. $\{4,3\}$ 3. $\{-7\}$ 5. $\{12,-1\}$ 7. $\{-4,8\}$ 9. $\{3,-12\}$ 11. $\left\{\dfrac{9}{2},-1\right\}$ 13. $\left\{\dfrac{1}{3},\dfrac{1}{2}\right\}$ 15. $\{5,-5\}$ 17. $\{4,-4\}$
19. $\{3,-3\}$ 21. $\{-1,-2\}$ 23. $\left\{\dfrac{3}{2},-4\right\}$ 25. $\{-6,-2\}$ 27. $w^2+8w-48=0$, width = 4 in., length = 12 in.

Trial Problems 5.3

1. $x=3, x=2$ 2. $x=-4, x=5$ 3. $x=-3, x=2$ 4. $x=-\dfrac{1}{3}, x=-2$ 5. $x=0, x=\dfrac{3}{2}$

Exercises 5.3

1. $\{-6,2\}$ 3. $\{-2\}$ 5. $\{3,-2\}$ 7. $\left\{\dfrac{-5+\sqrt{73}}{4},\dfrac{-5-\sqrt{73}}{4}\right\}$ 9. $\left\{1,-\dfrac{4}{3}\right\}$ 11. $\left\{\dfrac{\sqrt{13}+1}{4},\dfrac{-\sqrt{13}+1}{4}\right\}$
13. 12 in., 8 in.   15. $l=17$ in., $w=9$ in.

Trial Problems 5.4

1. $a=4, b=3, c=5$ 2. $a=3, b=-7, c=0$ 3. $a=\dfrac{1}{2}, b=0, c=-8$ 4. 64 5. $x=\dfrac{5\pm\sqrt{13}}{6}$

Exercises 5.4

1. $\{4,1\}$ 3. $\{6,-1\}$ 5. $\left\{\dfrac{3+\sqrt{5}}{2},\dfrac{3-\sqrt{5}}{2}\right\}$ 7. $\left\{\dfrac{5+\sqrt{13}}{6},\dfrac{5-\sqrt{13}}{6}\right\}$ 9. $\{0,5\}$ 11. $\{5\}$ 13. $\{0.83,-0.08\}$
15. $\{62.23, 0.49\}$ 17. 3 seconds 19. $I=\dfrac{E\pm\sqrt{E^2-4PR}}{2R}$

Trial Problems 5.5

1. $x<3$ 2. $x=-\dfrac{5}{2}$ 3. $x<4$ 4. $-3<x<-2$ 5. $-2<x<3$

Exercises 5.5

1. $-5<x<3$ 3. $x>5$ or $x<4$ 5. $-3<x<-\dfrac{3}{2}$ 7. $-4<x<3$ 9. $-3<x<5$ 11. $3\le x\le 5$
13. $x\le -1$ or $0<x\le 8$ 15. $t>\dfrac{5+\sqrt{65}}{4}$ sec

Review Exercises 5.7

1. [5.1]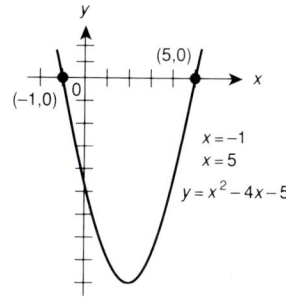

3. [5.1]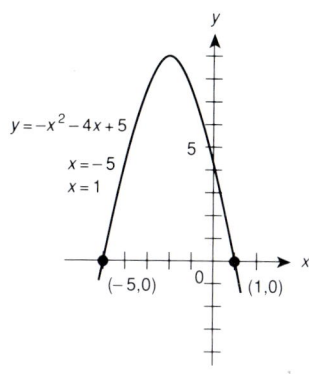

5. [5.1]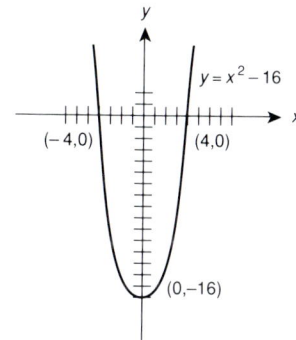

7. [5.2] $\{2,6\}$ **9.** [5.2] $\{-5\}$ **11.** [5.2] $\{-3,9\}$ **13.** [5.3] $\{3,4\}$ **15.** [5.3] $\left\{\dfrac{\sqrt{13}-3}{2}, \dfrac{-\sqrt{13}-3}{2}\right\}$
17. [5.3] $\left\{-\dfrac{3}{5}, 1\right\}$ **19.** [5.4] $\{5,1\}$ **21.** [5.4] $\left\{\dfrac{\sqrt{5}-5}{2}, \dfrac{-\sqrt{5}-5}{2}\right\}$ **23.** [5.4] $\{1,5\}$ **25.** [5.5] $x < \sqrt{17} - 1$
27. [5.5] $x > 3$ or $x < -2$ **29.** [5.5] $4 \leq x \leq 5$ **31.** [5.5] $0 < t < 3$

Test 5.8

1. [5.1] $x = 2, x = -3$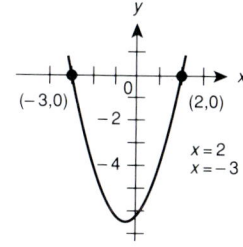

2. [5.1] $x = \pm 4$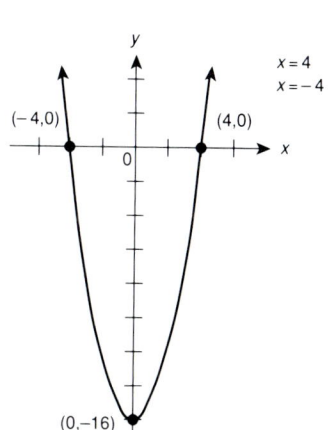

3. [5.2] $x = -1, x = \dfrac{9}{2}$ **4.** [5.2] $a = \pm 7$ **5.** [5.3] $x = -9, x = 3$ **6.** [5.3] $x = \dfrac{1 \pm \sqrt{6}}{5}$

7. [5.4] $z = \dfrac{7 \pm \sqrt{97}}{4}$ **8.** [5.4] $x = \dfrac{5 \pm \sqrt{13}}{6}$ **9.** [5.5] $x > 2$ or $x < -1$ **10.** [5.5] $-3 < x < 4$

11. [5.5] $0 < t < 2$ **12.** [5.5] $t > \dfrac{7}{2}$

Cumulative Review—Chapters 3, 4, and 5—5.9

1. [3]

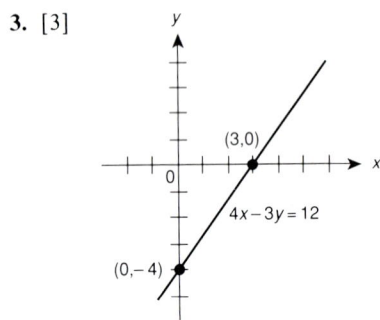

2. [3]

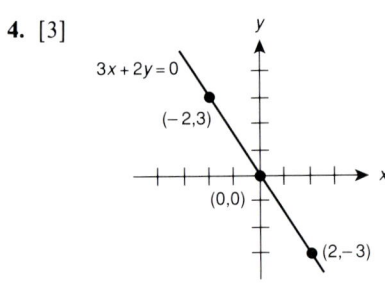

3. [3]

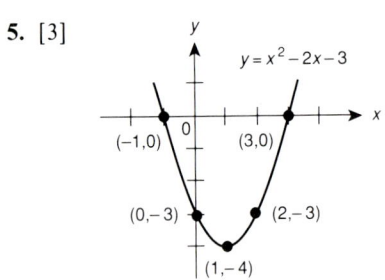

4. [3]

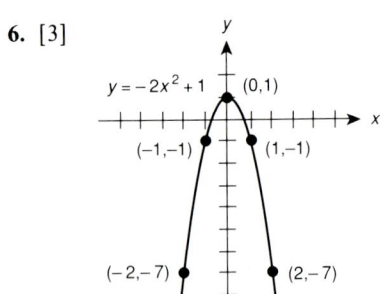

5. [3]

6. [3]

7. [3] 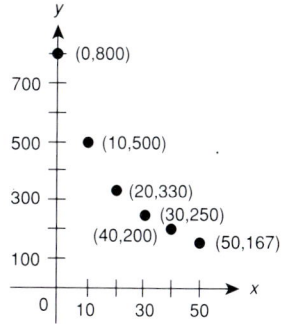   **8.** [3]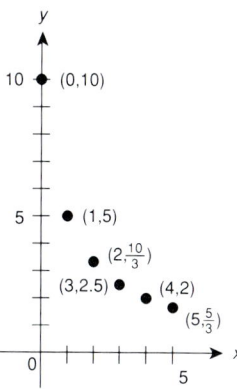

9. [3] $f(x) = 3x^2 - 2$ **10.** [3] $f(4) = 64$
11. [3] $f(x + h) = (x + h)^2 - 4(x + h) + 5 = x^2 + (2h - 4)x + h^2 - 4h + 5$
12. [4]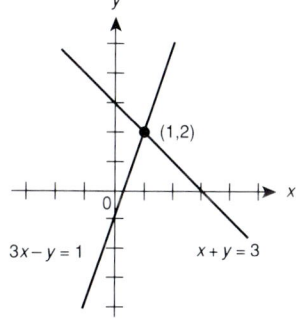

13. [4] $x = 500, y = 2000$ or $(500, 2000)$ **14.** [4] No intersection   **15.** [4] Length $= 7$ cm, width $= 5$ cm
16. [4] $p = 20, y = 1000$ **17.** [4] $x = 2, y = 1$ **18.** [4] $x = 2, y = 3$ **19.** [4] 5
20. [4] $x = -\dfrac{8}{15}, y = -\dfrac{13}{15}, z = \dfrac{22}{15}$
21. [5]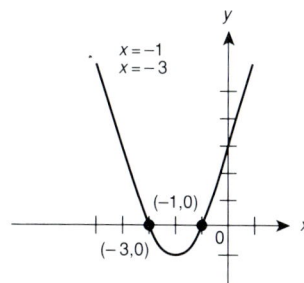

22. [5] $x = \pm 2$ **23.** [5] $y = -\dfrac{3}{2}, y = 1$ **24.** [5] $x = \dfrac{1 \pm \sqrt{13}}{4}$ **25.** [5] $x = \dfrac{3}{2}, x = 2$
26. [5] $x > 2$ or $x < \dfrac{2}{3}$ **27.** [5] $x > 20$ cm

Chapter 6 Answers

Trial Problems 6.1

1. $26°25' = \left(26 + \dfrac{25}{60}\right)° = 26.4167°$

   $\boxed{26}\ \boxed{+}\ \boxed{25}\ \boxed{\div}\ \boxed{60}\ \boxed{=}$

2. $A + B + C = 180$
   $93 + 62 + C = 180$
   $C = 180 - 93 - 62 = 25°$

3. $a^2 + b^2 = c^2$, where $a = 16.5$, $b = 18.3$
   $(16.5)^2 + (18.3)^2 = c^2$
   $c = \sqrt{(16.5)^2 + (18.3)^2} = 24.6$

   $\boxed{16.5}\ \boxed{x^2}\ \boxed{+}\ \boxed{18.3}\ \boxed{x^2}\ \boxed{=}\ \boxed{\sqrt{x}}$

4. $a^2 + b^2 = c^2$, where $a = b$ and $c = 14.3$
   $a^2 + a^2 = (14.3)^2$
   $2a^2 = (14.3)^2$
   $a = \sqrt{\dfrac{(14.3)^2}{2}} = 10.1$

   $\boxed{14.3}\ \boxed{x^2}\ \boxed{\div}\ \boxed{2}\ \boxed{=}\ \boxed{\sqrt{x}}$

5. $\dfrac{b'}{a'} = \dfrac{b}{a}$, where $a = 11.0$, $b = 14.8$, and $a' = 18.2$

   $\dfrac{b'}{18.2} = \dfrac{14.8}{11.0}$

   $b' = \dfrac{14.8 \times 18.2}{11.0} = 24.5$

   $\boxed{14.8}\ \boxed{\times}\ \boxed{18.2}\ \boxed{\div}\ \boxed{11}\ \boxed{=}$

Exercises 6.1

1. $30.3333°$ 3. $68.3667°$ 5. $118.3875°$ 7. $73.7°$ 9. $116°40'$ 11. $40°$ 13. $55°50'$ 15. $43°37'$ 17. $100°$
19. $74°$ 21. $b' = \dfrac{144}{7}$ in., $c' = \dfrac{192}{7}$ in.   23. $a' = 10.3$ in., $c' = 12.2$ in.   25. 169 27. 45 in.   29. 24.0 in.
31. 51.2 in.   33. $1\dfrac{1}{2}$ in.   35. 530.4 yd

Trial Problems 6.2

1. $\tan A = \dfrac{a}{b}$, where $A = 26°$ and $a = 18.3$

   $\tan 26° = \dfrac{18.3}{b}$

   $b = \dfrac{18.3}{\tan 26°} = 37.5$

   $\boxed{18.3}\ \boxed{\div}\ \boxed{26}\ \boxed{TAN}\ \boxed{=}$

2. $\tan A = \dfrac{a}{b}$, where $A = 48°$ and $b = 16.1$

   $\tan 48° = \dfrac{a}{16.1}$

   $a = 16.1 \tan 48° = 17.9$

   $\boxed{16.1}\ \boxed{\times}\ \boxed{48}\ \boxed{TAN}\ \boxed{=}$

Answer Key—Chapter 6

3. $\tan A = \dfrac{a}{b}$, where $a = 11.3$ and $b = 22.4$

$\tan A = \dfrac{11.3}{22.4}$, $A = 26.8°$

11.3 [÷] 22.4 [=] [INV] [TAN]

4. $\tan B = \dfrac{b}{a}$, where $a = 0.43$ and $B = 67°$

$\tan 67° = \dfrac{b}{0.43}$

$b = 0.43 \tan 67° = 1.01$

0.43 [×] 67 [TAN] [=]

5. $\tan B = \dfrac{b}{a}$, where $a = 11.0$ and $b = 16.7$

$\tan B = \dfrac{16.7}{11.0}$, $B = 56.6°$

16.7 [÷] 11.0 [=] [INV] [TAN]

Exercises 6.2

1. 4.67 in. **3.** 31.03 in. **5.** 57.42 yd **7.** 86.70 in. **9.** 16.65 cm **11.** 6.92 m **13.** $B = 73°47'15''$, $a = 42.45$ in. **15.** $A = 61.3°$, $B = 28.7°$ **17.** 56.25 ft **19.** 71.3 m **21.** 53.6 ft **23.** 23.9° **25.** 18.4° **27a.** 76.5° **b.** 13.5° **c.** 51.4 ft **29.** 24.95 ft **31.** 43.35 ft **33.** 2.54 cm **35.** 37.7° **37.** 6° **39.** 37.4° **41.** 3 ohms

Trial Problems 6.3

1. $\sin A = \dfrac{a}{c}$, where $a = 17.9$ and $A = 65.3°$

$\sin 65.3° = \dfrac{17.9}{c}$

$c = \dfrac{17.9}{\sin 65.3°} = 19.7$

17.9 [÷] 65.3 [SIN] [=]

2. $\cos B = \dfrac{a}{c}$, where $a = 17.9$ and $c = 25.8$

$\cos B = \dfrac{17.9}{25.8}$, $B = 46.1°$

17.9 [÷] 25.8 [=] [INV] [COS]

3. $\cos A = \dfrac{b}{c}$, where $b = 14.3$ and $c = 25.9$

$\cos A = \dfrac{14.3}{25.9}$, $A = 56.5°$

14.3 [÷] 25.9 [=] [INV] [COS]

4. $\cot A = \dfrac{1}{\tan A} = \dfrac{1}{\tan 22.6°} = 2.4023$

22.6 [TAN] [1/x]

5. $P = EI \cos \theta$

$P = (110)(20)(\cos 60°) = 1100$

110 [×] 20 [×] 60 [COS] [=]

Exercises 6.3

1. $A = 34.8°, B = 55.2°$ **3.** $B = 58°, a = 8.6$ in., $b = 13.7$ in. **5.** $B = 29°, b = 81.5$ in., $c = 168.1$ in.
7. $A = 45.2°, B = 44.8°, c = 169.1$ ft **9.** $A = 57°35', a = 6.9$ cm, $b = 4.4$ cm **11.** $B = 73°30', b = 2.44$ in., $c = 29.05$ in. **13a.** $\cot\theta = 1.3270, \sec\theta = 1.2521, \csc\theta = 1.6616$ **b.** $\cot\theta = 0.0945, \sec\theta = 10.6261, \csc\theta = 1.0045$ **c.** $\cot\theta = 0.5187, \sec\theta = 2.1718, \csc\theta = 1.1265$ **15a.** 1167 watts **b.** 1429 watts **c.** 697 watts
17. angle $XYZ = 22.54°$, angle $WZY = 67.46°$, angle $ZXY = 66.93°$ **19.** 49° **21.** 136° **23.** 1.12 in. **25.** 2.1°
27. 23.2° **29.** 714.5 watts **31.** 5.20 in. **33.** 1.91 cm **35.** 27.8° **37.** 922 ft **39.** 282 ft

Trial Problems 6.4

1. $\sin\theta = 0.8480$ $\csc\theta = \dfrac{1}{\sin\theta} = 1.1792$

122 [SIN]   122 [SIN] [1/x]

2. $\dfrac{\sin A}{a} = \dfrac{\sin B}{b}$, where $A = 43°, B = 68°$, and $b = 14$

$\dfrac{\sin 43°}{a} = \dfrac{\sin 68°}{14}$

$a = \dfrac{14\sin 43°}{\sin 68°} = 10.3$

14 [×] 43 [SIN] [÷] 68 [SIN] [=]

3. $\dfrac{\sin A}{a} = \dfrac{\sin B}{b}$, where $B = 62.3°, A = 66.8°$, and $a = 16.5$

$\dfrac{\sin 66.8°}{16.5} = \dfrac{\sin 62.3°}{b}$

$b = \dfrac{16.5\sin 62.3°}{\sin 66.8°} = 15.9$

16.5 [×] 62.3 [SIN] [÷] 66.8 [SIN] [=]

4. $\dfrac{\sin C}{c} = \dfrac{\sin B}{b}$, where $C = 85.3°, c = 18.6$, and $B = 62.3°$

$\dfrac{\sin 85.3°}{18.6} = \dfrac{\sin 62.3°}{b}$

$b = \dfrac{18.6\sin 62.3°}{\sin 85.3°} = 16.5$

18.6 [×] 62.3 [SIN] [÷] 85.3 [SIN] [=]

5. $\dfrac{\sin A}{a} = \dfrac{\sin C}{c}$, where $C = 105°, a = 19.6$, and $c = 28.3$

$\dfrac{\sin A}{19.6} = \dfrac{\sin 105°}{28.3}$

$\sin A = \dfrac{19.6\sin 105°}{28.3}, A = 42°$

19.6 [×] 105 [SIN] [÷] 28.3 [=] [INV] [SIN]

Exercises 6.4

1. $\sin \theta = 0.9659$, $\cos \theta = -0.2588$
   $\tan \theta = -3.7321$, $\cot \theta = -0.2679$
   $\sec \theta = -3.8637$, $\csc \theta = 1.0353$
5. $C = 30°$, $b = 9.54$ in., $c = 5.08$ in.
9. $A = 89°10'$, $a = 244.5$ ft, $b = 201.1$ ft
13. No such triangle
17. $BC = 1448$ ft, $AD = 1159$ ft, $CD = 1634$ ft
21. $AC = 175.3$ ft, $AD = 203.6$ ft
    $BC = 196.6$ ft, $BD = 168.4$ ft

3. $\sin \theta = 0.9903$, $\cos \theta = -0.1392$
   $\tan \theta = -7.1154$, $\cot \theta = -0.1405$
   $\sec \theta = -7.1853$, $\csc \theta = 1.0098$
7. $C = 73.5°$, $a = 4.58$ cm, $c = 6.65$ cm
11. $B = 45.4°$, $C = 69.6°$, $c = 14.5$ cm
15. 4.7 in.
19. 19.4 in.
23. 3.05 A

Trial Problems 6.5

1. $c = \sqrt{a^2 + b^2 - 2ab \cos C}$, where $a = 17$, $b = 14$, and $C = 98°$
   $c = \sqrt{(17)^2 + (14)^2 - 2(17)(14)(\cos 98°)} = 23$

   17 $\boxed{x^2}$ $\boxed{+}$ 14 $\boxed{x^2}$ $\boxed{-}$ 2 $\boxed{\times}$ 17 $\boxed{\times}$ 14 $\boxed{\times}$ 98 $\boxed{\cos}$ $\boxed{=}$ $\boxed{\sqrt{x}}$

2. $b = \sqrt{a^2 + c^2 - 2ac \cos B}$, where $c = 18.6$, $a = 13.3$, and $B = 46.3°$
   $b = \sqrt{(13.3)^2 + (18.6)^2 - 2(13.3)(18.6)(\cos 46.3°)} = 13.5$

   13.3 $\boxed{x^2}$ $\boxed{+}$ 18.6 $\boxed{x^2}$ $\boxed{-}$ 2 $\boxed{\times}$ 13.3 $\boxed{\times}$ 18.6 $\boxed{\times}$ 46.3 $\boxed{\cos}$ $\boxed{=}$ $\boxed{\sqrt{x}}$

3. $a = \sqrt{c^2 + b^2 - 2cb \cos A}$, where $c = 0.45$, $b = 0.86$, and $A = 25.5°$
   $a = \sqrt{(0.45)^2 + (0.86)^2 - 2(0.45)(0.86)(\cos 25.5°)} = 0.49$

   0.45 $\boxed{x^2}$ $\boxed{+}$ 0.86 $\boxed{x^2}$ $\boxed{-}$ 2 $\boxed{\times}$ 0.45 $\boxed{\times}$ 0.86 $\boxed{\times}$ 25.5 $\boxed{\cos}$ $\boxed{=}$ $\boxed{\sqrt{x}}$

4. $a^2 = b^2 + c^2 - 2bc \cos A$, where $a = 14.3$, $b = 16.7$, and $c = 20.4$
   $\cos A = \dfrac{b^2 + c^2 - a^2}{2bc}$
   $\cos A = \dfrac{(16.7)^2 + (20.4)^2 - (14.3)^2}{2(16.7)(20.4)}$, $A = 43.9°$

   16.7 $\boxed{x^2}$ $\boxed{+}$ 20.4 $\boxed{x^2}$ $\boxed{-}$ 14.3 $\boxed{x^2}$ $\boxed{=}$ $\boxed{\div}$ 2 $\boxed{=}$ $\boxed{\div}$ 16.7 $\boxed{=}$ $\boxed{\div}$ 20.4 $\boxed{=}$ $\boxed{\text{INV}}$ $\boxed{\cos}$

5. $\cos C = \dfrac{a^2 + b^2 - c^2}{2ab}$, where $a = 0.72$, $b = 0.85$, and $c = 0.98$
   $\cos C = \dfrac{(0.72)^2 + (0.85)^2 - (0.98)^2}{2(0.72)(0.85)}$, $C = 76.8°$

   0.72 $\boxed{x^2}$ $\boxed{+}$ 0.85 $\boxed{x^2}$ $\boxed{-}$ 0.98 $\boxed{x^2}$ $\boxed{=}$ $\boxed{\div}$ 2 $\boxed{=}$ $\boxed{\div}$ 0.72 $\boxed{=}$ $\boxed{\div}$ 0.85 $\boxed{=}$ $\boxed{\text{INV}}$ $\boxed{\cos}$

Exercises 6.5

1. $a = 15.7$ in., $B = 38.8°$, $C = 61.2°$ **3.** $b = 22.7$ ft, $A = 82.4°$, $C = 37.6°$ **5.** $c = 167.6$ ft, $A = 14.3°$, $B = 98.7°$
7. $a = 29.9$ cm, $B = 55°40'$, $C = 43°55'$ **9.** $b = 7.85$ m, $A = 85°56'$, $C = 33°39'$ **11.** $C = 67°20'$, $A = 76°55'$,
$B = 35°45'$ **13.** 640 ft   **15.** 6.35 mi   **17.** 144.5 m   **19.** 16.0 in., 11.4 in.   **21.a.** 749.58 ft   **b.** 128.47 ft   **c.** 680.38 ft

Review Exercises 6.7

1. [6.3] $\sin \theta = 0.6018$, $\cos \theta = 0.7986$
$\tan \theta = 0.7536$, $\cot \theta = 1.3270$
$\sec \theta = 1.2521$, $\csc \theta = 1.6616$

3. [6.3] $\sin \theta = 0.5344$, $\cos \theta = 0.8453$
$\tan \theta = 0.6322$, $\cot \theta = 1.5818$
$\sec \theta = 1.1831$, $\csc \theta = 1.8714$

5. [6.1] 9.57 cm   **7.** [6.1] 3.04 in.   **9.** [6.2] 24.2°   **11.** [6.3] 57.5°   **13.** [6.3] 12.7   **15.** [6.4] $B = 43°$, $b = 7.89$,
$c = 11.22$ **17.** [6.4] $B = 75.8°$, $C = 68.1°$, $c = 4.9$ or $B = 104.2°$, $C = 39.7°$, $c = 3.4$ **19.** [6.4] $A = 26°48'$,
$B = 87°52'$, $c = 12.9$ **21.** [6.5] $A = 50.7°$, $B = 33.6°$, $C = 95.7°$ **23.** [6.1] 20.2 cm   **25.** [6.3] 825 watts
27. [6.3] 450 ft   **29.** [6.4] 156.2 lb   **31.** [6.4,6.5] $AD = 33.4$ ft, $AC = 91.9$ ft, $BC = 57.7$ ft, $BD = 70.2$ ft, $CD = 91.6$ ft
33. [6.3] 109.7 ft   **35.** [6.3] 17.6   **37.** [6.4] 5.0   **39.** [6.4] yes   **41.** [6.4] no

Test 6.8

1. [6.1] $39°25'$ **2.** [6.1] $71°5'$ **3.** [6.1] 7.1 cm   **4.** [6.1] 175.0 in.   **5.** [6.1] 8.0 in., 13.2 in.
6. [6.3] sine is 0.5736, cosine is 0.8192   **7.** [6.3] sine is 0.5248, cosine is 0.8515, tangent is 0.6163   **8.** [6.3] 261 ft
9. [6.1] 20.9 cm   **10.** [6.2] 36.5°   **11.** [6.2] 61.5°   **12.** [6.3] 19.9   **13.** [6.3] 25.7   **14.** [6.4] $b = 5.0$ cm, $c = 10.0$ cm
15. [6.4] No such triangle   **16.** [6.5] $A = 57.9°$, $B = 75.5°$ **17.** [6.5] $A = 70°20'$, $c = 16.9$ **18.** [6.3] 145 watts
19. [6.5] 81.7 kg   **20.** [6.4,6.5] $BD = 91.8$, $BC = 63.8$, $CD = 116.3$

Chapter 7 Answers

Trial Problems 7.1

1. $41°40' = \left(41\frac{40}{60}\right)$ degrees $= 41.6667$ degrees

$41.6667° = \frac{41.6667 \times \pi}{180} = 0.727$ radians

$\boxed{41.6667} \; \boxed{\times} \; \boxed{\pi} \; \boxed{\div} \; \boxed{180} \; \boxed{=}$

2. 6.17 radians $= \frac{6.17 \times 180}{\pi} = 354$ degrees

$\boxed{6.17} \; \boxed{\times} \; \boxed{180} \; \boxed{\div} \; \boxed{\pi} \; \boxed{=}$

3. $\theta = \frac{s}{r}$, where $r = 6.3$ and $s = 7.2$

$\theta = \frac{7.2}{6.3} = 1.14$ radians

4. $v = r\omega$, where $r = 1.42$ and $\omega = 3.2$
$v = 14.2(3.2) = 45.44$ ft/min

5. $A = \dfrac{1}{2}r^2\theta$, where $r = 3.5$ and $\theta = \dfrac{\pi}{8}$

$A = \dfrac{1}{2}(3.5)^2\left(\dfrac{\pi}{8}\right) = 2.41$ sq in.

0.5 × 3.5 x^2 × π ÷ 8 =

Exercises 7.1

1. 0.61   **3.** 1.16   **5.** 0.93   **7.** 0.754   **9.** 22.5   **11.** 75   **13.** 149.5   **15.** 615.6   **17.** 1.96 radians   **19.** 3.78 radians
21. $r = 2\dfrac{2}{3}$ **23.** 6.55 in.   **25.** 3.67 in.   **27.** 8.70 in.   **29.** 7.77 in./min   **31.** 6.2 rad/sec   **33.** 11.4 ft
35. 0.7 ft or 8.4 in.   **37.** 4.03 cm   **39.** 1.9°   **41.** 7.4 in.   **43.** 436 mi   **45.** 363 sq in.   **47.** 2,945 sq mi   **49.** 2.8 sq in.
51. $v = 9948$ mi/hr, $\omega = 2.1$ rad/hr   **53.** 69 rad/sec, 660 rpm   **55.** 0.061 rad/sec   **57.** 3.84 rad/sec

Trial Problems 7.2

1. π ÷ 6 = +/− COS 1/x

$\sec\left(-\dfrac{\pi}{6}\right) = 1.1547$

2. $\cos\left(\dfrac{17\pi}{4}\right) = \cos\left(\dfrac{16+1}{4}\pi\right)$

$= \cos\left(4\pi + \dfrac{\pi}{4}\right)$

$= \cos\left(\dfrac{\pi}{4}\right) = \dfrac{\sqrt{2}}{2}$

3. 25 × π ÷ 6 = + π ÷ 3 = SIN × 30 =

Display 30, $y = 30$

4. π × 18 ÷ 12 = SIN × 1 400 = + 10000 =

Display 8600, $y = 8600$

5. 0.632 × 2 = SIN × 240 x^2 = ÷ 32 x^2 =

Display 53.623462, $y = 53.623$

Exercises 7.2

1.

TABLE 7.2A

t	$\sin t$	$\cos t$	$\tan t$	$\cot t$	$\sec t$	$\csc t$
0	0	1	0	undefined	1	undefined
$\frac{\pi}{6}$	$\frac{1}{2}$	$\frac{\sqrt{3}}{2}$	$\frac{\sqrt{3}}{3}$	$\sqrt{3}$	$\frac{2\sqrt{3}}{3}$	2
$\frac{\pi}{4}$	$\frac{\sqrt{2}}{2}$	$\frac{\sqrt{2}}{2}$	1	1	$\sqrt{2}$	$\sqrt{2}$
$\frac{\pi}{3}$	$\frac{\sqrt{3}}{2}$	$\frac{1}{2}$	$\sqrt{3}$	$\frac{\sqrt{3}}{3}$	2	$\frac{2\sqrt{3}}{3}$
$\frac{\pi}{2}$	1	0	undefined	0	undefined	1
$\frac{2\pi}{3}$	$\frac{\sqrt{3}}{2}$	$-\frac{1}{2}$	$-\sqrt{3}$	$-\frac{\sqrt{3}}{3}$	-2	$\frac{2\sqrt{3}}{3}$
$\frac{3\pi}{4}$	$\frac{\sqrt{2}}{2}$	$-\frac{\sqrt{2}}{2}$	-1	-1	$-\sqrt{2}$	$\sqrt{2}$
$\frac{5\pi}{6}$	$\frac{1}{2}$	$-\frac{\sqrt{3}}{2}$	$-\frac{\sqrt{3}}{3}$	$-\sqrt{3}$	$-\frac{2\sqrt{3}}{3}$	2
π	0	-1	0	undefined	-1	undefined
$\frac{7\pi}{6}$	$-\frac{1}{2}$	$-\frac{\sqrt{3}}{2}$	$\frac{\sqrt{3}}{3}$	$\sqrt{3}$	$-\frac{2\sqrt{3}}{3}$	-2
$\frac{5\pi}{4}$	$-\frac{\sqrt{2}}{2}$	$-\frac{\sqrt{2}}{2}$	1	1	$-\sqrt{2}$	$-\sqrt{2}$
$\frac{4\pi}{3}$	$-\frac{\sqrt{3}}{2}$	$-\frac{1}{2}$	$\sqrt{3}$	$\frac{\sqrt{3}}{3}$	-2	$-\frac{2\sqrt{3}}{3}$
$\frac{3\pi}{2}$	-1	0	undefined	0	undefined	-1
$\frac{5\pi}{3}$	$-\frac{\sqrt{3}}{2}$	$\frac{1}{2}$	$-\sqrt{3}$	$-\frac{\sqrt{3}}{3}$	2	$-\frac{2\sqrt{3}}{3}$
$\frac{7\pi}{4}$	$-\frac{\sqrt{2}}{2}$	$\frac{\sqrt{2}}{2}$	-1	-1	$\sqrt{2}$	$-\sqrt{2}$
$\frac{11\pi}{6}$	$-\frac{1}{2}$	$\frac{\sqrt{3}}{2}$	$-\frac{\sqrt{3}}{3}$	$-\sqrt{3}$	$\frac{2\sqrt{3}}{3}$	-2

3. -0.5298 **5.** -7.6966 **7.** 2.1344 **9.** 0.8163 **11.** 0 **13.** -1 **15.** 0 **17.** $\sqrt{2}$ **19.** 100 **21.** 20 **23.** 0.5
25. 4689 **27.** 54.4 **29.** 3851

Trial Problems 7.3

1. amplitude $= |a| = |1| = 1$

period $= \left|\dfrac{2\pi}{b}\right| = \left|\dfrac{2\pi}{\frac{3}{2}}\right| = \dfrac{4\pi}{3}$

2. amplitude $= |a| = |1| = 1$

period $= \left|\dfrac{2\pi}{b}\right| = \left|\dfrac{2\pi}{3.5}\right| = \dfrac{4\pi}{7}$

3. amplitude $= |a| = |3.5| = 3.5$

period $= \left|\dfrac{2\pi}{b}\right| = \left|\dfrac{2\pi}{1}\right| = 2\pi$

4. amplitude $= |-1.4| = 1.4$

period $= \left|\dfrac{2\pi}{b}\right| = \left|\dfrac{2\pi}{3}\right| = \dfrac{2\pi}{3}$

5. amplitude $= |3.8| = 3.8$

period $= \left|\dfrac{2\pi}{b}\right| = \left|\dfrac{2\pi}{\frac{\pi}{4}}\right| = 8$

Exercises 7.3

1.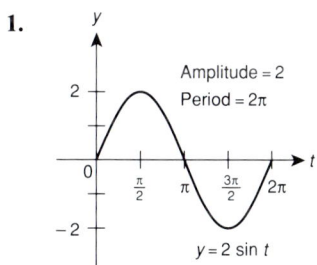

3.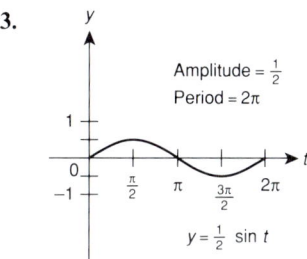

5.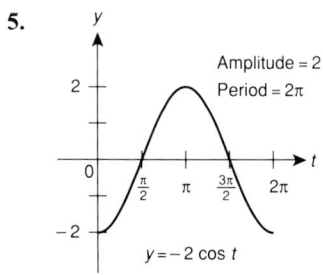

7.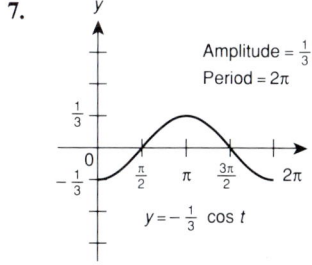

9.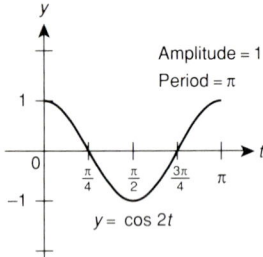

11.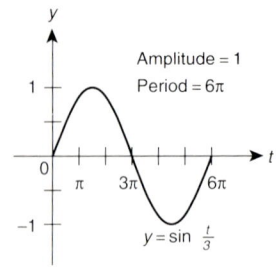

13.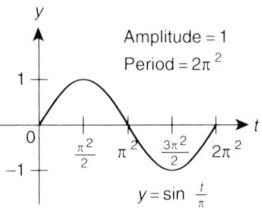

15.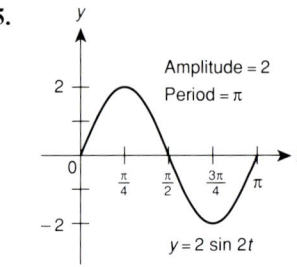

17.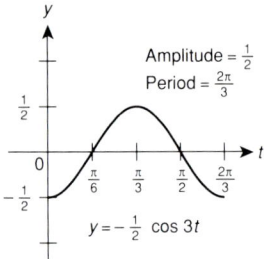

19.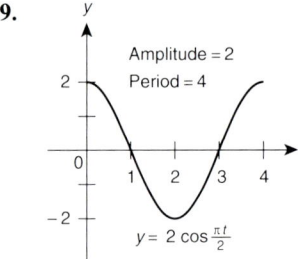

21.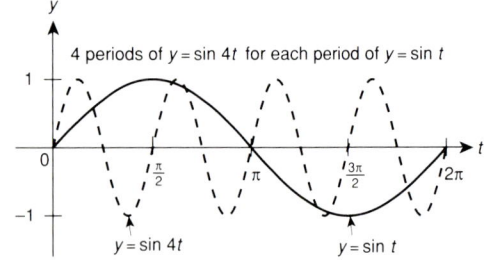

23.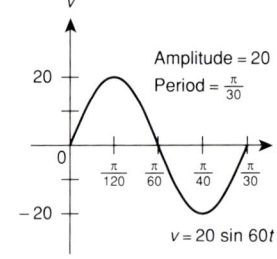

25.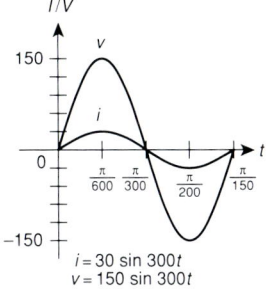

27.

t	$\sin t$
0.00	0.00
0.04	0.03999
0.08	0.0799
0.12	0.1197
0.16	0.1593
0.20	0.1987

29.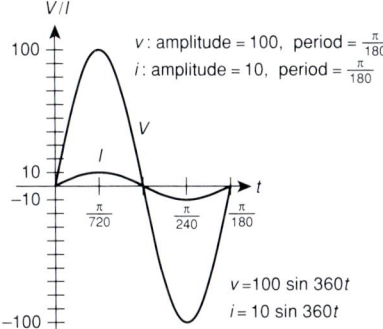

31.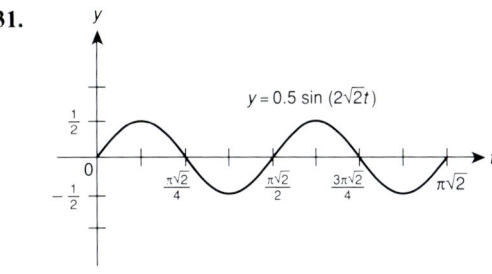

33.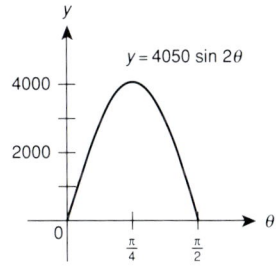

Trial Problems 7.4

1. period $= \left|\dfrac{2\pi}{b}\right| = \left|\dfrac{2\pi}{1}\right| = 2\pi$

phase shift $= -\dfrac{c}{|b|} = -\dfrac{-\frac{\pi}{8}}{1} = \dfrac{\pi}{8}$

2. period $= \left|\dfrac{2\pi}{b}\right| = \left|\dfrac{2\pi}{1}\right| = 2\pi$

phase shift $= -\dfrac{c}{|b|} = -\dfrac{\frac{\pi}{3}}{1} = -\dfrac{\pi}{3}$

3. period $= \dfrac{2\pi}{|2|} = \pi$

phase shift $= -\dfrac{c}{|b|} = -\dfrac{-\frac{\pi}{6}}{|2|} = \dfrac{\pi}{12}$

4. period $= \left|\dfrac{2\pi}{-3}\right| = \dfrac{2\pi}{3}$

phase shift $= -\dfrac{c}{|b|} = -\dfrac{-\frac{\pi}{12}}{|-3|} = \dfrac{\pi}{36}$

5. period $= \dfrac{2\pi}{|40|} = \dfrac{\pi}{20}$

phase shift $= -\dfrac{\frac{\pi}{6}}{|40|} = -\dfrac{\pi}{240}$

Exercises 7.4

1.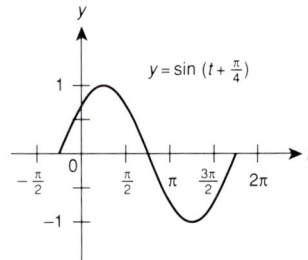

3.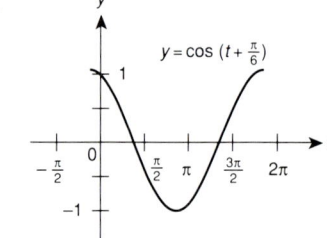

5.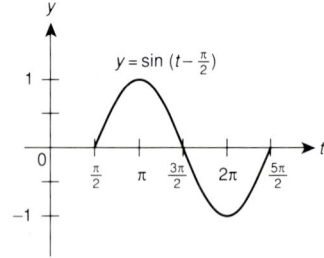

7.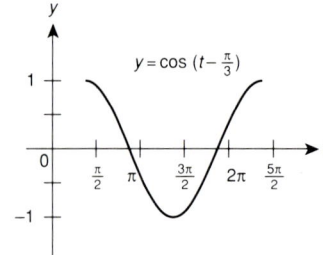

Answer Key—Chapter 7 1019

9.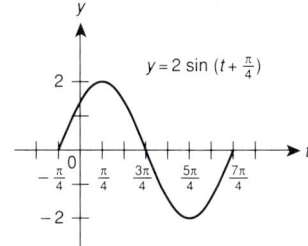

11.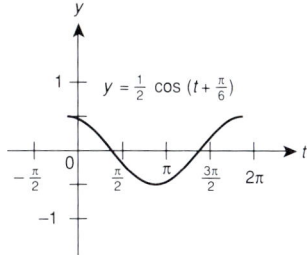

13.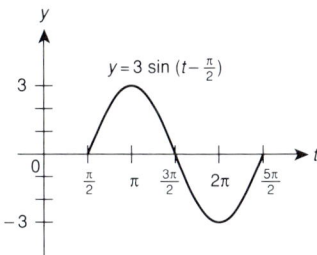

15.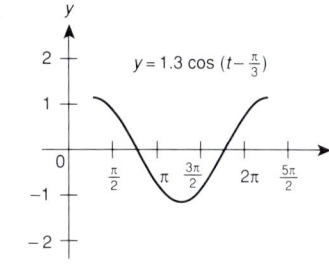

17.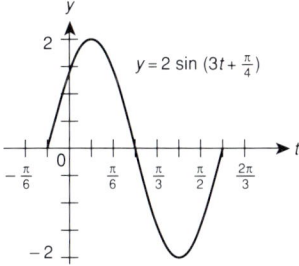

19.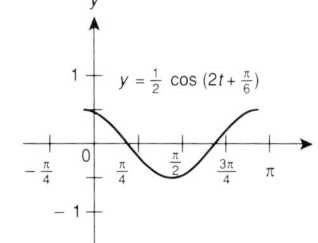

21.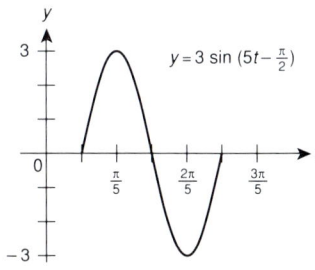

23.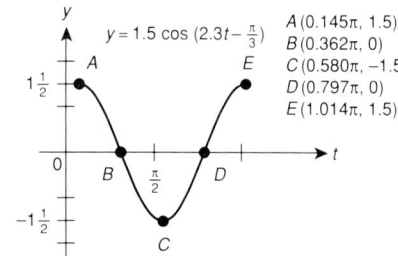

25.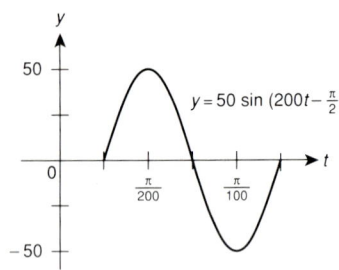

27.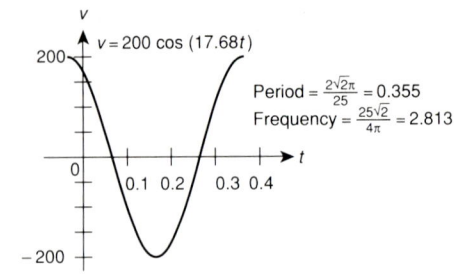

29.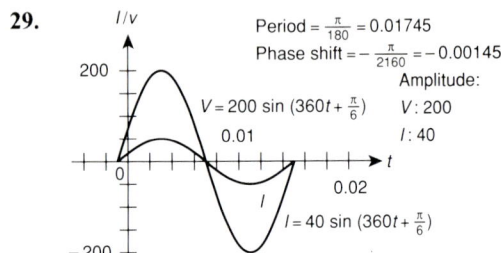

31. Maximum = 10
Period = $\dfrac{\pi}{100}$
Phase shift = 0 for i
Maximum = 200
Phase shift = $-\dfrac{\pi}{2}$
Period = $\dfrac{\pi}{100}$ for v

33. Amplitude = 1.4
$f = 150$

Trial Problems 7.5

1. $\tan 3t$ is undefined if $\cos 3t = 0$.
$\cos 3t = 0$ for $3t = \dfrac{\pi}{2}, \dfrac{3\pi}{2}, \dfrac{5\pi}{2}$
$t = \dfrac{\pi}{6}, \dfrac{\pi}{2}, \dfrac{5\pi}{6}$

2. $\cot \dfrac{t}{2}$ is undefined if $\sin \dfrac{t}{2} = 0$.
$\sin \dfrac{t}{2} = 0$ if $\dfrac{t}{2} = 0, \pi, 2\pi$
$t = 0, \pi, 4\pi$

3. $\dfrac{1}{2} \sec 2t$ is undefined if $\cos 2t = 0$.
$\cos 2t = 0$ if $2t = \dfrac{\pi}{2}, \dfrac{3\pi}{2}, \dfrac{5\pi}{2}$
$t = \dfrac{\pi}{4}, \dfrac{3\pi}{4}, \dfrac{5\pi}{4}$

4. $\csc \dfrac{t}{3}$ is undefined if $\sin \dfrac{t}{3} = 0$.
$\sin \dfrac{t}{3} = 0$ if $\dfrac{t}{3} = 0, \pi, 2\pi$
$t = 0, 3\pi, 6\pi$

5. $1.5 \tan 4.5t$ is undefined if $\cos 4.5t = 0$.
$\cos 4.5t = 0$ if $4.5t = \dfrac{\pi}{2}, \dfrac{3\pi}{2}, \dfrac{5\pi}{2}$
$t = \dfrac{\pi}{9}, \dfrac{\pi}{3}, \dfrac{5\pi}{9}$

Exercises 7.5

1.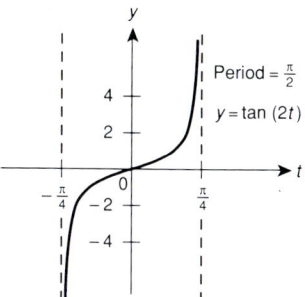

3.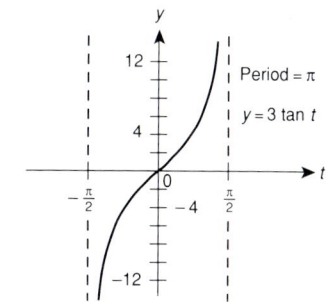

5.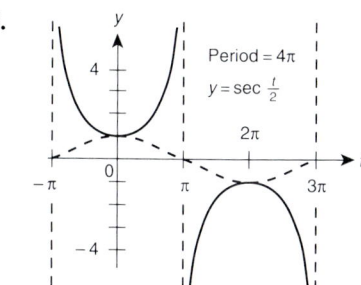

7.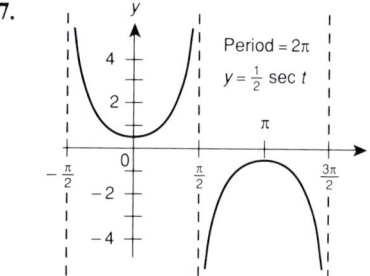

9.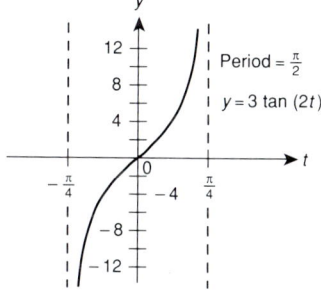

11.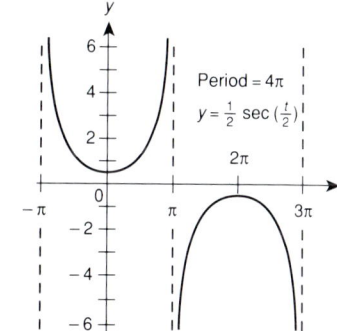

13.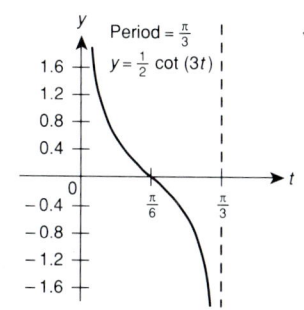

15.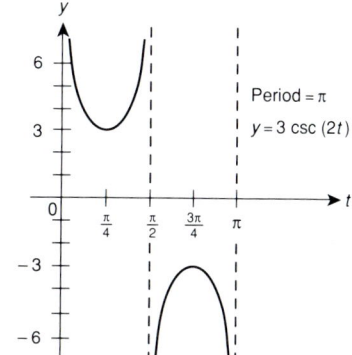

Trial Problems 7.6

1. Since the maximum and minimum values of sin t are 1 and -1, maximum $y = 2 - (-1) = 3$ and minimum $y = 2 - 1 = 1$.

2. Since the maximum and minimum values of cos t are 1 and -1, maximum $y = 3 + 1 = 4$ and minimum $y = 3 - 1 = 2$.

3. Since t can be made as large or as small as desired, and $-1 \leq \cos 2t \leq 1$, y has no maximum or minimum.

4. The maximum values of $2000 \cos\left(\dfrac{\pi t}{12}\right)$ are 2000 and -2000, maximum $y = 8000 + 2000 = 10{,}000$ and minimum $y = 8000 - 2000 = 6000$.

5. Since t has no maximum and minimum values, the function has no maximum and minimum values.

Exercises 7.6

1.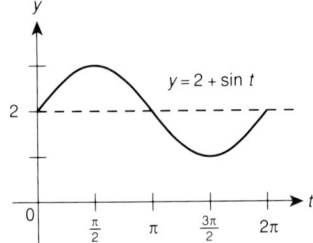

3.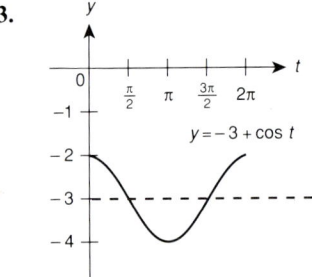

5.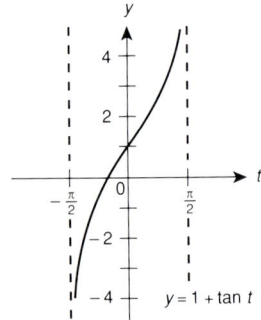

7.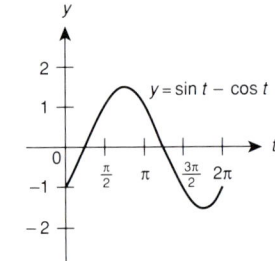

9.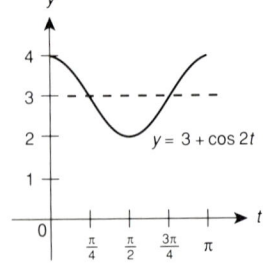

11.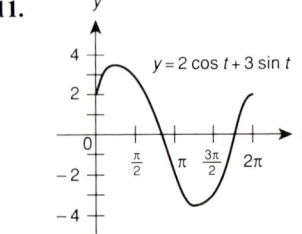

13.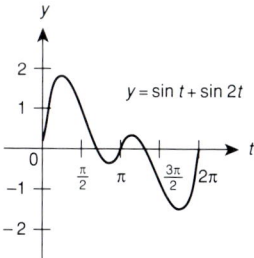

15.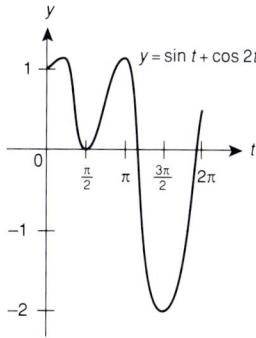

17.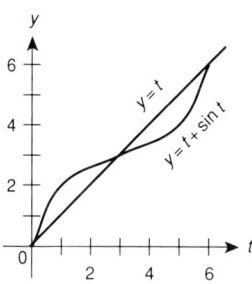

19.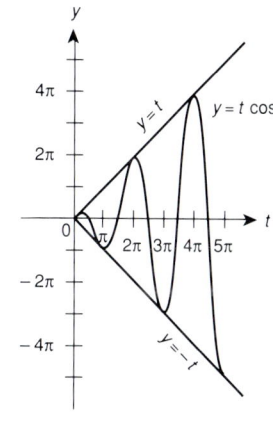

21.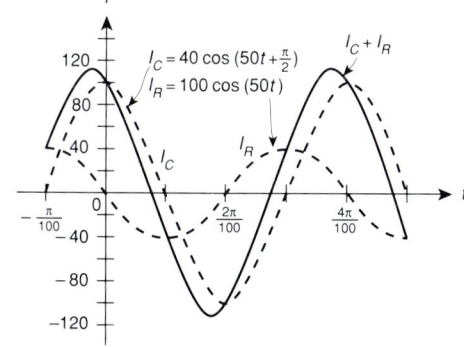

23.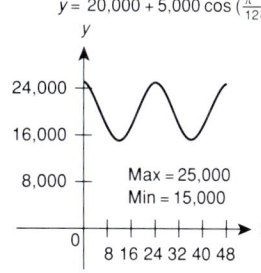

25.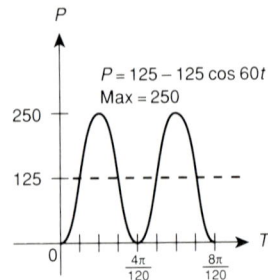

27.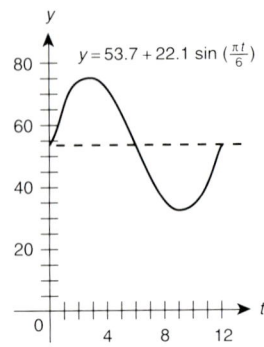

29.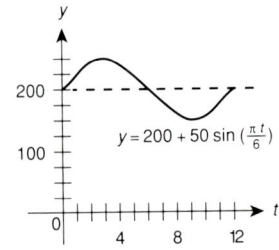

Trial Problems 7.7

1. 0.4632 [+/−] [INV] [COS]

Display: 2.0523988
$\cos^{-1}(-0.4632) = 2.0524$

2. 0.5617 [INV] [SIN]

Display: 0.5964391
$\sin^{-1}(0.5617) = 0.5964$

3. 2 [×] [π] [÷] 3 [=] [INV] [TAN]

Display: 1.1253388
$\tan^{-1}\left(\dfrac{2\pi}{3}\right) = 1.1253$

4. [π] [÷] 4 [=] [INV] [SIN] [TAN]

Display: 1.2688364
$\tan\left(\sin^{-1}\left(\dfrac{\pi}{4}\right)\right) = 1.2688$

5. 0.5 [INV] [SIN] [SIN]

Display: 0.5
$\sin(\sin^{-1}(0.5)) = 0.5$

Exercises 7.7

1. 1.0552 **3.** 2.0535 **5.** Does not exist **7.** 1.0802 **9.** −1.5442 **11.** 0.2649 **13.** 0.5069 **15.** 1.1928 **17.** 0.5224
19. 0.8520 **21.** 1.2688 **23.** 0.7232

25.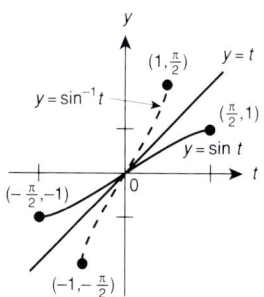

27.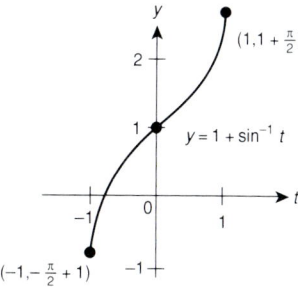

29.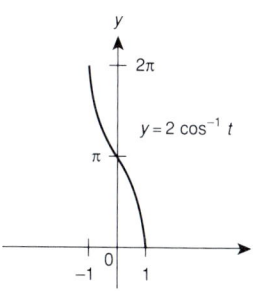

31.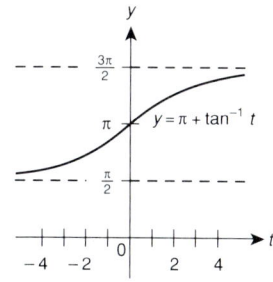

35. 0.654 radians or 37.5°   **37.** 0.524 radians or 30°   **39.** $\dfrac{\pi}{4}$ radians or 45°   **41.** 0.31 radians or 17.7°

Review Exercises 7.9

1. [7.1] 1.07   **3.** [7.1] 60   **5.** [7.1] 185.6   **7.** [7.1] 1.42 radians   **9.** [7.1] 2.5 cm   **11.** [7.1] $\dfrac{3.3}{\pi}$ ft
13. [7.1] 4.0 rad/sec   **15.** [7.2] -1 **17.** [7.2] -1 **19.** [7.2] -0.4755 **21.** [7.2] -20 **23.** [7.2] 27.94
25. [7.3] 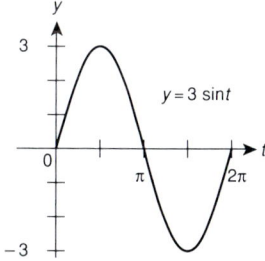   **27.** [7.3]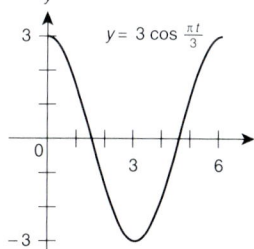

29. [7.3]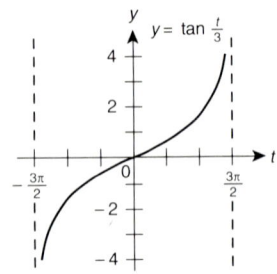

31. [7.4]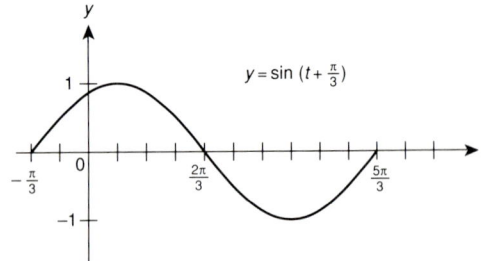

33. [7.4]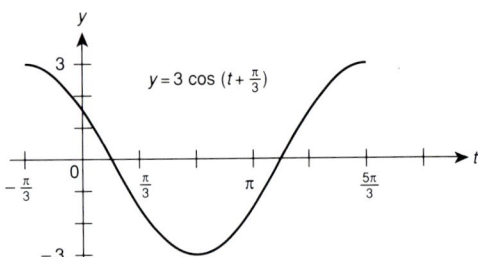

35. [7.4]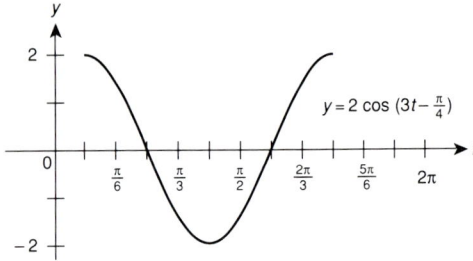

37. [7.4]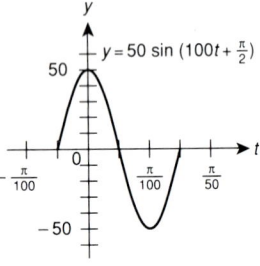

39. [7.6]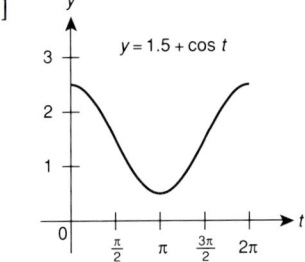

41. [7.6]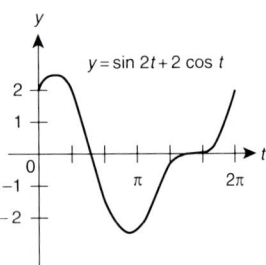

43. [7.7] 3.31

45. [7.7] 1.27   **47.** [7.7] -0.694 **49.** [7.7] $\dfrac{\pi}{4}$ **51.** [7.7]   **53.** [7.6] 8,800, 11,200

55. [7.6] 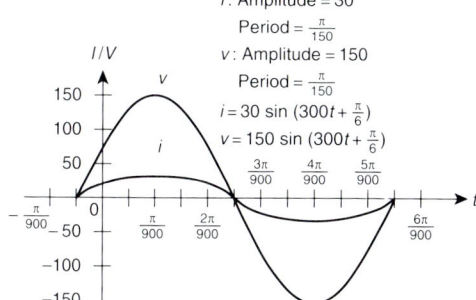   **57.** [7.7] 33.2°

Test 7.10

1. [7.1] 0.93   **2.** [7.1] 107.7   **3.** [7.1] 2.6 cm   **4.** [7.1] 2.97 radians   **5.** [7.1] 13.75 ft   **6.** [7.1] 46 rad/sec   **7.** [7.2] 1
8. [7.2] $-\dfrac{\sqrt{2}}{2}$ **9.** [7.2] -0.22 **10.** [7.2] 0   **11.** [7.2] 50.4   **12.** [7.2] 2.06

13. [7.3] 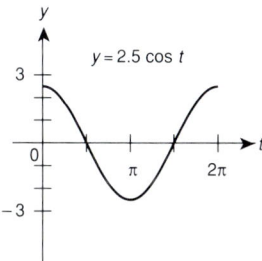   **14.** [7.3]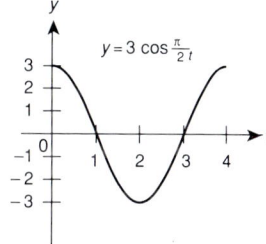

15. [7.5] 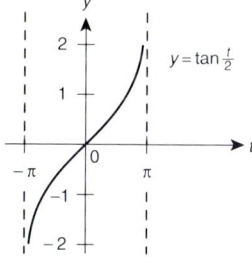   **16.** [7.4]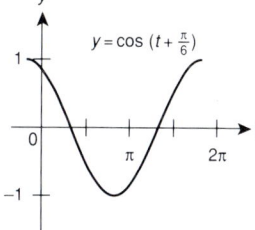

17. [7.4] 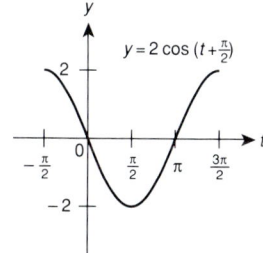   **18.** [7.4]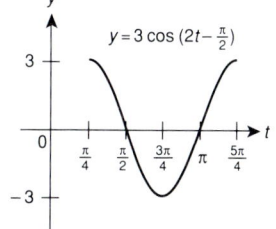

19. [7.4]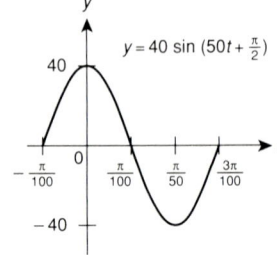

20. [7.6]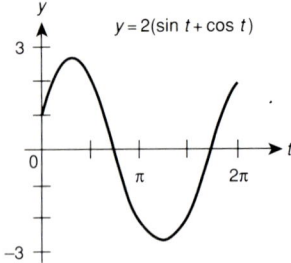

21. [7.6]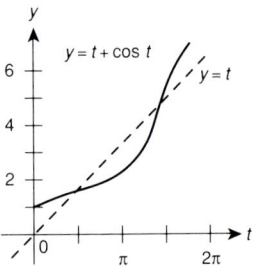

22. [7.7]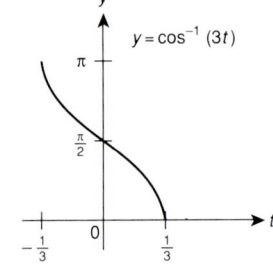

23. [7.2] 34.64

24. [7.2] -0.23 in.

Chapter 8 Answers

Trial Problems 8.1

1. $(\tan t)(\csc t)(\sec t) = \left(\dfrac{\sin t}{\cos t}\right)\left(\dfrac{1}{\sin t}\right)\left(\dfrac{1}{\cos t}\right)$

$= \dfrac{1}{\cos^2 t} = \sec^2 t$

2. $\dfrac{\cos t}{\cot t} = \dfrac{\cos t}{\dfrac{\cos t}{\sin t}} = \dfrac{\cos t \sin t}{\dfrac{\cos t}{\sin t} \sin t} = \dfrac{\cos t \sin t}{\cos t} = \sin t$

3. $\dfrac{\tan^2 t}{1 + \tan^2 t} = \dfrac{\dfrac{\sin^2 t}{\cos^2 t}\cos^2 t}{\left(1 + \dfrac{\sin^2 t}{\cos^2 t}\right)\cos^2 t} = \dfrac{\sin^2 t}{\cos^2 t + \sin^2 t} = \dfrac{\sin^2 t}{1} = \sin^2 t$

4. $\dfrac{\sin t + \sin t \cot^2 t}{\csc^2 t} = \dfrac{\sin t + \sin t \left(\dfrac{\cos^2 t}{\sin^2 t}\right)}{\dfrac{1}{\sin^2 t}}$

$= \left[\sin t + \sin t \left(\dfrac{\cos^2 t}{\sin^2 t}\right)\right](\sin^2 t)$

$= \sin^3 t + \sin t \cos^2 t$
$= \sin t(\sin^2 t + \cos^2 t)$
$= \sin t(1) = \sin t$

Exercises 8.1

1. $\tan t$ **3.** $\csc t$ **5.** $\sec t$ **7.** $\tan t$ **9.** $\cot t$ **11.** 1 **13.** $\sin t$ **15.** $\cos t$ **17.** 1 **19.** $2 \sec t$ or $\dfrac{2}{\cos t}$

Trial Problems 8.2

1. $\cos(u + v) = \cos u \cos v - \sin u \sin v$

$= \dfrac{\sqrt{2}}{2} \dfrac{2\sqrt{5}}{5} - \dfrac{\sqrt{2}}{2} \dfrac{\sqrt{5}}{5}$

$= \dfrac{2\sqrt{10}}{10} - \dfrac{\sqrt{10}}{10} = \dfrac{\sqrt{10}}{10}$

2. $\cos(u - v) = \cos u \cos v + \sin u \sin v$

$= \dfrac{\sqrt{2}}{2} \dfrac{2\sqrt{5}}{5} + \dfrac{\sqrt{2}}{2} \dfrac{\sqrt{5}}{5}$

$= \dfrac{2\sqrt{10}}{10} + \dfrac{\sqrt{10}}{10} = \dfrac{3\sqrt{10}}{10}$

3. $\cos 2u = 2 \cos^2 u - 1$

$= 2\left(\dfrac{\sqrt{2}}{2}\right)^2 - 1 = 1 - 1 = 0$

4. $\cos \dfrac{u}{2} = \sqrt{\dfrac{1 + \cos u}{2}}$

$= \sqrt{\dfrac{1 + \sqrt{2}/2}{2}} = \sqrt{\dfrac{2 + \sqrt{2}}{4}} = \dfrac{\sqrt{2 + \sqrt{2}}}{2}$

5. $\cos 2v = 2 \cos^2 v - 1$

$= 2\left[\dfrac{2\sqrt{5}}{5}\right]^2 - 1 = \dfrac{8}{5} - 1 = \dfrac{3}{5}$

Exercises 8.2

1. $\dfrac{\sqrt{6} - \sqrt{2}}{4}$ 3. $-\dfrac{\sqrt{2} - \sqrt{6}}{4}$ 5. $\dfrac{\sqrt{2 - \sqrt{2}}}{2}$ 7. $-\dfrac{\sqrt{2 - \sqrt{2}}}{2}$ 9. $\dfrac{4}{5}$ 11. $\dfrac{16}{65}$ 13. $-\dfrac{7}{25}$ 15. $\dfrac{2\sqrt{5}}{5}$ 17. $\dfrac{\sqrt{10}}{4}$

19. $\dfrac{\sqrt{15}}{5}$ 21. $\dfrac{2\sqrt{10}}{5}$ 23. $\sin 2t = -\dfrac{120}{169}$, $\cos 2t = \dfrac{119}{169}$, $\tan 2t = -\dfrac{120}{119}$ 25. $\cos \dfrac{t}{2} = -\dfrac{2\sqrt{13}}{13}$,

$\sin \dfrac{t}{2} = \dfrac{3\sqrt{13}}{13}$, $\tan \dfrac{t}{2} = -\dfrac{3}{2}$ 33. $\cos 6t = 2\cos^2 3t - 1$ 37. $\dfrac{\sqrt{2 + \sqrt{2 + \sqrt{2}}}}{2}$ 41. $3.62 \cos(20t - 0.488)$

43. $f(t)$ $y = 3\cos 2t + 4\sin 2t$ 45. i $i = 1.5\cos 20t + 2\sin 20t$

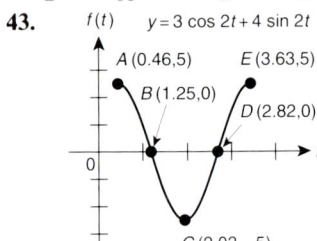

A (0.46, 5)   E (3.63, 5)
B (1.25, 0)
D (2.82, 0)
C (2.03, −5)

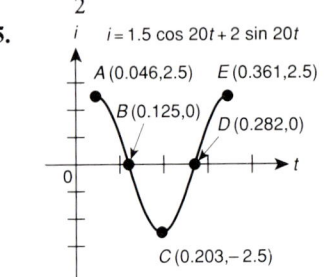

A (0.046, 2.5)   E (0.361, 2.5)
B (0.125, 0)
D (0.282, 0)
C (0.203, −2.5)

47. $y = 0.32(\cos 2t - 38.7°)$

Trial Problems 8.3

1. $\sin(u + v) = \sin u \cos v + \cos u \sin v$

$$= \dfrac{\sqrt{2}}{2} \dfrac{2\sqrt{5}}{5} + \dfrac{\sqrt{2}}{2} \dfrac{\sqrt{5}}{5}$$

$$= \dfrac{2\sqrt{10}}{10} + \dfrac{\sqrt{10}}{10} = \dfrac{3\sqrt{10}}{10}$$

2. $\sin(u - v) = \sin u \cos v - \cos u \sin v$

$$= \dfrac{\sqrt{2}}{2} \dfrac{2\sqrt{5}}{5} - \dfrac{\sqrt{2}}{2} \dfrac{\sqrt{5}}{5}$$

$$= \dfrac{2\sqrt{10}}{10} - \dfrac{\sqrt{10}}{10} = \dfrac{\sqrt{10}}{10}$$

3. $\sin 2v = 2 \sin v \cos v$

$$= 2 \dfrac{\sqrt{5}}{5} \dfrac{2\sqrt{5}}{5} = \dfrac{4(5)}{25} = \dfrac{4}{5}$$

4. $\sin \dfrac{v}{2} = \sqrt{\dfrac{1 - \cos v}{2}} = \sqrt{\dfrac{1 - \dfrac{2\sqrt{5}}{5}}{2}} = \sqrt{\dfrac{5 - 2\sqrt{5}}{10}}$

5. $\tan \dfrac{v}{2} = \dfrac{1 - \cos v}{\sin v} = \dfrac{1 - \dfrac{2\sqrt{5}}{5}}{\dfrac{\sqrt{5}}{5}} = \dfrac{5 - 2\sqrt{5}}{\sqrt{5}} = \dfrac{5\sqrt{5} - 10}{5} = \sqrt{5} - 2$

Exercises 8.3

1. $\dfrac{\sqrt{2}+\sqrt{6}}{4}$ 3. $\sqrt{\dfrac{2-\sqrt{3}}{2}}$ 5. $\sqrt{\dfrac{2-\sqrt{3}}{2}}$ 7. $\sqrt{2}+1$ 9. $\dfrac{3}{5}$ 11. $\dfrac{63}{65}$ 13. $\dfrac{24}{25}$ 15. $\dfrac{\sqrt{5}}{5}$ 17. $\dfrac{63}{16}$ 19. $-\dfrac{24}{7}$

21. $\dfrac{1}{2}$ 23. $\dfrac{\sqrt{6}}{6}$ 25. $\dfrac{\sqrt{5}}{5}$ 27. $\dfrac{\sqrt{30}}{5}$ 37. $\sin 8t = 2(\sin 4t)(\cos 4t)$ 39. $\dfrac{\sqrt{2-\sqrt{2+\sqrt{2}}}}{2}$

43. $\sec\dfrac{t}{2} = \dfrac{\sqrt{2-2\cos t}}{2}$

Trial Problems 8.4

1. Use $u = \dfrac{\pi}{8}$, $v = \dfrac{\pi}{8}$, in theorem 8.13.

$$\sin u \cos v = \dfrac{\sin(u+v) + \sin(u-v)}{2}$$

$$\sin\dfrac{\pi}{8}\cos\dfrac{\pi}{8} = \dfrac{\sin\dfrac{\pi}{4} + \sin 0}{2} = \dfrac{\dfrac{\sqrt{2}}{2}+0}{2} = \dfrac{\sqrt{2}}{4}$$

2. Use $u = \dfrac{\pi}{8}$, $v = \dfrac{\pi}{8}$ with theorem 8.14.

$$\sin\dfrac{\pi}{8}\sin\dfrac{\pi}{8} = \dfrac{\cos 0 - \cos\dfrac{\pi}{4}}{2} = \dfrac{1-\dfrac{\sqrt{2}}{2}}{2} = \dfrac{2-\sqrt{2}}{4}$$

3. $\cos\dfrac{\pi}{8}\cos\dfrac{\pi}{8} = \dfrac{\cos\dfrac{\pi}{4}+\cos 0}{2} = \dfrac{\dfrac{\sqrt{2}}{2}+1}{2} = \dfrac{\sqrt{2}+2}{4}$

4. $2\cos\left[\dfrac{t+\pi}{2}\right]\cos\left[\dfrac{t-\pi}{2}\right] = \cos u + \cos v = \cos t + \cos\pi = \cos t - 1$ (where $u = t$, $v = \pi$)

5. $2\cos\left[\dfrac{t+\pi}{2}\right]\sin\left[\dfrac{t-\pi}{2}\right] = \sin u - \sin v = \sin t - \sin\pi = \sin t$ (where $u = t$, $v = \pi$)

Exercises 8.4

1. $\dfrac{\sin 10t - \sin 2t}{2}$ 3. $\dfrac{\cos\dfrac{t}{2}+\cos\dfrac{t}{6}}{2}$ 5. $\dfrac{\cos(-4)-\cos 12t}{2}$ 7. $\dfrac{\sin(8t+2\pi)}{2}$ 9. $\dfrac{1}{2}\cos\left(4t-\dfrac{\pi}{6}\right)$

11. $2\cos(4t)(\sin t)$ 13. $-2\sin\left(\dfrac{5t}{2}\right)\sin\left(\dfrac{3t}{2}\right)$ 15. $2\sin\left(\dfrac{5t-5}{2}\right)\cos\left(\dfrac{t+1}{2}\right)$ 17. $2\cos(4t)\cos(-1)$

19. $\sqrt{3}\sin\left(50t+\dfrac{\pi}{2}\right)$ 21. $2\sin(4t)\cos t$ 23. $2\cos\left(\dfrac{5t}{2}\right)\cos\left(\dfrac{3t}{2}\right)$ 25. $2\cos\left(\dfrac{5t-5}{2}\right)\sin\left(\dfrac{t+1}{2}\right)$

27. $-2\sin(4t)\sin(-1)$ 33. $\dfrac{1}{2}+\dfrac{\sqrt{3}}{4}$

35.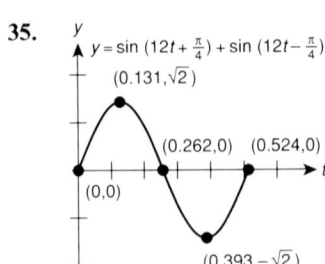

37.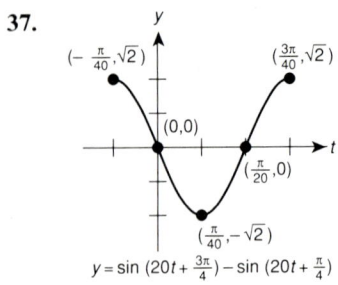

39. $-1000 \cos\left(200t + \dfrac{7\pi}{6}\right)$ **41.** $125 \sin\left(2\omega t + \dfrac{25\pi}{9}\right)$ **43.** $24(-\cos(40t + 2b) + 1)$

49. $i = a[\sin(2k_1\pi t) + \dfrac{b}{2}\cos(2(k_1 - k_2)\pi t) - \dfrac{b}{2}\cos(2(k_1 + k_2)\pi t)]$

Trial Problems 8.5

1. $\sin 3t = 1$ if $3t = \dfrac{\pi}{2}$

$$t = \dfrac{\pi}{6} \text{ or } 30°$$

2. $\cos \pi t = 0$ if $\pi t = \dfrac{\pi}{2}$

$$t = \dfrac{1}{2}$$

3. $2 \cos t + 1 = 0$ if $2 \cos t = -1$

$$\cos t = -\dfrac{1}{2}$$

$$t = \dfrac{2\pi}{3} \text{ or } 120°$$

4. $\sin^2 t - 2 \sin t + 1 = 0$

$(\sin t - 1)^2 = 0$

$\sin t - 1 = 0$

$\sin t = 1$

$$t = \dfrac{\pi}{2}$$

5. Use theorem 8.17 on the left-hand side of the equation with $u = 3t$ and $v = t$:

$$\sin 3t + \sin t = \sin\left(\frac{3t+t}{2}\right)\cos\left(\frac{3t-t}{2}\right)$$
$$= 2\sin 2t \cos t.$$

The equation becomes:
$$2\sin 2t \cos t = \cos t$$
$$2\sin 2t \cos t - \cos t = 0$$
$$\cos t(2\sin 2t - 1) = 0$$

$$\cos t = 0 \quad\bigg|\quad 2\sin 2t - 1 = 0$$
$$t = \frac{\pi}{2} \quad\bigg|\quad \sin 2t = \frac{1}{2}$$
$$\quad\bigg|\quad 2t = \frac{\pi}{6}$$
$$\quad\bigg|\quad t = \frac{\pi}{12}$$

The smallest positive solution is $t = \frac{\pi}{12}$.

Exercises 8.5

1. $\frac{\pi}{6} + 2k\pi, \frac{5\pi}{6} + 2k\pi$ **3.** $\frac{\pi}{4} + k\pi$ **5.** $\frac{\pi}{4} + \frac{k\pi}{2}$ **7.** $\frac{1}{6} + 2k, \frac{11}{6} + 2k$ **9.** $t = \frac{\pi}{2} + k\pi, \frac{\pi}{4} + k\pi$
11. $\frac{\pi}{4}, \frac{3\pi}{4}, \frac{5\pi}{4}, \frac{7\pi}{4}$ **13.** $\frac{7\pi}{6}, \frac{11\pi}{6}$ **15.** $\frac{\pi}{2}, \pi$ **17.** $\frac{\pi}{3}, \frac{5\pi}{3}, 0, \pi, 2\pi$ **19.** $0, \frac{\pi}{2}, \pi, \frac{3\pi}{2}, 2\pi$ **21.** $\frac{\pi}{6}, \frac{5\pi}{6}, \frac{7\pi}{6}, \frac{11\pi}{6}$
23. $\frac{\pi}{6}, \frac{5\pi}{6}$ **25.** $\frac{\pi}{4}, \frac{5\pi}{4}$ **27.** $0, \frac{\pi}{3}, \frac{5\pi}{3}, 2\pi$ **29.** $\frac{\pi}{12}, \frac{5\pi}{12}, \frac{13\pi}{12}, \frac{17\pi}{12}, \frac{\pi}{2}, \frac{3\pi}{2}$ **31.** No solutions
33. $\frac{\pi}{4}, \frac{5\pi}{4}, 0.2450, 3.3866$ **35.** $\frac{k\pi}{4}$ **37.** $-\frac{\pi}{8}, \frac{3\pi}{8}, \frac{7\pi}{8}$ **39.** $-\frac{\pi}{6} - \frac{1}{3}, \frac{\pi}{2} - \frac{1}{3}, \frac{7\pi}{6} - \frac{1}{3}, \frac{11\pi}{6} - \frac{1}{3}$
41. $\frac{\pi}{30}, \frac{\pi}{15}, \frac{\pi}{10}$ **43.** $\frac{\pi}{150}, \frac{\pi}{60}, \frac{2\pi}{75}$ **45.** 4.881×10^{-4} **47.** $39.2°$

Review Exercises 8.7

1. [8.1] 1   **3.** [8.1] 1   **5.** [8.1] 1   **7.** [8.1] $\sin t + 1$ **9.** [8.1] $\sin t \cos t$ **21.** [8.2,8.3] $\frac{2\sqrt{13}}{13}$ **23.** [8.2,8.3] $\frac{12}{13}$
25. [8.2,8.3] $-\frac{12}{5}$ **27.** [8.4] $2(\cos(10t) + \cos(2t))$ **29.** [8.4] $\frac{1}{2}(\sin(6t) - \sin(2t))$ **31.** [8.4] $-2\sin(3t)\sin(1)$
33. [8.4] $2\cos(3t)\cos t$ **35.** [8.5] $\frac{3\pi}{4}, \frac{5\pi}{4}$ **37.** [8.5] $\frac{\pi}{2}, \frac{7\pi}{6}, \frac{11\pi}{6}, \frac{5\pi}{2}, \frac{19\pi}{6}, \frac{23\pi}{6}$ **39.** [8.5] $\frac{\pi}{2}, \pi, \frac{3\pi}{2}, \frac{5\pi}{2}, 3\pi, \frac{7\pi}{2}$
41. [8.5] $\frac{\pi}{3}, \pi, \frac{5\pi}{3}$ **43.** [8.5] $0, \frac{\pi}{3}, \frac{5\pi}{3}, 2\pi$ **45.** [8.5] $\frac{\pi}{50}$ **47.** [8.5] 1.6667×10^{-3} **49.** [8.3] 0.3598
51. [8.3] 2.6677   **53.** [8.3] 0.8520   **55.** [8.2] 0.03062   **57.** [8.2] $y = 3.14\cos\left(\frac{\pi t}{20} - 1.01\right)$

Test 8.8

1. [8.1] $\cos t$ **2.** [8.1] $\sin^2 t$ **3.** [8.1] 1 **4.** [8.1] $\dfrac{1}{\sin^2 t \cos^2 t}$ **5.** [8.1] $\sin t$ **11.** [8.2,8.3] $\dfrac{\sqrt{10}}{5}$ **12.** [8.2,8.3] $\dfrac{3\sqrt{6}+2}{10}$
13. [8.2,8.3] $\dfrac{2\sqrt{6}}{5}$ **14.** [8.2,8.3] $\sqrt{\dfrac{10+3\sqrt{10}}{20}}$ **15.** [8.2,8.3] $\dfrac{3}{4}$ **16.** [8.4] $\cos 6t + \cos 2t$ **17.** [8.4] $\dfrac{\sin 6t + \sin 2t}{2}$
18. [8.4] $-2\sin(2t)\sin(1)$ **19.** [8.4] $2\cos\left(\dfrac{9t}{2}\right)\cos\left(\dfrac{3t}{2}\right)$ **20.** [8.5] $\dfrac{5\pi}{4}, \dfrac{7\pi}{4}$ **21.** [8.5] $\dfrac{7\pi}{8}, \dfrac{15\pi}{8}, \dfrac{23\pi}{8}, \dfrac{31\pi}{8}$ **22.** [8.5] $\dfrac{\pi}{2}$
23. [8.5] $\dfrac{\pi}{6}, \dfrac{5\pi}{6}, \dfrac{7\pi}{6}, \dfrac{11\pi}{6}$ **24.** [8.5] $0, \dfrac{\pi}{4}, \dfrac{\pi}{2}, \pi, \dfrac{5\pi}{4}$ **25.** [8.5] $\dfrac{7\pi}{6}, \dfrac{11\pi}{6}$ **26.** [8.5] $\dfrac{3\pi}{100}, \dfrac{7\pi}{100}$ **27.** [8.5] 0.098 sec
28. [8.2] $i = 1.37\cos(8t + 1.40)$

Cumulative Review—Chapters 6, 7, and 8 8.9

1. [6] 10.75 **2.** [6] 25.2, 16.8 **3.** [6] cosine is 0.6157, tangent is 1.2799 **4.** [6] sine is 0.3846, cosine is 0.9231, tangent is 0.4167 **5.** [6] 50.4° **6.** [6] 27.5 **7.** [6] 380 **8.** [6] $b = 11.0, c = 13.9$ **9.** [6] No such triangle
10. [6] $A = 41.8°, B = 51.0°$ **11.** [7] 1.42 **12.** [7] 81.4 **13.** [7] 21 rad/sec **14.** [7] 0.42 **15.** [7] 2.15
16. [7] 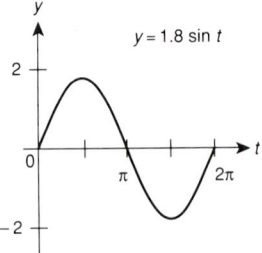 **17.** [7]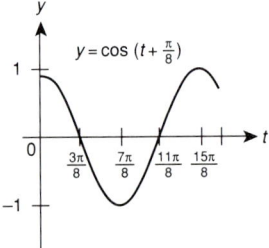

18. [7] 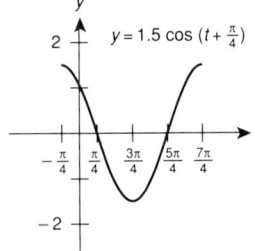 **19.** [7]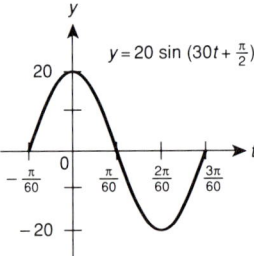

20. [7]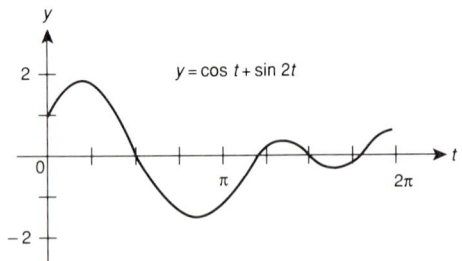

21. [8] 1   **22.** [8] $\sin t$
23. [7]

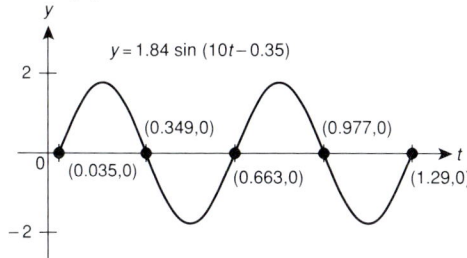

24. [8] $\dfrac{\sqrt{15}}{4}$ **25.** [8] $\dfrac{4 - 3\sqrt{15}}{20}$ **26.** [8] $\dfrac{7}{25}$ **27.** [8] $\dfrac{\sqrt{10}}{4}$ **28.** [8] 3   **29.** [8] $\dfrac{\pi}{6}, \dfrac{11\pi}{6}$ **30.** [8] $\dfrac{7\pi}{12}, \dfrac{5\pi}{4}, \dfrac{23\pi}{12}$
31. [8] $0, \dfrac{\pi}{2}, \dfrac{3\pi}{2}, 2\pi$ **32.** [8] $\dfrac{\pi}{6}, \dfrac{3\pi}{2}, \dfrac{5\pi}{6}$ **33.** [8] 0.034, 10.036

Chapter 9 Answers

Trial Problems 9.1

1. $|V_x| = |V| \cos \theta = 8.6 \cos 48° = 5.8$
$|V_y| = |V| \sin \theta = 8.6 \sin 48° = 6.4$

2. $|V_x| = 6.0 \cos 32° = 5.1$
$|V_y| = 6.0 \sin 32° = 3.2$

3. $|V_x| = 2.34 \cos 1.31 = 0.60$
$|V_y| = 2.34 \sin 1.31 = 2.26$

4. $\alpha = 180° - 41.3° = 137.8°$
$|V + W| = \sqrt{V^2 + W^2 - 2VW \cos \alpha}$
$ = \sqrt{(110)^2 + (220)^2 - 2(110)(220) \cos 137.8°}$
$ = 310$

5. $\alpha = 3.14 - 2.34 = 0.80$ radians
$|V + W| = \sqrt{(0.53)^2 + (0.76)^2 - 2(0.53)(0.76) \cos 0.80}$
$ = 0.55$

Exercises 9.1

1. $V_1 = V_3 = V_7$ **3.** V_1 and V_2, V_2 and V_3, V_7 and V_2
5. $-V_1$

7. $2V_1$

9. $-3V_1$

11. $V_1 + V_2$, V_2, V_1

13.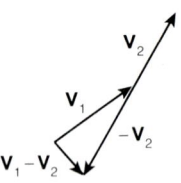

15.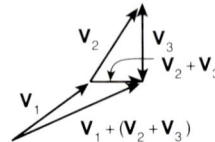

17.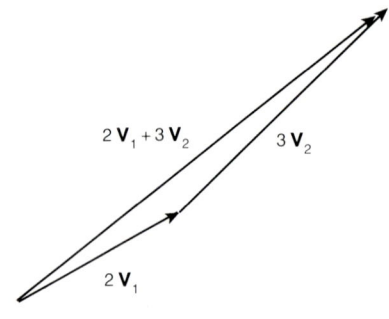

19. 2.6, 5.8   **21.** 4.44, 4.65   **23.** 1.35, 3.82   **25.** 492   **27.** 7.26   **29.** 67.7   **31.** 11.7, 22.1   **33.** 0.42, 40.2°
35. 6.6, 3.2   **37.** 204 mi   **39.** 13.1 lb   **41.** 6.4, 44.4°   **43.** 13.8, 44.7°   **45.** 687 ft-lb

Trial Problems 9.2

1. $(2 + 3j) - (5 - 5j) = (2 - 5) + (3j - (-5j))$
$$= -3 + 8j$$

2. $(2 + j)(3 - j) = (2)(3) + (3 - 2)j - j^2$
$$= 6 + j - j^2$$
$$= 6 + j + 1$$
$$= 7 + j$$

3. $\dfrac{2 + j}{3 - j} = \dfrac{(2 + j)(3 + j)}{(3 - j)(3 + j)}$
$$= \dfrac{6 + 5j + j^2}{9 - j^2}$$
$$= \dfrac{5 + 5j}{10}$$
$$= \dfrac{1}{2} + \dfrac{1}{2}j$$

4. $x = \dfrac{-2 \pm \sqrt{2^2 - 4(1)(6)}}{2(1)}$
$$= \dfrac{-2 \pm \sqrt{-20}}{2} = \dfrac{-2 \pm 2\sqrt{-5}}{2} = -1 \pm j\sqrt{5}$$

5. $j^{15} = j^{12}j^3 = (j^4)^3(j^3) = (1)^3(-j) = -j$

Exercises 9.2

1. $9 - 8j$ **3.** $3 + 10j$ **5.** $5 - 8j$ **7.** -2 **9.** $18 + j$ **11.** $5 - 12j$ **13.** 17 **15.** 13 **17.** $56 + 2j$
19. $\dfrac{1}{2} - \dfrac{1}{2}j$ **21.** $\dfrac{6}{13} - \dfrac{9}{13}j$ **23.** $\dfrac{5}{13} + \dfrac{12}{13}j$ **25.** $\dfrac{1}{2} + \dfrac{1}{2}j$ **27.** j **29.** 1 **31.** $2j$ **33.** $-2\sqrt{3}$ **35.** $-j\sqrt{6}$
37. -1 **39.** -1 **41.** -1 **43.** $-\dfrac{1}{2} \pm \dfrac{j\sqrt{3}}{2}$ **45.** $1 \pm j\sqrt{5}$ **47.** $-\dfrac{3}{5} \pm \dfrac{j\sqrt{11}}{5}$ **49.** $-\dfrac{3}{7} \pm \dfrac{2j\sqrt{3}}{7}$ **51.** $-2 + 3j$
53a. $2j$ **b.** $512j$ **55.** $9 - 12j$ **57.** $4.77 + 2.34j$ **59.** $7.02 + 3.10j$ **61.** $6.3 + 5.4j$

Trial Problems 9.3

1. $r = \sqrt{a^2 + b^2} = \sqrt{(\sqrt{3})^2 + (-1)^2} = \sqrt{4} = 2$
$\tan \theta = \dfrac{-1}{\sqrt{3}} = -\dfrac{\sqrt{3}}{3}$
$\theta = 330°$, since (a,b) is in the fourth quadrant.
$\sqrt{3} - j = 2(\cos 330° + j \sin 330°)$

2. Since $\cos \dfrac{\pi}{2} = 0$ and $\sin \dfrac{\pi}{2} = 1$,
$2\left(\cos \dfrac{\pi}{2} + j \sin \dfrac{\pi}{2}\right) = 2(0 + j) = 2j$

3. $uv = (2)(4)\left[\cos\left(\dfrac{\pi}{3} + \dfrac{\pi}{6}\right) = j \sin\left(\dfrac{\pi}{3} + \dfrac{\pi}{6}\right)\right]$
$= 8\left(\cos \dfrac{\pi}{2} + j \sin \dfrac{\pi}{2}\right)$

4. $\dfrac{u}{v} = \dfrac{2}{4}\left[\cos\left(\dfrac{\pi}{3} - \dfrac{\pi}{6}\right) + j \sin\left(\dfrac{\pi}{3} - \dfrac{\pi}{6}\right)\right]$
$= \dfrac{1}{2}\left(\cos \dfrac{\pi}{6} + j \sin \dfrac{\pi}{6}\right)$

5. $u^2 = \left[2\left(\cos \dfrac{\pi}{3} + j \sin \dfrac{\pi}{3}\right)\right]\left[2\left(\cos \dfrac{\pi}{3} + j \sin \dfrac{\pi}{3}\right)\right]$
$= 4\left[\cos\left(\dfrac{\pi}{3} + \dfrac{\pi}{3}\right) + j \sin\left(\dfrac{\pi}{3} + \dfrac{\pi}{3}\right)\right]$
$= 4\left(\cos \dfrac{2\pi}{3} + j \sin \dfrac{2\pi}{3}\right)$

Exercises 9.3

1.

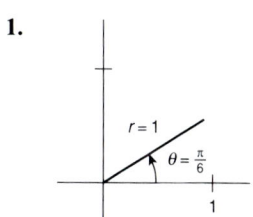

3.

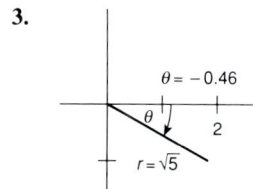

5.

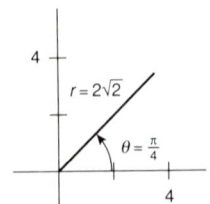

7.

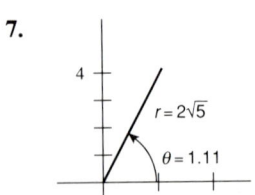

9.

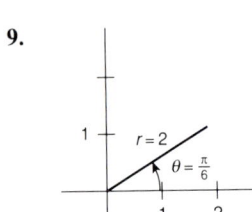

11. $\dfrac{\sqrt{2}}{2} + \dfrac{j\sqrt{2}}{2}$ **13.** -2 **15.** $-\dfrac{7}{2} + \dfrac{7j\sqrt{3}}{2}$ **17.** $-1 + j$ **19.** $1 - 3j$ **21.** $24\left(\cos\dfrac{7\pi}{12} + j\sin\dfrac{7\pi}{12}\right)$ **23.** $0.15\ cjs\ 60°$

25. $0.0864\ cjs\ 2.08$ **27.** $1.5\left(\cos\dfrac{\pi}{12} + j\sin\dfrac{\pi}{12}\right)$ **29.** $2\ cjs\ 30°$ **31.** $1.78\ cjs\ 0.5$ **33.** $0.732 + 2.732j$

35. $0.183 + 0.683j$ **37.** $2 - 2j\sqrt{3}$ **39.** $-\dfrac{1}{4} - \dfrac{1}{4}j$ **41.** $26.9\ \underline{/50°}$ **43.** $35.2\ \underline{/68°}$ **45.** $6.7\ \underline{/40°}$

Trial Problems 9.4

1. $[2(\cos(0.6) + j\sin(0.6))]^5 = 2^5[\cos(5)(0.6) + j\sin(5)(0.6)] = 32(\cos 3.0 + j\sin 3.0)$

2. $\left[\sqrt{2}\left(\cos\dfrac{\pi}{6} + j\sin\dfrac{\pi}{6}\right)\right]^4 = (\sqrt{2})^4\left(\cos\dfrac{4\pi}{6} + j\sin\dfrac{4\pi}{6}\right) = 4\left(\cos\dfrac{2\pi}{3} + j\sin\dfrac{2\pi}{3}\right)$

3. $(1 + j)^4 = \left[\sqrt{2}\left(\cos\dfrac{\pi}{4} + j\sin\dfrac{\pi}{4}\right)\right]^4$

$= (\sqrt{2})^4\left(\cos\dfrac{4\pi}{4} + j\sin\dfrac{4\pi}{4}\right)$

$= 4(\cos\pi + j\sin\pi)$
$= 4(-1 + 0j) = -4$

4. $(1 + j)^{1/2} = \left[(\sqrt{2})^{1/2}\left(\cos\dfrac{\pi}{4} + j\sin\dfrac{\pi}{4}\right)\right]^{1/2}$

$= \sqrt[4]{2}\left[\cos\left(\dfrac{1}{2}\dfrac{\pi}{4}\right) + j\sin\left(\dfrac{1}{2}\dfrac{\pi}{4}\right)\right]$

$= \sqrt[4]{2}\left(\cos\dfrac{\pi}{8} + j\sin\dfrac{\pi}{8}\right)$

5. $j^{1/2} = \left(\cos\dfrac{\pi}{2} + j\sin\dfrac{\pi}{2}\right)^{1/2}$

$= \cos\dfrac{\pi}{4} + j\sin\dfrac{\pi}{4}$

$= \dfrac{\sqrt{2}}{2} + j\dfrac{\sqrt{2}}{2}$

Exercises 9.4

1. 1   **3.** 4096   **5.** $\cos 5.28 + j\sin 5.28$ or $0.54 - 0.84j$ **7.** $27(\cos 5.61 + j\sin 5.61)$ or $21.1 - 16.83j$ **9.** $-8j$
11. $-128 - 221.7j$ **13.** $0.002 - 0.104j$ **15.** $0.125j$ **17.** $\frac{\sqrt{3}}{2} + \frac{3}{2}j, -\frac{3}{2} + \frac{\sqrt{3}}{2}j, -\frac{\sqrt{3}}{2} - \frac{3}{2}j, \frac{3}{2} - \frac{\sqrt{3}}{2}j$
19. $\sqrt{2} + j\sqrt{2}, -\sqrt{2} - j\sqrt{2}, 1.93 - 0.52j, -1.93 + 0.52j, 0.52 - 1.93j, -0.52 + 1.93j$
21. $2 + j\sqrt{2}, -0.91 + 1.78j, -1.98 - 0.31j, -0.31 - 1.98j, 1.78 - 0.91j$
23. $1.37 + 0.37j, -0.37 + 1.37j, -1.37 - 0.37j, 0.37 - 1.37j$ **25.** $1.57 + 0.69j, -1.38 + 1.01j, -0.19 - 1.71j$
27.

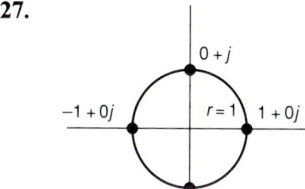

29.

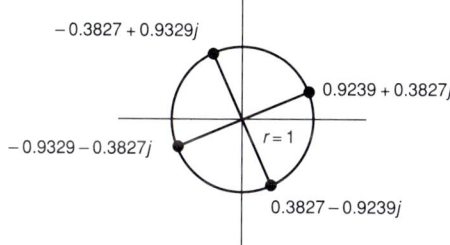

31. $j, -\frac{\sqrt{3}}{2} - \frac{1}{2}j, \frac{\sqrt{3}}{2} - \frac{1}{2}j$ **33.** $1.55 + 0.64j, -1.55 - 0.64j$

Review Exercises 9.6

9. [9.1] 2.6, 2.7   **11.** [9.1] 528, 334   **13.** [9.1] 213   **15.** [9.1] 151 newtons   **17.** [9.1] 55.4 lb   **19.** [9.1] 0.3936, -0.48
21. [9.2] $7 - 3j$ **23.** [9.2] $1 + 5j$ **25.** [9.2] $11 - 10j$ **27.** [9.2] $\frac{3}{13} - \frac{2}{13}j$ **29.** [9.2] $\frac{4}{13} + \frac{7}{13}j$ **31.** [9.2] j
33. [9.2] $3.3 + 0.16j$ **35.** [9.2] $0.66 + 0.93j$ **37.** [9.3] $3\sqrt{2}\left(\cos\frac{7\pi}{4} + j\sin\frac{7\pi}{4}\right)$ **39.** [9.3] $\cos\frac{\pi}{4} + j\sin\frac{\pi}{4}$
41. [9.3] $-\frac{\sqrt{2}}{2} - \frac{\sqrt{2}}{2}j$ **43.** [9.3] $1 + j$ **45.** [9.3] $2\,\text{cjs}\,\frac{3\pi}{2}$ or $-2j$ **47.** [9.3] $\sqrt{3}\,\text{cjs}\,\frac{7\pi}{12}$ **49.** [9.4] 4096
51. [9.4] $\sqrt[8]{2}\,\text{cjs}\,\frac{\pi}{16}, \sqrt[8]{2}\,\text{cjs}\,\frac{9\pi}{16}, \sqrt[8]{2}\,\text{cjs}\,\frac{17\pi}{16}, \sqrt[8]{2}\,\text{cjs}\,\frac{25\pi}{16}$

1040 ANSWER KEY—CHAPTER 10

Test 9.7

5. [9.1] 9.7, 14.3   **6.** [9.1] 506, 239   **7.** [9.1] 372   **8.** [9.1] 131 newtons   **9.** [9.2] $5 - 2j$ **10.** [9.2] $1 + 3j$
11. [9.2] $5 - j$ **12.** [9.2] $27.92 - 7.04j$ **13.** [9.2] $\frac{1}{10} + \frac{3}{10}j$ **14.** [9.2] $\frac{7}{17} - \frac{6}{17}j$ **15.** [9.3] $3\sqrt{2}\left(\cos\frac{\pi}{4} + j\sin\frac{\pi}{4}\right)$
16. [9.3] $2\left(\cos\frac{5\pi}{3} + j\sin\frac{5\pi}{3}\right)$ **17.** [9.3] $-\frac{\sqrt{3}}{2} + \frac{1}{2}j$ **18.** [9.3] $-\frac{1}{2} + \frac{\sqrt{3}}{2}j$ **19.** [9.3] $4\,cjs\,\frac{5\pi}{6}$ **20.** [9.3] $0.25\,cjs\,\frac{\pi}{2}$
21. [9.4] 1   **22.** [9.4] $-8 - 8j\sqrt{3}$ **23.** [9.4] 16   **24.** [9.4] $2\,cjs\,\frac{\pi}{8},\,2\,cjs\,\frac{5\pi}{8},\,2\,cjs\,\frac{9\pi}{8},\,2\,cjs\,\frac{13\pi}{8}$

Chapter 10 Answers

Trial Problems 10.1

1. $\left(-2,\frac{1}{4}\right)\left(-1,\frac{1}{2}\right)(0,1)\,(1,2)\,(2,4)$ **2.** $(-2,9)\,(-1,3)\,(0,1)\,\left(1,\frac{1}{3}\right)\left(2,\frac{1}{9}\right)$
3. $(0,1)\,(1,10)\,(2,100)\,(3,1000)$
4. $\left(2,\frac{1}{4}\right)\left(1,\frac{1}{2}\right)(-1,2)\,(-2,4)$
5. Decreasing

Exercises 10.1

1. 4   **3.** $\sqrt{2} \approx 1.414$ **5.** $\frac{1}{2}$ **7.** -1 **9.** 0.6
11.

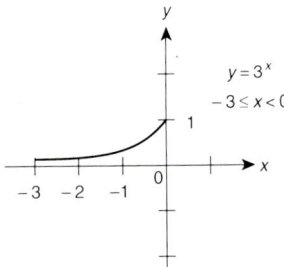

13.

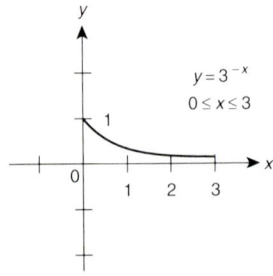

15.

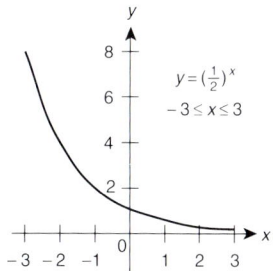

17. 40,000   **19.** 13.86 years

Trial Problems 10.2

1. $\log_5 25 = 2$ **2.** $\log_r t = s$ **3.** $2^3 = 8$ **4.** $10^2 = 100$ **5.** $x = 2$

Exercises 10.2

1. $2 = \log_3 9$ **3.** $\frac{1}{4} = \log_{16} 2$ **5.** $-\frac{1}{3} = \log_8 \frac{1}{2}$ **7.** $0 = \log_{10} 1$ **9.** $-1 = \log_{10} 0.1$ **11.** $-3 = \log_a \frac{1}{8}$
13. $2 = \log_x 100$ **15.** $x = \log_{10} 100$ **17.** $3 = \log_2 x$ **19.** $y = \log_3 9$ **21.** $b = \log_a 8$ **23.** $x = \log_2 y$ **25.** $x = \log_b y$
27. $s = \log_r \frac{1}{2}$ **29.** $x + y = \log_a s$ **31.** $36 = 6^2$ **33.** $1000 = 10^3$ **35.** $1 = 10^0$
37. $0.01 = 10^{-2}$ **39.** $x = 4^3$ **41.** $\frac{1}{8} = a^{-3}$ **43.** $100 = x^2$ **45.** $100 = 10^x$ **47.** $x = 2^3$ **49.** $9 = 3^y$ **51.** $8 = a^b$
53. $y = 2^x$ **55.** $y = b^x$ **57.** $\frac{1}{2} = r^s$ **59.** $s = a^{x+y}$ **61.** 2   **63.** 2   **65.** 1   **67.** 1   **69.** 0   **71.** 8   **73.** 2   **75.** -3
77. 1.5   **79.** $\frac{1}{2}$ **81.** $10^{0.4(J-H)} = \frac{B_1}{B_2}$

Trial Problems 10.3

1. $\log_a 3 + \log_a x$ **2.** $\log_b x - \log_b a$ **3.** $3 \log_x b$ **4.** $\log_b rs$ **5.** $\log_b \frac{u}{v}$

Exercises 10.3

1. $\log_b 2 + \log_b x$ **3.** $\log_b 3 + \log_b x + \log_b y$ **5.** $\log_b y - \log_b z$ **7.** $3 \log_b x$ **9.** $\frac{1}{3} \log_b x$ **11.** $\frac{3}{2} \log_b x$
13. $\log_{10} 2\pi - \frac{1}{2} \log_{10} g$, $\log_{10} 2 + \log_{10} \pi - \frac{1}{2} \log_{10} g$ **15.** $\log_b xy$ **17.** $\log_b \frac{x^3}{y^2}$ **19.** $\log_b \sqrt[3]{\frac{xy}{z^2}}$ **21.** $a = \frac{2}{3} b$ **23.** 7
25. -4

Trial Problems 10.4

1. 0.7226   **2.** 2.4771   **3.** -2.3979 **4.** 2.50   **5.** 22.8

Exercises 10.4

1. 0.516535 **3.** -1.910095 **5.** 1.144729 **7.** -1.440962 **9.** 1000 **11.** 65.463617 **13.** 3.141593 **15.** 1.003219
17. -0.41 **19.** 2.2 **21.** 3.56 **23.** 0.08 **25.** -0.205

Trial Problems 10.5

1. $x = \dfrac{\log 3}{\log 5} = 0.68261$ **2.** $x = \dfrac{-1}{\log 6} = -1.2851$ **3.** $x = 10{,}004$ **4.** $x = \dfrac{4}{3}$ **5.** $x = 7, x = -2$

Exercises 10.5

1. $x = \dfrac{\log 2}{\log 3}$ **3.** $x = \dfrac{\log 5 - \log 3}{\log 4}$ **5.** $x = \dfrac{-\log 8}{\log 7}$ **7.** $x = \dfrac{\log 5}{\log 6}$ **9.** $x = 10^{-b - \log n}$ or $x = \dfrac{10^{-b}}{n}$
11. $x = 20$ or -5 **13.** $36\sqrt{6}$ **15.** $x = 2.43$ **17.** $x = 1$ **19.** $x = 3$ **21.** $\dfrac{-1}{0.000174} \log \dfrac{A}{A_0}$
23. 196,986,822.7, calculator solution: 196,986,822.8

Review Exercises 10.7

1. [10.1]

[Graph of $y = 3^x$, $-2 \le x \le 2$]

3. [10.2] $\log_2 8 = 3$ **5.** [10.2] $\log_{49} 7 = \dfrac{1}{2}$ **7.** [10.2] $\log_{64} \dfrac{1}{4} = -\dfrac{1}{3}$ **9.** [10.2] $5^2 = 25$
11. [10.2] $10^{-2} = 0.01$ **13.** [10.2] $e^3 = x$ **15.** [10.2] 2 **17.** [10.2] $\dfrac{1}{2}$ **19.** [10.2] 2 **21.** [10.2] 9 **23.** [10.2] 2
25. [10.2] $x = 2$ **27.** [10.2] $a = 5$ **29.** [10.3] $\log_b 2x$ **31.** [10.3] $\log \dfrac{y}{z}$ **33.** [10.3] $\log_b \dfrac{x^3}{y^5}$ **35.** [10.3] 5
37. [10.3] 7 **39.** [10.3] 1.512951 **41.** [10.4] -0.458866 **43.** [10.4] 10,000 **45.** [10.4] 42.521082 **47.** [10.4] -0.34
49. [10.4] -0.17 **51.** [10.5] $\dfrac{\log 5}{\log 4}$ **53.** [10.5] $\dfrac{\log 5 - \log 2}{\log 3}$ **55.** [10.5] $\dfrac{\log 3}{\log 5}$ **57.** [10.5] $10^{-5 - \log n}$ or $x = \dfrac{1}{10^5 n}$
59. [10.5] $L = 50.00$ **61.** [10.5] 368.05175 **63.** [10.5] 44,800,000 **65.** [10.5] 8.7 g/cm²

Test 10.8

1. [10.1] $y = 2$ **2.** [10.1]

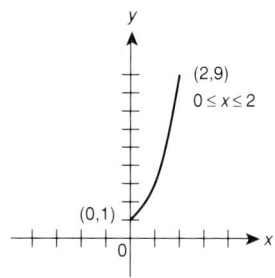

3. [10.1] 24,000   **4.** [10.2] $\log_3 x = y$ **5.** [10.2] $x^y = 16$ **6.** [10.2] $x = 5$
7. [10.2] $x = 2.7$ **8.** [10.3] $\log 5 + \log x$ **9.** [10.3] $\log_b \dfrac{x^5}{y^3}$ **10.** [10.3] $\dfrac{13}{4}$ **11.** [10.4] 0.585686
12. [10.4] $x = 4.2503$ **13.** [10.4] -0.04 **14.** [10.5] $\log_4 \dfrac{3}{5} = x$ **15.** [10.5] $x = 8.1$

Chapter 11 Answers

Trial Problems 11.1

1. Zero, one, or two   **2.** Zero, one, two, three, or four   **3.** One   **4.** Market equilibrium   **5.** None

Exercises 11.1

1.
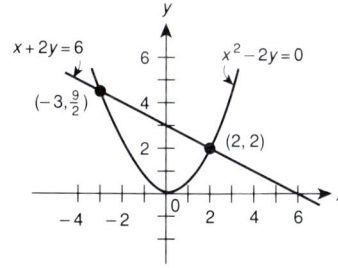

3.
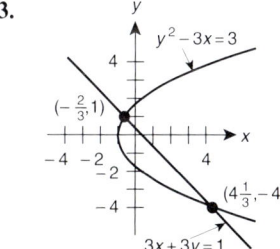

5.
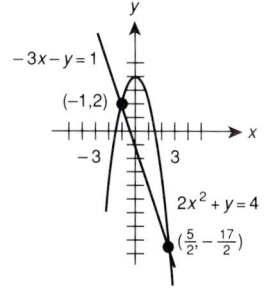

7.
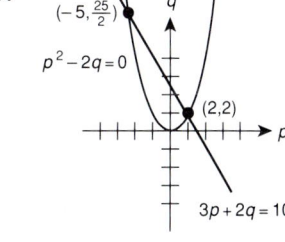

9.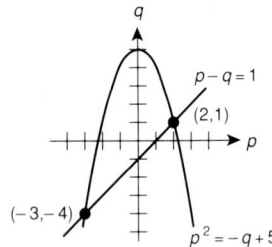

11.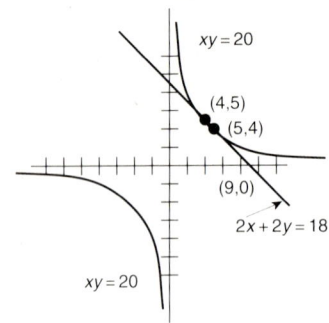

13.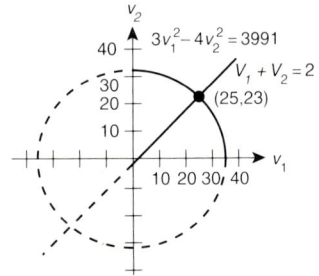

Trial Problems 11.2

1. Four **2.** Zero **3.** An infinite number **4.** $(1,2)\ (-3,10)$ **5.** $\left(\dfrac{4}{3}, 0.1249\right)$

Exercises 11.2

1. $\left(-5, \dfrac{25}{2}\right)$, $(2,2)$ **3.** $(3,3), (-3,-3)$ **5.** $(3, 1+\log_6 2), (3, 2-\log_6 3)$

7. 10 and 16 or -10 and -16 **9.** 12 and -8 **11.** $\sqrt[3]{5}$ and $\sqrt[3]{25}$ **13.** $L = 4$ cm, $W = \dfrac{7}{2}$ cm **15.** $v_1 = 27, v_2 = 23$

Trial Problems 11.3

1. $x = 64$ **2.** $x = 256$ **3.** $x = 125$ **4.** $x = 27$ **5.** $x = -2$

Exercises 11.3

1. $x = 64$ **3.** $x = -2$ **5.** $x = -\dfrac{1}{3}$ **7.** $y = -27$ **9.** $t = 17$ **11.** $s = -\dfrac{1}{2}$ or $s = 2$ **13.** $x = \dfrac{49}{16}$ **15.** $x = 5$

17. $x = -1$ or $x = -\dfrac{3}{4}$ **19.** $y = \dfrac{9}{25}$ **21.** $l = 12, w = 5$ **23.** $r = \dfrac{d^2 - h^2}{2h}$ **25.** $r = \dfrac{GM}{v^2}$

Review Exercises 11.5

1. [11.1]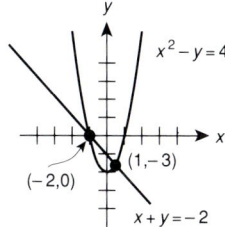

3. [11.1] $\left(2, \dfrac{4}{3}\right)$ **5.** [11.3] $r = \sqrt{\dfrac{-h^2 \pm \sqrt{h^4 + \dfrac{4A^2}{\pi^2}}}{2}}$ **7.** [11.2] 9 and 4

9. [11.3] $x = 5$, sides are 3, 3, and 11 **11.** [11.3] $h = \dfrac{0.0009\, Q^2}{A^2}$ **13.** [11.3] $L = 3.27\sqrt[3]{\dfrac{SAD}{P}}$

Test 11.6

1. [11.1] $p = 3$, $q = 2$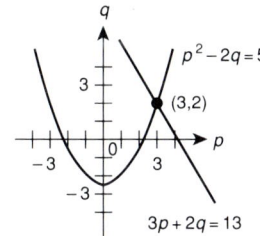

2. [11.1] $(-4, -1)$, $(4, -1)$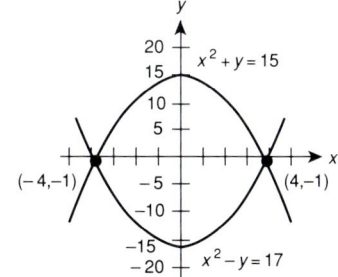

3. [11.2] $(x,y) = (4,3), \left(-7, \dfrac{39}{2}\right)$ **4.** [11.2] $(x,y) = (2, \sqrt{2}), (2, -\sqrt{2}), (-2, \sqrt{2}), (-2, -\sqrt{2})$ **5.** [11.2] 12 and 9
6. [11.3] $x = 3$ **7.** [11.3] 25 and 9

1046 Answer Key—Chapter 11

Cumulative Review—Chapters 9, 10, and 11   11.7

1. [9] Horizontal component is 5.9, vertical component is 4.1   **2.** [9] 586   **3.** [9] 21 pounds   **4.** [9] $5 + 2j$
5. [9] $56 - 2j$ **6.** [9] $1 - \frac{12}{5}j$ **7.** [9] $x = \frac{-3 \pm j\sqrt{7}}{2}$ **8.** [9] $\cos\frac{\pi}{6} + j\sin\frac{\pi}{6}$ **9.** [9] $\frac{\sqrt{2}}{2} + \frac{\sqrt{2}}{2}j$
10. [9] $15\cos\frac{7\pi}{12} + j\sin\frac{7\pi}{12}$ **11.** [9] $\cos 2\pi + j\sin 2\pi$ or $1 + 0j$ **12.** [9] $-4\sqrt{2} - 4\sqrt{2}j$
13. [9] $2\left(\cos\frac{3\pi}{20} + j\sin\frac{3\pi}{20}\right), 2\left(\cos\frac{11\pi}{20} + j\sin\frac{11\pi}{20}\right), 2\left(\cos\frac{19\pi}{20} + j\sin\frac{19\pi}{20}\right), 2\left(\cos\frac{27\pi}{20} + j\sin\frac{27\pi}{20}\right),$
$2\left(\cos\frac{35\pi}{20} + j\sin\frac{35\pi}{20}\right)$ **14.** [10] $y = 4$ **15.** [10] $y = \frac{5}{2}$ **16.** [10] $\log_2 y = x$ **17.** [10] $a^b = 8$ **18.** [10] 2
19. [10] $x = 3^2$ or 9   **20.** [10] $\frac{2}{3}\log_b x$ **21.** [10] $\log_b x^8$ **22.** [10] 5   **23.** [10] 0.37254   **24.** [10] 41.495
25. [10] 1.51   **26.** [10] $x = \frac{\log 5 - \log 3}{\log 4}$ **27.** [10] $x = 3$ **28.** [10] $x = 3, 5$
29. [11]

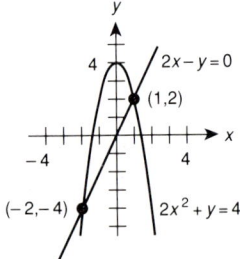

30. [11] $(-4,3), (4,3), (-3,-4), (3,-4)$ **31.** [11] $(3,1), (-3,-1)$ **32.** [11] $x = 5$ **33.** [11] $x = 7$

Chapter 12 Answers

Trial Problems 12.1

1. $[-6,5)$ **2.** $[-5,\infty)$ **3.** $x < 4$ **4.** $-4 < x \le 6$
5. $3x + 5 < 20,$
   $3x + 5 - 5 < 20 - 5,$
      $3x < 15,$
       $x < 5$
6. $7 \le 2x - 3 \le 11,$
   $7 + 3 \le 2x - 3 + 3 \le 11 + 3,$
      $10 \le 2x \le 14,$
       $5 \le x \le 7$

Exercises 12.1

1. $(-2,8)$ **3.** $[-5,-3]$ **5.** $(-2,\infty)$ **7.** $[4,\infty)$ **9.** $(-\infty,-7)$ **11.** $(-\infty,3]$ **13.** $[3,7]$

15.
```
   -1    3
 ←—(————]—→
     -1 < x < 3
```

17.
```
   -5    2
 ←—[————]—→
    -5 ≤ x ≤ 2
```

19.
```
   -2    3
 ←—(————]—→
    -2 < x ≤ 3
```

21.
```
         5
 ←———————]—→
        x ≤ 5
```

23.
```
         3
 ←———————)—→
         x < 3
```

25.
```
         3
 ←———————(—→
         x > 3
```

27.
```
     2
 ←—[————→
     x ≥ 2
```

29. $x \leq 4$ **31.** $x \leq \dfrac{15}{2}$ **33.** $x \leq 11$ **35.** $x \leq \dfrac{33}{2}$ **37.** $x < 5$ **39.** $x \geq \dfrac{1}{2}$ **41.** $\dfrac{11}{2} \leq x \leq \dfrac{23}{2}$ **43.** $-\dfrac{2}{3} \leq x \leq 6$
45. $-5 \leq x \leq 5$ **47.** $x < -7$ **49.** $x \leq 458$ **51.** $5 \leq L \leq 20$ **53.** $2.73 \leq I \leq 10.91$ **55.** $550 \leq P \leq 1650$
57. $32 \leq F \leq 41$ **59.** $6.55 \leq t \leq 9$ **61.** $0.2125 \leq g \leq 0.3625$ **63.** $30 \leq v \leq 60$ **65.** $17.8 \leq R_T \leq 21.3$

Trial Problems 12.2

1.
```
       --------0++++++++ (x+2)
       ----0++++++++++++ (x+3)
       ←—+—+—+—+—+—+—+—→
          -3  -2
           x < -3 or x > -2
```

2.

3.
```
       ----0++++++++++++ (x-1)
       ---------------0++++ (x-6)
       ←—+—+—+—+—+—+—+—→
           1           6
            x ≤ 1 or x > 6
```

4.

5.
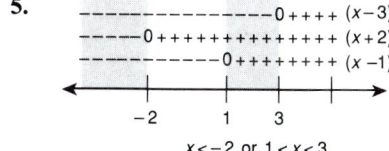

Exercises 12.2

1. $x < -2$ or $x > 3$ **3.** $-2 < x < 4$ **5.** $x \leq \dfrac{1}{2}$ or $x \geq \dfrac{3}{2}$ **7.** $-\dfrac{4}{5} \leq x \leq \dfrac{2}{3}$ **9.** $x \leq -3$ or $x \geq 4$

11. $-3 < x < -2$ **13.** No solutions   **15.** $x \leq \dfrac{5 - \sqrt{5}}{2}$ or $x \geq \dfrac{5 + \sqrt{5}}{2}$ **17.** All real numbers

19. $x < 2$ or $3 < x < 4$ **21.** $-3 \leq x \leq 2$ or $x \geq 4$ **23.** $-\dfrac{1}{3} < x < \dfrac{1}{2}$ **25.** $x < -2$ or $x > 2$

27. $x < -2$ or $0 < x < 1$ **29.** $-4 < x < -1$ or $0 < x < 2$ **31.** $-3 < x < -1$ or $1 < x < 2$ or $x > 4$

33. $\dfrac{5}{3} < x < 3$ **35.** $-3 < x < 1 - \sqrt{5}$ or $1 < x < 1 + \sqrt{5}$ **37.** $8 < t < 11$ **39.** $0 \leq n \leq 3$

41. $0 \leq I \leq 1$ or $3 \leq I \leq 4$ **43.** $x \leq -1$ or $x \geq 2$ **45.** $0 \leq x \leq 1.91$

Trial Problems 12.3

1. $x - 3 = 5$ or $x - 3 = -5$
     $x = 8$ or $x = -2$

2. $2x - 3 = x + 4$ or $2x - 3 = -(x + 4)$
     $x = 7$ $2x - 3 = -x - 4$
                            $3x = -1$
                            $x = -\dfrac{1}{3}$

3. $-4 \leq 2x - 3 \leq 4$
     $-1 \leq 2x \leq 7$
     $-\dfrac{1}{2} \leq x \leq \dfrac{7}{2}$

4. $5x - 4 < -17$ or $5x - 4 > 17$
     $5x < -13$ $5x > 21$
     $x < -\dfrac{13}{5}$ $x > \dfrac{21}{5}$

5. $-2 < \left(\dfrac{x - 6}{x - 3}\right) < 2$

$\dfrac{x - 6}{x - 3} > -2$ and $\dfrac{x - 6}{x - 3} < 2$

$\dfrac{x - 6}{x - 3} + 2 > 0$ $\dfrac{x - 6}{x - 3} - 2 < 0$

$\dfrac{x - 6 + 2(x - 3)}{x - 3} > 0$ $\dfrac{x - 6 - 2(x - 3)}{x - 3} < 0$

$\dfrac{3x - 12}{x - 3} > 0$ $\dfrac{-x}{x - 3} < 0$

$\dfrac{3(x - 4)}{x - 3} > 0$

-------0+++++++++ (x−3)   -----------0++++ (x−3)
---------0+++++++ (x−4)   ++++++0--------- (−x)
    3 4   0   3

($x < 3$ or $x > 4$)   and   ($x < 0$ or $x > 3$)
The solution set is $x < 0$ or $x > 4$.

Exercises 12.3

1. $x = -8$ or $x = 8$ **3.** $x = 0$ **5.** $x = -\dfrac{1}{2}$ **7.** $x = \dfrac{2}{3}$ or $x = \dfrac{6}{5}$ **9.** $x = 1$ **11.** $x = 11$ or $x = -3$
13. $x = 4$ or $x = 1$ **15.** $-4 < x < 4$ **17.** $x < -5$ or $x > 5$ **19.** All real numbers   **21.** $-3 < x < 11$
23. $x < -5$ or $x > -1$ **25.** $-\dfrac{3}{2} \leq x \leq \dfrac{9}{2}$ **27.** $x < -\dfrac{3}{4}$ or $x > \dfrac{13}{4}$ **29.** $x < \dfrac{17}{9}$ or $x > \dfrac{23}{11}$
31. $\dfrac{9}{5} \leq x \leq \dfrac{7}{3}$ and $x \neq 2$ **33.** $x < \dfrac{13}{4}$ **35.** $-6 \leq x \leq -\dfrac{4}{5}$ and $x \neq -\dfrac{5}{3}$ **37.** $-\dfrac{1}{3} < x < 17$ **39.** $[39.7, 41.9]$
41. $|m - x| \geq 1.2$

Trial Problems 12.4

1.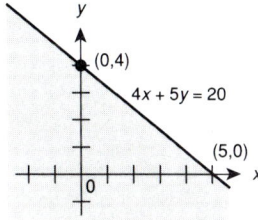

2.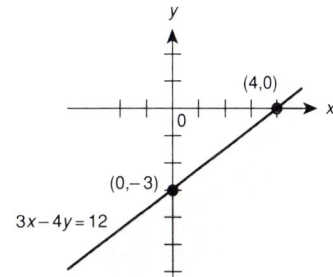

3.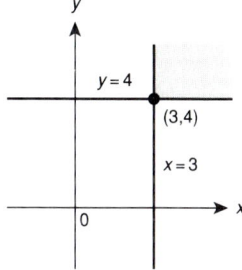

4.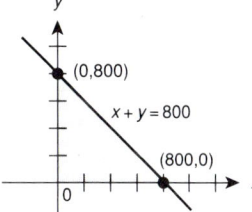

5.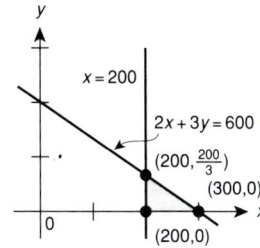

Exercises 12.4

1.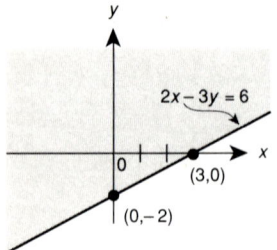

3.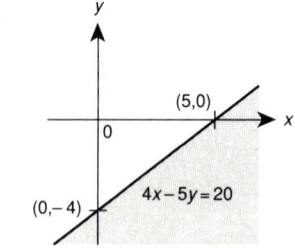

5.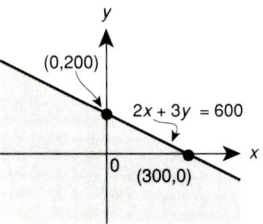

7.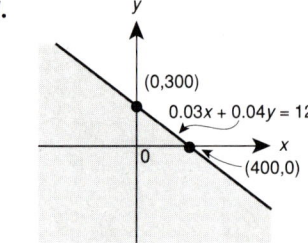

9.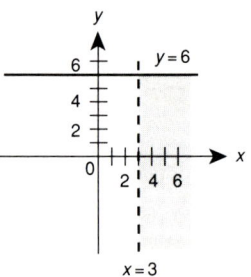

11.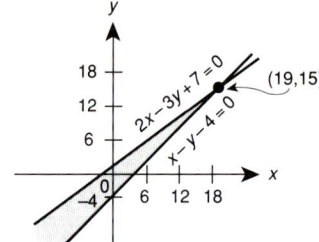

13.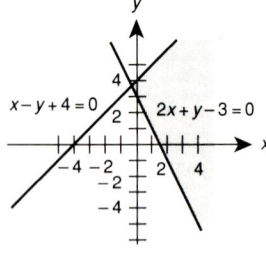

15.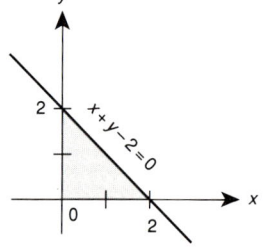

17.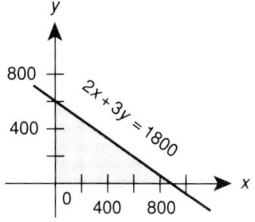

19.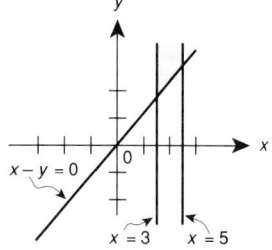

21.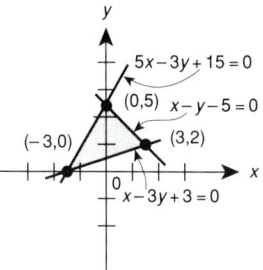

23.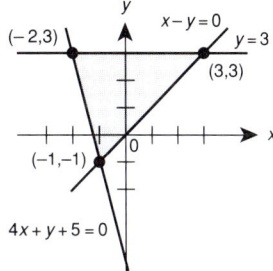

25.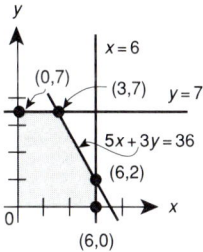

27.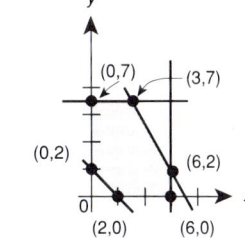

29.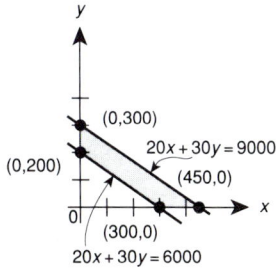

31.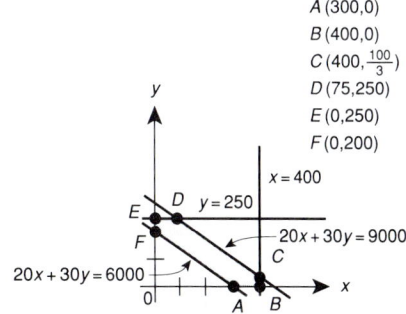

33.

Constraints:
$300x + 600y \leq 30{,}000$
$500x + 700y \leq 42{,}000$
$5x + 4y \leq 400$

$A(80, 0)$
$B(\frac{224}{3}, \frac{20}{3})$
$C(\frac{140}{3}, \frac{80}{3})$
$D(0, 50)$

35.

$A(80, 0)$
$B(\frac{224}{3}, \frac{20}{3})$
$C(\frac{140}{3}, \frac{80}{3})$
$D(40, 30)$
$E(30, 30)$
$F(30, 0)$

Trial Problems 12.5

1. At A, $f(300, 0) = 300 - 0 = 300$
   At B, $f(900, 0) = 900 - 0 = 900$
   At C, $f(1000, 300) = 1000 - 300 = 700$
   At D, $f(500, 600) = 500 - 600 = -100$
   At E, $f(0, 200) = 0 - 200 = -200$
   Maximum $= 900$

2. At A, $f(300, 0) = 600 + 0 = 600$
   At B, $f(900, 0) = 1800 + 0 = 1800$
   At C, $f(1000, 300) = 2000 + 900 = 2900$
   At D, $f(500, 600) = 1000 + 1800 = 2800$
   At E, $f(0, 200) = 0 + 600 = 600$
   Maximum $= 2900$

3. At A, $f(300, 0) = 90 + 0 = 90$
   At B, $f(900, 0) = 270 + 0 = 270$
   At C, $f(1000, 300) = 300 + 150 = 450$
   At D, $f(500, 600) = 150 + 300 = 450$
   At E, $f(0, 200) = 0 + 100 = 100$
   Maximum $= 450$

4. At A, $f(300, 0) = 600 + 0 = 600$
   At B, $f(900, 0) = 1800 + 0 = 1800$
   At C, $f(1000, 300) = 2000 - 900 = 1100$
   At D, $f(500, 600) = 1000 - 1800 = -800$
   At E, $f(0, 200) = 0 - 600 = -600$
   Minimum $= -800$

5. At A, $f(300, 0) = 150 + 0 = 150$
   At B, $f(900, 0) = 450 + 0 = 450$
   At C, $f(1000, 300) = 500 + 180 = 680$
   At D, $f(500, 600) = 250 + 360 = 610$
   At E, $f(0, 200) = 0 + 120 = 120$
   Minimum $= 120$

Exercises 12.5

1. max $= 4$, min $= -6$
3. max $= 902$, min $= -598$
5. max $= 90$, min $= -70$
7. max $= 550$, min $= -1250$
9. max $= -2$, min $= -10$
11. max $= 1$, min $= -38$
13. max $= 5$, min $= 0$
15. max $= 137$, min $= -3$
17. max $= 7.5$, min $= -1.9$
19. max $= 384.5$, min $= 35.3$
21. max $= 5$, min $= -\frac{63}{4}$
23. max $= 1700$, min $= 200$
25. max $= 348$
27. max $= 1000$
29. max $= 63{,}000$ at $(5, 12)$
31. $1.6\ T$ of Type I, $7\ T$ of Type II
33. $x = 5$, $y = 10$, $f = 19{,}000$

Review Exercises 12.7

1. [12.1] [2, 3]
3. [12.1] $(-\infty, 4)$
5. [12.1] $-6 < x < -2$
7. [12.1] $3 \leq x < 7$

9. [12.1]

$x > 2$

11. [12.1] $x \le \dfrac{1}{2}$ **13.** [12.1] $x \le 14$ **15.** [12.1] $\dfrac{7}{3} \le x \le \dfrac{10}{3}$ **17.** [12.1] $0 < x < 1$ **19.** [12.2] $2 \le x \le 4$
21. [12.2] $x < \dfrac{-5 - \sqrt{5}}{2}$ or $x > \dfrac{-5 + \sqrt{5}}{2}$ **23.** [12.2] $x \le -3$ or $-1 \le x \le 2$ **25.** [12.2] $x < -2$ or $0 \le x \le 1$
27. [12.3] $x = 3.5$ or $x = 0.5$ **29.** [12.3] $-9 \le x \le 9$ **31.** [12.3] $-0.5 < x < 6.5$ **33.** [12.3] $x < -8$ or $x > 2$
35. [12.3] $\dfrac{9}{14} < x < \dfrac{7}{10}$, $x \ne \dfrac{2}{3}$

37. [12.4]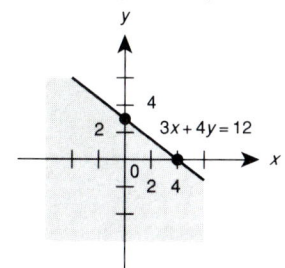

39. [12.4]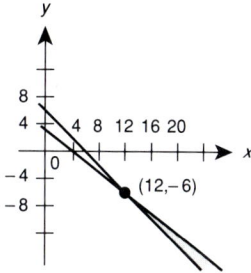

41. [12.4]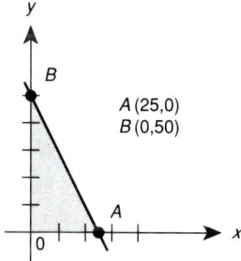

43. [12.4]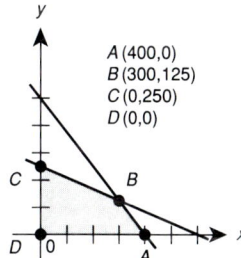

45. [12.5] max = 4875

Test 12.8

1. [12.1] [5,6] **2.** [12.1] [−6,∞) **3.** [12.1] $x \le -2$ **4.** [12.1] $3 \le x \le 7$ **5.** [12.1] $x > 2$ **6.** [12.1] $-2 < x < 3$
7. [12.2] $x \le 4$ **8.** [12.2] $x > 20$ **9.** [12.2] $4 \le x \le 6$ **10.** [12.2] $x \le 1$ or $x \ge 6$
11. [12.2] $x < -4$ or $-1 < x < 2$ **12.** [12.2] $0 < x < \dfrac{1}{2}$ or $x > 3$ **13.** [12.3] $-4 < x < 1$
14. [12.3] $x \ge 4$ or $x \le -5$ **15.** [12.3] $-12 < x < 12$ **16.** [12.3] $4, -5$

1054 ANSWER KEY—CHAPTER 12

17. [12.4]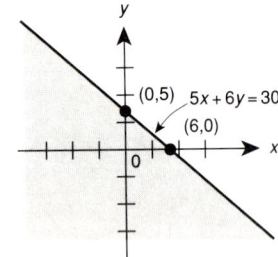

18. [12.4]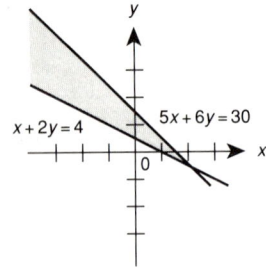

19. [12.4]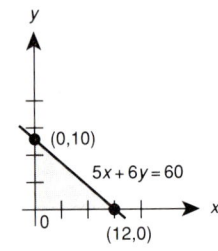

20. [12.4]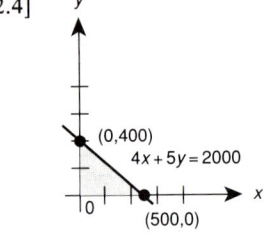

21. [12.4]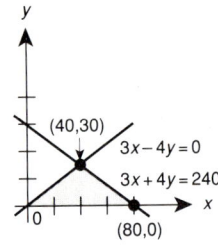

22. [12.5] $f = 85$ **23.** [12.5] 3 tables, 8 chairs

Chapter 13 Answers

Trial Problems 13.1

1. $[1 \quad 5 \quad 3] + [1 \quad 6 \quad 2] = [(1 + 1) \quad (5 + 6) \quad (3 + 2)]$
$= [2 \quad 11 \quad 5]$

2. $5\begin{bmatrix}1\\3\end{bmatrix} - 2\begin{bmatrix}-6\\-2\end{bmatrix} = \begin{bmatrix}5\\15\end{bmatrix} + \begin{bmatrix}12\\4\end{bmatrix} = \begin{bmatrix}17\\19\end{bmatrix}$

3. $4\begin{bmatrix}6\\-2\\2\end{bmatrix} + 7\begin{bmatrix}5\\6\\-3\end{bmatrix} = \begin{bmatrix}24\\-8\\8\end{bmatrix} + \begin{bmatrix}35\\42\\-21\end{bmatrix} = \begin{bmatrix}59\\34\\-13\end{bmatrix}$

4. $\begin{bmatrix}1 & 2 & 5\\6 & -3 & 1\\5 & 1 & -3\end{bmatrix} + \begin{bmatrix}1 & 4 & 6\\-3 & 0 & 1\\0 & 2 & 3\end{bmatrix} = \begin{bmatrix}(1+1) & (2+4) & (5+6)\\(6-3) & (-3+0) & (1+1)\\(5+0) & (1+2) & (-3+3)\end{bmatrix} = \begin{bmatrix}2 & 6 & 11\\3 & -3 & 2\\5 & 3 & 0\end{bmatrix}$

Answer Key—Chapter 13

5. $[x \; y] + 2[3 \; 6] = [5 \; 2]$
$[x \; y] + [6 \; 12] = [5 \; 2]$
$[x + 6 \; y + 12] = [5 \; 2]$
$x + 6 = 5 \rightarrow x = -1$
$y + 12 = 2 \rightarrow y = -10$

Exercises 13.1

1. $\begin{bmatrix} 7 & 5 & 0 \\ 6 & 9 & -3 \\ 4 & 1 & 2 \end{bmatrix}$
3. $\begin{bmatrix} 8 \\ 8 \\ 9 \end{bmatrix}$
5. $[0 \; -10 \; 9]$

7. $\begin{bmatrix} -2 & -2 & -3 \\ 0 & -2 & 5 \end{bmatrix}$
9. $\begin{bmatrix} 7 & 1 & 4 & 19 \\ 11 & 42 & 8 & 15 \\ 47 & 3 & -12 & -18 \end{bmatrix}$

11. $x = 2, y = 4, z = 6$ **13.** $x = 5, y = 2$ **15.** $x = 0, y = 3, z = -2$

17a. $\begin{bmatrix} 1500 & 1600 & 1800 & 50 & 2200 \\ 4000 & 1800 & 1200 & 3150 & 5600 \end{bmatrix}$

b. $\begin{bmatrix} 2875 & 1150 & 1380 & 115 & 1150 \\ 2300 & 1380 & 1035 & 978 & 2990 \end{bmatrix}$

c. $\begin{bmatrix} -1375 & 450 & 420 & -65 & 1050 \\ 1700 & 420 & 165 & 2172 & 2610 \end{bmatrix}$

Trial Problems 13.2

1. $\begin{bmatrix} ① & -2 & | & -1 \\ 2 & 1 & | & 3 \end{bmatrix} \rightarrow \begin{bmatrix} 1 & -2 & | & -1 \\ 0 & ⑤ & | & 5 \end{bmatrix} \rightarrow \begin{bmatrix} 1 & -2 & | & -1 \\ 0 & ① & | & 1 \end{bmatrix} \rightarrow \begin{bmatrix} 1 & 0 & | & 1 \\ 0 & 1 & | & 1 \end{bmatrix}$
$x = 1, y = 1$

2. $\begin{bmatrix} ① & 2 & | & 8 \\ 2 & -1 & | & 1 \end{bmatrix} \rightarrow \begin{bmatrix} 1 & 2 & | & 8 \\ 0 & ⑤ & | & -15 \end{bmatrix} \rightarrow \begin{bmatrix} 1 & 2 & | & 8 \\ 0 & ① & | & 3 \end{bmatrix} \rightarrow \begin{bmatrix} 1 & 0 & | & 2 \\ 0 & 1 & | & 3 \end{bmatrix}$
$x = 2, y = 3$

3. $\begin{bmatrix} ① & 1 & 1 & | & 2 \\ 2 & -1 & 2 & | & 7 \\ 3 & -2 & 1 & | & 7 \end{bmatrix} \rightarrow \begin{bmatrix} 1 & 1 & 1 & | & 2 \\ 0 & ③ & 0 & | & 3 \\ 0 & -5 & -2 & | & 1 \end{bmatrix} \rightarrow \begin{bmatrix} 1 & 1 & 1 & | & 2 \\ 0 & ① & 0 & | & -1 \\ 0 & -5 & -2 & | & 1 \end{bmatrix}$
$\begin{bmatrix} 1 & 0 & 1 & | & 3 \\ 0 & 1 & 0 & | & -1 \\ 0 & 0 & ② & | & -4 \end{bmatrix} \rightarrow \begin{bmatrix} 1 & 0 & 1 & | & 3 \\ 0 & 1 & 0 & | & -1 \\ 0 & 0 & ① & | & 2 \end{bmatrix} \rightarrow \begin{bmatrix} 1 & 0 & 0 & | & 1 \\ 0 & 1 & 0 & | & -1 \\ 0 & 0 & 1 & | & 2 \end{bmatrix}$
$x = 1, y = -1, z = 2$

4. $\begin{bmatrix} \boxed{1} & 2 & -2 & | & 10 \\ 3 & -1 & 1 & | & 2 \\ 5 & 2 & 3 & | & 13 \end{bmatrix} \rightarrow \begin{bmatrix} 1 & 2 & -2 & | & 10 \\ 0 & \boxed{-7} & 7 & | & -28 \\ 0 & -8 & 13 & | & -37 \end{bmatrix} \rightarrow \begin{bmatrix} 1 & 2 & -2 & | & 10 \\ 0 & \boxed{1} & -1 & | & 4 \\ 0 & -8 & 13 & | & -37 \end{bmatrix}$

$\begin{bmatrix} 1 & 0 & 0 & | & 2 \\ 0 & 1 & -1 & | & 4 \\ 0 & 0 & \boxed{5} & | & -5 \end{bmatrix} \rightarrow \begin{bmatrix} 1 & 0 & 0 & | & 2 \\ 0 & 1 & -1 & | & 4 \\ 0 & 0 & \boxed{1} & | & -1 \end{bmatrix} \rightarrow \begin{bmatrix} 1 & 0 & 0 & | & 2 \\ 0 & 1 & 0 & | & 3 \\ 0 & 0 & 1 & | & -1 \end{bmatrix}$

$x = 2, y = 3, z = -1$

Exercises 13.2

1. $x = 1, y = 4, z = 5$ **3.** $x = 1, y = -1, z = 2$ **5.** $x = -1, y = -2, z = -4$ **7.** $x = 20, y = 30, z = 40$
9. $x = 1, y = 1, z = 3, w = 1$ **11.** $x = 10, y = 20, z = 30, w = 40$ **13.** $x = 30, y = 30, z = 40$
15. $I_1 = 2.0, I_2 = 0.759, I_3 = -0.815$ **17.** $V_1 = 7.71, V_2 = 14.92, V_3 = 9.72$
19. 5600 Type M, 8200 Type N, 4800 Type P **21.** 400 Type I, 300 Type II, 100 Type III
23. 200 Type A, 300 Type B, 400 Type C **25.** $x^2 + y^2 - 6x + 2y - 15 = 0$ **27.** $1.63x^3 - 1.4x^2 - 7.63x + 5.4$

Trial Problems 13.3

1. $a_{11} = -1(3) + 0(2) + 1(-3) = -6$
The product matrix is $[-6]$.

2. $a_{11} = 3(-1) = -3, a_{12} = 3(0) = 0, a_{13} = 3(1) = 3$
$a_{21} = 2(-1) = -2, a_{22} = 2(0) = 0, a_{23} = 2(1) = 2$
$a_{31} = 3(-1) = -3, a_{32} = 3(0) = 0, a_{33} = 3(1) = 3$

The product matrix is $\begin{bmatrix} -3 & 0 & 3 \\ -2 & 0 & 2 \\ -3 & 0 & 3 \end{bmatrix}$.

3. $a_{11} = 1(1) + 4(3) = 13, a_{12} = 1(2) + 4(-4) = -14$
The product matrix is $[13 \quad -14]$.

4. The matrices cannot be multiplied.

5. $a_{11} = 1(0) + 4(1) = 4, a_{12} = 1(1) + 4(0) = 1$
$a_{21} = 3(0) + 2(1) = 2, a_{22} = 3(1) + 2(0) = 3$

The product matrix is $\begin{bmatrix} 4 & 1 \\ 2 & 3 \end{bmatrix}$.

Exercises 13.3

1. $[24]$

3. $[4]$

5. $\begin{bmatrix} 3 & 12 & 9 \\ 6 & 24 & 18 \\ -1 & -4 & -3 \end{bmatrix}$

7. $\begin{bmatrix} -2 & 11 \\ -14 & -4 \end{bmatrix}$

9. $\begin{bmatrix} 2 & 1 \\ -4 & -2 \end{bmatrix}$

11. $\begin{bmatrix} 17 \\ 19 \end{bmatrix}$

13. Cannot be multiplied

15. $\begin{bmatrix} 0 & 0 & 0 \\ 0 & 0 & 0 \\ 0 & 0 & 0 \end{bmatrix}$

17. $\begin{bmatrix} 1 & 5 & 2 \\ 6 & 5 & 1 \\ 0 & 4 & 7 \end{bmatrix}$

19. $\begin{bmatrix} 8425 \\ 18,670 \\ 9308 \end{bmatrix}$

21. $\begin{bmatrix} 1 & 0 & 0 & 0 \\ 0 & 1 & 0 & 0 \\ 0 & 0 & 1 & 0 \\ 0 & 0 & 0 & 1 \end{bmatrix}$

23. $\begin{bmatrix} 155 \\ 325 \\ 330 \end{bmatrix}$

25a. $\begin{bmatrix} 65 & 60 & 43 \\ 68 & 82 & 40 \\ 58 & 72 & 30 \\ 84 & 76 & 53 \end{bmatrix}$

b. [69.17 38.62 71.55]
c. [963.49 982.66 576.64]

27. [0.34 0.355 0.305]

Trial Problems 13.4

1. $\begin{bmatrix} \fbox{1} & -1 & | & 1 & 0 \\ 2 & 2 & | & 0 & 1 \end{bmatrix} \to \begin{bmatrix} 1 & -1 & | & 1 & 0 \\ 0 & \fbox{4} & | & -2 & 1 \end{bmatrix} \to \begin{bmatrix} 1 & -1 & | & 1 & 0 \\ 0 & \fbox{1} & | & -\frac{1}{2} & \frac{1}{4} \end{bmatrix} \to \begin{bmatrix} 1 & 0 & | & \frac{1}{2} & \frac{1}{4} \\ 0 & 1 & | & -\frac{1}{2} & \frac{1}{4} \end{bmatrix}$

The inverse is $\begin{bmatrix} \frac{1}{2} & \frac{1}{4} \\ -\frac{1}{2} & \frac{1}{4} \end{bmatrix}$

2. There is no inverse.

3. $\begin{bmatrix} \fbox{1} & -2 & 2 & | & 1 & 0 & 0 \\ 0 & 1 & 0 & | & 0 & 1 & 0 \\ -2 & -2 & 4 & | & 0 & 0 & 1 \end{bmatrix} \to \begin{bmatrix} 1 & -2 & 2 & | & 1 & 0 & 0 \\ 0 & \fbox{1} & 0 & | & 0 & 1 & 0 \\ 0 & -6 & 8 & | & 2 & 0 & 1 \end{bmatrix}$

$\to \begin{bmatrix} 1 & 0 & 2 & | & 1 & 2 & 0 \\ 0 & 1 & 0 & | & 0 & 1 & 0 \\ 0 & 0 & \fbox{8} & | & 2 & 6 & 1 \end{bmatrix} \to \begin{bmatrix} 1 & 0 & 2 & | & 1 & 2 & 0 \\ 0 & 1 & 0 & | & 0 & 1 & 0 \\ 0 & 0 & \fbox{1} & | & \frac{1}{4} & \frac{3}{4} & \frac{1}{8} \end{bmatrix}$

$\to \begin{bmatrix} 1 & 0 & 0 & | & \frac{1}{2} & \frac{1}{2} & -\frac{1}{4} \\ 0 & 1 & 0 & | & 0 & 1 & 0 \\ 0 & 0 & 1 & | & \frac{1}{4} & \frac{3}{4} & \frac{1}{8} \end{bmatrix}$

The inverse is $\begin{bmatrix} \frac{1}{2} & \frac{1}{2} & -\frac{1}{4} \\ 0 & 1 & 0 \\ \frac{1}{4} & \frac{3}{4} & \frac{1}{8} \end{bmatrix}$

Exercises 13.4

1. $\begin{bmatrix} -\frac{1}{5} & \frac{2}{5} \\ \frac{3}{5} & \frac{4}{5} \end{bmatrix}$

3. $\begin{bmatrix} \frac{2}{7} & -\frac{5}{14} \\ \frac{1}{7} & \frac{1}{14} \end{bmatrix}$

5. $\begin{bmatrix} \frac{20}{9} & \frac{10}{9} \\ \frac{25}{18} & \frac{35}{18} \end{bmatrix}$

7. $\begin{bmatrix} 0 & -1 & 1 \\ -1 & 1 & 2 \\ 1 & 0 & -2 \end{bmatrix}$

9. $\begin{bmatrix} -5 & 6 & -7 \\ 10 & -11 & 13 \\ -1 & 1 & -1 \end{bmatrix}$

11. $\begin{bmatrix} 1 & 0 & 0 \\ 0 & -1 & 0 \\ -1 & 0 & 1 \end{bmatrix}$

13. $\begin{bmatrix} -1 & -1 & -1 \\ 4 & 5 & 0 \\ 0 & 1 & -3 \end{bmatrix}$

15. $\begin{bmatrix} 1 & 2 & 3 & 1 \\ 1 & 3 & 3 & 2 \\ 2 & 4 & 3 & 3 \\ 1 & 1 & 1 & 1 \end{bmatrix}$

17. $\frac{1}{19}\begin{bmatrix} -4 & 11 & -6 \\ 1 & 2 & -8 \\ 7 & -5 & 1 \end{bmatrix}$

19. $\frac{1}{42}\begin{bmatrix} 6 & -1 & 5 & 2 \\ 24 & -18 & 6 & -6 \\ 90 & -29 & 19 & -26 \\ 216 & -71 & 61 & -68 \end{bmatrix}$

21. $x = 1, y = 4, z = 2$ 23. $x = 5, y = -1, z = -3$ 25. $x = 0, y = 0, z = 2, w = 2$

27. $\begin{bmatrix} 47{,}777.8 \\ 48{,}611.1 \end{bmatrix}$

Trial Problems 13.5

1. $\begin{vmatrix} 2 & 4 & 3 \\ 0 & -6 & 2 \\ 0 & -5 & 1 \end{vmatrix} = 2\begin{vmatrix} -6 & 2 \\ -5 & 1 \end{vmatrix} = 2((-6)(1) - (2)(-5)) = 2(-6 + 10) = 8$

2. $\begin{vmatrix} 4 & 0 & 0 \\ 1 & -5 & 2 \\ 3 & 1 & 6 \end{vmatrix} = 4\begin{vmatrix} -5 & 2 \\ 1 & 6 \end{vmatrix} = 4((-5)(6) - (2)(1)) = 4(-30 - 2) = -128$

3. $\begin{vmatrix} 1 & 1 & 1 \\ 4 & 1 & 2 \\ -3 & 0 & 2 \end{vmatrix} = -1\begin{vmatrix} 4 & 2 \\ -3 & 2 \end{vmatrix} + 1\begin{vmatrix} 1 & 1 \\ -3 & 2 \end{vmatrix}$

$= -1(8 - (-6)) + 1(2 - (-3))$
$= -1(14) + 1(5) = -9$

4. $\begin{vmatrix} 2 & -3 & 1 \\ 1 & 4 & 2 \\ 0 & 3 & -1 \end{vmatrix} = 2\begin{vmatrix} 4 & 2 \\ 3 & -1 \end{vmatrix} + (-1)\begin{vmatrix} -3 & 1 \\ 3 & -1 \end{vmatrix}$

$= 2(-4 - 6) - 1(3 - 3)$
$= 2(-10) - 1(0) = -20$

Exercises 13.5

1. -2 3. 9 5. 3 7. 0 9. -1.83 11. -9 13. 26 15. -654 17. 21 19. 16.5 21. Yes   23. Yes

Trial Problems 13.6

1. $\begin{vmatrix} 1 & 1 & 1 \\ 0 & 1 & -2 \\ 0 & 0 & 4 \end{vmatrix} = (1)(1)(4) = 4$

2. $\begin{vmatrix} 1 & -4 & 3 \\ 0 & 1 & 3 \\ 0 & 2 & 4 \end{vmatrix} = \begin{vmatrix} 1 & -4 & 3 \\ 0 & 1 & 3 \\ 0 & 0 & -2 \end{vmatrix} = (1)(1)(-2) = -2$

3. $\begin{vmatrix} 4 & -4 & 0 \\ 2 & 2 & 0 \\ 1 & 3 & -6 \end{vmatrix} = \begin{vmatrix} 8 & 0 & 0 \\ 2 & 2 & 0 \\ 1 & 3 & -6 \end{vmatrix} = 8(2)(-6) = -96$

4. $\begin{vmatrix} 1 & 4 & -3 \\ 2 & 6 & -1 \\ -3 & 2 & 4 \end{vmatrix} = \begin{vmatrix} 1 & 4 & -3 \\ 0 & -2 & 5 \\ 0 & 14 & -5 \end{vmatrix} = \begin{vmatrix} 1 & 4 & -3 \\ 0 & -2 & 5 \\ 0 & 0 & 30 \end{vmatrix} = (1)(-2)(30) = -60$

Exercises 13.6

1. 8 **3.** 8 **5.** −8 **7.** $x = 1, y = -1, z = 2$ **9.** 12 **11.** 12 **13.** 48 **15.** 12 **17.** $x = 1, y = 4, z = 1, w = 2$
19. $|A| = 12, |B| = -27, |AB| = -324 = (12)(-27)$ **21.** $2e^{6x}$

Review Exercises 13.8

1. [13.1] $\begin{bmatrix} -14 & 15 \\ 14 & 20 \\ 19 & 11 \end{bmatrix}$ **3.** [13.2] $x = y = z = -2$ **5.** [13.2] $x = 1, y = -1, z = 2, w = 2$

7. [13.2] $x = 10, y = 20, z = 30$ **9.** [13.3] $\begin{bmatrix} 0 & 1 & -1 \\ 3 & 6 & -1 \\ 5 & 8 & 0 \end{bmatrix}$ **11.** [13.3] $\begin{bmatrix} 5 & 17 & -15 \\ 6 & 19 & -17 \\ -3 & -10 & 9 \end{bmatrix}$

13. [13.3] $\begin{bmatrix} -6 & 12 & 19 \\ 1 & 11 & 18 \\ 10 & 1 & -13 \end{bmatrix}$ **15.** [13.4] $\begin{bmatrix} 7 & -8 & 5 \\ -4 & 5 & -3 \\ 1 & -1 & 1 \end{bmatrix}$

17. [13.4] $\begin{bmatrix} 0 & 1 & 2 \\ 2 & -1 & 1 \\ -1 & 1 & 0 \end{bmatrix}$ **19.** [13.4] $\begin{bmatrix} 17 & -8 & 15 \\ -10 & 5 & -9 \\ 3 & 0 & 2 \end{bmatrix}$

21. [13.4] $x = 4, y = 1, z = 1$ **23.** [13.4] $x = -10, y = 20, z = 30$ **25.** [13.5] 60 **27.** [13.5] 37
29. [13.5] $x = 1.1, y = 2.14, z = 3.58$ **31.** [13.5] $x = 1, y = 0, z = 0, w = 2$

33a. [13.1] $\begin{bmatrix} 2400 & 4800 & 7200 \\ 4200 & 9600 & 2400 \end{bmatrix}$

b. [13.1] $\begin{bmatrix} 1700 & 3400 & 5100 \\ 2975 & 6800 & 1700 \end{bmatrix}$

c. [13.2] $\begin{bmatrix} 81,760 \\ 82,960 \end{bmatrix}$ **e.** [13.2] $\begin{bmatrix} 64,496 \\ 70,516 \end{bmatrix}$

d. [13.2] $\begin{bmatrix} 98,112 \\ 99,552 \end{bmatrix}$ **f.** [13.2] $\begin{bmatrix} 249,368 \\ 253,028 \end{bmatrix}$

Test 13.9

1. [13.1] $\begin{bmatrix} 8 \\ 4 \end{bmatrix}$
2. [13.1] $\begin{bmatrix} 0 & 6 \\ 2 & 1 \end{bmatrix}$
3. [13.2] $x = 1, y = 2, z = -1$
4. [13.2] $x = y = z = -1$
5. [13.1] $\begin{bmatrix} -1 & 23 & -6 \\ -1 & -4 & 4 \\ 2 & -21 & 2 \end{bmatrix}$
6. [13.3] $\begin{bmatrix} 2 & -43 & 16 \\ 3 & -65 & 24 \\ -6 & 127 & -47 \end{bmatrix}$
7. [13.4] $\begin{bmatrix} 3 & -2 & 4 \\ 2 & 1 & 2 \\ 5 & 3 & 5 \end{bmatrix}$
8. [13.4] $\begin{bmatrix} 1 & -1 & 0 \\ 3 & 2 & 2 \\ -1 & -1 & -1 \end{bmatrix}$
9. [13.4] $\begin{bmatrix} 1 & 2 & -7 & 7 \\ 0 & 1 & -2 & 4 \\ 0 & 0 & 1 & 0 \\ 0 & 0 & 0 & 1 \end{bmatrix}$
10. [13.3] Can't be calculated
11. [13.4] $x = 1, y = 2, z = 1$
12. [13.4] $x = 6, y = 6, z = 1$
13. [13.5,13.6] 0
14. [13.5,13.6] -88
15. [13.5,13.6] Infinite number of solutions
16. [13.5,13.6] $x = 30, y = 40, z = 50$

Cumulative Review—Chapters 12 and 13   13.10

1. [12] $(-2,4)$
2. [12] $(-\infty,3.5)$
3. [12] $x > -3$
4. [12] $-2 \le x < 0$
5. [12] $x > \dfrac{3}{2}$
6. [12] $x \le 7$
7. [12] $x \le -4$ or $x \ge 2$
8. [12] $x < 0$ or $\dfrac{2}{3} < x < 1$
9. [12] $-3 \le x \le 4$
10. [12] $x < -\dfrac{2}{3}$ or $x > 4$

11. [12]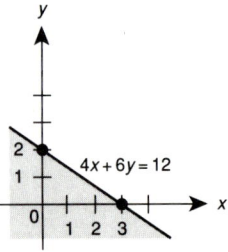

12. [12]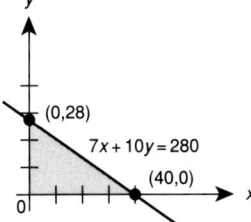

13. [12]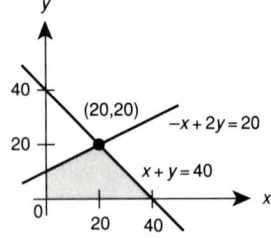

14. [12]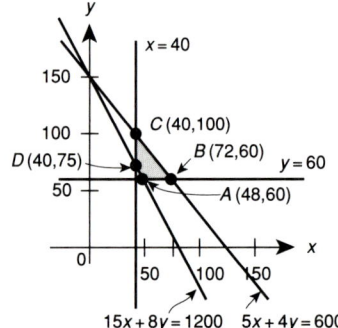

15. [12] 100 **16.** [12] 396

17. [12] $\begin{bmatrix} 24 & 1 \\ 28 & -3 \\ 26 & 10 \end{bmatrix}$

18. [13] $x = -3, y = -2, z = -4$

19. [13] $\begin{bmatrix} -2 & -1 & -4 \\ -2 & -5 & 6 \\ 3 & 3 & -3 \end{bmatrix}$

20. [13] $\begin{bmatrix} 1 & 4 & -3 \\ -2 & -5 & 4 \\ 1 & 3 & -2 \end{bmatrix}$

21. [13] $\begin{bmatrix} -2 & -1 & 1 \\ 0 & 1 & 2 \\ -1 & 1 & 3 \end{bmatrix}$

22. [13] $\begin{bmatrix} -5 & 8 & 0 \\ -6 & 3 & 8 \\ -10 & 0 & 10 \end{bmatrix}$

23. [13] $\begin{bmatrix} -5 & 8 & 0 \\ -6 & 3 & 8 \\ -10 & 0 & 10 \end{bmatrix}$

24. [13] $\begin{bmatrix} 1 & 0 & 0 \\ 0 & 1 & 0 \\ 0 & 0 & 1 \end{bmatrix}$

25. [13] 1 **26.** [13] 1 **27.** [13] $x = -2, y = 1, z = -2$

Chapter 14 Answers

Trial Problems 14.2

1. $\begin{array}{r|rrrr} 2 & 1 & 0 & -2 & -5 \\ & & 2 & 4 & 4 \\ \hline & 1 & 2 & 2 & -1 \end{array}$

$Q = x^2 + 2x + 2, \quad R = -1$

2. $\begin{array}{r|rrrrr} 2 & 1 & 0 & 0 & 0 & -16 \\ & & 2 & 4 & 8 & 16 \\ \hline & 1 & 2 & 4 & 8 & 0 \end{array}$

$Q = x^3 + 2x^2 + 4x + 8, \quad R = 0$

3. $\begin{array}{r|rrrrr} -3 & 1 & 0 & -3 & -1 & -6 \\ & & -3 & 9 & -18 & 57 \\ \hline & 1 & -3 & 6 & -19 & 51 \end{array}$

$Q = x^3 - 3x^2 + 6x - 19, \quad R = 51$

4. $\begin{array}{r|rrrrr} 2 & 2 & -1 & -6 & 4 & -8 \\ & & 4 & 6 & 0 & 8 \\ \hline & 2 & 3 & 0 & 4 & 0 \end{array}$

$Q = 2x^3 + 3x^2 + 4, \quad R = 0$

Exercises 14.2

1. $x^2 - x + 1, \quad R = 0$ **3.** $-3x^3 + 3x^2 - 3x + 8, \quad R = -20$ **5.** $x^2 - 5x + 1, \quad R = 0$ **7.** $6x^2 - 3x + 3, R = 0$ **9.** $x^3 + 3x^2 + 10x - 28, \quad R = 83$ **11.** $x^3 + 3x^2 + 4x + 12, \quad R = 40$ **13.** $-x^2 - 9x + 4, \quad R = 0$
15. $x^5 + x^4 + x^3 + x^2 + x + 1, \quad R = 0$ **17.** $(x - 1)(2x^3 - 4x^2 - x + 3)$ **19.** $(2x - 1)(2x^3 - 6x^2 + 2x - 2)$
21. $x - 3$ **23.** $3.23x^2 + 4.11x + 7.36, \quad R = 0$

Trial Problems 14.3

1. The possible roots are $\pm 1, \pm 2, \pm 3,$ and ± 6.

$$P(1) = 1 + 8 + 13 + 6 = 28 \neq 0$$
$$P(-1) = -1 + 8 - 13 + 6 = 0; \; -1 \text{ is a root}$$

$$\begin{array}{r|rrrr} -1 & 1 & 8 & 13 & 6 \\ & & -1 & -7 & -6 \\ \hline & 1 & 7 & 6 & 0 \end{array}$$

$$x^3 + 8x^2 + 13x + 6 = (x+1)(x^2 + 7x + 6)$$
$$= (x+1)(x+1)(x+6)$$

The zeros are -1 and -6.

2. The possible roots are ± 1 and ± 2.

$$P(1) = 1 - 4 + 5 - 2 - 2 \neq 0$$
$$P(-1) = 1 + 4 + 5 + 2 - 2 \neq 0$$
$$P(2) = 16 - 32 + 20 - 4 - 2 \neq 0$$
$$P(-2) = 16 + 32 + 20 + 4 - 2 \neq 0$$

There are no integer zeros.

3. The possible roots are $\pm 1, \pm 2, \pm \dfrac{1}{2}$.

$$P(1) = 2 - 1 - 4 + 2 \neq 0$$
$$P(-1) = -2 - 1 + 4 + 2 \neq 0$$
$$P(2) = 16 - 4 - 8 + 2 \neq 0$$
$$P(-2) = -16 - 4 + 8 + 2 \neq 0$$
$$P\left(\dfrac{1}{2}\right) = \dfrac{1}{4} - \dfrac{1}{4} - 2 + 2 = 0$$
$$P\left(-\dfrac{1}{2}\right) = -\dfrac{1}{4} - \dfrac{1}{4} + 2 + 2 \neq 0$$

$\dfrac{1}{2}$ is the only rational zero.

4. The possible roots are $\pm\dfrac{1}{2}, \pm\dfrac{1}{3}, \pm\dfrac{1}{4}, \pm\dfrac{1}{6}, \pm\dfrac{1}{8}, \pm\dfrac{1}{12}, \pm\dfrac{1}{16}, \pm\dfrac{1}{24}, \pm\dfrac{1}{32}, \pm\dfrac{1}{48}, \pm\dfrac{1}{96},$ and ± 1.

$$P\left(\dfrac{1}{2}\right) = 12 - 4 - 3 + 1 \neq 0, \quad P\left(-\dfrac{1}{2}\right) = -12 - 4 + 3 + 1 \neq 0$$
$$P\left(\dfrac{1}{3}\right) = \dfrac{32}{9} - \dfrac{16}{9} - 2 + 1 \neq 0, \quad P\left(-\dfrac{1}{3}\right) = -\dfrac{32}{9} - \dfrac{16}{9} + 2 + 1 \neq 0$$
$$P\left(\dfrac{1}{4}\right) = \dfrac{3}{2} - 1 - \dfrac{3}{2} + 1 = 0; \quad \dfrac{1}{4} \text{ is a root.}$$

$$\begin{array}{r|rrrr} \dfrac{1}{4} & 96 & -16 & -6 & 1 \\ & & 24 & 2 & -1 \\ \hline & 96 & 8 & -4 & 0 \end{array}$$

$$P(x) = \left(x - \frac{1}{4}\right)(96x^2 + 8x - 4)$$
$$= (4x - 1)(24x^2 + 2x - 1)$$
$$= (4x - 1)(4x + 1)(6x - 1)$$

The zeros are $\frac{1}{4}$, $-\frac{1}{4}$, and $\frac{1}{6}$.

Exercises 14.3

1. 1, 2, 3 **3.** −4, 2, 3 **5.** $\frac{1}{2}$, 3 **7.** $\frac{2}{3}$, $\frac{3}{4}$, 4 **9.** $\frac{1}{2}$ **11.** $\frac{1}{2}$, $\frac{2}{3}$ **13.** $\frac{2}{3}$, $\frac{-1 \pm j\sqrt{3}}{2}$ **15.** 2, $\frac{5 \pm j\sqrt{23}}{4}$
17. 2, 3, $\frac{-1 \pm j\sqrt{3}}{2}$ **19.** $\frac{3}{2}$, $-\frac{1}{3}$, $\frac{3 \pm \sqrt{5}}{2}$ **21.** 1, $\frac{3}{2}$, $-\frac{1}{3}$, $\frac{3 \pm \sqrt{5}}{2}$ **23.** 1, $-1 \pm \sqrt{2}$ **25.** 20 **27.** 30 hours
29. 3, 5, 6 **31.** 6 **33.** 3 **35.** 8

Trial Problems 14.4

1.

x	2	2.5	2.3	2.2	2.1	2.05
$P(x)$	−1	5.625	2.567	1.248	0.061	−0.485

The zero is 2.1.

2.

x	−4	−5	−4.5	−4.7	−4.6	−4.65
$P(x)$	8	−7	2.375	−0.883	0.824	−0.001

The zero is −4.6.

3.

x	1	2	1.5	1.3	1.35
$P(x)$	−3	11	2.125	−0.333	0.228

The zero is 1.3.

4.

x	−1	−2	−1.5	−1.3	−1.2	−1.21	−1.205
$P(x)$	−1	3	1.375	0.473	−0.008	0.041	0.016

The zero is 1.20.

Exercise 14.4

1. 2.5 **3.** 3.4 **5.** −4.3 **7.** 0.5 **9.** −1.4 **11.** −1.9, 0.3, 1.5 **13.** −2.4, −1.3, 1.7 **15.** −0.6, 1.6 **17.** −1.6, 1.4
19. 3.85 **21.** 20.4 **23.** 2.4

Review Exercises 14.6

1. [14.2] $x^2 + 2x + 3$, $R = 4$ **3.** [14.2] $x^2 + 3x - 2$, $R = 5$ **5.** [14.2] $2x^2 + 10x + 20$, $R = 72$
7. [14.2] $3x^3 - 12x^2 + 31x - 66$, $R = 143$ **9.** [14.2] $2.7x^2 - 3.49x + 6.753$, $R = 0$ **11.** [14.3] −1, 2, 3
13. [14.3] $\frac{2}{3}$ **15.** [14.3] $\frac{1}{3}$, 1, 2 **17.** [14.3] 1, 2 **19.** [14.3] $\frac{1}{2}$, $\frac{1 \pm \sqrt{5}}{2}$ **21.** [14.3] $\frac{1}{2}$, 1, $\frac{5}{3}$
23. [14.4] 0.4, 1.7, 2.6 **25.** [14.4] 1.4, 2.3 **27.** [14.3] 10 by 30 by 5 or 33 by 13 by 3.5 **29.** [14.3] 512
31. [14.3] 15, 16, 17 **33.** [14.4] 10° **35.** [14.4] 1° or 4°

Test 14.7

1. [14.2] $3x^2 + 14x + 51$, $R = 211$ 2. [14.2] $2x^2 - 5x + 9$, $R = -15$ 3. [14.2] $4x^3 + 4x^2 + 6x + 6$, $R = 1$
4. [14.2] $5x^3 - 7x^2 + 10x - 19$, $R = 37$ 5. [14.2] $2.4x^2 + 6.38x + 8.256$, $R = 8.3072$ 6. [14.3] $-\dfrac{1}{2}, -1$
7. [14.3] $4, -1, -2$ 8. [14.3] $\dfrac{1}{2}$ 9. [14.3] $-3, \dfrac{1}{2}, \dfrac{5}{2}$ 10. [14.3] $-3, -2, -1, 3$ 11. [14.3] $2, \sqrt{3}, -\sqrt{3}$
12. [14.3] $1, \dfrac{-7 \pm \sqrt{33}}{2}$ 13. [14.3] $-\dfrac{1}{2}, \dfrac{3}{4}, \dfrac{-1 \pm \sqrt{5}}{2}$ 14. [14.3] $2, 2 \pm \sqrt{7}$ 15. [14.3] $\dfrac{3}{2}, -1 \pm j$
16. [14.4] -0.9 17. [14.4] 1.76 18. [14.4] about 0.8 in. or about 12.65 in.

Chapter 15 Answers

Trial Problems 15.1

1. $10, 12$ 2. $32, 64$ 3. $16, -32, 64$ 4. $a_n = 4n - 3$ 5. $a_2 = 3$
$a_3 = 9$
$a_4 = 27$

Exercises 15.1

1. $11, 13, 15$ 3. $13, 16, 19$ 5. $15, 18, 21$ 7. $11, 14, 17$ 9. $26, 37, 50$ 11. $\dfrac{3}{9}, \dfrac{3}{11}, \dfrac{3}{13}$ 13. x^9, x^{11}, x^{13}
15. $\dfrac{5}{6}, \dfrac{6}{7}, \dfrac{7}{8}$ 17. $2n + 1$ 19. $3n - 2$ 21. $3n$ 23. $n^2 - n$ 25. $n^2 + 1$ 27. $\dfrac{3}{2n - 1}$ 29. x^{2n-1} 31. $\dfrac{n}{n+1}$
33. $2, 4, 8$ 35. $5, 8, 11$ 37. $1, -1, 1$ 39. $24, 60, 96, 132, 168$ 41. $28, 25, 22, 19$

Trial Problems 15.2

1. $[2(1) - 1] + [2(2) - 1)] + [2(3) - 1] = 1 + 3 + 5 = 9$
2. $(0^2 + 0) + (1^2 + 1) + (2^2 + 2) + (3^2 + 3) = 0 + 2 + 6 + 12 = 20$
3. $(3 + 4) + (4 + 4) + (5 + 4) = 7 + 8 + 9 = 24$ 4. $\sum_{n=1}^{4} 2n$ 5. $\sum_{n=1}^{4} (4n - 3)$

Exercises 15.2

1. $2 + 5 + 8 + 11$ 3. $2 + 3 + 5 + 9$ 5. $6 + 13 + 22 + 33 + 46$ 7. $-\dfrac{1}{3} + \dfrac{1}{6} - \dfrac{1}{9} + \dfrac{1}{12} - \dfrac{1}{15}$
9. $\sum_{n=1}^{4} (2n + 1)$ 11. $\sum_{n=0}^{4} (3n + 1)$ 13. $\sum_{n=0}^{3} (3n - 1)$ 15. $\sum_{n=1}^{4} \dfrac{2}{2n - 1}$ 17. $\sum_{n=1}^{4} x^{2n+1}$ 19. $\sum_{n=0}^{4} y^{2n}$
21. $60, 58, 56, 54, 52$ 23. $15, 13, 11, 9, 7, 5, 3, 1$

Trial Problems 15.3

1. $a_5 = 11$ **2.** $a_4 = -7$ **3.** $d = 2$ **4.** $a_1 = -1$ **5.** $s_5 = 37.5$

Exercises 15.3

1. -16 **3.** 2 **5.** -2 **7.** 60 **9.** 12 days **11.** 1600 ft **13.** 44,400

Trial Problems 15.4

1. $a_4 = 24$ **2.** $a_4 = -\dfrac{1}{4}$ **3.** $r = \pm\dfrac{3}{2}$ **4.** $a_1 = 1$ **5.** $S_4 = 60$

Exercises 15.4

1. 162 **3.** 3 **5.** 1 **7.** 31 **9.** 32 million **11.** $\dfrac{5}{4096}$ grams **13.** 0.0125%

Trial Problems 15.5

1. 3 **2.** 6 **3.** $\dfrac{8}{3}$ **4.** $\dfrac{2}{9}$ **5.** $\dfrac{35}{99}$

Exercises 15.5

1. $\dfrac{27}{2}$ **3.** 24 **5.** 3 **7.** $\dfrac{9}{20}$ **9.** $\dfrac{3}{405}$ **11.** $\dfrac{1}{3}$ **13.** $\dfrac{31}{99}$ **15.** $\dfrac{2408}{999}$ **17.** $\dfrac{29}{225}$ **19.** 30 ft **21.** 16 hours
23. 900 meters

Trial Problems 15.6

1. 120 **2.** 72 **3.** 1440 **4.** $m(m-1)(m-2)$ **5.** $3k(3k-1)(3k-2)$

Exercises 15.6

1. 5040 **3.** 7 **5.** 70,560 **7.** 35 **9.** 66 **11.** 1 **13.** $n(n-1)(n-2)$ **15.** $2n(2n-1)(2n-2)$
17. $2[n(n-1)(n-2)]$ **19.** $(n+1)n(n-1)$ **21.** $(n-2)(n-3)(n-4)$ **23.** 1 7 21 35 35 21 7 1
25. $16a^4 + 32a^3b + 24a^2b^2 + 8ab^3 + b^4$ **27.** $243a^5 - 810a^4b + 1080a^3b^2 - 720a^2b^3 + 240ab^4 - 32b^5$
29. $c^6 + 12c^5 + 60c^4 + 160c^3 + 240c^2 + 196c + 64$ **31.** $\dfrac{x^4}{16} - \dfrac{3x^3y}{2} + \dfrac{27x^2y^2}{2} - 54xy^3 + 81y^4$ **33.** $10r^3s^2$
35. $\dfrac{35}{8}x^8y^{12}$ **37.** $10x^2y^3$ **39.** $3360a^6b^4$ **41.** $35x^2y^{3/2}$ **43.** $\dbinom{n}{j} = \dfrac{n!}{j!(n-j)!}, \dbinom{n}{n-j} = \dfrac{n!}{(n-j)![n-(n-j)]!}$
$= \dfrac{n!}{(n-j)!j!} = \dfrac{n!}{j!(n-j)!}$ **45.** 0.9227

Review Exercises 15.8

1. [15.1] 11, 13, 15   **3.** [15.1] 8, −16, 32   **5.** [15.1] $2n + 3$ **7.** [15.1] $(-1)^n(2)^{n-1}$ or $(-1)(-2)^{n-1}$ **9.** [15.1] 2, 4, 8
11. [15.1] 6, 10, 14   **13.** [15.2] $[3(1) - 1] + [3(2) - 1] + [3(3) - 1] + [3(4) - 1] = 2 + 5 + 8 + 11 = 26$
15. [15.2] $[2^0 - 1] + [2^1 - 1] + [2^2 - 1] + [2^3 - 1] = 0 + 1 + 3 + 7 = 11$ **17.** [15.2] $\sum_{n=2}^{5}(2n+1)$
19. [15.2] $\sum_{n=1}^{4} x^{2n-1}$ **21.** [15.3] 17   **23.** [15.3] −2   **25.** [15.3] 60   **27.** [15.4] 162   **29.** [15.4] 2   **31.** [15.4] 1
33. [15.4] 31   **35.** [15.5] 9   **37.** [15.5] 36   **39.** [15.6] 362,880   **41.** [15.6] 504   **43.** [15.6] 210   **45.** [15.6] 28
47. [15.6] $(k-1)(k-2)(k-3)$ **49.** [15.6] $(2k+1)(2k)(2k-1)$ **51.** [15.6] $a^4 + 8a^3b + 24a^2b^2 + 32ab^3 + 16b^4$
53. [15.6] $\dfrac{1}{a^5} - \dfrac{5}{a^4b} + \dfrac{10}{a^3b^2} - \dfrac{10}{a^2b^3} + \dfrac{5}{ab^4} - \dfrac{1}{b^5}$ **55.** [15.6] $80x^3y^2$ **57.** [15.6] $5x^3y$ **59.** [15.6] 5062.5 or 5062

Test 15.9

1. [15.1] 17, 21, 25   **2.** [15.1] $\dfrac{16}{3}, \dfrac{32}{3}, \dfrac{64}{3}$ **3.** [15.1] $a_n = 3n - 2$ **4.** [15.1] 12, 48, 192   **5.** [15.2] $3 + 5 + 7 + 9$
6. [15.2] $2 + 6 + 12 + 20$ **7.** [15.2] $\sum_{i=1}^{5}(3i-1)$ **8.** [15.2] $\sum_{n=0}^{3} \dfrac{1}{y^n}$ **9.** [15.3] 39   **10.** [15.3] 7   **11.** [15.3] 5
12. [15.3] 6   **13.** [15.3] 5050   **14.** [15.4] $\dfrac{1}{54}$ **15.** [15.4] ± 2 **16.** [15.4] 5   **17.** [15.4] $s_4 = 200$ **18.** [15.4] $a_1 = 1$
19. [15.5] 18   **20.** [15.5] $\dfrac{2}{9}$ **21.** [15.6] 5040   **22.** [15.6] 21
23. [15.6] $x^5 + 10x^4y + 40x^3y^2 + 80x^2y^3 + 80xy^4 + 32y^5$ **24.** [15.6] $16{,}800x^5y^4$ **25.** [15.6] $1792xy^3$

Cumulative Review—Chapters 14 and 15   15.10

1. [14] $x^3 + 2x^2 - x + 6$, $R = 6$ **2.** [14] $3x^2 - 8x + 14$, $R = -10$ **3.** [14] −2   **4.** [14] $-2, \dfrac{1}{2}, \dfrac{3}{2}$
5. [14] $1, \dfrac{-3 \pm \sqrt{33}}{2}$ **6.** [14] $\dfrac{1}{2}$ **7.** [14] $\dfrac{1}{2}, \dfrac{-1 \pm \sqrt{17}}{2}$ **8.** [14] $-\dfrac{1}{3}, \dfrac{1}{2}, \dfrac{-1 \pm j\sqrt{11}}{2}$ **9.** [14] 2.2
10. [14] −2.65   **11.** [15] $3n$ **12.** [15] $2n - 4$ **13.** [15] 6, 18, 54   **14.** [15] 5, 13, 29   **15.** [15] $\sum_{j=1}^{4}(4j-1)$
16. [15] $\sum_{j=0}^{3} x^{4j+1}$ **17.** [15] $-1 + 0 + 1 + 2$ **18.** [15] $1 + 4 + 10 + 28$ **19.** [15] 16   **20.** [15] −2   **21.** [15] 64
22. [15] 40   **23.** [15] 48   **24.** [15] 12   **25.** [15] 840   **26.** [15] 364   **27.** [15] $(2k-3)(2k-4)(2k-5)$
28. [15] $(n+3)(n+2)(n+1)$ **29.** [15] $32a^5 - 80a^4b + 80a^3b^2 - 40a^2b + 10ab^4 - b^5$
30. [15] $\dfrac{16}{x^4} + \dfrac{32y}{x^3} + \dfrac{24y^2}{x^2} + \dfrac{8y}{x} + y^4$ **31.** [15] $1080x^3y^2$ **32.** [15] $240x^2y$

Chapter 16 Answers

Trial Problems 16.2

1. $d = \sqrt{(3-2)^2 + (-1-4)^2} = \sqrt{1^2 + 5^2} = \sqrt{26}$ 2. $y = mx + b$, where $m = -\dfrac{3}{4}$ and $b = -3$,
$y = -\dfrac{3}{4}x - 3$
3. $y - y_1 = m(x - x_1)$, where $m = -2$, $x_1 = -3$, and $y_1 = 2$
$y - 2 = -2(x - (-3))$
$y - 2 = -2(x + 3)$
$y - 2 = -2x - 6$
$2x + y + 4 = 0$
4. $y - y_1 = \left(\dfrac{y_2 - y_1}{x_2 - x_1}\right)(x - x_1)$, where $x_1 = 3$, $y_1 = -6$, $x_2 = 4$, and $y_2 = -2$
$y - (-6) = \left(\dfrac{-2 - (-6)}{4 - 3}\right)(x - 3)$
$y + 6 = 4(x - 3)$
$y + 6 = 4x - 12$
$-4x + y + 18 = 0$
5. Since the form of the equation is $x = k$, the equation is $x = 4$.

Exercise 16.2

1. $\sqrt{10}$ 3. $\sqrt{261}$ 5. $\sqrt{(h-5)^2 + (k-2)^2}$ 7. $x - 2y + 8 = 0$ 9. $2x + y + 6 = 0$ 11. $x + 2y + 6 = 0$
13. $2x - y - 1 = 0$ 15. $x - 2y + 7 = 0$ 17. $3x + y + 11 = 0$ 19. $x + 3y + 5 = 0$ 21. $x + 3y - 20 = 0$
23. $5x - 2y - 1 = 0$ 25. $2x - 3y - 1800 = 0$ 27. $y = -5000t + 40{,}000$ 29. $v = \dfrac{5t + 1365}{3}$
31. $h = -60t + 4500$ 33. $f = 0.35w$ 35. $R = 20.6$ 37. $r = 5.6$ ohms   39. $L = 0.004t + 14.96$
41. $i = 0.08v + 0.02$ 43. $122°$ 45. 27.5 mi/gal   47. $2x - 3y + 4 = 0$ 49. $5x - 4y + 22 = 0$

Trial Problems 16.3

1. $(x + 2)^2 + (y + 3)^2 = 25$
$(x - (-2))^2 + (y - (-3))^2 = 5^2$
center $(-2, -3)$, radius $= 5$
2. $(x^2 - 4x + \underline{}) + (y^2 - 10y + \underline{}) = -4$
$(x^2 - 4x + 4) + (y^2 - 10y + 25) = 25$
$(x - 2)^2 + (y - 5)^2 = 5^2$
center $(2, 5)$, radius $= 5$

3. $(x^2 + 12x + \underline{}) + (y^2 - 16y + \underline{}) = 0$
$(x^2 + 12x + 36) + (y^2 - 16y + 64) = 36 + 64$
$(x + 6)^2 + (y - 8)^2 = 100$
$(x - (-6))^2 + (y - 8)^2 = 10^2$
center $(-6, 8)$, radius $= 10$
4. $(x - 0)^2 + (y - (-5))^2 = 4^2$
$x^2 + (y + 5)^2 = 16$
5. Radius $=$ distance between $(0,0)$ and $(4,2)$
$= \sqrt{4^2 + 2^2} = \sqrt{18}$
Equation: $x^2 + y^2 = 18$

Answer Key—Chapter 16

Exercises 16.3

1. $(0,0)$, $r = 7$ 3. $(0,0)$, $r = 3\sqrt{2}$ 5. $(0,2)$, $r = 5$ 7. $(-3,0)$, $r = 3\sqrt{2}$ 9. $(2,3)$, $r = 9$
11. $(-2,-4)$, $r = 4$ 13. $\left(-\dfrac{3}{2}, \dfrac{1}{2}\right)$, $r = \dfrac{5}{3}$ 15. $x^2 + y^2 = 9$ 17. $x^2 + y^2 = 5$
19. $(x - 3)^2 + (y - 2)^2 = 16$ 21. $(x + 2)^2 + y^2 = 9$ 23. $(x - 2)^2 + (y + 3)^2 = 49$ 25. $(x + 3)^2 + (y + 5)^2 = \dfrac{1}{4}$
27. $(x + 2)^2 + (y + 4)^2 = 7$ 29. $(1,4)$, $r = \sqrt{10}$ 31. $(-3,-4)$, $r = 4$ 33. $(2,-1)$, $r = 3$
35. $(-5,3)$, $r = 2$ 37. $\left(-3, \dfrac{1}{3}\right)$, $r = \sqrt{2}$ 39. $x^2 + y^2 = 13$ 41. $(x - 3)^2 + (y + 2)^2 = 18$
43. $(x - 3)^2 + (y - 1)^2 = 5$ 45. $\left(x + \dfrac{3}{2}\right)^2 + \left(y - \dfrac{15}{4}\right)^2 = \dfrac{925}{16}$ 47. $x^2 + y^2 = 144$, $(x - 11)^2 + y^2 = \dfrac{1}{64}$
49. $(x - 14)^2 + (y + 3)^2 = 144$ 51. $y = 220 + \sqrt{(110)^2 - x^2}$, $y = -220 - \sqrt{(110)^2 - x^2}$
53. $(x - 6)^2 + y^2 = 1$, $(x + 6)^2 + y^2 = 1$, $x^2 + (y - 6)^2 = 1$, $x^2 + (y + 6)^2 = 1$

Trial Problems 16.4

1. $x^2 = 4(2)y$, $p = 2$
   Vertex: $(0,0)$, Focus: $(0,2)$, Directrix: $y = -2$
2. $(y - 2)^2 = x + 1$
   $(y - 2)^2 = 4\left(\dfrac{1}{4}\right)(x - (-1))$, $h = -1$, $k = 2$, $p = \dfrac{1}{4}$
   Vertex: $(h,k) = (-1,2)$, Focus: $(h + p, k) = \left(-\dfrac{3}{4}, 2\right)$, Directrix: $x = h - p$ or $x = -\dfrac{5}{4}$
3. $(y + 2)^2 = -\dfrac{1}{3}(x - 12)$
   $(y - (-2))^2 = -4\left(\dfrac{1}{12}\right)(x - 12)$, $h = 12$, $k = -2$, $p = \dfrac{1}{12}$
   Vertex: $(h,k) = (12,-2)$, Focus: $(h - p, k) = \left(\dfrac{143}{12}, -2\right)$, Directrix: $x = h + p$ or $x = \dfrac{145}{12}$
4. $h = 0$, $k = 0$, $p = 2$
   $(y - k)^2 = -4p(x - h)$
   $(y - 0)^2 = -4(2)(x - 0)$
   $y^2 = -8x$
5. $(x - h)^2 = 4p(y - k)$ $h = 1$, $k = 2$
   $(x - 1)^2 = 4p(y - 2)$
   Substitute $x = -2$, $y = 5$, and solve for p.
   $9 = 4(p)3$
   $p = \dfrac{3}{4}$
   $(x - 1)^2 = 4\left(\dfrac{3}{4}\right)(y - 2)$
   $(x - 1)^2 = 3(y - 2)$

Exercises 16.4

1.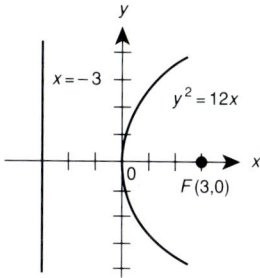

3.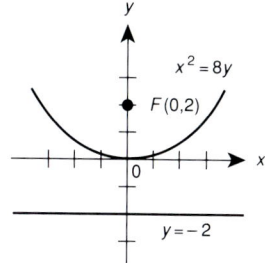

5.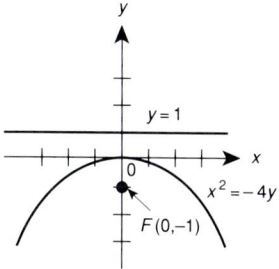

7.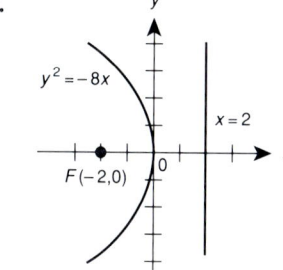

9.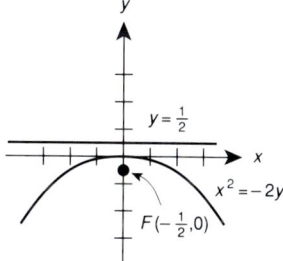

11.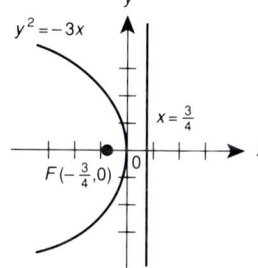

13.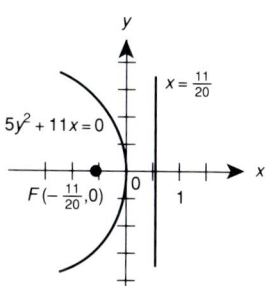

15. $y^2 = 16x$ **17.** $y^2 = -12x$ **19.** $x^2 = -8y$ **21.** $y^2 = 20x$ **23.** $x^2 = -12y$ **25.** $y^2 = \frac{4}{3}x$ **27.** $x^2 = 12y$

29. $(y - 2)^2 = 18(x - 3)$ **31.** $(y - 5)^2 = \frac{9}{5}(x - 1)$

33.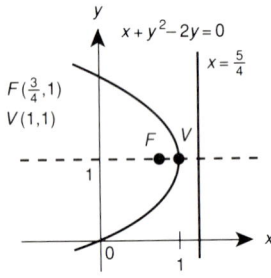

35.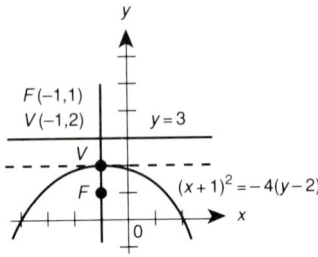

37.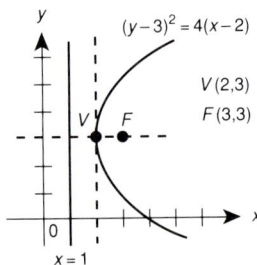

39. $x^2 = \frac{32{,}000}{7}y$, $\;-800 \le x \le 800$ **41.** $(0, 1.6)$ $x^2 = 6.4g$, $\;-4 \le x \le 4$ **43.** $x^2 = -\frac{720}{17}y$, $\;-60 \le x \le 60$

45. 2304 ft   **47.** about 14 ft

49.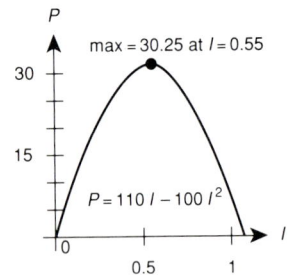

51.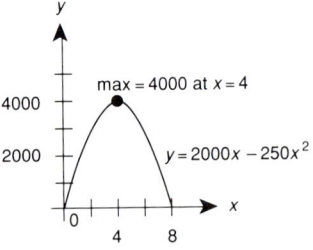

53. 13.5 ft

Trial Problems 16.5

1. $4x^2 + 36y^2 = 144$
   $\dfrac{x^2}{6^2} + \dfrac{y^2}{2^2} = 1$, $h = 0$, $k = 0$, $a = 6$, $b = 2$
   Center: $(0,0)$, Vertices: $(-a,0)$ and $(a,0)$ or $(-6,0)$ and $(6,0)$, Foci: $(-c,0)$ and $(c,0)$, where $c^2 = a^2 - b^2$
   $c = \sqrt{36 - 4} = \sqrt{32} = 4\sqrt{2}$
   $(-4\sqrt{2}, 0)$ and $(4\sqrt{2}, 0)$

2. $\dfrac{(x-3)^2}{4} + \dfrac{(y+2)^2}{9} = 1$, $h = 3$, $k = -2$, $a = 3$, $b = 2$
   Center: $(h,k) = (3,-2)$, Vertices: $(h, k-a)$, $(h, k+a)$ or $(3,-5)$, $(3,1)$, Foci: $(h, k-c)$, $(h, k+c)$, where $c^2 = a^2 - b^2$
   $c = \sqrt{3^2 - (2)^2} = \sqrt{5}$
   $(3, -2 - \sqrt{5})$, $(3, -2 + \sqrt{5})$

3. $\dfrac{x^2}{a^2} + \dfrac{y^2}{b^2} = 1$, where $a = 4$ and $b = 2$
   $\dfrac{x^2}{16} + \dfrac{y^2}{4} = 1$

4. $\dfrac{(x-h)^2}{b^2} + \dfrac{(y-k)^2}{a^2} = 1$, where $h = -2$ and $k = 3$
   The length of the major axis is $9 - (-3) = 12$. Thus, $a = \dfrac{12}{2} = 6$.
   $\dfrac{(x+2)^2}{b^2} + \dfrac{(y-3)^2}{36} = 1$.
   Substitute $x = 2$, $y = 3$, and solve for b.
   $\dfrac{16}{b^2} + \dfrac{0}{36} = 1$ or $b = 4$
   $\dfrac{(x+2)^2}{16} + \dfrac{(y-3)^2}{36} = 1$

Exercises 16.5

1. $V_1(-5,0)$
   $V_2(5,0)$
   $F_1(-3,0)$
   $F_2(3,0)$
   $\dfrac{x^2}{25} + \dfrac{y^2}{16} = 1$

3. $\dfrac{x^2}{144} + \dfrac{y^2}{121} = 1$
   $V_1(-12,0)$
   $V_2(12,0)$
   $F_1(-\sqrt{23}, 0)$
   $F_2(\sqrt{23}, 0)$

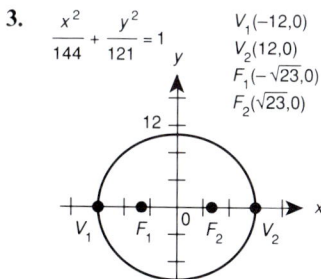

Answer Key—Chapter 16

5. $\dfrac{x^2}{49} + \dfrac{y^2}{25} = 1$
$V_1(-7,0)$
$V_2(7,0)$
$F_1(-\sqrt{24},0)$
$F_2(\sqrt{24},0)$

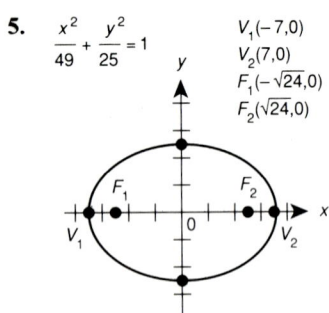

7. $V_1(0,-\sqrt{5})$
$V_2(0,\sqrt{5})$
$F_1(0,-1)$
$F_2(0,1)$
$\dfrac{x^2}{4} + \dfrac{y^2}{5} = 1$

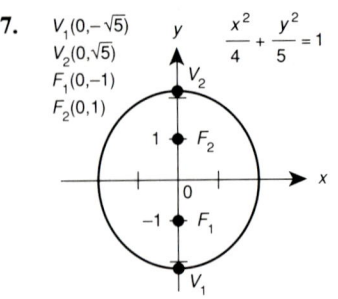

9. $V_1(0,-2\sqrt{3})$
$V_2(0,2\sqrt{3})$
$F_1(0,-\sqrt{2})$
$F_2(0,\sqrt{2})$
$\dfrac{x^2}{10} + \dfrac{y^2}{12} = 1$

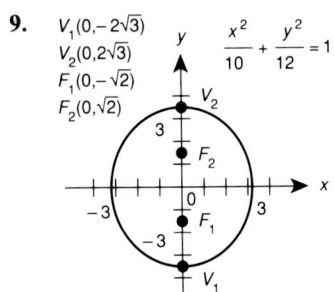

11. $V_1(-\tfrac{1}{2},0)$
$V_2(\tfrac{1}{2},0)$
$F_1(-\tfrac{\sqrt{5}}{6},0)$
$F_2(\tfrac{\sqrt{5}}{6},0)$
$\dfrac{x^2}{\tfrac{1}{4}} + \dfrac{y^2}{\tfrac{1}{9}} = 1$

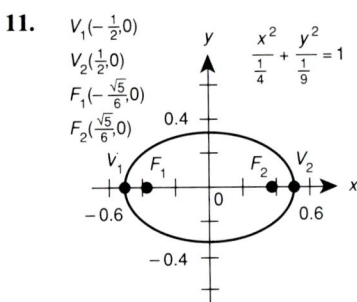

13. $V_1(0,-2)$
$V_2(0,2)$
$F_1(0,-\tfrac{\sqrt{35}}{3})$
$F_2(0,\tfrac{\sqrt{35}}{3})$
$\dfrac{x^2}{\tfrac{1}{9}} + \dfrac{y^2}{4} = 1$

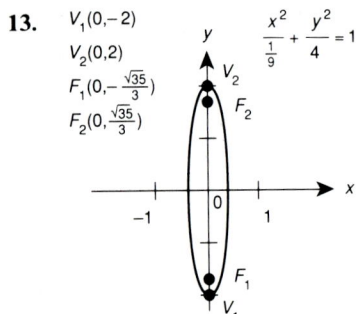

15. $\dfrac{x^2}{2^2} + \dfrac{y^2}{3^2} = 1$

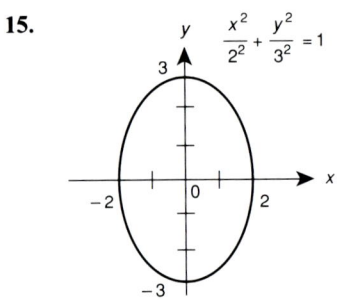

17. $\dfrac{x^2}{5^2} + \dfrac{y^2}{3^2} = 1$

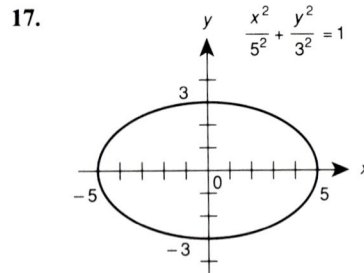

19. $\dfrac{x^2}{3^2} + \dfrac{y^2}{4^2} = 1$

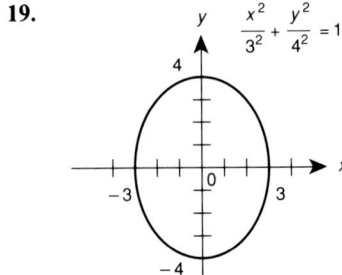

21.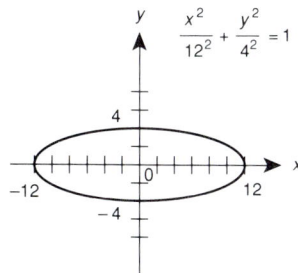

23.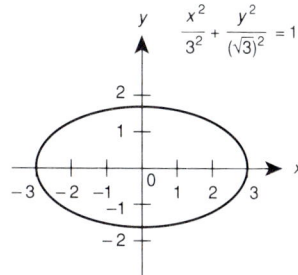

25.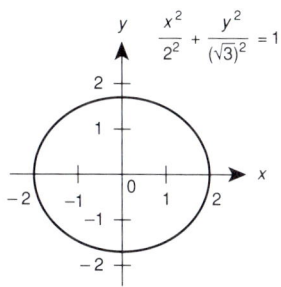

27.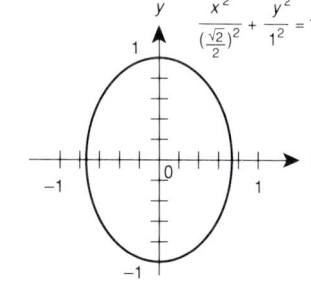

29. $\dfrac{x^2}{(4.5)^2} + \dfrac{y^2}{2^2} = 1$ or $\dfrac{x^2}{2^2} + \dfrac{y^2}{(4.5)^2} = 1$ **31.** $\dfrac{x^2}{7} + \dfrac{y^2}{16} = 1$ **33.** $\dfrac{x^2}{36} + \dfrac{y^2}{64} = 1$ **35.** $\dfrac{x^2}{16} + \dfrac{7y^2}{64} = 1$

37. $\dfrac{(x-4)^2}{25} + \dfrac{(y-3)^2}{4} = 1$

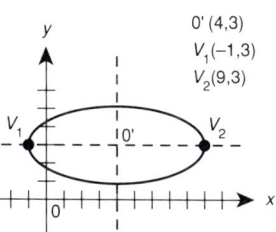

39. $\dfrac{(x-2)^2}{\frac{9}{4}} + \dfrac{(y+1)^2}{4} = 1$

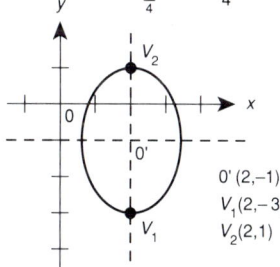

41. $\dfrac{(x-1)^2}{4} + \dfrac{(y+4)^2}{9} = 1$

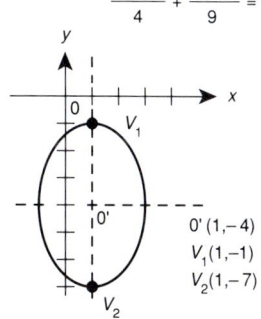

43. $\dfrac{x^2}{(1250)^2} + \dfrac{y^2}{(300)^2} = 1$, $\dfrac{x^2}{(1200)^2} + \dfrac{y^2}{(259)^2} = 1$, $A = 75{,}000\pi$ sq ft **45.** $\dfrac{x^2}{(12)^2} + \dfrac{y^2}{(10)^2} = 1$, $y \geq 0$

47. $\dfrac{x^2}{(70.4)^2} + \dfrac{y^2}{(70.1)^2} = 1$ **49.** 11.2 ft   **51.** 12π **53.** 16 ft   **55.** 4 ft from the ends of the major axis

57. $(0, -2.00), (0, 2.00)$

Trial Problems 16.6

1. $x^2 - y^2 = 9$

$\dfrac{x^2}{3^2} - \dfrac{y^2}{3^2} = 1$, $h = 0$, $k = 0$, $a = 3$, $b = 3$

Vertices: $(-a, 0), (a, 0) = (-3, 0), (3, 0)$, Foci: $(-c, 0), (c, 0)$ where $c^2 = a^2 + b^2$

$c = \sqrt{a^2 + b^2} = \sqrt{18} = 3\sqrt{2}$

$(-3\sqrt{2}, 0)$ and $(3\sqrt{2}, 0)$

Asymptotes: $y = \dfrac{b}{a}x$ and $y = -\dfrac{b}{a}x$ or $y = x$ and $y = -x$

2. $\dfrac{(x+2)^2}{9} - \dfrac{(y+2)^2}{4} = 1$, $h = -2$, $k = -2$, $a = 3$, $b = 2$

Vertices: $(h-a, k), (h+a, k) = (-5, -2), (1, -2)$

Foci: $(h \pm c, k)$, where $c = \sqrt{a^2 + b^2} = \sqrt{9 + 4} = \sqrt{13}$

$(-2 - \sqrt{13}, -2), (-2 + \sqrt{13}, -2)$

Asymptotes: $y - k = \pm \dfrac{b}{a}(x - h)$

$y + 2 = \dfrac{2}{3}(x + 2)$ and $y + 2 = -\dfrac{2}{3}(x + 2)$

3. $\dfrac{y^2}{a^2} - \dfrac{x^2}{b^2} = 1$, where $a = 2$, $c = 4$, and $b = \sqrt{c^2 - a^2}$

$b = \sqrt{16 - 4} = \sqrt{12}$

$\dfrac{y^2}{4} - \dfrac{x^2}{12} = 1$

4. $\dfrac{x^2}{a^2} - \dfrac{y^2}{b^2} = 1$, where $a = 4$

$\dfrac{x^2}{16} - \dfrac{y^2}{b^2} = 1$

Substitute $x = 8$, $y = 3$, and solve for b^2.

$\dfrac{64}{16} - \dfrac{9}{b^2} = 1$, $b^2 = 3$

$\dfrac{x^2}{16} - \dfrac{y^2}{3} = 1$

Exercises 16.6

1.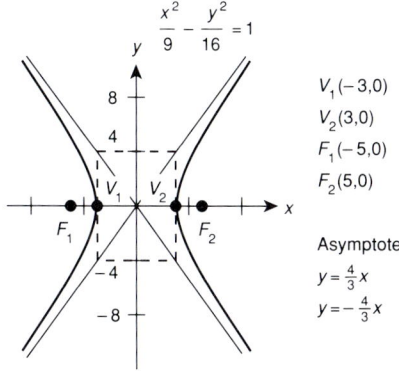

$\dfrac{x^2}{9} - \dfrac{y^2}{16} = 1$

$V_1(-3,0)$
$V_2(3,0)$
$F_1(-5,0)$
$F_2(5,0)$

Asymptotes:
$y = \dfrac{4}{3}x$
$y = -\dfrac{4}{3}x$

3.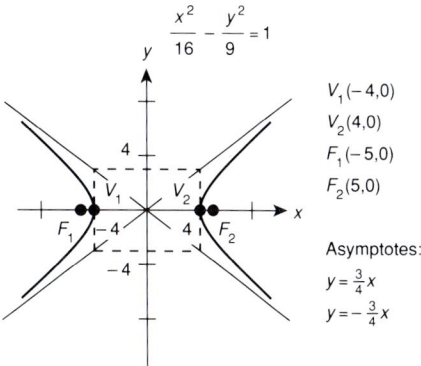

$\dfrac{x^2}{16} - \dfrac{y^2}{9} = 1$

$V_1(-4,0)$
$V_2(4,0)$
$F_1(-5,0)$
$F_2(5,0)$

Asymptotes:
$y = \dfrac{3}{4}x$
$y = -\dfrac{3}{4}x$

5.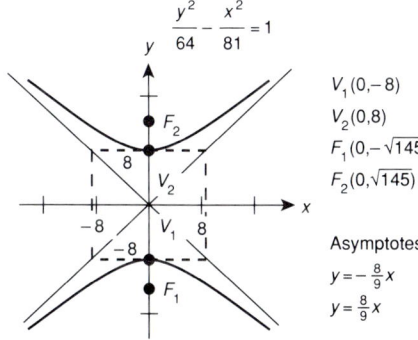

$\dfrac{y^2}{64} - \dfrac{x^2}{81} = 1$

$V_1(0,-8)$
$V_2(0,8)$
$F_1(0,-\sqrt{145})$
$F_2(0,\sqrt{145})$

Asymptotes:
$y = -\dfrac{8}{9}x$
$y = \dfrac{8}{9}x$

7.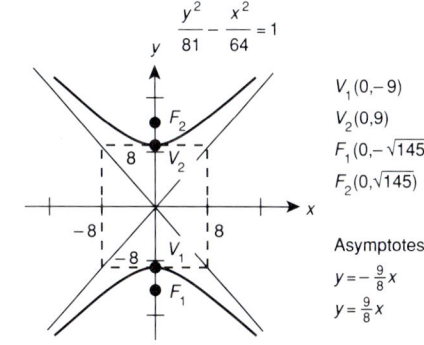

$\dfrac{y^2}{81} - \dfrac{x^2}{64} = 1$

$V_1(0,-9)$
$V_2(0,9)$
$F_1(0,-\sqrt{145})$
$F_2(0,\sqrt{145})$

Asymptotes:
$y = -\dfrac{9}{8}x$
$y = \dfrac{9}{8}x$

9.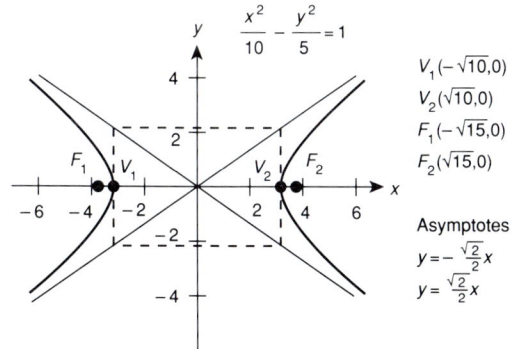

$\dfrac{x^2}{10} - \dfrac{y^2}{5} = 1$

$V_1(-\sqrt{10},0)$
$V_2(\sqrt{10},0)$
$F_1(-\sqrt{15},0)$
$F_2(\sqrt{15},0)$

Asymptotes:
$y = -\dfrac{\sqrt{2}}{2}x$
$y = \dfrac{\sqrt{2}}{2}x$

11.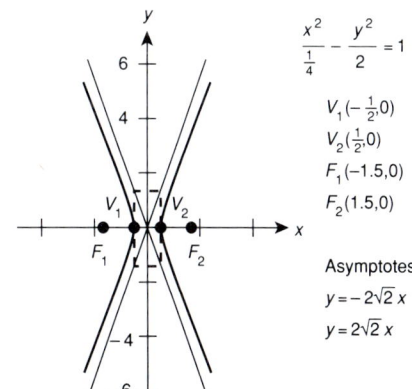

$\dfrac{x^2}{\frac{1}{4}} - \dfrac{y^2}{2} = 1$

$V_1(-\tfrac{1}{2},0)$
$V_2(\tfrac{1}{2},0)$
$F_1(-1.5,0)$
$F_2(1.5,0)$

Asymptotes:
$y = -2\sqrt{2}\,x$
$y = 2\sqrt{2}\,x$

13.

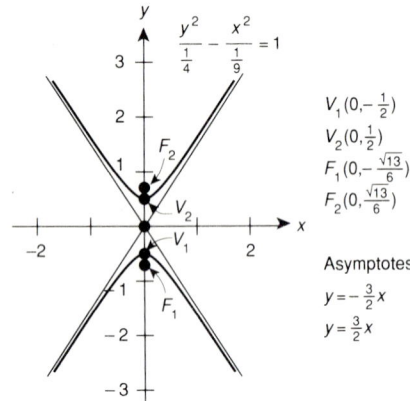

$$\frac{y^2}{\frac{1}{4}} - \frac{x^2}{\frac{1}{9}} = 1$$

$V_1(0, -\frac{1}{2})$
$V_2(0, \frac{1}{2})$
$F_1(0, -\frac{\sqrt{13}}{6})$
$F_2(0, \frac{\sqrt{13}}{6})$

Asymptotes:
$y = -\frac{3}{2}x$
$y = \frac{3}{2}x$

15.

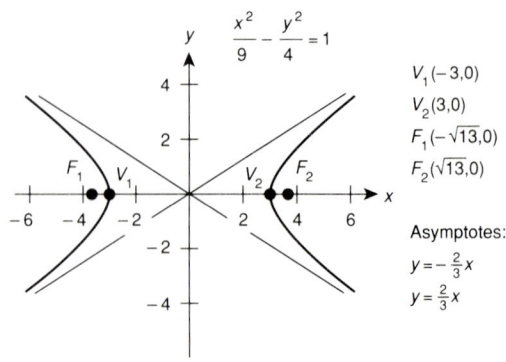

$$\frac{x^2}{9} - \frac{y^2}{4} = 1$$

$V_1(-3, 0)$
$V_2(3, 0)$
$F_1(-\sqrt{13}, 0)$
$F_2(\sqrt{13}, 0)$

Asymptotes:
$y = -\frac{2}{3}x$
$y = \frac{2}{3}x$

17.

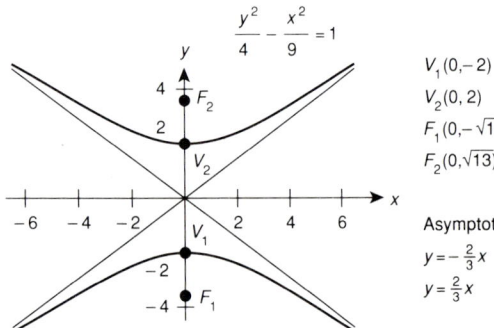

$$\frac{y^2}{4} - \frac{x^2}{9} = 1$$

$V_1(0, -2)$
$V_2(0, 2)$
$F_1(0, -\sqrt{13})$
$F_2(0, \sqrt{13})$

Asymptotes:
$y = -\frac{2}{3}x$
$y = \frac{2}{3}x$

19.

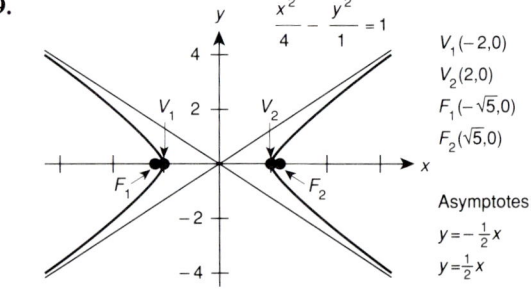

$$\frac{x^2}{4} - \frac{y^2}{1} = 1$$

$V_1(-2, 0)$
$V_2(2, 0)$
$F_1(-\sqrt{5}, 0)$
$F_2(\sqrt{5}, 0)$

Asymptotes:
$y = -\frac{1}{2}x$
$y = \frac{1}{2}x$

21.

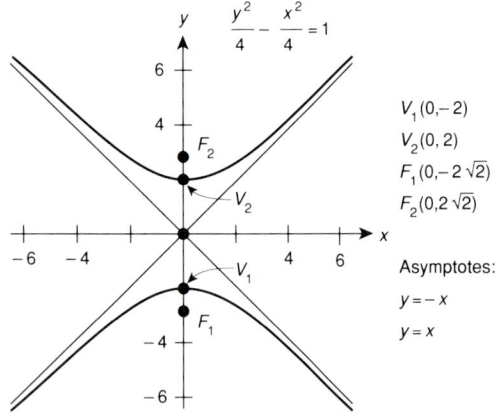

$$\frac{y^2}{4} - \frac{x^2}{4} = 1$$

$V_1(0, -2)$
$V_2(0, 2)$
$F_1(0, -2\sqrt{2})$
$F_2(0, 2\sqrt{2})$

Asymptotes:
$y = -x$
$y = x$

23.

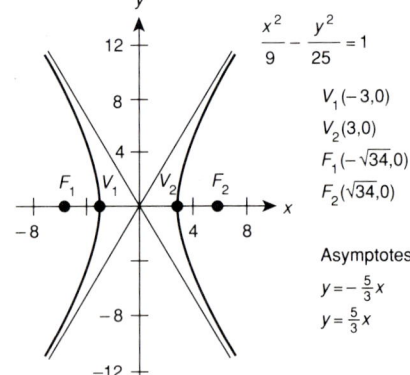

$$\frac{x^2}{9} - \frac{y^2}{25} = 1$$

$V_1(-3, 0)$
$V_2(3, 0)$
$F_1(-\sqrt{34}, 0)$
$F_2(\sqrt{34}, 0)$

Asymptotes:
$y = -\frac{5}{3}x$
$y = \frac{5}{3}x$

25.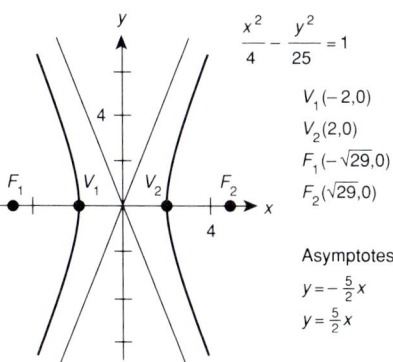

$$\frac{x^2}{4} - \frac{y^2}{25} = 1$$

$V_1(-2,0)$
$V_2(2,0)$
$F_1(-\sqrt{29},0)$
$F_2(\sqrt{29},0)$

Asymptotes:
$y = -\frac{5}{2}x$
$y = \frac{5}{2}x$

27.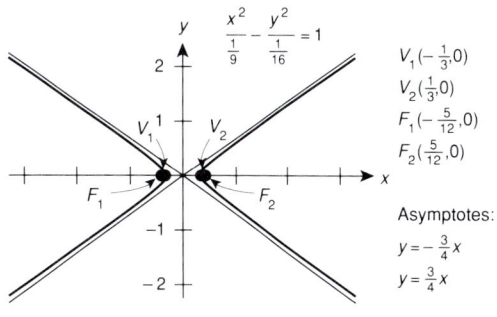

$$\frac{x^2}{\frac{1}{9}} - \frac{y^2}{\frac{1}{16}} = 1$$

$V_1(-\frac{1}{3},0)$
$V_2(\frac{1}{3},0)$
$F_1(-\frac{5}{12},0)$
$F_2(\frac{5}{12},0)$

Asymptotes:
$y = -\frac{3}{4}x$
$y = \frac{3}{4}x$

29.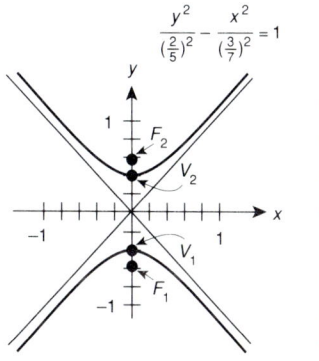

$$\frac{y^2}{(\frac{2}{5})^2} - \frac{x^2}{(\frac{3}{7})^2} = 1$$

$V_1(0,-\frac{2}{5})$
$V_2(0,\frac{2}{5})$
$F_1(0,-\frac{\sqrt{421}}{35})$
$F_2(0,\frac{\sqrt{421}}{35})$

Asymptotes:
$y = -\frac{14}{15}x$
$y = \frac{14}{15}x$

31. $\dfrac{x^2}{9} - \dfrac{y^2}{7} = 1$ **33.** $\dfrac{y^2}{9} - \dfrac{25x^2}{81} = 1$ **35.** $\dfrac{x^2}{5} - \dfrac{y^2}{20} = 1$ **37.** $\dfrac{x^2}{4} - \dfrac{y^2}{1} = 1$ **39.** $\dfrac{x^2}{1} - \dfrac{y^2}{2} = 1$

41.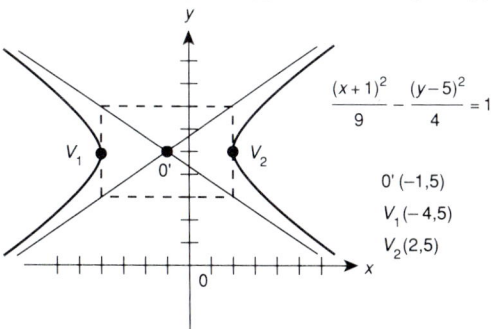

$$\frac{(x+1)^2}{9} - \frac{(y-5)^2}{4} = 1$$

$0'(-1,5)$
$V_1(-4,5)$
$V_2(2,5)$

43.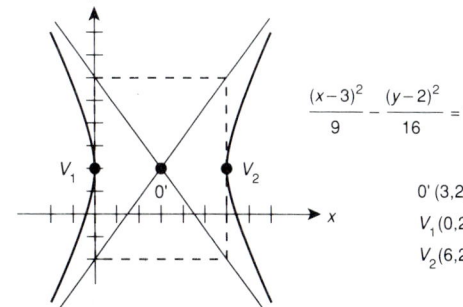

$$\frac{(x-3)^2}{9} - \frac{(y-2)^2}{16} = 1$$

$0'(3,2)$
$V_1(0,2)$
$V_2(6,2)$

45. $\dfrac{(x-7)^2}{64} - \dfrac{(y+2)^2}{36} = 1$

47.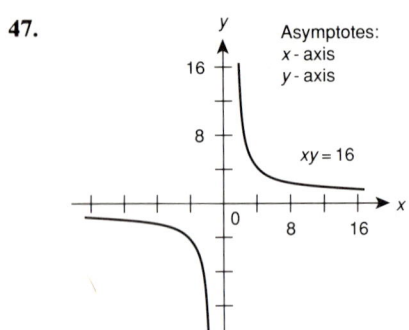

49.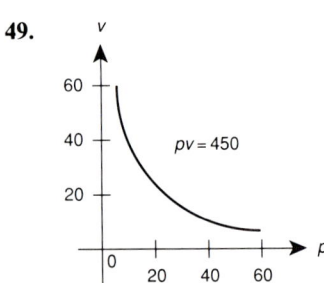

51.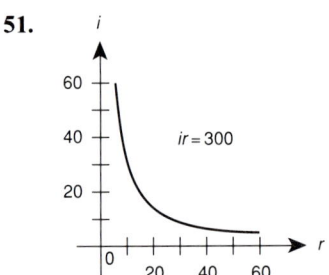

55.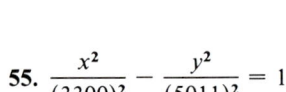
$\dfrac{x^2}{(3300)^2} - \dfrac{y^2}{(5011)^2} = 1$

Trial Problems 16.7

1. $x = r\cos\theta = 3\cos\dfrac{7\pi}{6} = -\dfrac{3\sqrt{3}}{2}$

$y = r\sin\theta = 3\sin\dfrac{7\pi}{6} = 3\left(-\dfrac{1}{2}\right) = -\dfrac{3}{2}$

2. $r^2 = x^2 + y^2,\ r = \sqrt{4^2 + (-4)^2} = \sqrt{32} = 4\sqrt{2}$

$\theta = \tan^{-1}\left(\dfrac{y}{x}\right) = \tan^{-1}(-1) = -45°$

Since θ is in the fourth quadrant, $\theta = 315°$ or $\dfrac{7\pi}{4}$

3. $x^2 - y^2 = 9$, where $x = r\cos\theta$ and $y = r\sin\theta$
$r^2\cos^2\theta - r^2\sin^2\theta = 9$
$r^2(\cos^2\theta - \sin^2\theta) = 9$

4. $r = \dfrac{4}{2 - \cos\theta}$
$2r - r\cos\theta = 4$
$2r = r\cos\theta + 4$
$2\sqrt{x^2 + y^2} = x + 4$
$4(x^2 + y^2) = (x + 4)^2$
$4x^2 + 4y^2 = x^2 + 8x + 16$
$3x^2 + 4y^2 - 8x = 16$

5.

θ	0	$\dfrac{\pi}{2}$	π	$\dfrac{3\pi}{2}$	2π
r	1	0	1	2	1

Exercises 16.7

1. (2,0)   **3.** (0,2)   **5.** $\left(\dfrac{\sqrt{2}}{2}, \dfrac{\sqrt{2}}{2}\right)$ **7.** $\left(-\dfrac{\sqrt{2}}{2}, -\dfrac{\sqrt{2}}{2}\right)$ **9.** (2,2)   **11.** $\left(\dfrac{\sqrt{3}}{4}, \dfrac{1}{4}\right)$ **13.** $\left(-\dfrac{5\sqrt{2}}{2}, \dfrac{5\sqrt{2}}{2}\right)$
15. (1.9319, 0.5167)   **17.** (2,0)   **19.** $\left(4, \dfrac{3\pi}{2}\right)$ **21.** $\left(3\sqrt{2}, \dfrac{\pi}{4}\right)$ **23.** $\left(2, \dfrac{\pi}{6}\right)$ **25.** $\left(2, \dfrac{7\pi}{4}\right)$ **27.** $r(\cos\theta + \sin\theta) = 5$
29. $r = 4$ **31.** $r^2 = \dfrac{9}{2\cos^2\theta - 1}$ **33.** $r^2 - 2r\cos\theta = 10$ **35.** $r = \dfrac{\sin\theta}{2\cos^2\theta}$ or $r = \dfrac{1}{2}\tan\theta\sec\theta$
37. $r = \dfrac{5}{\sqrt{1 + 3\sin^2\theta}}$ **39.** $x^2 + y^2 = 4$ **41.** $x = 2$ **43.** $x - y = 0$ **45.** $x^2 - 4x + y^2 = 0$ **47.** $3x - 4y = 7$
49. $16y^2 = 120x + 225$ **51.** $3x^2 + 4y^2 + 8x - 16 = 0$
53.

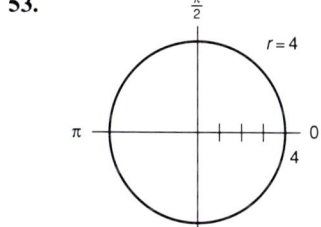

55.

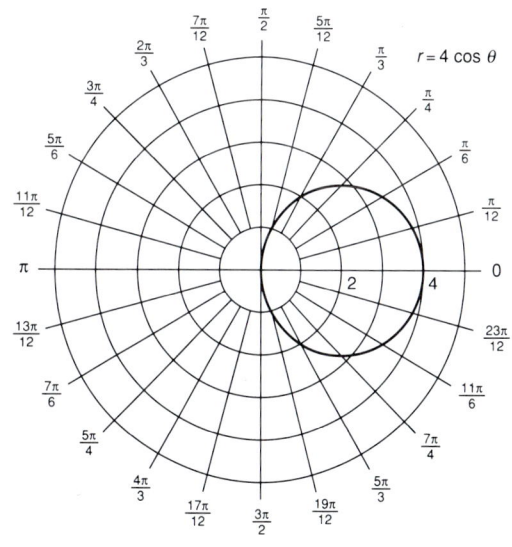

57.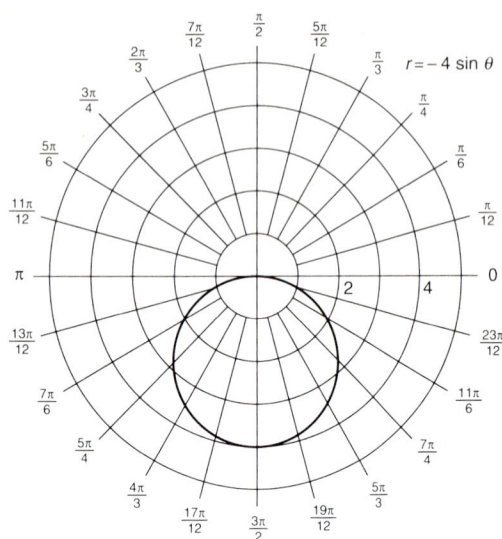

59.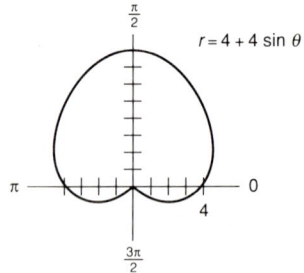

61.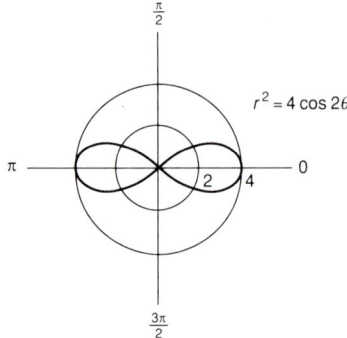

63.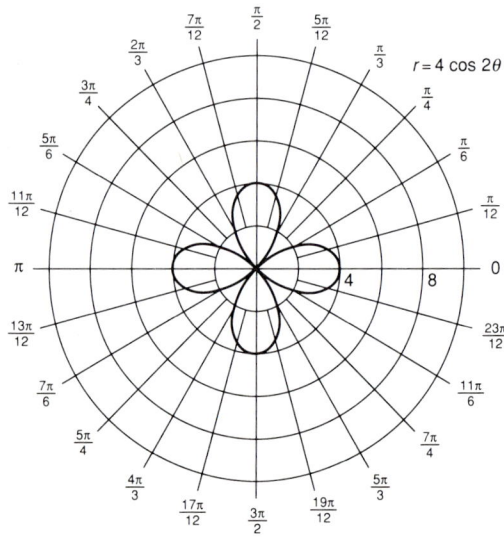

65.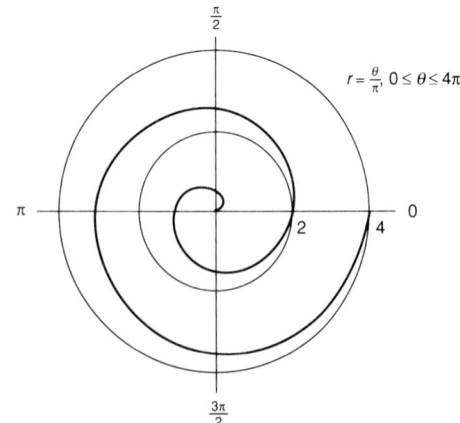

67.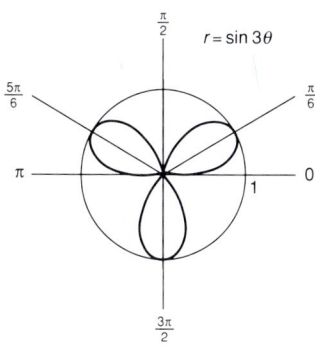

69. elliptical

Review Exercises 16.9

1. [16.2] $\sqrt{74}$ **3.** [16.2] $5\sqrt{5}$ **5.** [16.2] $\dfrac{1}{2}$ **7.** [16.2] $\dfrac{3}{5}$ **9.** [16.2] $2x + y - 5 = 0$ **11.** [16.2] $2x + y + 2 = 0$
13. [16.2] $7x + 5y + 1 = 0$ **15.** [16.2] $2x + 3y - 22 = 0$ **17.** [16.2] $3x - 2y + 9 = 0$ **19.** [16.3] $x^2 + y^2 = 12.25$
21. [16.3] $(x - 3)^2 + (y - 2)^2 = 16$ **23.** [16.3] $(x - 3)^2 + y^2 = 40$ **25.** [16.5] $\dfrac{x^2}{16} + \dfrac{y^2}{7} = 1$
27. [16.5] $\dfrac{x^2}{16} + \dfrac{7y^2}{64} = 1$ **29.** [16.5] $\dfrac{(x - 2)^2}{9} + \dfrac{(y + 1)^2}{16} = 1$ **31.** [16.4] $y^2 = 8(x - 3)$
33. [16.4] $y^2 = x - 2$ or $(x - 2)^2 = y$ **35.** [16.4] $(x - 3)^2 = -9(y - 2)$ **37.** [16.6] $\dfrac{x^2}{1} - \dfrac{y^2}{4} = 1$
39. [16.6] $\dfrac{x^2}{9} - \dfrac{y^2}{7} = 1$ **41.** [16.6] $\dfrac{(y - 2)^2}{9} - \dfrac{(x + 1)^2}{4} = 1$

43. [16.3]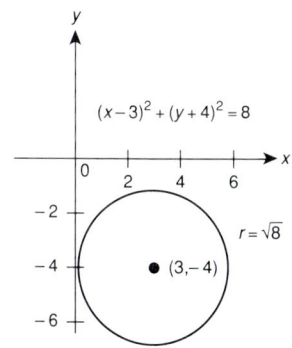

45. [16.5]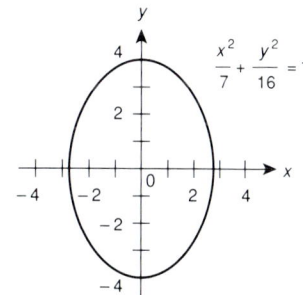

47. [16.5]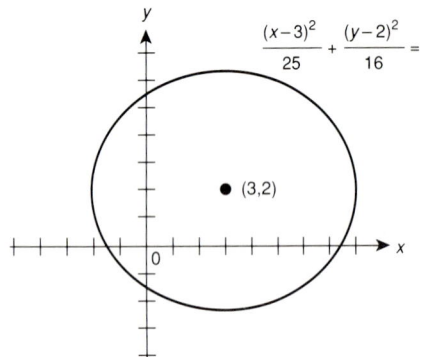

49. [16.4]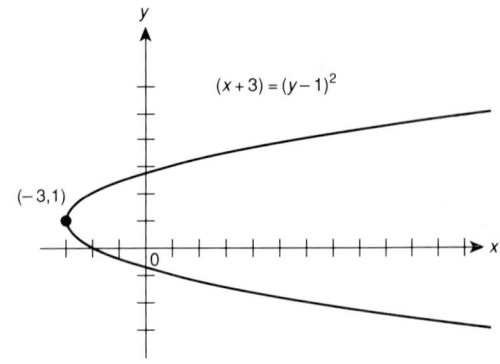

51. [16.6]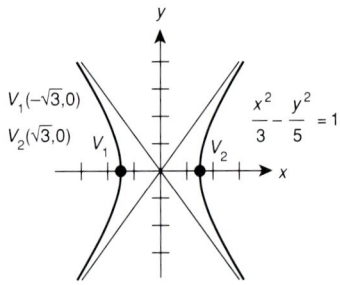

53. [16.6]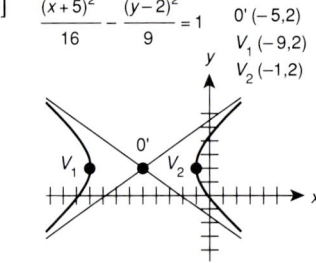

55. [16.7]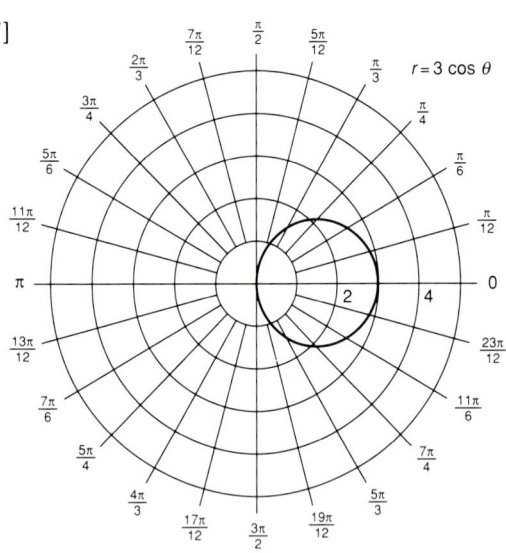

57. [16.7]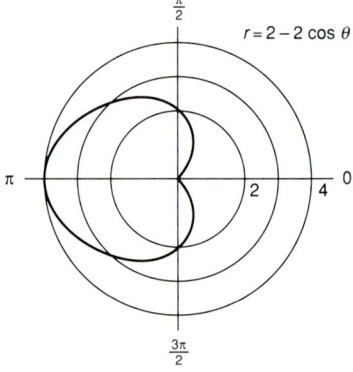

59. [16.7]

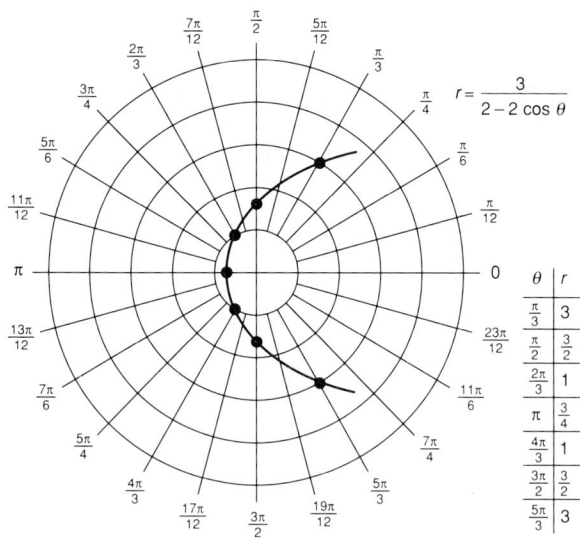

Test 16.10

1. [16.2] $\sqrt{5}$ **2.** [16.2] $8\sqrt{2}$ **3.** [16.2] -2 **4.** [16.2] $\dfrac{2}{5}$ **5.** [16.2] $x + 2y - 8 = 0$ **6.** [16.2] $3x + 2y = 0$

7. [16.3] $(x - 2)^2 + (y - 2)^2 = 9$ **8.** [16.3] $x^2 + (y - 3)^2 = 25$ **9.** [16.5] $\dfrac{x^2}{9} + \dfrac{y^2}{8} = 1$

10. [16.5] $\dfrac{(x - 4)^2}{9} + \dfrac{(y - 2)^2}{1} = 1$ **11.** [16.4] $y^2 = 2(x - 2)$ or $(x - 2)^2 = 2y$

12. [16.4] $(x - 3)^2 = \dfrac{1}{3}(y - 3)$ or $(y - 3)^2 = 9(x - 3)$ **13.** [16.6] $\dfrac{x^2}{4} - \dfrac{y^2}{12} = 1$ **14.** [16.6] $\dfrac{(x - 3)^2}{4} - \dfrac{(y - 2)^2}{9} = 1$

15. [16.3]

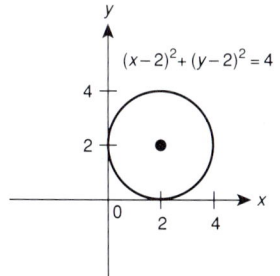

16. [16.5]

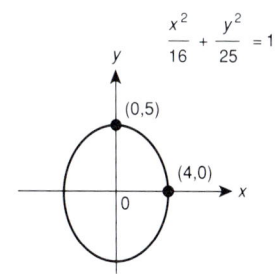

17. [16.5] $\dfrac{(x-2)^2}{2}+\dfrac{(y-3)^2}{9}=1$

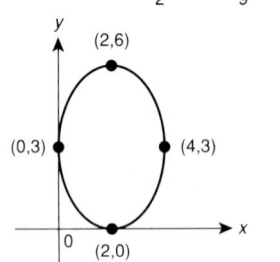

18. [16.4]

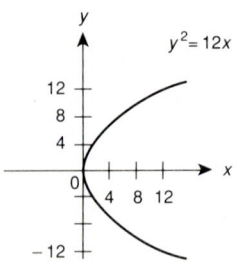

19. [16.4] $(x-2)^2=-4(y+3)$

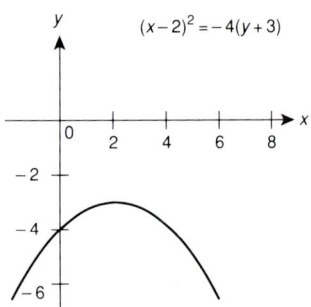

20. [16.6] $\dfrac{(x-1)^2}{16}-\dfrac{(y-3)^2}{9}=1$

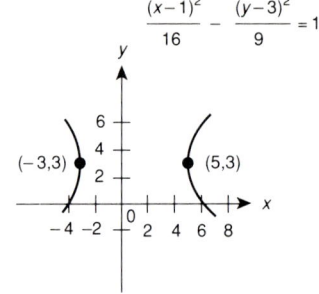

21. [16.7]

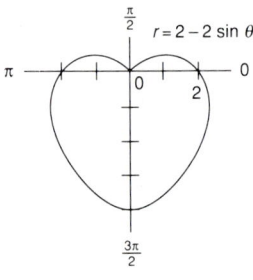

22. [16.7]

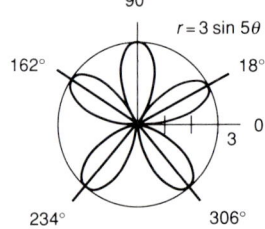

Chapter 17 Answers

Trial Problems 17.1

1.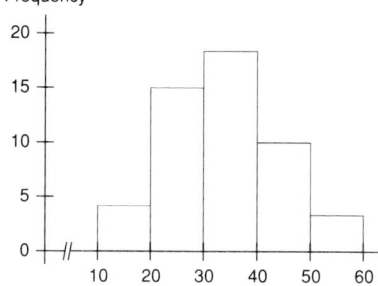

2.

Class	Frequency	Relative Frequency
10–19.9	4	0.08
20–29.9	15	0.30
30–39.9	18	0.36
40–49.9	10	0.20
50–59.9	3	0.06

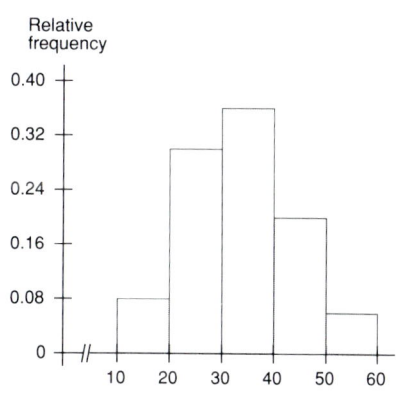

3–4.

Class	Frequency	Relative Frequency
40–49.9	3	0.075
50–59.9	10	0.250
60–69.9	12	0.300
70–79.9	10	0.250
80–89.9	3	0.075
90–99.9	2	0.050

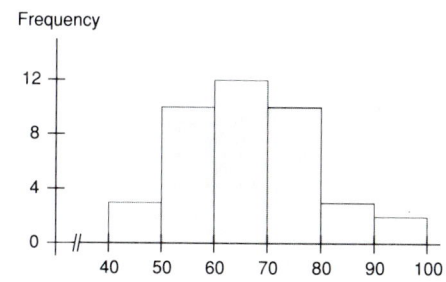 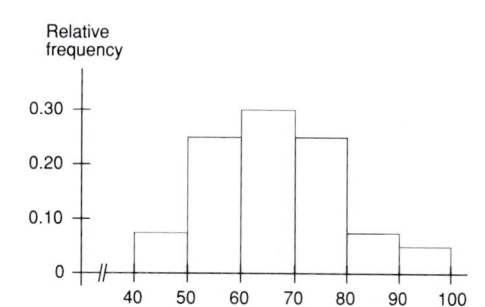

Exercises 17.1

1.

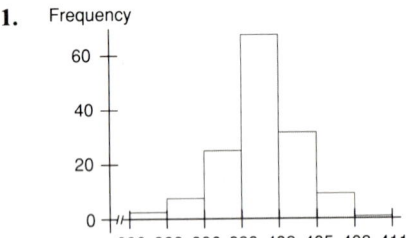

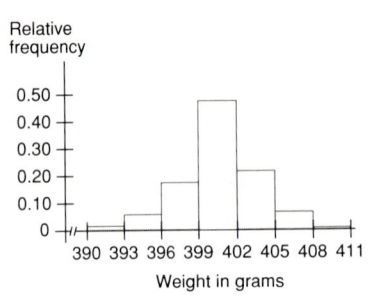

3a.

Interval	f	Relative Frequency
1–5	5	0.102
6–10	8	0.163
11–15	3	0.061
16–20	5	0.102
21–25	3	0.061
26–30	2	0.041
31–35	4	0.082
36–40	0	0.000
41–45	2	0.041
46–50	1	0.020
51–55	4	0.082
56–60	1	0.020
61–65	5	0.102
66–70	1	0.020
71–75	0	0.000
76–80	2	0.041
81–85	2	0.041
86–90	1	0.020

b.

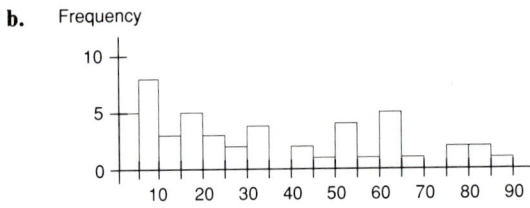

c.

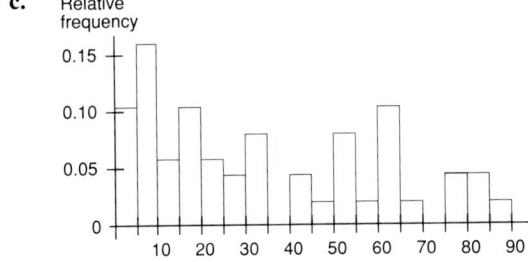

5a.

Interval	f	Relative Frequency
40–49	1	0.02
50–59	7	0.14
60–69	7	0.14
70–79	3	0.06
80–89	6	0.12
90–99	10	0.20
100–109	5	0.10
110–119	4	0.08
120–129	2	0.04
130–139	3	0.06
140–149	0	0.00
150–159	2	0.04

b.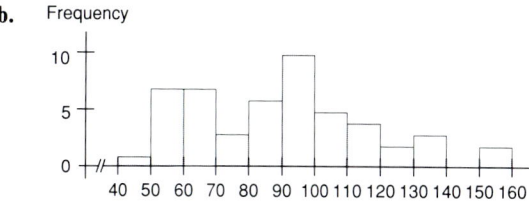

c.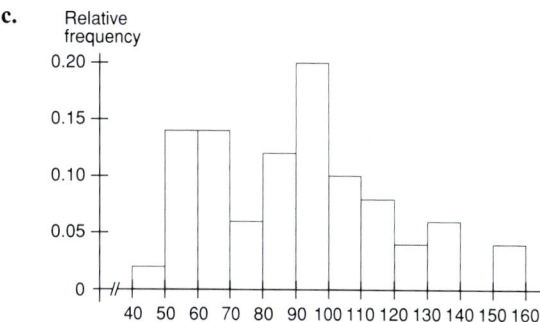

7a.

Diameter	f	Relative Frequency
10.1–10.5	6	0.150
10.6–11.0	5	0.125
11.1–11.5	1	0.025
11.6–12.0	1	0.025
12.1–12.5	2	0.050
12.6–13.0	2	0.050
13.1–13.5	1	0.025
13.6–14.0	6	0.150
14.1–14.5	4	0.100
14.6–15.0	7	0.175
15.1–15.5	5	0.125

b.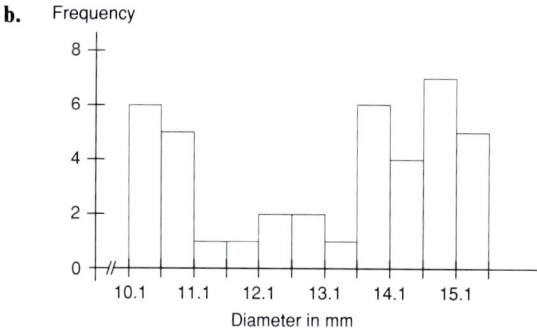

c.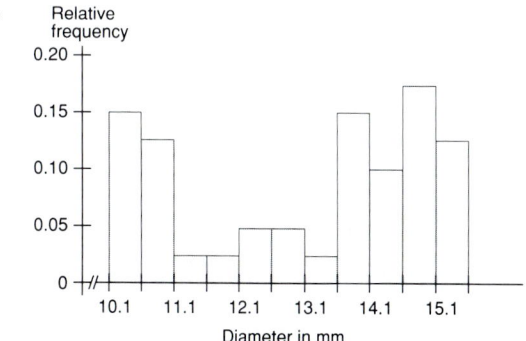

9a. Frequency

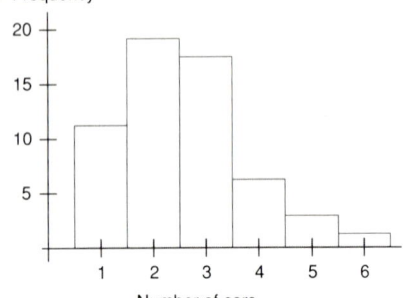

Number of cars

b. Relative frequency

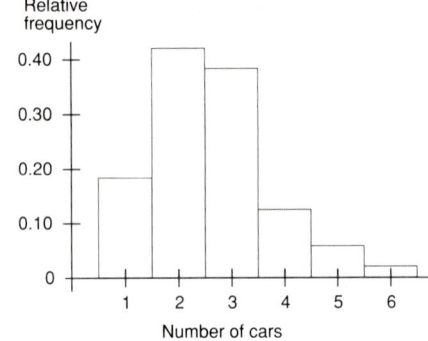

Number of cars

Trial Problems 17.2

1. Since $n = 20$, the median is the average of the tenth and eleventh terms.

$$\text{median} = \frac{3.9 + 4.0}{2} = 3.95$$

$$\text{mean} = \frac{\Sigma x}{n} = \frac{81.5}{20} = 4.075$$

2. Q_1 = average of fifth and sixth terms = $\frac{2.7 + 3.0}{2} = 2.85$

Q_2 = median = 3.95

Q_3 = average of fifteenth and sixteenth terms = $\frac{4.7 + 5.6}{2} = 5.15$

3.

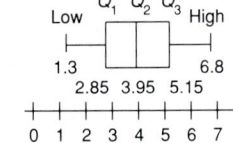

4.

x	f	xf
30	6	180
35	9	315
40	10	400
45	12	540
50	11	550
55	8	440
60	7	420
	63	2845

$$\bar{x} = \frac{\Sigma xf}{\Sigma f} = \frac{2845}{63} = 45.2$$

5.

Class	f	x	xf
10–19.9	4	14.95	59.80
20–29.9	15	24.95	374.25
30–39.9	18	34.95	629.10
40–49.9	10	44.95	449.50
50–59.9	3	54.95	164.85
	50		1677.50

$$\bar{x} = \frac{\Sigma xf}{\Sigma f} = \frac{1667.50}{50} = 33.55$$

Exercises 17.2

1. Modes: 20, 22, Median: 22, Mean: 22.2   3. Modes: 12, 15, 19, Median: 18.5, Mean: 21.45
5. $Q_1 = 5.3$, $Q_2 = 9.3$, $Q_3 = 14.75$

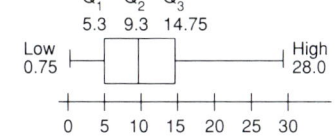

7. $Q_1 = 22.9$, $Q_2 = 36.35$, $Q_3 = 42.7$ 9. 12.51   11. 13.00   13. 11.03

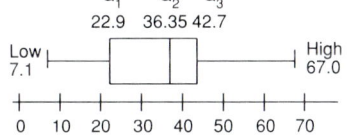

Trial Problems 17.3

1. $s = \sqrt{\dfrac{84}{6}} = 3.74$ 2. $s = \sqrt{\dfrac{14(672) - (84)^2}{(14)(13)}} = 3.59$ 3. $s = \sqrt{\dfrac{38(565,700) - (4590)^2}{(38)(37)}} = 17.46$

4.

x	x^2
17.1	292.41
20.2	408.04
35.3	1246.09
18.6	345.96
19.4	376.36
35.2	1239.04
26.1	681.21
19.0	361.00
20.4	416.16
26.3	691.69
237.6	6057.96

$$s = \sqrt{\dfrac{10(6057.96) - (237.6)^2}{(10)(9)}} = 6.77$$

5.

x	f	xf	x^2f
15	5	75	1125
16	12	192	3072
17	18	306	5202
18	20	360	6480
19	12	228	4332
20	9	180	3600
21	4	84	1764
	80	1425	25,575

$$s = \sqrt{\frac{80(25{,}575) - (1425)^2}{(80)(79)}} = 1.56$$

Exercises 17.3

1. $s^2 = 14.94$, $s = 3.86$ **3.** $s^2 = 28.08$, $s = 5.30$ **5.** $s^2 = 675.77$, $s = 26.0$ **7.** $s^2 = 0.027$, $s = 0.165$
9. $s^2 = 1.52$, $s = 1.23$ **11.** 30.61 to 45.09 **13.** 66.48 to 83.76 **15.** 132.4 to 140.0 **17.** 95.4%
19a. 24.99 to 27.75 **b.** 23.61 to 29.13 **c.** 22.23 to 30.51

Trial Problems 17.4

1. $r = \dfrac{30(6{,}810) - (33)(95)}{\sqrt{30(21{,}485) - (33)^2}\sqrt{30(16{,}820) - (95)^2}}$

$= \dfrac{201{,}165}{\sqrt{643{,}461}\sqrt{495{,}575}} = 0.356$

2. $y - \bar{y} = \dfrac{r(s_y)}{s_x}(x - \bar{x})$, where $\bar{y} = 68.3$, $\bar{x} = 60.4$, $s_y = 16.8$, $s_x = 21.4$, $r = 0.83$

$y - 68.3 = \dfrac{0.83(16.8)}{21.4}(x - 60.4)$

$y - 68.3 = 0.65(x - 60.4)$ or $y = 0.65x + 29.0$

3. $m = \dfrac{10(4325) - 180(335)}{10(5805) - (180)^2} = \dfrac{-17{,}050}{25{,}650} = -0.665$

$b = \dfrac{335 - (-0.665)(180)}{10} = 45.47$

$y = mx + b = -0.665x + 45.47$

x	y	xy	x^2	y^2
1	19	19	1	361
2	30	60	4	900
3	48	144	9	2304
4	49	196	16	2401
5	62	310	25	3844
6	71	426	36	5041
7	83	581	49	6889
28	362	1736	140	21,740

4. $r = \dfrac{7(1736) - 28(362)}{\sqrt{7(140) - (28)^2} \sqrt{7(21{,}740) - (362)^2}}$

$= \dfrac{2016}{\sqrt{196}\ \sqrt{21{,}136}} = 0.99$

5. $m = \dfrac{7(1736) - (28)(362)}{7(140) - (28)^2}$

$= \dfrac{2016}{196} = 10.29$

$b = \dfrac{362 - (10.29)(28)}{7} = 10.55$

$y = mx + b = 10.29x + 10.55$

Exercises 17.4

1. $r = 0.97$
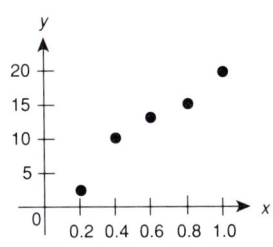

3. $r = 0.88$
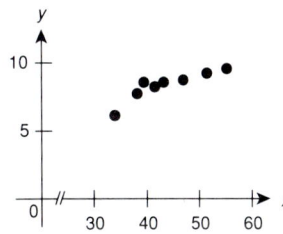

5. $r = -0.99$
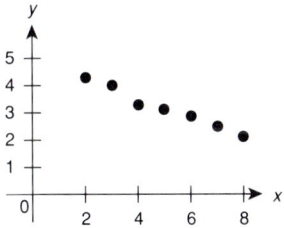

7. $r = 0.99$
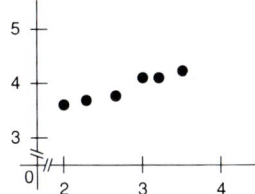

9a. $y = 4.96 - 0.357x$ **b.** 1.75
11a. $y = 2.58 + 0.486x$ **b.** 4.39
13a. $y = 1.064 + 0.9885x$ **b.** 45.55   **15a.** $y = 101.4 + 0.029x$ **b.** 104.3
17a. $y = 23.46 + 0.051x$ **b.** $201.96

Trial Problems 17.5

1.

x	y	$x^* = \sqrt{x}$	x^*y	$(x^*)^2$
10	23.1	$\sqrt{10}$	73.05	10
20	28.8	$\sqrt{20}$	128.80	20
30	33.2	$\sqrt{30}$	181.84	30
40	36.9	$\sqrt{40}$	233.38	40
50	40.1	$\sqrt{50}$	283.55	50
60	43.1	$\sqrt{60}$	333.85	60
70	45.8	$\sqrt{70}$	383.19	70
	251.0	42.62	1617.66	280

$$m = \frac{7(1617.66) - (42.62)(251)}{7(280) - (42.62)^2} = 4.36$$

$$b = \frac{251 - 4.36(42.62)}{7} = 9.31$$

$$y = mx^* + b = 4.36\sqrt{x} + 9.31$$

2.

x	y	$x^* = \dfrac{1}{x}$	x^*y	$(x^*)^2$
2	125	0.500	62.50	0.2500
4	75	0.250	18.75	0.0625
8	50	0.125	6.25	0.0156
10	45	0.100	4.50	0.0100
20	35	0.050	1.75	0.0025
40	30	0.025	0.75	0.0006
	360	1.050	94.50	0.3412

$$m = \frac{6(94.5) - (1.05)(360)}{6(0.3412) - (1.05)^2} = 200$$

$$b = \frac{360 - 200(1.05)}{6} = 25$$

$$y = \frac{m}{x} + b = \frac{200}{x} + 25$$

3.

x	y	$y^* = \ln y$	x^2	xy^*
0.5	12.8	2.5494	0.25	1.2747
1.0	16.5	2.8034	1.00	2.8034
1.5	21.2	3.0540	2.25	4.5810
2.0	27.2	3.3032	4.00	6.6064
3.0	44.8	3.8022	9.00	11.4066
3.5	57.5	4.0518	12.25	14.1813
11.5		19.5640	28.75	40.8534

$$m = \frac{6(40.8534) - 11.5(19.5640)}{6(28.75) - (11.5)^2} = 0.50$$

$$b^* = \frac{19.5640 - (0.50)(11.5)}{6} = 2.3023333$$

Since $b^* = \ln b$, then $b = e^{b^*} = e^{2.3023333} = 10.0$

$y = me^{bx} = 10.0e^{0.50x}$

Answer Key—Chapter 17

Exercises 17.5

1. $y = \dfrac{211.9}{x} + 13.7$ **3.** $y = 3.42\sqrt{x} + 6.85$ **5.** $y = 65.3x^2 + 13.7$ **7a.** $y = \dfrac{186.6}{x} + 1.95$ **b.** 7.3
9a. $y = 9.43(\ln x) + 44.69$ **b.** 65.4 **11.** $y = e^{(-0.083x + 5.33)}$ **13a.** $y = 90.14e^{0.0081x}$ **b.** 130 **15.** $y = 2.19x^{-1.23}$

Review Exercises 17.7

1a. [17.1]

Pounds	f	Relative Frequency
113–114.9	4	0.100
115–116.9	11	0.275
117–118.9	11	0.275
119–120.9	10	0.250
121–122.9	3	0.075
123–124.9	1	0.025

b. [17.1]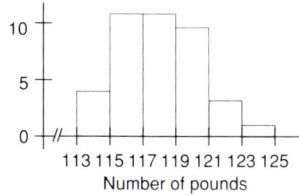

c. [17.1]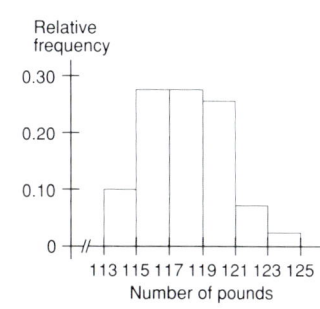

3. [17.2, 17.3] Mean = 2.39, $s = 6.03$ **5.** [17.2, 17.3] Mean = 18.99, $s = 1.43$ **7.** [17.3] 14.2 to 25.0
9a. [17.3] 78.9 to 91.5 **b.** [17.3] 72.6 to 97.8 **c.** [17.3] 66.3 to 104.1 **11a.** [17.4] 0.84 **b.** [17.4] $y = 14.8 + 0.66x$
13a. [17.4] $y = 860.5 + 0.018x$ **b.** [17.4] 861.4 **15a.** [17.5] $y = -8.33(\ln x) + 57.84$ **b.** [17.5] 24.6
17a. [17.5] $y = 79.3e^{0.00443x}$ **b.** [17.5] 103

Test 17.8

1a. [17.1]

Class	Frequency	Relative Frequency
10–19.9	1	0.03
20–29.9	4	0.13
30–39.9	9	0.30
40–49.9	8	0.27
50–59.9	5	0.17
60–69.9	3	0.10

b. [17.1] Frequency

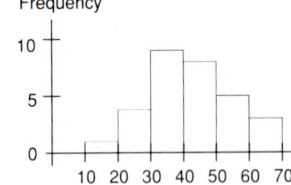

c. [17.1] Relative frequency

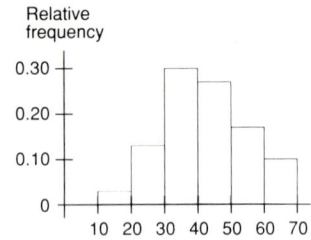

d. [17.2] 40.75   **e.** [17.2] 30.6, 40.75, 52.3   **f.** [17.2]

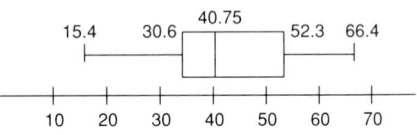

g. [17.3] 166.55   **h.** [17.3] 12.91   **2.** [17.2, 17.3] $\bar{x} = 3.83$. $s = 1.93$ **3.** [17.2, 17.3] $\bar{x} = 4.375$, $s = 1.32$
4. [17.3] 75% of data in (106.8, 185.6)   **5.** [17.3] 62.8 to 82.0   **6a.** [17.4] 0.968   **b.** $y = 2.37x - 30.3$ **7a.** [17.4] 0.991
b. [17.4] $y = 0.827x + 9.53$ **8a.** [17.5] 0.999   **b.** $y = 4.17x + 17.46$ **9a.** [17.5] $y = 2.60\sqrt{x} + 14.57$, **b.** [17.5] 29.95
10a. [17.5] $y = 3.50e^{-0.0557x}$,   **b.** 0.50

Cumulative Review—Chapters 16 and 17   17.9

1. [16] $\sqrt{26}$ **2.** [16] -1 **3.** [16] $y - 2 = \frac{1}{3}(x + 3)$ **4.** [16] $(x - 1)^2 + (y + 3)^2 = 16$ **5.** [16] $\frac{x^2}{12} + \frac{y^2}{16} = 1$

6. [16] $\frac{(x - 2)^2}{9} + \frac{(y - 3)^2}{4} = 1$ **7.** [16] $y^2 = 2(x - 4)$ or $(x - 4)^2 = 2y$ **8.** [16] $\frac{x^2}{9} - \frac{y^2}{16} = 1$

9. [16]

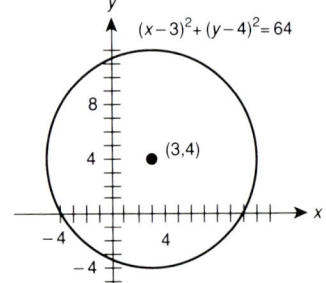

10. [16]

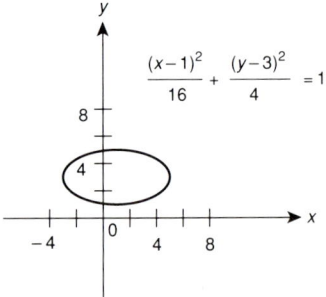

11. [16]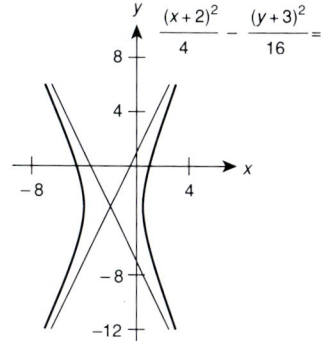

12. [16]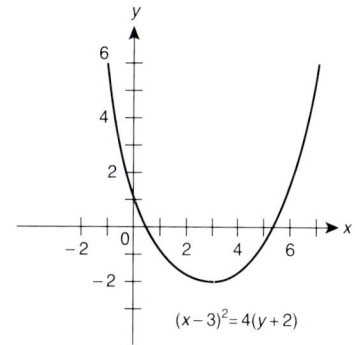

13. [16]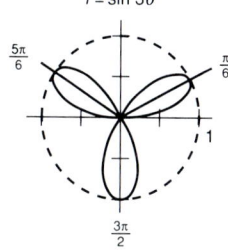

14a. [17]

Interval	Frequency
40–49.9	2
50–59.9	6
60–69.9	9
70–79.9	8
80–89.9	3
90–99.9	2

b. [17]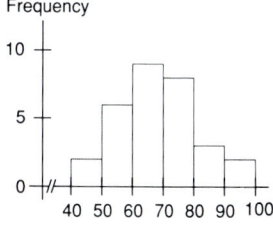

c. [17] 67.72 **d.** [17] 13.55 **15.** [17] Mean $= 40.1$, $s = 12.3$ **16a.** [17] 0.998, **b.** [17] $y = 9.753 + 0.353x$
17. [17] $y = 8.99e^{-2.77x}$

Chapter 18 Answers

Trial Problems 18.1

1. $\lim_{x \to 2} x = 2$ **2.** $\lim_{x \to 2} x^2 = 2^2 = 4$ **3.** $\lim_{x \to 0} \dfrac{x}{2} = \dfrac{0}{2} = 0$ **4.** $\lim_{x \to 1} \dfrac{3}{x} = \dfrac{3}{1} = 3$ **5.** $\lim_{x \to -1} 125 = 125$

Exercises 18.1

1. 0   **3.** $\dfrac{4}{5}$ **5.** $\dfrac{7}{4}$ **7.** 3   **9.** 0   **11.** -23 **13.** 0   **15.** $-\dfrac{1}{8}$ **17.** 8   **19.** does not exist (dne)   **21.** dne   **23.** -4
25. 337.57   **27.** 191.1 m

Trial Problems 18.2

1. $f(x) = x$ is continuous for all x.

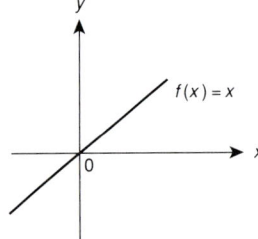

2. $f(x) = x + 2$ is continuous for all x.

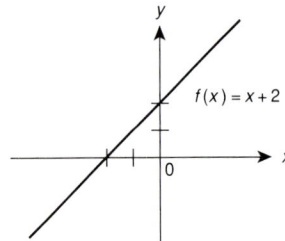

3. $f(x) = x - 2$ is continuous for all x.

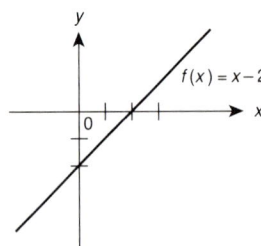

4. $f(x) = 4$ is continuous for all x.

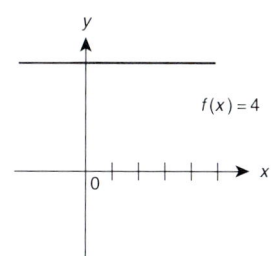

5. $f(x) = \sqrt{x}$ is continuous for all $x \geq 0$.

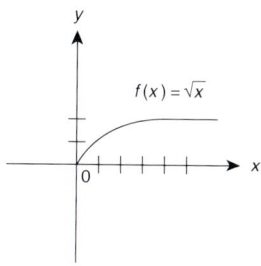

Exercises 18.2

1. discontinuous at $x = -2, 2$ **3.** discontinuous at $x = 0$ **5.** continuous for $x \geq 4$ **7.** continuous for all x
9. continuous for all x **11.** no current when $E = 0$ **13.** continuous at $x = 200$

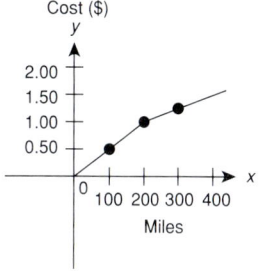

Trial Problems 18.3

1. $\lim\limits_{x \to \infty} \dfrac{1}{x} = 0$ **2.** $\lim\limits_{x \to \infty} \dfrac{1}{x^2} = 0$ **3.** $\lim\limits_{x \to -\infty} \dfrac{1}{x^2} = 0$ **4.** $\lim\limits_{x \to \infty} \dfrac{1}{1 + \dfrac{1}{x}} = \dfrac{1}{1 + 0} = 1$ **5.** $\lim\limits_{x \to 3} \dfrac{1}{x - 3} = \infty$

Exercises 18.3

1. $\dfrac{1}{2}$ **3.** dne   **5.** 2   **7.** dne   **9.** $-\dfrac{7}{3}$ **11.** dne   **13.** 0   **15.** 2   **17.** 0   **19.** $2x$ **21. a.** $+\infty$ **21. b.** 0   **23.** 0

Trial Problems 18.4

1. $f(x) = 2x$
$$\frac{f(x + \Delta x) - f(x)}{\Delta x} = \frac{2(x + \Delta x) - 2x}{\Delta x}$$
$$= \frac{2x + 2\Delta x - 2x}{\Delta x}$$
$$= \frac{2\Delta x}{\Delta x}$$
$$= 2$$

2. $f(x) = x^2$
$$\frac{f(x + \Delta x) - f(x)}{\Delta x} = \frac{(x + \Delta x)^2 - x^2}{\Delta x}$$
$$= \frac{x^2 + 2x\Delta x + \overline{\Delta x}^2 - x^2}{\Delta x}$$
$$= \frac{2x\Delta x + \overline{\Delta x}^2}{\Delta x}$$
$$= 2x + \Delta x$$

3. $f(x) = 4$
$$\frac{f(x + \Delta x) - f(x)}{\Delta x} = \frac{4 - 4}{\Delta x}$$
$$= \frac{0}{\Delta x}$$
$$= 0$$

4. $y = x + 1$
$$\frac{f(x + \Delta x) - f(x)}{\Delta x} = \frac{(x + \Delta x) + 1 - (x + 1)}{\Delta x}$$
$$= \frac{x + \Delta x + 1 - x - 1}{\Delta x}$$
$$= \frac{\Delta x}{\Delta x}$$
$$= 1$$

5. $y = 1 - x$
$$\frac{f(x + \Delta x) - f(x)}{\Delta x} = \frac{[1 - (x + \Delta x)] - (1 - x)}{\Delta x}$$
$$= \frac{1 - x - \Delta x - 1 + x}{\Delta x}$$
$$= -\frac{\Delta x}{\Delta x}$$
$$= -1$$

Exercises 18.4

1. 1 **3.** 5 **5.** 3 **7.** 3 **9.** 2 **11.** $2a$ **13.** $3a^2$ **15.** 0 **17.** 0 **19.** -1

Trial Problems 18.5

1. $f(x) = x, f'(x) = 1$ **2.** $y = x + 1, \dfrac{dy}{dx} = 1$ **3.** $y = 1 - x, y' = -1$ **4.** $f(x) = x^2, f'(x) = 2x, f'(1) = 2(1) = 2$
5. $g(x) = x^2 + x + 2, g'(x) = 2x + 1, g'(0) = 2(0) + 1 = 1$

Exercises 18.5

1. 0 **3.** $-2x$ **5.** 2 **7.** 6 **9.** -3 **11.** $-\dfrac{2}{x^3}$ **13.** $-\dfrac{8}{a^3}$ **15.** -2 **17.** $\dfrac{-1}{(x + 1)^2}$ **19.** $\dfrac{-2x}{(x^2 + 1)^2}$ **21.** $\dfrac{1}{2\sqrt{x}}$
23. $v = 2t - 2, -1.8$ **25.** If f is differentiable at $x = a$, then f is continuous at $x = a$.

Trial Problems 18.6

1. $y = (x + 1)(x - 1)$
$= x^2 - 1$
2. $y = (2x^2)^3$
$= 2^3(x^2)^3$
$= 8x^6$
3. $y = (x^2 + 1)^2$
$= (x^2 + 1)(x^2 + 1)$
$= x^4 + 2x^2 + 1$
4. $y = 3x(x^2 + 2)$
$= 3x^3 + 6x$
5. $y = \dfrac{x^2 + 3x - 10}{x - 2}$
$= \dfrac{(x - 2)(x + 5)}{(x - 2)}$
$= x + 5$

Exercises 18.6

1. $y' = 6x^2$ **3.** $y' = 18x^2 - 10x + 1$ **5.** $y' = 2x - 1$ **7.** $y' = -\dfrac{8}{x^3} - \dfrac{1}{x^2}$ **9.** $f'(x) = 324x^3$
11. $f'(x) = -\dfrac{2}{(x + 1)^3}$ **13.** $f'(x) = \dfrac{4x^3 - 7x^2 - 3}{x^2}$ **15.** $\dfrac{dy}{dx} = 7x^6 + 6x^5$ **17.** $f'(x) = x^2 + x$
19. $f'(x) = 52x^3 - 18x^2 + 1$ **21.** $y' = 6(2x^3 - 1)$ **23.** $y' = 12x^2(x^3 - 1)$
25. $f'(x) = -24x^7 + 108x^5 + 10x^4 - 12x^3 - 36x^2 + 2$ **27.** $y' = -\dfrac{2x}{(x^2 - 1)^2}$ **29.** $y' = \dfrac{-2x^3 - 21x^2 - 6}{(x^3 + x + 1)^2}$
31. $y' = \dfrac{-12x^2 + 15}{(4x^2 - 3x + 5)^2}$ **33.** $f'(s) = -1 + 8s - 20s^2$ **35.** $h'(r) = 18r^5 - 21r^2 + 4r$ **37.** $S'(x) = 2 - \dfrac{1}{2x^2}$
39. $f'(x) = -\dfrac{3x^2 + 2x + 1}{(1 + x + x^2 + x^3)^2}$ **41.** $T'(s) = -\dfrac{4}{3^4 s^5}$ **43.** $P'(v) = 24v^2 - 40v + 12$ **45.** $f'(x) = \dfrac{1}{3\sqrt[3]{x^2}}$
47. a. $y' = \dfrac{2 - 3x}{x^3}$ **b.** $y' = \dfrac{-2x^2 - 4}{x^5}$ **49.** $4x - 5y = -13$ **51.** $f'(5) = -10$ **53.** $f'(2) = 3$ **55.** $f'(1) = -9$
57. $f'(2) = -\dfrac{20}{9}$ **59.** $\dfrac{dV}{dt} = k$ **61.** $\dfrac{dD}{dh} = \dfrac{k}{2\sqrt{h}}$ **63.** $\dfrac{dC}{da} = k_1 - \dfrac{k_2}{a^2}$ **65.** $v' = \dfrac{2x - 1}{b - a}$ **67.** $\dfrac{dP}{dr} = \dfrac{E^2(R^2 - r^2)}{(R^2 + 2Rr + r^2)^2}$

Trial Problems 18.7

1. $y = (x + 1)^2$
   $y' = 2(x + 1)^1(1)$
   $= 2x + 2$

2. $f(x) = (x + 1)^{-1}$
   $f'(x) = -(x + 1)^{-2}(1)$
   $= -\dfrac{1}{(x + 1)^2}$

3. $y = (x + 1)^{1/2}$
   $y' = \dfrac{1}{2}(x + 1)^{-1/2}(1)$
   $= \dfrac{1}{2\sqrt{x + 1}}$

4. $g(x) = (x + 1)^{-1/2}$
   $g'(x) = -\dfrac{1}{2}(x + 1)^{-3/2}(1)$
   $= -\dfrac{1}{2(x + 1)^{3/2}}$

5. $y = \sqrt[3]{2x + 3} = (2x + 3)^{1/3}$
   $y' = \dfrac{1}{3}(2x + 3)^{-2/3}(2)$
   $= \dfrac{2}{3}(2x + 3)^{-2/3}$ or $\dfrac{2}{3(2x + 3)^{2/3}}$

Exercises 18.7

1. $y' = 6(2x + 5)^2$ 3. $f'(x) = -40(8x - 7)^{-6}$ 5. $y' = \dfrac{3(1 - 3x^2)}{(x^2 + 1)^3}$

7. $y' = -2(4x^5 - 3x^3 + 2x)^{-3}(20x^4 - 9x^2 + 2)$ 9. $f'(x) = \dfrac{2x + 3}{2\sqrt{x^2 + 3x + 2}}$ 11. $k'(x) = \dfrac{8x^2}{\sqrt[3]{(8x^3 + 27)^2}}$

13. $r'(s) = -\dfrac{3}{2(3s - 5)^{3/2}}$ 15. $f'(x) = 16\sqrt[4]{x^2 + 9}(4x + 7)^3 + \dfrac{x(4x + 7)^4}{2(x^2 + 9)^{3/4}}$ 17. $f'(t) = \dfrac{(5t + 1)^4(35t + 2)}{2\sqrt{(5t + 1)^5}}$

19. $y' = -\dfrac{1}{2\sqrt{1 - x}}$ 21. $y = \dfrac{2 - 3x}{2\sqrt{1 - x}}$ 23. $y' = \dfrac{-4x}{(2x + 1)^2\sqrt{4x + 1}}$ 25. $864x - y = 1647$ 27. $x = 0$

29. $m = \dfrac{1}{2}$ 31. $\dfrac{ds}{dr} = \dfrac{2\pi r^2 + \pi h^2}{\sqrt{r^2 + h^2}}$ 33. $\dfrac{dI}{dL} = \dfrac{-Rk^2L}{[k^2 + (kL)^2]^{3/2}}$

Trial Problems 18.8

1. $y = x^2$
$$\frac{\Delta y}{\Delta x} = \frac{f(x + \Delta x) - f(x)}{\Delta x}$$
$$= \frac{(x + \Delta x)^2 - x^2}{\Delta x}$$
$$= \frac{x^2 + 2x\Delta x + \overline{\Delta x}^2 - x^2}{\Delta x}$$
$$= \frac{2x\Delta x + \overline{\Delta x}^2}{\Delta x}$$
$$= 2x + \Delta x$$

2. $y = x^2$, $\lim_{\Delta x \to 0} \frac{\Delta y}{\Delta x} = \lim_{\Delta x \to 0} (2x - \Delta x) = 2x$ **3.** $y = x^2$, $y' = 2x$ **4.** $y = x^2$, $f'(x) = 2x$, $f'(3) = 2(3) = 6$

5. $y = \frac{1}{x} = x^{-1}$, $y' = -x^{-2}$, $y' = -\frac{1}{x^2}$, $y' = -\frac{1}{2^2}$ at $x = 2$, $y' = -\frac{1}{4}$

Exercises 18.8

1. 1 **3.** $2x$ **5.** $\frac{x^2 - 1}{x^2}$ **7.** 4 **9.** -16 **11.** $\frac{1}{2}$ **13.** -0.25, decreasing **15. a.** 3 **b.** 11 **c.** 19 **d.** 27 **17.** 12 **19.** 12 **21.** $\frac{9}{5}$ **23.** 0.148

Trial Problems 18.9

1. $y = x$, $dy = f'(x) dx = 1 \cdot dx = dx$ **2.** $y = x^2$, $dy = 2x \, dx$ **3.** $y = x^2$, $dy = 2x \, dx = 2(1)\left(\frac{1}{2}\right) = 1$
4. $y = x + 1$, $dy = 1 \cdot dx = dx$ **5.** $y = x^2 + 1$, $dy = 2x \, dx = 2(0)(0.1) = 0$

Exercises 18.9

1. $dy = 2 \, dx$ **3.** $dy = \frac{3 \, dx}{2\sqrt{3x - 2}}$ **5.** $dy = \frac{3 - 8x - 3x^2}{(x^2 + 1)^2} dx$ **7.** $dy = -\frac{1}{x^2} dx$ **9.** $dy = -0.06$ **11.** $dy = 0.075$
13. $dw = -1.3$ **15.** $dA = \frac{1}{4}$ **17.** $dF = -0.004 \, gm_1 m_2$ **19.** $dE = 11.9$

Trial Problems 18.10

1. $y = 2x^2$, $y' = 4x$, $y'' = 4$ **2.** $y = 2x^2$, $y' = 4x$, $y'' = 4$, $y''' = 0$ **3.** $y = 2x^{-2}$, $y' = -4x^{-3}$, $y'' = 12x^{-4}$
4. $x + y = 0$, $1 + y' = 0$, $y' = -1$ **5.** $2x + 3y = 0$, $2 + 3y' = 0$, $y' = -\frac{2}{3}$

Exercises 18.10

1. $y'' = 24x - 20$ 3. $y'' = 6x^{-5}$ 5. $y'' = 0$ 7. $y'' = \dfrac{8}{(2x+3)^3}$ 9. $y'' = 108(3x+2)^2$

11. $f'(x) = 24x^7, f''(x) = 168x^6$ 13. $k'(s) = \dfrac{1}{3}s^{-2/3} - 4s^{-3}, k''(s) = -\dfrac{2}{9}s^{-5/3} + 12s^{-4}$

15. $g'(x) = -3(2-9x)^{-2/3}, g''(x) = -18(2-9x)^{-5/3}$ 17. $y' = -\dfrac{2}{3}$ 19. $y' = \dfrac{8x-1}{3}$ 21. $y' = \dfrac{x}{y}$ 23. $y' = \dfrac{2x}{3y^2}$

25. $y' = \dfrac{-2x}{2y+1}$ 27. $y' = \dfrac{-1-y}{x+1}$ 29. $y' = \dfrac{x^2}{y^2}$ 31. $y' = -\dfrac{3y-2x}{2y-3x}$

33. $f'(x) = 6x^5 - 12x^3 + 6x^2 - 1, f''(x) = 30x^4 - 36x^2 + 12x, f'''(x) = 120x^3 - 72x + 12, f^{iv}(x) = 360x^2 - 72,$
$f^v(x) = 720x, f^{vi}(x) = 720$ 35. 3 37. $\dfrac{dy}{dx} = -\dfrac{x}{y}$ 39. $S' = 15t^2, S'' = 30t$

Review Exercises 18.12

1. [18.2] 1 3. [18.2] 2 5. [18.2] 16 7. [18.2] $3x^2$ 9. [18.2] dne 11. [18.3] discontinuous at $x = 0$ and $x = 2$
13. [18.7] $y' = 35x^4 - 1$ 15. [18.7] $y' = 30x(3x^2 - 1)^4$ 17. [18.7] $g'(t) = \dfrac{1}{\sqrt{2t+3}}$ 19. [18.7] $f'(w) = \dfrac{6w}{5(3w^2)^{4/5}}$
21. [18.7] $G'(x) = -4(x^2 - x^{-2})(x + x^{-3})$ 23. [18.7] $F'(x) = (x^5 + 1)^3(x + 2)^2(23x^5 + 40x^4 + 3)$
25. [18.7] $q'(x) = \dfrac{8x^2 - 24}{(x^2 + 4x + 3)^2}$ 27. [18.11] $y' = \dfrac{-y - 2x}{2y + x}$ 29. [18.5] -15
31. [18.11] $y' = \dfrac{-x - y}{x - y}$ and $y'' = \dfrac{2x^2 + 4xy - 2y^2}{(x-y)^3}$ 33. [18.9] $\dfrac{dA}{dr} = 4\pi$ 35. [18.7] $\dfrac{d\tau}{dR} = \dfrac{25}{(5+R)^2}$
37. [18.7] $\dfrac{dP}{dL} = \dfrac{\pi}{8}$ 39. [18.7] $\dfrac{d^2E}{dr^2} = \dfrac{2k}{r^3}$ 41. [18.9] $\dfrac{dP}{dx} = \dfrac{-10}{(x+4)^2}$ 43. [18.8] $\dfrac{dh}{dr} = \dfrac{-2r^2 - h^2}{rh}$
45. [18.9] $\dfrac{dD}{dv} = 1.7$ 47. [18.6] $\dfrac{dL}{dt} = \dfrac{2}{3}t\sqrt{L}$

Test 18.13

1. [18.2] 2 2. [18.2] $\dfrac{5}{4}$ 3. [18.3] continuous for $x \geq 2$ 4. [18.3] discontinuous at $x = 3$ 5. [18.2] dne 6. [18.2] 4
7. [18.2] $\dfrac{3}{4}$ 8. [18.7] 1 9. [18.7] $y' = 2(x-3)$ 10. [18.7] $f'(x) = 6x^2 - 10$ 11. [18.7] $f'(x) = 6x - 1$
12. [18.7] $y' = \dfrac{3 + 16x - 4x^2}{(4x^2 + 3)^2}$ 13. [18.8] $y' = 12(4x - 6)^2$ 14. [18.8] $f'(x) = \dfrac{-4x}{\sqrt{1 - 4x^2}}$
15. [18.9] $\dfrac{dI}{dR} = \dfrac{-50}{(R-10)^2} = -0.08$ 16. [18.10] $ds = (9t^2 - 8)dt, ds = 0.3, s = 2.3$
17. [18.11] $f'(x) = 8x, f''(x) = 8$ 18. [18.11] $f'(x) = -\dfrac{1}{2\sqrt{4-x}}, f''(x) = -\dfrac{1}{4(4-x)^{3/2}}$ 19. [18.11] $y' = -\dfrac{x}{y}$
20. [18.11] $y' = \dfrac{-1-y}{x-1}$

Chapter 19 Answers

Trial Problems 19.1

1. $y = \ln x$, $y' = \dfrac{1}{x}$ **2.** $y = \ln 2x$, $y' = \dfrac{1}{2x}(2) = \dfrac{1}{x}$ **3.** $y = \ln\left(\dfrac{1}{2}x\right)$, $y' = \dfrac{1}{\frac{1}{2}x}\left(\dfrac{1}{2}\right) = \dfrac{1}{x}$

4. $y = \ln(x + 1)$, $y' = \dfrac{1}{x + 1}$ **5.** $y = \ln(1 - x)$, $y' = \dfrac{1}{1 - x}(-1) = \dfrac{1}{x - 1}$

Exercises 19.1

1. $y' = \dfrac{1}{x}$ **3.** $f'(x) = \dfrac{9}{9x + 4}$ **5.** $f'(x) = \dfrac{3}{3x - 1}$ **7.** $y' = \dfrac{12}{3x - 2}$ **9.** $r'(x) = \dfrac{6x - 2}{3x^2 - 2x + 1}$

11. $f'(x) = 1 + \ln x$ **13.** $y' = \dfrac{1}{x \ln x}$ **15.** $g'(x) = \dfrac{2x}{x^2 - 1}$ **17.** $y' = 2x + \dfrac{1}{x}$ **19.** $y' = \dfrac{2}{1 - x^2}$

21. $y' = \dfrac{1}{2x\sqrt{2 + \ln x}}$ **23.** $y'' = -\dfrac{1}{x^2}$ **25.** $y'' = -\dfrac{2}{x^2}$ **27.** $y'' = \dfrac{1}{x^2}$ **29.** $y'' = \dfrac{-9}{(1 - 3x)^2}$ **31.** $y'' = -\dfrac{1}{x^2}$

33. $2x - y = e$ **35.** $y' = \dfrac{xy \ln y - y^2}{xy \ln x - x^2}$ **37.** $\dfrac{dt}{dx} = \dfrac{1}{c_1(c_2 - x)(c_3 - x)}$ **39.** $\dfrac{di}{dt} = \dfrac{2}{5}$

41. $\dfrac{dA}{dT} = \dfrac{6a}{(a - T)(T - a^2 + aT)}$

Trial Problems 19.2

1. $y' = 6e^{6x}$ **3.** $y' = -e^{-1/3x}$ **5.** $f'(x) = e^x - e^{-x}$ **7.** $h'(x) = 2^x \ln 2 + 10xe^{-2x} - 5e^{-2x}$ **9.** $y' = xe^x + e^x$
11. $y' = \dfrac{e^{\sqrt{x}}}{2\sqrt{x}}$ **13.** $f'(x) = 2xe^{x^2 + 1}$ **15.** $y' = -2x^3 e^{-x^2} + 2xe^{-x^2}$ **17.** $y' = \dfrac{2e^x}{(e^x + 1)^2}$ **19.** $y' = 1$

Exercises 19.2

1. $y' = 6e^{6x}$ **3.** $y' = -e^{-1/3x}$ **5.** $f'(x) = e^x - e^{-x}$ **7.** $h'(x) = 2^x \ln 2 + 10xe^{-x} - 5e^{-x}$ **9.** $y' = xe^x + e^x$
11. $y' = \dfrac{e^{\sqrt{x}}}{2\sqrt{x}}$ **13.** $f'(x) = 2xe^{x^2 + 1}$ **15.** $y' = -2x^3 e^{-x^2} + 2xe^{-x^2}$ **17.** $y' = \dfrac{2e^x}{(e^x + 1)^2}$ **19.** $y' = 1$
21. $f'(x) = (1 + \ln x)e^{x \ln x}$ **23.** $y'' = e^x$ **25.** $y'' = xe^x + 2e^x$ **27.** $x - y = -1$ **29.** $\dfrac{dT}{dt} = -3.35e^{-0.067t}$
31. $\dfrac{dV}{dt} = 10e^{-0.1t}$ **33.** $\dfrac{dP}{dt} = kP$

Trial Problems 19.3

1. $y = \sin 2x$, $y' = 2\cos 2x$ **2.** $y = \cos 2x$, $y' = -2\sin 2x$ **3.** $y = \tan 2x$, $y' = 2\sec^2 2x$
4. $y = \cot 2x$, $y' = -2\csc^2 2x$ **5.** $y = \sec 2x$, $y' = 2\sec 2x \tan 2x$

Exercises 19.3

1. $y' = 8\cos(8x - 3)$ **3.** $y' = -\dfrac{1}{2}\sin\dfrac{x}{2}$ **5.** $y' = -x\sin x^2$ **7.** $y' = -6x\sin 3x^2$ **9.** $y' = -2(3^{-2x})\ln 3 \cos 3^{-2x}$
11. $y' = 2x\sec^3 4x(6x\tan 4x + 1)$ **13.** $y' = e^x(\sec^2 x + \tan x)$ **15.** $y' = -18(3x - 2)\csc(3x - 2)^2\cot(3x - 2)^2$
17. $y' = \dfrac{12\sin^3\sqrt{3x+2}\cos\sqrt{3x+2}}{\sqrt{3x+2}}$ **19.** $y' = \cos 2x$ or $\cos^2 x - \sin^2 x$ **21.** $y' = 4\cos(2x - 1)$
23. $g'(x) = -2\sin x^2 \sin 2x + 2x\cos 2x \cos x^2$ **25.** $y' = \dfrac{3}{x}\cos 3x - \dfrac{1}{x^2}\sin 3x$ **27.** $y' = \dfrac{3\sec 3x \tan 3x}{2\sqrt{\sec 3x}}$
29. $f'(x) = \dfrac{-4\sin 4x(1 + \cot 3x) + 3\cos 4x \csc^2 3x}{(1 + \cot 3x)^2}$ **31.** $r'(x) = 2\sec 2x(\sec 2x - \tan 2x)$ **33.** $y'' = -\sin x$
35. $y'' = 2\sec^2 x \tan x$ **37.** $y'' = x\cos x + \sin x$ **39.** $m = 3\sqrt{2}$ **41.** $m = 1$ **43.** $\dfrac{dv}{dt} = 256\pi^2\cos(8\pi t)$
45. $\dfrac{di}{dt} = -1200\sin\left(120t + \dfrac{\pi}{6}\right)$

Trial Problems 19.4

1. $y = \sin^{-1} 2x$, $y' = \dfrac{1}{\sqrt{1 - (2x)^2}}(2) = \dfrac{2}{\sqrt{1 - 4x^2}}$ **2.** $y = \cos^{-1} 2x$, $y' = -\dfrac{1}{\sqrt{1 - (2x)^2}}(2) = -\dfrac{2}{\sqrt{1 - 4x^2}}$
3. $y = \tan^{-1} 2x$, $y' = \dfrac{1}{1 + (2x)^2}(2) = \dfrac{2}{1 + 4x^2}$ **4.** $y = \sec^{-1} 2x$, $y' = \dfrac{1}{2x\sqrt{(2x)^2 - 1}}(2) = \dfrac{1}{x\sqrt{4x^2 - 1}}$
5. $y = \sin^{-1} x + \cos^{-1} x$, $y' = \dfrac{1}{\sqrt{1 - x^2}} + \left(-\dfrac{1}{\sqrt{1 - x^2}}\right) = 0$

Exercises 19.4

1. $y' = \dfrac{1}{2\sqrt{x}\sqrt{1 - x}}$ **3.** $f'(x) = \dfrac{1}{2x^2 + 2x + 1}$ **5.** $g'(x) = \dfrac{6x}{\sqrt{6x - 9x^2}} + 2\sin^{-1}(3x - 1)$ **7.** $y' = \dfrac{1}{x^2 + 1}$
9. $h'(x) = \dfrac{2x^3}{1 + x^4} + 2x\tan^{-1}x^2$ **11.** $y' = \dfrac{-6(\cos^{-1} 3x + 1)}{\sqrt{1 - 9x^2}}$ **13.** $f'(x) = \dfrac{1}{(\cos^{-1}x)^2\sqrt{1 - x^2}}$
15. $g'(x) = \dfrac{-6(3x + 1)}{\sqrt{1 - (3x + 1)^4}}$ **17.** $y' = \dfrac{3x^2 e^{\sin^{-1}x^3}}{\sqrt{1 - x^6}}$ **19.** $y' = e^x$ **21.** $y' = \dfrac{(ye^x - 2x - \sin^{-1}y)\sqrt{1 - y^2}}{x - e^x\sqrt{1 - y^2}}$
23. $y'' = \dfrac{x}{(1 - x^2)^{3/2}}$ **25.** $y'' = \dfrac{-2x}{(1 + x^2)^2}$ **27.** $dy = 0.01$ **29. a.** $\dfrac{d\phi}{dR} = \dfrac{-x}{R^2 + x^2}$ **b.** $\dfrac{d\phi}{dx} = \dfrac{R}{R^2 + x^2}$
31. $d\theta \approx -86.21$ radians/hour

Review Exercises 19.6

1. [19.1] $y' = \dfrac{4x}{x^2 - 1}$ **3.** [19.2] $f'(x) = 3(e^{x-2})^3$ **5.** [19.3] $y' = \dfrac{2\cos x[\ln(2 + \sin x)]}{2 + \sin x}$

7. [19.2] $g'(x) = \dfrac{2e^{2x}(x^2 - x + 1)}{(x^2 + 1)^2}$ **9.** [19.3] $g' = \dfrac{-\sec^2\sqrt{2 - x}}{2\sqrt{2 - x}}$ **11.** [19.4] $y' = -1$ **13.** [19.1] $y' = 2 + 2\ln(2x)$

15. [19.2] $f'(x) = \dfrac{e^{2x} + 1 - 2xe^{2x}\ln x}{x(e^{2x} + 1)^2}$ **17.** [19.2] $h'(x) = 2xe^{-x^2}(1 - x^2)$ **19.** [19.2] $F'(x) = \dfrac{1}{2\sqrt{x}}$

21. [19.3] $y' = 3\sec x (\sec x + \tan x)^3$ **23.** [19.4] $y' = \dfrac{2x}{\sqrt{x^4 - 1}} + 2x\sec^{-1}x^2$

25. [19.4] $r'(x) = \dfrac{5e^x}{\sqrt{1 - 25x^2}} + e^x\sin^{-1}5x$ **27.** [19.2] $y' = 3\cos 3xe^{\sin 3x}$ **29.** [19.3] $f'(x) = \dfrac{-2 - 2\sec^2 x \tan x}{(2x + \sec^2 x)^2}$

31. [19.3] $h'(x) = -e\sin x(\cos x)^{e-1}$ **33.** [19.4] $G'(x) = \dfrac{e^{\tan^{-1}x}}{1 + x^2}$ **35.** [19.4] $y' = \dfrac{(1 - x^2) - 2x\sqrt{1 - x^2}\cos^{-1}x}{\sqrt{1 - x^2}(\cos^{-1}x)^2}$

37. [19.4] $y' = 2(\tan x + \tan^{-1}x)\left[\sec^2 x + \dfrac{1}{1 + x^2}\right]$ **39.** [19.2] $y' = \dfrac{-y(1 + x)}{x(1 - ye^{-y})}$ **41.** [19.2] $y' = -\dfrac{y}{x}$

43. [19.3] $y'' = -x\cos x - 2\sin x$ **45.** [19.2] $y'' = \dfrac{-2xe^{-x} - e^{-x} + x^2e^{-x}\ln x}{x^2}$ **47.** [19.3] $f''(x) = -\sec^2 x$

49. [19.2] $m = -2(e + 1)$ **51.** [19.3] $\dfrac{ds}{dt} = ak\cos kt - bk\sin kt$ **53.** [19.2] $f'(t) = -e^{-kt}(k + 1 - kt)$

55. [19.3] $P = -20\sin 2t$ **57.** [19.3] $x = y\cos\theta$ and $\dfrac{dx}{d\theta} = -20$ or $x = 10\cot\theta$, or $x = -10\csc^2\theta$

59. [19.2] $\dfrac{dV}{dt} \approx -4.09$ (the value of $100 is decreased by $4.09 during the fourth year)

Test 19.7

1. [19.1] $y' = \dfrac{1}{2 + 3x}$ **2.** [19.1] $f'(x) = 1 + \ln 3x$ **3.** [19.2] $g'(x) = e^x + e^{-x}$ **4.** [19.2] $y' = e^{-x}(1 - x)$

5. [19.3] $f'(x) = \cos(2x - 3)$ **6.** [19.3] $h'(x) = -\tan x$ **7.** [19.3] $y' = 2x\sec^2 x^2$

8. [19.3] $f'(x) = e^x\csc x(1 - \cot x)$ **9.** [19.4] $y' = \dfrac{2x}{\sqrt{1 - x^4}}$ **10.** [19.4] $g'(x) = \dfrac{2}{1 + (2x - 1)^2}$

11. [19.4] $r'(x) = -\dfrac{1}{x\sqrt{1 - (\ln x)^2}}$ **12.** [19.4] $y' = \dfrac{x}{(x^2 - 1)\sqrt{x^2 - 2}}$ **13.** [19.3] $f''(x) = -\csc^2 x$

14. [19.3] $y'' = -2e^x\sin x$ **15.** [19.3] $y' = \dfrac{\cos x - e^x}{\sin y}$ **16.** [19.3] $y' = \dfrac{ye^{xy}}{(\sec^2 y - xe^{xy})}$ **17.** [19.2] $m = 2e + 1$

18. [19.3] $\dfrac{ds}{dt} = 6\cos 3t - 6\sin 2t$ **19.** [19.2] ≈ -8.187 tons   **20.** [19.4] $dW = \dfrac{1}{2}$

Chapter 20 Answers

Trial Problems 20.1

1. $y = x^4 + 2x^3 - 5x^2 + x - 10$, $y' = 4x^3 + 6x^2 - 10x + 1$ **2.** $x^2 + 2x + y^2 = 12$, $2x + 2 + 2yy' = 0$, $y' = -\dfrac{x+1}{y}$ **3.** $y = e^{3x}$, $y' = 3e^{3x}$ **4.** $y = \ln \sqrt{x} = \ln x^{1/2}$, $y' = \dfrac{1}{\sqrt{x}}\left(\dfrac{1}{2}x^{-1/2}\right) = \dfrac{1}{2x}$
5. $y = \tan \dfrac{x}{3}$, $y' = \dfrac{1}{3}\sec^2 \dfrac{x}{3}$

Exercises 20.1

1. $x + y = -1$ **3.** Tangent Line: $6x + y = -3$, Normal Line: $x - 6y = -19$ **5.** $37x + y = -25$, $x - 37y = -1815$
7. $2x - y = 1$, or $18x - y = 81$ **9.** $x = \pm 2$ **11.**

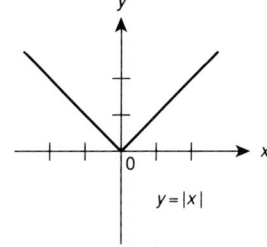

$y = |x|$

a. No horizontal tangent line **b.** No vertical tangent line **c.** No tangent line at $(0,0)$ **13. a.** $2x - y = 2$
b. $3x - 2y = 3$ **c.** $x - y = 1$ **15.** $x + 2y = 56$

Trial Problems 20.2

1. $f(x) = x^3 - \dfrac{5}{2}x^2 + 2$, $f'(x) = 3x^2 - 5x$ **2.** $y = \dfrac{x}{x^2 + 2}$, $y' = \dfrac{(x^2 + 2)1 - x(2x)}{(x^2 + 2)^2} = \dfrac{2 - x^2}{(x^2 + 2)^2}$
3. $g(x) = \sin x$, $g'(x) = \cos x$ **4.** $h(t) = t\sqrt{3-t} = t(3-t)^{1/2}$, $h'(t) = t\left[\dfrac{1}{2}(3-t)^{-1/2}(-1)\right] + (3-t)^{1/2}(1)$
$= (3-t)^{-1/2}\left[-\dfrac{t}{2} + (3-t)\right] = \dfrac{3(2-t)}{2\sqrt{3-t}}$ **5.** $y = e^{1/2x}$, $y' = \dfrac{1}{2}e^{1/2x}$

Exercises 20.2

1. Function is increasing on $(-\infty,\infty)$. No max. and min. values.

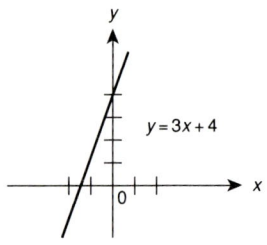

$y = 3x + 4$

3. Decreasing on $(-\infty,\infty)$. No max. and min. values

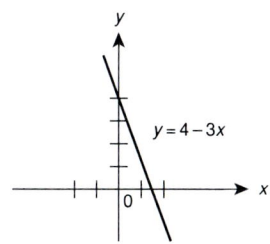

5. $y = \dfrac{x}{x^2 + 4}$ Decreasing on $(-\infty,-2)$ and $(2,\infty)$, increasing on $(-2,2)$, min. at $x = -2$, max. at $x = 2$

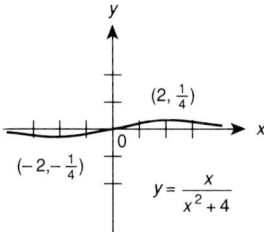

7. Increasing on $(0,\infty)$, min. at $x = 0$.

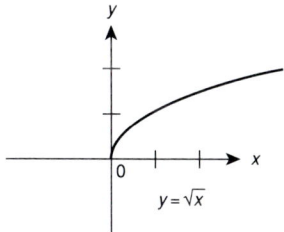

9. Increasing on $(-\infty,\infty)$

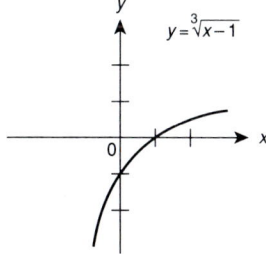

11. Decreasing on $(-\infty, 2)$, increasing on $(2, \infty)$, min. at $x = 2$.

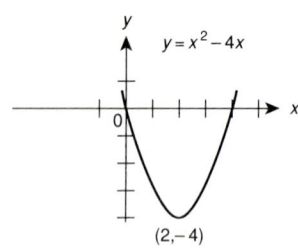

13. Increasing on $(-\infty, -\sqrt{3})$ and $(\sqrt{3}, \infty)$, decreasing on $(-\sqrt{3}, \sqrt{3})$, max. at $x = -\sqrt{3}$, min. at $x = \sqrt{3}$.

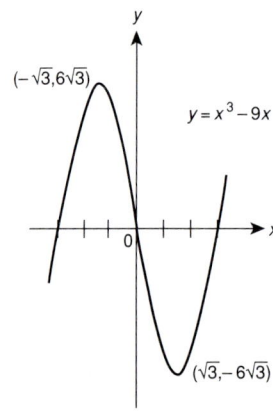

15. Max. at $x = \dfrac{\pi}{2}$, min. at $x = \dfrac{3\pi}{2}$ **17.** Max. at $x = \dfrac{\pi}{2}$ **19.** Max. at $x = e$ **21.** Min. at $x = \dfrac{1}{2}$

23. $f'(x)$ does not change signs around $x = 1$.   **25. a.** Lowest 1 week after dumping of organic waste   **b.** Highest 1 week before dumping of organic waste   **27.** Max. power when $R_2 = R_1$ **29.** $x = \$20$

Trial Problems 20.3

1. $y = (x^2 + 3)^2$, $y' = 2(x^2 + 3)(2x) = 4x^3 + 12x$, $y'' = 12x^2 + 12$ **2.** $y = \sin 2x$, $y' = 2\cos 2x$, $y'' = -4\sin 2x$

3. $f(x) = x^2 + \dfrac{1}{x} = x^2 + x^{-1}$, $f'(x) = 2x - x^{-2}$, $f''(x) = 2 + 2x^{-3} = 2 + \dfrac{2}{x^3}$ **4.** $g(x) = 10xe^{x/10}$, $g'(x) = 10x\left[\dfrac{1}{10}e^{x/10}\right] + e^{x/10}[10] = xe^{x/10} + 10e^{x/10}$, $g''(x) = x\left[\dfrac{1}{10}e^{x/10}\right] + e^{x/10}[1] + e^{x/10} = \dfrac{x}{10}e^{x/10} + 2e^{x/10}$

5. $w = \ln(x - 1)$, $w' = \dfrac{1}{x - 1} = (x - 1)^{-1}$, $w'' = -(x - 1)^{-2} = -\dfrac{1}{(x - 1)^2}$

Exercises 20.3

1. Max. at $x = -\dfrac{2}{3}\sqrt{3}$, Min. at $x = \dfrac{2}{3}\sqrt{3}$ **3.** No max., No min. **5.** Min. at $x = 2$, No max.

7. Min. at $x = \dfrac{1}{2}$, No max. **9.** Max. at $x = 0$, No min. **11.** Cannot determine by SDT

13. Max. at $x = \dfrac{\pi}{2}$, Min. at $x = \dfrac{3\pi}{2}$ **15.** Min. at $x = -1$, No max. **17.** Max. at $x = \dfrac{\pi}{4}$, Min. at $x = \dfrac{5\pi}{4}$

19. $x = \dfrac{1}{2}$ is not a critical number. **21.** 100 **23.** $t \approx 2.5$ min. **25.** Max. power $P = 40$.

Trial Problems 20.4

1. $h(t) = t^3 - \dfrac{5}{2}t^2$, $h'(t) = 3t^2 - 5t$, $h''(t) = 6t - 5$ **2.** $f(x) = x^4 + 1$, $f'(x) = 4x^3$, $f''(x) = 12x^2$

3. $y = x^{1/3} + 1$, $y' = \dfrac{1}{3}x^{-2/3}$, $y'' = -\dfrac{2}{9}x^{-5/3}$ **4.** $r(t) = 8t - t^3$, $r'(t) = 8 - 3t^2$, $r''(t) = -6t$

5. $g(x) = \dfrac{x}{1 + x^2}$, $g'(x) = \dfrac{(1 + x^2)(1) - x(2x)}{(1 + x^2)^2} = \dfrac{1 - x^2}{(1 + x^2)^2}$, $g''(x) = \dfrac{(1 + x^2)^2[-2x] - (1 - x^2)[2(1 + x^2)(2x)]}{(1 + x^2)^4}$
$= \dfrac{2x(x^2 - 3)}{(1 + x^2)^3}$

Exercises 20.4

1. Min. at $x = 0$, concave up on $(-\infty, \infty)$.

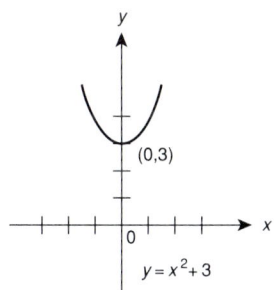

3. Max. at $x = 1$, min. at $x = -1$, concave up on $(-\infty, 0)$, concave down on $(0, \infty)$, point of inflection at $x = 0$.

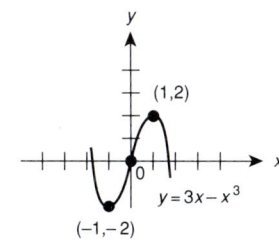

5. Min. at $x = -3$, concave up on $(-\infty, -2)$ and $(0, \infty)$, concave down on $(-2, 0)$, point of inflection at $x = -2$ and $x = 0$.

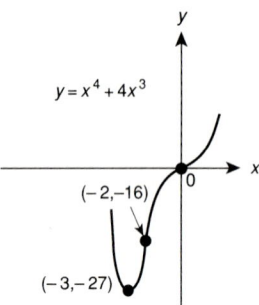

7. Concave down on $(-\infty, 0)$, concave up on $(0, \infty)$, point of inflection at $x = 0$.

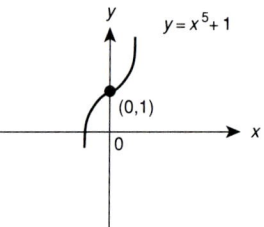

9. Max. at $x = 1$, min. at $x = -1$, concave down on $(0, \sqrt{3})$ and $(-\infty, -\sqrt{3})$, concave up on $(\sqrt{3}, \infty)$ and $(-\sqrt{3}, 0)$, points of inflection at $x = 0, \pm\sqrt{3}$.

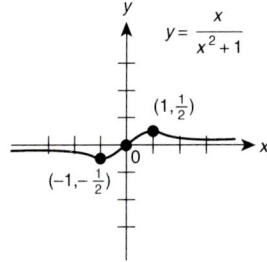

11. Min. at $x = -1$, concave up on $(-\infty, 0)$ and $(2, \infty)$, concave down on $(0, 2)$, point of inflection at $x = 0, 2$.

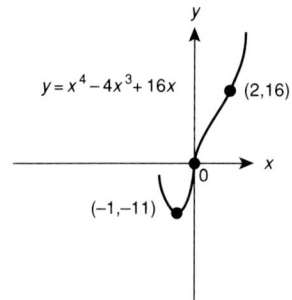

13. Concave up on $(-\infty,0)$, concave down on $(0,\infty)$, point of inflection at $x = 0$.

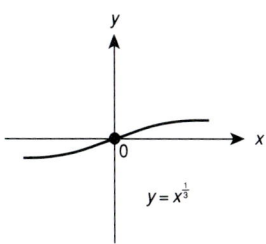

15. Max. at $x = -2$, min. at $x = 0$, concave down on $(-\infty,-1)$, concave up on $(-1,\infty)$, point of inflection at $x = -1$

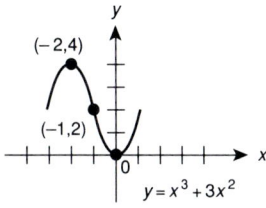

17. Concave down on $(0,\infty)$, concave up on $(-\infty,0)$, point of inflection at $x = 0$.

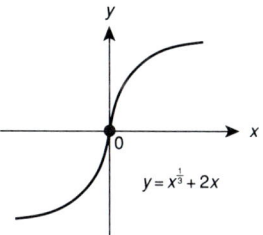

19. a. No   **b.** Concave up if $a > 0$ **c.** Concave down if $a < 0$ **d.** No

21. Max. at $x = 2$.

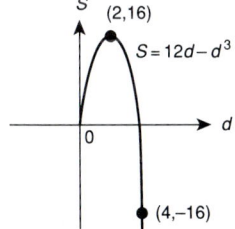

23. Max. at $t = 2$, point of inflection at $t = \dfrac{1}{2}$.

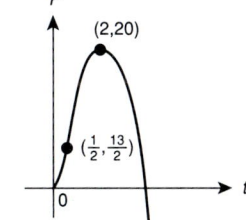

Trial Problems 20.5

1. $y = 3x^2 - 5x + 6$, $y' = 6x - 5$, $y'' = 6$ 2. $g(t) = 500t - 16t^2$, $g'(t) = 500 - 32t$, $g''(t) = -32$
3. $p = 32 \ln x + 5$, $p' = \dfrac{32}{x} = 32x^{-1}$, $p'' = -32x^{-2} = -\dfrac{32}{x^2}$ 4. $r(x) = 3 \sin 2x$, $r'(x) = 6 \cos 2x$, $r''(x) = -12 \sin 2x$
5. $y = x^2 + \sqrt{x}$, $y' = 2x + \dfrac{1}{2}x^{-1/2} = 2x + \dfrac{1}{2\sqrt{x}}$, $y'' = 2 - \dfrac{1}{4}x^{-3/2} = 2 - \dfrac{1}{4x^{3/2}}$

Exercises 20.5

1. 256   3. $3\dfrac{1}{4}$ 5. $\dfrac{26}{3}$ 7. 15,625 ft   9. $r = 0.6$ 11. $p = 8$ 13. 6 in.   15. $x = e^{-1/2}$ 17. $w = 4\dfrac{1}{3}$
19. $r\sqrt{2}$ by $\dfrac{r\sqrt{2}}{2}$ 21. 2 in.   23. max. at $x = \dfrac{\pi}{2}$

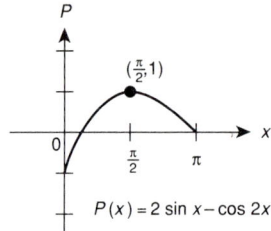

$P(x) = 2 \sin x - \cos 2x$

Trial Problems 20.6

1. $s = 7t^2 + 2t$, $s' = 14t + 2$, $s'' = 14$ 2. $s = 6t^{2/3}$, $s' = 4t^{-1/3}$, $s'' = -\dfrac{4}{3}t^{-4/3}$ 3. $s(t) = 2 \cos \pi t$, $s'(t) = -2\pi \sin \pi t$, $s''(t) = -2\pi^2 \cos \pi t$ 4. $s(t) = t^2 e^{-t}$, $s'(t) = t^2[-e^{-t}] + e^{-t}[2t]$, $s''(t) = t^2[e^{-t}] + (-e^{-t})[2t] + e^{-t}[2] + 2t[-e^{-t}] = t^2 e^{-t} - 4te^{-t} + 2e^{-t} = e^{-t}(t^2 - 4t + 2)$ 5. $s(t) = t^2 + \ln t$, $s'(t) = 2t + \dfrac{1}{t}$, $s''(t) = 2 - \dfrac{1}{t^2}$

Exercises 20.6

1. $v = 15$, $a = 12$ 3. $v = -6$, $a = 0$ 5. $v = \dfrac{1}{3}$, $a = 0$ 7. $v = 2$, $a = -6$ 9. $v = 15$, $a = 24$ 11. $v = 2t + 2$
13. $v = 96$ ft/sec, $a = 32$ 15. $v(0) = 0$, $v(1) = 15$, $v(4) = 30$, $v(9) = 45$ 17. $v_0 = 500$ ft/sec   19. 64
21. $v \approx 3.16$, $a \approx 0.05$ 23. $v(1) = 0$, $a(1) = 3$, $v(2) = \dfrac{8}{3}$, $a(2) = \dfrac{22}{9}$, $v(4) = \dfrac{36}{5}$, $a(4) = \dfrac{54}{25}$
25. a. $v = -6 \sin 2t$, $a = -12 \cos 2t$ b. $v = 2\pi \cos \pi t$, $a = -2\pi^2 \sin \pi t$

Trial Problems 20.7

1. $y = \dfrac{4+x}{4-x}$, $y' = \dfrac{(4-x)(1) - (4+x)(-1)}{(4-x)^2} = \dfrac{8}{(4-x)^2}$ 2. $y = x + \sqrt{x}$, $y' = 1 + \dfrac{1}{2}x^{-1/2} = 1 + \dfrac{1}{2\sqrt{x}}$
3. $y = \dfrac{6}{x^2} = 6x^{-2}$, $y' = -12x^{-3} = -\dfrac{12}{x^3}$ 4. $y = \sqrt{25-x^2}$, $y' = \dfrac{1}{2}(25-x^2)^{-1/2}(-2x) = -\dfrac{x}{\sqrt{25-x^2}}$
5. $y = \pi x^2$, $y' = 2\pi x$

Answer Key—Chapter 20

Exercises 20.7

1. 16 in./sec   **3.** ≈ 0.10 ft/sec   **5.** $\frac{4}{3}$ times the length   **7.** ≈ 2.1 ft/sec   **9.** ≈ 10.74 ft/sec   **11.** ≈ −1.88 cm/min
13. $2\pi r$ equals the circumference of the circle   **15.** ≈ 0.22 cm/min   **17.** ≈ 0.028 cm/sec   **19.** 1.6 m/sec

Review Exercises 20.9

1. [20.1] Tangent: $y = -3x$, Normal: $y = \frac{1}{3}x$

3. [20.2] Increasing: $(-\infty, 1)$ and $(2, \infty)$, Decreasing: $(1, 2)$

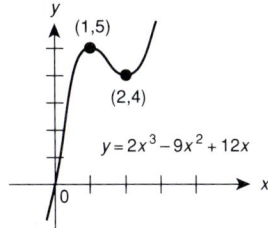

5. [20.2] Increasing: $\left(\frac{3}{2}, \infty\right)$, Decreasing: $\left(-\infty, \frac{3}{2}\right)$

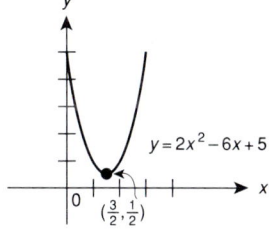

7. [20.2] Decreasing: $(-\infty, +\infty)$

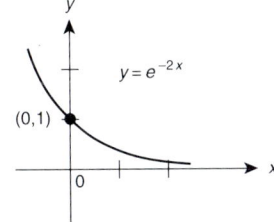

9. [20.2] Increasing: $\left(0, \frac{\pi}{2}\right)$ and $\left(\frac{3\pi}{2}, 2\pi\right)$, Decreasing: $\left(\frac{\pi}{2}, \frac{3\pi}{2}\right)$

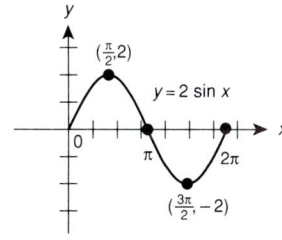

11. [20.5] Maximum: (3,18), minimum: $\left(-\dfrac{1}{3}, -\dfrac{14}{27}\right)$

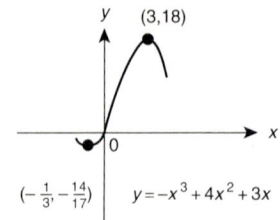

13. [20.5] Max. at $x = 1$, min. at $x = 2$, concave down on $\left(-\infty, \dfrac{3}{2}\right)$, concave up on $\left(\dfrac{3}{2}, \infty\right)$, point of inflection at $x = \dfrac{3}{2}$.

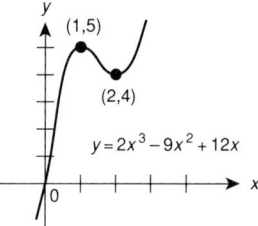

15. [20.5] Min. at $x = \dfrac{1}{4}$, concave up on $(-\infty, 0)$ and $\left(\dfrac{1}{6}, \infty\right)$, concave down on $\left(0, \dfrac{1}{6}\right)$, point of inflection at $x = 0, \dfrac{1}{6}$.

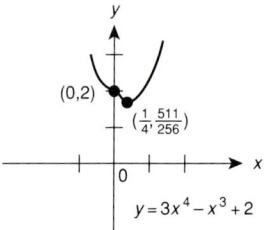

17. [20.5] Min. at $x = -\dfrac{1}{3}$, max. at $x = 3$, concave down on $\left(\dfrac{4}{3}, \infty\right)$, concave up on $\left(-\infty, \dfrac{4}{3}\right)$, point of inflection at $x = \dfrac{4}{3}$.

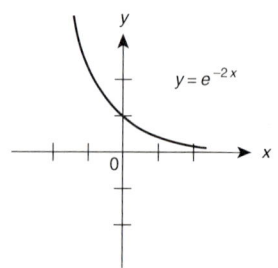

19. [20.5] Max. at $x = 0$, concave down on $\left(-\frac{\sqrt{3}}{3}, \frac{\sqrt{3}}{3}\right)$, concave up on $\left(-\infty, -\frac{\sqrt{3}}{3}\right)$ and $\left(\frac{\sqrt{3}}{3}, \infty\right)$, point of inflection at $x = -\frac{\sqrt{3}}{3}, \frac{\sqrt{3}}{3}$.

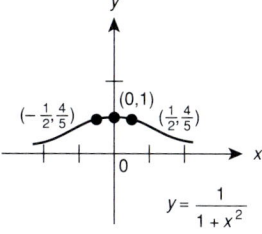

21. [20.5] Concave up on $(-\infty, -1)$, concave down on $(-1, \infty)$.

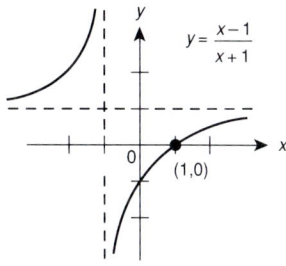

23. [20.5] Concave down on $(0, \infty)$, min. at $x = 0$.

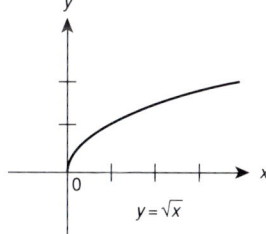

25. [20.5] Concave down on $(0, \infty)$.

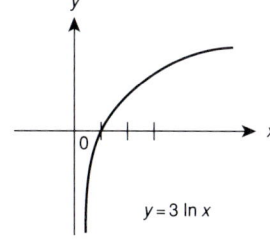

27. [20.4] $14x + y = 0$

29. [20.5] $\frac{40}{3}, \frac{20}{3}$ **31.** [20.5] $r = 2$ and $h = 4$ **33.** [20.7] $i = 2.75$ **35.** [20.5] $p \approx 0.53$ ft/lb at $t = 4.85$

37. [20.5] $t = 1$ **39.** [20.1] $13x + 2y = 2\tan^{-1}\left(\frac{3}{2}\right) + 39$ **41.** [20.1] $2x + 2e^{1/4}y = 2e^{1/2} + 1$ **43.** [20.5] $r = 10$

45. [20.6] $a = -2$ **47.** [20.7] ≈ 357.78 mi/hr **49.** [20.7] ≈ 2.36 ft/min

Test 20.10

1. [20.1] $y = 2x + 1$ **2.** [20.1] $x - 2y = -3$ **3.** [20.1] $y = 1$

4. [20.2] Min. at $x = \frac{2}{\sqrt{3}}$, max. at $x = -\frac{2}{\sqrt{3}}$.

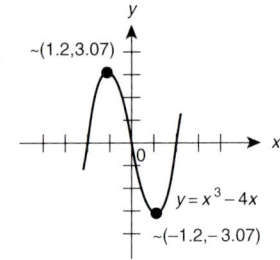

5. [20.3] Max. at $x = \frac{2 - \sqrt{7}}{3}$, min./max. at $x = \frac{2 + \sqrt{7}}{3}$.

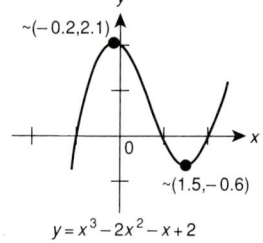

6. [20.4] Max. at $t = 2$, concave down on $(0,4)$.

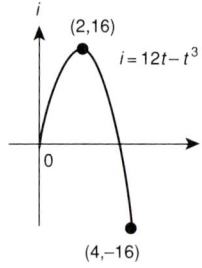

7. [20.5] $t = \dfrac{\pi}{2}$ **8.** [20.6] $v = 3$, $a = 2.75$ **9.** [20.7] -7.5, ≈ -5.9, ≈ -3.7

10. [20.5] Max. at $(0,2)$, min. at $(0,-2)$.

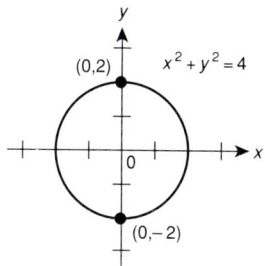

Cumulative Review—Chapters 18, 19, and 20   20.11

1. [18] 0   **2.** [18] $-\dfrac{1}{6}$ **3.** [18] $\dfrac{1}{3}$ **4.** [18] $-\infty$ **5.** [18] Continuous for $x > 4$ **6.** [18] Discontinuous at $x = 0,1$

7. [18] 5   **8.** [18] 0   **9.** [18] 1   **10.** [18] 6   **11.** [18] $-2\sqrt{2}$ **12.** [18] $\dfrac{2\sqrt{3}}{3}$ **13.** [18] $f'(x) = 16x^3 - 4$

14. [18] $-10(2x)^{-6}$ **15.** [18] $6x - 2$ **16.** [18] $\dfrac{-2}{(3x-2)^2}$ **17.** [18] $\dfrac{8x - 3x^2}{(4 - 3x)^2}$ **18.** [18] $\dfrac{1}{\sqrt{2x}}$ **19.** [18] $\dfrac{2x}{x^2 - 5}$

20. [18] $\dfrac{-2(x-1)}{x+1}$ **21.** [18] $x + 2x \ln x$ **22.** [18] $2e^{2x} - 2e^{-2x}$ **23.** [18] $2e^x(x+1)$ **24.** [18] $\dfrac{2x^2 e^{x^2} - e^2}{x^2}$

25. [18] $6x \cos 3x^2 - 6x^2 \sin 2x^3$ **26.** [18] $2xe^x \cos(x^2 - 1) + e^x \sin(x^2 + 1)$ **27.** [18] $\dfrac{3 \tan 3x - 9x \sec^2 3x}{\tan^2 3x}$

28. [18] $\dfrac{2}{\sqrt{1 - (2x+1)^2}}$ **29.** [18] $\dfrac{1}{x(1+x^2)} - \dfrac{\tan^{-1} x}{x^2}$ **30.** [18] $-\dfrac{e^x}{\sqrt{1-x^2}} + e^x \cos^{-1} x$ **31.** [18] $dy = 0.7$

32. [18] $dy = 2.4$ **33.** [18] $y' = \dfrac{x}{y}$ **34.** [18] $y' = 1$ **35.** [18] $y' = \dfrac{e^x - y + 1}{x - 1}$ **36.** [18] $y' = \dfrac{\cos x \cos y}{\sin x \sin y - 2y}$

37. [18] $y'' = 36x^2 + 4$ **38.** [18] $f''(x) = \dfrac{18}{(3x-2)^3}$ **39.** [18] $f''(x) = -(3 - 2x)^{-3/2}$ **40.** [18] $y'' = \dfrac{1}{x^2}$

41. [18] $g''(x) = -e^{-x}$ **42.** [18] $y'' = -\sin x - \cos x$ **43.** [18] $\dfrac{dI}{dR} = -0.006$, decreasing

44. [20] Tangent: $7x - y = 6$, Normal: $x + 7y = 58$ **45.** [20] Maximum: $(-3,28)$, Minimum: $\left(2, -\dfrac{41}{3}\right)$

46. [20] Maximum: $(0,0)$, Minimum: $\left(1, -\dfrac{1}{2}\right)$

47. [20] Max. at $x = 3$, Min. at $x = 0$, concave up on $\left(-\infty, \frac{3}{2}\right)$, concave down on $\left(\frac{3}{2}, \infty\right)$, point of inflection at $x = \frac{3}{2}$.

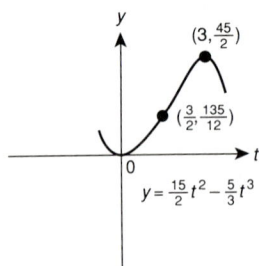

48. [20] Increasing: (1, 100,000), Decreasing: (100,000, 380,000). Max. votes: 100,000

49. [20] $\frac{dy}{dt} \approx 0.29$ at $t = 5$ yrs, $\frac{dy}{dt} \approx 0.17$ at $t = 10$ yrs, y is increasing on $(0, \infty)$

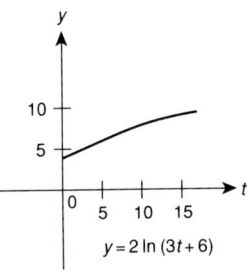

50. [20] $v = 250$ k/min., $a = 0$ **51.** [20] Altitude is 32 ft., velocity is 12 ft/sec and object is going up. acceleration is -4 ft/sec² **52.** [18] $4k(6000)^3$ **53.** [18] 1199 **54.** [20] Max. at $x = 1$, min. at $x = -3$.

55. [20] $390,000 **56.** [18] $\frac{dy}{dx} \approx 90.9$ at $x = 1$, $\frac{dy}{dx} \approx 60.7$ at $x = 5$, $\frac{dy}{dx} \approx 8.2$ at $x = 25$

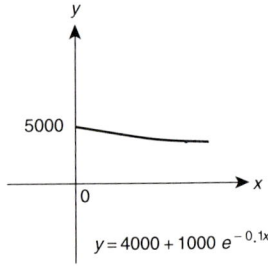

57. [18] $\lim_{x \to +\infty} T(x) = 300$

Maximum number of worker-hours needed for the project is 300.

58. [18] Continuous on $0 \leq t < 2$ and $2 \leq t \leq 6$, discontinuous at $t = 2$, not differentiable at $t = 2$

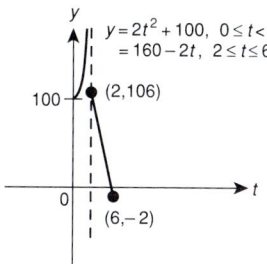

59. [20] Minimum average productivity was 837.5 units per day, and occurred at $t = 37.5$ months.

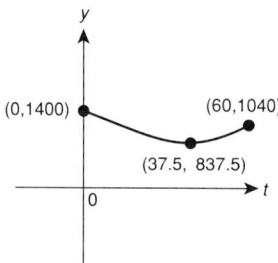

Chapter 21 Answers

Trial Problems 21.1

1. $f'(x) = 2, f(x) = 2x + c$ **2.** $f'(x) = x, f(x) = \dfrac{x^2}{2} + c$ **3.** $f'(x) = 2x, f(x) = \dfrac{2x^2}{2} + c = x^2 + c$
4. $f'(x) = x^7, f(x) = \dfrac{x^8}{8} + c$ **5.** $f'(x) = 6x^2 + 2, f'(x) = \dfrac{6x^3}{3} + 2x + c = 2x^3 + 2x + c$

Exercises 21.1

1. $\dfrac{d}{dx}\left(\dfrac{1}{2}x^4 + c\right) = (4)\left(\dfrac{1}{2}\right)x^3 + 0 = 2x^3$ **3.** $\dfrac{d}{dx}(-5x^{-1} + c) = (-1)(-5)x^{-2} + 0 = 5x^{-2}$
5. $\dfrac{d}{dx}(-5x + c) = -5 + 0 = -5$ **7.** $x^6 + c$ **9.** $x^2 - 7x + c$ **11.** $\dfrac{4}{7}x^{7/4} + c$ **13.** $\dfrac{3}{7}x^{7/4} + c$ **15.** $\dfrac{1}{x^3} + c$
17. $x^2 + c$ **19.** $-\dfrac{2}{x} + c$ **21.** $x^3 + x^2 + c$ **23.** $\dfrac{x^3}{3} + 3x^2 + 9x + c$ **25.** $\sin x + c$ **27.** $x + c$ **29.** $5x + c$
31. $(6x + 1)^{3/2} + c$ **33.** $3x^2 - 2y = -1$ **35. a.** $v = \dfrac{2}{3}t^{3/2} + v_0$ **b.** $s = \dfrac{4}{15}t^{5/2} + v_0 t + s_0$ **37.** $w = \dfrac{s^3}{3}\cos\theta$
39. $P(x) = 100x - 0.02x^2$

Trial Problems 21.2

1. $\int x\, dx = \dfrac{x^2}{2} + c$ **2.** $\int (x+2)\,dx = \dfrac{x^2}{2} + 2x + c$ **3.** $\int x^{-2}\,dx = \dfrac{x^{-1}}{-1} + c = -\dfrac{1}{x} + c$

4. $\int x^2 \cdot x^{1/2}\,dx = \int x^{5/2}\,dx = \dfrac{x^{7/2}}{\frac{7}{2}} + c = \dfrac{2}{7}x^{7/2} + c$ **5.** $\int \dfrac{1}{x^2}\,dx = \int x^{-2}\,dx = \dfrac{x^{-1}}{-1} + c = -\dfrac{1}{x} + c$

Exercises 21.2

1. $\dfrac{x^4}{4} - 7x + c$ **3.** $\dfrac{2}{5}x^{5/2} + x^2 + x + c$ **5.** $\dfrac{3}{5}x^{5/3} + c$ **7.** $-\dfrac{1}{2x^2} + c$ **9.** $x^2 - \dfrac{1}{x} + c$ **11.** $\dfrac{x^3}{3} + \dfrac{x^2}{2} - 2x + c$
13. $\dfrac{4}{5}r^5 - \dfrac{4}{3}r^3 + r + c$ **15.** $8s^2 - \dfrac{4}{7}s^{7/2} + c$ **17.** $\dfrac{x^5}{5} - \dfrac{4}{3}x^3 + 4x + c$ **19.** $\dfrac{10}{3}x^{-1/5} + c$
21. $x + \dfrac{x^3}{3} + \dfrac{x^5}{5} + \cdots + c$, divide $1 - x^2$ into 1, first. **23.** $s = t^3 + t^2 + s_0$ **25.** $y = k\left(\dfrac{x^6}{6} + \dfrac{1350}{4}x^4 - \dfrac{7000}{3}x^3\right) + c$
27. $y = 3x^2 - 5x + c$

Trial Problems 21.3

1. $\int (x+1)^2\,dx = \dfrac{(x+1)^3}{3} + c$ **2.** $\int 2x(x^2+1)\,dx = \dfrac{(x^2+1)^2}{2} + c$

3. $\int (x+1)^{1/2}\,dx = \dfrac{(x+1)^{3/2}}{\frac{3}{2}} + c = \dfrac{2}{3}(x+1)^{3/2} + c$ **4.** $\int 2x(x^2+1)^{1/2}\,dx = \dfrac{(x^2+1)^{3/2}}{\frac{3}{2}} + c = \dfrac{2}{3}(x^2+1)^{3/2} + c$

5. $\int \dfrac{1}{(x+1)^2}\,dx = \int (x+1)^{-2}\,dx = \dfrac{(x+1)^{-1}}{-1} + c = -\dfrac{1}{x+1} + c$

Exercises 21.3

1. $\dfrac{1}{15}(2+3x)^5 + c$ **3.** $\dfrac{1}{15}(x^3-1)^5 + c$ **5.** $4\sqrt{1+x^2} + c$ **7.** $-\dfrac{1}{3(1+x^3)} + c$ **9.** $\dfrac{15}{8}(1+x^2)^{4/3} + c$
11. $2(x-3)^{7/2} + c$ **13.** $-\dfrac{3}{8}(1-x^2)^{4/3} + c$ **15.** $-\dfrac{1}{2(z^2+2z-3)} + c$ **17.** $-\dfrac{2}{1+\sqrt{x}} + c$ **19.** $-\dfrac{1}{9s} + c$
21. yes, yes **23.** $\dfrac{1}{7-t} + c$ **25.** $A(t) = \dfrac{1}{2}\ln^2(t+1)$ **27.** $R(T) = 0.0005\,(3T^2+1)^{2/3} + c$
29. $M = -2e^{(0.1\ln 2)x} + 2$

Trial Problems 21.4

1. $\int_0^1 2\,dx = 2x\Big|_0^1 = 2(1) - 2(0) = 2$ **2.** $\int_0^1 2x\,dx = \frac{2x^2}{2}\Big|_0^1 = x^2\Big|_0^1 = 1^2 - 0^2 = 1$

3. $\int_0^1 2x^2\,dx = \frac{2x^3}{3}\Big|_0^1 = \frac{2 \cdot 1^3}{3} - \frac{2 \cdot 0^3}{3} = \frac{2}{3}$ **4.** $\int_0^4 \sqrt{x}\,dx = \int_0^4 x^{1/2}\,dx = \frac{x^{3/2}}{\frac{3}{2}}\Big|_0^4 = \frac{4^{3/2}}{\frac{3}{2}} - \frac{0}{\frac{3}{2}} = \frac{2}{3} \cdot 8 = \frac{16}{3}$

5. $\int_1^2 2x(x^2+1)dx = \frac{(x^2+1)^2}{2}\Big|_1^2 = \frac{(4+1)^2}{2} - \frac{(1+1)^2}{2} = \frac{25}{2} - \frac{4}{2} = \frac{21}{2}$

Exercises 21.4

1. $5x^2 + 15$ **3.** $\frac{3}{4}x^{4/3} + \frac{1}{4}$ **5.** $(4x^2 + 3x + 1)^3$ **7.** $-2\sqrt{3-x} + 2(1 + \sqrt{2})$ **9.** $-\frac{1}{8(x-5)^2} + \frac{33}{8}$ **11.** $\frac{31}{5}$

13. Cannot be done at this time.   **15.** $\frac{4}{25}$ **17.** 1   **19.** $\frac{2}{3}$ **21.** -2 **23.** $C(x) = x^2 - 12x + 136$, \$2036

25. $v = 4t^2 + 4$ **27.** 156   **29.** $P(t) = 60t + t^2 - \frac{1}{8}t^3$

Trial Problems 21.5

1. $\int e^x\,dx = e^x + c$ **2.** $\int e^{-x}\,dx = -e^{-x} + c$ **3.** $\int 2^x\,dx = \frac{1}{\ln 2} \cdot 2^x + c$ **4.** $\int \frac{2}{x}\,dx = 2\,|\ln|\,x + c$

5. $\int \frac{2x}{x^2 - 1}\,dx = \ln|x^2 - 1| + c$

Exercises 21.5

1. $\frac{1}{5}e^{3x} + c$ **3.** $\frac{4}{5}\ln|x| + c$ **5.** $\frac{3}{\ln 5}5^{x/3} + c$ **7.** $-\frac{1}{3}\ln|2 - 3r| + c$ **9.** $\frac{1}{6}e^{2x^3} + c$ **11.** $\frac{1}{2}\ln|v^2 - 4| + c$

13. $-\frac{3}{3^{2x}\ln 3} + c$ **15.** $\ln|x| + \frac{1}{2x^2} + c$ **17.** $\frac{(\ln x)^2}{2} + c$ **19.** $\frac{x^2}{2} + \frac{1}{2}\ln|x^2 - 1| + c$ **21.** $\ln|e^x + e^{-x}| + c$

23. $\frac{3}{2}\ln\left(\frac{8}{5}\right)$ **25.** $\frac{1}{2}e^3(e^2 - 1)$ **27.** $\frac{21}{16\ln 2}$ **29.** $\frac{1}{4}e^{4x} + c$ **31.** $\frac{1}{5}(1 - e^{-5})$ **33.** $\frac{1}{2\ln 2}$ **35.** $y = e^{x+2} - e^3$

37. $s = \ln 49$ cm   **39.** $T = 300(e^{3/100} - 1)°$

Trial Problems 21.6

1. $\int 2\sin 2x\,dx = -\cos 2x + c$ **2.** $\int \tan 2x\,dx = \frac{1}{2}\int 2\tan 2x\,dx = \frac{1}{2}\ln|\sec 2x| + c$

3. $\int \sec x \tan x\,dx = \sec x + c$ **4.** $\int \sin x \cos x\,dx = \frac{\sin^2 x}{2} + c$ **5.** $\int \frac{1}{\sqrt{1 - t^2}}\,dt = \sin^{-1} t + c$

Exercises 21.6

1. $-\frac{1}{3}\cos 3x + c$ **3.** $\frac{3}{4}\sin 4x + c$ **5.** $-2\ln\left|\cos\frac{\theta}{2}\right| + c$ **7.** $\frac{1}{5}\cot 5x + c$

9. $-\frac{1}{2}\ln|\cos 2\theta| + c$ or $\frac{1}{2}\ln|\sec 2\theta| + c$ **11.** $-\frac{1}{9}\cot 3r + c$ **13.** $-\cos(\ln x) + c$ **15.** $2\ln\left|\sec\frac{\theta}{2} + \tan\frac{\theta}{2}\right| + c$

17. $\frac{1}{8}\sin^4(2x) + c$ **19.** $-\frac{1}{4\sin^2(2x)} + c$ **21.** $-\ln|1 - \sin\theta| + c$ **23.** $\frac{1}{7}\sec 7x + c$ **25.** $\sin^{-1}\left(\frac{x}{3}\right) + c$

27. $\frac{1}{18}\tan^{-1}\left(\frac{x^3}{6}\right) + c$ **29.** $-\frac{1}{e^{\tan\theta}} + c$ **31.** $\frac{\sin^3(2x)}{3} + c$ **33.** $\frac{1}{6}\tan^{-1}\left(\frac{2x}{3}\right) + c$ **35.** $\frac{1}{5}\sec^{-1}\left(\frac{3x}{5}\right) + c$

37. $\frac{1}{\sqrt{21}}\tan^{-1}\left(\sqrt{\frac{3}{7}}x\right) + c$ **39.** $\frac{\sqrt{2}}{12}$ **41.** $\frac{1}{2}\sin 1$ **43.** $\frac{2}{3}\sin 6$ **45.** $\frac{\sqrt{30}}{3}\tan^{-1}\left(\frac{\sqrt{30}\,t}{30}\right) + c$

47. $s = \frac{1}{4} - \frac{\cos^4 t}{4}$ **49.** $\sin^{-1}\left(\frac{x}{A}\right) = \sqrt{\frac{k}{m}}\,t + \sin^{-1}\left(\frac{x_0}{A}\right)$

Review Exercises 21.8

1. [21.3] $\frac{2}{5}x^5 + c$ **3.** [21.3] $\frac{(x+1)^3}{3} + c$ **5.** [21.3] $\frac{r^2}{2} - \frac{1}{r} + c$ **7.** [21.3] $\frac{1}{8}(1 + 2z)^4 + c$

9. [21.4] $\frac{1}{2}\sqrt{1 + v^4} + c$ **11.** [21.6] $\frac{e^{2x} + 2xe^x - 1}{e^x} + c$ **13.** [21.6] $\ln|1 - e^{-t}| + c$ **15.** [21.6] $\frac{1}{10}e^{10x} + c$

17. [21.6] $\frac{3x^2}{2\ln 3} + c$ **19.** [21.7] $\frac{1}{3}\tan^{-1}(3r) + c$ **21.** [21.6] $\frac{1}{2}\ln(v^2 + 2) + c$ **23.** [21.7] $\frac{\sin^6\theta}{6} + 0$

25. [21.6] $\frac{(\ln x)^5}{5} + c$ **27.** [21.6] $2e^{\sqrt{x}} + c$ **29.** [21.4] $\frac{2}{3}\sqrt{t^3 + 8} + c$ **31.** [21.6] $-\frac{e^{-ax}}{a} + c$

33. [21.6] $\frac{2^{2x-1}}{\ln 2} + c$ or $\frac{2^{2x}}{2\ln 2} + c$ **35.** [21.6] $2\ln(1 + r^2) + c$ **37.** [21.7] $-\frac{1}{2}e^{\cos 2x} + c$ **39.** [21.7] $\frac{1}{3}\sec 3\theta + c$

41. [21.7] $\frac{1}{3}\tan v^3 + c$ **43.** [21.4] $\frac{1}{8}\ln|4x^2 + 25| - \frac{1}{5}\tan^{-1}\left(\frac{2x}{5}\right) + c$ **45.** [21.7] $\frac{\sin 2\theta}{2} + \frac{\sin^4 2\theta}{8} + c$

47. [21.5] 10 **49.** [21.5] $\frac{1}{4}$ **51.** [21.5] 5 **53.** [21.5] $\sqrt{2} - \sqrt{3}$ **55.** [21.5] $\frac{1}{2}\tan^{-1}\left(\frac{1}{2}\right)$ **57.** [21.5] e

59. [21.5] $\frac{2}{3}(2 - \sqrt{2})$ **61.** [21.5] $\frac{\sqrt{2}}{12}$ **63.** [21.5] $\frac{1}{8} + \frac{\pi}{6}$ **65.** [21.5] $\frac{3}{2\ln 2}$ **67.** [21.5] 4 **69.** [21.5] $\frac{e}{3}(e^2 - 1)$

71. [21.5] π^3 **73.** [21.5] $\int_a^c f'(x)dx = f(c) - f(a) = f(c) - f(a) + f(b) - f(b) = [f(b) - f(a)] + [f(c) - f(b)] = \int_a^b f'(x)\,dx + \int_b^c f'(x)\,dx$ **75.** [21.6] $s \approx 266.85$ ft **77.** [21.7] $w = \ln\left|\frac{2}{1 + \cos x}\right|$ **79.** [21.3] $v = \frac{t^3}{3} - t^2$

81. [21.5] $s = \frac{\pi}{6} \approx 0.5236$ ft **83.** [21.5] $\frac{\pi}{16}$

ANSWER KEY—CHAPTER 22 1123

Test 21.9

1. [21.2] $\dfrac{x^4}{4} + c$ 2. [21.2] $\dfrac{(x+1)^3}{3} + c$ 3. [21.2] $-\cos x + c$ 4. [21.2] $-\dfrac{1}{x} + c$ 5. [21.2] $\dfrac{x^2}{2} + \dfrac{2x^3}{3} + c$
6. [21.2] $2\sqrt{x} + c$ 7. [21.2] $\dfrac{(1+3x)^5}{15}$ 8. [21.2] $-\dfrac{2}{x-1} + c$ 9. [21.2] $\dfrac{2}{9}(x^3+2)^{3/2} + c$ 10. [21.2] $\dfrac{2}{3}\ln|x| + c$
11. [21.2] $-\dfrac{1}{2}e^{-2x} + c$ 12. [21.2] $\dfrac{1}{2}\ln|t^2 - 4| + c$ 13. [21.2] $-\cos 3x + c$ 14. [21.2] $-\ln|1 - \sin t| + c$
15. [21.2] $\dfrac{1}{4}\tan^{-1}\left(\dfrac{x}{4}\right) + c$ 16. [21.5] $\dfrac{4}{15}$ 17. [21.5] 0   18. [21.2] $y = x^2 - 16$
19. [21.2] $s = 2e^{1/2t} + \sin t + s_0$ 20. [21.2] $L = \dfrac{1}{2}\ln 2$ lumens

Chapter 22 Answers

Trial Problems 22.1

1. $\displaystyle\int x^2 \, dx = \dfrac{x^3}{3} + c$ 2. $\displaystyle\int \dfrac{1}{x} \, dx = \ln|x| + c$ 3. $\displaystyle\int \sin 2x \, dx = \dfrac{1}{2}\int 2\sin 2x \, dx = -\dfrac{1}{2}\cos 2x + c$
4. $\displaystyle\int e^{-2x} \, dx = -\dfrac{1}{2}\int -2e^{-2x} \, dx = -\dfrac{1}{2}e^{-2x} + c$ 5. $\displaystyle\int \dfrac{x}{x^2+1} \, dx = \dfrac{1}{2}\int \dfrac{2x}{x^2+1} \, dx = \dfrac{1}{2}\ln(x^2+1) + c$

Exercises 22.1

1. $x \sin x + \cos x + c$ 3. $\dfrac{x 2^x}{\ln 2} - \dfrac{2^x}{(\ln 2)^2} + c$ 5. $x \tan^{-1} x - \dfrac{1}{2}\ln(1 + x^2) + c$ 7. $-\dfrac{1}{2}x^2 e^{-x^2} - \dfrac{1}{2}e^{-x^2} + c$
9. $\dfrac{2}{3}x(x+1)^{3/2} - \dfrac{4}{15}(x+1)^{5/2} + c$ 11. $\dfrac{(\ln x)^2}{2} + c$ 13. $\dfrac{x\cos(\ln x) + x\sin(\ln x)}{2} + c$ 15. $-xe^{-x} - e^{-x} + c$
17. $-\dfrac{1}{x}\ln x - \dfrac{1}{x} + c$ 19. $-\dfrac{1}{3}$ 21. $2(\sqrt{x}\sin\sqrt{x} + \cos\sqrt{x}) + c$ 23. $y = -2x\sqrt{1-x} - \dfrac{4}{3}(1-x)^{3/2} + \dfrac{4}{3}$
25. $q = \dfrac{1}{2}t\sin 2t + \dfrac{1}{4}\cos 2t - \dfrac{1}{4}$ 27. a. $M \approx 5.97$ b. $M \approx 3.77$

Trial Problems 22.2

1. $\displaystyle\int \cos\dfrac{x}{2}\, dx = 2\int \dfrac{1}{2}\cos\dfrac{x}{2}\, dx = 2\sin\dfrac{x}{2} + c$ 2. $\displaystyle\int \csc^2 2x \, dx = \dfrac{1}{2}\int 2\csc^2 2x \, dx = -\dfrac{1}{2}\cot 2x + c$
3. $\displaystyle\int \tan 3x \, dx = \dfrac{1}{3}\int 3\tan 3x \, dx = \dfrac{1}{3}\ln|\sec 3x| + c$ 4. $\displaystyle\int \dfrac{5}{1+x^2} \, dx = 5\tan^{-1} x + c$
5. $\displaystyle\int \dfrac{1}{\sqrt{4-x^2}} \, dx = \sin^{-1}\dfrac{x}{2} + c$

Exercises 22.2

1. $\dfrac{x}{2} + \dfrac{\sin 2x}{4} + c$ **3.** $\sin\theta - \dfrac{\sin^3\theta}{3} + c$ **5.** $\dfrac{x}{8} - \dfrac{\sin 4x}{32} + c$ **7.** $\dfrac{\sin^4\theta}{4} + c$ **9.** $-\cot x - x + c$ **11.** $-\dfrac{\cos^4\theta}{4} + c$

13. $\tan\theta - \cot\theta + c$ **15.** $\dfrac{\sec 2x}{2} + c$ **17.** $\ln|2 + \tan x| + c$ **19.** $-\ln|x - \sqrt{x^2 - 16}| + c$ **21.** $-\dfrac{\sqrt{9 + x^2}}{9x} + c$

23. $\dfrac{1}{3}(x^2 - 36)^{3/2} + c$ **25.** $\dfrac{x}{4\sqrt{4 + x^2}} + c$ **27.** $\dfrac{\pi}{8}$ **29.** $-\sqrt{4 - x^2} + \sin^{-1}\left(\dfrac{x}{2}\right) + c$

31. $y = \dfrac{3}{2}x + \dfrac{1}{4}\sin 2x + 2$ **33. a.** $\underbrace{\int \sec^2 x \tan x \, dx}_{} = \underbrace{\int \tan x \sec^2 x \, dx}_{u \quad du} = \dfrac{u^2}{2} + c_1 = \dfrac{\tan^2 x}{2} + c_1$

b. $\int \sec^2 x \tan x \, dx = \int \underbrace{\sec x}_{u} \underbrace{\sec x \tan x \, dx}_{du} = \dfrac{u^2}{2} + c_2 = \dfrac{\sec^2 x}{2} + c_2$

and $\dfrac{\sec^2 x}{2} + c_2 = \dfrac{\tan^2 x}{2} + c_1$, where $c_2 - c_1 = -\dfrac{1}{2}$ **35.** $w = \dfrac{4\pi}{3} - 2\sqrt{3}$ **37.** $q = \sqrt{3} \approx 1.73$

Trial Problems 22.3

1. $\displaystyle\int \dfrac{3}{x - 2} \, dx = 3\ln|x - 2| + c$ **2.** $\displaystyle\int x^{-2} \, dx = \dfrac{x^{-1}}{-1} + c = -\dfrac{1}{x} + c$

3. $\displaystyle\int \dfrac{6}{(x-1)^2} \, dx = 6\int (x-1)^{-2} \, dx = \dfrac{6(x-1)^{-1}}{-1} + c = -\dfrac{6}{x - 1} + c$

4. $\displaystyle\int \dfrac{x}{x^2 + 4} \, dx = \dfrac{1}{2}\int \dfrac{2x}{x^2 + 4} \, dx = \dfrac{1}{2}\ln(x^2 + 4) + c$ **5.** $\displaystyle\int \dfrac{1}{x^2 + 4} \, dx = \dfrac{1}{2}\tan^{-1}\dfrac{x}{2} + c$

Exercises 22.3

1. $\ln\left|\dfrac{(x - 2)^3}{(x + 4)^2}\right| + c$ **3.** $-\dfrac{1}{x + 1} + 5\ln|x + 1| - 3\ln|x - 5| + c$ **5.** $\ln\left|\dfrac{x - 3}{x - 2}\right| + c$

7. $-\dfrac{1}{12}\ln(x^2 + 6) + \dfrac{1}{\sqrt{6}}\tan^{-1}\left(\dfrac{x}{\sqrt{6}}\right) + \dfrac{1}{6}\ln|x| + c$ **9.** $\dfrac{2}{x} + 3\ln|x| - 3\ln|x + 1| + c$

11. $\dfrac{1}{x} + 2\ln|x| - \dfrac{3}{5}\ln|5x + 3| + c$ **13.** $-\dfrac{1}{x + 2} + 2\ln|x + 2| + 3\ln|x - 2| + c$

15. $\dfrac{1}{2}\ln(x^2 + 1) + 2\tan^{-1}x + 3\ln|x + 1| + 3\ln|x - 1| + c$ **17.** $4\ln|x - 4| + \ln|x + 1| + c$

19. $\dfrac{1}{2}\ln|x^2 - 4| + c$ **21.** $x - \ln|x^2 + x + 1| + c$ **23.** $q = \dfrac{3}{4}\ln|t + 3| + \dfrac{1}{4}\ln|t - 1| - \dfrac{3}{4}\ln 5$

25. Methods will vary.

Trial Problems 22.4

1. $\int \dfrac{dx}{x(1+x^2)} = \dfrac{1}{2} \ln \left| \dfrac{x^2}{1+x^2} \right| + c$ (formula #5)　　2. $\int \dfrac{dx}{x\sqrt{x^2+1}} = -\ln \left| \dfrac{1+\sqrt{x^2+1}}{x} \right| + c$ (formula #13)

3. $\int \sin 3x \cos 2x \, dx = -\dfrac{\cos 5x}{10} - \dfrac{\cos x}{2} + c$ (formula #33)　　4. $\int \dfrac{dx}{x \ln x} = \ln|\ln x| + c$ (formula #36)

5. $\int \dfrac{dx}{1+e^x} = x - \ln(1+e^x) + c$ (formula #40)

Exercises 22.4

1. $-\dfrac{1}{2x} + \dfrac{3}{4} \ln \left| \dfrac{2+3x}{x} \right| + c$　　3. $\dfrac{1}{2\sqrt{10}} \ln \left| \dfrac{\sqrt{5}+\sqrt{2}x}{\sqrt{5}-\sqrt{2}x} \right| + c$　　5. $\sqrt{4x^2+9} - 3 \ln \left| \dfrac{3+\sqrt{4x^2+9}}{2x} \right| + c$

7. $\dfrac{1}{3}x^3 \left(\ln |x| - \dfrac{1}{3} \right) + c$　　9. $\sqrt{5+x^2} + c$　　11. $-\dfrac{\cos 8x}{16} - \dfrac{\cos 4x}{8} + c$　　13. $-\dfrac{\sin 4x}{8} + \dfrac{\sin 2x}{4} + c$　　15. $-\dfrac{1}{15}$

17. $2\sqrt{5+3x} + \dfrac{5}{\sqrt{5}} \ln \left| \dfrac{\sqrt{5+3x} - \sqrt{5}}{\sqrt{5+3x} + \sqrt{5}} \right| + c$　　19. $\dfrac{1}{2} \ln |\ln x^2| + c$　　21. $\sqrt{4x^2+9} - 3 \ln \left| \dfrac{3+\sqrt{4x^2+9}}{2x} \right| + c$

23. $q = 2 \ln \left| \dfrac{t + \sqrt{t^2+100}}{10} \right|$　　25. $w = \ln(\ln t) - \ln(\ln 2) = \ln \left(\dfrac{\ln t}{\ln 2} \right)$　　27. $t = \dfrac{k}{2} \ln(x^2+2) + c$

Review Exercises 22.6

1. [22.3] $-2 \ln |x-1| + \ln |x-2| + \ln |x-3| + c$　　3. [22.4] $\dfrac{1}{2}[x\sqrt{x^2+1} + \ln |x + \sqrt{x^2+1}|] + c$

5. [22.4] $\dfrac{x}{25\sqrt{25-x^2}} + c$　　7. [22.3] $\dfrac{1}{x} - 2 \ln |x| + 2 \ln |x-1| + c$　　9. [22.3] $\ln |x| - \dfrac{1}{2} \ln(x^2+4) + c$

11. [22.1] $\dfrac{e^x(2 \sin 2x + \cos 2x)}{5} + c$　　13. [22.3] $\dfrac{x^2}{2} + \dfrac{4}{3} \ln |x+2| + \dfrac{2}{3} \ln |x-1| + c$　　15. [22.4] $\dfrac{-\sqrt{36-x^2}}{36x} + c$

17. [22.1] $-x^2 \cos x + 2x \sin x + 2 \cos x + c$　　19. [22.1] $\dfrac{1}{2}x^2 e^{x^2} - \dfrac{1}{2}e^{x^2} + c$

21. [22.1] $x \ln(x^2 + x) - 2x + \ln |x+1| + c$　　23. [22.1] $x \ln |1+x| - x + \ln |1+x| + c$

25. [22.4] $2 \ln \left| \dfrac{2 - \sqrt{4-x^2}}{x} \right| + \sqrt{4-x^2} + c$　　27. [22.2] $\dfrac{1}{4} \sin^4 x - \dfrac{1}{6} \sin^6 x + c$

29. [22.1] $x \sec x - \ln |\sec x - \tan x| + c$　　31. [22.4] $\dfrac{1}{8}x\sqrt{4x^2+25} - \dfrac{25}{16} \ln |2x + \sqrt{4x^2+25}| + c$

33. [22.2] $\dfrac{1}{6} \sin^2 3x + c$　　35. [22.1] $s = \dfrac{\pi}{4} + \dfrac{1}{2} \approx 1.29$ cm　　37. [22.1] $y = xe^x - e^x + 1$

39. [22.3] $t = -3 \ln |1-x| - \ln |2-x| + c$　　41. [22.4] $q = \dfrac{9}{2} \sin^{-1} \left(\dfrac{t}{3} \right) - \dfrac{t\sqrt{9-t^2}}{2}$

Test 22.7

1. [22.1] $te^t - e^t + c$ 2. [22.1] $x^2 \sin x - 2 \sin x + 2x \cos x + c$ 3. [22.1] $xe^{-x} - e^{-x} + c$ 4. [22.1] $\dfrac{1}{2} \ln^2 x + c$

5. [22.2] $\dfrac{x}{8} - \dfrac{\sin 4x}{32} + c$ 6. [22.2] $\ln|\csc x - \cot x| + \cos x + c$ 7. [22.4] $\dfrac{t\sqrt{9-t^2}}{2} + \dfrac{9 \sin^{-1}\left(\dfrac{t}{3}\right)}{2} + c$

8. [22.1] $3 \ln|x-1| - \dfrac{1}{x-1} + c$ 9. [22.4] $\dfrac{1}{\sqrt{2}} \ln \left| \dfrac{\sqrt{2+3x} - \sqrt{2}}{\sqrt{2+3x} + \sqrt{2}} \right| + c$

10. [22.1] $v = 4 \ln|t-4| + \ln|t+1| - 4 \ln 4$

Chapter 23 Answers

Trial Problems 23.1

1. $\displaystyle\int_0^1 (x^2 + 1)dx = \dfrac{x^3}{3} + x \Big|_0^1 = \dfrac{1}{3} + 1 - 0 = \dfrac{4}{3}$ 2. $\displaystyle\int_0^4 \sqrt{x}\, dx = \int_0^4 x^{1/2}\, dx = \dfrac{2}{3} x^{3/2} \Big|_0^4 = \dfrac{2}{3}(8) - 0 = \dfrac{16}{3}$

3. $\displaystyle\int_1^3 \dfrac{1}{x} dx = \ln x \Big|_1^3 = \ln 3 - \ln 1 = \ln 3$ 4. $\displaystyle\int_0^{\pi/2} \cos x\, dx = +\sin x \Big|_0^{\pi/2} = \sin \dfrac{\pi}{2} - \sin 0 = 1$

5. $\displaystyle\int_0^2 (x+1)^2\, dx = \int_0^2 (x^2 + 2x + 1)dx = \dfrac{x^3}{3} + x^2 + x \Big|_0^2 = \dfrac{8}{3} + 4 + 2 - 0 = \dfrac{26}{3}$

Exercises 23.1

1. $A = \dfrac{1}{6}$

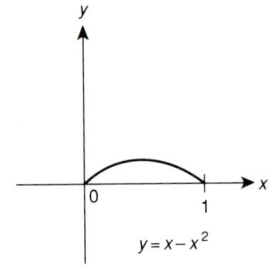

$y = x - x^2$

3. $A = \dfrac{1}{2}$

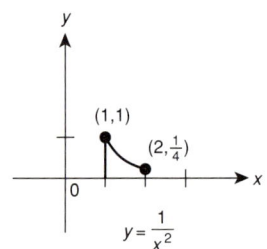

$(1,1)$

$(2, \tfrac{1}{4})$

$y = \dfrac{1}{x^2}$

5. $A = 76$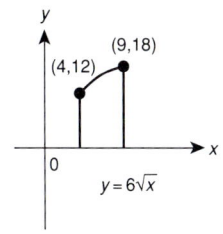

7. $A = A_1 + A_2 = \dfrac{1}{2}$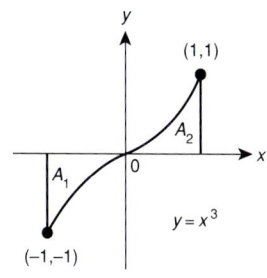

9. $A = 8$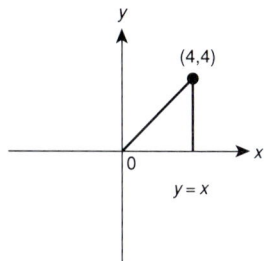

11. $A = \dfrac{1}{2} \ln 2$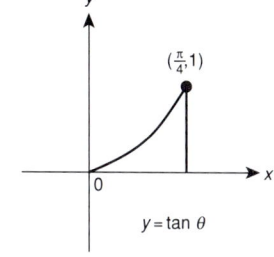

13. $A = e - 1$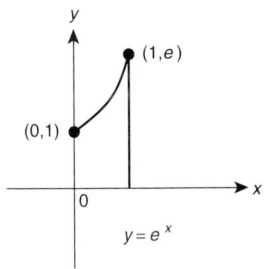

15. $A = 2$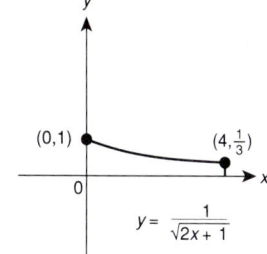

17. $A = 4\dfrac{1}{3}$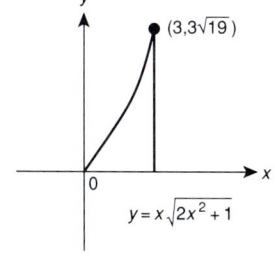

19. $A = \dfrac{2\pi}{3}$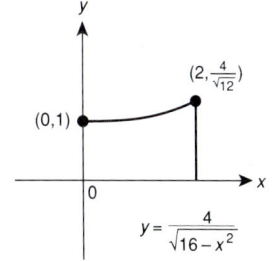

21. $A = 6\frac{1}{4}$

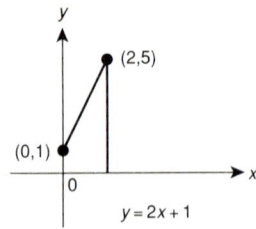

23. $A = \frac{1}{4}$

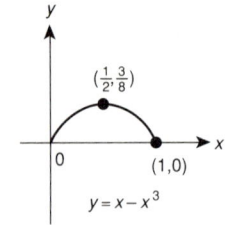

25. $A = 12\frac{2}{3}$

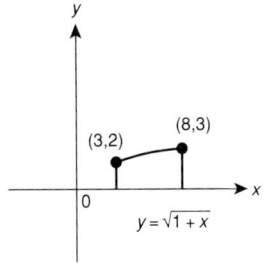

27. $A = 1\frac{1}{2}$

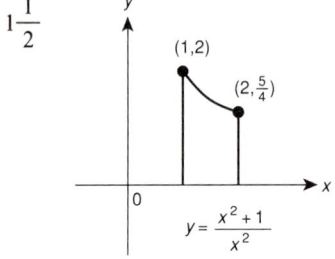

29. $A = 32\frac{1}{3}$

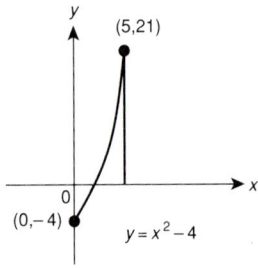

31. a. $\frac{1}{12}$ b. $4\frac{1}{4}$ c. $57\frac{1}{6}$ d. $26\frac{2}{3}$

33. $5\frac{1}{3}$ cm²

35. 51 complete mowers

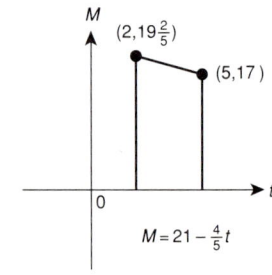

Answer Key—Chapter 23

Trial Problems 23.2

1. $\int_{-1}^{1}(1-x^2)dx = x - \dfrac{x^3}{3}\bigg|_{-1}^{1} = \left(1 - \dfrac{1}{3}\right) - \left(-1 + \dfrac{1}{3}\right) = 1\dfrac{1}{3}$ **2.** $\int_{0}^{1}(x^3 + x)dx = \dfrac{x^4}{4} + \dfrac{x^2}{2}\bigg|_{0}^{1} = \dfrac{1}{4} + \dfrac{1}{2} - 0 = \dfrac{3}{4}$

3. $\int_{0}^{1}(e^x + 1)dx = e^x + x\bigg|_{0}^{1} = e + 1 - e^0 + 0 = e$

4. $\int_{1}^{2} \ln x \, dx = x \ln x - x \bigg|_{1}^{2} = (2 \ln 2 - 2) - (1 \ln 1 - 1) = 2 \ln 2 - 1$

5. $\int_{0}^{1} \dfrac{x}{x^2 + 1} dx = \dfrac{1}{2}\int_{0}^{1} \dfrac{2x}{x^2 + 1} dx = \dfrac{1}{2} \ln(x^2 + 1)\bigg|_{0}^{1} = \dfrac{1}{2} \ln 2$

Exercises 23.2

1. $A = \dfrac{4}{3}$

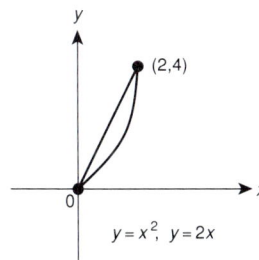

$y = x^2, \; y = 2x$

3. $A = \sqrt{2} - 1$

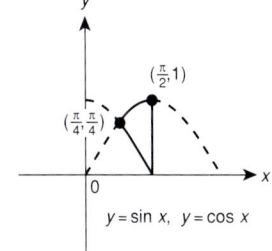

$y = \sin x, \; y = \cos x$

5. $A = \dfrac{2}{3}$

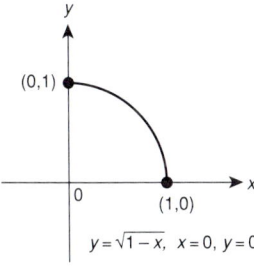

$y = \sqrt{1-x}, \; x = 0, \; y = 0$

7. $A = \dfrac{9}{2}$

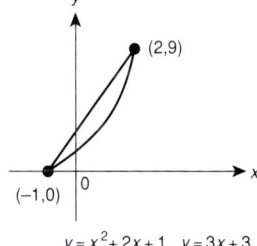

$y = x^2 + 2x + 1, \; y = 3x + 3$

9. $A = \dfrac{500}{27}$

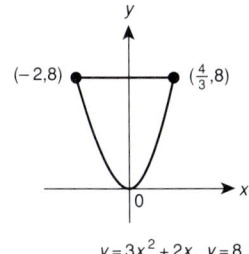

$y = 3x^2 + 2x, \; y = 8$

11. $A = \dfrac{128}{15}$

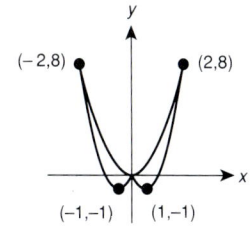

$y = x^4 - 2x^2, \; y = 2x^2$

13. $A = \dfrac{128}{5}$

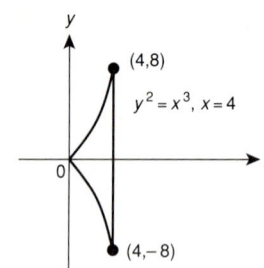

15. $A = \dfrac{1}{2}$

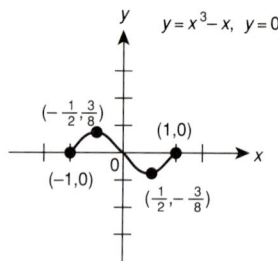

17. $A = 20\dfrac{5}{6}$

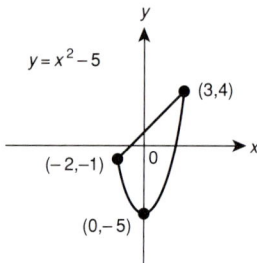

19. $A = \dfrac{44}{3}$

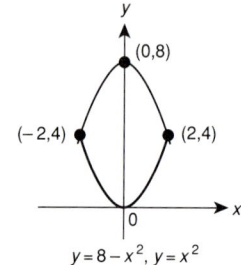

21. $e - \dfrac{5}{3}$

23. $\ln\left(\dfrac{2+\sqrt{2}}{2-\sqrt{2}}\right)$

25. $4\ln 4 - 3$

27. $\dfrac{1}{2}\ln 10$

Trial Problems 23.3

1. $\displaystyle\int_0^2 \pi(\sqrt{x})^2\,dx = \int_0^2 \pi x\,dx = \dfrac{\pi x^2}{2}\bigg|_0^2 = 2\pi$

2. $\displaystyle\int_0^\pi \sin^2 x\,dx = \int_0^\pi \left(\dfrac{1}{2} - \dfrac{1}{2}\cos 2x\right)dx = \dfrac{1}{2}\left(x - \dfrac{1}{2}\sin 2x\right)\bigg|_0^\pi = \dfrac{1}{2}\left(\pi - \dfrac{1}{2}\cdot 0\right) - 0 = \dfrac{\pi}{2}$

3. $\displaystyle\int_0^1 e^{-2x}\,dx = -\dfrac{1}{2}e^{-2x}\bigg|_0^1 = -\dfrac{1}{2}e^{-2} - \left(-\dfrac{1}{2}e^0\right) = \dfrac{1}{2} - \dfrac{1}{2}e^{-2}$

4. $\displaystyle\int_0^{\pi/4} \tan^2 x\,dx = \int_0^{\pi/4}(\sec^2 x - 1)dx = \tan x - x\bigg|_0^{\pi/4} = \tan\dfrac{\pi}{4} - \dfrac{\pi}{4} - 0 = 1 - \dfrac{\pi}{4}$

5. $\displaystyle\int_0^1 (1-x^2)^2\,dx = \int_0^1 (1 - 2x^2 + x^4)dx = x - \dfrac{2x^3}{3} + \dfrac{x^5}{5}\bigg|_0^1 = 1 - \dfrac{2}{3} + \dfrac{1}{5} - 0 = \dfrac{8}{15}$

Exercises 23.3

1. $\dfrac{117\pi}{3}$ 3. 8π 5. $\dfrac{128\pi}{7}$ 7. $\dfrac{\pi^2}{2}$ 9. $\dfrac{\pi}{2}(e^2 - 1)$ 11. $\dfrac{\pi}{2}[e^{-2} - e^{-4}]$ 13. $\dfrac{\pi^2}{4}$ 15. $\dfrac{\pi}{3}\tan^{-1}\left(\dfrac{1}{3}\right)$ 17. 2π 19. $\dfrac{3\pi}{4}$
21. $\dfrac{\pi}{3}$ 23. $\dfrac{3\pi}{5}$ 25. 12π 27. $\dfrac{106\sqrt{2}}{5}\pi$ 29. $\dfrac{162}{5}\pi$

Trial Problems 23.4

1. $\displaystyle\int_0^3 x(5-x)\,dx = \int_0^3 (5x - x^2)\,dx = \dfrac{5x^2}{2} - \dfrac{x^3}{3}\bigg|_0^3 = \dfrac{45}{2} - 9 = \dfrac{27}{2}$

2. $\displaystyle\int_0^4 2x\sqrt{x}\,dx = \int_0^4 2x \cdot x^{1/2}\,dx = 2\int_0^4 x^{3/2}\,dx = 2 \cdot \dfrac{2}{5} \cdot x^{5/2} = \dfrac{128}{5}$

3. $\displaystyle\int_0^\pi x\sin x\,dx = \sin x - x\cos x\bigg|_0^\pi = (\sin\pi - \pi\cos\pi) - (\sin 0 - 0\cdot\cos 0) = 0 - \pi(-1) - 0 = \pi$

4. $\displaystyle\int_1^3 xe^x\,dx = xe^x - e^x\bigg| = (3e^3 - e^3) - (1\cdot e - e) = 2e^3$

5. $\displaystyle\int_1^2 2x(1+x^2)^{-1}\,dx = \int_1^2 \dfrac{2x}{1+x^2}\,dx = \ln(1+x^2)\bigg|_1^2 = \ln(1+4) - \ln(1+1) = \ln 5 - \ln 2 = \ln\dfrac{5}{2}$

Exercises 23.4

1. 9π 3. $\dfrac{37}{3}\pi$ 5. 2π 7. $\dfrac{16\pi}{3}$ 9. $\pi\ln\dfrac{5}{2}$ 11. $\dfrac{96}{5}\pi$ 13. $\pi\sin 1$ 15. $8\pi\left(\dfrac{3}{4} - \sec^{-1}\dfrac{5}{4}\right)$ 17. $\dfrac{\pi}{3}$ 19. $\dfrac{32}{3}\pi$
21. $\dfrac{20}{3}\pi$

Trial Problems 23.5

1. $\displaystyle\int_0^5 \dfrac{3}{2}x\,dx = \dfrac{3}{2}\cdot\dfrac{x^2}{2}\bigg|_0^5 = \dfrac{3}{2}\cdot\dfrac{25}{2} - 0 = \dfrac{75}{4}$

2. $\displaystyle\int_0^5 1000\pi(10-x)\,dx = 1000\pi\left(10x - \dfrac{x^2}{2}\right)\bigg|_0^5 = 1000\pi\left(50 - \dfrac{25}{2}\right) = 37{,}500\pi$

3. $\displaystyle\int_4^9 8x^{-1/2}\,dx = 8\cdot\dfrac{2}{1}x^{1/2}\bigg|_4^9 = 16(3-2) = 16$

4. $\displaystyle\int_0^1 \left(\dfrac{y}{2}\right)^2(10-y)\,dy = \int_0^1 \dfrac{1}{4}y^2(10-y)\,dy = \dfrac{1}{4}\int_0^1 (10y^2 - y^3)\,dy = \dfrac{1}{4}\left(\dfrac{10}{3}y^3 - \dfrac{y^4}{4}\right)\bigg|_0^1 = \dfrac{1}{4}\left(\dfrac{10}{3} - \dfrac{1}{4}\right) - 0 = \dfrac{37}{48}$

5. $\displaystyle\int_3^7 \dfrac{1000}{x}\,dx = 1000\ln x\bigg|_3^7 = 1000(\ln 7 - \ln 3) = 1000\ln\dfrac{7}{3}$

Exercises 23.5

1. 48 in.-lbs   **3.** -1.2 **5.** 800 in.-lbs   **7.** $40{,}500\pi$ ft-lbs   **9.** $\approx 156{,}250\pi$ ft-lbs

Trial Problems 23.6

1. $\displaystyle\int_0^9 x\sqrt{x}\,dx = \int_0^9 x^{3/2}\,dx = \frac{2}{5}x^{5/2}\Big|_0^9 = \frac{2}{5}(3^5) - 0 = \frac{486}{5}$

2. $\displaystyle\int_0^1 (e^{2x})^2\,dx = \int_0^1 e^{4x}\,dx = \frac{1}{4}e^{4x}\Big|_0^1 = \frac{1}{4}(e^4 - e^0) = \frac{1}{4}(e^4 - 1)$

3. $\displaystyle\int_0^1 x(2 - x^2)\,dx = \int_0^1 (2x - x^3)\,dx = x^2 - \frac{x^4}{4}\Big|_0^1 = 1 - \frac{1}{4} - 0 = \frac{3}{4}$

4. $\displaystyle\int_{-1}^0 e^{-2x}\,dx = -\frac{1}{2}e^{-2x}\Big|_{-1}^0 = -\frac{1}{2}(e^0 - e^2) = \frac{e^2 - 1}{2}$

5. $\displaystyle\int_0^1 \sqrt{1 - x}\,dx = \int_0^1 (1 - x)^{1/2}\,dx = -\frac{2}{3}(1 - x)^{3/2}\Big|_0^1 = -\frac{2}{3}[(1 - 1)^{3/2} - (1 - 0)^{3/2}] = -\frac{2}{3}(-1) = \frac{2}{3}$

Exercises 23.6

1. Centroid: $\left(\dfrac{3}{2}, \dfrac{6}{5}\right)$

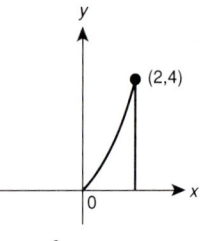
$y = x^2$, x-axis, $x = 2$

3. Centroid: $\left(\dfrac{8}{3}, \dfrac{4}{3}\right)$

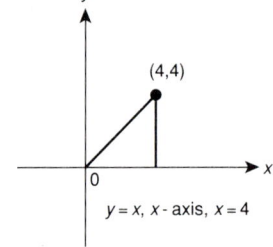
$y = x$, x-axis, $x = 4$

5. Centroid: $\left(\dfrac{1}{\ln 2}, \dfrac{1}{4\ln 2}\right)$

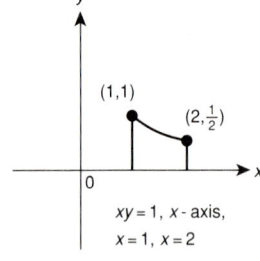
$xy = 1$, x-axis, $x = 1$, $x = 2$

7. Centroid: $\left(0, \dfrac{8}{5}\right)$

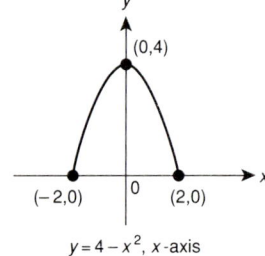
$y = 4 - x^2$, x-axis

9. Centroid: $\left(\dfrac{9}{4},\dfrac{27}{5}\right)$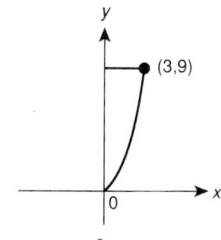

   $y=x^2$, y-axis, $y=9$

11. Centroid: $\left(\dfrac{8}{5},2\right)$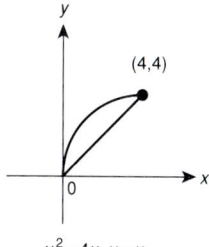

    $y^2=4x$, $y=x$

13. Centroid: $\approx (0.30, 0.26)$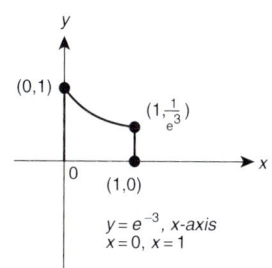

    $y=e^{-3}$, x-axis
    $x=0$, $x=1$

15. Centroid: $\left(\dfrac{2}{5},\dfrac{3}{8}\right)$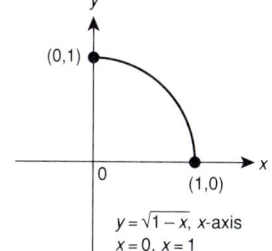

    $y=\sqrt{1-x}$, x-axis
    $x=0$, $x=1$

17. $\approx (0.66,0)$ 19. $\left(\dfrac{\pi}{2},0\right)$

21. $\left(0,\dfrac{9}{4}\right)$

23. $\left(0,\dfrac{27}{8}\right)$

Trial Problems 23.7

1. $\displaystyle\int_0^4 x^2\sqrt{x}\,dx = \int_0^4 x^2\cdot x^{1/2}\,dx = \int_0^4 x^{5/2}\,dx = \dfrac{2}{7}x^{7/2}\Big|_0^4 = \dfrac{2}{7}\cdot 2^7 - 0 = \dfrac{256}{7}$

2. $\displaystyle\int_0^1 x^2(4-x^{1/3})dx = \int(4x^2 - x^{7/3})dx = \dfrac{4x^3}{3} - \dfrac{3}{10}x^{10/3}\Big|_0^1 = \dfrac{4}{3} - \dfrac{3}{10} - 0 = \dfrac{31}{30}$

3. $\displaystyle\int_0^1 ye^{y^2}\,dy = \dfrac{1}{2}\int_0^1 2ye^{y^2}\,dy = \dfrac{1}{2}e^{y^2}\Big|_0^1 = \dfrac{1}{2}(e-1)$

4. $\displaystyle\int_0^1 xe^x\,dx = xe^x - e^x\Big|_0^1 = (1\cdot e - e) - (0\cdot e^0 - 1) = 1$ 5. $\displaystyle\int_1^8 y^2\cdot y^{1/3}\,dy = \int y^{7/3}\,dy = \dfrac{3}{10}y^{10/3}\Big|_1^8 = \dfrac{3}{10}(2^{10}-1)$

Exercises 23.7

1. $I_x = \dfrac{243}{5}$, $I_y = \dfrac{4374}{7}$ **3.** $I_x = \dfrac{768}{5}$, $I_y = \dfrac{32}{3}$ **5.** $I_y = \dfrac{1}{3}(e-1)$ **7.** $I_y = \dfrac{64}{15}$ **9.** $I_x = \dfrac{64}{3}\pi$ **11.** $I_y = \dfrac{32}{3}\pi$
13. $I_x = 4\pi$

Trial Problems 23.8

1. $\displaystyle\int_0^3 \sqrt{7}\,dx = \sqrt{7}x\Big|_0^3 = 3\sqrt{7}$ **2.** $\displaystyle\int_0^2 2x\sqrt{4-x^2}\,dx = -\dfrac{2}{3}(4-x^2)^{3/2}\Big|_0^2 = -\dfrac{2}{3}(0-8) = \dfrac{16}{3}$

3. $\displaystyle\int \dfrac{\ln x}{x}\,dx = \int \ln x\left(\dfrac{1}{x}\,dx\right) = \dfrac{(\ln x)^2}{2} + c$ **4.** $\displaystyle\int \dfrac{dx}{x^2+1} = \int \dfrac{dx}{1+x^2} = \tan^{-1}x + c$

5. $\displaystyle\int \dfrac{k}{(1+t)^2}\,dt = k\int(1+t)^{-2}\,dt = k\dfrac{(1+t)^{-1}}{-1} = -\dfrac{k}{1+t} + c$

Exercises 23.8

1. $s = 4\sqrt{2}$ **3.** $s = 2\sqrt{5}$ **5.** $s = \dfrac{38}{3}$ **7.** $s = 4$ **9. a.** $P \approx 0.93$ **b.** $P = 0.24$ **c.** $P \approx 0.46$ **11.** $k = \dfrac{3}{8}$
13. $F = 500$ lbs **15.** $F = 41{,}250$ lbs **17.** $F = 2{,}560$ lbs **19.** e **21.** 1 **23.** ∞ or dne **25.** dne **27.** e **29.** 0
31. dne

Review Exercises 23.10

1. [23.1] 8 **3.** [23.1] 2 **5.** [23.1] $2\ln 2 - \dfrac{3}{4}$ **7.** [23.1] $\dfrac{8}{3}$ **9.** [23.1] $\ln 3$ **11.** [23.1] $\ln\sqrt{3}$ **13.** [23.1] 4

15. [23.1] 2 **17.** [23.1] $\dfrac{1}{3} - \dfrac{\sqrt{3}}{8}$ **19.** [23.2] $\dfrac{14}{3}$ **21.** [23.2] $\dfrac{1}{3}(2\sqrt{2}-1)$ **23.** [23.2] $\dfrac{5}{12}$ **25.** [23.2] 1

27. [23.2] $\tan^{-1} 2 \approx 1.1$ **29.** [23.3] $\dfrac{2\pi}{3}$ **31.** [23.3] **a.** $\dfrac{32}{5}\pi$ **b.** 8π **33.** [23.3] 8π **35.** [23.4] $\dfrac{124}{5}\pi$

37. [23.5] 9 in.-lbs **39.** [23.5] $235{,}200\pi$ ft-lbs **41.** [23.6] $\left(3,\dfrac{6}{5}\right)$ **43.** [23.6] $(0,0)$ **45.** [23.7] $I_x = \dfrac{32}{3}$, $I_y = \dfrac{1536}{5}$

47. [23.8] $s = \sqrt{10} - \sqrt{2} + \ln\dfrac{3}{1+\sqrt{10}} - \ln\dfrac{1}{1+\sqrt{2}}$ **49.** [23.8] $F = 703\dfrac{1}{8}$ lbs **51.** [23.8] $P = \dfrac{2}{7}$

53. [23.8] $P = \dfrac{15}{16}$ **55.** [23.8] dne **57.** [23.8] $\dfrac{1}{2}$ **59.** [23.8] 1 **61.** [23.8] $\dfrac{1}{2e}$ **63.** [23.8] 1 **65.** [23.8] $\dfrac{1}{e}$

Test 23.11

1. [23.1] $A = \dfrac{9}{4}$ **2.** [23.2] $A = e^4 - \dfrac{19}{3}$ **3.** [23.3] $V = \dfrac{26\pi}{81}$ **4.** [23.4] dne **5.** [23.5] $W = 3033\dfrac{1}{3}$
6. [23.6] $\left(0,\dfrac{\pi}{8}\right)$ **7.** [23.7] $I_x = \dfrac{32}{7}$, $I_y = \dfrac{16}{5}$ **8.** [23.8] $\dfrac{1}{2}\ln 3$
9. [23.8] The total output of the mine over 4 years was 725.2 tons. **10.** [23.8] $P \approx 0.25$

Cumulative Review—Chapters 21, 22, and 23   23.12

1. [21] $x^6 + c$
2. [21] $x^3 - 2x + c$
3. [21] $\frac{2}{5}x^{5/3} + c$
4. [21] $\frac{8}{3}x^{3/2} + c$
5. [21] $e^x + 2x + c$
6. [21] $\sin x + c$
7. [21] $\ln |x| + c$
8. [21] $\frac{1}{3}(x + 2)^3 + c$
9. [21] $\frac{3}{5}x^{5/3} + \frac{1}{2}x^2 - 2x + c$
10. [21] $-\frac{1}{3x^3} + c$
11. [21] $8x^2 - \frac{4}{7}x^{7/2} + c$
12. [21] $\frac{3}{2}s^{2/3} + c$
13. [21] $\frac{1}{15}(2 + 3t)^5 + c$
14. [21] $-\frac{3}{2(1 + x^2)^2} + c$
15. [21] $\frac{1}{6}(r^4 - 2)^{3/2} + c$
16. [21] $-\frac{1}{4}\left(1 + \frac{1}{t}\right)^4 + c$
17. [21] $\frac{211}{5}$
18. [21] $\frac{4095}{6144}$
19. [21] $\frac{4}{25}$
20. [21] -2
21. [21] $\ln(x^2 + 2) + c$
22. [21] $\frac{(\ln t)^2}{2} + c$
23. [21] $\ln |x^4 - 3x + 5| + c$
24. [21] $\ln(e^x + 1) + c$
25. [21] $\frac{1}{3}e^{3x} + c$
26. [21] $\frac{1}{5}\ln \frac{3}{2}$
27. [21] $\frac{12}{(25)^2 \ln 5}$
28. [21] $-\frac{4}{3}(e^{-3} - 1)$
29. [21] $-\frac{1}{6}\cos 6x + c$
30. [21] $3 \ln \left|\sec \frac{\theta}{3}\right| + c$
31. [21] $-\frac{1}{3}(1 + \cos x)^3 + c$
32. [21] $\frac{1}{3}\sec 3x + c$
33. [21] $\frac{1}{3}\sin 1$
34. [21] $\frac{1}{2}\sqrt{2}$
35. [22] $\frac{x^2}{2}\ln 2x - \frac{x^2}{4} + c$
36. [22] $\frac{(\ln r)^2}{2} + c$
37. [22] $-\frac{2}{3}x\sqrt{1 - x} - \frac{4}{3}\sqrt{1 - x} + c$
38. [22] $\frac{e^x}{2}(\cos x + \sin x) + c$
39. [22] $\frac{x}{2}\sqrt{4 - x^2} + 2 \sin^{-1}\left(\frac{x}{2}\right) + c$
40. [22] $-\frac{1}{4}\sin^3 t \cos t + \frac{3}{8}t - \frac{3}{16}\sin 2t + c$
41. [22] $\ln |\csc x - \cot x| + \cos x + c$
42. [22] $-\ln \left|\frac{\sqrt{1 + x^2} + 1}{x}\right| + c$
43. [22] $-\frac{6}{5}\ln |x + 3| + \frac{11}{5}\ln |x - 2| + c$
44. [22] $-\frac{1}{5}\ln |x + 3| + \frac{1}{5}\ln |x - 2| + c$
45. [23] $\frac{8}{3}$
46. [23] $\frac{e^2 - e^{-2}}{2}$
47. [23] $17\frac{1}{3}$
48. [23] $A = \frac{9}{2}$

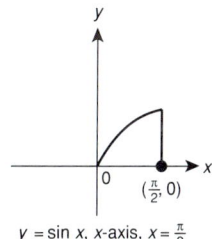

$y = 2 - x^2, y = x$

49. [23] $A = 1$

$y = \sin x,$ x-axis, $x = \frac{\pi}{2}$

50. [23] $A = 2 \ln 4 - \frac{3}{2}$

$y = \ln 2x, y = 0, x = 2$

51. [23] $\dfrac{\pi^2}{4}$ **52.** [23] $2\pi(e^2 - e^{-2})$ **53.** [23] $\dfrac{32\pi}{5}$ **54.** [23] $\dfrac{27\pi}{2}$ **55.** [23] $\dfrac{\pi}{2}(e^2 + 1)$ **56.** [23] π

57. [23] $W = 2000 \ln \dfrac{5}{2} \approx 1832.6$ ft-lbs **58.** [23] $W = 84{,}375\pi$ ft-lbs **59.** [23] $\left(\dfrac{8}{5}, \dfrac{3}{4}\right)$ **60.** [23] $\left(\dfrac{e^2 - 3}{2(e^2 - 1)}, 0\right)$

61. [23] $I_x = \dfrac{6144}{11}, I_y = \dfrac{1024}{9}$ **62.** $S = \sqrt{5} + \dfrac{1}{2}\ln(2 + \sqrt{5})$ **63.** [23] $F = 421\dfrac{7}{8}$ lbs **64.** [23] 1

65. [21] $0.6\sqrt{0.02t + 100}$ **66.** [21] $6.4 thousands **67.** [23] $A = 125{,}000$, $125{,}000 is the total savings over 25 years.
68. [23] $1{,}000{,}000 **69.** [21] $y = x^3 - 2x^2 + 3$ **70.** [21] $s(t) = 98t - 4.9t^2 + 200$, 690 ft **71.** [23] $P \approx 0.7$
72. [23] $1{,}250{,}000

Index

Abscissa, 113
Absolute inequality, 475
Absolute value, 6
  in inequalities, 475
Acceleration, 765, 842
Accuracy of a number, 6
Addition,
  of algebraic expressions, 34
  of complex numbers, 388
  of fractions, 95
  of matrices, 502
  of polynomials, 33
  of signed numbers, 7
  of vectors, 379
Addition property of equality, 10
Additive inverse, 11
Additive inverse properties, 88
Amplitude of a sine curve, 278, 288
Analytic geometry, 617–78
Angle, 202
  acute, 230
  central, 254
  of elevation, 219
  obtuse, 229
  phase, 249, 384
  of refraction, 228
  right, 204
  standard position, 202
  taper, 216
Angular velocity, 258, 324
Antiderivative, 856–59
Antilogarithm, 429
Applied maxima and minima, 838–40
Approximate roots of polynomial equations, 575
Arc length, 324, 964–65
Area,
  of a circular sector, 256
  under a curve, 928–33
  between two curves, 936–39
Argument of a complex number, 398
Arithmetic mean, 694
Arithmetic progression, 592

Arithmetic sequence, 592
Associative law, 11, 505, 524
Asymptote, 295, 415, 654
Augmented coefficient matrix, 508
Augmented matrix, 530
Average rate of change, 772
Average velocity, 842
Axes, coordinate, 113
Axis,
  of an ellipse, 643
  of a hyperbola, 653
  of a parabola, 632
  polar, 666
  of symmetry, 632

Base,
  of exponents, 14
  of logarithms, 419
Binomial, 31
Binomial coefficient, 606
Binomial expansion, 607
Binomial theorem, 605, 607
Bivariate data, 711
Box-and-whisker diagram, 693
Boyle's Law, 775, 847

Calculator operation,
  angles, 203
  antilogarithms, 429
  combined operations, 203
  inverse trigonometric functions, 213–41
  logarithms, 429
  powers, 21, 417
  reciprocals, 225
  triangle solution, 214–44
  trigonometric functions, 213–41
Capacitance, 292
Cardioid, 668
Center, 643, 653
Center of mass, 953–59
Central tendency, 690–96
Centroid formulas, 956

Centroids, 955
Chain rule, 768–70
Change of base property, 430
Charles' Law, 767
Chebyshev's theorem, 704
Circle, 627
Coefficient matrix, 508
Coefficient of correlation, 713
Column matrix, 517
Column vector, 517
Common difference, 592
Common logarithm, 428
Common ratio, 597
Commutative law, 11, 505
Complementary angles, 204
Completing the square, 182, 629
Complex numbers, 388
  standard form of, 388
  subtraction of, 389
Complex plane, 395
Complex roots, 393
Components of a vector, 381
Composite curves, 301
Concave,
  downward, 832
  upward, 832
Concavity, 832
Conditional inequalities, 192
Congruent, 52
Congruent triangles, 52
Conic sections, 618
Conjugate of a complex number, 390
Constant, 5
  derivative of, 760
  of integration, 857
  of proportionality, 56
Constant function, 414
Consumption matrix, 534
Continuity, 742–45
Continuous, 742
Converting between degrees and radians, 254
Converting coordinates, 397

Coordinate systems, 113
    Cartesian, 113
    polar, 386
    rectangular, 113
Correlation, 713
Corresponding angles, 208
Cosecant,
    of an angle, 225
    derivative of, 814
    graph, 276
    integration of, 885
    inverse, 312
Cosine, 220
    of an angle, 221
    double angle formula, 338
    function, 336
    graph, 276
    half angle formula, 339
    inverse, 311
    sum rule, 337
Cosines, law of, 239
Cotangent,
    of an angle, 225
    derivative of, 804
    graph, 297
    integration of, 885
    inverse, 311
Coterminal angles, 204
Counting numbers, 2
Cramer's Rule, 158
Critical number, 822
Cross-product rule, 50
Curve fitting,
    nonlinear, 720–26
Curve sketching, 832–36
Cycle of a curve, 277, 290
Cylindrical shell, 821, 827, 947

Decimal, repeating, 2
Decreasing function, 416, 821, 827
Definite integral, 874
Degree,
    measure of an angle, 202
    of a monomial, 33
    of a monomial in a variable, 33
    of a polynomial, 33
Demand curve, 443
DeMoivre's theorem, 404
Denominator,
    common, 97
    rationalizing, 27
Dependent system of equations, 140
Derivative, 756
    of a constant, 760
    of exponential functions, 797–99

    of implicit functions, 780–84
    of inverse trigonometric functions, 807–11
    of logarithmic functions, 792–95
    of a polynomial, 761
    of a power, 760
    of a product, 762
    of a quotient, 763
    as a rate of change, 772–74
    sum rule, 762
Descartes' rule of signs, 570
Determinant, 156
    higher order, 538
    properties, 542
    second order, 156
    third order, 538
    triangular form, 544
Deviation, standard, 480, 700
Diagonal, 211, 237
Diagonal of a determinant, 156, 544
Difference,
    common, 593
    of two fractions with different
        denominators, 96
    of two squares, 75
Differential, 777
Differentiation, 756
    implicit, 780–84
Dimensions,
    of a matrix, 500
    of the product matrix, 520
Directrix of a parabola, 633
Direct variation, 56
Discontinuous, 743
Discriminant, 189
Distance formula, 619
Distributive law, 11, 524
Division,
    of algebraic expressions, 39
    of complex numbers, 391
    of fractions, 92
    of a polynomial, 39
    of radicals, 24
    of signed numbers, 8
    synthetic, 557
Division rules for real numbers, 8
Domain, 130, 273
Double-angle formulas, 336, 338, 348, 352
Double inequality, 465
Double negative property, 11
Double reciprocal property, 12
Double roots, 175

e, 416
Electrical circuits, 215, 249, 292–93, 318,
    383, 386

Element,
    of a determinant, 538
    of a matrix, 501
Elimination by addition or subtraction,
    150–54
Ellipse, 643
Empirical rule, 705
Equal complex number, 388
Equal matrices, 501
Equal to, 5
Equal vectors, 378
Equation,
    of a circle, 628
    of a normal curve, 706
Equations,
    absolute value, 475
    exponential, 431
    graphic solution of, 142–44, 442–44
    of higher degree, 556–79
    involving fractions, 101–4
    linear, 43–45, 619
    logarithmic, 431
    polar, 664
    polynomial, 564
    quadratic, 123, 174, 388
    with radicals, 452
    simultaneous, 140
    systems of linear, 142–68, 507, 533, 547
    systems of quadratic, 413
    trigonometric, 363
Equilibrium, 153, 444
Equivalent fractions, 87
Even functions, 268
Expanded forms of a series, 589
Expansion by minors, 538
Exponential equations, 431
    derivative of, 777–79
    integral of, 877
Exponential functions, 414
    common, 797
    natural, 797
Exponents, 14–18
    laws of, 424
External demand matrix, 535
Extracting roots, 179
Extreme point, 828
Extremes, 49
Extremes-means rule, 51

Factorial, 605
Factoring, 74–84
    by grouping, 81
Factoring quadratic polynomials, 75
Factor theorem, 565
Feasible solution, 484, 490

Finite sequences, 584
First derivative, 782
First derivative test, 822
Foci, 653
Focus, 633
    of an ellipse, 643
    of a hyperbola, 653
    of a parabola, 633
Formulas,
    cosine function, 336
    sine function, 336
    sum and difference, 336
    tangent function, 346
Fractional equations, 101
Fractions, 84–89
    algebraic, 85
    equivalent, 87
Frequency, 290, 681
    distribution, 683
    histogram, 684
Functions, 130
    circular, 262
    derivative of, 757
    exponential, 414
    integration of, 856–926
    inverse, 214, 308
    limit of, 738–51
    logarithmic, 419
    quadratic, 123, 174, 188
    trigonometric, 262
    zeros of a polynomial, 564–79
Fundamental Theorem of Calculus, 874

General term of a sequence, 585
Geometric progression, 596
Graph, 4
    of an exponential function, 415
    of a function, 127
    of an inverse trigonometric function, 310–16
    of a linear equation, 116
    of a logarithmic function, 419
    of a polar function, 664
    of a quadratic function, 123
    of trigonometric functions, 274–306
    $y = a \cos(bt + c)$, 286
    $y = a \sin(bt + c)$, 286
Graphic representation of complex numbers, 395
Graphic solution of equations, 142–44, 442–44
Graphic solution of inequalities, 465
Greater than, 5, 462
Grouped data, 694

Half-angle formulas, 336, 339, 348, 352
Higher ordered derivatives, 780, 782
Histogram, 684–87
Hyperbola, 653
Hypotenuse, 205

Identity, 11
    matrix, 522
    trigonometric, 330
Imaginary axis, 395
Imaginary number, 388
Impedance, 383, 392
Implicit differentiation, 780
Implicit function, 780
    derivative of, 780–84
Improper integral, 969–70
Inconsistent system of equations, 140
Increasing-decreasing curves, 821
Increasing function, 416, 821, 827
Increment, 776
Indefinite integral, 872
Index, 24
    of a radical, 24
    of refraction, 228
    of summation, 589
Inequalities, 462
    absolute value, 475
    algebraic solution of, 462
    conditional, 192
    graphic solution, 465
    linear, 462
    quadratic, 192–96
    with two variables, 482
Infinite sequence, 584
Infinite series, 601
Infinity, 745
Inflection point, 832
Initial conditions, 874
Initial point of a vector, 378
Initial side of an angle, 203
Instantaneous rate of change, 772
Instantaneous velocity, 842
Integer, 2
Integral, 856
    definite, 874
    improper, 969–70
    indefinite, 872
    sign, 856
Integrand, 856
Integration, 856
    basic rules of, 864
    constant of, 857
    of exponential forms, 877
    of inverse trigonometric functions, 886
    of logarithmic functions, 778

    by partial fractions, 909–15
    by parts, 896–900
    of powers, 864, 866
    sum rule, 864
    of trigonometric functions, 882, 885
    by trigonometric substitutions, 901–9
Intercept, 621
Internal demand matrix, 534
Inverse function, 308
Inverse of a matrix, 529
Inverse trigonometric functions, 214, 310–16
    derivatives of, 807–11
    integration of, 886
Inverse variation, 58
Inverted ratios rule, 51
Irrational number, 4
Isosceles triangle, 210

Joint variation, 59

Kelvin's law, 767

Law,
    of cosines, 239
    of sines, 231
Least common denominator, 97
Least common multiple, 97
Least-squares line, 711–18
Legs, 205
Lemniscate, 671
Length of arc, 964–65
Less than, 5, 462
Limit, 738–51
    e as a, 792
    of a function, 738
    infinite, 745–51
    of an infinite series, 601
    of a sine, 801
    of the slope of a secant line, 753
Limits of integration, 874
Linear equation, 43, 45, 116, 140, 619
    in one variable, 43
    in two variables, 45
Linear inequality, 462, 482
Linear programming, 487, 490
Linear velocity, 258, 324
Literal number, 5
Location theorem, 575
Logarithm,
    of a power, 424
    of a product, 424
    of a quotient, 424
Logarithmic equations, 431

Logarithmic function, 419
  derivative of a, 792–95
  integral of, 877
Logarithms,
  base of, 10, 428
  other bases of, 419
  natural, 428
  properties of, 423–27
Lower limit of an integral, 874

Magnitude of a vector, 378
Main diagonal of a determinant, 544
Major axis of an ellipse, 643
Market equilibrium, 149
Market equilibrium point, 443
Matrix, 156, 500
  determinant of a, 338
  elements, 501
  identity, 522
  inverse, 529
  square, 529
  zero, 505
Matrix equation, 502
Maxima and minima, 838–40
Maximizing and minimizing a linear
    function, 489
Maximum points, 822
Mean proportional, 52
Means, 49
Mean using grouped data, 695
Measures of central tendency, 690
Measures of dispersion, 699–709
Mechanical advantage, 122
Median, 691
Minimum points, 822
Minor, 538
Minor axis of an ellipse, 643
Minor of an element of a determinant, 538
Minute (angle measure), 202
Mode, 690
Modulus, 398
Modulus of a complex number, 398
Moment of area, 955
Moment of inertia, 960–64
Monomial, 33
Multiplication,
  of algebraic expressions, 36
  of complex numbers, 389
  of fractions, 91
  of matrices, 517
  of polynomials, 36
  of radicals, 24
  scalar, 379, 503
  of signed numbers, 7

Multiplication property of equality, 10
Multiplicative inverse, 12

Natural logarithms, 428
Natural numbers, 2
Negative exponent, 16
Newton's law, 779, 815
Nonlinear curve fitting, 720–26
Nonlinear equations, 127
Nonlinear inequalities, 468
Normal distribution, 705
Normal line, 819
nth power, 14
nth roots of a complex number, 405
nth term of a sequence, 585
Number,
  complex, 388
  imaginary, 388
  irrational, 4
  natural, 2
  rational, 2
  real, 4

Objective function, 490
Obtuse angles, 229
Odd function, 268
One-to-one function, 308
Opens down, 123
Opens up, 123
Ordered pair, 112
Order of determinants, 156
Order of operations, 6
Ordinate, 113
Organizing data, 680
Origin, 113

Parabola, 123, 632
Parallelogram, 237, 241
Partial fractions, 909–15
Particular solution of an integral, 872
Parts,
  integration by, 896–900
Pascal triangle, 607
Perfect square, 24
Perfect square trinomial, 181
Period of a curve, 275, 279, 288
Periodic function, 275
Phase angle, 224, 318, 326, 384, 401
Phase shift, 287–88
Pivot,
  element, 509
  operation, 509
  position, 509

Point of inflection, 832
Point-slope form of a line, 622
Polar axis, 666
Polar coordinates, 396, 664
Polar equation, 664
Polar form, 398
Polar form of a complex number, 398
Pole, 666
Polygonal set, 484
Polynomial, 33
  derivative of, 760
  equations, 564
Positive integers, 2
Power, 14
  of a complex number, 404
  derivative of, 760
  integration of, 864, 866
  rule for integration, 862
    variation of, 866
Precision, 6
Primary diagonal, 156
Principal root, 24, 31
Probability density function, 966
Probability of an event, 967
Product,
  of complex numbers, 389, 399
  derivative of, 762
  of fractions, 91
  of matrices, 517–26
  rule, 762
  of sines and cosines, 357
  special, 70
  of the sum and difference of two terms, 72
Progression,
  arithmetic, 592
  geometric, 596
Properties,
  of the cosine function, 276
  of determinants, 542
  of inequalities, 462
  of matrix multiplication, 524
  of similar triangles, 53
  of the sine function, 275
  of a zero matrix, 505
Proportion, 49
Proportion solving strategy, 57
Pure waves, 290
Pythagorean theorem, 207

Quadrant, 113
Quadratic equation, 123, 174, 388
Quadratic formula, 187, 388
Quadratic inequality, 192–95
Quartile, 692

Quotient,
 of complex numbers, 391
 derivative of, 763
 of fractions, 92
 rule, 763
 rules, 87

Radian, 254
Radical form, 25
Radicals, 24
 division of, 24
 equations, 452
 multiplication of, 24
Radicand, 24
Range, 130, 682
Rate, 49
Ratio, 48
 common, 597
Rational,
 exponents, 31
 expression, 85
 number, 2
Rationalizing a denominator, 27
Rational roots of polynomial equations, 564
Rational Zero theorem, 566
Reactance, 383
Real axis, 395
Real number line, 4
Real numbers, 4, 10–12
 multiplication rules of, 7
Reciprocal, 12
Rectangular coordinate system, 113
Recursion formula, 586
Refraction, 228
Region of feasible solutions, 490
Related rates, 838, 845–47
Relative frequency, 683
Relative frequency histogram, 684
Relative maxima and minima, 838
Remainder theorem, 565
Remove a common factor, 74
Repeating decimal, 2
Resistance, 220
Resultant of two vectors, 241, 380
Right angle, 204
Right triangle, 205
Right triangle ratios, 221
Roots, 174
 of complex numbers, 405
 double, 175
 of linear equations, 43–47
 of quadratic equations, 174–90
 rational, 564–73
Rose curve, 669
Row matrix, 517

Row vector, 517
Rules,
 of exponents, 18
 for integration, 886
 for square roots, 25

Scalar, 503
Scalar multiples, 379
Scalar multiplication, 379, 503
Scalar quantity, 378
Scatter diagram, 711
Scientific notation, 19
Secant, 225
 derivative of, 804
 graph, 298
 integration of, 885
 inverse, 312
 line, 753
Second (angle measure), 202
Secondary diagonal, 156
Second derivative, 780, 782
Second derivative test, 827–30
Sector of a circle, 255
Sequence,
 arithmetic, 592
 finite, 584
 infinite, 584
Series, 589
Signed numbers, 7
Significant digits, 6
Signs,
 laws of, 7
 of trigonometric functions, 267
Similar, 53
Similar or like terms, 33
Similar triangles, 53, 207
Simple harmonic motion, 290
Sine, 220
 of an angle, 221
 derivative of, 802
 double-angle formula, 348
 graph, 275
 half-angle formula, 348
 integration of, 885
 inverse, 310
 sum formula, 348
Sines, law of, 231
Sinusoidal curve, 277
Slope, 620
 of a tangent line, 752–56
Slope-intercept form of a line, 621
Solid of revolution, 941
 volume of, 941
  by cylindrical shells, 947

Solution of,
 inequalities, 463–82
 linear equations, 43–47
 quadratic equations, 177–91, 388
 a system of equations by determinants, 156, 547
 a system of linear equations, 142–68, 507, 533, 547
 a system of quadratic equations, 413
Solution set, 177
Solve a proportion, 51
Solving,
 linear equations, 43
 quadratic equations,
  by completing the square, 181
  by factoring, 177
  by graphing, 174
  with the quadratic formula, 186
 quadratic inequalities, 192
 radical equations, 452
 similar triangles, 209
 systems of,
  equations using matrices, 507
  three linear equations in three unknowns, 161
  two linear equations,
   by addition or subtraction, 150
   by determinants, 156
   by graphing, 142
   by substitution, 145
 trigonometric equations, 367
Special factoring formulas, 81
Special products, 70
Spiral curve, 670
Square of a binomial, 70
Square root, 23, 24
 rules, 25
Standard deviation, 480, 700
Standard equation,
 of a circle, 628
 of an ellipse, 645
 of a hyperbola, 655
 of a parabola, 633
Standard form, 88, 178
 of a complex number, 388
Standard normal curve, 706
Standard position of an angle, 202
Statistics, 679–735
Stefan's law, 853
Substitution, elimination by, 145–49
Subtracting,
 algebraic expressions, 34
 complex numbers, 389
 fractions, 96
 matrices, 505

polynomials, 33
signed numbers, 7
vectors, 380
Sum,
   derivative of, 761
Sum and difference of two cubes, 83
Summation sign, 589
Sum of two fractions,
   with different denominators, 96
   with the same denominator, 95
Sums, differences, and products of sines and cosines, 357
Sums of sines and cosines, 359
Supplementary angles, 205
Supply curve, 443
Symmetry, 268
Synthetic division, 557
System of equations,
   algebraic solutions of, 446
   dependent, 140
   inconsistent, 140
   independent, 140
   linear, 43, 140
   quadratic, 448

Table of integrals, 916–20
Tangent, 212
   of an angle, 212
   derivative of, 803
   double-angle formula, 353
   graph, 295
   half-angle formula, 352
   integration of, 885
   inverse, 312
   line, 752, 818
   sum formula, 351
Tangent ratio, 212
Taper angle, 216
Tapered, 216
Terminal point of a vector, 378
Terminating decimals, 2
Terms, 33
   of a sequence, 585
   similar, 33

Theorem of Pythagoras, 206
Transcendental functions,
   derivatives of, 792–816
Transformation, 720
Transformed equation, 720
Transposing, 43
Transverse axis of a hyperbola, 653
Trial and error, 76
Triangle,
   congruent, 52
   isosceles, 210
   Pascal, 607
   right, 205
   similar, 53, 207
   solution of, 214–44
Triangular form of a determinant, 544
Trigonometric equations, 336, 363
Trigonometric form of a complex number, 398
Trigonometric functions, 263
   of angles, 212, 220
   derivative of, 801–6
   graphs of, 274–306
   integration of, 882, 885
   inverse, 308, 807
   for negative numbers, 267
   of negative numbers, 267
Trigonometric identities, 330
Trigonometric ratios, 220
Trigonometric ratios for obtuse angles, 230
Trigonometric substitution, 901–7
Trinomial, 33
Two-point form of the equation of a line, 623

Unit circle, 262
Upper limit of an integral, 874

Variable, 5
Variance, 700
Variance and standard deviation for grouped data, 702
Variation, 56–58
   combined, 59
   joint, 59

Variation in sign, 570
Vector,
   magnitude of, 378
   quantity, 378
Vectors, 378
   subtraction of, 380
Velocity,
   angular, 258
   average, 842
   instantaneous, 842
   linear, 258
Verifying a trigonometric identity, 331
Vertex,
   of an ellipse, 643
   of a hyperbola, 653
   of a parabola, 633
Vertical asymptote, 295
Vertical line test, 130
Vertices, 643, 653
Voltage, 292
Volumes of geometric figures,
   by integration, 944–50

Whole number, 2
Work, 952–54

x-axis, 113
x-coordinate, 113
x-intercept, 119

y-axis, 113
y-coordinate, 113
y-intercept, 119

Zero,
   exponent, 15
   factor property, 11
   of a function, 174, 566
   matrix, 505
   operations with, 10, 11
   of a polynomial, 566
Zeros, 174

Complex Numbers

Rectangular Form: $a + bj$, where $j^2 = -1$

Polar Form: $a + bj = r(\cos\theta + j\sin\theta)$, where $r = \sqrt{a^2 + b^2}$, $\tan\theta = \dfrac{b}{a}$, $a = r\cos\theta$ and $b = r\sin\theta$

Products and quotients: $(r\ cjs\ \alpha)(s\ cjs\ \beta) = rs(cjs(\alpha + \beta))$

$$\dfrac{r\ cjs\ \alpha}{s\ cjs\ \beta} = \dfrac{r}{s}\ cjs(\alpha - \beta)$$

Powers and Roots: $(r\ cjs\ \theta)^n = r^n\ cjs(n\theta)$

$$(r\ cjs\ \theta)^{1/n} = \sqrt[n]{r}\left(cjs\left(\dfrac{\theta + 2\pi k}{n}\right), k = 0, 1, 2, \ldots, (n-1)\right)$$

Analytic Geometry

Line: slope $= \dfrac{y_2 - y_1}{x_2 - x_1}$ Equation: $y = mx + b$, $y - y_1 = m(x - x_1)$

$$y - y_1 = \left(\dfrac{y_2 - y_1}{x_2 - x_1}\right)(x - x_1)$$

Circle: $(x - h)^2 + (y - k)^2 = r^2$

Ellipse: $\dfrac{(h - h)^2}{a^2} + \dfrac{(y - k)^2}{b^2} = 1$, $a > b$ (Major axis horizontal)

$\dfrac{(x - h)^2}{b^2} + \dfrac{(y - k)^2}{a^2} = 1$, $a > b$ (Major axis vertical)

Parabola: $(y - k)^2 = 4p(x - h)$ (Horizontal axis)
$(x - h)^2 = 4p(y - k)$ (Vertical axis)

Hyperbola: $\dfrac{(x - h)^2}{a^2} - \dfrac{(y - k)^2}{b^2} = 1$ $\dfrac{(y - k)^2}{a^2} - \dfrac{(x - h)^2}{b^2} = 1$

Statistics

Arithmetic Mean: $\bar{x} = \dfrac{\Sigma x}{n}$ Grouped Data: $\bar{x} = \dfrac{\Sigma xf}{n}$

Standard Deviation: $s = \sqrt{\dfrac{n\Sigma x^2 - (\Sigma x)^2}{n(n-1)}}$ $s = \sqrt{\dfrac{n\Sigma x^2 f - (\Sigma xf)^2}{n(n-1)}}$

Least Squares Line: $y = mx + b$, where

$$m = \dfrac{n\Sigma xy - (\Sigma x)(\Sigma y)}{n\Sigma x^2 - (\Sigma x)^2},\ b = \dfrac{\Sigma y - m(\Sigma x)}{n}$$

Coefficient of Correlation:

$$r = \dfrac{n(\Sigma xy) - (\Sigma x)(\Sigma y)}{\sqrt{n\Sigma x^2 - (\Sigma x)^2}\ \sqrt{n\Sigma y^2 - (\Sigma y)^2}}$$